T0204039

SAFETY *and* HEALTH *in* CONFINED SPACES

Neil McManus, CIH, ROH
NorthWest Occupational Health and Safety
North Vancouver, B.C.
Canada

CRC Press is an imprint of the
Taylor & Francis Group, an **informa** business

CRC Press
Taylor & Francis Group
6000 Broken Sound Parkway NW, Suite 300
Boca Raton, FL 33487-2742

First issued in paperback 2019

ISBN-13: 978-1-56670-326-0 (hbk)
ISBN-13: 978-0-367-40024-8 (pbk)

Library of Congress Cataloging-in-Publication Data

Catalog record is available from the Library of Congress

Visit the Taylor & Francis Web site at
http://www.taylorandfrancis.com

and the CRC Press Web site at
http://www.crcpress.com

Preface

Confined spaces are the sites of many fatal and nonfatal industrial accidents. Given appropriate conditions, potentially any space in which people work could be or could become a confined space.

We know very little about accidents involving confined spaces. Publicly available information exists primarily as a result of investigations of fatal accidents. NIOSH (National Institute for Occupational Safety and Health), OSHA (Occupational Safety and Health Administration), and MSHA (Mine Safety and Health Administration) reports are the main resources for research in this area in North America.

According to the most recent statistics provided by NIOSH, engulfment by loose materials caused 65% of the fatalities in workspaces normally called confined spaces, as well as excavations, trenches, and ditches. While excavations, trenches, and ditches normally are not considered to be confined spaces in regulations and standards, engulfment is engulfment, regardless of the semantics about location and type of material. Atmospheric hazards caused about 56% and engulfments about 34% of the fatalities in workspaces recognized as confined spaces. Drowning, electrocution, fall from heights, process-related accidents, plus other causes constituted the remaining 10%. The overwhelming cause of the very few process-related accidents was failure of live steam (utility) lines during work of an unrelated nature.

The most glaring deficiency identified in the reports from NIOSH, OSHA, and MSHA was lack of knowledge. Workers and supervision routinely failed to recognize the hazardous conditions that existed or could develop in these workspaces. This problem continues and will continue, despite the enactment of comprehensive legislation in industrialized countries.

Confined spaces differ from normal workspaces. One reason for this is the role of boundary surfaces. Containment by boundary surfaces amplifies or magnifies the severity of hazardous conditions. The relationship between the individual, boundary surfaces, and the source of energy or contamination or other hazardous condition is the major factor in the onset, outcome, and severity of accidents in confined spaces.

Another important factor is the function of the structure within which the space exists. Some structures act as a protective barrier between energy sources, such as machinery, and the surroundings. Other structures contain processes. Others contain or store substances and materials. Many of the conditions that occur in these spaces are unique and are not encountered in ordinary workspaces. Confined spaces also include temporary structures, such as excavations and ditches.

A disturbing number of fatal accidents occurred in spaces that lay outside conventional definitions of confined spaces. Without accommodation of these realities by standard-setters and regulators, accidents involving this type of space will continue to occur. Some examples include:

- **A mould used to manufacture 150-L (40-gal U.S.) plastic containers:** the victim inserted head and shoulders into the mould (50 cm diameter by 80 cm deep) and was overcome by vapor from the perchloroethylene that he used to wipe the surfaces.
- **A waist-high paint mixing pot:** the victim bent into the pot and was overcome by vapor from the methylene chloride that he used to clean the interior.
- **The shaft of a dumbwaiter:** the victim opened the door to determine the location of the car and was struck by it.
- **A sand-mixing machine:** the victim reached into a side hatch to repair a bearing and was struck by the blades when the drive motor started unexpectedly.

- **"Empty" 200-L drums:** ignition of the hazardous atmosphere contained in these drums from residual contents caused about 16% of the fatal welding and cutting accidents.
- **An open-topped degreaser:** the victim reached in over the top of the degreaser (2 m × 3 m × 1.5 m high) to recover a part that had fallen in and was overcome by the solvent-rich, oxygen-deficient atmosphere.
- **The cabinet of an abrasive blasting machine:** the operator walked in through the access door to retrieve a part that had fallen and was asphyxiated by the nitrogen atmosphere used to inert the interior.
- **An office:** a carpet layer sealed off the ventilation system, as well as gaps in walls and doors, to contain solvent vapors from adhesive used to anchor carpet tiles. He was overcome by toluene vapor.

Traditional definitions for confined spaces have focused on structures, not conditions. This then begs the questions: what exactly is a confined space and how should management of hazardous conditions occur?

The test for effectiveness of a particular definition is simple: does it encompass **all** of the workspaces in which the hazardous conditions peculiar to confined spaces could exist or could develop? This definitely is not the case with definitions used in present regulatory and consensus standards. Standards and regulations that fail to encompass the unusual workspaces in which people work leave open the door for future tragedies.

Work in confined spaces generally occurs during construction, inspection, maintenance, modification, and rehabilitation. This work is nonroutine, short in duration, nonrepetitive, and unpredictable in scheduling (often occurring during off-shifts and weekends). Hazardous conditions in confined spaces follow several themes: confinement of individuals, confined atmospheres, confined energy, and the amplified expression of health, safety, biological, and ergonomic hazards.

A major difficulty with the management of hazardous conditions in confined spaces is the fluid nature of the problem. A seemingly minor change or error or oversight in preparation of the space, selection, or maintenance of equipment or work activity can change the status of conditions from innocuous to life-threatening.

The term "confined space" itself has contributed to the problem. The term strikes fear into management because of the considerable administrative and technical burdens imposed by regulations onto workspaces that receive this label. The incentive not to label a particular space as a confined space, therefore, is considerable. This reality has distorted the manner in which confined spaces should be viewed.

Conditions in some confined spaces are life-threatening at all times. On the other hand, conditions in many other confined spaces are life-threatening only under unusual circumstances. Surely, the determinant that should govern the nature of the response to a particular circumstance is the level of concern appropriate to a particular condition and not the worst case. Yet, the typical depiction of work practices conveyed in articles in trade magazines and other sources and in advertising for products is the latter.

Hazard assessment is a device for identifying potential and actual hazardous conditions and assessing the level and acceptability of risk. Hazard assessment is a difficult process. Many of the conditions that can produce acute exposure or traumatic injury are difficult to recognize and assess. Hazard assessment is semiquantitative at best and intuitive at worst. Yet, as indicated by real-world accidents, the tolerance for error in judgment is very small.

The operating cycle of many confined spaces is divisible into the following segments: the undisturbed space, preentry preparation, prework inspection and work activity. A hazard assessment performed for each provides the best way to maximize control over conditions. The undisturbed space is the status quo established between closure at the end of one work cycle and the start of preparation for the next. Preentry preparation is the sum of activities undertaken to ready the space for entry. Activities undertaken during preentry preparation should minimize the risk of entry and performing work through hazard elimination, or at the least, control. Prework inspection is the initial entry into the space. (Prework inspection is a requirement in some jurisdictions.) The purpose for prework inspection is to ensure that the space is safe for the start of work. Work activity describes the individual tasks to be performed by entrants. Hazards that remain at the start of work activity or are created by it dictate the nature of possible accidents for which emergency response is required.

Performing the hazard assessment for each segment of work is essential to the process. A hazardous condition eliminated during preentry preparation could reappear as a result of work activity.

A decision about whether the risks associated with entry and work are acceptable is essential to the process. If control of conditions can be assured, the decision is not difficult to make. The less the level of perceived control, the greater becomes the need for contingencies. The only other alternative is to prohibit the entry.

This decision highlights a question that deserves to be asked, but is easily overlooked: should entry occur at all? This question receives little attention in discussion about confined spaces.

Consensus and regulatory standards have delegated the onus for hazard assessment and specification of control measures to the qualified person. The qualified person is deemed capable by education and/or specialized training and experience of anticipating, recognizing, and evaluating hazardous conditions and specifying control measures and/or protective actions. That is, the qualified person is expected to know what to do in the context of a particular situation.

Two models for the on-site management of hazardous conditions in confined spaces have evolved: the entry permit system and the on-site qualified person. Clear lines of authority, responsibility, and accountability are required under either system.

The entry permit summarizes actions and tests performed and indicates the need for precautionary measures. The permit also specifies procedures to follow and conditions under which entry and work can proceed. The entry permit also acts as the summary for the complete hazard assessment.

The permit system works best where hazardous conditions are known from previous experience and control measures have been tried and proven effective. The permit system enables apportioning of scarce expert resources in an efficient manner. The limitations of the permit system arise where previously unrecognized hazards are present. If the qualified person is not readily available, these can remain unaddressed.

The on-site qualified person provides readily available expertise for evaluation of conditions and control measures. A major advantage of this approach is the ability to respond to unanticipated situations on short notice. Following evaluation of the space and implementation of control measures, the qualified person issues a certificate. The certificate indicates tests performed and conditions under which the work can proceed. The certificate also indicates the status of the space in standardized language, such as "safe for workers."

This approach is ideally suited to operations that have numerous confined spaces or where conditions or the configuration of spaces can undergo rapid change. The limitation of this approach is the knowledge base of the qualified person and the skill and thoroughness in its application.

The preceding discussion highlights some of the challenges posed by confined spaces. There is a serious need to unify the themes that exist within this subject area. This subject area is fragmented by industrial sector. Little, if any, crossover occurs between them. This is most unfortunate, since hazard management models that work in one industrial sector may work just as well in another. The blockage to intersectoral application of these models is regulatory. Application of narrowly focused hazard management models across a broad spectrum of industry does not necessarily benefit all applications. Open discussion and consideration are essential to ensure that organizations can utilize the hazard management model that best suits their needs.

In order to achieve this objective, the entire spectrum of the subject of confined spaces must be brought forth for discussion. Only in this manner will the needless tragedies that continue to occur provide some useful lessons for the future.

Writing a book on confined spaces involves some difficult creative decisions because this subject does not progress from A to Z in linear fashion. The stylistic decisions are fundamental: either refer continually to material in other sections and avoid repetition or repeat some material to maintain continuity of thought. This book uses the latter approach in the hope that this will be easier to use for the greatest number of readers. The casual reader should be able to begin in many areas with enough peripheral information to be able to put the topic of interest into perspective with other issues.

The Author

Neil McManus is a practicing industrial hygienist with 20 years of broad-spectrum service "in the trenches" to workers and management. He is certified in the comprehensive practice of industrial hygiene by the American Board of Industrial Hygiene and by the Canadian Registration Board of Occupational Hygienists. Mr. McManus is a member of the American Conference of Governmental Industrial Hygienists, American Industrial Hygiene Association, British Occupational Hygiene Society, Health Physics Society, and the Marine Chemist Association. He has been an active volunteer in committees in the industrial hygiene profession and in the local community, and has written numerous articles and short publications. Mr. McManus has an M.Sc. in radiation biology and an M.Eng. in occupational health and safety engineering, as well as a B.Sc. in chemistry and a B.Ed. specializing in chemistry and biology.

Mr. McManus became interested in maintenance activities in confined spaces during turnarounds in an oil refinery in the early 1980s and followed this through subsequent employment in electrical generation, railway operations, and ship construction. He has taught courses on confined space hazards and made numerous presentations at meetings and conferences on this theme. He is a long standing member and former Chair of the Confined Spaces Committee of AIHA.

Contributors

Robert E. Brown, Jr., CIH, CHMM
Dexter Corporation
Pittsburg, CA

Richard P. Garrison, Ph.D., CIH, CSP
Associate Professor of Industrial Health
School of Public Health
Department of Environmental and Industrial Health
Ann Arbor, MI

John Gill, CIH
Springfield, MA

Gilda Green
Training by Design, Inc.
North Vancouver, BC, Canada

Robert E. Henderson
Director of Marketing
Biosystems Inc.
Middletown, CT

Michael S. Krupka, CIH
Medfield, MA

David T. Matthews, CIH
Clayton Environmental Consultants
Kennesaw, GA

David G. McCarthy
Belmont, MA

Contents

Acknowledgments

I wish to thank the following individuals for contributions to this book that otherwise would go unrecognized. Don Tardiff, Senior Safety Officer, Vancouver Shipyards, for his assistance on emergency response and first aid; Bob Henderson for his exuberance and encouragement; my wife, Gilda Green, for her indulgence in this undertaking and for producing the graphics, and Taffy, our yellow Labrador Retriever, for her patience, when all she wanted was some time to chase a stick.

Dedication

This book is dedicated to the memory of those people who either lost their lives or were seriously injured in accidents involving confined spaces. May we learn from these tragedies.

The way things were. This depiction shows the way that level gauging and other tasks involving confined spaces were performed, and perhaps still are. (Photograph entitled "Man with Rope" used by permission of Kistler Morse Systems, Bothell, WA.)

1 Atmospheric Hazards and Fatal Accidents in Confined Spaces

CONTENTS

INTRODUCTION

A recurring site of both fatal and nonfatal accidents in industry is the confined space. Confined spaces occur across the spectrum of industry. Potentially any structure in which people work could be or could become a confined space.

The term "confined space" normally is used to indicate danger in a particular structure or workspace. What the term confined space actually describes is hazardous conditions that can occur in a workspace, rather than the workspace itself. The enigma of confined spaces is that under some conditions a particular workspace may pose no extraordinary hazard. Yet, following seemingly minor change, the conditions become life threatening. On the other hand, the interior of some structures poses serious hazards under almost all conditions. The dangers are easily recognizable and these spaces receive due recognition and attention. In other spaces, hazardous conditions develop only under a few circumstances. The hazards that develop can be equally serious, yet receive no recognition. This, then, explains the difficulty in managing the hazards posed by confined spaces. The hazardous conditions may be transient and subtle and, therefore, difficult to recognize and to address. Also, the attributes of the space can exacerbate the hazards.

In the U.S., accidents involving confined spaces claim an estimated 200 victims per year across industry, agriculture, and the home. Sometimes these accidents produce multiple fatalities. In accidents involving hazardous atmospheres, individuals sometimes attempt to assist those in distress. The highly stressful conditions under which these actions occur subject the intervenor to considerably greater risk than the initial victim. Would-be rescuers comprise a sizable proportion of the victims in this type of accident.

Accidents associated with confined spaces or confined hazardous atmospheres differ from those occurring in normal workspaces. A minor error or oversight in preparation of the space, selection or maintenance of equipment, or work activity can produce a fatal outcome. The tolerance for mistakes is very small. Accidents involving atmospheric hazards in confined spaces are more likely to cause multiple fatalities than those occurring in normal workplace situations.

While most of the victims understandably are tradespeople who work in confined spaces, they also include engineering and technical people and supervisors and managers. Safety supervisors and superintendents also have died in these accidents. First aid attendants and emergency response personnel in the fire and ambulance services have died in rescue attempts.

Until recently, the industrial hygiene profession has had little formal involvement with the characterization, assessment, and management of hazards posed by confined spaces and confined hazardous atmospheres. The industrial hygiene literature of the past and present contains few technical articles on this subject (Garrison and McFee 1986). Until recently, the industrial hygienist could obtain little information or training in this area, reflecting the perspectives of this profession.

No doubt, a number of factors contributed to this situation. Possible influences may have been priorities in other areas imposed by legislation, analytical difficulties, and the philosophy of industrial hygiene practice. The problems posed by confined spaces and confined atmospheres are difficult to solve.

One possible influence in this situation may be the modes (production and nonproduction) in which industry operates. Production mode implies relatively invariate or steady-state conditions day after day. This could occur for many months or even years. Predictability and repeatability characterize tasks that occur during production mode. Nonproduction modes include construction, inspection, maintenance, modification and rehabilitation. Entire segments of industry and organizational units within individual organizations operate in nonproduction mode. Activities occurring during nonproduction mode usually are neither repetitive nor is the outcome fully predictable. Variability characterizes these activities and the conditions under which they are carried out.

Work occurring during nonproduction modes can be short in duration and highly hazardous, and often occurs irregularly, typically during off-shifts. Complicating matters further is time pressure to complete the work in orderly fashion. Time pressure extends to all aspects of this work, including hazard characterization and assessment. The first line of defence against workplace hazards in these situations usually is personal protection, supplemented by portable ventilating equipment. Complicating things further, the process and the individual are forced to accommodate to the capabilities offered by equipment that is available on site. Delivery of additional equipment may not be possible until after completion of the work. The tolerance for error in decision making in these situations likewise is very small.

The nature of work also influences the likelihood of encounters with confined spaces. In some segments of industry, encounters are infrequent. In others they are routine daily occurrences. Similarly, occupation also influences the likelihood of encounters with confined spaces. Some workers encounter confined spaces rarely, while others work in confined spaces all the time.

Confined spaces pose multidisciplinary problems. In many organizations, safety "owned" the problem. This "ownership" seemed logical, since the hazards associated with this type of work were perceived to be safety related. The safety profession thus assumed leadership in this area. To illustrate, safety professionals almost completely constituted the membership of technical committees and research groups active in this area (ANSI 1977, 1989, 1995; NIOSH 1979; NFPA 1993).

Nevertheless, evidence concerning the true nature of hazards that occur in confined spaces convincingly argues for significant involvement by the industrial hygiene profession.

Marine chemists face the problems of entry and work in confined spaces daily. Marine chemists are specialists in recognition, assessment, and control of hazards in confined spaces (Keller 1982, Willwerth 1994). Marine chemists to this point have focused on the maritime and shipyard industry. Hazards receiving attention from marine chemists include potential fire and explosion situations, as well as hygiene and safety-related hazards. (Marine chemists can seek certification through the Marine Chemist Qualification Board. The latter is administered by the National Fire Protection Association. Some of the Certified Marine Chemists also are Certified Industrial Hygienists and Certified Safety Professionals.)

CONFINED HAZARDOUS ATMOSPHERES: MAGNITUDE OF THE PROBLEM

In 1979 NIOSH published a systematic analysis about causal factors associated with fatal accidents in confined spaces in general industry (NIOSH 1979). These data provided evidence arguing for significant involvement by the industrial hygiene profession in this area. Data describing the development and control of hazardous conditions in confined spaces still are not readily available.

In beginning an examination of this subject, one needs to comprehend the full meaning and implications of the term "confined space." To the uninitiated, a confined space perhaps suggests a structure that limits access or egress, or possibly restricts mobility. While correct as far as it goes for some circumstances, this notion is overly restrictive, because it focuses exclusively on the relationship of the person to the structure. It ignores the relationship of atmospheric or environmental conditions to the structure. It also fails to acknowledge the possible presence of physical, chemical, biological, ergonomic, mechanical, process, and other hazards. At least 36 hazardous conditions can exist in confined spaces.

A point needing considerable emphasis here is that the purpose for many structures or much equipment labeled as confined spaces is to contain a process or condition. Entry was neither presumed nor intended by designers of these structures or manufacturers of this equipment. Entry, therefore, can defeat the purpose for containment by enabling contact with the process or condition. An implication of the function of the structure as a barrier is that conditions in confined spaces can be considerably different than those encountered in normal workspaces. These conditions can develop prior to opening the structure, during preparation for entry, or as a result of work activity.

No definition for the term, confined space, is yet universally accepted. This lack of agreement has impeded action to address hazards associated with these workspaces. The meaning attached to this term varies according to jurisdiction through standards and legislation. (See Appendix A for further information on standards and guidelines on confined spaces.) Another factor is that the meaning attached to confined space appears to have undergone evolutionary change.

The Division of Safety Research of the National Institute for Occupational Safety and Health (NIOSH) commissioned the first quantitative analysis of accidents that occurred in confined spaces (NIOSH 1978). Analysis of the circumstances surrounding accidents occurring in confined spaces was one purpose for the NIOSH-sponsored study. Another was to identify factors common to these events. The NIOSH definition for the term confined space contained the following elements:

- Limited openings for entry and exit by design
- Unfavorable natural ventilation that could contain or produce dangerous air contaminants
- Not intended for continuous employee occupancy

This definition limits the focus of concern almost exclusively to atmospheric hazards.

Data used in the study were developed from accident reports spanning the period 1974 to 1977. These reports documented events involving confined spaces that led to medical attention and lost-time injury or death. Associated with the 276 events selected for the study were 234 injuries and 193 fatalities. The ratios of serious injuries per accident and fatalities per accident were 0.8:1 and 0.7:1, respectively. The ratio of serious lost-time injuries per fatality was 1.2:1. Accidents occurring in confined spaces were considerably more severe than those occurring in other work environments.

Of particular interest to industrial hygienists was the role of hazardous atmospheric conditions in these accidents. Toxic and asphyxiating (oxygen-deficient) atmospheres were the leading cause of the accidents examined during this study. The next ranking cause was explosion/fire. All of the accidents attributed to these causes involved hazardous atmospheric conditions. No single causative agent dominated across industry. Hazardous atmospheric conditions were the leading cause in 23 of the 46 industrial classes. In 65% of the events examined during this study, the hazardous atmosphere existed prior to entry. In the remaining 35%, it developed after entry occurred. In the latter cases, work activity or accidental entry of gas produced the hazardous atmosphere.

One of the more important outcomes from the study was identification of the serious consequences of attempts to perform rescue. Would-be rescuers were injured or killed in 76% of attempts. The ratio of injuries per person rescued was 2.6:1 and 6:1 for deaths per person rescued. These statistics illustrate the tragic consequences that have arisen from the urge to assist someone in distress. The rescue attempt often put the would-be rescuer at far greater risk than the original victim. The study identified lack of preparedness for rescue as an extremely serious oversight.

More recently, the Occupational Safety and Health Administration (OSHA) published a series of analytical reports on fatal accidents. Each report focused on a specific type of workspace or work environment and provided a summary for each accident that was used in the investigation. The themes of several of the reports — fires and explosions, lockout and tagout, grain handling, toxic and asphyxiating atmospheres in confined spaces, welding and cutting, and shipbuilding and repair — are relevant for further discussion (OSHA 1982a, b, 1983, 1985, 1988, 1990).

The definition for confined workspace used by OSHA in these investigations contained the following criteria:

- An enclosed or partially enclosed workspace
- Limited means of entry and exit
- Subject to accumulation of toxic or flammable contaminants
- May develop an oxygen deficiency
- Not intended for continuous employee occupancy

This definition, although differing slightly from that utilized by NIOSH, still focused on atmospheric hazards. Neither definition referred to the breadth of hazardous conditions that exist or can develop in confined spaces.

The first OSHA report examined the relationship between fatal accidents and fires and explosions in confined workspaces (OSHA 1982a). Associated with the 50 accidents selected for the study during the period 1974 to 1979 were 76 fatal injuries. Heat and the rapid release of energy generated by the fire and/or explosion were the major causative agents. Multiple fatalities occurred in many of these accidents. Actions of entrants and faults in equipment provided the source of ignition in most cases.

The second OSHA report examined fatal accidents involving energized equipment (OSHA 1982b). The study grouped the accidents by causative agent: machines and conveyors, vehicles and equipment, and electrical equipment. Associated with the 83 cases selected for the study during the period 1974 to 1980 were 83 fatalities. Although the involvement of confined spaces was not considered as a factor during this study, many of the workspaces would satisfy current generally recognized criteria for inclusion.

The OSHA study of grain handling reported on 105 fatal accidents that occurred during the period 1977 to 1981 (OSHA 1983). Associated with these accidents were 126 fatal injuries. Although the involvement of confined spaces was not considered as a factor during this study, many of the workspaces would satisfy current, generally recognized criteria for inclusion.

The fourth OSHA report examined the relationship between fatal accidents, confined workspaces, and toxic and asphyxiating (oxygen-deficient) atmospheres (OSHA 1985). Associated with the 122 accidents selected for the study during the period 1974 to 1982 were 173 fatalities. A hazardous atmosphere was a causative factor in the vast majority of the accidents. Multiple fatalities occurred in many accidents. Many of the victims were would-be rescuers. In several cases the only victim was the would-be rescuer. The report emphasized that would-be rescuers were ill-prepared to carry out this task. Events documented in this report again illustrate the powerful urge to assist someone in distress, regardless of consequence to the intervenor.

The fifth report examined the relationship between fatal accidents and welding and cutting (OSHA 1988). Associated with the 217 accidents selected for the study during the period 1974 to 1985 were 262 fatal injuries. Ignition of flammable substances caused 48% of the accidents. The fuel sources in most cases were contents in containers on which the work occurred. In 22% of the accidents, the welder was working inside a workspace that could be considered a confined space. Misuse of oxygen occurred in 1% of the accidents.

The OSHA study of shipbuilding and repairing reported on 151 fatal accidents that occurred during the period 1974 to 1984 (OSHA 1990). Associated with these accidents were 176 fatal injuries. Ignition of flammable substances caused 12% of the accidents. The fuel sources in most cases were the contents of compartments and containers and mists from spray painting or coating. In 36% of the accidents, the victim was working inside a workspace that could be considered a confined space. Misuse of oxygen occurred in 2% of the accidents.

The authors of the OSHA reports repeatedly lamented about the lack of completeness of the database used in these studies. This occurred because the OSHA system for coding causal factors in accidents contained no reference to the parameters under study. Also, data were not available from all of the U.S. states. The authors also stated their belief that the accidents selected for inclusion in the study represented only a fraction of those that actually occurred. Despite these shortcomings, the summaries from individual accidents contained in these reports represent a valuable resource for further research.

NIOSH initiated the Fatality Assessment and Control Evaluation (FACE) in 1982 to identify factors that increase the risk of work-related fatal injury (NIOSH 1994). While previous studies utilized only written reports or forms as the source of information, FACE incorporated on-site visits. These visits provided the opportunity for firsthand investigation, including interviews with witnesses and observation of worksites. FACE also used a team approach. The intent of the team approach was to provide continuity and consistency from one investigation to another. This also afforded the opportunity for in-depth analysis and more precise reconstruction of accidents. FACE currently focuses on accidents involving falls from heights, contact with sources of electrical energy, entry into confined spaces, and contact with machinery.

Summaries of the FACE investigations have appeared in Operations Forum and the NIOSHTIC electronic database. These and other sources led NIOSH to estimate that approximately 200 persons in the U.S. die annually due to accidents in confined spaces (Reese and Mills 1986).

NIOSH (1994) recently published a summary of work by FACE. The FACE report provides an update of trends that are occurring. FACE investigated 423 accidents involving the deaths of 480 workers from 1983 to 1993. Of these, 70 involved confined spaces and caused 109 deaths. In 25 of these accidents, multiple deaths occurred. In performing its investigations, NIOSH utilized the definition of confined space mentioned previously. This definition focuses solely on atmospheric hazards.

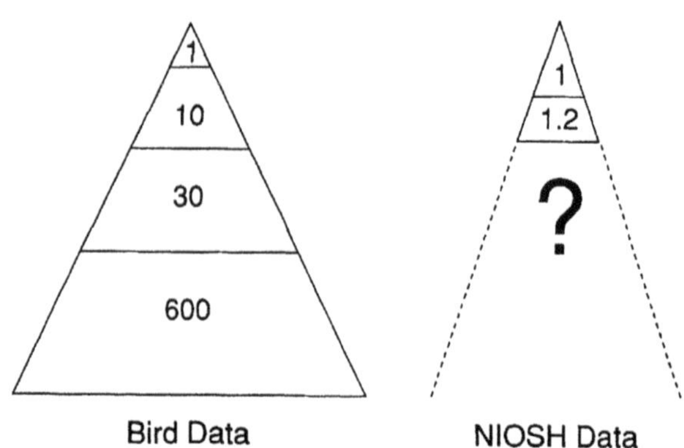

FIGURE 1.1 The incident/accident triangle applied to confined spaces. This figure compares the data from the Bird study of accidents in general industry to those involving confined spaces. The Bird study summarized the data in the ratios 1:10:30:600; that is, one serious or major injury: 10 minor injuries: 30 property damage accidents: 600 incidents involving no visible injury or damage. The corresponding data for incidents and minor accidents involving confined spaces are not available. (Adapted from Bird and Germain, 1990.)

Major findings of the FACE report parallel those reported previously by OSHA. Most of the victims failed to consider the workspace a confined space and underestimated its hazards. Atmospheric testing did not occur. Procedures for entry were not available or were not used at a large proportion of the sites. In many situations the procedures were ineffectual or had no practical meaning. Compounding the situation was the failure of the entrants to follow procedures.

A compilation of accident summaries in mining was prepared by the Mine Safety and Health Administration (MSHA) (MSHA 1988). MSHA documented 44 fatalities from 38 accidents that occurred during the period 1980 to 1986. MSHA (1994) subsequently reported that it had recorded 57 fatalities in accidents involving confined spaces at the end of 1993. MSHA noted that the majority of the fatalities resulted from the collapse of bridged and caked material in bins, silos, and hoppers.

Additional information concerning events that occur in confined spaces is not readily available. This lack of information arises from limitations in reporting systems. Normally, reporting systems address only serious accidents, those that cause lost-time injury or fatality. Not reported or made available for public review are the many nonserious incidents and so-called "near misses" that also occur. Taken together, incidents and minor and serious accidents comprise a spectrum. At the one extreme is the most insignificant of incidents. At the other is the fatal accident. Incidents and accidents of varying severity span the gap between these extremes. Most events that comprise this spectrum are nonserious.

The classic accident severity model cited in loss prevention texts was formulated from an extensive study of incidents and accidents reported by industry. For every reported serious or disabling injury, this model predicts 10 minor injuries, 30 property damage accidents, and 600 incidents having no visible injury or damage (Bird 1990). The data provided in the Bird model provide the basis for comparison with NIOSH-sponsored research mentioned previously (NIOSH 1978). According to this research, serious, lost-time accidents compared to fatalities occur in the ratio 1.2:1. This ratio is considerably smaller than the value (10:1) utilized in the Bird model for accident occurrence in general industry. The decrease in the ratio suggests a considerable narrowing of the triangle for accidents involving confined spaces. This would suggest that the tolerance between a minor incident and major accident involving a confined space is considerably smaller than for those occurring in general industry. Thus, the margin of safety between an event, a serious incident, and a near-miss or fatal accident that occurs in a confined space could be very small.

The fatal accident summaries in the OSHA, MSHA, and NIOSH reports provide the starting point for obtaining additional information about these accidents (OSHA 1982a, b, 1983, 1985,

TABLE 1.1
Summary of Data Presented in OSHA and MSHA Reports

Action/Condition	Period	Accidents	Fatalities	F/A
Grain handling	1977–1981	105	126	1.2
Lockout/tagout	1974–1980	83	83	1.0
Fire/explosion	1974–1979	50	76	1.5
Toxic/asphyxiating	1974–1982	122	173	1.4
Welding/cutting	1974–1985	217	262	1.2
Shipbuilding/repair	1974–1984	151	176	1.2
Mining	1980–1986	38	44	1.2

1988, 1990; MSHA 1988; NIOSH 1994). Table 1.1 summarizes the data presented in the OSHA and MSHA reports. The NIOSH report is not included at this point as the format is different.

A note of caution is needed before entering into overly enthusiastic use of these data. The OSHA reports spanned partially overlapping time-frames of varying length. Some of the accidents were utilized in more than one report: for example, welding and cutting and fire and explosion accidents in shipbuilding and repair. This is a natural outcome from the overlap between workspace and work environment specificity. Further, not all of the accidents believed to have occurred were available to the researchers who created these documents. That said, this is the extent of the information base that exists from which any analysis can occur.

Accidents in categories such as fire and explosion are focused by event, compared to those in shipbuilding and repair that report on one segment of industry. While accidents due to welding and cutting, fire and explosion, and toxic and asphyxiating atmospheres occurred in shipbuilding and repair, so also did other types. Similarly, while accidents due to welding and cutting, fire and explosion, and toxic and asphyxiating atmospheres occurred during grain handling, so also did accidents due to lockout and engulfment.

The ratio of fatalities to accidents provides a measure of the severity of accidents occurring in each category. The ratio of fatalities to accidents exceeded unity in all categories except lockout and tagout. That is, more than one fatal injury occurred during many of these accidents.

OSHA examined the role of confined spaces in only some of the reports listed in Table 1.1. Many of the workspaces described in the remaining reports also would fit current generally accepted criteria for confined spaces. In other situations the accident involved a hazardous atmosphere confined within a structure during work on external surfaces.

Table 1.2 illustrates the potential role of confined spaces in the OSHA and MSHA reports (OSHA 1982a, b, 1983, 1985, 1988, 1990). In order to make this evaluation, classification of the workspaces was performed according to current generally accepted criteria for confined spaces. This classification reflects the interpretation of information provided in the anecdotal summary for each accident. To the extent possible in this and subsequent analysis, the use of accident summaries that appeared in more than one report was minimized. The most prominent use of data from other sources occurred in the report on shipbuilding and repair.

According to this analysis, confined workspaces and/or confined hazardous atmospheres were associated with fatal accidents contained in each of the OSHA reports. The smallest association occurred in welding and cutting.

The ratio of fatalities to accidents increased in grain handling and accidents involving toxic and asphyxiating atmospheres compared to values presented in Table 1.1. This would suggest an increase in severity for accidents occurring in confined spaces compared to the overall group.

Data presented in the preceding tables indicate the relative involvement of confined spaces and/or confined hazardous atmospheres in these accidents. Of greater interest in the context of this

TABLE 1.2
Role of Confined Spaces/Confined Atmospheres in Fatal Accidents

	Accidents			Fatalities			
Action/condition	A_C	A_T	%	F_C	F_T	%	F/A
Grain handling[a]	58	105	55	78	126	62	1.3
Lockout/tagout[a]	29	83	35	29	83	35	1.0
Fire/explosion[a]	50	50	100	76	76	100	1.5
Toxic[a]	55	55	100	80	80	100	1.5
Asphyxiating[a]	46	46	100	71	71	100	1.5
Other[a]	19	21	90	19	22	86	1.0
Total[b] (T + A + O)	120	122	98	170	173	98	1.4
Welding/cutting[c]							
Interior	36	164	22	44	190	23	1.2
Exterior	58	164	35	72	190	38	1.2
Shipbuilding/repair[a,d]	31	31	100	36	36	100	1.2
Mining[a]	26	38	68	32	44	73	1.2

Note: A_C is fatal accidents associated with confined spaces/confined atmospheres; A_T is total accidents; F_C is fatalities associated with confined spaces/confined atmospheres; F_T is total fatalities.

[a] Data reflect review of summaries of individual accidents.

[b] Total is the sum of Toxic + Asphyxiating + Other.

[c] The data on welding and cutting accidents used in this table reflect analysis of individual accident summaries contained in the OSHA (1988) report. The report claims that this abridged group (164 cases and 190 associated fatalities) reflects the total (217 cases and 262 associated fatalities).

[d] Accident summaries utilized in other reports are not used in these statistics.

chapter are those accidents associated only with hazardous atmospheric conditions. Table 1.3 indicates the role of hazardous atmospheric conditions in accidents provided in the OSHA and MSHA reports (OSHA 1982a, b, 1983, 1985, 1988, 1990; MSHA 1988).

The first section of Table 1.3 presents the number of accidents and fatalities associated with hazardous atmospheric conditions.

Accidents associated with hazardous atmospheric conditions occurred in all categories except lockout and tagout and mining. The most severe accidents, expressed as the ratio of fatalities to accidents, occurred during grain handling. Most of these accidents involved fires and explosions. Fires and explosions and accidents involving toxic and asphyxiation hazards ranked next in severity. Welding and cutting accidents were the least severe.

Percent occurrence indicates the relative importance of hazardous atmospheric conditions to the total number of accidents associated with confined workspaces and/or confined atmospheres. Adverse atmospheric conditions are important in a minor fraction of grain handling accidents. Although they are a minor fraction of the accidents, these caused half of the fatalities associated with confined spaces in this category.

The OSHA and MSHA reports utilized in creating Tables 1.1 to 1.3 contain overlapping time frames of different lengths. This format prevents direct comparison between categories. Table 1.4 presents the data normalized to a time frame of 10 years. This technique necessitates extrapolation

TABLE 1.3
Role of Hazardous Atmospheric Conditions in Fatal Accidents

Action/Condition	Accidents	Fatalities	F/A	Percent occurrence A	Percent occurrence F
Grain handling	20	39	2.0	34	50
Lockout/tagout	0	0	0.0	0	0
Fire/explosion	44	67	1.5	88	88
Toxic	55	80	1.5	100	100
Asphyxiating	46	71	1.5	100	100
Other	0	0	0.0	0	0
Welding/cutting[a]					
Interior	24	30	1.3	67	68
Exterior	58	72	1.2	100	100
Shipbuilding/repair[b]	13	18	1.4	42	50
Mining	0	0	0.0	0	0

Note: A is fatal accidents associated with hazardous atmospheric conditions; F is fatalities occurring during these accidents.

[a] The data on welding and cutting accidents used in this table reflect analysis of individual accident summaries contained in the OSHA (1988) report. The report claims that this abridged group (164 cases and 190 associated fatalities) reflects the total (217 cases and 262 associated fatalities).

[b] Means accident summaries repeated in other reports are not used in these statistics.

of data contained in most categories and interpolation in the case of welding and cutting and shipbuilding and repair. Of necessity this comparison will be imperfect since the OSHA report on welding and cutting provided an incomplete group of accident summaries.

Assuming the validity of these calculations, hazardous atmospheric conditions associated with toxic and asphyxiating atmospheres would be involved in the greatest number of both accidents and fatalities. The smallest number of accidents and fatalities would occur in shipbuilding and repair. This results from the fact that these accidents are covered under the other categories.

The relative importance of other hazards of work in confined spaces becomes apparent from data presented in the second group of columns in Table 1.4. These include engulfment by bulk materials and entanglement by equipment and other hazardous conditions. An examination of these data forms the focus of the next chapter. The importance of hazardous atmospheric conditions during grain handling diminishes considerably compared to other hazards that occur in confined spaces.

The third group of columns in Table 1.4 provides the projection for all accidents in the category. According to these calculations, the number of fatal accidents and injuries associated with grain handling clearly would exceed those in the other categories. The next most hazardous activity would be welding and cutting. The least hazardous category from this perspective would be mining.

The dominance of hazardous atmospheric conditions as a causative agent in these tables is apparent. Of the 542 projected fatal accidents involving confined workspaces and/or confined atmospheres, hazardous atmospheric conditions would be a causative agent in 327 or 60%. Excluding exterior welding and cutting, hazardous atmospheric conditions would be a factor in 263 of the 478 or 55% of the fatal accidents occurring in confined spaces.

TABLE 1.4
Data Projected from OSHA Reports of Fatal Accidents

	Projected number					
	Hazardous atmosphere		Confined spaces		All causes	
Action/condition	A	F	A	F	A	F
Grain handling	40	78	116	156	210	252
Lockout/tagout	0	0	41	41	119	119
Fire/explosion	73	112	83	127	83	127
Toxic	61	89	61	89		
Asphyxiating	51	79	51	79		
Other	0	0	21	21		
Total[a] (T + A + O)	112	168	133	189	136	192
Welding/cutting[b]						
Interior	26	34	40	51		
Exterior	64	83	64	83		
Total	90	117	104	134	217	262
Shipbuilding/repair	12	16	28	33	137	160
Mining	0	0	37	46	54	63
Totals	**327**	**491**	**542**	**726**	**956**	**1,175**

Note: A is fatal accidents; F is fatalities occurring during these accidents.

[a] Total is the sum of Toxic + Asphyxiating + Other.

[b] The data on welding and cutting accidents used in this table reflect analysis of individual accident summaries contained in the OSHA (1988) report. The report claims that this abridged group (164 cases and 190 associated fatalities) reflects the total (217 cases and 262 associated fatalities).

At least 41 persons die annually in the U.S. in confined spaces through hazardous atmospheric conditions according to these calculations. At least 49 persons die annually when welding on the exterior of structures and containers confining a hazardous atmosphere is included. At least 64 persons die in accidents involving confined spaces. The estimates provided here are consistent with those obtained by other investigators (NIOSH 1994). The fact that the NIOSH data are so similar to those obtainable from the OSHA reports suggests that the fatality rate has not changed in the intervening years.

The NIOSH report on fatal accidents also introduced statistics from the National Traumatic Occupational Fatalities (NTOF) database (NIOSH 1994). The NTOF database provides comprehensive data on fatal injuries arising from work in confined spaces, as well as other situations. Cause is assigned from the death certificate. Table 1.5 provides summary statistics for confined spaces for the decade 1980 to 1989, from the NTOF data.

In creating these statistics, NIOSH stressed the incompleteness of the information base from which they were compiled. Comparison with the data in Table 1.4 suggests that a decrease in fatalities from accidents involving atmospheric hazards has occurred. Further, the accidents have become less severe. The severity in Table 1.4 for accidents occurring in confined spaces is 1.3 fatality per accident. (This excludes welding and cutting on structures that confine a hazardous atmosphere.) The severity in Table 1.5 is 1.1 fatality per accident.

TABLE 1.5
NTOF Statistics on Confined Spaces for 1980 to 1989

Statistic	Accidents	Fatalities
Atmospheric hazards		373
Engulfment in loose materials		227
Accidents with other causes (drowning, engulfment in sludge and manure)		70
Accidents in confined spaces	585	670
Accidents with 1 victim	428	
Accidents with 2 victims	61	
Accidents with 3 victims	9	
Accidents with 4 victims	2	
Accidents involving cave-ins (trenches, excavations, ditches)	572	606

DEVELOPMENT OF HAZARDOUS ATMOSPHERIC CONDITIONS

The hazardous atmospheric condition could exist prior to opening the space for entry or it could develop during the performance of work activity. This aspect could have major importance in determining the strategy for protective action. The temporal relationship between development of the atmospheric hazard and onset of the accident, therefore, is an important one.

The NIOSH-sponsored report on confined space accidents (1978) indicated that the hazardous condition existed prior to entry in 65% of events. The OSHA reports (1982a, 1985, 1988) provide an additional opportunity to investigate this question. The updated report on fatal accidents involving confined spaces published by NIOSH in 1994 also provides accident summaries from which additional information can be obtained (NIOSH 1994).

For the purposes of this analysis, an atmospheric hazard existing prior to the start of work activity qualifies as a preexisting hazard. An atmospheric hazard produced by the actions of someone in the space at the onset of the accident qualifies as work activity. Table 1.6 summarizes results of this examination. The data presented here apply only to accidents associated with hazardous atmospheric conditions. The data reflect subjective estimation of the situation described in the accident summary.

Accidents in the first three categories in the table (toxic and asphyxiating atmospheres, and fires and explosions) occurred inside the confined workspace. Accidents associated with welding and cutting occurred inside or outside, as noted. Some accidents occurred following reentry. Previous work created the hazardous condition.

Preexisting conditions were an important factor in accidents involving toxic and asphyxiating atmospheres and exterior welding and cutting. Work activity was important only in fires and explosions and welding and cutting that occurred inside the confined space.

Applying these percentages to the numerical projections presented in Table 1.4 produces a weighted average of 73%. That is, in 73% of all accidents involving hazardous atmospheric conditions, the hazardous condition probably would exist prior to the start of work.

The FACE data in the NIOSH report suggests the same trends as the OSHA data. The FACE data strongly suggest that recognition of the potential danger inherent in these situations still is not occurring.

TABLE 1.6
Temporal Development of Hazardous Atmospheric Conditions

Condition	Percent of fatal accidents				
	TA	AA	F/E	W/C[a]	
				Int	Ext
OSHA data					
Preexisting	84	76	48	46	100
Work activity	16	24	52	54	0
NIOSH (FACE) data					
Preexisting	70	100	Insufficient information		
Work activity	30	0			

Note: TA is toxic atmosphere; AA is asphyxiating (oxygen-deficient) atmosphere; F/E is flammable or explosive atmosphere; W/C is welding and cutting; Int is work occurring inside the confined space; and Ext is work occurring on the exterior of a structure or container.

[a] The data on welding and cutting accidents used in this table reflect analysis of accident summaries contained in the OSHA report (1988). The report claims that this abridged group (164 cases and 190 associated fatalities) reflects the total (217 cases and 262 associated fatalities).

NATURE OF ADVERSE ATMOSPHERIC CONDITIONS

The preceding section illustrated the importance of preexisting conditions as causative agents in accidents associated with confined workspaces and confined atmospheres. The next important aspect of this problem is the nature of the hazardous atmosphere.

Again, the accident summaries contained in the OSHA and NIOSH reports (OSHA 1982a, 1985, 1988; NIOSH 1994) provide an important starting point for exploring this subject. The information base for this enquiry is severely limited, since it originated from post hoc analysis and interpretation of preexisting conditions.

This lack of information in the OSHA reports is understandable, considering the difficult circumstances under which it was obtained. The criteria used to designate the hazardous agent do not appear to be consistent. Some overlap between toxic and asphyxiating atmospheres appears to have occurred. However, no attempt was made here to reclassify the individual accidents.

The FACE summaries in the NIOSH report provided considerably more reliable information about atmospheric conditions at the time of the accident. This resulted from the availability of real-time analytical instruments to emergency response personnel and accident investigators.

Table 1.7 summarizes data contained in the OSHA and NIOSH reports.

The data presented in Table 1.7 suggest that limited classes of contaminants are involved in this type of accident. Contaminants not assigned to a class were associated with a small proportion of the accidents. In the case of the NOTF data, oxygen deficiency was attributable to methane and inert gases, for a total of 36% of the fatalities.

An interesting feature of the data is the similarity between fires and explosions and welding and cutting accidents. Fires and explosions occurred inside the confined workspace, while welding and cutting accidents predominantly occurred on the exterior of a container or structure that confined

TABLE 1.7
Contaminants of Hazardous Atmospheres

Contaminant	Percent of Accidents	Percent of Fatalities
OSHA Data		
Toxic atmospheres[a]	54	55
Hydrogen sulfide		38
Carbon monoxide		15
Chlorinated solvent vapors		16
Organic solvent vapors		12
Fuel vapors		7
Undetermined/other		12
Asphyxiating atmospheres[a]	46	45
Oxygen deficiency		25
Nitrogen/process gas displacement		37
Fuel gas displacement		31
Welding gases		3
Other		4
Fire/explosion[b]		
Organic solvent vapors	32	25
Fuel vapors	30	37
Welding gases	23	21
Natural gas	7	12
Other	8	5
Welding/cutting[c]		
Organic solvent vapors	33	29
Fuel vapors	39	40
Welding gases	20	21
Natural gas	1	3
Other	7	7
NIOSH (FACE) Data		
Toxic atmospheres[b]	45	52
Hydrogen sulfide	8	8
Carbon monoxide	8	10
Chlorinated solvent vapors	8	7
Organic solvent vapors	8	7
Fuel vapors	—	—
Undetermined/other	13	20
Asphyxiating atmospheres[b]	49	42
Oxygen deficiency	37	34
Nitrogen/process gas displacement	8	4
Fuel gas displacement	4	4
Welding gases	—	—
Other	—	—
NIOSH (NOTF) Data		**100**
Hydrogen sulfide		14
Carbon monoxide		7
Methane		10
Inert gases		9
Sewer gases		7
Oxygen deficiency		17

TABLE 1.7 (continued)
Contaminants of Hazardous Atmospheres

Contaminant	Percent of Accidents	Percent of Fatalities
Not specified		20
Other		16

[a] The data in this section originated in summary tables provided in the OSHA report (1985). The potential complexity of the contaminated atmospheres and the lack of analytical data on composition prevented further analysis.
[b] The data used in this section reflect analysis of individual accident summaries contained in the report (NIOSH 1994).
[c] The data on welding and cutting used in this table reflect analysis of individual accident summaries contained in the OSHA report (1988). The report claims that this abridged group (164 cases and 190 associated fatalities) reflects the total (217 cases and 262 associated fatalities).

a hazardous atmosphere. These data argue strongly for inclusion of confined hazardous atmospheres within the domain of consideration of the term "confined space." Including confined hazardous atmospheres within the domain of confined spaces adds a dimension for addressing these accidents that is conspicuous by its absence.

The complexity of composition of these atmospheres imposes an obvious limitation on the analysis undertaken. Accident summaries contained in the OSHA reports provided little analytical data to substantiate judgments made about atmospheric composition. In fact, the classification assigned for several of the accidents appears to contradict the anecdotal information. Identifying individual atmospheric contaminants would have required investigative testing. The lack of testing is most unfortunate, although understandable, given the circumstances.

ETIOLOGY OF ACCIDENTS INVOLVING HAZARDOUS ATMOSPHERIC CONDITIONS

Accidents associated with confined spaces and confined atmospheres are complex events. They include elements from several areas: technical, procedural, and social. The technical and procedural elements, and the challenges that they pose, are more readily apparent than the social ones.

The focus of the remainder of this chapter is to develop information about the fundamental elements that comprise these accidents. This information could provide the basis for a model to describe the etiology of these accidents and to identify and highlight areas of deficiency. Identification of deficiencies could provide the basis for strategies for ameliorating the situation. This information will form the basis of the approach taken in subsequent chapters.

TECHNICAL ELEMENTS

Technical elements refer to the underlying aspects of the situation and the environment. Technical elements examined here include:

- Temporal aspects
- Characteristics of the space
- Conditions at time of entry

The summaries of individual accidents contained in the OSHA reports (OSHA 1982a, 1985, 1988), provide comprehensive data about these aspects of the accidents. The NIOSH report (1994) on fatal accidents in confined spaces provides supplemental data on this subject. This will be utilized, where beneficial to the discussion. The data used in the following tables pertain only to accidents involving atmospheric hazards.

Temporal Aspects

Temporal aspects refer to the time of day, day of the week, and month of the year. Table 1.8 shows occurrence of the accidents as a function of time of day. Initial inspection of the data indicated that times recorded by investigators tended to be hourly or half-hourly. Tabularizing the data according to conventional record points such as the hour or half hour could introduce a bias. Accordingly, this table presents data in half-hour intervals that begin and end on the quarter hour.

The data suggest several strong trends about the temporal occurrence of fatal accidents associated with hazardous atmospheric conditions. Most of the accidents occur during the day shift (normally 8:00 to 16:00). More of the remaining accidents occur during the afternoon shift (normally 16:00 to 24:00) than the night shift (normally 00:00 to 8:00). Of course, the data also reflect the relative occurrence of activity within the different shifts.

Establishing the reason for the trend is a point of minor importance. The fact that the vast majority of these accidents occurs during the normal workday makes these situations more amenable to correction. That is, the organizational resources capable of addressing the hazards represented by these situations are available during the hours of greatest need.

The data show no consistency in occurrence of these accidents during the day. Accidents involving work inside confined spaces (TA and AA) are more likely to occur during the afternoon. Accidents involving exposure to toxic atmospheres are most likely to occur during midafternoon (13:45 to 15:14). Accidents involving exposure to asphyxiating (oxygen-deficient) atmospheres are most likely to occur during late afternoon (15:15 to 16:14). Fires and explosions (F/E and W/C) are more likely to occur in the morning. Accidents occurring inside the confined space (F/E) are most likely to occur during midmorning (9:15 to 10:44). Welding and cutting accidents are most likely to occur during mid- to late morning (10:15 to 11:44). Most of the welding and cutting accidents occurred external to a structure confining a hazardous atmosphere.

The second temporal factor in occurrence of these accidents is the day of the week. Table 1.9 summarizes the data provided in the summaries of individual accidents.

The data indicate that most of the fatal accidents occur during the normal workweek, Monday to Friday. The proportion of accidents occurring on the weekend ranges from 8 to 23%. The weighted average for weekends is about 17%. Fatal accidents involving toxic and asphyxiating atmospheres and fires and explosions peak during midweek. Welding and cutting accidents peak on Friday. There is no apparent reason for these trends. Of course, these findings also reflect the level of activity that occurs in the different days of the workweek.

Again, the reason for the trend is a point of minor importance. The fact that the vast majority of the accidents occurs during the normal workweek should make these situations more amenable to correction.

The third temporal factor is month of the year. The climate in most parts of the U.S. (the origin of the data) undergoes distinct seasonal variations. Many of the fatal accidents documented in the OSHA reports occurred in outdoor locations, and therefore were potentially subject to this influence (OSHA 1982a, 1985, 1988). Climatic factors, such as extreme cold and heat, could have influenced the occurrence and outcome of some of these situations. However, summaries of individual accidents did not always clearly indicate location.

Table 1.10 summarizes occurrence of these accidents by month of the year.

TABLE 1.8
Temporal Occurrence of Fatal Accidents: Time of Day

	Percent of fatal accidents			
Time interval	**TA**	**AA**	**F/E**	**W/C**[a]
00:15–00:44				
00:45–1:14				
1:15–1:44			2	
1:45–1:14				
2:15–2:44				
2:45–3:14				
3:15–3:44		2	3	
3:45–4:14				
4:15–4:44	2			
4:45–5:14	2	3		1
5:15–5:44		5		
5:45–6:14			2	2
6:15–6:44	2	2		
6:45–7:14				
7:15–7:44	2		5	2
7:45–8:14	4			
Total for shift	**12**	**12**	**12**	**5**
8:15–8:44	7	4		1
8:45–9:14	6	7	3	
9:15–9:44	2	6	12	3
9:45–10:14	4	3	9	7
10:15–10:44	7	3	12	9
10:45–11:14		2		6
11:15–11:44	2	2		11
11:45–12:14	4	2	2	3
Halfshift total	**26**	**28**	**42**	**43**
12:15–12:44	4	6		5
12:45–13:14	2	8	9	9
13:15–13:44	4	6	2	6
13:45–14:14	7		5	5
14:15–14:44	9		7	3
14:45–15:14	5		3	3
15:15–15:44		13	2	2
15:45–16:14	4	13	5	1
Total for shift	**61**	**74**	**75**	**77**
16:15–16:44	7	3	2	2
16:45–17:14	4	3	2	2
17:15–17:44	4	2		1
17:45–18:14	5	4		1
18:15–18:44				
18:45–19:14	2			
19:15–19:44	2			
19:45–20:14	1			
20:15–20:44			3	
20:45–21:14			2	1
21:15–21:44			2	1
21:45–22:14				1

TABLE 1.8 (continued)
Temporal Occurrence of Fatal Accidents: Time of Day

Time interval	Percent of fatal accidents			
	TA	AA	F/E	W/C[a]
22:15–22:44				1
22:45–23:14	2		2	
23:15–23:44	1			
23:45–00:14		2		7
Total for shift	**27**	**14**	**13**	**18**

Note: TA is toxic atmosphere; AA is asphyxiating (oxygen-deficient) atmosphere; F/E is flammable or explosive atmosphere; W/C is welding and cutting.

[a] The data on welding and cutting accidents used in this table reflect analysis of accident summaries contained in the OSHA report (1988). The report claims that this abridged group (164 cases and 190 associated fatalities) reflects the total (217 cases and 262 associated fatalities).

TABLE 1.9
Temporal Occurrence of Fatal Accidents: Day of the Week

Day	Percent of fatal accidents			
	TA	AA	F/E	W/C[a]
Sunday	6	13	—	3
Monday	13	10	17	14
Tuesday	20	27	17	14
Wednesday	15	23	26	19
Thursday	19	13	14	14
Friday	17	8	19	25
Saturday	10	6	7	11
Workweek	**84**	**81**	**93**	**86**
Weekend	**16**	**19**	**7**	**14**

Note: TA is toxic atmosphere; AA is asphyxiating (oxygen-deficient) atmosphere; F/E is flammable or explosive atmosphere; and W/C is welding and cutting.

[a] The data on welding and cutting accidents used in this table reflect analysis of accident summaries contained in the OSHA report (OSHA 1988). The report claims that this abridged group (164 cases and 190 associated fatalities) reflects the total (217 cases and 262 associated fatalities).

Data from OSHA 1982a, 1985, and 1988.

TABLE 1.10
Temporal Occurrence of Fatal Accidents: Month of the Year

Month	Percent of fatal accidents			
	TA	AA	F/E	W/C[a]
January	13	2	12	5
February	5	4	12	13
March	5	4	9	3
April	9	17	5	14
May	7	13	5	8
June	19	13	5	7
July	15	15	—	8
August	5	8	6	12
September	5	4	14	5
October	7	10	19	10
November	3	6	6	10
December	7	4	7	5

Note: TA is toxic atmosphere; AA is asphyxiating (oxygen-deficient) atmosphere; F/E is flammable or explosive atmosphere; and W/C is welding and cutting.

[a] The data on welding and cutting accidents used in this table reflect analysis of accident summaries contained in the OSHA report (1988). The report claims that this abridged group (164 cases and 190 associated fatalities) reflects the total (217 cases and 262 associated fatalities).

Each category appears to be independent of the others. Within categories there appears to be seasonal influence. Accidents involving toxic and asphyxiating atmospheres peaked during warmer months. The lowest number occurred during the cooler months. By contrast, the greatest number of fires and explosions in confined workspaces occurred during the early autumn or the coldest months of the winter. Few of these accidents occurred during the warmer months. Welding and cutting accidents showed no pattern of occurrence.

Characteristics of the Space

Potentially important technical aspects in the etiology of these accidents are their functional and dimensional characteristics. Dimensional characteristics have not been explored previously in the literature on this subject.

Table 1.11 lists functional characteristics/work locations for confined spaces mentioned in the OSHA reports.

The data in Table 1.11 indicate that a large proportion of fatal accidents associated with confined work environments and confined atmospheres occurred in relatively few classes of workspace. While this observation is encouraging, the presence of a large proportion of unspecified "other" workspaces is cause for considerable concern.

The NTOF data provided in the NIOSH report (1994) indicate that tanks were the most common workspace for fatal accidents involving atmospheric hazards. These were followed by sewers, pits, and silos. The remaining accidents occurred across diverse structures, including: vats, wells, bins, pipes, trenches, and kilns. The occurrence of limited types of work environments creates the temptation to

TABLE 1.11
Confined Spaces Associated with Fatal Accidents

	Percent of fatal accidents			
Location	**TA**	**AA**	**F/E**	**W/C[a]**
Processing plant/animal matter	4	2		
Dip tank/degreaser	11			
Storage tank/compartment/vessel	33	47	75	8
Asphalt tanks				4
Hydrocarbon storage tanks				14
Fuel tanks				18
Transport tanks (truck, rail)				7
Chemical storage/reaction vessel				3
205-L (55 U.S. gal) drums				16
Wastewater treatment	31	21	5	
Trench			5	
Vault		6		
Fabrication				3
Other	21	24	15	27

Note: TA is toxic atmosphere; AA is asphyxiating (oxygen-deficient) atmosphere; F/E is flammable or explosive atmosphere; and W/C is welding and cutting.

[a] The data on welding and cutting accidents used in this table reflect analysis of accident summaries contained in the OSHA report (1988). The report claims that this abridged group (164 cases and 190 associated fatalities) reflects the total (217 cases and 262 associated fatalities).

Data from OSHA 1982a, 1985, and 1988.

identify these for the purposes of generating awareness and creating training materials. However, this approach fails to address the large proportion of accidents that occurred in the "other" work environments. By narrowing perspective, this approach overly restricts the process of awareness building.

Accidents associated with hazardous atmospheric conditions (TA and AA) occurred in a relatively narrow range of work environments. These included tanks of various types and workspaces associated with wastewater treatment. Diptanks and degreasing formed a small, but recognizable, category. Workspaces assigned to the "other" category formed a significant fraction of the total.

A large proportion of fires and explosions occurred in tanks, compartments, and other containers. A similar proportion of welding and cutting accidents involved tanks and containers of various types. This observation again reinforces the similarity between these types of accidents. A significant proportion of the welding and cutting accidents involved fuel tanks and 205-L (55 U.S. gal, 45 U.K. gal) drums. These containers are widely known for storing flammable and combustible liquids. Containers assigned to the "other" category also constituted a significant fraction of the workplaces.

The dimensional characteristics of confined workspaces are an important, but unknown aspect of the knowledge base about this subject. Complicating this inquiry, however, is the fact that confined spaces occur in different shapes. The only parameter common to all is volume.

Some of the accident summaries contained in the OSHA reports (1982a, 1985) provided the volume of the workspace or the basis for calculating this information. Table 1.12 summarizes information that is available.

TABLE 1.12
Volume of Confined Spaces

Volume (m^3)	Percent of fatal accidents TA	AA	F/E
0 to 4.9	31	33	—
5.0 to 9.9	17	6	13
10.0 to 19.9	17	17	—
20.0 to 49.9	7	22	38
50.0 to 99.9	10	11	25
Greater than 100	18	11	24
Accidents included in table	52	38	17

Note: TA is toxic atmosphere; AA is asphyxiating (oxygen-deficient) atmosphere; F/E is flammable or explosive atmosphere.

The data suggest that considerably more than half of the fatal accidents associated with toxic and asphyxiating atmospheres occurred in workspaces having a volume less than 20 m^3. Similarly, the data suggest that an appreciable proportion of these accidents occurred in workspaces having a volume less than 10 m^3. Accidents associated with fires and explosions occurred in larger workspaces, according to these data. While the data may reflect the true nature of these workspaces, the possibility of biases introduced by the incompleteness of the database also must be acknowledged.

Conditions Present at Time of Entry

Conditions encountered by the entrants or created during work activity are another important aspect in the etiology of these accidents. This information can provide important insight about the ability of entrants and would-be rescuers to recognize impending danger. Cues that could provide warning include visual appearance and odor.

The main resources for this enquiry are the anecdotal summaries contained in the OSHA reports (OSHA 1982a, 1985). This aspect was not pursued during the original investigation of these accidents. Extracting this information therefore requires careful reading of the original events. The data presented in the following Table 1.13 reflect subjective interpretation of the conditions as described in the accident summary. (The anecdotal comments did not always describe the exact nature of the conditions.)

The data suggest that both visual and olfactory cues about potential danger were provided to a varying extent to persons entering these workspaces. The extent to which these cues could provide warning depended heavily on the circumstances.

Visual cues were provided by materials contained in the space. These included both residual materials and materials in transit through it. The presence of materials that could provide visual cues depended on the previous history of the space.

Olfactory cues sometimes accompanied the visible materials or were provided by gases or vapors resident in the space. In many cases, the gases or vapors provided no olfactory cue, even at the hazardous concentrations that were present. The presence of detectable substances in the air at the entrance to the confined space depended on the previous history of the space.

A high percentage of the toxic atmospheres provided an indication about their presence to entrants in the form of an odor. Conversely, an even higher percentage of asphyxiating (oxygen-deficient) atmospheres provided no warning in the form of an odor to the entrants.

TABLE 1.13
Conditions in the Confined Space at Time of Entry

Condition	Percent of accidents		
	TA	AA	F/E
Visual			
Clean	31	56	68
Water/wastewater	22	15	2
Other liquid	20	9	18
Sludge	20	11	5
Other material	7	9	7
Olfactory			
Odor	69	19	43
No odor	31	81	57
Speed of Action of the Hazardous Atmosphere			
Slow acting	27	2	
Rapid acting	73	98	

Note: TA is toxic atmosphere; AA is asphyxiating (oxygen-deficient) atmosphere; and F/E is flammable or explosive atmosphere.

The rate of action of the hazardous atmosphere is another extremely important consideration. In the context of these circumstances, the only measure of this parameter is subjective. For the purposes of this investigation, the term "rapid acting" means a response that occurs up to 10 min after exposure to the hazardous atmosphere. Slow-acting responses required a timespan longer than 10 min. Owing to the circumstances, this investigation was compelled to rely on subjective information provided in the anecdotal summaries. For this reason, there is considerable potential for bias. However, the large spread between the values lends some credence to their validity.

According to this analysis, the hazardous atmospheres encountered in these situations acted rapidly. This was especially true for asphyxiating atmospheres. Although not documented in the table, many accident summaries indicated that would-be rescuers collapsed more rapidly than did the person originally overcome by the hazardous atmosphere.

As indicated by these observations, the hazards posed by asphyxiating atmospheres encountered in these environments pose a deadly combination: no warning properties and rapid action. The lack of an odor arouses no suspicion or concern about the need to pause for consideration. In addition, the hazardous atmosphere is likely to act rapidly. Atmospheres containing toxic substances are more moderate in their action. Some indicated their presence through an odor and were slower acting than oxygen-deficient atmospheres.

Procedural Elements

The process of work occurs within a framework containing technical and procedural elements, as well as social ones. The process follows a well-defined sequence. Initially, a person in authority recognizes the need to perform an activity. This person assigns responsibility to others for carrying out the required tasks in an organized manner. Workers later appear at the worksite in a predetermined sequence and perform their assigned tasks. During and upon completion of each task, the person in authority assesses the quality and completeness of the work. The preceding sequence is generic and occurs during all types of work, including activities associated with confined spaces.

The timing and conduct of activities are controlled by supervisors. Of course, any attempt to characterize the almost infinite number of combinations of activities would be pointless. The studies of fatal accidents published by OSHA (1982a, 1985, 1988) provide general insights into the

TABLE 1.14
Reasons for Entry or Action Involving Confined Spaces

	Percent of fatalities			
Reason for Entry/activity	**TA**	**AA**	**F/E**	**W/C[a]**
Normal job activity	62	64	87	82
Unusual job activity	—	—	7	2
Attempting a rescue	26	30	—	1
Unauthorized entry	5	3	6	—
Not work-related	4	1	—	2
Undetermined	4	2	—	13

Note: TA is toxic atmosphere; AA is asphyxiating (oxygen-deficient) atmosphere; F/E is flammable or explosive atmosphere; and W/C is welding or cutting.

[a] The data on welding and cutting used in this table reflect the full range of accidents (217) and fatalities (262) covered in the OSHA report. The involvement of adverse atmospheric conditions as a causative factor in these accidents is not known.

Data from OSHA 1982a, 1985, and 1988.

fundamental nature of these interactions. The NIOSH study of fatal accidents in confined spaces provides supplemental information (NIOSH 1994). Information provided in the NIOSH study will be utilized when beneficial to the discussion.

The following areas of enquiry will examine procedural elements:

- Situational aspects
- Activities associated with fatal accidents
- Underlying causes of fatal accidents
- Preventive measures

Situational Aspects

The reason for the entry is crucial to gaining an understanding about possible procedural problems. Table 1.14 summarizes information contained in the OSHA reports (OSHA 1982a, 1985, 1988). The table presents data about fatalities rather than accidents, because of the difficulty in interpreting information provided in the summaries. Fatalities, however, included original entrants as well as would-be rescuers. In some cases, entrants were the only fatalities. In others, the only fatalities were would-be rescuers. In still others, fatalities included both original entrants and would-be rescuers. For purposes of this discussion, would-be rescuers were individuals who responded after the onset of the accident. They included the original entrants, as well as individuals not normally associated with the activity or involved in it.

Fatal accidents summarized in the first three columns (toxic, asphyxiating, and flammable or explosive atmospheres) occurred inside the confined workspace. According to these data, situations unrelated to normal work were associated with up to 13% of the fatalities. Persons attempting a rescue comprised 26 to 30% of the fatalities. Some would-be rescuers, therefore, may have died while attempting to rescue individuals who entered the confined space for inappropriate reasons (unauthorized entry or entry not work related). The FACE data provided in the NIOSH report (1994) for selected accidents indicate that rescue was the reason for entry for 36% of the fatalities. A later section examines the subject of rescue more closely.

TABLE 1.15
Fatal Accidents and Normal Activities in Confined Spaces

Normal activity	Percent of fatal accidents			
	TA	AA	F/E	W/C[a]
Inspecting, testing, checking	9	30	9	
Retrieving objects	5	4		
Preparation — start/finish	5		5	
Installing, adjusting, repairing, replacing	22	35	16	
Cleaning, unblocking, scraping	49	26	9	
Abrasive blasting, painting	7	2	23	
Welding, cutting, joining		2	32	100
Other	3	1	6	

Note: TA is toxic atmosphere; AA is asphyxiating (oxygen-deficient) atmosphere; F/E is flammable or explosive atmosphere; and W/C is welding and cutting.

[a] The data on welding and cutting used in this table reflect analysis of individual accident summaries contained in the OSHA report (1988). The report claims that this abridged group (164 cases and 190 associated fatalities) reflects the total (217 cases and 262 associated fatalities).

Most of the fatal welding and cutting accidents occurred outside the confined space. The ignition of a flammable or explosive confined atmosphere caused most of these accidents. The pattern of work activity preceding these accidents was similar to that preceding fires and explosions. That such a high percentage of these accidents should occur during normal activity strongly suggests the failure to recognize the hazard or to appreciate the severity of its consequences.

Table 1.14 indicates that a high proportion of fatal injuries occurred during activities considered to be part of a normal job description. The nature of these "normal" activities is the next element needed to gain understanding about the nature of work performed in confined spaces.

Accident summaries contained in the OSHA reports (1982a, 1985, 1988) provide a starting point for investigating this question. Table 1.15 summarizes tasks identified in accident summaries contained in the OSHA reports. They reflect subjective estimation of the situation described in the accident summary.

Fatal accidents described in the first three columns (toxic and asphyxiating, and flammable or explosive atmospheres) occurred within the confined space. Most of the welding and cutting accidents occurred on the exterior of a container or structure that confined a hazardous atmosphere. The incidence of fatal accidents by activity varied according to the atmospheric hazard. A different pattern characterizes each atmospheric condition.

The highest percentage of fatal accidents involving toxic atmospheres occurred during cleaning and its associated tasks. This observation is understandable because of the disruptive nature of these activities. Cleaning and associated primary activities prepare the interior of the confined space for secondary activities such as modification of the existing configuration or installation of new equipment. Walls, horizontal surfaces, and depressed areas often retain residues of previous contents. Activities involved in cleaning and surface preparation disturb these residues and can liberate volatiles. These results strongly suggest that residual materials and cleaning products and processes are a major source of exposure to toxic substances.

The highest percentage of fatal accidents involving asphyxiating (oxygen-deficient) atmospheres occurred during inspection and modification. Modification also is associated with a significant

TABLE 1.16
Fatal Injuries and Normal Activities in Confined Spaces

	Percent of fatalities			
Normal activity	**TA**	**AA**	**F/E**	**W/C**[a]
Inspecting, testing, checking	13	31	6	
Retrieving objects	6	6		
Preparation — start/finish	5		4	
Installing, adjusting, repairing, replacing	16	39	18	
Cleaning, unblocking, scraping	49	22	12	
Abrasive blasting, painting	8	1	17	
Welding, cutting, joining		1	28	100
Other	3		15	

Note: TA is toxic atmosphere; AA is asphyxiating (oxygen-deficient) atmosphere; F/E is flammable or explosive atmosphere; and W/C is welding and cutting

[a] The data on welding and cutting used in this table reflect analysis of individual accident summaries contained in the OSHA report (1988). The report claims that this abridged group (164 cases and 190 associated fatalities) reflects the total (217 cases and 262 associated fatalities).

proportion of accidents involving toxic atmospheres. Processes utilized in these activities can generate hazardous contaminants. In addition, disturbance of residual contents also can contribute to the hazardous atmosphere. Many fatal accidents involving oxygen-deficient atmospheres occurred in outwardly clean workspaces that provided no warning to prospective entrants.

A surprisingly large percentage of fatal accidents occurred during inspection or the checking and testing of some characteristic within the confined space. This finding is especially important in workspaces having oxygen-deficient atmospheres. Inspection often occurs under outwardly clean conditions. The absence of visual or olfactory cues such as often occurred prior to accidents in these environments would arouse no suspicion and would provide no cue about the need to assess conditions.

Entry for retrieval of objects, while claiming few victims, remains an ever-present motivator.

Accidents associated with fires and explosions and welding and cutting were considerably more task oriented than those occurring in the other situations. Activities preceding the fire or explosion either exacerbated preexisting hazardous conditions or created them. Invariably, the source of ignition resulted from some aspect of the activity.

The preceding section examined the relationship between normal activities and fatal accidents. Accident summaries contained in the OSHA reports (1982a, 1985, 1988) also provide the basis for examining the relationship between fatal injuries and these activities. Additional information is provided in the NIOSH report.

This information provides the basis for estimating the relative severity of accidents associated with different types of activity. Data contained in Table 1.16 apply only to accidents associated with hazardous atmospheric conditions. The data reflect subjective estimation of the situation described in the accident summary.

The first three columns — toxic, asphyxiating, and flammable or explosive atmospheres — summarize fatal injuries that occurred inside confined workspaces. Most welding and cutting occurred during work on the exterior of a container or structure that confined a hazardous atmosphere.

As demonstrated in Table 1.16, the distribution of fatalities varies by activity within an atmospheric condition. The pattern of distribution of fatal injuries and fatal accidents, while similar, is not identical.

TABLE 1.17
Relative Severity of Accidents Involving Confined Spaces

	Ratio of Fatalities to Accidents			
Normal activity	TA	AA	F/E	W/C[a]
Inspecting, testing, checking	2.0	1.6	1.0	
Retrieving objects	1.7	2.0		
Preparation — start/finish	1.3		1.5	
Installing, adjusting, repairing, replacing	1.1	1.8	1.9	
Cleaning, unblocking, scraping	1.4	1.3	2.0	
Abrasive blasting, painting	1.5	1.0	1.2	
Welding, cutting, joining		1.0	1.4	1.2
Other	3.0		3.3	

Note: TA is toxic atmosphere; AA is asphyxiating (oxygen-deficient) atmosphere; F/E is flammable or explosive atmosphere; and W/C is welding and cutting.

[a] The data on welding and cutting used in this table reflect analysis of individual accident summaries contained in the OSHA report (OSHA 1988). The report claims that this abridged group (164 cases and 190 associated fatalities) reflects the total (217 cases and 262 associated fatalities).

The FACE data from the NIOSH report (1994) provide additional useful information. The report does not differentiate between type of atmospheric hazard in this analysis. The original presentation of the data included entry for rescue and entry for unknown reasons. Removing this information provides the means to make a direct comparison with the data in Table 1.16. Entry for repair and maintenance led to 64% of the fatalities. Inspection led to 16%. Retrieval of objects led to 6% of fatalities, and entry to dislodge material, 7%. Construction activity caused 7% of fatalities.

The data presented in Tables 1.15 and 1.16 provide the basis for estimating the relative severity of accidents occurring during these activities. Table 1.17 presents ratios of fatalities to accidents. A ratio greater than unity indicates accidents that cause multiple fatalities. The greater the ratio, the more serious the accidents were likely to have been.

Interpretations placed on these data warrant considerable caution, since the number of fatalities and accidents in several of the categories was very small. Thus, a minor change in either the number of accidents or the number of fatalities in a particular category could produce a substantial change in the ratio.

A second caveat concerns accidents in which multiple fatalities occurred. The victims could have included only entrants, would-be rescuers, or a combination of both groups. As presented in Table 1.17, the data do not indicate the relative occurrence of each type of accident. A third caveat concerns the possibility that a single, very serious accident that caused a large number of fatalities could have biased the ratio for the group.

Bearing these caveats in mind as possible limitations, the data do provide an indication about the relationship between work activity and atmospheric conditions.

Activities having the highest ratio of fatalities per accident in toxic and asphyxiating atmospheres were inspection, testing and checking, and retrieval. This observation is unexpected, since these activities normally would not cause major disruption of residual materials. By contrast, tasks such as cleaning, that normally are expected to disrupt residual contents, had a lower ratio of fatalities per accident. This anomaly suggests lack of awareness about hazards or preparedness for entry and rescue by workers performing inspections, testing, checking, and retrieval. The ratios of

TABLE 1.18
Assigned Causes of Fatalities in Confined Space Accident

Immediate cause	Percent of fatal accidents		
	TA	AA	F/E
Procedural			
Failure to test prior to entry	24	25	14
Inadequate training in use of testing equipment	2		
Improper test procedure	5		
Failure to follow entry procedures	13	32	
Entry against orders	7	5	5
Entry for unknown or invalid reason	5	5	
Faulty procedure	7	5	25
Failure to follow work/safety procedures	9	7	
Dangerous work practices or shortcuts	5	5	
Improper respirator selection	9		
Respirator or training in use not available	5		
Failure to use respirator	5		
Equipment, material, or facility			
Failure of supplied air system	2		
Respirator problem	2		
Faulty isolation		5	18
Ungrounded equipment		11	11
Unapproved lighting equipment			16
Other			11

Note: TA is toxic atmosphere; AA is asphyxiating (oxygen-deficient) atmosphere; and F/E is flammable or explosive atmosphere.

fatalities to accidents for other categories within this column are relatively constant. Toxic atmospheres appear to cause more serious accidents than asphyxiating atmospheres, based on this measure of severity.

Cleaning and associated activities produce the most severe fires and explosions. The severity of fires and explosions associated with other activities is equivalent to that reported for toxic and asphyxiating atmospheres. Welding and cutting accidents are slightly less severe than those associated with other activities.

Assigned Immediate Cause

A primary goal of accident investigation is to determine immediate cause. Immediate cause provides a focal point for targeting corrective action. OSHA assigned an immediate cause to accidents documented in its reports (OSHA 1982a, 1985). Table 1.18 summarizes data as reported by OSHA. These data pertain only to accidents associated with hazardous atmospheric conditions.

After examining accidents that occurred in confined workspaces containing toxic and asphyxiating atmospheres, OSHA commented that 95% resulted from procedural deficiencies or problems with materials, equipment, or facilities. The analysis of fires and explosions occurring in confined spaces produced almost the same conclusion. Approximately 93% of the accidents were directly attributable to deficiencies in operating procedures, problems with equipment, materials, or facilities; or environmental conditions.

The data presented in this table apply only to accidents involving hazardous atmospheres. Restricting the selection to these accidents provides a slightly sharper focus than that provided in the OSHA reports. The predominant characteristic indicated by the data was inadequate control of activities occurring in confined spaces. This expressed itself through failure to appreciate the potential hazards and failure to address them through an organized approach.

Table 1.18 did not incorporate data from the OSHA report (1988) on welding and cutting. The organization of this report differed from the others. Unfortunately, it did not provide a sufficiently detailed breakdown to establish causation. This report did comment that procedural deficiencies caused 80% of the incidents, that problems with equipment, materials, or facilities accounted for 11%, that environmental conditions accounted for 4%, and that 5% occurred for other reasons. These results are consistent with those contained in the other reports (OSHA 1982a, 1985). The report on welding and cutting also commented about issues raised in the other reports. These included the lack of precaution against fires and explosions involving the contents of containers and materials originating from other sources.

Preventive Measures

The OSHA reports repeatedly commented about the failure by entrants to act in a manner that would insure their safety. Prominent among the deficiencies were the lack of formal testing prior to entry, flawed or poorly conceived procedures, and the failure to use protective measures.

The OSHA reports (1982a, 1985) commented only very briefly about the inadequacies of atmospheric testing and ventilation of the confined space or confined atmosphere. The brevity of the comment does not correlate with the importance of these activities on the outcome of these accidents. Previous discussion has examined the relative importance of preexisting atmospheric hazards compared to those that developed during the course of the work.

The role of atmospheric testing and ventilation were not addressed in the OSHA reports. The accident summaries do provide sufficient information to permit an examination of this question (OSHA 1982a, 1985, 1988). The data presented in Table 1.19 apply only to accidents associated with hazardous atmospheric conditions. The data reflect subjective estimation of situations described in the accident summary.

The data indicate that atmospheric testing occurred in almost none of the situations in which it was needed. Even when attempted, the test almost always was carried out incorrectly.

The appropriate atmospheric test, when performed correctly, can identify potentially hazardous situations. This concept applies equally to the atmosphere in the confined space in which the work occurs, and to containers and structures that confine a hazardous atmosphere. Testing also can identify the development of a hazardous atmosphere during the progress of work.

The use of ventilation equipment was almost nonexistent. When such equipment was used, it usually was employed incorrectly. This is essentially the same situation as observed for atmospheric testing. Ventilation of confined workspaces is intended to control the concentration of airborne contaminants. This control can occur by introducing clean air into the contaminated workspace or by removing contaminated air from it. Ventilation of the confined workspace, properly conducted, prior to entry could have prevented many of these accidents. Ventilating during occupancy could have reduced the potential severity of the hazardous atmosphere resulting from work activity.

Social Elements

Social interactions play an extremely important role in the dynamics of work. These interactions are equally important in the dynamics of accidents that occur in confined spaces. Analysis that concentrates only on the technical aspects of this subject easily can ignore the important dynamics represented by the social elements. To fail to appreciate the social aspects of these accidents is to overlook a dimension of fundamental importance.

TABLE 1.19
Procedural Aspects of Accidents Involving Confined Spaces

	Percent of accidents			
Procedural aspect	**TA**	**AA**	**F/E**	**W/C**[a]
Atmospheric testing				
No test	94	100	100	98
Improper/incorrect test	4			2
Correct/appropriate test	2			
Ventilation				
No ventilation of the workspace	94	94	100	
Improper ventilation		4		
Lack of preparation/ventilation				97
Improper ventilation				1
Correct/appropriate ventilation	6	2		2

Note: TA is toxic atmosphere; AA is asphyxiating (oxygen-deficient) atmosphere; F/E is flammable or explosive atmosphere; and W/C is welding and cutting.

[a] The data on welding and cutting used in this table reflect the case studies contained in the OSHA report (1988) (164 cases and 190 associated fatalities). The report claims that this abridged group reflects the total (217 cases and 262 associated fatalities).

While people who work in confined workspaces usually are unrelated in the familial sense, their actions under stressful situations often reflect close-knit social bonding. Some people, co-workers and other bystanders, sometimes take charge during accident situations and assume the extremely dangerous role of rescuer. Would-be rescuers in these situations often act in a manner that shows complete disregard for their own personal safety. These individuals willingly assume unreasonable risks in situations clearly recognizable as hazardous. The events and resulting tragedies documented by the NIOSH- and OSHA-sponsored reports provide ample evidence to support this contention (NIOSH 1978, 1994; OSHA 1985). Addressing this situation will require a thorough understanding of the social dynamics that are operating.

This section will examine the following elements of the social aspects of work in confined spaces:

- Demographics of the workforce involved with confined workspaces
- Rescue during an accident situation

Demographics

The demographics of the population that enters and works in confined spaces is an important starting point for examining this subject. Unfortunately, little is known, due to the limited research performed on this topic. Accident summaries contained in the OSHA reports (1982a, 1985, 1988) provide a starting point for enquiry about accident situations. However, the information that they contain is relevant only to the question of who has died in confined spaces. They provide no information about the population of workers who enter these workspaces.

The following tables summarize data contained in the OSHA reports. Most of the fatalities in welding and cutting accidents occurred during work on the exterior of structures. The data apply only to hazardous atmospheric conditions.

TABLE 1.20
Demographics of Victims of Fatal Accidents: Age

Age Range	Percent of fatalities			
	TA	AA	F/E	W/C[a]
15 to 19	15	7	6	6
20 to 24	20	23	14	12
25 to 29	17	19	22	24
30 to 34	17	7	8	21
35 to 39	15	12	17	9
40 to 44	7	9	6	12
45 to 49	2	9	8	6
50 to 54	5	9	14	7
55 to 59		6	5	2
60 to 64	2			1
Number used in calculations	81	86	64	66

Note: TA is toxic atmosphere; AA is asphyxiating (oxygen-deficient) atmosphere; F/E is flammable or explosive atmosphere; and W/C is welding and cutting.

[a] The data on welding and cutting used in this table reflect the case studies contained in the OSHA report (1988) (164 cases and 190 associated fatalities). The report claims that this abridged group reflects the total (217 cases and 262 associated fatalities).

Age is an important consideration in the design of training and communications programs. Age of victims is summarized in Table 1.20.

Up to 2/3 of the fatal injuries occurred in the 20 to 40 age group, according to these data. About half the fatal injuries occurred in the 20 to 35 age group. Interestingly, the very young, who are the least experienced persons in the workforce, and older workers, who are the most experienced, were least likely to have been victims. These observations are virtually identical for each of the three types of hazardous conditions. Data from FACE investigations contained in the NIOSH report provide a similar picture (NIOSH 1994).

Fatal accidents associated with toxic and asphyxiating atmospheres tend to occur in younger workers. Fatal accidents involving fires and explosions and welding and cutting tend to occur in older workers.

Another important demographic in this problem is the occupation of people who enter and work in confined spaces. At this time there is no information concerning this subject. Table 1.21 indicates the occupations of individuals who entered the space prior to the onset of the accident.

Data presented in Table 1.21 suggest that the occupations of persons who enter confined spaces span a wide spectrum. The categories contain representation from all occupational groups within the workplace hierarchy. Representation in the three atmospheric classes is approximately the same. As would be expected, the doers — laborers and tradespeople — dominated the occupational categories. Supervisory/managerial and technical personnel were minor participants.

Workers entering confined spaces possess diverse educational and technical skills. A program of education and training for these people must recognize and accommodate to this diversity.

Rescue

Rescue of accident victims from confined spaces has generated concern in all investigative reports on this subject (NIOSH 1978, 1979, 1994; OSHA 1985). These reports commented that would-be

TABLE 1.21
Demographics of Work in Confined Spaces: Occupation

Occupation	Percent of fatal accidents		
	TA	AA	F/E
Management/Supervisory			
Supervisor/foreman	11	12	12
Superintendent/manager	3	12	
Trades			
Plumber/pipefitter	5	10	
Painter	3	2	16
Welder	24		
Maintenance	13	10	6
Operator	10	6	
Technical	7	10	2
Laborer	36	35	32
Other	12	3	8

Note: TA is toxic atmosphere; AA is asphyxiating (oxygen-deficient) atmosphere; and F/E is flammable or explosive atmosphere.

Data from OSHA 1982a and 1985.

TABLE 1.22
Rescue from Confined Spaces

Characteristic	Percent of Accidents	
	TA	AA
Rescue attempted	76	72
Rescue not attempted	24	28
Successful rescue of original victim	9	2

rescuers constituted a considerable proportion of the fatalities. Yet, the aspects of rescue remain to be defined. Rescue is both a social and a technical issue.

Table 1.22 summarizes information contained in the OSHA report on toxic and asphyxiating atmospheres (OSHA 1985). Of course, the data available for examination can provide only a limited view of this subject. First responders acted in one of two ways. (First responders were bystanders to the original event or were first on the scene to discover the accident.) Either they entered to retrieve the victim or did not attempt an entry but sought assistance. These actions define for the purposes of Table 1.22 whether or not rescue was attempted.

Rescue attempts occurred in three out of four accident situations. Successful rescue of the initial victim occurred in only a small fraction of these attempts. Asphyxiating (oxygen-deficient) atmospheres were more severe than toxic atmospheres, according to the data. This highlights the rapidity of action of oxygen deprivation compared to chemical intoxication.

Fundamental to the confined space problem are the demographics of rescue. That is, who went into the confined space to attempt the rescue? The accident summaries contained in both the OSHA and NIOSH reports on fatal accidents in confined workspaces provide a source of information (OSHA 1985, NIOSH 1994). Table 1.23 summarizes information from the OSHA report.

TABLE 1.23
Demographics of Rescue: Occupation of Would-be Rescuers

Occupation	Percent of Fatal Accidents			
	TA	Change	AA	Change
Management/supervisory				
Supervisor/foreman	17	6	15	3
Superintendent/manager	9	6	4	(8)
Trades				
Plumber/pipefitter	2	(3)	6	(4)
Painter	2	(1)		(2)
Maintenance	11	(2)	6	(4)
Operator	12	2	15	9
Technical	3	(4)	9	(1)
Laborer	26	(10)	21	(14)
Other	18	6	24	21

Note: TA is toxic atmosphere; AA is asphyxiating (oxygen-deficient) atmosphere; and ( ) indicates a negative value.

Data from OSHA 1985.

Within categories of atmospheric hazard, the spectrum of occupations of would-be rescuers was approximately the same. Would-be rescuers belonged to all groups within the occupational hierarchy.

An interesting comparison can be made between occupations of persons who enter confined workspaces to work and persons who enter to act as rescuers. Table 1.21 presented information about the occupation of entrants. The columns in Table 1.23 headed "change" indicate differences between the two tables. These differences indicate change in involvement by occupational groups during emergency response. The data indicate an increase in the involvement of supervisory and management personnel. This may reflect a taking charge during the process of the accident, perhaps as an extension of the normal modus operandi of this group.

The participation of "other" persons is an important aspect of the rescue process. This participation is particularly striking during accidents involving asphyxiating atmospheres. In many cases these "others" had no involvement in the work being performed and no reason to be familiar with the risks that they were assuming.

The final link in this chain is the relationship between occupation and fatal injury (Table 1.24).

Fatal injuries described in the first two columns (toxic and asphyxiating atmospheres and fires and explosions) occurred inside the confined space. Most fatal injuries in welding and cutting accidents occurred during work on the exterior of a container or structure that confined a hazardous atmosphere.

Laborers and trades workers formed the largest single group of fatalities during these accidents. These workers perform most of the tasks that occur in these workspaces.

Laborers perform less defined tasks than trades workers. They often handle bulk residual materials. Sometimes toxic vapors or gases effuse from these materials when they are disturbed. Laborers often are the first group to enter confined spaces. They prepare the workspace for the entry of other groups, such as trades workers and technical personnel.

In situations involving rescue attempts, the management and supervisory group forms a considerably larger proportion of victims than when rescue is not attempted. Under normal circumstances these workers entered the confined workspace to inspect, to troubleshoot, and to supervise.

TABLE 1.24
Fatalities Associated with Confined Spaces by Occupation

Occupation	Percent of fatalities		
	TA/AA	F/E	W/C[a]
Manager/superintendent/foreman	16	8	9
Trades			
Mechanic	7	9	
Painter	3	12	
Welder	3	13	49
Electrician	1		
Plumber	1		
Sandblaster	1	1	
Pipefitter/fitter		9	6
Mason	1	1	
Operator	12	1	
Technical	3	4	
Laborer	28	25	36
Other	24	17	

Note: TA/AA is toxic atmosphere/asphyxiating (oxygen-deficient) atmosphere; F/E is flammable or explosive atmosphere; and W/C is welding and cutting.

[a] The data on welding and cutting used in this table reflect the case studies contained in the OSHA report (1988) (164 cases and 190 associated fatalities). The report claims that this abridged group reflects the total (217 cases and 262 associated fatalities).

Data from OSHA 1982a, 1985, and 1988.

As mentioned in previous discussion, this group seems to take charge during the emergency and lead the rescue attempt.

Almost half the victims of the welding and cutting accidents were identified as being welders. These workers appeared to be unaware of the hazards posed by residual contents of containers. The involvement of such a high proportion of individuals in this one trade would suggest an opportunity for correction through training and certification programs. In many of these accidents, however, unskilled workers or workers employed in other trades operated the welding equipment. This represents an uncontrolled use of welding equipment. The fact that anyone with or without formal training can use this equipment complicates the task of providing training about these hazards.

More detailed examination of the data contained in the accident summaries discloses a frustrating picture (OSHA 1985). Rescue was attempted in 41 accidents (76%) involving toxic atmospheres and 34 (74%) involving asphyxiating atmospheres. At least one fatality occurred in each accident. In each of these situations, someone attempted to save the victim(s). For the rescue attempt to have been successful, someone also must have died. That person was either a victim of the original accident or a would-be rescuer. In situations involving multiple fatalities, the victims could have been the original entrants, rescuers, or some combination of these individuals. Table 1.25 summarizes this information. The data apply only to situations in which rescue was attempted.

The data in Table 1.25 provide a disturbing picture about emergency response during these accidents. The data on successful rescues refer to situations in which rescue was attempted rather

TABLE 1.25
Characteristics of Attempted Rescues Involving Confined Spaces

Characteristic	Percent of Attempts	
	TA	AA
Successful rescue of original victim	12	3
Death of original victim(s)	88	97
Death of rescuer(s)	44	50
Death of rescuer(s) only	12	3
Death of rescuer(s) and entrant(s)	32	47
Lack of use of respiratory protection by rescuers	78	91
Incorrect/improper use of respiratory protection by rescuers	12	
Correct use of respiratory protection by rescuers	10	9

TABLE 1.26
Ratios of Fatalities per Accident during Rescue Attempts

Ratio Statistic	Ratio	
	TA	AA
Not attempted		
Fatally injured entrants/accident	1.3	1.3
Unsuccessful attempts		
Fatally injured entrants/attempt	1.1	1.0
Fatally injured rescuers/attempt	1.2	1.4
Fatally injured (entrants + rescuers)/attempt	1.1	1.1
Successful attempts		
Entrant saved/successful attempt	1.2	1.0
Fatally injured rescuers/entrant saved	1.4	1.0

than the number of accidents, since rescue was not attempted in all cases. The success rate, therefore, was very small compared to the number of attempts. Would-be rescuers died in approximately half of the attempts at rescue. The proportion of situations in which only would-be rescuers died paralleled those involving successful rescue of original victims. In these cases, the original victim survived the accident.

The data suggest two main themes. Not only were the attempts to rescue the original victims highly unsuccessful, they were extremely costly to the would-be rescuers.

The second half of Table 1.25 examines the preparedness of would-be rescuers for this activity. The main indicator of preparedness was use of respiratory protective equipment by would-be rescuers during these attempts. In a large proportion of rescue attempts, would-be rescuers either used no respiratory protective equipment or used it incorrectly. They were unprepared to carry out this task. Doubtless, these deficiencies cost the lives of many would-be rescuers.

The previous tables have provided insight about the lack of success of rescue attempts. Table 1.26 details the human costs of these actions.

Table 1.26 attempts to detail the human costs of the decision to attempt or not to attempt a rescue from a confined space. Unfortunately, the small number of accidents used in calculating the ratios in some categories represents a serious potential limitation on the validity of these data. A small change in the number used in the calculation could produce a large change in the ratio. However limited, the data do provide a starting point for further discussion.

In situations in which the first responder did not attempt rescue, approximately 1.3 entrants died per accident. A successful rescue attempt reduced this value somewhat. From the perspective of the original victim, attempts at rescue are worthwhile. However, the cost to rescuers (fatally injured rescuers per entrant saved) can exceed the cost of not attempting a rescue. Overall, the cost of rescue is unacceptably high.

CONFINED SPACES — A PROBLEM OF ATMOSPHERIC HAZARDS

The work of investigators at NIOSH and OSHA has established the consequential role of hazardous atmospheric conditions in accidents involving confined spaces. Hazardous atmospheres confined within containers similarly play a consequential role in welding and cutting accidents. Accidents involving confined spaces occur across a wide spectrum of industry. Investigation reported in this chapter elaborated on the technical and procedural, as well as social characteristics of these accidents. This examination has revealed deficiencies common to many of these events. By correcting these deficiencies, considerable reduction in occurrence and severity of these accidents likely would occur.

Hazardous conditions predominating in accidents investigated in this chapter were caused by toxic substances, oxygen deficiency and enrichment, and the sudden release of energy during an explosion or fire. In most cases the hazardous condition developed prior to entry. The hazardous condition also could develop during work activity, the result of emissions from residual materials and processes employed.

An important factor highlighted in the accident summaries was unpredictability. Accident summaries repeatedly demonstrated the importance of this factor. What was predictable and familiar through many repetitions suddenly became an unknown quantity. Addressing the unpredictables that have occurred during these accidents in a proactive program is a formidable task.

To this point, discussion has avoided comment about the meaning of the term "confined space" and has relied on definitions utilized by the organizations that provided the analytical data. Data selected for inclusion in the analytical reports cited in the previous section reflected boundary conditions imposed by the investigators. The literature on this subject contains many definitions for the term "confined space." (Refer to Appendix A for more information.) Most of these reflected the needs of regulators or standard-setting groups, rather than enabling management and control of all of the hazardous conditions that could be present or could develop.

Over the years, the term "confined space" has been used to describe an environment in which an increasing number of hazardous conditions can occur. The meaning incorporated into the term has expanded greatly beyond that projected by the words "confined space." In effect, the term "confined space" has become jargon or shorthand or code. Jargon is useful only to people familiar with the full meaning intended for the term. However, the people most needing to understand the full scope of " confined space" are the least likely to appreciate it. This gap in understanding has been a major impediment to gaining control over the hazards, atmospheric and otherwise.

Assuming that "confined space" is overly restrictive, what then would evoke the necessary recognition? Obviously, the term must broaden perceptions beyond a cramped workspace that restricts personal movement. Yet, at the same time, the new term must retain the link to the traditional; otherwise confusion would reign. An unrelated term would lose the power of recognition achieved by the original. Finally, the term should communicate the concept of atmospheric confinement or entrapment, since this is the major hazard.

A phrase that bridges the gap and retains historical recognition is "**confined space/confined atmosphere**." "Confined space/confined atmosphere" conveys a fuller description of the hazardous conditions that can occur in these particular workspaces. This also informs about the need for improved precautionary measures for work occurring on the exterior of containers and structures that confine hazardous atmospheres. Confined space/confined atmosphere provides a constant reminder about conditions that constitute a major hazard — atmospheric contamination.

As a working definition, the term "confined space/confined atmosphere" would apply to workspaces inside which one or more of the following hazardous conditions could be present or could develop:

- Personal confinement
- Unstable interior condition
- Flowable solid materials or residual liquids or sludges
- Release of energy through uncontrolled or unpredicted motion or action of equipment
- Atmospheric confinement
  - Toxic substances
  - Oxygen deficiency/enrichment
  - Flammable/combustible atmosphere
- Chemical, physical, biological, ergonomic, mechanical, process, and safety hazards

This definition is intended to create the broadest possible description of situations in which the preceding conditions can occur or can develop. It also intends to emphasize concern about the ability of minor modifications in workspaces to create these conditions. For example, during renovation of office spaces, individuals sometimes block off the ventilation to prevent migration of contaminants into occupied areas. Such modifications could permit hazardous accumulation of solvent vapors during painting or the application of carpet or flooring adhesives. While the intent of the action is to minimize disturbance to others, deliberate confinement of vapors could produce acute overexposure. By their design and construction, office spaces normally are intended for continuous human occupancy. Modification using structurally trivial materials, such as sheet plastic and duct tape, is all that is required to create the conditions for atmospheric confinement.

Another issue for consideration is size. Fatal accidents continue to occur in structures large enough to insert the head, shoulders, and arms. A recently reported accident involved an individual who was required to clean the inside of metal molds used to form plastic containers (McCammon and McKenzie 1996). The mold in which the victim was found measured 49.5 cm (19.5 in) in diameter by 81 cm (32 in) deep. He was overcome by perchloroethylene, the solvent used to clean the interior. The practice was to apply perchloroethylene to a cloth and to wipe the interior of the mould.

In some legislation, for example, the OSHA Standard on confined spaces in general industry, such a structure is not defined as a confined space (OSHA 1993). Yet, the circumstances in this accident are typical of what occurs during fatal accidents in confined spaces. This publication does not consider size of the space to be an issue.

SUMMARY

This chapter has reported on factors intrinsic to fatal accidents involving confined workspaces and structures that cause atmospheric confinement. Research indicates that a major causative agent in these accidents is hazardous atmospheric conditions. Hazardous atmospheric conditions arise from the presence of toxic substances, oxygen deficiency and enrichment, and flammables and combustibles. Accidents associated with confined workspaces occurred during entry and work activities. Accidents similar to these also occurred during work on the exterior of containers and structures that confine flammable or explosive atmospheres. The hazardous condition was more likely to have developed prior to entry or the start of work activity.

Confined workspaces and containers and structures that confine hazardous atmospheres are deceptively innocuous. The hazardous condition is unlikely to provide warning to the senses.

Persons entered these spaces apparently without concern for the potential danger. Seemingly minor errors in judgment produced disproportionate and catastrophic consequences. These consequences appear to be considerably more severe than what occurs in accident situations involving normal workplaces. A sizable proportion of the victims were would-be rescuers. Many of these individuals were not involved in the original event. Would-be rescuers sometimes perished while the original victim survived. The urge to assist a person in distress is a powerful motivator for human actions, regardless of the consequences. Actions taken during rescue attempts demonstrated an obvious disregard for the danger in these situations. Understanding these behaviors is essential to ensuring appropriate response during these situations.

Providing safe systems for entry and work in confined spaces/confined atmospheres is the only way to reduce the human costs of these tragedies.

This chapter also has introduced the term "confined space/confined atmosphere." This expanded term intends to associate atmospheric confinement with the concept of the confined space. The term "confined space" fails to convey the broad nature of the hazards inherent in these workspaces. As a working definition, "confined space/confined atmosphere" should apply to workspaces inside which one or more of the following hazardous conditions could be present or could develop:

- Personal confinement
- Unstable interior condition
- Flowable solid materials or residual liquids or sludges
- Release of energy through uncontrolled or unpredicted motion or action of equipment
- Atmospheric confinement
  - Toxic substances
  - Oxygen deficiency/enrichment
  - Flammable/combustible atmosphere
- Chemical, physical, biological, ergonomic, mechanical, process, and safety hazards

REFERENCES

American National Standards Institute: *American National Standard Safety Requirements for Working in Tanks and Other Confined Spaces* (Standard Z117.1-1977). New York: American National Standards Institute, 1977. 24 pp.

American National Standards Institute: *American National Standard Safety Requirements for Confined Spaces* (Standard Z117.1-1989). Des Plaines, IL: American Society of Safety Engineers/American National Standards Institute, 1989. 24 pp.

American National Standards Institute: *American National Standard Safety Requirements for Confined Spaces* (Standard Z117.1-1995). Des Plaines, IL: American Society of Safety Engineers/American National Standards Institute, 1995. 24 pp.

Bird, F.E., Jr. and G.L. Germain: *Practical Loss Control Leadership.* Rev. ed. Loganville: Institute Press, 1990. pp. 20–21.

Garrison, R.P. and D.M. McFee: Confined spaces — A case for ventilation. *Am. Ind. Hyg. Assoc. J. 47*: 708–714 (1986).

Keller, C.: Firesafety in the shipyard. *Fire J. 76:* 60–66 (1982).

McCammon, J. and L. McKenzie: Workplace fatality related to perchloroethylene exposures. *Appl. Occup. Environ. Hyg. 11:* 156–157 (1996).

Mine Safety and Health Administration: Think "Quicksand": Accidents Around Bins, Hoppers and Stockpiles, Slide and Accident Abstract Program. Arlington, VA: U.S. Department of Labor, Mine Safety and Health Administration, National Mine Health and Safety Academy, 1988.

Mine Safety and Health Administration: Hazard Information Alert: Confined Space Fatalities — 57 since 1980. Arlington, VA: U.S. Department of Labor, Mine Safety and Health Administration, 1994. 3 pp.

National Fire Protection Association: *NFPA 306 — Control of Gas Hazards on Vessels*, 1993 ed. Quincy, MA.: National Fire Protection Association, 1993. 15 pp.

National Institute for Occupational Safety and Health: Search of Fatality and Injury Records for Cases Related to Confined Spaces (NIOSH Pub. No. 10947). San Diego, CA: Safety Sciences, 1978.

National Institute for Occupational Safety and Health: Criteria for a Recommended Standard — Working in Confined Spaces (DHEW/PHS/CDC/NIOSH Pub. No. 80-106). Cincinnati, OH: National Institute for Occupational Safety and Health, 1979. 68 pp.

National Institute for Occupational Safety and Health: Worker Deaths in Confined Spaces (DHHS/PHS/CDC/NIOSH Pub. No. 94-103). Cincinnati, OH: National Institute for Occupational Safety and Health, 1994. 273 pp.

Occupational Safety and Health Administration: Selected Occupational Fatalities Related to Fire and/or Explosion in Confined Work Spaces as Found in OSHA Fatality/Catastrophe Investigations. Washington, D.C.: U.S. Department of Labor, Occupational Safety and Health Administration (U.S. DOL/OSHA), 1982a. 76 pp.

Occupational Safety and Health Administration: Selected Occupational Fatalities Related to Lockout/Tagout Problems as Found in Reports of OSHA Fatality/Catastrophe Investigations. Washington, D.C.: U.S. Department of Labor, Occupational Safety and Health Administration (U.S. DOL/OSHA), 1982b. 113 pp.

Occupational Safety and Health Administration: Selected Occupational Fatalities Related to Grain Handling as Found in Reports of OSHA Fatality/Catastrophe Investigations. Washington, D.C.: U.S. Department of Labor, Occupational Safety and Health Administration (U.S. DOL/OSHA), 1983. 150 pp.

Occupational Safety and Health Administration: Selected Occupational Fatalities Related to Toxic and Asphyxiating Atmospheres in Confined Work Spaces as Found in Reports of OSHA Fatality/Catastrophe Investigations. Washington, D.C.: U.S. Department of Labor, Occupational Safety and Health Administration (U.S. DOL/OSHA), 1985. 230 pp.

Occupational Safety and Health Administration: Selected Occupational Fatalities Related to Welding and Cutting as Found in Reports of OSHA Fatality/Catastrophe Investigations. Washington, D.C.: U.S. Department of Labor, Occupational Safety and Health Administration (U.S. DOL/OSHA), 1988. 225 pp.

Occupational Safety and Health Administration: Selected Occupational Fatalities Related to Ship Building and Repairing as Found in Reports of OSHA Fatality/Catastrophe Investigations. Washington, D.C.: U.S. Department of Labor, Occupational Safety and Health Administration (U.S. DOL/OSHA), 1990. 195 pp.

OSHA: Permit-required confined spaces for general industry; final rule, *Fed. Regist. 58*: 9 (14 January 1993). pp. 4462–4563.

Reese. C.D. and G.R. Mills: Trauma epidemiology of confined space fatalities and its application to intervention/prevention now. In *The Changing Nature of Work and Workforce*. Proc. Third Joint U.S.-Finnish Science Symposium, Frankfort, Kentucky, October 22–24, 1986. Cincinnati, OH: National Institute for Occupational Safety and Health, 1986. pp. 65–67.

Willwerth, E.: Maritime confined spaces. *Occup. Health Safe. 63:* 39–44 (1994).

2 Nonatmospheric Hazards and Fatal Accidents in Confined Spaces

CONTENTS

INTRODUCTION

A recurring site of both fatal and nonfatal accidents in industry is the confined space. Potentially any structure in which people work could be or could become a confined space. The hazards in confined spaces depend on the conditions that exist at the time of entry or that develop as a result of work activity.

The first chapter explored the role of atmospheric hazards in accidents involving confined spaces. Until recently, the importance of atmospheric hazards in these accidents was largely unknown (NIOSH 1979). This changed with the publication of the NIOSH (National Institute for Occupational Safety and Health) criteria document and subsequent investigative reports by both NIOSH and OSHA (Occupational Safety and Health Administration) (NIOSH 1979, 1994; OSHA 1982a, 1982b, 1983, 1985, 1988, 1990). While the theme of these documents was not always oriented to confined spaces, accidents involving confined spaces were reported in each.

These documents drew needed attention to the role of atmospheric hazards in these accidents. In so doing, however, they left unexplored the role of other hazardous conditions. The approach taken in these documents regarding other hazardous conditions typically was to exhort elimination or control. The documents provided approaches for eliminating or controlling the more obvious of these, mainly mechanical, electrical, and process hazards. Other hazardous conditions were not mentioned or discussed.

This chapter explores the role of nonatmospheric hazardous conditions in accidents involving confined spaces. As with atmospheric hazards, the only sources of information are reports on fatal accidents. NIOSH recently published a report on fatal accidents in confined spaces that provided information about both atmospheric and nonatmospheric hazardous conditions (NIOSH 1994). MSHA (Mines Safety and Health Administration) also published a report on fatal accidents (MSHA 1988). This focused on accidents resulting from fluid behavior of bulk solids. Some of these accidents occurred in confined spaces.

The enigma of confined spaces is that under some circumstances the conditions in a particular workspace may pose no extraordinary hazard, yet, following seemingly minor change, the conditions become life threatening. On the other hand, the interior of some structures poses serious hazards under almost all circumstances. That is to say, the entire purpose for the structure is to contain hazardous conditions. Entry, without eliminating or at the very least controlling, these conditions, would be fatal. The dangers in situations such as these are easily recognizable. These spaces usually receive due recognition and preparation prior to entry. In other situations, the hazardous conditions develop only under limited circumstances. The hazards that develop can be equally serious, yet fail to receive recognition. Further, the attributes of the space exacerbate the hazardous condition. The hazardous condition may be transient and subtle, and therefore difficult to recognize and to address. This, then, explains some of the difficulty in the management of hazardous conditions in many confined spaces.

Nonatmospheric hazardous conditions are similar in many ways to atmospheric hazards. Some provide a warning that is detectable by the senses. Instrumental methods exist for detecting and estimating the magnitude of some of these hazards. More typically, however, awareness and knowledge develop over time through information passed from one individual to another. The content of this information is not intuitive. It must be learned, often from firsthand experience, and then passed from person to person. Continuity of communication between past, present, and future generations of workers about the hazards of a site is critical to ensuring safety, but, loss of this information easily could become a casualty of downsizing.

Much of industry and industrial equipment is concerned with the storage, transfer, and transport, and use or conversion of energy. (Energy cannot be created or destroyed, but can be converted from one form to another.) The actual outcome of the endeavor may be production of a tangible item or substance that may seemingly have nothing to do with energy. Nevertheless, production of products and substances involves energy, and energy has everything to do with the hazardous conditions that can occur in confined spaces. Viewed in this context, the sole purpose of many industrial structures is to confine processes and activities that involve energy. Entry into these structures before the energy is discharged and maintained at nonhazardous levels represents a preaccident condition.

Accidents associated with confined spaces differ from those occurring in normal workspaces. A minor error or oversight in preparation of the space, selection or maintenance of equipment, or work activity can produce a fatal outcome. The tolerance for mistakes is very small.

While most of the victims of accidents in confined spaces understandably are tradespeople, they also include engineering and technical people and supervisors and high-level managers. Rescue is not an issue in accidents involving nonatmospheric hazardous conditions. Energy dissipation occurs during the accident; this does not impact on the rescue attempt.

The modes in which industry operates (production and nonproduction) may influence the occurrence and outcome of accidents involving nonatmospheric hazardous conditions. Production mode implies relatively invariate or steady-state conditions day after day. These could occur for many months or even years. Predictability and repeatability tend to characterize conditions that occur within this framework. Nonproduction modes include construction, inspection, maintenance, modification, and rehabilitation. Activities occurring within this mode usually are not repetitive, nor is the outcome fully predictable. Variability characterizes these activities and the conditions under which they are carried out. Work occurring during nonproduction modes can be short in duration and highly hazardous and can occur unpredictably, often during off-shifts. These situations often are nonroutine and nonrepetitive.

The nature of work also influences the likelihood of encounters with confined spaces. In some segments of industry encounters are infrequent. In others, they are routine daily occurrences. Similarly, occupation also influences the likelihood of encounter with confined spaces. Some workers encounter confined spaces rarely, while others work in confined spaces all the time.

NONATMOSPHERIC HAZARDOUS CONDITIONS: MAGNITUDE OF THE PROBLEM

The main sources of information on nonatmospheric hazardous conditions in confined spaces are reports published over the last 15 years. The Division of Safety Research of NIOSH commissioned the first quantitative analysis of accidents that occurred in confined spaces (NIOSH 1978). The NIOSH definition of the term "confined space" used in this investigation contained the following elements:

- Limited openings for entry and exit by design
- Unfavorable natural ventilation that could contain or produce dangerous air contaminants
- Not intended for continuous employee occupancy

The NIOSH definition excluded accidents related to nonatmospheric hazardous conditions.

More recently, OSHA published a series of analytical reports on fatal accidents. Each report focused on a specific type of work environment or activity. The themes of several of the reports — fires and explosions in confined spaces, lockout and tagout, grain handling, toxic and asphyxiating atmospheres in confined spaces, welding and cutting, and shipbuilding and repair — are relevant for further discussion (OSHA 1982a, 1982b, 1983, 1985, 1988, 1990). Each report, excluding that on welding and cutting, provided a summary for every accident that was used in the investigation.

The definition for confined workspace used by OSHA in these investigations contained the following criteria:

- An enclosed or partially enclosed workspace
- Limited means of entry and exit
- Subject to accumulation of toxic or flammable contaminants
- May develop an oxygen deficiency
- Not intended for continuous employee occupancy

This definition differed slightly from that utilized by NIOSH. Neither referred to nonatmospheric hazardous conditions that exist or can develop in confined spaces. Regardless of the definition, the OSHA reports are useful because of the inclusion of the accident summaries. The accident summaries indicate that some of the accidents would be considered to have occurred in confined spaces under criteria generally accepted at this time.

The first OSHA report (1982a) examined the link between fatal accidents and fires and explosions in confined workspaces. Associated with the 50 accidents selected for the study during the period 1974 to 1979 were 76 fatal injuries. Heat and the energy generated by the fire and/or explosion were the major causative agents. Multiple fatalities occurred in many of these accidents. Some of the accidents involved process hazards.

The second OSHA report (1982b) examined fatal accidents involving energized equipment. Associated with the 83 cases selected for the study during the period 1974 to 1980 were 83 fatalities. Although the involvement of confined spaces was not considered as a factor during this study, many of the workspaces would satisfy criteria generally accepted at this time.

The OSHA study of grain handling reported on 105 fatal accidents that occurred during the period 1977 to 1981 (OSHA 1983). Associated with these accidents were 126 fatal injuries. Although the involvement of confined spaces was not considered as a factor during this study, many of the workspaces would satisfy criteria generally accepted at this time.

The fourth OSHA report (1985) examined the relationship between fatal accidents in confined workspaces and toxic and asphyxiating (oxygen-deficient) atmospheres and other causes. While most of the accidents involved hazardous atmospheric conditions, some were caused by other hazards. Associated with the 122 accidents selected for the study during the period 1974 to 1982 were 173 fatalities.

The fifth report published by OSHA (1988) examined the relationship between fatal accidents and welding and cutting. Associated with the 217 accidents selected for the study during the period 1974 to 1985 were 262 fatal injuries. In 22% of the accidents, the welder was working inside a workspace that would be considered a confined space according to criteria generally accepted at this time.

The OSHA study of shipbuilding and repairing (1990) reported on 151 fatal accidents that occurred during the period 1974 to 1984. Associated with these accidents were 176 fatal injuries. In 36% of the accidents, the victim was working inside a workspace that could be considered a confined space according to criteria generally accepted at this time.

The authors of the OSHA reports repeatedly commented about the lack of completeness of the data used in these studies. This occurred because the OSHA system for coding causal factors in accidents contained no reference to the parameters examined in these studies. Also, data were not available from all of the U.S. states. Despite these shortcomings, the summaries from individual accidents contained in these reports represent a valuable resource for further enquiry.

During the same period, NIOSH initiated the Fatality Assessment and Control Evaluation (FACE) program to investigate fatal accidents (NIOSH 1994). The purpose of FACE was to identify factors that increase the risk of work-related fatal injury. While previous studies utilized only written reports or forms as the source of information, FACE included on-site visits. These visits provided the opportunity for firsthand observation and investigation of the worksite by a team of specialists and interviews with witnesses. The intent of the team approach was to provide continuity and consistency from one investigation to another and the opportunity for more precise reconstruction of accidents. FACE currently focuses on accidents involving falls from heights, contact with sources of electrical energy, entry into confined spaces, and contact with machinery.

FACE investigated 423 accidents involving the deaths of 480 workers from 1983 to 1993. Of these, 70 involved confined spaces and caused 109 deaths. Multiple deaths occurred in 25 of these accidents. The NIOSH report provides a more recent view of trends that are occurring than is obtainable from the OSHA reports.

Summaries of the FACE investigations have appeared in Operations Forum and the NIOSHTIC electronic database. These and other sources led NIOSH to estimate that approximately 200 persons in the U.S. die annually due to accidents in confined spaces (Reese and Mills 1986).

A compilation of summaries of accidents involving confined spaces in mining was undertaken by MSHA (MSHA 1988). MSHA documented 44 fatalities from 38 accidents that occurred during the period 1980 to 1986. MSHA subsequently reported (1994) that by the end of 1993 there had been 57 fatalities in confined spaces. MSHA noted that the majority of the fatalities resulted from collapse of bridged and caked material in bins, silos, and hoppers.

Table 1.1 in Chapter 1 summarized the data presented in the OSHA and MSHA reports (OSHA 1982a, b, 1983, 1985, 1988, 1990; MSHA 1988). As mentioned, a note of caution is needed before entering into overly enthusiastic use of these data. These reports spanned partially overlapping time frames of varying length. Some of the accidents were utilized in more than one report. For example, accidents involving welding and cutting and fire and explosion reappeared in the report on shipbuilding and repair. This is a natural outcome from the overlap between workspace specificity and work environment.

TABLE 2.1
Role of Nonatmospheric Hazardous Conditions in Fatal Accidents

Action/Condition	Accidents	Fatalities	F/A	Relative occurrence % of total* A	F
Grain handling	38	39	1.0	36	31
Lockout/tagout	29	29	1.0	35	35
Fire/explosion	6	9	1.5	12	12
Toxic	0	0	0.0	0	0
Asphyxiating	0	0	0.0	0	0
Other	19	19	1.0	16	11
Welding/cutting[a]					
Interior	12	14	1.2	7	7
Shipbuilding/repair	18	18	1.0	58	50
Mining	26	32	1.2	68	73

Note: A is fatal accidents associated with nonatmospheric hazardous conditions; and F is fatalities occurring during these accidents.

[a] The data on welding and cutting accidents used in this table reflect analysis of individual accident summaries contained in the OSHA report (1988). The report claims that this abridged group (164 cases and 190 associated fatalities) reflects the total (217 cases and 262 associated fatalities).

Table 1.2 illustrated the potential role of confined spaces in the OSHA and MSHA reports (OSHA 1982a, b, 1983, 1985, 1988, 1990; MSHA 1988). In order to make this comparison, classification of the workspaces was performed according to broader criteria that included hazardous nonatmospheric conditions. To the extent possible in this and subsequent analysis, use of data that appeared in more than one report was minimized.

Data presented in the preceding tables indicated the relative involvement of confined spaces and/or confined atmospheres in these accidents. Confined workspaces were involved in fatal accidents contained in each of the OSHA reports, according to this analysis. Table 2.1 indicates the magnitude of the role of nonatmospheric hazardous conditions in accidents provided in the OSHA and MSHA reports.

The first columns of Table 2.1 present the actual number of accidents and fatalities, based on review of the summaries of individual accidents. The most severe accidents, expressed as the ratio of fatalities to accidents, occurred during explosions (releases of steam). The greatest fraction of the total number of accidents and fatalities occurred in mining.

The OSHA reports utilized in creating Table 2.1 utilized overlapping time frames of different lengths. This variation in format prevents direct comparison between categories. Table 2.2 provides the data normalized to a common time frame of 10 years. This technique necessitates extrapolation of data contained in most categories and interpolation in the case of welding and cutting and shipbuilding and repair. Of necessity, this comparison will be imperfect, since the OSHA report on welding and cutting provided an incomplete group of accident summaries. Assuming the validity of these calculations, nonatmospheric hazardous conditions would be associated with 214 of the 542 (39%) accidents involving confined spaces. Associated with these accidents were 232 fatal

TABLE 2.2
Data Normalized from OSHA Reports of Fatal Accidents

	Projected number					
	Nonatmospheric hazardous conditions		Confined spaces		All causes	
Action/condition	**A**	**F**	**A**	**F**	**A**	**F**
Grain handling	76	78	116	156	210	252
Lockout/tagout	41	41	41	41	119	119
Fire/explosion	10	15	83	127	83	127
Toxic	—	—	61	89		
Asphyxiating	—	—	51	79		
Other	21	21	21	21		
Total	21	21	133	189	136	192
Welding/cutting[a]						
Interior	13	15	40	51		
Exterior	0	0	64	83		
Total	13	15	104	134	217	262
Shipbuilding/repair	16	16	28	33	137	160
Mining	37	46	37	46	54	63
Totals	**214**	**232**	**542**	**726**	**956**	**1,175**

Note: A is fatal accidents; and F is fatalities occurring during these accidents.

[a] The data on welding and cutting accidents used in this table reflect analysis of individual accident summaries contained in the OSHA report (OSHA 1988). The report claims that this abridged group (164 cases and 190 associated fatalities) reflects the total (217 cases and 262 associated fatalities).

injuries, or 1.1 fatal injuries per accident. This frequency is considerably lower than the 1.5 fatal injuries per accident for accidents involving atmospheric hazards (Table 1.4). Accidents involving nonatmospheric hazardous conditions are considerably less consequential than those involving atmospheric hazards.

The NIOSH report on fatal accidents (1994) provided statistics from the National Traumatic Occupational Fatalities (NTOF) database, as well as the FACE program. In creating these statistics, NIOSH stressed the incompleteness of the information base from which they were obtained. The NTOF database provides comprehensive data on fatal injuries arising from work in confined spaces, as well as other situations. Cause is assigned from the death certificate. Table 2.3 provides summary statistics for confined spaces for the decade 1980 to 1989. According to these statistics, engulfment involving loose materials is the second ranking cause of fatalities in confined spaces. In compiling these statistics, NIOSH did not incorporate engulfment involving collapse of trenches, excavations, and ditches. Considered from the perspective that these workspaces also are confined spaces, engulfment from all causes easily would become the first ranking cause of fatal injury.

DEVELOPMENT OF NONATMOSPHERIC HAZARDOUS CONDITIONS

The temporal relationship between development of the hazardous condition and onset of the accident is an important one. Hazardous conditions could exist prior to opening the space for entry or they could develop during the performance of work activity.

TABLE 2.3
NTOF Statistics on Confined Spaces for 1980 to 1989

Statistic	Accidents	Fatalities
Atmospheric hazards		373
Engulfment in loose materials		227
Accidents with other causes (drowning, engulfment in sludge and manure)		70
Accidents in confined spaces	585	670
Accidents with 1 victim	428	
Accidents with 2 victims	61	
Accidents with 3 victims	9	
Accidents with 4 victims	2	
Accidents involving cave-ins (trenches, excavations, ditches)	572	606

TABLE 2.4
Temporal Development of Hazardous Conditions

	Percent of fatal accidents							
Condition	GH	L/T	F/E	CS	W/C[a]	S/R	M	NIOSH
Pre-existing	95	21	0	53	8	0	100	67
Work activity	5	79	100	47	92	100	0	33

Note: GH is grain handling; L/T is lockout/tagout; F/E is fires and explosions; CS is hazardous nonatmospheric conditions in confined spaces; W/C is welding and cutting (interior work only); S/R is shipbuilding and repair; M is mining; and NIOSH is data from the FACE study.

[a] The data on welding and cutting accidents used in this table reflect analysis of accident summaries contained in the OSHA report (1988). The report claims that this abridged group (164 cases and 190 associated fatalities) reflects the total (217 cases and 262 associated fatalities).

The OSHA, MSHA, and NIOSH reports provide an opportunity to investigate this question (OSHA 1982a, b, 1983, 1985, 1988, 1990; MSHA 1988; NIOSH 1994). For the purposes of this analysis, hazardous conditions existing prior to the start of work qualify as preexisting hazards. Hazardous conditions produced by the actions of someone in the space at the onset of the accident qualify as work activity. Table 2.4 summarizes this examination. The data presented here reflect subjective estimation of the situations described in the accident summaries.

The data suggest that nonatmospheric hazardous conditions occurring in specific work environments in most situations were highly polarized between preexisting conditions and those created by work activity. That is, the hazardous condition highly depended on the work environment and/or the work activity.

NATURE OF NONATMOSPHERIC HAZARDOUS CONDITIONS

The preceding section illustrates the importance of preexisting conditions and work activity as causative agents in accidents associated with confined workspaces. Circumstances highlighted by these data should be amenable to correction once the nature of these circumstances are understood.

TABLE 2.5
Hazardous Nonatmospheric Conditions in Confined Spaces

	Percent of fatal accidents							
Condition	**GH**	**L/T**	**F/E**	**CS**	**W/C**[a]	**S/R**	**M**	**NIOSH**
Engulfment	82	—	—	42	—	—	100	40
Entanglement	18	100	—	5	8	6	—	7
Process hazard	—	—	100	11	—	6	—	13
Electrical	—	—	—	16	42	39	—	20
Unstable interior Condition	—	—	—	5	50	—	—	—
Fall from height	—	—	—	11	—	49	—	20
Other	—	—	—	10	—	—	—	—

Note: GH is grain handling; L/T is lockout/tagout; F/E is fires and explosions; CS is hazardous nonatmospheric conditions in confined spaces; W/C is welding and cutting (interior work only); S/R is shipbuilding and repair; M is mining; and NIOSH is data from the FACE study.

[a] The data on welding and cutting used in this table reflect analysis of individual accident summaries contained in the OSHA report (1988). The report claims that this abridged group (164 cases and 190 associated fatalities) reflects the total (217 cases and 262 associated fatalities).

Table 2.5 summarizes hazardous conditions described in the accident summaries contained in the OSHA, MSHA and NIOSH reports on fatal accidents (OSHA 1982a, 1982b, 1983, 1985, 1988, 1990; MSHA 1988; NIOSH 1994). The data presented here reflect subjective estimation of the situations described in the accident summaries.

The data in Table 2.5 suggest that limited classes of hazardous conditions are involved in these accidents. This, of course, reflects the focus of particular reports toward types of work, but also the nature of work activity that occurs in some industrial sectors. Taken together, the data presented in Table 2.4 and 2.5 suggest that engulfment is highly associated with hazardous conditions that developed prior to the entry and start of work. Conversely, the other hazardous conditions identified in Table 2.5 are highly associated with work activity.

The data in Table 2.5 also can be expressed numerically. This would provide an estimate of the relative number of fatal accidents in each category. Table 2.6 provides the data normalized to a common time frame of 10 years. This technique necessitates extrapolation of data contained in most categories and interpolation in the case of welding and cutting and shipbuilding and repair. Of necessity, this comparison will be imperfect since the OSHA report on welding and cutting provided an incomplete group of accident summaries. The NIOSH data from the FACE study are not included, as these represented accidents specifically selected for investigation.

According to the data, 210 accidents due to nonatmospheric hazardous conditions would occur in confined spaces in a decade. Engulfment would be the most serious nonatmospheric hazardous condition, causing slightly more than half of these accidents. This result is consistent with the NTOF data provided in the NIOSH report and presented in Table 2.3 (NIOSH 1994). The estimate provided in Table 2.6 is considerably smaller than that provided by the NTOF data. This could reflect improvements in reporting about the nature of this problem.

Engulfment is followed by entanglement, process and electrical hazards, falls from heights, and unstable interior conditions as causative agents.

TABLE 2.6
Hazardous Nonatmospheric Conditions in Confined Spaces

	Projected number of fatal accidents							
Condition	**GH**	**L/T**	**F/E**	**CS**	**W/C[a]**	**S/R**	**M**	**Total**
Engulfment	62	—	—	9	—	—	37	108
Entanglement	14	41	—	1	1	1	—	58
Process hazard	—	—	10	2	—	1	—	13
Electrical	—	—	—	3	4	6	—	13
Unstable interior Condition	—	—	—	1	5	—	—	6
Fall from height	—	—	—	2	—	8	—	10
Other	—	—	—	2	—	—	—	2

Note: GH is grain handling; L/T is lockout/tagout; F/E is fires and explosions; CS is hazardous nonatmospheric conditions in confined spaces; W/C is welding and cutting (interior work only); S/R is shipbuilding and repair; and M is mining.

[a] The data on welding and cutting accidents used in this table relfect analysis of individual accident summaries contained in the OSHA report (1988). The report claims that this abridged group (164 cases and 190 associated fatalities) reflects the total (217 cases and 262 associated fatalities).

Data from OSHA 1982a, b, 1983, 1985, 1988, 1990; MSHA 1988; and NIOSH 1994.

ETIOLOGY OF ACCIDENTS INVOLVING CONFINED SPACES

Accidents associated with confined spaces are complex events. They include technical, procedural, and social elements. The focus of the remainder of this chapter is to develop information about the fundamental elements that comprise these accidents. This will form the basis for strategies taken in subsequent chapters to ameliorate the situation.

TECHNICAL ELEMENTS

Technical elements describe the situation and the environment in which these accidents occurred. Technical elements discussed here include:

- Temporal aspects
- Description of the space
- Other aspects
- Rescue

The summaries of individual accidents provided in the OSHA and MSHA reports contain useful information about these aspects of the accidents (OSHA 1982a, b, 1983, 1985, 1988, 1990; MSHA 1988). The NIOSH report on fatal accidents in confined spaces provides supplemental data on this subject (NIOSH 1994). This will be utilized, where beneficial to the discussion.

Temporal Aspects

Temporal aspects refer to the time of day, day of the week, and season of the year. Table 2.7 illustrates occurrence of accidents in major categories as a function of time of day. Initial inspection of the data indicated that times recorded by investigators tended to be hourly or half-hourly. Tabularizing the data according to conventional record points such as the hour or half hour could introduce a bias. Accordingly, this table presents data in half-hour intervals that begin and end on the quarter hour. Sufficient records for listing by the half hour are available only for engulfment and entanglement accidents. For the other accidents, the listing is made by shift or partial shift.

The data suggest that most of the accidents occur during the day shift (normally 8:00 to 16:00). Most of the day-shift accidents occur during the morning. A greater fraction occur during the afternoon shift (normally 16:00 to 24:00) than the night shift (normally 00:00 to 8:00). Of course, these findings also reflect the level of activity that occurs in the different shifts. These observations are essentially the same as for fatal accidents caused by atmospheric hazards.

Establishing the reason for the trend is a point of minor importance. The fact that the vast majority of the accidents occurs during the normal workday should make these situations more amenable to correction. That is, these are the normal hours of work of health and safety advisors. These individuals have the specialized knowledge needed to address the hazards posed by these situations.

The second line of inquiry under temporal aspects is occurrence of these accidents during the workweek. Table 2.8 summarizes the data provided in the summaries of individual accidents. This analysis focuses on the most common causes of fatal accidents.

The data indicate that most of the fatal accidents occur during the normal workweek. The proportion occurring on weekends ranges around 10% except for falls. The larger proportion of fatal accidents that occurred on weekends in this case may be an artifact from the small number of accidents (n = 13). Of course, these findings also reflect the level of activity that occurs in the different days of the workweek. These observations are essentially the same as for fatal accidents caused by atmospheric hazards.

Again, the reason for the trend is a point of minor importance. The fact that the vast majority of the accidents occurs during the normal workweek should make these situations more amenable to correction.

The third temporal factor is season of the year. The climate in most parts of the U.S. (the site of origin of the data) undergoes distinct seasonal variations. Many of the accidents occurred in outdoor locations and, therefore, were potentially subject to this influence. Climatic factors, such as extreme cold and heat, could have influenced the occurrence and outcome of these situations. However, summaries of individual accidents did not always clearly indicate location.

Table 2.9 provides data about the temporal occurrence of these accidents by season of the year. There were insufficient data to provide a breakdown by month. In addition, engulfments that occurred in grain handling were separated from those in mining, since there may be seasonal components in each. Engulfments in mining include those occurring in other industrial sectors. According to the data, seasonal differences exist in the occurrence of accidents in the different categories. Differences also exist within categories, such as engulfment. The pattern of engulfment accidents in grain handling differs from that in mining and general industry. While the pattern of occurrence of these accidents appears not to be random, it may reflect seasonal activity in the industries whose accidents were recorded here.

Description of the Space

Important technical aspects of confined spaces are their functional characteristics. Table 2.10 lists functional characteristics/work locations of confined spaces in which fatal accidents occurred (OSHA 1982a, b, 1983, 1985, 1988, 1990; MSHA 1988; NIOSH 1994). The number of accidents in each column is projected to a common period of 10 years, so that horizontal, as well as vertical, comparisons are enabled. This analysis focuses on the most common causes of fatal accidents.

TABLE 2.7
Temporal Occurrence of Fatal Accidents: Time of Day

	Percent of fatal accidents				
Time Interval	**Engulf**	**Entangle**	**Process**	**Electrical**	**Falls**
00:15–00:44					
00:45–1:14					
1:15–1:44					
1:45–2:14					
2:15–2:44					
2:45–3:14		3			
3:15–3:44					
3:45–4:14					
4:15–4:44		3			
4:45–5:14					
5:15–5:44	5				
5:45–6:14		4			
6:15–6:44					
6:45–7:14					
7:15–7:44					
7:45–8:14	2				
Total for shift	**7**	**10**	**14**		
8:15–8:44	3	3			
8:45–9:14	2				
9:15–9:44	10				
9:45–10:14	3	10			
10:15–10:44	13	13			
10:45–11:14	5	10			
11:15–11:44	7	3			
11:45–12:14	8	10			
Halfshift Total	**51**	**49**	**43**	**44**	**36**
12:15–12:44					
12:45–13:14	2	3			
13:15–13:44	3				
13:45–14:14	6				
14:15–14:44	2	3			
14:45–15:14	8	3			
15:15–15:44	2	10			
15:45–16:14	5				
Total for shift	**73**	**74**	**57**	**66**	**71**
16:15–16:44	8				
16:45–17:14	7	6			
17:15–17:44					
17:45–18:14		6			
18:15–18:44					
18:45–19:14					
19:15–19:44	3				
19:45–20:14		4			
20:15–20:44					
20:45–21:14					
21:15–21:44					
21:45–22:14					
22:15–22:44					

TABLE 2.7 (continued)
Temporal Occurrence of Fatal Accidents: Time of Day

	Percent of fatal accidents				
Time Interval	**Engulf**	**Entangle**	**Process**	**Electrical**	**Falls**
22:45–23:14					
23:15–23:44	2				
23:45–00:14					
Total for shift	**20**	**16**	**29**	**44**	**29**

TABLE 2.8
Temporal Occurrence of Fatal Accidents: Day of the Week

	Percent of fatal accidents				
Day	**Engulf**	**Entangle**	**Process**	**Electrical**	**Falls**
Sunday	—	8	—	—	23
Monday	18	27	11	25	15
Tuesday	24	11	11	19	8
Wednesday	20	8	33	13	15
Thursday	18	22	34	6	23
Friday	10	19	—	31	16
Saturday	10	5	11	6	—
Workweek	**90**	**87**	**89**	**94**	**77**
Weekend	**10**	**13**	**11**	**6**	**23**

TABLE 2.9
Temporal Occurrence of Fatal Accidents: Season of the Year

	Percent of fatal accidents					
	Engulf					
Month	**Grain**	**Mining**	**Entangle**	**Process**	**Electrical**	**Falls**
Spring	33	28	13	22	38	38
Summer	31	9	25	56	44	31
Fall	21	25	30	11	6	15
Winter	15	38	32	11	12	16

The data suggest that some accidents occur with high frequency in certain workspaces. For example, a high proportion of engulfments occurred in bins and chutes, with smaller proportions in hoppers and silos. A high proportion of entanglements occurred in a broad classification of locations called "mechanical equipment." A high proportion of process accidents occurred in rooms and vaults. Electrocutions occurred primarily in tanks/containers and vaults. Similarly, a high proportion of falls occurred in tanks.

Taken together, the data suggest that the location of most accidents involving nonatmospheric hazards will be bins and chutes, a broad range of sites called mechanical equipment, and tanks and containers. Less frequent sites include rooms, vaults, silos, and hoppers. Of course, the least frequent sites must not be ignored.

TABLE 2.10
Confined Spaces Associated with Fatal Accidents

Location	Number of fatal accidents					
	Engulf	Entangle	Process	Electrical	Fall	Total
Tank/container	3	1		7	9	20
Process vessel	2	1	2	1		6
Bin	67	1				68
Hopper	8	1				9
Silo	9				1	10
Chute	29					29
Trench	2		2			4
Pit	2	5				7
Room	3		6	2	1	12
Vault		1	6	5		12
Tunnel	2		2			4
Riser pipe					2	2
Mechanical equip.		21				21
Auger		5				5

From OSHA 1982a, b, 1983, 1985, 1988, 1990; MSHA 1988; and NIOSH 1994.

TABLE 2.11
Confined Spaces Associated with Fatal Accidents

Location	Percent of Fatal Accidents					
	Engulf	Entangle	Process	Electrical	Fall	Total
Tank/container	2	3		47	69	10
Process vessel	2	3	11	7		3
Bin	53	3				33
Hopper	6	3				4
Silo	7				8	5
Chute	22					14
Trench	2		11			2
Pit	2	13				3
Room	2		33	13	8	6
Vault		3	33	33		6
Tunnel	2		12			2
Riser pipe					15	1
Mechanical equipment		58				10
Auger	14					2

From OSHA 1982a, b, 1983, 1985, 1988, 1990; MSHA 1988; and NIOSH 1994.

Another way of examining the preceding data is to compare by percent occurrence, as presented in Table 2.11. Expressing the data in this manner normalizes the information and permits comparison across categories.

Percentages highlight different aspects about the data within columns, since each column is normalized. Cross-comparison suggests that tanks and compartments are the location of most electrocutions and falls. Similarly, vaults are the site of process accidents and electrocutions. Bins and chutes are associated almost exclusively with engulfments, and mechanical equipment, augers, and pits, with entanglements.

The "Total" column provides the global occurrence of accidents by location. When considered globally, bins, chutes, tanks, and containers are the workspaces in which most (2/3) accidents involving nonatmospheric hazardous conditions occur. The observation of limited types of work environments creates the temptation to focus on these for the purposes of generating awareness and creating training materials. However, to reinforce concerns expressed in the previous chapter, this approach fails to address the vacuum created by not identifying the other work environments in which accidents occurred. By narrowing perspective, this approach overly restricts the process of awareness-building.

Other Aspects

Data in the preceding tables indicate that accidents involving nonatmospheric hazardous conditions differ from those associated with hazardous atmospheres. That is, these hazardous conditions tend to occur in specific workspaces. There is little overlap between workspaces and hazardous conditions. All that is common in most cases is that these hazardous conditions occur in confined spaces. By contrast, the various conditions characterized as atmospheric hazards could occur simultaneously in the same space. For this reason, situational aspects of accidents involving nonatmospheric hazardous conditions will be examined in isolation from each other.

This discussion will utilize information contained in the anecdotal summaries provided in the OSHA, MSHA, and NIOSH reports on fatal accidents (OSHA 1982a, b, 1983, 1985, 1988, 1990; MSHA 1988, NIOSH 1994). This area of inquiry was not pursued during the original investigation of these accidents. Extracting this information requires careful reading of events and reflects subjective judgment.

Table 2.12 summarizes information on the elements of fatal accidents involving engulfment. Table 2.12 provides the number of fatal accidents projected to the time-frame of 10 years.

The data suggest that 108 accidents (excluding the NIOSH data from the FACE study) would occur due to engulfment over the period of a decade. About 57% of these accidents would occur during grain handling. The NTOF data in the NIOSH report on fatal accidents indicate that 124 of the 227 fatalities (55%) due to engulfment also occurred during grain handling (NIOSH 1994). (The NIOSH report provided no information about the accidents that corresponded to this data.) Reiterating the discussion following Table 2.3, which summarizes the NTOF data, engulfment would be the leading cause of fatal injury in confined spaces, when trenches and other excavations are considered in this category.

The data in Table 2.12 suggest a mixed pattern for the occupations of the victims of these accidents. The greatest variation occurred in mining; the least, in grain handling. The nature of the work performed at the time of the accident also varied. Work in grain handling mostly was part of the normal routine. In mining and general industry, work activity at the time of the accident mostly was unusual compared to the normal routine. In most circumstances the victim entered the space to start or to speed up the flow of the material. Cleanout of residual material occurred in some situations. In these cases, almost all of the contents of the space were removed previously by gravity. In most circumstances, the contents of the space were not flowing at the time of entry. Entry into spaces where contents were flowing was more likely during grain handling.

Contents usually are drawn from the bottom of the space. The bottom geometry of the space was either flat, with one or more draw-off openings, or tapered. During storage, coalescence of material sometimes occurs. This results from the presence of moisture or biological action. This can result in caking on vertical surfaces of the structure and bridging across the horizontal plane of the material. Material that has not coalesced can flow from the under side of the bridge, thus creating a hollow. Also, "rat holes" may be present. These are vertical channels in the coalesced material. "Rat holes" permit some flow to occur. This flow may hide the existence of a bridge.

TABLE 2.12
Elements of Fatal Accidents Involving Engulfment

	Number of fatal accidents			
Element	**Grain handling**	**Confined spaces**	**Mining**	**NIOSH**
Occupation of victim				
Laborer	50	6	11	2
Trades person			3	
Equipment operator			11	
Supervisor	5		6	1
Owner/manager/farmer	7	1		1
Other		2	6	2
Work activity				
Normal	40	3	11	3
Unusual	22	6	26	3
Reason for entry				
Cleanout	16	2	3	1
Start flow	20	4	14	4
Improve flow	12	3	6	
Other	4		10	1
Unknown	10		4	
Interior condition at entry				
Contents not flowing	34	6	31	5
Contents flowing	28	3	6	1
Reason for engulfment				
Bridge collapse (horizontal)	20	4	21	4
Cake collapse (vertical)	6	1	4	
Flow induced	24	4	10	2
Not known	12		2	

Many of the accidents resulted from the collapse of a bridge during attempts to open a channel for flow. Some accidents resulted following dislodging of material adhering to vertical surfaces.

Another significant cause of these accidents was engulfment by flowing material. In some cases, this process began as soon as the victim made contact with the contents of the space. In other cases, this process began with the collapse of a bridge or caked material. In most cases, the flowing material buried the victim. In some circumstances, the victim suffocated despite not being buried above chest level.

Table 2.13 summarizes information on fatal accidents involving entanglement (OSHA 1982b, 1983, 1988). The data from different reports were combined and projected to a time frame of 10 years.

The data suggest that over 80% of the victims normally worked with the equipment. Most of the work activity could be considered part of the normal routine. Entry typically occurred for cleaning, repair, inspection, or adjustment. In 35% of the accidents, entry occurred during operation of the equipment. Data in Table 2.11 indicate that 85% of these accidents would occur within the confines of mechanical equipment, augers, and machine pits.

The column "factors in entry during operation" provides some insight into the actions and motivation of the victims. In most of these situations, the victim deliberately entered the space while the hazardous condition was present. This occurred despite the existence of protective measures in many situations. These ranged from verbal instructions to written procedures and training to lockout/tagout systems, all intended to warn about and prevent the consequences of this action.

TABLE 2.13
Elements of Fatal Accidents Involving Entanglement

Element	Projected number
Occupation of victim	
Laborer	17
Trade	16
Supervisor	4
Manager	1
Other	2
Work activity	
Normal	24
Unusual	16
Reason for entry	
Clean	17
Repair	7
Adjust	3
Inspect	5
Other	3
Unknown	5
Reason for entanglement	
Entry during operation	14
Accidental activation	25
Equipment failure	1
Factors in entry during operation	
Intent to perform some action	8
Unintended entry	3
Reason for entry not known	3
Factors in accidental activation	
Miscommunication	12
Control remote from equipment	1
Control activated accidentally	10
Reason for entry not known	8

Data from OSHA 1982b, 1983, and 1988.

The column "factors in accidental activation" provides insight into why equipment was started while the victim was inside. Since more than one factor may have been operative, the total is greater than the number of accidents. Miscommunication was the most important factor in these situations. Miscommunication led to misperception about the location of the victim and status of occupancy of the equipment. In most situations, formal control procedures, such as lockout/tagout, were not utilized. However, one accident occurred despite the correct use of a lockout procedure.

Redundancy of control also was a factor in accidental activation. In many situations, safety depended on a single level of control. Activation, either accidental or deliberate, caused the accident. One accident occurred when a manual override was used inappropriately. Depressing the control opened a sliding gate. The off position of the control released the gate to the closed position as it was intended to do. Most of the accidents occurred because of activation through a single level of control. Controls were activated inadvertently or inappropriately or through mistaken identity. Multiple levels of control, however, do not necessarily provide greater security. One accident occurred following the deliberate activation of three levels of control.

Table 2.14 summarizes information on fatal accidents involving electrocution. The data from different reports were combined and projected to a time frame of 10 years.

TABLE 2.14
Elements of Fatal Accidents Involving Electrocution

Element	Projected number
Occupation of victim	
Laborer	5
Electrician	4
Welder	7
Other trade	1
Work activity	
Normal	16
Unusual	1
Reason for entry	
Welding	8
Electrical repair	2
Electrical installation	2
Clean	3
Paint	1
Other	1
Nonwelding Accidents	
Voltage	
110	5
220–1,000	2
>1,000	2
Factors at time of accident	
Water	3
Perspiration	3
Bare conductors	4
Ineffective ground	3
Current leakage	3
Work while active	4
Equipment failure	4
Circuit not tested	5
Reason for electrocution	
Existing energized circuit	5
Accidental activation	2
Exposed an energized circuit	2
Welding Accidents	
Factors at time of accident	
Water	3
Perspiration	5
Bare conductors	3
Equipment failure	1
Reason for electrocution	
Ineffective isolation	6
Body contact with electrode	2
Not known	1

Data from OSHA 1985, 1988, and 1990.

The data indicated that electrocutions occurred during three distinct activities: electrical work, accidental contact with energized conductors during nonelectrical work, and welding. Welding was considered separately during this analysis, as welding accidents are fundamentally different from other accidents involving electricity. Slightly less than half of the electrocutions occurred during welding or welding-related activity. Almost all of the victims were performing normal work at the time of these accidents. Almost 2/3 would be presumed to be knowledgeable about electrical hazards through trades training.

More than half of the nonwelding accidents involved electrical contact that was not anticipated, for example, during cleaning, painting, or other activity. More than half of these accidents involved contact with normal household voltage, 110 to 120V. More than half of the electrocutions occurred because of contact with a circuit that was energized prior to entry and the start of work. Exposure of energized but protected circuits as a result of work activity occurred in a small minority of cases. Water or perspiration was a factor in 2/3 of these accidents. Bare conductors or ineffective grounding were present in more than 3/4 of these cases. The data also suggest that lack of testing for live electrical circuits and lack of inspection of equipment and conductors were important factors in these accidents.

Almost half of these accidents occurred during welding activity. Electrical welding processes utilize both AC and DC (alternating and direct current), low voltage, high current equipment. Inherent in electrical welding processes is the requirement for the welder to remain electrically isolated. The data suggest that water and/or perspiration was a critical factor in all of these accidents. Water and perspiration can increase the conductivity of insulating materials and the skin. Maintaining an isolated condition is especially critical when the welder is working inside a conductive structure. Conductive metals, such as steel, stainless steel, and aluminum, are the materials of construction of many industrial structures.

Electrical energy was a causative agent in fatal accidents other than electrocutions, namely fires and explosions. Table 2.15 summarizes information obtained from the OSHA reports on fires and explosions in confined spaces and on shipbuilding and repair (OSHA 1982a, 1990). The data from different reports were combined and projected to a time frame of 10 years.

The other fatal accidents caused by electrical energy were explosions and fires. The data suggest that about 70% of the victims in these accidents were performing normal work. Their occupations would suggest that only a small proportion of the victims would have received formal training in electrical hazards. A high proportion of the victims were performing tasks in which electricity was integral to some aspect of the work. All of the nonwelding accidents involved electricity at household voltages, 110 to 120V. Almost all of the ignition sources in the nonwelding accidents were either lightbulbs or arcing electrical equipment or conductors. The lightbulbs either shattered at the time of ignition or acted as hot point sources. The electrical circuit was active at the time of the accident in almost all cases.

Another type of accident involving nonatmospheric hazards in confined spaces was caused by process hazards. Process hazards were a minor cause of accidents in confined spaces. Table 2.16 summarizes information about the elements of accidents involving process hazards. The number of accidents available for analysis is limited. The data are projected to a time frame of 10 years.

The data suggest that most of the situations encountered by the victims were within normal activity. Most of the accidents involved pressurized systems containing steam. Sulfur trioxide easily could have been some other process chemical. Steam systems seem to be especially prone to sudden failure in service. Most of the work activity in these situations did not directly involve equipment in the process system. That is, the entry occurred for a purpose unrelated to involvement with the process system that failed.

These accidents typify the concern for hazards in process systems in confined spaces that remain active during entry and work. That is, the system could fail during occupancy and discharge the contents into the space. The victim then is unable to escape from the space before being affected by the process emission.

TABLE 2.15
Other Fatal Accidents Involving Electrical Energy

Element	Projected number
Occupation of victim	
Laborer	6
Welder	4
Painter	10
Other trade	5
Other	2
Work activity	
Normal	19
Unusual	8
Reason for entry	
Welding	8
Repair	2
Installation	2
Clean	5
Paint	9
Other	1
Type of accident	
Explosion	21
Fire	6
Ignition source	
Lightbulb	16
Electric arc	2
Welding arc	8
Other	1
Voltage	
110	19
Factors at time of accident	
Bare conductors	3
Equipment failure	3
Hot point source	13
Reason for accident	
Existing energized circuit	16
Accidental activation	1
Exposed an energized circuit	1
Other	1

Data from OSHA 1982a and 1990.

Table 2.17 summarizes data from the OSHA reports on fatal accidents due to falls from heights (OSHA 1985, 1990). The data were combined due to limited numbers and projected to a time-frame of 10 years.

The data suggest that the victims almost exclusively were performing normal work activity. This comprised a broad range of tasks. The data suggest that the use of fall arrest equipment or other strategies to protect against falls from heights could have prevented many of these accidents. A concern identified by the investigators was use of "home-made" equipment. Failure of this equipment was a factor in several of these accidents.

The last major causative agent in these accidents was instability of the interior structure of the confined space. That is, movement of the structure or some part of the structure caused the accident.

TABLE 2.16
Elements of Fatal Accidents Involving Process Hazards

Element	Projected number
Occupation of victim	
Laborer	2
Maintenance	6
Other trade	2
Supervisor	1
Work activity	
Normal	8
Unusual	3
Causative agent	
Steam	9
Sulfur trioxide	2
Reason for entry	
Repair	3
Installation	2
Adjust	2
Inspect	3
Clean	1
Factors at time of accident	
Work while active	5
Equipment failure	11
Reason for accident	
Existing energized circuit	9
Exposed an energized circuit	2

Data from OSHA 1982a and 1985.

Table 2.18 summarizes data from the OSHA reports on fatal accidents (OSHA 1985, OSHA 1988). The data were combined due to limited numbers and projected to a time frame of 10 years.

The data suggest that accidents involving unstable interior structural conditions were equally likely to occur during routine work as compared to unusual activity. Work activity associated with installation or removal of parts of the structure or equipment created the unstable condition in almost all situations. Welding or flame cutting created the unstable condition in almost all cases. Failure to support the structure in the unstable condition or preventing contact with it occurred in all cases. The victims apparently failed to recognize the imminent danger in the situation to which they were party.

Rescue

Rescue of accident victims from confined spaces has generated concern in all investigative reports on this subject (NIOSH 1978, 1979, 1994; OSHA 1985). These reports commented that would-be rescuers constituted a considerable proportion of the victims in accidents caused by atmospheric hazards. This subject remains to be explored in accidents involving nonatmospheric hazards. Rescue is both a social and a technical issue.

Social interactions play an extremely important role in the dynamics of accidents that occur in confined spaces. While people who work in confined workspaces usually are unrelated in the familial sense, their actions under stressful situations often reflect close-knit social bonding. Sometimes individuals, co-workers, and other bystanders take charge during accident situations and

TABLE 2.17
Elements of Fatal Accidents Involving Falls from Heights

Element	Projected number
Occupation of victim	
Laborer	5
Painter	4
Welder	1
Other trade	2
Supervisor	3
Work activity	
Normal	15
Unusual	1
Reason for entry	
Welding	2
Repair	1
Installation/removal	2
Clean	3
Paint	5
Unknown	1
Factors at time of accident	
Water	2
Failure to use fall arrest	10
Inappropriate equipment	5
Failure to use other personal protective equipment	1
Reason for accident	
Solvent narcosis	1
Equipment failure	5
Other	6
Not known	3

Data from OSHA 1985 and 1990.

assume the extremely dangerous role of would-be rescuer. Would-be rescuers in these situations often act in a manner that shows complete disregard for their own safety. These individuals willingly assume unreasonable risks in situations clearly recognizable as hazardous.

Table 2.19 summarizes information on rescue contained in the OSHA, MSHA, and NIOSH reports (OSHA 1982a, b, 1983, 1985, 1988, 1990; MSHA 1988; NIOSH 1994). Of course, the data available for examination can provide only a limited view of this subject. First responders acted in one of two ways. (First responders were bystanders to the original event or were first to discover the accident.) Either they attempted to retrieve the victim or did not attempt an entry but sought assistance. These actions define for the purposes of Table 2.19 whether or not rescue was attempted.

According to the data, rescue during the progress of the accident was attempted in a small minority of situations. In most cases, the first person on the scene either deactivated the energy source or withdrew to obtain further assistance. Engulfment accidents and accidents due to process hazards posed the greatest risk to would-be rescuers, since these individuals often were present with the victim at the time of the accident. None of the would-be rescuers perished in the attempt. This contrasts sharply with accident situations involving atmospheric hazards. The hazard exposed by the onset of the accident was directly visible to would-be rescuers in circumstances involving nonatmospheric hazards. This did not appear to have been the case in many of the accidents that involved atmospheric hazards.

TABLE 2.18
Elements of Accidents Involving Unstable Interior Conditions

Element	Projected number
Occupation of victim	
Laborer	2
Welder	2
Other trade	1
Work activity	
Normal	3
Unusual	2
Reason for entry	
Installation/removal	4
Other	1
Action that caused instability	
Welding/flame cutting	4
Other	1
Reason for accident	
Work activity	3
Inherent instability	1
Failure to support structure	5
Other	1

Data from OSHA 1985 and 1988.

TABLE 2.19
Rescue from Confined Spaces

	Percent of accidents				
Characteristic	Engulf	Entangle	Process	Electrical	Fall
Rescue attempted	22	0	14	12	0
Rescue not attempted	78	100	86	88	100
Successful rescue	0	0	0	0	0
Rescuer fatally injured	0	0	0	0	0

Data from OSHA 1982a,b, 1983, 1985, 1988, 1990; MSHA 1988; and NIOSH 1994.

When rescue attempts occurred following entanglement, electrical, and process accidents, the energy source was deactivated. The rescuers were not subjected to the risk of entry experienced by the victim.

CONFINED SPACES — A PROBLEM OF NONATMOSPHERIC HAZARDS

The work of investigators at NIOSH, OSHA, and MSHA has established the consequential role of nonatmospheric hazardous conditions in accidents involving confined spaces. These accidents have occurred across a wide spectrum of industry. Investigation reported in this chapter elaborated on the technical, procedural, and social aspects of these accidents. This examination has revealed deficiencies common to many of these events. By correcting these deficiencies, considerable reduction in the occurrence and severity of these accidents likely would occur.

Accident situations investigated in this chapter were characterized by the (often) sudden and rapid release or conversion of energy. In some cases the hazardous condition developed prior to entry and the start of work. In other situations the hazardous condition developed as a result of work activity. Release or conversion of energy was expressed through engulfment, entanglement and entrapment, and mechanical, process, and other safety hazards.

Accident summaries in the OSHA, MSHA, and NIOSH reports repeatedly demonstrated the importance of unpredictability as a factor in these accidents. What was predictable and familiar through many repetitions and iterations suddenly and inexplicably became an unknown quantity. Addressing the unpredictables that cause these accidents to occur in a proactive manner is a formidable undertaking.

Over the years, the term "confined space" has been used to describe an environment in which an increasing number of hazardous conditions can occur. Current standards and legal statutes on this subject provide many definitions for the term confined space. (Refer to Appendix A for more information on current standards and legal statutes on this subject.) Most of these reflect the needs of regulators or standard-setting groups, rather than managers and technical professionals responsible for addressing problems. Discussion in this chapter has indicated that a definition for the term "confined space" that favors recognition of atmospheric hazards to the exclusion of others would be inadequate.

The meaning incorporated into the term has expanded greatly beyond what the words "confined space" can express. In effect, "confined space" has become jargon or shorthand or code. Jargon is useful only to cognoscenti, an elite of individuals who are familiar with the full meaning intended for the term. The people most needing to appreciate the full scope of the term "confined space" are the least likely to recognize it. Further complicating this problem is that a single term is used to describe hazardous conditions that vary from space to space and from one point in time to another in a specific space. This gap in understanding has been a major impediment to achieving control over the hazards, atmospheric and otherwise.

One of the greatest difficulties with confined spaces is creating a descriptive term containing only a small number of words that will enable recognition of the full complement of hazardous conditions. Hazardous conditions can include those already documented in the OSHA, MSHA, and NIOSH reports, as well as others not yet documented or less severe.

The previous chapter suggested the term "confined space/confined atmosphere" as a means of broadening recognition of hazardous conditions. Whereas the basis of action of atmospheric hazards is the acute toxicity of chemical and biological substances, the basis of action of other hazardous conditions is traumatic physical injury caused by rapid release of energy. This can occur through physical, ergonomic, mechanical, process, safety, and other hazards. A confined space could contain at least 36 different hazardous conditions. A more descriptive and inclusive term would be **confined space/confined atmosphere/confined energy**.

As a working definition, the term "confined space/confined atmosphere/confined energy" could apply to workspaces inside which one or more of the following hazardous conditions could be present or could develop:

- Personal confinement
- Unstable interior condition
- Flowable solid materials or residual liquids or sludges
- Release of energy through uncontrolled or unpredicted motion or action of equipment
- Atmospheric confinement
  - Toxic substances
  - Oxygen deficiency/enrichment
  - Flammable/combustible atmosphere
- Chemical, physical, biological, ergonomic, mechanical, process, and safety hazards

This definition intends to create the broadest possible description of situations in which the preceding conditions can occur or can develop. The definition also intends to emphasize concern about the ability of minor modifications in workspaces to create these conditions.

SUMMARY

This chapter has reported on factors intrinsic to fatal accidents caused by nonatmospheric hazardous conditions in confined workspaces and structures. These conditions arise from the storage of energy in equipment and contents. Accidents associated with nonatmospheric hazards occurred during entry and work activities. Conditions that could cause engulfment are likely to develop prior to the entry and start of work. Conversely, the other hazardous conditions, such as entanglement, electrocution, process hazards, falls from heights, and unstable interior structure are highly associated with work activity.

Nonatmospheric hazards are deceptively innocuous. The hazardous condition is unlikely to provide warning to the senses. Persons entered these spaces apparently without concern for the danger. Seemingly minor errors in judgment produced disproportionate and catastrophic consequences. These consequences appear to be considerably more severe than what results from accident situations involving normal workplaces.

Rescue was not an issue in these accidents. No rescuers were injured or killed. This outcome was fundamentally different from what occurred in accidents involving atmospheric hazards. The hazardous condition was immediately recognizable to respondents once the accident occurred.

Providing safe systems for entry and work in confined spaces/confined atmospheres is the only way to reduce the human costs of these tragedies.

This chapter also has introduced the term "confined space/confined atmosphere/confined energy." This expanded term intends to broaden recognition of hazards that can exist or develop in these workspaces. As a working definition, the term "confined space/confined atmosphere/confined energy" could apply to workspaces inside which one or more of the following hazardous conditions could be present or could develop:

- Personal confinement
- Unstable interior condition
- Flowable solid materials or residual liquids or sludges
- Release of energy through uncontrolled or unpredicted motion or action of equipment
- Atmospheric confinement
  - Toxic substances
  - Oxygen deficiency/enrichment
  - Flammable/combustible atmosphere
- Chemical, physical, biological, ergonomic, mechanical, process, and safety hazards

REFERENCES

Mine Safety and Health Administration: Think "Quicksand": Accidents Around Bins, Hoppers and Stockpiles, Slide and Accident Abstract Program. Arlington, VA: U.S. Department of Labor, Mine Safety and Health Administration, National Mine Health and Safety Academy, 1988.

Mine Safety and Health Administration: Hazard Information Alert: Confined Space Fatalities — 57 since 1980. Arlington, VA: U.S. Department of Labor, Mine Safety and Health Administration, 1994. 3 pp.

National Institute for Occupational Safety and Health: Search of Fatality and Injury Records for Cases Related to Confined Spaces (NIOSH Pub. No. 10947). San Diego, CA: Safety Sciences, 1978.

National Institute for Occupational Safety and Health: Criteria for a Recommended Standard — Working in Confined Spaces (DHEW/PHS/CDC/NIOSH Pub. No. 80-106). Cincinnati, OH: National Institute for Occupational Safety and Health, 1979. 68 pp.

National Institute for Occupational Safety and Health: Worker Deaths in Confined Spaces (DHHS/PHS/CDC/NIOSH Pub. No. 94-103). Cincinnati, OH: National Institute for Occupational Safety and Health, 1994. 273 pp.

Occupational Safety and Health Administration: Selected Occupational Fatalities Related to Fire and/or Explosion in Confined Work Spaces as Found in OSHA Fatality/Catastrophe Investigations. Washington, D.C.: U.S. Department of Labor, Occupational Safety and Health Administration (U.S. DOL/OSHA), 1982a. 76 pp.

Occupational Safety and Health Administration: Selected Occupational Fatalities Related to Lockout/Tagout Problems as Found in Reports of OSHA Fatality/Catastrophe Investigations. Washington, D.C.: U.S. Department of Labor, Occupational Safety and Health Administration (U.S. DOL/OSHA), 1982b. 113 pp.

Occupational Safety and Health Administration: Selected Occupational Fatalities Related to Grain Handling as Found in Reports of OSHA Fatality/Catastrophe Investigations. Washington, D.C.: U.S. Department of Labor, Occupational Safety and Health Administration (U.S. DOL/OSHA), 1983. 150 pp.

Occupational Safety and Health Administration: Selected Occupational Fatalities Related to Toxic and Asphyxiating Atmospheres in Confined Work Spaces as Found in Reports of OSHA Fatality/Catastrophe Investigations. Washington, D.C.: U.S. Department of Labor, Occupational Safety and Health Administration (U.S. DOL/OSHA), 1985. 230 pp.

Occupational Safety and Health Administration: Selected Occupational Fatalities Related to Welding and Cutting as Found in Reports of OSHA Fatality/Catastrophe Investigations. Washington, D.C.: U.S. Department of Labor, Occupational Safety and Health Administration (U.S. DOL/OSHA), 1988. 225 pp.

Occupational Safety and Health Administration: Selected Occupational Fatalities Related to Ship Building and Repairing as Found in Reports of OSHA Fatality/Catastrophe Investigations. Washington, D.C.: U.S. Department of Labor, Occupational Safety and Health Administration (U.S. DOL/OSHA), 1990. 195 pp.

Reese, C.D. and G.R. Mills: Trauma epidemiology of confined space fatalities and its application to intervention/prevention now. In *The Changing Nature of Work and Workforce*. Proc. Third Joint U.S.–Finnish Science Symposium, Frankfort, KY, October 22–24 1986. Cincinnati, OH: National Institute for Occupational Safety and Health, 1986. pp. 65–67.

3 Toxic and Asphyxiating Hazards in Confined Spaces

CONTENTS

INTRODUCTION

Confined spaces are inherently dangerous workplaces and the sites of many fatal and nonfatal accidents. The hazards in confined spaces arise from many factors: the geometric shape that forms the space, internal configuration, bulk and residual contents, active and passive chemical and physical processes, and mechanical and other equipment. The geometric shapes that enclose many confined spaces have the ability to cause atmospheric and personal confinement.

The National Institute for Occupational Safety and Health (NIOSH) and the Occupational Safety and Health Administration (OSHA) published a series of reports that examined fatal accidents associated with various workplace environments (NIOSH 1979, 1994; OSHA 1982a, b, 1983, 1985, 1988, 1990).

A major contribution of the NIOSH and OSHA reports was providing an estimate of the severity of atmospheric hazards in these accidents (NIOSH 1979, 1994; OSHA 1982a, 1985). Categories identified in the NIOSH and OSHA reports included toxic gases and vapors, oxygen deficiency and enrichment, and flammable and explosive atmospheres. Specific toxic agents included carbon dioxide, carbon monoxide and hydrogen sulfide, and vapors from hydrocarbon and halogenated hydrocarbon solvents. Flammable and explosive atmospheres resulted from fuel gases and vapors, vapors from hydrocarbon solvents, welding gases, and suspensions of combustible dusts. Analysis reported in the first chapter suggests that these categories apply to other situations involving atmospheric hazards (OSHA 1983, 1988). In most cases, the hazardous atmosphere had developed prior to the entry into the confined workspace or the start of work.

The OSHA report on confined spaces (1985) listed the toxic chemical agents. In decreasing importance these were: hydrogen sulfide, halogenated organic solvents, carbon monoxide, and organic solvents and fuels. Causes of asphyxiating atmospheres included oxygen deficiency, atmospheric displacement by process gases such as nitrogen, hydrogen, and carbon dioxide, and atmospheric displacement by fuel gases and welding gases. Atmospheric hazards that caused fires and explosions during work occurring inside confined workspaces were similar to those associated with welding and cutting accidents (OSHA 1982a, 1988). In decreasing importance these were: vapors from fuels and organic solvents, and welding and natural gases. Of course, any contaminant capable of causing a fire or explosion also could pose an inhalation hazard.

Accidents involving toxic and asphyxiating atmospheres resulted from acute (brief) exposure to rapidly acting chemical agents. Chemical agents exert acute effects by two main actions: asphyxiation and anesthesia. **Asphyxiation** is the condition of oxygen insufficiency and buildup of carbon dioxide in blood and tissues. **Anesthesia** is loss of sensation and depression of mental function produced by action of chemical agents on the nervous system (Dinman 1978).

As indicated in the OSHA report (1985), brief exposure to a hazardous atmosphere in a confined space can produce fatal consequences. This chapter will focus on atmospheric hazards found in confined spaces.

THE STANDARD ATMOSPHERE

The starting point for any discussion about the respiratory system and respiration is the atmosphere and our relationship with it. The atmosphere varies slightly according to where one lives. This variation reflects weather conditions, altitude, geographic location, and the presence of air pollutants.

Humans, other animals, and plants have evolved to utilize one or more components of the atmosphere. Humans require oxygen for respiration, and therefore survival. Accompanying oxygen in the normal atmosphere are several other gases, as indicated in the following table (after Weast 1987). Some are toxic, yet cause no deleterious effect at concentrations of normal occurrence. At elevated concentrations they have the potential to cause injury and death. Other gases are physiologically inert and exert no influence under normal conditions.

The term "standard atmosphere" refers to a dry atmosphere at sea level, having total pressure of 760 mmHg (101.325 kPa) and temperature of 15°C (Weast 1987). The main factors influencing the relationship between the atmosphere and respiration are composition and pressure. The major composition of the atmosphere remains nearly uniform to an altitude of 15,000 m. Only the total pressure and therefore the pressure of individual components decreases with increasing altitude. This means that the concentration of individual gases does not change in the range of altitudes

TABLE 3.1
Components of the Standard Dry Atmosphere

Component	Composition (ppm)	Partial pressure (mmHg) (re: sea level)
Major		
Nitrogen	780,900	593
Oxygen	209,500	159
Minor		
Argon	9,340	7
Carbon dioxide	335	0.3
Neon	18	0.01
Helium	5	0.004
Methane	2	0.002
Krypton	1	0.001
Hydrogen	0.5	
Nitrous oxide	0.5	
Xenon	0.09	
Ozone	0.01	
Total	1,000,000	760

within which humans live and work. The minor composition of the atmosphere, that containing air pollutants, varies according to geographic location.

In discussions about the atmosphere, observed readings are converted to sea level. This corrects for latitude and altitude. Atmospheric pressure also reflects weather conditions. These can increase or decrease the corrected pressure above or below that of the standard atmosphere. The real-world atmosphere, of course, also contains water vapor. Water vapor displaces other gases, causing slight reduction in their partial pressures and hence concentrations.

Atmospheric pressure and, hence, the partial pressure of oxygen are the major determinants in human survival. Atmospheric pressure decreases by 50% for each increase in elevation of 5,500 m (de Treville 1988). Thus, under normal conditions, elevation is the major variable affecting the pressure exerted by the atmosphere and its major components.

HUMAN RESPIRATION

Human respiration occurs at two levels: physiological and biochemical. Physiological respiration involves the process of gas exchange in the lungs and transport of oxygen and carbon dioxide through the body. Biochemical respiration occurs within cells of the body.

Respiration is the general name given to a process that incorporates a number of elements, including (Olishifski and Benjamin 1988):

- Breathing — movement of the chest–lung complex to ventilate the alveolar spaces
- External respiration — exchange of gases between the alveolar airspaces and the blood
- Internal respiration — exchange of gases between blood and the cells
- Cellular respiration — mitochondrial oxidation/reduction for utilization of food energy

Collectively, the acute effects of the asphyxiating and toxic gases and vapors identified in the studies of accidents that occur in confined spaces can interfere with respiration at all stages.

TABLE 3.2
Atmospheric Conditions in Different Environments

	Component							
Atmospheric Condition	**A**	**B**	**C**	**D**	**E**	**F**	**G**	**H**
Temperature (°C)	15	25	37	37	37	37	37	37
Relative Humidity (%)	0	50	100	100	100	NA	NA	NA
Atmospheric Component	**Partial pressure (mmHg) (referenced to sea level)**							
Nitrogen	593	584	557	562	566	566	566	566
Oxygen	159	157	149	116	100	95	40	30
Carbon dioxide	0.3	0.3	0.3	28	40	40	46	50
Other	8	7	7	7	7	7	7	7
Water	0	12	47	47	47	47	47	47
Total	760	760	760	760	760	755	706	700

After Selkurt (1982) and Davis (1979).

HUMAN RESPIRATORY SYSTEM

The human respiratory system is a complex, yet extremely effective, air-handling and air-conditioning system (Bouhuys 1974, Armstrong et al. 1958). Structures in the nose near the entrance to the respiratory system rapidly condition incoming air from extremes of temperature and humidity to the constant values of 37°C and 100% RH (relative humidity), respectively.

Table 3.2 compares atmospheric conditions and those present in various parts of the respiratory system.

Condition A represents the composition of dry air. Air of this composition is rarely encountered by an individual, except possibly in compressed breathing air specified for use at very low temperatures (CSA 1993). The dew point of this air is set very low to prevent freeze-up in regulators of respiratory protective equipment. Condition B represents the situation typically encountered in an office environment at sea level. The partial pressure of water vapor in the saturated atmosphere (100% relative humidity) at that temperature is 24 mmHg. Condition C is the composition of inspired air following entry into the respiratory system and humidification. Air contained within the respiratory system is fully saturated with water vapor. Condition D is the composition of gas expired by the respiratory system during normal breathing under saturated conditions. Condition E is the composition of alveolar gas at the site of gas exchange. Condition F is the composition of arterial blood that leaves the lung. Condition G is the composition of mixed venous blood that enters the lung. Condition H is the composition of blood in the tissues. Note the reduced total pressure of gases in blood compared to normal atmospheric pressure.

The normal points of entry into the human respiratory system are the nose and the mouth. These are also the points of discharge. A healthy person normally breathes through the nose during quiet periods. Mouth breathing usually occurs during periods of exertion and stress. Mouth breathing also could be expected during the highly stressful period of an unplanned rescue attempt following an accident. Obstructions or swollen structures in the nasal passages also may necessitate mouth breathing.

The nose is an effective temperature conditioning and humidifying structure. The effectiveness of the nose in performing these functions is related to its internal structure. Interior folds in the nasal passages produce a large surface area (about 160 cm^2) within a relatively small volume (20 mL). A sticky, moist and warm mucosal layer covers these surfaces (Bouhuys 1974).

The following example illustrates the effectiveness of the nose as an air conditioning structure (Schmidt-Nielsen et al. 1970). Consider dry air at a temperature of 0°C breathed into the nose during a 24-hour period. The body must heat the air to 37°C, the normal internal temperature. Saturation to 100% relative humidity at this temperature requires evaporation of 420 g of water. Heating and humidification require expenditure of energy, about 2,200 calories. The energy expended is almost the total amount of metabolic heat produced by a resting person in a day. Clearly, conservatory mechanisms must be operative to prevent loss of this energy. One mechanism for energy conservation is countercurrent heat exchange. Incoming cold air cools the nasal surfaces and evaporates moisture. Expired heated and humid air warms and moistens these surfaces. Water vapor in the warm moist air condenses on the cool surfaces.

In a hot environment, humidified air leaves the nose at a temperature of 37°C, carrying with it saturated water vapor. In this case this process contributes to heat loss by the body.

Mouth breathing bypasses the nose and its air conditioning structures. Resistance to airflow through the mouth is about half that of the nose. During quiet breathing, the nasal passages contribute nearly half the total resistance to flow in the respiratory system (Bouhuys 1974).

From the nose and the mouth, air passes through the nasal pharynx. The pharynx is a common path for both food and air. This structure is thought to facilitate interaction between the senses of taste and smell. The pharynx splits to form the larynx and the esophagus. The esophagus leads to the stomach. The larynx continues the path of air into the respiratory system. The larynx acts as a valve to prevent entry of food into the trachea. The larynx also is a major site of flow resistance during breathing (Bouhuys 1974).

Immediately below the larynx is the trachea. The trachea and its subsequent subdivisions function as air-conducting tubes. The trachea divides to form the right and left bronchi. Each bronchus enters the external envelope of the lung (the pleural membranes). The bronchi divide and subdivide into smaller and smaller tubes, the bronchioles. The smallest branches of the bronchioles contain outpockets called alveoli. The alveoli are the sites of gas exchange. For this reason, these bronchioles are known as respiratory bronchioles. The walls of successive branches of the respiratory bronchioles contain an increasingly greater proportion of alveoli. Subsequently, the bronchioles become alveolar ducts. The latter terminate in clusters of alveoli. The walls of the alveoli are only two cells thick. Gases not reacting chemically with the membranes and contents of these cells pass readily across this barrier.

To facilitate its function in air conduction and to prevent collapse, the internal structure of the lung requires structural support. Cartilage in the form of C-shaped rings reinforces the large conductive airways starting at the larynx. These structures create the necessary primary internal framework. This framework also provides a point of attachment for other structural elements. Small bronchioles contain no cartilage (Bouhuys 1974).

The walls of the large airways contain epithelial cells, smooth muscle, cartilage, and glands arranged in a framework of connective tissue. The arrangement of the network of smooth muscle permits the airways to change in length and girth. The epithelium is composed of ciliated and goblet cells, together with columnar epithelial and basal cells. The bronchioles contain a simpler cuboidal cell epithelium having fewer goblet cells (Bouhuys 1974).

Between the basal membrane of the epithelium and the cartilage of the larger airways are seromucinous glands. Some cells of the seromucinous glands produce serous liquid, while others produce mucus. Ducts from these cells discharge onto the epithelial surface. These glands are absent in bronchioles. The mucous produced by the glands and goblet cells coats the epithelial surface. The number of goblet cells decreases in smaller airways. Bronchioles normally contain few or no goblet cells. A thin fluid layer covers the mucosal surface of terminal bronchioles. A surfactant liquid (dipalmitoyl lecithin) covers the surfaces of the alveoli (Bouhuys 1974).

TABLE 3.3
Structural Elements in the Respiratory System

Structural element	Region T/B	Br	RB	Al
Cartilage	+	+	–	–
Smooth muscle	+	+	–	–
Helical elastic fibers	–	–	+	+
Ciliated epithelium	+	+	+	–
Serous liquid	+	+	+	–
Mucous	+	+	–	–
Surfactant layer	–	–	–	+

Note: + = present; – = absent.

T/B = tracheobronchial; Br = bronchioles; RB = respiratory bronchioles; and Al = alveoli.

Cilia extend from the brush border of epithelial cells. Each brush border contains approximately 200 cilia (hair-like projections). Ciliary fibers extend approximately 5 to 7 μm into the serous mucus (watery layer) and wave in a coordinated manner. The cilia bend rapidly in the direction of the pharynx and recoil slowly in the opposite direction. The gel layer of mucous floats on top of the serous liquid, closest to the interior of the airway. Debris becomes embedded in the mucous. The waving action of the cilia transports mucous and embedded debris toward the pharynx. The ciliated epithelium extend from the pharynx to the terminal bronchioles, but not into the alveoli (Satir 1961).

Toxic agents can inhibit ciliary action (Vander et al. 1990). Smoke from a single cigarette can immobilize cilia for several hours. Prolonged inhibition can cause lung infection or airway obstruction.

Table 3.3 summarizes the occurrence of various structural elements occurring in the respiratory system.

Researchers have proposed physical models to describe the internal structure of the lung. For example, the Weibel model proposed 16 branches in the respiratory tree, starting from the trachea and ending in the alveoli (Weibel 1963). The cross-sectional area of the air passages increases as a function of airway generation number or distance into the airway.

According to the Weibel model, the human respiratory system contains about 130,000 branches. The cross-sectional area of the trachea is approximately 2.5 cm^2. By comparison, the combined cross-sectional area of the terminal bronchioles is approximately 185 cm^2. The Weibel model proposes that there are 300 million alveoli, 150 million per lung. The area of the alveolar respiratory surfaces increases from about 30 m^2 at rest to approximately 100 m^2 at deepest inspiration. The surface area decreases by approximately one third from inspiration to expiration (von Hayek 1960). The area of respiratory surfaces for reference man is 75 m^2 and 66 m^2 for reference woman (ICRP 1975). The surface area in reference man is roughly the size of a tennis court, about 80 times greater than that of the skin.

The alveolar surfaces make intimate contact with the pulmonary capillaries. Gases diffuse across the alveolar epithelial cell and the capillary endothelial cell. Thus, the diffusion path between the alveolar airspace and the fluid-filled interior of the pulmonary capillary is as small as two cells thick (Bouhuys 1974). The length of the diffusion path varies from less than 0.4 μm to more than 2 μm (Schulz 1962).

The pulmonary capillary circulation contains about 75 to 100 mL of blood. The surface of the pulmonary capillaries covers an estimated area of 50 m^2 or about 30 times the surface area of the body. As much as 30 L/min of blood can pass through this network without exceeding colloidal osmotic pressure at which pulmonary edema occurs (Comroe et al. 1962).

BREATHING MECHANICS

Breathing is the periodic act of drawing air into the respiratory system and expelling air contained within it. Changes in the volume of the chest cavity external to the lungs induce air motion into and out of the air passages. The coordinated action of expanding and contracting the chest cage constitutes the breathing mechanism.

The lungs are elastic structures (Comroe et al. 1962). They expand elastically when expansive forces are applied and recoil passively when they are released. Lung tissue also has viscous properties. These produce a delay in deformation in response to applied stress.

In cross-section, the interior of the lung resembles a sponge. Noncollapsing structural elements form the internal framework. The interior structure of the lungs is enclosed by the pleural membranes. Pleural membranes also line the interior of the chest cavity. The tendency of the lung surface to adhere to the chest wall and the internal structure assists in preventing collapse of the open air passages. As a result, almost no free space exists between the pleural membranes. The remaining volume is occupied by lubricating fluids.

The chest cage also is an elastic structure. Contraction of the diaphragm, the external intercostal muscles, and certain accessory muscles causes expansion. This expansion occurs against the resistance of various elastic forces. The latter cause the contraction that occurs during the relaxation phase of the breathing cycle (Comroe et al. 1962, Bouhuys 1974).

The internal structure of the lung contains elastic fibers. They are attached to fibers of collagenic tissue. The elastic fibers form a helical structure around the alveolar ducts and the alveoli. The structural arrangement of these fibers produces elastic properties. During inspiration, the airways stretch and elongate. During this process, the helical fibers surrounding the alveoli and alveolar ducts uncoil elastically. The reverse occurs during contraction (Pierce and Ebert 1965).

The fluid film covering the alveolar surfaces exhibits viscous properties. The expansive forces must perform work to overcome this resistance. This also contributes to the elastic properties of the alveolar region (Bachofen et al. 1970, Moran Campbell et al. 1984).

During expiration, the chest cage returns passively to the smaller resting volume. The muscles of the chest cage perform so-called negative work as they lengthen. The helical fibers surrounding the alveolar ducts and the alveoli contract and the helix recoils. This action reduces the length of the alveolar ducts and the surface area of the respiratory surfaces. In this systematic way, the alveolar walls distend and retract without folding. The film covering the alveolar surface also contributes to the elastic recoil, especially at higher lung volumes.

The action of filling or emptying air passages in the respiratory system requires a driving force (Comroe et al. 1962, Bouhuys 1974). The lungs neither fill nor empty themselves. The force creating air motion results from change in the volume of the respiratory passages. This process occurs only through the actions of the breathing apparatus. The respiratory muscles and the skeletal framework create the changes in pressure (force/area) necessary to induce air to flow into the respiratory passages. Elastic recoil and viscous forces decrease the volume, thereby exerting the pressure needed to expel air from the respiratory passages. Air motion within the respiratory passages occurs only during changes in interior (intrapulmonic) pressure. Intrapulmonic pressure fluctuates continuously during the breathing cycle. Change in intrapulmonic pressure occurs only during motion of the chest cage. Intrapulmonic pressure is less than atmospheric pressure during inspiration. It equals atmospheric pressure at the peak point of inspiration. Intrapulmonic pressure exceeds atmospheric pressure during expiration and equalizes again at the end point of expiration. Intrapulmonic pressure also equalizes whenever a person deliberately ceases breathing movements. At that point airflow also ceases.

The change in alveolar pressure that occurs during this process at rest is less than 1 mmHg (Vander et al. 1990).

PULMONARY FUNCTION

Breathing is the outward response of the body to demands for replenishment of oxygen and elimination of carbon dioxide. The rate and depth of breathing reflect a number of influences, both subconscious and conscious, that may interact simultaneously. Subconscious pathways respond to the level of physical activity undertaken by the person, and physical and emotional stress experienced by the person. Within limits, breathing activity also responds to consciously imposed actions such as hyperventilation and breath-holding. Other factors such as gender, age, atmospheric conditions, and health status also influence the depth and rate of breathing.

Understanding the characteristics of human respiration is extremely important from the perspective of exposure to real-world atmospheric conditions. Parameters of pulmonary function influence important decisions, including, for example:

- Defining conditions of respiratory impairment due to oxygen deficiency or exposure to respiratory intoxicants
- Specification of delivery rates for supplying air through respirators
- Specification of air quality inside the facepiece of the respirator
- Specification of quality of compressed and atmospheric breathing air
- Defining personal fitness to wear a respirator

Spirometry is one of a number of techniques used to assess pulmonary function (Comroe et al. 1962, Shigeoka 1983, Cotes 1983). Spirometry uses noninvasive techniques to provide a volume/time record. This record, the spirogram, permits measurement of breathing parameters including lung volumes, capacities, and volumetric flow rates. These parameters provide information about normal and abnormal pulmonary function. Following are the primary parameters of pulmonary function (Figure 3.1).

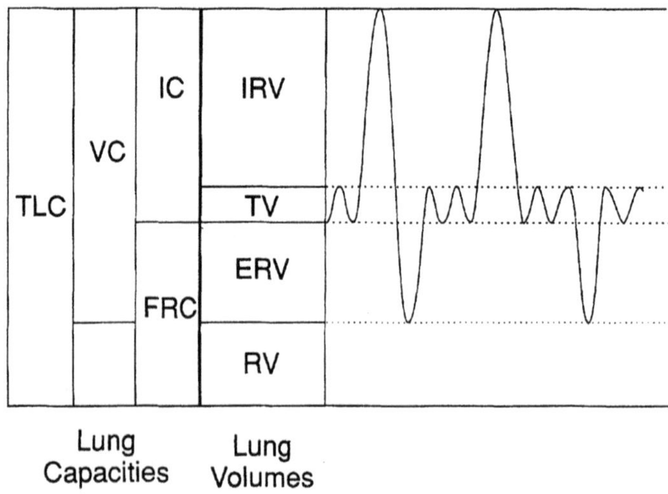

FIGURE 3.1 Lung volumes and capacities corresponding to the spirometric tracing. TLC: total lung capacity; vc: vital capacity; RV: residual volume; IC: inspiratory capacity; FRC: functional residual capacity; IRV: inspiratory reserve volume; TV: tidal volume; ERV: expiratory reserve volume. (Adapted from Comroe et al. 1962.)

TABLE 3.4
Respiratory Capacity for Reference Man and Woman

Volume classification	Adult males (L)	Adult females (L)
Total lung capacity	5.6	4.4
Functional residual capacity	2.2	1.8
Vital capacity	4.3	3.3
Anatomic dead space	0.16	0.13

After ICRP 1975.

Tidal volume is the volume of air exchanged during each respiratory cycle consisting of inspiration and expiration. Tidal volume is controlled involuntarily and reflects various factors that affect breathing, including rate of work.

Inspiratory reserve volume is the maximum amount of air inspired following exhalation of a normal tidal volume.

Expiratory reserve volume is the maximum amount of air voluntarily and forcefully exhaled following inspiration of a normal tidal volume.

Residual volume is the volume of air remaining in the lung following maximum voluntary, forced expiration. This quantity cannot be measured directly.

The following secondary parameters are derived from manipulation of these quantities.

Functional residual capacity is the volume of air remaining in the lungs at resting expiratory level (FRC = ERV + RV).

Vital capacity is the maximum volume of air that can be expelled from the lungs by forceful effort following maximum inspiration (VC = ERV + TV + IRV).

Inspiratory capacity is the volume of air inhaled during a forced voluntary inhalation to inflate the lungs to their maximum extent. This quantity represents the maximum volume of air that the subject can inhale. The starting point of this inspiration is the end of normal resting expiration (IC = TV + IRV).

Total lung capacity is the estimated volume of air contained in the lung at maximum inflation. Maximum inflation occurs at the end point of maximum forced inspiration, a voluntary action. Total lung capacity cannot be measured directly (TLC = RV + ERV + TV + IRV).

The following tables summarize respiratory parameters for reference man and reference woman (ICRP 1975) (Tables 3.4 and 3.5). Reference man is a 70 kg male whose other dimensions are proportionate to this mass. Reference woman is a 58 kg female configured in a parallel manner. Obviously, not all persons in the adult population conform to the dimensions of these individuals. Hence, caution is advised when applying physiologic and morphometric data for reference man and reference woman to specific individuals, since a range should be expected. Vital capacity depends on height, age, race, and gender.

One parameter not mentioned to this point is anatomic dead space: the volume of the conducting airways in the respiratory tree. This volume varies from approximately 150 mL to 300 mL, depending on the state of inflation of the lungs (Bouhuys 1974). As mentioned previously, the airways are elastic. They elongate and swell during each inspired breath, and contract and shorten during each expiration. The anatomic dead space varies according to the point of measurement in the respiratory cycle.

A volume of air remains in the airways and airspaces at all times during the breathing cycle (Comroe et al. 1962, Bouhuys 1974, Vander et al. 1990). This applies even during forced maximum exhalation. The expiratory force applied during forced maximum exhalation cannot collapse the respiratory passages to expel all of the air.

After each normal breath, the lungs contain the functional residual capacity (FRC), about 2.2 L of air (ICRP 1975). During each inhalation, the tidal volume (TV) enters the respiratory tree and mixes with the air contained inside (FRC + TV). During each exhalation, the tidal volume is expelled. The outcome from this process is that air drawn into the alveoli during each breathing cycle does not have the same composition as the air outside the body. The presence of unexpelled air in the respiratory passages introduces a potential inefficiency to the process of gas exchange. Replenishment of air in the alveoli involves mixing of inspired air with that remaining from the preceding exhalation. The effectiveness of the replenishment depends on the extent to which expiration minimizes the volume of residual air and inspiration maximizes the volume of outside air. The extent to which mixed alveolar air approaches the composition of air found outside the body influences the efficiency of the process of gas exchange.

In normal males, the average tidal volume is 450 to 600 mL. During normal resting ventilation, the anatomic dead space contains approximately 150 mL of air at the end of expiration. Inspiration of 450 mL will increase the alveolar volume by 450 mL (Figure 3.2). The first 150 mL to enter the alveolar spaces is the air remaining in the airways (anatomic dead space) at the end of the previous exhalation. This air has the same composition as alveolar air. Hence, this addition to the air in the alveolar spaces does not alter the composition. The next 300 mL to enter the alveolar space is the inspired air containing outside air. This addition alters the composition of alveolar gas. The remaining 150 mL of the inspired air fills the airways. No gas exchange occurs in the airways. This air is expired first in the exhaled breath, and other than humidification and heating or cooling, is not modified by the body (Vander et al. 1990).

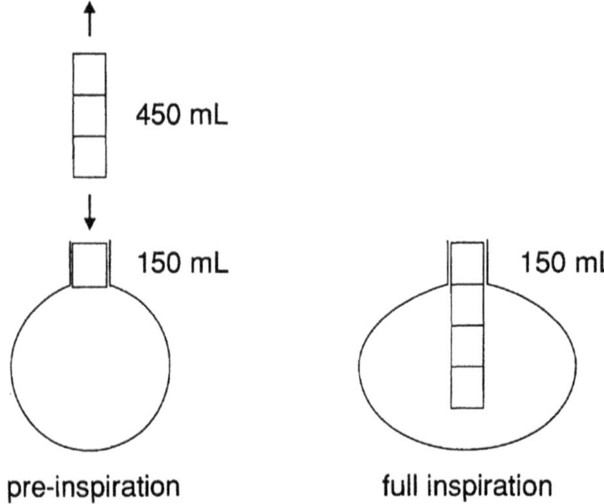

FIGURE 3.2 Normal ventilation: (a) Pre-inspiration, (b) Full Inspiration. Normal inspiration adds 450 mL of air to the 150 mL of gas remaining in the air passages at the end of exhalation. This produces an approximate total volume of 600 mL. Of this, the alveoli receive 300 mL of new air plus 150 mL of residual air for a total of 450 mL. The volume of inspired air varies. Normal exhalation removes 450 ml of gas from the lung, leaving behind 150 mL in the airspaces. During hypoventilation (shallow breathing) only 150 mL of air may enter the airways. The alveoli receive poorly mixed residual gas from the dead space. (Adapted from Comroe et al. 1962.)

TABLE 3.5
Respiratory Requirements for Reference Man and Reference Woman

	Respiratory volume	
Activity classification	**Adult males (L)**	**Adult females (L)**
8-h Workshift, light activity	9,600	9,100
8-h Nonoccupational activity	9,600	9,100
8-h Resting	3,600	2,900
Total	22,800	21,100

After ICRP 1975.

The forward velocity of air inside the terminal bronchioles and alveolar spaces is nearly zero. In this region, gases move primarily by diffusion, rather than by bulk flow. The dimensions of the terminal alveoli determine the length of travel of gas molecules and, hence, the effectiveness of diffusion. According to the Weibel lung model, the average length of the pathway from the respiratory bronchiole to the terminal alveolus is less than 5 mm (Weibel 1963). Using a 7-mm long cylinder for comparison, gas concentrations decrease to 16% of their initial value by diffusion alone after 0.38 seconds (Bouhuys 1974). The rapidity with which the concentration changes indicates that diffusion in this region is a highly important and effective mechanism in gas transport.

During the breathing cycle, the composition of alveolar air changes continuously. The composition of blood perfusing through pulmonary capillaries changes less dramatically. During expiration, the concentration of carbon dioxide in alveolar air increases, while the concentration of oxygen decreases. The total gas pressure in the alveolus remains the same. The partial pressure of oxygen decreases to a minimum at the beginning of inspiration. As fresh air dilutes alveolar gas during inspiration, the partial pressure of oxygen increases, while the partial pressure of carbon dioxide decreases (Comroe et al. 1962).

The pattern of airflow and gaseous exchange in the lung has been optimized in one practical application — the artificial ventilator. Normal peak airflow occurs about one third of the way through inspiration. If peak flow occurred later in the breathing cycle, alveolar air would contain more CO_2. Similarly, more CO_2 would be removed in the same expired volume. When peak flow occurs later, the same gas exchange rate could be maintained with 6% less alveolar ventilation. Thus, the same exchange of gas can occur with less ventilation by delaying expiration and hastening inspiration. This combination produces a gasping type of breathing (Bouhuys 1974).

Table 3.5 summarizes daily respiratory requirements for reference man and woman for the average workday.

According to this model for an average workday, 42% of the air consumed by reference males and 43% of that consumed by reference females is breathed during the workshift. Oxygen consumption could increase during performance of more strenuous activity.

GAS EXCHANGE

The exchange of gases between alveolar air and blood in pulmonary capillaries is the essential normal function of the lung. The amount of exchange depends on the alveolar ventilation rate and the flow of blood through pulmonary capillaries (perfusion of the lungs), diffusivity through cellular membranes, and solubility in blood.

Henry's law describes the relationship between gas or vapor and liquids with which they are in contact (Reid et al. 1987). The quantity of a gas dissolved in a liquid at equilibrium is proportional to the partial pressure of the gas above the liquid. For each gas, there is an individual Henry's constant. The value of the constant depends on a number of factors, including: temperature, pH, and interactions between molecules of the gas and the solvent.

Henry's law describes the equilibrium situation. There are two possible nonequilibrium situations that also must be considered. The first involves contact between a solvent containing no gas or a weak solution and gas-rich atmosphere. Gas will dissolve into the solvent or weak solution until equilibrium is attained or another factor intervenes. The converse situation involves contact between a solution containing dissolved gas and an atmosphere containing no gas or a concentration less than the equilibrium value. Gas will effuse from the solution into the gas-lean atmosphere until equilibrium again is attained or another factor intervenes. Both of these processes occur in the lung and the tissues as part of gas exchange.

The relationship between atmospheric and other gases and body fluids, such as blood and extra- and intracellular fluids is a critical part of the process of transport and respiration. These considerations represent a direct application of Henry's law. Oxygen diffuses into the liquid part of the blood in the lung and is transported to regions having lower concentration. This process occurs because the partial pressure of atmospheric oxygen exceeds the equilibrium partial pressure of dissolved oxygen. Carbon dioxide diffuses into the liquid part of the blood from the tissues and effuses into airspaces in the lung. The latter process occurs because the partial pressure of dissolved gas exceeds the equilibrium partial pressure of atmospheric gas.

Gases and vapors that do not react with components of tissue or cellular fluids pass freely across the membrane barrier in both directions. Gases and vapors diffuse in response to a pressure gradient — from an area of high partial pressure to an area of low partial pressure. The difference in partial pressure between alveolar air and the blood determines the net direction of flow. Gases and vapors will diffuse across the membrane barrier into or from a particular volume of blood until equilibration occurs (partial pressures become equal), or the flow has reached the end of the alveolar-capillary contact (Comroe et al. 1962, Bouhuys 1974).

Under normal conditions, the partial pressure of oxygen in alveolar air is greater than that in blood entering the pulmonary capillaries. At the same time, the partial pressure of oxygen in tissue capillaries is greater than that in tissue fluids and greater in tissue fluids than in cells of the body. Conversely, the partial pressure of carbon dioxide is higher in the cells than in the intercellular fluids, higher in the intercellular fluids than blood flowing through tissue capillaries, and higher in pulmonary capillaries than in alveolar air (Bouhuys 1974).

During the breathing cycle, the alveolar partial pressure of oxygen increases from a minimum of 97.9 mmHg to a maximum of 101.5 mmHg. The corresponding alveolar partial pressure of carbon dioxide changes from 40.8 mmHg to 38.2 mmHg. However, these changes in partial pressure do not correspond exactly to the inspiratory and expiratory motions of the chest. The change in alveolar partial pressure is not the same for the two gases. Metabolism consumes more oxygen than the amount of carbon dioxide produced. This means that a greater amount of oxygen is exchanged per unit time than carbon dioxide. The relative amount of carbon dioxide produced and oxygen taken up depends on metabolic activity, i.e., work (Comroe et al. 1962).

Oxygen tension of mixed venous blood entering the pulmonary capillaries is 40 mmHg. Oxygen tension of oxygenated blood in the pulmonary veins is 100 mmHg (Figure 3.3). This is identical to the partial pressure of oxygen in the alveolar space. The normal time spent in the pulmonary capillary bed is 0.75 s. The oxygen tension increases to almost 100 mmHg in 0.35 s or less. This is less than half of the normal transit time. This efficiency provides redundancy for situations that are less than ideal (Comroe et al. 1962). In normal individuals, only during the most strenuous of exercise when blood flow through the capillaries is extremely rapid is there insufficient time for complete equilibration. This may not be the case in individuals whose lung and circulatory function is compromised by disease, age, obesity, or lack of physical conditioning. The combination of the

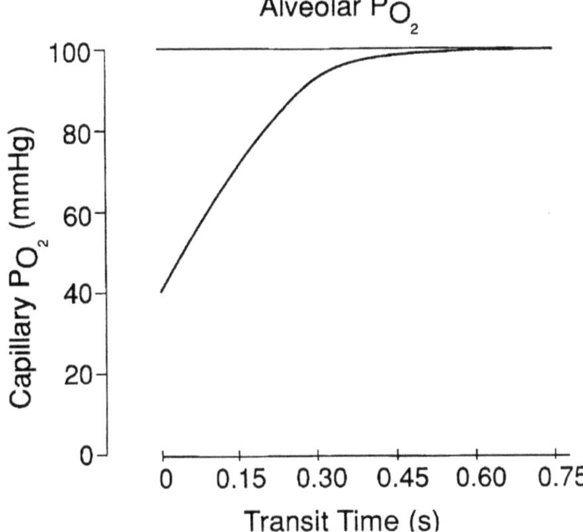

FIGURE 3.3 Oxygenation of hemoglobin in pulmonary capillaries under normal conditions. The partial pressure of oxygen increases rapidly from 40 to 100 mmHg during passage of deoxygenated blood through the pulmonary capillaries (0.75 s). This rapid increase provides a reserve for less than ideal conditions. (Adapted from Comroe et al. 1962.)

stress induced by the situation, coupled with these factors, easily could provide the required conditions for insufficiency in gas exchange.

This process is affected by the diffusing capacity of the pulmonary capillaries and other factors. On a micro scale, this process is very complex. Ventilation of the alveoli occurs only during inspiration. On the other hand, blood flow and gas exchange occur continuously. Imbalance between the rate of ventilation and perfusion causes inefficient exchange between alveolar airspaces and the blood.

Diffusion through cellular membranes does not limit gas exchange. The rate of uptake or clearance of a gas or vapor depends on solubility in blood, the alveolar ventilation rate, and the perfusion rate. The factor limiting the importance of the alveolar ventilation rate compared to the perfusion rate is solubility of the gas or vapor in the blood (Farhi 1967). Clearance of a relatively insoluble gas or vapor depends almost exclusively on the perfusion rate. The alveolar ventilation rate has little effect. For example, the rate of clearance from the blood of xenon, a relatively insoluble gas, depends mostly on the perfusion rate. Oxygen also behaves as a relatively insoluble gas. The rate of uptake of oxygen is perfusion limited (Bouhuys 1974).

The rate of clearance of a relatively soluble gas or vapor depends almost exclusively on the alveolar ventilation rate. The perfusion rate has little effect. Clearance of the relatively soluble vapor, diethyl ether, increases dramatically with increasing alveolar ventilation at constant perfusion rate. The rate of clearance is little affected by the perfusion rate at constant alveolar ventilation rate (Farhi 1967). The rate at which carbon dioxide leaves the blood is largely determined by the rate of alveolar ventilation. Carbon dioxide behaves as a soluble gas.

The ratio of partition coefficients of oxygen and carbon dioxide is about 1:10. Carbon dioxide diffuses 20 times more readily than oxygen through the pulmonary membranes (Bouhuys 1974).

The ratio of the alveolar ventilation rate to the perfusion rate provides a useful measure of efficiency. The alveolar partial pressure and the partial pressure in venous blood of relatively soluble gases are nearly equal at ratios less than one. The alveolar partial pressure of less soluble gases is much less than that in mixed venous blood, even when the ratio is very low. In a direct comparison between an insoluble gas and a more soluble gas for each value of the ratio, the insoluble gas has a lower alveolar pressure relative to the next venous pressure than the more soluble gas (Farhi 1967).

Blood leaving the alveoli contains nitrogen in direct proportion to the alveolar partial pressure of nitrogen. No net exchange between gas and blood normally occurs because nitrogen from atmospheric air saturates the tissues of the body (Moran Campbell et al. 1984).

OXYGEN TRANSPORT IN BLOOD

Oxygen enters the fluid of the blood by diffusion and forms a simple solution in the plasma. However, this solution rapidly saturates. Oxygen surplus to the solvation potential of liquid blood diffuses into red blood cells (erythrocytes) and binds to the iron atom of hemoglobin. The amount of oxygen that remains dissolved in the liquid component of blood is small (about 1.5% of that bound to hemoglobin). Dissolved oxygen is extremely important in the transfer between air and hemoglobin, and hemoglobin and tissue fluids (Bouhuys 1974).

Hemoglobin is a tetramer composed of four molecular units. Each of the four subunits can bind an oxygen molecule. The binding site is the iron-containing heme group. The heme group lies in a depression on the surface of the hemoglobin molecule. During the binding process, the oxygen atom fits into the depression, yet does not increase the volume of the molecule (Fenn 1971). Binding causes a slight change in the radius of the iron atom. This change, in turn, causes slight movement of the peptide chains of the hemoglobin subunit near the heme group. Associated with this movement is weakened attraction between the subunits (Perutz 1970).

Oxygen binds to heme groups successively rather than simultaneously. This implies that all hemoglobin molecules in a group will be saturated to the same approximate extent. Binding to the first two or three heme groups produces several changes. These include conformational changes around unfilled heme pockets in the remaining subunits and disruption of molecular attraction between subunits. The result is an increase in affinity for oxygen by unoccupied heme groups. The shape of the dissociation curve for hemoglobin reflects these changes. Oxygen binding to the first heme group increases the affinity for oxygen of the others (Bouhuys 1974).

The presence of certain ionic species in the cytoplasm in the erythrocyte further affects the affinity of hemoglobin for oxygen. Organic phosphates (in particular, 2,3-diphosphoglycerate or 2,3-DPG) play an important role. These compounds bind to the hemoglobin molecule and appear to stabilize the deoxygenated form. This process increases the ability of hemoglobin to bind oxygen (Benesch and Benesch 1969, Benesch et al. 1971).

The affinity of hemoglobin for oxygen also is pH dependent. A lower pH decreases the affinity of hemoglobin for oxygen and promotes unloading of oxygen. This appears as a shift to the right by the dissociation curve. The dissociation curve is a graphical representation of the affinity between oxygen and hemoglobin. While the curve retains its shape, the displacement indicates decreased oxygen affinity at the same partial pressure. Hence, oxygen binds to a smaller percentage of the available sites. This is the situation in the tissues. Hydrogen ions present in abundance in this environment bind to the hemoglobin molecule. Coincidentally, this process also increases the ability of hemoglobin to bind carbon dioxide. Thus, deoxygenated hemoglobin binds more carbon dioxide than oxygenated hemoglobin (Bouhuys 1974, Harper et al. 1977).

Myoglobin, a protein found in muscle, also binds oxygen. Myoglobin is structurally similar to one of the monomeric units of hemoglobin. That is, the molecule is a protein containing a heme group arranged into a depression on the surface and an iron atom. The binding of oxygen to myoglobin follows the predictions of the law of mass action. As predicted, the relationship between oxygen partial pressure and the degree of oxygenation is hyperbolic (Bouhuys 1974). At partial pressures less than 60 mmHg, myoglobin binds more oxygen than hemoglobin. Thus, in muscle tissue, hemoglobin releases oxygen and myoglobin binds it. This oxygen storage mechanism is especially important to functioning of the myocardium (the heart muscle) (Ayers et al. 1969).

Nonlung storage of oxygen in normal males is approximately 1 L. Nonlung storage of carbon dioxide is approximately 17 L. The blood stores practically all of the oxygen. The bulk of the carbon dioxide is contained in tissue fluids as bicarbonate ions. Oxygen storage responds to a change in ventilatory volume or composition of inspired air within 2 min. Carbon dioxide storage responds within 15 min. A given change in oxygen level produces a far greater response than the same change in carbon dioxide (Moran Campbell et al. 1984).

CELLULAR RESPIRATION

In the tissues, oxygen molecules migrate in response to a pressure gradient. This migration occurs from an area of high partial pressure in the plasma to one of low partial pressure in the interior of the cells. Oxygen dissociates from hemoglobin and dissolves into the plasma. Diffusion then occurs through the capillary membranes into the extracellular fluid and from the extracellular fluid through the cell membrane. Within the cell, oxygen molecules diffuse through the cytoplasm and into the mitochondria. The mitochondrion, site of cellular respiration, is the final destination for up to 90% of the oxygen entering the cell (Bouhuys 1974).

The mitochondrion is a double-walled cylindrical enclosed structure formed by two membranes, one inside the other. The inner membrane contains a series of folds or in-pocketings, called *cristae*. The cristae greatly increase the surface area. Respiratory enzymes are located on the cristae. Isolated mitochondria require a minimum oxygen tension of 0.5 mmHg. The critical intracellular partial pressure of oxygen for intact cells is estimated to be 3.5 mmHg (Jobsis 1964).

CARBON DIOXIDE TRANSPORT

The main products of cellular respiration are carbon dioxide and water. The blood is the route of transport of carbon dioxide from tissues to the lungs, the point of excretion. The partial pressure of carbon dioxide in the atmosphere (0.27 mmHg) is insignificant compared to that present in alveolar air (approximately 40 mmHg) (Comroe et al. 1962, Bouhuys 1974).

The transport of carbon dioxide from the cells to the lungs is a complex process. First, carbon dioxide must enter the blood. This occurs by diffusion from cells into interstitial spaces and from the interstitial spaces into capillary plasma. This occurs in response to the pressure gradient. The tension of carbon dioxide in the cells is higher than that in mixed venous blood (46 mmHg) (Comroe et al. 1962).

Transport of carbon dioxide occurs in a number of ways (Comroe et al. 1962, Bouhuys 1974, Moran Campbell et al. 1984, Harper et al. 1977). Some CO_2 (about 10%) dissolves in the plasma to form a simple solution. A very small amount reacts slowly with water to form carbonic acid, H_2CO_3. Some of the acid dissociates into H^+ and HCO_3^- ions. Equation 3.1 summarizes these reactions:

$$CO_2 + H_2O \rightleftharpoons H_2CO_3 \rightleftharpoons H^+ + HCO_3^- \quad (3.1)$$

Plasma buffering controls the concentration of H^+. A reaction also occurs between dissolved CO_2 and the amino group of plasma proteins to form carbamino compounds:

$$CO_2 + \text{R-NH}_2 \rightleftharpoons \text{R-NHCOOH} \quad (3.2)$$

Most of the CO_2 entering the blood through tissue capillaries diffuses into erythrocytes. Three reactions can occur within this environment. Some CO_2 remains unreacted as gas dissolved in simple solution within the erythrocyte. Some CO_2 combines with amino groups of hemoglobin to form carbamino compounds. This reaction occurs very rapidly. Parts of the hemoglobin molecule buffer this reaction. Dissociation of oxygen from the hemoglobin molecule simultaneously facilitates the binding of carbon dioxide. This occurs because conversion of oxyhemoglobin to the reduced form through dissociation of oxygen produces a weaker acid. Reduced hemoglobin can accept additional H^+ with little change in pH.

Some CO_2 combines with water within the erythrocyte to form carbonic acid, as indicated above. In this case, the reaction is catalyzed by carbonic anhydrase, an enzyme that is concentrated within the erythrocyte. Hence, the conversion is rapid. Excess HCO_3^- ions diffuse from the erythrocyte into the plasma. This process maintains ionic equilibrium between the erythrocyte and the plasma. Chloride ions maintain electrical neutrality by migrating from the plasma into the erythrocyte.

The plasma contains considerably greater CO_2 (in all forms) than the erythrocyte. In fact, the plasma transports more than 60% of CO_2 added to capillary blood. Chemical reactions occurring within the erythrocyte provide practically all the additional bicarbonate ion transported in the plasma.

Just as the amount of oxygen carried by the blood is related to the partial pressure of oxygen to which the blood is exposed, the amount of CO_2 carried in blood is related to the partial pressure of carbon dioxide in the blood.

In the pulmonary capillaries, diffusion of carbon dioxide into the alveolar airspaces rapidly decreases the partial pressure of CO_2 in blood. The rapid decrease reverses the processes of storage described above. Dissociation of carbonic acid into water and carbon dioxide is a slow process. This process cannot occur rapidly enough without catalysis to make available the carbon dioxide needed for rapid transfer into the alveolar airspace. Carbonic anhydrase reversibly catalyzes the rapid dissociation of H_2CO_3 to form H_2O and CO_2. The CO_2 rapidly diffuses from the cytoplasm of the erythrocyte into the plasma and from the plasma, across the alveolar membrane. This reaction produces an imbalance of HCO_3^- ions between the erythrocyte and the plasma and a consequent influx of bicarbonate from the plasma. Chloride ions migrate back into the plasma to balance the H^+ ions. Loss of CO_2 from the plasma during this process also reverses the protein-binding mechanism for molecular CO_2 (Harper et al. 1977).

CONTROL OF BREATHING

Breathing is the action that draws air into the lungs for the purpose of gas exchange. Breathing usually occurs under subconscious control. However, within certain limits, a person can consciously and deliberately influence the rate and depth of breathing. Despite efforts exerted through the conscious mind to stop breathing, the subconscious asserts control and forces breathing to occur.

Breathing occurs at a base rate that is subject to feedback. The brain stem contains the respiratory centers. The precise arrangement and action of the controlling system are not yet known. Networks of nerve cells in the brain stem (respiratory centers) initiate involuntary rhythmic breathing. Coordination of muscle activity involved in breathing occurs within the spinal column. The respiratory center, as well as the coordinating centers, monitors results of motor output. A number of mechanisms can modify inspiration. These influences include both chemical and nervous stimuli. Expiration is a passive process (Bouhuys 1974).

Chemoreceptor cells located centrally near the respiratory centers and peripherally in carotid and aortic bodies influence the rate and depth of breathing. Of the two types of chemoreceptors, the central are by far the more important. Central chemoreceptors respond to the partial pressure of CO_2 through change in the concentration of H^+ ions in extracellular fluid in the brain. Peripheral chemoreceptors respond to the partial pressure of O_2 and CO_2, and to the H^+ concentration in arterial blood. Nervous control originates from higher centers in the midbrain and from cerebral centers. In most circumstances, the concentration of carbon dioxide in arterial blood regulates the depth and rate of breathing (Vander et al. 1990).

Stimulation of chemoreceptors in the respiratory centers depends on pH and partial pressure of carbon dioxide in the interstitial fluid of the brain (Leusen 1972, Mitchell et al. 1963, Pappenheimer et al. 1965). These quantities depend on the composition of blood perfusing the medulla and cerebrospinal fluid (CSF). When the composition of CSF is kept constant, large ventilation increases can occur through increased partial pressure of CO_2. Carbon dioxide readily penetrates the blood–brain barrier. Thus, the composition of CO_2 in the CSF readily follows that in arterial blood. The chemoreceptor cells appear to be equally accessible to changes in CSF and the blood. Thus, response of the central chemoreceptors reflects pH and partial pressure of CO_2 in both CSF and blood.

Chemoreceptors also are located in the carotid bodies near the bifurcation of the common carotid artery and the arch of the aorta (Bouhuys 1974). These constitute the peripheral chemoreceptors. The partial pressure of oxygen sensed by cells of the carotid bodies is equivalent to that present in the arteries. Peripheral chemoreceptors respond to a decrease in partial pressure of oxygen and pH and an increase in arterial partial pressure of carbon dioxide.

Both central and peripheral chemoreceptors respond to localized changes in the partial pressure of carbon dioxide. The peripheral chemoreceptors act rapidly but exert little influence. On the other hand, the central chemoreceptors act more slowly, but produce a greater response (Bouhuys 1974).

EXERCISE AND WORK

Exercise and work create considerable demands on the respiratory and cardiovascular systems. The body responds to the demands of exercise and work through increase in cardiac output and alveolar ventilation. Increased ventilation and other compensatory mechanisms maintain alveolar and arterial partial pressures of oxygen and arterial partial pressure of carbon dioxide and pH at normal levels, despite the considerable increase in pulmonary blood flow (Comroe et al. 1962).

Conditions associated with exercise and work include lower partial pressure of oxygen and elevated partial pressure of carbon dioxide, lower pH, and increased temperature in the tissues. All of the preceding factors favor dissociation of oxyhemoglobin and delivery of oxygen to the tissues (Figure 3.4). Increased dissociation in the tissues lowers the partial pressure of venous oxygen far below the level present in the resting individual (Comroe et al. 1962, Bouhuys 1974). The impact of these changes on control of ventilation is not known (Vander et al. 1990).

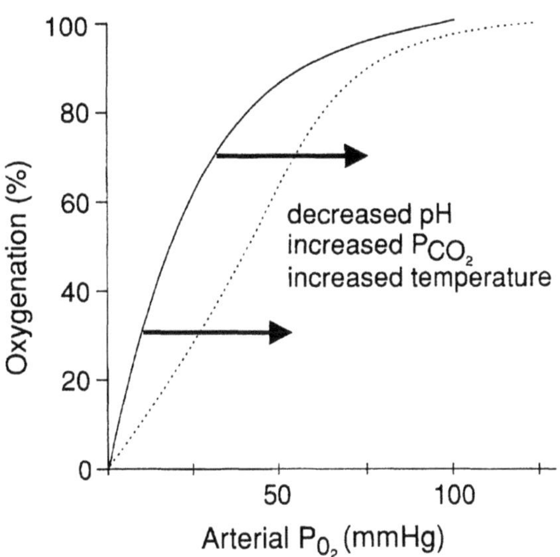

FIGURE 3.4 Oxygen dissociation from hemoglobin. Because of the shape of the dissociation curve, hemoglobin remains about 90% saturated despite a drop in partial pressure in blood of 40 mmHg. The hemoglobin dissociation curve shifts to the right following a decrease in pH, an increase in the partial pressure of carbon dioxide, or increase in temperature. (Adapted from Bouhuys 1974.)

TABLE 3.6
Minute Volumes Under Various Conditions

Activity Classification	Minute Volume Adult males (L/min)	Minute Volume Adult females (L/min)
Sleep	6.0	
Resting	7.5	6.0
Light activity	20.0	19.0
Medium work	29.2	
Medium heavy work	43.0	26.0
Heavy work	59.5	
Maximum work	132	
Maximum breathing capacity	160	

After NIOSH (1976a) and ICRP (1975).

During exercise, the residence time of blood in pulmonary capillaries may decrease from 0.75 to 0.30 seconds. Oxygen transfer across the alveolar membrane normally is 90% complete in 0.45 s. In 0.30 s, the transfer is 75% complete. Despite the considerable reduction in residence time, the transfer process is affected only slightly. Arterial partial pressure of oxygen need not decrease even during maximum work and arterial partial pressure of carbon dioxide need not increase. In the normal person this adaptation to increased demand for oxygen and increased need to excrete carbon dioxide is very effective (Comroe et al. 1962, Bouhuys 1974).

Table 3.6 illustrates the influence of work demands on lung ventilation. Ventilatory flow rate or minute volume is an important measure in respiratory physiology.

The attainable upper limits of minute volume are extremely important, as these illustrate potential air consumption rates at the height of emotional and physical stress during accident situations. Persons carrying out rescue activities in a hostile atmosphere that also could be thermally stressful operate under considerable physical and emotional duress. At high rates of ventilation, the impact and rate of onset of intoxication or oxygen deficiency is considerably greater than under normal conditions of work.

These limits of performance also have major impact on delivery specifications for supplied air respirators. Supplied air respirators must deliver sufficient air to meet the respiratory demand of the wearer. As demonstrated above, respiratory demand varies according to level of activity. Design parameters for delivery of air to supplied air respirators must consider respiratory demands in highly stressful situations, since these represent probable applications for this equipment. Insufficient delivery will impede the ability of the wearer to sustain a rate of activity. In extreme cases, the wearer may even remove the facepiece in a hostile atmosphere in an attempt to gain more air.

BREATH HOLDING

Entry into a contaminated atmosphere while breath holding sometimes occurs. This practice is especially tempting when the entry is anticipated to last for only a "few seconds" and the preparations needed for carrying out a safe procedure would require several minutes. Failure to return to the clean air within the span of the breath hold means that the entrant must breathe the air contained within the space. This could place the entrant in considerable jeopardy.

A person voluntarily can breath hold for periods usually lasting less than a minute. The duration of this interval depends on several factors: motivation, level of physical exertion and emotional stress, initial lung volume, and alveolar partial pressure of oxygen and carbon dioxide at the beginning of the breath hold. Voluntary hyperventilation prior to initial inspiration can prolong breath holding.

During breath holding, alveolar ventilation stops. Gas exchange in the alveolar airspaces continues, however, but proceeds at a rapidly decreasing rate as the alveolar–arterial pressure gradient decreases. The decrease in alveolar partial pressure of oxygen is much greater than the increase in alveolar partial pressure of carbon dioxide. This imbalance occurs partly because of the greater solubility of carbon dioxide in blood and tissues (Bouhuys 1974). Both the O_2 and CO_2 stimuli influence the point at which a person can no longer resist the urge to breathe. A person quietly breathing room air prior to breath holding commonly reaches the point of endurance when alveolar partial pressure of O_2 decreases to about 70 mmHg and CO_2 is about 50 mmHg (Kellogg 1964).

The major concern about breath holding and entry into confined spaces is obligatory inspiration of contaminated or oxygen-deficient air. This would occur if the entrant does not succeed in vacating the contaminated atmosphere prior to obligatory breathing. Obligatory breathing following prolonged breath holding is deep and rapid. This situation can lead to replacement of oxygen-deficient air from the respiratory airspaces by air of even lesser quality. Coupled with obligatory breathing, this could lead to rapid inspiration of contaminated or oxygen-deficient air. Inspiring an atmosphere low in oxygen and high in carbon dioxide can enable the person to hold the breath again, even though the alveolar gas composition has not improved (Fowler 1954).

At the end of the respiratory cycle, the alveolar partial pressure of oxygen and carbon dioxide (100 mmHg and 40 mmHg, respectively) approximate those in the pulmonary capillaries. Inspiring gas of the same composition would affect only the volume, not the partial pressures of alveolar gases. Oxygen uptake continues at the end of the breathing cycle. Even without enrichment, alveolar oxygen still has a higher partial pressure than mixed venous blood. However, without enrichment, transfer of oxygen from alveolar air eventually would cease, thus leading to decreasing saturation of hemoglobin and arterial anoxemia. The accompanying decrease in exchange of carbon dioxide into alveolar air would lead to respiratory acidosis (Comroe et al. 1962).

HYPOVENTILATION

Hypoventilation exists when ventilation of the alveoli is insufficient to supply the metabolic needs of the body (Comroe et al. 1962). The term "hypoventilation syndrome" originally was applied to obese individuals who were anoxemic because of insufficient alveolar ventilation. However, hypoventilation can occur for many reasons: uneven distribution of air in the lung, impairment of diffusion, depression of respiratory centers and neural transmission, injury, disease, obesity, age-related processes, and drugs. With each breath providing insufficient fresh air, alveolar and arterial partial pressure of oxygen decreases, arterial and alveolar partial pressures of carbon dioxide increase, and pH decreases.

The significance of breathing pattern on alveolar ventilation is illustrated in Table 3.7. This table illustrates the impact of various combinations of breathing depth and rate. Note that minute ventilation (tidal volume × frequency) remains constant in all cases at 6,000 mL/min. That is, the same volume of air is breathed per minute in each case.

As illustrated, the greater influence on effectiveness of alveolar ventilation is exerted by depth of breathing, rather than frequency.

Shallow breathing, regardless of frequency, may replace only the air contained in the conducting airways and may be incapable of replacing that in the alveolar spaces. Regardless of the quality of air external to the body, unconsciousness would occur in several minutes. This situation easily could occur during an accident in a confined space containing a mildly contaminated or slightly oxygen-deficient atmosphere. The combination of the stress induced by the situation, coupled with a respiratory system compromised by disease, obesity, age, or lack of physical conditioning easily could provide the required conditions for ventilatory insufficiency.

The second example illustrates normal breathing depth and rate.

TABLE 3.7
Breathing Pattern and Alveolar Ventilation

Tidal volume (mL)	Anatomic dead space (mL)	Net ventilation (ml/breath)	Frequency (breath/min)	Alveolar ventilation rate (ml/min)
150	150	0	40	0
500	150	350	12	4,200
1,000	150	850	6	5,100

After Vander et al. 1990.

The third example illustrates the effect of deep, slow breathing. This example also highlights the significance of tidal volume on alveolar ventilation. This breathing pattern would produce the greatest alveolar ventilation rate.

Taken at face value, the phenomenon highlighted in these examples could explain one of the perverse aspects of accidents involving confined spaces. There are several examples in which the entrant was overcome by atmospheric conditions and survived, while would-be rescuers who later entered the space perished (OSHA 1985). The initial victim may have survived through the combination of deep, slow, highly efficient breathing, while the would-be rescuer perished through shallow, rapid, highly inefficient breathing. However, inspired air enters the alveoli in a cone- or spike-shaped front, rather than a square front (Comroe et al. 1962). The net result is that alveolar ventilation occurs even when the tidal volume is 60 to 70 mL and anatomical dead volume is 150 mL.

Restrictive and obstructive lung diseases both can reduce forced vital capacity (Shigeoka 1983). As a result, the subject becomes breathless on exertion. This leads to shallow, rapid breathing. Restrictive disease processes include interstitial fibrosis, pleural scarring, rib cage abnormality, and muscle weakness. Restrictive disease could lead to low rate of gas transfer and an increase in ventilation relative to consumption of oxygen during exercise (Cotes 1983). The latter increases demands on the lung and heart. This increase in breathlessness also could occur due to obesity. Obesity disproportionately increases the uptake of oxygen and ventilation during the performance of a task. Obstructive processes include arthritis, bronchitis, emphysema (destruction of alveolar walls), asthma, bronchiectasis (chronic dilation of the bronchi or bronchioles, uniformly or in bulbous enlargements), and tumor formation. Obstructive processes involve narrowing of the airways. This traps air in the lung because there is insufficient time during exhalation for emptying to occur. An asthmatic attack is a special concern, since this can be triggered by nonspecific stressors. Hypoventilation also could affect individuals lacking in physical conditioning during the highly demanding and stressful conditions of an accident situation.

HYPERVENTILATION

Hyperventilation refers to alveolar ventilation in excess of that needed to maintain normal arterial partial pressure of oxygen and carbon dioxide (Comroe et al. 1962). Hyperventilation can occur for many reasons: anxiety, injury, disease, hypoxemia (low partial pressure of oxygen in the blood), mechanical overventilation during use of respirators and rebreathers, hypotension (low blood pressure), triggers of pulmonary reflexes, acidosis, hormones, and drugs.

Hyperventilation has little impact on hemoglobin saturation. This is an outcome of the shape of the saturation curve. Increasing alveolar ventilation from 4.3 L/min to 7.5 L/min, for example, increases hemoglobin saturation from 97.4 to 98.8%. Alveolar partial pressure of oxygen increases from 104 to 122 mmHg, while alveolar partial pressure of carbon dioxide decreases from 40 to 23 mmHg. Arterial pH increases from 7.40 to 7.56.

Hyperventilation can have more serious consequences where toxic gases and vapors are present. Increasing alveolar ventilation increases the rate of absorption.

HYPEROXIA (OXYGEN ENRICHMENT)

Hyperoxia is the condition resulting when the partial pressure of oxygen exceeds that found at sea level (20.9% or 159 mmHg in dry air). Hyperoxia can occur at normal pressures from contact with enriched mixtures or pure oxygen and through pressurization of atmospheres having normal composition. Breathed continuously, oxygen contained in a hyperbaric or enriched atmosphere acts as a toxic agent (Behnke 1978). Despite this, enrichment has been utilized successfully in surgical procedures and hyperbaric oxygen therapy. Two applications of the latter include transport of inert gases from tissues during decompression and use as a therapeutic agent during decompression sickness (Davis 1979).

Pressurized atmospheres are used in specialized work environments: tunnels and caissons and shallow diving. Tunnels and caissons require pressurization to prevent collapse or to prevent entry of water or gases from soils and rock. Diving requires pressurized air to counteract pressure exerted by water. Total pressures utilized in such operations range from less than 2 atmospheres (1,520 mmHg) to more than 4 atmospheres (3,040 mmHg). Corresponding partial pressures of oxygen range from 318 to 637 mmHg (Behnke 1978).

Enrichment of atmospheres at normal pressures occurred during fatal accidents involving confined spaces (OSHA 1985). Oxygen enrichment occurred following leakage from valves and supply lines of compressed gas systems, as well as use of oxygen as a coolant. Other possible sources include processes involving hydrogen peroxide or other peroxides that utilize or generate oxygen.

An important aspect of oxygen enrichment is the increased flammability of clothing and other combustible materials, including the skin (OSHA 1985). The risk of fire is greatly enhanced under these conditions. Oxygen is believed to adhere to clothing and other materials when sprayed as a gas at normal pressure. In hyperbaric environments at a pressure of 3 atmospheres (2,280 mmHg), fire burns at twice the normal rate (Behnke 1978).

Hyperoxia has little impact on hemoglobin saturation. Increasing alveolar partial pressure beyond normal values increases hemoglobin saturation insignificantly. This is an outcome from the dynamics of the saturation process as reflected in the saturation/partial pressure curve. At partial pressures considerably greater than those found in normal atmospheres, oxygen exerts both acute and chronic toxic effects. Some of these effects may be due to increased total pressure or the diluent (gas, nitrogen or helium). This discussion will focus on effects and conditions most relevant to confined spaces.

Table 3.8 indicates the toxic activity of oxygen at elevated partial pressures.

Oxygen toxicity is exerted in the lungs, central nervous system, and the eyes, although it is probably toxic to all organs at sufficient partial pressure (Piantadosi 1991). Generally, the rate of onset is a hyperbolic function of the inspired partial pressure (Clark and Lambertson 1971a, Clark and Lambertson 1971b). Sensitivity of the central nervous system to the toxic effects of oxygen is considerably greater than the that of the pulmonary system. Tolerance time to elevated partial pressures of pure oxygen atmospheres ranges from several minutes to 2 hours. Toxic action of hyperbaric oxygen atmospheres is greatly enhanced by exercise and elevated levels of carbon dioxide (Yarborough 1947). This translates into reduced tolerance time. Individual tolerance varies widely (Donald 1947).

Oxygen toxicity is expressed through production of reactive intermediates, such as the superoxide anion O_2^- and the hydroxyl radical (OH) (Freeman and Crapo 1982). The superoxide anion is highly reactive toward biological molecules. Normally, these species are removed by enzymatic action and reaction by free radical scavengers, such as reduced glutathione. During hyperoxia, production of reactive oxygen metabolites greatly increases and may exceed capacity of scavengers to remove them. Tissue injury and subsequent effects in both brain and lungs appear to be related to increased metabolism (Mayevsky 1984).

The setting of exposure limits involving oxygen enrichment is considered in Appendix C.

TABLE 3.8
Toxic Action of Oxygen

Atmospheric pressure		
Total (mmHg)	Oxygen (mmHg)	Comments
760	159	Sea level
	400	Respiratory irritation
	760	Throat irritation; no systemic effects provided that exposure is brief
	1,520	Tracheal irritation, slight burning on inhalation; tolerance increased when periods of oxygen interspersed with air; reduced vital capacity develops
	>1,520	Signs and symptoms of oxygen poisoning: tingling of fingers and toes, visual disturbances, acoustic hallucinations, confusion, muscle twitch, nausea, vertigo, possible convulsions
	>2,280	Nervous signs and symptoms twitching, vertigo, anxiety, paresthesia in toes and fingers, nausea, convulsive seizures

Data from Yarborough 1947, Donald 1947, Dukes–Dobos and Badger 1977, and Behnke 1978.

HIGH ALTITUDE

The atmosphere of habitable areas above sea level contains the same relative concentration of gases. The total pressure, and hence the partial pressures of individual components, including oxygen, decreases with increasing altitude (de Treville 1988, Lahiri et al. 1972, Davis 1979). Acclimatization from sea level to high level can require weeks or even months. This discussion will consider acute effects of transition to high altitude, as these are more likely to be comparable to events that occur in confined spaces.

Travel by large numbers of unacclimatized individuals to high altitudes has increased considerably over the last 3 decades (Hultgren 1992). The transient population of ski resorts in the U.S. is estimated at one million. Most of these individuals reside near sea level. This phenomenon adds another dimension to the study of hypoxia (oxygen deficiency). Travel characteristically entails rapid ascent, often within several hours, a brief stay at altitude, and rapid descent. Travel activities can include skiing, backpacking, trekking, and hiking. All of these involve strenuous exercise.

Table 3.9 summarizes characteristics of the atmosphere at different altitudes encountered during travel.

Moderate altitude includes many commonly visited and well-inhabited regions of the world. Mild discomfort may occur in susceptible individuals.

The zone of high altitude begins at 8,000 ft (2,440 m). The latter is generally regarded as the threshold above which altitude-related illness occurs. At this altitude, the arterial partial pressure of oxygen is 60 mmHg. Corresponding hemoglobin saturation relative to sea level is 92%. At higher altitudes, hemoglobin saturation decreases rapidly. At 14,000 ft (4,270 m), arterial partial pressure is 46 mmHg; arterial oxygen saturation is 82%.

Altitude illness and high altitude pulmonary edema are extremely rare at ski lodges below 7,000 ft (2,135 m), yet occur with low frequency at lodges located at 9,000 ft (2,745 m). Ski areas are located at higher levels. The important factor seems to be related to sleep.

Very high altitudes are easily accessible to trekkers and climbers. Rapid ascent to these levels is accompanied by high incidence of severe medical problems, including fatalities. The upper level, 18,000 ft (5,490 m) is the limit for prolonged stay. Prolonged stay above this altitude results in deterioration, not acclimatization. This, coincidentally, also is the limit for permanent habitation.

Most people who ascend rapidly to altitudes above 10,000 feet (3,050 m) experience some form of altitude effect. At this altitude, total atmospheric pressure is 530 mmHg and the partial

TABLE 3.9
Altitudes Encountered during Travel

Altitude		Atmospheric pressure		Oxygen level	
ft	m	Total (mmHg)	Oxygen (mmHg)	(%)	Comments
0	0	760	159	20.9	Sea level dry reference atmosphere
5,000	1,525	636	133	17.5	Moderate altitude
to 8,000	2,440	570	120	15.8	
8,000	2,440	570	120	15.8	High altitude
to 14,000	4,270	456	95	12.5	
14,000	4,270	456	95	12.5	Very high altitude
to 18,000	5,490	390	82	10.8	
18,000	5,490	390	82	10.8	Extreme altitude
to 29,028	8,850	249	52	6.8	

Data from Hultgren 1992.

pressure of oxygen is 111 mmHg. Symptoms include breathlessness, heart palpitations, headache, nausea, fatigue, and impairment of mental processes (Vander et al. 1990). These symptoms are similar to those quoted for similar pressures in the table describing oxygen deficiency. These effects disappear during the course of several days, although maximum physical capacity remains reduced.

The first response of a person acclimatized to sea level upon arrival at high altitude is increased ventilation at rest and during work. Ventilation increases to compensate for acute hypoxia. Hyperventilation increases the partial pressure of O_2 and decreases the partial pressure of CO_2. The increase in alveolar partial pressure of O_2 continues during the period of acclimatization. Acclimatization requires weeks or even months to accomplish. Thus, acclimatization results in increased alveolar partial pressure of O_2 at the cost of increased ventilation and decreased alveolar partial pressure of CO_2. Decrease in alveolar and arterial partial pressure of carbon dioxide initially increases pH in blood and cerebrospinal fluid (Bouhuys 1974, Lahiri 1972, Lahiri et al. 1972, Davis 1979). The increase in pH modifies the oxygen–hemoglobin binding relationship. This results in increased hemoglobin saturation beyond what would be predicted, based solely on consideration of partial pressure. Also, hemoglobin binds oxygen more tightly at higher pH and releases less to the tissues for a given decrease in arterial partial pressure (Bellingham et al. 1970).

Despite the increase in pH, the hemoglobin dissociation curve for healthy humans shifts to the right within 24 to 36 h after arrival at high altitudes (3,000 m or more). This shift promotes unloading of oxygen from hemoglobin, thus increasing its availability to body tissues. This increase reverts to normal upon return to sea level. Associated with this effect is an increase in the level of 2,3-DPG in red blood cells. When long-term high altitude residents travel to sea level, the reverse occurs. That is, the level of 2,3-DPG decreases and oxygen affinity of hemoglobin increases. Increased 2,3-DPG formation appears to be part of an adaptive response to high altitudes (Lenfant et al. 1968, Lenfant and Sullivan 1971).

There are many genetic variants of hemoglobin in humans. Some lead to disease, whereas others represent adaptation to environmental conditions, such as high altitude (Bouhuys 1974). Many variants have higher or lower affinity for oxygen than "normal" hemoglobin (Stamatoyannopoulos et al. 1971). In general, the higher the affinity for oxygen, the higher the capacity.

Residents of high altitudes ventilate less than newly acclimatized lowlanders during exercise or in hypoxic conditions (Lahiri et al. 1972). Highlanders native to 2,900 m or higher tolerate hypoxia better than acclimatized lowlanders and apparently can work harder.

TABLE 3.10
Effect of Brief Exposure to Oxygen-Deficient Atmospheres

Oxygen concentration (%)	Tidal volume (mL)	Breathing frequency (breath/min)	Minute volume (L/min)	Alveolar ventilation rate (L/min)
20.9	500	14	7.0	4.9
18.0	500	14	7.0	4.9
16.0	536	14	7.5	5.4
12.0	536	14	7.5	5.4
10.0	593	14	8.3	6.2
8.0	812	16	13	10.4
6.0	—	—	18	—
5.2	—	—	22	—
4.2	933	30	28	23.2

Data from Comroe et al. 1962.

An important effect demonstrated by travel to high altitude is a progressive decrease in maximum exercise capacity and maximum oxygen consumption and decrease in maximum heart rate (West et al. 1983). This decrement occurs even at moderate altitudes and led to increases in times of 5 to 10% for distance races in the Mexico Olympics. The altitude of Mexico City is 7,350 ft (2,240 m) (Grover et al. 1986).

Decreased performance capacity could have important significance in accidents that occur in confined spaces. This could be especially significant in oxygen-deficient atmospheres during rescue attempts. The rescuer operates under extreme physical and emotional duress. Decreased performance capacity considerably increases the risk of exceeding one's limits under such circumstances.

HYPOXIA (OXYGEN DEFICIENCY)

A condition that mimics the effects of hypoventilation in normal individuals is exposure to an atmosphere containing less than the normal partial pressure of oxygen. In the occupational setting, hypoxia is produced by asphyxiants. Asphyxiants interfere with the supply or use of oxygen in the body. Asphyxiants include both simple asphyxiants and chemical asphyxiants. Simple asphyxiants include: acetylene, argon, ethane, ethylene, hydrogen, helium, methane, neon, nitrogen, propane, and propylene (ACGIH 1994). Chemical asphyxiants will be discussed in a later section.

Simple asphyxiants are physiologically inert; that is, they do not affect biochemical processes. They dilute or displace the normal atmosphere, so that the resultant partial pressure of oxygen is insufficient to maintain oxygen tensions at levels needed for normal tissue respiration. The areas of the body considered most sensitive to oxygen deprivation are the brain and myocardium (heart muscle). Brain cells perish in 3 to 5 minutes under conditions of complete hypoxia. Damage sustained by these oxygen-sensitive tissues is not reversible upon restoration of the atmosphere (Ayers et al. 1969, Davis 1979).

The physiological effects of brief exposure (8 to 10 min) to oxygen-deficient atmospheres in resting subjects are summarized in Table 3.10 (Comroe et al. 1962).

The characteristic response to hypoxemia induced by breathing an oxygen-deficient atmosphere is an increase in depth (tidal volume) and frequency of breathing. This is a direct response to triggering of oxygen chemoreceptors in the carotid and aortic bodies by the decrease in arterial partial pressure. These receptors are somewhat insensitive and not immediate in their response. Atmospheric oxygen concentration must decrease to 16% (sea level) prior to the initiation of

TABLE 3.11
Effect of Oxygen Partial Pressure on Alveolar Gas Exchange

Atmospheric Oxygen (%)	Alveolar partial pressure (mmHg)	Capillary partial pressure		Partial pressure gradient		Hemoglobin saturation	
		Start (mmHg)	End (mmHg)	Start (mmHg)	End (mmHg)	Start (%)	End (%)
20.9	101	40	100	61	1	75	97
14.0	57	32	51	25	6	58	84
12.0	44	27.5	40	16.5	4	53	75

Data from Comroe et al. 1962.

response. Said another way, the delay in increasing the depth and frequency of breathing in these situations appears to correlate with decrease in hemoglobin saturation to the steep part of the curve.

The apparent delay in response could be construed as an emergency response when hypoxemia becomes severe. This may not be the case, since there is a similar delay in the onset of more rapid and deeper breathing following the start of vigorous exercise, such as running, from resting status. Hypoxemia is capable of causing increased respiration in normal individuals. However, hypoxemia greater than that seen in most patients with chronic pulmonary disease is required before breathing in normal individuals is stimulated conspicuously (Comroe et al. 1962).

The extent of saturation of hemoglobin reflects partial pressure of oxygen in the blood. Many normally occurring situations, including changing metabolic status from rest to vigorous exercise, rapid ascent to high altitude, and cardiac or pulmonary insufficiency are characterized by reduced alveolar and therefore arterial partial pressure of oxygen.

Decrease of arterial partial pressure from 100 to 60 mmHg would cause only a 10% decrease in hemoglobin saturation. Hyperventilation by a normal person at sea level produces little change in hemoglobin saturation for this reason (Vander et al. 1990). At arterial partial pressures less than 50 mmHg, saturation of hemoglobin decreases rapidly. Oxygen tension in tissue capillaries is 40 mmHg. Oxygen dissociates from the hemoglobin molecule and enters into physical solution in the plasma whenever the oxygen tension in the plasma decreases. Thus, as fast as oxygen diffuses from the plasma into tissues through the capillaries, it is replenished by oxygen dissociating from the hemoglobin (Bouhuys 1974). Oxygenated hemoglobin gives up large quantities of oxygen under these conditions.

Another aspect in exposure to reduced levels of oxygen (normal resting subjects) is transfer from alveolar spaces into blood. This is illustrated in Table 3.11 (Comroe et al. 1962).

As mentioned previously, the normal time spent in the pulmonary capillary bed is 0.75 s. Under normal conditions, the oxygen tension increases to almost 100 mmHg in 0.35 s or less. This results from the steepness of the pressure gradient across the capillary membranes. The change in pressure gradient with time as blood perfuses through the capillary is hyperbolic and reaches equilibrium asymptotically.

At low levels of oxygen, for example, 12%, oxygen tension in incoming blood decreases to 27.5 mmHg. The pressure gradient is linear and less steep than that at higher concentrations. The rapid increase in saturation early in the passage through the capillary no longer occurs. Instead, saturation increases proportionate to distance along the capillary. In addition, there is a net partial pressure difference between alveolar air and blood at the end of the capillary due to lack of equilibration.

Complicating this situation is the impact of exercise and work. Exercise decreases the time spent by blood in the pulmonary capillaries. This would further reduce saturation in an individual breathing a reduced level of oxygen. To a first approximation (this could be influenced by change

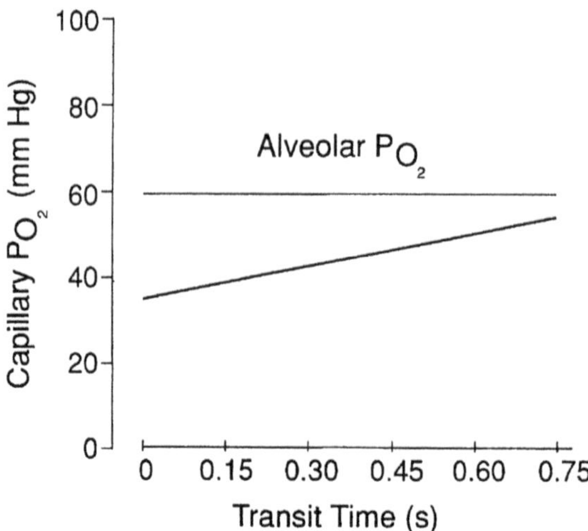

FIGURE 3.5 Oxygenation of hemoglobin in pulmonary capillaries under oxygen-deficient conditions (12 to 14% oxygen). Under oxygen-deficient conditions oxygenation occurs at a slower rate throughout the duration of travel through the pulmonary capillaries. Equilibration of partial pressure of oxygen in the alveoli and blood does not occur. (Adapted from Comroe et al. 1962.)

in pH), the remaining pressure gradient could be estimated using reduced transit time as a fraction of normal and the linear increase of saturation with time in the capillary. For example, in 14% oxygen and 0.30 s for transit time in place of 0.75 s, partial pressure in blood would increase from 32 to 40 mmHg. Saturation of hemoglobin would increase from 58 to 75%. Reducing the level of exercise so that the transit time increases to 0.45 s would provide only marginal increase in partial pressure in blood from 40 to 45 mmHg. Partial pressure of 45 mmHg corresponds to saturation of 80%. Saturation would increase during passage through the pulmonary capillaries from 58 to 80% (an increase from 75 to 80%).

Oxygen deficiency is a major concern in the occupational setting and the subject of many standards and regulations (Figure 3.5). Table 3.12 summarizes the effects of acute exposure to oxygen deficient atmospheres as commonly reported.

The rate of onset of the symptoms presented in this table depends on many factors, including breathing rate, work rate, temperature, emotional stress, age, and individual susceptibility (Timar 1983). These factors can exacerbate the effects of an oxygen-deficient atmosphere and influence the onset, course, and outcome of accidents that occur under these conditions in confined spaces.

Loss of consciousness is a key outcome in an oxygen-deficient atmosphere (Figure 3.6). At a concentration of 5% oxygen at sea level, unconsciousness in inactive subjects begins after about 12 seconds, or about 2 breaths of air (Davis 1979, Miller and Mazur 1984). For slight increases in concentration, the duration of consciousness for inactive subjects increases rapidly to about 30 s at 6.5% oxygen. For active or active and highly stressed subjects, loss of consciousness would occur at higher concentrations. High activity and high stress is the likely state of a would-be rescuer during an accident situation.

A number of stressors that reflect the metabolic demand for gas exchange can modify the breathing pattern. Feedback about arterial partial pressures of oxygen and carbon dioxide and hydrogen ion concentration provides the information. Under most conditions, ventilation rate regulates arterial oxygen and carbon dioxide tensions within narrow limits. Oxygen deprivation also can become regulating. This occurs when the oxygen content of the inspired gases is reduced to nearly half that in air at sea level (approximately 11%). Hence, under normal circumstances regulation of breathing occurs by bodily requirements to control carbon dioxide tension. However, the concentration of oxygen in an oxygen-deficient atmosphere may become the regulator of

TABLE 3.12
Effects of Acute Exposure to Oxygen-Deficient Atmosphere

Effect	Atmospheric Oxygen (dry air, sea level) Concentration (%)	Pressure (mmHg)
No symptoms	16 to 20.9	122 to 159
Increased heart and breathing rate, some incoordination, increased breathing volume, impaired attention and thinking	16	122
Abnormal fatigue upon exertion, emotional upset, faulty coordination, impaired judgment	14	106
Very poor judgment and coordination, impaired respiration that may cause permanent heart damage, nausea, and vomiting	12	9
Nausea, vomiting, lethargic movements, perhaps unconsciousness, inability to perform vigorous movement or loss of all movement, unconsciousness followed by death	<10	<76
Convulsions, shortness of breath, cardiac standstill, spasmatic breathing, death in minutes	<6	<46
Unconsciousness after one or two breaths	<4	<30

Data from NIOSH 1976, Miller and Mazur 1984, ANSI 1992, and CSA 1993.

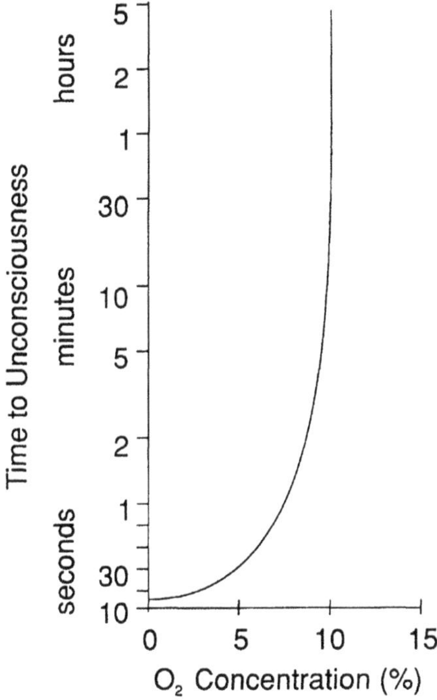

FIGURE 3.6 Approximate time to unconsciousness versus oxygen concentration for seated subjects at sea level. This curve resulted from a compilation of data from several sources. (Adapted from Miller and Mazur 1984.)

breathing. Elevated levels of carbon dioxide (30,000 to 70,000 ppm) produce the following effects: increase in tidal volume, breathing rate, and minute ventilation (Bouhuys 1974).

Healthy people live long and active lives at high altitudes where arterial saturation ranges from 85 to 95%. Few patients with cardiopulmonary disease have arterial oxygen saturation less than

85%. The lower limit of arterial oxygen saturation compatible with moderately active existence depends on the abruptness with which the hypoxemia develops, compensatory mechanisms, and other limiting factors in the disease process. Hemoglobin saturation in persons with congenital heart disease may be less than 80% without causing disability. On the other hand, an asthmatic may sustain adequate alveolar gas exchange and arterial saturation only by extreme effort. Persons with emphysema may experience disability, despite the fact that arterial saturation is 90 to 95% (Comroe 1962).

The setting of exposure limits involving oxygen deficiency is considered in Appendix C.

CHEMICAL ASPHYXIANTS

Chemical asphyxiants interfere with the supply or use of oxygen in the body. Chemical asphyxiants exert their toxicological influence through several mechanisms:

- Preventing uptake of oxygen by the blood
- Interfering with the transport of oxygen from the lungs to the tissues
- Preventing normal oxygenation of the tissues, even though the blood is well oxygenated

Chemical asphyxiants produce anemic hypoxia and histotoxic hypoxia. Anemic hypoxia implies partial or total lack of availability of hemoglobin for transport of oxygen by red blood cells. Chemical asphyxiants that affect hemoglobin include carbon monoxide, aniline, dimethyl aniline, and toluidine. Histotoxic hypoxia results from the action of agents that block or interfere with enzymes involved in cellular respiration. Reduction in consumption of oxygen at the cellular level can lead to increased saturation of hemoglobin in arterial and venous blood. Chemical asphyxiants that act in this manner include: cyanogen, hydrogen cyanide, hydrogen sulfide, and nitrites (Dinman 1978, NIOSH 1976).

The preceding information indicates that carbon monoxide and hydrogen sulfide normally are classified as chemical asphyxiants, rather than toxic agents. OSHA assigned accidents involving these substances into the category of toxic atmosphere, rather than asphyxiating atmosphere, in the report on atmospheric hazards in confined workspaces (OSHA 1985). This discrepancy in classification implies that asphyxiation could have occurred in a much higher proportion of accidents than ascribed by OSHA. That is, OSHA may have underestimated the role of asphyxiants as the causative agent.

HYDROGEN SULFIDE

The substance implicated most frequently in fatal accidents occurring in confined spaces is hydrogen sulfide (OSHA 1985). Hydrogen sulfide is produced mainly through two main processes: biological decay of sulfur-containing organic materials and the action of acids on inorganic sulfides. Hydrogen sulfide also occurs widely in petroleum and natural gas (NIOSH 1977).

Table 3.13 summarizes the physical and chemical properties of hydrogen sulfide.

The hazards posed by hydrogen sulfide during workplace exposure are a direct reflection of its physical, chemical, and toxic properties. The gas density of hydrogen sulfide gas is somewhat greater than that of air. Thus, a concentrated cloud produced in a depressed area will tend to remain at the point of generation. Dispersion of a concentrated cloud generated in an unconfined area by convective currents and diffusion also could be hindered. This characteristic of hydrogen sulfide is a major concern in confined workspaces, since the concentration required for producing acute toxic effects is relatively low. Hydrogen sulfide forms explosive mixtures in air. This characteristic could be a concern in some confined workspaces; however, the concentration required for flammability is very high relative to that capable of producing acute toxic effects. Hydrogen sulfide also is colorless, and thus provides no visual warning about its presence.

TABLE 3.13
Chemical and Physical Properties of Hydrogen Sulfide

Molecular formula	H_2S
Molecular weight	34.1
Gas density (Air = 1)	1.19
Solubility in water @ 20°C	2.9% by weight
Autoignition temperature	250°C
Explosive range in air	4.5% to 45.2%
Color	Colorless
Reaction in water	Ionizes in water; high pH increases solubility by increasing ionization
Relevant chemistry	Reacts with many metal ions to form insoluble sulfides; acidification of solutions containing sulfide salts generates hydrogen sulfide
Odor	Rotten eggs (when detectable)

After Macaluso (1969).

Hydrogen sulfide is quite soluble in pure water. This results partly from the polar nature of the molecule and the similarity in shape that to water. Solubility in water also is enhanced because molecules of H_2S ionize to form an acid solution containing hydrosulfide and sulfide ions. Solubility is greatly enhanced in alkaline solutions (pH >7) due to increased ionization. The enhanced solubility of hydrogen sulfide in alkaline solutions is extremely important to the hazard potential of this gas. Rapid decrease in pH, by direct acidification or other means, reverses the equilibrium to form undissociated H_2S molecules. This process can exceed the solubility of hydrogen sulfide and can lead to effusion of gas from solution into the atmosphere.

De Morbis Artificum Diatriba contains the first reported reference about occupational exposure to hydrogen sulfide (Ramazzini 1964). In industrialized North America, poisoning by hydrogen sulfide has become an important concern only in the last 70 years (NIOSH 1977).

NIOSH and other reviewers of the toxicological literature have concluded that effects produced by exposure to hydrogen sulfide are primarily acute in nature (NIOSH 1977, ACGIH 1991). At concentrations between 50 and 500 ppm, hydrogen sulfide acts primarily as an irritant of the respiratory system and the eyes. It also can induce olfactory fatigue and nausea. Pulmonary edema and pneumonia may accompany prolonged exposures at concentrations exceeding 250 ppm. At low concentrations, effects on the eyes predominate. Conjunctivitis and keratitis are common. Eye effects are generally acute in nature. No reports of lasting eye damage were found. At high concentrations, in the range of 500 to 1,000 ppm, hydrogen sulfide acts primarily as a systemic poison, causing unconsciousness and death through respiratory paralysis.

Brief exposures to moderate concentrations, 140 mg/m^3 (100 ppm), commonly cause conjunctivitis and keratitis. At concentrations exceeding 280 mg/m^3 (140 ppm), hydrogen sulfide can cause unconsciousness, respiratory paralysis, and death. Hydrogen sulfide paralyzes the respiratory center of the brain and also the olfactory nerve. At concentrations of 350 to 700 mg/m^3 (500 to 1,000 ppm), hydrogen sulfide causes rapid unconsciousness and death through respiratory paralysis. Higher concentrations rapidly deaden the sense of smell. Individual case histories have shown that other disorders have resulted from exposure to hydrogen sulfide. These include: cardiovascular, nervous system, and gastrointestinal disorders. No conclusive support exists for adverse health effects from repeated long-term exposure at low concentration (NIOSH 1977).

Hydrogen sulfide blocks cellular oxidation at the center of the brain that controls respiration. This effect occurs through interference with the function of oxidative enzymes, mainly cytochrome oxidase. The respiratory center becomes extremely hypoxic. This frequently results in a nearly simultaneous respiratory arrest and hypoxic seizure (Kurt 1983).

TABLE 3.14
Effects of Exposure to Hydrogen Sulfide

Toxicological Effect	Concentration (ppm)
Odor Threshold	0.003 to 0.02
Rotten egg odor	Up to 30
Sickly sweet odor	30 to 100
Olfactory fatigue	>100
Olfactory fatigue	Prolonged exposure <100
Eye irritation, conjunctivitis, corneal erosion	<20 ppm/<8 h
Burning eyes, headache, shortness of breath, loss of appetite, weight loss, dizziness	15–25
Acute conjunctivitis, pain, lachrymation, photophobia, keratoconjunctivitis	50 ppm/1 hour
Loss of consciousness, arm cramps, low blood pressure	230 ppm/20 min
Unconsciousness, low blood pressure, pulmonary edema, convulsions, hematuria, death	>1,000 ppm/>1 min
Coma after single breath	1,000 to 2,000
Pulmonary edema	250, prolonged exposure

Data from NIOSH 1977.

The hydrosulfide anion, HS^-, forms a complex with methemoglobin (a form of hemoglobin in which the iron atom has been oxidized to the III oxidation state rather than the II state that is normally present). The resulting complex is known as sulfmethemoglobin. The production of methemoglobin with nitrite solution has been used successfully to resuscitate victims severely poisoned by hydrogen sulfide (Smith and Gosselin 1966, Smith et al. 1977).

Table 3.14 summarizes toxicological effects caused by hydrogen sulfide (NIOSH 1977).

The data presented in this table indicate that hydrogen sulfide is acutely toxic, to the point that a single breath can produce collapse and coma. Death can occur rapidly following even brief exposure under these conditions. The concentration of hydrogen sulfide at which this response can occur is not high in relative terms.

At concentrations less than 10 ppm, acute exposure is likely to produce nasal discomfort due to the characteristic odor of rotten eggs. It is detectable at concentrations in the range of 2 ppb by some people. However, at higher concentrations hydrogen sulfide has a sickly sweet odor, and at still higher concentrations, no odor, due to instantaneous olfactory fatigue.

Because hydrogen sulfide possesses the characteristic odor of rotten eggs, one can easily be lulled into relying on smell as an indicator of concentration. Olfactory fatigue rapidly occurs at 100 to 150 ppm. The lack of odor tends to create a false sense of security. A person suffering from olfactory fatigue could easily mistake the apparent lack of odor as indicating an environment free from hydrogen sulfide. This is the reason that air purifying respirators cannot be used for protection against H_2S.

The toxicological properties of hydrogen sulfide form the basis for judgments made by standard-setting agencies. Table 3.15 summarizes these limits.

Emergency Response Planning Guidelines (ERPGs) are prepared by the Emergency Response Planning Committee of the American Industrial Hygiene Association (AIHA 1995). They are intended to provide estimates of concentration ranges above which one could reasonably anticipate observing adverse acute effects in most individuals. That is, they are considered thresholds above which there would be an unacceptable likelihood of observing the defined effect. ERPGs contain no safety factors. They are intended for planning and emergency response, rather than exposure guidelines.

ERPG-1 is the maximum airborne concentration below which it is believed that nearly all individuals could be exposed for up to 1 h without experiencing other than mild transient adverse health effects or perceiving a clearly defined objectionable odor. ERPG-2 is the maximum airborne

TABLE 3.15
Exposure Standards — Hydrogen Sulfide

Exposure standard	Concentration (ppm)
Emergency Response Planning Guideline (ERPG-1)	0.1
Emergency Response Planning Guideline (ERPG-2)	30
Emergency Response Planning Guideline (ERPG-3)	100
Threshold Limit Value-Time Weighted Average	10
Threshold Limit Value-Short Term Exposure Limit	15
Recommended Exposure Limit-Ceiling	10
Immediately Dangerous to Life and Health	300
Lethal Concentration (50%), estimated	952

Compiled from Alexeeff et al. 1989, NIOSH 1990, ACGIH 1994, and AIHA 1995.

concentration below which it is believed that nearly all individuals could be exposed for up to 1 h without experiencing or developing irreversible or other serious health effects or symptoms that could impair their abilities to take protective action. ERPG-3 is the maximum airborne concentration below which it is believed that nearly all individuals could be exposed for up to 1 h without experiencing or developing life-threatening health effects.

Threshold Limit Values (TLVs) are recommended by the TLV Committee of the American Conference of Governmental Industrial Hygienists (ACGIH 1994). TLVs are guidelines, not fine lines, between safe and dangerous conditions. TLVs are believed to represent conditions under which nearly all workers may be repeatedly exposed day after day without adverse health effects.

TLV-TWA is the time-weighted concentration for a normal 8-h workday and 40-h workweek to which nearly all workers may be repeatedly exposed, day after day, without adverse effect. TLV-STEL is a 15-min time-weighted average that should not be exceeded at any time during a workday, even if the 8-hour TWA is within the TLV-TWA. Up to four excursions at the TLV-STEL may occur per day, provided that each is separated by at least 1 hour and the TLV-TWA is not exceeded.

Recommended Exposure Levels (RELs) and Immediately Dangerous to Life or Health limits (IDLHs) were prepared by the National Institute for Occupational Safety and Health as part of the Standards Completion Project (NIOSH 1990). REL-TWA is the time-weighted average concentration for up to a 10-h workday during a 40-h workweek. A short-term exposure limit (REL-STEL) is a 15-min time-weighted average that should not be exceeded at any time during a workday. A ceiling level (REL-Ceiling) should not be exceeded at any time. IDLH was defined for the purpose of selecting respiratory protection. IDLH represents the presumed maximum concentration from which a person could escape in 30 min in the event of respirator failure without experiencing any escape-impairing or irreversible health effects.

In listing and discussing these standards, there is the underlying notion that some represent limits of human tolerance to atmospheric conditions. Mention of these values in this discussion is not intended to condone their use as acceptable limits for unprotected exposure, since most clearly exceed legal limits. Rather, their use is intended to illustrate levels that individuals may have exceeded during work activities or voluntary rescue attempts that occurred during fatal accidents.

The limits of greatest interest from the perspective of acute exposure that could be tolerated during gross overexposure are the ERPG-3 and IDLH. These levels provide some insight regarding conditions of gross overexposure that could be tolerated during fatal accidents in which hydrogen sulfide was implicated (OSHA 1985, NIOSH 1994). The ERPG-3 and IDLH levels are considerably higher than concentrations measured in these spaces during follow-up investigation.

TABLE 3.16
Chemical and Physical Properties of Carbon Monoxide

Molecular formula	CO
Molecular weight	28.0
Gas density (air = 1)	0.968
Solubility in Water (20°C)	0.004% by weight
Autoignition temperature	609°C
Explosive range in air (by volume)	12.5% to 74.2%
Color	Colorless
Reaction in water	None
Relevant chemistry	None
Odor	Odorless

After NAPA (1970).

Carbon Monoxide

Carbon monoxide is one of the most studied of chemical hazards. Its toxic effects have been appreciated since the discovery of fire. The consequences of exposure have been acknowledged in the earliest of writings on health hazards.

Carbon monoxide is the product of incomplete combustion of carbon-containing materials. Internal combustion engines, gasoline, diesel, and alternate fuelled, and fuel-powered space heaters are important sources of carbon monoxide that have had roles in fatal accidents involving confined spaces (OSHA 1985, NIOSH 1994). Sources in process operations, such as refineries, chemical plants, and steel production and foundries also could be involved.

Table 3.16 summarizes the chemical and physical properties of carbon monoxide.

The hazards posed by carbon monoxide during workplace exposure are a direct reflection of its physical, chemical, and toxic properties. The gas density of carbon monoxide gas is slightly less than that of air. Thus, a concentrated cloud will tend to remain at the point of generation until dispersed by convection currents and diffusion. This property of carbon dioxide is responsible for many fatal and nonfatal overexposures and is a major concern in poorly ventilated workspaces. Carbon monoxide forms flammable and explosive mixtures in air. This characteristic could be of major concern in some confined workspaces; however, the concentration required for flammability is very high relative to that capable of producing acute toxic effects. Carbon monoxide also is colorless, and thus provides no visual warning about its presence. However, at the low concentrations at which carbon monoxide exerts its toxic effects, this property likely would be of little importance.

The main toxic action of carbon monoxide in the body is binding to hemoglobin in the erythrocyte (red blood cell). Smaller amounts bind to myoglobin, a muscle-bound protein having a structure similar to a single subunit of hemoglobin. The remainder of CO binds with cytochrome oxidase, cytochrome P-450, and the hydroperoxidases (ACGIH 1991).

Carbon monoxide reacts with hemoglobin in a manner similar to that of oxygen. The molecule fits into pockets in the hemoglobin molecule formed by the heme group that are normally occupied by oxygen. The mode of chemical bonding is similar to, but stronger than that of oxygen. This also occurs less rapidly than does bonding with oxygen. Despite this, the resulting carbon monoxide–hemoglobin complex is 220 to 290 times more stable than the oxygen–hemoglobin complex. The binding of CO to hemoglobin also occurs in four steps, as does the process involving oxygen (NIOSH 1972).

The response of the hemoglobin association/dissociation curve to intoxication by carbon monoxide is displacement to the left. This occurs because the affinity of hemoglobin for CO is much greater than that for oxygen. The amount of oxygen in physical solution in the blood remains near normal during carbon monoxide intoxication. Not only is oxygen storage by the blood lowered during exposure to CO, but displacement of the dissociation curve to the left decreases the amount of oxygen on oxyhemoglobin that is available to the tissues. Both mechanisms effectively lower the partial pressure of oxygen in tissue spaces and hence, can create generalized tissue hypoxia (Bartlett 1968, Lilienthal 1950, Dinman 1968, Lundwig and Blakemore 1957).

Carbon monoxide is a normal product of the catabolism of heme, the ring structure that holds the iron atom. Production of carbon monoxide in the body by endogenous processes results in a level of carboxyhemoglobin normally ranging from 0.5 to 0.8%. The great affinity of hemoglobin for carbon monoxide provides the body an expedient method for isolation and removal of carbon monoxide produced endogenously (Lundwig and Blakemore 1957, Coburn 1967).

Myoglobin has approximately 16% of the oxygen-carrying capacity of an equal quantity of hemoglobin. Oxymyoglobin is a major supplier of oxygen for cellular respiration during heavy work by muscle. Myoglobin accounts for only 20% of the body's total capacity for carbon monoxide. The formation of carboxymyoglobin is analogous to that of carboxyhemoglobin. The affinity constant of myoglobin for carbon monoxide, however, is only 40 times greater than for oxygen (Coburn 1967, Rossi-Fanelli and Antonini 1958).

The heart muscle (myocardium) is particularly at risk from binding of carbon monoxide to myoglobin. Binding of carbon monoxide reduces the number of sites available for oxygen storage and supply. The heart responds to generalized tissue hypoxia by increasing blood supply. The heart increases supply by an increase in both rate and volume per contraction. This increase in activity also requires increased supply of oxygen to the myocardium. Under hypoxic conditions, two mechanisms potentially could provide this increase: increased blood flow (vascular dilation) and/or increased oxygen extraction by the tissues. The reality is that the increase in supply is provided only by an increase in coronary circulation (Ayers et al. 1969).

While peripheral tissues normally extract only 25% of the oxygen content of perfusing arterial blood during resting conditions, the myocardium extracts 75%. This highly efficient extraction leaves mixed venous blood in the myocardial circulation only 25% saturated. Under normal conditions, this process maintains oxygen tension in the myocardium at a higher level than would be present in other muscle tissue. This situation ensures continual aerobic metabolism in the myocardium, even under hypoxic duress. The presence of carboxyhemoglobin alters this process. The leftward shift of the oxyhemoglobin association/dissociation curve reduces the amount of oxygen available to the tissues. The only alternative to maintaining oxygen levels is increased coronary circulation (Ayers et al. 1969).

A person with coronary heart disease is especially sensitive to this situation. One outcome of coronary heart disease is reduced coronary circulation. Binding of carbon monoxide by myocardial myoglobin and reduced oxygen tension in myocardial capillaries due to carboxyhemoglobin may cause a serious deficiency in oxygen supply to the myocardium. A person with diminished coronary circulation, therefore, may constantly be near the point of myocardial hypoxia. Experiments involving coronary heart patients have demonstrated the gravity of this situation (Ayers et al. 1970).

Table 3.17 summarizes the effects of exposure to carbon monoxide (NIOSH 1972).

The data presented in this table indicate that intoxication produced by carbon monoxide depends on the level of carboxyhemoglobin in the blood, rather than airborne concentration. Further, carbon monoxide is not acutely toxic in the manner of hydrogen sulfide; a single breath of highly contaminated air cannot produce collapse and coma. Accumulation of carboxyhemoglobin is a time-dependent process. This observation is extremely important from the context of the dynamics of fatal accidents that have occurred in confined spaces (OSHA 1985, NIOSH 1994).

Studies of controlled exposure to carbon monoxide provide insight about the kinetics of formation of carboxyhemoglobin (Stewart and Peterson 1970). Unconsciousness is a recognizable end point in many of the accidents involving confined spaces or atmospheric confinement. The concentration of carboxyhemoglobin that produces coma is a minimum of 50%. Accumulation of

TABLE 3.17
Effects of Exposure to Carbon Monoxide

Effect/condition	Carboxyhemoglobin Concentration (%)
Odor recognition	Odorless
Carboxyhemoglobin normally present in blood	0.5 to 0.8
Impaired cardiovascular function — patients with cardiovascular disease	3 to 5
Carboxyhemoglobin present in blood of smokers	3 to 10
Slight headaches	10 to 20
Effects on visual function	>20
Coma with intermittent convulsions	50 to 60
Death	70 to 80

Data from NIOSH 1972.

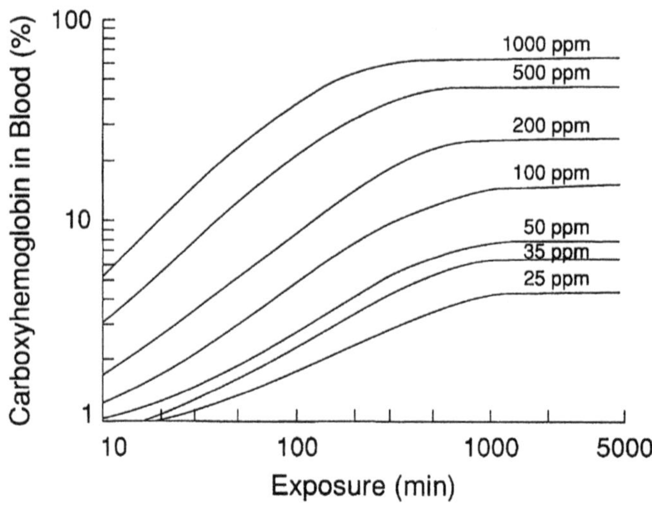

FIGURE 3.7 Absorption of carbon monoxide by hemoglobin in blood at normal atmospheric pressure. (Adapted from Stewart and Peterson 1970.)

carboxyhemoglobin to this level would require exposure to 1,000 ppm for approximately 180 min. For this same level of saturation to be reached in a shorter period of time, exposure to a considerably higher concentration of carbon monoxide would be required (Figure 3.7).

Tobacco smokers are the most heavily exposed nonindustrial segment of the population. Carboxyhemoglobin saturation in the blood of this group ranges from 3 to 20%. The mean for a one pack per day smoker is 5 to 6% (Kurt 1983).

Carbon monoxide is odorless. Thus, it provides no warning about its presence. This situation complicates the selection and use of respirators where CO may be present. There are two negative outcomes from this situation in confined spaces: failure to use a respirator and use of a respirator inappropriate to the circumstances. Both factors contributed to fatal accidents reported by OSHA and NIOSH (OSHA 1985, NIOSH 1994). Failure to use a respirator could occur from the perceived lack of hazard because of the absence of a warning odor. Without testing to evaluate the composition of the atmosphere, the possibility of inappropriate selection is ever-present. The latter also excludes the ability to establish the true contributory value of carbon monoxide to atmospheric contamination in these work environments.

TABLE 3.18
Exposure Standards — Carbon Monoxide

Exposure Standard	Concentration (ppm)
Threshold Limit Value — Time-Weighted Average	25
Recommended Exposure Limit — Time-Weighted Average	35
Recommended Exposure Limit — Ceiling	200
Immediately Dangerous to Life and Health	1,500
Lethal Concentration (50%), estimated	5,207

Compiled from Alexeeff et al. 1989, NIOSH 1990, and ACGIH 1994.

The toxicological properties of carbon monoxide form the basis for judgments made by standard-setting agencies. The following table summarizes these criteria (Alexeeff et al. 1989, NIOSH 1990, ACGIH 1994).

Threshold Limit Values (TLVs) are recommended by the TLV Committee of the American Conference of Governmental Industrial Hygienists (ACGIH 1994). TLVs are guidelines, not fine lines, between safe and dangerous conditions. TLVs are believed to represent conditions under which nearly all workers may be repeatedly exposed day after day without adverse health effects. TLV-TWA is the time-weighted concentration for a normal 8-h workday and 40-h workweek to which nearly all workers may be repeatedly exposed, day after day, without adverse effect.

Recommended Exposure Levels (RELs) and Immediately Dangerous to Life or Health limits (IDLHs) were prepared by the National Institute for Occupational Safety and Health as part of the Standards Completion Project. REL-TWA is a time-weighted average concentration for up to a 10-h workday during a 40-h workweek. A ceiling (REL-Ceiling) should not be exceeded at any time. IDLH was defined for the purpose of selecting respiratory protection. IDLH represents the presumed maximum concentration from which a person could escape in 30 min in the event of respirator failure without experiencing any escape-impairing or irreversible health effects.

In listing and discussing these standards, there is the underlying notion that some represent limits of human tolerance to atmospheric conditions. Mention of these values in this discussion is not intended to condone their use as acceptable limits for unprotected exposure, since most clearly exceed legal limits. Rather, their use is intended to illustrate levels that individuals may have exceeded during work activities or voluntary rescue attempts that occurred during fatal accidents.

The recommended limit of greatest interest from the perspective of acute exposure that could be tolerated during gross overexposure is the IDLH (immediately dangerous to life or health). This level provides some insight regarding conditions of gross overexposure that could be tolerated during fatal accidents in which carbon monoxide was implicated (OSHA 1985, NIOSH 1994). The IDLH limit is considerably higher than concentrations measured in these spaces during follow-up investigation.

OTHER CHEMICAL SUBSTANCES

Accident investigations reported by OSHA and NIOSH also documented the involvement of chemical substances in fatal accidents occurring in confined spaces (OSHA 1985, NIOSH 1994). Substances mentioned most frequently included carbon dioxide and halogenated and nonhalogenated organic solvents. The following section discusses the actions of these toxic agents.

CARBON DIOXIDE

One of the chemical substances identified frequently during investigation of fatal accidents involving confined spaces was carbon dioxide (OSHA 1985, NIOSH 1994). Carbon dioxide is ubiquitous in the atmosphere. It normally occurs at a concentration of 350 ppm.

Carbon dioxide is a by-product of human respiration, internal combustion engines, and other fuel-powered equipment and industrial chemical and biological processes. It also is produced for industrial use, mostly by recovery from industrial processes of which it is a by-product. These processes include: lime production, ammonia synthesis, synthesis gas (carbon monoxide and hydrogen), catalytic oxidation of benzene and ethylene oxide, fermentation of biological substances, and combustion of fuels (NIOSH 1976b).

Solidified carbon dioxide (dry ice) is used extensively for the rapid refrigeration and freezing of food products. Solid carbon dioxide is used as an abrasive for removal of paint. Liquid carbon dioxide is used as a refrigerant. Other specialized applications utilize the pressure generated during vaporization of liquid to gas. Gaseous carbon dioxide is used for various purposes: a fire extinguishant, to carbonate beverages, as a shield gas in welding, and as an inerting blanket in process operations. Carbon dioxide is a raw material in the production of some chemical products: urea, sodium carbonate, sodium bicarbonate, and sodium salicylate.

The following table summarizes the chemical and physical properties of carbon dioxide.

TABLE 3.19
Physical and Chemical Properties of Carbon Dioxide

Molecular formula	CO_2
Molecular weight	44.0
Gas density (Air = 1)	1.53
Solubility in water (20°C)	0.14% by weight
Autoignition temperature	Not applicable
Explosive range (by volume)	Not applicable
Color	Colorless
Reaction in water	Ionizes in water; high pH enhances ionization and increases solubility
Relevant chemistry	Reacts with many metal ions to form insoluble salts; acidification of solutions containing carbonate and bicarbonate salts generates carbon dioxide
Odor	Odorless

After NIOSH 1976b.

The hazards posed by carbon dioxide during workplace exposure directly reflect its physical, chemical, and toxic properties.

The gas density of carbon dioxide gas is somewhat greater than that of air. Thus, a concentrated cloud produced in a depressed area will tend to remain at the point of generation. A concentrated cloud generated in an unconfined area will tend to flow to the lowest point until dispersing by convection and diffusion. This tendency would be exacerbated when the temperature of the carbon dioxide is less than ambient. Carbon dioxide neither forms flammable or explosive mixtures in air nor supports combustion under normal circumstances. Carbon dioxide also is colorless, and thus provides no visual warning about its presence.

Carbon dioxide is slightly soluble in pure water (more than carbon monoxide but considerably less than hydrogen sulfide). Solubility is enhanced because molecules of CO_2 react with water to form an acid (carbonic acid) that ionizes in solution. Solubility is greatly enhanced in alkaline solutions ($pH > 7$) due to ionization of the acid to form bicarbonate and carbonate ions. The enhanced solubility of carbon dioxide, carbonate, and bicarbonate salts in alkaline solutions is extremely important to the hazard potential of this gas. Rapid decrease in pH of these solutions, by direct acidification or other means, causes formation of CO_2 molecules. This process can lead to effusion of gas from solution into the atmosphere.

Carbon dioxide is a normal product of human metabolism. Transfer from the blood into the airspaces of the lung is the major route for the elimination of metabolic carbon dioxide. The partial pressure of carbon dioxide in alveolar airspaces normally is 40 mmHg (53,000 ppm) at sea level.

Despite the high concentration in the alveolar airspaces and tolerance by the body at this and higher concentrations, carbon dioxide does exert toxic effects. It appears to play a major role in the complex atmospheric conditions that characterize some accident situations.

Normally, carbon dioxide acts as a respiratory stimulant in the neurological control of breathing. To some extent, carbon dioxide also is instrumental in the control of cerebral blood flow and local vasodilation. Effects of exposure to elevated levels of carbon dioxide on the respiratory and central nervous systems occur rapidly. This happens because of the high permeability of the membranes of the blood–brain and blood–cerebrospinal fluid barriers to carbon dioxide. In addition, the solubility of carbon dioxide in tissue fluids is approximately 20 times that of oxygen. Changes in concentration are readily identified by the central chemoreceptors. Response is virtually immediate (Comroe et al. 1962, Bouhuys 1974).

Exposure to elevated levels produces contradictory effects: stimulation of the respiratory center, mild narcotic effects, respiratory arrest, and asphyxiation. The occurrence of a particular effect depends on the concentration and duration of exposure. Among the first of the toxicological properties of carbon dioxide to be exploited was its ability to anesthetize. Carbon dioxide was used as an anesthetic during animal surgery (Cutting 1969). This effect also was explored in human experiments involving psychiatric patients (Friedlander and Hill 1954).

Both the central and peripheral chemoreceptors are sensitive to changes in local arterial partial pressure of carbon dioxide. Peripheral chemoreceptors act rapidly but contribute only a small portion of the total ventilatory drive. Central chemoreceptors react more slowly, but provide about 80% of the ventilatory drive. When healthy persons are exposed to air containing 30,000 to 70,000 ppm of carbon dioxide, tidal volume, breathing rate, and minute ventilation increase. The increased ventilation rate reduces the increase of arterial partial pressure of CO_2. It cannot prevent the marked increase in alveolar partial pressure of CO_2 when the tension of inspired CO_2 is higher than the initial alveolar partial pressure of CO_2. At still higher levels, the narcotic effect of CO_2 reduces ventilation and causes rapid retention of CO_2 in blood and tissues (Bouhuys 1974).

Experiments reviewed by NIOSH indicated the occurrence of behavioral effects and depression of the central nervous system after short exposures in the range of 75,000 ppm. Exposure of longer duration lasting hours and days produced these effects at lower concentrations. Graded increases in concentration at lower levels produce immediate and significant effects. These include graded increases in respiratory minute volume and ventilatory rate. These effects are illustrated in Table 3.20. Respiratory minute volume gradually increased from an average resting rate of 7 L/min at 350 ppm to 76 L/min at 104,000 ppm. Tidal volume increased correspondingly from 440 mL to 2,500 mL.

The neuroendocrine system also responds rapidly to an increase in the level of carbon dioxide. The adrenal glands increase production of corticosteroids and catecholamines in response to inhalation of carbon dioxide at elevated levels. This response occurs within 10 to 15 minutes (Sechzer et al. 1960).

The most far-reaching effects of exposure to carbon dioxide involve modification of the acid–base balance and electrolyte chemistry of the blood. These changes generally occur following prolonged exposure lasting several days. Prolonged exposure causes increased urinary excretion of bicarbonate. Renal excretion of ammonium ions and titratable acid provides for removal of H^+ ions in exchange for the reabsorption of Na^+ ions. The sodium ions maintain ionic balance with bicarbonate ions in the plasma. Respiratory mechanisms act to stabilize carbon dioxide concentration. Renal mechanisms act to stabilize bicarbonate concentration. Acclimation, or tolerance, by activation of compensatory mechanisms is an outcome of long-term exposure to elevated levels. In general, the greater the physiologic imbalance, the greater the compensatory response (NIOSH 1976b).

Table 3.21 summarizes effects of exposure to carbon dioxide (after NIOSH 1976b).

TABLE 3.20
Effect of Exposure to Carbon Dioxide

Concentration (%)	Tidal volume (mL)	Breathing frequency (breath/min)	Minute volume (L/min)	Alveolar ventilation rate (L/min)
0.03	440	16	7	4.6
1.0	500	16	8	5.6
2.0	560	16	9	6.6
4.0	823	17	14	11.3
5.0	1,300	20	26	22.5
7.6	2,100	28	52	47.9
10.4	2,500	35	76	69.0

After Comroe et al. 1962.

TABLE 3.21
Effects of Exposure to Carbon Dioxide

Effect/Condition	Concentration (ppm)
Normal atmosphere	350
Alveolar airspace	53,000
No noticeable effect	5,500 (5 h)
No measurable effects	15,000 (long-term)
Slight effect (normal oxygen content), weakly narcotic, reduced hearing acuity, increased blood pressure and pulse	30,000
Respiratory volume doubled	40,000
Respiratory volume redoubled	50,000
Headache, restlessness, dizziness	75,000 (7 to 15 min)
Increase in heart rate and blood pressure, shortness of breath, throbbing headaches, dizziness, vertigo, poor memory, inability to concentrate, photophobia	76,000
Unconsciousness	>100,000 (prolonged)
Unconsciousness	110,000 (<1 min)
Unconsciousness	300,000 (25 s)

After NIOSH 1976b.

The data presented in this table indicate that carbon dioxide can exhibit acute toxicity at extremely high concentrations. A single breath can produce collapse and unconsciousness. Of course, this also could be related to low oxygen concentration.

Real-world exposure to high concentrations, such as mentioned in Table 3.21, likely would occur only in specific situations. These could include process operations where a cloud of carbon dioxide escapes from containment. Another example would be a bioreactor or storage container or structure containing organic debris where fermentation or respiration by microbial action can occur. Even photosynthesizing plants produce an excess of carbon dioxide in the absence of sunshine. These conditions could result in atmospheric displacement, thereby producing an oxygen-deficient atmosphere. The net physiological result could be due to the combined effect of the high level of carbon dioxide and the oxygen deficiency. Acute exposure to lower concentrations of carbon dioxide also would occur in an oxygen-deficient atmosphere. Lower concentrations would increase the breathing rate, thereby enhancing the effect of the oxygen deficiency. Coupled with the presence of other toxic agents, as discussed previously, this could become the deadly combination needed to produce a fatal outcome. The increased breathing rate could increase the rate of oxygen depletion and, at the same time, increase the rate of exposure to other inhalation hazards.

Exercise is a factor in the severity of response following exposure to increased concentrations of carbon dioxide. At concentrations below 28,000 ppm, physically fit subjects tolerate strenuous exercise (180 Watts) without detectable stress. At or above this concentration, however, subjects begin to experience respiratory exhaustion, headaches, and pain in the intercostal muscles. At 39,000 ppm, subjects reported mild to severe frontal headaches, although these did not interfere with performance. At 52,000 ppm, strenuous exercise caused mental confusion, impaired vision, and collapse (Menn et al. 1968, 1970; Craig et al. 1970).

Elevated levels of carbon dioxide may have been involved in a higher proportion of fatal accidents in confined spaces than previously recognized. Rescue activity during these accidents occurred under highly stressful, physically demanding conditions. The presence of elevated levels of carbon dioxide in the contaminated atmosphere may have compounded the effects of other substances. The physical exertion and emotional stress could have put the rescuer under considerably greater duress than the inactive victim. This may explain why so many would-be rescuers perished in situations in which the original victim survived.

Generally, results of animal studies into the effects of exposure to carbon dioxide parallel those obtained through human experimentation and industrial experience. One animal experiment that may have considerable implication for humans examined rapid removal from high levels of carbon dioxide to normal conditions. The animals were exposed for 2 hours to a mixture containing 30% carbon dioxide and 70% oxygen. At the end of this period, the concentration of carbon dioxide was increased to 40% for an additional 2 hours. At that point, some animals were instantaneously exposed to air. These animals developed cardiac arrhythmias within 6 minutes. Most developed ventricular fibrillation and died within 10 min after this sudden change. By contrast, animals returned gradually to a normal atmosphere survived and showed no signs of cardiac dysfunction (Brown and Miller 1952).

Normal actions taken during rescue of victims from confined spaces parallel the protocol in these experiments. The rescuer attempts to relocate the victim to an uncontaminated atmosphere as quickly as possible. Assuming that the preceding results can be extrapolated to humans, rapid return of an individual exposed to a high level of carbon dioxide to a normal atmospheric environment may exacerbate the situation.

The toxicological properties of carbon dioxide form the basis for judgments made by standard-setting agencies. Table 3.22 summarizes these criteria (ACGIH 1994, NIOSH 1990).

Threshold Limit Values (TLVs) are recommended by the TLV Committee of the American Conference of Governmental Industrial Hygienists. TLVs are guidelines, not fine lines, between safe and dangerous conditions. TLVs are believed to represent conditions under which nearly all workers may be repeatedly exposed day after day without adverse health effects.

TLV-TWA is the time-weighted concentration for a normal 8-h workday and 40-h workweek to which nearly all workers may be repeatedly exposed, day after day, without adverse effect. TLV-STEL is a 15-min time-weighted average which should not be exceeded at any time during a workday, even if the 8-h TWA is within the TLV-TWA. Up to four excursions at the TLV-STEL may occur per day, provided that each is separated by at least 1 hour and the TLV-TWA is not exceeded.

Recommended Exposure Levels (RELs) and Immediately Dangerous to Life or Health limits (IDLHs) were prepared by the National Institute for Occupational Safety and Health as part of the Standards Completion Project. REL-TWA is a time-weighted average concentration for up to a 10-h workday during a 40-h workweek. A short-term exposure limit (REL-ST) is a 15-min time-weighted average that should not be exceeded at any time during a workday. A ceiling (REL-Ceiling) should

TABLE 3.22
Exposure Standards — Carbon Dioxide

Effect	Concentration (ppm)
Threshold Limit Value — Time-Weighted Average	5,000
Threshold Limit Value — Short-Term Exposure Limit	30,000
Recommended Exposure Limit — time-weighted average	5,000
Recommended Exposure Limit — short term	30,000
Immediately Dangerous to Life and Health	50,000

Compiled from ACGIH 1994 and NIOSH 1990.

not be exceeded at any time. IDLH was defined in the NIOSH Standards Completion Project for the purpose of selecting respiratory protection. IDLH represents the presumed maximum concentration from which a person could escape in 30 min in the event of respirator failure without experiencing any escape-impairing or irreversible health effects.

The limit of greatest interest that could be tolerated during gross overexposure is the IDLH. This level provides some insight regarding conditions that would have been required to produce fatal outcomes in accidents in which carbon dioxide may have been implicated (OSHA 1985, NIOSH 1994).

Organic and Halogenated Solvents

The OSHA and NIOSH reports on fatal accidents associated with confined workspaces and toxic atmospheres implicated organic solvents in an appreciable percentage of situations (OSHA 1985, NIOSH 1994). Halogenated solvents were the subject of specific concerns (NIOSH 1989, Novak and Hain 1980).

Many solvents produce narcotic and anesthetic effects. Narcotics and anesthetics primarily cause simple anesthesia, namely, depression of bioelectrical activity in the central and peripheral nervous systems. The action of anesthetics, of course, is well known from medical applications in anaesthesiology (Dinman 1978). Usually, anesthesia and narcosis occur without serious systemic effects, unless the dose is massive. Depending on the concentration present, the depth of anesthesia ranges from mild, to complete loss of consciousness and death. In accidents involving exposure to very high concentrations, death may occur due to asphyxiation.

Nearly all organic solvents found in the industrial environment possess narcotic properties. Anesthetics and narcotics include aliphatic alcohols, aliphatic ketones, acetylene hydrocarbons, and ethers (Andrews and Snyder 1991). The fact that a wide range of substances produces similar effects, despite differences in chemical structure, suggests a common pathway of interaction between molecules of the solvent and the sensitive cells of the nervous system. The actual mechanism of interaction leading to narcosis and anesthesia is not known. The rapidity of onset and depth of effect depend on concentration.

Effects caused by exposure to high levels of solvent vapors typically include: headache, nausea, lack of coordination, disorientation, euphoria, giddiness, confusion progressing to unconsciousness, paralysis, convulsion, and death from respiratory arrest or cardiovascular failure (Browning 1965). The rapidity of onset of the symptoms strongly suggests that the solvent and not its metabolites is the causative agent. Recovery of the majority of persons from the effects produced by nonfatal exposures usually is rapid following removal from the source of exposure.

The ability of solvent vapors to enter the body depends on their solubility in lipids. Entry into the body is contingent on traversing cellular membranes that are composed of lipid–protein layers. The rate at which a solvent distributes to organs of the body from the blood depends on the blood-to-air partition coefficient. Substances having a high blood to air partition coefficient, such as diethyl ether, pass into organs from the blood at a slow rate. Halothane, an anesthetic having a low blood-to-air partition coefficient, distributes to organs more rapidly (Andrews and Snyder 1991).

CONFINED SPACE ACCIDENTS AND ATMOSPHERIC HAZARDS

Of considerable interest to the industrial hygienist is the composition of the contaminated atmosphere associated with accidents that occur in confined spaces. This information would provide the key to better understanding about the hazardous nature of these workplace environments. Ultimately, this information would determine the nature and scope of the response needed to address and manage these conditions. The OSHA and NIOSH reports on fatal accidents occurring in confined spaces provide the main source of data about this subject (OSHA 1985, NIOSH 1994). Both reports provided descriptive summaries of individual accidents. These make possible further speculation about the composition of the atmosphere present at the time of the accident. Sometimes the summary alluded to the presence of more than one hazardous substance. Unfortunately, these summaries provide little or no analytical data about the composition of the hazardous atmospheres involved in the accidents. This aspect definitely requires further study. Elucidating the composition of these atmospheres is important in the extreme.

Anecdotal information in the accident summaries provides some indication that these atmospheres are more complex than originally described. Further, there are discrepancies between the progression of events that actually occurred and what could be expected based on controlled studies of the toxic agents implicated in the accidents. That is, the outcomes produced by some of these substances under controlled conditions differed from what was observed during accidents attributed to them. This discrepancy is extremely important. It may indicate the involvement of more or different substances and greater complexity in real-world hazardous atmospheres than previously believed.

Considerable similarity exists in the progression of events in individual accident situations. During a typical accident, the victim usually is affected by the atmospheric condition either at the time of entry or soon afterward. This individual collapses and may yell for help, or is discovered soon afterward by someone outside the space. The discoverer or some other individual nearby undertakes the role of would-be rescuer and enters the space without respiratory protection. The would-be rescuer possibly succeeds in transferring the victim from the interior of the space to the access opening after expending considerable physical effort. The conditions overcome the would-be rescuer, who then collapses. The would-be rescuer often collapses more rapidly than the victim. Additional would-be rescuers may suffer the same fate as the first. These events all occur prior to response by individuals equipped appropriately for the rescue. Either the victim, a would-be rescuer, or both are fatally injured during this process.

During a real-world accident, entrants often collapse either immediately or shortly after initial contact with the hazardous atmosphere. This action suggests the presence of a rapidly acting, acutely hazardous condition. The rapid onset of debilitation under real-world conditions contrasts with the slower action of many substances, including carbon monoxide and organic solvents, as discussed previously.

The onset of coma following exposure to carbon monoxide occurs when carboxyhemoglobin saturation exceeds 50 to 60% (NIOSH 1972). Saturation to the 50% level by an atmosphere containing 1,000 ppm requires approximately 180 min (Stewart and Peterson 1970). This time sequence is much too slow to account for the rapid onset of unconsciousness observed during actual accident situations. This discrepancy suggests that carbon monoxide alone was not the causative agent in these accidents.

Hydrogen sulfide can cause rapid collapse when inspired in high concentration. Yet, in many accidents in which hydrogen sulfide was implicated, and in which air sampling subsequently occurred, concentrations typically were in the range of 50 ppm (OSHA 1985, NIOSH 1994).

Concentrations in this range are sufficiently high to cause only eye irritation, not rapid collapse. However, the test results possibly were not reliable or dilution could have occurred following the accident.

In high concentrations, solvent vapors can cause rapid collapse. This response is consistent with situations in which exposure to high concentrations of solvent vapors did occur. However, solvents were implicated in only a small proportion of the fatal accidents described by OSHA and NIOSH.

Oxygen deficiency also can cause rapid collapse. Collapse can occur after one or two breaths of atmospheres containing less than 4% oxygen (Miller and Mazur 1984). Atmospheres deficient in oxygen contain other gases that maintain total pressure at ambient levels. Carbon dioxide stimulates breathing at concentrations up to 70,000 ppm (7%). Thus, elevated levels of carbon dioxide could stimulate inhalation of other contaminants present in the same contaminated atmosphere. At the same time under this circumstance, this atmosphere also could produce impairment because of the oxygen deficiency (Figure 3.5).

Atmospheres encountered in confined environments likely are complex mixtures of several of the substances discussed in this chapter. Considerable investigative work is needed to elucidate this question further.

SUMMARY

This chapter has considered the relationship of the respiratory system to the atmosphere which humans must breathe for survival. The human organism survives and thrives in atmospheric conditions of relatively narrow tolerance. The predominant factor controlling human survival is the partial pressure of oxygen.

The form and function of the respiratory system are predicated on the transfer of oxygen into the circulatory system and delivery to individual cells of the body, and the removal of carbon dioxide. In the course of this process, the respiratory system may be exposed to toxic agents and conditions of oxygen enrichment or deficiency. These situations can develop within confined spaces.

The focus of this chapter was an examination of the effects of the toxic substances and asphyxiating conditions identified in studies of fatal accidents occurring in confined spaces. The predicted action of several of these agents does not appear to correspond with the progression of real-world accidents in which they were implicated. This lack of correspondence suggests that the atmospheric conditions present during many of these accidents may have been more complex than previously considered. This uncertainty potentially complicates actions needed to ensure competent management and control of these work environments.

REFERENCES

Alexeeff, G.V., M.J. Lipsett, and K.W. Kizer: Problems associated with the use of immediately dangerous to life and health values for estimating the hazard of accidental chemical releases. *Am. Ind. Hyg. Assoc. J. 50:* 598–605 (1989).

American Conference of Governmental Industrial Hygienists: *Documentation of the Threshold Limit Values and Biological Exposure Indices*. 6th ed. vol. 1. Cincinnati, OH: American Conference of Governmental Industrial Hygienists, 1991. 865 pp.

American Conference of Governmental Industrial Hygienists: *1994–1995, Threshold Limit Values for Chemical Substances and Physical Agents and Biological Exposure Indices*. Cincinnati, OH: American Conference of Governmental Industrial Hygienists, 1994. 119 pp.

American Industrial Hygiene Association: *Emergency Response Planning Guidelines*. Fairfax, VA: American Industrial Hygiene Association, 1995.

American National Standards Institute: *American National Standard Practices for Respiratory Protection*. (ANSI Z88.2-1992) New York: American National Standards Institute. 1992.

Andrews, L.S. and R. Snyder: Toxic effects of solvents and vapours. In *Casarett and Doull's Toxicology, The Basic Science of Poisons*, 4th ed. Amdur, M.O., J. Doull, and C.D. Klaassen (Eds.). New York: Macmillan. 1991. pp. 681–722.

Armstrong, H.G., A.C. Burton, and G.E. Hall: The physiological effects of breathing cold atmospheric air. *J. Aviat. Med. 29*: 593–597 (1958).

Ayres, S.M., H.S. Mueller, J.J. Gregory, S. Gianelli, Jr., and J.L Penny: Systemic and myocardial hemodynamic responses to relatively small concentrations of carboxyhemoglobin (COHb). *Arch. Environ. Health 18*: 699–704 (1969).

Ayres, S.M., S. Giannelli, and H. Mueller: Effects of low concentrations of carbon monoxide. *Ann. N.Y. Acad. Sci. 174*: 268–293 (1970).

Bachofen, H., J. Hildebrandt, and M. Bachofen: Pressure-volume curves of air- and liquid-filled excised lungs — surface tension *in situ*. *J. Appl. Physiol. 29*: 422–431 (1970).

Bartlett, D., Jr.: Pathophysiology of exposure to low concentrations of carbon monoxide. *Arch. Environ. Health. 16*: 719–727 (1968).

Behnke, A.R., Jr.: Physiological effects of abnormal atmospheric pressures. In *Patty's Industrial Hygiene*. 3rd rev. ed., vol. 1. *General Principles*. New York: John Wiley & Sons, 1978. pp. 237–274.

Bellingham, A.J., J.C. Detter, and C. Lenfant: The role of hemoglobin affinity for oxygen and red cell 2,3-diphosphoglycerate in the management of diabetic ketoacidosis. *Trans. Assoc. Am. Physicians 83:* 113–120 (1970).

Benesch, R. and R.E. Benesch: Intracellular organic phosphates as regulators of oxygen release by haemoglobin. *Nature 221*: 618–622 (1969).

Benesch, R.E., R. Benesch, R. Renthal, and W.B. Gratzer: Cofactor binding and oxygen equilibria in haemoglobin. *Nature (New Biol.) 234*: 174–176 (1971).

Bouhuys, A.: *Breathing; Physiology, Environment and Lung Disease*. New York: Grune & Stratton, 1974. pp. 25–233.

Brown, E.B., Jr. and F. Miller: Ventricular fibrillation following a rapid fall in alveolar carbon dioxide concentration. *Am. J. Physiol. 189*: 56–60 (1952).

Browning, E.: *Toxicity and Metabolism of Industrial Solvents*. New York: Elsevier, 1965.

Canadian Standards Association: *Selection, Use, and Care of Respirators*. (Z94.4-93) Rexdale, ON: Canadian Standards Association. 1993. 103 pp.

Clark, J.M. and C.J. Lambertsen: Pulmonary oxygen toxicity: a review. *Pharmacol. Rev. 23*: 37–133 (1971a).

Clark, J.M. and C.J. Lambertsen: Rate of development of pulmonary O_2 toxicity in man during O_2 breathing at 2.0 ATA. *J. Appl. Physiol. 30*: 739–752 (1971b).

Coburn, R.F.: Endogenous carbon monoxide production and body carbon monoxide stores. *Acta Med. Scandinav. Suppl. 472*: 269–282 (1967).

Comroe J.H., Jr., R.E. Forster, II, A.B. DuBois, W.A. Briscoe, and E. Carlsen: *The Lung, Clinical Physiology and Pulmonary Function Tests*, 2nd ed. Chicago: Year Book Medical Publishers, 1962. 390 pp.

Cotes, J.E.: Lung function tests. In *Encyclopaedia of Occupational Health and Safety*, 3rd. rev. ed., vol. 2, L–Z. Geneva: International Labour Organization, 1983. pp. 1250–1256.

Craig, F.N., W.V. Blevins, and E.G. Cummings: Exhausting work limited by external resistance and inhalation of carbon dioxide. *J. Appl. Physiol. 29*: 847–851 (1970).

Cutting, W.: *Handbook of Pharmacology*, 4th. ed. New York: Appleton-Century-Crofts, 1969. Chaps. 54–55.

Davis, J.C.: Abnormal pressure. In *Patty's Industrial Hygiene*, vol. 3, 2nd. ed. New York: John Wiley & Sons, 1979. pp. 525–542.

de Treville, R.T.P.: Occupational medical considerations in the aviation industry. In *Occupational Medicine — Principles and Practical Applications*, 2nd ed. Chicago: Year Book Medical Publishers, 1988. pp. 909–923.

Dinman, B.D.: Pathophysiologic determinants of community air quality standards for carbon monoxide. *J. Occup. Med. 10*: 446–456 (1968).

Dinman, B.D. The mode of entry and action of toxic materials. In *Patty's Industrial Hygiene*. vol. 1, General Principles, 3rd rev. ed., G.D. Clayton and F.E. Clayton, Eds. New York: John Wiley & Sons, 1978. pp. 135–164.

Donald, K.W.: Oxygen poisoning in man. *Br. Med. J. 1:* 717–722 (1947).

Dukes-Dobos, F.N. and D.W. Badger: Atmospheric variations. In Occupational Diseases: A Guide to Their Recognition, rev. ed. DHEW (NIOSH) Pub. No. 77-181. Washington, D.C.: U.S. Government Printing Office (DHEW/PHS/CDC/NIOSH), 1977. pp. 497–520.

Farhi, L.E.: Elimination of inert gas by the lung. *Respir. Physiol. 3*: 1–11 (1967).

Fenn, W.O.: Partial molar volumes of oxygen and carbon monoxide in blood. *Respir. Physiol. 13*: 129–140 (1971).

Fowler, W.S.: Breaking point of breath-holding. *J. Appl. Physiol. 6*: 539–545 (1954).

Friedlander, W.J. and T. Hill: EEG changes during administration of carbon dioxide. *Dis. Nerv. Syst. 15*: 71–75 (1954).

Freeman, B.A. and J.D. Crapo: Free radicals and tissue injury. *Lab. Invest. 47*: 412–426 (1982).

Grover, R.F., J.V. Weil, and J.T. Reeves: Cardiovascular adaptation to exercise at high altitudes. *Exerc. Sport Sci. Rev.* 14: 269 (1986).

Harper, H.A., V.A. Rodwell, and P.A. Mayes, Eds.: *Review of Physiological Chemistry*, 16th ed. Los Altos, CA: Lange Medical, 1977. pp. 587–595.

Hultgren, H.N.: High-altitude medical problems. In *Scientific American Medicine.* New York: Scientific American, 1994. pp. 1–16.

International Commission on Radiological Protection: *Report of the Task Group on Reference Man.* Oxford: Pergamon Press. 1975.

Jobsis, F.F.: Basic processes in cellular respiration. In *Handbook of Physiology, Section 3: Respiration,* vol. 1. Washington, D.C.: American Physiological Society, 1964. pp. 63–124.

Kellogg, R.H.: Central chemical regulation of respiration. In *Handbook of Physiology, Section 3. Respiration,* vol. 1. Washington: American Physiological Society, 1964. pp. 507–534.

Kurt, T.L.: Chemical asphyxiants. In *Environmental and Occupational Medicine.* Boston: Little, Brown, 1983. pp. 289–299.

Lahiri, S., J.S. Milledge, and S.C. Sorensen: Ventilation in man during exercise at high altitude. *J. Appl. Physiol. 52*: 766–769 (1972).

Lahiri. S: Dynamic aspects of regulation of ventilation in man during acclimatization to high altitude. *Respir. Physiol. 16*: 245–258 (1972).

Lenfant, C., J. Torrance, E. English, C.A. Finch, C. Reynafarje, J. Ramos, and J. Faura: Effect of altitude on oxygen binding by hemoglobin and on organic phosphate levels. *J. Clin. Invest. 47:* 2652–2656 (1968).

Lenfant, C. and K. Sullivan: Adaptation to high altitude. *N. Engl. J. Med. 284:* 1298–1309 (1971).

Leusen, I.: Regulation of cerebrospinal fluid with reference to breathing. *Physiol. Rev. 52*: 1–56 (1972).

Lilienthal, J.L., Jr.: Carbon monoxide. *Pharmacol. Rev.* 2: 324–354 (1950).

Lundwig, G.D. and W.S. Blakemore: Production of carbon monoxide by hemin oxidation. *J. Clin. Invest. 36*: 912 (1957).

Macaluso, P.: Hydrogen sulphide. In *Encyclopedia of Chemical Technology*, 2nd rev. ed. vol. 19. New York: Interscience, 1969. pp. 375–389.

Mayevsky, A.: Brain oxygen toxicity. In *Proc. 8th Symp. Underwater Physiology.* Bachrach, A.J. and M.M. Matzen (Eds.). Bethesda, MD: Undersea Medical Society, 1984. pp. 69–89.

Menn, S.J., R.D. Sinclair, and B.E. Welch: Response of Normal Man in Graded Exercise in Progressive Elevations of CO_2. Rep. No. SAM-TR-68-116. Brooks Air Force Base, TX: U.S. Air Force, Aerospace Medical Division (AFSC), USAF School of Aerospace Medicine. 1968. pp. 1–16.

Menn, S.J., R.D. Sinclair, and B.E. Welch: Effect of inspired PCO_2 up to 30 mmHg on response of normal man to exercise. *J. Appl. Physiol. 28*: 663–671 (1970).

Miller, T.M. and P.O. Mazur: Oxygen deficiency hazards associated with liquified gas systems: derivation of program controls. *Am. Ind. Hyg. Assoc. J. 45:* 293–298 (1984).

Mitchell, R.A., H.H. Loescheke, W.H. Massion, and J.W. Severinghaus: Respiratory responses mediated through superficial chemosensitive areas on the medulla. *J. Appl. Physiol. 18*: 523–533 (1963).

Moran Campbell, E.J., C.J. Dickinson, J.D.H. Slater, C.R.W. Edwards, and R. Sikora, Eds.: *Clinical Physiology,* 5th ed. Oxford: Blackwell Scientific, 1984. pp. 96–153.

National Air Pollution Control Administration: *Air Quality Criteria for Carbon Monoxide.* (HEW/NAPCA) Pub. No. AP-62. Washington: U.S. Government Printing Office. 1970.

National Institute for Occupational Safety and Health: Criteria for a Recommended Standard... Occupational Exposure to Carbon Monoxide. (DHEW/HSMHA/NIOSH) Pub. No. 73-11000. Cincinnati, OH: U.S. Government Printing Office. 1972.

National Institute for Occupational Safety and Health: A Guide to Industrial Respiratory Protection, by J.A. Pritchard. (DHEW/PHS/CDC/NIOSH Pub. No. 76-189). Cincinnati, OH: National Institute for Occupational Safety and Health, 1976a. 150 pp.

National Institute for Occupational Safety and Health: Criteria for a Recommended Standard... Occupational Exposure to Carbon Dioxide. (DHEW/PHS/CDC/NIOSH) Pub. No. 76-194. Cincinnati, OH: U.S. Government Printing Office. 1976b. 141 pp.

National Institute for Occupational Safety and Health: Criteria for a Recommended Standard... Occupational Exposure to Hydrogen Sulphide. (DHEW/PHS/CDC/NIOSH) Pub. No. 77-158. Cincinnati, OH: Government Printing Office. 1977. 112 pp.

National Institute for Occupational Safety and Health: Criteria for a Recommended Standard — Working in Confined Spaces. (DHEW/PHS/CDC/NIOSH Pub. No. 80-106). Cincinnati, OH: National Institute for Occupational Safety and Health, 1979. 68 pp.

National Institute for Occupational Safety and Health: NIOSH Alert, Request for Assistance in Preventing Death from Excessive Exposure to Chlorofluorocarbon 113 (CFC-113). (DHHS/PHS/CDC?NIOSH Pub. No. 89-109). Cincinnati, OH: National Institute for Occupational Safety and Health, 1989.

National Institute for Occupational Safety and Health: NIOSH Pocket Guide to Chemical Hazards. (DHHS/NIOSH Pub. No. 90-117). Cincinnati, OH: DHHS/PHS/CDC/NIOSH, 1990. 245 pp.

National Institute for Occupational Safety and Health: Worker Deaths in Confined Spaces. (DHHS/PHS/CDCP/NIOSH Pub. No. 94-103). Cincinnati, OH: National Institute for Occupational Safety and Health, 1994. 273 pp.

Novak, J.J. and J.R. Hain: Furniture stripping vapour inhalation fatalities: two case studies. *Appl. Occup. Environ. Hyg. 5*: 843–847 (1990).

Occupational Safety and Health Administration: Selected Occupational Fatalities Related to Fire and/or Explosion in Confined Work Spaces as Found in OSHA Fatality/Catastrophe Investigations. Washington, D.C.: U.S. Department of Labor, Occupational Safety and Health Administration (U.S. DOL/OSHA), 1982a. 76 pp.

Occupational Safety and Health Administration: Selected Occupational Fatalities Related to Lockout/Tagout Problems as Found in Reports of OSHA Fatality/Catastrophe Investigations. Washington, D.C.: U.S. Department of Labor, Occupational Safety and Health Administration (U.S. DOL/OSHA), 1982b. 113 pp.

Occupational Safety and Health Administration: Selected Occupational Fatalities Related to Grain Handling as Found in Reports of OSHA Fatality/Catastrophe Investigations. Washington, D.C.: U.S. Department of Labor, Occupational Safety and Health Administration (U.S. DOL/OSHA), 1983. 150 pp.

Occupational Safety and Health Administration: Selected Occupational Fatalities Related to Toxic and Asphyxiating Atmospheres in Confined Work Spaces as Found in Reports of OSHA Fatality/Catastrophe Investigations. Washington, D.C.: U.S. Department of Labor, Occupational Safety and Health Administration (U.S. DOL/OSHA), 1985. 230 pp.

Occupational Safety and Health Administration: Selected Occupational Fatalities Related to Welding and Cutting as Found in Reports of OSHA Fatality/Catastrophe Investigations. Washington, D.C.: U.S. Department of Labor, Occupational Safety and Health Administration (U.S. DOL/OSHA), 1988. 225 pp.

Occupational Safety and Health Administration: Selected Occupational Fatalities Related to Shipbuilding and Repairing as Found in Reports of OSHA Fatality/Catastrophe Investigations. Washington, D.C.: U.S. Department of Labor, Occupational Safety and Health Administration (U.S. DOL/OSHA), 1990.

Olishifski, J.B. and G.S. Benjamin: The lungs. In *Fundamentals of Industrial Hygiene*, 3rd. ed. Chicago, IL: National Safety Council, 1988. pp. 31–45.

Pappenheimer, J.R., V. Fencl, S.R. Heisey, and D. Held: Role of cerebral fluids in control of respiration as studied in unanesthetized goats. *Am. J. Physiol. 208*: 436–450 (1965).

Perutz, M.P.: Stereochemistry of cooperative effects of haemoglobin. *Nature 228*: 726–739 (1970).

Piantadosi, C.A.: Physiological effects of altered barometric pressure. In *Patty's Industrial Hygiene*, 4th ed., vol. 1, part A. New York: John Wiley & Sons, 1991. pp. 329–359.

Pierce, J.A. and R.V. Ebert: Fibrous network of the lung and its change with age. *Thorax 20*: 469–476 (1965).

Ramazzini, B.: *Diseases of Workers — De Morbis Artificum Diatriba*. W.C. Wright (trans.) New York: C. Hafner Publishing, 1964. pp. 549.

Reid, R.C., J.M. Prausnitz, and B.E. Poling: *The Properties of Gases & Liquids*. 4th ed. New York: McGraw-Hill, 1987. pp. 332–337.

Rossi-Fanelli, A. and E. Antonini: Studies of the oxygen and carbon monoxide equilibrium of human myoglobin. *Arch. Biochem. Biophys. 77*: 478–492 (1958).

Satir, P.: Cilia. *Sci. Am. 204*: 108–116 (1961).

Schmidt-Nielsen, K., F.R. Hainsworth, and D.E. Murrish: Counter-current heat exchange in the respiratory passages: effect on water and heat balance. *Respir. Physiol. 9*: 263–276 (1970).

Schulz, H.: Some remarks on the submicroscopic anatomy and pathology of the blood. Air pathway in the lung. In *Ciba Foundation Symp. Pulmonary Structure and Function.* A.V.S. DeReuck and M. O'Connor, Eds. Boston: Little, Brown & Co. 1962. pp. 205–214.

Sechzer, P.H., L.D. Egbert, H.W. Linde, D.Y. Cooper, R.D. Dripps, and H.L. Price: Effect of CO_2 inhalation on arterial pressure, ECG and plasma catecholamines and 17-OH corticosteroids in normal man. *J. Appl. Physiol. 13*: 454–458 (1960).

Selkurt, E.E.: Respiration. In *Basic Physiology for the Health Sciences,* 2nd. ed. E.E. Selkurt, Ed. Boston: Little, Brown, 1982. pp. 324–379.

Shigeoka, J.W.: Pulmonary function testing. In *Environmental and Occupational Medicine.* Boston: Little, Brown, 1983. pp. 99–112.

Smith, R.P. and R.E. Gosselin: On the mechanism of sulphide inactivation by methemoglobin. *Toxicol. Appl. Pharmacol. 8:* 159–172 (1966).

Smith, L., H. Kruszyna, and R.P. Smith: The effect of methemoglobin on the inhibition of cytochrome c oxidase by cyanide, sulphide or azide. *Biochem. Pharmacol. 26:* 2247–2250 (1977).

Stamatoyannopoulos, G., A.J. Bellingham, C. Lenfant, and C.A. Finch: Abnormal hemoglobins with high and low oxygen affinity. *Annu. Rev. Med. 22:* 221–234 (1971).

Stewart, R.L. and M.R. Peterson: Experimental human exposure to carbon monoxide. *Arch. Environ. Health. 21*: 154–164 (1970).

Timar, M.: Hypoxia and anoxia. In *Encyclopaedia of Occupational Health and Safety,* 3rd. rev. ed., vol. 1 A–K, Geneva: International Labour Organization, 1983. pp. 1093–1096.

Vander, A.J., J.H. Sherman, and D.S. Luciano: *Human Physiology: The Mechanisms of Body Function,* 5th ed. New York: McGraw-Hill, 1990. pp. 427–469.

von Hayek, H.: *The Human Lung.* New York: Hafner (trans. by V.E. Krahl), 1960. p. 202.

Weast, R.C., Ed.: *CRC Handbook of Chemistry and Physics.* Boca Raton, FL: CRC Press, 1987. pp. F143–F150.

Weibel, E.R.: *Morphometry of the Human Lung.* Berlin: Springer-Verlag. 1963.

West, J.B., S.J. Boyer, D.J. Graber et al.: Maximal exercise at extreme altitudes on Mount Everest. *J. Appl. Physiol. 55:* 688 (1983).

Yarborough, O.D., W. Welham, E.S. Brinton, and A.R. Behnke: *Experimental Diving Unit Report No. 1.* Washington, D.C., 1947.

4 Ignitable and Explosive Atmospheric Hazards

CONTENTS

INTRODUCTION

Fires and explosions are abrupt traumatic events. They are a major cause of injury and death during accidents involving confined spaces and confined hazardous atmospheres (NIOSH 1979, 1994; OSHA 1982, 1983, 1985, 1988, 1990). Fires and explosions caused a relatively large proportion of fatal injuries per accident compared to other situations.

A large proportion of these accidents occurred while the victims performed normal job activities: welding and associated activities, painting and associated activities, cleaning and associated activities, and mechanical work. The most traumatic occurred during cleaning and associated activities and mechanical work. Unauthorized activity was not a major component in these accidents.

The victims of fires and explosions almost exclusively were tradespeople. Workers having the job title "welder" constituted almost half the victims of welding and cutting accidents occurring externally to a hazardous confined atmosphere. The accident investigations noted that these individuals did not necessarily possess formal training and certification as welders. The remainder of the victims were pipefitters, laborers or unspecified tradespeople.

The immediate cause of accidents occurring inside confined workspaces was attributed to the following deficiencies: failure to test the condition of the atmosphere, faulty procedure, faulty isolation of the workspace, use of ungrounded and unapproved equipment, and lack of ventilation. Substances that acted as the fuel in fires and explosions occurring inside and outside confined spaces were essentially the same: vapors from organic solvents and fuels, welding gases, natural gas, and other substances. Sources of fuel commonly encountered in these situations included:

- Substances resident in the space at the start of work
  - Incompletely purged gases
  - Liquids and sludges
  - Finely powdered solids
  - Coatings and linings
- Substances introduced into the space
  - Solvents, thinners, cleaning agents, and degreasers
  - Coatings
  - Process gases (welding and cutting)
- Substances produced during the work activity
  - Aerosols
    - Sanding, grinding, chipping
    - Turbulent agitation
    - Spraying
  - Gases/vapors
    - Action of acids on metals
    - Thermal decomposition

In about half the of the fatal accidents, the hazardous atmosphere existed at the time of initial entry. The entrants failed to appreciate or failed to act on sensory warnings or visual cues provided by the conditions. In the remaining situations, the hazardous atmosphere developed as a result of the activity. Again, the entrants failed to appreciate the danger or failed to act on sensory or visual cues. Sources of ignition commonly associated with these accidents included:

- Open flames
- Cigarettes and matches or lighters
- Hot surfaces (such as gas and propane heaters)
- Sparks from faulty electrical equipment
- Buildup of static electricity
- Welding, cutting, burning, gouging equipment

Accidents involving confined hazardous atmospheres contained similar elements. The container often was "empty." Residues that remain on surfaces inside an "empty" container sometimes can form an atmosphere that can burn or explode. In these situations the victim failed to appreciate or act on dangers posed by contents. No testing of the atmosphere in the container occurred.

Work occurring on the exterior of containers during these accidents typically involved welding and cutting. Heat applied to the exterior surface degrades interior coatings and linings and may vaporize contents and pressurize the internal atmosphere. Sufficient application of heat could cause ignition or detonation. In none of the welding and cutting accidents did the fatal situation develop from overexposure to the contents following release. Fatal injury resulted only from the effects of the fire or explosion.

Work involving hazardous confined atmospheres routinely occurs aboard ships that carry flammable and combustible materials and in chemical process operations. However, fires and explosions occur infrequently in these industries. This reflects intervention by the National Fire Protection Association (NFPA) and American Petroleum Institute. The Marine Chemist certification program administered by NFPA is one such example (Keller 1982). The Marine Chemist is a technical expert in the hazards of confined spaces. The Marine Chemist inspects, assesses, and recommends measures to eliminate or control hazards in these environments. This approach is worth broadening to other areas of industry.

This chapter examines the relationship between confined spaces and fires and explosions that occur in them. That is, how the unique environment of the confined space exacerbates the severity of fire and explosion.

COMBUSTION

Combustion is an oxidative chemical process. It produces energy as heat and often as light. Combustion is called *fire* when the oxidative process occurs fast enough to be self-sustaining. Combustion that produces a sudden and violent release of energy is called an *explosion* (Meyer 1989). Combustion differs from slow oxidative processes, such as *rusting*. The difference is the rate of release of heat. Temperature near the surface of slow oxidation increases only slightly — never more than 1°C above that of the surroundings. Combustion occurs so rapidly that heat is generated faster than dissipated. Temperature increases by hundreds and often thousands of °C above the surroundings. Heating is so intense that light is emitted (Friedman 1991).

Fire occurs in situations where the fuel and oxidizer (usually oxygen in air) have not undergone thorough mixing. As a result, combustion proceeds slowly and potentially is delayed by the need for mixing of fuel and oxidizer. Combustion rate is controlled by the rate at which mixing occurs, rather than the rates of chemical reactions occurring within the flame. Conditions conducive to explosion require thorough mixing of oxidizer and fuel prior to ignition. Burning rate is much greater during explosions than in fire situations (Drysdale 1991).

Combustion occurs in two modes: flaming (including explosions) and surface (including glow and deep-seated glowing embers) (Haessler 1986). These modes occur singly or in combination, but are not mutually exclusive.

Flammable liquids and gases burn in flaming mode only, as do plastics that melt and vaporize on heating. Flaming mode occurs only when the fuel is present in vapor state. Vapor formation can occur through vaporization from liquid, or through melting and vaporization or pyrolysis of solids. Pyrolysis is the process of breakdown of complex molecular structures into smaller pieces. Pyrolysis products include ignitable gases and vapors.

Surface combustion occurs directly on the surface of the material as with pure carbon and readily oxidizable metals and nonmetals. Flaming and surface combustion occur during the burning of coal; sugars and starches; wood, straw, and vegetable matter; and certain plastics that do not melt. Early stages of combustion of these materials involve thermal decomposition and destructive distillation.

Considered from simplest perspectives, fire is a sustaining chemical reaction involving three components: fuel (a substance capable of being oxidized under the conditions available), an oxidizer, and a source of energy. The energy source initiates the reaction (ignites the fuel) and sustains and maintains the reaction. Energy produced by oxidizing the fuel may enable the fire to be self-sustaining, or an external source (such as a lighted match) may be necessary. Removing any of these elements extinguishes the fire.

In chemical terms, an oxidizer is a substance capable of accepting electrons during a chemical reaction. Oxidizers usually are gases, although they need not be. The oxidizer most commonly encountered in fire situations is oxygen. Gaseous oxidizers that occur in industrial applications include: halogens and halogen oxides, nitric oxide and nitrogen dioxide, oxygen, and ozone. Some liquid and solid substances readily release oxygen during chemical or thermal decomposition. Liquid oxidizers include hydrogen peroxide, per- and peroxyacids, per- and peroxycompounds, and nitro-compounds, among others. Solid oxidizers include peroxides, hydroperoxides and peroxidates, hypohalites, halites, halates and perhalates, metal oxides, metal peroxides, chromates, permanganates, and nitrates and nitrites, among others (Breatherick 1990).

Historically, the components considered essential for the sustenance of fire — oxidizer, fuel, and heat — were represented as a triangle: the fire triangle. All three components must be present simultaneously in a mutually beneficial relationship. The fire triangle is a valid representation of flameless surface combustion (including glow and deep seated glowing embers). However, other situations indicate that the process is more complex. That this is true, is attributed by the ability of magnesium, aluminum, and calcium to "burn" in nitrogen, or zirconium in carbon dioxide. These processes occur only under highly specific conditions. This is an important caveat, since nitrogen and carbon dioxide are commonly used as extinguishing agents or fire suppressants under normal conditions. Further, some materials such as hydrazine, nitromethane, hydrogen peroxide, and ozone, when heated, will decompose, emitting light and heat (Haessler 1986).

Flaming combustion is mediated through the elements contained in the fire triangle plus chemically reactive species called *free radicals* (Haessler 1974). A free radical is a molecular fragment having one or more unpaired electrons, but no electrical charge. Free radicals are formed in many kinds of chemical processes. Free radicals are highly reactive, and therefore short-lived. Free radicals actively seek to form chemical bonds having paired electrons. This occurs as part of the process of combustion.

The burning of methane provides an illustration of the importance of free radicals in combustion. The combustion of methane occurs through a series of elementary reactions. The product of one reaction is a reactant in one or more others. The reaction mechanism thus is a series of elementary steps. The reaction mechanism for the combustion of more complex organic substances is considerably more elaborate (Meyer 1989).

Free radicals are fundamental to the process of flaming combustion. Destroy the free radicals and flaming combustion ceases. That is, the fire is extinguished and the explosion prevented. This technique is effective for flaming combustion, but not surface (glow) combustion. Extremely effective extinguishing agents, such as halogenated ethanes and methanes, utilize this principle. These substances readily form free radicals through loss of a halogen atom. The free radicals formed from this decomposition unite with intermediates of combustion, which are also free radicals, thereby interrupting and terminating the process. Free radicals add a fourth component to the three-sided fire triangle. The result is the so-called fire tetrahedron (Haessler 1974).

Combustion produces heat. Heat of combustion is the energy released during the reactions that constitute combustion. Heat released during combustion is essential to sustaining the combustion reaction. Heat activates molecules of both fuel and oxidizer by raising their internal energy. For combustion to be self-sustaining, the rate of production of activated molecules must exceed the rate of decay back to the unexcited state (Meyer 1989). Once ignition has occurred, combustion will continue until one of the following has occurred:

- All available fuel or oxidant has been consumed
- The flame is extinguished by cooling
- The number of reactive species has been reduced by some means below sustaining levels (Drysdale 1991).

Combustion of substances containing carbon results in the formation of two possible gaseous products: carbon monoxide or carbon dioxide. Carbon monoxide, the product of incomplete combustion, can undergo further oxidation to carbon dioxide. Carbon dioxide is the product of complete combustion. Unburned substances can form smoke: solid particles containing elemental carbon, sludge, or ash.

MEASURES OF COMBUSTION

The study of combustion is complicated by the different states in which matter can exist. Different measures of combustion apply to matter in its different forms: gas, pure liquid and solution, bulk and powdered solids, and aerosols (airborne solid and liquid particulates). Liquids that can burn and solutions containing them themselves do not burn or explode. Combustion occurs only in the vapor space above the liquid. A vapor–air mixture having appropriate characteristics of concentration must form before combustion can occur. A similar process must occur in the vapor space above volatile solids.

Molecules of gas or vapor are constrained in their motion only by the boundaries of the containment and are free to intermingle. Given sufficient time, molecules of a gas or vapor introduced into a container of air as a discrete cloud will migrate to form a uniform mixture. Airborne dusts and liquid aerosols possess considerably enhanced surface area at which vapor formation can occur compared to bulk solids and liquids.

Combustion of gases and vapors occurs only within a range of concentration in air (the ignitable range). Ignition occurs only above a specific temperature (the ignition temperature).

GASES

Gas is the physical state of a substance that takes the shape and volume of the container in which it is confined. The term *gas* refers to a substance that exists exclusively in the gaseous state at ambient temperature and pressure (21°C and 101 kPa or 760 mmHg). The National Fire Protection Association (NFPA) considers a flammable liquid to have a vapor pressure not exceeding 40 psia at 100°F (275 kPa or 2,070 mmHg at 38°C) (NFPA 1991a). Conversely, by this definition, a gas would have a vapor pressure exceeding 40 psia at 100°F (275 kPa or 2,070 mmHg at 38°C).

NFPA considers a gas that will burn in the normal concentration of oxygen in air to be flammable or combustible. Gases that do not burn in any concentration of air or oxygen are considered by NFPA to be nonflammable. Some of these gases support combustion of other substances, while others suppress combustion. Gases that suppress combustion are called inerting gases. By this definition, under normal consideration, inerting gases include: carbon dioxide and nitrogen, as well as the chemically inert gases, helium, neon, argon, krypton, xenon and radon. However, both carbon dioxide and nitrogen will support combustion of some metals under specific circumstances (Lemoff 1991).

Flammable gases burn in air in the same manner as the vapors from flammable and combustible liquids. Flammable gases burn only within a range of concentration in air (the flammable range) and ignite only above a certain temperature (the ignition temperature). A fire involving a flammable gas is an aborted explosion. In outdoor situations, fires usually occur. Massive release and partial confinement by structures instead could lead to an explosion. The intent of continuously burning pilot flames in gas-burning equipment, such as ovens, water heaters, boilers, and furnaces is to cause a gas fire instead of permitting development of conditions that could lead to a more destructive explosion.

Volatile Liquids and Solids

A key property of some liquid and solid substances is the ability to form vapors that can be ignited. Vaporization and condensation occur continuously at the surface of volatile liquids and solids. At any particular temperature, an equilibrium establishes between vaporization and condensation, given sufficient time. As temperature increases, vaporization increases relative to condensation, thus producing an ever-increasing concentration of vapor. At a critical temperature, the concentration of vapor is sufficient to be ignitable (burst into flame) by an energetic ignition source. When heated slowly in the absence of a source of ignition, the vapor will self-ignite at a higher critical temperature, assuming that breakdown of the substance does not occur.

Flash Point

The minimum temperature of the liquid or solid at which there is sufficient vapor in air to be ignited by an energetic source is the *flash point*. At this temperature, the flame propagates away from the source of ignition. This is an important descriptive distinction, since at lower temperatures the vapor will burn only in the zone immediately surrounding the source of ignition. At higher temperatures and corresponding increase in concentration, the flame propagates (spreads) away from the source of ignition. Flash points of many common substances are available from standard sources (NFPA 1991a).

Practical measurement of flash point is a difficult process. Replicating measurements requires meticulous attention to detail and procedure, and use of standardized apparatus (Wierzbicki and Palladino 1994). The measured value depends on the specific conditions created by each type of testing apparatus and the environment in which it is used. The configuration of the interface between the chamber in which the sample is held and the airspace in which the ignition source is situated reflects two fundamental design philosophies. Closed-cup instruments contain an enclosed chamber having a small opening in the top for escape of vapor. Methods using these instruments attempt to simulate flammability in a semiclosed environment, such as the vent on a container. Open-cup instruments provide much larger surface area for emission of vapor. Methods using these instruments attempt to simulate flammability in open environments, such as an open-topped container. The Cleveland tester is completely open, whereas the other testers have openings ranging between 5.4% to 6.4% of the surface area of liquid. Flash points measured using open-cup testing equipment generally are higher than those measured using the closed-cup units. For most liquids, the open-cup flash point is 10 to 20% higher in degrees Fahrenheit than the closed-cup value (NFPA 1991a). Reproducibility of results from the closed-cup testing instruments lies in the following ranges: Setaflash: 3.3 to 5°C; Pensky-Martens: 3.5 to 8.5°C; and Tag: 2.2 to 3.3°C (Fujii and Hermann 1982).

Flash point testing introduces its own hazards. These include: burn hazard from the flash and potential for overexposure to vapor and combustion products.

Apparatus for determining flash point are specified in standards issued by the American Society for Testing and Materials (ASTM), International Standards Organization (ISO), and the European Community.

With reference to ASTM, the Tag Closed Tester is used for testing liquids, except for certain viscous and film-forming liquids having a flash point below 200°F (93°C) (ASTM 1982a). The Pensky Martens Closed Tester is intended for testing liquids having a flash point of 200°F (93°C) or higher and certain viscous and film-forming liquids (ASTM 1980). The most recently developed flash point tester is the Setaflash Closed Tester (ASTM 1982b, 1981). This is used for testing paints, enamels, varnishes, and related products and components. The Setaflash Closed Tester is intended for testing liquids having a flash point between 32°F and 230°F (0°C and 110°C). The Cleveland Open Tester is sometimes used for liquids having high flash points (ASTM 1978a). The Tag Open Tester used for liquids having low flash points provides data more representative of conditions in open tanks (ASTM 1982c).

TABLE 4.1
NFPA-Recommended Methods for Determining Flash Point

Condition/description	Test Method
• Liquids having viscosity less than 45 SUS (Saybolt Universal Seconds) at 100°F (38°C) and flash point below 200°F (93°C) (in those countries which use the Abel or Abel-Pensky closed-cup tests as an official standard, they will be accepted as equal to the Tag Closed-Cup Method)	ASTM D56-79
• Aviation turbine fuels	ASTM D3828-81
• Liquids having flash points in the range of 32°F (0°C) to 230°F (110°C)	ASTM D3278-82
• Viscous and solid chemicals	ASTM E502-74
• Liquids having viscosity of 45 SUS (Saybolt Universal Seconds) or more at 100°F (38°C) or a flash point of 200°F (93°C) or higher	ASTM D-93-79

From NFPA 1990.

In the interest of reproducibility, the National Fire Protection Association recommends use of the testing protocols listed in Table 4.1.

Flash point provides an indication about several things: ability or inability of vapors to burn, relative level of concern appropriate to ignitability, and susceptibility to ignition. Flash point is a key indicator in consideration about the hazard posed by liquids that can burn. Flash point is a predictor about the potential for vapor formation. The lower the flash point relative to the temperature at which the liquid or solid is used, the more likely that vapor will be present.

Data tabulated in reference sources have practical limitations. These apply to pure substances. Commercial products are seldom as highly purified (NFPA 1991a). This limitation highlights a broader concern about flash point. Actual experience from real world use of chemical products bears little relationship to properties determined in the controlled environment of a laboratory testing instrument.

The most important application of flash point is comparison of ignitability (ability to burn) between substances. In order to optimize the utility of these comparisons, benchmarks are required, since flash point temperatures vary from very low to very high. Benchmarks provide reference points for putting individual values into perspective. The most commonly used system for classification was created by the National Fire Protection Association (NFPA 1991b). This system has been adopted in various forms by many jurisdictions. The NFPA classification system is presented in Table 4.2.

The division of classes is somewhat arbitrarily based on temperatures found in normal situations. Class I refers to temperatures that could occur at some time during the year. The cut-off of 100°F (38°C) also is body temperature. Stated another way, volatile liquids, such as gasoline, having flash points considerably below room temperature (–45°F or –43°C) pose a considerable ignitability hazard. Given an energetic source of ignition, gasoline vapor is easily ignitable at room temperature. The same also applies to the vapors of other industrial solvents in common use. Substances having flash points in Class I have the potential to be very hazardous in normal circumstances of use. Substances having flash points in Class II require moderate heating to generate sufficient vapor for ignition. Moderate heating of industrial solvents other than in process situations involving closed systems is not common. Fires involving substances having flash points in this region are less likely to occur in normal circumstances. Class III refers to substances requiring considerably more heating than provided by ambient sources to generate sufficient vapor for ignitability.

Fujii and Hermann summarized historical models and developed models using regression analysis for estimating flash point (Fujii and Hermann 1982). Closed-cup flash points for pure substances in a number of chemical families can be modeled effectively by Equation 4.1.

TABLE 4.2
NFPA Classification of Flammability and Combustibility

Classification	Flash point	Boiling point
Flammable liquid:	Flash point below 100°F (38°C) and vapor pressure not exceeding 40 psia (275 kPa or 2,070 mmHg) at 100°F (38°C)	
Class IA	<73°F (23°C)	<100°F (38°C)
Class IB	<73°F (23°C)	≥100°F (38°C)
Class IC	73°F (23°C) ≤ FP < 100°F (38°C)	
Combustible liquid:	Flash point at or above 100°F (38°C)	
Class II	100°F (38°C) ≤ FP < 140°F (60°C)	
Class IIIA	140°F (60°C) ≤ FP < 200°F (93°C)	
Class IIIB	≥200°F (93°C)	

Note: NFPA defines liquid as a fluid having a vapor pressure not exceeding 40 psi absolute (275 kPa or 2,070 mmHg) and fluidity greater than that of 300 penetration asphalt (a unit of viscosity).

TABLE 4.3
Values of Regression Constants for Calculating Flash Point

	Regression Constant	
Chemical Family	**a**	**b**
Alkenes	3.097	0.424
Ethers	3.056	0.357
Ketones	3.033	0.381
Aldehydes	2.924	0.443
Acetates	2.976	0.380
Esters	2.948	0.385
Alcohols	2.953	0.323
Phenols	2.953	0.323
Amines	3.077	0.322
Acids	2.777	0.491

After Fujii and Hermann 1982.

$$\frac{1}{T_{FP}} \times 10^3 = a + b\left(\log\ VP_{25}\right) \tag{4.1}$$

where
T_{FP} = the flash point in °K;
VP_{25} = the vapor pressure in mmHg at 25°C; and
a and b are constants.

The values of a and b are tabulated in Table 4.3.

It was determined that alkanes and aromatics are modeled more suitably by the following quadratic equation:

$$\frac{1}{T_{FP}} = 2.996 + 0.324\left(\log VP_{25}\right) + 0.074\left(\log VP_{25}\right)^2 \tag{4.2}$$

Vapor pressure at the specified temperature can be estimated: predictive equations and resources that provide graphical or tabular data (Boublik et al. 1973, Reid et al. 1987, Weast 1968).

Mixtures are not well modeled by the preceding equations.

Closed-cup flash point of hydrocarbon mixtures (mineral spirits and aromatic solvents, as used in paint formulations) can be approximated from their boiling points using the formula (Fujii and Hermann 1982):

$$t_{FP} = 0.70\, t_B - 72.7 \tag{4.3}$$

where
t_{FP} = closed-cup flash point,°C;
t_B = initial boiling point,°C.

The vapor pressure of hydrocarbon mixtures at the flash point can be approximated using Trouton's rule (Butler et al. 1956).

$$\log VP = 2.881 + 4.60 \times \left(1 - T_B/T_{FP}\right) \tag{4.4}$$

where
VP = the vapor pressure of the mixture in mmHg;
T_B = the boiling point in °K; and
T_{FP} = the flash point temperature in °K.

Liquids occurring in mixtures may be completely miscible, partly miscible, or completely immiscible in each other. In an ideal solution containing completely miscible liquids, interactions between molecules of solute and solvent are the same as solute–solute and solvent–solvent interactions. Raoult's law provides a basis for predicting vapor pressure of components of ideal solutions in equilibrium situations. According to Raoult's law, the partial pressure of a vapor above a solution is equal to the product of the mole fraction of the liquid in the solution times the partial pressure of the pure liquid at the same temperature. Raoult's law applies, for example, to solutions of volatile hydrocarbons such as butane and propane in a nonvolatile liquid, such as a hydrocarbon oil (Treybal 1980). Raoult's law holds for a component that approaches 100% in solution. For mixtures where the mole fraction of the organic material in solution is 0.8 or more, Raoult's law holds with an error of 7% or less, except in extremely unusual cases (Perry et al. 1984). Physical and chemical interactions between components of the mixture affect their individual vapor pressures. Most solutions are nonideal at some composition. In nonideal solutions, interactions between solute and solvent differ in magnitude from solvent–solvent and solute–solute interactions. This can result in greater or lesser vapor pressure for individual components than that predicted by Raoult's law (Moore 1962). In the case of complete immiscibility, the vapor pressure of the mixture is the sum of the vapor pressures of each component. Partial miscibility produces complex relationships.

Henry's law applies to dilute nonideal solutions at equilibrium. Henry's law states that the partial pressure of a volatile component above a solution equals the mole fraction in the solution times a constant. Henry's law applies to dilute solutions of flammable gases, such as methane, hydrogen, ethane, and ethylene in water.

Chapter 5 more fully explores the implications of solutions that obey Raoult's law and Henry's law.

Mixtures containing both flammable and nonflammable liquids complicate the concept of ignitable. Ignitability can be suppressed by addition of a volatile nonignitable substance to a flammable/combustible one. The mixture as initially formulated may be neither flammable nor combustible. This occurs because vapors from the flammable and nonflammable components contribute to the total vapor pressure of the mixture. An example of this situation and its potential ramifications is a blend containing carbon tetrachloride and gasoline. Carbon tetrachloride is more volatile than some of the components in gasoline. On standing in an open container, the carbon tetrachloride will evaporate more rapidly than the gasoline. The residual fluid thus becomes enriched in hydrocarbons. Over time, the residual liquid will exhibit a high flash point that progressively decreases. The flash point of the final 10% of the original mixture approximates that of the heavier fractions of the gasoline (Sly 1991).

A mixture of flammable/combustible components represents a variation on this concept. Unequal evaporation would lead to enrichment by the more volatile component(s). This would result in a decrease or increase in the flash point, depending on the ignitability of the more volatile components. An example of this situation is a high-flash nonvolatile mixture (such as motor oil) that contains a volatile low-flash substance (such as toluene). Preferential evaporation by the volatile substance produces an unexpectedly low flash point for the mixture. Used motor oil containing a small percentage of toluene is an example of such a mixture. This type of mixture easily could arise as a waste liquid in a garage operation. Toluene can vaporize readily from this mixture. The flash point of the mixture would be the temperature at which sufficient toluene vapor is present for ignition. This temperature would be considerably less than the flash point for the oil. Waste oil products containing small amounts of solvent can constitute major fire hazards. Another example would be the complex sludge mixtures that occur in some confined spaces.

Appreciable evaporation and consequent change in ignitability can occur under conditions involving prolonged exposure of solvent blends to atmosphere. Examples include solvents used in dip tanks and degreasers and waste solvents whose history is not known. Fractional evaporation tests conducted at room temperature in open vessels provide a means to evaluate the fire hazard of these mixtures. Flash point can be determined on fractions containing 10, 20, 40, 60, and 90% of the original sample. The results of such tests indicate the ignitability grouping into which the liquid should be placed. The open-cup test method may give a more reliable indication of the ignitability hazard than closed-cup methods in such situations (NFPA 1990).

Flammable organic substances can occur in aqueous solutions or emulsions. Vapor from the organic component still can pose an ignitability hazard. Examples of such mixtures include spent caustic solution containing traces of gasoline, or commercial products containing aqueous solutions of alcohols. Raoult's law can be used to estimate the closed-cup flash points of these mixtures (Johnston 1974). This method assumes that Raoult's law applies and that the vapor space in the container above the liquid is occupied by water, solvent vapor, and air. The temperature of the mixture must be sufficient to generate the vapor concentration at the flash point in dry air. This temperature is greater than the flash point of the pure liquid. For substances having high flash points, the contribution of water vapor to the mixture in the vapor space becomes critical at elevated temperatures. At elevated temperatures, water vapor can suppress vapor formation by the organic component, so that an ignitable mixture cannot develop.

Decreasing total atmospheric pressure decreases the flash point relative to sea level. For example, the flash point of toluene measured in Denver, CO (altitude = 1.6 km or 1 mi; standard atmospheric pressure = 627 mmHg or 83.6 kPa) is 1°C less than the value at sea level (Bodurtha 1980). This results from increased vaporization relative to condensation as total pressure decreases. Increasing the total atmospheric pressure increases the flash point. This property has practical significance. The fire hazard associated with flammable solvents increases slightly with increasing altitude or decreased atmospheric pressure and decreases slightly with depth, as in a deep mine. Another example is a confined space under negative pressure. In the real world, these effects are negligible.

Mist

Mist is a suspension of droplets in air. Mist can develop by condensation of vapor or by active processes, such as spraying and spray painting, pouring, aerating, grinding and burning, sieving, splashing, and nebulization, among others. Another source of liquid aerosols is foam. The bursting of bubbles in a froth propels liquid droplets into the air. These processes increase surface area by breaking the mass into smaller particles.

Droplets dramatically increase the surface area at which vaporization can occur. Vapor formation occurs at the surface of each droplet. This can provide conditions for ignitability at temperatures below the flash point. This property depends on particle size. The concept of flash point is predicated on passive generation of vapor above the surface of a volatile liquid or solid. Insufficient concentration of vapor for ignition exists at temperatures below the flash point.

Processes that aerosolize liquids dramatically increase ignitability by loading the air with an ignitable mixture at a temperature below the flash point of the liquid. This is an especial concern with combustible liquids. The lower ignitable limit for mists containing fine droplets plus accompanying vapor occurs at a concentration of approximately 48 g/m^3 at 0°C and 101 kPa (760 mmHg), regardless of the flash point of the liquid. This concentration of small droplets (smaller than 10 μm) corresponds to a very dense mist (Burgoyne 1963, Zabetakis 1965). A 100-Watt bulb is visible only for a few centimeters (inches) at this concentration (Burgoyne 1957).

Dust

Most finely divided combustible solids are hazardous (Schwab 1991). Deposits of combustible dust on beams, machinery, and other surfaces are subject to flash fires and explosions. Two types of fires occur in dust deposits: smoldering (surface) and flaming (Lees 1980).

Smoldering involves very slow combustion. Smoldering results from restricted air access and heat loss. Dust layers 2 mm and thicker can sustain smoldering for periods measured in years. Smoldering may give no readily detectable effects, such as smoke or odor. Upon reaching the surface of a layer of dust, a smoldering fire can burst into flame. Smoldering rate commonly is determined from rate of travel along a dust "train" of stated dimensions. Typical smoldering rates for dusts are approximately 5 cm/h (wood) and 20 cm/h (coal) in a bed having a depth of 1 cm. Magnesium has an extremely high rate of 14 m/h. Characteristics of dust that influence the type of fire include volatile content, melting point, and particle size. Detection of dust fires is difficult. The effects of smoldering are difficult to detect with the sensors normally used in fire protection systems.

In a flaming fire, heat from the flame causes volatilization of volatile substances from the surface of dust particles. The flame may spread across the surface of the dust while smoldering occurs underneath. Heat produced in a smoldering fire liberates volatile compounds from the dust. If sufficient volatile material is present, the flame may travel rapidly across the surface. Otherwise, the smoldering rate determines the rate of propagation over the surface. Large particles reduce packing density, thereby permitting airflow and flame propagation through the dust. Combustion occurs rapidly. Large particles smoulder poorly.

IGNITABILITY LIMITS

Flammable gases and vapor from flammable and combustible liquids and solids can be ignited by an energetic source of ignition within specific limits of concentration (Figure 4.1). Under equilibrium conditions, these concentrations correspond to a range of temperature in air or oxygen. The lower limit of concentration at which ignition by an energetic source can occur is the **lower flammable limit** (**LFL**). Under equilibrium conditions for volatile liquids and solids, this occurs

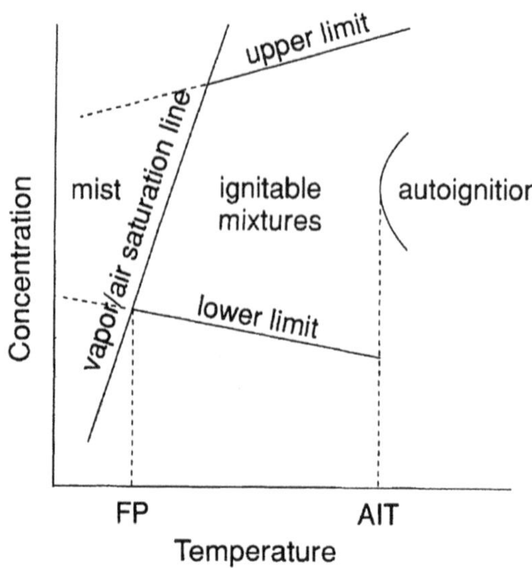

FIGURE 4.1 Ignitability limits in air as a function of temperature and constant initial pressure. An important inclusion in this diagram is the transition from ignitable mists (droplets and vapor) to ignitable vapor. The vapor–air mixture is saturated and igitable only along the vapor–air saturation line. Ignitable mixtures to the right of the air–vapor saturation line and between the lower and upper limits are unsaturated. Heterogeneous mixtures, such as sprays, mists, and foams, can ignite at temperatures at which homogeneous mixtures normally would be nonignitable. FP is the flash point temperature. AIT is the autoignition temperature (Adapted from Zabetakis 1965.)

at the flash point. A flame does not propagate away from the source of ignition at concentrations below the lower flammable limit. The upper limit of concentration is the **upper flammable limit (UFL)**. The difference between these concentrations is the **flammable range**. Values of the LFL and UFL for many chemical substances are available from standard sources (NFPA 1991a).

The lower flammable limit is often referred to as the lower explosive limit (LEL), the upper flammable limit as the upper explosive limit (UEL), and the flammable range as the explosive range. Strictly speaking, these terms are not identical. Explosive mixtures will burn. Not all ignitable mixtures will explode. The difference rests with the amount of energy needed to cause an explosion, rather than a fire, and the limits of concentration within which the mixture will explode.

Stated in popular terms, a mixture whose concentration is less than the lower flammable limit is too "lean" to burn or explode. A mixture whose concentration is greater than the upper flammable limit is too "rich." Flammability limits for selected substances are provided in Table 4.4.

While commonly measured in volume percent, when expressed on a mass/concentration basis, LFLs for hydrocarbons generally lie in the range 40 to 50 g/m^3 of air at 0°C and 101.325 kPa. Consequently, ventilation rates needed to reduce concentrations of equal masses of different hydrocarbons to a specified percent of the lower limit are approximately the same (Bodurtha 1980). On a mass-concentration basis, UFLs increase with increasing molecular weight within a homologous series (Zabetakis 1965). Alcohols and other oxygen-containing compounds have higher LFL values when expressed on a mass-concentration basis than hydrocarbons. Hydrogen has a much lower value. Other substances having low molecular weight have slightly lower values.

The lack of uniformity in LFL values across chemical classes is a major disincentive to the calibration of testing instruments by one substance for assessment of hazardous atmospheres containing other substances. This lack of uniformity becomes an especial concern in the assessment of unknowns and mixtures of different classes of substances.

TABLE 4.4
Flammability Limits of Selected Substances

Substance	Volume units Lower	Volume units Upper	Mass units Lower	Mass units Upper
	%	%	g/m³	g/m³
Normal paraffins				
Methane	5.0	15.0	38	126
Ethane	3.0	12.4	41	190
Propane	2.1	9.5	42	210
Butane	1.8	8.4	48	240
Pentane	1.4	7.8	46	270
Hexane	1.2	7.4	47	310
Heptane	1.05	6.7	47	320
Octane	0.95	—	49	—
Nonane	0.85	—	49	—
Decane	0.75	—	48	—
Unsaturated hydrocarbons				
Ethylene	2.7	36	35	700
Propylene	2.4	11	46	210
Butene-1	1.7	9.7	44	270
cis-Butene-2	1.8	9.7	46	270
Isobutylene	1.8	9.6	46	260
3-Methyl-butene-1	1.5	9.1	48	310
Propadiene	2.6	—	48	—
1,3-Butadiene	2.0	12	49	320
Aromatic hydrocarbons				
Benzene	1.3	7.9	47	300
Toluene	1.2	7.1	50	310
Ethylbenzene	1.0	6.7	48	340
Xylenes	1.1	6.4	53	320
Alicyclic hydrocarbons				
Cyclopropane	2.4	10.4	46	220
Cyclobutane	1.8	—	46	—
Cyclopentane	1.5	—	48	—
Cyclohexane	1.3	7.8	49	320
Ethylcyclobutane	1.2	7.7	46	310
Cycloheptane	1.1	6.7	49	310
Methylcyclohexane	1.1	6.7	49	310
Alcohols				
Methanol	6.7	36	103	810
Ethanol	3.3	19	70	480
n-Propanol	2.2	14	60	420
n-Butanol	1.7	12	57	—
Ethers				
Dimethylether	3.4	27	72	760
Diethylether	1.9	36	64	1,880
Ethylpropylether	1.7	9	68	390
Diisopropylether	1.4	7.9	60	290
Divinylether	1.7	27	54	1,160

TABLE 4.4 (continued)
Flammability Limits of Selected Substances

Substance	Volume units Lower %	Volume units Upper %	Mass units Lower g/m³	Mass units Upper g/m³
Esters				
Methyl formate	5.0	23.0	142	800
Ethyl formate	2.8	16.0	95	630
n-Butyl formate	1.7	8.2	79	410
Methyl acetate	3.2	16.0	106	630
Ethyl acetate	2.2	11.0	88	510
n-Propyl acetate	1.8	8.0	83	400
n-Butyl acetate	1.4	8.0	73	450
n-Amyl acetate	1.0	7.1	65	440
Methyl propionate	2.4	13.0	97	580
Ethyl propionate	1.8	11.0	83	510
Aldehydes and ketones				
Acetaldehyde	4.0	36	82	1,100
Propionaldehyde	2.9	14	77	420
Paraldehyde	1.3	—	78	—
Acetone	2.6	13	70	390
Methylethyl ketone	1.9	10	62	350
Methylpropyl ketone	1.6	8.2	63	340
Diethylketone	1.6	—	63	—
Methylbutyl ketone	1.4	8.0	64	390
Sulfur compounds				
Hydrogen sulfide	4.0	44	63	1190
Carbon disulfide	1.3	50	45	3,400
Methyl mercaptan	3.9	22	87	600
Ethyl mercaptan	2.8	18	80	610
Dimethyl sulfide	2.2	20	62	690
Other				
Hydrogen	4.0	75	3.7	270
Carbon monoxide	12.5	74	145	858

From NIOSH 1990, Zabetakis 1965, and NFPA 1991a.

The vapor composition at the lower flammable limit can be estimated from the vapor pressure at the flash point (closed-cup) using the following relationships (Sly 1991):

$$\text{LFL} = \frac{\text{VP}}{\text{P}_{\text{atmospheric}}} \times 100\% \text{ v/v} \tag{4.5}$$

LFL = percent vapor by volume at atmospheric pressure
VP = vapor pressure at flash point temperature.

LFL and UFL for mixtures can be estimated using Le Chatelier's rule. Le Chatelier's rule applies to mixtures whose components obey Raoult's law (Le Châtelier 1891, Bodurtha 1980). The following equation may be used to calculate the LFL and UFL. These equations must be used with discretion, particularly with chemically dissimilar substances.

$$FL = \frac{100\%}{\frac{C_1}{FL_1} + \frac{C_2}{FL_2} + \ldots + \frac{C_n}{FL_n}} \tag{4.6}$$

FL = the flammability limit (LFL or UFL) calculated for the mixture.
C_1 = the percent by volume of component 1 in the mixture.
FL_1 = the flammability limit (LFL or UFL) for component 1 in air.

Limits of flammability are not a fundamental parameter of combustion. They depend upon many variables, including: the surface-to-volume ratio of the reaction chamber, flow direction, and velocity (NFPA 1991c). Elevated temperature and pressure affect flammability limits. These effects mostly concern process operations involving liquids and vapors confined in tanks, pipes, reactors, fractionating towers, and other process equipment during operation, rather than preparation for entry.

The general effect of increasing temperature is to decrease the LFL and increase the UFL, thus broadening the flammable range (Zabetakis 1965). This outcome is very specific to equipment and conditions. An increase in temperature also can cause a previously nonflammable mixture to become flammable by increasing vapor formation. A decrease in temperature can cause a previously flammable mixture to become nonflammable by decreasing vapor formation (Drysdale 1991). The following equations provide a means for estimating flammability limits at elevated temperatures (Bodurtha 1980).

$$LFL_{t\,°C} = LFL_{25°C} - \left(0.8\,LFL_{25°C} \times 10^{-3}\right)(t - 25) \tag{4.7}$$

$$UFL_{t\,°C} = UFL_{25°C} + \left(0.8\,UFL_{25°C} \times 10^{-3}\right)(t - 25) \tag{4.8}$$

Variations in pressure at normal atmospheric levels have little effect on flammability limits. The effect of larger pressure changes is specific to each mixture. Decreasing the pressure below atmospheric can increase the LFL and reduce the UFL until the flammability limits coincide, thus rendering the mixture nonflammable. Increasing the pressure above atmospheric levels can widen the flammable range. The effect is more marked on the UFL than the LFL. However, in some cases, an increase in pressure narrows the flammability range (Zabetakis 1965).

One aspect about flash point and flammability that cannot be overemphasized is the problem posed by "empty" containers. "Empty" containers were involved in 16% of the fatal welding and cutting accidents reported by OSHA (1988). Methods for the safe handling of "empty" containers are the subject of several safety standards (Nat. Safety Council DataSheet 432 1956, NFPA 327 1993a, ANSI/AWS 1988). No empty container should be presumed to be clean or safe. Containers that have held hazardous substances can be made safe by following appropriate measures.

An "empty" container easily can confine a flammable atmosphere under subambient, ambient, and superambient conditions of temperature. The governing factors are temperature of the surroundings and flashpoint of the liquid. All that is required for formation of a flammable or explosible atmosphere is sufficient residual liquid on interior surfaces and time. Time is needed for the formation of the vapor–air mixture. Containers also may be lined or may contain sludge. When heated, coatings or liners or sludges could release vapors or liquids trapped between the coating or liner and the container wall.

Table 4.5 outlines conditions that could exist in "empty" containers prior to the application of heat.

Heating a container holding only vapor that is within or above the flammable range creates the potential for fire or explosion. Heat applied from a cutting torch creates a hot spot on the surface.

TABLE 4.5
Conditions in "Empty" Containers

Contents	Flammability	Comments
Gas or vapor	Nonflammable	• Concentration below flammable range • Concentration above flammable range
Gas or vapor	Flammable	• Concentration within flammable range
Liquid	Nonflammable	• Vapor concentration below flammable range • Vapor concentration above flammable range
Liquid	Flammable	• Vapor concentration within flammable range

This pressurizes the gaseous contents in a sealed container and would cause outflow from any opening, including that created by the torch. Other less dramatic heat sources capable of creating hot spots that could ignite vapors include saws and grinders.

The flammability of the atmosphere inside a container holding both liquid and vapor depends on the temperature of the containment and the contents. Flammability could change with the season since radiant heating from the intense summer sun could sufficiently heat a high-flash liquid to create a flammable mixture. On the other hand, heat from the summer sun could put the vapor concentration from a low-flash liquid, such as gasoline, above the flammable range. The reverse could apply under winter conditions. The key to assessing this concern is the interrelationship between vapor pressure and temperature and flash point.

The relationship between a liquid and its vapor in a container at any point in time reflects two possible conditions: equilibrium or nonequilibrium. Equilibrium is a precise condition in which the rate of vaporization and rate of condensation are constant. Equilibrium mixtures can develop in the vapor space of all containers, from large and small tanks, to small-sized containers used for everyday use, to "empty" containers that contain just sufficient liquid to form a flammable mixture. Attainment of equilibrium is fully possible only in closed containers where vapor is prevented from escaping and temperature remains constant. Equilibrium is unlikely to be attained in containers in which the vapor is not fully confined. This is especially problematic for low flash liquids in summer conditions, since they are volatile enough for the vapor to escape from containment.

Flash point and flammability limits are measured under equilibrium conditions. Disturbance of equilibrium occurs following raising or lowering the temperature or changing the liquid content in the container. The relationship between temperature and vapor concentration depends on the kinetics of attainment of equilibrium. Where equilibration is rapid, the lag time in reaching the new condition would be short. This situation is less problematic for liquids having high volatility.

The question then becomes whether the vapor mixture that develops in nonequilibrium situations can be ignited. The answer is related to the concentration of vapor in the region in which the energetic source of ignition is located. The mixture should be ignitable, provided that the concentration exceeds the LFL and lies within the flammable range. Conditions under which a nonequilibrium mixture develops are extremely common. To summarize, the flash point and flammable range are measured under equilibrium conditions. This then begs the question whether equilibrium is needed to have an ignitable mixture. The answer is no.

SOURCES OF IGNITION

Discussion

Fires and explosions occurring in industry usually involve a source of ignition, rather than slow heating and self-ignition. Sources of ignition are many and varied. They may be obvious or unobtrusive. Recognizing that they are present is critical to addressing the hazard that they pose.

Sources of ignition can be considered as related to an existing process, introduced as part of a maintenance activity, or associated with the actions of the participants (Lees 1980).

Sources related to existing processes include:

- Open flames (burners, heaters, furnaces, and flare stacks)
- Hot soot (ships and railway engines)
- Unlagged surfaces (process equipment and piping)
- Distressed machinery (faulty or damaged bearings and seals)
- Static electricity (bulk materials handling)
- Compression (Diesel effect)
- Abrasion (buffing, sanding, scraping)

Sources introduced as part of maintenance activities include:

- Electric arc (welding, cutting, burning, electrical equipment)
- Hot metal and slag (welding, cutting, grinding, and burning)
- Engines/exhaust systems (vehicles and mobile equipment)
- Static electricity (abrasive blasting, spray painting, steam cleaning)

Sources unrelated to the activity include:

- Open flames (cigarettes, matches and lighters)
- Static electricity (nylon and static-generating fabrics in clothing, nails in soles and heels of footwear, tramp metal)

The preceding sources are obvious to the trained observer. Raising awareness about less obvious sources is critical to addressing this problem. All fire-producing and spark-producing sources should be considered as potential sources of ignition.

Ignition sources in the industrial environment must be capable of heating the vapor to its ignition temperature in the presence of the surrounding air. Under many conditions, the ignition source also must provide sufficient energy to volatilize the fuel. Hot surfaces and flames can be a source of ignition if large enough and hot enough. The smaller the surface, the hotter it must be to cause ignition. Also, the vapor from the liquid or solid must remain close to the source of ignition to enter the flammable range prior to being carried off by convection currents (Sly 1991).

Chemical Energy

Rapidly generated chemical energy can act as a source of ignition. Rapid generation of chemical energy can occur during different classes of chemical reaction under appropriate conditions. These include: reduction–oxidation (redox) reactions of which combustion reactions and reactions by chemical explosives are subgroups, acid–base neutralization, hydration of salts, and formation of solutions. Runaway chemical reactions have caused considerable devastation because of rapid release of energy.

Situations that provide the conditions for rapid generation of energy through redox reactions include metallic smears and pyrophoric iron sulfide. Rusty steel smeared with metallic aluminum or aluminum paint or magnesium is an example of a metallic smear. Struck by a heavy metallic object, a strong spark results. Energy that creates the spark is partly mechanical and partly chemical. The mechanical component arises from the sudden impact. The chemical component arises from the release of chemical energy during the oxidation–reduction process. Rapid oxidation of aluminum and reduction of iron oxide is the thermite reaction. Thermite welding is used industrially to bond steel. Heat evolved during the reaction melts the newly formed iron.

Pyrophoric iron sulfide scale forms in the reaction between hydrogen sulfide and iron in exposed steel. This reaction is a concern in industrial situations where Sulfur and sulfides, including hydrogen sulfide, are generated or otherwise potentially present. Under dry, warm conditions, the scale may glow red and act as a source of ignition. Pyrophoric iron sulfide is best removed by dampening the surface and scraping to remove the scale. No attempt should be made to remove it by scraping prior to dampening (Lees 1980).

Mechanical Energy

Mechanical energy is the source of ignition for a significant number of fires each year. Frictional heating and friction sparks, and heat of compression are the source of ignition in most of them (Drysdale 1991).

Frictional heating results from conversion of energy used to overcome resistance to motion when two solids are rubbed together. The significance of this hazard reflects availability of mechanical energy and the rate of heat generation and dissipation.

Friction sparks result from the sudden impact of two hard surfaces. One surface usually is metal. In workplace situations, friction sparks often arise in a casual manner. Examples include dropping of steel tools onto a concrete floor or machinery (or piping), ricocheting of tramp metal inside a grinding mill, and impact between shoe nails and a concrete floor. Heat, generated by impact or friction, initially heats a metal particle. The freshly exposed surface may oxidize at the elevated temperature. The heat of oxidation may increase the temperature of the particle until it is incandescent. The extent of this process depends on the ease of oxidation and the heat of combustion of the particle. The incandescence temperature of most metals is well above the ignition temperature of ignitable materials. The ignition potential of a spark, however, depends on total heat content. Thus, the practical danger from mechanical sparks is limited by their small size and low heat content. They cool quickly and start fires only under highly favorable conditions.

Special tools made from copper–beryllium and other alloys are designed to minimize the generation of sparks. Such tools cannot, however, wholly eliminate the danger of spark generation in hazardous locations, because a spark may be produced under several conditions. Concern about the hazard of ignition of vapors or gases by friction sparks has been de-emphasized somewhat by reports, such as API 2214 (API 1989). API 2214 indicates that use of nonsparking hand tools in place of steel tools provides no benefit in preventing explosion of hydrocarbons. Aluminum tools also provide little benefit. Hit against iron oxide (rust) with sufficient force, an aluminum tool could initiate a thermite reaction. Leather, plastic, and wooden tools are free from the friction spark hazard. However, they may not be practical. Nickel, monel, and bronze pose a very slight spark hazard. Stainless steel has a much lower spark hazard than ordinary tool steel (Drysdale 1991).

Heat of compression is the heat released when a gas is compressed. Heat of compression has found practical application in the operation of the diesel engine. In a diesel, air is compressed in the cylinder by motion of the piston, and then an oil spray is injected into it. The heat contained in the compressed air is sufficient to vaporize and ignite the oil spray.

Electrical Energy

Electrical energy converted to heat is an important source of ignition energy (Drysdale 1991). This conversion can occur through several mechanisms.

Resistive heating occurs during flow of current through a conductor. The power required for current flow is the resistance of the conductor times the square of the current. Less power is needed for passage of current through conductors, such as copper and silver, that have low resistivity. Heat generated by filaments in incandescent bulbs and infrared heaters is an example of resistive heating.

Dielectric heating results from distortion of atomic or molecular structure caused by externally applied time- and direction-varying electric potential. A rapidly varying external potential can cause considerable heating of a dielectric (a good insulator).

Induction heating occurs in a conductor subjected to a fluctuating or alternating magnetic field. This also occurs in a conductor moved across the lines of force of a magnetic field. Rapidly changing or alternating potentials produce heat due to mechanical and electrical distortion of the atomic or molecular structure. Heating increases with the frequency of alteration. Microwave heating is an application of this principle.

Leakage current also is a source of heat. Leakage current is current flow through insulators that are subjected to substantial voltages. The heat produced by leakage current becomes important only in unusual situations involving breakdown of insulation.

Arcing is a phenomenon of current electricity. Arcing occurs when current flow in an electric circuit is interrupted. Arcing is especially severe when motor or other inductive circuits are involved. The temperature of the arc is very high. In some circumstances the arc may melt the conductor and scatter the molten metal. Energy released by an intrinsically safe electrical circuit during a fault condition such as arcing must be less than that needed to ignite the hazardous atmosphere in which the circuit is located.

Prevention of unexpected discharge of electrical energy is paramount to protecting people and facilities. Complementing the NFPA classification of flash point is the classification of the atmospheric hazard by substance in the U.S. National Electrical Code. This classification includes gases, vapors, and dusts. Article 500 divides fire and explosion hazards into the following classes, divisions and groups (Table 4.6).

Class I locations are subdivided into four groups. Groups contain chemical substances having similar properties. Class II locations are subdivided into three groups. Division 2 locations normally are not hazardous. They apply to areas that become hazardous in the event of accidental discharge of flammable materials from confined systems.

Explosion-proof equipment generally is required for Class I, Division 1 locations. Explosion-proof electrical equipment is not gastight. The cast-metal electrical enclosure must prevent escape of flames and operate below the ignition temperature of the flammable material in the ambient environment. Essentially, the chief purpose of an explosion-proof enclosure is to prevent initiation of a fire or explosion in the ambient atmosphere.

Normally, nonsparking equipment or apparatus that has make-or-break contacts immersed in oil or hermetically sealed is used in Division 2 areas. This equipment can cause ignition only if it malfunctions at the same time as a flammable concentration develops. This event is unlikely, considering the tiny probability of simultaneous occurrence of electrical failure and release of flammable materials. When nonsparking equipment is not available, explosion-proof equipment or electrical apparatus contained in explosion-proof housings must be used.

Alternative approaches exist for addressing these requirements (Elcon 1989, NFPA 1993c). The first is to locate electrical equipment in less hazardous or nonhazardous areas. The second is to pressurize the electrical system with noncontaminated air or a nonreactive gas. This system must be fail-safe. A third approach is to use intrinsically safe equipment. Intrinsic safety means that the electrical equipment is designed to release insufficient electrical or thermal energy under normal or abnormal conditions to ignite a specific hazardous atmosphere. Abnormal conditions include accidental damage to field-installed wiring, failure of electrical components, over-voltage, adjustment and maintenance operations, and other similar conditions. The low energy requirements limit the use of intrinsically safe equipment to low-power devices, such as process-control instrumentation and communication equipment.

Lightning also is a form of current electricity (Frydenlund 1966, Whitehead 1983, Davis 1991). The destructive potential of lightning, of course, is well known. Less well appreciated is the fact that current electricity is considered to be able to flow without the guidance provided by wires

TABLE 4.6
Hazardous Locations — U.S. National Electrical Code

Classification/Description

Class I: Locations in which flammable gases or vapors are or may be present in quantities sufficient to produce explosive or ignitable mixtures.

Division 1: Location in which hazardous concentrations of flammable gases or vapors exist continuously, intermittently, or periodically under normal operating conditions

or

in which concentrations of such gases or vapors may exist frequently because of repair or maintenance operations or because of leakage

or

in which breakdown or faulty operation of equipment or processes might release hazardous concentrations of flammable gases or vapors, and might also cause simultaneous failure of electrical equipment.

Division 2: Locations in which volatile flammable liquids or flammable gases are handled, processed, or used, but in which they will normally be confined within closed containers or closed systems, from which they can escape only in case of accidental rupture or breakdown of such containers or systems, or in case of abnormal operation of equipment

or

in which hazardous concentrations of gases or vapors are normally prevented by positive mechanical ventilation, and which might become hazardous through failure or abnormal operation of the ventilating equipment

or

which are adjacent to a class I, division 1 location and to which hazardous concentrations of gases or vapors might occasionally be communicated unless such communication is prevented by adequate positive pressure ventilation from a source of clean air and effective safeguards against ventilation failure are provided.

Class II: Locations that are hazardous because of the presence of combustible dust.

Division 1: Location in which combustible dust is or may be in suspension in the air continuously, intermittently, or periodically under normal operating conditions in quantities sufficient to produce explosive or ignitable mixtures

or

where mechanical failure or abnormal operation of machinery or equipment might cause such explosive or ignitable mixtures to be produced and might also provide a source of ignition through simultaneous failure of electric equipment, operation of protection devices, or from other causes

or

in which combustible dusts of an electrically conducting nature may be present.

Division 2: Location in which combustible dust will not normally be in suspension in the air or will not likely be thrown into suspension by the normal operation of equipment or apparatus in quantities sufficient to produce explosive or ignitable mixtures,

but

where deposits or accumulations of such combustible dust may be sufficient to interfere with the safe dissipation of heat from electrical equipment or apparatus

or

where such deposits or accumulations of combustible dust on, in, or in the vicinity of, electric equipment might be ignited by arcs, sparks, or burning material from such equipment.

Class III: Locations that are hazardous because of the presence of flammable fibers and flyings.

From NFPA 1991d,e, 1992a, and 1993b.

from source to destination. Lightning involves the flow of 2,000 to 200,000 amperes of current (median 30,000 amperes) at a potential difference of 10 million to 100 million volts for periods lasting for several microseconds. Lightning is most likely to strike the object in a group that is: the best conductor, nearest the approaching cloud, or the most prominent. The lightning bolt generally follows a metallic path to ground. Fulfilling this mandate may entail jumps from one source of metal to another in direct and indirect paths. The lightning stroke in effect is an arc between two plates of a capacitor, each having the opposite charge. Charge separation within clouds leads to induction of the opposite charge in the ground under the cloud. Lightning is believed to move downward to ground through a leader that creates an ionized path. The leader involves movement of only a small current. The major part of the discharge current is carried in the return stroke from earth to the charged cloud through the ionized path.

Protection against lightning strikes is the subject of several standards (API 1991, NFPA 1992b). Maintaining lightning protection during turnaround work on structures containing confined spaces or confined hazardous atmospheres is essential to protecting health and safety. Similar to protection against static electrical discharges, lightning protection relies on bonding and grounding (Davis 1991). Turnaround activities raise the possibility of disrupting the continuous path from point of contact on a structure to ground and creating isolated conductors. This could create an isolated conductor and lead to arcing to complete the path to earth.

Static electricity is a potential source of ignition wherever flammable substances are present. Static electricity is believed to be the cause of many apparently mysterious explosions that have occurred in process plants (Lees 1980). Static electricity is a surface effect produced by the contact and separation of dissimilar materials. During the action of separation, one surface acquires a positive electrical charge, the other a negative. If the materials are good conductors, the charges move freely and neutralize at last contact before separation. If one or both of the materials are poor conductors, the charges do not move freely, and the bodies retain charge after separation. A charged body can induce charge and charge separation in a neutral body. Charge induced in a conductor by an oppositely charged insulator could be considerably more consequential than the inducing charge that accumulated on the insulator.

Static electricity is characterized by low current and high voltage. The increase in potential caused by contact and separation easily can exceed 10,000 V (volts). Process materials, equipment, and the human body all can become charged. Accumulation of charge increases the strength of the associated electrical field. Accumulation beyond a critical value leads to electrostatic discharge. This discharge is similar in some respects to lightning. Electrostatic discharge through the air can occur by several mechanisms: complete electrical breakdown (spark discharge) or partial breakdown (corona discharge). A spark discharge occurs in a very small fraction of a second and gives a short, sharp, crackling sound. A corona discharge occurs over a longer time and may give a faint glow and a hissing sound. Both corona and spark discharges are capable of igniting a flammable mixture. The corona discharge usually is less hazardous than a spark discharge. Static arcs normally do not produce sufficient heat to ignite ordinary combustible materials such as paper. Some, however, are capable of igniting flammable vapors and gases and clouds of combustible dust. The role of static electricity in dust explosions is discussed in a later section.

The ability of an electrostatic spark to cause ignition depends on the delivery of energy to the flammable mixture (Scarbrough 1991). The minimum voltage needed to produce a spark across the shortest measurable gap of 0.01 mm under ideal conditions is about 350 V. Quenching by associated equipment and the surrounding atmosphere precludes ignition under this condition. However, incendive sparks induced across gap lengths in the range of 0.5 mm to 1.5 mm (which exceed the quenching distance), with voltages in the range of 1,500 to 5,000 V can produce ignition under ideal conditions. Most hydrocarbon mixtures require 0.25 mJ (millijoules) of energy for ignition under ideal conditions. Ignition energy is discussed in the next section. Ionization potential of air is about 30,000 V/cm.

Sparks from good conductors are more incendive than those from poor conductors or insulators. Charge flows along the contiguous surface of a conductor during discharge. On the other hand, charge flows only from a localized region of an insulator during discharge. For this reason, discharge from insulators or poor conductors usually is not implicated in fire and explosion.

Discharge occurs more readily when one, or both, of the oppositely charged objects are pointed (Lees 1980). Thus, a metal object protruding into a tank and electrically bonded to it can act as a discharge path. Objects in process operations that can act as discharge paths include: grounded probes, dipsticks and ullage tapes, tank-washing machines, and metallic objects that float on the surface of the liquid.

Actions that introduce pointed conductors into containers holding flammable liquids are especially hazardous. Activities during which this can occur include level gauging and drum sampling. Discharge between the charged liquid and the conductive object may occur in two ways. First, the object may act as a discharge path from liquid to ground directly or via the person. Alternatively, discharge may occur from the liquid to the tank. The hazard of charged liquids is reduced dramatically through use of antistatic additives or time allowance for charge relaxation. Nonconductive ullage tapes and permanently fixed and grounded sounding pipes that extend to within a few centimeters of the bottom of the tank reduce the hazard from level gauging.

Relative, as compared to absolute, humidity, influences the buildup and stability of electrostatic charge (Scarbrough 1991). Most fires attributed to static electricity occur indoors in winter when relative humidity is less than 30%. Electrostatic charge tends to dissipate when the relative humidity is kept above 50%. Under the latter condition, the surface conductivity increases sufficiently to prevent static accumulation. To illustrate, the surface conductivity of plate glass increases 1,000-fold when relative humidity increases from 20% to 50%. Humidification is effective only when surfaces are held at room temperature. Surfaces heated above room temperature, as in the case of dry textiles or other materials moved over heated surfaces, would be unaffected by humidification.

Many industrial processes involve contact, movement, and separation of poorly conducting materials (NFPA 1988). These processes may involve solid, liquid, or gaseous phases singly or mixed. Examples include:

- Fluid handling: pipeline flow, tank filling, agitation in process vessels
- Dust and powder handling: grinding, sieving, pneumatic conveying
- Sprays and mists: steam cleaning, steam leaks, spray washing, spray painting
- Moving equipment: conveyor belts, bucket elevators, web-handling of rolled materials

Each of the preceding processes can generate sufficient static electricity to ignite a flammable vapor.

Static effects in liquids can usefully be considered in terms of the classical theory of the electrical double layer at an interface (Lees 1980). According to this model, a layer of positive ions and a layer of negative ions accumulate at the interface between two immiscible liquids or the liquid and the wall of a container. Movement of the liquid produces unequal distribution of the ions and a resultant electrostatic charge. Situations causing charge separation include: pipeline flow, especially of immiscible liquids such as oil and water, settling of water droplets through oil, splash filling of tanks involving free fall, agitation of liquids in tanks, and splashing of oil droplets on the side of a tank. Filters, valves, and other constrictions in pipelines are points of generation of high levels of electrostatic charge.

The extent of charge separation depends on the resistivity of the liquid. Appreciable charge separation without immediate recombination can occur in liquids high in resistivity. Static electricity is a major concern in the handling of these liquids. Liquids having high resistivity include gasoline, kerosene, naphtha, benzene, and other white oils. Liquids having low resistivity include water, ethanol, and crude oil.

Most industrial dusts and powders are poor electrical conductors. Production and transport of these substances tend to generate static electricity. Operations in which this occurs include micronizing, grinding, mixing, sieving, gas filtration, pneumatic conveying, and mechanical transfer. The rate of generation is influenced by conductivity of the material, turbulence, interfacial surface area between materials, and presence of impurities. High generation rates are common in dispersing operations and when materials are mixed, thinned, combined, or agitated. Hazardous sparking can occur between poorly conductive materials and the agitator blade in a mixer or the conductive fill pipe in a ball or pebble mill.

In piping systems, the generation and accumulation of static electricity are functions of the materials, flow rate, velocity, and pipe dimensions (Gregg 1996). In filling operations, high rates of flow, turbulence in splashing or free-falling liquids, and powder fines contribute to charge accumulation. Disconnection of hoses and valves can result in hazardous discharge. Filters can generate levels of electrostatic charge 200 times higher than in pipe alone. The most serious mistake in powder and dust handling is failure to bond and ground equipment made from conductive material.

Process gases also can cause problems with static electricity (Scarbrough 1991). Steam, carbon dioxide, compressed air, and other process gases can generate static charge under certain conditions. The electrostatic hazard from steam arises from the formation of liquid water droplets through condensation. Strong electrification can occur at the orifice during escape of steam containing droplets of condensed water. When escaping under high pressure from an orifice as a liquid, carbon dioxide immediately changes to gas and solid (snow). This process can result in accumulation of static charge on the discharge device and the receiving container.

Carbon dioxide should not be used for the rapid inerting of flammable atmospheres by injection under high pressure for this reason. Compressed air or other process gas containing solid or liquid impurities also can produce strong electrification on escaping from an orifice. Bombardment of a conductive body by gas contaminated with dust, mist, scale, and metallic oxides can produce strong electrification of conductive fittings that are not bonded and grounded. Gas contaminated in this manner should not be used for purging or cleaning.

Static generated in this manner also can charge a nearby conductor that is insulated from ground (Lees 1980, Gregg 1996). Conductors insulated from ground can include:

- Tramp metal located inside vessels containing liquids and powders, screens
- Metal rims on nonconductive drums
- Probes
- Thermometers
- Spray nozzles
- High-pressure cleaning equipment
- Wire netting around lagging and insulated metal containers

Another aspect of static control is the type of container in which substances are handled and stored. Containers may be classified in order of increasing electrostatic hazard:

- Made from conducting material and grounded
- Made from conducting material and insulated from ground
- Made from insulating material

Containment in process plants usually involves grounded metal structures. A large electrostatic charge still can accumulate in the liquid and discharge through the containment to ground. A metal drum containing mixed oily waste that is sitting on a wooden pallet is an example of a metal container insulated from ground. In this case, a large charge can accumulate in the liquid and on the container. A metal sampling tube could provide a discharge path to ground. A similarly large charge may accumulate in a liquid held in a container made from an insulating material, such as plastic.

The conductive material of containers, piping, and other equipment provides a means to control the generation and buildup of static electricity (Scarbrough 1991). The main technique is bonding and grounding.

Bonding involves electrical connection between conducting objects. Electrostatic charging currents are small. Accumulation of a hazardous amount of charge through unimpeded flow to ground can generally be prevented by ensuring that resistance in the bonding system is low. The presence of paint, grease and oil, corrosion products, and rust on conductive elements can increase resistance, thus impeding flow and permitting charge accumulation. Bonding across flanged joints assures better electrical continuity than use of bolts that pass through the flange. This is important to consider during procedures that require opening of flanges to isolate lines from vessels.

Grounding drains away charge that develops or accumulates on objects under control of the bonded system. Grounding requires an electrical connection between a conducting object and ground. Ground means the earth. Bonding maintains the connected objects at the same electrical potential. Grounding a bonded system maintains the electrical potential of the system to the same level as the ground. Grounding in normal circumstances involves electrical connection to water pipes that pass through the ground or to rods of conductive metal that are driven into the ground. Grounding is usually effected using copper strips or wire.

No part of the electrical current-carrying system should ever be used as a ground (Gregg 1996). Fires have occurred where the ground for static control was tied into the neutral of the electrical system. Similarly, water pipes are poor candidates for grounds, as are underground systems equipped with cathodic protection. (Cathodic protection involves use of a sacrificial metal that corrodes to protect the metal of the system.) Disconnection for maintenance or modification renders the grounding capabilities of these systems inoperative.

The concept of grounding and equalization of potential usually receives little attention beyond this point in general discussion. Establishing an effective ground can be difficult to achieve in some circumstances. Poor soil conductivity and high resistivity can deter establishment of an effective ground. An ineffective ground could have disastrous consequences where high voltages and currents are involved. This situation can affect the safety and effectiveness of lightning protection systems and isolation of conductors involved in high voltage electrical transmission. A ground rod literally can become a missile, being propelled into the air by energy released during an ineffective grounding episode.

Equalization of potential is equally important in establishing an effective ground. Four-legged grazing animals, such as cattle, are considerably more sensitive to differences in ground potential than human bipeds because of the spacing between the hind legs and the mouth. These animals will refuse to graze where they detect even a slight difference in potential.

Precautions against charge accumulation on machinery can take several forms (Scarbrough 1991). In some cases, grounding is sufficient to prevent accumulation of electrostatic charge on moving machinery. Grounding is not sufficient for conveyor belts and transmission belts. These can accumulate considerable charge. The usual solution is to increase conductivity of the material of the belt by incorporating additives during manufacture or by use of dressing compound during operation. Charge-scavenging combs made from conductive metals can provide a discharge path.

Static eliminators based on ionization are used to prevent charge accumulation in many applications. The charge either drains to ground through ionized air or is neutralized by ions from the air. Air can be ionized by heat, ultraviolet light, electrical discharge, or radioactive substances. Equipment used for static elimination itself must not act as a source of ignition (Lees 1980).

IGNITION ENERGY

Ignition is the process of initiating self-sustained combustion. Ignition requires energy. The energy needed for ignition varies from substance to substance and depends on conditions. In general, molecules of both fuel and oxidizer require activation before they can undergo chemical reaction. Activation requires energy. The energy needed for activation can be obtained in two ways: an external source or an increase in the internal energy of the mixture as reflected by its temperature. As the temperature of the mixture increases, the amount of supplemental energy needed from the external source decreases. At a sufficiently high temperature, the mixture ignites spontaneously. No supplemental source of ignition is required. This process is called **self-** or **autoignition**. Ignition occurring through supply of energy from an external source — a flame, spark, or glowing object (ember) — is called **piloted ignition**. The temperature of the fuel–air mixture at which piloted ignition can occur is considerably lower than the autoignition temperature (Drysdale 1991).

The energy required to ignite a flammable mixture varies according to composition. This passes through a minimum in the middle of the flammable range (Zabetakis 1965). The minimum ignition energy is the minimum energy needed from an external source, usually a spark discharge to cause ignition. These values are provided in Table 4.7.

Flammable mixtures of hydrogen and acetylene require only 0.02 mJ to cause ignition. This compares to about 0.25 mJ required to ignite other hydrocarbons. These quantities are extremely small. Excellence in design of equipment is the reason that acetylene and hydrogen cause so few problems in everyday use. Equipment handling these gases contains features to control static and other electrical discharges.

The energy stored by the human body following acquisition of electrostatic charge ranges around 10 mJ. This assumes a potential of 10,000 V and capacitance of 200 pF (picofarads). Much higher voltages can develop in a dry environment. This buildup depends on clothing worn, humidity, materials of construction, and grounding (Lees 1980). A person charged to 15 kV could provide a discharge of 22.5 mJ. This level of energy is within the range produced by an ordinary spark plug (20 to 30 mJ). Thus, commonplace sparks and arcs can ignite flammable vapor and gas mixtures with energy to spare. More energy is required to ignite dusts (Bodurtha 1980).

Tabulated values for minimum ignition energies from spark discharges reflect optimized conditions, for example, a fixed gap between electrodes and fixed electrodes vs. breaking a contact. A spark gap of 2.5 mm requires less energy than a wider gap. The minimum ignition energy for sparks generated by broken electrical contacts is considerably higher than for fixed electrodes. Oxygen concentration and pressure also influence minimum ignition energy. Increasing the oxygen concentration and total pressure reduces the minimum ignition energy.

IGNITION TEMPERATURE

Ignition temperature of a substance, solid, liquid, or gaseous, is the minimum temperature at which self-sustained combustion in the absence of any source of ignition and independent of the heating or heated element can occur.

Ignition temperature depends on the conditions of measurement. Changing these conditions can have significant effect on the ignition temperature. Some of the variables known to affect ignition temperature include: percentage composition of the vapor– or gas–air mixture, shape and size of the space in which the ignition occurs, rate and duration of heating, kind and temperature of the ignition source, catalytic or other material that may be present, and oxygen concentration. Additional complicating factors include differences in test methods, size and shape of ignition chambers, composition of ignition chambers, method of heating and ignition source, rate of heating, residence time, and method of flame detection. That ignition temperatures are affected by the test method is not surprising. For this reason, measured values are considered only as estimates (NFPA 1991a).

Ignition is a complex process involving several stages. The National Fire Protection Association has identified the following benchmarks as definable during ignition. **Reaction threshold** is the lowest temperature at which any reaction of the sample or its decomposition products occurs for any fuel-to-air ratio. **Pre-flame reaction** is a slow, nonluminous gas-phase reaction of the sample or its decomposition products with the oxidant with which it is in contact. **Preflame reaction threshold** (RTT) is the lowest temperature at which exothermic gas phase reactions are observed for a particular system. **Cool-flame ignition** is a relatively slow, self-sustaining, barely luminous gas phase reaction between the sample or its decomposition products and an oxidant. Cool flames are visible only in a darkened area. **Cool-flame reaction threshold** (CFT) is the lowest temperature at which cool-flame ignitions are observed for a particular system. **Hot-flame ignition** is a rapid, self-sustaining, sometimes audible gas-phase reaction of the sample or its decomposition products

TABLE 4.7
Minimum Energy for Ignition of Substances and Mixtures

Substance	Energy (mJ)	Substance	Energy (mJ)
	Mixtures in Air		
Agricultural dusts		Chemicals	
Range	25–320	Acetone	1.15
Typical	40	Benzene	0.2
		Carbon disulfide	0.009
Carbonaceous		Cyclohexane	0.22
Asphalt	40	Diethyl ether	0.19
Carbon black	180	Diisopropyl ether	1.14
Charcoal	20	Neohexane	0.25
Coals	30–60	Dimethyl ether	0.29
		Ethyl acetate	1.42
Metal dusts		Furan	0.22
Aluminum	10–50	Heptane	0.24
Iron	20	Hexane	0.24
Magnesium	20–80	Methanol	0.14
Manganese	80–320	Methyl ethyl ketone	0.53
Titanium	10–40	*i*-Propyl alcohol	0.65
Zinc	100	Sulfur	15
		Tetrahydrofuran	0.54
Plastic Dusts			
Methylmethacrylate	15–20	Gases	
Nylon	20	Acetylene	0.017
Phenolic resin	10–25	Butane	0.25
Polycarbonate	25	Ethane	0.24
Polyethylene	10–30	Ethylene	0.07
Polypropylene	25–30	Ethylene oxide	0.06
Polystyrene	15	Hydrogen	0.017
Polyurethane foam	15	Hydrogen sulfide	0.068
Rayon	240	Methane	0.28
Urea formaldehyde	80	Propane	0.25
	Mixtures in Oxygen		
Chemicals		Gases	
Diethyl ether	0.0012	Acetylene	0.0002
		Ethane	0.0019
		Ethylene	0.0009
		Hydrogen	0.0012
		Methane	0.0027
		Propane	0.0021

After Frankel 1991 and Schwab 1991.

with an oxidant. A readily visible yellow or blue flame usually accompanies the reaction. **Autoignition temperature** (AIT) is now defined as the hot-flame reaction threshold temperature. **Catalytic reaction** is a relatively fast, self-sustaining, energetic, sometimes luminous, sometimes audible reaction that occurs as a result of the catalytic action of any substance upon the sample or its products of decomposition in admixture with an oxidant (NFPA 1991c).

Heretofore, ignition temperatures reported in the literature corresponded roughly to the AIT. Future reporting of ignition data likely will include CFT and RTT. Both temperatures are lower than AIT and are very significant factors in assessing the overall risk of autoignition in a given system. Cool flames are self-sustaining, exothermic ignition reactions. Under proper circumstances they may act as the source or point of interaction of more energetic hot-flame reactions. Pre-flame reactions have the capability under adiabatic or near adiabatic conditions to elevate the temperature of a fuel–air mixture to the point where cool- or hot-flame ignition may occur ("Adiabatic conditions" means change occurs without gain or loss of heat).

As part of the process of broadening the reporting criteria, NFPA and the American Society for Testing and Materials (ASTM) have altered methods for determining ignition. ASTM E 659, Standard Test Method for Autoignition Temperatures of Liquid Chemicals, has replaced ASTM D 2155 (ASTM 1978b). An earlier method, ASTM D 2883, Test Method for Reaction Threshold Temperature of Liquid and Solid Materials, provided for the study of autoignition phenomena at reduced as well as elevated pressures (ASTM 1983). Federal Test Method Standard 791b, Method 5050, also a current standard, provides for the measurement of autoignition properties in the same terms used by the ASTM methods. Former ASTM standards (ASTM D 286 and ASTM D 2155) provided only for visual detection of flame. As a result, ignition temperatures obtained by these methods were the minimum temperatures at which hot-flame ignitions occurred. The current methods (ASTM D 2883, ASTM E 659 and FTM Standard 791b, Method 5050) employ thermoelectric flame detection systems. These permit detection of nonluminous or barely luminous reactions that were difficult or impossible to detect by the older procedures (NFPA 1991c).

SPONTANEOUS (SELF) HEATING

Exposed to the normal atmosphere, practically all organic substances undergo oxidation, even at ambient temperatures. The rate of oxidation usually is so slow that heat of combustion is transferred to surroundings as rapidly as it is produced. The usual result is no appreciable increase in temperature of the substance (Drysdale 1991).

Under certain conditions, heat is retained in the material. Spontaneous or self-heating results in an increase in temperature of the material without input from the surroundings. If enough energy remains in the material, spontaneous heating can produce spontaneous ignition. Several factors control the severity of the process: nature of the substance, rate of heat generation, air supply, and geometry and insulating properties of the immediate surroundings (Drysdale 1991, Lees 1980).

Bulk materials in process, storage, or transport may undergo spontaneous heating. Examples include bulk materials handled in process driers, stored in piles in warehouses, or transported in large containers, such as ships. Well known is the spontaneous combustion of coal stored in piles or bunkers (Lees 1980). Self-heating is a significant source of ignition of dust. Dust in a pile has high surface area and controlled air circulation, both of which favor self-heating. Storage of a large quantity of dust at a high initial temperature creates the greatest hazard. A typical example is the discharge of a hot dusty product from a drier into a hopper (Bodurtha 1980).

Spontaneous heating is not likely to occur in volatile substances having access to the atmosphere. Vaporization removes the material, as well as the accumulating heat. Spontaneous heating is restricted to materials, such as oils, that have low volatility and combustible dusts.

Sufficient air must be available for oxidation, yet heat must not be removed by convection. Tightly packed materials might provide the necessary insulation, yet impede air movement. Predicting the occurrence of spontaneous heating under a given set of conditions is very difficult because of the many possible interactions between air supply and insulation (Drysdale 1991).

Substances undergoing oxidation first form intermediate products of oxidation. These may catalyze further oxidation. For example, oil that has become rancid oxidizes faster than fresh oil, due to the presence of intermediate products of oxidation.

Supplemental heat provided by other sources can promote the process by increasing the rate of oxidation. An additional source of supplemental heat is bacterial metabolism. This is a common cause of heating in agricultural crops stored in silos or wood chips stored in piles. Continued heating beyond temperatures generated by bacterial action occurs due to chemical oxidation. Moisture promotes bacterial activity. However, evaporation of the moisture removes heat. Agricultural products having a high content of oxidizable oils, such as cornmeal feed, linseed, rice bran, and pecan meal are susceptible to spontaneous heating.

A concern of major importance in process operations arising from self-heating is a lagging fire (Lees 1980). Lagging is the insulation applied to piping and process equipment. Lagging on plant equipment frequently becomes impregnated with oils and other liquids. Heat transferred from the equipment to the lagging can cause preheating of the liquid and spontaneous combustion. The temperature which can be attained in the lagging depends on geometry and temperature of the underlying equipment. Typical leakage points at which a lagging fire can occur include pumps, flanged joints, and sample and drain points.

The most important factor in a lagging fire is the liquid that impregnates the lagging. Significant self-heating requires several conditions:

- Liquid having low volatility
- Intrinsic reactivity such as unsaturated chemical bonds
- Low fire resistance — fire-resistant liquids containing antioxidants or hydraulic fluids containing water can become combustible if preferential loss of the antioxidant or water occurs in the warm lagging
- Favorable leakage rate
- Thickness of lagging — the relationship between pipe temperature and lagging thickness is the reverse of that required for heat insulation
- Time — time needed to attain critical temperature may be considerable, possibly months. Frequently, there is an induction period because of the presence of natural antioxidants in many materials.
- Insulation — a good insulating material has low thermal conductivity and porous structure of low density. Surface area, porosity, and heat retention favor self-heating.
- Protective coverings — impervious coatings, such as cement finishes, bituminous materials, or aluminum foil greatly reduce diffusion of oxygen into the insulation.

Several precautions can be taken against lagging fires. The most obvious is to prevent leakage into the lagging. This requires not only a high standard of operation and maintenance, but also the action of knowledgeable personnel. Additional measures include leaving bare known points of leakage and protecting lagging at critical points with metal collars. Another approach is to use insulating materials less prone to fire. This type of solution often introduces additional cost. A final alternative is not to insulate.

OXYGEN ENRICHMENT

The principal oxidizer in common experience is oxygen. Oxygen occurs in the normal atmosphere at a concentration of 20.9% and partial pressure of 21 kPa (159 mmHg) at sea level. Oxygen in process situations can occur at concentrations and pressures considerably higher than those encountered under ambient conditions. Oxygen sometimes is used as a process gas, for example, in steelmaking and the bleaching of pulp. Processes could involve transport within pressurized lines. Small quantities of compressed oxygen, relative to process applications, are used, sometimes with disastrous consequences, in medical applications and in oxy-fuel torches.

The likelihood of ignition and the rate of flame propagation of a combustible are greatly influenced by the oxygen content of the atmosphere. Oxygen is a nonflammable gas, meaning it

TABLE 4.8
Effect of Oxygen Enrichment on Combustibility/Flammability

Pressure (mmHg)		
Oxygen	**Total**	**Comments**
160	760	Normal atmosphere, sea level, dry air
Range	760	Decrease in autoignition temperature of hydraulic fluids with increase in partial pressure of oxygen
Range	760	Decrease in autoignition temperature of lubricants with increase in partial pressure of oxygen from less than normal through 760 mmHg
236	760	Increase in combustibility in oxygen/nitrogen mixture of materials (fabrics, paper, polymers) that did not burn in normal atmosphere
258	760	Considerable increase in flame spread rate in combustible materials (fabrics and polymers)
319	760	Decrease in ignition temperature of combustible fabrics and sheeting
760	760	Slight decrease in autoignition temperature of most hydrocarbon fuels, solvents, and anesthetic gases; broadening of flammable range by increase in upper flammable limit

Data from Hugget et al. 1965, Johnson and Woods 1966, Kuchta et al. 1967, Kuchta and Cato 1968, Frankel 1991.

does not burn. In general, oxygen enrichment significantly increases the fire hazard of materials, although this is not guaranteed (Frankel 1991). Oxygen enrichment widens flammability limits by decreasing the LFL and increasing the UFL (Zabetakis 1965). The LFL in oxygen is not markedly affected, since the concentration of oxygen in the normal atmosphere already exceeds combustion requirements. The UFL increases markedly in oxygen-rich atmospheres, tending to be above 50%.

The fire hazard in an oxygen-enriched atmosphere is significantly greater than that in a normal atmosphere (Frankel 1991). This is due in part to the reduction in minimum energy needed for ignition and the greater rate of flame spread. That is, combustible materials ignite more easily and burn more rapidly in an oxygen-enriched atmosphere. Generally, ignition energy decreases with increasing oxygen concentration. The rate of flame spread increases with the increase in the oxygen concentration at constant pressure or with the increase in the total pressure at constant percentage of oxygen (increased oxygen partial pressure).

The autoignition temperature (AITs) of most hydrocarbons tends to be slightly less in oxygen than in air. In the case of lubricants and hydraulic fluids, this decrease is significant. These products pose significantly greater fire hazard in oxygen-enriched atmospheres than under normal conditions. Minimum ignition energy decreases significantly in an oxygen-enriched atmosphere. Ignition occurs more readily in an oxygen-enriched atmosphere.

In an enriched atmosphere, even at normal pressure, oxygen adsorbs onto fabrics and the skin, thus increasing combustibility. In general, ignition temperature and flame resistance are lower in an oxygen-enriched atmosphere. OSHA documented fatal accidents in which oxygen enrichment occurred through inadvertent or deliberate release of pressurized oxygen (OSHA 1985). The resulting fires indicated that risk of combustibility is greatly enhanced, even at normal atmospheric pressure. Almost all materials will burn in a pure oxygen environment. This would change the NFPA classification of some materials from nonflammable to flammable/combustible. This situation can seriously challenge presumptions about safety in selection of materials for use in oxygen service.

Table 4.8 summarizes the effects on combustibility and flammability caused by exposure of substances, fabrics, and polymers to an enriched oxygen atmosphere.

Lubricants and hydraulic fluids are the most sensitive of the types of substances for which information is available. In the case of lubricants, this sensitivity changes from oxygen deficiency through normal concentrations through oxygen enrichment. The lowest of the tested partial pressures corresponded to a concentration of 31% oxygen relative to the sea level dry atmosphere.

OXYGEN DEFICIENCY

Many process environments contain little, if any, oxygen. However, even in these circumstances, the concentration of oxygen could exceed combustion requirements. At the LFL, oxygen concentration definitely exceeds combustion requirements. Decreasing the oxygen content of a flammable mixture decreases heat generation. There is a minimum concentration of oxygen below which flame will not propagate.

Depletion of oxygen by dilution is a technique exploited in fire and explosion methods (Bodurtha 1980, Zabetakis 1965, Wysocki 1991). Ignitability is suppressed by addition of inert gases such as nitrogen, carbon dioxide, or steam (Zabetakis 1965). In general, carbon dioxide reduces the height of the flammability envelope compared to nitrogen per unit of volume. Many mixtures can be rendered nonflammable by adding inerting gases, such as carbon dioxide and nitrogen. For most flammable mixtures, the inerting concentrations of carbon dioxide and nitrogen are 28% and 42%, respectively. The lesser quantity of carbon dioxide required for fire suppression reduces the oxygen level to a lesser extent. The greater safety and efficiency in fire suppression is the basis for use of carbon dioxide in fire extinguishers. Steam, carbon dioxide, and nitrogen are used to purge and ventilate process vessels prior to and after opening. The use of steam and carbon dioxide requires great care and consideration where flammable/combustible substances are present because of the potential for buildup of static charges. Condensate from steam and solid carbon dioxide formed during high pressure release to atmosphere are potential sources of static electricity. Steam is not recommended for inerting in process or marine environments (NFPA 1993d, API 2015 1993).

IGNITABILITY DIAGRAMS

The relationship between composition of air–vapor mixtures and ignitability often is a major concern regarding work involving confined spaces. One approach to addressing these concerns is provided by the ignitability diagram (Figure 4.2).

This provides information concerning ignitability of a vapor–air mixture for a specific temperature–pressure combination (Zabetakis 1965, AGA 1975). These diagrams also include information about the effect caused by the addition of inerting agents, such as steam, carbon dioxide, nitrogen, and helium. This information can be shown on a triangular plot, but more customarily by rectangular coordinates, since the sum of all components in a mixture is 100%. These diagrams apply to mixtures containing flammable or combustible vapors and gases. The maximum concentration of vapor is the concentration at the saturation vapor pressure for the specific temperature of interest. The concentration of oxygen in an oxygen–vapor–inert mixture is given by:

$$\text{Oxygen}\ (\%) = 100\% - \text{flammable/combustible}\ (\%) - \text{inert}\ (\%) \tag{4.9}$$

The corresponding concentration of oxygen in a mixture containing air is given by:

$$\text{Air}\ (\%) = 100\% - \text{flammable/combustible}\ (\%) - \text{inert}\ (\%) \tag{4.10}$$

$$\text{Oxygen}\ (\%) = 0.21 \times \text{Air}\ (\%) \tag{4.11}$$

While this information strictly applies only to systems that have equilibrated and are well mixed, the data do provide an indication about how conditions could develop in nonequilibrated systems.

These diagrams contain an envelope that encloses all of the possible compositions of ignitable mixtures in air or oxygen. Bordering one side of the envelope is the fuel composition axis. Defining

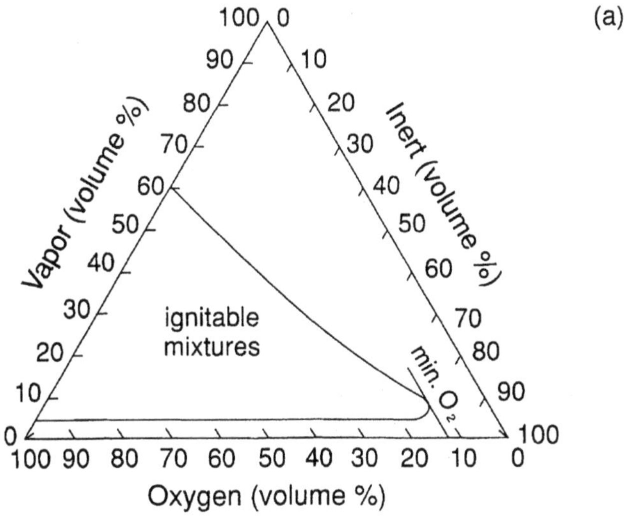

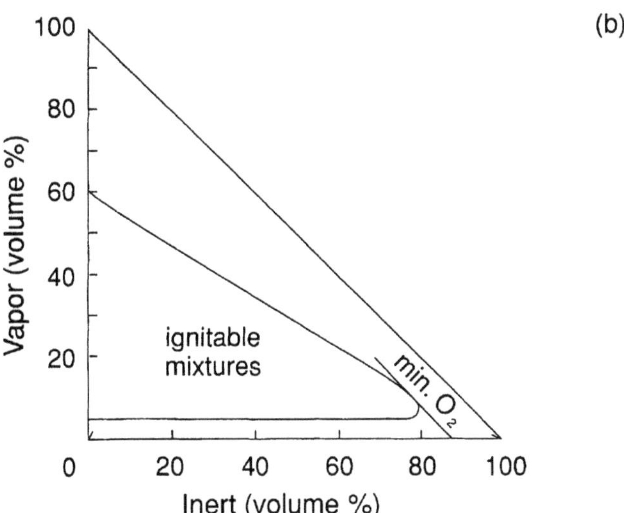

FIGURE 4.2 Ignitability diagrams: (a) triangular coordinates; (b) rectangular coordinates. Flammability diagrams provide a means of predicting ignitability of mixtures of gases and vapors following changes in composition. These diagrams apply only when homogeneous conditions exist in the mixture. (Adapted from Zabetakis 1965.)

the width of the envelope are the LFL and UFL. The diagram also indicates the minimum concentration of oxygen (air) needed to support combustion as a tangent to the curve that forms the envelope. External to the envelope within the boundaries of the diagram are the compositions of other possible mixtures.

Addition of one or more of the original components results in an increase in volume and change in composition. Adding one of the components yields all compositions along the line joining the starting composition to the vertex of the triangle at 100% composition (fuel, diluent, or oxygen [air], if displayed). This occurs as the mixture enriches in that component at the expense of the others. Addition of two of the components is considered stepwise, as the addition of each sequentially. Where oxygen (air) is added and is not listed as an axis in the diagram, 0% diluent or inert is used as the vertex for 100% oxygen.

Similar use of construction lines can be applied to predict the composition of mixtures under real-world situations. In any situation where the composition line passes through the envelope that encloses ignitable mixtures, an ignitable mixture will develop at some point during the change. Some applications for these diagrams include:

- Purging a container holding an air–vapor or an inert–vapor mixture with air or inert gases such as steam, nitrogen, or carbon dioxide
- Effect of hot water washing and purging with air on formation of ignitable mixtures
- Escape of vapor or a vapor–inert or an air–vapor–inert gas mixture from a container
- Maximum vapor to inert ratio in nonignitable mixtures

These diagrams can be modified to suit conditions other than ambient.

The minimum oxygen concentration (MOC) needed for combustion and other quantities can be measured experimentally or estimated for pure substances containing carbon, hydrogen, and oxygen, using the following equations (Jones 1938):

$$C_nH_xO_y + \left(n + \frac{x}{4} - \frac{y}{2}\right)O_2 \rightarrow n\,CO_2 + \frac{x}{2}H_2O \tag{4.12}$$

$$\text{Oxygen} = \left(n + \frac{x}{4} - \frac{y}{2}\right) \text{ mol/mol of fuel} \tag{4.13}$$

$$\text{Air} = \frac{100\%}{20.95\%} \times \left(n + \frac{x}{4} - \frac{y}{2}\right) \text{ mol/mol of fuel}$$

$$= 4.77\left(n + \frac{x}{4} - \frac{y}{2}\right) \text{ mol/mol of fuel} \tag{4.14}$$

$$C_{st} = \frac{100}{4.77\,n + 1.19\,x - 2.38\,y + 1} \%\ v/v \tag{4.15}$$

$$\text{LFL} \approx \frac{55}{4.77\,n + 1.19\,x - 2.38\,y + 1} \%\ v/v \text{ at } 25°C \tag{4.16}$$

$$\text{MOC} = \text{LFL} \times \left(n + \frac{x}{4} - \frac{y}{2}\right) \%\ v/v \tag{4.17}$$

C_{st} = the stoichiometric concentration. The line joining C_{st} on the fuel axis to 0% oxygen intersects the MOC and is used in creating the flammability diagram. The lower flammable limit for organic substances in air is about 55% of the stoichiometric concentration in air.

MOC = minimum oxygen concentration. The intersection of MOC and the C_{st} line approximates the outer limit of the flammability envelope.

Similar, but more complex equations are also available for halogenated hydrocarbons (Zabetakis 1965).

For combustion of a mixture of discrete ingredients, MOC is given by:

$$\text{MOC} = \frac{\text{LFL}_{\text{mixture}} \times \left[\left(n + \frac{x}{4} - \frac{y}{2}\right)_1 + \left(n + \frac{x}{4} - \frac{y}{2}\right)_2 + \ldots + \left(n + \frac{x}{4} - \frac{y}{2}\right)_N\right]}{N} \tag{4.18}$$

$LFL_{mixture}$ = the lower flammable limit for the mixture calculated using Le Chatelier's rule (Equation 3.6)

N is the number of discrete substances.

EXPLOSION

An explosion is an event characterized by the sudden release of energy (Cruice 1991). Explosions result from chemical and physical processes. Detonation of an explosive or rapid combustion of a cloud of flammable vapor or gas are examples of explosions resulting from chemical processes. Rupture of a container caused by overpressure or rapid vaporization and uncontrolled escape of a liquid or vapor are examples of explosions resulting from physical processes. The effects produced by an explosion occur due to at least one of the following processes: production of gases, rapid expansion of gases, and motion of projectiles.

One of the main characteristics of an explosion is the shock wave or blast wave. The shock wave is characterized by the rapid change in atmospheric pressure caused by rapid expansion of gases. The shape of the pressure profile depends on the type of explosion. The shock wave produced during an explosion moves outward at subsonic or supersonic velocities. Again, this depends on the type of explosion. A subsonic shock wave results from a deflagration, one type of explosion involving gas/vapor–air mixtures. During a deflagration, the pressure equalizes at the speed of sound following the overpressure. The pressure drop across the flame front is relatively small. Supersonic propagation of the flame from the source of ignition occurs during a detonation explosion. During a detonation, the rate of pressure equalization at the flame front is less than the rate of propagation. This results in considerable pressure drop across the flame front (Zabetakis 1965).

During an explosion, atmospheric pressure rises almost instantaneously to a peak and gradually decreases to the starting level. As the shock wave travels outward, the height of the peak of the pressure wave decreases at the shock front with increasing distance. The shock wave in air usually is referred to as a "blast wave" because it may be accompanied by a strong wind. The peak wind velocity behind the shock front depends on the peak overpressure. Overpressures generated by the blast wave can injure or kill people and damage or destroy equipment and buildings. The overpressure in the shock wave is followed by a region of negative pressure, or underpressure. The underpressure usually is quite weak and usually does not exceed about 100 kPa (760 mmHg) gauge. The high-pressure components of the shock wave move outward at higher velocities. Initially, the shock wave from a detonation travels at supersonic speed. As the intensity of the wave subsides, it becomes sonic (Lees 1980, Bodurtha 1980). Table 4.9 provides information about the effects of blast-produced overpressures.

Ordinary buildings can be demolished with relatively slight overpressures, as indicated from the table. Not surprisingly, even reinforced concrete buildings have been demolished by these types of explosions.

PHYSICAL EXPLOSIONS

Physical explosions result from physical events, rather than rapid chemical reactions (Cruice 1991). Mechanical means or other physical events are the source of all of the high-pressure gas.

Vapor Explosions

A vapor explosion results from rapid heating (superheating) and rapid vaporization of a liquid in a confined geometry (Bodurtha 1980). Superheating results from contact between a hot liquid and a cold liquid. The energy source is the sensible heat plus the heat of fusion of the hot liquid. A liquid may become superheated under hydrostatic pressure or in a boiling regime that favors

TABLE 4.9
Effects of Overpressure

Overpressure (kPa)	Structural element	Failure condition
	Physical effects	
3.4 to 6.9	Glass windows	Shattering, occasional frame failure
6.9 to 14	Corrugated asbestos siding	Shattering
	Corrugated steel or aluminum siding	Connection failure + buckling
	Wooden exterior siding panels; standard house construction	Failure at main connections; panels blown in
14 to 21	Concrete or cinder block wall; 200 to 300 mm thick; not reinforced	Shattering of the wall
48 to 55	Brick wall; 200 to 300 mm thick; not reinforced	Shearing and flexure failures
>70	Structure	Total destruction
1930	Ground	Crater formation
	Physiological effects	
15	Knockdown	Knockdown threshold
34	Eardrum	Threshold of rupture
207 to 255	Lung	Threshold of damage
240	Human body	Threshold of fatal injury
345	Human body	50% fatalities
450	Human body	99% fatalities

Data from Robinson 1944, Glasstone 1962, after Cruice 1991.

superheating. Normally the release of superheat energy occurs without great violence through generation of nucleated bubbles, as in normal heat transfer to a boiling liquid. In some cases, the release of vapor is considerably more violent.

The cold liquid generally is water. The hot liquid often is molten metal. Molten salts also can produce vapor explosions if they are quenched or come into contact with water. The usual mode of occurrence follows from a spill of molten metal onto a puddle of water on the floor. A vapor explosion also can occur when water is poured onto the melt. A layer of water at the bottom of a tank of hot oil can produce vapor in the same manner. This has produced boilover, slopover, and frothover during some serious fire situations involving burning liquids (Sly 1991).

Water may act as the hot liquid if it comes into contact with a relatively cold liquid having a low boiling point. Some examples include contact between chlorodifluoromethane or LNG (liquified natural gas) and water. In the latter case, the superheat energy may be released explosively, although without ignition.

Another type of vapor explosion occurs when a hot pressurized liquid is suddenly depressurized. The bulk liquid may undergo a spontaneous nucleation with sudden vaporization and development of a shock wave (Bodurtha 1980).

Containment Rupture

Containment rupture occurs when the internal pressure exceeds the strength of the containment (Cruice 1991). Overpressure can result from the heating of liquids and gas/vapor in the container. Vessel failure generally occurs at four times the allowed working pressure. Failure occurs at the weakest point. This can result in expulsion of projectiles or flinging open of walls. Gas release is extremely rapid under these circumstances. The pressure wave and projectiles from the vessel are highly directional. Damage potential is approximated by the product of the volume of gas and the pressure at time of failure. Damage potential can be assessed in units of TNT-equivalents. This is the amount of TNT needed to produce a shock wave of equivalent intensity.

Boiling Liquid Expanding Vapor Explosions

Another important type of physical explosion is the boiling liquid expanding vapor explosion (BLEVE). This occurs when a container is heated so that the containment ruptures and discharges a stream of boiling liquid. In the lower pressure environment outside the containment, the liquid vaporizes. This is especially critical where the boiling point of the liquid is low, as in the case of liquified gases (superheated liquids), compared to ambient temperatures. The aftereffects of a BLEVE depend on the flammability of the liquid in the vessel. In all cases, the initial explosion may generate a blast wave and missiles. If the liquid is flammable, a fire always occurs. The boiling liquid vaporizes and forms a vapor cloud giving rise to a second explosion (Lees 1980). The definition of the term, BLEVE, indicates that mechanisms for the release of energy are strictly physical, including the effects of flying missiles and blast. Combustion following release of a flammable or combustible liquid is a secondary occurrence (Cruice 1991).

In most BLEVEs the failure of containment originates in the metal of the vapor space (Lemoff 1991). The metal stretches and thins. A longitudinal tear develops and progressively lengthens to a critical length. At this point, the metal fatigues from the heat of the impinging fire and pressure exerted by contained superheated liquid. The failure propagates at sonic velocity in both longitudinal and circumferential directions. As a result, the container often ruptures in two or more places.

In a fire situation, a BLEVE will occur if one of two conditions exist. First is the lack of a pressure relief device or presence of an undersized one. The second condition is impingement of the fire on the metal above the liquid in the container. The magnitude of a BLEVE depends on the quantity of liquid that vaporizes when the container fails and the weight of the pieces. Most BLEVEs involving liquefied gases occur when the container is half to three quarters full. Under these conditions, pieces can be propelled up to 1 km.

All liquefied gases are stored in containers at temperatures above their boiling points at normal temperatures and pressures. This pressure ranges from less than 7 kPa for some cryogenic gases to more than a thousand kPa for noncryogenic liquefied gases. Container failure reduces the internal pressure to atmospheric. Rapid vaporization of a portion of the liquid occurs. The amount is related to the difference in temperature between the liquid at the instant of container failure and the normal boiling point. For many liquefied flammable gases, this can result in vaporization of a considerable fraction of the liquid. The liquid remaining is refrigerated by "self-extraction" of heat and cooled to near its normal boiling point.

Accompanying the vaporization is the liquid-to-vapor expansion. This expansion provides the energy for explosion of the container, atomization of the remaining liquid, and rapid mixing of the vapor and air that results in the characteristic fireball upon ignition. Many atomized droplets burn as they fly through the air.

Chemical Explosions

Chemical explosions result from the evolution of gaseous products during a rapid chemical reaction (Cruice 1991). The products differ substantially from the reactants.

Explosives

An explosive is a substance or mixture of substances that when subjected to the appropriate stimulus, undergoes an exceedingly rapid self-propagating reaction (Porter 1991). The products of the reaction usually include stable gases, heat, and a blast wave. The explosive release of these gases occurs in microseconds. The blast wave results from the action of heat on the gaseous products. Commercial explosives have been developed to exploit the properties of instability inherent in some materials in a controlled (read safe) manner.

The process undergone by an explosive following detonation is a redox (reduction–oxidation) reaction in which the explosive acts as both oxidizing and reducing agent. Gases produced during decomposition of explosives can include: carbon dioxide, carbon monoxide, nitrogen, nitrogen oxides, oxygen, steam, and sulfur dioxide (Meyer 1989). The oxygen is generated during the decomposition of the solid or liquid material that constitutes the explosive and does not originate from the atmosphere. Further, atmospheric oxygen is unnecessary for the decomposition. The gases produced following the detonation expand to 10,000 times the original volume of the material at the elevated temperature and pressure of the reaction.

Primary or initiating high explosives include compounds that are inherently unstable: mercury fulminate (cyanate), lead styphnate (salt of 2,4-dihydroxy-1,3,5-trinitrobenzene), and lead azide. More stable explosive compounds include organic and inorganic nitrates, amine and nitro compounds, and their metal salts.

Flammable Gas/Vapor Explosions

Fire occurs in situations where fuel and oxidizer are initially unmixed. Mixing of fuel and oxidizer is controlled by the combustion process itself. Burning rates are restricted primarily by the supply of fuel and oxidizer, rather than the rate of elementary chemical reactions occurring within the flames. These reactions generally occur so fast that they immediately consume all available fuel and oxidizer. The basic gas-phase process usually occurs along thin flame sheets, called "diffusion flames." These separate regions rich in fuel vapor from regions rich in oxidizer. Fuel vapor and oxidizer diffuse toward the flame sheet. Combustion products and heat, in turn, diffuse away from the flame sheet (Drysdale 1991).

Explosions involving flammable mixtures generally occur only where fuel and oxidizer have mixed intimately before ignition (Cruice 1991). As a consequence, combustion occurs very rapidly. Two types of mixtures can result: homogeneous, uniform mixtures and heterogeneous, nonuniform mixtures. The components of a homogeneous mixture are intimately and uniformly mixed. A small sample is representative of the entire mixture. Examples of substances able to form homogeneous mixtures include gases and vapors. Other types of flammable or combustible mixtures are heterogenous. Examples include mists, foams, and dusts. Gas– and vapor–air mixtures have both flammability limits (deflagration) and explosive limits. Explosive limits depend upon the initiating stimulus and the environment. They usually are slightly narrower than flammability limits. The most intense explosions of vapor–air mixtures occur near the middle of the flammable range. Explosions involving flammable vapor–air mixtures most frequently occur in confined spaces.

Explosions resulting from a chemical process are classified as deflagrations or detonations. (Bodurtha 1980). The violence of gas or vapor explosions depends upon the nature of the substance, the quantity of the mixture, and the enclosure confining it.

A deflagration is an exothermic reaction that propagates from the burning gases to unreacted material by conduction, convection, and radiation. The combustion rate during a deflagration is less than the velocity of sound. A detonation is an exothermic reaction characterized by the presence of a supersonic shock wave. The shock wave establishes and maintains the reaction. The reaction zone propagates at a rate greater than the velocity of sound. The principal mechanism of heating is shock compression. The temperature increase is directly associated with the intensity of the shock wave, rather than thermal conduction.

The pressure rise associated with a deflagration in an unvented vessel is approximately 8 atmospheres (800 kPa) (Zabetakis 1965, Bodurtha 1980). The increase may be as much as 20 atmospheres (2,000 kPa) in fuel–oxygen systems. The corresponding pressure rise during a detonation is approximately 40 atmospheres (4,000 kPa).

Ignition energy required for a detonation is considerably greater than for a deflagration. If a sufficiently energetic source is present, detonation can occur immediately upon ignition (Zabetakis 1965). The occurrence of a detonation during a process accident is likely associated with special circumstances (Lees 1980).

Explosions in Contained Systems

While not common, fires and explosions have occurred in process equipment, such as gas compression systems (Burgoyne and Craven 1973, Perlee and Zabetakis 1963, Anon. 1959). The causes relate to the process of compression or to operation of the compressor. Compression of gas or vapor–air mixtures in the presence of air can lead to explosion. Air should be purged from systems handling flammable gases or vapors with inert gas to prevent this occurrence. Intakes must be located carefully to avoid drawing in air containing hydrocarbon vapors. Special care is required in the compression of endothermic gases such as acetylene. Another potentially dangerous condition can occur when a centrifugal pump containing a flammable liquid and an air pocket is run deadheaded. Adiabatic compression of the gas pocket may develop temperatures above the autoignition temperature or cause decomposition of the liquid.

Another cause of compressor explosions is the compressor itself. Generally, the type of compressor involved in these explosions is an oil-lubricated reciprocating compressor. Carbonaceous deposits in the vicinity of final-stage outlet valves and oil films appear to be the principal sources of combustibles. Excessive carbon deposits may result from an improper oil feed rate.

A suspension of finely divided drops of combustible or flammable liquid in air can form in gas compression systems from condensation of saturated vapor or atomization of liquid by mechanical forces (Lees 1980). Explosions in compressed air systems are essentially oil mist explosions. An explosion occurs because of the presence of oil mist and condensed oil in the system. Often the compressor has a record of faulty operation and high outlet temperature prior to the explosion. The risk of explosion is not completely linked to abnormally high oil usage.

The explosion that occurs in a compressed air line having a thin film of flammable oil is unique. Typically, the air line ruptures at intervals along its length. The explosion is a detonation. Dispersing the oil film into a mist by some primary shock or explosion creates the conditions for a more powerful secondary explosion. High outlet temperature at the compressor can vaporize and ignite the oil. Sudden release at high pressure can cause simultaneous formation of mist and ignition. Carbonaceous residues can undergo self-heating and ignite. Oil film explosions have occurred in the compressed air starter systems of large diesel engines. In these cases, the explosion generally appears to have been initiated by the diesel engine, rather than the air compressor.

Crankcase explosions in engines and compressors involve oil mists. The suspension of oil mist in a crankcase during normal operation originates from two sources: mechanical spray and condensed mist from the lubricated parts. Lubricated parts are hotter than the average temperature in the crankcase. Serious overheating accelerates mist formation, a condition often referred to as "smoke." Ventilation and injection of inert gas are measures to suppress flammability.

A flame arrester, or flame trap, is a device to prevent the passage of a flame along a pipe or duct (Bodurtha 1980). Most flame arresters are an assembly of narrow passages through which gas or vapor can flow, but which are too small to allow the passage of flame. Desirable properties of a flame arrester include large, free, cross-sectional area, low resistance to flow, freedom from blockage, high capacity to absorb the heat from the flame, and the ability to withstand mechanical and explosive shock. Typical applications include vents on storage tanks, piping systems that supply fuel gas to burners, pipelines and flare stacks, and storage cabinets. Flame arresters are also used on exhausts of internal combustion engines working in atmospheres with a flammability hazard and on crankcases of small internal combustion engines.

Vapor Cloud Explosions

Previous discussion has centered on gases and vapors confined in experimental apparatus or process or industrial equipment. Gases and vapors that have escaped from containment also can pose considerable explosion hazard. The more familiar cause of explosion involves ignition of flammable mixtures that have escaped from containment as a cloud. The majority of vapor cloud explosions documented in recent years resulted from ignition of heavier-than-air vapors. These tend to stratify near the ground and resist dispersion. Release could occur from vent stacks above ground level, pumps, valves, and other sources at ground level or below grade in pits, pump rooms, sumps, chemical sewers, and other containments. Flammable concentrations from releases of dense vapors at the ground seldom extend to appreciable heights. Less dense gases rise in the atmosphere as a result of buoyancy and generally do not accumulate at low levels. Vapor cloud explosions usually result from leaks of flashing liquids — liquids stored under pressure at temperatures above their atmospheric boiling points. These liquids vaporize rapidly at the higher temperature and lower pressure outside containment (Kletz 1977). The source of ignition can be a considerable distance from the point of release.

Rate of release, rather than quantity, is the primary criterion for assessing the potential hazard of continuous releases. Concentration remains constant once steady state is achieved. High concentrations can occur at large distances from the release point of instantaneous emissions. These concentrations are short-lived at any fixed location because of the brevity of the emission. An unconfined vapor cloud explosion or turbulent wind could disperse more concentrated sections of the cloud to form puffs. A puff expands as it moves downwind. This could dilute unreacted vapor previously above the upper flammable limit into the flammable range. Total quantity of vapor in the puff is the critical factor (Bodurtha 1980).

When ignited, the flammable mixture burns rapidly and produces heat rapidly. The heat is absorbed by the fuel, combustion products, and other gases in the vicinity of the flame. Confinement by structures restricts the ability of the heated gases to expand, thereby increasing pressure and the severity of the explosion. Oxygen enrichment further increases explosion pressure.

Accumulation of gas or vapor in a structure is affected by the rate of emission, density of the gas/vapor, and ventilation rate. Classic laws of diffusion are inapplicable under actual conditions because the combination of extremely slow release rates and airtight structures is seldom encountered. The flammable mixture occupied less than 25% of the structure in most explosions. In addition, most flammable mixtures are approximately 90% air. The density of the mixture is approximately the same as the density of air, regardless of the density of the gas or vapor. Hence, density of the gas or vapor is seldom a significant factor in gas explosions involving structures (Lemoff 1991).

Dust Explosions

The hazard of a dust explosion exists wherever ignitable dusts are handled. Historically, industries particularly affected by dust explosions have been flour milling, grain processing and storage, and coal mining and processing (Lees 1980). Stringent regulations that control releases into the environment have introduced new opportunities for dust explosions. There is a growing trend toward installation of dry collectors, such as baghouses and electrostatic precipitators, to handle these materials, which can include large quantities of combustible dust. Explosions and fires have occurred in these installations (Bodurtha 1980).

A dust explosion can occur only when the dust is dispersed in air. Transition from a dust fire to a dust explosion or vice versa can occur. Burning particles from an explosion may act as the source of ignition for a fire of other flammable materials (Lees 1980).

The ability of detonation explosions to occur in normal industrial situations has not been established. Combustion in a dust explosion is very rapid. The flame speed is comparable with that in gas deflagrations. Maximum explosion pressures are often close to theoretical values calculated assuming no heat loss during the explosion. Most of the evidence for detonation relates to coal mines, where a strong ignition source initiates the explosion. Industrial plants generally have weak ignition sources. The general practice for protection against dust explosions in industrial plants is to assume the occurrence of deflagration, rather than detonation. This procedure has proved to be satisfactory in practice (Palmer 1973).

An industrial dust explosion often involves two stages: a primary and a secondary explosion. The shock from the primary explosion dislodges settled dust from horizontal surfaces or ruptures dust handling and storage equipment in adjacent areas. Burning material produced during the primary explosion acts as the ignition source for the newly formed, unexploded cloud. The quantity of dust in the secondary explosion often exceeds that in the primary one. Secondary explosions often are more destructive than the primary one. The possibility of a highly destructive secondary explosion makes dust explosions rather unpredictable (Lees 1980). Venting dust-collecting equipment inside a building could create serious repercussions in the progress of a primary explosion. Where feasible, dust handling and processing equipment should be installed outdoors or at least vented outside through as short a duct as possible (Bodurtha 1980).

As with flammable gases and vapors, dusts explode within a specific range of concentration (Dept. of the Interior 1954, 1962, 1965). Generally, the lower explosible limit (LEL) of fine solid organic materials in air is about 20 g/m^3. This loading in air is described as a very dense fog.

Particle size is extremely important in this determination. The test specified by the U.S. Bureau of Mines uses dust that passes through a 200-mesh screen (74-μm particle size or smaller). Dusts coarser than 200 mesh have high LEL. The effect is negligible for finer dusts (Dept. of the Interior 1961). In general, dusts having particle size larger than 500 μm do not explode (Dept. of the Interior 1968). Sample purity, oxygen concentration, strength of ignition source, turbulence of dust cloud, and uniformity of dispersion also effect the LEL (Schwab 1991). Upper explosible limits (UEL) for dust clouds have not been determined. This results from experimental difficulties. However, this information would have limited applicability. UELs are estimated to be about 4 kg/m^3 (Bodurtha 1980). Dust suspensions in industrial equipment often occur above the UEL. Using the UEL as a design parameter is not generally practical, since the LEL is the concentration of prime interest (Lees 1980). The most violent dust explosions occur at concentrations slightly above that required for reaction with all of the oxygen in the atmosphere. At lower concentrations, less heat is generated and smaller peak pressure develops. At higher concentrations, absorption of heat by unreacted dust apparently reduces explosion pressures (Schwab 1991).

The minimum spark ignition energy (MIE) of combustible dusts is about 100 times that of flammable vapors or gases. MIE is the minimum spark energy from a condenser required to produce flame propagation 100 mm or longer in the test apparatus. MIEs are determined at the most easily ignitable concentration, generally 5 to 10 times the LEL for the dust. Consequently, ignition energies for marginally explosible concentrations are much higher than published MIEs. MIE values are affected by many factors, including particle size and the presence of flammable or combustible vapors. Dusts wetted by flammable or combustible solvent or suspended in a vapor–air mixture can be ignited more easily. Sources in industrial settings having sufficient energy to ignite dust with an MIE >500 mJ are uncommon (Dept. of the Interior 1960).

Ignition temperature of a dust cloud is determined in a Godbert-Greenwald furnace. The ignition temperature measured by this determination is the hot-surface temperature in the furnace, not the lower (and unknown) temperature of the dust–air mixture. Thus, ignition temperature, while not the autoignition temperature, is similar to the hot-surface ignition temperature for vapors and gases. Natural substances tend to have lower ignition temperatures than synthetic products (Dept. of the Interior 1961, 1960). Ignition temperatures required for dust explosions are much lower than the temperatures and energies available in most common sources of ignition. As with flammable vapors and gases, dust clouds can be ignited by common sources of high temperature: open flames, lightbulbs, smoking materials, electric arcs, hot filaments of infrared heaters, friction sparks from grinding and other processes, high pressure steam lines and other hot surfaces, welding and cutting operations, and other common sources of heat. Not surprisingly, dust explosions have been caused by all common sources of ignition (Schwab 1991).

Trouble lamps on extension cords or bare bulbs are extremely hazardous in dusty locations. Ignition of dust by electric lightbulbs has often been reported. Bulbs overheat when coated with dust. The hot glass can ignite the dust before the filament burns out. Moreover, a loose bulb presents a double hazard, as it may arc as well as produce heat (LeVine 1972). Hot surfaces also result from distress in machinery such as pumps and motors. Another important source of ignition is mechanical sparks. Dust may block equipment and cause overloading. Sparks can arise from foreign materials such as tramp metal in rotating machinery (Lees 1980).

Moisture in dust particles raises the ignition temperature of the dust cloud. Heat is required to vaporize the moisture. Moisture in air surrounding a dust particle has no significant effect on the course of a deflagration once ignition has occurred. There is a direct relationship between moisture content and minimum energy required for ignition, lower explosibility limit, maximum pressure, and rate of pressure increase. With spark ignition, the moisture content in dust needed to prevent ignition varies from 16% for coal to 35% for paper (Dept. of the Interior 1964a). In practice, moisture is not an explosion preventive, since most ignition sources provide more than enough heat to vaporize the moisture. In order for moisture to prevent ignition, the dust would have to be so damp that a dust cloud could not form (Schwab 1991).

Static electricity is a source of ignition for dusts, accounting for 10% of dust ignitions (Eichel 1967). Electric charges are generated on dust particles when dust clouds are produced and conveyed through transfer equipment. The particles, often highly insulating, can retain charge for considerable time (Jones and King 1991). Pneumatic transport produces the highest level of charge. Thus, airborne dust can become the source of electrostatic charge, an ignition source. Strong electric charges can develop in dust clouds. This is a concern in industrial situations only in dust clouds at least the size of a small house. The density of electric charge in the dust cloud at the bottom of a chute in a receiver, such as a silo, can be relatively high. An electrical discharge from the charged powder to a grounded pointed object in the silo could cause an explosion involving dust or associated flammable vapors. Protrusions into the tops of containers receiving dusty bulk solids should be avoided for this reason (Cooper 1953).

Sparking between an insulated electrical conductor charged by the dust and nearby grounded equipment (capacitive discharge) has caused the vast majority (90%) of static-related dust explosions (Glor 1985). A spark from an isolated conductor rapidly dissipates the entire stored energy in a single discharge. The voltage needed for discharge is approximately 350 V; 100 V is used as a design standard to provide a margin of safety (Gibson 1979). Sparks from a charged insulator are relatively weak. They contain only a portion of the energy stored on the insulator, since discharge occurs only locally. They also can be relatively long in duration.

Other types of electrostatic discharge can occur in powder handling systems (Glor 1985). While these thus far have had little impact on the occurrence of dust explosions, some types of discharge have sufficient energy to ignite flammable vapors. Brush discharges can occur between a conductor having a radius of curvature in the range 5 to 50 mm (0.2 to 2 in.) and either another conductor or a charged insulating surface. Brush discharges, while unlikely to ignite dust, can ignite flammable vapors (Maurer 1979). Corona discharge involves electrodes having a radius of curvature less than 5 mm (Glor 1985). These include sharp edges, protrusions such as threaded bolts, welded seams, and brackets inside vessels. Corona discharge is too weak to ignite gases and vapors. Propagating brush (Lichtenberg) discharges occur when a thin insulating layer inside the vessel, such as a lining, that is backed by a conducting layer, becomes charged. These are the most energetic of all discharges and the most potentially destructive. The thin insulating layer acts as a broadly distributed capacitor. For this reason, nonconductive linings inside vessels, bins, and pipes must be avoided. Maurer discharge is a form of brush discharge. It occurs along the conical surface of the powder during filling operations, in conjunction with sliding or avalanching of particles (Maurer 1979). Maurer discharges possess sufficient energy to ignite a dust explosion. Lightning-like discharges emanating from the dust cloud suspended above the pile to a grounded vessel wall are considered to be possible, but have not yet been reported.

Accordingly, grounding of all dust-handling equipment, either directly or by bonding to grounded equipment, is essential. This includes conveyor belts and pneumatic conveying systems. Potentially dangerous electric charging occurs in plastic pipe when highly resistive dusts are transported (Bodurtha 1980).

Low humidity promotes static buildup because of the decrease in moisture content and conductivity of the particles of dust. This condition is associated with the higher frequency of fires and dust explosions in winter than in summer (Kissel et al. 1973).

Addition of an inert gas to a dust-handling system (transfer lines, hoppers, and silos) can render dust clouds nonexplosible. The main gases used for this purpose are nitrogen, carbon dioxide, and flue gas. In selecting a suitable inert gas, care must be taken to ensure that it is not reactive with the dust. Certain metallic dusts, for example, can react with carbon dioxide or nitrogen under specific conditions. Helium and argon would be suitable inerts in such situations (Schwab 1991). The concentration of inert gas must reduce oxygen below the minimum needed to support combustion. Inerting gases vary in their ability to suppress combustion. For example, successful inerting using carbon dioxide requires reduction of oxygen to 11%, and 8% using nitrogen (Dept. of the Interior 1964b). The appropriate margin of safety depends on the dust and on the conditions, but normally would be a further reduction of at least 2%. Thus, the system should maintain the oxygen concentration below 9% when the minimum needed to support combustion is 11%. Concentrations reported in the literature normally are measured at ambient temperature. The appropriate concentration of inert gas for protection against ignition at high temperatures is considerably higher. This also requires confirmatory testing (Lees 1980). Minimum ignition energy and ignition temperature increase as oxygen concentration decreases. Inerting down to the limiting oxygen concentration of 11% may not be essential if a strong source of ignition is not present. Partial inerting by carbon dioxide and/or the water vapor generated in a dust dryer may be entirely satisfactory (Bodurtha 1980).

Careful design and testing of a system relying on inerting as a mode of control are essential. Often this type of plant is totally enclosed. Dust is added and removed through valves. These permit escape of only a small amount of inert gas. Where recycle occurs, preventing accumulation of fines not removed by gas-cleaning equipment is critical. Dead spots and points of in-leakage of air are possible. Multiple points of injection usually are required. Reliance on inerting as a control requires monitoring of oxygen concentration and mechanisms for shutdown of dusty operations when malfunction occurs.

Inert dust also is used for inerting. Limestone dust is utilized on the floor of passageways in coal mines (Lees 1980). The inert powder reduces combustibility by absorbing heat from the source of ignition. However, the amount of airborne inert powder needed to prevent an explosion usually is considerably higher than concentrations normally found or tolerated. To be effective, the inert dust must constitute at least 65% of the total airborne concentration (Dept. of the Interior 1964b).

Dust explosions are considered to be more destructive than those involving flammable gases and vapors (Cruice 1991). The destructiveness of a dust explosion depends on a number of factors, although primarily on the rate and duration of pressure increase. Other contributing factors include: maximum pressure, confinement, concentration of oxygen, and quantity of dust compared to the quantity of heat generated during combustion.

The rate of pressure increase is an important consideration in the design of explosion vents (Schwab 1991). This factor largely determines the area required for the vent and its practicality. Dust explosions produce gases. Heat raises the temperature of air and combustion gases. Rapid expansion of these gases produces the pressure that will be exerted destructively on surrounding enclosures unless enough vent area is provided.

Closely associated with rate of pressure increase and maximum pressure as indicators of destructiveness is the duration of excess pressure. The area under the time–pressure curve determines the total impulse exerted. Total impulse rather than instantaneous force determines the amount of destruction. This partly explains why dust explosions, which generally have slower rates of pressure rise than gas and vapor explosions, are more destructive. The destructiveness in many dust explosions does not reach full potential, because the dust is not uniformly dispersed throughout the cloud. A dust explosion in real-world situations rarely occurs under optimum conditions.

Particle size has a profound effect on several characteristics of dust explosions. The lower explosible limit, ignition temperature, and energy necessary for ignition decrease as particle size decreases. Fine dust is much easier to ignite than coarse dust. Rate of pressure increase and maximum pressure are greater for fine dusts. The exposed surface per unit weight increases as particle size decreases. Also, fine dust disperses more easily and remains in suspension longer. Particle size also affects the rate of pressure increase during the explosion. Explosions involving fine dusts have a higher rate of pressure increase.

Unique strategies, procedures, and equipment are required for dust and powder handling because they are solid materials (Lees 1980). Equipment that produces powders or dusts by size reduction is prone to dust explosions. Grinding elements or foreign bodies, such as tramp metal, can act as sources of ignition. Minimizing quantities of material where dust suspensions may occur is advisable. Several small quantities pose less hazard than one large one. Overdesigning or oversizing a process unit, such as a hopper or a mill, creates large void spaces in which a dust cloud can form. This also provides the possibility for long free fall. Removing dust from suspension as soon as possible reduces the possibility and severity of ignition. Removing suspended dust locally is preferable to transferring it long distances along ducts to a central cleaning unit. Design that minimizes accumulation of settled dust is highly preferable to reliance on housekeeping. Startup, shutdown, or fault conditions are critical periods during plant operations. Conditions that otherwise provide acceptable control can deteriorate rapidly.

Routine materials handling poses potential challenges with dusts and powders. Scoops used for manual transfer of material should be grounded plastic or metal to prevent formation of a static spark. The ground wire requires attention and inspection. Other devices for minimizing the potential for generating static sparks include grounded wristlets and anklets. These are worn by the person performing the transfer. Water sprays provide a simple means of dust suppression (Bodurtha 1980).

In process operations, dust and powder handling occurs by manual methods or by mechanical or pneumatic conveyors. Manual methods should be restricted mostly to bagging or cleanup. Mechanical and other equipment must be chosen carefully to minimize potential for static generation (Lees 1980).

Mechanical conveying equipment can include screw and drag-link conveyors, belt conveyors, and bucket elevators. Screw and drag-link conveyors are effective for dusts because of the minimal free volume. The return leg of the drag-link conveyor usually requires explosion protection. A belt conveyor is suitable for dusts only if enclosed and protected against explosion. This equipment is particularly prone to generate static. The bucket elevator poses severe hazards unless safeguarded. All mechanical conveying methods involve the risk of overheating due to mechanical failure. Pneumatic conveying utilizes two methods: low volume/high pressure air and high volume/low pressure air. Pneumatic systems are particularly prone to static buildup. Systems must be bonded and grounded to minimize this possibility.

Separation of dust from air or gas is another area of concern. From the perspective of dust explosions, the critical factor in selecting a technology for separation is free volume in the equipment. Settling chambers are unsuitable because of large free volume. Cyclones are widely used for this purpose. (The cyclone is an air-purifying device that utilizes the difference in angular momentum between particles and the airstream for separation.) In the cyclone, the concentration of dust increases from the center to the wall. At some distance from the center, the dust concentration could lie within the explosible range. The baghouse is another common separatory device. An explosible concentration may arise during mechanical shaking to clean the bags. The electrostatic

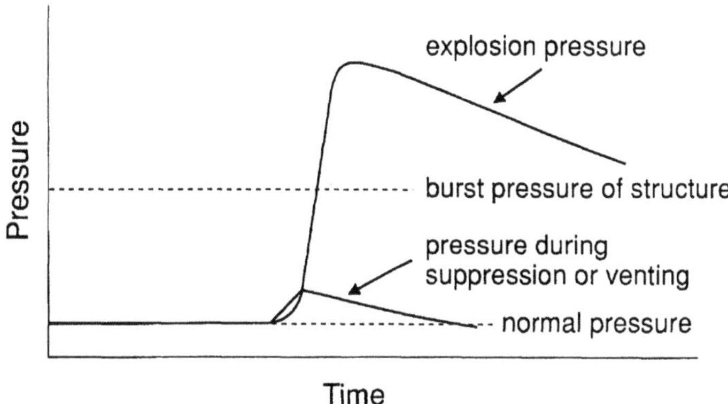

FIGURE 4.3 Pressure change in a structure during an explosion and the role of venting or suppression. Venting requires discharge of the explosion from the structure. Suppression involves detection of the developing explosion and rapid injection of a suppressant. The suppressant stops further development of the explosion by stopping the propagation phase. (Adapted from Chatrathi 1995.)

precipitator uses ionization for separating dust from gases. The inlet concentration of dust normally is less than the explosible limit. Mechanical rapping of the electrodes to loosen deposited material can create an explosible concentration within the unit. Electrostatic precipitators are not generally suitable for removal of flammable dusts.

Drying is another important part of many processes. In some processes, the liquid evaporated from the solid is flammable. Given appropriate conditions of concentration, temperature, and pressure, the vapor could exist in the flammable range and thereby could constitute a potential fire or explosion hazard. Ventilation and inerting are the main methods used for maintaining the concentration of vapor below the lower flammable limit. Low feed rate can lead to overheating of the feedstock. Overheating also can occur when residual material is exposed to hot air on start-up. Hot product is more prone to self-heating.

Screening, classifying, and other methods of size classification and separation, as well as mixing and blending, tend to create dust clouds. Operations should be enclosed and preferably run under a slight vacuum. A silo or storage hopper also is prone to formation of dust clouds.

Vacuum cleaners and mobile sweepers used for spill cleanup and general housekeeping are reservoirs of settled dust. Although rare, explosions have occurred in this equipment. All metal parts, including wands and wire in flexible hoses, should be grounded to minimize the chance of ignition. Normal techniques for explosion protection often are not feasible with vacuum cleaners. Minimizing accumulation of powder by prior washdown and shovelling is preferable to overreliance on vacuum cleaning in these circumstances (Bodurtha 1980).

Another means of protecting systems that handle flammable dusts, as well as gases and liquids, is explosion protection (Senecal 1991, Garzia and Guaricci 1995). Explosion protection systems utilize several strategies. They all act to prevent buildup of the pressure wave from the explosion and in so doing, to prevent damage and destruction (Figure 4.3). The first system releases buildup of explosion pressure in a controlled manner by providing an escape path from the equipment. This involves the use of venting panels (Figure 4.4A). These burst open under slight overpressure in the equipment and permit the pressure wave and flame to vent into an area where damage cannot occur. The second system relies on the closure of rapidly acting valves or discharge of a chemical suppressant to isolate the explosion to one part of the system (Figure 4.4B). The third strategy utilizes one or more extinguishers to discharge suppressant into vulnerable parts of the system (Figure 4.4C; Figure 4.5). The suppressant typically is a dry chemical extinguishant, water, or a halogenated methane or ethane. The chemical extinguishant acts by removing heat from the flame front and by interfering with the chain reaction.

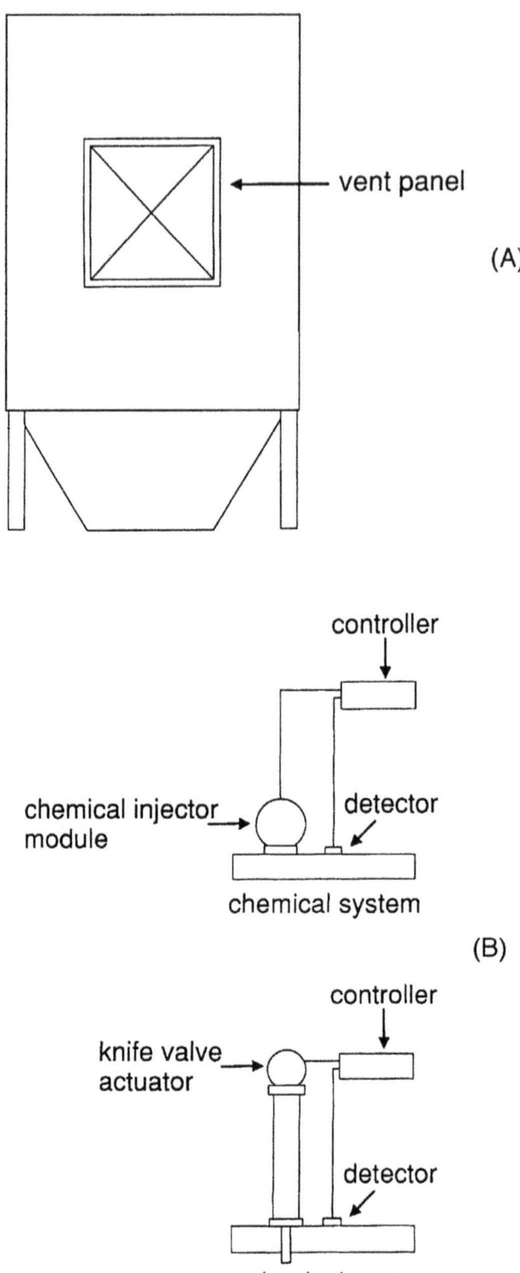

FIGURE 4.4 Explosion protection systems. (A) Venting: Venting utilizes a structure that fails quickly and reliably at a predetermined pressure. This permits the flame and pressure shock wave to dissipate outside the structure. (B) Isolation: Isolation systems contain a detector that senses buildup of pressure and a control unit that activates a rapidly acting mechanical or chemical blocking device. The mechanical device physically blocks the path of the explosion through the action of a high-speed knife valve. The chemical device injects a suppressant into the path of the explosion. (C) Suppression: Suppression systems contain multiple detectors and suppressant injector modules. These act in coordinated fashion to protect equipment and structures. Suppressants include dry chemical substances, as well as water. The suppressant chemically interferes with the chain reaction and removes heat. (Adapted from Garzia and Guaricci 1995.)

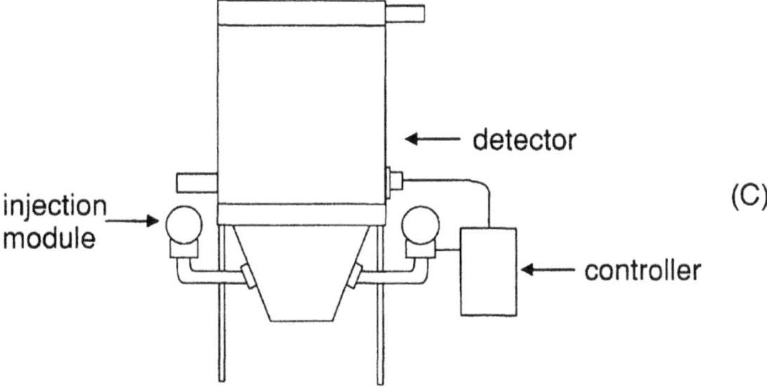

FIGURE 4.4 (continued)

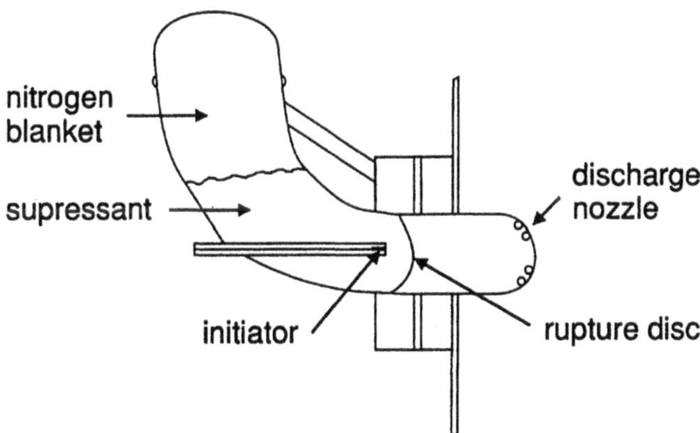

FIGURE 4.5 Suppressant injector module. The suppressant discharges from the module under pressure from the nitrogen blanket after the initiator causes the rupture disc to open. Discharge occurs within milliseconds. (Adapted from Staggs 1995.)

The suppression systems employ one or more detectors and discharge units. These systems act to contain the explosion within milliseconds of the increase in pressure in the system. The detectors and discharge devices are located inside the structure to be protected.

LOSS OF CONTROL

The first chapter in this publication introduced a theme that will recur throughout: loss of control. Control is the entirety of all inputs and outputs needed to operate an industrial facility in safety. Loss of control and the manifestations that demonstrate it were antecedent in all accidents reported in the OSHA publications (OSHA 1982, 1983, 1985, 1988). Loss of control refers to the loss of predictability over actions and conditions within an organization. Fires and explosions involving confined spaces and atmospheres confined in containers are one inevitable outcome of this deficiency.

Loss of control is most likely to occur during nonproduction activities: construction, commissioning, maintenance, decommissioning, and demolition. Production is the only period within which conditions and actions, the inputs and outputs of control, can be predicted within reasonable limits. Nonproduction work, especially maintenance, involves highly variable activities of short duration.

These activities typically involve work inside containers and enclosures that often are inaccessible. These boundaries enclose spaces whose conditions under normal operations could be lethal. The interior of these spaces is precisely designed. Change of interior or exterior conditions as a result of entry could cause deleterious outcomes later. These changes could be seemingly as minor as: addition of a welded bracket or a hanger to support a scaffold, unnoticed damage to detector heads or discharge openings in explosion suppression systems, addition of an isolated conductive probe in a detection system, substitution of a piece of pipe, failure to ensure contiguity of an electrical bond, and so on. Process systems are complex configurations. The knowledge required for recognition about the implications of a seemingly insignificant change easily could be beyond the level of those who make them.

Many organizations routinely utilize outside contractors to perform nonproduction activities. This workforce usually has little knowledge about the equipment, structures, processes, and inherent hazards in worksites in which they will spend a brief period. Loss of control over conditions and circumstances is a logical outcome of this situation. This situation received considerable attention by OSHA in its deliberations in preparing the confined spaces standard for general industry (OSHA 1993). This standard defines host and contractor relationships with the view to maintaining control in entry situations.

SUMMARY

This chapter has provided background information about fire and explosion. Fire and explosion are major killers in accidents involving confined spaces and confined hazardous atmospheres. Conditions conducive to causing a fire or explosion can develop rapidly in these workspaces following subtle flaws in work practices or technique.

Many properties — flash point, ignition temperature, range of flammability or explosibility, etc. — can be used to estimate the hazard posed by a product or material. Most are measured under strictly controlled conditions in the laboratory. Properties measured under such conditions are narrowly focused and may not reflect the hazard of a given real-world situation. Additional factors to consider include ignition temperature, minimum ignition energy, rate of evaporation of liquid or solid, rate of diffusion of vapor, and reactivity. These factors determine the relative susceptibility of a flammable or combustible to ignition. Flash point has proven to be a workable real-world indicator of relative flammability hazards. That is, low flash point means potentially high risk and high flash point, potentially low risk.

Ignitability and explosibility have profound impacts upon the safe conduct of work involving confined spaces and confined hazardous atmospheres. Minor changes in conditions can have profound consequences.

REFERENCES

American Gas Association: *Purging Principles and Practice,* 2nd ed. Arlington, VA: American Gas Association, 1975. 180 pp.

American Petroleum Institute: *Spark Ignition Properties of Hand Tools,* 3rd ed. (API Publ. 2214). Washington, D.C.: American Petroleum Institute, 1989. 5 pp.

American Petroleum Institute: *Protection Against Ignitions Arising Out of Static, Lightning, and Stray Currents,* 5th ed. (API Publ. RP2003). Washington, D.C.: American Petroleum Institute, 1991. 39 pp.

American Petroleum Institute: *Safe Entry and Cleaning of Petroleum Storage Tanks,* 5th ed. (API Std. 2015). Washington, D.C.: American Petroleum Institute, 1993. 60 pp.

American Society for Testing and Materials: *Standard Test Method for Flash and Fire Points by Cleveland Open Cup* (ASTM D 92-78). Philadelphia, PA: American Society for Testing and Materials, 1978a. 6 pp.

American Society for Testing and Materials: *Standard Test Method for Autoignition Temperature of Liquid Chemicals* (ASTM E 659-78). Philadelphia, PA: American Society for Testing and Materials, 1978b. 7 pp.

American Society for Testing and Materials: *Standard Test Methods for Flash Point by Pensky-Martens Closed Tester* (ASTM D 93-80). Philadelphia, PA: American Society for Testing and Materials, 1980. 16 pp.

American Society for Testing and Materials: *Standard Test Methods for Flash Point by Setaflash Closed Tester* (ASTM D 3828-81). Philadelphia, PA: American Society for Testing and Materials, 1981. 8 pp.

American Society for Testing and Materials: *Standard Test Method for Flash Point by Tag Closed Tester* (ASTM D 56-82). Philadelphia, PA: American Society for Testing and Materials, 1982a. 7 pp.

American Society for Testing and Materials: *Standard Test Methods for Flash Point of Liquids by Setaflash Closed Tester* (ASTM D 3278-82). Philadelphia, PA: American Society for Testing and Materials, 1982b. 12 pp.

American Society for Testing and Materials: *Standard Test Method for Flash Point of Liquids by Tag Open-Cup Apparatus* (ASTM D 1310-82). Philadelphia, PA: American Society for Testing and Materials, 1982c. 9 pp.

American Society for Testing and Materials: *Standard Test Method for Reaction Threshold Temperature of Liquid and Solid Materials* (ASTM D 2883-83). Philadelphia, PA: American Society for Testing and Materials, 1984. 13 pp.

American Welding Society: *Recommended Safe Practices for the Preparation for Welding and Cutting of Containers That Have Held Hazardous Substances* (ANSI/AWS F4.1-88). Miami, FL: American National Standards Institute/American Welding Society, 1988. 4 pp.

Anonymous: Maintenance Notes. *Maintenance*: December 1959. pp. 11–12.

Bodurtha, F.T.: *Industrial Explosion Prevention and Protection.* New York: McGraw-Hill, 1980. 167 pp.

Boublik, T., V. Fried, and E. Hala: *The Vapor Pressures of Pure Substances.* New York: Elsevier Science, 1973.

Breatherick, L.: *Breatherick's Handbook of Reactive Chemical Hazards,* 4th ed. London: Butterworth, 1990. 2,003 pp.

Burgoyne, J.H.: Mist and spray explosions. *Chem. Eng. Progr. 53:* 121M-124M (1957).

Burgoyne, J.H.: Flammability of mists and sprays. In *Proc. 2nd Symp. Chemical Process Hazards With Special Reference to Plant Design.* London: Institute of Chemical Engineers Symposium Series 15, 1973. pp. 1–5.

Burgoyne, J.H. and A.D. Craven: Fire and explosion hazards in compressed air systems. In *Proc. Chemical Engineering Progress 7th Loss Prevention Symp.* New York: 1973. pp. 79–87.

Butler, R.M., G.M. Cooke, G.C. Lukk, and B.G. Jameson: Prediction of flash points of middle distillates. *Industr. Eng. Chem. 48*: 808–812 (1956).

Chatrathi, K: Explosion suppression validation. *Safety Technol. News* (a Publication of Fike Corporation) 7: Spring, 1995. 7 pp.

Cooper, W.F.: Practical evaluation of electrostatic hazards. *Br. J. Appl. Phys. 4:* Suppl. 2, S71–S77 (1953).

Cruice, W.: Explosions. In *Fire Protection Handbook,* 17th ed. Cote, A.E. and J.L. Linville (Eds.). Quincy, MA: National Fire Protection Association, 1991. pp. 1-56 to 1-71.

Davis, N.H., III: Lightning protection systems. In *Fire Protection Handbook,* 17th ed. Cote, A.E. and J.L. Linville (Eds.). Quincy, MA: National Fire Protection Association, 1991. pp. 2-293 to 2-304.

Department of the Interior: Laboratory Explosibility Study of American Coals, by I. Hartmann, M. Jacobson, and R.P. Williams (Report of Investigation, 5052). Pittsburgh, PA: U.S. Department of the Interior/Bureau of Mines, 1954.

Department of the Interior: Laboratory Equipment and Test Procedures for Evaluating Explosibility of Dusts, by H.G. Dorsett, Jr., M. Jacobson, J. Nagy, and R.P. Williams (Report of Investigation, 5624). Pittsburgh, PA: U.S. Department of the Interior/Bureau of Mines, 1960.

Department of the Interior: Explosibility of Agricultural Dusts, by M. Jacobson, J. Nagy, A.R. Cooper, and F.J. Ball (Report of Investigation, 5753). Pittsburgh, PA: U.S. Department of the Interior/Bureau of Mines, 1961.

Department of the Interior: Explosibility of Dusts Used in the Plastics Industry, by M. Jacobson, J. Nagy, A.R. Cooper, and F.J. Ball (Report of Investigation, 5971). Pittsburgh, PA: U.S. Department of the Interior/Bureau of Mines, 1962.

Department of the Interior: Pressure Development in Laboratory Dust Explosions, by J. Nagy, A.R. Cooper, and J.M. Stupar (Report of Investigation, 6561). Pittsburgh, PA: U.S. Department of the Interior/Bureau of Mines, 1964a.

Department of the Interior: Preventing Ignition of Dust Dispersions by Inerting, by J. Nagy, H.G. Dorsett, and M. Jacobson (Report of Investigation, 6543). Pittsburgh, PA: U.S. Department of the Interior/Bureau of Mines, 1964b.

Department of the Interior: Float Coal Hazard in Mines: A Progress Report, by J. Nagy, D.W. Mitchell, and E.M. Kawenski (Report of Investigation, 6581). Pittsburgh, PA: U.S. Department of the Interior/Bureau of Mines, 1965.

Department of the Interior: Dust Explosibility of Chemicals, Drugs, Dyes and Pesticides, by H.G. Dorsett, Jr. and J. Nagy (Report of Investigation, 7132). Pittsburgh, PA: U.S. Department of the Interior/Bureau of Mines, 1968.

Drysdale, D.D.: Chemistry and physics of fire. In *Fire Protection Handbook*, 17th ed. Cote, A.E. and J.L. Linville (Eds.). Quincy, MA: National Fire Protection Association, 1991. pp. 1-42 to 1-55.

Eichel, F.G.: Electrostatics. *Chem. Eng.:* March 13, 1967. pp. 153–167.

Elcon Instruments: *Introduction to Intrinsic Safety.* Annapolis, MD: Elcon Instruments, 1989. 110 pp.

Frankel, G.J.: Oxygen-enriched atmospheres. In *Fire Protection Handbook*, 17th ed. Cote, A.E. and J.L. Linville (Eds.). Quincy, MA: National Fire Protection Association, 1991. pp. 3-160 to 3-169.

Friedman, R.: Theory of fire extinguishment. In *Fire Protection Handbook*, 17th ed. Cote, A.E. and J.L. Linville (Eds.). Quincy, MA: National Fire Protection Association, 1991. pp. 1-72 to 1-82.

Frydenlund, M.M.: Modern lightning protection. *Fire J. 60:* 10–15 (1966).

Fujii, A. and E.R. Hermann: Correlation between flash points and vapour pressures of organic compounds. *J. Safety Res. 13*: 163–175 (1982).

Garzia, H. and D. Guaricci: How to protect your drying process from explosions. *Powder Bulk Eng. 9:* 53–64 (1995).

Gibson, N.: Electrostatic hazards in filters. *Filtr. Separ. 16:* 382–386 (1979).

Glasstone, S.: Effects of Nuclear Weapons, rev. ed. Washington, D.C.: U.S. Atomic Energy Commission, 1962.

Glor, M.: Hazards due to electrostatic charging of powders. *J. Electrostat. 16:* 175–181 (1985).

Gregg, B.: Generation and control of static electricity. *Plant Serv. 17(6):* 83–87, 1996.

Haessler, W.M.: *Extinguishment of Fire,* rev. ed. Quincy, MA: National Fire Protection Association, 1974.

Haessler, W.: Theory of fire and explosion control. In *Fire Protection Handbook*, 16th ed. Cote, A.E. and J.L. Linville (Eds.). Quincy, MA: National Fire Protection Association, 1986. pp. 4-42 to 4-47.

Hugget, C. et al.: Effects of 100% Oxygen at Reduced Pressure on the Ignitability and Combustibility of Materials (SAM-TR-65-78). Brooks Air Force Base, TX: 1965.

Johnson, J.E. and F.J. Woods: Flammability in Unusual Atmospheres. Part 1--Preliminary Studies of Materials in Hyperbaric Atmospheres Containing Oxygen, Nitrogen, and/or Helium (NRL Report 6470). Washington, D.C.: Naval Research Laboratory, 1966.

Johnston, J.C.: Estimating flash points for organic aqueous solutions. *Chem. Eng. 81(25):* 122, 1974.

Jones, G.W.: Inflammation limits and their practical application in hazardous industrial operations. *Chem. Revs. 22:* 1-26 (1938).

Jones, T.B. and J.L. King: *Powder Handling and Electrostatics.* Chelsea, MI: Lewis Publishers, 1991. 103 pp.

Keller, C.: Firesafety in the shipyard. *Fire J. 76:* 60–66 (1982).

Kissell, F.N., A.E. Nagel, and M.G. Zabetakis: Coal mining explosions: seasonal trends. *Science 179:* 891–892 (1973).

Kletz, T.A.: Unconfined vapour cloud explosions. In *Proc. Chemical Engineering Progress 11th Loss Prevention Symp.,* Houston, TX, 1977. pp. 50–58.

Kuchta, J.M. et al.: *Flammability of Materials in Hyperbaric Atmospheres (USDI/BM Final Report 4016).* Pittsburgh, PA: Explosive Research Center, 1967.

Kuchta, J.M. and R.J. Cato: Review of Ignition and Flammability Properties of Lubricants (Technical Report AFAPL-TR-67-126). Wright Air Force Base, OH: Air Force Aero Propulsion Laboratory, 1968.

Le Châtelier, H.: Estimation of firedamp by flammability limits. *Ann. Mines 19:* 388–395 (1891).

Lees, F.P.: *Loss Prevention in the Process Industries,* vol. 1. London: Butterworths, 1980. pp. 477–634.

Lemoff, T.C.: Gases. In *Fire Protection Handbook*, 17th ed. Cote, A.E. and J.L. Linville (Eds.). Quincy, MA: National Fire Protection Association, 1991. pp. 3-63 to 3-82.

LeVine, R.Y.: Electrical safety in process plants ... Classes and limits of hazardous areas. *Chem. Eng. 79:* 51–58 (1972).

Maurer, B.: Discharges due to electrostatic charging of particles in large storage silos. *Ger. Chem. Eng. 2:* 189–195 (1979).

Meyer, E.: *Chemistry of Hazardous Materials*, 2nd ed. Englewood Cliffs, NJ: Prentice-Hall, 1989. 509 pp.

Moore, W.J.: *Physical Chemistry,* 3rd ed. Englewood Cliffs, NJ: Prentice-Hall, 1962. pp. 117–140.

National Fire Protection Association: *Recommended Practice on Explosion Venting* (NFPA 68-1983). Quincy, MA: National Fire Protection Association, 1983. p. 8.

National Fire Protection Association: *Recommended Practice on Static Electricity* (NFPA 77-1988). Quincy, MA: National Fire Protection Association, 1988. 30 pp.

National Fire Protection Association: *Standard System for the Identification of the Fire Hazards of Materials* (NFPA 704-1990). Quincy, MA: National Fire Protection Association, 1990. 12 pp.

National Fire Protection Association: *Fire Protection Guide to Hazardous Materials,* 10th ed. Quincy, MA: National Fire Protection Association, 1991a. 477 pp.

National Fire Protection Association: *Basic Classification of Flammable and Combustible Liquids* (NFPA 321-1991). Quincy, MA: National Fire Protection Association, 1991b. 7 pp.

National Fire Protection Association: *Fire Hazard Properties of Flammable Liquids, Gases and Volatile Solids* (NFPA 325M-1991). Quincy, MA: National Fire Protection Association, 1991c. 94 pp.

National Fire Protection Association: *Recommended Practice for Classification of Class II Hazardous (Classified) Locations for Electrical Installations in Chemical Process Areas* (NFPA 497B-1991). Quincy, MA: National Fire Protection Association, 1991d. 18 pp.

National Fire Protection Association: *Classification of Gases, Vapors, and Dusts for Electrical Equipment in Hazardous (Classified) Locations* (NFPA 497M-1991). Quincy, MA: National Fire Protection Association, 1991e.

National Fire Protection Association: *Recommended Practice for Classification of Class I Hazardous Locations for Electrical Installations in Chemical Plants* (NFPA 497A-1992). Quincy, MA: National Fire Protection Association, 1992a.

National Fire Protection Association: *Lightning Protection Code* (NFPA 780-1992). Quincy, MA: National Fire Protection Association, 1992b. 44 pp.

National Fire Protection Association: *Cleaning or Safeguarding Small Tanks and Containers* (NFPA 327-1993). Quincy, MA: National Fire Protection Association, 1993a. 9 pp.

National Fire Protection Association: *National Electrical Code* (NFPA 70-1993). Quincy, MA: National Fire Protection Association, 1993b. 389 pp.

National Fire Protection Association: *Purged and Pressurized Enclosures for Electrical Equipment* (NFPA 496-1993). Quincy, MA: National Fire Protection Association, 1993c. 19 pp.

National Fire Protection Association: *Control of Gas Hazards on Vessels* (NFPA 306-1993). Quincy, MA: National Fire Protection Association, 1993d. 15 pp.

National Institute for Occupational Safety and Health: *Criteria for a Recommended Standard— Working in Confined Spaces* (DHEW/PHS/CDC/NIOSH Pub. No. 80-106). Cincinnati, OH: National Institute for Occupational Safety and Health, 1979. 68 pp.

National Institute for Occupational Safety and Health: *Pocket Guide to Chemical Hazards* (DHHS (NIOSH) Pub. No. 90-117). Cincinnati, OH: DHHS/PHS/CDC/NIOSH, 1990. 245 pp.

National Institute for Occupational Safety and Health: *Worker Deaths in Confined Spaces* (DHHS/PHS/CDCP/NIOSH Pub. No. 94-103). Cincinnati, OH: National Institute for Occupational Safety and Health, 1994. 273 pp.

National Safety Council: *Cleaning Small Containers That Have Held Combustibles* (Data Sheet 432). Chicago, IL: National Safety Council, 1956. 2 pp.

Occupational Safety and Health Administration: Selected Occupational Fatalities Related to Fire and/or Explosion in Confined Work Spaces as Found in OSHA Fatality/Catastrophe Investigations. Washington, D.C.: U.S. Department of Labor, Occupational Safety and Health Administration (U.S. DOL/OSHA), 1982. 76 pp.

Occupational Safety and Health Administration: Selected Occupational Fatalities Related to Grain Handling as Found in Reports of OSHA Fatality/Catastrophe Investigations. Washington, D.C.: U.S. Department of Labor, Occupational Safety and Health Administration (U.S. DOL/OSHA), 1983. 150 pp.

Occupational Safety and Health Administration: Selected Occupational Fatalities Related to Toxic and Asphyxiating Atmospheres in Confined Work Spaces as Found in Reports of OSHA Fatality/Catastrophe Investigations. Washington, D.C.: U.S. Department of Labor, Occupational Safety and Health Administration (U.S. DOL/OSHA), 1985. 230 pp.

Occupational Safety and Health Administration: Selected Occupational Fatalities Related to Welding and Cutting as Found in Reports of OSHA Fatality/Catastrophe Investigations. Washington, D.C.: U.S. Department of Labor, Occupational Safety and Health Administration (U.S. DOL/OSHA), 1988. 225 pp.

Occupational Safety and Health Administration: Selected Occupational Fatalities Related to Ship Building and Repairing as Found in Reports of OSHA Fatality/Catastrophe Investigations. Washington, D.C.: U.S. Department of Labor, Occupational Safety and Health Administration (U.S. DOL/OSHA), 1990.

OSHA: Permit-Required Confined Spaces for General Industry; Final Rule, *Fed. Regist. 58*: 9, 4462–4563 (1993).

Palmer, K.N.: *Dust Explosions and Fires.* London: Chapman and Hall, 1973.

Perlee, H.E. and M.G. Zabetakis: Compressor and Related Explosions (U.S. Bureau of Mines Report of Investigation, 8187). Washington, D.C.: U.S. Department of the Interior, 1963.

Perry, J.H., D.W. Green, and J.D. Maloney (Eds.): *Perry's Chemical Engineer's Handbook*, 6th ed. New York: McGraw-Hill, 1984.

Porter, S.J.: Explosives and blasting agents. In *Fire Protection Handbook*, 17th ed. Cote, A.E. and J.L. Linville (Eds.). Quincy, MA: National Fire Protection Association, 1991. pp. 3-92 to 3-100.

Reid, R.C., J.M. Prausnitz, and B.E. Poling: *The Properties of Gases and Liquids*, 4th ed. New York: McGraw-Hill, 1987. 741 pp.

Robinson, C.S.: *Explosions: Their Anatomy and Destructiveness.* New York: McGraw-Hill, 1944.

Scarbrough, D.R.: Control of electrostatic ignition sources. In *Fire Protection Handbook*, 17th ed. Cote, A.E. and J.L. Linville (Eds.) Quincy, MA: National Fire Protection Association, 1991. pp. 2-284 to 2-292.

Schwab, R.F.: Dusts. In *Fire Protection Handbook*, 17th ed. Cote, A.E. and J.L. Linville (Eds.) Quincy, MA: National Fire Protection Association, 1991. pp. 3-133 to 3-142.

Senecal J.A.: Explosion prevention and protection. In *Fire Protection Handbook*, 17th ed. Cote, A.E. and J.L. Linville (Eds.). Quincy, MA: National Fire Protection Association, 1991. pp. 6-184 to 6-201.

Sly, O.M., Jr.: Flammable and combustible liquids. In *Fire Protection Handbook*, 17th ed. Cote, A.E. and J.L. Linville (Eds.). Quincy, MA: National Fire Protection Association, 1991. pp. 3-43 to 3-53.

Staggs, W.A.: Flushmount, telescoping dispersion nozzle for food and pharmaceutical application. *Safety Technol. News* (Publication of Fike Corporation) 7: Spring, 1995. 7 pp.

Treybal, R.E: *Mass-Transfer Operations*, 3rd ed. New York: McGraw-Hill, 1980. 784 pp.

Weast, R.C. (Ed.): *Handbook of Chemistry and Physics*, 49th ed. Cleveland, OH: The Chemical Rubber Co., 1968. pp. D-105 to D-142.

Whitehead, E.R.: Lightning. In *Encyclopaedia of Occupational Health and Safety*, 3rd rev. ed., vol. 2 (L-Z), Parmeggiani, L (Ed.) Geneva: International Labour Organisation, 1983. pp. 1231–1234.

Wierzbicki, V. and D. Palladino: Standard flash-point testing methods. *Environ. Testing Anal. 3:* 30–37 (1994).

Wysocki, T.J.: Carbon dioxide and application systems. In *Fire Protection Handbook*, 17th ed. Cote, A.E. and J.L. Linville (Eds.). Quincy, MA: National Fire Protection Association, 1991. pp. 5-232 to 5-240.

Zabetakis, M.G.: Flammability Characteristics of Combustible Gases and Vapours (Bull. 627). Washington, D.C.: U.S. Department of the Interior/Bureau of Mines, 1965. 121 pp.

5 Atmospheric Confinement: The Role of Boundary Surfaces

CONTENTS

INTRODUCTION

Analytical studies performed by the National Institute for Occupational Safety and Health (NIOSH) provided the first quantitative evidence for the dominant role of atmospheric hazards in accidents associated with confined spaces (Safety Sciences 1978, NIOSH 1979). The reports published by the Occupational Safety and Health Administration (OSHA) provided the basis for establishing similarities between fires and explosions in confined spaces, with similar accidents occurring during welding external to containers and structures (OSHA 1982, 1988). Establishing similarities between these accidents is significant, since they usually are not considered to be related. Analysis of the OSHA data presented in the first chapter demonstrated the similar ranking of atmospheric hazards present during both types of accident. Further, the relative occurrence of contaminants in each ranked group was almost identical.

The first chapter also identified other characteristics of accidents involving atmospheric contamination (OSHA 1983, 1985, 1990). In most cases, the hazardous atmosphere developed prior to opening the space to prepare for entry and the start of work. In only a minor fraction of the accidents did the activity itself generate the hazardous atmosphere. The atmospheric hazard most likely to be produced by human activity was ignitability or explosibility.

In almost all situations examined by NIOSH and OSHA, management and workers failed to recognize the importance of boundary surfaces formed by structures in trapping atmospheric contaminants. They failed to appreciate the potential danger resulting from atmospheric confinement and the relationship between minor change in geometry and the level of hazard. Confinement by boundary surfaces is the factor that enables development of hazardous atmospheric conditions. Enabling personnel to recognize conditions conducive to atmospheric confinement is an outcome of utmost importance.

The occurrence in a workspace of conditions conducive to atmospheric confinement results from many factors: introduction of a source of contamination to an area not designed or intended to accommodate it, inappropriate modifications to control equipment, disablement of ventilation systems, and so on. The generation of a hazardous condition reflects the relationship between the source of contamination and the extent of confinement.

To illustrate, a typical room in an office building easily can confine atmospheric contaminants to hazardous levels. Confinement would be exacerbated by blocking supply and exhaust vents in the ventilation system and sealing leakage paths, such as gaps around doors and windows. Such situations have occurred during renovation of isolated sections in occupied office buildings. Individuals applying solvent-based flooring adhesives and coatings have sealed off rooms in an attempt to prevent migration of vapors into occupied areas.

As a result of these modifications, an ordinary office could provide the conditions of confinement to enable considerable overexposure to the solvent vapor. Under these circumstances, considering the room to be a confined space does not seem to be unreasonable. (This discussion is not intended to challenge the meaning imparted to the term "confined space" as used in the legal parlance of a particular jurisdiction. Rather, the term "confined space" is used here in the generic sense to demonstrate the role of geometry in atmospheric confinement.)

The frustrating aspect in the circumstance described here is that upon dissipation of the solvent vapors and removal of the sealing materials, the room reverts to its normal status: a workspace intended for occupancy by virtue of its ventilation, lighting, furniture, and other accoutrements. The circumstance described here, that changes a controlled condition into an uncontrolled one, illustrates how a minor change to the status quo can lead to a consequential or even fatal outcome.

The effects of minor change in circumstances and conditions can stymie attempts by organizations to identify and inventory confined spaces. No list can encompass all of the possible situations. The omissions and exceptions could form the basis for potential catastrophes. For an organization that has a large number of workspaces, such a list would be cumbersome in the extreme. In addition, this approach eliminates the benefit of developing recognition skills in the individuals who have direct involvement with these workspaces.

Educational materials on the subject of confined spaces generally attempt to foster recognition skills through examples of familiar structures, such as sewers and tanks. This approach forces the learner to extrapolate from the specifics provided by these examples to the specifics of the actual circumstance. An alternate approach is to build awareness about the elements that create the hazard: confinement of emissions by boundary surfaces, and the relationship between the source, the boundary surface, and the affected individual. This strategy is used in this chapter.

Boundary surfaces and emission sources are present in all confined workspaces that have atmospheric hazards. The relationship between these elements and the occupants is the factor that enables contaminants to accumulate to hazardous levels.

The intent of this chapter is to develop the relationship between workspaces as geometric entities containing boundary surfaces and atmospheric confinement. Discussion will explore the relationship between geometry and confinement, existing knowledge about confinement in actual confined spaces and the generation of contaminated atmospheres, and tools for predicting hazard. How these boundaries restrict visual communication between the attendant and entrants and complicate rescue operations is discussed in Chapter 15.

ATMOSPHERIC CONFINEMENT

Discussion

The industrial hygiene literature contains little information on the subject of confined spaces and atmospheric confinement. Similarly, information about the process by which atmospheric contamination develops in these workspaces also is sparse. By way of contrast, the literature has focused on evaluation and control of situations recognized as potentially hazardous. This, of course, is rightly so, since emphasis should be placed on prevention. However, lack of attention to processes involved in accumulation of contamination has severely constrained understanding about this subject. This, in turn, has hampered development of a knowledge base about the development of hazardous conditions in confined spaces.

Gases and vapors were the airborne contaminants implicated in fatal accidents investigated by NIOSH and OSHA (NIOSH 1979, 1994; OSHA 1982, 1983, 1985, 1988, 1990). This observation reflects the relationship between mode of entry into the body and acute toxic action of substances in the gaseous state. Other possible agents include dusts, mists, and fumes. The most likely mode of action of these agents would be irritation or corrosive attack of the skin, eyes, mucous membranes in the nose and throat, and the lower respiratory system. Irritation and corrosive attack were not mentioned as causal factors in any of the accidents investigated by NIOSH and OSHA. Irritation or corrosive attack, however, could be a factor in less consequential situations. Irritation and corrosive attack usually provide a strong sensual warning. Lack of sensual warning was a major factor in the failure to recognize the seriousness of the danger posed by hazardous atmospheres containing gases and vapors.

Factors mentioned in the anecdotal summaries of accidents investigated by NIOSH and OSHA regarding development of the hazardous atmosphere included:

Characteristics of the source

- Point source vs. area source
- Fixed source vs. mobile source
- Single phase (gas/vapor) vs. multiphases (vapor + droplets)
- Single source vs. multiple sources

Number of sources

Quantity of contaminant present in the air and on surfaces

Emission rate

Passive mechanisms by which contaminant becomes airborne

- Desorption
- Vaporization
- Sublimation

Active mechanisms by which contaminant becomes airborne

- Welding processes
- Thermal plumes/convection
- Sprays

- Jets
- Turbulent mixing
- Aeration
- Chemical reaction
- Mechanical action

Factors influencing airborne concentration

- Vapor pressure
- Gas/vapor density
- Density and settling characteristics of aerosols
- Evaporation rate
- Air-to-solution partition coefficient
- Solvent-to-solvent partition coefficient
- Temperature
- Extent of equilibration/disequilibration
- Agitation/turbulence/aeration/fluidization
- Time

Passive mechanisms that deplete atmospheric oxygen

- Surface oxidation (rust formation)
- Absorption
- Adsorption
- Chemical reaction

Active mechanisms that deplete atmospheric oxygen

- Respiration
- Fermentation
- Combustion
- Displacement by other gases
  - Purging
    - Nitrogen
    - Carbon dioxide
    - Helium
  - Process gases
  - Shield gases (welding)
    - Argon
    - Helium
    - Carbon dioxide

Mechanisms by which atmospheric contaminants accumulate

- Convection
- Diffusion

Geometry and boundary conditions imposed by the workspace

Ventilation

- Wind
- Convective air motion
- Mechanically induced air motion

None of the factors presented in the preceding list is unique to confined spaces or atmospheric confinement. They can apply to any industrial situation involving work with or production of toxic agents. Interaction of these factors with barrier surfaces is the element that differentiates conditions that develop in confined spaces from those occurring in normal workspaces. This interaction magnifies the individual or collective consequence of these factors and increases the danger of working in these environments.

Analysis presented in the first chapter suggests that a large proportion of the hazardous conditions involved in fatal accidents developed in the undisturbed space (NIOSH 1979, 1994; OSHA 1982, 1983, 1985, 1988, 1990). A smaller proportion of the hazardous conditions developed as a direct result of actions taken during preentry preparation or work activity. Common industrial chemicals and naturally occurring substances were identified as the causative agents in these accidents. The chemical, physical, and toxicological properties of these substances normally do not inspire undue concern. Thus, substances that can cause tragedy in confined spaces generally pose no special hazard under conditions of routine encounter.

Empirical Studies

This section focuses on information that has been published about atmospheric conditions in confined spaces.

Following a fatal accident involving a magnesium chloride brine evaporator (16 m high × 3.7 m diameter), Melcher et al. reported on a technical investigation to determine the cause of the oxygen deficiency that caused the tragedy. The vessel had been washed, vented, and drained, but remained closed for several days prior to the entry (Melcher et al. 1987). The follow-up investigation determined that the oxygen content in the vessel decreased from normal levels to 1% in 24 h. Equation 5.1 was derived to describe the kinetics of oxygen depletion in the vessel.

$$C = 21\,e^{-\left(\frac{\tau-3}{6.344}\right)} \tag{5.1}$$

C = concentration in %
τ = time in hours

The follow-up laboratory investigation determined that surface scale (magnetite) increases oxygen depletion more than 10-fold over that of clean metal. The tests showed that formation of ferric oxide (Fe_2O_3) over the first 24 h was not a factor. The results also showed formation of Iowaite, a magnesium–iron hydroxy complex, which contains a high proportion of oxygen, on highly scaled surfaces. The scale retains a large amount of $MgCl_2$ solution. The scale becomes black, indicating formation of magnetite, Fe_3O_4, the primary corrosion product. Highly scaled surfaces show a 10-fold increase in oxygen depletion compared to clean metal. This study also showed that other salt solutions, such as KCl, NaCl, $CaBr_2$, and $MgSO_4$, in conjunction with high humidity also can cause hazardous oxygen depletion.

Slack compared levels of hydrogen sulfide in residual fuel oil against those in the vapor space of storage tanks and sample containers (Slack 1988). Liquid temperatures typically were 60°C. Temperatures in the vapor space at the point of measurement were near ambient. He found an approximate vapor phase to liquid phase ratio of 50:1 (ppm_v: ppm_w), based on concentrations measured in both phases. A power function provided the best fit to the data. Variance increased with increasing concentration.

As expected, venting and product transfer decreased concentrations in both the liquid and vapor phases. Product transfer and agitation of the liquid accelerated the decrease. Concentrations were somewhat higher in samples taken near the bottom of the liquid compared with those taken from the surface.

Concentrations in the vapor space were essentially homogeneous from the liquid level to the top of the storage compartment.

Mignone et al. sampled in manholes in two sewerage systems for oxygen, combustible gases, carbon monoxide, hydrogen sulfide, and volatile organic compounds (VOCs) (Mignone et al. 1990). Both sites (an industrial and a domestic sewer) were chosen for their depth (>5 m), large volume of flow (>12 ML/day), and location remote from major roadways. VOCs from the industrial sewer were sampled at 1-h intervals. Those from the domestic sewer were sampled at 4-h intervals. Samplers were positioned 1.2 m above the water, the assumed height of the breathing zone of a worker in a typically bent-over or crouched position. Sampling at each location occurred on seven 1-day intervals between September and March.

Results indicated periods of sudden change in the composition of the atmosphere in both sewer systems. During these episodes, the level of carbon monoxide, hydrogen sulfide, and VOCs was observed to rise slightly from the background level of zero. The greatest change occurred in the sewer that receives domestic sewage. Excursions in the level of carbon monoxide best seemed to reflect the presence of vehicles outside the manhole.

Decreases in oxygen and increases in carbon monoxide levels were the most frequently observed events. The lowest concentration of oxygen, 15.5%, occurred during the period of lowest flow in the domestic sewer. Decreases in oxygen sometimes correlated with the presence of flammable/combustibles. However, the increase in level of flammable/combustibles was too small to account for the decrease in oxygen. Carbon dioxide, which also may have been involved in these excursions, was not measured. The contaminant responsible for decreasing the level of oxygen was not identified. Oxygen levels were observed to decrease rapidly in pulses lasting around 10 min. (Ten minutes is more than sufficient time for the onset of a fatal accident involving an oxygen deficiency. The occurrence of unpredictable episodes involving low levels of oxygen illustrates the difficulty that would ensue in estimating conditions during a follow-up accident investigation.)

Fingas et al. examined the fate of volatile liquids in a model sewer (Fingas et al. 1988). They constructed the model from a 104-m length of a 3.8-cm (1.5-in.) diameter pipe. The pipe descended at a constant slope of 0.0048 to a sump. Water supplied from a constant head tank provided a constant flow rate. Fuel "spills" were introduced at a head box containing a concrete pad designed to resemble a street drain. The model also contained simulated manholes or ports along the pipe at approximately logarithmic intervals.

Gasoline spilled into the model sewer produced two vapor peaks, whereas diesel fuel produced only one. The first vapor peak from the evaporation of gasoline formed as it rode along on the water. This peak moved at approximately the same speed as the water. It contained the higher boiling fractions of gasoline. The second peak originated from evaporation of gasoline near the point of entry and consisted of low boiling fractions. This peak moved slowly through the system, as its only driving force was the weak friction with the underlying water. Diesel fuel produced significantly less vapor than gasoline.

These authors also investigated the influence of dispersants on vapor formation in sewer systems. Dispersants increased the rate of volatilization of fuel already in the system and thus increased the amount of vapor. Dispersants apparently decrease the size of droplets of fuel. Use of dispersants in this manner in real-world situations potentially could increase the fire hazard.

Groves and Ellwood investigated atmospheric conditions in sealed vertical farm silos during anaerobic conversion of green crops to silage (Groves and Ellwood 1989). They measured conditions in the airspace at varying heights above the silage. Silage production evolves carbon dioxide, nitric oxide, and nitrogen dioxide and consumes oxygen. The carbon dioxide displaces the atmosphere to produce an oxygen-deficient condition. These studies showed rapid development of a stratified layer of gases, mostly carbon dioxide initially and later nitric oxide, above the last load of material. The stratified layer of gases was about 1 to 2 m thick and highly oxygen deficient. It developed rapidly within 12 h after the addition of crop materials to the top of silage remaining in the silo.

This buildup occurs regardless of whether the silo is sealed or top hatches are open. Dispersion to form uniform concentrations with depth does occur in sealed silos, but the process requires months. The resulting atmospheric condition, while uniform in concentration, can be highly oxygen deficient. This study also showed development of local pockets of high contamination and low levels of oxygen in low spots. Contamination persisted in low spots after the silos were opened to permit natural ventilation.

Glass described work on a survey of exposure of water reclamation workers in the U.K. to hydrogen sulfide (Glass 1990). This study examined three job classifications: site workers, mobile workers, and tanker truck drivers. Only 14% of the routine exposure samples exceeded the limit of detection (0.1 ppm) for the 8-h shift. None exceeded the occupational exposure standard of 10 ppm. Also, Glass determined levels in potential sources of exposure. These ranged to over 400 ppm in narrow, deep desludging chambers on primary sedimentation tanks. Lesser but significant levels were found in desludging sections in other tanks, on top of digesters, in wet wells, and during tanker loading and discharge. Hydrogen sulfide generation in sewage is slow for the first two days and then rises rapidly. Rate of generation is temperature dependent (Baumgartner 1934).

Groves and Ellwood continued their investigation of agricultural hazards with an investigation of gases emitted during the handling of liquid manure. The liquid is collected through drains into pits and tanks for storage until it is spread onto the fields. Mixing and emptying operations are considered to be especially problematic (Groves and Ellwood 1991). To put the magnitude of this problem into perspective in today's intensive farming operations, the average production of waste per day is 41 L from dairy cows, 15 L from pigs, and 0.114 L from laying hens (Grundey 1980, Ministry of Agriculture 1980).

Agitation of the slurry before and during emptying can cause rapid release of large quantities of gases. Ammonia, carbon dioxide, and hydrogen sulfide comprise the bulk of these emissions (Groves and Ellwood, 1991). The rapid release of gases on agitation results from the following effects: altering the mass kinetics of the system by increasing the surface area of the liquid phase, releasing gases entrapped in bubbles in the more viscous portions of the slurry, and disruption of the surface layer (Donham et al. 1982). The quantity of gases released partly depends on the fluidity of the slurry (Noren 1977).

The main hazard during mixing and transfer was found to be high transient concentrations of hydrogen sulfide. The concentration rose rapidly after the start of mixing and decreased slowly (Groves and Ellwood 1991). This could range as high as 500 ppm at a height just above the level of the slurry, although 100 to 200 ppm was more typical. Carbon dioxide levels rose to 10,000 to 20,000 ppm at the same locations. Oxygen concentrations decreased to the extent of increase in carbon dioxide.

The most extensive source of information on conditions that can develop in confined spaces is the NIOSH summary of investigations under the Fatality Assessment and Control Evaluation (FACE) Program (NIOSH 1994). The purpose of FACE was to investigate fatal accidents in detail, using a multidisciplinary team. The quality of information gathered during the FACE investigations reflected the greater awareness about confined spaces that developed in the 1980s and availability of monitoring equipment. FACE, however, provided a narrow view of the subject, since its focus was situations in which fatal accidents had occurred. By necessity, a delay always preceded the detailed investigation. Atmospheric conditions almost certainly changed prior to the investigative proceedings.

Data contained in the FACE summaries are provided in Table 5.1. These are listed to conditions judged most likely to represent those at the time of the accident.

Despite delays in assessing conditions following fatal accidents, these results provide some indication about the potentially lethal conditions that can exist in confined spaces. In some situations, the results confirm the apparent stratification demonstrated in other studies.

TABLE 5.1
Conditions in Confined Spaces Following Fatal Accidents

FACE reference	Type of space	Comments
86-13	Fermentation tank	1,200 gal (5 m^3) Carbon dioxide: 48% Oxygen: 6%
89-46	Manure pit	Closed top, 24 ft × 20 ft × 11 ft (150 m^3) Methane: 2% of LFL Hydrogen sulfide: 18 ppm Oxygen: 20.2%
86-37	Vault (valve pit) Potable water	Standing water, rusting metal 6 ft × 8 ft × 10 ft (14 m^3) Conditions in similar, previously undisturbed vault: Oxygen 3%
87-06	Vault (valve pit) Potable water	Standing water, wood waste, bacteria 12 ft × 6 ft × 8 ft (16 m^3) Oxygen: 7%
91-17	Vault (valve pit) Potable water	6 ft diameter × 7 ft deep (6 m^3) Standing water Oxygen: 2% at bottom
86-48	Stormwater drainage pit	12 ft (4 m) deep Oxygen: <5% at bottom
86-54	Sanitary sewer	15 ft (5 m) deep Oxygen: 6% at the bottom Methane: 20% of LFL at the bottom
87-45	Sanitary sewer	15 ft (5 m) deep Sludge present Oxygen: 7% at bottom
89-28	Sanitary sewer	4 ft diameter × 10 ft deep (4 m^3) Tannery waste and water present Hydrogen sulfide: >200 ppm
88-44	Sanitary sewer, new construction	15 ft (5 m) deep Standing water, mud Oxygen profile: 20.0% at 2.1 m below the surface to 14.0% at 2.7 m, 6.5% at 3.4 m, and 4.0% at 4.0 m
84-13	Sanitary sewer new construction	Gasoline-powered pump Carbon monoxide: 600 ppm
87-67	Sanitary sewer new construction	7 ft (2 m) deep Water present Oxygen: 7% at 2 m
88-01	Wastewater treatment plant manhole	5 ft diameter × 8 ft (4 m^3) Oxygen: 11%
85-44	Wastewater treatment plant, sludge distribution chamber	8 ft × 9 ft × 9 ft (18 m^3) Sludge present Hydrogen sulfide: >500 ppm Methane: 10% of LFL

HAZARDOUS CONDITIONS IN THE OPERATIONAL CYCLE OF THE CONFINED SPACE

Analysis of fatal accidents presented in the first chapter and empirical data from the preceding section suggest that the operational cycle of confined spaces should be divided into four distinct segments, based on hazard generation:

- The undisturbed space
- Preentry preparation
- Prework inspection
- Work activity

A hazardous condition in a confined workspace can develop during any of the segments in its operational history. The undisturbed space encompasses the period following the last closure to the start of actions to prepare for entry and work. Preentry preparation describes activities and actions taken to prepare for entry. Prework inspection is the initial entry into the space to assess the safety of conditions. Work activity includes actions undertaken during the course of work that occurs in the space.

The Undisturbed Space

Adverse atmospheric conditions generated in the undisturbed space arose through passive mechanisms or through mechanisms associated with processes. Contamination resulted from residual contents, from entry of substances from the surroundings (soil, water, and air) into the space, or from biological or chemical activity occurring in the space. There are numerous examples of sources of residual material in process equipment — bottoms; sludges; material trapped in lines, valves, and pumps; and so on. These materials may be volatile. They may absorb oxygen. They may decompose or be decomposed to form other substances. These materials can form extended sources on interior surfaces as pools of undrained materials and sludges that remain in low-lying areas. Gases and vapors may be trapped in some spaces or enter from surrounding soils and water through the walls of subgrade porous containments or broken lines. Subsident clouds of gas or vapor also may enter openings of below-grade spaces. This could be especially important where the temperature of a buoyant gas or vapor is much less than ambient.

The atmospheric hazard in the undisturbed space usually develops over an extended period. This results from the nature of the process that produced the contaminant or the manner by which it entered the space and lack of dispersive mechanisms other than diffusion. The latter could produce a concentration gradient that decreases with distance from the source. The resulting hazard can be inhalation or fire related. However, the hazard is more likely to be inhalation related because of the high concentration needed for ignitability.

Preentry Preparation

Preentry preparation refers to all actions taken to prepare the space for entry. These can include draining, flushing and cleaning, atmospheric testing, purging, inerting, and ventilating, breaking connections and opening lines, and so on.

Preparation of process equipment for entry often involves use of gases. One application is to displace gaseous contents from the interior of the space on the basis of density. The displacing gas and the resident gas form stratified layers. Hydrogen and nitrogen can be used to displace denser gases and vapors downward. Hydrogen is not an inert gas and, therefore, it must be displaced from the vessel prior to opening and ventilating with air. Hydrogen is used as a feedstock in some processes and in some nonprocess equipment to maintain a reducing atmosphere. Carbon dioxide is effective for upward displacement of atmospheres having high air content or methane and ethane. Stratification also can be a hinderance when used inappropriately, since the stratifying gas may remain in the space.

Another application of gases is to remove the contents of the space prior to admitting air. Inert gases can be used to remove residual vapors and evaporate liquid contents. Use of inert gases offers the advantage of avoiding contact between the contents of the space and gases of the atmosphere. Mixing air with flammable gases and vapors during ventilation of a space can create a flammable atmosphere or reactive mixture. The mixture of air and vapor or gas in the space passes from concentrations of inflammability above the upper flammable limit into the flammable range during the dilution.

From the perspective of entry, establishment of an inert atmosphere usually is not the goal in itself, but merely a step in the process of providing a safe atmosphere for the worker to enter. An inert atmosphere would be lethal, and could be entered only after taking the most stringent of precautions, including complete reliance on air-supplied respiratory protective equipment. In some situations, entry into a space containing a deliberately inerted atmosphere may be required. Inerting would be necessary when safety could not be achieved by other means of control. Inerting of spaces adjacent to the one in which work is to occur could be required when the work to be performed could generate a hazardous condition in those spaces. When inerting is used, the composition of the atmosphere must be monitored and maintained under careful control.

Gases commonly used for purging and inerting include nitrogen, carbon dioxide, combustion products, and steam. Putting the space under vacuum also can be used where the strength of the enclosure permits. This is not practical for entry, however.

Steam has the advantage of condensing on cooling. Air is automatically drawn into the space. Condensate should be constantly removed from the space, as removal of a material in the vapor state is very slow. Steaming, as a method of cleaning as well as inerting, should continue as long as condensate escaping from the vessel shows any sign of impurity. Steam cleaning must be used with considerable care due to the potential for buildup of static electricity. In addition, steam cleaning followed by closure of a vessel prior to thorough equilibration with the surroundings could lead to collapse as a result of the negative pressure created during condensation and cooling (Lees 1980).

Prework Inspection

The purpose of the prework inspection is to determine the unanticipated and the unexpected. The prework inspection is a mandatory part of some confined space programs. For example, the operator of a process unit may perform an inspection following preparation of the space and prior to the start of work. Also, the prework inspection has been a fundamental part of shipyard safety programs since the early 1920s (Keller 1982). The prework inspection is a mandatory part of some legislative approaches to work involving confined spaces (OSHA 1994).

Work Activity

Work activity refers to work occurring following sanctioned entry into the confined space or from work occurring externally to equipment or a vessel that confines a hazardous atmosphere. Work activity created the hazardous atmospheric condition in a minor but significant fraction of the fatal accidents investigated by NIOSH and OSHA (NIOSH 1979, 1994; OSHA 1982, 1985, 1988). That is, the hazard would not have developed had the activity not occurred. The manner of using substances, actions that changed existing conditions, and the boundary conditions of the workspace strongly influenced development of the hazardous condition.

Sources of contamination arising from work activity include both point and extended sources. Point sources are a common feature of work activity. The purpose of many activities, such as spray painting, is to distribute material from the point source onto a surface having extended area. This involves passage of droplets of volatile material through an ever-increasing volume of space. Welding and cutting create a plume that disseminates into a larger volume of space from the point source. Development of the hazard from these processes occurs through the process of dissemination, as well as the extended surface that results. Processes that disperse from point sources into large air volumes involve jets, plumes, puffs, and sprays. The most potentially hazardous point sources produce hazardous atmospheric conditions through the interaction of several factors:

- Energetics of the release of the substance
- Potential for the substance to become or remain airborne
- Toxicity or flammability
- Build-up of electrostatic charge

Dissemination of the substance from the point of origin can occur passively or energetically. Passive dissemination involves minimal expenditure of energy. Energetic dissemination involves the expenditure of energy to disseminate the substance from the point of origin to some distant point in space. The process often utilizes high pressure air or other pressurized gas as a propellant. Energetic dissemination rapidly increases the surface area of source material by breaking it into smaller pieces. This process also distributes the material into the volume of the confined workspace.

There are many work activities and processes that involve dissemination from point sources to form distributed sources. Some examples include:

- Metalworking: brazing, welding, burning, cutting, grinding
- Spraying or brushing products containing volatile solvents
- Exhaust gas emissions from heaters, engine exhaust systems, and other combustion sources
- Escape of gases and vapors from valves, pumps, and flanges, and leakage from containment

A common example of a point source is welding. Welding is commonly carried out in confined spaces. Welding generates intense heat and a plume of gases and particulate matter. As the plume rises, the envelope of the plume radiates outward. The plume rises until deflected by a boundary surface or until the contents have lost their positive buoyancy through cooling. The boundary surfaces of a confined space are more likely to prevent dissipative motion upward than in normal workspaces. The boundary forces the plume to spread horizontally along its surface and eventually downward. The rate of production of contaminants, the proximity of the boundary surface to the source, and the geometry and volume of the structure dictate the rate at which contaminated air accumulates in the airspace.

Steam poses an underappreciated hazard as a point source in confined spaces. Steam is used to loosen scale and to heat sludges during cleaning activities. Released from the nozzle, the volume of the steam increases, while pressure decreases to that in the space (assumed here to be ambient). Expansion and forward momentum of the cloud of steam can displace atmospheric gases from its path. These gases escape from openings of the space. Atmospheric displacement can produce an oxygen-deficient condition in the space under these conditions.

A similar situation can result from use of helium, argon, carbon dioxide, and blends of these gases during gas-shielded welding. Accumulation of these gases within the boundaries of the space can displace the atmosphere, creating an oxygen-deficient condition.

Compressed oxygen that escapes from oxyfuel, cutting and burning equipment, is an ever-present potential hazard in these environments. Oxygen enrichment more frequently occurs following deliberate release of oxygen for personal cooling. This has resulted from the mistaken belief that this can be done safely. Accidental leakage through faulty or incompletely shut valves is another route of escape of oxygen and other process gases into the confined workspace.

Some work activities generate extended sources. Extended sources often originate from activities involving point sources, such as spraying. These usually involve application of materials that coat some or all of a surface. The result of this action is a dramatic increase in the surface area from which volatile components of the material can evaporate compared to that of the product in the original container. Vapor formation at extended sources has caused fatal accidents in confined workspaces and many nonfatal overexposures to solvent vapors. Activities that create extended sources during work in confined spaces include:

- Painting and coating
- Solvent washing
- Abrasive blasting that exposes a contaminated surface or a surface capable of undergoing chemical reaction
- High pressure cleaning that exposes volatile surfaces
- Disturbance of sludges to expose volatile surfaces

BOUNDARY SURFACES

Previous attention directed to confined spaces has focused predominantly on partly or completely enclosed structures having limited access or egress. The traditional concept of the confined space, however, does not satisfactorily account for inhalation hazards resulting from atmospheric confinement by partly enclosed structures. That is to say, under some circumstances, geometrically simple structures can trap contaminants efficiently enough to create hazardous atmospheric conditions. These structures may not restrict personal movement, or access/egress, in any obvious way.

The traditional concept of the confined space is an enclosed structure having limited access or egress. Such a structure is eminently capable of creating a hazardous atmosphere through confinement. The emphasis on enclosure means that less enclosed structures housing vigorously active sources would not be considered to be confined spaces, despite the obvious hazard. The criterion of the traditional view for acceptance as a confined space is based solely on the geometry of the structure. The relationship of the structure to the source, toxic properties of the contaminant, and work location of affected individuals is not considered. The difficulty with the traditional concept is rationalizing the occurrence of fatal accidents in less enclosed structures where vigorous release of contamination has occurred.

This discussion takes the view that assignment of status as a confined space must consider the ability of a structure to confine in context with the characteristics of the source and proximity of potentially affected individuals. This approach emphasizes the continuous increase in complexity of structures relative to their role in atmospheric confinement. The confined space that fulfills the traditional concept would fit near one end of this continuum. This approach also accommodates the view that at some point the geometric complexity of a particular workspace in context with the sources and potentially affected individuals exceeds the threshold for acceptance as a confined space. The difficulty with a continuum is to determine where the threshold of concern should lie.

Recent legislation on this subject addressed these concerns to some extent. The OSHA confined space standard for general industry took the approach of creating a broadly based general definition for confined spaces (OSHA 1993). Within the envelope of situations encompassed by the definition, some are assessed to pose minimal hazard (nonpermit confined space). Others (permit-required confined space) are considered capable of causing serious injury or death. Corrective measures are required in the latter case to eliminate or control hazards. Work activity that could create hazardous conditions also must be considered in the assignment of classification.

The OSHA confined space standard for shipyard employment considered all enclosed spaces to pose a potential hazard of serious injury or death and requires a hazard assessment by a competent person prior to the start of work (OSHA 1994). Hazards discovered during this process must be eliminated prior to entry for the start of work. The nature of the work to be performed also must be considered in the elimination or control of hazards.

Both standards rely on definitions for entry into the space that are narrower than that espoused by the continuum concept. For example, the OSHA standard for general industry considers that the space must be capable of being entered bodily (OSHA 1993). This would exclude chambers into which only the head, shoulders, and arms could be inserted. Yet, fatal accidents have occurred in spaces this small (McCammon and McKenzie 1996). Under the OSHA concept, entry occurs once any part of the body crosses the plane of the opening of the space. This concept of entry generated considerable discussion during hearings prior to adoption of the standard.

The boundary surfaces of a workspace define the limits of motion of contaminants, hence the extent to which atmospheric confinement can occur. The geometry of ordinary workspaces imposes less stringent boundary conditions than exist in confined spaces. This results from the closer proximity of the source(s) to potentially affected individuals because of the presence of boundary surfaces. Boundary surfaces accelerate entry of contaminants into the breathing zone of affected individuals. The relationship between boundary surfaces, source(s), and potentially affected individuals determines the kinetics of entry of contaminants into the breathing zone.

The habitats, workspaces, and workplaces of the industrialized world are a marriage between form and function. Architects and industrial designers utilize three-dimensional geometry as the vehicle for achieving this goal. Under appropriate conditions, virtually any shape provides a boundary capable of causing atmospheric confinement. Of considerable importance is the fact that the ability to confine the atmosphere within a geometric shape does not in any way imply restriction of access or egress. The only relevant issues are the efficacy of the shape in confining the atmosphere and the potential for exposure. Atmospheric confinement in and of itself is sufficient to provide conditions capable of causing fatal accidents by overexposure, asphyxiation, fire, or explosion.

Simple geometric shapes confine or restrict only the atmosphere. More complex shapes also can restrict movement of personnel. Shapes of varying complexity occur throughout industry and architecture. Listing all possible examples of atmospheric confinement by these shapes is impossible. One means of exploring the role of boundary surfaces is to start from simplest principles and progress stepwise to the more complex.

The least confining boundary surface is none at all. This would occur in a situation in which person works while suspended in free space. Work on a skeletal structure, such as the steel frame of a building under construction is an example of a situation where boundary surfaces are not present. The task, such as welding on the side of an upright beam, could generate an airborne contaminant. Contaminated air would emit upward from the source. The only potential boundary surfaces in this situation are other structural members. These would exert minimal or minor influence dispersion of the contaminant or its concentration. Either they are too remote from the person or their surface area is too small to pose significant resistance to movement of the contaminant. The concentration of the contaminant decreases with increasing distance through unhindered migration. Entry into the breathing zone may not occur. Accumulation cannot occur because boundary surfaces are absent. The only constraints on migration of the contaminant are the physical laws governing transport. The contaminant disperses as fast as it is generated. Of course, this situation does not preclude overexposure of the person along the direct path of emission due to proximity to the source, its rate of generation, and dispersal.

Any surface that interferes with the motion of airborne contaminants constitutes a boundary. These surfaces can be planar, angular, or curved. The boundary alters and restricts motion of the airborne contaminant. When present in the vicinity of the breathing zone, the boundary surface can increase exposure by altering motion of airborne contaminants. Redirection may lead to accumulation of contaminants by preventing dispersive motion. The net impact depends on the characteristics of the source, and the proximity of the source and the person to the barrier.

The simplest boundary surfaces that can alter motion of contaminated air are planes: ceilings, floors, walls, and oblique surfaces. A ceiling can trap a contaminant plume that rises due to heating. Vapors emitted from a coating applied to a ceiling can accumulate at this level. Work performed near a ceiling under these conditions carries considerable potential for exposure. Similarly, vapor from a solvent having high vapor density can accumulate during work performed at floor level. This situation also could occur following a spill that creates a pool or application of a coating. Inclined planar surfaces have long been utilized in buildings to direct buoyant or subsident air in a desired direction. Heated contaminated air flows along the inclined surface while seeking the highest level in a structure. Older buildings housing processes that generate considerable heat were designed with tapered roof structures to channel the flow of heated (contaminated) air toward the highest point in the center of the building. The flow of cool air down the surface of a hill into a valley demonstrates the influence of a boundary surface on subsident air.

A semicylinder is a geometrically simple boundary surface that can effectively trap airborne contaminants. The curved roof of a covered walkway is an example of this type of structure. Buoyant heated air produced within the structure is trapped in the airspace at the top of the curve. Accumulation can occur rapidly. The longer the structure, the greater the propensity for accumulation. Openings at the ends of the structure provide the only route of escape. The same process can occur in subsident vapors or gases in structures used to conduct water and other liquids.

An additional consideration with preformed structures such as a semicylindrical trough is migration of contaminants over long distances. A liquid flowing in a trough induces motion of the air above it. Contaminated air can migrate for considerable distances from the point of generation within a structure having suitable shape and a flowing liquid.

Structures containing two planar or curvilinear surfaces form more complex boundary surfaces. The orientation of these surfaces can be either parallel or intersecting. The most familiar boundary surface formed by the intersection of two planar surfaces is the corner. Oriented in the appropriate direction, a corner forms an effective collector of contaminated air. The inverted "V"-shaped roof is an example of such a structure. A "V"-shaped trench is another example. The corner formed by the intersection of a wall and a floor or ceiling or even two walls can provide the conditions for confinement. The body of a person working near these surfaces forms an additional boundary surface. The channel formed by these boundary surfaces restricts motion of contaminated air and also can localize and direct flow. A potential consequence of this geometry is the ability of contaminants to migrate far beyond the point of generation.

Structures containing parallel surfaces also can trap and channel contaminated air. Critical factors include: orientation in space, distance between the two surfaces, surface area, and the distance of the source from the plane of the surfaces. An example of such a structure is a large crawl space or ceiling space having large surface area and limited height.

Workspaces, such as those discussed here, normally are not considered to be confined spaces. Yet, atmospheric confinement can occur readily in these structures. The only difference between these workspaces and those that gain unchallenged recognition as confined spaces is perception. Fatal accidents have occurred in the simple structures already mentioned (OSHA 1985). Conventional wisdom would argue that conditions of atmospheric confinement can occur only within highly restrictive geometries. These examples illustrate, however, that atmospheric confinement can occur under simple boundary surfaces.

Two curvilinear surfaces join to form a cylinder. The cylinder is the simplest boundary surface having limited access and egress. The walls form a continuous barrier, resulting in potential personal and atmospheric confinement. Airborne contaminants easily can accumulate within this type of structure, as for example, in a cylindrical tunnel, pipeline, or sewer. A horizontal cylinder is the simplest structure in which both buoyant and subsident vapors or gases can accumulate simultaneously. Critical parameters for atmospheric confinement in simple cylinders include length and diameter. A vertically oriented cylinder can restrict motion of contaminated air having neutral buoyancy.

Joining a curved and a planar surface forms a covered trough. Inverted examples include the Quonset hut and railway tunnel. Airborne contaminants easily can accumulate in the roofspace and at the floor of these structures. This geometry also can be exploited to prevent entry of contaminants into the breathing zone. Oriented vertically, these structures can trap contaminated air having neutral buoyancy.

Three planar or curvilinear surfaces join to form a structure that is triangular in cross section. The structure can have a roof and two walls; a floor and two walls; or a floor, wall, and roof. These structures behave similarly to cylinders. The V-shaped boundary surfaces can enhance buildup of contaminants compared to the rounded structure of the cylinder.

Three planar surfaces joined together at a common point form a pyramid. A pyramid-shaped roof is extremely effective for trapping contaminated air. The cone and hemispherical roof, both having a continuous curvilinear boundary, are variations on the pyramidal shape. These structures, similarly, are very efficient in atmospheric confinement. In all of these examples, the only escape path for contaminated air is under the edges of the open bottom. Also, the corners of a pyramid are extremely effective in channeling contaminated air toward the apex.

A confining structure formed by two curved surfaces and a plane is the open-topped or open-bottomed cylinder. An open-topped tank or cylindrical diked structure are examples. An open-topped cylinder can effectively confine atmospheric contaminants that are denser than air. This geometry is the simplest capable of producing virtually complete atmospheric confinement. The ability to confine in this case depends on the diameter of the base and height of the walls, as well as the density of the contaminant.

The least confining structure containing four planar surfaces is a four-sided pyramid. Like the three-sided pyramid, this structure is extremely effective at confining and channeling atmospheric contaminants. A fourth planar or curvilinear surface can produce a structure completely enclosed by boundary surfaces. Adding a fourth planar surface to a three-sided pyramid produces a completely enclosed structure. An enclosed pyramid is the simplest geometric shape containing planar surfaces that is capable of complete atmospheric and personal confinement. A structure capable of complete atmospheric and personal confinement would satisfy requirements by any definition for inclusion as a confined space.

Combining four curvilinear surfaces produces an approximation of a sphere. Another combination having four curved surfaces is a cylinder having rounded ends. A silo is a cylinder having a flat bottom joined to three curved surfaces.

Five planar surfaces produce several geometric shapes: a vertical pit, horizontal shaft, or a bottomless structure having a flat top. These shapes are similar to an open-bottomed or open-topped cylinder and so can trap atmospheric contaminants effectively.

The most commonly encountered six-sided structures are cubic or rectangular boxes. These form the majority of structures found in industry. Examples include building envelopes and interior substructures, such as rooms. This geometry also forms the interior of equipment and storage structures.

The geometries discussed in this section, including even the simplest, provide boundary surfaces capable of atmospheric confinement. Atmospheric confinement can lead to fatal buildup of contaminants or depletion of oxygen. The more completely the workspace encloses the source of contamination, the more effective its ability to confine the atmosphere. The geometric shapes discussed here occur throughout industry in an infinite number of structures. The most notable regarding traditional interest in confined spaces are structures that form process vessels, tanks, pits, and shafts. As this section illustrates, there is a need to reconsider the assumption that a geometrically complex shape is needed to create the conditions of atmospheric confinement. The term "confined space" and the image of personal confinement implied by this term are too restrictive to encompass the potential for atmospheric confinement.

VENTILATION

The second factor affecting concentration of airborne contaminants in confined workspaces is ventilation. The design of normal workplaces usually incorporates supply and exhaust ventilation systems. At the very least, the design should address requirements for human respiration, and beyond that, comfort and control of airborne contaminants (ASHRAE 1989). Workspaces capable of atmospheric confinement either lack ventilation entirely or are poorly ventilated. The less restrictive boundary conditions and better ventilation found in normal workspaces together act to maintain the concentration of contaminants within acceptable limits during normal activity. Development of hazardous conditions does not occur, or does not occur as rapidly.

Ventilation of confined spaces is discussed in detail in later chapters.

TIME

The third factor involved in the development of hazardous atmospheric conditions or depletion of oxygen is time. Time controls the process. Time is the factor that permits an extremely slow process to proceed to a hazardous end point. In terms of hazard potential, this end point need not be the ultimate completion of the process. The appropriate end point is generation of a suitably hazardous atmosphere. A considerable period of quiescence preceding entry, as mentioned in many accident summaries contained in the OSHA report, satisfies this requirement (OSHA 1985).

PROPERTIES OF GASES AND VAPORS

The physical behavior of atmospheric contaminants is critical to understanding the mechanisms behind the accumulation of contaminants or depletion of oxygen. The contaminants involved in these situations are gases and vapors.

Gases and vapors are formless fluids in the gaseous state. The term "gas" refers to a substance that exists exclusively in the gaseous state at ambient temperature and pressure (21°C and 101 kPa or 760 mmHg). The National Fire Protection Association (NFPA) considers a flammable liquid to have a vapor pressure not exceeding 40 psia at 100°F (275 kPa or 2,070 mmHg at 38°C) (NFPA 1991). Conversely, by this definition, a gas would have a vapor pressure exceeding 40 psia at 100°F (275 kPa or 2,070 mmHg at 38°C). Vapor is the gaseous form of substances that exist as solids or liquids at ambient conditions of temperature and pressure. Vapor forms at the surface of volatile substances.

A gas or vapor completely fills the container in which it resides. Lacking boundary surfaces, the open atmosphere has no means to confine a gas or vapor, hence the widespread dispersion of pollutants from point sources across continents and into the upper atmosphere. Bulk air motion due to wind and mechanical ventilation, motion of objects and people, and convection due to heating and the diffusion act to randomize the distribution of molecules of gas and vapor in workspaces.

Gases boil at temperatures below ambient. Energy requirements for the process limit the rate of vaporization. Pressure exerted by gaseous substances when in equilibrium with the liquid form usually is greater than atmospheric pressure. Pressure exerted by vapor in equilibrium with the liquid form usually is less than atmospheric. Vapor pressure equals atmospheric pressure at the boiling point of substances that exist in the liquid state under ambient conditions. While the pressure inside a vented container will remain equal to atmospheric pressure, the gas will boil and completely displace the air. A liquid will evaporate without boiling. A gas or vapor confined in the headspace above the liquid similarly will pressurize an unvented container. To convert gases into liquids requires compression, removal of heat, or both. To convert vapor into the liquid form requires either compression or removal of heat.

Gases can be categorized according to physical properties (Lemoff, 1991). **Compressed gases** exist solely in the gaseous state at all pressures and normal temperatures. The pressure in the container depends on the quantity of gas, the temperature, and the volume of the container. Under containment, **liquefied gases** exist partly in the liquid state and partly in the gaseous state. Vaporization of liquid pressurizes the container. The temperature of the liquid primarily controls pressure of the gas. **Cryogenic gases** exist as liquids at atmospheric pressure, but at temperatures far below ambient.

Physical properties of gases provide important cues about their behavior outside containment. The density of the gas at the temperature of interest is the important factor. Instruments provide the only means to define the extent of the cloud formed by an odorless and colorless gas when outside containment. Liquefied gases extract the heat needed for vaporization from the air. This sudden cooling condenses water vapor and produces a visible mist that roughly defines the area occupied by the gas. However, ignitible gas–air mixtures can extend beyond the boundaries of the visible fog. These can pose considerable hazard since they are not visible. These gases are always denser than air at normal temperatures because of their low temperature. The low temperature, and the associated water fog, cause normally lighter-than-air gases to hug the ground for some distance. Liquefied gases often vaporize rapidly following contact with the air or ground. They generally will not exist in the liquid phase once unconfined. Exceptions to this rule are gases having low vapor pressures. Low air and ground temperature also will inhibit vaporization. Cryogenic liquefied gases will form a pool if leakage into a localized area occurs with sufficient rapidity.

The relationship of gas and vapor clouds to surrounding air is an important concern. The relationship of pressure to volume for "ideal" gases is given by Boyle's Law. That is, for constant temperature, the product of pressure × volume = constant (Metz 1976).

$$P_1V_1 = P_2V_2 = \text{constant} \tag{5.2}$$

Charles' Law provides the relation between volume of an "ideal" gas and its absolute temperature. That is, for constant pressure, volume/absolute temperature = constant (Metz 1976).

$$\frac{V_1}{T_1} = \frac{V_2}{T_2} = \text{constant} \tag{5.3}$$

$T_1 = (t_1 + 273)$; t_1 is measured in°C.
$T_2 = (t_2 + 273)$; t_2 is measured in°C.

Combining Boyle's and Charles' Laws creates the relationship for predicting the pressure inside a container holding an "ideal" gas or vapor at different temperatures, when V remains constant ($V_2 = V_1$).

$$\frac{P_1V_1}{T_1} = \frac{P_2V_2}{T_2} = \text{constant} \tag{5.4}$$

The density of a gas that acts as an "ideal" gas varies with temperature and pressure according to the following relationship (Hering, 1989):

$$\frac{\rho_1T_1}{P_1} = \frac{\rho_2T_2}{P_2} = \text{constant} \tag{5.5}$$

ρ = density (metric units).

This equation provides the means to calculate gas density at temperatures and pressures different from ambient.

A more convenient comparison is based on gas specific gravity (GSG). Gas density is the mass per unit volume of pure gas. Gas specific gravity is the ratio of the density of gas to the density of air under the same conditions of temperature and pressure. Equal volumes of ideal gases at the same temperature and pressure contain the same number of molecules. Where volumes are the same, only total masses differ. Total mass is the product of the number of molecules × molecular weight. Since the volumes are the same, the numbers of molecules are identical in both samples; hence, the only difference is due to molecular weight. The GSG at the temperature of interest is $MW_{gas}/29$ compared to air.

$$\rho_{gas} = \frac{n \times (MW)_{gas}}{V} \tag{5.6}$$

$$\rho_{air} = \frac{n \times (MW)_{air}}{V} = \frac{n \times 29}{V} \tag{5.7}$$

$$GSG_{gas} = \frac{\rho_{gas}}{\rho_{air}} = \frac{(MW)_{gas}}{29} \tag{5.8}$$

ρ = density;
n = number of molecules;
MW = molecular weight; and
GSG = gas specific gravity.

The relationship between GSG and temperature and pressure parallels the relationship between gas density and temperature and pressure. Gas specific gravity provides a way of predicting the behavior of gas clouds at varying temperatures that are released into air at constant fixed temperature, say 20°C, and normal atmospheric pressure. The behavior of these clouds has major relevance to atmospheric confinement by barrier surfaces.

$$(GSG)_2 = (GSG)_{20°C} \times \frac{293}{T_2} \tag{5.9}$$

Table 5.2 provides GSG for gases having importance in confined spaces at different temperatures at normal atmospheric pressure. The table contains data only for those conditions likely to be encountered under normal conditions of entry and work.

To a first approximation, data presented in Table 5.2 can be used to predict the buoyancy characteristics of a cloud of gas released into ambient air by molecular weight and temperature. A gas cloud having the same temperature as another can exhibit different buoyancy simply because of the difference in weight of its molecules. The highlighted column under the temperature 20°C illustrates differences in specific gravity due only to molecular weight. Similarly, clouds of the same gas at different temperatures exhibit different buoyancy due to thermal expansion or contraction.

The horizontal row of values for air indicates the change in GSG due to temperature. Change in specific gravity with temperature is much smaller than would be expected from the ratio of temperatures expressed in °C. The small change results from use of the Kelvin scale of temperature

TABLE 5.2
Buoyancy Characteristics of Gases at Different Temperatures

		Gas Specific Gravity (re 20°C) Temperature (°C)							
Substance	MW	–100	–50	–20	0	**20**	50	100	200
Hydrogen	2	0.12	0.09	0.08	0.08	**0.07**			
Helium	4	0.24	0.18	0.16	0.15	**0.14**			
Methane	16	0.93	0.72	0.64	0.59	**0.55**			
Ammonia	17	*	*	0.68	0.63	**0.59**	0.54	0.46	0.37
Steam	18							0.49	0.38
Acetylene	26	*	1.18	1.04	0.97	**0.90**	0.82	0.71	0.56
Carbon monoxide	28	1.64	1.27	1.12	1.04	**0.97**	0.88	0.76	0.60
Nitrogen	28	1.64	1.27	1.12	1.04	**0.97**	0.88	0.76	0.60
Air	29	1.69	1.31	1.16	1.07	**1.00**	0.91	0.79	0.62
Oxygen	32	1.86	1.45	1.27	1.18	**1.10**	1.00	0.86	0.68
Hydrogen sulfide	34	*	1.54	1.35	1.26	**1.17**	1.06	0.92	0.72
Argon	40	2.34	1.81	1.60	1.48	**1.38**	1.25	1.08	0.85
Carbon dioxide	44					**1.52**	1.38	1.19	0.94
Propane	44					**1.52**	1.38	1.19	0.94
Methyl mercaptan	48					**1.66**	1.51	1.30	1.03
Butane	58					**2.00**	1.81	1.57	1.24
Sulfur dioxide	64					**2.21**	2.00	1.74	1.37
Chlorine dioxide	68					**2.34**	2.12	1.84	1.45
Chlorine	71					**2.45**	2.22	1.92	1.52

Notes: * Ammonia boils at –33°C; acetylene boils at –83°C; hydrogen sulfide boils at –60°C.

and calculations involving the ratio of small changes in large numbers. Change in specific gravity due to molecular weight is the dominant factor.

A cloud of pure gas having a GSG value somewhat greater than 1.0 would be expected to flow to ground and outward from the point of contact to the lowest attainable elevation. The latter could be the interior of a tank, sump, trench, or other structure that is located below grade. The flow of cool air down the slope of a mountainside into a valley at night or of cold air onto the floor of a room from a walk-in freezer are practical examples of this phenomenon. The GSG of a cloud of pure gas having a temperature somewhat less than 20°C that is released into air would be expected to increase to its value at ambient temperature as the cloud warmed up.

GSG values calculated for temperatures that are less than 0°C provide an indication of behavior of gas clouds under winter conditions or self-cooling during vaporization of a liquid pool or release of chilled gas as from a refrigeration system.

A cloud of pure gas having a GSG value somewhat less than 1.0 would be expected to rise from the point of emission to the highest attainable point. This could include the roofspace of a building or the workspace. The rise of engine exhaust to the roofspace of a tunnel, or a welding plume into the roofspace of a building, or diesel exhaust or woodsmoke to form flat layers under cold winter conditions are practical examples of this phenomenon. The GSG of a cloud of pure gas having a temperature somewhat greater than 20°C that is released into air would be expected to decrease to its value at ambient temperature as the cloud cooled. Diesel exhaust and wood smoke under cold winter conditions form horizontal layers for this reason.

GSG values calculated for temperatures that are greater than 50°C provide an indication of behavior of gas clouds as in a heated plume, such as from engine exhaust or from welding, burning or cutting. Gas clouds at temperatures markedly greater than the reference value of 20°C would be expected to lose energy rapidly to the surroundings.

Strong heating or cooling does not markedly influence the behavior of gases having very small GSG, such as hydrogen or helium, or very large values such as sulfur dioxide, chlorine dioxide, or chlorine. Strong cooling can influence the behavior of gases having reference GSG values of 0.5 and larger. Methane, for example, normally is considered to rise in air. Strong external cooling or self-cooling could increase the GSG to greater than 1.0, in which case methane would exhibit neutral buoyancy or even subsident behavior. Gases that have GSG values around 1.0 could become strongly subsident or strongly buoyant following strong cooling or heating, respectively.

PROPERTIES OF LIQUIDS AND LIQUID MIXTURES

The behavior of clouds formed by the vaporization of liquids is equally important to that of gases. The properties of vapors can be determined by direct measurement and by predictive equations, the latter reflecting the results of empirical work.

ASTM Method D 323-82 provides a means for determination of vapor pressure of liquids, such as gasoline and other volatile petroleum products (ASTM 1982).

The usual basis for predicting the behavior of a vapor cloud in air is a comparison between the ratio of molecular weight of vapor and the average molecular weight of air (MW/29) and unity. This comparison assumes that the vapor is pure and contains no air. Yet, the vapor space above a liquid in a container seldom contains only vapor, since the vapor pressure is almost always less than the atmospheric pressure. (When vapor pressure equals atmospheric pressure, the liquid begins to boil. This can occur only by heating the liquid or by transporting the container to a suitably high altitude where the atmospheric pressure equals the vapor pressure.) This means that the vapor space in a container also contains air and that the "vapor" is actually a mixture of vapor and air. How well mixed is the mixture, is a function of air motion inside the container.

Vapor-air specific gravity (VASG) is the equivalent to the GSG. That is, this is the ratio of the density of the air–vapor mixture to the density of air under identical conditions of temperature and pressure. The following equation calculates the specific gravity of an air–vapor mixture resulting from vaporization of a liquid at equilibrium temperature and pressure conditions compared to an equal volume of air at the same conditions. The first term in the equation is the contribution of the vapor to the specific gravity of the mixture. The second term is the contribution of the air (Drysdale 1991).

$$\begin{aligned} \mathrm{VASG} &= \frac{\mathrm{VP} \times \mathrm{VSG}}{\mathrm{P}} + \frac{\mathrm{P} - \mathrm{VP}}{\mathrm{P}} \\ &= \frac{\mathrm{VP}(\mathrm{VSG} - 1) + \mathrm{P}}{\mathrm{P}} \end{aligned} \tag{5.10}$$

where

VASG = the specific gravity of the air–vapor mixture at the temperature of interest;
VP = the vapor pressure of the substance at the temperature of interest;
VSG = the specific gravity of pure vapor (MW/29); and
P = the atmospheric pressure at the temperature of interest.

This equation incorporates changes due to the increase in vapor pressure with temperature.

Vapor pressure at the specified temperature can be estimated by several methods: predictive equations and resources that provide graphical or tabular data (Boublik et al. 1973, Reid et al. 1987, Weast 1968). Reid et al. discuss various mathematical equations used for estimating vapor pressure at different temperatures and provide tabular data (Reid et al. 1987). The focus of the discussion that follows is equations that can be used for predicting conditions in which people normally work. Following are two equations used to predict vapor pressure.

TABLE 5.3
Vapor Pressure for Selected Substances

Substance	MW	(mmHg) Temperature (°C)						
		−20	−10	0	10	**20**	30	40
Benzene	78	*	*	*	45	**75**	119	182
Toluene	92	OR	OR	OR	12	**22**	36	58
Methanol	32	7	15	30	55	**92**	163	264
Acetone	58	22	41	70	116	**180**	282	419
Chloroform	119	OR	34	59	98	**160**	238	353

Notes: * Benzene melts at 5.5°C; OR means outside the range of the calculation.

From Reid et al. 1987, Himmelblau 1989, NIOSH 1990.

Antoine's equation provides a means for predicting vapor pressure of pure substances over a specified, but limited range of temperature. The Antoine equation should not be used for extrapolated temperatures.

$$\ln VP = A - \frac{B}{T + C} \tag{5.11}$$

A, B, C = constants specific to the substance
VP = calculated in bars (1 bar = 1 atm = 760 mmHg = 101 kPa)
T = °K

The **Wagner equation** is an alternate to the Antoine equation. It provides more accurate results over a wider range of temperatures.

$$\ln\left(VP/P_c\right) = \frac{A\tau + B\tau^{1.5} + C\tau^3 + D\tau^6}{1 - \tau} \tag{5.12}$$

$$\tau = 1 - \frac{T}{T_c} \tag{5.12a}$$

A, B, C, D = constants specific to the substance
P_c = the critical pressure in bars
VP = calculated in bars
T_c = the critical temperature in °K

Table 5.3 provides vapor pressure for equilibrium mixtures of vapor in the vapor space in a container. A container could include an undisturbed confined space. The liquid source could arise from wetted interior surfaces or from pools of liquid. Table 5.3 provides data at ambient pressure and different temperatures. These temperatures correspond to conditions to be expected inside a container during normal occupancy. These conditions assume a well-mixed atmosphere. Data were calculated using the preceding equations.

Table 5.3 provides vapor pressures calculated for a range of temperatures for a representative group of substances. For comparison against substances not listed, sources provide vapor pressures at standard reference temperatures (NIOSH 1990). Many substances have vapor pressures similar to those in Table 5.3 at 20°C.

TABLE 5.4
Vapor-Air Specific Gravity for Selected Substances

		Temperature (°C)						
Substance	VSG	−20	−10	0	10	**20**	30	40
Benzene	2.69	*	*	*	1.10	**1.17**	1.26	1.40
Toluene	3.17	OR	OR	OR	1.03	**1.06**	1.10	1.17
Methanol	1.10	1.00	1.00	1.00	1.01	**1.01**	1.02	1.03
Acetone	2.00	1.03	1.05	1.09	1.15	**1.24**	1.37	1.55
Chloroform	4.10	OR	1.14	1.24	1.40	**1.65**	1.97	2.44

Notes: * Benzene melts at 5.5°C; OR means outside the range of the calculation; VSG is the specific gravity of pure vapor (MW/29).

From Reid et al. 1987, Himmelblau 1989, NIOSH 1990.

These vapor pressures could be converted into concentrations by application of Equation 5.13. This equation gives the equilibrium concentration of vapor in a container at ambient conditions of temperature and pressure (Olishifski 1988).

$$\text{Concentration} = \frac{\text{VP (mmHg)}}{760 \text{ mmHg}} \times 1{,}000{,}000 \text{ ppm} \tag{5.13}$$

In this equation, concentration is expressed in units of parts per million (ppm). Multiplying by 100% in place of 1,000,000 ppm would calculate in units of percent. There is little to be gained in converting the vapor pressures listed in Table 5.3 to concentrations, since the temperatures of interest are near the flash points. Concentrations of vapor in the range of the flash point are substantial. (Flash points are expressed in the percent range. See Chapter 4 for more information.)

Table 5.4 provides VASG for the same substances, calculated using Equation 5.10. This calculation assumes that total pressure remains constant.

These results indicate that the specific gravity of vapor–air mixtures is considerably less than predicted simply by calculating the ratio of the molecular weight of the substance to the average molecular weight of air (MW/29). The vapor–air clouds are denser than air at the same temperature because the molecules of vapor have molecular weights greater than the molecules that constitute the air. The relative contribution of a vapor to the specific gravity depends on volatility of the substance and molecular weight. The densest vapor–air mixtures at a particular temperature result from highly volatile substances that have large molecular weight.

Table 5.5 provides VASG for the substances calculated relative to air at 20°C compared to air of the same temperature as the vapor–air mixture. Equation 4.9 was used in the calculation. In this case, 20°C is the new temperature. These data provide an indication about the behavior of a vapor–air mixture generated at the initial temperature when released into air at ambient temperature.

Cold vapor mixtures of methanol and acetone could become buoyant when released into warmer air. Similarly, the heated vapors could become more subsident when released into cooler air.

The deviation of magnitude of GSG or VASG from unity is expected to determine the driving force behind vertical motion.

Air parcels are subject to vertical buoyant forces that arise from density differences between the parcel and the surrounding air (Moran and Morgan 1989). An air parcel will rise or sink until it reaches air of equivalent density. In the atmosphere, cooling occurs on expansion of an air parcel as it rises, and heating due to compression as it settles. Heating and cooling are unlikely to be factors in this discussion, since the distances are very small.

TABLE 5.5
Vapor-Air Specific Gravity for Selected Substances

		(re 20°C) Temperature (°C)						
Substance	**VSG**	−20	−10	0	10	**20**	30	40
Benzene	2.69	*	*	*	1.06	**1.17**	1.30	1.50
Toluene	3.17	OR	OR	OR	0.99	**1.06**	1.14	1.25
Methanol	1.10	0.86	0.90	0.93	0.98	**1.01**	1.05	1.10
Acetone	2.00	0.89	0.94	1.02	1.11	**1.24**	1.42	1.66
Chloroform	4.10	OR	1.02	1.16	1.35	**1.65**	2.04	2.61

Notes: * Benzene melts at 5.5°C; OR means outside the range of the calculation; VSG is the specific gravity of pure vapor (MW/29).

Some perspective about flow of air masses having different specific gravities is needed. To illustrate, cool air readily flows outward and downward into a room from a refrigerator. The temperature in a typical refrigerator is 7.5°C (280.5 °K) and 21°C (294 °K) in a room. The GSG of air in the refrigerator is 1.05 times that of air in the room. The atmospheric pressure differences that cause daily sea and land breezes in coastal areas during the summer are approximately 20 mb (millibars) and range from 990 to 1,010 mb. The GSG of air at the higher pressure compared to air at the lower pressure is 1,010/990 = 1.02. Diethylether vapor will flow visibly from the opening of the container down a "U"-shaped trough under conditions of ambient temperature and pressure. The VASG of diethylether vapor at 20°C and 760 mmHg calculated using Equations 4.10 and 4.12 is 1.90. This is considerably greater than the values determined in the previous cases, where subsidence also is apparent. In the case of the diethylether, the vapor must escape through the opening in the top of the container prior to settling.

Of considerable importance as well, is the emission of vapor from mixtures. The vapor pressure of hydrocarbon mixtures can be approximated using Trouton's rule (Butler et al. 1956).

$$\log \mathrm{VP} = 2.881 + 4.60 \times \left(1 - \frac{T_B}{T}\right) \tag{5.14}$$

VP = the vapor pressure of the mixture in mmHg;
T_B = the boiling point in °K;
T = the temperature of interest in °K.

Total vapor pressure of a hydrocarbon mixture is useful to the extent of gauging the importance of all contributors to the vapor. Of greater concern is the contribution made by individual substances, as reflected in standards for occupational exposure and fire hazard. Liquid components of mixtures may be completely miscible, partly miscible, or completely immiscible in each other. Their behavior may be characterized as ideal or nonideal. In an ideal solution, the components are completely miscible. Interactions between molecules of solute and solvent are the same as solute–solute and solvent–solvent interactions. Further, the volume of the solution is the sum of the volumes of the constituents and the heat of formation of the solution is zero. The total pressure of components above the solution is the sum of the partial pressures exerted by the components (Barrow 1966).

Raoult's law provides a basis for predicting vapor pressure of components in the airspace above ideal solutions in equilibrium situations. Raoult's law applies to components in mixtures whose mole fraction approaches unity and for mixtures whose chemical behavior is similar (Himmelblau

1989). According to Raoult's law, the partial pressure of a vapor above a solution is equal to the product of the mole fraction of the liquid in the solution times the partial pressure of the pure liquid at the same temperature.

$$p_i = VP_i x_i \tag{5.15}$$

p_i = the pressure of component i in the gas phase;
VP_i = the vapor pressure of pure component i at the temperature of interest;
x_i = the mole fraction of component i in the solution.

The mole fraction, x_i, of component i in the solution is given by the following equation:

$$x_i = \frac{n_i}{n_1 + n_2 + \ldots + n_n} \tag{5.16}$$

where n_1, n_2, …, n_n are the number of moles of substance 1, substance 2, …, substance n in the mixture.

$$x_1 + x_2 + \ldots x_n = 1 \tag{5.17}$$

The mole fraction, X_i, of component i in the vapor phase is given by the following equation:

$$X_i = \frac{p_i}{p_1 + p_2 + \ldots + p_n} \tag{5.18}$$

where p_1, p_2, …, p_n are the partial pressures exerted by substance 1, substance 2, …, substance n in the vapor.

$$X_1 + X_2 + \ldots + X_n = 1 \tag{5.19}$$

The vapor phase will be richer relative to the liquid in the more volatile component.

Raoult's law applies, for example, to solutions of volatile hydrocarbons such as butane and propane, or benzene, toluene, or xylene in a nonvolatile liquid, such as a hydrocarbon oil (Treybal 1980).

As an application of Raoult's law, consider a tank holding waste oil that is contaminated by a commercial grade of toluene. The toluene contains a small proportion of benzene as a contaminant. Raoult's law provides a means to predict the concentrations of toluene and benzene in the vapor space of the tank. The atmospheric conditions in the tank are ambient, 20°C and 760 mmHg (101 kPa). Vapor pressures of toluene and benzene (22 and 75 mmHg, respectively) were obtained from Table 5.3. The vapor pressure of the oil is considered negligible under these conditions.

The contribution of toluene to the vapor–air mixture depends on the mole fraction in the oil–toluene mixture. To calculate the number of moles of toluene, the mole fraction in solution is multiplied by the total number of moles. In this case, 1 m^3 of solution weighs about 950 kg, that being almost completely due to the oil. Thus, the total number of moles is approximately equal to the number of moles of oil, 950 kg/400 g/mole = 2,375 moles. (The molecular weight of typical blended motor oil is about 400 g/mole. The molecular weights of light viscosity oils, such as light hydraulic oil, light gear oil, and turbine oil, and medium viscosity oils, such as heavier hydraulic and gear oils, are approximately 350 and 450 g/mole, respectively.) The weight of toluene in solution

TABLE 5.6
Equilibrium Concentrations of Contaminants of Waste Oil

Liquid fraction			Air–vapor mixture		
Mole	Concentration		Partial pressure	Concentration	
Fraction	ml/L	g/m³	mmHg	ppm	g/m³
			Toluene		
0.1	25.2	21,850	2.2	2,895	10.9
0.01	2.5	2,185	0.22	290	1.1
0.001	0.25	219	0.022	29	0.1
0.0001	0.025	22	0.0022	2.9	0.01
			Benzene		
0.001	0.21	185	0.075	99	0.3
0.0001	0.021	18.5	0.0075	9.9	0.03
0.00001	0.0021	1.9	0.00075	1.0	0.003

is the (mole fraction of toluene) × 2,375 × (molecular weight of toluene). The contribution of benzene in turn depends on the level of contamination of the toluene. The density of liquid toluene and benzene are 0.867 and 0.879 g/ml, respectively.

Table 5.6 demonstrates the impact of the addition of small quantities of commercial grade toluene to waste motor oil. These calculations suggest that a small level of contamination of a nonvolatile fluid by a volatile one can produce high levels of airborne contamination in the vapor space. Commercial grades of toluene can contain small levels of benzene. Attainment of equilibrium in the airspace in a tank depends on the volume of oil, the volume of the airspace above the liquid, and contact time. Attainment of equilibrium would deplete the liquid initially put into the tank as the contaminants partition between oil and air. These calculations suggest that depletion of oil into the airspace would be a small fraction of the starting level.

Raoult's law holds for a component that approaches 100% in solution. That is, the vapor pressure of the solvent tends to follow the ideal vapor pressure curve as the limit of infinite dilution is approached (Barrow 1966). For mixtures where the mole fraction of the organic material in solution is 0.8 or more, Raoult's law holds with an error of 7% or less, except in extremely unusual cases (Perry 1984). Physical and chemical interactions between components of the mixture affect their individual vapor pressures. Most solutions are nonideal at some composition. In nonideal solutions, interactions between solute and solvent differ in magnitude from solvent–solvent and solute–solute interactions. This can result in greater or lesser vapor pressure for individual components than that predicted by Raoult's law (Moore 1962). In the case of complete immiscibility, the vapor pressure of the mixture is the sum of the vapor pressures of each component. Partial miscibility produces complex relationships.

Henry's law describes the equilibrium between gases and vapors in very dilute liquid solution and the gaseous phase above it. The vapor pressure of the solute varies linearly with concentration in the region of composition in which Henry's law applies (Barrow 1966).

Henry's law states that the partial pressure of a volatile component in the airspace above a solution equals the mole fraction in the solution times a constant.

$$p_i = x_i H_i \tag{5.20}$$

p_i = the pressure of component i in the gas phase;
x_i = the mole fraction of component i in the liquid phase;
H_i = the Henry's law constant for component i.

TABLE 5.7
Equilibrium Concentrations of Hydrogen Sulfide above Water

Solution		Air-gas mixture		
Mole fraction	Concentration (g/m^3)	Partial pressure (Pa)	Concentration ppm	Concentration g/m^3
0.001	1,890	55,200	545,000	763
0.0001	189	5,520	54,500	76
0.00001	18.9	552	5,450	7.6
0.000001	1.9	55	545	0.76
0.0000001	0.19	6	55	0.076

Henry's law applies to dilute solutions of flammable gases, such as methane, hydrogen, ethane, and ethylene in water. Values of H depend on the system and the temperature. In general, up to a total pressure of 5 atm (500 kPa or 3,800 mmHg), the value of p_i in equilibrium with x_i is independent of total pressure. If Henry's law applies to a system, then it usually applies for values of p_i up to 1 atm (101 kPa or 760 mmHg) (Geankoplis 1972).

Henry's law likely would apply to work situations involving confined spaces where total atmospheric pressure is 1 atm (101 kPa or 760 mmHg). Regardless of whether Henry's law applies, the relationship between gases and vapors dissolved in fluids, such as residual liquids and sludges, can be determined from direct observation. Henry's constants are available from various sources (Geankoplis 1972, Heinsohn 1991).

As an application of Henry's law, consider a tank containing an aqueous solution of hydrogen sulfide. The Henry's constant for dissolved hydrogen sulfide is 5.52×10^7 N/m^2 (Pa) or 4.15×10^5 mmHg at ambient temperature. Normal atmospheric pressure is 101,325 Pa. Table 5.7 presents possible combinations of airborne concentration and dissolved gas. To calculate the number of moles of H_2S, the mole fraction in solution is multiplied by the total number of moles. In this case, 1 m^3 of solution weighs about 1,000 kg, that being almost completely due to water. Thus, the total number of moles is approximately equal to the number of moles of water, 5.56×10^4. The weight of H_2S in solution is the (mole fraction) $\times 5.56 \times 10^4 \times$ (MW).

These calculations suggest that a very small quantity of hydrogen sulfide dissolved in water can produce a lethal airborne concentration in the equilibrium condition. This has major ramifications where water contains dissolved sulfide salts. Rapid acidification of such solutions would upset the ionic–molecular equilibrium.

Hydrogen sulfide is readily obtained from the action of acids on metal chalconides (complex chain-like structures) (Cotton and Wilkinson 1988). Hydrogen sulfide dissolves in water to form a 0.1-M solution at atmospheric pressure. The dissociations are given by the following equations:

$$H_2S + H_2O \rightleftharpoons H_3O^+ + HS^- \qquad pKa = 6.88 \pm 0.02 \tag{5.21}$$

$$HS^- + H_2O \rightleftharpoons H_3O^+ + S^{2-} \qquad pKa = 14.15 \pm 0.05 \tag{5.22}$$

Hydrogen sulfide behaves as a weak acid. Essentially, only HS^- ions are present in aqueous solution, owing to the small second dissociation constant. While S^{2-} is present in concentrated alkaline solutions, it cannot be detected below 8 *M* NaOH, owing to the formation of the HS^- ion in the following reaction.

$$S^{2-} + H_2O \rightleftharpoons HS^- + OH^- \qquad K \approx 1 \tag{5.23}$$

TABLE 5.8
Equilibrium Data for the Methanol–Water System

Liquid phase — mole fraction	Vapor phase			
	Partial Pressure (mmHg)	Conc. (ppm)	Partial pressure (mmHg)	Conc. (ppm)
0.0	**0.0**	0	**0**	0
0.05	**25.0**	32,900	**50**	65,800
0.10	**46.0**	60,500	**102**	134,000
0.15	**66.5**	87,500	**151**	199,000

Note: Temperature: 39.9C/59°C.

Only the alkali and alkaline earth elements form sulfides that appear to be mainly ionic and dissolve in water. Sulfides of other elements are extremely insoluble. Addition of acid to a sulfide solution could form H_2S and evolve gas from the solution. Organic and biochemical substances are potential sources of bound sulfur. This can be released through chemical and enzymatic action.

Another method for predicting behavior of vapors above solutions for which Henry's constants may not exist is empirical data. Table 5.8 provides data for the methanol–water system (after National Research Council 1929). As provided by the reference source, only the mole fraction and partial pressures were included (highlighted entries). Other entries were calculated, using equations provided previously in this chapter. The density of methanol is 0.791 g/mL.

These calculations suggest that small quantities of dissolved methanol produce large concentrations of airborne methanol relative to exposure and flammability standards. The concentration of methanol in water in usual units such as mL/L cannot be calculated without information about the mixing relationship between the two liquids.

The data should make available opportunities for interpolation and limited extrapolation. Partial pressure curves at different temperatures form a family of curves. Some approximations should be possible, especially for dilute solutions.

Together, the methods outlined in this section provide a basis for predicting the concentration of substances at equilibrium. However, equilibrium is an end point in a process that begins the moment a volatile substance appears on a surface inside a container. This substance can result from loss of containment, as in the case of a spill or from application of a coating onto a large surface area by brushing, rolling or spraying, or as a result of incomplete drainage.

ASTM Method D 1901-67 and D 3539-76 are laboratory methods for determining the rate of evaporation of volatile liquids (ASTM 1967, ASTM 1981). These methods have been applied to organic lacquer solvents and thinners, paint and varnish, hydrocarbon thinners, cleaners, naphtha, mineral spirits, and insecticide spray base oils. ASTM Method D 1901-67 determines the time needed for complete evaporation of a thin film of solvent from a panel of sheet metal. ASTM Method D 3539-76 determines the change in weight of a paper disk wetted by the liquid in an airstream of controlled temperature, humidity, and flow. Both methods use reference solvents for comparison.

Evaporation of liquids can occur in the open, in a building that possesses normal ventilatory flow, or under stagnant conditions within the confines of an enclosed structure. Any source of vapor can have important consequences for the atmosphere inside a confined space.

Evaporation of spilled liquids from a pool on the ground or the surface of water has been studied for many years. The surface temperature of the pool is the enigmatic quantity in this process (Daggupaty 1990). It is a function of many inputs: net solar radiation (outdoor), heat transfer between pool surface and the ground and container, heat transfer between liquid and air, and evaporative cooling of the liquid. Multicomponent liquids add to the complexity. The difficulty in modeling multicomponent systems is expressing vapor pressure as a function of the changing composition of the liquid (Stiver and McKay 1984). As a liquid evaporates, the more volatile materials evaporate preferentially, thus leading to a decrease in the vapor pressure of the remaining mixture.

The U.S. Environmental Protection Agency uses several equations for modeling evaporation and exposure (Hummel et al. 1996). In one example the value of G, the generation or evaporation rate, is calculated from the following equations. These equations are applicable where air is presumed to flow horizontally across the surface of the pool of liquid. The equation is applicable to liquids having low vapor pressure (less than 35 torr or 5% of the ambient pressure).

$$G = \frac{8.79 \times 10^{-5} \times MW^{0.833} \times VP \times \left(\frac{1}{29} + \frac{1}{MW}\right)^{0.25} \times v^{0.5} \times SA}{T^{0.05} \times z^{0.5} \times P^{0.5}} \tag{5.24}$$

G = generation rate in g/s;
MW = molecular weight of the evaporating substance;
VP = vapor pressure in atm;
v = air velocity in cm/s;
SA = surface area in cm^2;
T = temperature in °K;
z = pool length along direction of airflow in cm; and
P = atmospheric pressure in atm.

The following equation based on mass balance is used to calculate concentration (EPA 1991a, Fehrenbacher and Hummel 1996).

$$C = \frac{\left(1.7 \times 10^5\right) \times T \times G}{MW \times Q \times k} \tag{5.25}$$

C = concentration of contaminant in ppm;
T = temperature of ambient air in °K;
G = generation rate in g/s;
MW = molecular weight;
Q = the ventilation rate in ft^3/min. Rate of general ventilation ranges from 500 to 10,000 ft^3/min (14.2 to 283 m^3/min). Typical value is 3,000 ft^3/min (85 m^3/min). Worst case is 500 ft^3/min (14.2 m^3/min). For outdoor situations with minimal structure, ventilation rate is estimated at 26,400 × v (ft^3/min), where v is wind speed in mi/h. Average wind speed is estimated to be 9 mi/h;
k = a dimensionless factor that expresses the extent of mixing of the displaced ventilated air. The mixing factor is a function of room size and location of air inlet and exhaust. The following values of k are used: best mixing (0.67 to 1), good (0.5 to 0.67), fair (0.2 to 0.5), poor (0.1 to 0.2) (ACGIH 1995).

Another approach to the problem of evaporation was provided by Nielsen et al (Nielsen et al. 1995). These authors noted that other models usually are correlations of data obtained in ducts or open airstreams or open-air chemical spills. These authors combined an expression for the mass transfer coefficient from laminar boundary layer theory with the equation for generalized mass transfer. The equation that they derived is intended for use in situations with low horizontal airflows

(0.1 to 0.7 m/s) found in many workplace environments. This equation produced excellent results when compared to other predictive equations. Following is the equation that they derived.

$$R = 0.662 \times D^{2/3} \times \varphi^{-1/6} \times V^{1/2} \times \left[\frac{\left(L^{3/4} - L_0^{3/4}\right)^{2/3}}{L - L_0} \right] \times \frac{\ln \frac{M^S}{M^\infty}}{\frac{M^S}{M^\infty} - 1} \times \frac{P}{p_{lm,air}} \times \frac{VP}{RT} \tag{5.26}$$

R = the specific evaporation rate of pure liquid in mol/m²/s;
D = the diffusion coefficient of the substance in air in m²/s;
φ = the kinematic viscosity of air, 1.5×10^{-5} m²/s;
V = the air velocity, 0.1 to 0.7 m/s;
L = the length of the surface of liquid in the direction of airflow in m;
L_0 = the possible starting length preceding the evaporating surface in the same plane in m; the term containing L is the configuration term;
M^S = the average molar mass of the air–vapor mixture at the evaporating surface in g/mole;
M^∞ = the average molar mass of the air–vapor mixture in the workroom air in g/mole;
P = the total pressure in Pa;
$p_{lm,air}$ = the log-mean partial pressure of air;
VP = the vapor pressure of the substance in Pa;
R = the gas constant in Pa m³/mol/°K)
T = the temperature of the liquid (and vapor, assumed constant) in °K.

$$p_{lm,air} \approx \frac{VP}{\ln \frac{P}{P - VP}} \tag{5.27}$$

The situation represented by the preceding equations addresses evaporation where air is moving above the evaporating surface. The opposite extreme is evaporation of liquid from a surface where cross-current air motion is not occurring. The air above the liquid is assumed to be stagnant. This could occur in a closed tank containing wetted interior surfaces or pooled liquid. Evaporation is assumed to occur by molecular diffusion.

Molecular diffusion of pure substances (gas or vapor) in a gas (air or other gas) is regulated by the molecular diffusion coefficient (D). The value of D is determined experimentally or may be obtained from reference sources (Treybal 1980, Reid et al. 1987, Feigley and Lee, 1987, Heinsohn, 1991). The molecular diffusion coefficient varies with temperature and pressure according to the following equation:

$$\frac{D_2 P_2}{\left(T_2\right)^{1.81}} = \frac{D_1 P_1}{\left(T_1\right)^{1.81}} \tag{5.28}$$

D = molecular diffusion rate in m²/s.

One approach to solving the question of evaporation in a stagnant environment is to consider a container having an open top through which vapors can escape (Heinsohn 1991). This approach incorporates the following assumptions: air above the opening does not enter the container, nor does it induce motion in the airspace; air does not dissolve in the volatile liquid. Also, air currents carrying off the vapor maintain the concentration at the opening at negligible levels. The concentration gradient is considered to exist only in the vertical direction. Evaporation rate is given by:

$$G = \frac{P \times D \times \ln\left(1 - \frac{VP}{P}\right) \times SA \times MW}{R \times T \times h} \tag{5.29}$$

G = generation or evaporation rate in g/h;
D = molecular diffusion rate in m^2/s;
VP = vapor pressure in kPa;
P = atmospheric pressure in kPa;
SA = surface area in m^2;
MW = molecular weight;
R = universal gas constant, 8.314 kJ/(kmol °K);
T = temperature in °K;
h = height in m.

This calculation assumes no back diffusion and condensation into the liquid. To a first approximation, this approach would describe the early part of the process in a completely closed container before condensation and pressure effects became significant. Vapor pressure of the volatile component is the dominant factor controlling the emission rate.

The preceding calculation applies to pure substances. Each volatile component in a mixture containing several volatile components will evaporate at a unique rate. As a result, the composition of the liquid will change because of preferential evaporation of the more volatile liquids. Consequently, the total vapor pressure decreases.

Evaporation is an important application of mass transfer concepts. Mass transfer describes the movement of matter at the molecular level resulting from a concentration gradient, rather than motion of bulk fluid due to mechanical work. The driving force for this movement is the difference in concentration between two spatial locations (Incropera and DeWitt 1985).

Air–liquid contact and sometimes air–solid contact are important mechanisms leading to the development of hazardous atmospheric conditions in confined spaces. Generally, this process occurs at ambient conditions of temperature and pressure. Initially, the contaminant is present in the liquid or solid phase as residual contents or sludge. Diffusive movement of the contaminant occurs within phases and across the phase boundary.

Vaporization and the opposing process, condensation, occur simultaneously at the surface of liquids. When a volatile surface is first exposed to air, vaporization occurs virtually uninfluenced by condensation as the vapor enters the atmosphere just above the volatile surface. At the same time, transport of vapor into the airspace of the container is occurring by diffusion. In an open-topped container, this creates and maintains a concentration gradient between the liquid–gas interface and the uncontaminated air at the top of the container. Equilibrium between vaporization and condensation cannot develop under these conditions.

In a closed container, accumulation of vapor in the airspace above the interface gradually occurs. As vapor accumulates, condensation becomes increasingly more important as an opposing factor. The vaporization rate gradually decreases and the condensation rate gradually increases. At equilibrium, both the vaporization rate and condensation rate are equal. In addition, the concentration of vapor at any point in the airspace above the volatile substance becomes constant.

The rate at which a component transfers from one phase to the other depends on the mass transfer or rate coefficient and the extent of departure of the system from equilibrium (concentration difference). Net transfer stops when equilibrium is attained.

An increase or decrease in temperature changes the rate of each process. As the temperature of the liquid in a container increases, vaporization increases. This, in turn, increases the total pressure exerted on the interior walls of the container. The escape of the gasoline vapor–air mixture from the filler pipe of a gas tank on a hot day at the filling station indicates the importance of temperature on vaporization.

Diffusion can occur in two modes: molecular diffusion and eddy or turbulent diffusion (Geankoplis 1972, Treybal 1980). Molecular diffusion occurs in fluids that are stagnant or undergoing laminar flow. Turbulent diffusion occurs when the fluid undergoes turbulent motion. In laminar flow the fluid flows in smooth streamlines. In turbulent flow, there are no orderly streamlines. Rather, large eddies or "chunks" of fluid of varying size move in seemingly random fashion. These eddies can move rapidly from one part of the fluid to another, including perpendicular to the direction of flow. As a result, the concentration of solute differs from one eddy packet to another. The rapid movement from one part of the fluid to another, combined with the transfer of relatively large amounts of solute per packet, enhances mass transfer from the interface. Eddy or turbulent diffusion is very rapid compared to molecular diffusion.

Adjacent to the gas–liquid (or solid) interface is a quiescent laminar layer in which mass transfer occurs by molecular diffusion. Adjacent to this layer is a transition region where both molecular and turbulent diffusion occur. This region is characterized by a gradual transition from one form of diffusion to the other.

Under conditions of turbulence, molecular diffusion is relatively unimportant. Mass transfer coefficients then become much more nearly alike for all components. Mass transfer coefficients regulate the rate at which equilibrium is approached. In the nonturbulent conditions of many confined spaces they control the time required for formation of the hazardous atmosphere. While mass transfer coefficients can be calculated for molecular diffusion, they cannot be computed for turbulent flow because the conditions cannot be described mathematically.

Roach first explored the role of turbulent diffusion in dispersion of contaminants in workplace situations (Roach 1981). Franke and Wadden have shown that turbulent diffusion is 1,000 to 10,000 times greater than molecular diffusion in workplaces, and hence is the operative mode in contaminant dispersion (Franke and Wadden 1985). (Molecular diffusion coefficients are in the range of 0.5 to 2×10^{-5} m^2/s) (Heinsohn 1991). Even the human body is a potential agent in the turbulent mixing of airborne contaminants. Flynn and Ljungqvist reviewed studies of airflow patterns around the human body (Flynn and Ljungqvist 1995). They reported that boundary layer separation leads to the formation of a wake region downstream from the body, similar to that behind a ship moving through the water. This could occur in an air stream or by motion of the body in stagnant air. This wake is characterized by eddies or vortices that entrain air into a reverse flow region near the body. That is, the wake is a mixing zone and the entrained airflow and contaminant generation rate determine exposure. Similar properties can be expected from the motion of other equipment, such as forklifts.

MODELING THE ENVIRONMENT IN CONFINED SPACES

Discussion

The ability to model the process by which atmospheric contamination develops would be a valuable asset for understanding the problem posed by confined hazardous atmospheres. A model is a mathematical expression that attempts to describe an empirical reality. A model could provide the basis for predicting the airborne concentration of a contaminant as a function of time and other variables. A model would permit an assessment of the impact of control measures, such as ventilation, and conditions, such as proximity to the source and mode of generation. The health hazard of a particular activity could be assessed by comparing, for example, maximum projected concentration against exposure standards and guidelines.

Models for assessing conditions that develop in confined spaces presently do not exist per se. Models developed for use in other areas of endeavor may be adaptable to the peculiarities of confined spaces. Sources of potentially compatible models include:

- Workplace emission models
- Indoor air quality models
- Chemical spills models
- Environmental transport and dispersion models

These will be discussed later in this chapter. The types of models developed in these areas include:

- Box models
- Dispersion models
- Fluid mechanical models

The box model treats the volume of the workspace or a designated portion of it as a box (Jayjock 1988, Jayjock 1991). The box is defined by the boundaries of the sampling points or by the walls of the space. The mass of the contaminant is assumed to be conserved as it is generated into the box and removed from it. The box model describes the emission rate within the box as a function of the upwind and in-box concentrations, box size, and advective velocity. The concentration of a contaminant in the box is equal to the upwind concentration plus the contribution due to internal sources less losses due to removal mechanisms. Removal from the box can occur by advection, physical transport out of the box, chemical transformation inside the box, and absorption/adsorption onto surfaces in the box. The box model assumes rapid mixing throughout, that the system is at steady state, and that air velocity and pollutant emission rate are constant.

Moss discusses the characteristics of calibration and exposure chambers that apply the box concept (Moss 1989). The need for precision in assuring conditions, such as instantaneous mixing and uniform, well-mixed atmospheres, is considerably greater than what would be required in other situations, since the box is small by comparison. Following instantaneous mixing of each incremental increase in contamination, an exponential buildup of concentration occurs. Moss discusses the consequences of poor mixing and other deficiencies, including leakage and plate-out, on concentration within the chamber.

In the context of confined spaces, box models would work well when the source is well distributed, as for example, during the application of a coating to the interior surfaces. The walls of the confined space could serve as the walls of the box. The box model would not work well where multiple point sources are present. The ability of contaminants to mix uniformly is seriously compromised. A mixing efficiency factor attempts to compensate for variable mixing under these conditions. This mixing efficiency factor serves two functions (Park and Garrison 1990). The first is to approximate the effectiveness of fresh air in mixing with contaminated air. The second is to apply a safety factor in an attempt to provide some level of overdesign in ventilation flow rate.

Another limitation involves a point source in a space of large volume and variable movement by the affected individual. The model also would not work well where stratification occurs, as in a tunnel where exhaust gases from mobile equipment stratify at the ceiling.

In a more complex version of the box model, Nicas has modeled a room as a two-zone system (Nicas 1996). Ventilation air enters the upper zone and room air and unmixed ventilation air exits from it. Occupants normally contact air in the lower zone. This model has possible application where stratification has occurred.

The limitations of the box model become the strengths of dispersion models (Jayjock 1988). A model that considers dispersion from point sources omnidirectionally by random air motion from a point source into infinite space is the eddy diffusion model. This model considers the dimensions of the volume around the source in which the majority of the contaminant would be measured, relative to the time frame of its removal by ventilation or other mechanism. Diffusion exerts its greatest influence on concentration within a short distance from the source.

This model requires a description of the spatial dispersion of an instantaneous release of a finite amount of contaminant during the time period in order to gain insight into the size of the affected volume. A technique to determine this using tracer gases has been described by Antonsson (Antonsson 1990). This utilizes the tracer gas to create a continuous twin source. The emission rate of the industrial source is determined from correlation with the controlled release of the tracer. The success of this approach is contingent on similar densities between the contaminant and the tracer and lack of confounders such as particulates in the contaminant airstream.

The speed with which the emitted contaminants move through space is directly proportional to the product of the change in concentration with distance from the source and molecular diffusivity (Fick's law of diffusion) (Wadden et al. 1989). In indoor spaces where air motion is approximately random, eddy diffusivity is more appropriate than molecular diffusion. Elevated sources are best described as a sphere and floor-level ones as a hemisphere.

The eddy diffusion model depends on randomized air motion and entrainment of contaminant. This would not be the case where directed airflows are employed. These would include supply ventilation through flexible ducts whose termination is close to the affected worker. In other situations flow from supply air from ceiling level diffusers also would distort dispersion. Similarly, exhaust ventilation located to remove contamination at the source also would distort the process. The eddy diffusion model would not be able to accommodate plumes from an energetic source, such as a welding plume.

A variant of the eddy diffusion model is the advection–diffusion model (Scheff et al. 1992). The basis of this model is diffusion in all directions and advection (directed air motion or cross-draft) in the x-direction. The advection–diffusion model is useful in cases where advection (directed air motion in the x direction caused by wind or mechanical induction) and diffusion (all directions) are both significant. This model can be solved for concentrations resulting from emissions into an infinite space with advection in the x direction, neglecting plume rise. The model also can accommodate two sources. The equation for the two-source model has four unknowns. It is exactly solvable if simultaneous measurements of concentration are available from four locations.

The previously mentioned models accommodate two sources of emission. In circumstances where many sources are present, the best approach may be the completely mixed space mass balance model (Wadden et al. 1994). This is a variant of the box model. This model assumes uniform mixing. A mixing efficiency factor is incorporated into the model. This model utilizes averaged data from air sampling, as well as volumetric airflow rates at entry and exit points.

In a subsequent article, Keil et al. compared emissions in a printing shop using two mass balance models: the experimental mass balance model and the completely mixed space model (Keil et al. 1997). The experimental mass balance model is most useful where air entry and exit (air balance) are easily identified and quantified. The completely mixed space model assumes that pollutant concentrations approach completely mixed conditions. This model is applicable when identification and measurement of entry and exit points is difficult. In this case, the mixing factor is a major unknown. The completely mixed space model also is more appropriate in airspaces having lower net air exchange rates. These result in a longer residence time for the pollutant.

Some of the models describe bulk flow of fluids. Bulk flow occurs following rupture of containers, especially those under pressure and pipelines. It also occurs in rooms and other workspaces under the influence of ventilation and motion of people and equipment. These models utilize concepts of fluid mechanics.

Fluid mechanics describes the behavior of fluids at rest and in motion (Hughes and Brighton 1967). Fluids include liquids, and gases and vapors. The study of fluids is concerned with macroscopic properties rather than behavior at the atomic or molecular level (Fox and McDonald 1985). A fluid is considered to be an infinitely divisible continuum, rather than an assemblage of atoms or molecules. A consequence of the continuum concept is that each property of the fluid is assumed to have a definite value at every point in space. The only force exerted by the fluid on the container is that due to pressure. A fluid remains at rest in the absence of shear (tangential) stresses. A fluid in equilibrium (at rest) therefore cannot sustain tangential or shear forces without deforming. A fluid deforms continuously under the application of a shear (tangential) force, no matter how small.

A fluid in motion deforms continuously when a shear stress is applied. Fluid motion can occur simultaneously in three dimensions.

Studies in air pollution and micrometeorology utilize fluid mechanical concepts to describe the behavior of gases and vapors in the outdoor environment (Perkins 1974, Bowne 1984). From the perspective of confined spaces and atmospheric confinement, airflow near boundary surfaces is an important consideration (The boundary layer is the region near the surface where velocity is less than 99% of that in the free air stream due to frictional effects) (Hughes and Brighton 1967).

Studies of stagnation and stratification of air are of considerable interest in the context of confinement of denser-than-air gases and gas and vapor mixtures. Gases and gas and vapor mixtures can originate from a stack as a denser-than-air plume, as a puff resulting from spurious emission from process equipment, as evaporated liquid from leaks or spilled liquid, or seepage from the ground. Pooling and confinement can occur in open channels, trenches, berms, subgrade structures, open pit mines, and so on. During pooling, dilution of the cloud through diffusion is greatly reduced or even eliminated. In nearly calm nighttime conditions, the time to remove the dense cloud from the valley can be many hours.

Briggs et al. performed wind tunnel experiments using simulated valleys and stratified dense gases to study entrainment and flushing by crosswinds (Briggs et al. 1990). The purpose for these studies was to model behavior of dense toxic gases and vapors arising from accidental releases into the stratified air of a valley during evening and nighttime hours and to quantify flushing (entrainment) by crosswinds. Flushing a large volume such as a valley would require many hours, according to their calculations. They predict that a modest release of 1,000 kg of chlorine into a holding pond 900 m^2 would require 40 min for evacuation by a wind of 1 m/s at 20°C.

Also of concern is contamination resulting from sprays and jets. Most well known to the industrial hygienist are sprays used in the application of coatings and spray washing and degreasing. Jets can arise from rupture of pipelines. These also include release points for pressurized gases used in welding, burning, and cutting. Jets result from release of liquid stored under pressure and at temperatures exceeding normal boiling point (Papadourakis et al. 1991). Material released under these conditions may emerge as a liquid and can flash to form a two-phase jet of liquid or droplets and vapor. Processes accompanying the formation of the jet can include evaporation, coalescence of droplets, liquid rain-out, vapor condensation, ambient humidity condensation, and reaction with released material.

Workplace Emission Models

Workplace emission models consider normal workplace conditions and the relationship of the source to the recipient of exposure.

Jayjock reviewed possible approaches for modeling inhalation exposures in the workplace (Jayjock 1988). He discussed the box model in some detail, pointing out strengths and limitations in application to the workplace environment. He also examined the role of dispersion models that incorporate eddy diffusion.

Jayjock further investigated a refinement to the box model through incorporation of a term for backpressure as a retardant to evaporation (Jayjock 1994). Backpressure retards evaporation due to the presence of vapor previously evaporated. The retarding effect of the contaminant is absent only at the beginning of the evaporative process. That is, the rate of evaporation decreases from its maximum value as the products of the evaporation accumulate. The effect of backpressure is especially important where the ratio of the area of the evaporating surface to the volume of the box is large. The effect of increasing ventilation rate under these circumstances is to increase evaporation, with little effect on airborne concentration. That is, the increased ventilation removes the backpressure and stimulates evaporation. This effect does not appear to be significant where the surface area of the evaporating source to the volume of the box is small.

The model predicts that dilution ventilation would be relatively ineffective in controlling exposure to sources having large vaporizing surface area-to-volume ratios. This outcome is especially troubling for confined spaces that contain large wetted surfaces of residual contamination or during the application of coatings. This suggests serious problems with the use of portable ventilation systems as a control strategy for vapors in large structures, such as tanks under these conditions.

In a later article, Jayjock et al. reported on investigations about the role of sinks and sources within the box (Jayjock et al. 1995). Sinks adsorb or absorb vapor. At the same time, they also can become additional sources through desorption. Processes that can occur in sinks include chemical, physical, or biological degradation. Adsorption is a measure of the amount of substance that is airborne. Desorption is a measure of the amount of substance that is in the sink.

Wadden et al. used a point source dispersion model (two-point eddy diffusion model) to determine emissions from two vapor degreasers (Wadden et al. 1989). This group also used this approach with chromium plating (Conroy et al. 1995). They used concentration patterns surrounding these units to develop mass emission rates and emission factors. Measurement of an emission factor puts area concentration data into a format suitable for estimating emissions from the same type of equipment at other sites. This also provides a quantitative basis for developing engineering controls.

Wadden et al. compared the utility of the two-point eddy diffusion model and the completely mixed space (box) model for estimating emission factors for several chemical processes (Wadden et al. 1991). Each is based on a mass balance for the contaminant in a specified volume: a semi-infinite sphere or hemisphere, or room volume. The point diffusion model is suited to workspaces having discrete sources. The completely mixed space (box) model is more suitable to a distributed layout, since this considers the workspace to be a completely mixed container. Each model incorporates some, but not all, of the major features that describe the spread of a contaminant from single or multiple sources. This investigation indicated that the two-point eddy diffusion model was more suitable for modeling vapor dispersion. Cross-drafts were believed to have exerted an unconsidered influence on the data.

Scheff et al. explored the use of the box model and the two-source advection–diffusion model to predict emissions from vapor degreasing units (Scheff et al. 1992). The two source advection–diffusion model is a variant of the two point eddy diffusion model that incorporates a cross-draft. The two models provided similar estimates for parameters quantifying emissions from the degreasers.

Wadden et al. utilized the completely mixed space mass balance model in a candy factory and printing shop (Wadden et al. 1994, Wadden et al. 1995). This is a variant of the box model. It is applicable in circumstances where many identical or nonidentical sources are present. This model assumes uniform mixing. A mixing efficiency factor is incorporated into the model. This model utilizes averaged data from air sampling, as well as volumetric airflow rates at entry and exit points.

Bjerre created a six-parameter box model to express average and maximum concentrations of solvent vapor in indoor air during the application of solvent-based products, such as paints and adhesives (Bjerre 1989). The model is valid for solvents that evaporate from the liquid without fractionation and at a constant rate. This model provides support for the results obtained by Jayjock (Jayjock 1994). The model assumes complete mixing of solvent vapors with the room air. The assumption that evaporation rate is independent of time restricts the solvent to either a pure substance, an azeotropic mixture, or a mixture of hydrocarbons that boil within a limited temperature range. No matter which solvent is used, a decrease in evaporation rate occurs during the final stage of drying when resistance to diffusion through the coating becomes comparable to the resistance in the gas film adhering to it. The model will be unsatisfactory in spray applications when the aerosol evaporates appreciably before deposition. The assumption of perfect mixing restricts application of the model to rooms of limited size and regular form.

Haberlin and Heinsohn used a sequential box model to predict time-varying concentrations at arbitrary points within a large tank during recoating (Haberlin and Heinsohn 1993). Work proceeds on wedge-shaped sections of the tank. The methodology used in this paper was the multicell well-mixed model (sequential box model). This model assumes that well-mixed conditions exist within each slice or wedge.

Park and Garrison developed multicellular models as part of their studies of ventilation in model confined spaces (Park and Garrison 1990). These models divide three-dimensional space into subspaces or cells and describe migration between the cells. The model presumes instantaneous uniform mixing of a contaminant within each cell. Movement of contaminant within the larger space results from interaction across cell boundaries.

These models were patterned after systems developed to model migration of water pollution in lakes and streams. This approach involved design of cell structures to accommodate the geometry and mass flow characteristics of three-dimensional space. This also involved approximation of dispersion coefficients to describe transport in addition to that resulting from mass flow between cells. The model utilizes a mass balance equation to predict contaminant dispersion as a function of time. The primary limitations of the multicellular method are associated with experimental approximations of flow patterns and dispersion coefficients. This model assumes that reactions of contaminants and buoyant effects inside the three-dimensional space are negligible.

Fontaine reported on the use of a computational fluid dynamics model to predict air velocities, contaminant concentrations, and air temperatures inside ventilated workspaces (Fontaine 1991). The basis of the model is a software simulation package, EOL. EOL predicts airflows, dispersions within ventilated premises, time evolution of concentrations, aerosol flows, and velocities in three dimensions. EOL also calculates velocity fields and captures efficiency of local exhaust systems. EOL considers a room to be constructed from parallelepipedic elements. It can accommodate laminar and turbulent airflows, natural and forced convection, and transport of gases and aerosols.

The behavior of aerosols also has generated the interest of modelers. Sutter and co-workers examined behavior of solid and liquid aerosols in a chamber of fixed volume (Sutter et al. 1982, Sutter 1983, Sutter and Halverson 1984). The chamber was equipped with several high-volume samplers that discharged into the chamber. Sample collection by these units amounted to vacuum-cleaning aerosols from the air in the chamber. The samplers were run until of 1% of the starting level of aerosol remained in the air. The results represent the average concentration over the period of collection. The first report examined the effect of overturning a beaker and allowing the contents, a fine powder or a solution, to drop onto the floor of the chamber (Sutter et al. 1982). The second report examined pressurized release of powder, as might happen through a pressure-relief system (Sutter 1983). This configuration was examined as a follow-up to an incident that had occurred. The third report examined the generation of liquid aerosols under equivalent conditions (Sutter and Halverson 1984). The chamber confined the plume that resulted from the pressurized releases.

Cowherd et al. described an apparatus for determining dustiness of materials (Cowherd et al. 1989a). This unit tips a container of powdered material onto a surface. Dust generated during this activity is collected onto a filter. They followed up with an investigation to determine the relationship between dustiness parameters determined from use of the testing unit and personal exposure in a simulated work situation (Cowherd et al. 1989b).

Cooper and Horowitz examined dust generation following the dropping of powder into an enclosed column and onto the floor of an open room (Cooper and Horowitz 1986). They compared the results obtained from air sampling against two models for predicting exposure following powder spills: the perfect mixing model (a box model) and the turbulent puff diffusion model (a dispersion model). Their results plus a subsequent letter of clarification indicated that the dust concentrations decreased with distance and adhered to the turbulent puff model (Cooper 1986).

Plinke et al. reported on factors involved in dust generation in a situation more typical of industrial experience: continuous fall of powdered material from a hopper through air onto a receiving surface (Plinke et al. 1991). They also reviewed models for dust generation published by other authors. The factors examined in this report were moisture content, drop height, and material flow. Dust generation increases with increasing drop height and decreases with increasing moisture content. Dust generation with rate of flow is specific to the material. In a follow-up report, Plinke et al. examined dust generation during the handling of granulated materials (Plinke et al. 1995). Cohesion due to material properties affected dust generation more than impaction arising from material handling. Size distribution of the granular material was not as important as either cohesion or impaction.

Nazaroff and Cass developed a mathematical model for indoor aerosols (Nazaroff and Cass 1989). This model predicts concentration and fate of particulate matter in indoor air. The model accounts for ventilation, filtration, deposition onto surfaces, direct emission, and coagulation. Initial applications were expected to be modeling of tobacco smoke, radon, and radon progeny, and soiling of industrially important surfaces by aerosols.

Indoor Air Quality (IAQ) Models

Interest and concern about the indoor environment spurred efforts to model the impact of design, choice of building materials, and effects of activity. Some of these have potential application to confined spaces.

A series of review articles by Grot and Lagus illustrates modeling concepts used with tracer gas techniques (Grot and Lagus 1991a,b,c,d). The word "contaminant" easily could be substituted for tracer gas. They point out that the central problem in modeling dispersion or migration of airborne contaminants is accurate prediction of the spatial and temporal variation of concentration for all processes that may affect the dispersion. In these models, there is one well-mixed zone into which air enters through known or unknown flow paths and from which it exits through known or unknown paths.

Following are the basic assumptions implicit in the derivation of equations used in these models. The tracer gas in the zone is well mixed and therefore can be represented by a concentration density independent of spatial position in the room. This assumption can be relaxed somewhat by assuming that the average concentration in the zone is the same as the average value of the tracer in air that flows out of the zone. The air in the room has uniform density which is independent of spatial position in the room.

Equations in Appendices D to H of ASHRAE Standard 62-1989 address a number of themes in indoor air quality, starting with physiological requirements for respiration (ASHRAE 1989). A simple mass balance equation gives the outdoor airflow rate needed to maintain the steady-state concentration of carbon dioxide below a given limit. Appendix E provides models for assessing the use of cleaned or recirculated air. The requisite amount of outdoor air depends on contaminant generation in the space, contaminant concentration in indoor and outdoor air, filter location, filtration efficiency, ventilation effectiveness, supply air circulation rate, and the fraction recirculated. Ventilation effectiveness depends on location of the supply outlet and return inlet, and the design and performance of the supply diffuser. Appendix F provides a model for calculating ventilation effectiveness. This model considers stratification of air between the zone of delivery above the level of the occupants and the occupied zone. A fraction of the supply air may bypass directly to the return inlet without mixing at the occupied level. The model in Appendix G provides the rationale for delaying operation of the ventilation system upon occupancy and utilizing existing air for dissipating and diluting contaminants. Appendix H provides a model for providing outdoor air to multizone systems.

The U.S. EPA has developed several models for assessing exposure to indoor air contaminants (USEPA 1989a, 1989b, 1991b, 1991c). EPA has investigated emissions from furnishings and building materials and use of chemical products indoors. As part of this activity, EPA produced models for predicting exposure to airborne contaminants. These are broadly classed as source emission and indoor air quality (transport) models. The latter are discussed here.

Indoor air quality (transport) models characterize the movement of air pollutants through defined indoor spaces. These models provide an estimate of the concentration of pollutants in a given microenvironment under specified conditions. These models have potential application to confined spaces when ventilation rates are severely restricted. Many industrial products have compositions similar to those mentioned in the models that are intended for consumer use.

Indoor Air Quality Model Version 1.0 (INDOOR) was written to run on MS-DOS-based microcomputers (USEPA 1989b). This model was designed to estimate the impact of various sources on air quality in a multiroom building. The model treats each room as a well-mixed chamber that contains pollutant sources and sinks. The model provides a wide variety of source terms including random on–off sources, such as cigarettes, sources such as heaters that operate for specified periods, steady-state sources such as mothballs, and sources such as floor wax, whose emissions decrease with time.

EXPOSURE Version 2 is the successor to Indoor Air Quality Model Version 1.0 (USEPA 1991b). Improvements offered by EXPOSURE over INDOOR are the ability to calculate individual exposure due to a given source and to model patterns of personal activity. The model permits estimation of exposure through use of products such as aerosol sprays and wood stains. The model calculates instantaneous and cumulative exposure.

A third model created by EPA, Indoor Air Quality Simulator for Personal Computers, provides concentration profiles to cover up to 62 days based on information entered on building layout, heating, ventilating and air-conditioning parameters, and sources and sinks (USEPA 1991c). This model can handle six contaminants simultaneously in single-story layouts having up to 20 rooms.

The National Institute of Standards and Technology (NIST) also produced computer models on air movement and contaminant migration (U.S. Dept. of Commerce 1991). NBSAVIS is a description processor for multizone buildings. It creates and edits a building description, calculates zone and opening data including leakage, and fan and contaminant data required to predict infiltration and internal air movement and perform an IAQ analysis. CONTAM88 performs dispersal analysis for multiple reactive and nonreactive contaminants and calculates airflows and dynamic and steady-state levels of indoor contaminants. Both of the preceding are macroscopic models.

NIST also is developing a microscopic flow model based on a fluid mechanical turbulence model. EXACT3 uses numerical methods to predict air movements in rooms. EXACT3 estimates local airflow fields around obstacles and models levels of contaminants, such as carbon dioxide in these situations (U.S. Dept. of Commerce 1990).

Chemical Spills Models

The dynamics of emission of chemical substances into the environment have been investigated for some time. These emissions occur following loss of containment or from deliberate releases from stacks and vents. There is considerable need to be able to predict the impact of these accidents on the plant, the surrounding community, and the environment. Behavior of spilled chemicals is a major concern in environmental emergency response and rehabilitation. Models addressing emission of industrial chemicals into the environment focus on the following areas: spills and releases in-plant and during transportation, and the fate of waste products following disposal by burial in the ground. These models derive from laboratory-scale experiments of evaporation to large-scale staged releases.

Spills and process emissions in the industrial environment can occur indoors or outdoors. The behavior of denser-than-air vapors and gases can create serious consequences for work occurring concurrently or subsequently in confined spaces downwind from these sources.

Fingas described the urgency and utility of computer models and databases on site for assessment of hazards and management of response (Fingas 1991a). These can be especially valuable in extreme cases when little information is known or available from conventional sources about the spilled chemical. In another article, Fingas surveyed computer models that were currently available and their suitability for use at spill sites (Fingas 1991b). Models for chemical spills describe aspects ranging from the microenvironment, involving the immediate area, to the macroenvironment, involving geographic distribution.

The Center for Chemical Process Safety reviewed models for estimating conditions that can develop following chemical accidents (Hanna and Drivas 1987). Source emission models are catalogued according to the following source conditions:

- Liquid or two-phase jet from pressurized tank
- Gas jet from pressurized tank
- Slowly evaporating pool
- Momentum jet and dense gas jet

Probable source conditions typically included tank rupture, pipe break, and venting of runaway reactions. These models attempt to describe very complex conditions.

Fluid discharge from a pressurized container or pipeline normally occurs in two ways (Lees 1980). The discharge will be liquid when the opening is below the liquid level, and vapor or a vapor–liquid mixture when above. For a given pressure difference, usually considerably more liquid or vapor–liquid mixture escapes than gas or vapor. Fluid having high momentum forms a turbulent momentum jet. Liquid having low momentum escapes as a liquid stream. Liquid having high momentum escapes as a high "throw" liquid jet. The circumstances of the release, indoor or open air, and the height greatly affect dispersion of the material.

Gases and vapors can form a puff or a plume. A puff results from an instantaneous or short-term release. A puff separates from the source when the emission ceases (Hanna and Drivas 1987). A plume results when the release occurs over a long duration. Pipeline breaks or liquid leaks from large storage tanks that subsequently evaporate generally produce plumes. During a pipeline break, some of the fluid (liquid at the initial operating pressure) may rapidly vaporize or flash to the gaseous phase as the pressure drops. As the pressure is relieved upstream, more liquid flashes to vapor.

Vaporization depends on the physical properties of the spilled liquid (Lees 1980). A liquid, such as acetone, evaporates relatively slowly and provides a steady and continuous source of vapor. The process depends on the vapor pressure of the liquid and wind flow across the surface of the pool. Heat transfer from the air and the ground provides the latent heat needed for vaporization and prevents cooling of the liquid unless the rate of evaporation is rapid due to the combination of high vapor pressure and wind speed.

Evaporation of superheated liquids occurs in two stages. Liquids become superheated due to the conditions of storage. Superheating occurs when a liquified gas is stored under pressure at ambient temperature or a liquid heated beyond its normal boiling point is stored under high pressure. When spilled, a portion flashes off. The remainder, cooled by removal of the latent heat, evaporates more slowly at its normal boiling point. The secondary stage usually is regarded as less important than the initial flash-off, particularly when a cloud of flammable gas may form.

Gases, such as methane, that are liquefied by refrigeration at low temperature and stored at atmospheric pressure, also evaporate in two stages. When spilled, this type of liquid evaporates rapidly at first and then more slowly as the surroundings are cooled.

Environmental Transport and Dispersion Models

Environmental transport and dispersion models focus on the ultimate fate of gases and vapors. Dispersion and dilution of a dense gas or vapor occur in several steps. Different physical processes occur in each. These processes can include source-controlled dilution, gravitational slumping, combined gravitational slumping and ambient flow interaction, and passive dispersion (Raj 1985).

Transport and dispersion of dense gases highly depend on local topography, including both natural and man-made features (Hanna and Drivas 1987). Dense gases flow like water, seeking low spots and deflecting around obstacles such as hills and buildings. This behavior also affects gases having neutral buoyancy to some extent. Dense gases released from an elevated stack or vent may slump down to ground level. Slumping poses considerable concern regarding potential for accumulation in subgrade structures.

Four basic approaches are used for modeling the dispersion of pure heavy gases: box models, similarity models, the intermediate approach, and numerical models (Raj 1985).

Box models consider only overall features of the dispersing cloud (or a plume), such as mean radius (or plume width), height, concentration, and temperature. These models presume uniform internal conditions across the cloud after a specified travel time or downwind distance. The principal phenomena considered in box models are gravitational slumping of the cloud, entrainment of ambient air, and turbulent dilution.

All box models regard the initial cloud to be describable by a right circular cylinder or rectangular box. Invariably, they assume that the dispersing vapor and air are ideal gases. Presuming the vapor and air to have the same molar specific heat results in conservation of volume following adiabatic mixing at any given pressure or temperature with ambient air. Final volume of the mixture is the sum of the initial volume of the gas and the volume of air added to it. As a result, buoyancy remains constant. This aspect has been assumed in almost all of the box models.

When the cloud is diluted and the effects of negative buoyancy are small, the cloud disperses essentially as a passive cloud. Dilution is mainly controlled by atmospheric turbulence. In this case, the concentration distributions tend to become Gaussian.

Some modelers distinguish between box models and so-called "similarity" models (Hanna and Drivas 1987). The Gaussian model is a similarity model because the crosswind distribution is always similar (Gaussian). Simple box or similarity models agree fairly well with the limited databases on the transport and dispersion of hazardous materials (Raj 1985). Yet, they fail to address the structure within the plume or cloud.

Intermediate models solve for spatial and temporal variation of concentration and other quantities in one or two horizontal dimensions within the dispersing cloud. "Shallow layer" models provide true representation of the gravity head, within which most of the dispersing mass appears to be located during the initial period of dispersion of a puff. However, they neglect the vertical gradients or simplify distributions in the vertical direction.

Numerical approaches solve the three-dimensional equations of motion (mass, momentum, and energy), together with certain recipes representing the turbulence field. The solution utilizes numerical schemes, and requires the representation of the entire flow field in the form of nodal grids. Each of the properties of the cloud is obtained at each "node" for each time.

At some travel time or downwind distance, the cloud or plume ceases to be influenced by the density perturbation (Hanna and Drivas 1987). Subsequent transport and dispersion proceeds as if the plume were neutrally buoyant. Most of the models make arbitrary assumptions about the transition point. At the transition point, most box or slab models change to a Gaussian plume or puff model. Mass fluxes should be conserved, maximum concentrations should be continuous, and the plume dimensions conserved, if possible.

The Gaussian equation should be used only at downwind distances (greater than 100 m) where initial jet effects have become unimportant and plume density has decreased to near ambient. The transition usually is accomplished by matching emission parameters with distribution parameters (e.g., cloud height and width) from the dense gas model with Gaussian parameters.

The transition from an initial jet model to a dense gas slumping model relies on the notion that once the dense gas plume reaches the surface, its evolution can be approximated by that of a plume from a ground level release with the same mass flux rate. Mass must be conserved. Conservation of other fluxes, such as enthalpy or momentum, is not considered.

The box model transits to a Gaussian dispersion model when the rate of expansion of the cloud radius by gravitational slumping is less than the rate of increase of the standard deviation of the Gaussian profile, consistent with the atmospheric stability (Raj 1985). To ensure "continuous" transition to the Gaussian model, the source at the transition point is assumed to be a cylindrical source with the concentration prevalent in the cloud at transition time.

TRANSPORT IN GROUNDWATER AND SOIL

Groundwater and soils are the potential residences for large volumes of liquid spilled accidentally or otherwise on or near the surface. Substances in spilled liquids can reappear in subgrade structures remote from the spill either as seepage or in vapor form. Hazardous concentrations of vapor can form in subgrade structures under conditions of confinement from only small quantities of contaminant. Soils and groundwater therefore can represent an important contributor to the hazardous atmosphere in some confined spaces.

A vertical profile through the soil contains several zones (Palmer and Johnson 1989a). The vadose zone extends between the surface and the water table and includes the capillary fringe at the edge of the saturated zone. The saturated zone lies below the vadose zone. As the layer potentially in initial contact with spilled material, the vadose zone is very important to the transport of contaminants. For this reason, the vadose zone often contains greater amounts of foreign organic matter and metal oxides than the saturated zone. Also, the unsaturated portion of the vadose zone can be a pathway for the transport of gases and volatile organics.

Substances are transported through soils by advection and dispersion. Advection is transport at the average velocity of the groundwater. Spreading in the longitudinal and transverse directions occurs through dispersion. Gases and vapors transported through the unsaturated zone of the vadose zone eventually may diffuse into the atmosphere. The key physical process that affects the transport of gases and vapors in the vadose zone is diffusion. This results from the large diffusion coefficient for gases (10^{-5} m^2/s compared to 10^{-9} m^2/s for solutes).

The involvement of organic substances and gases with the aqueous phase is more complex (Palmer and Johnson 1989b). Some organic contaminants are volatile and can partition between the liquid phase and the vapor phase. Liquids that do not readily dissolve in water but can exist in a separate fluid phase are known as nonaqueous phase liquids (NAPLs). Generally, these are subdivided into two classes: those that are less dense than water (LNAPLs), and those that are more dense (DNAPLs). Most LNAPLs are hydrocarbon fuels, such as gasoline, heating oil, kerosene, jet fuel, and aviation gas. Most DNAPLs are chlorinated hydrocarbons, such as 1,1,1-trichloroethane, carbon tetrachloride, chlorophenols, chlorobenzenes, tetrachloroethylene, and PCBs.

NAPLs move through geologic media. They can displace water and air. Because water is the wetting phase for both air and NAPLs, it tends to line the surfaces of pores and to cover sand grains. The NAPL, the nonwetting phase, tends to move through the central portion of the pore. Neither the water nor the NAPL occupies the entire pore. As the fraction of the pore space occupied by the NAPL increases, a corresponding decrease occurs in the fraction of water within the pore space. A point is reached where the water phase is effectively immobile and there is no significant flow of water. The opposite condition, immobilization of the NAPL, also can occur. These immobile fractions of NAPL cause great concern because they cannot be removed from the pores except by simple dissolution in flowing groundwater. Thus, NAPLs that enter the subsurface can remain for decades and can appear in groundwater that enters subgrade structures over that period.

As a spilled LNAPL enters the unsaturated zone, it flows through the central portion of unsaturated pores. If the amount of product released is small, the product flows until residual saturation is reached. This creates a three-phase system consisting of water, product, and air within the vadose zone. Infiltrating water dissolves the components within the LNAPL and carries them to the water table. The dissolved contaminants then form a plume that emanates from the area of the residual product. Many of the components commonly found in LNAPLs are volatile and, as a result, can partition into the soil air and be transported by molecular diffusion to other parts of the aquifer. As the vapors diffuse into adjacent soil areas, they partition back into the water phase and spread the contamination over a wider area. If the surface is porous, the vapors may diffuse across the surface boundary and into the atmosphere. A relatively impermeable boundary covering the surface will prevent mass transfer.

Flow from a large spill may reach the top of the capillary fringe. Being less dense than the water in the saturated zone, the LNAPL tends to float on top of the capillary fringe, and spread laterally. This can depress the water table and permit the product to accumulate in the depression. After the spill has been controlled, the LNAPL continues to flow under the influence of gravity within the vadose zone. Groundwater passing through this area of residual saturation and water infiltrating from the surface dissolves components within the residual product, thus creating a contaminant plume.

DNAPLs can have great mobility in the subsurface as a result of their relatively low solubility, high density, and low viscosity. They do not readily mix with water and therefore remain as a separate phase. The relatively high density of these substances provides the driving force that can carry them deep into an aquifer.

Spills involving a small amount of DNAPL flow through the unsaturated zone until reaching residual saturation in the vadose zone. Infiltrating water can dissolve the residual DNAPL or the vapors and transport them to the water table, thus creating a dissolved chemical plume within the aquifer. Under appropriate conditions, the DNAPL accumulates at the interface with the groundwater and forms a pool.

If even larger amounts of DNAPL are spilled, the DNAPL can, in principle, penetrate to the bottom of the aquifer, forming pools in depressions. If the impermeable boundary slopes, the DNAPL will flow down the dip of the boundary. This direction can be upgradient from the area of the original spill. The DNAPL also can flow along bedrock troughs, which may be oriented differently from the general direction of groundwater flow. This flow along low permeability boundaries can spread contamination in directions that would not be predicted on the basis of hydraulics.

SUMMARY

Inherent in many of the geometric shapes used in industry is the capability of atmospheric confinement. Confinement to potentially hazardous levels can occur in the least complex of structures under appropriate conditions. "Appropriate conditions" describes the relationship between the boundary surface, the source(s) of emission of the contaminant, and the affected worker(s). Structures not ordinarily considered to be confined spaces demonstrate ample capability for atmospheric confinement.

The ability to describe and predict conditions in confined spaces through mathematical models is essential to gaining greater understanding about these workspaces. Mathematical models created for application in other endeavors offer potential for use in confined spaces. Little has been published to characterize the sources of atmospheric contamination. What becomes apparent through rudimentary calculations is that very small quantities of contaminant can produce highly hazardous atmospheric conditions.

REFERENCES

American Conference of Governmental Industrial Hygienists: *Industrial Ventilation: A Manual of Recommended Practice*, 22nd ed. Cincinnati, OH: ACGIH, 1995. pp. 2-1 to 2-16.

American Society of Heating, Refrigerating and Air-Conditioning Engineers: *Ventilation for Acceptable Indoor Air Quality* (ASHRAE 62-1989). Atlanta, GA: ASHRAE, 1989. 26 pp.

American Society for Testing and Materials: *Standard Test Method for Relative Evaporation Times of Halogenated Organic Solvents and Their Admixtures* (D 1901-67). Philadelphia, PA: American Society for Testing and Materials, 1967. 2 pp.

American Society for Testing and Materials: *Standard Test Method for Evaporation Rates of Volatile Liquids by Shell Thin-film Evaporometry* (D 3539-76, Reapproved 1981). Philadelphia, PA: American Society for Testing and Materials, 1981. 11 pp.

American Society for Testing and Materials: *Standard Test Method for Vapor Pressure of Petroleum Products (Reid Method)* (D 323-82). Philadelphia, PA: American Society for Testing and Materials, 1982. 7 pp.

Antonsson, A-B.: A new method for measuring emission rates from point and surface sources. *Am. Ind. Hyg. Assoc. J. 51:* 352–355 (1990).

Barrow, G.M.: *Physical Chemistry*, 2nd ed. New York: McGraw-Hill, 1966. pp. 576–611.

Baumgartner, W.H.: Effect of temperature and seeding on hydrogen sulphide formation in sewage. *Sewage Works J. 6:* 399–412 (1934).

Bjerre, A.: Assessing exposure to solvent vapour during the application of paints, etc. — model calculations versus common sense. *Ann. Occup. Hyg. 33:* 507–517 (1989).

Boublik, T., V. Fried, and E. Hala: *Vapor Pressures of Pure Substances.* New York: Elsevier Science, 1973.

Bowne, N.E.: Atmospheric dispersion. In *Handbook of Air Pollution Technology.* Calvert, S. and H.M. Englund (Eds.). New York: John Wiley & Sons, 1984. pp. 859–891.

Briggs, G.A., R.S. Thompson, and W.H. Snyder: Dense gas removal from a valley by crosswinds. *J. Haz. Mater. 24:* 1–38 (1990).

Butler, R.M., G.M. Cooke, G.C. Lukk, and B.G. Jameson: Prediction of flash points of middle distillates. *Industr. Eng. Chem. 48*: 808–812 (1956).

Conroy, L.M., R.A. Wadden, P.A. Scheff, J.E. Franke, and C.B. Keil: Workplace emission factors for hexavalent chromium plating. *Appl. Occup. Environ. Hyg. 10:* 620–627 (1995).

Cooper, D.W.: Letter to the editor. *Am. Ind. Hyg. Assoc. J. 47:* A-780 to A-781 (1986).

Cooper, D.W. and M. Horowitz: Exposures from indoor powder releases: models and experiments. *Am. Ind. Hyg. Assoc. J. 47:* 214–218 (1986).

Cotton, F.A. and G. Wilkinson: *Advanced Inorganic Chemistry*, 5th ed. New York: John Wiley & Sons, 1988. pp. 500–501.

Cowherd, C., M.A. Grelinger, P.J. Englehart, R.F. Kent, and K.F. Wong: An apparatus and methodology for predicting the dustiness of materials. *Am. Ind. Hyg. Assoc. J. 50:* 123–130 (1989a).

Cowherd, C., M.A. Grelinger, and K.F. Wong: Dust inhalation exposures from the handling of small volumes of powders. *Am. Ind. Hyg. Assoc. J. 50:* 131–138 (1989b).

Daggupaty, S.M.: A source strength model for accidental release of hazardous substances. In *Proc. 7th Technical Seminar on Chemical Spills, June 4 to 5, 1990, Edmonton, Alberta* (EN 40-327/1990). Ottawa, ON: Minister of Supply and Services Canada (Environment Canada/Technology Development Branch/Environmental Protection/Conservation and Protection). pp. 55–60.

Donham, K.J., L.W. Knapp, L.W. Monson, and K. Gustafson: Acute toxic exposure to gases from liquid manure. *J. Occup. Med. 24:* 142–145 (1982).

Drysdale, D.D.: Chemistry and physics of fire. In *Fire Protection Handbook*, 17th ed. Cote, A.E. and J.L. Linville (Eds.). Quincy, MA: National Fire Protection Association, 1991. pp. 1-42 to 1-55.

Fehrenbacher, M.C. and A.A. Hummel: Evaluation of the mass balance model used by the environmental protection agency for testing inhalation exposures to new chemical substances. *Am. Ind. Hyg. Assoc. J. 57:* 526–536 (1996).

Feigley, C. and B.M. Lee: Determination of sampling rates of passive samplers for organic vapours based on estimated diffusion coefficients. *Am. Ind. Hyg. Assoc. J. 48:* 873–876 (1987).

Fingas, M.F., K.A. Hughes, and A.M. Bobra: Volatile liquids in sewers: behaviour and countermeasures. In *Proc. 5th Technical Seminar on Chemical Spills, February 9 to 11, 1988, Montréal, Québec* (EN 40-327/1988). Ottawa, ON: Minister of Supply and Services Canada (Environment Canada/Technology Development and Technical Services Branch/Environmental Protection/Conservation and Protection), 1988. pp. 91–111.

Fingas, M.F.: New chemicals: ways to set up input data for running forecast models. In *Proc. Workshop on Risk Assessment of Accidental Pollution Related to the Maritime Transport of Harmful Substances*, NATO Committee on The Challenges of Modern Society. Brest, France: Centre de Documentation de Recherche et d'Experimentations sur les Pollutions Accidentelles des Eaux, 1991a. 8 pp.

Fingas, M.F.: Choice of PC models for operational purposes. In *Proc. Workshop on Risk Assessment of Accidental Pollution Related to the Maritime Transport of Harmful Substances*, NATO Committee on The Challenges of Modern Society. Brest, France: Centre de Documentation de Recherche et d'Experimentations sur les Pollutions Accidentelles des Eaux, 1991b. 8 pp.

Flynn, M.R. and B. Ljungqvist: Review of wake effects on worker exposure. *Ann. Occup. Hyg. 39:* 211–221 (1995).

Fontaine, J.R., R. Braconnier, R. Rapp, and J.C. Sérieys: EOL: A Computational Fluid Dynamics Software Designed to Solve Air Quality Problems. Paper presented at Ventilation '91, 3rd Int. Symp. Ventilation for Contaminant Control, Cincinnati, OH, September 16 to 20, 1991.

Fox, R.W. and A.T. McDonald: *Introduction to Fluid Mechanics*, 3rd ed. New York: John Wiley & Sons, 1985. pp. 1–94.

Franke, J.R. and R.A. Wadden: Eddy Diffusivities Measured Inside a Light Industrial Building, Poster #107. Paper presented at American Industrial Hygiene Conference, Las Vegas, NV, May 23, 1985.

Geankoplis, C.J.: *Mass Transport Phenomena.* Columbus, OH: C.J. Geankoplis (distributed and sold by the Ohio State University Bookstores, Columbus, OH), 1972. pp. 313–347.

Glass, D.C.: Assessment of the exposure of water reclamation workers to hydrogen sulphide. *Ann. Occup. Hyg. 34:* 509–519 (1990).

Grot, R.A. and P.L. Lagus: Application of tracer gas analysis to industrial hygiene investigations. *Ind. Hyg. News.* May, 1991a.

Grot, R.A. and P.L. Lagus: Evaluation of ventilation systems using tracer gas methods. *Ind. Hyg. News.* July, 1991b.

Grot, R.A. and P.L. Lagus: Airborne hazardous substance assessment by tracer gas methods. *Ind. Hyg. News.* September, 1991c.

Grot, R.A. and P.L. Lagus: Airflow and contaminant migration modelling. *Ind. Hyg. News.* December, 1991d.

Groves, J.A. and P.A. Ellwood: Gases in forage tower silos. *Ann. Occup. Hyg. 33:* 519–535 (1989).

Groves, J.A. and P.A. Ellwood: Gases in agricultural slurry stores. *Ann. Occup. Hyg. 35:* 139–151 (1991).

Grundey, K.: *Tackling Farm Waste.* Ipswich, Suffolk, UK: Farming Press, 1980.

Haberlin, G.M. and R.J. Heinsohn: Predicting solvent concentrations from coating the inside of bulk storage tanks. *Am. Ind. Hyg. J. 54:* 1–9 (1993).

Hanna, S.R. and P.J. Drivas: *Guidelines for Use of Vapour Cloud Dispersion Models.* New York: Centre for Chemical Process Safety of the American Society of Chemical Engineers, 1987. pp. 4–60.

Heinsohn, R.J.: *Industrial Ventilation: Engineering Principles.* New York: John Wiley & Sons, 1991. 699 pp.

Hering, S.V (Ed.): *Air Sampling Instruments for Evaluation of Atmospheric Contaminants*, 7th ed. Cincinnati, OH: American Conference of Governmental Industrial Hygienists, 1989. 612 pp.

Himmelblau, D.M.: *Basic Principles and Calculations in Chemical Engineering*, 5th. ed. Englewood Cliffs, NJ: Prentice Hall, 1989. 735 pp.

Hughes, W.F. and J.H. Brighton: *Schaum's Outline Series. Theory and Problems of Fluid Dynamics.* New York: McGraw-Hill, 1967. pp. 75–105.

Hummel, A.A., K.O. Braun, and M.C. Fehrenbacher: Evaporation of a liquid in a flowing airstream. *Am. Ind. Hyg. Assoc. J. 57:* 519–525 (1996).

Incropera, F.P. and D.P. DeWitt: *Fundamentals of Heat and Mass Transfer*, 2nd ed. New York: John Wiley & Sons, 1985. pp. 711–752.

Jayjock, M.A.: Assessment of inhalation exposure potential from vapours in the workplace. *Am. Ind. Hyg. Assoc. J. 49:* 380–385 (1988).

Jayjock, M.A.: Back Pressure Modeling of Indoor Air Concentrations from Volatilizing Sources. *Am. Ind. Hyg. Assoc. J. 55:* 230–235 (1994).

Jayjock, M.A., D.P. Doshi, E.H. Nungesser, and W.D. Shade: Development and evaluation of a source/sink model of indoor air concentrations from isothiazolone-treated wood used indoors. *Am. Ind. Hyg. Assoc. J. 56:* 546–557 (1995).

Keil, C.B., R.A. Wadden, P.A. Scheff, J.E. Franke, and L.M. Conroy: Determination of multiple source volatile organic compound emission factors in offset printing shops. *Appl. Occup. Environ. Hyg. 12:* 111–121 (1997).

Keller, C.: Firesafety in the shipyard. *Fire J. 76:* 60–66 (1982).

Lees, F.P.: *Loss Prevention in the Process Industries*, vols. 1 and 2. London: Butterworths, 1980.

Lemoff, T.C.: Gases. In *Fire Protection Handbook*, 17th ed. Cote, A.E. and J.L. Linville (Eds.). Quincy, MA: National Fire Protection Association, 1991. pp. 3-63 to 3-82.

McCammon, J. and L. McKenzie: Workplace fatality related to perchloroethylene exposures. *Appl. Occup. Environ. Hyg. 11:* 156–157 (1996).

Melcher, R.G., C.E. Crowder, J.C. Tou, and D.I. Townsend: Oxygen depletion in corroded steel vessels. *Am. Ind. Hyg. J. 48:* 608–612 (1987).

Metz, C.R.: *Theory and Problems of Physical Chemistry.* New York: McGraw-Hill, 1976. 424 pp.

Mignone, A.T. Jr., E.C. Beckhusen, K. O'Leary, and M. Gochfeld: Temporal variation in oxygen and chemical concentration in a confined workspace: the wastewater manhole. *Appl. Occup. Environ. Hyg. 5:* 428–434 (1990).

Ministry of Agriculture, Fisheries and Food: *Slurry Handling, Useful Facts and Figures* (Booklet 2356). London, U.K.: Ministry of Agriculture, Fisheries and Food, 1980.

Moore, W.J.: *Physical Chemistry,* 3rd ed. Englewood Cliffs, NJ: Prentice-Hall, 1962. pp. 117–140.

Moran, J.M. and M.D. Morgan: *Meteorology: The Atmosphere and The Science of Weather,* 2nd ed. NY: Macmillan, 1989. pp. 300–302.

Moss, O.R.: Sampling in calibration and exposure chambers. In *Air Sampling Instruments for Evaluation of Atmospheric Contaminants,* 7th ed., Hering, S.V (Ed.). Cincinnati, OH 45211-4438: American Conference of Governmental Industrial Hygienists, 1989. p. 157–162.

National Fire Protection Association: *Fire Protection Guide to Hazardous Materials,* 10th ed. Quincy, MA: National Fire Protection Association, 1991. 477 pp.

National Institute for Occupational Safety and Health: *Search of Fatality and Injury Records for Cases Related to Confined Spaces* (NIOSH Pub. No. 10947). San Diego, CA: Safety Sciences, 1978.

National Institute for Occupational Safety and Health: Criteria for a Recommended Standard — Working in Confined Spaces (DHEW/PHS/CDC/NIOSH Pub. No. 80-106). Cincinnati, OH: National Institute for Occupational Safety and Health, 1979. 68 pp.

National Institute for Occupational Safety and Health: NIOSH Pocket Guide to Chemical Hazards (DHHS/NIOSH Pub. No. 90-117). Cincinnati, OH: DHHS/PHS/CDC/NIOSH, 1990. 245 pp.

National Institute for Occupational Safety and Health: Worker Deaths in Confined Spaces: A Summary of NIOSH Surveillance and Investigative Findings (DHHS/PHS/CDCP/NIOSH Pub. No. 94-103). Cincinnati, OH: National Institute for Occupational Safety and Health, 1994. 273 pp.

National Research Council: *International Critical Tables,* vol. 3. New York: McGraw-Hill, 1929.

Nazaroff, W.W. and G.R. Cass: Mathematical modeling of indoor aerosol dynamics. *Environ. Sci. Technol. 23:* 157–166 (1989).

Nicas, M.: Estimating exposure intensity in an imperfectly mixed room. *Am. Ind. Hyg. Assoc. J. 57:* 542–550 (1996).

Nielsen, F., E. Olsen, and A. Fredenslund: Prediction of isothermal evaporation rates of pure volatile organic compounds in occupational environments — a theoretical approach based on laminar boundary layer theory. *Ann. Occup. Hyg. 39:* 497–511 (1995).

Noren, O.: Noxious gases and odours. In *Animal Wastes.* Taigainides, E.P (Ed.). Amsterdam, The Netherlands: Elsevier (Applied Science Publishers), 1977. pp. 111–129.

Occupational Safety and Health Administration: Selected Occupational Fatalities Related to Fire and/or Explosion in Confined Work Spaces as Found in OSHA Fatality/Catastrophe Investigations. Washington, D.C.: U.S. Department of Labor, Occupational Safety and Health Administration (U.S. DOL/OSHA), 1982. 76 pp.

Occupational Safety and Health Administration: Selected Occupational Fatalities Related to Grain Handling as Found in Reports of OSHA Fatality/Catastrophe Investigations. Washington, D.C.: U.S. Department of Labor, Occupational Safety and Health Administration (U.S. DOL/OSHA), 1983. 150 pp.

Occupational Safety and Health Administration: Selected Occupational Fatalities Related to Toxic and Asphyxiating Atmospheres in Confined Work Spaces as Found in Reports of OSHA Fatality/Catastrophe Investigations. Washington, D.C.: U.S. Department of Labor, Occupational Safety and Health Administration (U.S. DOL/OSHA), 1985. 230 pp.

Occupational Safety and Health Administration: Selected Occupational Fatalities Related to Welding and Cutting as Found in Reports of OSHA Fatality/Catastrophe Investigations. Washington, D.C.: U.S. Department of Labor, Occupational Safety and Health Administration (U.S. DOL/OSHA), 1988. 225 pp.

Occupational Safety and Health Administration: Selected Occupational Fatalities Related to Toxic and Asphyxiating Atmospheres in Shipyards and Ship Building as Found in Reports of OSHA Fatality/Catastrophe Investigations. Washington, D.C.: U.S. Department of Labor, Occupational Safety and Health Administration (U.S. DOL/OSHA), 1990.

OSHA: Confined and Enclosed Spaces and Other Dangerous Atmospheres in Shipyard Employment; Final Rule, *Fed. Regist. 59*: 141 (25 July 1994). pp. 37816–37863.

OSHA: Permit-Required Confined Spaces for General Industry; Final Rule, *Fed. Regist. 58*: 9 (14 January 1993). pp. 4462–4563.

Olishifski, J.B.: Methods of evaluation. In *Fundamentals of Industrial Hygiene*, 3rd. ed. Plog, B.A (Ed.). Chicago, 1988. pp. 397–416.

Palmer, C.D. and R.L. Johnson: Physical processes controlling the transport of contaminants in the aqueous zone. In Transport and Fate of Contaminants in the Subsurface (EPA/625/4-89/019). Cincinnati, OH: U.S. Environmental Protection Agency, Centre for Environmental Research Information and Robert S. Kerr Environmental Research Laboratory, 1989a. pp. 5–22.

Palmer, C.D. and R.L. Johnson: Physical processes controlling the transport of non-aqueous phase liquids in the subsurface. In Transport and Fate of Contaminants in the Subsurface (EPA/625/4-89/019). Cincinnati, OH: U.S. Environmental Protection Agency, Centre for Environmental Research Information and Robert S. Kerr Environmental Research Laboratory, 1989b. pp. 23–27.

Papadourakis, A., H.S. Caram, and C.L. Barner: Upper and lower bounds of droplet evaporation in two-phase jets. *J. Loss Prev. Process Ind. 4:* 93–101 (1991).

Park, C. and R.P Garrison: Multicellular model for contaminant dispersion and ventilation effectiveness with application for oxygen deficiency in a confined space. *Am. Ind. Hyg. Assoc. J. 51:* 70–78 (1990).

Perkins, H.C.: *Air Pollution*. New York: McGraw-Hill, 1974. pp. 145–219.

Perry, J.H., D.W. Green, and J.D. Maloney (Eds.): *Perry's Chemical Engineer's Handbook*, 6th ed. New York: McGraw-Hill, 1984.

Plinke, M.A.E., D. Leith, D.B. Holstein, and M.G. Boundy: Experimental examination of factors that affect dust generation. *Am. Ind. Hyg. Assoc. J. 52:* 521–528 (1991).

Plinke, M.A.E., D. Leith, M.G. Boundy, and F. Löffler: Dust generation from handling powders in industry. *Am. Ind. Hyg. Assoc. J. 56:* 251–257 (1995).

Raj, P.K.: Summary of heavy gas spills modelling research. In *Proc. Heavy Gas (LNG/LPG) Workshop, January 29 to 30, 1985, Toronto, ON*, Portelli, R.V (Ed.). Downsview, ON: Concord Scientific Corporation, 1985. pp. 51–75.

Reid, R.C., J.M. Prausnitz, and B.E. Poling: *Properties of Gases and Liquids*, 4th ed. New York: McGraw-Hill, 1987. 741 pp.

Roach, S.A.: On the role of turbulent diffusion in ventilation. *Ann. Occup. Hyg. 24:* 105–132 (1981).

Scheff, P.A, R.L. Friedman, J.E. Franke, L.M. Conroy, and R.A. Wadden: Source activity modelling of freon emissions from open-top vapour degreasers. *Appl. Occup. Environ. Hyg. 7:* 127–134 (1992).

Slack, D.J.: Hydrogen sulphide in residual fuel oil and storage tank vapour space. *Am. Ind. Hyg. Assoc. J. 49:* 205–206 (1988).

Stiver, W. and D. McKay: Evaporation rates of hydrocarbons and petroleum mixtures. *Environ. Sci. Technol. 18:* 834–840 (1984).

Sutter, S.L., J.W. Johnston, and J. Mishima: Investigation of accident-generated aerosols: releases from free fall spills. *Am. Ind. Hyg. Assoc. J. 43:* 540–543 (1982).

Sutter, S.L.: Powder aerosols generated by accidents: pressurized release experiments. *Am. Ind. Hyg. Assoc. J. 44:* 379–383 (1983).

Sutter, S.L. and M.A. Halverson: Aerosols generated by accidents: pressurized liquids release experiments. *Am. Ind. Hyg. Assoc. J. 45:* 227–230 (1984).

Treybal, R.E: *Mass-Transfer Operations*, 3rd ed. New York: McGraw-Hill, 1980. 784 pp.

U.S. Department of Commerce: A Numerical Method for Calculating Indoor Airflows Using a Turbulence Model, by T. Kurabuchi, J.B. Fang, and R.A. Grot (Pub. No. NISTIR 89-4211). Gaithersburg, MD: U.S. Department of Commerce/National Institute of Standards and Technology, 1990.

U.S. Department of Commerce: User Manual NBSAVIS. CONTAM88. A User Interface for Air Movement and Contaminant Dispersal Analysis in Multizone Buildings, by R.A. Grot (Pub. No. NISTIR 4585). Gaithersburg, MD: U.S. Department of Commerce/National Institute of Standards and Technology/Building and Fire Research Laboratory, 1991.

U.S. Environmental Protection Agency: Report to Congress on Indoor Air Quality. Volume II: Assessment and Control of Indoor Air Pollution (EPA/400/1-89/001C). Washington, D.C.: U.S. Environmental Protection Agency/Indoor Air Division/Office of Air and Radiation, 1989a. pp. 2-1 to 2-37.

U.S. Environmental Protection Agency: Project Summary. Indoor Air Quality Model Version 1.0, by L.E. Sparks (EPA/600/S8-88/097). Research Triangle Park, NC: U.S. Environmental Protection Agency/Air and Energy Engineering Research Laboratory, 1989b.

U.S. Environmental Protection Agency, Office of Toxic Substances: A Manual for the Preparation of Engineering Assessments, vol. 1: CEB Engineering Manual (CEB/ETD/OTS). Washington, D.C.: U.S. Environmental Protection Agency, 1991a.

U.S. Environmental Protection Agency: EXPOSURE Version 2. A Computer Model for Analyzing the Effects of Indoor Air Pollutant Sources on Individual Exposure, by L.E. Sparks (EPA-600/8-91-013). Research Triangle Park, N.C.: U.S. Environmental Protection Agency/Air and Energy Engineering Research Laboratory, 1991b.

U.S. Environmental Protection Agency: Indoor Air Quality Simulator for Personal Computers, by P.A. Lawless and M.K. Owen (EPA/600/S8-91-014). Research Triangle Park, N.C.: U.S. Environmental Protection Agency/Air and Energy Engineering Research Laboratory, 1991c.

Wadden, R.A., P.A. Scheff, and J.E. Franke: Emission factors for trichloroethylene vapour degreasers. *Am. Ind. Hyg. Assoc. J. 50:* 496–500 (1989).

Wadden, R.A., J.L. Hawkins, P.A. Scheff, and J.E. Franke: Characterization of emission factors related to source activity for trichloroethylene degreasing and chrome plating processes. *Am. Ind. Hyg. Assoc. J. 52:* 349–356 (1991).

Wadden, R.A., D.I. Baird, J.E. Franke, P.A. Scheff, and L.M. Conroy: Ethanol emission factors for glazing during candy production. *Am. Ind. Hyg. Assoc. J. 55:* 343–351 (1994).

Wadden, R.A., P.A. Scheff, J.E. Franke, L.M. Conroy, M. Javor, C.B. Keil, and S.A. Milz: VOC emission rates and emission factors for a sheetfed offset printing shop. *Am. Ind. Hyg. Assoc. J. 56:* 368–376 (1995).

Weast, R.C (Ed.): *Handbook of Chemistry and Physics*, 49th Ed. Cleveland, OH: The Chemical Rubber Co., 1968. pp. D-105 to D-142.

6 Nonatmospheric Hazardous Conditions: The Role of Confined Energy

CONTENTS

INTRODUCTION

Nonatmospheric hazardous conditions cause many fatal accidents in confined spaces. According to reports published by the National Institute for Occupational Safety and Health (NIOSH), the Occupational Safety and Health Administration (OSHA), and the Mine Safety and Health Administration (MSHA), leading nonatmospheric hazards include engulfment, entanglement, process and electrical hazards, falls from heights, and unstable interior conditions (NIOSH 1979, 1994; OSHA 1982a,b, 1983, 1985, 1988, 1990; MSHA 1988). Statistics compiled by NIOSH indicate that engulfment by loose materials is the second ranking cause of fatalities in accidents in these workspaces (NIOSH 1994). In compiling these statistics, NIOSH did not include engulfment resulting from collapse of trenches, excavations, and ditches. Trenches, excavations, and ditches normally are not considered to be confined spaces. However, engulfment is engulfment. The mechanics of engulfment in a trench, excavation, or ditch are similar to what occurs in a structure, such as a silo or hopper. When considered without the imposition of regulatory constraints, engulfment easily would become the first ranking cause of fatal accidents in these workspaces.

Chapter 2 examined the association between nonatmospheric hazards and fatal accidents in confined spaces. The data suggest that engulfment is highly associated with hazardous conditions that developed prior to entry and the start of work. Conversely, the other hazardous conditions (entanglement, process and electrical hazards, falls from heights, and unstable interior conditions) are highly associated with work activity.

Engulfment occurs through two main mechanisms: sudden release of material from walls of storage structures and downward suction by motion of material. Some accidents resulted following dislodging of material adhering to vertical surfaces. Injury occurred during the sudden release of large coalesced fragments. Engulfment occurred during release of trapped flowable material. The other significant cause of engulfment was downward suction by the vortex created by the flowing material.

The data on entanglement presented in Chapter 2 suggest that the victims normally worked with the equipment in which the accident occurred and that most of the work activity could be considered routine. Entry typically occurred for cleaning, repair, inspection, or adjustment. In some cases, entry occurred during operation of the equipment. A high proportion of these accidents occurred within the confines of mechanical equipment, augers, and machine pits.

In most of these situations the victim deliberately entered the space while the hazardous condition was present. This occurred despite the existence of protective measures ranging from verbal instructions to written procedures and training, to lockout/tagout systems.

Miscommunication was the most important factor in start-up of equipment while the victim was inside. Miscommunication led to misperception about the location of the victim and occupancy within the boundary surfaces of the equipment. In many situations safety depended on a single level of protection: an ON/OFF control switch. Activation of control switches occurred inadvertently or inappropriately or through mistaken identity. Redundancy or multiple levels of protection, however, did not necessarily provide greater security.

Electrocutions occurred during three distinct activities: electrical work, accidental contact with energized conductors during nonelectrical work, and welding. Almost all of the victims were performing normal work at the time of these accidents. During more than half of the nonwelding accidents, electrical contact occurred during cleaning, painting, or other activity. In more than half of the accidents the voltage was 110 to 120 V, normal household voltage. In more than half of the electrocutions, the circuit was energized prior to entry and the start of work. Accidental exposure to energized, protected circuits during work activity occurred in a small minority of cases. Water or perspiration was a factor in 2/3 of these accidents. Bare conductors or ineffective grounding were factors in more than 3/4 of these situations. The data also suggest that lack of testing for live electrical circuits and lack of inspection of equipment and conductors were important factors in these accidents.

Almost half of the electrical accidents occurred during welding. The data suggest that water and/or perspiration was a critical factor in all of these accidents. Water and perspiration can increase the conductivity of insulating materials and the skin. Maintaining an isolated condition is especially critical when the welder is working inside a structure fabricated from conductive materials. Conductive metals, such as aluminum, steel, or stainless steel are the materials of construction of many industrial enclosures and structures that form the boundary surfaces of confined spaces.

Most process-related accidents occurred during normal activity. Most of the accidents resulted from failure of pressurized systems containing steam. Other process chemicals were involved in a minor proportion of these accidents. Utility steam systems seem especially prone to sudden in-service failure. Most of the work activity in the accidents studied was unrelated to the process system that failed.

These accidents exemplify the concern expressed over the years for utility and process systems that remain active during entry and work — namely, some component of the system fails during occupancy and discharges process chemicals into the space. The victim then is unable to escape from the space before being injured by the discharged substance.

Accidents involving unstable interior structural conditions were as likely to occur during routine work as during unusual activity. Work activity associated with installation or removal of the structure or equipment (welding or flame cutting) created the unstable condition in almost all situations. Failure to support the structure in the unstable condition or to prevent contact with it occurred in all cases. The victims apparently failed to recognize the imminent danger of the situation.

ENERGY MOVEMENT AND STORAGE IN EQUIPMENT AND SYSTEMS

Discussion

Rapid release of stored energy was a cause in all of the accident situations summarized in the previous section. Controlled storage of energy has been utilized in equipment for many centuries. The catapult provides an early example. Rotation of a rope winch flexes the hurling arm. A locking mechanism permits loading of the spoon with a rock and prevents release of the drum until the appropriate moment. At that point, releasing the trigger causes virtually instantaneous transformation of stored energy from potential to kinetic. The ability of this energy to hurl rocks a considerable distance suggests the inherent ability of these units to injure or kill the individuals who operated and serviced them.

Experience would have indicated that the appropriate time to service these units was when the hurling arm was not under tension. The rationale for this is that the unit was not storing energy and that pieces could be secured readily to prevent unexpected or unintended movement. Under tension, motion of the hurling arm is not completely predictable. The ropes and the locking mechanism could fail. The timber in the hurling arm could split. Accidental release or failure of the trigger mechanism could occur. Unexpected release of the tensioned hurling arm when an individual was working in the mechanism would have caused serious or fatal injury.

The catapult operates through a sequence of actions that store energy in a controlled manner for release at a specific moment. Because the unit is fully open and operated manually, the sequence and consequence of storing energy are fully observable.

The guillotine is another example of equipment that stores energy for rapid conversion from one form to another. Some maintenance entails work on the innards of the mechanism that lifts the blade. In order to do this, let us say that the worker must insert head, shoulders, and arms across the path of the blade. Of course, the blade must be positioned in the "up" position in order to provide access. Guillotines, like most equipment, were made for operation, not for maintenance. Work on the lift mechanism can occur in safety, so long as the rope that lifts the blade is tied off. However, unforeseen actions, such as a burn-through caused by a lighted torch, or the misdirected swing of a sword or axe, could change the circumstances immediately. Again, the potential for these types of accidents is obvious because the system is fully exposed and its actions observable.

A wind-up alarm clock is a more modern example of a device that stores energy by a similar mechanism. The alarm mechanism is essentially similar to that in the catapult. The alarm can be armed and triggered at will. Increasing tension in the spring and the end of rotational travel provides feedback to the operator.

A major difference exists between the wind-up clock and the catapult and guillotine. The action produced by rotating the spring-winding keys is not visible. The case of the clock shields the internal mechanism from view. Mechanisms in wind-up clocks fail. The hands fail to turn or the alarm mechanism fails to function. The cause for these failures is not observable.

The only way to determine the cause of the problem is to enter the shell of the clock or to dismantle it. Of course, no one can enter the shell of a normal wind-up alarm clock. However, the interior of large clocks, such as "Big Ben" in London, is accessible. Removing the shell of the wind-up alarm clock is a simple matter of removing several machine screws. This exposes a framework containing many intermeshing gears and two flat metal springs. Imagine the shock as the mainspring rapidly unwinds and flies into a curious young face after loosening one screw too many. Stored energy can release rapidly and create a lasting impression. Instructions accompanying wind-up clocks provided no guidance to the budding young mechanic about how to fix them. Clocks, as with other machines, are intended to perform an operating function, not to be disassembled by curious young minds.

There are additional examples of energy storage systems in common experience. For example, capacitors in television sets and video monitors store electrical energy even after the unit is unplugged. This energy is released in an instant by bridging between certain contacts using a screwdriver. The automobile cooling system retains considerable energy for some time after the engine is shut down. Removing the cap from the radiator releases thermal energy and built-up pressure. This process occurs despite locking out through removal of the key from the ignition and locking of the door.

Energy utilization in machines and systems is not always consistent. For example, air-braking systems in trucks and trailers operate completely opposite to those in trains. In trucks, air pressure releases the brakes. In trains, air pressure applies the brakes. The hydraulic system in a bulldozer operates by controlling fluid pressure on both sides of the piston. Changing the pressure on one side relative to the other causes movement of the piston.

Understanding how the movement, storage, and transformation of energy occurs in systems and equipment is essential for safe conduct of work in and around them. In the simplest of machines, the workings were visible to the operator and maintenance person. Actions and consequences of actions that could produce injury were readily observable. As a result, the measures needed to prevent these actions could be perceived, implemented, and assessed at little risk of injury.

In more sophisticated equipment, the outer enclosure hides actions performed by internal subsystems from the observer. Complicating matters further, sensors and feedback circuits control operation of these subsystems. The user is several levels removed from the action and control of modern equipment. Without sophisticated analytical tools and procedures, the operator or maintenance person can do little to ascertain the energy status of the equipment. Working on or inside equipment under these conditions is akin to attempting to defuse a newly exposed, unexploded WWI bomb or an explosive device planted by terrorists. There is no assurance about the ability to perform this type of work in safety. As a result, demolitions experts attempt to recover these devices using robotic equipment and to detonate them under controlled conditions. The luxury afforded by this approach is not available to persons who must dismantle and service industrial equipment that is located in, or that forms a confined space.

Mechanical Systems (Machines)

In the academic context, a "machine" is a mechanical device that harnesses forces and controls the direction of expression in order to produce required work (Olivo and Olivo 1984). Mechanical devices change the magnitude, direction, or intensity of forces, or the speed resulting from them. Mechanical devices can be as simple as the screwdriver, wrench, or hammer, and as complex as a passenger aircraft. In the industrial context, a machine usually is an assembly of subsystems, each of which can contain simpler subsystems composed of simple mechanical devices. Simple mechanical devices and the subsystems found within industrial equipment can store, transform, and transfer energy.

One of the simplest mechanical devices is the lever (Walton 1968). The lever is a rigid piece, capable of turning about a point or fulcrum, by which force applied at one point is transmitted or modified at a second. That is, the lever amplifies force applied at a point. Small force applied through a large distance from the fulcrum produces the effect of a large force applied through a small distance. This effect (mechanical advantage) is fully exploited in machines. Mechanical advantage of force is the ratio of the forces balanced by the lever. Mechanical advantage of distance is the ratio of the distances at which the forces balance on the lever (Olivo and Olivo 1984). Grouping together several simple levers with linkages produces the compound lever. The compound lever provides the same mechanical advantage in a much shorter distance. Angled lever arms and complex designs can provide high mechanical advantage in a compact package.

Motion in a machine is either translational (straight line or curved path) or rotational (Raczkowski 1979). Work performed by mechanical systems results from the action of forces or torques. Torque is the tendency to move around a point. Expressed mathematically, torque is the product of the force × distance from the point. Machine elements must carry loads, for example, static and dynamic forces. Resulting forces or torques balance input loads.

Power (energy) transmission equipment is the medium through which practically all motion occurs in industrial equipment (power is the rate of use of energy [Nelson 1983]). Power (energy) transmission equipment refers to specific parts of a machine assembly that function in the mechanical transmission. By far the most common form of power (energy) transmission is transfer of rotary motion from one shaft to another. This occurs axially and shaft to adjacent shaft. Axial transmission requires a clutch or coupling. Transfer of rotary motion between adjacent shafts occurs through physical contact.

Rotation of one disc causes rotation of another with which it is in contact. While in contact, the surfaces travel equal distances at equal surface speeds. This also applies to power (energy) transmission devices, such as belts and chains, since they are in contact with the moving surfaces. That is, the belt or chain travels equal distances at the same speed as the surfaces of the discs. Where the discs are unequal in size, in order to maintain continuity, the shaft of the smaller disc must rotate faster than that of the larger disc.

Belt drives provide quiet, compact, and resilient power (energy) transmission. Belts are used singly or in multiples, depending on service requirements. Belt types include "V"-belts, flat belts, and positive drive (gear) belts. In the case of the V-belt, transmission of power (energy) occurs at the plane of contact between the sides of the V-belt and the sloping walls of the sheave (pulley). Applications in power (energy) transmission involving flat belts are mostly historical. However, flat belting is widely used in conveying systems. Positive drive belts contain molded teeth. The interface between the teeth of the belt and those of the pulley is the plane of power (energy) transmission. Positive drive belts offer the quiet operation and flexibility of the belt with the power-transfer interface of gears and chains.

A gear is a disc or wheel that contains teeth around the periphery. A rack is a linear version of a gear. Gears provide positive transfer of power (energy) by meshing with the teeth on another gear. Alignment is critical for proper performance. Lubrication and freedom from the presence of foreign materials are key determinants in the life of gears. For this reason, enclosures are used to house gears and to provide access to lubricants. At the same time, the housing prevents observation of the complexity of the gear train.

A gear is equivalent to a cluster of evenly spaced levers mounted on a common fulcrum (Walton 1968). As a result, a small gear connected to the powered shaft connected to a large gear on the driven shaft would provide a considerable increase in torque. Increasing torque is a major incentive for use of gears and gear trains.

Chain drives consist of one or more driven sprockets and an endless chain. The links of the chain mesh with the teeth of the sprocket. This maintains a positive speed ratio between the powered and driven sprockets because they do not slip or creep. The principal advantages of chains are simplicity, economy, efficiency, and adaptability. The most widely used chain for power (energy) transmission is the roller chain (bicycle chain). Roller chains are manufactured in single and multiple strands. The inverted tooth chain is similar in concept to the positive drive belt. The links contain inverted teeth. These engage the teeth in a cut tooth wheel. This design provides the flexibility and quiet of operation of belts, the positive action and durability of gears, and the convenience and efficiency of chains.

Enclosures that surround and shield against the actions of complex machines form the boundary surfaces of many confined spaces. Mechanical power (energy) transmission subsystems in this equipment may contain no shielding or guarding. This is most likely the case where entry into the enclosure during operation is considered to be unrealistic or prohibited or prevented by interlock safety devices. Entanglement is especially possible around exposed mechanical power (energy) transmission equipment where the protection afforded by the enclosure is not respected or is circumvented. Power (energy) transmission equipment does not store mechanical energy. The energy content of the equipment dissipates rapidly when the source is shut down. However, energy storage can occur in equipment to which power (energy) transmission equipment is connected.

Energy storage in mechanical systems occurs in a number of ways. Gravity is the most widespread mechanism. Parts in machines that move as part of normal operation, such as levers, knives, hammers, and many others, can stop in a position from which downward motion can occur under appropriate conditions. As well, loss of anchorage during dismantlement can lead to pendular motion of many types of functional and structural parts of machines.

Rotational inertia is another important mechanism. A rotating mass will continue to move after removal of the energy source. The flywheel and pendulum are examples of energy storage devices in mechanical systems that utilize rotational inertia. Energy storage in rotation is calculated using the moment of inertia (Juvinall and Marshek 1991). Energy stored in a flywheel is proportional to the square of the rotational speed, the length and density of the material, and the fourth power of the diameter. In the case of a ring-type flywheel, energy depends on the difference between the inner and outer diameters raised to the fourth power. The flywheel in some machines can store considerable residual energy.

Springs are another type of energy storage device. Springs are elastic members. They exert forces or torques and absorb energy which is stored and later released (Juvinall and Marshek 1991). Springs are classed according to function as controlled-action, variable-action, or static springs (Jensen and Helsel 1996). Controlled-action springs have a constant range of action or well-defined function for each cycle of operation. Valve springs in an engine and die and switch springs are examples of controlled-action springs. Variable-action springs have variable range of action. This reflects the varying conditions imposed on them. Suspension, clutch, and cushion springs are examples of variable-action springs. Static springs exert comparatively constant pressure or tension between parts. Packing or bearing pressure, antirattle, and seal springs are examples.

Classification of springs also reflects the type of spring. Type of spring is governed by function, shape, application, or design. A compression spring is an open-coiled helix. The helix can contain cylindrical or flat material. A compression spring offers resistance to compressive force. An extension spring offers resistance to a pulling force. An extension spring is a close-coiled helical spring. Torsion springs exert force along a circular path. Torsion springs include helical and bar springs. A power spring is a flat metal coil spring that is wound on an arbor and usually confined to a case or drum. The springs in a mechanical clock are examples of power springs. Flat springs contain flat metal formed in a manner to oppose force in the desired direction when deflected in the opposite direction. The leaf spring is an example of a series of flat springs nested together and arranged to provide approximately uniform distribution of stress throughout the length. Individual levers also can act as springs under certain circumstances.

Mechanical systems containing the components mentioned here can convert, store and transmit, and translate energy within equipment. This energy can release slowly in the case of the flywheel or rapidly in the case of a compressed, stretched, flexed, or twisted spring. Energy released from these sources in this manner, in conjunction with the compounding action of lever systems, can produce unexpected movement in machine parts. These unexpected movements have caused fatal injuries to individuals working in confined spaces containing this equipment.

Fluid Power Systems

Fluid power systems provide flexibility in speed and precise control of motion, while utilizing few moving parts.

Fluid power systems that utilize a liquid as the medium for energy transfer are referred to as hydraulic systems (Stewart 1976). The simplest hydraulic system consists of a tank or reservoir containing oil, a pump, tubing or piping, and a load, such as a motor or cylinder. The pump exerts force on the oil. Pressure exerted by the oil against walls of the downstream containment causes some action to occur. Actions could include longitudinal movement of a piston in a cylinder or rotation of the rotor in a motor. Depressurized oil then returns to the reservoir. A bypass valve controls the quantity of oil diverted from the load.

The volume of liquid fluids does not change with applied pressure. Compressed air can undergo further compression until liquefaction of individual gases occurs. Volume decreases and heat is generated. Liquid fluids also act as a lubricant of moving parts. When completely filled with liquid fluid, movement of a piston in a cylinder can be controlled precisely.

Fluid power systems that utilize a gas are referred to as pneumatic systems. While the gas usually used in pneumatic systems is compressed air, other gases, such as nitrogen, are possible. While compressed air in such systems can be obtained from cylinders, a dedicated or plant-system compressor is the usual source. Pressure exerted by the compressed air can be used to move the piston in a cylinder against a load or to rotate the rotor of a motor. Pneumatic devices present no spark hazard in an explosive atmosphere, nor an electrical shock hazard.

A fundamental concept in fluid power is the relationship between pressure, force, and surface area, as expressed in Pascal's law (Vickers 1992). A confined fluid against which a force is applied exerts the same force perpendicularly on equal areas of the surface of the containment. Pressure is a defined quantity: force/surface area. The force exerted against walls of the containment is pressure × surface area. In a static system containing two interconnected cylinders of different areas, a small force exerted over a small area and transmitted through the fluid equals a large force distributed over a proportionately larger area.

Fluid machines are divided into two groups: velocity (dynamic) type and positive displacement (pressure) type (Stewart 1976). In the velocity (dynamic) fluid machine, action between a mechanical part and the fluid produces appreciable change in the velocity of the fluid. The centrifugal and axial flow compressors are examples of the velocity (dynamic) type of machine. In the positive displacement (pressure) type of machine, volumetric change or displacement occurs. Pressure is developed primarily by displacement action. Positive displacement fluid machines produce both reciprocating (back and forth) and rotary motion.

Pneumatic and hydraulic cylinders are familiar applications of fluid power. In those used in bulldozers and excavators, fluid pressure is applied on both sides of the piston. This enables a single piston to perform work in both directions. Disconnecting the fluid supply to one side of the piston does not inactivate the piston, however, since pressure exists on the other.

Various types of pressure regulation are used in these systems. In one type, intake valves in the compressor remain open during the suction and compression strokes. Hence, no compression occurs in the cylinder. Another type closes the intake line. This prevents air from entering the compressor. In the third type, the speed of the drive decreases or the drive shuts off. A compressor controlled in the latter manner could reactivate at any moment.

Certain subsystems within fluid power systems carry important implications for service work. The first is the pressure booster (intensifier) (Figure 6.1). The pressure booster increases fluid pressure. The pressure boosting fluid and the pressure boosted fluid can be liquid or gas. The pressure booster is the fluid power analogue of the electrical transformer. The pressure booster functions through the relationship between surface area and fluid pressure. That is, a low pressure exerted on a large surface area becomes a high pressure exerted by a small surface area. In the practical application of this principle, pressure applied by a fluid to a large piston becomes much

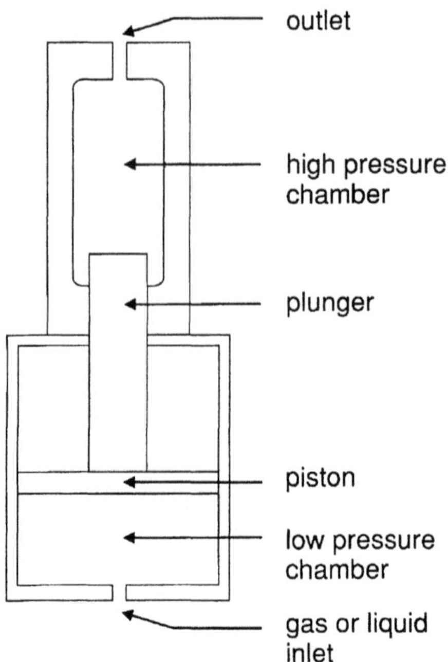

FIGURE 6.1 Pressure intensification in hydraulic systems. Pressure boosters use air or hydraulic pressure acting on a piston of large surface area to exert pressure on hydraulic fluid through movement of an interconnected piston of small surface area. High pressure circuits are often difficult to recognize. (Adapted from Stewart 1976.)

higher pressure applied by a smaller piston. Equation 6.1 provides the relationship between pressure and surface area.

$$P_1A_1 = P_2A_2 \tag{6.1}$$

P_1 is the pressure applied to piston of surface area, A_1. P_2 is the pressure applied by piston of surface area, A_2. The units of pressure and area on both sides of the equation must be consistent.

Pressure increases of 50 to 1 or higher are achievable. Boosters can provide much higher pressure than high pressure pumps under reduced heat production. Also, low pressure valves can be used in the low pressure side of the system. This means that equipment containing low pressure components can operate at very high pressure. The existence of high pressure in the system may not be evident from casual observation of low pressure components.

Another important subsystem is the low power subsystem. Low power subsystems are used to actuate high power outputs. The pilot valve is a small valve whose operation requires low effort. Fluid output from pilot valves is used to control the position and hence the throughput of larger valves.

An important application of pilot valves is the servo control system. The function of servo control systems is to maintain operation of large systems within fixed parameters, such as speed, pressure, and temperature. Servo systems can include hydraulic or pneumatic components or a combination of both. Servo systems receive input from sensors that measure the parameter and also provide error sensing and correction. Pilot valves and low pressure circuits provide the physical means to control operation of large-scale equipment.

Fluid power systems also can store energy (Figure 6.2). The air receiver or storage tank in a pneumatic system is an energy storage device. The air receiver dampens pulsations created by intermittent discharge of the compressor and provides reserve capacity for peak demand. The

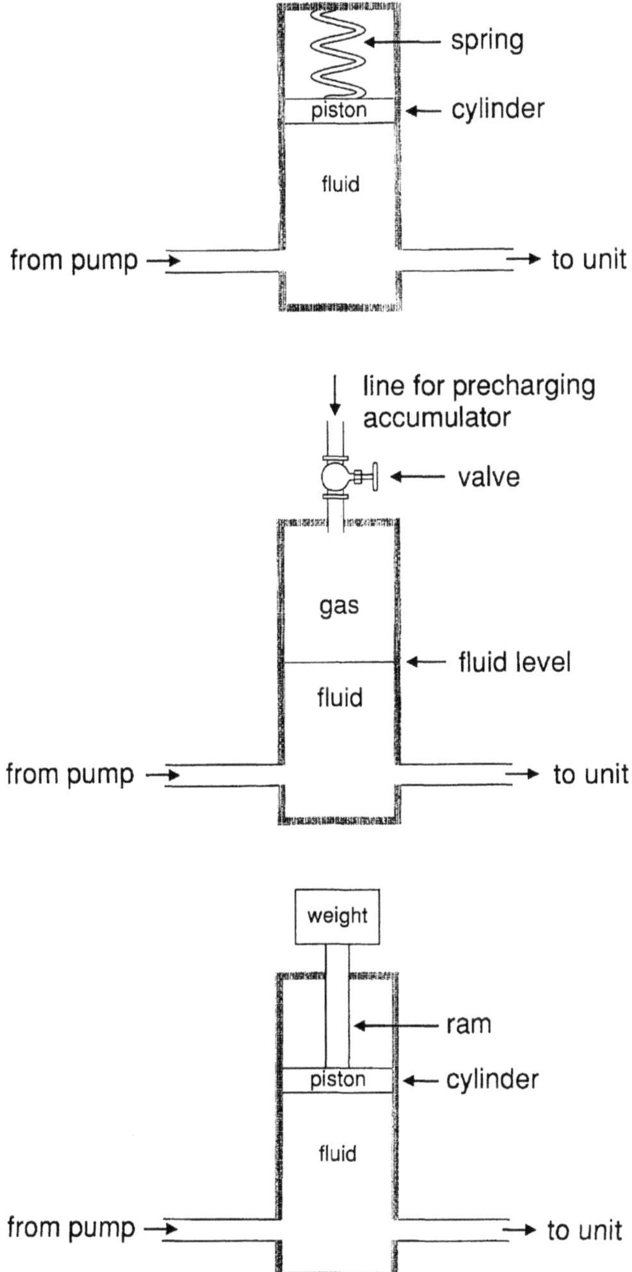

FIGURE 6.2 Energy storage in hydraulic systems. Pressure is exerted on hydraulic fluid by the following sources: a spring (6.2A), compressed gas (6.2B), or a weight (6.2C). The result is that pressure in the fluid may remain following deactivation of the hydraulic pump. (Adapted from Stewart 1976.)

volume of the receiver determines the demand that can be met. There is no limit to the size of a receiver tank. Large receivers provide stability to a system. Hence, a receiver tank can continue to supply air to a system after shutdown of the compressor.

The equivalent of the receiver tank in hydraulic systems is the accumulator. The accumulator provides emergency power following failure (shutdown) of the normal supply and maintains constant delivery pressure without necessitating continuous operation of the pump. Accumulator designs include weight- and spring-loaded pistons and chambers containing oil and pressurized gas or air. These devices can store large volumes of fluid under pressure.

Electrical Systems

Electrical equipment includes analogues to equipment described in pneumatic and hydraulic and mechanical systems. Electrical equipment can boost and store energy.

Transformers convert AC (alternating current) voltage from one level to another, up or down (Palmquist 1973). Transformers contain a rectangular ring of thin stacked iron plates or strips (laminations) around which are wound isolated or nonisolated coils of wire. Transformers utilize the principle of induction to impress a voltage in the output side. The winding receiving the impressed voltage is the primary. The winding receiving the induced voltage is the secondary. Ideal transformers follow the relationship:

$$V_{secondary}N_{primary} = V_{primary}N_{secondary} \tag{6.2}$$

V refers to voltage; N is the number of turns in the coil; primary refers to the input side, secondary refers to the output side.

Capacitors are energy storage devices (Palmquist 1973). The capacitor is constructed from two conductors held in close proximity, and separated by a nonconducting material (dielectric). The larger the surface area of the conductors, the greater the energy storage. Each conductor carries electrical charge opposite that on the other. The conductors can accumulate charge to the level at which the insulating property of the dielectric fails and charge transfer occurs. Charge storage (capacity or capacitance) increases as the distance between the conductors decreases. Voltage on the conductors also decreases as the space decreases. Capacitance also depends on the dielectric and the shape of the conductor. Electrical energy stored in the capacitor following the charging phase is released into the circuit once the circuit is opened. This energy is capable of sustaining operation of equipment, as well as spark generation.

Capacitors also are found in resistance welders, induction heaters, power stabilizers, and conveyor-drive power supplies (UAW-GM National Joint Committee on Health and Safety 1985). Large power transformers can act as capacitors and store energy in some cases.

Capacitance also can develop in unusual situations. Isolated conductors in a grounded metallic raceway can store charge between the conductors and ground (Palmquist 1973). The conductors and raceway must be grounded to prevent severe shock potential related to charge accumulation. Similar charge accumulation can occur between power line conductors and conductors and ground. This storage of charge occurs through the process of induction. Induction occurs when a conductor moves across a magnetic flux so as to cut the lines of force. The conductor may be a straight wire, coil of wire, or a solid block of metal. Maximum induction occurs when the lines of force are cut perpendicularly.

Capacitance also can result in metal structures in which static electricity can be generated (Jones and King 1991). A metal structure and a nongrounded metal surface in close proximity to it and separated from it by a dielectric material, such as a gasket, can form a capacitor.

Some industrial facilities contain DC systems that utilize energy from storage batteries. Batteries usually are considered as part of backup systems. In some industrial facilities, banks of batteries are integral to the process. Electrical generation systems utilize batteries to produce the magnetic field in the rotor. Other electrical installations also utilize DC current provided from batteries.

Electrical systems are inherently different from hydraulic and pneumatic systems. Unlike hydraulic and pneumatic systems, where leakage of liquid and gas is readily detected, leakage from electrical systems is not obvious. Leakage of electrical current can cause fatal accidents. Detecting leakage of electrical energy is the function of insulation testing and ground fault circuit interrupters (GFCIs) at 110 V and equipment leakage circuit interrupters (ELCIs) at higher voltages.

Testing for insulation leakage in new systems should occur prior to energizing. This enables correction of faults and determination of later deterioration due to dirt and moisture. Deterioration due to dirt and moisture generally is considered to cause larger leakage currents. Insulation testing exposes defects not found in any other manner (Palmquist 1973). Insulation testing could have prevented some of the fatal electrical accidents documented in the OSHA reports on confined spaces (OSHA 1982b, 1985, 1988, 1990).

Stray electrical currents and water are potent factors in electrocutions. This was illustrated numerous times in accidents involving confined spaces (OSHA 1985, 1990). An energized conductor that is exposed to water can produce currents in the water (Smoot and Bentel 1964). The magnitude of the current will depend upon the shape and the size of the conductor–water contact surface, the conductivity of the water, and the resistance in the current path to ground (Novotny and Priegel 1974). The water in these situations can include sweat on the skin, as well as water or other conductive liquid that is residual in the space. As well, the surfaces of the space could be constructed from conductive metals, such as steel and aluminum.

The electric arc is a special example of the power inherent in the rapid discharge of electrical energy (Bernstein 1991a). Depending on its size and available energy, the electric arc can act as a source of ignition at currents less than 15 A. The temperature of an electric arc at these current levels ranges from 2,000°C to 4,000°C (3,600°F to 7,200°F) (Cobine 1958). Such arcs can cause damage if sufficient energy is available in them (Baitinger 1995).

The voltage that can initiate an arc in air depends on the shape of the electrodes and the waveform. A typical value is 30 kV/cm (Cobine 1958). A fundamental property of arcs is that no matter how small the gap between two conductors, the voltage must exceed 300 V. Breakdown across an air gap cannot develop at smaller voltages. An arc, however, can develop at much lower voltages when electrodes initially in contact are separated. After being established, only 20 V/cm is needed to sustain an arc (Golde 1973). Further, the arc established in this way can exist over a large air gap. Voltage needed to sustain the arc depends primarily on arc length and is practically independent of current.

To illustrate, an arc welder with an 80-V open circuit voltage cannot initiate an arc across an air gap (Bernstein 1991a). Separating the electrodes after touching initiates the arc. This voltage is sufficient to maintain the arc.

Protection strategies for electrical circuits involve three main approaches: grounding, double insulation, and ground fault circuit interrupters.

Grounding is the routing of unwanted current to the ground or earth (Palmquist 1973, OSHA 1993). Grounding is a concern where static or current electrical sources are present. Grounding is performed for the following reasons:

- Protection from overvoltage from inadvertent crossing of primary and secondary leads to transformers or high voltage and low voltage lines
- Dissipation of lightning or static charges or other types of surge voltages
- Rendering noncurrent-carrying parts of an electrical system at zero potential relative to ground

Grounding is effected through direct connection of a circuit of low resistance between the object and the ground. The grounding conductor is analogous to a safety valve (Bernstein 1991a). While not needed for normal operation, the ground conduction system is an essential safety feature in the event of a circuit failure. Under normal operating conditions, the ground conductor carries no current. The ground conductor provides a return path for the electricity in the event of insulation failure or misconnection. Sufficient current flow in the energized conductor will actuate overcurrent devices protecting the ungrounded conductor. The equipment should then deenergize.

Equipment grounding ensures that all exposed metal surfaces will be maintained at ground potentials, even during a fault condition (Bernstein 1991a). A low-resistance internal fault in contact

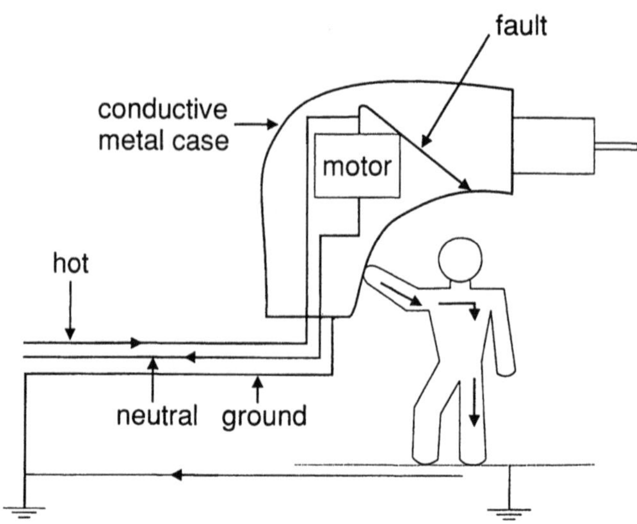

FIGURE 6.3 Current leakage path in a grounded circuit. A person can provide an alternate leakage path to ground following development of a fault that leaks to the grounded metal case of a tool. (Adapted from OSHA 1993.)

with an exposed metal surface probably will cause the overcurrent protection device to trip if the equipment is properly grounded. This disconnects power from the equipment and removes the hazard. The overcurrent device operates because of the low-resistance path provided by the series combination of the internal fault and the equipment ground. A properly installed equipment ground in a fault condition involving high internal resistance will tend to maintain exposed metal surfaces at or near ground potential. In the latter case, the high internal resistance limits flow of the fault current to a level below that necessary to activate the overcurrent device. Thus, even though disconnection of the power does not occur, surfaces of the equipment still are safe to touch.

Without equipment grounding, a fault that enables an energized conductor to contact exposed metal parts may not cause the overcurrent device to operate (Bernstein 1991a). Activation of the overcurrent device would depend upon the existence of a path to ground of sufficiently low resistance from the exposed metal parts. A situation involving faulted equipment resting on a dry wooden table likely would not satisfy this requirement. The path to ground from the exposed metal parts likely would have too great a resistance to activate the overcurrent protection device. The equipment could continue to function normally with its exposed metal parts at a potential of 120 V. Simultaneously touching the exposed metal and a grounded object could provide the current pathway leading to a lethal shock (Figure 6.3).

Grounding relies on conduction by nonconductive materials, such as soil (Palmquist 1973). Resistivity is the measure of conduction of soil. Resistivity of soil depends on composition, moisture content, salts and mineral content, and temperature. Dry, washed, sand and gravel is a very poor conductor. Wet, salt-containing fills — ashes, cinders, and brine wastes — should be excellent conductors. Testing is the only means to establish the efficacy of a given ground. Grounding in unsatisfactory situations can be improved by use of longer electrodes, use of additional parallel electrodes, or through addition of chemical additives to the soil near the electrode. Increase in depth beyond 3 m (10 ft) produces little tangible benefit. Reduction of ground resistance through use of chemical treatment is not considered a permanent solution. Of the three possible techniques, driving additional ground rods is the preferred route.

Fuses and circuit breakers are overcurrent devices (Bernstein 1991a). These protect the wire insulation of the ungrounded conductor from overheating and damage because of overcurrent. An overcurrent device functions through an inverse current–time relationship: the higher the current, the faster the device acts. However, overcurrent devices offer no protection from electrical shock, or from fires caused by high resistance or arcing faults. Leakage current above which humans cannot "let-go" of an energized circuit is about 0.05 A (50 mA). This level is well below the usual activation level of overcurrent devices. As well, they offer no protection from fires caused by the heat of an arc, overheating of a small conductor or component, or the heat generated by a loose terminal or connection. Fires can be started at currents below 15 A. Hence, the function of an overcurrent device is to protect wiring, nothing more.

Assured grounding is one technique sanctioned in regulatory statutes on electrical safety. (In the U.S., see for example, 29 CFR 1926.404 on wiring design and protection and 1926.405 on wiring methods, components, and equipment for general use, for specific guidelines on ground fault protection and confined space entry (OSHA 1996). In this statute, assured grounding is called the Assured Equipment Grounding Conductor Program.) The concept behind assured grounding is a program of continuous testing and recordkeeping. Testing includes tests of ground conductors for continuity and for correct attachment of the grounding connector to each receptacle, attachment plug, and cord set. Tests must occur before first use, before return to service after repairs, before reuse after any incident that could have caused damage, and routinely at intervals not exceeding 3 months. One or more individuals trained to recognize safety hazards and who have the authority to take corrective action must constantly monitor the program.

Manufacturers of electrical equipment have devised ways to simplify the monitoring requirements. One example is the ground continuity monitor (Anonymous 1995). This device uses the glow from a lamp to indicate continuity. Failure of the lamp to glow indicates one or more of the following conditions: reversed polarity, open hot, hot on neutral/hot unwired, open neutral, or hot and ground reversed. These conditions apply within the cord, jobsite receptacle, or branch wiring.

Double insulation is a design feature that increases the safety of electrical tools and equipment (Bernstein 1991a). The term "double insulation" means that insulation between internal energized conductors and any possible point of external contact consists of two insulating systems that are physically separated (UL 1983). Functional insulation is the insulation necessary for the normal operation of the unit. Functional insulation is the insulation on windings of a motor or a transformer. Protective insulation provides protection against electric shock following failure of functional insulation. Protective and functional insulation are independent of each other. A protective enclosure in a portable tool manufactured from an insulating material is an example of protective insulation (Bernstein 1991a). Exposed metal parts, such as the exposed metal chuck on a drill, have internal protective insulation. Double-insulated appliances are required to have only a two-bladed plug (Figure 6.4).

Double insulation is not sanctioned as a protective measure in regulatory statutes on electrical safety. There is no method to assure that the integrity, and hence performance of the insulation, remains satisfactory throughout the life of the product (Carbone 1994). Wetting of the insulation and introduction of foreign substances can cause deterioration of performance. Wetting, for example, can create a continuous conductive path between live conductors inside the tool and the outside surface.

Figure 6.5 illustrates the third method for assuring safety in the use of electrical equipment, the ground fault circuit interrupter (GFCI) (Bernstein 1991a, Roberts, 1996).

GFCIs are available only for household voltages (110 to 120 V). The equivalent of the GFCI at higher voltages is the equipment leakage circuit interrupter (ELCI). The GFCI interrupts power following low current flow to ground through a path other than a grounded conductor. The GFCI compares the current in the ungrounded conductor to that in the grounded conductor (Palmquist 1973, Roberts 1996). The GFCI is actually a current transformer or zero sequence transformer. The principle on which the GFCI operates is that the voltage produced in the sense coil wound on a magnetic core that encircles all conductors in a circuit is proportional to the difference in current in the conductors. That is, the magnetic flux is zero when the current is balanced. The GFCI detects low-level ground faults and opens the circuit at currents equal to or exceeding the preset value of 6 mA (UL 1985). This is considerably less than current flow needed to trigger a conventional circuit

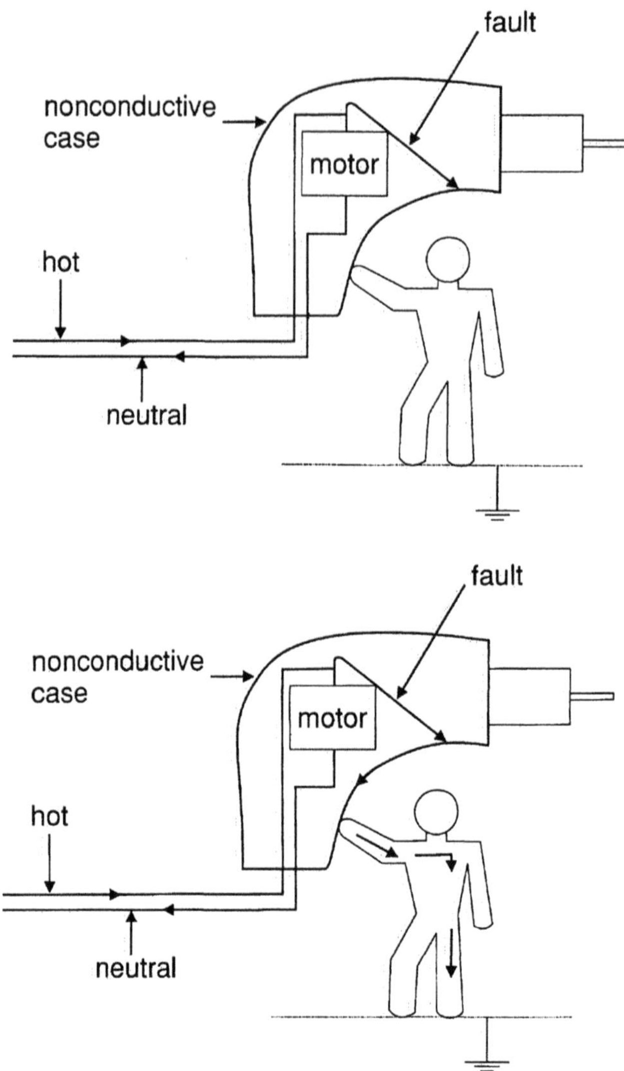

FIGURE 6.4 (A) Current leakage protection in a double-insulated tool. The extra layer of insulation provided by the case of a double-insulated tool acts as a barrier to transmission of leakage current through limbs of the user to ground. (B) Leakage path to ground in a double-insulated tool. Double-insulated tools are used with the understanding that current leakage cannot occur. Current leakage paths can develop under the rugged conditions that exist in some industrial settings. Possible situations include degradation of insulation, damage to the case (crackage or puncture), and intrusion of conductive materials.

breaker operating at 15 to 20 A. The rate of interruption of flow of current in the circuit is inversely proportional to the imbalance in current. At 6 mA, the GFCI will deenergize the circuit within 0.025 second. The ELCI operates at much higher current flows.

The trip times for the GFCI are designed to be in the safe range for human response (Bernstein 1991a). By tripping at 6 mA, an individual will be able to let go of the circuit and not receive a lethal shock. The shock may be painful, but would not be lethal.

GFCIs are sanctioned in regulatory statutes as a protective device in electrical circuits. (In the U.S., see for example, 29 CFR 1926.404 on wiring design and protection and 1926.405 on wiring methods, components, and equipment for general use, for specific guidelines on ground fault protection and confined space entry [OSHA 1996].)

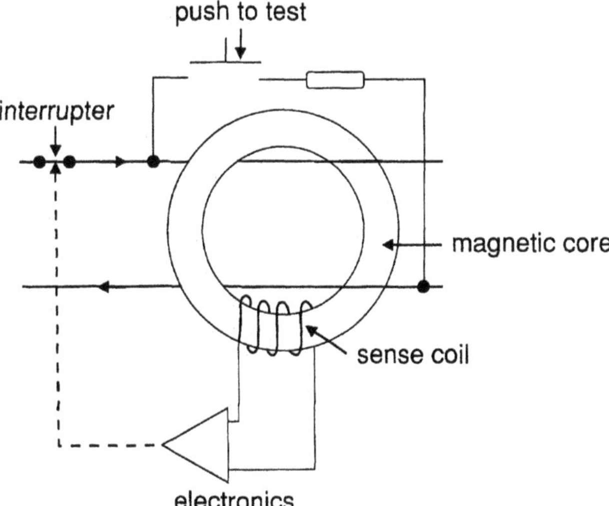

FIGURE 6.5 Ground fault circuit interrupter. The ground fault circuit interrupter (GFCI) utilizes the electrical principle of induction. The conductors pass through a magnetic core around which is wrapped the sense coil. Imbalance in the current flowing in the conductors induces a magnetic field in the magnetic core and an electrical current in the sense coil. Currents above the set-point of the electronics cause the circuit to open. (Adapted from Greenwald 1991.)

GFCIs mandated for use in industrial applications differ from those approved for use in the home (Anonymous 1995). The difference is the open neutral protection required in industrial applications. Opening or lifting the line side neutral at a panel creates the condition called an open neutral. The domestic GFCI lacks the power required for its operation and no longer provides protection. Protection against the open neutral condition occurs when both the hot and neutral circuits on the load side of the GFCI are disengaged following opening of the neutral or hot lines on the line side (UL 1985). In addition, portable industrial GFCIs also provide protection against reverse phasing conditions, while GFCI breakers do not. Reverse phasing occurs when the neutral and phase conductors are reversed at the terminals in the receptacle. Protection is achieved by switching both conductors when a ground fault is detected.

Low voltage equipment is the fourth approach to safety in the use of electrical equipment in confined spaces (Anonymous 1995). Thus far, lighting is the principal industrial application of low voltage equipment. Low voltage systems operate at 6 or 12 V AC through transformers that convert 110 to 120 V AC. Some designs shield the low voltage side from the high voltage. The transformer may be located upstream from the light or in the assembly. These units are produced for use in hazardous and nonhazardous locations.

Low voltage lighting is sanctioned in regulatory statutes as protective devices in conductive and hazardous locations. In the U.S., see for example, 29 CFR 1926.404 on wiring design and protection and 1926.405 on wiring methods, components, and equipment for general use, for specific guidelines on protection using low voltage equipment (OSHA 1996). OSHA mandates use of hand lamps operating at 12 V or less. Hand lamps designed for use in hazardous locations are specially designed units.

Hazardous Effects of Electrical Energy

In the U.S., about 10% of fatal occupational injuries result from electrocution. This translates to about 700 lives lost per year (Centers for Disease Control 1984).

TABLE 6.1
Effects of 60 Hz AC Electrical Current

Current (mA)	Reaction
1	Perception of faint tingle
5	Slight shock; disturbing, not painful. Average individual can let go of energized conductor. Injury can occur from involuntary reaction.
6–25 (women) 9–30 (men)	Painful; muscular control lost; let-go range
30 (men)	Respiratory paralysis, stoppage of breathing
50–150	Extreme pain, respiratory arrest, severe muscle contractions. Individual cannot let go. Death possible.
75 (men)	Fibrillation threshold 0.5%
250 (men)	Fibrillation threshold 99.5%
1,000–4,300	Ventricular fibrillation, muscle contraction and nerve damage after short duration. Death likely.
> 10,000	Cardiac arrest, severe burns, probable death

After Lee 1971 and OSHA 1986.

The usual cause of death due to contact with electrical energy is cardiac arrest (Bernstein 1991b). The most prevalent electrical contact is 60 Hz. The effects of normal 60-Hz electrical energy are related to the root mean square of the current and the duration. Voltage is important through its involvement in Ohm's law with impedance (resistance) of the body for calculating current flow. Current flow through the body occurs when different parts contact surfaces having different potentials (voltages). Impedance varies with the part of the body. Impedance rather than circuit resistance determines current flow in AC circuits. Table 6.1 summarizes the effects of electrical current at different levels.

The threshold of perception for a finger tapping contact at 60 Hz is approximately 0.2 mA. The perceptual level for half of a group of men is 0.36 mA, and 0.24 mA for half of a group of women (Kahn and Murray 1966). Further studies at Underwriters Laboratories indicated that no uncontrolled startle reaction resulted when the shock current was below 0.5 mA, even though perception of a shock might occur. Women seemed to have greater sensitivity to shock than men (Smoot and Stevenson 1968). As a result, the threshold level for a startle reaction was established to be 0.5 mA.

Current at a level somewhat higher than the threshold of perception can cause the hand involuntarily to close and grasp the electrified contact that the palm or fingers were touching (Bernstein 1991b). The individual cannot "let go" and remains trapped unless the power is turned off or he/she is physically removed from contact with the circuit. Shocks at the let-go current level are quite painful, although not usually lethal. The let-go current threshold can be considered as a "go-no-go" situation. Once immobilized or frozen to a circuit, an individual either will break the contact and live or will not be able to break the contact. If the contact is not broken, the person's skin contact resistance may decrease and increase the current to the lethal level, ultimately causing death. To break contact and become free of the circuit, the individual must overcome involuntary muscular contraction while enduring the painful shock.

The let-go current for women is lower than that for men. Dalziel and Massoglia found that 0.5% of women could not let go at 6 mA, whereas 0.5% of men could not let go at 9 mA (Dalziel and Massoglia 1956). As a result, 6 mA was set as the safe value for let-go current because shocks at this level will not freeze individuals to energized circuits. The ground fault circuit interrupter detects current imbalances greater than 6 mA between supply and return currents and disconnects the circuit in sufficient time to prevent electrocution (UL 1985).

Asphyxia can occur when the passage of current through the chest cavity causes constant contraction of the chest muscles (Cabanes 1985). The individual cannot breathe or let go of the circuit.

Respiratory arrest can occur when current passes through the respiratory center (Lee 1966). The respiratory center in the medulla is situated slightly above a horizontal line extending from the back of the throat. Thus, shocks from the head to a limb or between two arms could lead to respiratory arrest (Bernstein 1991b). Shocks at current levels above those that cause ventricular fibrillation can produce respiratory arrest, even though the primary path is not through the respiratory center (Hodgkin et al. 1973).

Interference with the normal control of heart muscle contraction is the usual cause of death in electrocution (Bernstein 1991b). Death will occur if the ventricles do not contract to circulate the blood. Lack of oxygenated blood causes irreversible brain damage within 3 to 6 min.

Ventricular fibrillation is the uncoordinated, asynchronous contraction of the ventricular muscle fibers. An electrical shock passing through or across the chest can cause ventricular fibrillation. The victim becomes unconscious in less than 10 seconds, because blood circulation stops. Cardiopulmonary resuscitation, promptly administered, can provide some circulation of oxygenated blood to the brain and heart until a defibrillator can be used (NIOSH 1986a). The only way to restore heart rhythm is to use a defibrillator. A defibrillator applies a pulse shock to the chest.

For shocks shorter than a cardiac cycle, the electrical current to cause fibrillation must be large and occur during the vulnerable period (Bernstein 1991b). Shocks longer than a cardiac cycle lower the shock threshold current by causing premature ventricular contractions. Using these concepts, the following safe current limits were proposed: 500 mA for shocks less than 0.2 seconds and 50 mA for shocks longer than 2 s (Biegelmeier and Lee 1980).

Asystole is cardiac standstill; the heart does not beat. Current flow above 1 A through the chest cavity can cause asystole. Such high currents are associated with high voltage. Hence, asystole rather than ventricular fibrillation often is the cause of death in accidents at voltages greater than 1,000 V (Bernstein 1991b). Unlike ventricular fibrillation, asystole may revert to a normal heart rhythm once the triggering voltage is removed. As a result, individuals have survived asystole following exposure to high voltage, high current shock, while others have died from ventricular fibrillation following a low current, 120-V shock. This is the basis for concern expressed in NIOSH Alert documents about 110- to 120-V electrical contacts (NIOSH 1986a, b).

Pulse and impulse-type shocks also can cause damage to the heart. These shocks have a time duration much shorter than the heart cycle. Pulse and impulse-type shocks result from capacitor, electrostatic, and lightning discharges. The hazard in an impulse-shock is due to the electrical energy in the discharge.

Any pulse shock with an energy content of 50 J probably is hazardous. Shocks below 0.25 J, while disagreeable, probably are not hazardous (Dalziel 1971). The annoying electrostatic discharge shock produced by walking across a carpet is about 10 mJ. Defibrillators have a maximum output in the range of 200 to 400 J (Bernstein 1991a).

Direct current levels that produce the perception or startle reaction are about three times the current required at 60 Hz. Similarly, the current needed for ventricular fibrillation is three times that for the 60 Hz for shocks lasting longer than 2 s. The value is the same for shocks less than 0.2 s. Direct current causes severe pain when the circuit is made or broken, but little while current is maintained. Let-go level for direct current is the magnitude when test subjects refuse to let go because of the anticipated severe jolt (Dalziel and Massoglia 1956).

Surface burns on the body can result from the heat produced by an arc in nearby equipment, heating at the point of contact by a high-resistance contact or large currents, or the splatter of molten particles from damaged conductors or equipment (Bernstein 1991b). A shock from contact with energized equipment can cause external or internal burns. The magnitude of the current and type and duration of contact determine the severity of the injury.

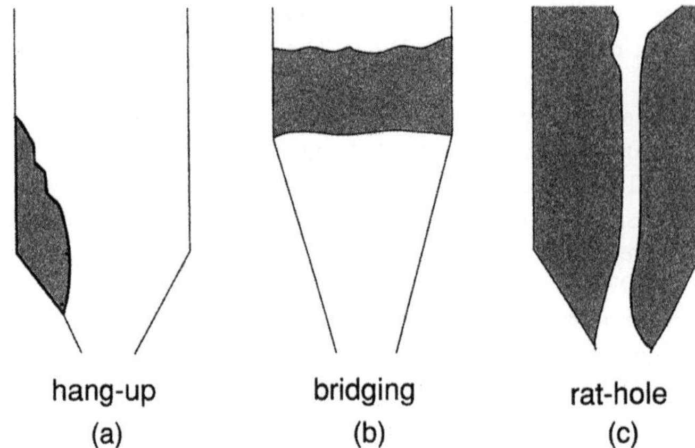

FIGURE 6.6 Modes of retention of flowable bulk materials in silos. These include hang-up on walls (6.6A), bridges across the material (6.6B), and ratholes (6.6C). (Courtesy of Northern Vibrator Manufacturing Ltd., Georgetown, ON.)

Flowable Bulk Materials

The storage and handling of flowable bulk materials poses problems of energy storage and dissipation similar to those experienced in mechanical, pneumatic, hydraulic, and electrical systems. During storage, coalescence of material sometimes occurs (Figure 6.6). The presence of flow-inducing products in the industrial marketplace is ample testimony to the widespread occurrence of this problem. Coalescence results from the presence of moisture or biological action. Coalescence can result in caking on vertical surfaces of the structure and bridging across the horizontal plane of the material. Bridging can occur at or below the surface of the material (Figure 6.6B). Material that has not coalesced can flow from underneath the bridge, thus creating a hollow. Also, "rat holes" may be present (Figure 6.6C). "Rat holes" are vertical channels in the coalesced material. "Rat holes," therefore, permit some flow to occur. This flow may hide the existence of a bridge. Many of the accidents in silos, hoppers, and other storage structures resulted from the collapse of a bridge during attempts to open a channel or to improve flow.

Flowable bulk solids are akin to soils and snow. In fact, soils and ground and pulverized rocks and minerals comprise a substantial portion of industrial flowable bulk solids. Soil is an aggregate of mineral grains that can be separated by gentle mechanical action (Terzaghi et al. 1996). Soils are characterized by index properties. Index properties subdivide into classes: soil grain properties and soil aggregate properties. The principal grain properties are size and shape. The principal aggregate properties are cohesiveness, relative density, and consistency.

Sand (particle size to 2 mm) is a cohesive aggregate of fragments of rock and minerals. Inorganic silt is a fine-grained soil that has little or no plasticity. The least plastic varieties (rock flour) contain equidimensional grains of quartz. The most plastic varieties (plastic silt) contain flake-shaped particles. Clay is an aggregate of microscopic and submicroscopic particles, resulting from chemical decomposition of rock.

Flowable bulk solids are the most difficult of the states of matter to handle (Wahl 1985). Failure of a material to flow by gravity frequently results from a combination of properties, such as density, compressibility, and hygroscopicity. Compressibility is computed from the aerated density and the packed density of the material. The dividing line between free-flowing granular and nonfree-flowing powder is about 20% compressibility. A higher percent indicates a powder that is not free-flowing and is likely to bridge. Hygroscopicity refers to the ability of a substance to absorb moisture. Hygroscopic substances will flow erratically or not at all when exposed to high humidity.

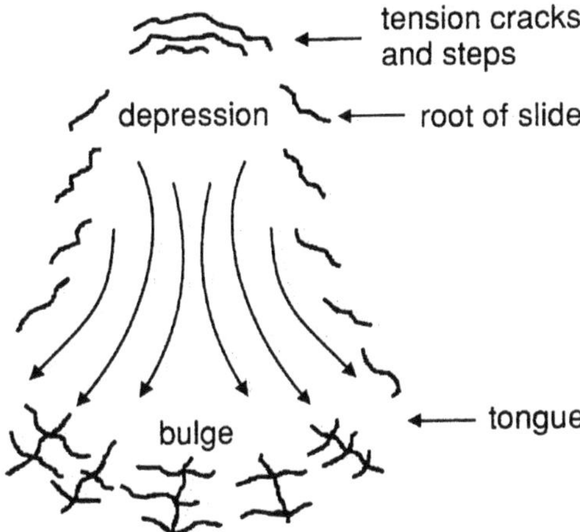

FIGURE 6.7 Typical slide in cohesive material. Almost every slide exhibits the same general characteristics: formation of tension cracks on the upper slope or beyond its crest. The upper part (root) subsides, whereas the lower part (tongue) bulges. (Adapted from Terzaghi et al. 1996.)

Four angular indicators of the free flow potential of a bulk material are the angle of repose, the angle of fall, the angle of slide, and the angle of spatula (Wahl 1985). The angle of repose is the included angle between the edge of a cone-shaped pile formed by falling material and the horizontal. The smaller the angle of repose, the more flowable the material. The angle of fall is the corresponding angle measured when the pile settles after a weight is dropped nearby on the same surface. The greater the difference between the two angles, the greater the free-flow potential. The angle of slide is the angle at which a material will slide down a surface under its own weight. To measure the angle of spatula, a spatula covered to the maximum is lifted from a pile of material. A weight is then dropped onto the same surface. The angle of spatula is the difference between the angle of the side of the pile before and after the dropping of the weight.

A classification scheme containing four divisions is used in the assessment of flowable bulk solids (Wahl 1985). These classes range from granular, free flowing materials to fibrous or flaky materials of low bulk density that tend to interlock and absorb vibration. Flowability is a measure of the tendency of the material to form a bridge. A bridge is a condensed layer of material that does not flow. The material under the bridge may flow away, potentially leaving behind a hollow. The bridge may support the weight of an individual, but could collapse at any time. Bridging and ratholing (formation of flow channels through a condensed mass of material) in storage hoppers and silos is subject to mathematical treatment (Carson and Johanson 1985).

Flowable bulk materials contained in a sloping pile or on the walls of a silo have a tendency to move downward and outward under the influence of gravity (Terzhagi et al. 1996). If counteracted by the shearing resistance of the material, the slope is stable; otherwise, a slide occurs (Figure 6.7). Downward movement of material is highly desirable during normal operation. This constitutes the normal operation of the silo in its function as a surge and storage structure. Under abnormal conditions the material does not flow. Often this is related to the presence of abnormal levels of moisture.

Flowable bulk materials adherent to the walls of structures contain potential energy. Hangup on walls and formation of bridges and other flow problems pose such problems that a whole industry has developed to correct them. This industry manufactures equipment to dislodge adherent material, as well as open channels to enable the flow of material. Sending workers into storage structures to perform the same functions, still is common practice. Dislodging hung-up material can require use of shovels, jackhammers, and even dynamite. In these situations, the purpose for the activity is to

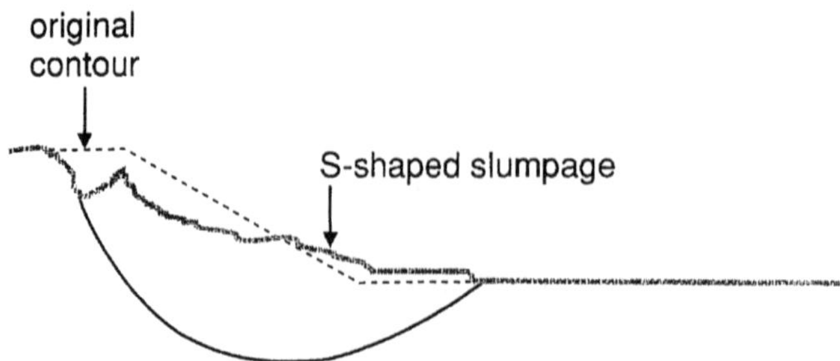

FIGURE 6.8 Cross-section through a typical slide of cohesive material. Note the "S" shape of the slumpage. (Adapted from Terzhagi et al. 1996.)

destabilize the structure of the material in order to induce flow, which, in turn, reduces the energy content. Techniques include undercutting the foot of the slope or digging an excavation with unsupported sides. These techniques mimic naturally occurring phenomena, including earthquakes. As fatal accidents that have occurred in these structures have shown, the moment of failure of the slope is not always predictable (OSHA 1983, MSHA 1988). Slumping can occur slowly or suddenly, with or without apparent provocation.

Parallels exist in soil mechanics with the mechanics of flow of flowable bulk solids that are constrained in a storage structure. Failure of a mass of material located beneath a slope is a slide (Figure 6.7). A slide involves downward and outward movement of the entire mass of material that participates in the failure. Almost every slide of cohesive material is preceded by formation of tension cracks on the upper part of the slope or beyond its crest (Terzaghi et al. 1996). During the slide, the upper part of the slide area (root) subsides, while the lower part (tongue) bulges. The shape of the bulge depends on the nature of the material. The plane of the slide forms an S-shaped curve (Figure 6.8). This may be constrained when the flow of material reaches the opposite side of the structure.

Experience indicates that a slope ratio in soils of 1.5 (horizontal:vertical) usually is stable (Terzaghi et al. 1996). Slopes having smaller ratios than this can occur in storage structures when the material hangs-up on the walls and refuses to flow. Certain types of soils also are less stable and require larger ratios. The ratio used in regulatory standards pertaining to trenches and excavations also could differ from these values. The variety of factors and processes that lead to slides are extraordinary and complex. As a result, excavations involving these materials require input from experienced civil engineers. Massive flows of material (flow slides) have occurred unexpectedly under natural conditions as well as during construction activities. The former can occur because of poor adhesion between layers of stratified material.

A slope underlain by a dry, cohesionless material, such as clean, dry sand, is stable regardless of height, provided that the angle between the slope and the horizontal is equal to or smaller than the angle of internal friction for the material in the loose state. The factor of safety is the tangent of the angle of friction divided by the tangent of the angle between the slope and the horizontal. Few natural soils and materials are cohesionless. In the case of cohesive soils and flowable bulk materials stored in silos, stable vertical slopes can exist, as demonstrated by the existence of hung-up material. Stability of a slope is a function of height.

Stability computations for slopes are based on moments of rotation about an imaginary point. Curvature of the surface of motion by the subsurface material resembles the arc of an ellipse. In stability computations, the arc of a circle is used as an approximation for the arc of the ellipse. Moments are calculated for the slice of material that tends to produce failure and the slice that tends to resist it. The moment is calculated using the center of mass for each slice and the distance along the perpendicular line drawn from the center of mass to the radius of the circle. In the simplest case, sliding resistance at equilibrium depends on the difference between the moments and the radius of the imaginary circle and the length of the surface of sliding.

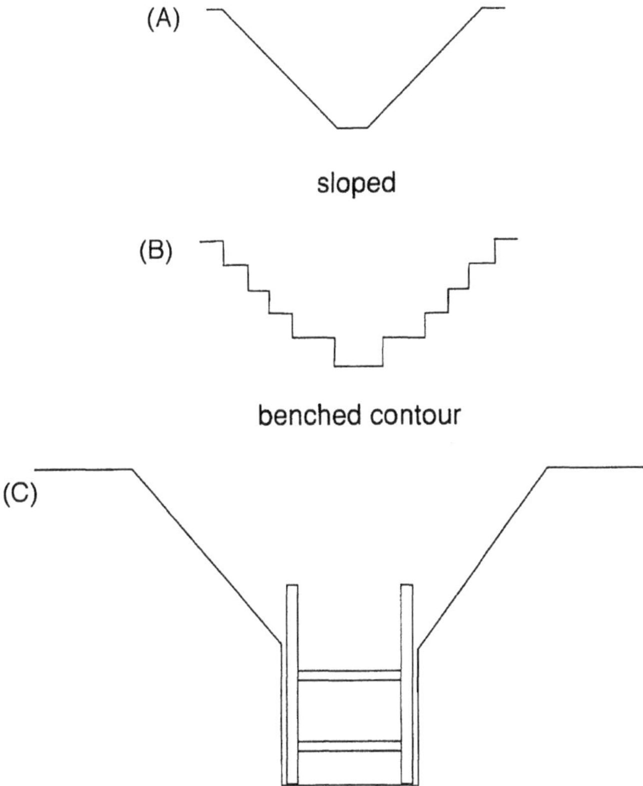

FIGURE 6.9 Excavation and trenching techniques: (A) Sloped Walls; (B) Benching; (C) Combination of sloping and shoring. Slope angle varies according to soil type. (Adapted from Yokel and Chung 1983.)

Protection of individuals who work in excavations and trenches is the function of sloping or benching, shoring or shielding, or a combination of both (Figure 6.9). Sloping involves the creation of an oblique surface. Benching involves a series of one or more steps having horizontal and vertical surfaces. Shoring is a cross-braced structure formed from horizontal and vertical planks (Figure 6.10A). Shielding is a manufactured structure that incorporates the framework and vertical protective sheeting (Figure 6.10B). The depth at which protective measures should occur is 1.5 m (5 ft), although ground conditions could necessitate these at even shallower depths (Yokel and Chung 1983, Krieger and Montgomery 1997). The nature of the precautions depends on depth and soil type.

OTHER SOURCES OF ENERGY IN CONFINED SPACES

The preceding sections focused on energy sources intrinsic to the space and materials contained in it. Other sources also release energy into these spaces. Some of these sources are associated with operation of the space and equipment inside it. Others are associated with work activity that occurs during entry. Some of these sources are familiar and readily recognized, while others are elusive.

Most of the energy provided by these sources is not sufficient to cause visible expression of acute injury following an overexposure of a workshift or less. Many of these agents act chronically. This is not to advocate, of course, any relaxation of standards during short-term work involving confined spaces. Rather, this is to say that lack of expression of acute injury is the reason that high-level exposures to these agents have not resulted in reported accidents involving confined spaces.

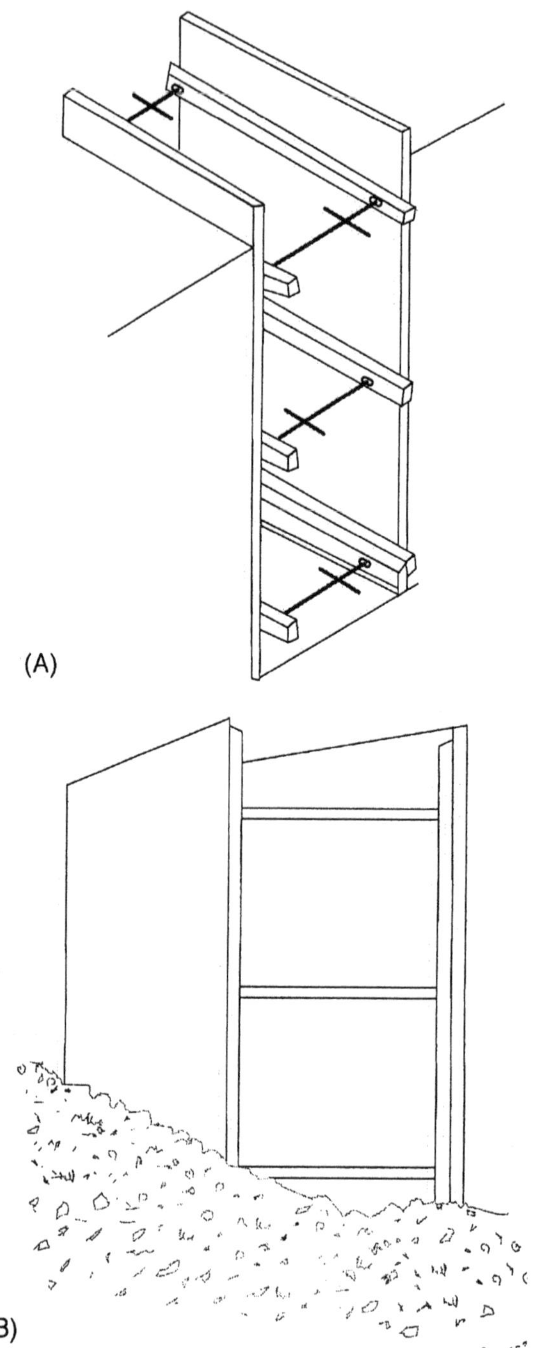

FIGURE 6.10 Shoring and shielding structures: (A) Shoring structure composed of horizontal and vertical planks; (B) Prefabricated shielding structure. (Adapted from Krieger and Montgomery 1997.)

Noise

The long-term effects of high levels of noise on human hearing and productivity are well known and are well described in other sources (Berger et al. 1986).

The main effect of acute exposure to noise is the temporary threshold shift (Hétu and Parrot 1978). The temporary threshold shift (TTS) is a temporary shift in the sensitivity of the auditory system to sound signals at different frequencies. The shift appears following noise exposure and disappears following a prolonged period of relative quiet, optimistically provided during sleep and leisure hours.

Hétu and Parrot examined TTS in a group of workers exposed during most of the workshift for a period of many years to noise from a bottling line (Hétu and Parrot 1978). Spectral data indicated that the band pressure level ranged from 85 to 89 dB across much of the sound spectrum. The average noise dose for the exposure period of 6 h was an L_{eq} of 96 dBA. This group had suffered considerable permanent loss, ranging from 4.5 to 15 dB across the speech range and about 40 dB in the 3- to 8-kHz region. The daily noise exposure produced a temporary threshold shift of 7 to 12 dB across the speech region. The losses incurred in the TTS reflected the ameliorating effect of a prolonged break during midday. Greatest loss occurred between the start of the shift and midday. Greatest initial shift occurred at 2 kHz.

Schmidek and Carpenter studied temporary threshold shift in chain saw operators who were engaged in thinning forested areas (Schmidek and Carpenter 1974). These authors studied temporary threshold shift at 3, 4, and 6 kHz 2 min after the end of the noise exposure (TTS_2). Their experience indicated that these frequencies were most sensitive to exposure to loud noise and produced temporary threshold shifts ranging from 13 to 21 dB. Greatest temporary threshold shift occurred at 4 kHz.

Temporary threshold shift has received considerable attention in past years (Ward 1969). Growth in TTS (measured in dB) is almost linear with the logarithm of time. TTS begins at noise levels of 80 dB and increases linearly to at least 130 dB. This means that TTS produced between 80 and 85 dB will be about the same as that produced between 90 and 95 dB. Considerably higher temporary threshold shifts than those mentioned above can occur. TTS depends on noise level and individual sensitivity. As indicated by observation, TTS is less for equal levels of noise occurring in low frequencies compared to high frequencies. TTS from noise energy concentrated in a narrow range of frequencies occurs at half to a full octave higher. Intermittency is another important factor in development of TTS. TTS is proportional to the fraction of the time that the noise is present. That is, short bursts are tolerated better than continuous noise.

The (U.S.) National Academy of Sciences/National Research Council Committee on Hearing Bioacoustics and Biomechanics (CHABA) examined the problem of temporary and permanent threshold shift (Kryter 1963). CHABA offered the view that the limits below which handicap in speech reception would not occur are:

- Not greater than 10 dB at frequencies up to 1 kHz
- 15 dB at 2 kHz
- 20 dB at 3 kHz and above

The temporary shift in hearing sensitivity compounds the effect of permanent losses incurred following previous exposure to loud noise plus that lost due to aging plus that due to use of hearing protection. The net result for an individual can be considerable loss of hearing sensitivity. The impact of this loss can result in the inability to hear warning signals.

Seshagiri and Stewart reported on the audibility of warning devices when masking sounds producing high noise are present (Seshagiri and Stewart 1992). This report was the outcome from an investigation of accidents resulting from the inability of workers to hear warnings from train whistles. In these and other situations, the workers were exposed to high levels of workplace noise.

The loss of hearing sensitivity can be especially important during work occurring in confined spaces where warning signals are utilized. Many confined spaces contain reflective surfaces. Reflective surfaces increase the contribution of reflected energy to the sound field. The sound pressure level in a confined space that is fully or partially enclosed receives contributions from the direct and reflected fields (Magrab 1975). The sound pressure level at any point, r, from a point source is given by:

$$L_p = L_w + 10 \times \log_{10}\left(\frac{Q}{4\pi r^2} + \frac{4}{R}\right) \quad (6.3)$$

$$R = \frac{S\alpha}{1-\alpha} \quad (6.4)$$

L_p = the sound pressure level, dB re 20 μPa;
L_w = the sound power level, dB re 1 pW;
Q = a geometry factor, 1, 2, 4, or 8;
r = the distance from the source, m;
R = the absorption constant of boundary surfaces;
S = the surface area, m^2;
∝ = the absorption coefficient for the boundary surfaces. In this case, the surfaces are assumed to be the same material; otherwise, the weighted average is used.

The net result from Equations 6.3 and 6.4 is that, in a structure constructed from highly reflective material, the sound pressure level approaches the sound power level.

A confined space can contain unusual sources of noise other than those introduced during work activity. Some devices used for level detection emit sound at supra audible frequencies, typically 40 to 50 kHz. These devices could remain operational through oversight or lack of comprehension of the hazard during entry and work activity.

Ultrasound, sound at frequencies above the normal audible range, has received little attention in the industrial hygiene literature (Wiernicki and Karoly 1985). Exposure to airborne ultrasound is considerably less hazardous than contact exposure, since air readily attenuates transmission. Most common industrial applications occur in the range 20 to 300 kHz. Sources utilize continuous and pulsed output. The main effects of high-energy exposure to ultrasound include: heating of tissue, mechanical disruption of membranes, formation of gas bubbles, and damage to tissues and organ systems. Many of these high-energy effects occur in the brain and central nervous system.

The basis for the Threshold Limit Value for frequencies above 20 kHz is prevention of possible hearing losses from subharmonics of those frequencies (ACGIH 1991). Subharmonics of these frequencies can cause temporary threshold shifts (Parrack 1966). Other effects that can occur in sensitive individuals include: nausea, headache, tinnitus (ringing in the ears), pain, tingling in the scalp and cheeks, dizziness, and fatigue. Almost all of the adverse health effects occur below 50 kHz (International Non-Ionizing Radiation Committee 1984).

Some of these effects could occur in individuals working in confined spaces containing sources of airborne ultrasound. More serious are the potential effects that could result from physical contact with these sources. Physical contact could occur accidentally or through deliberate grasping.

IONIZING RADIATION

Sources of ionizing radiation other than background levels that occur in confined spaces have two main origins: naturally occurring and man-made.

Naturally occurring sources of ionizing radiation result from substances containing compounds of uranium, radium, and thorium (Shapiro 1981). These substances occur in low concentrations in the earth's crust and at substantially higher concentrations in certain minerals, such as the monazite sands, phosphate-containing rock, and pitchblende (Metzger et al. 1980, Hewson and Fardy 1993). Industrial processes, such as combustion of coal and oil, can concentrate compounds containing these elements to considerably higher concentrations (de Santis and Longo 1984, Cohen 1985). The net result is that deposits containing high concentrations of these substances can form on the inside (boundary) surfaces of processing equipment. Such deposits develop during processing of phosphates in mineral extraction and the production of fertilizer and in the processing of oil and natural gas (Wilson and Scott 1992). Airborne dust from these deposits can cause significant exposure to the lungs during work in these structures. Uranium, radium, and thorium decay through a series of unstable isotopes (Shapiro 1981). These isotopes emit alpha, beta, and gamma radiation in varying amounts. Radiation from these substances is unlikely to produce symptoms of acute exposure.

Radon, the heaviest of the inert gases, is an exception to the other naturally occurring radioactive substances, which are solids (Shapiro 1981). Radon-222, the most prevalent form of radon, originates from the radiological decay of radium-226. Radon can diffuse into confined spaces from surrounding soil or from materials of construction containing radium-226. Radon-222 can accumulate in spaces in which air exchange is low. Radon-222 has a half-life of 3.8 days. This means that half of the original number of atoms disappears in 3.8 days. Radon acts chronically. High levels, as encountered in uranium mining, are a causative agent in lung cancer. There is no evidence to support any concern about injury from acute exposure, as might occur during entry into a poorly ventilated space.

Gamma-emitting sources have been used industrially for many years in level and thickness gauges, as well as for radiographic nondestructive testing. When used as a level gauge, the source is mounted externally to the vessel. The beam passes through the walls of the vessel and the substance under measurement to a detector. The extent of attenuation is an indication of level, or thickness, in the case of a thickness gauge. The source contains a shutter that is to be closed and locked out during entry and servicing of the vessel. Damage to the shutter mechanism has occurred on occasion in corrosive environments. In these circumstances, the shutter failed to close when the arm was moved to the closed position. In other cases involving poor housekeeping, burial of the source by powder has occurred. In these situations, the source was undetectable to outside contractors.

Industrial radiography utilizes gamma sources to produce an image on photographic film. The source is located on one side of the object to be examined, and the film on the other. Prevention of unnecessary exposure of bystanders and the radiographer is a function of the skill used in planning the exposure and bystander control.

In these situations, in the absence of a confirmatory check using a gamma-detecting radiation survey instrument, entry and work could occur under conditions of full irradiation.

In the previous instances, the source is external to the enclosure. In many other applications, the enclosure is designed to confine energy from the source. The concept of using shielding for confinement of energy of radiological significance has formed the basis for radiation protection for many years. Application of this concept ranges from the walls that shield the diagnostic X-ray machine in the dental office to the thick, high density concrete used to shield high energy gamma sources used in irradiation of food and sterilization of equipment. In the former case, there are no constraints to prevent the operator from entering the field of the X-rays during operation of the equipment. In the latter case, interlocks on doors shut down or otherwise deactivate the source at time of unsanctioned entry.

These safeguards, however, are not foolproof, as the recent accident in San Salvador would indicate (IAEA 1990). In that accident, three workers entered the controlled zone of an irradiation facility to dislodge a jammed cobalt-60 source array. All three workers later developed acute irradiation syndrome from the exposure. Two sustained amputations of the leg. The third man died from causes unrelated to the accident. In parallel to fatal accidents that have occurred in conventional confined spaces, this accident resulted from many causes. These included defeat of safety systems, damaged equipment, lack of radiation monitoring, and miscommunication resulting from inadequate translation of information from English to Spanish.

The acute effects of overexposure to ionizing radiation affect all organs and systems of the body (Cember 1969). The acute effects are classified into three syndromes in order of increasing dose: hematopoietic (blood-forming elements) syndrome, gastrointestinal syndrome, and central nervous system syndrome. Effects that are common to all of these syndromes include: nausea and vomiting, malaise and fatigue, increased temperature, and blood changes. The appearance of symptoms depends on the dose and the syndrome. None of the symptoms is immediate, as would be the case with some types of chemical overexposure.

Additional radiation accidents have resulted from mishandling of abandoned cesium-137, iridium-192, and cobalt-60 sources (Maletskos and Lushbaugh 1991). These accidents have caused the deaths of several persons. Much of the injury resulted from direct handling of the source material.

NONIONIZING RADIATION

Nonionizing radiation, gamma radiation, and X-rays all are forms of electromagnetic energy. Electromagnetic energy is carried by waves of different length. The distinction between ionizing and nonionizing radiation is the ability to ionize (remove electrons from) atoms and molecules. Electromagnetic waves classed as nonionizing radiation carry insufficient energy to cause ionization. These wavelengths are subdivided, sometimes artificially, by the ability to cause other types of biological effects.

ULTRAVIOLET REGION

The ultraviolet (UV) region of the electromagnetic spectrum contains the highest energies of nonionizing radiation. These energies can cause excitation of atoms and molecules, but not outright ionization. UV energy has considerable application in industrial processes. As a result, UV sources are used in confined spaces, as well as other environments. A major industrial application of UV energy is the polymerization of monomers by activation of photosensitive initiators (Moss 1980). Applications of this process include the curing of inks, adhesives, and wood and metal coatings. The UV sources used in these processes produce significant output in the 300- to 400-nm region. This is within the region (180 to 400 nm) that produces biological effects (INIRC 1985a). Energy in the UV region may not be visible to the eye. Workers who enter confined spaces containing UV sources may not be aware of the presence of this energy. The only indication may be emission of blue-coloured light and fluorescing of certain types of clothing. In such circumstances, warning devices and interlocks should be present at entrances to the space. There is no guarantee, however, that these devices will function as intended or that tampering to defeat them cannot occur.

The critical targets for UV energy are the skin and the eye These structures absorb energy in this region of the electromagnetic spectrum. The most pertinent example of exposure to UV from an uncontrolled source occurred in a school gymnasium containing a damaged mercury vapor lamp (Andersen, 1980). A mercury vapor lamp contains an inner lamp enclosed in quartz and an outer glass envelope. The inner lamp continues to function when the glass is damaged. The layer of glass serves to absorb the UV energy. Without the intact glass envelope, the lamp emits considerable UV energy. In general lighting situations, the glass envelope of mercury vapor lamps effectively absorbs UV energy (McKinlay 1992). High intensity discharge lamps, such as are used in photopolymerizers and sun lamps, can emit considerable UV energy.

The gymnasium incident resulted in severe corneal photokeratitis (inflammation of the cornea), conjunctivitis (inflammation of membranes in the eye), and erythema (redness similar to sunburn in the skin). Photokeratitis is a delayed effect. It produces the sensation of sand in the eyes. The delay in onset is proportional to the dose. Erythema is the most conspicuous short-term change in the skin following UV exposure. The severity of erythema reflects genetic differences. The impact of UV sources in some confined spaces is exacerbated by the reflective interiors and barrier surfaces.

Another major source of UV energy during work in confined spaces is welding. Welding emissions cover the range from the ultraviolet to the infrared (Hinrichs 1980). In general, UV emissions from electric welding processes decrease in the following order: gas metal arc welding, shielded metal arc welding, gas tungsten arc welding, and plasma arc welding. Welding arcs produce emissions at wavelengths below 300 nm. As a consequence, these are sources of ozone.

Visible Region

The energies contained in a narrow band of the electromagnetic spectrum are detectable by the retina in the eye. Visible light only recently has received recognition as being hazardous. "Blue light" in the region 400 to 550 nm can cause photochemical injury to the retina (Sliney 1990). The discovery of photochemical damage from blue light provided an explanation for unanswered questions about effects, such as eclipse blindness (Sliney 1983). The duration of exposure must exceed 10 seconds. Injury appears several hours to 2 days after the exposure (Sliney 1992). Blue-light retinal injury can result from viewing an intense source for a short time or a less intense one for a longer time. Energy levels capable of causing blue-light injury can vary more than 1,000-fold. High intensity discharge lamps are potential sources of blue-light hazard (McKinlay 1992). Welding arcs are another potential source of exposure to blue light. Reflections from boundary surfaces of structures can cause indirect exposure to these sources.

Infrared Region

Infrared sources span the range from 760 nm to 1 mm (1,000,000 nm) (McKinlay 1992). Infrared energy from 770 to 1,400 nm is focused onto the retina. At sufficient intensity, a retinal burn can occur (Sliney 1983). The cornea absorbs infrared energy at wavelengths longer than 2,000 nm. Injury to the retina at wavelengths above 550 nm in the visible and infrared regions occurs principally from thermal injury (Sliney 1992). However, this appears likely only from pulsed sources. Conventional sources of infrared energy, such as molten glass and metal, apparently do not produce the irradiance necessary to cause either acute or chronic effects beyond "dry eye" (Sliney 1983). Incandescent (heated filament) lamps pose no hazard to the retina. However, infrared heat sources that provide no strong visual stimulus could pose a retinal hazard during prolonged staring (McKinlay 1992).

Sources of infrared energy could be present in confined spaces. Provision of a strong visual or other sensual warning inside the space is unimportant in situations where boundary surfaces serve as shielding. In such circumstances, warning devices and interlocks should be present at entrances to the space. There is no guarantee, however, that these devices will function as intended, or that tampering to defeat them cannot occur. The impact of IR sources in some confined spaces is exacerbated by the reflective interiors and barrier surfaces.

Microwave and Radiofrequency Region

Microwave energy ranges from 1 mm to 10 m (Wilkening 1991). Radiofrequency radiation ranges from 1 m to 3 km (Stuchly 1992). Radiofrequency and microwave energies represent an artificial division of the same continuum.

The outer surface of the skin absorbs energy at wavelengths less than 3 mm. Wavelengths ranging from 3 mm to 10 cm penetrate 1 to 10 mm. Wavelengths of 10 to 20 cm penetrate deeply enough that the potential for damage to internal organs is a consideration. The body is transparent to wavelengths greater than 500 cm.

Of the many possible effects in this region of the electromagnetic spectrum investigated over the years, most relate to heating of the exposed subject (Cleary 1980). The photon energy at microwave frequencies is too low to ionize molecules, regardless of the number absorbed (Wilkening 1991). As a result, ionization effects are excluded. However, nonthermal effects are considered possible.

Some individuals can "hear" microwave pulses (Frey 1962). The "sound" is sensed as a buzzing or ticking, depending on pulse rate. The current explanation for this effect is induction of thermoelastic waves in the head. To date, there is no evidence of injury from this effect. Microwave energy also can cause cataracts under special conditions of exposure (Wilkening 1991).

Some level-sensing devices that are used in storage hoppers, bin, tanks, and so on, emit RF energy. High levels of RF energy are utilized in other industrial applications, such as sealers, welding, and heating processes. In such circumstances, warning devices and interlocks should be present at entrances to the space. There is no guarantee, however, that these devices will function as intended or that tampering to defeat them cannot occur.

Extremely Low Frequency Region

Extremely low frequencies (ELF) have wavelengths longer than 3 km (Stuchly 1992). The most important applications of energies at these frequencies are electrical power generation, distribution, and utilization (50 or 60 Hz). The effects of the electric and magnetic fields generated at these frequencies have received considerable attention in recent years. These fields can induce currents in the human body. However, the currents induced by power frequency fields are very small.

The body and other materials readily perturb the electric field component. Hence, shielding is readily achieved. However, the body perturbs the magnetic field component to a minimal extent. The body is essentially transparent to the magnetic field. Providing portable shielding to nullify the magnetic field from an isolated conductor would be extremely difficult, if not technically impossible. To date, no short-term effect has been characterized in fields to which humans normally are exposed in industry or the home (Bernhardt and Matthes 1992).

Lasers

"Laser" is an acronym for "light amplification by stimulated emission of radiation" (WHO 1982, ACGIH 1990). Lasers have undergone widespread utilization in industrial applications (Court and Courant 1992). Applications include alignment, welding, cutting, drilling, heat treatment, distance measurement, entertainment, advertisement, and surgery (INIRC 1985b). Some industrial applications can produce exposure to highly intense beams through direct viewing, as well as reflected paths.

Adverse health effects from laser radiation are particularly a concern from 400 to 1,400 nm (visible to near infrared) (INIRC 1985b). Retinal damage can occur in this range. Biological effects can occur across the optical spectrum of wavelengths from 180 nm to 1 mm. The biological effects induced by optical radiation at any wavelength are essentially the same for coherent (laser) and noncoherent sources. The increased concern with lasers results from the intensities and irradiances that are produced. As well, lasers are highly monochromatic (single wavelength). The eye and skin are the critical organs for laser exposure, as for other types of optical radiation (ICNIRP 1996).

Damaging levels of laser energy are utilized in many industrial applications, including welding and cutting, heating processes, optical processes, electronic processes, photochemical processes, and so on. Provision of a strong visual or other stimulus as a warning inside spaces where boundary surfaces serve as shielding is unimportant. In such circumstances, warning devices and interlocks should be present at entrances to the space. There is no guarantee, however, that these devices will function as intended or that tampering to defeat them cannot occur.

Work activity in the confined space also may involve use of laser processes. This could include welding and cutting, distance sensing and alignment equipment, photochemical processes, and so on. Reflections from boundary surfaces could exacerbate exposure to laser energy.

HOT SURFACES AND HEAT-RELATED INJURY

Many confined spaces contain heated surfaces. These include surfaces of process equipment, as well as equipment and processes used during work activity. Contact between the unprotected skin and heated surfaces can cause thermal burns. The extent of a burn injury is affected by the temperature of the surface and duration of contact. Surface temperatures above 45 to 50°C can cause burn injury. Duration of contact is subject to the location on the body on which contact occurs. Contact with skin containing few sensory nerves could occur without warning longer than contact with skin in the face and neck that contains many sensory nerves. The nature of the contact also influences the outcome. Contact can occur through grasping the hot surface, through momentary brushing or grazing against it, or through prolonged contact following an uncontrolled action, such as a fall onto the surface.

Depth, size, and location determine severity of a burn injury (Wilkerson 1985). The terms "partial" and "full thickness" burns have replaced the older "first," "second," and "third degree" burns. First and second degree burns now are considered partial thickness burns. Partial thickness burns are less destructive. Under the old terminology, a first degree burn produced only redness of the skin and no death of tissue. Second degree burn damaged the upper layers of the skin and produced blisters. Recovery could occur without the need for grafting. Third degree burns now are classed as full-thickness burns. Third degree burns require skin grafts. Deeper burn injury can include underlying muscle, as well as the skin.

Until recently, few individuals survived full-thickness burns covering more than 50% of the skin surface area (Wilkerson, 1985). This applied even when treatment occurred in specialized burn centers. Under proper care, few burns covering less than 15 to 20% are lethal.

The more commonly occurring outcome involving thermal energy is heat stress. Heat stress is the impact of the total net heat load on the body (ACGIH 1991). Heat stress causes an increase in the core temperature of the body. An increase to 41.1°C (106°F) is almost certainly fatal unless cooling and medical attention are administered promptly. The Threshold Limit Value was set to prevent the deep body temperature from rising above 38°C (100.4°F). Considerable compensatory action occurs with even a seemingly slight increase in core temperature above 37°C (98.6°F). This temperature represents consensus among work physiologists for minimizing risk of heat illness.

The contribution to the heat load from the external environment is a serious concern in confined spaces under some conditions. The materials of construction of the boundary surfaces can absorb and then transmit solar energy to the airspace inside the structure. They also can retain heat produced by processes used during work activity.

The surface solar intensity (insolation) at noon in early July in the northern U.S. from radiation at wavelengths less than 3.5 μm is about 3.3 $J/cm^2/s$ (Moran and Morgan 1989). This was measured by a pyranometer, a standard instrument used for measuring the intensity of solar radiation striking a horizontal surface. Insolation varies with time of day, month of the year, latitude, and presence of clouds.

Heat transfer from boundary surfaces into the airspace occurs by the standard mechanisms: conduction, convection, and radiation. Heat transfer to the air inside the space can increase the interior temperature to a level considerably above that outside.

COLD SURFACES AND COLD RELATED INJURY

Cold is a serious workplace hazard in many parts of the world. Fatal accidents involving humans almost always have resulted from accidental exposure to low environmental air temperatures or immersion in water (ACGIH 1991). The most important life threatening aspect from exposure to cold is a decrease in the deep core temperature of the body. Cold stress is impact of loss of heat from the body or a portion, such as feet, hands, limbs, or head. The Threshold Limit Value for cold stress was set to prevent the deep core temperature from decreasing below 36°C (96.8°F). Considerable compensatory action occurs with even a seemingly slight decrease in core temperature from 37°C (98.6°F). Prolonged exposure to cold air or immersion in cold water at temperatures above freezing can cause this condition.

Some confined spaces contain cold surfaces. These can result from refrigeration processes used to maintain temperature, but more probably from weather conditions. For the many confined spaces located in the outside environment, weather conditions raise serious concerns. Mobility is restricted in small spaces. Restricted mobility can lead to reduced ability to maintain body heat through physical activity. Barrier surfaces can provide a shield from the wind, but they cannot protect against the effects of low temperatures. Ventilation used in the space to control airborne contaminants can exacerbate heat loss by convection.

Cold surfaces can damage unprotected skin. The most familiar examples are the "freezer burn" on the skin of a frozen chicken or turkey. Superficial or deep tissue freezing of human skin occurs at temperatures less than –1°C (30°F) (ACGIH 1991). Unprotected contact with air should not occur at wind chill equivalent temperature of –32°C (–26°F). Contact frostbite with surfaces can occur at temperatures less than –7°C (19°F). Rapid freezing can occur. Severe freezing injuries also can occur from contact with liquids, such as gasoline, alcohol, or some solvents that remain liquid at temperatures far below 0°C (Wilkerson 1986a). Liquids at these temperatures can absorb considerable heat.

Frostbite results from the formation of ice crystals in the fluids and underlying soft tissues of the skin (McDonald 1986). Frostbite damage results mostly from disruption of cellular activity caused by extraction of water and obstruction of blood supply. Obstruction of the blood supply results from leakage of serum into the tissues (Wilkerson 1986a). Frostbite is ranked according to extent of damage. First degree is freezing without blistering or peeling. Second degree includes blistering and peeling. Third degree includes freezing and possibly death of skin and deeper tissues (McDonald 1986). The most common sites of frostbite are the extremities: toes, fingers, earlobes, and the nose (Wilkerson 1986a).

In addition to frostbite, damage to tissue can occur from immersion injury (Wilkerson 1986b). Trench foot (immersion foot) results from prolonged wetting of the feet in near-freezing temperatures. Prolonged cooling damages the tissues. Vasoconstriction also occurs to preserve heat in the core of the body. Greatest damage occurs to the nerves. This causes pain, prickling, or tingling sensations (paresthesia) or the total anesthesia that may result. Damage also occurs to the skin and other tissues.

SUMMARY

Nonatmospheric hazardous conditions that can occur in confined spaces cover a wide range. Some can be potentially life threatening, while others pose the potential for causing injury. The OSHA, NIOSH, and MSHA reports on fatal accidents that occur in confined spaces provide the basis for determining which of the nonatmospheric conditions pose the greatest hazard.

A major theme with nonatmospheric hazards is the storage, conversion, and transport of energy. The boundary surfaces of many confined spaces are the shields used to protect outsiders from the transformation of energy through the motion of machinery and stored materials and utilization of energy in processes and process equipment.

Some nonatmospheric conditions involving energy result from work that occurs in the space. Prime sources include welding and processes that involve use of lasers.

REFERENCES

American Conference of Governmental Industrial Hygienists: *A Guide for Control of Laser Hazards*, 4th ed. Cincinnati, OH: American Conference of Governmental Industrial Hygienists, 1990. 73 pp.

American Conference of Governmental Industrial Hygienists: *Documentation of the Threshold Limit Values and Biological Exposure Indices*, 6th ed. Cincinnati, OH: American Conference of Governmental Industrial Hygienists, 1991.

Andersen, F.A.: Sodium and mercury vapor lamps. In *Non-Ionizing Radiation. Proc. Topical Symp., November 26–28, 1979, Washington, D.C.* Cincinnati, OH: American Conference of Governmental Industrial Hygienists, 1980. pp. 239–243.

Anonymous: *Safety First, Ground Fault Protection and Confined Space Lighting.* Willoughby OH: Ericson Manufacturing Company, 1995. 7 pp. [pamphlet]

Baitinger, W.F.: PC issues are hot in the electrical utility business. *Safety Prot. Fab. 4(1):* 23–25 (1995).

Berger, E.H., W.D. Ward, J.C Morrill, and L.H. Royster (Eds.): *Noise and Hearing Conservation Manual,* 4th ed. Akron, OH: American Industrial Hygiene Association, 1986. 592 pp.

Bernhardt, J.H. and R. Matthes: ELF and RF Electromagnetic Sources. In *Non-Ionizing Radiation. Proc. 2nd Int. Non-Ionizing Radiation Workshop, Vancouver, BC, Canada, 1992 May 10 to 14,* Greene, M.W (Ed.). London, England: International Radiation Protection Association, The Institution of Nuclear Engineers, 1992. pp. 59–75.

Bernstein, T.: Electrical systems, terminology and components — relationship to electrical and lightning accidents and fires. In *Electrical Hazards and Accidents, Their Causes and Prevention,* Greenwald, E.K. (Ed.). New York: Van Nostrand Reinhold, 1991a. pp. 1–27.

Bernstein, T.: Physiological effects of electricity — relationship to electrical and lightning death and injury. In *Electrical Hazards and Accidents, Their Causes and Prevention,* Greenwald, E.K. (Ed.). New York: Van Nostrand Reinhold, 1991b. pp. 28–49.

Biegelmeier, G. and W.R. Lee: New considerations on the threshold of ventricular fibrillation for AC Shocks at 50–60 Hz. *Proc. IEEE 127(2):* 103–110 (1980).

Cabanes, J.: Physiological effects of electric currents on living organisms, more particularly humans. In *Electrical Shock Safety Criteria.* Bridges, J.E., G.L. Ford, I.A. Sherman, and M. Vainberg (Eds.). New York: Pergamon Press, 1985. pp. 7–24.

Carbone, F.: *Double Insulated vs. GFCI Protection.* Willoughby, OH: Ericson Manufacturing Company, 7/20/94. 1 pp. [unaddressed memo]

Carson, J.W. and J.R. Johanson: Design of bins and hoppers. In *Materials Handling Handbook,* 2nd ed., Kulwiec, R.A. (Ed.). New York: John Wiley & Sons, 1985. pp. 901–940.

Cember, H.: *Introduction to Health Physics.* Oxford, U.K.: Pergamon Press, 1969. 420 pp.

Centers for Disease Control: Leading work-related diseases and injuries — United States. *Morbid. Mortal. Weekly Rep. 33:* 3–5 (1984).

Cleary, S.: Biological effects of low intensity microwave and radiofrequency radiation exposure. In *Non-Ionizing Radiation. Proc. Topical Sym., November 26 to 28, 1979, Washington, D.C.* Cincinnati, OH: American Conference of Governmental Industrial Hygienists, 1980. pp. 71–74.

Cobine, J.D.: *Gaseous Conductors.* New York: Dover Publishing, 1958.

Cohen, B.L.: Comparison of radiological risks from coal burning and nuclear power. *Health Phys. 48:* 342–343 (1985).

Court, L. and D. Courant: Laser: characteristics and emissions. In *Non-Ionizing Radiation. Proc. 2nd Int. Non-Ionizing Radiation Workshop, Vancouver, BC, Canada, 1992 May 10 to 14,* Greene, M.W. (Ed.). London, England: International Radiation Protection Association, The Institution of Nuclear Engineers, 1992. pp. 289–306.

Dalziel, C.F. and F.P. Massoglia: Let-go currents and voltages. *AIEE Trans. 75 Part II:* 49–56 (1956).

Dalziel, C.F.: Deleterious effect of electric shock. In *Handbook of Laboratory Safety,* 2nd ed., Steere, N.V. (Ed.). Boca Raton, FL: CRC Press, 1971. pp. 521–527.

Frey, A.H.: Human auditory system response to modulated electromagnetic energy. *J. Appl. Physiol. 17:* 689–692 (1962).

Golde, R.H.: *Lightning Protection.* New York: Chemical Publishing.

Hétu, R. and J. Parrot: Field evaluation of noise-induced temporary threshold shift. *Am. Ind. Hyg. Assoc. J. 39:* 301–311 (1978).

Hewson, G.S. and J.J. Fardy: Thorium metabolism and bioassay of mineral sands workers. *Health Phys. 64:* 147–156 (1993).

Hinrichs, J.F.: A bright spot in welding. In *Non-Ionizing Radiation. Proc. Topical Symposium, November 26 to 28, 1979, Washington, D.C.* Cincinnati, OH: American Conference of Governmental Industrial Hygienists, 1980. pp. 245–247.

Hodgkin, B., O. Langworthy, and W.B. Kouwenhoven: Effect on Breathing. *IEEE Trans. Power Appar. Sys. PAS-92(4):* 1388–1391 (1973).

International Atomic Energy Agency: *The Radiological Accident in San Salvador.* Vienna, Austria: International Atomic Energy Agency, 1990. 94 pp.

International Non-Ionizing Radiation Committee of the International Radiation Protection Association: Interim guidelines on limits of human exposure to airborne ultrasound. *Health Phys. 46:* 969–974 (1984).

International Non-Ionizing Radiation Committee of the International Radiation Protection Association: Guidelines on limits of exposure to ultraviolet radiation of wavelengths between 180 nm and 400 nm (incoherent optical radiation), *Health Phys. 49:* 331–340 (1985a).

International Non-Ionizing Radiation Committee of the International Radiation Protection Association: Guidelines on limits of exposure to laser radiation of wavelengths between 180 nm and 1 mm. *Health Phys. 49:* 341–359 (1985a).

International Commission on Non-Ionizing Radiation Protection: Guidelines on limits of exposure to laser radiation of wavelengths between 180 nm and 1,000 µm. *Health Phys. 71:* 804–819 (1996).

Jensen, C. and J.D. Helsel: *Fundamentals of Engineering Drawing,* 4th ed. Westerville, OH: Glencoe/McGraw-Hill, 1996. 480 pp.

Jones, T.B. and J.L. King: *Powder Handling and Electrostatics.* Chelsea, MI: Lewis Publishers, 1991. 103 pp.

Juvinall, R.C. and K.M. Marshek: *Fundamentals of Machine Component Design,* 2nd ed. New York: John Wiley & Sons, 1991. 804 pp.

Kahn, F. and L. Murray: Shock free electric appliances. *IEEE Trans. Ind. Gen. Appl. IGA-2(4):* 322–327 (1966).

Krieger, G.R. and J.F. Montgomery (Eds.): *Accident Prevention Manual for Business and Industry, Engineering & Technology,* 11th ed. Itasca, IL: pp. 74–76.

Kryter, K.D.: Exposure to steady-state noise and impairment to hearing. *J. Acous. Soc. Am. 35:* 1515 (1963).

Lee, W.R.: Death from electric shock. *Proc. IEEE 113(1):* 144–148 (1966).

Lee, R.H.: Electrical safety in industrial plants. *IEEE Spectrum,* June 1971.

Magrab, E.B.: *Environmental Noise Control.* New York: John Wiley & Sons, 1975. pp. 176–214.

Maletskos, C.J. and C.C. Lushbaugh: The Goiânia Radiation Accident. *Health Phys. 60:* 1 (1991).

McDonald, O.F.: Cold stress: how to deal with it. *Natl. Safety Health News:* February 1986. pp. 18–24.

McKinlay, A.: Optical Radiation. In *Non-Ionizing Radiation. Proc. 2nd Int. Non-Ionizing Radiation Workshop, Vancouver, BC, Canada, 1992 May 10 to 14,* Greene, M.W. (Ed.). London, England: International Radiation Protection Association, The Institution of Nuclear Engineers, 1992. pp. 227–251.

Metzger, R., J.W. McKlveen, R. Jenkins, and W.J. McDowell: Specific activity of uranium and thorium in marketable rock phosphate as a function of particle size. *Health Phys. 39:* 69–75 (1980).

Mine Safety and Health Administration: Think "Quicksand": Accidents Around Bins, Hoppers and Stockpiles, Slide and Accident Abstract Program. Arlington, VA: U.S. Department of Labor, Mine Safety and Health Administration, National Mine Health and Safety Academy, 1988.

Moran, J.M. and M.D. Morgan: *Meteorology: The Atmosphere and The Science of Weather. 2nd ed.* New York: Macmillan, 1989. pp. 27–55.

Moss, C.E.: Radiation hazards associated with ultraviolet radiation curing processes. In *Non-Ionizing Radiation. Proc. Topical Sym., November 26 to 28, 1979, Washington, D.C.* Cincinnati, OH: American Conference of Governmental Industrial Hygienists, 1980. pp. 203–209.

National Institute for Occupational Safety and Health: Criteria for a Recommended Standard — Working in Confined Spaces (DHEW/PHS/CDC/NIOSH Pub. No. 80-106). Cincinnati, OH: National Institute for Occupational Safety and Health, 1979. 68 pp.

National Institute for Occupational Safety and Health: NIOSH Alert, Request for Assistance in Preventing Fatalities of Workers Who Contact Electrical Energy (DHHS(NIOSH) Pub. No. 87-103. Cincinnati, OH: DHHS/PHS/CDC/NIOSH, 1986a. 6 pp.

National Institute for Occupational Safety and Health: NIOSH Alert, Request for Assistance in Preventing Electrocutions Due to Damaged Receptacles and Connectors (DHHS(NIOSH) Pub. No. 87-100. Cincinnati, OH: DHHS/PHS/CDC/NIOSH, 1986b. 6 pp.

National Institute for Occupational Safety and Health: Worker Deaths in Confined Spaces (DHHS/PHS/CDC/NIOSH Pub. No. 94-103). Cincinnati, OH: National Institute for Occupational Safety and Health, 1994. 273 pp.

Nelson, C.A.: *Millwrights and Mechanics Guide.* New York: The Bobbs-Merrill, 1983. 1032 pp.

Novotny, D.W. and G.R. Priegel: *Electrofishing Boats.* Tech. Bull. No. 73. Madison, WI: State of Wisconsin Department of Natural Resources, 1974.

Occupational Safety and Health Administration: Selected Occupational Fatalities Related to Fire and/or Explosion in Confined Work Spaces as Found in OSHA Fatality/Catastrophe Investigations. Washington, D.C.: U.S. Department of Labor, Occupational Safety and Health Administration (U.S. DOL/OSHA), 1982a. 76 pp.

Occupational Safety and Health Administration: Selected Occupational Fatalities Related to Lockout/Tagout Problems as Found in Reports of OSHA Fatality/Catastrophe Investigations. Washington, D.C.: U.S. Department of Labor, Occupational Safety and Health Administration (U.S. DOL/OSHA), 1982b. 113 pp.

Occupational Safety and Health Administration: Selected Occupational Fatalities Related to Grain Handling as Found in Reports of OSHA Fatality/Catastrophe Investigations. Washington, D.C.: U.S. Department of Labor, Occupational Safety and Health Administration (U.S. DOL/OSHA), 1983. 150 pp.

Occupational Safety and Health Administration: Selected Occupational Fatalities Related to Toxic and Asphyxiating Atmospheres in Confined Work Spaces as Found in Reports of OSHA Fatality/Catastrophe Investigations. Washington, D.C.: U.S. Department of Labor, Occupational Safety and Health Administration (U.S. DOL/OSHA), 1985. 230 pp.

Occupational Safety and Health Administration: Controlling Electrical Hazards (OSHA 3075, Rev.). Washington, D.C.: U.S.DOL/OSHA, 1986. 12 pp.

Occupational Safety and Health Administration: Selected Occupational Fatalities Related to Welding and Cutting as Found in Reports of OSHA Fatality/Catastrophe Investigations. Washington, D.C.: U.S. Department of Labor, Occupational Safety and Health Administration (U.S. DOL/OSHA), 1988. 225 pp.

Occupational Safety and Health Administration: Selected Occupational Fatalities Related to Ship Building and Repairing as Found in Reports of OSHA Fatality/Catastrophe Investigations. Washington, D.C.: U.S. Department of Labor, Occupational Safety and Health Administration (U.S. DOL/OSHA), 1990. 195 pp.

Occupational Safety and Health Administration: Ground-Fault Protection on Construction Sites (OSHA 3007). Washington, D.C.: U.S. Department of Labor, Occupational Safety and Health Administration (U.S. DOL/OSHA), 1993 (reprint). 18 pp.

Occupational Safety and Health Administration: *OSHA CD-ROM* (OSHA A95-4). Washington, D.C.: U.S.DOL/OSHA, 1995.

Olivo, C.T. and T.P. Olivo: *Fundamentals of Applied Physics*, 3rd ed. Albany, NY: Delmar Publishers, 1984. 500 pp.

Palmquist, R.E.: *Electrical Course for Apprentices and Journeymen.* Indianapolis, IN: Howard W. Sams, 1973. 438 pp.

Parrack, H.O.: Effects of airborne ultrasound on humans. *Int. Audiol. 5:* 294–308 (1966).

Raczkowski, G.: *Principles of Machine Dynamics.* Houston, TX: Gulf Publishing, 1979. 104 pp.

Roberts, E.W.: *Overcurrents and Undercurrents: All About GFCIs.* Mystic, CT: Reptec, 1996. 145 pp.

de Santis, V. and I. Longo: Coal energy vs. nuclear energy: a comparison of the radiological risks. *Health Phys. 46:* 73–84 (1984).

Schmidek, M. and P. Carpenter: Intermittent noise exposure and associated damage risk to hearing of chain saw operators. *Am. Ind. Hyg. Assoc. J. 35:* 152–158 (1974).

Seshagiri, B. and B. Stewart: Investigation of the audibility of locomotive horns. *Am. Ind. Hyg. Assoc. J. 53:* 726–735 (1992).

Shapiro, J.: *Radiation Protection, A Guide for Scientists and Physicians*, 2nd. ed. Cambridge, MA: Harvard University Press, 1981. 480 pp.

Sliney, D.H.: Biohazards of ultraviolet, visible and infrared radiation. *J. Occup. Med. 25:* 203–206 (1983).

Sliney, D.H.: Ultraviolet radiation and the eye. In *Light, Lasers, and Synchrotron Radiation*, Grandolfo, M. et al. (Eds.). New York: Plenum Press, 1990. pp. 237–245.

Sliney, D.H.: Measurements and bioeffects of infrared and visible light. In *Non-Ionizing Radiation. Proc. 2nd Int. Non-Ionizing Radiation Workshop, Vancouver, BC, Canada, 1992 May 10 to 14*, Greene, M.W (Ed.). London, England: International Radiation Protection Association, The Institution of Nuclear Engineers, 1992. pp. 252–267.

Smoot, A.W. and C.A. Bentel: Electric shock hazard of underwater swimming pool light fixtures. *IEEE Trans. Power Apparat. Sys. 83:* 945–964 (1964).

Smoot, A. and J. Stevenson: *Report on Investigation of Reaction Currents* (Subject 965-1). Melville, NY: Underwriters Laboratories, 1968.

Stewart, H.L.: *Pneumatics and Hydraulics.* Indianapolis, IN: Theodore Audel, 1976. 486 pp.

Stuchly, M.A.: Introduction to bioelectromagnetics. In *Non-Ionizing Radiation. Proc. 2nd Int. Non-Ionizing Radiation Workshop, Vancouver, BC, Canada, 1992 May 10 to 14*, Greene, M.W (Ed.). London, England: International Radiation Protection Association, The Institution of Nuclear Engineers, 1992. pp. 17–31.

Terzaghi, K., R.B. Peck, and G. Mesri: *Soil Mechanics in Engineering Practice*, 3rd ed. New York: John Wiley & Sons, 1996. 549 pp.

UAW-GM National Joint Committee on Health and Safety: *Lockout.* Detroit, MI: UAW-GM Human Resources Center, 1985.

Underwriters Laboratories: *Standard for Safety, Double Insulated Systems for Use in Electrical Equipment* (UL 1097). Northbrook, IL: Underwriters Laboratories, 1983.

Underwriters Laboratories: *Standard for Safety, Ground-Fault Circuit Interrupters* (UL 943). Northbrook, IL: Underwriters Laboratories, 1985.

Vickers Training Center: *Vickers Industrial Hydraulics Manual.* Rochester Hills, MI: Vickers, 1992.

Wahl, R.C.: Properties of bulk solids. In *Materials Handling Handbook,* 2nd ed., Kulwiec, R.A. (Ed.). New York: John Wiley & Sons, 1985. pp. 882–900.

Walton, H.: *The How and Why of Mechanical Movements.* New York: Popular Science Publishing, 1968. 297 pp.

Ward, W.D.: Effects of Noise on Hearing Thresholds. In *Noise as a Public Health Hazard,* Ward, W.D. and J.E. Fricke (Eds.). Washington, D.C.: American Speech and Hearing Association, 1969. pp. 40–48.

Wiernicki, C. and W.J. Karoly: Ultrasound: biological effects and industrial hygiene concerns. *Am. Ind. Hyg. Assoc. J. 46:* 488–496 (1985).

Wilkening, G.M.: Nonionizing Radiation. In *Patty's Industrial Hygiene,* 4th ed., vol. 1, Part B, *General Principles,* Clayton, G.D. and F.E. Clayton (Eds.). New York: John Wiley & Sons, 1991. pp. 657–742.

Wilkerson, J.A. (Ed.): *Medicine for Mountaineering,* 3rd. ed. Seattle, WA: The Mountaineers, 1985. 438 pp.

Wilkerson, J.A.: Frostbite. In *Hypothermia, Frostbite and Other Cold Injuries,* Wilkerson, J.A., C.C. Bangs, and J.S. Hayward (Eds.). Seattle, WA: The Mountaineers, 1986a. pp. 84–95.

Wilkerson, J.A.: Other localized cold injuries. In *Hypothermia, Frostbite and Other Cold Injuries,* Wilkerson, J.A., C.C. Bangs, and J.S. Hayward (Eds.). Seattle, WA: The Mountaineers, 1986b. pp. 96–101.

Wilson, A.J. and Scott, L.M.: Characterization of radioactive petroleum piping scale with an evaluation of subsequent land contamination. *Health Phys. 63:* 681–685 (1992).

World Health Organization: *Lasers and Optical Radiation* (Environmental Health Criteria 23). Geneva, Switzerland: World Health Organization, 1982. 154 pp.

Yokel, F.Y. and R.M. Chung: Proposed standards for construction practice in excavation. *Prof. Safety 28:* September, 1983. pp. 34–39.

7 Hazard Management for Confined Spaces

CONTENTS

INTRODUCTION

Previous chapters focused on causative agents and factors that underlie fatal accidents that occur in confined spaces. Developing this picture is a frustrating quest, since the information that is available has resulted from personal tragedy. Many accidents occurred in surroundings or under circumstances that do not fit conventional descriptions for the term "confined space." Accidents have occurred in structures whose purpose is to isolate and prevent contact with hazardous energy sources. Accidents have occurred during work on containers or structures that confined a hazardous

TABLE 7.1
Reasons for Entry into Confined Spaces

Operational

- Preinspection to determine safety for entry by others
- Inspection of the space
- Inspection of equipment
- Cleaning
- Adjusting/modifying/maintaining equipment
- Modifying the structure
- Increase/improve flow

Nonoperational

- Curiosity
- Misadventure
- Defiance of organizational policy
- Rescue

atmosphere. The work activity caused a fire or explosion. These situations were very similar to fires and explosions that occurred during work activity in confined spaces.

Later discussion elaborated on the technical nature of hazardous conditions that can develop in confined spaces. These can be very different from what is encountered in normal workspaces. These sections established the role of geometry as an exacerbating factor in personal confinement, atmospheric confinement, and confinement of hazardous energy.

This analysis suggests the need to reconfigure the term "confined space." The term "confined space/confined atmosphere/confined energy," while cumbersome, is more inclusive of the hazardous conditions that can occur in these workspaces.

A major goal of preceding chapters was to identify and explore the factors and conditions associated with a typical accident in a confined space, and from these, to build a descriptive model. A model provides the basis for describing the phenomenon and for prescribing measures to gain control over the process. The following section summarizes observations determined previously.

CONFINED SPACES: WHAT WE KNOW

The data from research studies clearly indicate that confined spaces pose a hazard to persons who enter and work in them (NIOSH 1979, 1994; OSHA 1982a,b, 1983, 1985, 1988, 1990; MSHA 1988). The corollary question is, "What is the need for people to work in such hazardous environments?" People enter and work in confined spaces for what usually are valid operational reasons. Conversely, people sometimes enter confined spaces for reasons beyond normal procedures.

Entry into confined workspaces usually occurs for the reasons summarized in Table 7.1.

An important general concept is the nature of work involving confined spaces. Work in confined spaces usually differs from production-oriented activity. This can be unpredictable in time, in occurrence, in duration, and location. By contrast, production-oriented activity is predictable and occurs during a much longer time frame. In addition, the conditions of work in confined spaces and confined atmospheres are less readily definable than those occurring during production activity.

Table 7.2 summarizes work situations mentioned in reports of fatal accidents. The temporary nature of work in these workspaces severely constrains the ability of an organization to engineer-out hazards. Some of the accident reports, however, indicated that entry was a routine activity, occurring at least once per shift into the same space. This situation highlights a potential use for entry analysis as a tool for assessing whether engineering control should be implemented.

TABLE 7.2
Characteristics of Work Involving Confined Spaces

Temporal Aspects	Duration of Activity
• Partial shift • Daily • Weekly • Monthly • Semiannually • Annually • Greater than annually	• Partial shift • Full shift • Multiple shift **Nature of Activity** • Start new work • Continue work in progress
Entry Aspects	**Workforce**
• Single • Multiple entry/reentry	• Single person • Group having constant composition • Group having variable composition

Work activity in these spaces creates complex interactions between many variables. The varying nature of these interactions is partly responsible for the need to create a comprehensive system for addressing them.

For the industrial hygienist, the issue of greatest concern is atmospheric hazard. Acutely toxic atmospheres are associated with approximately two of every three fatal accidents that occur in these workspaces (excluding excavation accidents). The occurrence of atmospheric hazards is a result of confinement by the structure. Atmospheric confinement can occur in structures considerably different from those conventionally defined as confined spaces. Structures that confine the atmosphere do not necessarily restrict movement during work activity, or access or egress.

Table 7.3 summarizes the causes of atmospheric contamination in confined workspaces. Fires or explosions result from the interaction between a flammable or explosive atmosphere and a source of ignition. The latter usually is introduced during the work activity.

Table 7.4 lists sources of ignition of fires and explosions inside confined spaces or external to a structure or container that confined a hazardous atmosphere mentioned in fatal accidents. Almost all of these sources reflect human activity in the space, rather than conditions produced by the process intended to occur in the space.

Nonatmospheric hazardous conditions cause a significant proportion of the accidents that occur in confined spaces. These accidents differ fundamentally from those involving hazardous atmospheres. The most common types of accidents involving nonatmospheric hazards include engulfment, entanglement, electrocution, failure of process equipment, fall from height, and unstable interior conditions. Table 7.5 lists major causes of these accidents. The moment that these accidents occur, the surroundings became stable, as if energy, like the spring in a mousetrap, has been released. Rescue was not an issue in these accidents. The would-be rescuer typically recognized the gravity of the situation and deactivated equipment, as needed, prior to beginning to act.

The authors of the NIOSH and OSHA reports commented that these accidents resulted from organizational and procedural deficiencies. Table 7.6 summarizes factors mentioned in the reports that enhanced the potential for onset or exacerbated the severity of these events.

Accidents involving confined spaces and confined atmospheres are complex events. Tables 7.1 through 7.6 present many of the factors that can contribute to a particular situation. A particular situation can involve one or more of these factors, as well as others not listed. Obviously, no simple description can encompass all situations. The complex nature of these situations considerably complicates the ultimate need: strategies for gaining control over them.

TABLE 7.3
Causes of Atmospheric Contamination

Toxic Substances/Chemical Asphyxiants

- Substances present prior to entry
- Substances enter during occupancy
- Substances used during work activity
  - Solvents
  - Spray painting
- Substances produced during work activity
  - Welding fumes and exhaust gases
  - Exhaust gases generated by combustion sources
  - Grinding
- Chemical reactions involving contents and introduced substances

Simple Asphyxiants

- Chemical reaction between contents and atmospheric oxygen
  - Oxidation of metals (rusting)
  - Welding processes
- Physical processes that deplete oxygen
  - Absorption
  - Adsorption
- Biological processes that deplete oxygen
  - Fermentation
  - Aerobic respiration
- Displacement of atmosphere by purge/shield gases
  - Nitrogen
  - Carbon dioxide
  - Helium
  - Argon
  - Steam

Oxygen Enrichment

- Accidental release of oxygen into the confined space
  - Welding/cutting
  - Process source
- Deliberate release of oxygen into the confined space for cooling

Flammable or Explosive Atmosphere

- Substances remaining prior to entry
- Substances entering during occupancy
- Substances used during work activity
  - Solvents
  - Spray painting
- Substances produced during work activity
  - Chemical reactions involving contents and introduced substances

CONFINED SPACES: RESPONDING TO THE CHALLENGE

Discussion to this point has focused on the past; that is, on what can be learned through analysis of recorded events. To paraphrase one observer, "… those who ignore lessons provided by past events are doomed to repeat them." Data provided in the NIOSH, OSHA, and MSHA reports represent lessons gained through human tragedy (NIOSH 1979, 1994; OSHA 1982a,b, 1983, 1985, 1988, 1990; MSHA 1988). These tragedies will continue to occur until the lessons that they provide

TABLE 7.4
Sources of Ignition of Fires and Explosions

- Welding, burning, cutting, grinding
- Current electricity
  Electrical tools
- Lightbulbs
  Static electricity
  Matches
  Cigarettes
  Heaters

TABLE 7.5
Causes of Accidents Involving Nonatmospheric Hazards

Engulfment

- Bridge collapse (horizontal)
- Cake collapse (vertical)
- Flow-induced

Entanglement

- Entry during operation
- Accidental activation
- Equipment failure

Electrocution

- Contact with existing energized circuit
- Contact following accidental activation
- Contact following exposure of an energized circuit
- Ineffective isolation
- Contact with electrode

Process Hazards

- Failure of existing energized circuit
- Exposure of an energized circuit

Falls from Heights

- Equipment failure
- Solvent narcosis
- Other

Unstable Interior Conditions

- Work activity
- Inherent instability of structure
- Failure to support structure
- Other

are implemented fully across industry. Failing to heed and apply the counsel that they provide would be doubly tragic.

The second part of this book focuses on strategies and actions for achieving safe conditions for work in confined workspaces. This goal is achieved by minimizing the risk of encountering a

TABLE 7.6
Organizational and Procedural Deficiencies in Management Systems

Policy

- Lack of organizational policy toward confined spaces
- Poorly elucidated organizational policy

Planning

- Lack of planning of entry and work activities
- Lack of planning of emergency response procedures
- Lack of planning for rescue

Organization

- Lack of accountability in supervision

Preparation

- Lack of equipment appropriate to the task
- Lack of rescue equipment
- Inappropriate equipment

Procedure

- Circumventing of organizational policy
- Lack of ventilation
- Habituation to faulty procedure
- Lack of procedure
- Refusal to follow procedure
- Failure to follow procedure
- Unusual procedure
- Impulsive decision making — e.g., someone drops something into the space and impulsively enters to retrieve it
- Defeat of safety devices
- Improper preparation — incomplete purging
- Failure to follow requirements of entry document
- Failure to use experience gained during near-misses

Training

- Inappropriate training
- Change of process without training to upgrade skills
- Inappropriate response when attempting to assist a person in distress following an accident
- General lack of explanation about hazards to suppress urges of curiosity/misadventure
- Lack of training in entry procedures
- Lack of technical knowledge about the potential hazard represented by the condition
- Lack of training in rescue procedures
- Failure to appreciate the seriousness of accidents occurring in confined spaces

Recognition

- Presence of the confined space/confined atmosphere
- Conditions that could arise due to entry or the work undertaken
- Conditions present or likely to develop

Testing

- Lack of testing to determine conditions
- Inappropriate testing procedure
- Equipment not calibrated/serviced
- Failure to test at the appropriate moment

Process Conditions

- Unusual chemical or physical process
- Unusual condition

hazardous condition. Minimizing the risk is accomplished by gaining control over hazardous conditions and enabling individuals through increased knowledge to act in a manner consistent with safe practice. Gaining this control requires thorough understanding about the factors present in a particular situation and exerting influence over them, combined with an organized, assertive approach.

Previous discussion indicated that lack of understanding about, and lack or loss of control over hazardous conditions were major factors in the etiology of these accidents. Lack or loss of control over hazardous conditions is fundamental to models for accident causation (Heinrich et al. 1980, Bird and Germain 1990). According to the ILCI (International Loss Control Institute) model, incidents are the inevitable result of loss of control. Loss of control leads to development of basic and immediate causes. These, in turn, lead to contact and the onset of the incident.

Basic or underlying causes (also called root, real, indirect, or contributing causes) are the factors that permit the development of substandard practices and conditions. Immediate causes are the "unsafe acts" or substandard actions and conditions that permit the occurrence of an accident. They usually are quite apparent. In the ILCI model, contact is the event during which the harm or damage occurs. The energy involved in the contact exceeds some critical threshold, thereby causing personal injury or property damage. Among the practical lessons to be learned from study of accident causation is that loss-producing events seldom result from a single cause. Many substandard practices and conditions may be present.

ILCI contends that loss of control by management is a critical factor in the causation of accidents (Bird and Germain 1990). Control is one of the four essential functions of management (the others are plan, organize, and lead). According to ILCI, the common reasons for lack or loss of control are:

- Inadequate or superficial program
- Inadequate program organization
- Inadequate compliance with standards

Certainly these propositions are consistent with findings in the NIOSH, OSHA, and MSHA reports (NIOSH 1979, 1994; OSHA 1982a, 1982b, 1983, 1985, 1988, 1990; MSHA 1988). Correcting the preceding deficiencies to eliminate these accidents is a critical responsibility of management.

In highlighting the complexity of accident causation, the ILCI model also illustrates opportunities for control (Bird and Germain 1990). These can occur at the stages in the sequence of an incident:

- Precontact
- Contact
- Postcontact

Precontact control includes preventive measures that avoid the risks and prevent the losses, and actions that reduce loss if and when contact occurs. Precontact produces the most beneficial results. This is the time to develop programs and establish standards and audit procedures. Contact control includes measures that reduce energy transfer during contact. These measures include substitution of less hazardous materials, engineering controls, and personal protective equipment. Minimizing the extent of energy transfer at time of contact reduces the severity of the incident. Measures taken postcontact reduce the impact of the incident. These include emergency response and rescue operations, and fire and explosion control. Postcontact controls do not prevent accidents; they minimize their impact. Exercising control through management is essential to reducing the risks of work involving confined spaces.

Management involves a multistage process: examine the situation, and investigate and analyze its elements; synthesize a model to describe their interactions, and devise measures to gain control. The management process must address emergency situations, as well as normal activity. This means anticipating what can go wrong, and then responding proactively to address these possibilities.

Auditing is essential to ensure that all elements of the management system are in place and functioning correctly. An audit identifies deficiencies and ensures that corrective action has occurred. A situation is not managed until all elements of the system are fully integrated.

The safety profession utilizes several technical approaches for evaluating the potential for accident situations. One is the system safety model. The term "system safety" was chosen to describe safety studies of complex systems. These studies involve formal systematic analyses for both hazard identification and control (Heinrich et al. 1980). A system is a set of interdependent components that are combined in such a way as to perform a given function under specified conditions. A system can be a machine, a machine plus human operator, or other combination of equipment, facilities, and people. The dimensions of the system reflect the scope of the activity. System safety is built around rational evaluation of potential hazards in all aspects of the system rather than a focus on areas that reflect statistics. System safety is active during the life cycle of the system, from concept through design, through construction, operation, and retirement.

An important contribution from system safety is the principle known as the "hazard control order of precedence." This principle states that control over each identified hazard should be provided by:

- Designing to minimize the occurrence and potential severity of effects
- Providing warning and protection devices in the design
- Including hazard controls in training, operating, and maintenance procedures

System safety considers accidents to have two underlying causes: energy transformation and energy deficiency. Transformation of energy from one form to another exceeds some threshold for damage. Energy deficiency occurs when energy needed to perform a vital function is not available at a critical moment and the accident occurs.

An important area of system safety is analysis (Capps 1980). Analysis identifies factors influencing the probability of unwanted events and determines the way in which they occur. The outcome is to develop preventive measures to reduce the probability of occurrence. System safety analysis may be performed *a priori* or *a posteriori*. An *a priori* analysis is undertaken before the unwanted occurrence. This attempts to discover factors leading to the unwanted occurrence from previous examples. An *a posteriori* analysis is performed after an unwanted occurrence has occurred. Both types of analysis are complimentary (Ho 1983). These analyses involve combinations of deductive and inductive reasoning. The deductive component relates to the cause of failure, the inductive, to its effect on the safety of a system (Heinrich et al. 1980).

The methods for determining the mechanism or logic behind the sequence of two or more events are the direct and reverse methods. The direct or inductive method (event tree analysis) starts with causes and attempts to foresee their effects. This technique starts from an initiating event and moves forward step by step through a hazard scenario. If the probabilities of intermediate events are known, the probability of the final event can be calculated. This method has the disadvantage of beginning with a single initiating event. As a result, this method cannot accommodate multiple initiating events and interactions. The direct method is suited to simulations. The value of this analysis largely depends on the relevance of the dysfunction (Ho 1983, Kletz 1985, Gressel and Gideon 1991).

The reverse method (fault tree analysis) starts with effects and works backward to the cause. The diagram resulting from application of the reverse method resembles a tree, hence, "fault tree analysis." The reverse method is often used in *a priori* analyses of tangible systems and *a posteriori* analysis of industrial accidents. Fault tree analysis starts with the "top" (or end) event and develops a logic tree showing the causes. Analysis of the fault tree identifies the smallest sequence of events that would result in the final event.

Probabilistic risk assessment is an extension of fault tree analysis. The probability of occurrence of the final event can be calculated from the probability for intermediate events. The quality of the analysis reflects the uncertainty in each of the probabilities. They typically range from unreliable to nonexistent. If reliable data could be obtained, this technique would be very effective.

Gressel and Gideon described several of the tools used in the system safety approach, starting with checklists, preliminary hazard analysis and "what if" analysis (Gressel and Gideon 1991).

A checklist is a series of questions specific to the situation of interest. Checklists can be used to identify deviations from expected or intended conditions. Personnel at the operating level with a modicum of training can use a checklist to recognize hazards and ensure compliance with standards. Implicit in use of checklists is that they are developed by professionals thoroughly familiar with the process and associated hazards. Checklists are best suited to situations where hazards have been identified and controlled.

Preliminary hazard analysis (PHA) is a means of assessing situations that are not fully characterized. Typically, a PHA is performed at the earliest stages of planning. Hazardous materials, equipment, and process conditions are listed. This list becomes the basis for identifying potential hazards. For each hazard, possible consequences are identified. For each possible consequence, corrective measures are derived. A PHA results in recommendations for reduction or elimination of hazards (Heinrich et al. 1980, Gressel and Gideon 1991).

A "what if" analysis examines the effects on a process of upset or malfunction of equipment. The initial questions usually result from findings of an earlier evaluation, such as PHA. Negative outcomes become the basis for reducing or eliminating the potential hazard. The ultimate effectiveness of this type of analysis depends on the quality of the questions.

Failure modes and effects analysis (FMEA), also known as failure modes, effects, and criticality analysis (FMECA), examines potential failure modes of process components, their causes and effects. FMEA can be used to assess qualitative risk of failure. FMEA can assess single failures only.

The hazard and operability study (HAZOP) is a rigorous and powerful technique for identifying complex failure scenarios involving multiple independent events. HAZOP considers small segments in all aspects of a process, including instrumentation. Deviations from normal operating conditions are evaluated by applying a series of guide words. The likelihood and consequence of the deviation are determined. These form the basis for recommendations for improvement or further study. HAZOP can be used to evaluate maintenance and operating procedures and management systems, as well as performance of process equipment (Knowlton 1981).

Another approach to improving human performance is the management oversight risk tree (MORT). MORT was developed as a synthesis of the best features of successful methods from system safety, quality assurance, and program management (Buys 1988). The model begins with the basic philosophy or theorems of accident occurrence and prevention. The next steps focus on managerial decision making through the following activities: collect data, analyze, select remedy, apply remedy, and monitor or measure results. The model represents an ongoing process. Following the monitoring step is a reevaluation to determine the need for fine-tuning.

Workplace conditions are the result of the complex interrelationship between people, machines, and the environment. An important premise held by the safety profession is that safety should be managed like any other business function. The basis for the role is economic, rather than moralistic. The outcome of the lack of safety — accidents, injuries, and death — is readily apparent to management and the workforce. The key to accountability is measurement of safety performance through standard indicators (Heinrich et al. 1980). Procedure development, training, motivation, and supervision all are critically important to the success of safety functions.

An important difference between the approach of the safety and industrial hygiene professions to the management of programs is worth highlighting (Crutchfield 1981). This difference could have important bearing on programs created to manage hazards in confined spaces. An important determinant of success in providing safe conditions of work is recognition and acceptance of responsibility for safety and industrial hygiene by line management. These programs should be implemented through line activities.

According to Crutchfield, industrial hygienists tend to focus on the relationship between machines and the environment. This approach tends to view people as passive receptors of contaminants. The outcome from this approach is that hygienists tend to adopt a solitary role and incorrectly assume responsibility for control of hazardous conditions. Also, the successes and failures in hygiene are less obvious. These determinants are based on compliance with intangibles, usually expressed through instrument-generated numbers.

A technical solution alone does not insure success in eliminating or controlling the situation that created the problem. Involvement by line management and the workforce is essential to success. This is one of the concepts of motivational control (Dawson et al. 1987). Without involvement and buy-in, the technical solution could remain unused, misused, or become subject to sabotage.

Motivational controls are concerned with the development and maintenance of general safety awareness. The principal elements of motivational control include:

- General objectives, culture, and organization and atmosphere of the organization
- Definitions of responsibility and authority
- Mechanisms of accountability and measurement of performance

The "climate" of safety awareness in an organization is believed to be crucial to success in hazard control. This climate substantively depends on the words and actions of senior site and company management. The extent of commitment of senior management is viewed as a crucial determinant to motivation throughout the organization (Cohen 1977).

Allocation of specific responsibilities for technical control is essential for the key individuals in the organization to have a clear understanding of expectations. Otherwise, what is done is likely to occur in haphazard fashion. Further, key elements may remain unattended because of lack of ownership. Authority is the essential partner of accountability. Motivational controls need to be positively structured into the patterns of rewards and sanctions that motivate management activity. Motivation to strive for health and safety must be built into the culture of the organization as a long-term priority (Dawson et al. 1987).

Motivational controls have more diffuse coverage than technical (engineering) controls. Enabling this involvement in the intangible world of hygiene is difficult. The technical issues are more complex and not related to the technical or trades training people normally receive.

Many conflicting influences operate on available funds. Demands can range from providing more lucrative compensation to investing in productivity enhancements. The direction of allocation often reflects the powerbase in the organization (Dawson et al. 1987). These conflicting influences demonstrate why commitment from senior management to health and safety is so important. This sets the mandate, for control of hazards extends beyond the specifics of legal and technical considerations. This also gives strength to the existence of the powerful ideal that is difficult to attain, given the realities of organizational life.

In managing the complex situation posed by confined workspaces, the most difficult question to answer is where to start. The fact to establish is whether these workspaces exist on the property and whether they are entered or could be entered by employees or outside contractors. This is the starting point for standards and legislation on this subject. (See Appendix A for further information about standards and guidelines.) The answer to this question determines whether further action is necessary.

A common definition for confined space is not universally accepted. For the purposes of this publication, the following definition will be used. As a working definition, the term "confined space/confined atmosphere/confined energy" could apply to workspaces inside which one or more of the following hazardous conditions could be present or could develop:

- Personal confinement
- Unstable interior condition
- Flowable solid materials or residual liquids or sludges

- Release of energy through uncontrolled or unpredicted motion or action of equipment
- Atmospheric confinement
  - Toxic substances
  - Oxygen deficiency/enrichment
  - Flammable/combustible atmosphere
- Chemical, physical, biological, ergonomic, mechanical, process, and safety hazards

The meaning of individual terms may be defined in the legislative framework of a jurisdiction. This definition represents 36 or more potentially hazardous conditions. Many could exist or could develop simultaneously and independent of each other. One important outcome from this definition is that the term "confined space/confined atmosphere/confined energy" encompasses a range of hazardous conditions. The end points of concern for individual hazardous conditions range from minimal to life threatening.

To illustrate, a workspace may possess attributes of geometry capable of confining contaminants. Inverted semicylinders, currently fashionable in architecture as roof structures on walkways and other types of shelters, are simple examples. These structures can confine smoke from cigarettes or vapors from the solvents used to clean them. Access into or egress from the spaces under these structures is unrestricted in the normal sense. Yet, under appropriate conditions, considerable overexposure could occur due rapid buildup of contaminants and low rate of removal.

A house normally is not considered to pose a hazard to occupants. Yet, a burning residential building into which fire fighters must enter to search for possible victims is a primary example of a potentially life threatening situation in which the atmosphere is acutely toxic and structural conditions are unknown, and possibly unstable. Victims and fire fighters alike face highly uncertain conditions presenting extreme risks.

A geometry capable of atmospheric confinement can be as simple as a planar deck of a building under construction. The boundary surface traps contaminants by preventing motion in one or more dimensions. Structures such as this do not restrict access or egress in any normally perceived manner. If the rate of generation exceeds the rate of removal, accumulation occurs. The opposite extreme is a structure having complex internal geometry that is entered through a small opening. Atmospheric conditions inside this type of structure could be completely unpredictable due to residual contents, or substances introduced or produced as a result of the activity.

Substances confined in the space could be relatively harmless or extremely toxic. In either case, the level of concern depends on concentration. The atmospheric condition could be small compared to regulated levels, to immediately dangerous to life or health or higher. Hazardous atmospheric conditions often present no warning to the senses, either visual or olfactory.

A PROGRAM FOR MANAGING HAZARDOUS CONDITIONS IN CONFINED SPACES

GENERAL

Management is the process that oversees the operation of an organization. Management exercises control, and provides vision, focus, and direction. Control can be sought over conditions, the sequence and conduct of events, and the actions of people. The absence or loss of control during activities involving confined spaces is potentially very serious. The consequences of this situation are readily apparent in the accident summaries provided in the NIOSH, OSHA, and MSHA reports (NIOSH 1979, 1994; OSHA 1982a,b, 1983, 1985, 1988, 1990; MSHA 1988). These accidents appeared to be unusually severe compared to those that occur in normal workplace environments. Seemingly minor errors led to major consequences.

Activities conducted in confined workspaces are essentially the same as those conducted in other environments. The manner in which they are conducted also is comparable. In both cases,

TABLE 7.7
Elements in the Hazard Management Model

- Goals and objectives
- Description of program
- Statement of management commitment
- Organization, responsibility, and accountability
- Program elements
- Specific strategies

management and supervisors focus on the orderly performance and completion of the task. The only difference is the workplace environment. It would appear that management practices that produce acceptable results in normal workplaces do not provide sufficient control over the hazardous conditions that occur in confined workspaces.

Work to be performed in confined spaces should be considered a three-way interaction: the worker to the workspace, the worker to the task, and the task to the workspace. Ordinary supervisory and management practices focus exclusively on the orderly performance and completion of the task. This approach fails to address the relationship of the worker to the surrounding environment. The disproportionate severity of accidents that occur in confined spaces indicates that management of the interaction between the worker and the surroundings is critical.

One approach for addressing this problem is to create a management model. The model for managing work in confined spaces is a variant of the general model for the practice of industrial hygiene. The model consists of a series of interconnected elements, each expanding on the previous one. The elements in the model are contained in Table 7.7. The model can be visualized as a group of increasingly larger wheels connected to a common axle. Each wheel represents an element within the model. Each wheel has a central hub, radiating spokes, and an outer rim.

The central hub represents the focus of a particular element and the interface to other elements. The spokes are reinforcing concepts necessary for articulation of the element. The rim of each wheel is a network that connects each of the spokes. Just as the wheel of a bicycle cannot function properly without all of its spokes, each theme cannot function unless all of the reinforcing concepts are in place. The axle links all of the elements.

In the model, the length and cross section of individual reinforcing concepts (spokes) symbolize the importance attached to that element in a particular situation. The magnitude of importance of a reinforcing concept ranges from minimal to extreme. As the length and cross section of the spokes (reinforcing concepts) increase, so also does the diameter of the central hub that supports them (the central focus of that element). The hazard potential of a particular situation thus is related to the length and cross section of the reinforcing elements in each of the wheels.

The hub of each successive wheel in the model, starting from the top, increases in diameter. This reflects the increasing complexity of each successive element and the number of reinforcing concepts. The axle joining successive elements also increases in diameter. In a metaphor from the communications industry, each of the spokes and the rims, as well as the fibrils that form the increasingly larger axle, can be considered to be a line of communication such as a strand of fiber optic cable. All concepts are interconnected, some more strongly than others. Naturally, as the focus of the elements changes from conceptualization to implementation, their complexity and size increase. At the level of full implementation, all details relevant to a particular organization must be considered.

The overall hazard potential represented by a particular situation reflects the outline formed by joining the reinforcing concepts of each element (spokes of each wheel). In most cases, the outlines are anything but circular. The wheels representing the normally perceived confined space have long, heavy spokes and large central hubs. In situations having low hazard, the wheels have short, lightweight spokes and a small central hub.

The management system should function consistently in all cases. The following sections describe the parts of the management model and offer guidance for implementing it.

Goals and Objectives

The first element in the program, goals and objectives, sets the tone and direction for all of the others (Toca 1981, Toca 1994). Without a clear statement of purpose, the program can become piecemeal and fragmented. Goals and objectives should be created by the stakeholders having overall responsibility and accountability for the program. Goals and objectives express the vision for the program and provide substantive concepts against which success or failure can be determined. Goals and objectives need not necessarily be published in this form. They form the underlying framework for other expressions of policy and direction.

Description of Program

The next element in the model is the description of program. The description of program presents the *modus vivendi* and *modus operandi* for the program in conceptual terms. This is needed for presentation to senior management to explain the need for the program, and its direction, and scope. Senior management approval for and commitment to the program are essential for its success (Bird and Loftus 1976, Findlay and Kuhlman 1980, Coletta 1982, Jensen 1984). Obtaining this support at an early stage will minimize interference in enabling creation of the program. The statement of program also provides evidence to senior management that the program will be created in a coherent manner that reflects prior planning and consideration of cost and change to the organization. Stating the *modus vivendi* and *modus operandi* for this program in clearly understood, unambiguous terms is critical for securing approval and ensuring its success. The statement of program should clearly declare purpose, direction, and scope.

Creating an inclusive, coherent statement of program requires some structure. As a guide, the probing questions, what, why, when, where, how, by whom, and for whom, provide the basis for framing the defining, supporting, and clarifying statements. Table 7.8 provides an example of the statement of a program that might be defined within an organization. A management system that systematically incorporates these themes will have the greatest chance for success. Successive elements in the management model describe the means with which to make this happen.

Organizational Concepts

The theme of the third element in the management model is organizational. This describes how the program will fit into or require modification of the existing structure of the organization. The organizational aspects should include at least the following components:

- Statement of organizational commitment from senior management
- Statement defining confined space
- Designation of personnel
- Assignment of responsibilities

The components stated here provide the foundation for establishing an organized and focused program. The statement defining confined space/confined atmosphere/confined energy provides a common definition for the term within the organization. The statements designating personnel and assigning responsibilities indicate through whom and how the program will function. These statements also express important dynamics, responsibilities, and interrelationships of persons participating within the program.

The most important outward starting point for participants in the program is an expression of commitment from senior management. A statement of support issued from this level can make the

TABLE 7.8
Management System for Confined Spaces

The purpose for this program is to prevent incidents and accidents that involve workspaces known as confined spaces and workspaces that because of their geometry are capable of confining a hazardous atmosphere.

Accidents involving confined workspaces can be severe. Conditions that cause these accidents often provide no warning to the senses. These hazardous conditions often can be detected only by using measuring equipment. Accidents in these workspaces are entirely preventable. The reason for creating this program is to ensure that appropriate precautions are taken to enable work to occur in confined spaces in safety.

This program applies to all workspaces deemed to be confined spaces and to all work involving situations in which atmospheric confinement, personal confinement, confinement of energy, and other safety hazards can occur. This program applies to all aspects of preparation, entry, and the conduct of work.

Fundamental elements within this program include organizational policies, practices, and procedures; provision of testing and protective equipment; education and training; and recordkeeping and auditing.

The program will be carried out by persons trained to an appropriate level of competence. A key individual will oversee creation and implementation of the program. Resources will be made available to carry out the routine functions of the program in operating units.

All persons required to enter confined spaces or who supervise those who enter them will participate in this program. Persons who work on structures that can confine a hazardous atmosphere or who supervise these persons also will participate in this program. All persons who working in proximity to confined atmospheres shall receive training.

difference between failure and success (Bird and Loftus 1976, Findlay and Kuhlman 1980, Coletta 1982, Jensen 1984).

Table 7.9 provides a sample statement of organizational commitment. The sample statement provided in Table 7.9 acknowledges the awareness of senior management about the dangers inherent in confined spaces and the need for a program to address them. A clear, unequivocal statement is critical for gaining active input and cooperation from personnel at all levels. The sample statement also directs the organization to take all necessary steps to minimize the hazards of work in confined spaces. This sends a clear message to intermediate management and lower-level supervisors.

A clear, unequivocal message to this group is important, since the usual focus of intermediate management is maximizing production and minimizing cost. This is a direct result of financial pressure to operate the organization in the most cost-effective manner possible. Intermediate management sometimes tends to lose sight of safety concerns and may need reminding about them. A strong statement from senior management enables the commitment of resources to ensure compliance with this policy.

People are central to the success of the program. A necessary part of a successful program is a clear description of the infrastructure through which it will function. This description must identify participants, their functions, responsibilities, and interrelationships — that is to say, who will do what to whom. Clarifying roles and relationships prior to implementation of the program is extremely important. This avoids discord once the program has begun and enables participants to maximize their effectiveness in full knowledge of the role of the others.

Appendix E provides an example of an organizational structure within the program. This outlines responsibilities and accountabilities.

Program Elements

The theme of the next element of the management model is the confined space program. The program achieves success by enabling employees to manage their interaction with the environment in these workspaces through provision of knowledge and equipment. The requirement for such a program often is stated in standards and legislation on confined spaces. However, little information on achieving this goal usually is provided in these documents.

TABLE 7.9
Sample Statement of Organizational Commitment

Senior management wishes to state its support for the confined space program. This program is an important and distinct element within the organizational health and safety program.

The following statement reiterates the organizational health and safety policy. Senior management believes that the greatest assets of this organization are its employees, without whose dedicated service it could not exist. The organization owes its employees the right to a safe and healthful workplace.

Senior management realizes that work associated with confined spaces involves inherently greater risk than that occurring in other workplace environments. As a working definition, the term "confined space" can be expanded to read: "confined space/confined atmosphere/confined energy." This term applies to workspaces inside which one or more of the following hazardous conditions could be present or could develop:

- Personal confinement
- Unstable interior condition
- Flowable solid materials or residual liquids or sludges
- Release of energy through uncontrolled or unpredicted motion or action of equipment
- Atmospheric confinement
  - Toxic substances
  - Oxygen deficiency/enrichment
  - Flammable/combustible atmosphere
- Chemical, physical, biological, ergonomic, mechanical, process, and safety hazards

(The meaning of individual terms may be defined in the legislative framework of a jurisdiction.) This definition represents 36 or more potentially hazardous conditions. Many could exist or could develop simultaneously and independent from each other.

We support all endeavors to reduce the risk of working in these workspaces to the minimum possible. Accordingly, a program dedicated to management of hazards associated with confined spaces will develop as a distinct element within the overall safety program of this organization.

In pursuit of this goal, all levels of management will conduct activities involving confined spaces in a manner that will minimize the risk to which employees will be subjected.

Management will take all steps and support all initiatives to minimize the risk of injury to its employees. In return, to ensure that this policy is realized and carried out, management expects constructive and active participation at all levels.

This initiative regarding confined spaces will be realized through a program administered by the Safety Department. The program will involve participation by managers of support and operating departments, and professionals, including industrial hygienists and safety officers from the Safety Department. At the operational level, the program will function through Qualified Persons. Work practices will incorporate all elements necessary to ensure success.

The only satisfactory measure of success in this program is an operation that is illness-, injury-, and fatality-free.

The elements presented here are the minimum needed to create a stand-alone program. These are a composite obtained from standards, guidelines, and legislation. The elements are interconnected. Each interacts with the others. An important consideration in proposing program elements is that they address both routine and emergency aspects of individual situations. They also should address contingencies and provide redundancy. Table 7.10 presents the elements of the program. As outlined to this point, the program is completely general. Specific programs created for use in organizations should consider inclusion of the preceding elements. Adaptation to reflect the circumstances within an organization will necessitate emphasizing some elements and de-emphasizing others.

The following sections briefly describe the preceding concepts. Most are discussed in greater depth in subsequent chapters and appendices.

Inventory

A practical starting point for implementing this program is an inventory to determine all confined spaces within the organization. This could provide invaluable information about the potential extent

TABLE 7.10
Elements of a Confined Space Program

- Inventory of confined spaces
- Identification of hazardous conditions
- Access control
- Hazard assessment
- Procedures for site preparation, and entry and work
- Atmospheric/environmental testing
- Ventilation
- Respiratory and other personal protective equipment, standby rescue/emergency equipment and other specialized equipment
- Communication
- Education and training
- Medical aspects
- Emergency preparedness and response
- Recordkeeping
- Interaction with contractors
- Audit and program review

and magnitude of the problem. In order to ensure consistency and comprehensiveness, the inventory should be conducted by a Qualified Person who is thoroughly familiar with structures and equipment within the organization. Inventories also can be performed by members of the Safety and Health Committee following familiarization with the legal description of confined space applicable in the jurisdiction.

An overly thorough inventory is more desirable than a superficial one. A superficial inventory could understate the extent of the problem. Trivial assignments always can be eliminated during review and further refinement. Missing assignments create the potential for lack of attention to a space.

This topic is discussed more thoroughly in Chapter 8.

Identification

Identification of confined spaces by signs or labels is one possible follow-up to the inventory. Identification can take many forms. One of the simplest is a sign stating that a space is a confined space. More complex forms utilize pictograms or other communicative devices. Use of identifying devices can have the positive outcome of raising the level of awareness among members of an organization, as well as outside contractors.

This subject is discussed more thoroughly in Chapter 8.

Security

Coincident with labeling is the question of security. Security of these workspaces against unauthorized entry is an important issue. Permitting unhindered access to a structure identified as a confined space invites disaster. The NIOSH, OSHA, and MSHA reports indicated that people enter these workspaces for many reasons, not all of which are consistent with expected work activity (NIOSH 1979, 1994; OSHA 1982a,b, 1983, 1985, 1988, 1990; MSHA 1988). Simple, easily performed modifications, such as the locking of doors using padlocks, putting bars across openings, and securely fastening covers make casual entry considerably more difficult. The need to overcome a securing device as simple as a padlock should force reflection about the reason for its use. Simple deterrence may be enough to avoid a tragedy.

A follow-up to the question of security prior to opening a space is security during preparation for entry and work activity. Establishing security to prevent casual entry by passersby during entry and work activity is absolutely essential. Hazard warning tapes and signs, while providing a minimal physical barrier, do pose a psychological barrier that mitigates against unwanted intrusion. More substantial barriers that isolate the access opening from the surrounding area provide an additional level of security. Unfortunately, even the use of barriers and warning signs will not stop a determined individual. The only way to prevent unauthorized entry during occupancy is to post a watch at the entrance.

Unauthorized entry during work activity is an additional concern. The solution is to ensure that the confined space is not left unattended during coffee and lunch breaks or at the end of the workshift. That is, no one should be able to enter the workspace for any reason after authorized personnel have left the area. The only way to ensure security of the work area is to close the point of entry or to post a watch.

This topic also is discussed in Chapter 9.

Hazard Assessment

Hazard assessment focuses on anticipating hazards that may arise in these environments, predicting their potential severity, assessing the level of concern, and judging its acceptability.

Fundamental to the success of the hazard assessment are the action and involvement of a key individual, the Qualified Person. The Qualified Person is expected to know through education, training, and experience what is required in the context of a particular situation.

The temporal aspects of the operational cycle form the framework for hazard assessment. Events that occur in confined spaces can be subdivided into the following parts:

- Undisturbed space
- Preentry preparation
- Prework inspection
- Work activity
- Emergency response

Development of hazardous conditions can occur during any part of the cycle, especially before initial entry. Anticipation of hazardous conditions in the undisturbed space leads to control during preentry preparation. Preentry preparation can create hazardous conditions, since the contents of the space could be disseminated to the surroundings. Anticipation and correction of this problem also occur during this step. A prework inspection occurs in some circumstances prior to the start of work. This can lead to identification of conditions not anticipated previously.

In the ideal situation, hazardous conditions are eliminated prior to entry. Work activity also can create hazardous conditions. These can be completely unrelated to what was anticipated and corrected during previous preparatory work. The purpose for hazard assessment for work activity is to control conditions that could arise during this aspect. Emergency response refers to accident situations still believed possible after all other precautions are taken. This defines potential situations for which emergency preparedness and response are necessary.

The task for the decision maker is to decide whether potential or actual risks inherent in the proposed work are acceptable. The decision maker must anticipate situations possibly unforeseen to the proponent that could arise during the project. The basis for making this decision is to integrate knowledge of process, hazardous conditions, and work activity to the specifics of a particular workspace at a particular time.

The next step is to decide whether the work can be conducted in a manner that is considered sufficiently safe. At some point the potential for injury may exceed the protection offered by control measures. The challenge for the decision maker is to determine the point at which the margin of safety is too small to accept. Or put another way, how safe is safe enough?

Safety is achieved through control — confidence in knowledge about what could happen, assurance that steps have been taken to minimize the possibility for this, and assurance that contingencies exist to address what still could happen. If control of the conditions can be assured, the decision is not difficult to make. If control of the situation cannot be assured, the decision then becomes difficult. The less the level of perceived control, the greater becomes the need to provide for contingencies. The only other alternative is to prohibit the entry.

Hazard assessment is a focus of Chapter 8.

Site Preparation, Entry, and Work

In the event that entry is considered feasible, the next step is to undertake site preparation, entry, and work procedures. These must be consistent with the hazard assessment. In the long term, this process can lead to formalized practices and procedures. Formal procedures are useful devices for achieving control in situations where conditions remain the same from entry to entry.

No matter how much energy and effort have been expended on review and planning, site preparation, entry, and work are all that matters. If matters have been considered correctly, everything will proceed without event. However, mismatch between the world of planning and the actual situation could occur. The extent of the mismatch directly reflects the foresight of planning and preparation and the experience and capability of participants.

Work activity involving confined spaces often is time-critical, as are most other aspects of maintenance. The space may be a critical component of a production process, or a ship in drydock. Unavailability could represent a considerable economic impact to the organization. The pressures created by time-criticality impose a considerable burden on all participants in the project — management, supervision, trades and technical people, and the Qualified Person.

The Qualified Person certainly has no immunity to these pressures. Assessing conditions and their impact in a rapidly changing environment, so that critical decisions can be made, is a difficult and stressful undertaking. Time-criticality has a special meaning to the Qualified Person when things do not proceed according to schedule or when something goes wrong or when the unexpected occurs. Whereas the focus of other participants in these situations is task oriented, seeking ways to correct the problem or to make up for lost time, the Qualified Person must focus on the implications of those proposals on the safety and well-being of the participants.

Perhaps the most difficult action required from the Qualified Person is making the decision that work must stop. This could be forced when conditions in the space exceed prescribed limits or some unanticipated event has occurred. The need to stop work is an acknowledgment that control is inadequate. Stopping work, or recommending evacuation or other action that alters work flow under these circumstances is an especially difficult decision.

Site preparation, entry, and work are the focus of Chapter 9.

Atmospheric/Environmental Testing

Testing is fundamental to assessing atmospheric and other hazards in confined spaces. As demonstrated in the NIOSH and OSHA reports, lack of testing was an important factor in many of the fatal accidents (NIOSH 1979, 1994; OSHA 1982a,b, 1985, 1988, 1990). Testing provides the basis for evaluating the relationship between the entrant and the environment in the workspace. It provides the warning not to enter without protection and the warning to leave when hazardous conditions develop. Atmospheric testing also provides the basis for selecting respiratory protection.

Atmospheric testing must be appropriate to the contaminant of interest. Use of incorrectly selected equipment is equivalent to not testing at all. It is no understatement that reliance on data provided by this equipment can become a matter of life and death. Use of incorrectly selected equipment has led to fatal accidents (OSHA 1985, 1990; NIOSH 1994). The entrants were lulled into believing results that were inappropriate to the situation.

No single instrument can assess all the of the contaminants that may be present in a particular confined space. There are many situations for which no dedicated instrument is available. In these cases, determining the most appropriate instrument for the circumstances requires knowledge about the composition of the contaminated atmosphere, as well as the capabilities of available instruments. Cross-sensitivity and loss of sensitivity due to the presence of interfering contaminants are confounding factors that must always be considered. Also, intrinsically safe equipment may be required.

Testing equipment also is available for assessing energy sources that may be present in the space. These include: noise and vibration, ionizing and nonionizing radiation sources, ultraviolet and visible energy, electrical current, air motion, and so on.

Atmospheric testing is the theme of Chapter 10.

Ventilation

Ventilation is a fundamental measure for controlling exposure to atmospheric hazards in confined workspaces. Confined workspaces are fundamentally different from normal environments. The proximity of boundary surfaces often creates conditions favorable to atmospheric confinement near the point of generation. This can lead to accumulation of airborne contaminants at a much greater rate than would occur under normal circumstances.

Airborne contamination can arise from residual contents, contaminants that enter from the surroundings or work activity. The characteristics of the source and its relationship to the position of the occupants and boundary surfaces govern the ventilation strategy. Ventilation can utilize natural forces of wind and convective air motion, as well as portable equipment.

The usual means of ventilating confined spaces is local supply or exhaust. This concept makes use of a duct connected to an air mover either to convey clean air into the space or to remove contaminated air from it. The transfer duct often passes through the portal used for access and egress. The transfer duct can be located in the immediate area in which the work is occurring. Both supply and exhaust have application in some circumstances.

The ability of contaminants to accumulate affects selection of the air mover and location of the duct used to supply or exhaust the air. Portable ventilating units must provide sufficient volumetric flow rate. Undersupply could lead to insufficient air motion. Oversupply could cause problems with processes such as shielded gas welding. Turbulence could destroy the shield around the weld provided by the gas. The source of supply air or location of discharge must be chosen carefully to avoid cross-contamination.

Ventilation is the theme of Chapter 11 and associated appendices.

Respiratory and Other Personal Protective Equipment

Respiratory and other personal protective devices may be required for entry and work under some circumstances. Compared to other working environments, the need to rely on personal protective equipment is greater in confined spaces. This results from the nature of work in these workspaces: infrequent and short in duration, and the impracticality of other means of control.

Specification of personal protective equipment for use in confined spaces can become a considerable challenge. Compounding the problem is the fact that the only means of access may be a small opening. Rapid egress during an emergency through such an opening by persons encumbered by excessive amounts of personal protective equipment could be impossible. This problem could compound injury produced by conditions in the space. Also, overburdening a person with protective equipment under these circumstances could cause physiological stress, such as heat stress, and psychological stress, such as claustrophobia.

The selection of personal protective equipment for use in these workspaces must reflect conditions against which protection is required and the necessity of the protected person to work in confining surroundings. Personal protective equipment selected for use in these workspaces should not compound the hazards of work. This problem must be considered carefully during hazard assessment and planning. Discovering these problems during entry and work activity is too late.

Respiratory and other personal protective equipment is the theme of Chapter 12.

Communications

Communications are an important but often unconsidered part of the strategy for providing safe conditions of work in confined spaces. Establishing and maintaining safe conditions requires defined and continuous communication between persons working inside the confined space and those working or attending outside. Communication can occur through various means, such as voice, two-way pager, or two-way radio. The means of maintaining communication need not be elaborate, just effective. The critical outcome is to enable persons inside the confined space to obtain assistance when needed. Use of intrinsically safe equipment may be required in some circumstances.

Communication issues are discussed in Chapter 9.

Education and Training

Education and training at all levels of the organization are another important part of the strategy for gaining control. This is especially important for persons required to work in confined workspaces or who authorize entry by others. Lack of training was a factor in virtually all of the accidents described in the NIOSH, OSHA, and MSHA reports (NIOSH 1979, 1994; OSHA 1982a, 1982b, 1983, 1985, 1988, 1990; MSHA 1988). Informed persons utilizing training about confined spaces could have influenced the outcome of many of these accidents. Increasing the level of knowledge of all persons working in these environments is essential. Without competent training, any program in hazard assessment and control is jeopardized.

Education and training require considerable effort and commitment by an organization. Acquiring competency requires ongoing instruction, reinforcement, practice, evaluation, and review.

Education and training are the subject of Chapter 13.

Emergency Preparedness and Response

Emergency preparedness bridges the gap between what is supposed to happen during an activity and what could happen. Determining what *could* happen is a critical part of hazard assessment. Although unpleasant to ponder because of the considerations involved, the cost of not performing this exercise is greater in the long run. Not thinking about accidents will not prevent their occurrence. However, considering the possibilities in advance provides the basis for minimizing the consequences of such events. The consequences of lack of such preparations were amply demonstrated in the NIOSH and OSHA reports (NIOSH 1979, 1994; OSHA 1985). Lack of preparation was a major factor in the inability of persons at the scene to respond appropriately to situations presented to them.

Consideration of possible accidents is the first part of emergency preparedness. The second part focuses on actions required to address the problem. An important consideration is the level of response that is realistically possible from within an organization. If the level of response is less than what is needed to address a realistic accident situation, reassessment is needed. The decision to respond during an emergency cannot be taken lightly, as demonstrated in the NIOSH and OSHA reports. The intent to use any rescue equipment necessitates a serious commitment to training.

Organizations can provide an initial response to accidents that minimizes risk to first responders. These efforts must occur from outside the confined space. Their main focus is to stabilize the victim and ameliorate conditions before arrival of professional first responders. Establishing procedures and lines of communication with providers of emergency response is essential for determining their needs and for familiarizing them with the organization and its activities.

Emergency response is the subject of Chapter 15.

Recordkeeping

Records ultimately define the existence of the program to inside and outside auditors. While viewed as nonproductive activity by some, records can provide important information. Records are the link with the past. They can provide valuable information about performance of testing equipment or the characteristics and behavior of atmospheric conditions in particular circumstances. They also can provide strategies for approaching future problems and data for accident investigation. Records should include:

- The inventory of confined spaces
- Hazard assessments
- Standardized entry procedures
- Dates of calibration and performance and repair of equipment
- Dates and nature of training of personnel
- Updating of procedures, calibration, retraining, equipment
- Entry documents
- Audit reports

Recordkeeping is integral to many of the activities discussed in later chapters.

Interaction with Contractors

Contractors perform an appreciable fraction of the work that occurs in confined spaces. This results from specialized expertise, as well as the current trend of downsizing. Contractors seldom know the concerns of a particular situation as well as the people who work there. The seriousness of this problem is recognized in current standards and legislation (ANSI 1995; OSHA 1993, 1994).

Interaction with contractors is discussed further in Chapter 9.

Audit and Program Review

Periodic review of the program is essential to ensure that it remains current and meets the needs of participants. Review could occur by calendar or at any time that conditions indicate that existing procedures provide inadequate protection.

Specific Strategies for Providing Safe Conditions of Work

Successive elements of the management model progress through two broad themes: conceptualization and implementation. The first elements emphasize concepts and strategies. The latter elements focus on the apparatus needed to bring the program to fruition within a particular organization. The theme of the last element in the management model is specific strategies. These are practical approaches for applying the model and customizing it to fit the characteristics and needs of a particular organization.

The need for organizational strategies to address these problems cannot be overstressed. Lack of organization was a major factor in many of the fatal accidents documented by NIOSH and OSHA (NIOSH 1979, 1994; OSHA 1982a,b, 1983, 1985, 1988, 1990; MSHA 1988). Creating strategies that address these problems in a practical way is a difficult challenge. The challenge is to integrate the expertise of existing health and safety resources into daily operations of line functions. This is least consequential where these resources are available. However, they are unavailable in the vast majority of organizations. By far, most organizations have fewer than 20 persons and are incapable of providing the necessary resources.

Hazard Management Models

Current standards and legislation provide models of approach to this problem. The OSHA Standards on confined spaces provide examples of organizational and philosophical directions that can be used in confined space programs (OSHA 1993, 1994). The approaches taken in these standards are completely different from each other and are worth examining. The present Canadian regulation in the federal jurisdiction provides an example of a hybrid to the approaches in the OSHA Standards (Canada Gazette 1992). (Standards and guidelines are described in Appendix A and summarized here.)

The OSHA Standard on confined spaces in general industry requires employers to identify confined spaces in their workplaces and to determine whether they are permit-required or nonpermit-required (OSHA 1993). A permit space is a confined space that has certain characteristics that make entry hazardous without special precautionary measures. The permit must be completed before entry is allowed into a permit-required confined space. Similarly, the employer must reevaluate nonpermit spaces following changes that might increase hazards to entrants above the threshold that defines a permit space.

The permit authorizes entry and documents measures required to ensure that employees can safely enter and work in the permit space. The permit also provides a concise summary of the entry procedure. Critical information that must appear on the OSHA version of the permit is contained in Table 7.11. (The OSHA Standard for general industry is one of the few standards on confined spaces that indicates content required in entry permits.)

TABLE 7.11
Information Required on OSHA-Compliant Entry Permit

- Identification of the space to be entered
- Purpose for the entry
- Date and authorized duration of the entry permit (this may not be actual time but may be completion of the task)
- List of authorized entrants (this may occur through a systematic tracking system)
- Current entry supervisor
- Hazards of the space
- Specific measures for isolating the permit space or controlling hazards prior to entry
- Acceptable entry conditions
  - Atmospheric conditions
  - Energy control measures
- Measures for obtaining acceptable entry conditions
- Initial and periodic test results corresponding to the specified entry conditions plus signature or initials of the tester and an indication about when the tests were performed
- Rescue and emergency services that can be summoned and the means for summoning them
- Communication procedures to be used by the attendant and authorized entrants during entry operations
- Equipment to be provided for compliance with requirements of the Standard
  - Personal protective equipment
  - Testing equipment
  - Communications equipment
  - Alarm systems
  - Rescue equipment
  - Other equipment
- Existence of any additional permits, such as hot work permits

The permit must be prepared after establishing safe entry conditions and prior to the entry. The entry supervisor must sign the permit. The permit must be available for review by all entrants at the time of entry. The duration of the permit is the period of time needed to complete the assigned task or job. As long as acceptable entry conditions are present, employees can safely enter and perform work in the permit space. The entry supervisor cancels the permit. "Entry supervisor" is a transferrable title; hence, someone other than the author of the permit may terminate entry and cancel the permit.

The OSHA Standard requires a written permit space program where entry of permit spaces occurs. The general purpose for the program is to ensure the existence of an infrastructure that underlies the issuance of entry permits. OSHA separated requirements for the program from the content of the permit in order to emphasize the inclusive nature of the system from preparation of the permit.

The Standard defines job titles for personnel who work in and around confined spaces. These include: authorized entrants, entry supervisor, attendant, testing personnel, and rescue and emergency services. The Standard mandates training to ensure that individuals attain the appropriate level of performance.

The entry supervisor is responsible for evaluating conditions in and around a permit space where entry is planned. The entry supervisor authorizes entry to begin or allows entry already underway to continue and oversees entry operations. This means prescribing necessary measures to protect personnel from hazards in the permit space and prohibiting or terminating entry, as deemed necessary. The entry supervisor must determine at appropriate intervals whether conditions within the space remain safe and cancel the permit whenever conditions are otherwise. The entry supervisor and the person who supervises the actual work could be one and the same or they could be different.

The attendant monitors entrants and conditions inside and outside the space, prevents entry of unauthorized persons, and summons rescue services in emergencies.

The OSHA Standard on confined and enclosed spaces in shipyard employment covers ships and ship sections, as well as anywhere within the confines of the property of the shipyard (OSHA 1994). OSHA considers the shipyard Standard to be a more restrictive subset of the Standard for general industry and even more comprehensive in its coverage of hazardous atmospheres. Work in confined spaces in a shipyard is highly complex and can involve many industrial processes. Processes whose hazards are highly interactive could be utilized in close proximity. To illustrate, poorly planned hot work on one side of a bulkhead could cause a fire during spray painting on the other. Complicating things further, these activities and the hazards inherent in them must be managed within the confines of tight production schedules.

The shipyard Standard requires the same approach for all spaces, confined or otherwise. This means initial testing and visual inspection, possibly followed by continuous ventilation and further testing. OSHA argues that the requirements of the shipyard Standard constitute an informal permit system. That is, it requires evaluation, tracking, and control measures that are equivalent in protection.

The foundation of the shipyard Standard is the shipyard competent person. The shipyard competent person completes the initial evaluation, and as appropriate, implements the first level of control. The competent person can determine that the atmosphere exceeds a regulatory limit and can recommend installation of ventilation to reduce the atmospheric concentration to within acceptable limits. If ventilation fails to control the level or other aspects of the situation exceed their capabilities, the competent person must seek advice from an expert: an NFPA (National Fire Protection Association), Certified Marine Chemist, a U.S. Coast Guard authorized person, or a Certified Industrial Hygienist. The latter then will provide assistance in the development and implementation of control measures. The shipyard competent person or outside expert records his/her findings in a certificate. This is posted at the entrance to the space.

The certificate utilizes specific terms to describe conditions that exist or are likely to exist in the space: Safe for Workers, Not Safe for Workers, Enter with Restrictions, Safe for Hot Work,

Not Safe for Hot Work. The only spaces that employees are permitted to enter are those certified as Safe for Workers. Entry is not permitted until that certification is granted. Shipyard employees are trained to remain outside spaces that do not have this certification.

OSHA mandates training to ensure that the shipyard competent person has attained a suitable level of performance and that workers and other participants are aware of their responsibilities.

These Standards provide distinctly different approaches for the management of work involving confined spaces. The permit system works best where hazardous conditions are known from previous experience and control measures have been tried and proven effective. The permit system enables scarce expert resources to be apportioned in an efficient manner. The limitations of the permit arise where previously unrecognized hazards are present or develop. If the expert is not readily available, these can remain unrecognized and unaddressed. The issuer of the permit should be independent of the supervisory hierarchy in order to avoid potential pressure to speed the performance of work.

The on-site expert provides hands-on expertise in the recognition, assessment, evaluation, and control of hazards. An added advantage is the ability to respond to concerns on short notice and to address unanticipated hazards. This approach is ideally suited to operations that have numerous confined spaces or where conditions or the configuration of spaces can undergo rapid change.

The Canada Occupational Safety and Health (COSH) regulations offer a hybridized approach (Canada Gazette 1992). The approach used in the COSH regulations begins with recognition that a particular workspace is a confined space. At this point the employer must appoint a Qualified Person. The Qualified Person assesses the physical and chemical hazards to which an entrant could be exposed, and specifies tests necessary to determine whether the entrant likely would be exposed. The Qualified Person then provides the results of the assessment in a signed and dated report to the employer. The employer then establishes procedures for entry, work, and egress. The employer also must establish a system of entry permits. Also, the employer must specify protective equipment and tools to be used by entrants and emergency equipment required by rescuers.

An important feature of the Canadian approach to workplace health and safety is the statutory involvement of the Safety and Health Committee or representative at all stages in the process. The employer must inform and consult with the Committee or representative about all aspects of the situation.

Just prior to entry and during occupancy of the confined space, the Qualified Person performs tests to verify compliance with regulated specifications. At the conclusion of the work the Qualified Person prepares a signed and dated report outlining the work performed and results obtained from the tests. Where the report indicates that the occupants have been in danger, the employer must revise the procedures.

In circumstances where conditions in the space or the nature of the work preclude compliance with the preceding specifications, the employer must establish emergency and evacuation procedures and provide other measures to ensure communication and immediate assistance to occupants. A Qualified Person trained in entry and rescue procedures must attend outside the confined space. In addition, additional trained help must be immediately available. The Qualified Person specifies equipment required and procedures to be followed in a signed and dated report to the employer. Also, the Qualified Person explains the procedures to entrants and obtains their acknowledgment through signature.

Performance of hot work introduces additional constraints. Where the confined space contains hazardous concentrations of flammable or explosive materials, the employer must provide a Qualified Person to maintain a fire watch in the area outside the space.

Each of the approaches described here offers advantages and disadvantages. One may offer the best combination to respond to the needs of a particular organization, where that flexibility is available. (Flexibility may not be available in some jurisdictions.) Regardless of the approach that is chosen, the *modus operandi* should be described in a formal manner.

Appendix E provides an example of an organization for a confined space program that designates personnel and assigns responsibilities.

Many organizations have available the services of health and safety professionals. These individuals, however, likely would be unavailable to attend at every entry. The only way to fill this gap is to designate subordinates. The role of the subordinates can fit into each of the models mentioned here. Adoption of this approach means that the only real difference between the models is the means of recording and communicating the hazard. In each case, a Qualified Person (Entry Supervisor, Competent Person, Qualified Person) will be present at the activity. Since none of the models specifies training required in these individuals, this opens the opportunity for the expert–subordinate approach.

The OSHA Standard for shipyard employment clearly requires the competent person to call for assistance when the situation is beyond the expertise of that individual. Similarly, by organizational policy, the entry supervisor or Qualified Person could be instructed to request assistance when the situation is beyond the capabilities of that individual.

The key to implementing any of these models is the hazard assessment. The hazard assessment must proceed through a repeatable, comprehensive format. This provides the greatest potential for consistent identification and elimination or control of hazardous conditions. The hazard assessment used in this publication contains 36 potentially hazardous conditions. This serves as an indicator of the breadth of enquiry that this assessment should include. A comprehensive hazard assessment can be utilized by a wide group of individuals.

This approach also provides a vehicle to small organizations that face two realities: lack of human resources and an abundance of contact with confined spaces. While the problem of human resources can never be resolved, use of existing resources in a more focused manner should be possible.

Individualization

Management of conditions associated with confined workspaces can take several directions. At the one extreme is the philosophy that can best be described as "one size fits all." By definition, this is the only strategy that is simultaneously applicable to all organizations and circumstances. Out of necessity, this approach must presume that the situation is worst-case at all times. Of course, this demands a response of appropriate magnitude. While this approach could be applied carte blanche to all situations identified as confined spaces by the criteria published here or elsewhere, the question of appropriateness arises. Certainly, this approach would require expenditure of considerable resources. The benefits of large-scale implementation definitely are open to question.

The view taken here, and in standards and legislation, is that organizations should be able to establish strategies to fit their individual needs. Individualization provides opportunity for innovation. However, any strategy proposed for a specific situation must conform to the general framework described in the preceding section. In addition, the strategy must satisfy the following questions:

- Have steps appropriate to ensure control over conditions during entry and work activity been taken?
- What events still can happen?
- What steps have been taken to address them?

Any organization taking this approach must prepare to document and justify its decisions.

Activities occurring in these workplace environments likely will involve interactions with either the physical surroundings or the atmospheric environment. Identifying these interactions and potential ramifications is critical to determining actions needed for exercising control.

Appendix F provides assistance in this matter through a large number of focused questions that direct attention to issues that may influence the outcome of entry and work involving these workspaces. These questions also focus on establishing potential consequences and actions needed for response.

SUMMARY

This chapter serves as the bridge between the two main parts of this publication. The first was concerned about the past; that is, what currently is known about work involving confined spaces and confined hazardous atmospheres. The second focuses on the future: gaining control over hazardous conditions associated with these environments through management strategies.

A formal program of hazard control is absolutely essential for gaining control over conditions associated with work in these workspaces. This chapter presents a model for hazard control through management of conditions. The model contains several elements around which the program is built. Each element contains supporting structures that provide additional detail about the theme. Each element is interconnected to the others.

REFERENCES

American National Standards Institute: *Safety Requirements for Confined Spaces* (ANSI Z117.1-1995). Des Plaines, IL 60018-2187: American Society of Safety Engineers/American National Standards Institute, 1995. 32 pp.

Bird, F.E. and Loftus, R.G.: *Loss Control Management.* Loganville, GA: Institute Press, 1976. 562 pp.

Bird, F.E., Jr. and G.L. Germain: *Practical Loss Control Leadership,* rev. ed. Loganville, GA: Institute Publishing, 1990.

Buys, J.R.: How to grow A SAF-T-TREE. *Safety Health:* May, 1988. pp. 56–59.

Capps, J.H.: Coal utilization technology: a safety analysis and review system. *Prof. Safety. 25:* 43–49 (1980).

Cohen, A.: Factors in successful occupational safety programs. *J. Safety Res. 9:* 168–178 (1977).

Coletta, G.C.: Managing your risk control program. *Prof. Safety. 27:* 20–26 (1982).

Canada Gazette: Part XI, Confined Spaces, *Canada Gazette, Part II. 126*: 3863–3876 (1992).

Crutchfield, C.D.: Managing occupational safety and health programs — an overview. *Am. Ind. Hyg. Assoc. J. 42:* 226–228 (1981).

Dawson, S., P. Poynter, and D. Stevens: How to secure an effective health and safety program at work. *Prof. Safety 32:* 32–41 (1987).

Findlay, J.V. and R.L. Kuhlman: *Leadership in Safety.* Loganville, GA: Institute Press, 1980. 197 pp.

Gressel, M.G. and J.A. Gideon: An overview of process hazard evaluation techniques. *Am. Ind. Hyg. Assoc. J. 52:* 158–163 (1991).

Heinrich, H.W., D. Petersen, and N. Roos: *Industrial Accident Prevention.* New York: McGraw-Hill, 1980. 468 pp.

Ho, T.: System analysis (safety): methods and analysis. In *Encyclopedia of Occupational Health and Safety, vol. 2.* Geneva: International Labour Office, 1983. pp. 2133–2137.

Jensen, E.W.: Effective communication with top management. *Prof. Safety. 29:* 30–33 (1984).

Kletz, T.A.: Eliminating potential process hazards. *Chem. Eng.:* April 1, 1985. pp. 48–68.

Knowlton, R.E.: *An Introduction to Hazard and Operability Studies. The Guide Word Approach.* Vancouver, BC: Chemetics International, 1981. 43 pp.

Mine Safety and Health Administration: Think "Quicksand": Accidents Around Bins, Hoppers and Stockpiles, Slide and Accident Abstract Program. Arlington, VA: U.S. Department of Labor, Mine Safety and Health Administration, National Mine Health and Safety Academy, 1988.

National Institute for Occupational Safety and Health: Criteria for a Recommended Standard — Working in Confined Spaces (DHEW/PHS/CDC/NIOSH Pub. No. 80-106). Cincinnati, OH: National Institute for Occupational Safety and Health, 1979. 68 pp.

National Institute for Occupational Safety and Health: Worker Deaths in Confined Spaces (DHHS/PHS/CDC/NIOSH Pub. No. 94-103). Cincinnati, OH: National Institute for Occupational Safety and Health, 1994. 273 pp.

Occupational Safety and Health Administration: Selected Occupational Fatalities Related to Fire and/or Explosion in Confined Work Spaces as Found in OSHA Fatality/Catastrophe Investigations. Washington, D.C.: U.S. Department of Labor, Occupational Safety and Health Administration (U.S. DOL/OSHA), 1982a. 76 pp.

Occupational Safety and Health Administration: Selected Occupational Fatalities Related to Lockout/Tagout Problems as Found in Reports of OSHA Fatality/Catastrophe Investigations. Washington, D.C.: U.S. Department of Labor, Occupational Safety and Health Administration (U.S. DOL/OSHA), 1982b. 113 pp.

Occupational Safety and Health Administration: Selected Occupational Fatalities Related to Grain Handling as Found in Reports of OSHA Fatality/Catastrophe Investigations. Washington, D.C.: U.S. Department of Labor, Occupational Safety and Health Administration (U.S. DOL/OSHA), 1983. 150 pp.

Occupational Safety and Health Administration: Selected Occupational Fatalities Related to Toxic and Asphyxiating Atmospheres in Confined Work Spaces as Found in Reports of OSHA Fatality/Catastrophe Investigations. Washington, D.C.: U.S. Department of Labor, Occupational Safety and Health Administration (U.S. DOL/OSHA), 1985. 230 pp.

Occupational Safety and Health Administration: Selected Occupational Fatalities Related to Welding and Cutting as Found in Reports of OSHA Fatality/Catastrophe Investigations. Washington, D.C.: U.S. Department of Labor, Occupational Safety and Health Administration (U.S. DOL/OSHA), 1988. 225 pp.

Occupational Safety and Health Administration: Selected Occupational Fatalities Related to Ship Building and Repairing as Found in Reports of OSHA Fatality/Catastrophe Investigations. Washington, D.C.: U.S. Department of Labor, Occupational Safety and Health Administration (U.S. DOL/OSHA), 1990. 195 pp.

OSHA: Permit-Required Confined Spaces for General Industry; Final Rule, *Fed. Regist. 58*: 9 (14 January 1993). pp. 4462–4563.

OSHA: Confined and Enclosed Spaces and Other Dangerous Atmospheres in Shipyard Employment; Final Rule, *Fed. Regist. 59*: 141 (25 July 1994). pp. 37816–37863.

Toca, F.M.: Program evaluation: industrial hygiene. *Am. Ind. Hyg. Assoc. J. 42:* 213–216 (1981).

Toca, F.M.: In search of IH excellence. *Occup. Hazards. 56:* 113–116 (1994).

8 Preentry Planning, Hazard Assessment, and Hazard Management

CONTENTS

INTRODUCTION

Success in the conduct of work involves planning and prior preparation, as well as action and coordination. This is true regardless of the nature of the workspace in which the work occurs. Certainly this is true for confined spaces. Viewed in the context of planning and work flow, confined spaces are merely a class or subclass of workspace. From a planner's perspective, the term "confined space" merely reflects the distinguishing characteristics of the workspace, not the manner in which work is planned and performed. The manner in which work in confined spaces is approached, therefore, should be no different in principle from the approach used in other environments.

Work is performed in confined spaces in much the same way as in any other workspace. Certainly this has been the case in the past, as reflected in fatal accidents that have occurred in confined spaces. The nature of work and causative factors involved in these fatal accidents form the focus of the first part of this publication. One of the recurrent themes in the accident summaries was the ordinary nature of everything that was involved (NIOSH 1979; OSHA 1982a,b, 1983, 1985, 1988). Nothing about the activities was unusual, nor about the way in which they were planned and executed. These events were the end product of actions that resulted from conscious decision making by well-meaning individuals about circumstances that, in their view, were completely normal.

These accidents clearly reflected faulty decision making. This resulted at least in part from deficiency in the knowledge of the decision maker. These accidents also reflected lack of attention to detail in the way in which the work was to be undertaken. Correcting these deficiencies is critical to preventing further accidents in these workspaces, rather than merely reacting to them once they are progressing.

Many organizations have proven over the years that entry and work involving confined spaces can proceed without incident. A critical element in the strategy employed in these organizations is planning. The advantage of the planning process is the disciplined approach that it forces to problem solving. In the context used here, planning encompasses two major activities: assessment and proactivity. The one naturally leads to the other. Both must occur in order to achieve success in the conduct of work involving confined spaces. That is to say, success results from a series of deliberate actions. Success is not random. The deliberateness of those actions reflect investigation, assessment, and informed decision making.

Confined spaces are complex environments. The input and decision making required for the planning of entry and work involving these workspaces must occur at a high level. The quality of this input and the resulting decision making make the difference between the successful effort, the partially successful one that misses some key element, and the token one that could expose everyone involved to high-level risk. Unfortunately, input and decision making at the level that is required are not readily available to the average organization or individual.

Chapter 7 presented an introductory discussion about hazard management during activities involving confined spaces. Successful management of hazards associated with confined spaces starts long before anyone enters and starts to work. The beginnings of this process are the recognition and assessment of real or potential hazards. In a specific situation this process could necessitate extraction and utilization of a small fragment of highly pertinent information from an extensive knowledge base. The facility that enables this information to be readily available and the ability to apply it to a specific situation highlight the difficulties and complexity inherent in hazard recognition and assessment. The problem with this situation is that someone, the "Qualified Person," must be available to or within an organization to facilitate and implement the process. Providing the tools to enable individuals possessing a broad range of skills and knowledge to attain the status of Qualified Person is a major challenge.

This intent of this chapter is to provide guidance in planning entry and work involving confined spaces. Subsequent chapters will discuss other aspects in the process: entry procedures, ventilation, atmospheric testing, training, personal protective equipment, emergency response, medical aspects, safety, and other issues.

ASSESSMENT AND PREPARATION: THE KEY ELEMENTS OF PLANNING

Planning presumes the existence of a problem — an imbalance, uncertainty, inequity, or unknown. The problem requires a solution. The first element in the process of problem solving is inquiry to discover the nature and extent of the problem. The initial inquiry could lead, in turn, to further inquiry as the situation unfolds. These inquiries can lead to possible solutions. Solutions offer options. The next step, therefore, is to examine the options to determine their individual strengths and weaknesses. Finally, the option offering the greatest strengths and fewest risks should become apparent. A formal decision-making assessment model should be used, in order to ensure consistency and discipline in the process (Kepner and Tregoe, 1981). Whether the "logical option" is chosen and implemented involves judgment by the decision maker.

A major difficulty regarding planning for work involving confined spaces has been identification of what requires doing, and therefore, planning. Previously, the elements and actions required within a program usually were learned by experience or handed down by experienced practitioners. This process provided no guideposts to individuals endeavoring to start out without a mentor. Also, the approach currently taken by standard-setting organizations and regulators to managing the conduct of work in these workspaces has not helped. These groups have chosen performance, as compared to specification, as the strategy for obtaining outcomes.

Performance-based standards and legal statutes specify end results or key outcomes. At the same time, they provide few, if any, details about the process for attaining them. In theory, this approach provides maximum flexibility for implementation. The reality is that this approach provides no guidance to the average individual required to make real-world decisions about complex technical issues. The background knowledge and experience needed to recognize, assess, and control real and potential hazards in these workspaces is extensive and must not be underestimated.

Fundamental to the concept of performance-based strategies is the action and involvement of a key individual, the "Qualified Person." The Qualified Person is expected to know through education, training, and experience what is required in the context of a particular situation. While defined in various ways in different documents, a relatively comprehensive description for the Qualified Person is someone designated by the employer as capable by education and/or specialized training of anticipating, recognizing, and evaluating exposure to hazardous substances or other unsafe conditions. Also, the Qualified Person must be capable of specifying control measures and/or protective actions necessary to ensure safety (Standards Australia 1995). Individuals presently satisfying the requirements of the Qualified Person for confined spaces are not employed by the average organization.

The level of performance required from Qualified Persons involved with confined spaces is extraordinary. The Qualified Person simultaneously must possess and be able to focus encyclopedic knowledge, skill, and experience in hazard assessment and decision making onto specific situations. Equally important are the quality and level of decision making. Quality and level of decision making involve similar characteristics. They depend on noncontrollable factors, such as the inherent capability and skill of the individual, and also controllable ones, such as knowledge and experience.

Specification-based strategies attempt to dictate actions required in a particular context and situation through documentation. The main limitation in this approach is the encyclopedic content that would be needed to address all situations and contexts. This, of course, would be impossible to provide. An additional problem with this approach would be providing this information in an accessible format.

Neither of the preceding approaches is satisfactory for real-world needs. Qualified persons, having the background requisites, should not need regulatory standards or guidelines, as they already know what is required. For these persons, these documents are little more than strategic outlines and are essentially superfluous to their needs. For everyone else, these documents describe a level of attainment, but provide no means for reaching it.

The difficulties mentioned above were recognized by the drafters of the U.S. Standard on confined spaces (OSHA 1993). The OSHA Standard identifies many of the performances required from Qualified Persons. (In the language of the OSHA Standard, everyone in an organization in which confined spaces are entered must become a "Qualified Person" to some extent. This varies according to context of activities.) This document contains detailed performance outcomes for which planning is required. Sources of additional performance outcomes are the standards and legal statutes in other jurisdictions (Standards Australia 1995; ANSI 1995; Canada Gazette 1987, 1992; OSHA 1994).

Performance outcomes vary somewhat from document to document. What is specified in one document in some detail may appear only as a word phrase in another. Lack of inclusion or lack of detail does not necessarily mean lack of need for the performance. A performance missing from one document that appears in another may be essential to the process. The omission simply may reflect the skill of one group or the detail orientation of another.

Performance outcomes form the basis for planning. A comprehensive list of performance outcomes is essential to the planning process, regardless of the jurisdiction. That is to say, a performance outcome that provides suitable detail to a considerable extent defines the activity or action needed to achieve it. At the very least, it indicates the existence of a problem for which a resolution involving planning is required.

Table B.1 (Appendix B — The Qualified Person) identifies performances associated with the conduct of work in confined spaces. Table B.1 is a composite of performance outcomes synthesized from the standards and legislative statutes of various jurisdictions (NIOSH 1979; Standards Australia 1995; ANSI 1995; Canada Gazette 1987, 1992; OSHA 1994). Not all performance outcomes necessarily apply to the circumstances within a particular organization. Knowledgeable individuals should be able to recognize those that apply. For those who are less sure, starting from the broad perspective offers greater assurance than underestimating the need.

The elements identified in Table B.1 indicate that programs for entry and work in confined spaces require involvement from a number of people and the generation and management of a sizable knowledge base. This is a considerable planning challenge. As indicated, planning for entry and work in confined spaces simply does not entail the acquisition of rudimentary training and some new equipment. Rather, this is a highly complex undertaking that requires development of a substantial multielement program.

Several of the preceding elements form the topic of subsequent chapters.

A STRATEGIC APPROACH FOR WORK INVOLVING CONFINED SPACES

Discussion

The most important questions for an organization whose operations involve entry and work in structures and equipment are:

- Are they confined spaces?
- What must be done to ensure safety during the work to be performed?

The answers to these questions will vary from organization to organization. Also, they are likely to vary within an organization from one point in time to another. Hazardous conditions encountered in confined spaces are not static by location or by time. Minor change can alter conditions from innocuous to hazardous.

The challenge facing standard-setting agencies and regulators has been to find a process for managing the infinite combination of hazardous conditions that occur in confined spaces. The variation in the approaches is evident in Appendix A. (Appendix A examines strategies utilized in various regulatory and consensus standards.) The U.S. Department of Labor, for example, uses two markedly different approaches in its standards on confined spaces (general industry and shipyard employment). Further, the term "confined space" varies from standard to standard.

The approach proposed here can be visualized as two onions. One represents the hazardous conditions posed by a workspace and the work to be performed in it. The other onion represents the actions needed to render the space safe before and during the work. The most hazardous of spaces and the preventive actions that they require would be represented by each onion containing all of its layers. Said another way, a life-threatening situation demands the highest level of response.

On the other hand, a situation posing minimal hazards needs minimal response. The average situation is likely to be far less serious than the worst case, yet deserving more attention than minimal response.

The challenge is to create a systematic, defensible process for removing as many layers from the onions as justified by a particular situation. This approach conserves resources, so that they can be utilized where most appropriate. The caveat with this concept is to ensure that mismatch does not occur. Underaddressing a hazardous situation could lead to a fatal accident. Overreacting to a nonhazardous situation also could lead to injury.

Another issue in starting a real-world program is the number of workspaces in an industrial complex. Preceding discussion provides an approach to consider. Spaces that are entered frequently or are more likely to be encountered should be addressed first. This enables initial effort to produce greatest results. Spaces identified in the initial list can be assigned to categories of decreasing concern: critical, serious, moderate, and low. This assessment should involve consideration of previous and present status of the workspace, frequency of entry, and the nature of work to be performed.

Human Resources

The resource that is likely to be the limiting factor in facilitating this process is the human resource. This situation can be exacerbated by the regulatory philosophy of the jurisdiction in which the organization is located. Table B.1 highlights an important difference in utilization of human resources between regulators.

One regulatory approach designates the employer as being solely responsible for all aspects of the safety and health program in the organization. This includes the confined space program as a particular example. Within the confines of this approach, the regulator acts as the regulator and watchdog. This approach puts considerable stress on technical resources available within the organization. This ignores the fact that operators of equipment (workers) and entrants (mostly workers) are likely to know far more about the operation and its idiosyncrasies than supervisors and higher level management. Yet, this approach deliberately excludes workers from participation in the process of program development and implementation.

A way to approach this challenge is to create a team composed of safety and industrial hygiene practitioners and engineering, operations, and maintenance personnel. The team could then undertake the implementation, utilizing the strengths of each of the participants. This option offers the advantages of involvement of technical personnel in areas of endeavor beyond normal interests and responsibilities. Involving these groups provides the potential for major benefits, since they are involved in the design, operation, and maintenance of equipment and structures that ultimately become workspaces. To some in these groups, entry into these spaces for maintenance, repair, and modification may never previously have been a consideration. Involvement in the process can produce positive benefits in future endeavors through awareness raising.

Involving designers in this process can provide additional benefits: improved accessibility in future structures and equipment and reduced need for entry. Obtaining involvement in this process requires visible commitment from senior management to hazard recognition and assessment during the design phase. This is one reason that visible commitment from senior management to all elements of the confined space program is vital to its success.

The disadvantage of the approach discussed here is that the vast majority of the people who work in confined spaces (i.e., workers) have no input into the process. This can lead to oversight, since the workforce often are more knowledgeable about the operation than supervisors and management. Certainly, they bring a unique perspective to the discussion.

These concerns were addressed in the approach utilized by regulators in other jurisdictions, the so-called internal responsibility system. The internal responsibility system mandates creation of the Health and Safety Committee. The Health and Safety Committee acts as a forum for raising and addressing concerns regarding the workplace. However, the Committee has a larger mandate. The employer is required to develop programs in consultation with the Health and Safety Committee. In its most superficial form, this process is merely a formality involving adoption of programs submitted for approval by management or labor.

Downsizing of supervisory and management ranks and reengineering of organizations is forcing more effective utilization of the talent resident within these Committees. This utilization can include extensive partnership in the development of programs, such as one on confined spaces. The fullest possible participation of all members can provide other benefits, including the satisfaction that involvement and cooperation brings.

In one example of active involvement, members were provided with a copy of the regulation on confined spaces and a hazard identification guide sheet. They were asked to identify potential confined spaces within an industrial complex. Their approach to this task included physical inspection, but additionally, interviews with operating and maintenance personnel. The Committee then invited input from a "Qualified Person" to facilitate further action. This action included assistance in developing entry and emergency response procedures, selection of testing instruments, protocols for air monitoring, and ventilation concepts.

Where workplace Health and Safety Committees exist, the expertise, talent, and viewpoint represented in these individuals is a resource that deserves to be utilized.

IDENTIFICATION

The first step in implementing a confined space program should be to develop an accurate, up-to-date inventory of all confined spaces. This should include location, characteristics, contents, hazardous conditions, and so on. Each workspace that is examined should generate an electronic or paper-based file. This can be used to store information for use during future enquiries. Workspaces involved in this enquiry may be shown to be confined spaces or not to be confined spaces. Workspaces that are shown to be confined spaces may or may not entered by employees of the organization. Workspaces not entered by employees of the organization may be entered by a contractor. The employer is required to determine confined spaces on the property and to inform employees and contractors about their existence.

The following tables present a form that can be used in identifying confined spaces. The form is broken into pieces to permit discussion about individual sections. The first part of the form begins by establishing the location and (unique) identity of the workspace under examination. **Date** establishes when the form was created. **Assessor** and **Qualification** establish who performed the assessment and what qualifies that person to do so. Many jurisdictions specify that a Qualified Person must perform these assessments. Some jurisdictions specify professional designations that establish the capability of an individual to perform this work, while others do not. Establishing formal qualifications of the individual who performs this work is extremely important, since the outcomes from these decisions can have life or death implications.

Basis establishes the approach taken by the Qualified Person in establishing the decision about the status of the workspace. The most comprehensive basis for establishing the status of a workplace is a personal inspection and examination of plans or drawings. This establishes firsthand the geometric configuration of the space, potential hazards, and its relationship to its surroundings. A less comprehensive approach would involve an examination of drawings or plans of the workspace. Lack of presence in the space by the Qualified Person represents potential loss of information to the process. Personal inspection is considered an extremely important part of strategy for hazard assessment and control in some standards (NFPA 1993, OSHA 1994). Plans or drawings of the space and its surroundings should become part of any permanent record. This has obvious importance for follow-up activities.

TABLE 8.1A
Identification Information

ABC Company
Confined Space — Identification and Hazard Assessment

1. Descriptive Information

Department:

Location:

Building/Shop:

Equipment/Space:

Part:

Date:	**Assessor:**
	Qualification:
	Basis:
	Date:
Previous Assmt:	**Assessor:**
	Qualification:

TABLE 8.1B

2. Assessment Information

Description:

Adjacent spaces:

Access/egress:

Function/use:

Contents:

Process:

Equipment:

Previous assessment establishes that a previous assessment had been performed, when, and by whom. Some jurisdictions require periodic assessment of workspaces deemed to be confined spaces (Canada Gazette 1992).

Table 8.1b provides the second section of the confined space identification and hazard assessment form. Assessment information forms the basis for determining whether a workspace is or is not a confined space.

TABLE 8.1C

3. Confined Space Identification

Geometry/design

Space can cause personal confinement:	**Yes** No **Possible**
Space has unstable interior condition:	**Yes** No **Possible**
Space contains flowable solids or liquids:	**Yes** No **Possible**
Release of energy into the space can occur:	**Yes** No **Possible**
Space can cause atmospheric confinement:	**Yes** No **Possible**

Response in the **highlighted** areas for one or more conditions indicates a potential or actual confined space for which a detailed hazard assessment is required.

Administrative aspects

Entry occurs or planned by Company personnel:	**Yes** No **Possible**
Entry occurs or planned by Contractor personnel:	**Yes** No **Possible**

Entry by a contractor into a confined space owned and controlled by the Company necessitates formal written contractual agreement of responsibilities.

Description is used to provide a physical description of the space. This could include its dimensions, location, and position relative to grade. Ideally, the identification and hazard assessment form should contain a diagram showing the space and other spaces adjacent to it. **Adjacent spaces** indicate spaces that are located beside, above, below, and across diagonals from the space. This concept is especially important in ship environments, where the adjacent space and its contents could pose a greater hazard than the space to be entered (NFPA 1993, OSHA 1994). However, this concept has universal applicability.

Access/egress specifies the points of entry and egress from the space. This section should list all possible points of access and egress. These may be apparent during the investigation to identify confined spaces, but not during later work. **Function/use** describes the function or use of the space for which this identification applies. A space could be used for many applications. The level of hazard associated with conditions could vary considerably from application to application. A single identification and hazard assessment would not be appropriate. **Contents** reflect the contents of the space at the time of the identification and hazard assessment. Contents of historical importance also should be listed. Previous contents are considered to be extremely important in some standards (NFPA 1993, OSHA 1994). Previous contents are a concern where residues either could react chemically with the newer contents or where these residues could cause exposure during subsequent entry. **Process** refers to a chemical, physical, or biological process that normally occurs in the space. These processes can be the source of residual liquids and solids, and gases and vapors. Continuance of the process during occupancy could create hazardous conditions. **Equipment** refers to equipment contained in the space. Operation of this equipment could create hazardous conditions to occupants.

Table 8.1c provides the third part of the identification and hazard assessment form. This section contains the framework on which a workspace is assessed. That framework could originate from a legal statute or consensus standard, depending on the jurisdiction. Decision making expressed in Table 8.1c draws on information provided in Table 8.1b.

Answering "yes" or "possible" to any of the statements in the section **Geometry/Design** would lead to classification of this workspace as a confined space. (The electronic version of this form has spaces under the headings into which "yes," "no," and "possible" can be typed.) The word **possible** is included here to accommodate situations where the evaluator is unsure about the answer. This means that every option must have an answer, even if it expresses uncertainty. In this case the evaluator must proceed with a **yes** or **possible** and cannot stop the process unless certain that the answer is **no**. This sets the logic to err on the side of caution.

The function of **Administrative aspects** is to enter into the record employers whose workers enter the space. Details pertaining to entry should be recorded under the appropriate statement. This also could include a statement that asserts the negative. Entry by a contractor into a confined space owned by an employer brings obligations under common law and obligations stated in statutes, examples being the OSHA standards (OSHA 1993, 1994).

The preceding example used the definition for a confined space utilized in this publication. The definition used in this publication considers a confined space to be a workspace that satisfies or could satisfy one or more of the following conditions:

- Personal confinement
- Unstable interior condition
- Flowable solid materials or residual liquids or sludges
- Release of energy through uncontrolled or unpredictable motion or action of equipment
- Atmospheric confinement
  - Toxic substances
  - Oxygen deficiency
  - Oxygen enrichment
  - Flammable or combustible atmosphere
- Chemicle, physical, biological, ergonomic, mechanical, process, and safety hazards

In this definition, the meaning of "toxic substances," "oxygen deficiency," "oxygen enrichment," and "flammable or combustible atmosphere" requires further specification according to standards that exist in a particular jurisdiction.

Based on the definition presented here, a particular workspace may or may not be a "confined space." Designation of a particular workspace as a confined space sets into motion the requirements of a particular jurisdiction.

As demonstrated in Appendix A, the definition of confined space differs from jurisdiction to jurisdiction and from standard to standard. Using the definition contained in the Canadian Occupational Safety and Health Regulations (federally regulated companies), the section on Geometry/Design would read:

Geometry/design

Space enclosed or partially enclosed:	**Yes** No **Possible**
and	
Access or egress restricted:	**Yes** No **Possible**
and	
Designed/intended for human occupancy except for the purpose of performing work:	Yes **No** **Possible**

Response in the **highlighted** areas for all three conditions (**yes** or **possible**, **yes** or **possible,** and **no** or **possible**) indicates a potential or actual confined space for which a detailed hazard assessment is required.

This section also can be modified to accommodate the definition for confined space used in the OSHA standard for general industry, as follows:

Geometry/design

Space has adequate size and configuration for employee entry:	**Yes** No **Possible**
Space has limited means of access and egress:	**Yes** No **Possible**
Space is not intended for continuous occupancy:	**Yes** No **Possible**

Response in the **highlighted** areas for all three conditions (**yes** or **possible**) indicates a potential or actual confined space for which a detailed hazard assessment is required.

The definition used in the OSHA standard is more complex than those used in previous examples, since OSHA requires further subdivision into permit-required confined space and non-permit-required confined space. These concepts will be addressed in the section on hazard identification and assessment.

The outcome of the process of identification should be a series of files (paper-based or electronic) on the workspaces that were examined. This compilation must be recognized for its static nature and the population of confined spaces for its dynamics. An inventory is a static entity within a dynamic organizational infrastructure. Dynamic entities change with time. This change must be acknowledged and accommodated at the outset in any inventory process. An inventory that remains frozen in time can produce a false sense of security and, ultimately, cynicism, when its limitations become apparent. Periodic inspection to identify new or previously unidentified or modified confined spaces is essential. An annual inspection is suggested until experience dictates otherwise. This process of reevaluation is a requirement in some jurisdictions (Canada Gazette 1992).

Labeling

A useful outcome from the inventory process is an opportunity to identify all of the confined spaces through some physical means. A label in conjunction with training can provide considerable benefit, since all employees then become familiar with the concept of confined spaces and their location in the physical infrastructure. Employees then can relate the term "confined space" to equipment and structures at their work location. This approach has distinct advantages, since the training becomes localized to familiar surroundings.

The downside to the process is that over time the label could disappear into a landscape filled with other labels. In this situation, a new label rapidly fades into the background. In addition, organizations that have many confined spaces potentially would have great difficulty in labeling all of them. Another reality for which constant awareness is required is that the population of confined spaces is dynamic and could change. Overdependence on labels means that confined spaces could be overlooked.

Where the decision to label is made, entry portals to the confined space should be clearly identified by permanent labeling or posting. A unique identifier, such as a name or code, is essential to eliminate ambiguity about the space or its location. This could become critical when entry into more than one space is occurring simultaneously. The label also could contain other information, such as known hazards and the authority controlling entry. Precautionary wording on signs may be regulated by statute.

HAZARD IDENTIFICATION AND ASSESSMENT

General Discussion

A number of issues deserve consideration before the decision to enter a particular workspace is made. These provide the purpose and necessity for preentry planning: to explore issues and alternatives prior to actual entry. Pursuing these considerations can expose unexpected hazards and deficiencies, as well as lead to unexpected opportunities. The more that can be learned about a particular workspace prior to the decision to enter, the greater the potential for management and control of the expected and unexpected hazards and for minimizing the risk of entry and work.

The key to successful preentry planning is to structure enquiry to generate relevant information in an accessible format. The urge to reach the decision point in a seemingly unproductive line of inquiry represents a major temptation that must be resisted. Certainly, the lack of discipline in evaluating conditions in confined spaces was a major factor in the fatal accidents documented by OSHA (OSHA 1982a, 1985, 1988).

TABLE 8.2
Possible Hazardous Conditions in Confined Spaces

- Hot work
- Atmospheric hazards
  - Oxygen deficiency
  - Oxygen enrichment
  - Biochemical/chemical
  - Fire/explosion
- Biological hazard
- Ingestion/skin contact hazard
- Physical agents
  - Noise/vibration
  - Heat/cold stress
  - Non/ionizing radiation
  - Laser
- Personal confinement
- Mechanical hazard
- Hydraulic hazard
- Pneumatic hazard
- Process hazard
- Safety hazards
  - Structural
  - Engulfment/immersion
  - Entanglement
  - Electrical
  - Fall
  - Slip/trip
  - Visibility/light level
  - Explosive/implosive
  - Hot/cold surfaces

Note: Each entry in Table 8.2 can be expanded to describe a much larger set of circumstances. These guide words can be used by the Qualified Person to describe the situation in a confined space at a particular point in time.

A structured approach to this problem starts with hazard identification. Hazard identification focuses on conditions in the workspace that satisfy or could satisfy or could contribute to the rubrics in the definition of confined space. These determine why a particular workspace is a confined space. Determining the extent of concern regarding these conditions is the role of hazard assessment. Hazard management focuses on work procedures and processes, and control of hazardous conditions associated with them. Hazard assessment and hazard management are cooperative and iterative processes.

The starting point in the process of hazard identification is to recognize the full scope of hazardous conditions that could be present in a confined space. These expand outward from the definition of the confined space. Expansion of these concerns is analogous to the decompression of computer files. In the case of confined spaces, the Qualified Person acts in the role of the decompression program. The definition used here can be expanded, as shown in Table 8.2.

The most complex and difficult element within the confined space program is hazard assessment. Hazard assessment focuses on predicting the severity of hazardous conditions that exist or may develop, and judging their acceptability. One of the most difficult tasks within hazard assessment is determining the potential impact of a particular situation. Some aspects of hazard assessment can be instrumentally based. For example, instrumental monitoring can generate a profile of

TABLE 8.3
Possible Benchmarks for Atmospheric Hazards

Oxygen

- Normal atmospheric composition: 20.9%
- Normal atmospheric pressure, sea level, dry air: 159 mmHg
- Oxygen deficiency (90% hemoglobin saturation): 110 mmHg
- Oxygen deficiency (regulatory limits): 19.5%, 18%
- Oxygen enrichment (regulatory limits): 22%, 23.5%

Toxic substances

- Threshold Limit Value
- 10 × (Threshold Limit Value) or 1000 ppm, whichever is less
- IDLH (immediately dangerous to life and health) or 10,000 ppm, whichever is less

Flammable/combustible substances

- Percentage of lower flammable limit
- Lower flammable limit

physical, chemical, and biological conditions. This could occur at the time of entry or during a prolonged period. The latter offers the potential for establishing the stability (or instability) of conditions. Instrument-based assessment offers the potential for ranking the anticipated level of hazard against some criterion. The benchmarks in Table 8.3 may provide suitable criteria for ranking atmospheric hazards. Values presented in this table are familiar benchmarks in occupational health and safety. As such, they are completely applicable to situations involving confined spaces and confined atmospheres. However, most of the hazardous conditions that may occur are not readily amenable to instrumental measurement. Instead, assessing their severity draws heavily on the experience and knowledge base of the assessor; hence the real meaning of the term "Qualified Person."

Every confined space deserves its "day in court." That is, prior to entry, consideration must be given to the actual or potential presence or development of hazardous conditions, as well as work plans and work processes that could interface with that workspace at a particular moment. Hazard assessment requires an examination of each of the elements that define the confined space in as much detail as is possible to document. This assessment could be appropriate each time that entry will occur, depending on factors to be discussed.

Hazard assessment also can provide a qualitative assessment of the level of concern to attach to a particular situation at a particular moment. Hazard assessment can establish differences in the level of concern between one confined space and another. This sense of perspective can assist in specifying the level of response. An important concept in the process of hazard assessment is the breadth of hazardous conditions encompassed within each of the rubrics. Like all ranges, this has two end points, one representing minimal consequence or concern and the other a life threatening situation. Real-world situations rarely exist at either end of the spectrum.

A situation at the lesser extreme may barely satisfy the criteria for inclusion under one of the rubrics. For example, the geometry of the workspace may possess the ability to confine contaminants. Inverted semicylinders or steeples, currently fashionable as roof structures on walkways and other types of shelters, are simple examples. Access into or egress from these workspaces is unrestricted in the normal sense, except that a ladder may be required. Under appropriate conditions, considerable overexposure could occur due to accumulation of contaminants.

TABLE 8.4
Assessment of Hazardous Conditions — Burning House

	Real or Potential Consequence				
Hazardous Condition	**Low**		**Moderate**		**High**
• Hot work	NA				
• Atmospheric hazards					
Oxygen deficiency					x
Oxygen enrichment	NA				
Biochemical/chemical					x
Fire/explosion					x
• Biological hazard	NA				
• Ingestion/skin contact			x		
• Physical agents					
Noise/vibration		x			
Heat/cold stress					x
Nonionizing/ionizing radiation			x		
Laser	NA				
• Personal confinement					x
• Mechanical hazard	NA				
• Hydraulic hazard	NA				
• Pneumatic hazard	NA				
• Process hazard	NA				
• Safety hazards					
Structural					x
Engulfment/immersion	NA				
Entanglement				x	
Electrical		x			
Fall				x	
Slip/trip					x
Visibility/light level					x
Explosive/implosive	NA				
Hot/cold surfaces					x

Note: In this table, the meaning of "toxic substance," "oxygen deficiency," "oxygen enrichment," and "flammable or combustible atmosphere" requires further specification according to standards that exist in a particular jurisdiction. NA means not applicable.

The enclosed space whose atmosphere is acutely toxic and whose interior condition is unknown and possibly unstable represents the opposite extreme. A burning residential building that fire fighters enter to search for possible victims is a primary example of such a situation. Victims and fire fighters alike face highly uncertain conditions presenting extreme risk of injury, intoxication, and asphyxiation.

Conditions in the confined space may satisfy one or more of the rubrics in the definition. The level of concern attached to each could differ considerably. Table 8.4 illustrates a way to express a level of concern for each hazardous condition considered in the hazard assessment, using the burning residential building as an example.

Table 8.4 provides a powerful means of indicating the level of concern attached to a particular confined space at a particular moment in its historical cycle.

The historical cycle of a confined space is an important consideration in hazard assessment and control. The key segments in the historical cycle of every confined space include:

- Undisturbed space
- Preentry preparation
- Prework inspection
- Work activity (emergency response)

The historical cycle can be short or long, repetitive or nonrepetitive. **Undisturbed space** represents the status quo established between the closure following one entry and the start of preparations for the next. This likely is the longest period in the historical cycle. **Preentry preparations** are the actions taken to render the space safe for entry and work. **Prework inspection** is practiced in some workplaces. During this period, someone, perhaps a process operator or shipyard competent person, examines the space to ensure that it is safe prior to entry and start of work by others. **Work activity** is the individual work task to be undertaken by entrants. **Emergency response** refers to events that can still happen, despite measures taken in the previous segments, for which action is required.

Fatal hazardous conditions can exist or develop during any of these segments. This was amply illustrated in the reports produced by OSHA (1982a, 1985, 1988). Performing the hazard assessment for each segment considerably simplifies the process because the focus changes. To illustrate, the level of concern about a particular hazardous condition could change dramatically following preentry preparation and during, or as a result of, an activity. For this reason, assessing a level of concern to a hazardous condition for all time based only on an appraisal of pre-opening conditions would be inappropriate. The only way to ensure that the level of concern and consequently the level of response for any segment are appropriate is to assess conditions prior to each. This need not occur literally in that sequence since experience and knowledge about processes and conditions can be applied to particular circumstances. The strategy used for decision making by the Qualified Person must take this concern into account. In some jurisdictions, an appraisal is required prior to each entry for this reason (Canada Gazette 1992).

Stopping the process of hazard assessment following assessment of the undisturbed space is a potentially useful strategic option. This could occur one step beyond designation of a workspace as a confined space. This approach provides a mechanism for allocating scarce technical resources by assigning a level of concern to each space with minimum effort. This also enables designation of those spaces needing greatest effort during preentry planning and preparation.

Assessing potential or actual hazardous conditions for each confined space/confined atmosphere is a time-consuming exercise. Yet, comprehensive assessment of all situations is the only way to define the magnitude of the problem facing an organization. Obviously, there is serious need for simplification, so that resources can be apportioned appropriately. Several strategies for simplification are found in technical standards and regulatory statutes. These are discussed in a later section.

Estimating levels of contamination by preentry conditions, as well as proposed work activities, *a priori* requires considerable knowledge and experience. Procedural and process controls must strive to maintain levels within the most stringent of these criteria. This goal should be achievable in almost all situations. Unpredictable situations, such as entry into burning buildings, should remain among the few circumstances for which assessment and prediction of conditions prior to entry are not readily possible.

The scope of work should provide enough detail about the activity to alert the Qualified Person about potential hazards. The task for the decision maker is to decide whether potential or actual hazards accompanying this activity are acceptable. In order to make this assessment, the decision maker must anticipate hazards and episodes possibly unforeseen to the proponent that could arise during the project. The basis for making this decision is integration of knowledge about the process and its hazards and their interaction with the geometry of a particular workspace.

The next step is to focus on possible injury and damage that could arise during episodic situations, and the adequacy of contingencies proposed to address them. The definition of confined space/confined atmosphere (or other appropriate definition) provides a starting point for considering the concepts involved in control of conditions.

The last part in the process is the critical one. The Qualified Person is faced with a decision between the following alternatives: to permit entry and the activity or not to permit entry and the activity. That is, are the risks associated with entry and work acceptable and are the hazards controlled sufficiently by the actions proposed?

The choice between the alternatives directly confronts the question about whether the work can be conducted in a manner that is considered sufficiently "safe." That is, is the risk perceived in the activity acceptable to the decision maker? At some point the potential for overexposure/injury may exceed the protection offered by control measures. The challenge for the Qualified Person is to determine the point at which the margin of safety is too small to accept. Or put another way, how safe is safe enough?

Lowrance (1976) explored this subject and pointed out that safety involves both objective measurement of risk and subjective judgment (professional judgment) about the acceptability of the risk compared to other influencing factors. Some of the influencing factors include laws, group and personal attitudes, and the mores of society. Lowrance concluded that "a thing is safe if its risks are acceptable." In this judgment, acceptable must be qualified against the best available information.

People are accustomed to taking risks as part of daily life. What is unfamiliar to people is the dual nature of risk (Findlay and Kuhlman 1980). People only expect risk to involve some element for gain. Risks, however, can be divided into two broad categories: speculative and nonspeculative. Speculative risks involve the possibility of either gain or loss. Decisions involving speculative risks may be correct, leading to a desired outcome. They may also be incorrect, resulting in a loss. Nonspeculative (pure) risks involve the possibility of either loss or no loss; there is no chance for gain. There is no incentive to take this type of risk. Yet, this is the type of risk associated with accidents involving confined spaces.

Table 8.5 illustrates a way of visualizing these concepts for a hypothetical situation. This assessment could be applied at any point in the process. For each point of judgment, there is a corresponding zone of uncertainty. The magnitude of the uncertainty compared to the size of the transition zone is the critical consideration.

Determining uncertainty in a quantitative manner, while highly desirable, is impossible for everything except measurable quantities, such as airborne concentrations. "Hazardous condition" could mean the TLV (Threshold Limit Value) or even the IDLH (immediately dangerous to life or health) concentration. This assignment reflects the judgment of the Qualified Person. In all other cases, the judgment point and the uncertainty represent "judgment call." To be quantitative would require a numerical expression of the gap between the safe and the hazardous condition and the uncertainty. This type of information does not exist. These assignments, therefore, always will represent the best guess of the Qualified Person.

The most desirable situation is a small uncertainty and a large transition zone between the judgment point and the hazardous condition. This case represents high confidence in the assignment of a situation toward the safe condition. In some situations, the gap between the judgment point and the hazardous condition may be very small. Assessment of even a small uncertainty in confidence can bridge the gap between the two positions. Similarly, a large uncertainty and a large transition zone also present the same problem. Situations such as these, that rest on the borderline between safe and potentially hazardous conduct of work, require careful consideration. Bridging the gap from the judgment point to the hazardous condition would require only a small change in circumstances. Reducing the level of concern and uncertainty reflected in a particular assessment by changing the approach to a problem increases the margin of safety.

Following the decision that entry and work can proceed safely, the next step is to devise procedures and to arrange for emergency response, as needed. Audit of the work process provides

TABLE 8.5
Assessment of Hazardous Conditions

Hazardous Condition	Safe	Hazardous
• Hot work	NA	
• Atmospheric hazards		
Oxygen deficiency	. . . . x	
Oxygen enrichment	NA	
Biochemical/chemical	. . . . x	
Fire/explosion		. . . . x . . . .
• Biological hazard	NA	
• Ingestion/skin contact	. x .	
• Physical agents		
Noise/vibration	NA	
Heat/cold stress	. . x . .	
Nonionizing/ionizing radiation	NA	
Laser	NA	
• Personal confinement	. . x . .	
• Mechanical hazard	NA	
• Hydraulic hazard	NA	
• Pneumatic hazard	NA	
• Process hazard	NA	
• Safety hazards		
Structural		. . . . x . . . .
Engulfment/immersion	NA	
Entanglement		. . . . . . x . . . . . .
Electrical	NA	
Fall		. . . . x . . . .
Slip/trip		. . . . . x . . . . .
Visibility/light level		. x .
Explosive/implosive	NA	
Hot/cold surfaces		. x .

Note: In this table, the meaning of "toxic substance," "oxygen deficiency," "oxygen enrichment," and "flammable or combustible atmosphere" requires further specification according to standards that exist in a particular jurisdiction. NA means not applicable. The dots ... represent the uncertainty assigned to a particular value of "x," the perceived or measured level of concern.

information about the effectiveness of control measures. This provides valuable input for future decision making.

Safety can best be assured through control. If control of conditions can be assured, the decision is not difficult to make. If control of conditions cannot be assured, the decision then becomes difficult. The less the level of perceived control, the greater becomes the need to provide for contingencies. The only other alternative is to prohibit the entry. Safety is achieved through control — confidence in knowledge about what could happen, assurance that steps have been taken to minimize the possibility for this, and assurance that contingencies exist to address what still could happen.

The Undisturbed Space

The first part of hazard assessment is concerned with conditions in the space prior to opening and preparation for entry.

This period begins with closure following previous work and ends when preparations for the next entry begin. This segment in the historical cycle usually is long compared to entry and work. During this period, the confined space may be undisturbed and possibly uninspected. Unless conditions in the space are carefully monitored and controlled, as in a process unit, for example, the Qualified Person may have little firm information about conditions in the space. The period during which the space remains undisturbed thus provides the least information and creates the greatest uncertainty.

Research presented in the first chapter indicates that the conditions that develop during the preopening period can be especially hazardous (OSHA 1985). Every confined space has a story to tell. That story, for example, may include one or more fatal accidents. Without investigation, that story may remain unknown. As much of that story as is possible to assemble must become known prior to preparing for entry in order for successful hazard management to occur.

Most confined spaces lie within the bounds of operating organizations. As such, they "belong" to some entity within the organization. This means that someone has responsibility for the space.

Active operations offer opportunity for discussion about on-going activity, materials currently and historically handled, availability of operating records and documentation, and the presence of operational utilities, alarms, etc. Because entry into confined spaces at operating locations can require extensive coordination with operations and maintenance personnel, this liaison is extremely important.

Personnel who operate the equipment or structure into and within which entry and work will occur are invaluable resources about its structure and interior conditions. Maintenance personnel, as well, could have experience in performing work activities in the space. The experience and knowledge and perspective of these individuals are invaluable to the process of hazard assessment. They can provide input regarding historical aspects about what was done and why, and how successfully. How successfully is a relative term, especially in organizations where industrial hygiene surveillance was not performed during this work.

Some confined spaces are located in abandoned or inoperative sites. These operations may have been shut down through war or strife, natural disaster, process obsolescence, insolvency, or even criminal abandonment. Information about these sites may be completely lacking. In operating organizations, downsizing through layoffs and early retirements similarly may have created a break in the continuity of information. With any of these circumstances, loss of the expertise and knowledge resident in former employees and operating and maintenance records considerably complicates hazard assessment. The starting point for hazard assessment in these situations is a thorough historical review of the site and its confined spaces.

Information regarding operations, materials handled, and maintenance may be nonexistent for abandoned or inactive sites. Hazardous conditions that could be exacerbated by the passage of time or discontinuity of knowledge include (among others):

- Accumulation or depletion of atmospheric contaminants due to lack of disturbance
- Loss of integrity of the internal structures due to incompatibility between contents and materials of construction
- Unknown or unexpected residual contents
- Bridging of flowable solids
- Flammable or toxic atmosphere
- Pressurized atmosphere

Any of these conditions could cause an accident during initial opening of the equipment or structure. The focus of hazard assessment under these conditions is to learn as much as possible about the status quo before disturbing it. This could have a profound influence on strategy used for assessing and opening the equipment or structure in order to prepare for entry.

Hazard assessment for the preentry period should begin with a review to determine what is known about the site, and just as importantly, what is not known. A review of documents maintained by regulatory or other public agencies at various levels may assist in providing this information. Insufficient information will necessitate use of elevated levels of protective equipment, as well as intensive evaluation of both physical and chemical hazards.

Appendix F provides questions for guiding the Qualified Person through this investigation. This investigation will not be pursued in detail here.

Identification of materials that were handled, stored, used, produced, or consumed in the space defines the need for information about their properties. This information should be available from material safety data sheets (MSDSs) or other information resources. Materials handled in the vicinity outside the space represent an additional concern. These could cause exposure during the work process or could react with substances removed from the space. A subsidiary concern with chemical products is the condition of storage, since this could destabilize the product or lead to formation of new substances through reaction with moisture, oxygen in the air, or substances stored in proximity.

There is the possibility that the identity of contents in the space is unknown. This possibility could delay any consideration about opening the space to prepare for entry until procedures have been devised for determining the identity, properties, and quantity of material. Abandoned, inactive, or uncharacterized sites deserve special consideration. This could include review of hazardous material inventories filed with government agencies.

A subsidiary concern with underground tanks, piping, and valves is leakage into the surrounding soil. Vapor emitted from this soil, once exposed, could create an exposure and possibly a flammability hazard. An equivalent concern with porous underground structures located near former landfills that handled garbage is buildup of methane or solvent vapors. These concerns highlight an important concept within hazard assessment: the surroundings. Opening the space to prepare for entry cannot occur until these subsidiary concerns have been addressed.

Confined spaces usually are considered only in terms of the space enclosed by some defining geometry. This space is viewed in isolation from its surroundings. Analysis in the first chapter of fatal accidents documented by OSHA (1982a, 1985, 1988) indicates that this approach can lead to a false sense of security. In reality, a confined space is inseparably connected to its surroundings, just as visual inspection would indicate. What is misperceived is that conditions or activities that occur in the surroundings do not impact on the confined space and vice versa. This observation forms the important link between the confined space inside which people work and the hazardous atmosphere confined in equipment or structures outside which people work. The consequences of misjudgment are the same.

NFPA 306 used by Marine Chemists in the practice of hazard assessment on ships and the OSHA standard on confined spaces in shipyard employment requires evaluation of conditions across the boundaries — planar, curved, and angular — that enclose the space of immediate interest (NFPA 1993, OSHA 1994). Knowledge gained from tragic experience indicates that concern must extend in all directions across the boundaries of the space. While this concept is foreign to land-based practitioners, it has considerable merit and deserves to be included in the hazard assessment. This can draw attention to areas around the confined space where activities unrelated to the entry could have major influence on its safety.

The undisturbed confined space could include residual contents and a related or unrelated residual atmosphere. Information provided by MSDSs and other resources forms the basis for learning about the status quo contained in the undisturbed space. The status quo in these situations usually represents stability. That is, conditions have remained the same, or changed so gradually that nothing is likely to happen with passage of several more days, weeks, or even a month, and deciding how to evaluate potentially hazardous situations that could arise once the status quo is disturbed by opening the space.

Another essential input in the assessment process is a detailed plan for the confined space. Plans should provide information about all aspects of the space and the equipment that it contains. Plans should provide information about:

- Layout, including access points
- Piping, flanges, valves and vents
- Mechanical equipment and controls
- Hydraulic and pneumatic equipment and controls
- Electrical circuits
- Structural detail
- Materials of construction

The information contained in these drawings is a vital resource for understanding about the space. This input may contain information that has been lost over the passage of time and changing of personnel. In addition, information contained in plans could form the basis for planning entry for initial assessment of conditions, ventilation strategy, strategy for performing the work, and so on.

The current physical condition of the worksite and the spaces of interest are major concerns. An evaluation of the integrity of the space for entry, as well as surrounding structures, should be made prior to entry operations, especially at inactive or abandoned worksites. In cases where structural integrity is the least bit questionable, review of all structures by a qualified Professional Engineer is highly recommended. Under no circumstances should spaces in or around structures of questionable integrity be entered. This review may require three-dimensional imaging to determine internal structure and configuration and materials of construction.

The last aspect of hazard assessment of the undisturbed space is to summarize the information obtained in the process. This summary should clearly indicate conditions that may exist in the space at the time that preentry preparations begin. This summary should be usable by anyone planning preentry preparations. Table 8.6 provides a format for identifying and characterizing potentially hazardous conditions found in confined spaces. Table 8.6 and following tables continue the numerical sequence begun in other tables in this chapter.

Table 8.6 describes conditions that an observer transported into the undisturbed space likely would encounter. These form the basis for preentry preparation. Subsequent tables continue the process through the sequence of steps.

The following sections will elaborate on decisions recorded in Table 8.6.

Oxygen deficiency

Oxygen deficiency can develop through biological action involving organisms resident on surfaces of the structure or in residual liquids or by chemical action involving residual contents or the material of construction of the structure.

Oxygen deficiency can develop in a stratified layer or throughout the space. The condition in a particular space depends on several factors: the process by which oxygen deficiency occurs, time, and turbulence. Atmospheric testing must occur throughout the atmosphere of the space to ensure that all possibilities are examined.

Oxygen enrichment

Oxygen enrichment can result from leakage from a cylinder of compressed oxygen or from chemical reactions that produce oxygen. Certain compounds such as peroxides, peracids, perhalogenates, and compounds used in explosives produce oxygen during chemical reaction or decomposition.

Oxygen enrichment can develop in a stratified layer or throughout the space. The condition in a particular space depends on several factors: the process by which oxygen enrichment occurs, time, and turbulence. Atmospheric testing must occur throughout the atmosphere of the space to ensure that all possibilities are examined.

TABLE 8.6
Hazard Assessment — Undisturbed Space

4. Hazard Assessment

Hazard assessment is required for all confined spaces into which entry is being considered or occurs. Hazard assessment is considered according to the following elements:

- Undisturbed space
- Preentry preparation
- Prework inspection
- Work procedures
- Emergency response

Some or all of the preceding elements may apply to a particular workspace. For elements judged inapplicable, the **no** response is completed in the following sections. Where **yes** or **possible** applies, this section should provide additional information for explanation and clarification.

4.1 Hazard Assessment — Undisturbed Space

The undisturbed space poses hazardous conditions requiring a hazard assessment.	**Yes** No **Possible**
Atmospheric hazards	
Oxygen deficiency	**Yes** No **Possible**
Oxygen enrichment	**Yes** No **Possible**
Biochemical/chemical	**Yes** No **Possible**
Fire/explosion	**Yes** No **Possible**
Biological hazard	**Yes** No **Possible**
Ingestion/skin contact hazard	**Yes** No **Possible**
Physical agents	
Noise/vibration	**Yes** No **Possible**
Heat/cold stress	**Yes** No **Possible**
Non-ionizing/ionizing radiation	**Yes** No **Possible**
Laser	**Yes** No **Possible**
Personal confinement	**Yes** No **Possible**
Mechanical hazard	**Yes** No **Possible**
Hydraulic hazard	**Yes** No **Possible**
Pneumatic hazard	**Yes** No **Possible**
Process hazard	**Yes** No **Possible**
Safety hazards	
Structural hazard	**Yes** No **Possible**
Engulfment/immersion	**Yes** No **Possible**
Entanglement	**Yes** No **Possible**
Electrical	**Yes** No **Possible**
Fall	**Yes** No **Possible**
Slip/trip	**Yes** No **Possible**
Visibility/light level	**Yes** No **Possible**
Explosive/implosive	**Yes** No **Possible**
Hot/cold surfaces	**Yes** No **Possible**

Hazardous conditions in the confined space may satisfy one or more of the rubrics in the definition. The following table summarizes concern during this activity in the confined space.

TABLE 8.6 (continued)
Hazard Assessment — Undisturbed Space

	Real or Potential Consequence		
	Low	Moderate	High
Hazardous Condition			
• Atmospheric hazards			
Oxygen deficiency			
Oxygen enrichment			
Biochemical/chemical			
Fire/explosion			
• Biological hazard			
• Ingestion/skin contact			
• Physical agents			
Noise/vibration			
Heat/cold stress			
Nonionizing/ionizing radiation			
Laser			
• Personal confinement			
• Mechanical hazard			
• Hydraulic hazard			
• Pneumatic hazard			
• Process hazard			
• Safety hazards			
Structural			
Engulfment/immersion			
Entanglement			
Electrical			
Fall			
Slip/trip			
Visibility/light level			
Explosive/implosive			
Hot/cold surfaces			

In this table, the meaning of "toxic substance," "oxygen deficiency," "oxygen enrichment," and "flammable or combustible atmosphere" requires further specification according to standards that exist in a particular jurisdiction. NA means not applicable.

Action Required

Biochemical/chemical hazard

Biochemical/chemical hazard indicates the presence or potential presence of some physical form of the chemical or biochemical substance — gas, vapor, mist, particulate — in air.

Contamination can develop in a stratified layer or throughout the space. The condition in a particular space depends on several factors: the process by which contamination occurs, time, and turbulence. Atmospheric testing must occur throughout the atmosphere of the space to ensure that all possibilities are examined. Grab-sampling methods provide discrete indication of conditions. A limited variety of person-portable real-time instruments containing chemical-specific sensors also exist in the marketplace.

Fire/explosion hazard

A fire/explosion hazard results from the presence or potential presence of an ignitable substance — gas, vapor, mist, or particulate. Formation of an ignitable mixture would require the presence of air or oxygen in the atmosphere of the space. Concentrations that can be ignited or exploded generally are much higher than those associated with the onset of toxicological effects.

Contamination can develop in a stratified layer or throughout the space. The condition in a particular space depends on several factors: the process by which contamination occurs, time, and turbulence. The possibility of an ignitable mixture indicates that atmospheric testing must occur throughout the atmosphere of the space to ensure that all possibilities are examined. The action of testing should not itself create an ignition hazard. This can occur during actions to open the space or in use of equipment that can accumulate and discharge static electricity.

Biological hazard

Biological hazards include airborne hazards, such as viable microorganisms, spores, cysts, or pollen and hazards resulting from aggressive animals. The latter could include rodents, insects, and spiders.

Ingestion/skin contact

Ingestion refers to the presence of a substance that can be hazardous in quantities ingestible during the course of normal work. Skin contact refers to substances that could cause a hazard following contact with the skin. Toxicological effects could include percutaneous absorption, irritation, allergic reaction, or destructive attack on the structure and integrity of the skin.

Noise/vibration

Noise and vibration can result from operation of equipment inside or adjacent to the space or from motion of contents. External sources of noise and vibration sometimes are used to promote flow and to prevent caking of flowable solids.

Heat/cold stress

Heat and cold stress can result from thermal conditions in the space that differ from conditions of comfort.

Nonionizing/ionizing radiation

Nonionizing radiation can include intense electrical and magnetic fields, and ultraviolet, visible, infrared, microwave, and radio frequency sources.

Potential sources of ionizing radiation include residues from naturally and artificially radioactive process materials, materials of construction that are radioactive, and sources that are present inside the space or whose beams are directed through the space. Radioactive beams that are directed through the space are used in level gauges or for irradiating the contents.

Laser

Equipment that produces laser energy may be present in the space.

Personal confinement

Personal confinement refers to the ability of the space, its portals, and interior geometry to confine an entrant. For some persons, claustrophobia could be a major consequence of confinement. Also, individuals whose dimensions are large could experience considerable difficulty in passing through small openings. This could create a considerable problem in situations where rapid evacuation is required. Complicated internal geometry can create considerable difficulty for extrication of an injured individual following an accident.

Mechanical hazard

Mechanical hazard results from unexpected or unpredictable operation of mechanical equipment located in the space. Mechanical equipment has moving parts such as blades or paddles, levers, screw conveyors, propellers or augurs, belts, chains, and so on. Unexpected motion of these parts can strike occupants and cause injury. Mechanical hazard includes discharge of stored energy through motion of parts.

Hydraulic hazard

Energy storage occurs in hydraulic systems. Release of energy stored in these systems can cause unexpected movement of machine parts.

Pneumatic hazard

Energy storage also occurs in pneumatic systems. Release of energy stored in these systems can cause unexpected movement of machine parts.

Process hazard

Process hazard refers to hazardous conditions that arise from chemical, physical, and biological processes that normally occur in the space.

Structural hazard

Structural hazard refers to hazardous conditions caused by instability of the internal structure of the space. Structural instability can result from incompatibility between contents and materials of construction. The hazard results from unexpected or unpredicted change in the status quo.

Engulfment/immersion

Engulfment/immersion considers the potential for residual contents — flowable solids or liquids — to engulf or immerse entrants. Engulfment can result from failure of caked materials on walls or of bridged flowable solids. Engulfment also can result from instability of walls and floors of trenches, excavations, shoring, and similar structures. These could form the boundary surfaces of the confined space.

Entanglement

Entanglement refers to the ability of structural features of the space and its interior equipment to entangle clothing or other equipment, such as lifelines.

Electrical hazard

Electrical hazard may result from exposure to or work on live electrical circuits in equipment that operates in the space.

Fall hazard

Fall hazard exists when vertical distances through which fall can occur could cause serious bodily injury. Fall hazard is defined in regulatory limits.

Slip/trip

Slip/trip refers to conditions that can cause slipping and tripping accidents. Conditions conducive to occurrence of this type of problem could include surfaces that are slippery due to residual contents or sloping angle. Tripping hazards could result from geometric features of the interior structure.

Visibility hazard/light level

Visibility hazard can result from presence of airborne particulates and mists or sprays. Light level refers to the absence of a source of illumination in the space.

Explosive/implosive hazard

Explosive/implosive hazard refers to physically explosive or implosive conditions caused by internal atmospheric pressure that differs from ambient. A small difference in pressure between surfaces can create considerable force. "Explosive" also refers to the presence of chemically explosive substances.

Hot/cold surfaces

Hot/cold surfaces refers to surface temperatures that differ markedly from ambient. These could cause burn injury, or if high enough, act as a source of ignition.

Action required

This section lists actions that are recommended for assessing, controlling, or more preferably, eliminating, hazards identified during this stage of the hazard assessment.

Actions that could be required include measurement of atmospheric conditions in the space during the undisturbed period. These measurements might establish, for example, that the oxygen level always remains normal. Establishing for the record that the oxygen level remains normal raises confidence that this would be the case at all times unless conditions changed dramatically. Proactivity of this kind increases certainty by establishing facts and reduces the uncertainty inherent in speculation about possibilities. The greater the extent of knowledge that can be established about conditions prior to preentry preparation, the less of the unknown that remains to be discovered and addressed. Investigating conditions throughout the undisturbed period represents one strategy for reducing the level of concern during hazard assessment. The value of this initiative in the overall process of hazard assessment and hazard management cannot be overstressed.

Hazardous conditions identified in the undisturbed space are addressed in subsequent segments of hazard assessment and hazard management. Most of the hazards identified in the undisturbed space likely will be controlled or eliminated during preentry preparation. It is possible that a hazardous condition may not be controllable. This could persist through remaining segments of the hazard assessment. Confinement caused by the geometry of the space is one such example. The ability of the space to cause confinement may not be ameliorated, regardless of actions that are taken.

Hazardous conditions that cannot be controlled or eliminated become the subject of study for determining emergency response and ultimately the decision about safe conduct of the proposed entry and work.

Preentry Preparation

Following evaluation of the characteristics of the undisturbed space and its contents, the next segment of hazard assessment considers preparation for entry and work. Entry preparation encompasses all actions taken to prepare the space for the prework inspection, or where this does not occur, the start of work. The start of preentry preparation is any action that changes the existing status quo; that is, the relation of the space and its contents to its surroundings.

Establishing contact between conditions in the interior of the space and those outside changes the status quo immediately. Establishing contact literally could occur through preparation to insert a small diameter probe to perform atmospheric testing. Entry preparation may be elementary or extensive. While the period occupied by preparation is brief relative to the overall cycle of quiescence–preparation–entry–work–post-entry, it can pose significant hazards.

Preentry preparation is the first instant of contact between a potentially hazardous condition and foreign equipment introduced for the purposes of testing, ventilating, and protecting entrants. The chemical, physical, and flammability characteristics of the atmosphere and residual contents in the undisturbed space, therefore, could influence selection of the type of equipment, mode of power, and materials of construction.

Entry preparations can involve a number of steps. These are described in varying levels of detail in different standards and guidance documents (NIOSH 1979, 1983; ANSI 1992, 1995; OSHA 1989, 1993, 1994; Canada Gazette 1992; API 1984, 1987, 1993; ASTM 1984; NSC 1985, 1986, 1987a, b; NFPA 1988, 1993, 1992; Standards Australia 1995; Jones and King 1991; Grund 1995). Actions performed during preparation for entry can include:

- Isolation of the space by closing and locking out valves and pumps in supply systems
- Isolating supply and drain lines by installing blanks, blinds, cups, and plugs
- Breaking open supply lines
- Draining residual contents
- Displacement of residual gases and vapors
- Removal of solids and sludges
- Flushing/washing to remove residues
- Purging with gases to dilute residual gases and vapors
- Flooding with water
- Filling with an inert gas to create an atmosphere that does not support combustion
- Opening access/egress points
- Ventilating with air
- Cutting, burning, and gouging of structural members
- Deactivation, de-energization and lockout of equipment and machinery
- Isolation of control systems to prevent reactivation

Each of these activities can create hazardous conditions. Proposed actions deserve careful attention to ensure that the preceding activities fulfill their objectives without creating additional hazards. As there are many types of confined spaces, preentry preparation could involve additional actions not mentioned here.

Many of the actions mentioned above involve interaction with systems that store, transport, and transform energy. Energy can occur in many forms: chemical, thermal, nuclear, electromagnetic, electrical, kinetic, and potential. Exposure to a burst of many of these forms of energy through an appropriate system of delivery could be fatal. While confined by the barrier surfaces of the space, this energy is controlled and isolated from contact with humans.

Entry into confined spaces often puts workers in close proximity to processes and mechanical, hydraulic, and pneumatic systems. Entry is made hazardous because there is no obvious way to determine the exact status of energy within a space and the equipment that contains it. This requires analysis. A second concern about energy in spaces is control of flow from one compartment to another; for example, from storage to transport to consumption. Human interaction with unexpected or uncontrolled flow of energy has caused many injuries and fatalities both inside and outside confined spaces.

Ensuring safety during entry and work in confined spaces therefore requires control over the generation, storage, transport, and consumption, as well as control of flow of energy in confined spaces. Control normally means removal of all forms of energy to achieve a "zero energy state." This would mean that nothing could happen. Mechanical equipment would be completely immobile. Achieving this end requires analysis of processes occurring in the space and flow of energy through equipment and control systems.

These concepts will be discussed further in Appendix G.

Table 8.7 provides a format for consideration of hazardous conditions potentially associated with preentry preparation. This table follows the format introduced in the previous section. The following sections explain the meaning of hazard assessment in the context of preentry preparation.

Hot work

Hot work is any type of activity or process that produces arcs, sparks, flames, heat, or other sources of ignition. Hot work can include welding, grinding, flame and mechanical cutting, gouging, soldering, brazing, burning, and drilling. Other sources of ignition include energetic chemical reactions, static electrical discharge, abrasive blasting and space heating, electrical short-circuits, tobacco smoking, lightning, engines and compressors, power tools, fixtures, electrical switches, and appliances that are not explosion-proof (ANSI 1995, NFPA 1993, NIOSH 1979, API 1993).

TABLE 8.7
Hazard Assessment — Preentry Preparation

4.2 Hazard Assessment — Preentry Preparation

Description: This section is used to provide a description of action(s) that occur during preentry preparation.

Entry preparation poses hazards requiring hazard assessment.	**Yes** No **Possible**

There are circumstances in which preentry preparation does not occur. In other situations, preentry preparation could be quite extensive. The details of a particular situation should be outlined here in sufficient detail to provide insight about actions that are required and why they are needed.

Hot work	**Yes** No **Possible**
Atmospheric hazards	
Oxygen deficiency	**Yes** No **Possible**
Oxygen enrichment	**Yes** No **Possible**
Biochemical/chemical	**Yes** No **Possible**
Fire/explosion	**Yes** No **Possible**
Biological hazard	**Yes** No **Possible**
Ingestion/skin contact hazard	**Yes** No **Possible**
Physical agents	
Noise/vibration	**Yes** No **Possible**
Heat/cold stress	**Yes** No **Possible**
Nonionizing/ionizing radiation	**Yes** No **Possible**
Laser	**Yes** No **Possible**
Personal confinement	**Yes** No **Possible**
Mechanical hazard	**Yes** No **Possible**
Hydraulic hazard	**Yes** No **Possible**
Pneumatic hazard	**Yes** No **Possible**
Process hazard	**Yes** No **Possible**
Safety hazards	
Structural hazard	**Yes** No **Possible**
Engulfment/immersion	**Yes** No **Possible**
Entanglement	**Yes** No **Possible**
Electrical	**Yes** No **Possible**
Fall	**Yes** No **Possible**
Slip/trip	**Yes** No **Possible**
Visibility/light level	**Yes** No **Possible**
Explosive/implosive	**Yes** No **Possible**
Hot/cold surfaces	**Yes** No **Possible**

Hazardous conditions in the confined space may satisfy one or more of the rubrics in the definition. The following table summarizes concern during this activity in the confined space.

The indication of intent to pursue hot work externally to a confined space containing or potentially containing flammables or combustibles either in the residual atmosphere or residual materials during preentry preparation is cause for major concern. Guidance for performing hot work is provided in a number of standards and regulations (NFPA 1988, 1992, 1993; AWS 1988; Standards Australia 1995; API 1984, 1987, 1993; OSHA 1993, 1994; Canada Gazette 1992; Jones and King 1991).

Hot work undertaken prior to opening the space could be perceived as not disturbing the status quo in any obvious way. However, hot work can act as a source of ignition to flammable/combustible materials both inside and external to the confined space. This situation again draws attention to the

TABLE 8.7 (continued)
Hazard Assessment — Preentry Preparation

	Real or Potential Consequence		
	Low	Moderate	High
• Hot work			
• Atmospheric hazards			
Oxygen deficiency			
Oxygen enrichment			
Biochemical/chemical			
Fire/explosion			
• Biological hazard			
• Ingestion/skin contact			
• Physical agents			
Noise/vibration			
Heat/cold stress			
Nonionizing/ionizing radiation			
Laser			
• Personal confinement			
• Mechanical hazard			
• Hydraulic hazard			
• Pneumatic hazard			
• Process hazard			
• Safety hazards			
Structural			
Engulfment/immersion			
Entanglement			
Electrical			
Fall			
Slip/trip			
Visibility/light level			
Explosive/implosive			
Hot/cold surfaces			

In this table, the meaning of "toxic substance," "oxygen deficiency," "oxygen enrichment," and "flammable or combustible atmosphere" requires further specification according to standards that exist in a particular jurisdiction. NA means not applicable.

Action Required

fact that a confined space usually is considered in isolation from its surroundings. The reality is that a confined space is inseparably connected to its surroundings, just as visual inspection would indicate. Activities that occur in the space directly impact on conditions in the surroundings and vice versa.

Surfaces in the area in which hot work is to occur may require cleaning to remove ignitable material or shielding to prevent heating. These considerations apply to both sides of the surface affected by the heat, since hot work occurring on one side could ignite material on the other. Nothing happens in isolation. This is the basis for one of the key concepts found in some standards: concern for adjacent spaces (NFPA 1993, OSHA 1994). This concept deserves to be applied to all confined spaces.

The approach to hazard assessment utilized here is to regard the hazardous condition represented by hot work as being greater than the sum of its contributing parts. This is the case because of the amplifying role of geometry in creating the hazard in the space. A person working in the tight quarters posed by many confined spaces would experience considerably greater difficulty in escaping from an accident caused by hot work. For this reason, as well as the need to consider additional protective measures, hot work is considered as an independent category.

Oxygen deficiency

Oxygen deficiency that exists in the undisturbed space could produce fatal consequences during preentry preparation. This can occur from the seemingly inconsequential action of putting the face across the plane of the opening of the space during a cursory inspection. In some jurisdictions, this action is not considered entry. Persons involved in preentry preparation of oxygen-deficient spaces must recognize the severity of this hazard and not expose the respiratory system to the atmosphere inside the space under any circumstances. In these circumstances, atmospheric testing for oxygen should occur immediately at the beginning of preentry preparation.

Oxygen deficiency can develop in a stratified layer or throughout the space. The condition in a particular space depends on several factors: the process by which oxygen deficiency occurs, time, and turbulence. Atmospheric testing must occur throughout the atmosphere of the space to ensure that all possibilities are examined.

Forced exchange of an oxygen-deficient atmosphere inside the space with the atmosphere outside the space must occur with care, since this could transfer the problem outside the space under conditions of poor mixing.

Inerting is a technique using a nonreactive gas, such as nitrogen, to reduce the concentration of oxygen to levels below which combustion cannot occur (API 1987). Inerting could be essential where residual contents continue to evolve flammable vapor or where flammable gas or vapor can seep into the space. Inerting also is required in spaces containing substances that react with oxygen in air. Substances requiring inerting include pyrophores, such as iron sulfide, and unregenerated catalysts, strong oxidizers, such as peroxides, that promote oxidation; and reactive substances that undergo a self-accelerating exothermic reaction when a critical temperature is reached.

During work under the inerting procedure, oxygen level must be maintained below 5% and monitored continuously. The inerting gas must enter the space at a sufficient rate to prevent in-leakage of air. In-leakage of air or use of contaminated inerting gas (for example, engine exhaust) could compromise the process. Mixing of the inerted atmosphere with outside air following escape or deliberate venting could produce a flammable atmosphere exterior to the space. This could be ignited by ignition sources located outside the space.

An inerted atmosphere cannot sustain life. This situation, therefore, necessitates the highest level of concern and precaution for persons entering the confined space.

Oxygen enrichment

Oxygen enrichment can develop in a stratified layer or throughout the space. The condition in a particular space depends on several factors: the process by which oxygen enrichment occurs, time, and turbulence in the atmosphere. Atmospheric testing must occur throughout the atmosphere of the space to ensure that all possibilities are examined.

Exchange of an oxygen-enriched atmosphere with the atmosphere outside the space must occur with care, since this oxygen could enrich the atmosphere outside the space under conditions of poor mixing. Ventilation of a space containing an oxygen-enriched atmosphere could severely tax the safe use of portable ventilating equipment and of materials external to the space. This is due to enhancement of combustibility of materials used in construction of air movers and duct and those present in the surroundings. Relatively minor enrichment seriously enhances combustibility. Most materials (including some metals) burn in oxygen. In addition, the conditions needed for ignition are considerably less stringent in an enriched atmosphere than are required under normal conditions (Frankel 1986, Turner 1987, Lowrie 1987).

Biochemical/chemical hazard

A substance that is airborne in the undisturbed space could pose a serious chemical hazard, one that could produce fatal consequences, during preentry preparation. This situation can occur from the seemingly inconsequential action of putting the face across the plane of the opening of a space during a cursory inspection. Persons involved in preentry preparation of spaces containing chemical or biochemical substances must recognize the potential severity of this hazard and not expose the respiratory system to the atmosphere inside the space under any circumstances. In these situations, atmospheric testing for these substances should occur immediately at the beginning of preentry preparation.

Contamination can develop in a stratified layer or throughout the space. The condition in a particular space depends on several factors: the process by which contamination occurs, time, and turbulence. Atmospheric testing must occur throughout the atmosphere of the space to ensure that all possibilities are examined. Grab-sampling methods provide discrete indication of conditions. A limited variety of person-portable, real-time instruments containing chemical-specific sensors also exist in the marketplace.

An important concern that deserves mention at this point is reliance on the sense of smell by many persons, and the lack of warning provided by many chemical substances. Reliance on the sense of smell to provide a warning as the covers on a sealed space are removed is virtually an instinctive behavior.

The subject of odor perception, of course, is much more complex and important than the casual coverage that it usually receives. To have an odor, a substance must be sufficiently water-soluble to be absorbed into the mucous layer lining the nasal passages. At the same time, it must be lipid-soluble enough to interact with receptors in olfactory tissue and have a chemical structure that interacts with an olfactory receptor (Norman and Cooper, 1994).

Sense of smell varies with age and gender. About one person in 50 has a decreased sense of smell (hyposmia). About one person in 500 has no sense of smell (anosmia). Both of these conditions occur when the olfactory epithelium becomes damaged. Depending on the extent of damage, these conditions may be temporary or permanent. Paralysis of smell caused by hydrogen sulfide is well known. Odor and odor threshold are addressed in a number of resources (Hellman and Small, 1974; Ruth, 1986; TRC Environmental Consultants, 1989).

Ventilating a space during preentry preparation transfers materials previously confined within the space to the exterior. The cloud formed in the surroundings under these conditions could constitute a serious exposure hazard to persons working outside the space near the discharge of the ventilator. Similarly, entry of hazardous materials from the surroundings could create the hazardous condition under some circumstances. This is an especial concern in process operations. This also could occur when the space is located below grade. For this reason, vapor density of contaminants known to be present in the surroundings merits special consideration. In some situations, this concern could extend to atmospheric conditions.

Fire/explosion hazard

Contamination can develop in a stratified layer or throughout the space. The condition in a particular space depends on several factors: the process by which contamination occurs, time, and turbulence. The possibility of an ignitable mixture indicates that atmospheric testing must occur throughout the atmosphere of the space to ensure that all possibilities are examined. The action of testing should not itself create an ignition hazard. This can occur during actions to open the space or in use of equipment that can accumulate and discharge static electricity.

A cloud formed in the surroundings external to the space during release of a vapor or gas from a pressurized space or removal by exhaust ventilation could contain an ignitable mixture. Similarly, introduction of air into the space during forced ventilation could form an ignitable mixture by displacement. This mixture could be present in the ventilation system. Bonding and grounding and use of equipment rated for use in hazardous atmospheres are essential under these circumstances.

Biological hazard

Forced ventilation could release biological hazards, such as viable microorganisms, viruses, spores, cysts, or pollen into the surrounding air during preentry preparation. As with chemical and biochemical hazards, this action could create hazardous conditions external to the space where none existed previously. This situation could be especially serious where the biological agents are highly hazardous to plants, animals, or humans. Complicating matters is the difficulty in performing real-time assessment of existence and migration of contamination.

Skin contact/ingestion hazard

Preentry preparation involving removal and transfer of solid and liquid materials from the space can result in skin contact and ingestion hazards. Removal and transfer of toxic substances can result in contamination of skin and clothing, instruments and work equipment, and the environment surrounding the space. Any one of these can become the source of contamination by ingestion or toxic action on the skin. Small quantities of highly toxic material transferred in this way could be highly significant. Also, chemically reactive substances can produce irritation and corrosion and sensitization reactions. Ingestion and skin contact could occur during handling and transfer of materials contained in the space.

Concepts of containment, contamination, and decontamination could be highly relevant during preentry preparation (NIOSH/OSHA/USCG/EPA, 1985). Initially, the hazardous substance is contained in the space. Faulty technique or catastrophic release during removal and transfer could expand the zone of contamination considerably beyond the confines of the space.

Noise/vibration hazard

Noise and vibration can result from operation of equipment inside or adjacent to the space or from motion of contents. Additional noise sources may be involved in preentry preparation.

Heat/cold stress

Preentry preparation occurs outside the space. Transfer of thermal conditions from the interior to the surroundings could cause heat/cold stress only under unusual conditions. Preentry preparation is the time to restore interior conditions to ambient, if possible. In some circumstances, this may not be possible.

Nonionizing/ionizing radiation

Preentry preparation is the time for deactivating sources of nonionizing and ionizing radiation that may be active in the space or directed through the space. These concerns should be addressed through deactivation/de-energization analysis and a lockout procedure. Where the hazard cannot be eliminated, assistance from a specialist in these matters should be solicited.

Laser

Preentry preparation is the time for deactivating sources of laser energy that may be active in the space or directed through the space. These concerns should be addressed through deactivation/de-energization analysis and a lockout procedure.

Personal confinement

Personal confinement is an inherent property of the geometry of the space, its internal configuration, and portals. Few options may be available during preentry preparation for ameliorating confinement. Complicated internal geometry and small access/egress portals can create considerably complicated extrication of an individual injured in an accident. For this reason, personal confinement may continue to occur as a potentially hazardous condition throughout the segments of hazard assessment. This situation necessitates considerable attention during planning for emergency response to ensure that appropriate measures are taken.

Mechanical hazard

Mechanical hazard results from unexpected or unpredictable discharge of stored energy through parts of equipment. The type of physical or mechanical process (agitation, aeration, mixing, fluidization, size reduction, dewatering, pressurization/depressurization, and so on) provides important cues about energy source, generation, storage, usage, and release. Process and instrumentation diagrams, operations manuals, and site reviews all can contribute to the knowledge base that is needed for performing a deactivation/de-energization analysis to create a lockout procedure.

Hydraulic hazard

Unexpected discharge of energy stored in hydraulic systems during preentry preparation poses considerable potential for injury. Process and instrumentation diagrams, operations manuals, and site reviews all can contribute to the knowledge base that is needed for performing a deactivation/de-energization analysis to create a lockout procedure.

Pneumatic hazard

Unexpected discharge of energy stored in pneumatic systems during preentry preparation poses considerable potential for injury. Process and instrumentation diagrams, operations manuals, and site reviews all can contribute to the knowledge base that is needed for performing a deactivation/de-energization analysis to create a lockout procedure.

Process hazard

Preparation to protect against process hazards focuses on preventing the occurrence of the process in the space during entry and work. Stopping the conditions that permit the process to occur is the first part of this action. This could require deactivation analysis, since computerized process control could be involved. In order to minimize hazard to entrants, contents must be removed from the space and prevented from reentry. This consideration also extends to process chemicals and wastes. Complete removal of contents could require flushing, washing, or high-pressure water jetting, chemical removal, or mechanical scouring, as well as purging and ventilating.

Isolation of the space to prevent entry of process chemicals or waste products involves severing connection between the space and the source of supply or removal. This can occur in several ways. One technique involves opening lines and inserting flanges, cups, or plugs to prevent flow. Another involves use of pairs of isolating valves having a third valve in between that opens to atmosphere (double block and bleed). The idea behind this concept is that the first valve stops the flow. The second valve drains leakage to atmosphere, and the third isolates the space. In this case, movement of the valves must be actively prevented by lockout devices. A variation of this approach is to lock out individual valves. The latter approach is much less secure since the connection between the space and line is still intact. Deactivation analysis and lockout procedure are required. These can occur simultaneously with activities directed at controlling other hazardous conditions.

Introduction of liquids and gases foreign to the process during preentry preparation can create problems of compatibility with residues and materials of construction of equipment or structures. Lines containing hazardous substances must be depressurized, drained, and possibly flushed with a cleaning agent prior to opening. Introduction of air into a space containing a flammable liquid can produce a flammable mixture both inside and outside the space.

Temperature difference between the residual atmosphere and the external atmosphere is an important concern. A residual atmosphere that is hotter than the air in the surroundings could condense into a mist on being removed from the space. Depending on the characteristics of the substance, this could create or enhance a fire hazard. The potential for production of static electrical charges increases with the presence of mist. Also, condensation exposes equipment to chemical and physical attack, as well as negative internal pressure.

The opposite situation, removing a cooler atmosphere from the space, could alter the vapor-to-liquid ratio in mixtures containing vapor and mist. Increased vaporization could put a mixture that was nonflammable into the flammable (explosive) range, thus potentially increasing the fire hazard.

Flammable and combustible atmospheres could result from activities during preentry preparation. The atmosphere generated during cleaning processes could simultaneously contain a mixture of gas or vapor, and dust or mist originating from a single source. Combustible solvents, for example, when present in air as a mixture of mist and vapor, pose considerably greater flammability hazard than when present solely as vapor. Solids, when finely divided and suspended in air as dusts, can pose considerable flammability or explosibility hazard.

Flooding using water or other suitable liquid is another option for eliminating fire hazard. Flooding eliminates contact between residues on the walls of structures and oxygen in air. Flooding can only be used under narrow circumstances, since the liquid could dissolve some of the residue, creating a potential problem for disposal. Incompatibility between the liquid used for flooding and the material of construction of the equipment or structure could exclude this option.

Chemical incompatibility between materials of construction of work equipment and instruments, and personal protective equipment used during preentry preparation and the components of the residual atmosphere can be a serious issue during selection. Incompatibility could lead to damage, as well as premature failure of components. Incompatibility can occur through the following actions:

- Solvent action
- Corrosive attack

A solvent is a substance that dissolves another substance. Solvent action is a concern during contact between fabrics and polymeric coatings and the residual atmosphere and materials.

Corrosion is a deterioration or destruction of materials brought about by chemical or electrochemical reduction–oxidation (redox). Metallic components are susceptible to certain types of corrosive action, while fabrics and polymeric coatings and substances are susceptible to others.

Corrosive attack is an important consideration during the design of equipment and structures. Materials of construction are chosen to resist corrosive attack by the contents. However, portable work equipment used to prepare for entry into confined spaces is intrusive to this environment. Materials used in construction of this equipment should not be presumed *a priori* to be equivalent to those of the existing structure in corrosion resistance.

Corrosives can be both acidic or basic (alkaline) (Meyer 1989). Acids and bases attack amphoteric metals such as aluminum and zinc. These could be present in fan casings and blades. Acidic substances or substances that react with water to form acids can cause extensive damage. Concentrated sulfuric acid extracts water from organic materials containing hydroxyl groups (OH). Fabrics and polymeric coatings used in flexible duct and protective clothing face potential attack. Less concentrated solutions of sulfuric acid can attack steel. Acids and to some extent, bases can hydrolyze the intermolecular linkages present in natural and synthetic polymers present in textiles and coatings.

Corrosives of greatest concern to the materials of construction of portable equipment are oxidizers or oxidizing agents. These include nitrates, chlorine, bromine, iodine, fluorine, hypochlorites, ferric chloride, peroxides such as hydrogen peroxide, chromic acid and chromates, permanganates, and ozone (Furr 1989). Oxidizers can react with alloys containing active metals such as aluminum, magnesium, and zinc and organic polymers.

Nitric acid is both an acid and an oxidizing agent (Meyer 1989). It attacks certain metals, such as zinc and aluminum, and even iron in steel, to form nitrogen oxides. Nitric acid also oxidizes organic materials. Fabrics and polymeric coatings also may be susceptible to attack by nitric acid.

Perchloric acid is another powerful oxidizing agent. Chlorosulfuric and perchloric acid can attack both metals and organic substances.

Electrochemical corrosion can occur when reactive metals such as aluminum, magnesium, and zinc found in lightweight alloys come into contact with metals that occur lower in the electrochemical series (Barrow 1966). Some organic compounds also may be sufficiently reactive to cause corrosion of these metals.

Extensive discussion about corrosion and reactivity is beyond the scope of this chapter. The reader is directed to references by Meyer and Breatherick for further information (Meyer 1989, Breatherick 1990). This subject is extremely important for the safe use of externally introduced equipment, especially in chemical process operations.

Structural hazard

Actions taken during preentry preparation could exacerbate an existing structural hazard or create one. Development of a structural hazard may not be recognized at the time that the critical action occurs. This situation is more likely to remain unrecognized when prework inspection is absent from the process of hazard assessment. In situations where prework inspection does not occur, consideration about the impact of actions performed during preentry preparation on structural stability and integrity is essential.

Advice from a structural, civil, or mechanical engineer about structural stability should be sought where this issue is a concern.

Engulfment/immersion

Since preentry preparation occurs outside the space, engulfment and immersion should be preventable by appropriate control of movement by personnel at the entry portals. Residual contents — flowable solids or liquids — certainly could pose serious safety hazards to entrants if not removed during preentry preparation. Later discussion provides options for removing these materials without the necessity for entry.

Engulfment also can occur due to instability of walls of trenches, excavations, shoring, and similar structures created during entry preparations. These could form the boundary of the confined space.

Entanglement hazard

Entanglement hazard refers to the ability of structural features of the space and its interior equipment to entangle clothing or other equipment, such as lifelines. Entanglement should not pose a concern during preentry preparation, since this is performed from outside the space.

Electrical hazard

Electrical hazard can result from work on live electrical circuits in equipment that operates in the space. Preentry preparation is the time for deactivating sources of electrical energy. These concerns should be addressed through deactivation/de-energization analysis and a lockout procedure.

Fall hazard

Preentry preparation should address fall hazards identified during the hazard assessment of the undisturbed space. However, access to the interior of the space may not be available at this time. Under circumstances involving entry and work in spaces whose interior characteristics are not known, a fall hazard could go unrecognized until the prework inspection, or if this does not occur, the start of work.

Slip/trip

Preentry preparation should address slip/trip hazards identified during the hazard assessment of the undisturbed space. However, access to the interior of the space may not be available at this time. Under circumstances involving entry and work in spaces whose interior characteristics are not known or where residual contents remain, a slip/trip hazard could go unrecognized until the prework inspection, or if this does not occur, the start of work.

Visibility hazard

Preentry preparation should eliminate a visibility hazard resulting from airborne particulates and mists or sprays. Provision for installation of lights should occur where absence of a source of illumination in the space occurs. Installation of lights in the space would be one of the first work procedures to occur after work in the space begins. Conditions could warrant use of equipment rated for use in hazardous atmospheres. The potential for fire or explosion as an influencing factor deserves careful consideration.

Explosive/implosive hazard

Identification of chemically explosive contents may not be possible until the prework inspection, or if this does not occur, initial entry for work.

Pressure inside the space is a fundamental concern. Pressure could mean pressure above or below atmospheric pressure. Unexpected opening or structural failure of a pressurized or depressurized space could pose serious hazards due to rapid equalization of pressure. Rapid equalization could cause structural failure. Controlled depressurization also raises concern regarding transfer of hazardous contents from the interior of the space to the exterior. Hazardous conditions could develop around the vent or opening. Similarly, repressurizing, which involves introducing the external atmosphere into the space, also could produce a hazardous condition through mixing of atmospheric oxygen with residual contents.

Hot/cold surfaces

Preentry preparation is the time for eliminating sources having abnormal temperatures. These concerns should be addressed through deactivation/de-energization analysis and a lockout procedure. This may not be possible in circumstances where temperature maintenance is critical. Insulation or personal protective equipment could be the only means for assuring safety under these circumstances.

Action required

Actions taken during preentry preparation attempt to eliminate or at the least control hazards in the space. The conclusion of this process could render the space equivalent to any normal workspace, except possibly for personal confinement.

Actions required could include measuring atmospheric conditions in the space during the initial contact between the internal and external atmospheres. These measurements should establish the potential severity of hazard posed by the internal atmosphere.

Prework inspection

Prework inspection is the briefest segment in the cyclic process of work involving confined spaces. It usually does not occur as a distinct phase in most situations, and in these cases is seamless in the overall process. However, prework inspection offers many benefits toward the safe conduct of work in confined spaces. Prework inspection offers the last opportunity for intercession by the Qualified Person prior to the start of work. A personal inspection of the confined space and testing of atmospheric conditions after preparations for entry have concluded can provide confirmation of success or the opportunity to discover the unexpected. Prework inspection by the Qualified Person is the only way to do this.

Prework inspection is mandatory in some jurisdictions (NFPA 1993, OSHA 1994). Prework inspection is a precondition to issuance of a certificate by a marine chemist (NFPA 1993). Prework inspection also occurs in process operations. The process operator inspects the space prior to granting clearance to tradespeople for entry and work. Prework inspection is especially important where hot work is to occur.

Prework inspection can be extremely hazardous, since this represents the first contact with conditions inside the space. Preentry preparation may be unable to eliminate all hazardous conditions in the space. These may be anticipated or unexpected at the time of the prework inspection. In addition, there also could be unrecognized hazards.

Table 8.8 provides a format for extending hazard assessment to prework inspection. This table follows the format introduced in the previous section.

The following sections explain the meaning of hazard assessment in the context of prework inspection.

Hot work

While hot work would not occur during a prework inspection, the prework inspection is critical to ensuring the safe performance of hot work. Firsthand visual inspection is essential where flammable/combustible substances are/were present in the space. These could include residues of materials held in the space, as well as coatings applied to its surfaces.

Surfaces in the area in which hot work is to occur may require additional cleaning to remove ignitable material, or shielding to prevent heating. These considerations apply to both sides of the surface affected by the heat, since hot work occurring on one side could ignite material on the other.

The intent to pursue hot work externally to a confined space containing or potentially containing flammables or combustibles either in the residual atmosphere or residual materials is cause for major concern. Hot work can act as a source of ignition to flammable/combustible materials both inside and external to the confined space at any time.

Oxygen deficiency

Ventilation during preentry preparation should have corrected any oxygen deficiency existing in the undisturbed space. The only remaining potential sources of oxygen-deficient air should be pockets of the original atmosphere not removed by ventilation or residues of material undergoing rapid reaction that consumes oxygen. Both of these conditions could be detected during a prework inspection through use of appropriate instruments. Discovery of a pocket of oxygen-deficient atmosphere should prompt immediate evacuation and continuance of ventilation.

Oxygen deficiency can develop in a stratified layer or throughout the space. The condition in a particular space depends on several factors: the process by which oxygen deficiency occurs, time, and turbulence. Atmospheric testing must occur throughout the atmosphere of the space to ensure that all possibilities are examined. Ideally, the instrument should actively sample the atmosphere to provide rapid indication about the presence of oxygen deficiency during initial occupancy of the space.

An inerted space containing oxygen at a level below 5% should be entered only under extreme caution, since this atmosphere cannot sustain life. This situation, therefore, necessitates the highest level of concern and precaution for persons entering the confined space under any circumstance.

Oxygen enrichment

Ventilation during preentry preparation should have corrected any oxygen enrichment existing in the undisturbed space. The only remaining potential sources of enriched air should be pockets of the original atmosphere not removed by ventilation or residues of material that release oxygen during decomposition. Both of these conditions could be detected during a prework inspection through use of appropriate instruments. Discovery of a pocket of oxygen-enriched atmosphere should prompt immediate evacuation and continuance of ventilation.

Oxygen enrichment can develop in a stratified layer or throughout the space. The condition in a particular space depends on several factors: the process by which oxygen enrichment occurs, time, and turbulence in the atmosphere. Atmospheric testing must occur throughout the atmosphere of the space to ensure that all possibilities are examined. Ideally, the instrument should actively sample the atmosphere to provide rapid indication about the presence of oxygen enrichment during initial occupancy of the space.

TABLE 8.8
Hazard Assessment — Prework Inspection

4.3 Hazard Assessment — Prework Inspection

Description: This section is used to provide a description of action(s) that occur during prework inspection.

Prework inspection is required as part of the entry preparation process. **Yes** No **Possible**

Prework inspection does not occur in most circumstances. In other situations, preentry preparation could be quite extensive. The details should be outlined here in sufficient detail to provide insight about a particular situation.

Hot work	**Yes** No **Possible**
Atmospheric hazards	
Oxygen deficiency	**Yes** No **Possible**
Oxygen enrichment	**Yes** No **Possible**
Biochemical/chemical	**Yes** No **Possible**
Fire/explosion	**Yes** No **Possible**
Biological hazard	**Yes** No **Possible**
Ingestion/skin contact hazard	**Yes** No **Possible**
Physical agents	
Noise/vibration	**Yes** No **Possible**
Heat/cold stress	**Yes** No **Possible**
Nonionizing/ionizing radiation	**Yes** No **Possible**
Laser	**Yes** No **Possible**
Personal confinement	**Yes** No **Possible**
Mechanical hazard	**Yes** No **Possible**
Hydraulic hazard	**Yes** No **Possible**
Pneumatic hazard	**Yes** No **Possible**
Process hazard	**Yes** No **Possible**
Safety hazards	
Structural hazard	**Yes** No **Possible**
Engulfment/immersion	**Yes** No **Possible**
Entanglement	**Yes** No **Possible**
Electrical	**Yes** No **Possible**
Fall	**Yes** No **Possible**
Slip/trip	**Yes** No **Possible**
Visibility/light level	**Yes** No **Possible**
Explosive/implosive	**Yes** No **Possible**
Hot/cold surfaces	**Yes** No **Possible**

Hazardous conditions in the confined space may satisfy one or more of the rubrics in the definition. The following table summarizes concern during this activity in the confined space.

Biochemical/chemical hazard

Ventilation during preentry preparation should have eliminated any chemical or biochemical hazard existing in the undisturbed space. The only remaining potential sources of contaminated air should be pockets of the original atmosphere not removed by ventilation or residues of volatile or friable material. Both of these conditions could be detected during a prework inspection by visual inspection and use of appropriate testing equipment. Selection of testing equipment is an important issue, since rapid notification about the presence of contamination is essential to ensure protection of the

TABLE 8.8 (continued)
Hazard Assessment — Prework Inspection

	Real or Potential Consequence		
	Low	Moderate	High
• Hot work			
• Atmospheric hazards			
Oxygen deficiency			
Oxygen enrichment			
Biochemical/chemical			
Fire/explosion			
• Biological hazard			
• Ingestion/skin contact			
• Physical agents			
Noise/vibration			
Heat/cold stress			
Nonionizing/ionizing radiation			
Laser			
• Personal confinement			
• Mechanical hazard			
• Hydraulic hazard			
• Pneumatic hazard			
• Process hazard			
• Safety hazards			
Structural			
Engulfment/immersion			
Entanglement			
Electrical			
Fall			
Slip/trip			
Visibility/light level			
Explosive/implosive			
Hot/cold surfaces			

In this table, the meaning of "toxic substance," "oxygen deficiency," "oxygen enrichment," and "flammable or combustible atmosphere" requires further specification according to standards that exist in a particular jurisdiction. NA means not applicable.

Action Required

person performing the inspection. Discovery of a pocket of contaminated atmosphere should prompt continuance of ventilation. Immediate evacuation also may be required.

Contamination can develop in a stratified layer or throughout the space. The condition in a particular space depends on several factors: the process by which contamination occurs, time, and turbulence. Atmospheric testing must occur throughout the atmosphere of the space to ensure that all possibilities are examined. Grab-sampling methods provide discrete indication of conditions. A limited variety of person-portable, real-time instruments containing chemical-specific sensors also exist in the marketplace. Ideally, the instrument should actively sample the atmosphere to provide rapid indication about the presence of contamination during initial occupancy of the space.

A hazardous condition resulting from off-gassing of residual materials is much more likely where removal was not undertaken during preentry preparation. Removal of residual contents during preentry preparation considerably reduces potential for exposure during entry and work.

Fire/explosion

Ventilation during preentry preparation should have eliminated any fire/explosion hazard existing in the undisturbed space. The only remaining potential sources of contaminated air should be pockets of the original atmosphere not removed by ventilation or residues of volatile or friable material. Both of these conditions could be detected during a prework inspection by visual inspection and use of appropriate testing equipment. Discovery of a pocket of contaminated atmosphere should prompt immediate evacuation and continuance of ventilation.

Contamination can develop in a stratified layer or throughout the space. The condition in a particular space depends on several factors: the process by which contamination occurs, time, and turbulence. The possibility of an ignitable mixture indicates that atmospheric testing must occur throughout the atmosphere of the space to ensure that all possibilities are examined. The action of testing should not itself create an ignition hazard. This can occur in use of equipment that can accumulate and discharge static electricity. Ideally, the instrument should actively sample the atmosphere to provide rapid indication about the presence of contamination during initial occupancy of the space.

A hazardous condition arising from residual materials is much more likely where removal was not undertaken during preentry preparation. Removal of residual contents during preentry preparation is essential to eliminate the hazardous condition during entry and work.

Biological hazard

Ventilation during preentry preparation should have eliminated any biological hazard existing in the undisturbed space. The only remaining potential sources of contaminated air should be pockets of the original atmosphere not removed by ventilation or residues of friable material. A visual inspection could determine only the latter condition. The lack of testing equipment for rapid measurement of airborne contamination could be a serious concern. The presence of contamination should dictate further cleaning measures or signal the need for personal protective equipment.

Without further action directed toward removal of residual materials, forced ventilation during work activity could cause release of biological hazards, such as viable microorganisms, viruses, spores, cysts, or pollen, into the air of the space. This situation could be especially serious where the biological agent is highly hazardous to humans.

Ingestion/skin contact hazard

Actions taken during preentry preparation should have attempted to remove bulk and residual quantities of substances that are hazardous by ingestion or skin contact. However, residues still could remain. A visual inspection provides the means for determining the extent of concern merited by this situation. The presence of contamination could dictate the need for further cleaning measures or use of personal protective equipment during work procedures.

Contact with liquid and solid residues during prework inspection can result in contamination of skin and clothing and instruments. Any one of these can become the source of contamination leading to ingestion and toxic action on the skin. Small quantities of highly toxic material could be highly injurious under these circumstances. Also, chemically reactive substances could produce corrosion of materials, as well as skin irritation and sensitization reactions.

Concepts of containment, contamination, and decontamination could be highly relevant during prework inspection (NIOSH/OSHA/USCG/EPA 1985). Contamination by the hazardous substance originates in the space. Faulty technique during removal of protective clothing and decontamination of instruments and equipment could expand the zone of contamination considerably beyond the confines of the space.

Noise/vibration

Noise and vibration can result from operation of equipment inside or adjacent to the space.

Heat/cold stress

Preentry preparation should attempt to moderate thermal conditions in the space to ambient. This may not be possible in all cases. Where this accommodation is not possible, conditions must be anticipated to enable the entrant to perform the prework inspection safely. Safe entry for inspection and work could require use of personal protective equipment.

Nonionizing/ionizing radiation

Effectiveness of the deactivation/de-energization analysis and lockout procedure for shielding or deactivating sources of nonionizing and ionizing radiation that may be active in the space or directed through the space should be tested prior to entry and during the prework inspection. Testing should include inspection of the lockout point and could involve testing inside the space with a suitable instrument. (The shutter-arm coupling on radioactive sources has failed on occasion in hostile environments where inspection and maintenance were not performed. This means that rotation and locking the shutter arm into the off position alone may not guarantee disablement.)

Where the hazard cannot be eliminated, assistance from a specialist in these matters should be solicited.

Laser

Effectiveness of the deactivation/de-energization analysis and lockout procedure for deactivating laser sources that may be active in the space or directed through the space should be tested during the prework inspection. Testing should include inspection of the lockout point and control circuits. Testing also could involve use of instruments under some conditions.

Personal confinement

As an inherent property of the geometry of the space, its internal configuration, and portals, personal confinement may continue to occur as a potentially hazardous condition throughout the stages of hazard assessment. Prework inspection is the first contact with the internal configuration of the space. Complicated internal geometry and small access/egress portals can considerably complicate extrication of an individual injured in an accident.

Mechanical hazard

Effectiveness of the deactivation/de-energization analysis and lockout procedure for deactivating mechanical equipment and energy sources that may be active in the space should be tested during the prework inspection. Testing should include inspection of the lockout point and control circuits. Testing also could involve use of instruments under some conditions.

The prework inspection could pose considerable hazard when the entrant must move around equipment inside the space. The safety of the entrant performing the inspection entirely depends on the effectiveness of the lockout procedure, but more especially the deactivation/de-energization analysis for deactivating mechanical equipment and energy sources. Unexpected or unpredictable discharge of energy stored in mechanical equipment could lead to motion of parts such as blades or paddles, levers, propellers or augurs, belts, chains, and so on.

Hydraulic hazard

Sudden discharge of energy stored in hydraulic systems could produce serious consequences during the prework inspection. Comments about mechanical hazards are equally applicable here.

Pneumatic hazard

Sudden discharge of energy stored in pneumatic systems could produce serious consequences during the prework inspection. Comments about mechanical hazards are equally applicable here.

Process hazard

Effectiveness of the deactivation/de-energization analysis and lockout and isolation procedure for isolating processes that normally occur in the space should be tested during the prework inspection.

Testing should include inspection of isolation points and control circuits. Testing also could involve use of instruments.

Prework inspection also provides the opportunity to determine the effectiveness of actions to remove residual contents.

Structural hazard

Actions taken during preentry preparation could exacerbate an existing structural hazard or create one. Development of a hazardous condition under these circumstances may not be recognized at the time that the critical action occurs. This situation can add to the risk of performing a prework inspection. In situations where prework inspection does not occur, consideration about the consequence of actions performed during preentry preparation on structural stability and integrity is essential.

Engulfment/immersion

Preentry preparation should include procedures for removing, shoring, or stabilizing materials capable of causing engulfment and immersion. The purpose for the prework inspection in this case is to confirm the effectiveness of these procedures.

Entanglement

Entanglement could pose a concern during the prework inspection, since this is the first entry into the space.

Entanglement hazards are an inherent property of the geometry of the space, its internal configuration, and that of equipment inside it. Few options may be available for ameliorating potential entanglement hazards.

Electrical hazard

Effectiveness of the deactivation/de-energization analysis and lockout procedure should be tested during the prework inspection. Testing should include inspection of isolation points and control circuits. Testing also could involve use of instruments under some circumstances.

Fall hazard

Fall could be a major hazard during prework inspection. While preentry preparation should have addressed fall hazards, this may not have occurred or have been possible. Under circumstances involving entry and work in spaces whose interior characteristics are not known, a fall hazard could remain unrecognized until the prework inspection, or if this does not occur, the start of work.

Installation of fall prevention and fall arrest equipment may be one of the first work activities following entry. A complete prework inspection may not be possible until installation of this equipment has occurred.

Slip/trip

Preentry preparation should address slip/trip hazards identified during the hazard assessment of the undisturbed space. Under circumstances involving entry and work in spaces whose interior characteristics are not known, or where residual contents remain, a slip/trip hazard could remain unrecognized until the prework inspection, or if this does not occur, the start of work.

In spaces where thorough removal of residual contents has not occurred, slip/trip may pose a serious and underappreciated safety hazard.

Visibility/light level

Purging and ventilation of the space during preentry preparation should have eliminated the visibility hazard resulting from airborne particulates and mists or sprays.

Installation of lights in the space could be one of the first work activities to occur. Conditions could warrant use of low voltage equipment or equipment rated for use in hazardous atmospheres.

This also would apply during prework inspection. The potential for fire or explosion as an influencing factor deserves careful consideration.

Explosive/implosive hazard

Identification of chemically explosive contents may not be possible until the prework inspection. Preentry preparation should have addressed concerns related to deviation of pressure inside the space from normal atmospheric values. The effect of pressurization or depressurization on the structural integrity of the space might not be recognized even after a personal inspection.

Hot/cold surfaces

Preentry preparation should attempt to eliminate hot or cold surfaces. This may not be possible in many circumstances. Where this is not possible, this condition must be anticipated to enable the entrant to perform the prework inspection safely. Safe entry for inspection and work could require use of personal protective equipment and application of insulated coverings.

Action required

The function of the prework inspection is to confirm the effectiveness of actions taken during preentry preparation to control or eliminate hazardous conditions. The conclusion from the prework inspection could be to consider the space equivalent to any normal workspace, except possibly for personal confinement.

However, the prework inspection could uncover hazardous conditions not anticipated during prior consideration or created by actions taken during preentry preparation. Actions required as a result of the prework inspection attempt to correct these hazardous conditions.

Actions required could include measuring atmospheric conditions in the space. This would confirm effectiveness of cleaning to remove residual materials and its impact on atmospheric quality. These measurements should establish the potential severity of the hazard posed by the internal atmosphere prior to the start of work. This approach increases certainty about conditions and reduces concern about possibilities.

Work Activity

The next segment of hazard assessment is concerned with individual procedures to be performed in the confined space. Assessment of the impact of work activities in the environment of the confined space is the most difficult task in the process of hazard assessment. This also is the aspect about which least is known. Work activities in confined spaces have received little attention over the years. **Preparing** the confined space for work traditionally has been the focus of attention during hazard assessment. **Work activity** in the confined space apparently has been assumed to pose the same level of hazard as work elsewhere.

Concern about hot work, for instance, focuses on fire and explosion, not on the airborne contaminants and physical agents generated during these processes. However, these substances and agents could pose considerably greater hazard than fire alone. Performed in confined spaces in the same manner as in open surroundings, so-called normal work activities have led to tragic consequences. These were amply illustrated in the OSHA reports on fatal accidents (OSHA 1982a, 1985, 1988).

Assessing the level of concern to apply to a work activity or process prior to its application in a confined workspace is extremely difficult. The restrictive geometry and small volume of many confined spaces magnify hazards associated with normal work practices. This is especially important where plume rise, dilution into a larger volume, and general ventilation are depended on to control exposure to airborne contaminants. Application of this approach to the restrictive geometry and small volume of many confined spaces is not practicable. The small roofspace above the occupied zone can rapidly be flooded by a heated plume, with resultant downward fumigation. Similarly, application of coatings by spraying creates a large surface area of volatile liquid relative to air volume, as well as an atmosphere of mist and vapor. These realities indicate the need to focus on contaminant generation and migration in the airspace.

Knowledge about the impact of work activities in the context of confined workspaces is essential for gaining control over these conditions. Unfortunately, there is little in the published literature to provide assistance. This puts reliance on the knowledge base and experience of the Qualified Person. Recognition of the need to develop an information base is implicit in the approach taken in some legislation (Canada Gazette 1992). This approach requires the Qualified Person to attend the entry into the confined space and to sample to determine effectiveness of control measures. Improved procedures for entry and work and development of a knowledge base for the Qualified Person should be important outcomes from this approach.

Information on process and contaminants should be readily accessible through published work in textbooks, from manufacturers, suppliers and trade associations, and material safety data sheets. Also, software-based exposure models discussed in Chapter 5 may provide assistance. These possibly could combine source terms provided by other sources.

Table 8.9 provides a format for consideration of hazardous conditions potentially associated with individual work activities. This table follows the format introduced in the previous section. By activities, from the variety of work activities and the potentially hazardous conditions that could accompany them, Table 8.9 is incomplete. The following sections explain the meaning of hazard assessment in the context of work activity.

Hot work

Hot work is any type of activity or process that produces arcs, sparks, flames, heat, or other sources of ignition. Hot work can include welding, grinding, flame and mechanical cutting, gouging, soldering, brazing, burning, and drilling. Other sources of ignition include energetic chemical reactions, static electrical discharge, abrasive blasting and space heating, electrical short-circuits, tobacco smoking, lightning, engines and compressors, power tools, fixtures, electrical switches, and appliances that are not explosion-proof (ANSI 1995, NFPA 1993, NIOSH 1979, API 1993).

The intent to pursue hot work inside or externally to a confined space containing flammables or combustibles, either in the residual atmosphere or residual materials, during a work procedure is cause for major concern. Guidance for performing hot work is provided in a number of standards, regulations, and guidance documents (NFPA 1993, 1992, 1988; AWS 1988, Standards Australia 1995, API 1984, 1987, 1993; OSHA 1993, 1994; Canada Gazette 1992; Jones and King 1991).

Hot work can act as a source of ignition to flammable/combustible materials both inside and external to the confined space.

Surfaces in the area in which hot work is to occur may require cleaning to remove ignitable material, or shielding to prevent heating. These considerations apply to both sides of the surface affected by the heat, since hot work occurring on one side could ignite material on the other. Nothing happens in isolation. This is the basis for one of the key concepts found in some standards: concern for adjacent spaces (NFPA 1993, OSHA 1994).

The approach to hazard assessment utilized here is to regard the hazardous condition represented by hot work as being greater than the sum of its contributing parts. This is the case because of the amplifying role of geometry in creating the hazard in the space. A person working in the tight quarters posed by many confined spaces would experience considerably greater difficulty in escaping from an accident caused by hot work. For this reason, plus the need to consider additional protective measures, hot work is considered as an independent category.

Oxygen deficiency

Ventilation during preentry preparation and follow-up measures during the prework inspection should have corrected any oxygen deficiency existing in the space. The only remaining potential sources of oxygen-deficient air should be residues of material undergoing rapid reaction that consumes oxygen. This condition could be monitored during a work procedure through use of appropriate instruments.

A process introduced as part of a work activity also could consume oxygen. This condition could be monitored during a work procedure through use of appropriate instruments. This situation potentially is controllable through use of ventilation.

Oxygen deficiency can develop in a stratified layer or throughout the space. The condition in a particular space depends on several factors: the process by which oxygen deficiency occurs, time, and turbulence. Continuous testing using a real-time instrument can provide the basis for ensuring the safety of a work activity. Ideally, the instrument should actively sample the atmosphere in the breathing zone or proximal to the source of contaminant to provide rapid indication about the oxygen deficiency during work activity in the space.

An inerted space containing oxygen at a level below 5% should be entered only under extreme caution, since this atmosphere cannot sustain life. This situation, therefore, necessitates the highest level of concern and precaution for persons entering the confined space under any circumstance. A Qualified Person should be present during this work to assure adherence to procedures and to assure control of the atmosphere in the space.

Inerting is a technique using a nonreactive gas, such as nitrogen, to reduce the concentration of oxygen to levels below which combustion cannot occur (API 1987). Inerting could be essential where residual contents continue to evolve flammable vapor or where flammable gas or vapor can seep into the space. Inerting also is required in spaces containing substances that react with oxygen in air. Substances requiring an inerted atmosphere include pyrophores, such as iron sulfide and unregenerated catalysts; strong oxidizers, such as peroxides, that promote oxidation; and reactive substances that undergo a self-accelerating exothermic reaction when a critical temperature is reached.

The inerting gas must enter the space at a sufficient rate to prevent in-leakage of air. In-leakage of air or use of contaminated inerting gas (for example, engine exhaust) could compromise the process. Mixing of the inerted atmosphere with outside air following escape or deliberate venting could produce a flammable atmosphere exterior to the space. This could be ignited by ignition sources located outside the space.

Oxygen enrichment

Ventilation during preentry preparation and follow-up measures during the prework inspection should have corrected any oxygen enrichment existing in the space. The only remaining potential source of enriched air should be residues of material that release oxygen during decomposition. This condition could be detected during a work procedure through use of appropriate instruments.

Equipment used during a work procedure, such as an oxy-fuel torch, could utilize oxygen as a process gas. Leakage from this equipment has occurred in confined spaces and led to tragic outcomes. This possibility must be considered as part of hazard assessment. This condition could be monitored during a work procedure through use of appropriate instruments. This situation also should be controllable through use of ventilation. The latter could be required to control contamination by combustion products and other emissions.

Accidental oxygen enrichment could compromise the safety of any activity occurring in the space from use of portable ventilation equipment to use of processes or even occupancy. This is due to enhancement of combustibility of materials by adherence of oxygen to surfaces. Relatively minor enrichment seriously enhances combustibility. Most materials (including some metals) burn in oxygen. As well, the conditions needed for ignition in an enriched atmosphere are considerably less stringent than what are required under normal conditions. The space should be evacuated following accidental oxygen enrichment and the situation assessed carefully prior to implementing any action, including continuation or resumption of ventilation.

TABLE 8.9
Hazard Assessment — Work Activity

4.4 Hazard Assessment — Work Activity

Complete a separate hazard assessment for each work procedure that creates distinct hazards.

4.4.x Work Procedure: title of work procedure

Description: This section is used to provide a description of action(s) that occur during the work procedure.

This hazard assessment considers potential hazards associated with the work activity in relation to those associated with the space.

Hot work	**Yes** No **Possible**
Atmospheric hazards	
Oxygen deficiency	**Yes** No **Possible**
Oxygen enrichment	**Yes** No **Possible**
Biochemical/chemical	**Yes** No **Possible**
Fire/explosion	**Yes** No **Possible**
Biological hazard	**Yes** No **Possible**
Ingestion/skin contact hazard	**Yes** No **Possible**
Physical agents	
Noise/vibration	**Yes** No **Possible**
Heat/cold stress	**Yes** No **Possible**
Nonionizing/ionizing radiation	**Yes** No **Possible**
Laser	**Yes** No **Possible**
Personal confinement	**Yes** No **Possible**
Mechanical hazard	**Yes** No **Possible**
Hydraulic hazard	**Yes** No **Possible**
Pneumatic hazard	**Yes** No **Possible**
Process hazard	**Yes** No **Possible**
Safety hazards	
Structural hazard	**Yes** No **Possible**
Engulfment/immersion	**Yes** No **Possible**
Entanglement	**Yes** No **Possible**
Electrical	**Yes** No **Possible**
Fall	**Yes** No **Possible**
Slip/trip	**Yes** No **Possible**
Visibility/light level	**Yes** No **Possible**
Explosive/implosive	**Yes** No **Possible**
Hot/cold surfaces	**Yes** No **Possible**

Hazardous conditions in the confined space may satisfy one or more of the rubrics in the definition. The following table summarizes concern during this activity in the confined space.

Oxygen enrichment can develop in a stratified layer or throughout the space. The condition in a particular space depends on several factors: the process by which oxygen enrichment occurs, time, and turbulence. Continuous testing using a real-time instrument can provide the basis for ensuring the safety of a work activity. Ideally, the instrument should actively sample the atmosphere in the breathing zone or proximal to the source of contaminant to provide rapid indication about the enrichment during work activity in the space.

Procedures and other measures specified under "Actions Required" should minimize the probability for such events, as well as their severity.

TABLE 8.9 (continued)
Hazard Assessment — Work Activity

	Real or Potential Consequence		
	Low	Moderate	High
• Hot work			
• Atmospheric hazards			
Oxygen deficiency			
Oxygen enrichment			
Biochemical/chemical			
Fire/explosion			
• Biological hazard			
• Ingestion/skin contact			
• Physical agents			
Noise/vibration			
Heat/cold stress			
Nonionizing/ionizing radiation			
Laser			
• Personal confinement			
• Mechanical hazard			
• Hydraulic hazard			
• Pneumatic hazard			
• Process hazard			
• Safety hazards			
Structural			
Engulfment/immersion			
Entanglement			
Electrical			
Fall			
Slip/trip			
Visibility/light level			
Explosive/implosive			
Hot/cold surfaces			

In this table, the meaning of "toxic substance," "oxygen deficiency," "oxygen enrichment," and "flammable or combustible atmosphere" requires further specification according to standards that exist in a particular jurisdiction. NA means not applicable.

Action Required

Biochemical/chemical hazard

Ventilation during preentry preparation and follow-up measures during the prework inspection should have eliminated any chemical or biochemical hazard existing in the space. The only remaining potential source of contaminated air would be residues of volatile or friable material. This condition possibly could be assessed during a work procedure by use of testing equipment. Selection of testing equipment is an important issue, since rapid notification about the presence of contamination is essential to ensure protection of the person working in the area. Eliminating the need for testing is highly desirable, as this would considerably simplify the work activity.

Contamination can develop in a stratified layer or throughout the space. The condition in a particular space depends on several factors: the process by which contamination occurs, time, and turbulence. Atmospheric testing must occur throughout the atmosphere of the space to ensure that all possibilities are examined. Grab-sampling methods provide discrete indication of conditions. A limited variety of person-portable, real-time instruments containing chemical-specific sensors also exist in the marketplace. Ideally, the instrument should actively sample the atmosphere in the breathing zone or proximal to the source of contaminant to provide rapid indication about the presence of contamination during work activity in the space.

A hazardous condition resulting from off-gassing of residual materials is much more likely where removal was not undertaken during preentry preparation.

Many work procedures utilize chemical materials and processes that emit newly formed products. A hazardous condition could arise under any number of circumstances. The Qualified Person must attempt to judge the outcome of this activity, erring on the side of prudence in the use of control measures, such as local exhaust ventilation and respiratory and other personal protective equipment.

Contaminants can occur in any of the forms — gas or vapor, dust, mist, or fume — encountered in the normal workplace. The influence of geometry on accumulation of contaminants is substantial. For example, the geometry of a structure can trap gas used in a process or accidental leakage from a delivery valve.

Applied to surfaces in a space by spraying, brushing or rolling, a volatile liquid rapidly can cover a large surface area in a space having relatively small volume. By comparison, the surface area of the opening in the container in which the product is stored is very small. Another measure of the same concept is the surface-area-to-depth-of-liquid ratio. Application of volatile liquids by spraying creates an additional atmospheric contaminant: liquid aerosols from which additional vaporization can occur. A confined space provides the optimum workplace conditions for vaporization and accumulation of vapor.

Process emissions sometimes can be monitored during a work procedure through use of instruments. Appropriateness to task is a major concern.

Ventilating a space during a work procedure transfers materials from the confined space to the exterior atmosphere. The cloud formed in the surroundings under these conditions could constitute a serious exposure hazard to persons working outside the space near the discharge of the ventilator. Similarly, entry of hazardous materials unrelated to the work procedure from the surroundings could create a hazardous condition under some circumstances. This is an especial concern in process operations where fugitive emissions could be present. This also could occur when the space is located below grade. For this reason, vapor density of contaminants known to be present in the surroundings merits special consideration. In some situations, this concern could extend to atmospheric conditions.

Procedures and other measures specified under "Actions Required" should minimize the probability for such events, as well as their severity.

An important concern that deserves reiteration at this point is reliance on the sense of smell by many persons and the lack of warning properties of many chemical substances.

The subject of odor perception, of course, is much more complex and important than the casual coverage that it usually receives. To have an odor, a substance must be sufficiently water soluble to be absorbed into the mucous layer lining the nasal passages. At the same time, it must be lipid soluble enough to interact with receptors in olfactory tissue and have a chemical structure that interacts with an olfactory receptor (Norman and Cooper 1994).

Sense of smell varies with age and gender. About one person in 50 has a decreased sense of smell (hyposmia). About one person in 500 has no sense of smell (anosmia). Both of these conditions occur when the olfactory epithelium becomes damaged. Depending on the extent of damage, these conditions may be temporary or permanent. Paralysis of smell caused by hydrogen sulfide is well known. Odor and odor threshold are addressed in a number of resources (Hellman and Small 1974; Ruth 1986; TRC Environmental Consultants 1989).

Fire/explosion hazard

Ventilation during preentry preparation and follow-up measures during the prework inspection should have eliminated any fire/explosion hazard existing in the space. The only remaining potential sources of contaminated air would be residues of volatile or friable material.

A hazardous condition arising from residual materials is much more likely where removal was not undertaken during preentry preparation. Removal of residual contents during preentry preparation is essential to eliminate the hazardous condition during work procedures.

The process utilized during the work procedure could contain sources of ignition. A source of ignition used in a confined space containing only residues of flammable/combustible material could be extremely hazardous.

Some activities introduce flammable/combustible materials into the space or generate them. Certain activities, such as spraying, in addition, can generate static electrical charge, a possible source of ignition.

Contamination can develop in a stratified layer or throughout the space. The condition in a particular space depends on several factors: the process by which contamination occurs, time, and turbulence. The possibility of an ignitable mixture indicates that atmospheric testing must occur throughout the atmosphere of the space to ensure that all possibilities are examined. The action of testing should not itself create an ignition hazard. This can occur in use of equipment that can accumulate and discharge static electricity. Ideally, the instrument should actively sample the atmosphere to provide rapid indication about the presence of contamination during work activity.

A cloud formed in the surroundings external to the space during removal of a vapor or gas could contain an ignitable mixture. Similarly, introduction of air into the space during forced ventilation could form an ignitable mixture by displacement. This mixture also could be present in the ventilation system. Bonding and grounding and use of equipment rated for use in hazardous atmospheres are essential under these circumstances.

The cloud formed in the surroundings under these conditions could constitute a serious exposure hazard to persons working outside the space.

Procedures and other measures specified under "Actions Required" should minimize the probability for such events, as well as their severity.

Biological hazard

Ventilation during preentry preparation and follow-up measures during the prework inspection should have eliminated any biological hazard existing in the space. The only remaining potential source of contamination would be residues of friable material.

The possibility of involvement of a biological vector in an unrelated work procedure is cause for concern. Without further action directed toward complete removal of residual materials, forced ventilation utilized during work activity could cause release of biological hazards, such as viable microorganisms, viruses, spores, cysts, or pollen into the air of the space. This situation could be especially serious where the biological agent is highly hazardous to humans.

The lack of testing equipment for rapid measurement of airborne biological contamination further complicates this situation.

The presence of residual contamination should dictate the need for further cleaning measures or use of personal protective equipment. Soliciting advice from a specialist familiar with microbiology should be considered in this situation.

Ingestion/skin contact hazard

Actions taken during preentry preparation and follow-up measures resulting from the prework inspection should have attempted to remove bulk and residual quantities of substances that are hazardous by ingestion or skin contact. However, residues still could remain despite all attempts at removal. The potential for contamination could dictate the need for protective clothing and other personal protective equipment during work procedures.

Contact with liquid and solid residues as well as materials used during work activity can result in contamination of skin and clothing, and instruments and work equipment. Any one of these could become the source of contamination leading to ingestion and toxic action on the skin. Small quantities of highly toxic material could be highly injurious under these circumstances. As well, chemically reactive substances could produce corrosion of materials, as well as skin irritation and sensitization reactions.

Work procedures can involve removal and transfer of unrelated solid and liquid materials from the space. This transfer also can result in skin contact and ingestion hazards.

Concepts of containment, contamination, and decontamination could be highly relevant during prework inspection (NIOSH/OSHA/USCG/EPA 1985). Contamination by the hazardous substance originates in the space. Faulty technique during removal of protective clothing and decontamination of instruments and work equipment could expand the zone of contamination considerably beyond the confines of the space.

Procedures and other measures specified under "Actions Required" should minimize the probability for such occurrences, as well as their severity.

Noise/vibration

Noise and vibration can result from operation of equipment inside or adjacent to the space during work activity. The level of noise will depend on the context of the activity and noise sources that are operating.

Heat/cold stress

Preparative activities would have attempted to moderate thermal conditions in the space to ambient. This may not be possible in all cases. Where this accommodation is not possible, these conditions must be anticipated to enable the entrant to perform the work activity safely. This could require use of personal protective equipment, as well as other supplies.

Heat and cold stress also could be influenced by weather conditions and location. Radiant energy from the sun incident on a steel container could considerably elevate the temperature inside. Damp, windy, or wet conditions could cause hypothermia in the unprotected.

Equipment and processes utilized during work activities could heat the interior of the space.

Procedures and other measures specified under Actions Required should minimize the probability for such occurrences, as well as their severity.

Nonionizing/ionizing radiation

Preentry preparation and follow-up activities may be sufficient to control sources of nonionizing or ionizing radiation in the space. Temporary shielding and other techniques also may be required.

Work processes or equipment may produce or contain sources of nonionizing radiation, including intense electrical and magnetic fields, and ultraviolet, visible, infrared, microwave, and radio frequency sources. Reflective surfaces present in many confined spaces can exacerbate irradiation by these sources.

Some processes utilize sources of ionizing radiation from naturally and artificially radioactive materials. The hazard arising from these materials depends on the isotope and strength, shielding, and mode of use. Conditions of use are provided in the license for the source. The beam from sources used in industrial radiography can penetrate the wall of the structure. The beam could be directed through the wall of the space from the exterior or vice versa. In either case, occupancy must be prevented during this work.

Procedures and other measures specified under "Actions Required" should minimize the probability for such occurrences, as well as their severity.

Laser hazard

Actions during preentry preparation and follow-up to the prework inspection should have ensured the effectiveness of the deactivation/de-energization analysis and lockout procedure for deactivating laser sources that may be active in the space or directed through it.

A work procedure may introduce equipment containing or utilizing a laser. The hazard arising from this equipment depends on the type and strength of the laser, shielding, and mode of use. Reflective surfaces present in many confined spaces can exacerbate irradiation by these sources.

Procedures and other measures specified under "Actions Required" should minimize the probability for such occurrences, as well as their severity.

Personal confinement

As an inherent property of the geometry of the space, its internal configuration, and portals, personal confinement may continue to occur as a potentially hazardous condition during work procedures. Complicated internal geometry and small access/egress portals can considerably complicate extrication of an individual injured in an accident.

Mechanical hazard

Effectiveness of the deactivation/de-energization analysis and lockout procedure for deactivating mechanical equipment and energy sources should have been confirmed during preentry preparation or the prework inspection.

Work activity could introduce mechanical equipment into the space. Operation of this equipment could pose considerable hazard because of the confines of the space. For example, failure of a grinding wheel inside a confined space poses considerably greater hazard than a similar failure in a normal work area. The walls of the space act as reflectors to rapidly moving objects. Evasive movement to avoid a flying object is constrained by walls and other obstacles.

Procedures and other measures specified under "Actions Required" should minimize the probability for such occurrences, as well as their severity.

Hydraulic hazard

Effectiveness of the deactivation/de-energization analysis and lockout procedure for deactivating mechanical equipment and energy sources should have been confirmed during preentry preparation or the prework inspection.

Work activity could introduce equipment containing hydraulic systems into the space. Procedures and other measures specified under "Actions Required" should minimize the hazard of using this equipment.

Pneumatic hazard

Effectiveness of the deactivation/de-energization analysis and lockout procedure for deactivating mechanical equipment and energy sources should have been confirmed during preentry preparation or the prework inspection.

Work activity could introduce equipment containing pneumatic systems into the space. Procedures and other measures specified under "Actions Required" should minimize the hazard of using this equipment.

Process hazard

Effectiveness of the deactivation/de-energization analysis and lockout and isolation procedure for isolating processes that normally occur in the space should have been confirmed during preentry preparation or the prework inspection.

Work activity often involves use of chemical processes. While the quantities of material used in these processes usually are small, the close proximity to the worker can produce a considerable exposure hazard.

Procedures and other measures specified under "Actions Required" should minimize the probability for such occurrences, as well as their severity.

Structural hazard

Actions taken during preentry preparation could exacerbate an existing structural hazard or create one. Development of a hazardous condition under these circumstances may not be recognized at the time that the critical action occurs. In situations where prework inspection does not occur, consideration about the impact of work procedures on structural stability and integrity is essential.

Advice from a professional engineer about structural stability should be sought where this issue is a concern.

Engulfment/immersion

An engulfment/immersion hazard could exist at the time of entry for performing work. Work procedures could exacerbate the hazard, depending on circumstances.

Entanglement

Entanglement hazards are an inherent property of the geometry of the space, its internal configuration, and that of equipment inside it. Few options may be available for ameliorating potential entanglement hazards.

Electrical

Effectiveness of the deactivation/de-energization analysis and lockout procedure should have been confirmed during preentry preparation or the prework inspection.

Work activity often introduces electrical equipment to the space. The safe use of this equipment is contingent on its integrity and performance. The integrity of electrical equipment can be destroyed rapidly during accident situations. For this reason, GFCIs (ground fault circuit interruptors) should be required on all electrical supply lines. Circumstances could merit use of low voltage equipment or equipment rated for use in hazardous atmospheres.

Procedures and other measures specified under "Actions Required" should minimize the probability for such occurrences and minimize their severity.

Fall

Fall hazard should be corrected as one of the first work activities following entry. Installation of fall prevention and fall arrest equipment should be a prerequisite for any work activity where fall can produce serious injury or exceeds regulated limits.

Slip/trip

Preentry preparation and follow-up measures to the prework inspection should address slip/trip hazards existing in the space. In spaces where thorough removal of residual contents has not occurred, slip/trip may pose a serious and underappreciated safety hazard.

Work activity can introduce slip and trip hazards. This is especially likely where hoses, lines, duct, pipe, and electrical cable could be present on the floor of the space. These objects can create a considerable housekeeping problem.

Procedures and other measures specified under "Actions Required" should minimize the probability for such occurrences, as well as their severity.

Visibility/light level

Purging and ventilation of the space during preentry preparation should have eliminated the visibility hazard resulting from airborne particulates and mists or sprays. However, an uncontrolled or poorly controlled process used during a work procedure can create a new problem.

Installation of lights in the space should be one of the first work activities to occur. Conditions could warrant use of equipment rated for use in hazardous atmospheres. The potential for fire or explosion as an influencing factor deserves careful consideration.

Explosive/implosive

Work activity that introduces explosive substances to the space deserves exceedingly careful consideration because of the consequences of a deliberately planned or accidental explosion. A small difference in pressure across surfaces can create considerable force. Rapid pressurization and depressurization following an explosion could cause structural failure.

Procedures and other measures specified under "Actions Required" should minimize the probability for such occurrences, as well as their severity.

Hot/cold surfaces

Preentry preparation should attempt to eliminate hot or cold surfaces. Surface temperatures that differ from ambient could cause burn, or if high enough, act as a source of ignition. This accommodation may not be possible in circumstances where temperature maintenance is critical. Insulation applied to the surface or personal protective equipment could be the only means for assuring safety under these circumstances. Where this is not possible, this condition must be anticipated to enable the entrant to perform the work activity safely.

Many processes that could be used during work activity involve controlled generation and application of heat. These processes may involve hot surfaces. Hot surfaces utilized in this manner can pose considerable hazard due to proximity of the surface to the worker.

Procedures and other measures specified under Actions Required should minimize the probability for such occurrences, as well as their severity.

Action required

Actions required intend to assess, control, or eliminate hazardous conditions that could occur or develop during work procedures.

Tables 8.10 and 8.11 provide examples of the application of these hazard assessment forms to actual situations.

Table 8.10 highlights the dilemma for hazard management of confined spaces. The steering compartment on a ship fits the concept of a confined space conveyed in many definitions. This space is entered regularly during the watch by engineering personnel. Many other compartments on a ship also would fit this description. Modification of the steering compartment to remove the designation as a confined space could affect the seaworthiness of the vessel. Treating entry into the steering compartment in the manner of other confined spaces would drastically alter operation of the ship, yet provide questionable benefit. The issue with the steering compartment is to provide a mechanism that can accommodate minor hazardous conditions associated with normal work, yet can address more hazardous situations, such as painting and welding.

Table 8.11 provides an example of how work activity can alter requirements for hazard management in a single space: the hoppers in a covered hopper railcar. Railcars undergo repainting, as well as repair of collision damage. The hoppers in a covered hopper car fit the description of a confined space. Requirements for hazard management differ considerably based on the nature of the work to be performed.

TABLE 8.10
Hazard Assessment for Work Involving a Steering Compartment

ABC Company
Confined Space — Identification and Hazard Assessment

1. Descriptive Information

Department: Marine Operations

Location: Waterfront City

Building/shop: Marine Ship

Equipment/space: Steering compartment

Part: Not applicable

Date: xx-yy-zz

Assessor: Hy Gienist

Qualification: aaa, bbb

Basis: personal inspection

Date: xx-yy-zz

Previous Assmt:

Assessor:

Qualification:

2. Assessment Information

Description: The steering compartment extends across the width of the stern of the vessel below the main deck. The space is closed to the atmosphere except through the access manways which normally remain closed. There is no mechanical ventilation.

Adjacent spaces: No. 3 W.B. Tanks P&S

Access/egress: Manways on P&S sides of the main deck

Function/use: The space houses steering gear and associated hydraulic pumps for the vessel. Steering machinery is situated on a platform above the frames (vertical structural members) and bilges.

Contents: Void spaces under the hydraulic pumps may contain a film or small amount of oil originating from the hydraulic pumps. There is no route for entry of other substances.

Process: Not applicable

Equipment: The space contains steering gear and associated hydraulic pumps.

3. Confined Space Identification

Geometry/design

Space can cause personal confinement:	**Yes**		
Space has unstable interior condition:		No	
Space contains flowable solids or liquids:		No	
Release of energy into the space can occur:			**Possible**
Space can cause atmospheric confinement:	**Yes**		

Response in the **highlighted** areas for one or more conditions indicates a potential or actual confined space for which a detailed hazard assessment is required.

TABLE 8.10 (continued)
Hazard Assessment for Work Involving a Steering Compartment

Administrative aspects

Entry occurs or planned by Company personnel:	**Yes**
Entry occurs or planned by Contractor personnel:	**Yes**

Entry by a contractor into a confined space owned and controlled by the Company necessitates formal written contractual agreement of responsibilities.

4. Hazard Assessment

4.1 Hazard Assessment — Undisturbed Space

The undisturbed space poses hazardous conditions requiring a hazard assessment. No

This space is entered several times during the watch. It is not undisturbed within the normal meaning of this term.

4.2 Hazard Assessment — Preentry Preparation

Description: Not applicable

Entry preparation poses hazards requiring hazard assessment. No

This space undergoes no preparation prior to entry. Entry is a normal occurrence during operation of the vessel.

4.3 Hazard Assessment — Prework Inspection

Description: Not applicable

Prework inspection is required as part of the entry preparation process. No

Prework inspection does not occur. Entry is a normal occurrence during operation of the vessel.

4.4 Hazard Assessment — Work Activity

Complete a separate hazard assessment for each work procedure that creates distinct hazards.

4.4.1 Work Procedure: routine inspection/minor repairs

Description: Mechanical personnel enter the space several times during the watch to examine the condition of the steering gear. Minor repairs involve use of hand tools, such as wrenches, pliers, and screwdrivers.

This hazard assessment considers potential hazards associated with the work activity in relation to those in the space and work practices encountered at the time of the inspection.

Hot work No

Atmospheric hazards

Oxygen deficiency **Possible**

Oxygen level in this space remains to be established on a long-term basis.

Oxygen enrichment No

Biochemical/chemical No

Fire/explosion No

Biological hazard No

Ingestion/skin contact hazard **Possible**

Hydraulic fluid may cause skin reaction.

Physical agents

Noise/vibration **Possible**

Space contains noise-producing equipment. Sound levels not determined. Use of noise-producing equipment could produce overexposure.

TABLE 8.10 (continued)
Hazard Assessment for Work Involving a Steering Compartment

Heat/cold stress	No
Nonionizing/ionizing radiation	No
Laser	No
Personal confinement	**Yes**

Workers enter through a manway in the main deck. The manway remains closed except during occupancy. The space is entirely enclosed under the deck. Movement is unrestricted except in the rudder hydraulics area and in the void spaces between the frames. Headroom is adequate for persons of average height, except in the entry walkway where protruding deck beams could cause head injury.

Mechanical hazard	No
Hydraulic hazard	**Possible**

Activation of hydraulic pumps or the rudder mechanism during inspection/servicing could cause injury. Pressure in hydraulic lines used to move the rudder could release suddenly during servicing. Rudder arm could move during change in hydraulic pressure across the piston.

Pneumatic hazard	No
Process hazard	No
Safety hazards	
Structural hazard	No
Engulfment/immersion	No
Entanglement	No
Electrical	**Possible**

Electrical hazard could exist during servicing of hydraulic pumps.

Fall	**Possible**

Fall into bilges is possible under some circumstances during servicing.

Slip/trip	**Possible**

Slip is possible if hydraulic oil spills onto deck plates. Trip is possible if tools and equipment are left on walkways.

Visibility/light level	No
Explosive/implosive	No
Hot/cold surfaces	No

Hazardous conditions in the confined space may satisfy one or more of the rubrics in the definition. The following table summarizes concern during this activity in the confined space.

TABLE 8.10 (continued)
Hazard Assessment — Routine Inspection/Minor Repairs

	Real or Potential Consequence		
	Low	**Moderate**	**High**
• Hot work	NA		
• Atmospheric hazards			
Oxygen deficiency	x		
Oxygen enrichment	NA		
Biochemical/chemical	NA		
Fire/explosion	NA		
• Biological hazard	NA		
• Ingestion/skin contact	x		
• Physical agents			
Noise/vibration		x	
Heat/cold stress	NA		
Nonionizing/ionizing radiation	NA		
Laser	NA		
• Personal confinement	x		
• Mechanical hazard	NA		
• Hydraulic hazard	x		
• Pneumatic hazard	NA		
• Process hazard	NA		
• Safety hazards			
Structural	NA		
Engulfment/immersion	NA		
Entanglement	NA		
Electrical	x		
Fall	x		
Slip/trip	x		
Visibility/light level	NA		
Explosive/implosive	NA		
Hot/cold surfaces	NA		

In this table, the meaning of "toxic substance," "oxygen deficiency," "oxygen enrichment," and "flammable or combustible atmosphere" requires further specification according to standards that exist in a particular jurisdiction. NA means not applicable.

Action Required

- Provide entrants with two-way radio communication with persons outside the space. Create a procedure to ensure consistency of communication under normal and emergency conditions.
- Ensure availability of first aid to persons working in the space and a means for summoning emergency services.
- Require wearing of hard hats.
- Measure and record oxygen levels randomly over a period of time (several months) to determine actual conditions in the space. Normal atmospheric level is 20.9%. Do not enter if level drops below 20.9%. Force-ventilate the space.
- Consult the material safety data sheet to determine whether skin contact with the hydraulic fluid is likely to cause a reaction. If so, provide protective gloves recommended in the MSDS when skin contact is likely.
- Require use of hearing protection during use of noise-producing equipment.
- Devise procedure for locking out hydraulic system and immobilizing the hydraulic ram to prevent unexpected activation of equipment or motion during servicing.
- Require use of fall arrest equipment or install barriers when fall into bilges could occur.

TABLE 8.11
Hazard Assessment for Work Involving a Covered Hopper Rail Car

ABC Company
Confined Space — Identification and Hazard Assessment

1. Descriptive Information

Department: Mechanical Repair

Location: Any City

Building/shop: Car Refurbishing

Equipment/space: Covered hopper car

Part: Hopper

Date: xx-yy-zz **Assessor:** Hy Gienist

Qualification: aaa, bbb

Basis: personal inspection

Date: xx-yy-zz

Previous Assmt: **Assessor:**

Qualification:

2. Assessment Information

Description: The covered hopper car contains two to four compartments. Compartments have sloping and vertical walls. Top access occurs through a circular or rectangular opening. Bottom access occurs through a horizontal sliding or pivoted gate. Interior surfaces of the hopper are coated.

Adjacent spaces: Adjacent hopper(s), end compartment

Access/egress: Opening in top of hopper, gate at bottom of hopper

Function/use: The space is used to hold flowable solids and slurries.

Contents: Cargo residues

Process: Not applicable

Equipment: Not applicable

3. Confined Space Identification

Geometry/design

Space can cause personal confinement:	**Yes**	
Space has unstable interior condition:		No
Space contains flowable solids or liquids:		No
Release of energy into the space can occur:		No
Space can cause atmospheric confinement:	**Yes**	

Response in the **highlighted** areas for one or more conditions indicates a potential or actual confined space for which a detailed hazard assessment is required.

Administrative aspects

Entry occurs or planned by Company personnel:	**Yes**	
Entry occurs or planned by Contractor personnel:		No

TABLE 8.11 (continued)
Hazard Assessment for Work Involving a Covered Hopper Rail Car

Entry by a contractor into a confined space owned and controlled by the Company necessitates formal written contractual agreement of responsibilities.

4. Hazard Assessment

4.1 Hazard Assessment — Undisturbed Space

The undisturbed space poses hazardous conditions requiring a hazard assessment. **Yes**

Atmospheric hazards

Oxygen deficiency **Possible**

Oxygen may be consumed by chemical action with cargo residues.

Oxygen enrichment No
Biochemical/chemical **Possible**

Substances in airborne dust could pose an exposure hazard.

Fire/explosion **Possible**

Dusts from some cargoes can form explosible mixtures.

Biological hazard **Possible**

Spores from organisms and dusts may accompany cargoes of agricultural origin. Assessing level of concern is extremely difficult, since these agents affect only sensitive individuals.

Ingestion/skin contact hazard **Possible**

Substances in cargo residues may cause skin irritation. Some substances pose an inhalation hazard.

Physical agents
Noise/vibration No
Heat/cold stress No
Nonionizing/ionizing radiation No
Laser No
Personal confinement **Yes**

The space is vertical in orientation. Walls are either vertical or sloping or curved. Access/egress is restricted to the openings at top and bottom of the hopper.

Mechanical hazard No
Hydraulic hazard No
Pneumatic hazard No
Process hazard No
Safety hazards
Structural hazard No
Engulfment/immersion No
Entanglement No
Electrical No
Fall **Yes**

Entry from the top and work on the top of the car would pose a fall hazard.

Slip/trip **Possible**

Cargo residues on surfaces could pose a slip hazard.

Visibility/light level **Yes**

TABLE 8.11 (continued)
Hazard Assessment for Work Involving a Covered Hopper Rail Car

The space contains no interior lighting.

Explosive/implosive	No
Hot/cold surfaces	No

Hazardous conditions in the confined space may satisfy one or more of the rubrics in the definition. The following table summarizes concern during this activity in the confined space.

Hazard Assessment — Undisturbed Space

Hazardous Condition	Real or Potential Consequence		
	Low	Moderate	High
• Hot work	NA		
• Atmospheric hazards			
Oxygen deficiency	x		
Oxygen enrichment	NA		
Biochemical/chemical	x		
Fire/explosion	x		
• Biological hazard	x		
• Ingestion/skin contact	x		
• Physical agents			
Noise/vibration	NA		
Heat/cold stress	NA		
Nonionizing/ionizing radiation	NA		
Laser	NA		
• Personal confinement		x	
• Mechanical hazard	NA		
• Hydraulic hazard	NA		
• Pneumatic hazard	NA		
• Process hazard	NA		
• Safety hazards			
Structural	NA		
Engulfment/immersion	NA		
Entanglement	NA		
Electrical	NA		
Fall	x		
Slip/trip	x		
Visibility/light level	x		
Explosive/implosive	NA		
Hot/cold surfaces	NA		

In this table, the meaning of "toxic substance," "oxygen deficiency," "oxygen enrichment," and "flammable or combustible atmosphere" requires further specification according to standards that exist in a particular jurisdiction. NA means not applicable.

Action Required

4.2 Hazard Assessment — Preentry Preparation

Description: Not applicable

Entry preparation poses hazards requiring hazard assessment. No

This space undergoes no general preparation prior to entry. Specific actions occur during preparation for work activity. Entry is a normal occurrence during servicing, repair, modification, and repainting.

TABLE 8.11 (continued)
Hazard Assessment for Work Involving a Covered Hopper Rail Car

4.3 Hazard Assessment — Prework Inspection

Description: This section is used to provide a description of action(s) that occur during prework inspection.

Prework inspection is required as part of the entry preparation process. No

This space undergoes no inspection prior to the start of work.

4.4 Hazard Assessment — Work Activity

Complete a separate hazard assessment for each work procedure that creates distinct hazards.

4.4.1 Work Procedure: Abrasive blasting to remove interior coating

Description: Abrasive blasting to remove the interior coating is performed in the shotblast booth in the paint shop. Interior blasting is performed from a platform that is installed inside the hopper either from the top or the bottom. The top-mounted platform is used in most cases. Abrasive blasting utilizes high pressure air from the plant air system.

This hazard assessment considers potential hazards associated with the work activity in relation to those in the space and work practices encountered at the time of the inspection.

Hot work No

Atmospheric hazards

Oxygen deficiency **Possible**

Oxygen may be consumed by chemical action with cargo residues. Compressed air used for abrasive blasting will displace the atmosphere in the space soon after blasting starts. Workers use supplied-air respiratory protection. This minimizes potential for exposure to oxygen-deficient conditions.

Oxygen enrichment No

Biochemical/chemical **Possible**

Substances in airborne dust from cargo and paint residues may exceed legal limits. Workers use supplied-air respiratory protection. This minimizes potential for inhalation of airborne contaminants.

Fire/explosion **Possible**

Dusts from some cargoes are explosible. Compressed air used for abrasive blasting creates high dust levels.

Biological hazard **Possible**

Spores from organisms and dusts may accompany cargoes of agricultural origin. Assessing level of concern is extremely difficult, since these agents affect only sensitive individuals. Compressed air used for abrasive blasting will displace the original atmosphere from the space soon after blasting starts. Workers entering to perform abrasive blasting wear supplied-air respiratory protection. This minimizes potential for inhalation of contaminants of biological origin.

Ingestion/skin contact hazard **Possible**

Substances in cargo and paint residues may cause skin irritation and pose an ingestion hazard. Workers wear skin protection because of abrasive blasting. Skin contact is unlikely to occur under these circumstances.

Physical Agents

Noise/vibration **Yes**

Abrasive blasting is a very noisy process.

Heat/cold stress No

Nonionizing/ionizing radiation No

Laser No

Personal confinement **Yes**

TABLE 8.11 (continued)
Hazard Assessment for Work Involving a Covered Hopper Rail Car

The space is vertical in orientation. The geometry of this space is confining. Walls either are vertical or sloping or curved. Access/egress is restricted to the opening at the top and bottom of the hopper.

Mechanical hazard	No	
Hydraulic hazard	No	
Pneumatic hazard	**Yes**	

Pressure of air used in abrasive blasting is high enough to cause injection injuries.

Process hazard	No	
Safety hazards		
Structural hazard	No	
Engulfment/immersion	No	
Entanglement	No	
Electrical	No	
Fall		**Possible**

Fall from the platform at the bottom of the hopper could occur.

Slip/trip	**Yes**

Tapering walls inside the space pose a slipping hazard. Potential for contact with the walls occurs only when work occurs on the platform at the bottom of the hopper.

Visibility/light level	**Yes**

The space contains no interior lighting. Particulate level during blasting inside the hopper creates a potential visibility hazard.

Explosive/implosive	No
Hot/cold surfaces	No

Hazardous conditions in the confined space may satisfy one or more of the rubrics in the definition. The following table summarizes concern during this activity in the confined space.

Hazard Assessment — Abrasive Blasting

	Real or Potential Consequence				
Hazardous Condition	**Low**		**Moderate**		**High**
• Hot work	NA				
• Atmospheric hazards					
Oxygen deficiency	x				
Oxygen enrichment	NA				
Biochemical/chemical	x				
Fire/explosion			x		
• Biological hazard	x				
• Ingestion/skin contact	x				
• Physical agents					
Noise/vibration				x	
Heat/cold stress	NA				
Nonionizing/ionizing radiation	NA				
Laser	NA				
• Personal confinement		x			

TABLE 8.11 (continued)
Hazard Assessment for Work Involving a Covered Hopper Rail Car

	Real or Potential Consequence			
Hazardous Condition	**Low**		**Moderate**	**High**
• Mechanical hazard	NA			
• Hydraulic hazard	NA			
• Pneumatic hazard			x	
• Process hazard	NA			
• Safety hazards				
Structural	NA			
Engulfment/immersion	NA			
Entanglement	NA			
Electrical	NA			
Fall		x		
Slip/trip	x			
Visibility/light level		x		
Explosive/implosive	NA			
Hot/cold surfaces	NA			

In this table, the meaning of "toxic substance," "oxygen deficiency," "oxygen enrichment," and "flammable or combustible atmosphere" requires further specification according to standards that exist in a particular jurisdiction. NA means not applicable.

Action Required

- Provide entrants two-way radio communication with persons outside the space. Create a procedure to ensure consistency of communication under normal and emergency conditions.
- Consider a visual indicator of occupancy in the abrasive blasting booth. An annunciator panel is a possible option.
- Ensure availability of first aid to persons working in the space and a means for summoning emergency services.
- Measure and record oxygen levels randomly over a period of time (several months) to determine actual conditions in the space. Normal atmospheric level is 20.9%. Do not enter if level drops below 20.9%. Force ventilate.
- Refer to industrial hygiene surveys for information about dust levels generated during this work and recommendations for control.
- Minimize potential for generating airborne dust from protective clothing by use of a HEPA-filtered vacuum cleaner. Vacuum cleaning around the head and neck, hands, feet, and zipper is strongly urged to minimize dust generation during removal of this clothing. Air blowing to remove dust must not be used.
- Require entrants to wash face and hands and, as necessary, head and neck prior to eating, drinking, or smoking, or leaving the site.
- Provide information to entrants about contaminants known to be present in the dust. Provide training in personal hygiene measures to minimize potential for internal contamination during this work.
- Provide hearing protection to entrants and training in its use.
- Instruct entrants about injection hazards posed by use of high pressure air.
- Provide portable lighting for use in the hopper.
- Test dust for explosibility to establish electrical rating of lighting.

4.4.2 Work Procedure: Spray painting interior of hopper

Description: Spray painting to apply the interior coating is performed in the spray painting booth in the paint shop. Interior painting is performed from a platform that is installed inside the hopper through the top loading hatch. Spray painting utilizes high pressure air from the plant air system.

This hazard assessment considers potential hazards associated with the work activity in relation to those in the space and work practices encountered at the time of the inspection.

TABLE 8.11 (continued)
Hazard Assessment for Work Involving a Covered Hopper Rail Car

Hot work No

Atmospheric hazards

Oxygen deficiency **Possible**

Oxygen may be consumed by chemical action with iron in exposed steel. Compressed air used for spray painting will displace original contents of the space soon after painting starts. Spray painter uses a supplied-air respirator.

Oxygen enrichment No

Biochemical/chemical **Possible**

Substances in paint solids, mist, and vapor may exceed airborne legal limits. Spray painter uses supplied-air respirator.

Fire/explosion **Possible**

Substances in paint mist and vapor may form combustible/explosive mixture.

Biological hazard No

Ingestion/skin contact hazard **Possible**

Substances in paint may cause skin irritation and ingestion hazard. Painter wears skin protection against the paint. Skin contact with paint should be minimal.

Physical agents

Noise/vibration **Yes**

Spray painting in confined spaces is potentially noisy.

Heat/cold stress No

Nonionizing/ionizing radiation No

Laser No

Personal confinement **Yes**

While the structure that is lowered into the hopper is confining, alternative approaches, such as entry from the bottom, are more so. The geometry of the hopper is very confining. Walls may be vertical or sloping or curved.

Mechanical hazard No

Hydraulic hazard **Yes**

Pressure of liquid used in spray painting is high enough to cause injection injury.

Pneumatic hazard **Yes**

Pressure of air used in spray painting is high enough to cause injection injury.

Process hazard No

Safety hazards

Structural hazard No

Engulfment/immersion No

Entanglement No

Electrical No

Fall No

Slip/trip No

Visibility/light level **Yes**

The space contains no interior lighting.

TABLE 8.11 (continued)
Hazard Assessment for Work Involving a Covered Hopper Rail Car

Explosive/implosive	No
Hot/cold surfaces	No

Hazardous conditions in the confined space may satisfy one or more of the rubrics in the definition. The following table summarizes concern during this activity in the confined space.

Hazard Assessment — Spray Painting Interior of Hopper

Hazardous Condition	Real or Potential Consequence		
	Low	Moderate	High
• Hot work	NA		
• Atmospheric hazards			
Oxygen deficiency	x		
Oxygen enrichment	NA		
Biochemical/chemical			x
Fire/explosion		x	
• Biological hazard	x		
• Ingestion/skin contact	x		
• Physical agents			
Noise/vibration		x	
Heat/cold stress	NA		
Nonionizing/ionizing radiation	NA		
Laser	NA		
• Personal confinement	x		
• Mechanical hazard	NA		
• Hydraulic hazard		x	
• Pneumatic hazard		x	
• Process hazard	NA		
• Safety hazards			
Structural	NA		
Engulfment/immersion	NA		
Entanglement	NA		
Electrical	NA		
Fall	NA		
Slip/trip	NA		
Visibility/light level	x		
Explosive/implosive	NA		
Hot/cold surfaces	NA		

In this table, the meaning of "toxic substance," "oxygen deficiency," "oxygen enrichment," and "flammable or combustible atmosphere" requires further specification according to standards that exist in a particular jurisdiction. NA means not applicable.

Action Required

- Provide entrants two-way radio communication with persons outside the space. Create a procedure to ensure consistency of communication under normal and emergency conditions.
- Ensure availability of first aid to persons working in the space and a means for summoning emergency services.
- Measure and record oxygen levels randomly over a period of time (several months) to determine actual conditions in the space. Normal atmospheric level is 20.9%. Do not enter if level drops below 20.9%. Force ventilate.

TABLE 8.11 (continued)
Hazard Assessment for Work Involving a Covered Hopper Rail Car

- Refer to industrial hygiene surveys for information about dust levels generated during this work and recommendations for control.
- Require entrants to wash face and hands and, as necessary, head and neck prior to eating, drinking, or smoking, or leaving the site.
- Provide information to entrants about contaminants known to be present in the paint. Provide training in personal hygiene measures to minimize potential for internal contamination during the work.
- Provide hearing protection to entrants.
- Instruct entrants about injection hazards posed by work with pressurized liquids.
- Provide portable lighting rated for use in hazardous atmospheres.
- Measure concentration of mist and vapor in the space during painting.

4.4.3 Work Procedure: Hot work in hoppers

Description: Hot work (grinding, air arcing, welding, burning, and cutting) in hoppers is performed in the car repair shop. This work decomposes and burns paint, as well as cargo residues. Entry occurs through the bottom and top of the hopper. The body of the car usually is rotated so that the openings are located at the side. This considerably improves access to work surfaces, but depending on orientation, can significantly reduce natural convective airflow through the openings of the hopper.

During repair work, the interior of the hopper usually is partially exposed due to removal of steel side sheets.

This hazard assessment considers potential hazards associated with the work activity in relation to those in the space and work practices encountered at the time of the inspection.

Hot work **Yes**

Hot work includes grinding, plasma and air arcing, welding, burning, and cutting. Hot work destroys coatings and burnable residual contents. These processes can generate combustion products, such as carbon dioxide, carbon monoxide, sulfur dioxide, and nitrogen oxides, and many decomposition products. Welding and related processes produce a plume that contains metals and metallic oxides, as well as other substances and gases. Hot work produces extremely hot surfaces. In the confinement of the space, these surfaces can pose considerably greater hazard than work in open conditions.

Atmospheric hazards

Oxygen deficiency **Possible**

Oxygen may be consumed by burning of paint or cargo residues. Oxygen deficiency can occur due to displacement by shield gases during welding.

Oxygen enrichment **Possible**

Accidental enrichment caused by leakage from an oxygen supply line to an oxy-fuel torch or other similar device could occur. This event should have low probability, provided that safety practices involving redundant or fail-safe measures are followed.

Biochemical/chemical **Yes**

High heat produces substances different from those in the paint or cargo residues. Hot work destroys coatings and burnable residual contents. These processes can generate combustion products, such as carbon dioxide, carbon monoxide, sulfur dioxide, and nitrogen oxides, and numerous decomposition products. Welding and related processes produce a plume containing metals and metallic oxides, as well as other substances and gases. Grinding produces metal particles and remnants of the wheel.

Fire/explosion **Possible**

Dusts of some cargoes are potentially explosible. Hot work of this type does not appear to be a cause of fire. Fuel gases used in oxy-fuel cutting could leak from equipment and accumulate in the space.

Biological hazard **Possible**

TABLE 8.11 (continued)
Hazard Assessment for Work Involving a Covered Hopper Rail Car

Spores from organisms and dusts may accompany cargoes of agricultural origin. Assessing level of concern is extremely difficult, since these agents affect only sensitive individuals.

Ingestion/skin contact hazard		**Possible**

Substances in some cargo residues may cause skin irritation or pose an ingestion hazard.

Physical agents			
Noise/vibration	**Yes**		

Some activities associated with hot work can generate considerable noise.

Heat/cold stress			**Possible**

Heat stress is possible during summer months due to entrapment of thermal energy within the structure.

Nonionizing/ionizing radiation	**Yes**		

Oxy-fuel and arc welding equipment are sources of ultraviolet, visible, and/or infrared energies. Reflection from surfaces of the structure increases the paths by which exposure can occur.

Laser		No	
Personal confinement	**Yes**		

The geometry of this space is highly confining. Walls may be vertical or sloping or curved. Access/egress is restricted to the opening at the bottom of the hopper and the opening at the top. Openings may be more accessible following rotation of the car.

Mechanical hazard	**Yes**		

Use of portable grinders and milling machines introduces the hazard from the shattered wheel and runaway equipment.

Hydraulic hazard		No	
Pneumatic hazard	**Yes**		

Pressure of air used in air-arcing is high enough to cause injection injury.

Process hazard		No	
Safety hazards			
Structural hazard		No	
Engulfment/immersion		No	
Entanglement		No	
Electrical	**Yes**		

Under hot conditions, perspiration can render the body considerably more conductive. Improper use of or defective electrical equipment can lead to electrocution.

Fall			**Possible**

Fall hazard can exist under some orientations of the body of the car.

Slip/trip	**Yes**		

Sloping walls pose a slipping hazard.

Visibility/light level			**Possible**

The space contains no interior lighting.

Explosive/implosive		No	
Hot/cold surfaces	**Yes**		

TABLE 8.11 (continued)
Hazard Assessment for Work Involving a Covered Hopper Rail Car

Hot surfaces created by hot work combined with the sloping surface of the interior of the hopper can pose serious burn hazard.

Hazardous conditions in the confined space may satisfy one or more of the rubrics in the definition. The following table summarizes concern during this activity in the confined space.

Hazard Assessment — Hot Work in Hopper

Hazardous Condition	Real or Potential Consequence		
	Low	Moderate	High
• Hot work		x	
• Atmospheric hazards			
Oxygen deficiency		x	
Oxygen enrichment	x		
Biochemical/chemical		x	
Fire/explosion	x		
• Biological hazard	x		
• Ingestion/skin contact	x		
• Physical agents			
Noise/vibration		x	
Heat/cold stress	x		
Nonionizing/ionizing radiation		x	
Laser	NA		
• Personal confinement		x	
• Mechanical hazard		x	
• Hydraulic hazard	NA		
• Pneumatic hazard	x		
• Process hazard	NA		
• Safety hazards			
Structural	NA		
Engulfment/immersion	NA		
Entanglement	NA		
Electrical	x		
Fall	x		
Slip/trip	x		
Visibility/light level	x		
Explosive/implosive	NA		
Hot/cold surfaces		x	

In this table, the meaning of "toxic substance," "oxygen deficiency," "oxygen enrichment," and "flammable or combustible atmosphere" requires further specification according to standards that exist in a particular jurisdiction. NA means not applicable.

Action Required

- Provide entrants with protective clothing and equipment suitable for hot work and failure of equipment.
- Instruct entrants to inspect tools and equipment carefully before starting work and to tag damaged items out of service.
- Provide entrants two-way radios or other suitable means to maintain communication with persons outside the space. Create a procedure to ensure consistency of communication under normal and emergency conditions.

TABLE 8.11 (continued)
Hazard Assessment for Work Involving a Covered Hopper Rail Car

- Provide an attendant at the entrance to the space who is equipped to assist the entrant in situations where work is occurring in a fully enclosed hopper or where the entrant is not readily visible to other workers in the shop.
- Ensure availability of first aid to persons working in the space and a means for summoning emergency services.
- Measure and record oxygen levels randomly over a period of time (several months) to determine actual conditions in the space. Normal atmospheric level is 20.9%. Do not enter or remain if level drops below 20.9% or if it exceeds 20.9%. Force ventilate.
- Remove gas-shielded welding equipment and oxy-fuel cutting torches from the space prior to leaving for breaks or other prolonged period. Close valves on supply cylinders. If equipment is mistakenly left in the space and/or the cylinder valves are left open, test for oxygen and flammables/combustibles prior to starting work. If contamination has occurred (oxygen greater than 20.9% or flammables/combustibles greater than 0% LFL [lower flammable limit]), prohibit entry. Either leave the space undisturbed until normal conditions are restored, or force ventilate using a portable ventilator. Ventilator and duct must be electrically bonded to the metal of the hopper and rated for use in hazardous atmospheres.
- Determine levels of airborne contaminants during hot work in hoppers. Provide entrants with NIOSH-approved, full-facepiece airline respirators equipped with 5-minute escape bottle until testing can confirm that conditions would accommodate use of other respiratory protection.
- Utilize the local exhaust system at all times during hot work in hoppers.
- Require entrants to wash face and hands and, as necessary, head and neck prior to eating, drinking, or smoking, or leaving the site.
- Provide training in personal hygiene measures to minimize potential for internal contamination during work activity.
- Provide information to entrants about contaminants known to be present in combustion and decomposition products.
- Provide hearing protection to entrants and instruction in its use.
- Provide education about heat-related illnesses and make available cold water for consumption during summer months.
- Instruct entrants about injection hazards posed by work with high pressurized air.
- Provide portable lighting and electrical equipment incorporating GFCIs (ground fault circuit interrupter).
- Utilize fall arrest equipment where fall can cause serious injury or fall distance exceeds legal limits.

EMERGENCY RESPONSE

Events occurring under emergency conditions involving confined spaces have cost the lives of many would-be rescuers (OSHA 1985). These tragedies have resulted from the basic human urge to assist those in distress. Persons unprepared and unprotected have entered hostile environments to rescue friends and co-workers. These acts of courage will continue no matter what is written or done to dissuade them.

Emergency preparedness and response are the subject of Chapter 15.

The greatest single action that could reduce the occurrence of these tragedies is to eliminate or at least control hazardous conditions in the confined space. This has formed the focus of the "Action Required" section of the hazard assessment process.

Accidents will continue to occur in confined spaces despite the taking of precautions. The amplifying effect of the boundary surfaces makes these spaces inherently more dangerous in which to work. There is no reason not to expect, though, that these accidents should differ in type from those that occur in ordinary workplaces. If preparatory and precautionary measures for work activity are adequate, the only remaining consideration should be access and egress.

Table 8.12 provides a format for consideration about emergency response needed for individual work procedures. This table follows the format used in previous sections. Table 8.12 provides a focused basis for considering measures needed to address individual elements.

TABLE 8.12
Emergency Response Assessment — Work Procedures

4.5 Emergency Response Assessment — Work Procedures

Complete a separate emergency response assessment for each work procedure that creates distinct hazards.

4.5.x Work Procedure: title of work procedure

Description: This section is used to provide a description of action(s) that occur during the work procedure.

This emergency response assessment considers potential hazards associated with the work activity in relation to those associated with the space.

Hot work	**Yes** No **Possible**
Atmospheric hazards	
Oxygen deficiency	**Yes** No **Possible**
Oxygen enrichment	**Yes** No **Possible**
Biochemical/chemical	**Yes** No **Possible**
Fire/explosion	**Yes** No **Possible**
Biological hazard	**Yes** No **Possible**
Ingestion/skin contact hazard	**Yes** No **Possible**
Physical agents	
Noise/vibration	**Yes** No **Possible**
Heat/cold stress	**Yes** No **Possible**
Nonionizing/ionizing radiation	**Yes** No **Possible**
Laser	**Yes** No **Possible**
Personal confinement	**Yes** No **Possible**
Mechanical hazard	**Yes** No **Possible**
Hydraulic hazard	**Yes** No **Possible**
Pneumatic hazard	**Yes** No **Possible**
Process hazard	**Yes** No **Possible**
Safety hazards	
Structural hazard	**Yes** No **Possible**
Engulfment/immersion	**Yes** No **Possible**
Entanglement	**Yes** No **Possible**
Electrical	**Yes** No **Possible**
Fall	**Yes** No **Possible**
Slip/trip	**Yes** No **Possible**
Visibility/light level	**Yes** No **Possible**
Explosive/implosive	**Yes** No **Possible**
Hot/cold surfaces	**Yes** No **Possible**

Hazardous conditions in the confined space may satisfy one or more of the rubrics in the definition. The following table summarizes concern during this activity in the confined space.

Hazard Assessment — Emergency Response

Hazardous Condition	Real or Potential Consequence		
	Low	Moderate	High
• Hot work			
• Atmospheric hazards			
Oxygen deficiency			
Oxygen enrichment			
Biochemical/chemical			
Fire/explosion			
• Biological hazard			

TABLE 8.12
Emergency Response Assessment — Work Procedures

Hazardous Condition	Real or Potential Consequence		
	Low	Moderate	High
• Ingestion/skin contact			
• Physical agents			
Noise/vibration			
Heat/cold stress			
Nonionizing/ionizing radiation			
Laser			
• Personal confinement			
• Mechanical hazard			
• Hydraulic hazard			
• Pneumatic hazard			
• Process hazard			
• Safety hazards			
Structural			
Engulfment/immersion			
Entanglement			
Electrical			
Fall			
Slip/trip			
Visibility/light level			
Explosive/implosive			
Hot/cold surfaces			

In this table, the meaning of "toxic substance," "oxygen deficiency," "oxygen enrichment," and "flammable or combustible atmosphere" requires further specification according to standards that exist in a particular jurisdiction. NA means not applicable.

Action Required

The following sections provide meaning to terms used in the hazard assessment in the context of emergency preparedness and response.

Hot work

Hot work can act as a source of ignition to flammable/combustible materials both inside and external to the confined space. It also could cause burn and other heat-related injuries.

Oxygen deficiency

Preparatory work, as well as ventilation and atmospheric testing during the work activity, as deemed necessary, should be capable of preventing oxygen deficiency from developing in the space under normal circumstances.

An inerted space containing oxygen at a level below 5% cannot sustain life. This situation, therefore, necessitates the highest level of precaution for persons entering the confined space to attempt rescue.

Oxygen enrichment

Preparatory work, as well as ventilation and atmospheric testing during the work procedure, as deemed necessary, should be capable of preventing oxygen enrichment from developing in the space under normal circumstances.

Biochemical/chemical

Preparatory work, including ventilation and testing, should have successfully eliminated or controlled the chemical or biochemical hazard from residual contents.

Ventilation may or may not be effective in controlling emissions from processes utilized during work procedures. Also, respiratory protection could be required. Emergency response must be prepared to address these conditions as a prerequisite for entry into the space.

Fire/explosion

Ventilation and other preparatory measures should have eliminated any fire/explosion hazard resulting from residual materials. A hazardous condition arising from residual materials is much more likely where removal was not undertaken during preentry preparation.

A process utilized during work activity could introduce flammable/combustible materials into the space or generate them, as well as sources of ignition. A source of ignition used in a confined space containing only residues of flammable/combustible material could be extremely hazardous.

A cloud formed in the surroundings external to the space during removal of a contaminant by exhaust ventilation could contain a flammable/explosive/explosible mixture.

These situations should have low probability, provided that safety practices involving redundant or fail-safe measures are followed. As the possibility for fire does exist despite these measures, careful attention is warranted during planning for emergency response to ensure that appropriate measures are taken.

Biological hazard

Ventilation and other preparatory measures should have eliminated any biological hazard existing in the space. Involvement of a biological vector in a situation that would prompt an emergency response is unlikely. The most probable incident would be an allergic reaction.

Ingestion/skin contact hazard

Preparatory activities should have attempted to remove bulk and residual quantities of substances that are hazardous by ingestion or skin contact. However, residues still could remain despite all attempts at removal.

Materials introduced into the space during work procedures also could produce toxic effects on the skin. The most probable toxic effect for which emergency response would be required would be irritation or corrosive attack on the skin. Consult the Material Safety Data Sheet.

Noise/vibration

Noise and vibration are unlikely to prompt an emergency situation.

Heat/cold stress

If addressed through preparatory activities and precautionary measures in work procedures, heat and cold stresses are unlikely to prompt an emergency situation.

Nonionizing/ionizing radiation

If addressed through preparatory activities and precautionary measures in work procedures, sources of nonionizing and ionizing radiation are unlikely to prompt an emergency situation.

Laser

If addressed through preparatory activities and precautionary measures in work procedures, laser sources are unlikely to prompt an emergency situation.

Personal confinement

As an inherent property of the geometry of the space, its internal configuration, and portals, personal confinement may continue to occur as a potentially hazardous condition during work procedures. Personal confinement could lead to claustrophobia in some individuals. Complicated internal geometry and small access/egress portals can considerably complicate extrication of an individual injured in an accident.

Mechanical hazard

Mechanical equipment introduced into the space by the work activity could pose considerable hazard because of the confines of the space. Portable tools could run away. Grinding wheels could shatter and the pieces could ricochet off surfaces. Despite procedures and precautionary measures to minimize the probability for accidents, as well as their severity, these events still could cause serious traumatic injury.

Hydraulic hazard

Equipment containing hydraulic systems introduced into the space by the work activity likely would pose little hazard when used as directed. Uncontrolled release of liquids under high pressure could cause injection injuries.

Pneumatic hazard

Equipment containing pneumatic systems introduced into the space by the work activity likely would pose little hazard when used as directed. Uncontrolled release of gases under high pressure could cause injection injuries.

Process hazard

If addressed through preparatory activities and precautionary measures in work procedures, process hazards are unlikely to prompt an emergency situation.

Structural hazard

Preparatory activities and precautionary measures in work procedures under the guidance of a Professional Engineer, as appropriate, should minimize the potential for involvement of structural hazards in emergency situations.

Accidents involving "struck by" and "struck against" situations are virtually impossible to prevent. Despite procedures and precautionary measures to minimize the probability for these accidents, as well as their severity, these events still could cause serious traumatic injury.

Engulfment/immersion

An engulfment/immersion hazard could exist at the time of work activity. Work procedures could exacerbate engulfment or immersion hazards.

Entanglement

If addressed through preparatory activities and precautionary measures in work procedures, entanglement is unlikely to prompt an emergency situation.

Entanglement is caused by the interaction between structural features of the space and its interior equipment, and clothing and other equipment, such as lifelines. Entanglement of lifelines could pose a concern during work activities where the entrant must have freedom to move throughout a space, and into and around equipment. The benefits offered by use of lifelines should not be outweighed by the consequences.

Electrical

Preparatory activities and precautionary measures in work procedures should minimize the potential for electrical accidents. The main concerns about electrocution result from use of equipment containing faulty wiring, and increased conductivity of the body caused by perspiration and other sources of moisture.

Fall

Preparatory activities and precautionary measures in work procedures should minimize the potential for fall-related accidents.

Accidents involving falls are virtually impossible to prevent. Despite procedures and precautionary measures to minimize the probability for these accidents, as well as their severity, falls could cause serious traumatic injury.

Slip/trip

Preparatory activities and precautionary measures in work procedures should minimize the potential for slip/trip-related accidents.

Accidents involving slips and trips are virtually impossible to prevent. Despite procedures and precautionary measures to minimize the probability for these accidents, as well as their severity, slips and trips still could cause serious traumatic injury.

Visibility/light level

If addressed through preparatory activities and precautionary measures in work procedures, visibility and light level are unlikely to prompt an emergency situation.

Explosive/implosive

If addressed through preparatory activities and precautionary measures in work procedures, use of explosives is unlikely to prompt an emergency situation.

Hot/cold surfaces

If addressed through preparatory activities and precautionary measures in work procedures, hot or cold surfaces are unlikely to prompt an emergency situation.

Action required

Actions required are those needed to respond to emergency situations believed able to occur during a particular work procedure.

HAZARD MANAGEMENT SYSTEMS

General Discussion

"Hazard management" is the term used to describe actions taken in an organized, progressive, and coherent manner to assess, control, or eliminate hazardous conditions. Hazard management is a direct outcome from hazard assessment. Hazard management is essential for assuring the safety of work involving confined spaces. Hazard management involves the management of both people and conditions. Focusing on one to the exclusion of the other could lead to failure or only partial success.

Hazard management is separate from project management. Hazard management focuses on actions needed to perform work in the manner that minimizes risks. Project management focuses on performing the work as efficiently as possible. Conflict between these elements is natural, although not necessary, because the focus of each is very different. The perception of the efficient conduct of work unfortunately could conflict with performance of work in a manner that minimizes hazards. Hazard management and project planning should be undertaken by different people. One individual, unless highly cognizant about these conflicts of interest, is unlikely to be able to perform both tasks in a manner that does justice to each.

Management of conditions associated with confined workspaces can take several directions. One extreme is expressed by the philosophy best described as "one size fits all." By definition, this is the only strategy that is simultaneously applicable to all circumstances. This approach presumes worst-case situations at all times. Of course, presumption of worst case demands a response of appropriate magnitude. While this approach could be applied to all confined spaces, the question of appropriateness arises. Certainly, this approach would probably waste resources.

Clearly, the response to two situations that differ considerably in level of hazard need not be the same. Assigning the same high level of effort to the two situations also would constitute a misapplication of resources. On the other hand, providing a median level of response to the two situations also would be inappropriate. This could mean overresponse in the one case and underresponse in the other. The most appropriate approach is to fit the response to the needs of the situation. On the surface, this approach would seem to be very inefficient, since considerable expenditure of time would be required. This need not be the case, as later discussion will indicate.

Individualization provides an opportunity for innovation to establish strategies for addressing specific situations. However, any strategy proposed for a specific situation must conform to the general framework described in the preceding section. In addition, the strategy must satisfy the following questions:

- Have appropriate steps been taken to ensure control over conditions during entry and the course of work?
- What unforeseen events can happen?
- What steps have been taken to address them?

Any organization taking this approach must prepare to document and justify its decisions.

A starting point for discussion about this question is the nature of work procedures in confined spaces. One or more activities may occur within a confined space during a particular entry and work situation. Describing the nature of these activities serves to illustrate the potential utility of individual hazard assessments. Table 8.13 documents the nature of work activities in confined spaces and illustrates that conditions can range from predictability and minimal uncertainty to unpredictability and high uncertainty. Transitions of this kind tax the ability of all but the most knowledgeable and experienced of experts. The preceding bluntly illustrates why accidents occur in confined spaces. These are exceedingly complex environments.

Predictability is the basis on which planning decisions can be made. The certainty that one situation is unpredictable is as important a finding as predictability in another. These conclusions can be used to direct planning. The extent of certainty or uncertainty mandates a type and level of response. The conclusion that a situation is predictable within limits of a reasonable level of doubt also can form the basis for planning. The more certain and defined that a situation can be made, the more precise and planned the action that it mandates can become.

Obviously, spaces possessing high potential severity of hazard that are entered frequently deserve to be addressed sooner (when that commodity must be rationed) than spaces posing low levels of hazard that are entered infrequently. Of course, decision making usually is not this clear-cut.

Attention first should be given to spaces that are entered frequently and that are believed to pose hazardous conditions. Devoting time and energy to a confined space that will not be entered for a year to come while others are entered routinely is counter productive. Also, focus first should go toward work procedures that occur frequently, rather than those that occur rarely. Entry for inspection creates little hazard, compared to other activities.

Hazard Classification

The ANSI Standard on confined spaces and legislation, such as the OSHA standard for general industry, requires classification of the overall hazard-posed conditions in the confined space (ANSI 1995, OSHA 1993). This classification determines the level of response needed to enter the space safely. The analytical process provided in the preceding section is compatible with these requirements.

The ANSI Standard uses two classes — **nonpermit confined space** and **permit-required confined space** — to govern the nature and extent of the response (ANSI 1995).

For nonpermit confined spaces, the Standard requires written procedures that specify measures and precautions for safe entry and work. These also would identify conditions that would revoke the status as a nonpermit confined space and require reevaluation. A nonpermit confined space would require control of atmospheric hazards within acceptable limits.

TABLE 8.13
Nature of Work Activities in Confined Spaces

Frequent entry, nonhazardous conditions, nonhazardous work represents the simplest situation. In this case, the conditions in the space, as well as the work to be performed, pose minimal hazard. This can be established through random measurement of atmospheric conditions, job hazard analysis, and consideration of previous experience. Frequency of entry establishes constancy of conditions with time.

Frequent entry, nonhazardous conditions, hazardous work introduces the possibility of hazardous conditions produced by the performance of work. This situation is very common and very difficult to assess. Conditions at time of entry can be completely nonhazardous. Yet, processes and activity that occur during work may create highly hazardous conditions. The difficulty in assessing this situation results from the interaction between the activity and geometry of the confined workspace.

Frequent entry, hazardous conditions, nonhazardous work is a common situation. In this case, the hazardous condition develops in the span of time between entries. Frequency of entry offers continuity of experience. Preparatory and precautionary measures may render conditions in the space nonhazardous prior to entry.

Frequent entry, hazardous conditions, hazardous work is a common situation. In this case, the hazardous condition develops in the span of time between entries. Frequency of entry offers continuity of experience. Preparatory and precautionary measures may render conditions in the space nonhazardous prior to entry. Conditions at time of entry can be completely nonhazardous. Yet, processes and activity that occur during work create highly hazardous conditions. The difficulty in assessing this situation results from the interaction between the activity and geometry of the confined workspace.

Infrequent entry, nonhazardous work introduces the potential for development of hazardous and unpredictable conditions in the span of time between entries. Any space that is entered infrequently must be treated with considerable concern. Assessment of conditions and entry preparation could involve considerable hazard, due to the uncertainty of conditions. Preparatory and precautionary measures may render conditions in the space nonhazardous prior to entry.

Infrequent entry, hazardous work is the situation about which least is known. This situation poses potential hazards during entry preparation and work. Any space that is entered infrequently must be treated with considerable concern. Assessment of conditions and entry preparation could involve considerable hazard due to the uncertainty of conditions. Preparatory and precautionary measures may render conditions in the space nonhazardous prior to entry. Conditions at time of entry can be completely nonhazardous. Yet, processes and activity that occur during work may create highly hazardous conditions. The difficulty in assessing this situation results from the interaction between the activity and geometry of the confined workspace.

TABLE 8.14
Classification of Confined Space (ANSI Criteria)

4.6 Classification of Confined Space

Evaluation of conditions in the space indicates that there is little potential for generation of hazards or the hazards have been eliminated by engineering control.	Yes **No**
Evaluation of the space indicates that there are actual or potential hazards.	**Yes** No

Response in a highlighted box indicates a **permit-required confined space** that requires written authorization for entry.

For permit-required confined spaces, the Standard requires issuance of a permit prior to all entries. The intent for the permit is to ensure systematic review of hazards prior to each entry, communication of this information to potential occupants, and acceptance by them of requirements to be followed. Changes in conditions or the nature of the work would terminate the permit.

Table 8.14 provides a format for assigning the classification to a particular confined space using the ANSI criteria. The ANSI Standard does not refer to work procedures in this classification. A

TABLE 8.15
Classification of Confined Space (OSHA Criteria)

4.6 Classification of Confined Space

The space contains or has a potential to contain a hazardous atmosphere.	**Yes** No
The space contains a material that has the potential for engulfing an entrant.	**Yes** No
The space has an internal configuration, such that an entrant could be trapped or asphyxiated by inwardly converging walls or by a floor which slopes downward and tapers to a smaller cross section.	**Yes** No
The space contains any other recognized serious safety or health hazard.	**Yes** No

Response in one or more highlighted box indicates a **permit-required confined space** that requires written authorization for entry.

work procedure that generates hazardous conditions could easily be performed in a nonpermit confined space without further consideration according to this decision scheme.

The OSHA Standard for general industry defined the term "confined space" indicating characteristics that are common to a broad range of confined spaces. OSHA took the view that certain characteristics make some types of confined spaces hazardous, while others lacking them are not. OSHA chose the terms **nonpermit confined space** (nonpermit space) and **permit-required confined space** (permit space) to distinguish the level of concern.

A nonpermit confined space (nonpermit space) is a confined space that does not contain or, with respect to atmospheric hazards, have the potential to contain any hazard capable of causing death or serious physical harm. The element that minimizes concern about hazards in these environments is the presence of natural or mechanical ventilation.

A permit-required confined space (permit space) is a confined space that poses or could pose health and safety problems. Health and safety problems could include atmospheric hazard, engulfment hazard, configuration hazard, or other recognized serious hazard.

Table 8.15 provides a format for assigning the classification to a particular confined space using the OSHA criteria. Like the ANSI Standard, the OSHA Standard does not refer to work procedures directly in this classification. However, the preamble and the Standard both provide indications that conditions specified in the permit could include those generated by the work procedure. However, the logic applied here does not address a work procedure that generates hazardous conditions and is performed in a nonpermit confined space.

The strength of this approach is to permit allocation of resources to situations that pose serious and potentially life threatening hazards. The weakness is that changes in a particular situation can occur and escape notice.

A controversial simplification in the OSHA Standard permits use of continuous forced ventilation in lieu of other measures to control atmospheric hazards. A permit is not required for entry into permit spaces whose hazards can be controlled by ventilation alone. Control of hazards, however, is not the same as elimination of hazards. Alternate ventilation procedures do not turn a permit-required confined space into a nonpermit space. Instead, they allow certain permit-required spaces to be entered without the need for a comprehensive program, including a fully comprehensive entry permit. These spaces, therefore, remain classified as permit-required, even though a less comprehensive "written certification" would suffice instead of the complete permit (Anonymous 1993). In effect, this creates two categories of permit-required confined space (Easter 1993). The first covers those spaces that pose not only atmospheric, but also other hazardous conditions. The second category applies to spaces that contain or potentially contain only atmospheric hazards. This would necessitate an alternate and less extensive program involving training, atmospheric testing, forced air ventilation, and an abbreviated permit and notification program. An attendant would not be required.

A confined space containing a hazardous environment would require initial entry to ascertain conditions and to apply any necessary measures to prepare it for further work. Such measures could include cleaning. Initial entry would require all precautions, as appropriate. Once cleaning and other preparatory activities were complete, the only remaining hazard would be atmospheric. The space then would revert to the condition discussed above.

OSHA determined that spaces in which all hazards have been eliminated can be reclassified as nonpermit spaces for as long as the hazards remain eliminated. Ventilation controls hazards; it does not remove them. Reclassification must be rigorous. This involves completion of a dated and signed certificate.

In the Criteria Document on confined spaces, NIOSH established a classification system containing three levels of concern (NIOSH 1979). **Class A** confined spaces are immediately dangerous to life and health (IDLH) due to oxygen deficiency, flammable or explosive atmosphere, or the presence of toxic agents. **Class B** confined spaces have the potential to cause injury and illness if preventive measures are not taken. **Class C** confined spaces require no specific modification of work procedures due to low hazard potential. Classification would precede precautionary measures specific to the workspace. The most hazardous condition that could occur during entry, work, or exit would determine the classification. A work practice that could increase existing hazards or generate additional ones would necessitate reevaluation to determine whether a change in classification would be warranted.

The American Petroleum Institute (API) published several consensus documents for work involving tanks that have contained petroleum products (API 1984, 1987, 1993). These cover various aspects of work on the exterior of containers, as well as entry and work in confined spaces. API also created a classification scheme for confined spaces. The approach taken in API 2015 to categorizing hazards leads to three possible outcomes.

Category I describes conditions under which entry is not permitted. These include excessive flammable vapors, low oxygen level, excessive hydrogen sulfide, airborne levels of toxic substances exceeding levels set by the employer and government, or unsafe physical conditions.

Category II would require use of full facepiece and positive-pressure respiratory protective equipment. Also, an entry permit must be issued. Category II describes conditions that could be life threatening.

Category III describes conditions in which entry could occur without the need for respiratory protective equipment.

A classification system provides the means to simplify the response to hazardous conditions into broad categories. This method can facilitate hazard management by applying similar precautions to all entries whose hazards match those covered by a particular category. A classification system is most valuable when persons other than trained industrial hygienists or safety professionals must administer entry and work on a local level. A classification system provides specific direction to a foreman or maintenance supervisor. This simplifies decision making from formulating many individual decisions to applying a standardized group of procedures. This approach can reduce the likelihood of error. A classification system also simplifies training.

However, a classification system cannot replace careful evaluation of each situation, nor can a rigid structure apply to all circumstances. The decision boundaries of a single class may not apply to a particular circumstance. This can lead to corruption of the borders of the classification as the edges blur. Also, a single rating may not apply to all activities carried out in a particular workspace or group of workspaces. An additional problem in applying a classification system is potential lack of comprehensiveness to the full range of hazardous conditions that may be present. A hazardous condition may not be included within the borders of the classification. Also, the classification may focus on the space prior to entry and not on work procedures that will occur.

Grouping

Thus far, discussion has focused on addressing the problems posed by individual confined spaces. The challenge facing an organization is to provide a collective response that satisfies the requirements of every confined workspace without sacrificing quality. At the same time, a practical impetus within the mandate to manage is to simplify and thus to eliminate unnecessary effort. Minimizing the number of tasks also improves the potential for success and acceptance of the process.

The collective information resident in the hazard assessments may indicate trends. For example, at one extreme of complexity, hazardous conditions increase from space to space. Each requires an ever-increasing response. At the other extreme, individual confined spaces contain similarities, perhaps in geometry, hazardous conditions, or other factors. The responses needed to address problems posed by these, the latter group, are similar and may even be identical.

The latter example suggests one possible means of simplification: grouping. Grouping relies on the concept that a generic response may be satisfactory for a number of situations and locations. Hazardous conditions inherent in individual members of the group may not be identical. However, there is nothing to suggest that a single response would not be appropriate or acceptable to satisfy the needs of all members in the group. Rather than individualizing the response to address the idiosyncracies of every situation, a generic response could be satisfactory. A generic response that meets the requirements of the group also would save considerable development time and energy.

Applying the concept of grouping requires careful and detailed analysis to ensure that the generic response is appropriate to the needs of all members. In a true group, the differences that distinguish individual members are relatively small; for example, location or geometry. Critical aspects such as hazard potential must be similar for all members.

The generic approach can offer significant advantages to an organization, starting with consistency of approach. This strategy minimizes requirements for training and generation of procedures. The negative side of such an approach is the possible drain on resources to acquire equipment that may exceed need in individual situations.

Organizations also may have a large number of confined spaces and work procedures that conform to the main group and a small number of outliers. A single response based on the needs of all spaces may not be practicable. The needs of the outliers may be considerably different than those of the main group. An approach to this situation is to retain the single response for the main group and to devise individualized solutions that accommodate the needs of the outliers.

In organizations having confined spaces of diverse character, the needs in individual situations may differ considerably. There may be no main group. A single response may not be practicable. Addressing confined spaces individually in situations such as this rapidly becomes impractical. Subdivision becomes a necessary approach. The challenge is to define the boundaries of each group. Two approaches are possible: defining subgroups around the hazards of individual members or defining levels of response.

The first approach utilizes the results of hazard assessments for individual confined spaces and work procedures. The response required for the group encompasses the minimum needs for all confined spaces in the group. The response must encompass the minimum needs for all hazardous conditions. The group response thus is greater than that needed for any individual member.

The second approach defines the level of response. Individual confined spaces and work procedures require less than the response provided to the group. This approach places less emphasis on hazard assessment of individual members. The intent is to ensure that the defined level of response exceeds the need of any member by a visible margin. The purpose for the visible margin is to allow for uncertainty between the perceived need and the response. The philosophy in this approach is to minimize the amount of time devoted to hazard assessment. The cost is overresponse in level of precaution.

This strategy can work as long as the hazardous situations are classifiable into relatively well-defined groups. A possible approach to creating these groups is to divide the hazard spectrum into regions requiring defined levels of response. This approach could provide the most efficient allocation of resources in some circumstances.

A note of caution in applying any type of rating scheme to a heterogeneous system is essential. Determining relativity between the different hazards that define confined spaces is critical. Otherwise, ratings that elicit response cannot be intercompared. The absolute magnitude of risk likely differs from one hazard to another. One possible means to resolve this difficulty is to utilize the Delphi technique (Linstone and Turoff 1975). Briefly, the Delphi technique focuses the knowledge of experts onto a particular question. A well-chosen panel containing as few as eight members can provide valid results. Through the process of consensus, the opinions of members of the panel can rapidly mould ideas into workable concepts. The Delphi technique could be a highly effective means to resolve this difficulty.

In order for this approach to function well, several factors must occur. First, differences between levels of response should be large. This reduces uncertainty in making the assignment during the hazard assessment. The number of situations requiring attention also should be large. Ideally, a particular level of response should be localized to specific areas that have dedicated operating personnel. This localizes training and response only to persons normally working in the area.

The graded approach also has advantages and disadvantages. This approach is more conserving of organizational resources than the others. The level of training can match the specific need. However, this approach creates responders having different levels of training. People migrate within organizations. Training applicable in one area may not be applicable to another. This situation would necessitate periodic training and upgrading of a very small groups of individuals.

The graded approach also creates considerable administrative requirements. There is the need to communicate the meaning of the different levels of hazard inherent in the level of response. This is critical to ensuring that response continues to reflect the hazard. A system built on differing levels of response easily could lose its integrity.

An additional problem with this approach is that the assignment of hazard must be considered to be dynamic. The hazard associated with a particular workspace could vary considerably, depending on conditions and the activity. Thus, the assignment of response also could change rapidly. This could result in a very complicated system.

Grouping confined spaces having similar hazards because of construction, planned activity, and the surrounding environment is specifically permitted under some jurisdictions (Canada Gazette 1992). A single hazard assessment could cover the group. The strength of this concept is the time saved in not repeating the exercise for every space in the group. The limitation is that conditions must be known with a high level of confidence.

Hazard Labeling Systems

Several of the strategies mentioned in the previous section utilized the concept of fixed level of response. That is, the level of response would provide protection against some collection of hazardous conditions. Inherent in a system that utilizes multiple levels of action is the need to communicate about their application. Ensuring the ability of workers to differentiate between groups or ratings to ensure the appropriate level of response in a particular case is a major concern.

One approach for addressing this problem is a scheme for hazard communication. A label mounted at the entrance to the confined workspace would indicate the nature and magnitude of a hazard known or suspected to be present. This label would be equivalent to a precautionary label affixed to a container holding a hazardous chemical. Elements that could appear on the label could include:

- Identity of the confined space
- Potential or actual hazardous conditions
- Seriousness of hazard

Hazard rating scales utilize several strategies for communicating relative severity. One approach uses a numerical scale. Another possibility is a word scale. Still another is the use of symbols, such as internationally recognized traffic symbols. Color can play an important role. Colors provide important cues to assist the process of recognition. The use of color requires caution because of color-blindness in the male population.

Potentially the most effective system is the one that builds upon familiar concepts that warn, such as traffic symbols. The more imprinted the meaning associated with the communicative device, the more effective will be the warning conveyed during conditions of high stress, such as emergencies. Under these conditions, familiar warnings to which we respond daily likely will produce more effective reaction than a system that uses unfamiliar communicators. To be most effective, this system must communicate at the level of instinct.

A rating system comprehended by all users can become a valuable tool for communicating the relative severity of hazards. A system of this type could become the common language of communication between occupational health and safety professionals and the lay workforce about hazards present or likely to be present in a workspace.

Implicit in the decision to implement a rating system is the need to ensure recognition and understanding of its concepts. Recognition and understanding will not occur without extensive education and training of persons potentially at risk. Although this task seems enormous, this requirement has been utilized previously. Education and training about chemical hazards, hazardous waste, and asbestos abatement represent several examples.

Hazard Reduction/Hazard Elimination

Reduction of the hazard associated with work in confined spaces is an area of primary importance. This should represent a major opportunity for development. Several techniques and technological approaches for hazard reduction currently exist. Most are based on removal of residual material and cleaning techniques that utilize equipment that is controlled from outside the space. Since the presence of residual toxic materials in confined spaces adds to the risk of entry, removal of these materials can dramatically reduce the potential for unnecessary exposure during work in the space. In addition, this approach is supported in legislation such as the OSHA Standards on confined spaces (OSHA 1993, 1994).

Use of remotely controlled equipment in a contaminated space can create its own problems, as mentioned previously. For example, compatibility between the residual materials and materials of construction of the equipment is the first concern. Contamination of surfaces of the equipment will necessitate decontamination or possibly even sacrifice. Decontamination itself could become an issue, depending on the toxicity, volatility, and disposal of the material. The material removed from the confined space may require treatment as a hazardous waste. Disposal in the appropriate location must be a consideration during this procedure. Use of remotely operated equipment in these circumstances therefore requires careful planning.

(For more in-depth information about hazard reduction/hazard elimination, consult Appendix H.)

The ultimate goal should be to eliminate the need for entry altogether.

TABLE 8.16
Examples of Intervening Regulations (U.S. Context)

Technical Concern	U.S. Standard
Hazard communication	Hazard Communication 29 CFR 1910.1200 29 CFR 1926.59
Personal protective equipment	Various sections 29 CFR 1910 29 CFR 1926
Safety requirements	Various sections 29 CFR 1910 29 CFR 1926
Process safety	Process Safety Management 29 CFR 1910.119
Decontamination	Hazardous Waste Operations and Emergency Response 29 CFR 1910.120
Lockout/tagout	Control of Hazardous Energy (Lockout/Tagout) 29 CFR 1910.147
Toxic substances	Various U.S. standards
Biological substances	Bloodborne Pathogens 29 CFR 1910.130

OTHER CONSIDERATIONS

Entry into confined spaces cannot be considered as an activity isolated from other realities. For example, entrants could contaminate themselves and their equipment. Further, preparation of the space for entry could require draining of hazardous materials from lines during isolation. In sufficient quantity without collection, these drainings could constitute environmental releases requiring reporting. Purging to remove gaseous components similarly could release into the atmosphere a considerable quantity of vapor.

In the context of the U.S., some of the intervening regulatory standards are provided in Table 8.16.

AN ORGANIZATIONAL DECISION POINT

The preceding discussion indicates that satisfying technical and legal requirements for work in confined spaces involves commitment of considerable organizational resources. Prior to embarking in this direction, an organization that presently has no such program needs to determine the answers to two key questions.

- Is entry into confined spaces necessary?
- Should the organization use its own people?

While these questions may seem trivial in organizations where these practices have occurred for many years, the answers are fundamental to future direction. The costs associated with upgrading present practices to meet increasingly stringent requirements of regulations and standards could outweigh those associated with other options. Other options include use of contractors that specialize in performing work in confined spaces and techniques for entry avoidance. These and other options likely will become more widely available and sophisticated as this field develops.

Because of the hazards associated with entry and work in confined spaces and the complexity of a program, serious consideration should be given to measures for entry avoidance. Entry avoidance approaches the problem of need to perform work without entry into the space. Whenever possible, entry avoidance should be the preferred alternative. Entry avoidance can be achieved by minimizing hazards during design, modifying the space or equipment it contains, or performing the work in alternate fashion.

REFERENCES

American National Standards Institute: *Safety Requirements for Lockout/Tagout of Energy Sources* (ANSI Z244.1-R1992). New York: American National Standards Institute, 1992.

American National Standards Institute: *Safety Requirements for Confined Spaces* (ANSI Z117.1-1995). Des Plaines IL 60018-2187: American Society of Safety Engineers/American National Standards Institute, 1995. 32 pp.

American Petroleum Institute: *Guidelines for Confined Space Work in the Petroleum Industry* (API Pub. 2217). Washington, D.C.: American Petroleum Institute, 1984. 5 pp.

American Petroleum Institute: *Guidelines for Work in Inert Confined Spaces in the Petroleum Industry* (API Pub. 2217A). Washington, D.C.: American Petroleum Institute, 1987. 7 pp.

American Petroleum Institute: *Safe Entry and Cleaning of Petroleum Storage Tanks* (API Pub. 2015, 5th ed.). Washington, D.C.: American Petroleum Institute, 1993. 60 pp.

American Society for Testing and Materials: *Standard Practice for Confined Area Entry* (ASTM D 4276-84, reapproved 1989). Philadelphia, PA: American Society for Testing and Materials, 1989. 3 pp.

American Welding Society: *Recommended Safe Practices for the Preparation for Welding and Cutting of Containers That Have Held Hazardous Substances* (ANSI/AWS F4.1-88). Miami, FL: American National Standards Institute/American Welding Society, 1988. 4 pp.

Anonymous: Reply to "Confined Spaces Clarification" by Gary Easter. *Occupational Hazards* May 1993 p. 9.

Barrow, G.M.: *Physical Chemistry,* 2nd ed. New York: McGraw-Hill, 1966. pp. 712–752.

Breatherick, L.: *Breatherick's Handbook of Reactive Chemical Hazards,* 4th ed. London: Butterworths, 1990. 2003 pp.

Canada Gazette: "Part IX, Confined Spaces," *Canada Gazette, Part II. 121*: (26 March 1987) pp. 1337–1364.

Canada Gazette: "Part XI, Confined Spaces," *Canada Gazette, Part II. 126*: (17 September 1992) pp. 3863–3876.

Easter, G.: "Confined Spaces Clarification." Letter to the Editor. *Occupational Hazards* May 1993 p. 9.

Findlay, J.V. and R.L. Kuhlman: *Leadership in Safety.* Loganville, GA: Institute Press, 1980. pp. 29–62.

Frankel, G.J.: Oxygen-Enriched Atmospheres. In *Fire Protection Handbook.* 16th ed. Cote, A.E. and J.L. Linville (Eds.). Quincy, MA: National Fire Protection Association, 1986. pp. 12-12 to 12-21.

Furr, A.K.: *CRC Handbook of Laboratory Safety,* 3rd ed. Boca Raton, FL: CRC Press, 1989. pp. 268–284.

Grund, E.V.: Lockout/Tagout: *The Process of Controlling Hazardous Energy.* Itasca, IL: National Safety Council, 1995. 429 pp.

Jones, T.B. and J.L. King: *Powder Handling and Electrostatics: Understanding and Preventing Hazards.* Chelsea, MI: Lewis Publishers, 1991. 103 pp.

Hellman, T.M. and F.H. Small: Characterization of the odor properties of 101 petrochemicals using sensory methods. *J. Air Poll. Control Assoc. 24:* 979–982 (1974).

Kepner, C.H. and B.B. Tregoe: *The New Rational Manager.* Princeton, NJ: Princeton Research Press, 1981. 224 pp.

Linstone, H.A. and M. Turoff: *The Delphi Method. Techniques and Applications.* Reading, MA: Addison-Wesley, 1975. pp. 1–12.

Lowrance, W.W.: *Of Acceptable Risk: Science and the Determination of Safety.* Los Altos, CA: William Kaufman, 1976. pp. 8–9.

Lowrie, R.: Materials for oxygen service. *Chem. Eng.* April 27, 1987. pp. 75–80.

Meyer, E.: *Chemistry of Hazardous Materials,* 2nd. ed. Englewood Cliffs, NJ: Prentice Hall, 1989. pp. 204–359.

National Fire Protection Association: *NFPA 77: Recommended Practice on Static Electricity* (1988 ed.). Quincy, MA: National Fire Protection Association, 1988. 30 pp.

National Fire Protection Association: *NFPA 328: Control of Flammable and Combustible Liquids and Gases in Manholes, Sewers, and Similar Underground Structures* (1992 ed.). Quincy, MA: National Fire Protection Association, 1992. 11 pp.

National Fire Protection Association: *NFPA 306: Control of Gas Hazards on Vessels* (1993 ed.). Quincy, MA. National Fire Protection Association, 1993. 15 pp.

National Institute for Occupational Safety and Health: Criteria for a Recommended Standard — Working in Confined Spaces (DHEW/PHS/CDC/NIOSH Pub. No. 80-106). Cincinnati, OH: DHEW(NIOSH), 1979. 68 pp.

National Institute for Occupational Safety and Health: Guidelines for Controlling Hazardous Energy During Maintenance and Servicing (DHHS (NIOSH) Pub. No. 83-125). Morgantown, WV: DHHS/PHS/CDC/NIOSH, 1983. 72 pp.

National Institute for Occupational Safety and Health/Occupational Safety and Health Administration/U.S. Coast Guard/Environmental Protection Agency: Occupational Safety and Health Guidance Manual for Hazardous Waste Site Activities (DHHS/PHS/CDC/NIOSH Pub. No. 85-115). Washington, D.C. U.S. Government Printing Office, October, 1985.

National Safety Council: *Confined Space Entry Control System for R & D Operations*, No. 12304-0704. Itasca IL: National Safety Council, 1985. 5 pp.

National Safety Council: *Inspecting and Cleaning Pipes and Sewers*, No. 12304-0577. Itasca IL: National Safety Council, 1986.

National Safety Council: *Atmospheres in Sub-Surface Structures and Sewers*, No. 12304-0550. Itasca IL: National Safety Council, 1987a. 5 pp.

National Safety Council: *Entry into Grain Bins and Food Tanks*, No. 12304-0663. Itasca IL: National Safety Council, 1987b. 6 pp.

Norman, C. and D. Cooper: Sniffing out trouble: the smell of danger. *Occup. Health Safety Canada 10:* May/June, 1994. pp. 64–65.

Occupational Safety and Health Administration: Selected Occupational Fatalities Related to Fire and/or Explosion in Confined Work Spaces as Found in OSHA Fatality/Catastrophe Investigations. Washington, D.C.: U.S. Department of Labor, Occupational Safety and Health Administration (U.S. DOL/OSHA), 1982a. 76 pp.

Occupational Safety and Health Administration: Selected Occupational Fatalities Related to Lockout/Tagout Problems as Found in Reports of OSHA Fatality/Catastrophe Investigations. Washington, D.C.: U.S. Department of Labor, Occupational Safety and Health Administration (U.S. DOL/OSHA), 1982b. 113 pp.

Occupational Safety and Health Administration: Selected Occupational Fatalities Related to Grain Handling as Found in Reports of OSHA Fatality/Catastrophe Investigations. Washington, D.C.: U.S. Department of Labor, Occupational Safety and Health Administration (U.S. DOL/OSHA), 1983. 150 pp.

Occupational Safety and Health Administration: Selected Occupational Fatalities Related to Toxic and Asphyxiating Atmospheres in Confined Work Spaces as Found in Reports of OSHA Fatality/Catastrophe Investigations. Washington, D.C.: U.S. Department of Labor, Occupational Safety and Health Administration (U.S. DOL/OSHA), 1985. 230 pp.

Occupational Safety and Health Administration: Selected Occupational Fatalities Related to Welding and Cutting as Found in Reports of OSHA Fatality/Catastrophe Investigations. Washington, D.C.: U.S. Department of Labor, Occupational Safety and Health Administration (U.S. DOL/OSHA), 1988. 225 pp.

OSHA: Control of Hazardous Energy Source (Lockout/Tagout); Final Rule, *Fed. Regist. 54*: 169 (1 September 1989). pp. 36644–36696.

OSHA: Permit-Required Confined Spaces for General Industry; Final Rule, *Fed. Regist. 58*: 9 (14 January 1993). pp. 4462–4563.

OSHA: Confined and Enclosed Spaces and Other Dangerous Atmospheres in Shipyard Employment; Final Rule, *Fed. Regist. 59*: 141 (25 July 1994). pp. 37816–37863.

Ruth, J.H.: Odor thresholds and irritation levels of several chemical substances: a review. *Am. Ind. Hyg. Assoc. J. 47:* A-142–A-151 (1986).

Standards Australia: *Safe Working in a Confined Space* (AS 2865-1995). North Sydney, New South Wales, Standards Association of Australia: 1995. 42 pp.

TRC Environmental Consultants: *Odor Thresholds for Chemicals with Established Occupational Health Standards.* Akron OH: American Industrial Hygiene Association, 1989. 90 pp.

Turner, K.B.: Oxygen safety. *Professional Safety.* January, 1987. pp. 13–16.

9 Logistical Considerations for Work Involving Confined Spaces

CONTENTS

INTRODUCTION

Preparation, entry, and work activity are the real world of confined spaces. No matter how much energy and effort are expended in review and planning, the real world is all that matters. If matters have been considered correctly, everything will proceed without event. However, if a mismatch between the world of planning and the real world occurs, the outcome and consequences could be completely unforeseen. The extent of the mismatch directly reflects the thoroughness of planning and preparation, and the experience and capability of participants.

The decision to proceed with the project ends the planning phase. Planning offers the advantage of the intellectual exercise. Any number of situations can be imagined and followed through. Error brings no consequence. While seemingly minor compared to other decisions in the corporate sphere, the decision to proceed with entry and work in a confined space can have major implications for those involved. This decision represents the acceptance of the perceived risk and the belief that the hazardous conditions can be eliminated or are controllable.

Research documented in the NIOSH, OSHA, and MSHA reports indicates that the critical period in the cycle of events involving confined workspaces is the entry (NIOSH 1979, 1994; OSHA 1982a, 1982b, 1983, 1985, 1988, 1990; MSHA 1988). Most of the accidents occurred soon after the victim was exposed to hazardous conditions within the confined space. Coincidentally, the investigations also revealed the lack of a formal process for regulating entry and work in these workspaces.

The conduct of work in confined spaces reflects the realities of nonproduction mode: nonrepetitiveness, unpredictability, and irregularity in occurrence and outcome, variability of conditions, and briefness in duration. As well, the work often is performed in field locations under construction conditions. The dynamics of interaction between people at the site and conditions is an important consideration. The focus of workers and supervision alike is completion of the task in orderly, timely fashion. As a result, individuals who work in these surroundings may be considerably more tolerant of adverse conditions than those who work in regular industrial workplaces.

ON-SITE SUPERVISION AND MANAGEMENT

The greatest influence over the direction, impact, and outcome of work in these situations is the tone set by site supervision and management. This tone is especially critical to safety concerns. Safety must receive more than lip service in these operations. Otherwise, the planning and other activity undertaken to prepare for work could be for nought, and the frequency of accidents and injuries could exceed what is considered readily preventable. Also implicit in an approach that does not contain a bona fide commitment to safe conduct of work could be infraction of legal statutes.

Management has responsibility and accountability for the conduct and outcome of work. Lack of capable management was a readily identifiable factor in many of the fatal accidents reported by NIOSH and OSHA (NIOSH 1979, 1994; OSHA 1982a,b, 1983, 1985, 1988, 1990; MSHA 1988). Typically, the management structure at any large site is a hierarchy that includes senior or general managers or superintendents, junior managers, and first or second line supervisors. At small sites, a supervisor or foreman may be the sole supervisory person present.

It is not uncommon for every member of this hierarchy to have originated from the lowest levels of the workforce. Through seniority and innate leadership skills, these individuals move through the ranks from workers to supervision. This change in status does not necessarily reflect any change in technical knowledge, attitude, or ability. These individuals may simply transfer responsibility for doing work tasks to responsibility for overseeing others who do them. The major focus remains the same: to ensure that the work is completed as expeditiously as possible. Anything that interferes with the straightest path from beginning to end is regarded as an intrusion. Intrusion could bring forth hostility. This is understandable, since these individuals bear bottom-line, day-to-day responsibility and accountability for completing the work on time. Their livelihood, promotability, and future prosperity all depend on achieving production. From the perspective of these individuals, this is the only criterion that satisfies and earns praise and respect from senior management.

Some senior managers progressed through the ranks. Others entered from professional backgrounds. Their main exposure to concerns about confined spaces likely is trade periodicals. A higher level of education or status within an organization provides no guarantee of a difference in attitude about the conduct of work involving confined spaces. Intellectual and physical distance from the situation can decrease sensitivity to concerns about the potentially hazardous nature of these workspaces. Senior management likely will enter the space only after the "dirty work" has been completed. An important distinguishing characteristic about senior management is that inspecting the progress of work is a small part of their jobs at the site.

In a number of fatal accidents documented by OSHA and NIOSH, a supervisory individual (foreman, supervisor, superintendent, manager) entered the confined space during the accident in an attempt to rescue the victim, a subordinate (OSHA 1985, NIOSH 1994). The superior seemed to take responsibility for the predicament into which the subordinate had fallen. These individuals were completely unprepared for this role.

An element underlying many of these situations was an apparent lack of outward concern about their own safety. The incompetent manner in which these actions were undertaken did not benefit the situation. The actions of these individuals were inappropriate in the circumstances and needlessly increased the level of tragedy. Given a different attitude by supervision and management toward confined spaces, these tragedies needn't have happened.

Hazard management in confined spaces is a difficult issue for site managers. On the one hand, managers likely are aware about the hazards posed by confined spaces from articles in periodicals. They know that working in the hazardous conditions that can occur in confined spaces poses serious moral, financial, and legal concerns. Of course, people in management are concerned about hazards in confined spaces and for legitimate reasons. On the other hand, their focus is production and completing the work as economically as possible. Management would prefer to "fix" the problem through a one-time expenditure, rather than be subjected to on-going, project-by-project concerns. From their perspective, this action would reduce the uncertainty in production planning.

REGULATORY FRAMEWORK FOR HAZARD MANAGEMENT

Many jurisdictions directly regulate entry into confined spaces. This also is the subject of a number of standards created by industry and other groups. (For further information on this subject, consult Appendix A.) These documents outline formal processes for identifying, assessing, and controlling hazards in confined spaces. In addition, these standards attempt to eliminate casualness in the entry and work process by introducing a disciplined formal approach.

This regulatory framework can play an extremely important role by mandating approaches to enable the safe conduct of work involving confined spaces. This best can occur by establishing in clear legal terms the requirement for and mandate of an individual responsible for health and safety at the worksite. This approach institutionalizes the requirement for health and safety to take precedence over production. Reference to regulated limits for exposure indicates that legal sanctions could be applied when these are exceeded. Tolerance of adverse conditions by workers at the site has no place as a criterion for considering them to be acceptable. Legal limits are considerably lower than tolerable concentrations.

Ultimately, this approach will enable the success and harmony of the on-site management team. Challenge that can arise to a less direct course of action would be stopped by the acquiescence that legal requirements must take precedence over seeming expedience.

What remains is to formalize the organizational and operational relationships within which this can happen. Legal statutes and standards on confined spaces mandate the existence and function of the "Qualified Person." (The Qualified Person is the subject of Appendix B.) Since the existence and function of the Qualified Person are mandated in law, this individual must become an essential member of the management team. Two examples of possible models are utilized in the OSHA Standards on confined spaces (OSHA 1993; 1994).

The OSHA Standard for general industry utilizes mission-oriented job titles and specification of performances required from individuals holding them (OSHA 1993). This approach creates a hierarchical structure that is directed solely toward the safe conduct of entry and work in the confined space.

The Standard requires employers to identify confined spaces in their workplaces and to determine whether they are permit-required or nonpermit-required. A permit space is a confined space that has certain characteristics that make entry hazardous without special precautionary measures. A permit must be completed before entry is allowed into a permit-required confined space. Similarly, the employer must reevaluate nonpermit spaces following changes that might increase hazards to entrants above the threshold that defines a permit space.

The Standard utilizes the entry permit as the entry control document. The permit authorizes entry and documents measures required to ensure that employees can safely enter and work in the permit space. The permit also provides a concise summary of the entry procedure. The permit must be prepared after establishing safe entry conditions and prior to the entry. The permit must be available for review by all entrants at the time of entry. The duration of the permit is the period of time needed to complete the assigned task or job. As long as acceptable entry conditions are present, employees can safely enter and perform work in the permit space.

The Standard identifies job titles of personnel who work in and around confined spaces and mandates training to ensure that individuals attain the appropriate level of performance.

The entry supervisor is responsible for evaluating conditions in and around a permit space where entry is planned. The entry supervisor signs the entry permit authorizing entry to begin. The entry supervisor allows entry already underway to continue and oversees entry operations. This means prescribing necessary measures to protect personnel from hazards in the permit space and prohibiting or terminating entry, as deemed necessary. The entry supervisor must determine at appropriate intervals whether conditions within the space remain safe and whether to cancel the permit whenever conditions are otherwise. "Entry supervisor" is a transferrable title; hence, someone other than the author of the permit may terminate entry and cancel the permit. The entry supervisor and the person who supervises the actual work could be one and the same or they could be different.

Authorized entrants are the persons authorized by the employer to enter the space. Authorized entrants can work in the space, test the atmosphere, supervise the entry, or provide rescue services. The attendant monitors authorized entrants and activities inside and outside the space to determine whether they can remain inside safely. The attendant orders occupants to leave the space when evidence of an uncontrolled hazard appears. The attendant watches for conditions not envisioned during preparation of the entry permit and for other prohibited conditions. The attendant works from a location outside the permit space. Testing personnel determine the condition of atmospheric hazards in the confined space. Rescue and emergency services respond to accidents and other emergency situations involving permit spaces.

The strength of this approach is that it creates a system of work focused in a systematic and comprehensive manner on the confined space. An individual normally at the site during the entry can serve as the entry supervisor. Everyone entering the space must receive a set level of training. This approach can work in small and large organizations.

The difficulty with the approach is its operational structure. This differs markedly from what is found in organizations. The latter is oriented toward the orderly completion of work tasks. The entry supervisor need not necessarily supervise the performance of work tasks. An outsider to the process, such as an individual with specialized expertise in safety and industrial hygiene, could be appointed as entry supervisor. This approach could create serious jurisdictional problems with functional supervision. On the other hand, supplementing the knowledge and skills of functional supervisors with those needed to become entry supervisors could be a daunting task. Also, this approach creates conditions for a perceived, if not real, conflict of interest between production and safety.

The OSHA Standard for shipyard employment takes a considerably different approach to hazard assessment and control (OSHA 1994). This Standard requires the same approach for all spaces, confined or otherwise. OSHA argues that the requirements of the shipyard Standard constitute an informal permit system. The Standard requires assessment of conditions at any time deemed appropriate during entry and work. Hazardous conditions must be assessed and eliminated or corrected to within acceptable limits prior to entry of any worker. Since the work itself can be the source of the hazard, the process of evaluation and correction must continue at all stages. When conditions exceed acceptable limits during work, work must stop and workers must vacate the space. Work can resume once conditions are restored to within acceptable limits.

The foundation of this approach is the shipyard competent person. The employer designates the shipyard competent person and provides training to attain a standard of performance. The competent person evaluates atmospheric conditions and inspects the space, and as appropriate, implements control through installation of ventilation. If ventilation fails to reduce the atmospheric concentration to within acceptable limits or other aspects of the situation exceed their capabilities, the competent person must seek advice from a higher level of expertise: an NFPA (National Fire Protection Association) Certified Marine Chemist, a U.S. Coast Guard authorized person, or a Certified Industrial Hygienist. The latter then will provide assistance in the development and implementation of control measures.

Results of tests made by the competent person and outside experts must be recorded by location and time and date, and posted in the immediate vicinity of the affected operation before work can start. A certificate must be issued, either by the competent person or the outside expert. The certificate provides an interpretation of conditions in the space in standardized language, along with results of testing. A certificate issued by the outside expert can contain instructions to be followed by the competent person and management. All spaces, therefore, must be certificated prior to the start of work.

The strength of this approach is that it incorporates requirements for hazard assessment and control into normal functional relationships. The competent person is part of a team that eliminates or controls hazards as a seamless part of performing work in these environments. There is nothing foreign about the working relationships between the participants.

One of the failings of the approach is that OSHA did not provide sufficient detail about qualifications and capabilities needed in the competent person. The competent person could be a junior person in the hierarchy who lacks the formal credentials needed to gain respect from management and labor. The reporting relationship of the competent person also is not specified or suggested. This could lead to considerable abuse of the individual holding this position. This would be a classic case of accountability before the law without the organizational clout needed to achieve results. There also is a serious potential for the competent person to attempt actions beyond the level of competence. A person with limited knowledge easily could step beyond that limit without realizing that this has occurred.

Other standards and legal statutes on confined spaces take an intermediate approach for defining the role of the Qualified Person. These specify only actions that must occur (NIOSH 1979, ANSI 1995, Canada Gazette 1992). This approach does not specify a job title for the Qualified Person, nor does it necessarily require the same individual to perform all of the functions. This approach provides maximum flexibility for integrating the functions of the Qualified Person into existing organizational structures. However, this approach does force organizational discipline, since assignment of responsibilities and accountabilities must be clear and unambiguous to ensure that they are performed appropriately.

The approach taken in the Canada Occupational Safety and Health (COSH) Regulations begins with recognition that a particular workspace is a confined space (Canada Gazette 1992). At this point, the employer must appoint a Qualified Person. The Qualified Person assesses the physical and chemical hazards to which an entrant could be exposed, and specifies tests necessary to determine whether the entrant likely would be exposed. The Qualified Person provides the results of the assessment in a signed and dated report to the employer. The employer then establishes procedures for entry, work, and egress. The employer also must establish a system of entry permits. As well, the employer must specify protective equipment and tools to be used by entrants and emergency equipment required by rescuers.

An important feature of the Canadian approach to workplace health and safety is the statutory involvement of the Safety and Health Committee or representative at all stages in the process. The employer must inform and consult with the Committee or representative about all aspects of the situation.

Just prior to entry and during occupancy of the confined space, the Qualified Person performs tests to verify compliance with regulated specifications. At the conclusion of the work, the Qualified Person prepares a signed and dated report outlining the work performed and results obtained from the tests. Where the report indicates that the occupants have been in danger, the employer must revise the procedures.

In circumstances where conditions in the space or the nature of the work preclude compliance with the preceding specifications, the employer must establish emergency and evacuation procedures and provide other measures to ensure communication and immediate assistance to occupants. A Qualified Person trained in entry and rescue procedures must attend outside the confined space. Further, additional trained help must be immediately available. The Qualified Person specifies equipment required and procedures to be followed in a signed and dated report to the employer. As well, the Qualified Person explains the procedures to entrants and obtains their acknowledgment through signature.

Performance of hot work introduces additional constraints. Where the confined space contains a hazardous concentration of a flammable or explosive material, the employer must provide a Qualified Person to maintain a fire watch in the area outside the space.

These Standards provide distinctly different approaches for management of work involving confined spaces. The approach using the entry supervisor works best where hazardous conditions are known from previous experience and where control measures have been tried and proven effective. This system enables scarce expert resources to be apportioned in an efficient manner. The limitations of this approach arise where previously unrecognized hazards are present or develop. If the expert is not readily available, these can remain unrecognized and unaddressed. The issuer of the permit should be independent of the supervisory hierarchy in order to avoid potential or real conflict of interest.

The on-site expert provides hands-on expertise in the recognition, assessment and evaluation, and control of hazards. An added advantage is the ability to respond to concerns on short notice and to address unanticipated hazards. This approach is ideally suited to operations that have numerous confined spaces or where conditions or the configuration of spaces can undergo rapid change.

Many organizations have available the services of health and safety professionals. These individuals, however, likely would be unavailable to attend at every entry. The only way to fill this gap is to designate subordinates. Subordinates fit into each of the models mentioned here. Adoption of this approach means that the only real difference between the models is the means of recording and communicating the hazard. In each case a Qualified Person (Entry Supervisor, Competent Person, Qualified Person) will be present at the activity. Since none of the models specifies training required in these individuals, this opens the opportunity for the expert-subordinate approach.

The OSHA Standard for shipyard employment clearly requires the competent person to call for assistance when the situation is beyond the expertise of that individual. Similarly, by organizational policy, the entry supervisor or Qualified Person could be instructed to request assistance when the situation is beyond the capabilities of that individual.

The key to implementing any of these models is the hazard assessment. The hazard assessment must proceed through a repeatable, comprehensive format. This provides the greatest potential for consistent identification, and elimination or control of hazardous conditions. The hazard assessment used in this publication contains 36 potentially hazardous conditions. This serves as an indicator of the breadth of enquiry that this assessment should include. A comprehensive hazard assessment produced by a high-level resource (industrial hygienist and safety professional) can be utilized by a wide group of individuals at distributed sites.

This approach also is consistent with the needs of small organizations. A knowledgeable consultant can prepare a comprehensive hazard assessment for situations known to be encountered. While the problem of human resources can never be resolved, use of existing expert resources in a more focused manner should be possible.

The presence of the on-site Qualified Person (Entry Supervisor, Competent Person, Qualified Person) and the input that this person provides still represents a potential problem to management and supervision. Project management is unaccustomed to working with technical advisors who have the power to stop work. Recommendations from these individuals are not always seen as helpful. This is especially true when management has a preconceived notion that conditions on a project are within acceptable (that is, personally tolerable) limits. Personally tolerable limits and legal limits, of course, usually have no relationship to each other. Personal tolerance according to this mindset is related to irritation and discomfort. One person's comfort easily could be another person's intolerance. To complicate matters further, standards, such as the Threshold Limit Values®, have become so small that preventing overexposure is almost impossible without complete adherence to best practice. There is no room at all for the antiquated attitudes expressed here.

The on-site Qualified Person deals with what project management perceives with considerable frustration as "black magic" or "witchcraft." The sources of this frustration are the numbers on the display of a "black box" or the change in color in a glass tube containing some exotic substance. These numbers represent an intangible that management does not understand and does not appreciate. Yet, these numbers have the potential to require shutting down the work. This situation represents a destabilizing influence of considerable and unacceptable magnitude to the goal of expeditious production. Critical situations could pit safety against expediency and bottom-line considerations. The way in which management moves during these conflicts ultimately sets the tone and credibility of other measures.

The preceding discussion indicates that the situation at the worksite could deteriorate rapidly without clear agreement about responsibilities and accountabilities. Site supervision and management must report through a single individual. That way, groups do not act in isolation from each other. The supervisor in charge must receive support from all participants, so that decision making can occur in orderly fashion. This aspect is essential for maintaining order during the height of a critical situation. A critical situation could include the need to formulate decisions and resolve disputes that have arisen because of differing points of view.

The supervisor in charge needs to possess specialized knowledge and training for managing hazardous conditions associated with confined spaces. An understanding about hazardous conditions will enable the relationship and cooperation with the on-site Qualified Person. A respectful relationship is essential for minimizing conflict and enabling problem solving. The relationship between the on-site Qualified Person and the supervisor in charge must have documented support from senior management. This is absolutely essential, so that input from the Qualified Person will be acted upon without hesitation and interference.

PREPARATORY AND WORK PROCEDURES

The focus of the previous chapter was a systematic approach for identification of hazardous conditions during the various phases of activity involving confined spaces. The intended outcome is action to eliminate or at least control hazardous conditions, so that work can proceed safely. This outcome also could include procedures to be used during these activities. These procedures could require sophistication, because the hazards could be subtle or highly dangerous. The approaches needed in these procedures may not be part of the knowledge base of management and supervision.

A number of resources offer assistance both generally and in specific areas. This section identifies the resource and briefly summarizes the area in which it can offer assistance.

Purging Principles and Practice, from the American Gas Association, discusses techniques for gas-freeing containers, vessels, and pipelines (AGA 1975). This book discusses the use of inert gases and water.

ANSI/AWS F4.1-88 recommends practices for preparation for welding and cutting of some types of containers that have held hazardous substances (ANSI/AWS 1988). Practices include preparing the container for cleaning, cleaning methods, and preparations for welding. Preparations include housekeeping, ventilation, pressure relief, testing, inerting, and water-filling.

ASTM Standard D 4276-84 provides recommendations for preparing vessels for entry (ASTM 1989). ASTM commented that these procedures are particularly suited for use in confined spaces containing halogenated organic solvents. They were not intended for use in all confined spaces.

Australian Standard AS 2865-1995 provides recommendations for performing cleaning tasks using several processes: waterjetting, steam cleaning, abrasive blasting, and chemical cleaning (Standards Australia 1995). AS 2865-1995 provides additional recommendations for performing hot work. The latter focus on fire and explosion prevention.

The NIOSH Criteria Document on confined spaces provided recommendations for work practices to prepare for entry into confined spaces (NIOSH 1979). These focus on purging, ventilating, fire prevention and protection, isolation, lockout/tagout, cleaning procedures, and safe work practices with tools and equipment.

NFPA 306 provides detailed requirements for preparing interior spaces to meet requirements for issuance of a Marine Chemist's Certificate (NFPA 1993). The predecessor of NFPA 306 (*Control of Gas Hazards on Vessels*), Appendix A of Regulations Governing Marine Fire Hazards, was first published in 1922. These procedures stress draining of cargo and residues, and cleaning and securing of valves in the position appropriate to the work to be performed. Cleaning includes flushing with water, blowing with steam, or inerting.

NFPA 327, *Standard Procedures for Cleaning or Safeguarding Small Tanks and Containers,* also has had a long history (NFPA 1975). This standard first appeared in 1922 as "Suggested Good Practices for Freeing Fuel Oil and Other Oil Tanks at Refineries, Tank Farms, Distilleries and Other Industrial Plants of Flammable and Explosive Vapors Previous to Entering for Making Repairs or Other Purposes." NFPA 327 is intended for use only with small tanks or containers that cannot be entered, but can confine a hazardous atmosphere. It provides procedures for removal of residual gases and vapors by displacement with water, displacement with air, displacement with inert gas, or by inerting. It also provides procedures for removal of residual liquids by steam cleaning and chemical cleaning.

National Safety Council Data Sheet 663 provides procedures for entry into grain bins and food tanks (NSC 1987). These focus on atmospheric hazards, fire and explosion prevention, ventilation, respiratory protection, protection against engulfment, and lockout.

API RP 1631 recommends practices for the interior lining of existing steel and fiberglass-reinforced plastic underground storage tanks (API 1992). These include methods for vapor freeing, removing sediment, and cleaning interior surfaces.

API RP 2003 describes precautionary measures for use against static electricity and stray currents (API 1991a).

API 2214 provides information about hazards attributed to use of ferrous and nonferrous hand tools in the petroleum industry (API 1989).

API 2009 recommends precautions during welding and cutting on equipment in and around petroleum operations (API 1988a).

API 2013 provides safety practices for cleaning mobile tanks that have contained flammable or combustible liquids (API 1991b).

API 2015 provides safety practices for the preparing, emptying, isolating, ventilating, atmospheric testing, cleaning, entry, and hot work on and recommissioning of storage tanks that have held flammable or combustible liquids (API 2015 1993). These include methods for control of ignition sources, isolation, ventilation, and vapor freeing.

API 2217A provides guidelines for safe work practices involving inerted confined spaces (API 2217A 1987).

API 2207 outlines safety precautions to be followed to prepare tank bottoms for hot work (API 1991c).

API 2219 outlines procedures to prevent accidents during operation of vacuum trucks used in petroleum service (API 2219 1986).

API 2027 outlines procedures to reduce risk of ignition during abrasive blasting of tanks used in hydrocarbon service whose vapor space may contain a flammable atmosphere (API 2027 1988b).

The outcome from use of these standards should be procedures that address (as appropriate to the situation):

- Draining, cleaning, and removal of residual contents
- Isolation
- Purging and ventilating
- Deenergization, deactivation, and lockout
- Testing
- Decontamination
- Control and disposal of spilled or released materials

These procedures are essential to the safe conduct of work. They also form the basis for describing preparation of the space that is reported in the entry documents.

ENTRY DOCUMENTS

A critical part of the entry process is communication of conditions to be encountered and precautions required against them. Legislation and standards on confined spaces require communication of the hazard assessment and its recommendations in written form to all persons affected by it. This provides a written record of the basis for decisions and reduces the potential for misunderstanding.

The entry document is the vehicle for communicating the results of the hazard assessment to persons who will work in the confined space. Entry documents have existed for many years, and effectively have become institutionalized. The success of the program outlined here reflects the skill with which the entry document is prepared, communicated, and its conditions carried out.

The Entry Permit

The entry permit has been a prominent feature of consensus and regulatory standards on confined spaces over the years. Some jurisdictions require a permit for all entries into confined spaces (Canada Gazette 1992). Others require entry permits under specific conditions (ANSI 1995, OSHA 1993).

Aside from the fact that the entry permit conveys permission to enter and work in the space, the anatomy and dynamics of these documents are worth discussing. Entry permits sometimes are produced in two versions: cold work and hot work permits. Table 9.1 is a composite from various sources of information required or provided in entry permits (NIOSH 1979, Standards Australia 1995, OSHA 1993). Sources also included industrial organizations. These are not identified to respect the privacy of these organizations.

TABLE 9.1
Information Provided in Entry Permits

Descriptive
- Department
- Location of space
- Permit prepared by
- Purpose for entry
- Description of work
- Start time
- End time
- Date of entry
- Time of entry
- Last contents of space

Isolation measures
- Blanking and/or disconnecting
- Electrical
- Mechanical

Hazardous work
- Burning
- Brazing
- Welding
- Open flame
- Painting
- Chemical cleaning
- Other

Hazards expected
- Corrosive materials
- Hot equipment
- Toxic materials
- Drains open
- Cleaning (specify type)
- Spark producing operations
- Spilled liquids
- Pressure systems
- Other

Vessel cleaned
- Deposits
- Method
- Inspection
- Neutralized with
- Purged
- Ventilated
- Continuing ventilation required

Lockout
- Electrical
- Mechanical
- Hydraulic/pneumatic
- Stored energy
- Chemical lines

Atmospheric testing
- Oxygen
- Flammability (calibrating gas, date)
- Toxic substance (calibrating vapor, date)
- Periodic/continuous
- Location
- Time

Performed by
- Acceptable entry conditions

Fire safety precautions
- Fire extinguisher
- Equipment to be specified

Protective equipment
- Respirators
- Protective clothing
- Head, hand, and foot protection
- Eye protection, faceshields
- Lifelines, harness, and retrieval equipment
- Explosion-proof or low voltage lighting
- Communications
- Emergency alarms
- Communication
- Equipment grounded and GFCI installed or low voltage
- Warning signs
- Testing

Work practices
- Buddy system
- Employee qualified
- Standby person qualified
- Communication procedure used by authorized entrants and attendants to maintain contact with each other and with Qualified Person

Training
- CPR
- First aid
- Emergency egress procedures
- Preparatory and other procedures discussed
- Fire extinguisher
- Air monitor

Authorizations
- Entry supervisor
- Supervisor who authorized entry
- Production supervisor
- Line supervisor
- Safety supervisor
- Evaluator

Personnel
- Entrants (authorized entrants)
  - Company personnel
  - Noncompany personnel
- Standby person
- Rescue
- Attendants
- Entry supervisor

Rescue and emergency services
- Equipment
- Numbers to call

Other necessary information
- Diagram of internal configuration of the space
- MSDSs for contents, materials to take in
- Bonding/grounding (e.g., spray painting equipment to structure of space

Compiled from NIOSH 1979, Standards Australia 1995, and OSHA 1993.

Additional Permits

Entry permits from organizations also were reviewed. These show the real-world implementation of the concept. Those examined here reflected the model provided by NIOSH (NIOSH 1979). These documents solicited varying levels of detail. The following observations were made about the group:

Hazard assessment/hazard control checklist/permit to work: These documents attempted to act as a combination hazard assessment, hazard control checklist, and permit to work.

Omission of hazardous conditions: The current permit can lead to omission of consideration about hazardous conditions. This can occur because of the inclusion on the permit of only a small number of hazardous conditions. At least 36 hazardous conditions could be present in confined spaces.

Hazard control: The permits use a check-box to indicate status of hazardous conditions. However, they provided no quantitative criteria against which to measure success, except for atmospheric hazards. Even these were limited to common contaminants. The only basis for decision making was the judgment of the creator of the permit.

Hazardous conditions produced by work activity: Aside from fire prevention/fire protection, the permits provided little information or consideration about hazardous conditions that could be generated by work activity. What information was provided was limited to typical safety hazards. No consideration was given to health hazards.

Qualified person: The permits did not contain a requirement for an individual to identify competence to be able to assess hazardous conditions or to recommend corrective action. Similarly, the permits provided no indication about anyone directly responsible for the content of the permit and the conduct of work in the space. Most permits used sign-offs by varying levels of supervision acknowledging review of the content of the permit. Some used sign-offs to indicate completion of individual tasks. Aside from day-to-day responsibility for conduct of the operation at the site, there was no indication in the sign-offs by senior supervision that they were technically qualified or capable of judging the validity of information provided in the permit.

Use of qualifiers and vague terminology: Many of the permits used qualifiers such as: "required" or "as required," "appropriate" or "as appropriate," "proper" or "properly," "equivalent," "etc.," "understands" or "knows," "adequate" or "adequately." These words convey judgmental qualities to performance of a task. The permits contain no antecedents or reference points to which these qualifiers referred. Similarly, many of the permits also used vagueness in technical terminology: ventilated, permit safe working, sewer basins protected, space checked for extreme heat, levels acceptable for human survival, explosive condition (no reference level), periodic test required, surfaces safe, procedures understood, fire protection (type not specified). The lack of antecedents or reference points means that no one, including the individual attempting to perform the instruction, can assess the level of performance or suitability of the task.

Generic use of "toxic": Many of the permits reviewed here used the work "toxic" in a generic way to indicate unspecified hazardous substances. The permits provided no space to indicate the name of the substance of concern. The test protocol also was not specified.

Calibrant for flammable/combustible tester: Methane and hydrocarbon vapors, such as pentane or hexane, are used for calibration of flammable/combustible sensors in direct reading instruments. Each substance has different ignition and combustion characteristics. Methane calibration could cause under-indication of hydrocarbon vapors, whereas hexane calibration could produce overestimation in a methane atmosphere. Of the two, overestimation would pose less of a concern. As well, the instrument should be calibrated with the substance of interest, where the contaminant is pure or a known mixture. Without indication about the method of calibration, there is no way to determine the validity of results.

Continuous vs. discrete testing (timing of discrete tests): Some of the permits specified discrete tests. These were to be performed at intervals specified by the author of the permit. There is no way to judge whether either the timing or frequency of testing would be relevant to the hazards of concern. With the availability of direct-reading instruments for many hazardous conditions, there is no reason for not performing continuous testing. However, there are many substances for which direct-reading instruments are not available. For some, there are discrete methods for rapid assessment, such as colorimetric detector tubes. The latter methods still pose the question about frequency of sampling. The timing for sampling still reflects judgment from a Qualified Person.

Information overload: The permits as reflected by the examples described here attempt to provide or solicit considerable information on one or at most two sides of paper. This information often contains exhortations to work safely or provides summaries of safe work practices. These documents provide so much information that the consumer is likely not to read everything in the manner intended. The intended audience for the information — first-line supervisors and tradespeople — appears not to have been considered in the document.

The preceding observations about entry permits recommended in standards and those in use in actual organizations indicate that these documents can rapidly become clumsy and bloated as they attempt to be all things to all people. In so doing, they could satisfy the needs of no one.

The function of an entry permit is to inform and to document. The entry permit would be considerably more effective a communication document as a summary of actions performed and indicating the need for precautionary measures. The permit can inform about measures taken to ensure inactivation and isolation, draining and purging, for example, by referring to completion of procedures for so doing. The permit can inform about other preparations for entry by referring to other actions. The permit can document conditions at the time permission to begin work is given and can specify tests where contamination may continue.

The key to successful implementation of the permit system is the hazard assessment. The hazard assessment serves as the antecedent or background documentation to the entry permit. In essence, using this approach, the entry permit reports on the outcome of hazard assessment, actions taken during preentry preparation, and documents remaining concerns that require attention. In this manner, the permit can act as a summary and refer to the larger enquiry that forms the basis of the hazard assessment. The permit then focuses attention on concerns that remain.

Table 9.2 presents a model for a permit that follows concepts developed in the preceding chapter (Chapter 8). Reference to legislative standards should occur to ensure that this would comply with requirements in a particular jurisdiction. The expectation in use of this permit is to provide **only** information relevant to a particular situation. The permit becomes a concise summary of the information contained in the hazard assessment. The hazard assessment provides full information for documentation purposes.

The signed permit is posted at the entry or portal to the space or as specified by the regulatory authority. It remains posted until the work is completed or is replaced by a new permit.

The signed permit remains in effect as long as there is no change in personnel or designated activity. Should there be a change in the authorizer, tester, or work crew, a new permit may be required. Should work not authorized on the permit be necessary, again, a new permit may be required.

The entry permit becomes a necessary record upon completion of the work. It must be retained for recordkeeping according to requirements of the regulatory authority.

TABLE 9.2
Sample Entry Permit

ABC Company
Confined Space — Entry Permit

1. Descriptive Information

Department:

Location:

Building/Shop:

Equipment/Space:

Part:

Date: **Assessor:**

Duration: **Qualification:**

2. Adjacent Spaces

Space:

Description:

Contents:

Process:

3. Prework Conditions

Atmospheric hazards

Oxygen deficiency **Yes** No **Controlled**
Concentration: (Acceptable minimum:)

Oxygen enrichment **Yes** No **Controlled**
Concentration: (Acceptable maximum:)

Biochemical/chemical **Yes** No **Controlled**
Substance Concentration (Acceptable standard:)

Fire/explosion **Yes** No **Controlled**
Substance concentration (Acceptable maximum:)

Biological hazard **Yes** No **Controlled**
Refer to corrective action.

Ingestion/skin contact hazard **Yes** No **Controlled**
Refer to corrective action.

Physical agents

Noise/vibration **Yes** No **Controlled**
Level: (Acceptable maximum: dBA)

Heat/cold stress **Yes** No **Controlled**
Temperature: (Acceptable range:)

Non-ionizing/ionizing radiation **Yes** No **Controlled**
Type Level (Acceptable maximum:)

Laser **Yes** No **Controlled**
Type Level (Acceptable maximum:)

TABLE 9.2 (continued)
Sample Entry Permit

Personal confinement **Yes** No **Controlled**
Refer to corrective action.

Mechanical hazard **Yes** No **Controlled**
Refer to procedure.

Hydraulic hazard **Yes** No **Controlled**
Refer to procedure.

Pneumatic hazard **Yes** No **Controlled**
Refer to procedure.

Process hazard **Yes** No **Controlled**
Refer to procedure.

Safety hazards

Structural hazard **Yes** No **Controlled**
Refer to corrective action.

Engulfment/immersion **Yes** No **Controlled**
Refer to corrective action.

Entanglement **Yes** No **Controlled**
Refer to corrective action.

Electrical **Yes** No **Controlled**
Refer to procedure.

Fall **Yes** No **Controlled**
Refer to corrective action.

Slip/trip **Yes** No **Controlled**
Refer to corrective action.

Visibility/light level **Yes** No **Controlled**
Level: (Acceptable range: lux)

Explosive/implosive **Yes** No **Controlled**
Refer to corrective action.

Hot/cold surfaces **Yes** No **Controlled**
Refer to corrective action.

For entries in **highlighted** boxes, provide additional detail and refer to protective measures. For hazards for which tests can be made, refer to testing requirements. Provide date of most recent calibration.

4.0 Work Procedure

Description:

Hot work **Yes** No **Possible**
Refer to protective measures.

Atmospheric hazards

Oxygen deficiency **Yes** No **Possible**
Refer to requirement for additional testing. Record results. Refer to requirement for protective measures.
Concentration: (Acceptable minimum:)

Oxygen enrichment **Yes** No **Possible**
Refer to requirement for additional testing. Record results. Refer to requirement for protective measures.
Concentration: (Acceptable maximum:)

TABLE 9.2 (continued)
Sample Entry Permit

Biochemical/chemical **Yes** No **Possible**
Refer to requirement for additional testing. Record results. Refer to requirement for protective measures.
Substance concentration: (Acceptable standard:)

Fire/explosion **Yes** No **Possible**
Refer to requirement for additional testing. Record results. Refer to requirement for protective measures.
Substance concentration: (Acceptable standard:)

Biological hazard **Yes** No **Possible**
Refer to requirement for protective measures.

Ingestion/skin contact hazard **Yes** No **Possible**
Refer to requirement for protective measures.
Physical agents:

Noise/vibration **Yes** No **Possible**
Refer to requirement for protective measures. Refer to requirement for additional testing. Record results.
Level: (Acceptable maximum: dBA)

Heat/cold stress **Yes** No **Possible**
Refer to requirement for protective measures. Refer to requirement for additional testing. Record results.
Temperature: (Acceptable range:)

Non-ionizing/ionizing radiation **Yes** No **Possible**
Refer to requirement for protective measures. Refer to requirement for additional testing. Record results.
Type Level: (Acceptable maximum:)

Laser **Yes** No **Possible**
Refer to requirement for protective measures.

Mechanical hazard **Yes** No **Possible**
Refer to requirement for protective measures.

Hydraulic hazard **Yes** No **Possible**
Refer to requirement for protective measures.

Pneumatic hazard **Yes** No **Possible**
Refer to requirement for protective measures.

Process hazard **Yes** No **Possible**
Refer to requirement for protective measures.

Safety hazards

Structural hazard **Yes** No **Possible**
Refer to requirement for protective measures.

Engulfment/immersion **Yes** No **Possible**
Refer to requirement for protective measures.

Entanglement **Yes** No **Possible**
Refer to requirement for protective measures.

Electrical **Yes** No **Possible**
Refer to requirement for protective measures.

Fall **Yes** No **Possible**
Refer to requirement for protective measures.

TABLE 9.2 (continued)
Sample Entry Permit

Slip/trip	**Yes** No **Possible**
Refer to requirement for protective measures.	
Visibility/light level	**Yes** No **Possible**
Refer to requirement for protective measures.	
Explosive/implosive	**Yes** No **Possible**
Refer to requirement for protective measures.	
Hot/cold surfaces	**Yes** No **Possible**
Refer to requirement for protective measures.	

For entries in **highlighted** boxes, provide additional detail and refer to protective measures. For hazards for which tests can be made, refer to testing requirements. Provide date of most recent calibration.

Protective measures

Personal protective equipment (specify)
Communications equipment and procedure (specify)
Alarm systems (specify)
Rescue Equipment (specify)
Ventilation (specify)
Lighting (specify)
Other (specify)

Testing requirements

Specify testing requirements and frequency.

Personnel

Entry supervisor
Originating supervisor
Authorized entrants
Testing personnel
Attendants

Entry Certificate

The entry certificate has been utilized primarily in the maritime industry. Its origins date back to the early 1920s. In that era the marine chemist was one of the few professionals trained in hazard assessment and control. Even if the protocols for testing were provided in entry documents, these would have provided little, if any, benefit to persons reading them. As a result, the approach and communication in the process became completely expert centered. The expert communicated the judgment about a particular space through standard phrases and provided recommendations for further action, as necessary.

Table 9.3 provides a generalized version of the Marine Chemist Certificate contained in NFPA 306 (after NFPA 1993).

TABLE 9.3
Entry Certificate

ABC Company
Confined Space — Entry Certificate

1. Descriptive Information

Department:

Location:

Building/shop:

Equipment/space:

Part:

Date: **Assessor:**

Duration: **Qualification:**

2. Adjacent Spaces

Space:

Description:

Contents:

Process:

3. Results of Tests

Contaminant	Result	Standard	Comments
Oxygen			
Flammables/combustibles			
Carbon monoxide			
Carbon dioxide			
Hydrogen sulfide			
Sulfur dioxide			
•			
•			
•			

4. Comments from Physical Inspection

5. Safety Designation

After NFPA 1993.

STANDARD SAFETY DESIGNATIONS

Safe for workers: means that in the space so designated: (a) the oxygen content of the atmosphere is at least 19.5% (or as regulated) by volume; and that (b) toxic materials in the atmosphere are within permissible concentrations; and that (c) the residues are not capable of producing toxic materials under existing atmospheric conditions while maintained as directed on the Certificate.

Not safe for workers: means that in the space so designated, the requirements of "safe for workers" have not been met.

Enter with restrictions: means that in the space so designated, entry for work may be made only if conditions of proper protective equipment, clothing, and time are as specified.

Safe for hot work: means that in the space so designated: (a) oxygen content of the atmosphere is at least 19.5% (or as regulated) by volume, with the exception of inerted spaces or where external hot work is to be performed; and that (b) the concentration of flammable materials in the atmosphere is below 10% (or as regulated) of the lower flammable limit; and that (c) the residues are not capable of producing a higher concentration than permitted by (b) above under existing atmospheric conditions in the presence of fire, and while maintained as directed on the Certificate; and further, that, (d) all adjacent spaces containing or having contained flammable or combustible materials have been cleaned sufficiently to prevent the spread of fire, or are satisfactorily inerted.

Not safe for hot work: means that in the space so designated, the requirements of "safe for hot work" have not been met.

The focus of the Marine Chemist Certificate is atmospheric hazards that can result in an inhalation or fire and explosion hazard. The consumer of the information must rely on the thoroughness of the inspection and the comprehensiveness of the knowledge of the inspector. The Certificate provides no indication about the manner in which the inspection was performed or about criteria that were examined.

This approach works in the confines of the shipyard where structures, functions, spaces, and hazardous conditions are relatively constant. The problem with the approach is that it provides no ability for follow-up or audit. The user of the information has no means to enquire about the information or the process used in developing it. This limits the utility of the information to the recollection of the inspector. As well, concerns in confined spaces are expanding. This publication, for example, considers at least 36 potentially hazardous conditions that could exist.

The Certificate has continuing viability as a reporting format when a formal means of hazard assessment, such as that advocated here, is employed. The Certificate becomes the vehicle for communicating the outcome of the hazard assessment in the language of the shipyard.

THE WORK ENVIRONMENT

Work activity involving confined spaces often is time-critical, as are most aspects of maintenance. The space may be a critical component of a production process. Unavailability of this component could represent a considerable economic impact to the organization. This comment applies regardless of whether the space is a process unit, a hydroelectric generator, a railway tunnel, a compartment on a ship or barge, or other such structure. The confined space could be located in a downtown street, an industrial complex, or in a remote location.

Downtime is very costly. The pressures created by time-criticality impose a considerable burden on all participants in the project — management, supervision, trades- and technical people, and the on-site Qualified Person. All sense the urgency to complete the work on time.

The on-site Qualified Person is especially vulnerable to these pressures. Assessing conditions and their impact in a rapidly changing environment, so that critical decisions can be made, is a difficult and stressful undertaking. Individual activities that involve exposure to hazardous agents often are short in duration and unpredictable in their actual, vs. planned, occurrence. Situations not anticipated during hazard assessment must be corrected.

Time-criticality has a special meaning to the on-site Qualified Person when things do not proceed according to schedule or when something goes wrong or the unexpected occurs. Whereas the focus of other participants in these situations is task oriented — seeking ways to correct the problem or to make up for lost time — the Qualified Person must focus on the implications of those proposals on the safety and well-being of the participants.

Perhaps the most difficult action required from the Qualified Person is making the decision that the work must stop. This could be forced when conditions in the space exceed prescribed limits. The need to stop work is an acknowledgment that control was inadequate. Recovering from this situation, so that the work can resume, could be very difficult and very costly. Minimizing the need to act in this manner is essential. This best can be achieved during hazard assessment through careful analysis and planning and preparation.

Stopping work or recommending evacuation or other action that alters work flow under these circumstances is an especially difficult decision. Assessing and evaluating proposed actions and determining ways to ameliorate the condition that motivated this decision also can be extremely difficult under these conditions. Determining the action needed to bring these situations back into control can tax the abilities of even the most experienced and knowledgable of individuals. Minimizing the need to act in this manner is essential, since decision making in these circumstances is extremely stressful and potentially more prone to error than what occurs during normal activity.

Safety is not foremost as a concern in most minds on a worksite. The focus is task oriented — getting the work done. This outlook should not be regarded overly harshly, since in any organization, only at most a few minds are permitted the luxury to consider in any detail the ramifications of actions. The Qualified Person, therefore, has an extremely important role to play in establishing and maintaining the safe conduct of work in these situations.

EXPOSURE ASSESSMENT AND EVALUATION

Evaluating acute or instantaneous exposures during work in confined spaces traditionally has posed difficulty. This difficulty extends from the need to determine and interpret and act on the consequences of exposure that occurs over a short period. To a large extent, these problems have been technology limited.

Uncertainties associated with the identity of contaminants and the occurrence, duration, and magnitude of exposure in these situations together pose considerable difficulties to anyone responsible for assessing these activities and specifying controls. Determining the temporal occurrence of excursion(s) requires in-depth knowledge about activities and processes. Coupled with this are the irregularities of scheduling and unpredictable progress within the scope of work. A successful approach to this problem begins with a detailed analysis of all activities, tasks, and procedures contained in the scope of work. Subsequent steps include assessment of hazards associated with each activity and evaluation by measurement to determine the need for and suitability of controls.

A major impediment to evaluating exposures under these conditions has been availability of technology advanced enough to perform the measurements. One of the major challenges in these work environments is assessment of acute exposure to substances whose identity may not be known.

Measuring the airborne concentration of most contaminants during short-term exposures is difficult, if not impossible. This difficulty reflects the shortcomings of available instruments and methods. This situation is changing rapidly, however. A major factor is the development of person-portable analytical instruments. These instruments provide capabilities from broad-spectrum to substance-specific. Some contain several substance-specific sensors. Many can provide continuous "real-time" output, and in some cases possess data-logging capabilities. (Dataloggers are electronic memory modules that record information at selected intervals within the sampling period.) These new tools provide the capability to assess the previously unmeasurable.

A serious concern in testing for ignitable atmospheres is that equipment used in the procedure or the testing equipment itself not act as a source of ignition. There are examples on record where tools used to open ports or manhole covers have created sparks. The sparks ignited vapor/gas mixtures. Similarly, static discharges from clothing through metal objects, such as instrument probes, can create sparks. Again, there is evidence that this sort of event can occur.

Guidelines and legislated standards used for assessing exposure primarily reflect the experience and needs of production-oriented industry. Threshold Limit Value — Time-Weighted Average (TLV-TWA®) and most legislated standards, for example, are based on day-after-day exposure for an 8-h workshift or 40-h work week (ACGIH 1995).

Exposure to toxic substances in production-oriented industry often occurs with some degree of consistency, and therefore predictability, day after day (NIOSH 1977). On the other hand, exposures occurring within entire segments of industry are neither uniform nor predictable nor necessarily long in duration relative to the 8-h shift.

Evaluation of exposure under these circumstances poses difficulty. One option is to prorate the exposure to a timebase of 8 h and compare against the TLV-TWA or its equivalent, as applicable. This depends on the mode of action of the chemical agent and duration of exposure, and not exceeding other limits where they exist (ACGIH 1995).

Other measures apply to exposures of very short duration. An approach for assessing single exposures lasting less than 15 minutes is to compare the time-weighted average to the Threshold Limit Value-Short Term Exposure Limit (TLV-STEL®), as available. The TLV-STEL is based on the time-weighted average concentration during the period of 15 min. Hence, the TLV-STEL would apply to an exposure lasting less than 15 min.

ACGIH recently deleted the STELs for many chemical substances. Another evaluative tool mentioned by ACGIH for chemical substances lacking an STEL is the short-term excursion. Short-term excursions should not exceed three times the TLV-TWA for more than 30 min during the work day. Under no circumstances, should the exposure exceed five times the TLV-TWA. The preceding approaches for assessing short-term exposure presuppose that the 8-h Time-Weighted Average exposure does not exceed the TLV-TWA. These approaches do not apply to substances in the TLV booklet carrying the designation TLV-Ceiling or TLV-C®. Exposure to these substances must not exceed the TLV-C at any time.

While the preceding approaches may satisfy the rules for application of the TLVs or equivalent standards, these are unacceptable to many regulators, and workers and their unions. In situations such as these, these groups will not accept any exceedence of the minimum standard (the TLV-TWA or its equivalent) for any length of time. This view recognizes several realities: the magnitude of the exposure is unpredictable and is unlikely to be known until after its occurrence, and exceedence of the lowest standard is a warning that must be heeded.

An additional concern that may arise is exposure during extended shifts. Extended shifts can result from deliberate planning or from ad hoc decisions taken when a crisis arises. As mentioned, most exposure standards are intended for 8-h shifts. These are inapplicable to workshifts of longer length (Brief and Scala 1986). Brief and Scala commented (1975) that little work on the pharmacokinetics of specific compounds has occurred since publication of their recommendation for modifying TLVs to accommodate longer shift lengths. This issue remains unresolved to this time. The Brief and Scala modification is the most simple to work with and also the most conservative.

LOGISTICAL CONSIDERATIONS

The following topics relate to the actual conduct of work on the site. Many could be considered as planning concerns. The focus really is a function of timing, since these could easily arise as a foreseen item during planning for which consideration could be given or an unforeseen one during the conduct of work.

Hazard Communication

Hazard communication is required for substances that will be present, generated, or used at the site and with which workers potentially will be in contact. This could apply to substances residual in the space, introduced through work activity and products of reaction between residual substances and processes used during work activity. This also could apply to exhaust gases and spurious emissions from surrounding processes. Hazard communication is coincident with the requirement for training concerning hazardous conditions that could be present in the space.

Training is the subject of Chapter 13.

Emergency First Aid

The approach taken in this publication is to ensure safe conditions of work through the orderly anticipation, recognition, evaluation, and elimination and control of hazardous conditions. Despite these actions, accident and injury still can occur in these workspaces. The capability and speed of the response could influence the outcome of the injury.

The types of injuries for which rapid response is needed include:

- Chemical splashes in the eye
- Foreign object in the eye
- Chemical contact with the skin
- Cuts and abrasions that induce serious bleeding
- Struck by and struck against accidents
- Slips, trips, and falls from heights
- Heat/cold stress injury

Eye injuries especially deserve rapid attention. The first seconds following the splash of a corrosive into the eye are critical to saving the sight (Jagger, 1988, Champagne 1994). This is especially important where chemical splash has occurred. Caustics are more consequential in this matter than acids. Growth of microorganisms in plumbed-in and portable eyewashes also is a concern (Tyndall et al. 1987, Bier and Sawyer 1990, Paszko-Kolva et al. 1991, 1995). In situations where rapid evacuation cannot occur, provision of emergency medical supplies inside the space should be considered.

For work that occurs in remote locations, a self-contained first aid/emergency response unit may be required. This could be a van outfitted with a bed, blankets, and first aid supplies, and other items for personal hygiene and protection. There is a definite need in these situations for a first-response vehicle that can serve as a comfort station, first aid station, work station for the First Aid Attendant and on-site Safety Officer, and stores for personal protective equipment needed in oversight. As well, a vehicle of this type could be especially valuable in cold weather for maintaining medical supplies; for example, eyewash supplies.

Emergency preparedness and response is the subject of Chapter 15.

Human Factors

The comfort of workers at the site has a profound impact on productivity and their personal well-being and safety. Human factors considerations include the timing and length of workshifts, and provision of sanitary facilities and comfort stations.

An example (a composite situation) serves to illustrate these concerns. The location was a field site 100 km (60 mi) from the nearest town. The town has doctors and a small hospital. The site was off-road about 5 km (3 mi). As originally conceived, the project was expected to be completed in 27 to 29 h. Work was to proceed nonstop. Laborers were to work 12-h shifts. Supervisory and technical personnel planned to work throughout the operation without a break. A contractor supplied heavy equipment and operators. The contractor planned to work nonstop until the work was completed.

Due to an unforeseen problem in scheduling, the start of the project was delayed 3 h. Instead of starting at 11:00, the work started at 14:00. Participants, including supervisors, technical personnel, and heavy equipment operators were accustomed to day-shift hours. Thus, the start of work was late in their normal workday. Daytime weather was hot and sunny with periods of rain. Nighttime weather was very cool, just above freezing, with periods of rain. Most of the personnel were unable to take shelter from the rain during downtime.

An additional delay occurred due to partial flooding of the site. The site was located in the path of seepage from a lake. The excavation created by the heavy equipment lowered the level of the grade below the water table. Pumps had to be located and a channel dug to remove the water.

The project was completed 42 h after the start.

As applied to supervisory and technical personnel, and the contractor's employees, the original plan contained no contingency for delay or emergency. As events unfolded, this decision put unrealistic demands on capabilities and performance of these individuals. From the perspective of exposure assessment, the most heavily exposed persons were the same individuals: supervisors and technical personnel and heavy equipment operators.

Most of the work involving operation of the heavy equipment (bulldozers and front-end loaders) occurred in close proximity to laborers and technical and supervisory personnel. This operation required intense concentration from the equipment operators. Lighting was nonideal, being provided by running lights on the units and portable generator-light sets. Also, pedestrians periodically passed close to the equipment. The latter activity necessitated gaining attention from the operator and waiting until the equipment had stopped operating. Long hours of this type of activity without breaks would be greatly fatiguing. These conditions were highly conducive to error. Safe operation in these conditions required human performance within narrow tolerances of perfection.

People work best with a schedule that incorporates formal break periods. Formal breaks for operators of equipment are a legal requirement in many jurisdictions. There is a very real tendency under the pressure of time-critical operations to continue working, oblivious to the onset of fatigue. Contingencies to address fatigue and the outcomes of fatigue were not considered in this project.

An additional concern arising from the long work schedule exemplified by this project is exposure assessment. Exposure considerations normally are based on shift lengths of 8 h. Interpretation of results from exposure during considerably extended workshifts is very difficult. Continuous long exposure to contaminants, even at very low levels, potentially was a factor in the fatigue experienced by the workforce.

In time-critical projects, the overall goal — timely completion — always must remain clearly in focus. Not having replacements for supervisory and technical personnel put timely completion of this project at risk. Loss of one or more of these individuals from uncontrollable fatigue or injury could have increased project time beyond even that which was required and created other lasting consequences. Accidental damage to critical equipment resulting from substandard performance from overly fatigued personnel could have produced the same result. The costs of failing to meet the goal because of unexpected events and failures can far outweigh any perceived savings through "short cuts."

Considerable useful information and even some wisdom can be gained from examining in some detail the processes and dynamics that occurred during this project. Considerable improvement in future projects can result by building on the lessons provided by this episode.

A major useful point to start is to acknowledge the limitations of human performance. By planning work within normally accepted work schedules, the project should progress faster, since the minds and bodies of all participants will be rested and not overly stressed. This approach provides contingency and redundancy in the event of injury and other unforeseen circumstances.

Provision of food and hot and cold refreshments also deserves consideration. This, however, becomes less of an issue when the first consideration (normal work schedules) is addressed. Weather can be an important consideration. Susceptibility to cold stress especially is exacerbated by rain and long periods of inactivity in cool temperatures. Having refreshments available to combat the effects of weather could be very important for maintaining well-being, and hence, productivity.

To illustrate, Wallin and Wright documented the impact of working conditions on performance in a methodological survey (Wallin and Wright 1986). This work illustrated the effects of situational aspects on the well-being of both white- and blue-collar workers. The results illustrated the impact of controllable and noncontrollable conditions. Stress reactions can occur because of overload, lack of influence over working conditions, conflict, and low quality of work. Overload and pace were mentioned as the leading causes of stress reactions. The result was tiredness, listlessness, cardiovascular involvement, and tendency to depression. This occurred in managers and others who held positions that provided varied work opportunities. The negative aspects of this work far outweighed the positive, according to this methodology.

The relationship between fatigue and accidents has been postulated since 1931 (Heinrich 1931). Heinrich postulated the role of fatigue in the sequence:

$$\text{overtime} \rightarrow \text{fatigue} \rightarrow \text{incidents} \rightarrow \text{injuries}$$

Research has not proven that this sequence models reality. This model appears to be too simplistic and may need qualifiers. These include: amount of overtime and recovery from fatigue, type of work, attitude toward overtime, pace of work and planning, and control and supervision (Gaunt 1980).

Many human physiological and psychological functions follow a 24-h cycle. This is related to but not necessarily controlled by the diurnal activity pattern (Luce 1970). Kleitman showed (1963) that performance is related to change in metabolic rate of chemical processes in the brain. Reaction time is closely related to body temperature. This relationship also exists in the processing of simple information for which the speed of response is the major variable (Hockey and Colquhoun 1972). The low point of performance during shift work for industrial workers occurs about 3:00 (am) (Bjerner et al. 1955). The diurnal temperature curve is a sine wave that peaks at 18:00 and troughs at 4:00 (Aschoff 1981) (Figure 9.1).

The timing of the workshift can cause the demands of performance to be completely out of phase with diurnal rhythms. This is especially the case where individuals work considerably longer than normal hours, as in the previous example. During the night, when body temperature drops, the individual attempts to remain awake. During the day, when the individual needs to sleep, body activity increases. Sleep deprivation can occur under these conditions. This can produce deleterious effects on performance (Coplen 1988). The change from day shift acclimatization to night shift acclimatization occurs over the period of a week (Colquhoun 1971). The sine wave flattens somewhat (Colligan and Tepas 1986). This suggests that the activity cycle has an influence on body temperature but does not control or regulate it.

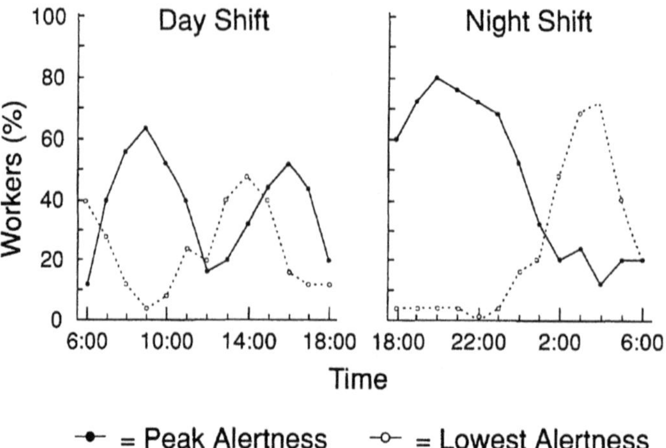

FIGURE 9.1 Effect of time within the workshift (12 h) on alertness. Peak, average, and lowest reported levels of alertness were lower during the night shift than during the day. (Adapted from Budnick et al. 1994.)

Communications Systems

Communications is an integral part of site safety. Fatal accidents documented by NIOSH, OSHA, and MSHA occurred because of the lack of communication or communication breakdown (NIOSH 1979, 1994; OSHA 1982a,b, 1983, 1985, 1988, 1990; MSHA 1988). To illustrate, individuals entered or remained inside confined spaces without informing others. Persons inside the space did not communicate or were unable to communicate with those attending outside. The person inside could not alert other people about the need for help or to modify their actions.

Legislation such as the OSHA Standard for general industry attempts to correct this situation by mandating the existence of the attendant (OSHA 1993). The attendant monitors activities inside and outside the space and determines whether authorized entrants can safely remain inside. The attendant orders entrants to leave the space when evidence of an uncontrolled hazard exists. These tasks require knowledge about hazards that may be present, modes of exposure, and signs and symptoms and consequences. The attendant watches for conditions not envisioned during preparation of the permit and for other prohibited conditions. As well, the attendant warns individuals not authorized to enter the space to stay away, and those who have entered, to leave. In the event of an emergency, the attendant summons rescue services and assists entrants in evacuation. The attendant may order the evacuation when unsafe or unpermitted conditions are present.

The attendant works from a location outside the permit space. This could include a control room that allows remote monitoring of entrants through television monitors, or by electronic monitors or radio communications.

Other legislation, such as the OSHA Standard on shipyard employment, did not specify the need for attendants during work in confined spaces (OSHA 1993). This Standard operates on the premise that hazards will be eliminated or controlled prior to the start of work. This premise may be satisfactory for atmospheric hazards, but it does not cover hazardous conditions that lead to accidents, such as, slips, trips, falls from heights, and "struck by" and "struck against." Ensuring a means of communication for individuals working in remote compartments is essential for enabling rapid response in these situations.

Depending on the site and its activities, communications can involve the following path:

entrant → attendant → entry supervisor → site supervisor → superintendent/manager

The entrant who discovers a hazardous condition may start communication through a chain that leads to the senior manager at the site. In sites where few people are working, the path of communication is much shorter. In large sites, multiple activities can occur simultaneously in the space. These could involve work by independent groups of company employees and by one or more contractors. This situation can lead to complications. Each group may have an entry supervisor. Work activity in one part of the space could have a negative impact on work occurring in another.

Resolving the conflict that could arise in this situation requires action from senior management at the site.

Breakdown in communication between entrants and the attendant can occur for several reasons (Ibbetson 1994). The most important operational reason is noise. Noise produced by equipment in the surroundings or inside the space can render communication very difficult. Audio warnings produced by instruments may go unrecognized. This problem is compounded when entrants cannot maintain direct visual contact with each other or with the attendant.

Table 9.4 describes communications options that are available. Other options include two-way pagers and avalanche locators. Their suitability for use in the industrial environment remains to be determined (Figures 9.2, 9.3, 9.4).

Each option in Table 9.4 offers strengths and limitations. These should be considered within the context of the circumstances of an individual situation before making a selection.

Fire/Explosion Protection

Where fire or explosion is a consideration, the most productive action is to eliminate the potential to the greatest extent possible. The means removing the fuel and/or sources of ignition. The principal oxidizer, the oxygen that we breathe, is controllable only under rigorous conditions. These could involve inerting (displacing the atmosphere from the structure by an inert gas) or filling the structure with water. Neither is a routine action. Since purging and ventilating can displace the internal atmosphere from the structure into the surroundings, a flammable/explosive mixture could develop. A flammable/explosive atmosphere also could develop inside the structure during this process.

Eliminating sources of ignition involves a rigorous process. This could necessitate measuring accumulation of static charge. Static accumulation is a concern where humidity levels are low. Bonding and grounding of portable equipment, such as air movers and duct, and vacuum cleaning equipment, to the structure may be required (Figure 9.5). Testing to ensure the effectiveness of bonding and grounding may be necessary.

Sources of thermal energy can be measured using remote sensing infrared thermometers (Figure 9.6). These units measure infrared emissions from heat sources. Some units use laser beams to pinpoint the source of the measurement. This technology provides an indication about whether equipment operating in the vicinity could be or could become a source of ignition. Some units also can detect cold sources.

The concentration of gas/vapor inside the space and in the surroundings also is critically important. Measuring concentrations of vapor in the flammable range could be very difficult; equipment normally is designed to measure levels lower than this.

Fire protection also includes portable fire protection equipment. Selection of fire protection equipment must follow review of specific requirements of the situation. Equipment provided at the site may represent only the initial response from a more extensive fire protection system. The latter may require time for activation — that is, for volunteers to become available from normal work duties.

Emergency preparedness and response is the subject of Chapter 15.

TABLE 9.4
Communication Systems for Use in Confined Spaces

Type of Product	Comments
Hand-held, two-way radio	Requires free hand to activate
	May not be approved for use in hazardous atmospheres
	Belt-mounted case adds extra bulk during passage through access/egress opening or small spaces
	Easy to contaminate, difficult to decontaminate
	May interfere with operation of monitoring instruments
	Difficult to hear in noisy environments
Noise-attenuating, two-way radio communication headset	Worn over the head or mounted to hardhat; microphone + speaker + noise-attenuating muffs
	Designed for use in high noise area
	Voice activated (VOX) or push to talk; push-to-talk units require free hand to operate
	May not be approved for use in hazardous atmospheres
	Easy to contaminate; difficult to decontaminate
	May interfere with monitoring instruments
	Some products use FM transceiver; limited range
In-ear receiver, throat microphone radio unit	Designed for use with respiratory protection
	Some noise attenuation
	Voice activated or push to talk; push-to-talk units require free hand to operate
	When worn inside encapsulating suit, protected from contamination; difficult to decontaminate
	May interfere with monitoring instruments
Tethered in-ear or headset intercom unit	Provides more assured signal than radio
	Units have rating for intrinsic safety
	Cable may snag and tear on sharp edges
	Cable requires decontamination; see previous comments regarding headset and in-ear units
In-facepiece microphone, external speaker	Permanently mounted unit
	Speaker easy to contaminate, difficult to decontaminate
	Voice activated or push to talk
Motion sensor/audio alarm	Senses motion of wearer; prealarm sounds if no motion after 20 seconds
	Sealed, watertight case
	Alarm only, no voice contact
	Belt-mounted case adds extra bulk during passage through access/egress opening or small spaces
	May be difficult for others, including the attendant, to hear in noisy environments
Air horn	Horn utilizes aerosol can of pressurized gas as a power source
	Manually activated, free hand required to operate
	Pressurized can could rupture in an accident situation
	Container adds extra bulk during passage through access/egress opening or small spaces
	Requires prearranged recognition of signal
	May be difficult to hear in noisy environments
	Requires free hand to activate

Electrical Systems

Inappropriate use or use of faulty 110 V equipment was a significant cause of fire and explosion and electrocution in confined spaces (OSHA 1982a, 1985, 1988, 1990; NIOSH 1994). These accidents resulted from faulty grounds, contact with bare conductors, use of unshielded lightbulbs in hazardous atmospheres, and inappropriately modified circuits, as well as other circumstances.

Preventing electrical accidents during work in confined spaces is a considerable challenge. Many confined spaces are constructed from conductive materials, such as steel or aluminum. The conductivity of these metals and the nonconductive materials that are used in construction of structures can be enhanced through wetting, especially by solutions containing salts. Concerns about moisture extend from low voltage/high current processes, such as welding, to 110 V systems.

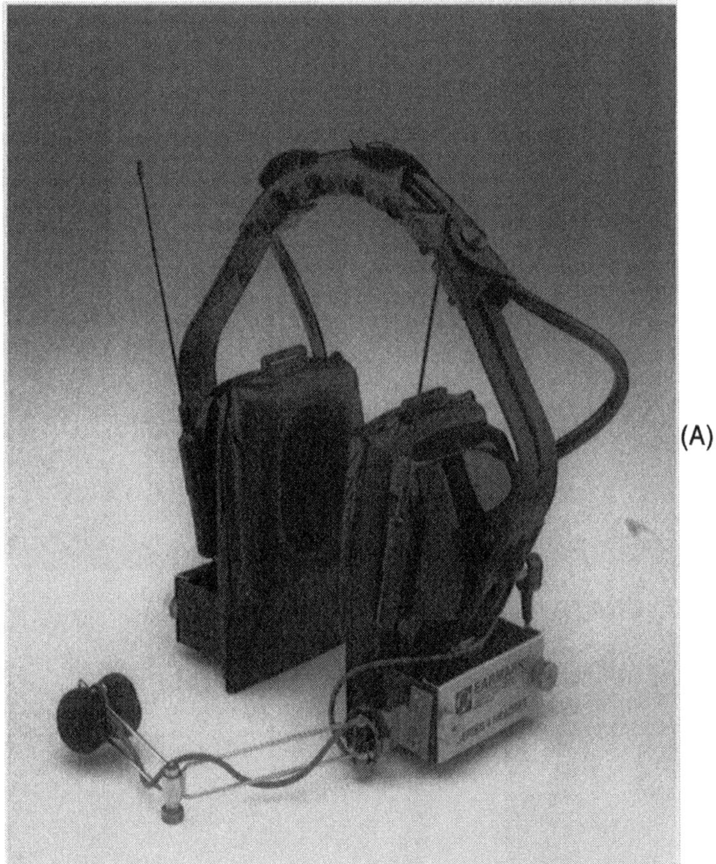

(A)

FIGURE 9.2 Communications equipment used in confined spaces: (A) Self-contained headset. Headset contains voice-activated transmitter/receiver and boom microphone. Noise-cancellation technology is available; (Courtesy of EARMARK, Hamden, CT.) (B) In-ear speaker and throat-mounted microphone connected to a standard two-way radio; (Courtesy of CON-SPACE COMMUNICATIONS LTD., Richmond, BC.) (C) In-ear speaker and choice of throat-mounted or surface-adhering microphone, and voice-activated radio; (Courtesy of EARMARK, Hamden, CT.) (D) Hard-wired intercom relay station. This unit relays radio communication signals from inside the space to the attendant located outside, who listens to a unit containing a loudspeaker and a microphone. (Courtesy of EARMARK, Hamden, CT.)

These voltages by far were the major causes of the accidents documented in the OSHA and NIOSH reports. Water from rain or washing of the interior of structures is a normal part of the work environment in many confined spaces.

This work often entails use of portable electrical equipment and lighting systems. These require temporary wiring. Wire in temporary systems often means plastic- or rubber-insulated extension "cords." This wire is highly vulnerable to damage, especially when used in close quarters. Wire at entry/exit portals usually is left on the floor where it can be stepped on, or crushed and cut by falling equipment and materials. This type of wire is ill-suited to the treatment that it can receive.

The most important precautionary measures for preventing electrical accidents are the simplest:

- Inspect wire and equipment for damage prior to and during use
- Position and protect temporary wiring against damage; locate overhead whenever possible
- Prohibit use of damaged equipment

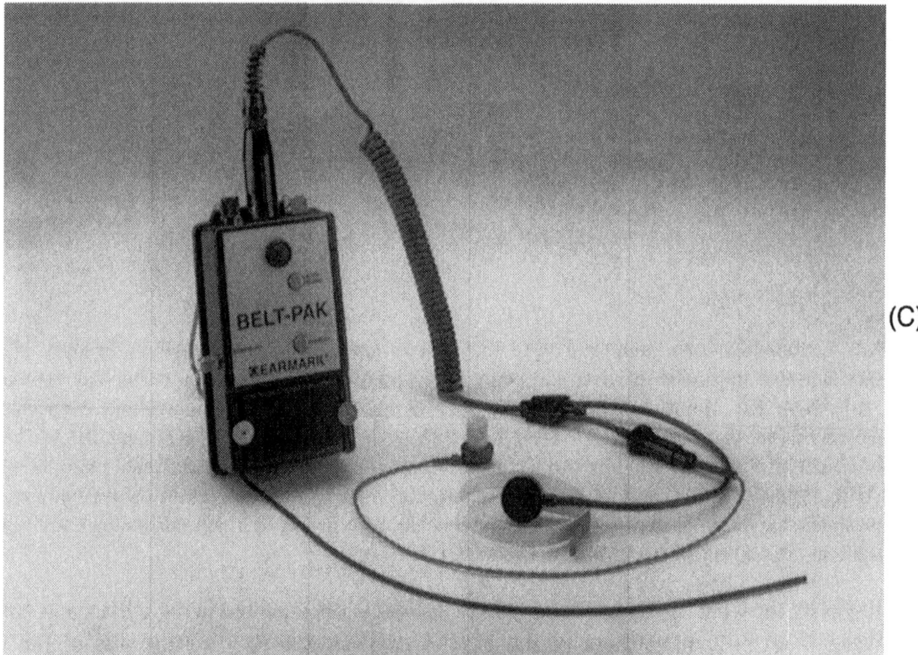

FIGURE 9.2 (continued)

- Forbid repairs or alterations to wire or equipment except those made by a qualified electrician
- Use low voltage equipment; if 110V equipment is used, incorporate a ground fault circuit interrupter (GFCI) in the circuit

These measures will provide some assurance against electrical accidents. Obtaining cooperation from workers and on-site supervision in this matter is essential. All individuals must be informed through formal training about the serious consequences that failure to follow these measures have had on others.

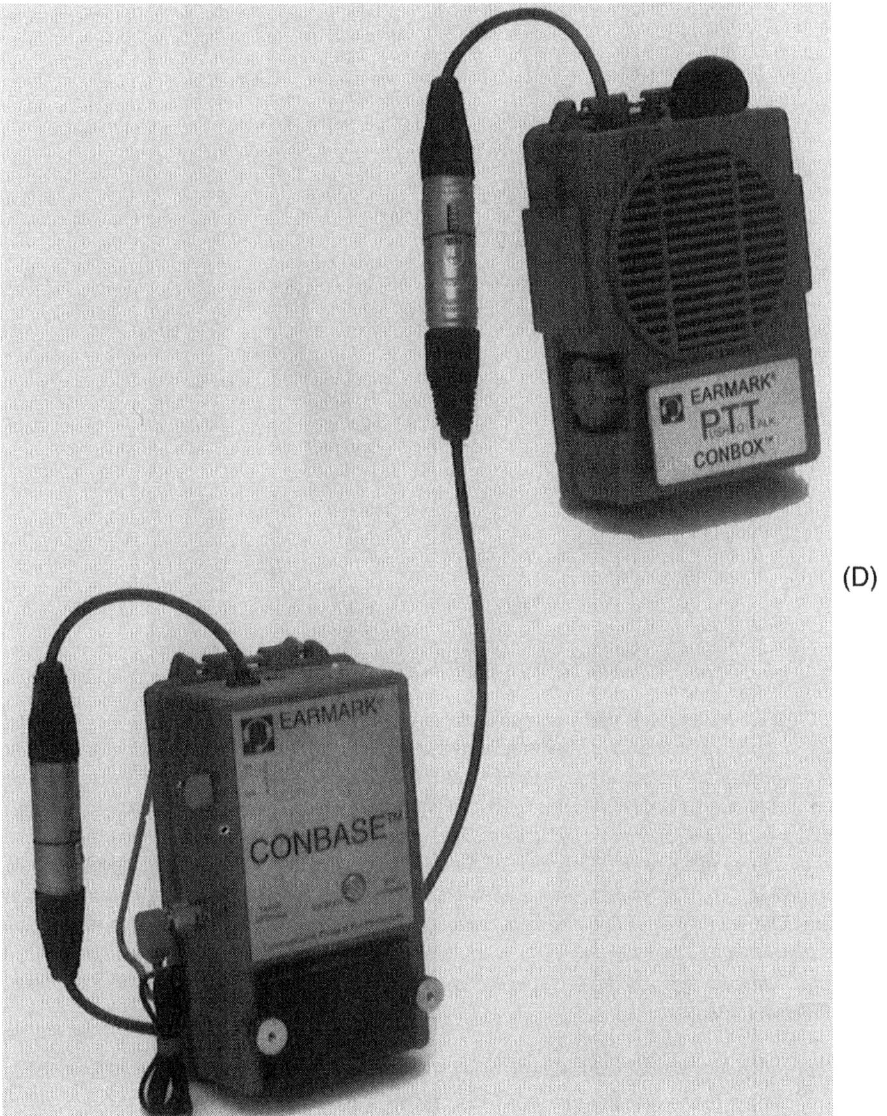

(D)

FIGURE 9.2 (continued)

Additional strategies to minimize the consequences of wiring or equipment failure operate on the premise that these circumstances can and will happen. Additional measures include:

- Fault-proofing equipment
- Detecting flaws in the circuit and stopping operation
- Equipment that operates at low enough voltage levels that will not produce injury
- Eliminating the need for electrical supply through use of battery- or air-powered equipment

Fault-proofing equipment is not an option except in the most unusual cases. Legislation imposes a duty of care on employers in the use of equipment in hazardous conditions. In the context of this discussion these could include moist or wet conditions, as well as atmospheres that are hazardous due to the presence or possible presence of flammable/explosive mixtures. Fault-proofing to withstand the

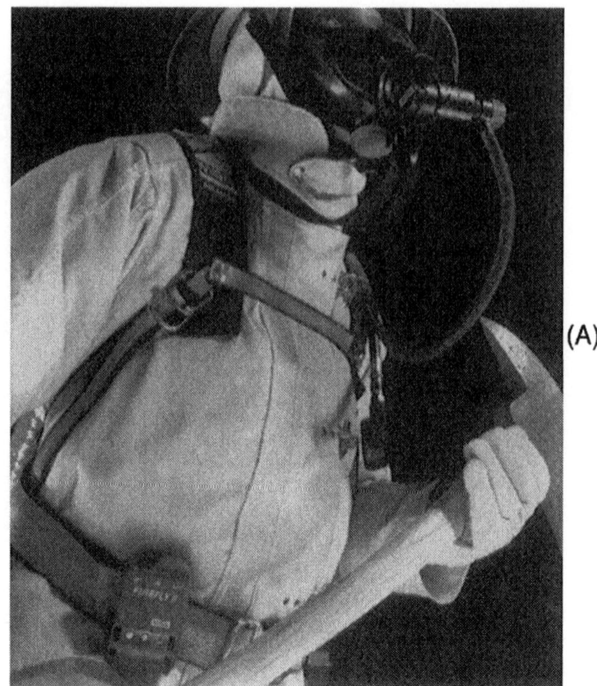

(A)

FIGURE 9.3 Worker status monitoring equipment: (A) This unit detects lack of motion and sounds a pre-alarm after 20 seconds of inactivity. Movement at that point resets the alarm (Courtesy of Mine Safety Appliances, Pittsburgh, PA). More advanced units combine a microcomputer with two-way radio communication between a base station and remote units; (B, C) The motion sensor initiates an alarm to the base station, as well as other units, after 90 seconds of inactivity. The user also can activate the alarm manually. This system can accommodate up to three users (Courtesy of Safety-Link Systems, a Division of Talkie Tooter Canada Ltd., Vancouver, BC); (D) This system uses a barcode scanner and swipe card reader and can accommodate up to 100 users. The user must acknowledge a computer-generated query from the base unit by pressing a button on the remote unit. The user can initiate an alarm and use the remote unit as a two-way radio. This system also can integrate air and other types of monitoring equipment. (Courtesy of Safe Environment Engineering, Valencia, CA.)

rigors of these conditions would include the ability to pass tests administered by independent testing laboratories. Meeting these conditions would be prohibitive.

A device for ensuring electrical safety is the ground fault circuit interruptor (GFCI or GFI) (Figure 9.7). A GFCI is a circuit breaker that detects overloads and shorts as do regular breakers, but also detects a ground fault (Bernstein 1991). A ground fault is a flow of current through an unintended path from the "live" wire to ground. The GFCI measures imbalance between current flowing into the circuit and current flowing out. The breaker opens the circuit when leakage is occurring. In this case, current is flowing to ground. Leakage can occur through damp or worn insulation, defective wiring, or cutting of the cord. The rapid opening of the breaker decreases the shock hazard. The GFCI will not protect against contacting the "live" wire and the neutral wire (short circuit) or two "live" wires. Protection is offered against contact with a "live" wire and a ground path. GFCIs monitor leakage in the range 4 to 6 mA. Testers are available for determining performance of GFCIs and to verify the amount of leakage.

AC sensors detect the presence of AC voltage in a circuit. This can occur through insulation. These sensors can locate blown fuses, defective breakers, breaks in circuitry, and neutral/live wires. This sensor can determine whether a piece of equipment is safe to contact for maintenance or other purposes.

Self-contained, battery-powered equipment is available for many applications. Battery life could be an issue with these tools. Many are not rated for use in hazardous atmospheres.

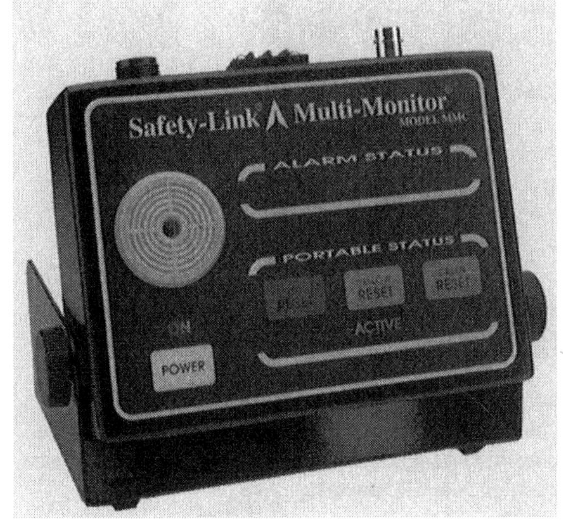

(B)

(C)

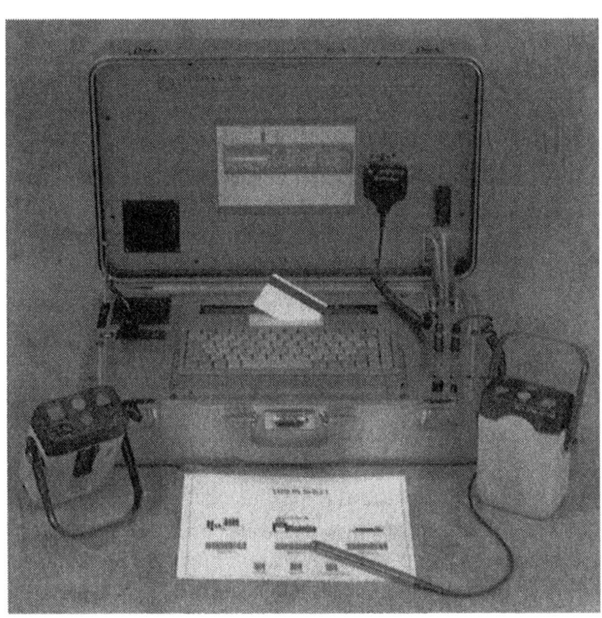

(D)

FIGURE 9.3 (continued)

FIGURE 9.4. Identification and status of participants. The tags on this board contain the photographs of workers who hold various positions. Positions on the board identify the entry supervisor, attendants, and entry workers. Each individual is responsible for placing the tag into the appropriate position on the board. Placement of the tag identifies who has entered the space. (Courtesy of Idesco Corporation, NY.)

Air-powered tools also are available. These do not pose an electrical hazard. Mist from oils used to lubricate these tools could cause an exposure problem.

Portable Lighting Systems

The criteria on which lighting systems are selected are to provide adequate levels of illumination in safety. Lighting systems are a major concern in confined spaces. Faulty or inappropriate lighting equipment was the cause of fatal fires and explosions and electrocutions in confined spaces (OSHA 1982a, 1985, 1990; NIOSH 1994).

Portable lighting is needed for specific applications involving confined spaces. The first is the prework inspection. Prework inspection is undertaken to ensure that preentry preparation has eliminated or at least controlled hazardous conditions in the space. Prework inspection is a mandatory requirement in some legislative approaches to confined spaces. Hazardous conditions could exist in the space at the time of the prework inspection.

Hazardous conditions may occur or may be produced during the work activity. The more obvious condition is a flammable/combustible atmosphere. This is caused by vaporization of residual materials or materials introduced into the space during work activity. Airborne dust and liquid aerosols can create an explosible atmosphere.

A flashlight designated for use in hazardous locations is designed so that in the case of breakage of the bulb, the reflector will de-energize the circuit and crush the filament. This eliminates possible contact between an energized circuit and a flammable/explosive atmosphere. Vapor-tight or gasketed lighting is required to withstand water vapor and noncombustible dust. Explosion-proof lighting is designed to be strong enough to withstand an internal explosion and not leak hot vapors, and operate at temperatures low enough not to ignite the surrounding atmosphere.

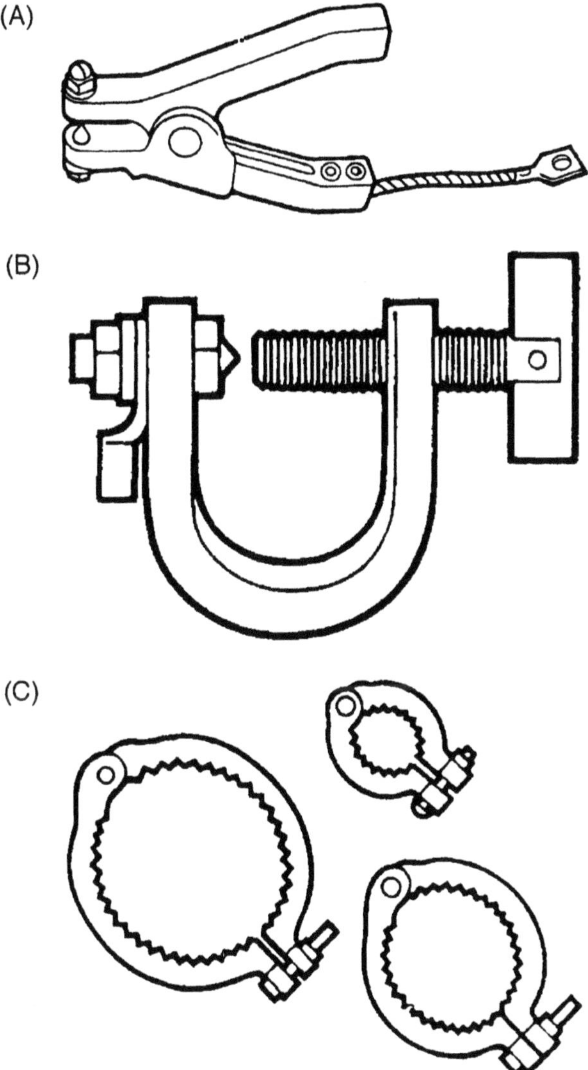

FIGURE 9.5 Bonding and grounding clamps. Bonding and grounding clamps are used to establish reliable electrical connection with surfaces. Clamps include: (A) jaw-type clamps; (B) "C"-clamps; and (C) circular pipe clamps. Bonding cable attaches to the clamps. (Courtesy of Stewart R. Browne Mfg. Co., Atlanta, GA.)

Table 9.5 describes (and Figures 9.8 and 9.9 illustrate) the various types of lighting systems that are available.

Each option in Table 9.5 offers strengths and limitations. These should be considered within the context of the circumstances of an individual situation before making a selection.

Hazard Removal Hierarchy/Techniques

Confined spaces may contain one, or any combination of many hazards to entry personnel. In order to insure safety of the entry team, the hazards must be removed prior to entry. Hazard removal techniques should be applied in order of priority.

The first priority for hazards management is flammable/combustible atmospheres. Even before opening the space, the potential for flammable atmospheres must be considered. Breaking open a

FIGURE 9.6 Infrared noncontact thermometer. Noncontact thermosensing devices can measure temperature of hot and cold surfaces. Sensing can occur typically to 3 m (10 ft). At that distance, target spot size is 400 mm (16 in). Some units have laser devices to assist in sighting on a target. (Courtesy of Raytek Corporation, Portable Products Division, Santa Cruz, CA.)

manway, or other opening, may result in sparking and potential ignition. Possible methods for managing potentially flammable atmospheres include:

- Ventilating the atmosphere within the space to reduce the concentration of flammable vapors to safe levels
- Inerting the atmosphere with a nonreactive gas, such as nitrogen or carbon dioxide

Further considerations about these methods are provided in other chapters.

Once concerns regarding flammable atmospheres have been addressed, the next order of priority is toxic materials/atmospheres within the space. Although no longer a physical hazard, there may be significant risk of dermal or respiratory exposure when working near any openings to the confined space. Possible methods for minimizing these risks include:

- Ventilation with remote exhaust (possibly through an appropriate scrubber)
- Use of personal protective equipment
- Removal of free material from the space
- Containment of vapor through the use of vapor-containing foam, water, or other liquid pad

Further considerations about these methods are provided in other chapters.

An additional consideration is the intended sequence for the work. A useful example relates to vessels that are to be demolished. Instead of entering and cleaning and then dismantling/cutting for removal, change the work process to dismantlement/cold-cutting followed by cleaning and/or demolition.

(A)

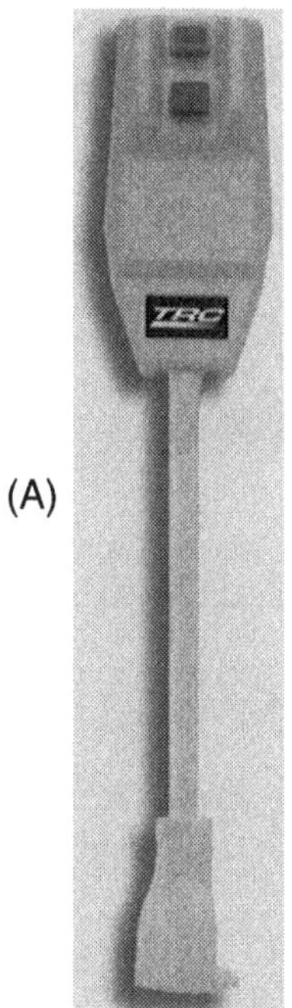

(B)

FIGURE 9.7 Ground fault circuit interrupters: (A) In-line plug-in cord set; (B) Quad box receptacle; (C) User attachable plug (120V); (D) User attachable plug (120/240V). GFCIs designed for use in industrial applications differ from those used in the home. (Courtesy of Technology Research Corporation, Clearwater, FL.)

If, after thorough review, there are no alternatives to entry, the next consideration is to modify the entry to eliminate or minimize risks. Some of the possible modifications include:

- Consider turning the vessel on edge and blocking it in place. Once the vessel is on its side, it can be entered in a safer fashion, and egress is greatly simplified. This is especially helpful with open-topped vessels (round or box type).
- Stabilize unstable structures and remove unstable sections prior to entry. When evaluating/modifying structures of questionable integrity, a structural engineer should be consulted.
- Enter from the side instead of the top when the option exists. (If the level of material is above the side opening, the material can be pumped down through the top opening, and the cover to the side opening subsequently removed.)
- Increase the number of openings to increase ventilation. This can be achieved by cutting holes (if the vessel/structure is scheduled for destruction/removal), or by breaking open additional flanges and manways, as available.
- Manage or control activities in the surroundings that are unrelated to the entry operation. Many activities that occur outside of the confined space, including such common activities

(C) 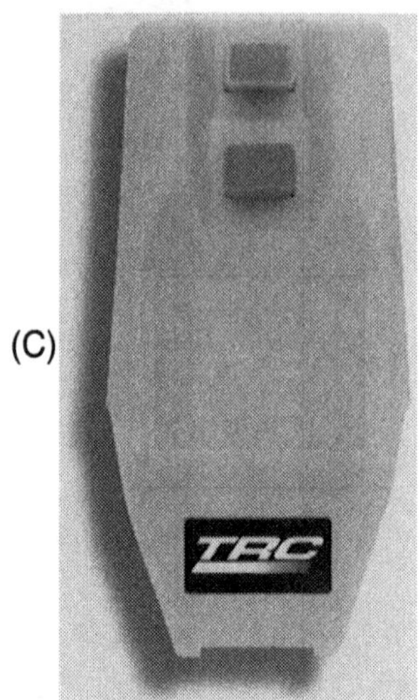

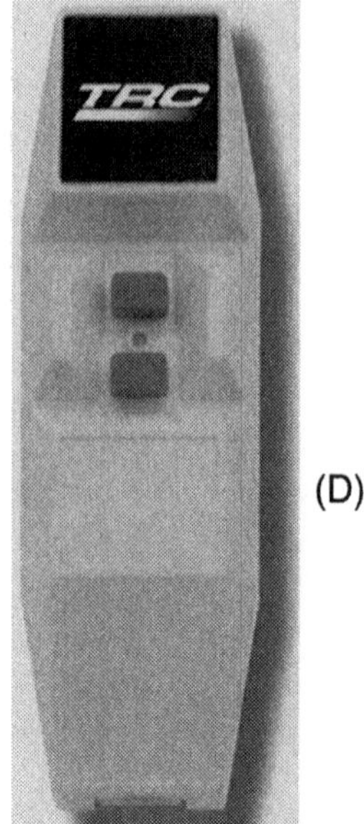 (D)

FIGURE 9.7 (continued)

TABLE 9.5
Temporary Lighting Systems

Description	Comments
Flashlights/lanterns	Some units rated for use in hazardous atmospheres Up to 100,000 candlepower available Hand required to hold in desired position
Portable light plants	Self-contained units, trailer or skid mounted; some powered by gasoline or diesel engines Exhaust gases can enter the space if location not considered carefully Not rated for use in hazardous atmospheres
110 V Hand lamps	Some units rated for use in hazardous atmospheres; these have shielded high impact lenses Incandescent and low wattage fluorescent lamps are available
110 V Flood lamps	Some units rated for use in hazardous atmospheres Some units rated for use in wet locations Incandescent and quartz lamps are available
12 V Hand lamps	Some units rated for use in wet locations Some units also rated for use in hazardous atmospheres Units contain 110 V feed to a 12 V transformer or connect directly to 12 V power sources

as material processing, heating, cooling, welding, combustion, etc., generate flammable, toxic, or oxygen-deficient atmospheres. These emissions also may include materials in storage, such as gases (nitrogen, oxygen, carbon dioxide, sulfur hexafluoride, etc.). Under appropriate conditions, these may enter the space. If these emissions cannot be controlled, either through ventilation or through cessation of activity, then the entry must be postponed.

- Consider entry during off-hours in cases where workforce density near the entry operation is high, and where hazardous conditions inside or outside the space could affect either the entry crew or outside personnel.
- Ensure that all lines and hoses are connected correctly. Outlets that do not have unique fittings can be cross-connected. Cross-connection has caused fatal accidents in confined spaces.

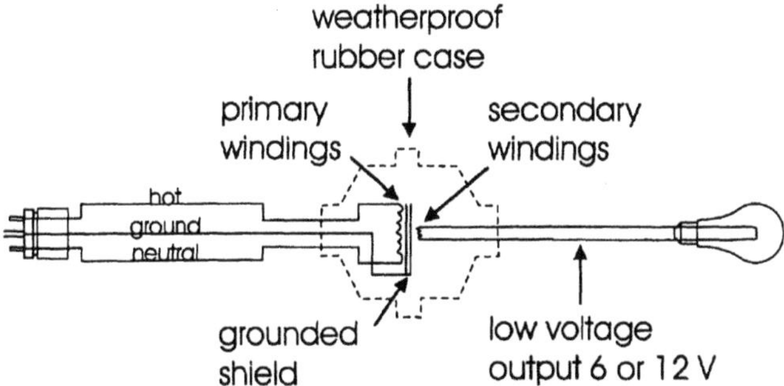

FIGURE 9.8 Low voltage circuitry. An in-line transformer converts 110 VAC to 6 VAC or 12 VAC (depending on design). Low voltage equipment is an AC system. A grounded metal shield between the primary and secondary windings is used to isolate the two voltages from each other. A break in the low voltage AC line is much less likely to cause a spark than a break in a 12V DC system. (Courtesy of Ericson Manufacturing Co., Willoughby, OH.)

Ventilation

Ventilation often is essential for the control of atmospheric hazards in confined spaces. Portable systems are the usual means of ventilating confined spaces.

A major concern about ventilating confined spaces is that the act of ventilating could create a hazard. The most critical period is the first opening of the space to air during preentry preparation. Up to that point, the atmosphere in the space was contained and undisturbed. Opening the space and beginning ventilation poses considerable potential for escalation of hazard. Where flammable/combustible or oxygen-enriched atmospheres may be present, ventilating equipment should be rated for this application, so that its use does not constitute a hazard.

The discharge from the ventilator must be directed well away from the work area. This is necessitated both by the need to keep the ventilated toxic and flammable vapors away from personnel on the entry team, and perhaps more importantly, the vapors could impact upon outside personnel and operations in the work area. The use of scrubbers/collectors may be indicated, depending upon specific circumstances.

Ventilation of confined spaces is the subject of Chapter 11 and Appendices I and J.

Waste Generation

Work involving the space is a potential source of contamination to which environmental legislation could apply.

Preparation of the space for entry could require draining of hazardous materials from lines during isolation. In sufficient quantity without collection, these drainings could constitute environmental releases requiring reporting. Purging to remove gaseous components similarly could release into the atmosphere a quantity of vapor that could require reporting.

Liquids available for cleaning and flushing the space during preentry preparation are numerous. Selection should depend upon the residue to be removed and the facilities available for removal. The most preferable course of action is to flush with the least hazardous substance. Sometimes this means use of several substances in succession, starting from most to least hazardous. Water is usually the least hazardous and most desirable flushing agent. Aqueous cleaning solutions followed by water, or organic solvents followed by aqueous solutions and then by water, may be appropriate for some situations.

Decontamination

Work in a space containing residual material or hazardous material to be removed (such as asbestos or lead-containing paint dust) creates the potential for contamination of persons and equipment.

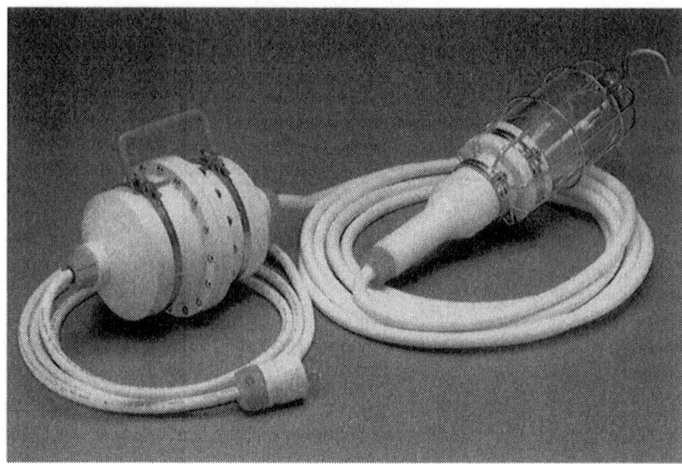

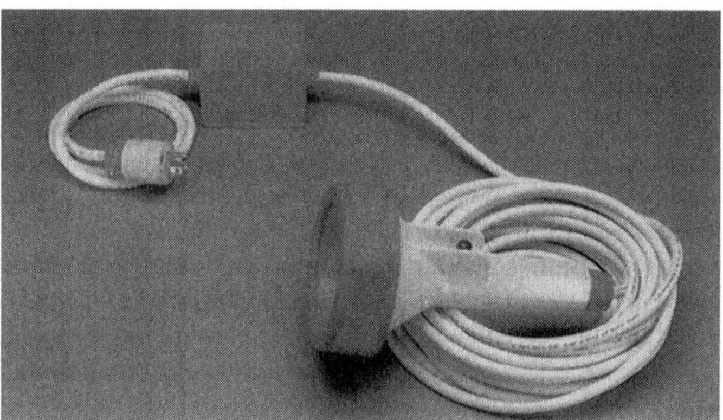

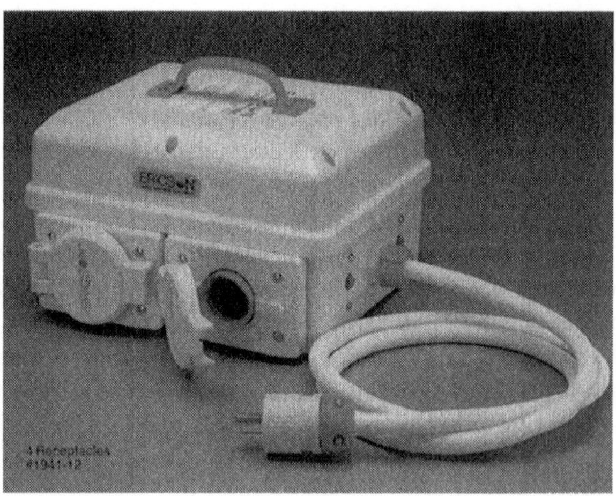

FIGURE 9.9 Low voltage electrical lighting equipment: (A) Trouble (Hand) lamp; (B) Spot or flood lamp; (C) Low voltage receptacle. The transformer for voltage conversion is located either in-line or in the receptacle into which the lamp plugs. (Courtesy of Ericson Manufacturing Co., Willoughby, OH.)

FIGURE 9.10 Decontamination. Decontamination is the physical and/or chemical removal of unwanted substances and materials from the surfaces of chemical protective clothing and equipment. Decontamination of some materials is greatly assisted through use of shower units, as illustrated. (Courtesy of Kappler Safety Group, Guntersville, AL.)

Where the potential for contamination can occur, the means to prevent contamination and to decontaminate also must occur. Otherwise, the potential for transfer of contamination offsite becomes a major concern.

Contamination is unwanted material that adheres to a surface. Surfacial contamination may be fixed or loose. Fixed material is removable only with great difficulty.

Personal contamination occurs on the surfaces of the body — hair, skin, eyes, and the interior of the respiratory system. Personal contamination may be fixed or loose or may penetrate through boundary surfaces. Penetration can occur by diffusion or following chemical attack.

Personal contamination was not mentioned as a concern in the fatal accidents documented in the NIOSH, OSHA, and MSHA reports (NIOSH 1979, 1994; OSHA 1982, 1983 1985, 1988, 1990; MSHA 1988). This included the both the victims and the survivors. The acute action of atmospheric and other hazardous conditions was the sole focus of these accidents. Contamination of personnel, clothing, and equipment was not considered as an issue, nor was it addressed.

Prevention of contamination of workers and equipment at the site is a concern that must be addressed.

Contamination of personnel and equipment can result from normal operations or breakdown of protective measures (Cember 1969). Prevention of personal contamination is the function of respiratory and skin protection, the latter in the form of clothing, gloves, and boots.

Contamination during work in confined spaces generally takes the form of loose residual solids, liquids, sludges and slurries, and solid and liquid aerosols. These can transfer to the skin and surfaces of clothing. Depending on circumstances, contamination can be radioactive or nonradioactive. Fixed contamination, unless radioactive, does not pose the level of concern of loose contamination.

The hazard posed by a chemically contaminated surface is difficult to assess. The most probable mechanism for human contamination occurs by inhalation following suspension into the air. Ingestion following skin contamination also could be an important consideration.

Decontamination is the process of removal of contamination from surfaces. Contamination of surfaces results from chemical and physical binding forces. Decontamination, therefore, involves chemical and physical methods. The nuclear industry was confronted by the contamination problem long before this became an issue in the chemical industry. As a result, considerable information on this subject is available (Ayres 1970). The extent and thoroughness of the process reflects the level of acceptable residual contamination and value attached to the item. (See Figure 9.10.)

The extent to which decontamination becomes an issue during the work phase depends on strategies selected during planning. Decontamination, therefore, is both a planning and an operational issue. The first issue is the nature of the contamination. This is determined by the nature of residual materials and work activity. Usual sources include contact with surfaces containing residual or newly applied materials. Contact also can occur with aerosols generated during spraying or similar activity.

The nature of the activity, expected contact, and properties of contaminants should determine the need and mode of selection of protective clothing and equipment, and for decontamination.

The need for decontamination can be simplified by the use of sacrificial layers. This can occur in several ways. One strategy is to wear layered protective clothing. The outer is an inexpensive sacrificial suit. That is, the suit is intended for disposal, rather than decontamination. The seams and weak points on such suits can be reinforced using duct tape prior to use. Lightweight suits tear easily under some conditions. The purpose for the inner is to provide a secondary barrier in the event of damage to the outer. The strategy may call for a second inner, a more expensive suit that provides a more resistant barrier, albeit at higher cost. The advantage to this strategy is that the outer layer (layers, as appropriate) are sacrificed to protect the inner. The outer layers become coated with grime and dirt. The innermost layer remains clean.

This strategy is most successful where the space has been cleaned prior to entry for work activity and where contamination is readily visible. Precleaning can considerably reduce operating time.

Protective equipment becomes contaminated. That is its sole function: to intercept contamination prior to contact with underclothing or the skin (Cember 1969). Bearing this in mind, this can form the strategy on which the selection is made.

Special consideration must be given to accident situations (Shapiro 1981). The objectives of remedial action in these situations are to:

- Minimize entry of contaminants into the body of the victim by any route, including the wounds
- Prevent spread of contamination from the area of the accident
- Remove contamination from the victim
- Begin decontamination of the area, as necessary

This approach can apply regardless of the type of contamination, radiological or chemical. Accidents in confined spaces involving radiological contamination likely involve small quantities. The situation involving chemical contamination is likely to be similar, unless substances having acute toxicity are involved. Normal decontamination procedures can lead to needless delay and spread of contamination. Remove contaminated clothing away from the wound. Decontamination of the wound should occur under the direction of a physician. This would involve use of mild soap or detergent and tepid water.

Protective clothing and equipment and contamination/decontamination are the subjects of Chapter 12. This chapter will elaborate on the themes mentioned here.

WORK COMPLETION

Upon completion of the work the space must be vacated and the entry process reversed, so that recommissioning can occur. The entry supervisor must verify that the space is clear and meets specifications. Work completion and recommissioning are as important as entry preparation.

Preparation for entry and work activity can necessitate modification to the internal configuration of the space or these can occur unintentionally. In some cases, these modifications could create hazardous conditions during reuse. For example, metal brackets welded onto smooth surfaces as temporary anchors could act as discharge points for static electricity. Bonding to other equipment during reassembly could be impaired because of accidentally painted or greased contact surfaces or installation of gaskets. Leakage from reconnected flanges can produce serious exposure problems during start-up. Recommissioning the space to preentry condition can be a very complex undertaking.

A number of hazardous conditions can occur during this process (Lees, 1980). The main hazards encountered in the petrochemical industry include:

- Mixing of air and hydrocarbons
- Contact between hot oil and water
- Water shots and water freezing
- Over- and underpressurizing of equipment
- Thermal and mechanical shock to equipment
- Corrosive and poisonous fluids
- Formation of pyrophoric iron sulfide

Start-up is the time when there is a much higher risk of presence or entry of unwanted materials or objects in the structure. These can include: air, water, tracked-in dirt, waste or unused materials related to the work activity, tramp metal, and forgotten or misplaced tools. Prolonged lack of operation can cause deterioration of seals and gaskets, as well as corrosion of metal surfaces. A formal system for start-up, backed up by proper documentation, is required. This is essential to ensure correct sequential removal of blanks and blinds, caps and plugs, lockout devices, reconnection of lines, and so on.

Purging of air and moisture from the structure, pumps, valves, and lines is an essential practice in some circumstances (AGA 1975, Lees 1980). Gases, such as steam, nitrogen, and carbon dioxide, or water are used. Steam also can bring the unit up to temperature. Heating removes unwanted liquid water through vapor formation. Heating the system using steam can create serious hazards, including water hammer, over- and underpressurizing, and thermal and mechanical shock.

Water (fluid) hammer is one example of a highly hazardous condition that can occur during recommissioning. Water (fluid) hammer is the banging or hammering resulting from the sudden contact of moving fluid with boundaries in partly filled piping or other equipment. Water (fluid) hammer can be highly destructive. One kind of water (fluid) hammer results from the sudden closure of a valve in a long pipeline filled with water. Another is caused when a slug of condensate is flung against the boundaries of pipe by steam pressure. Typically, the steam line is not used for some time, but steam pressure is maintained. An additional requirement is nonfunctioning steam traps and drains. Care in making large alterations in steam flow during start-up is essential. Precautions include checks to minimize condensate accumulation, and slow operation of steam traps and valves.

Contacting hot oil or other immiscible liquid with water can cause rapid boiling and formation of foam, a condition known as boilover or slopover. This condition typically occurs when oil with a water layer is heated, or hot oil is permitted to contact water, or vice versa.

Leak testing is another essential part of the process. This may occur once the structure and associated equipment have undergone purging. Vacuum equipment would receive a leak test under pressure and a subsequent vacuum test.

Process systems contain subsystems that shut down the operation when a particular parameter exceeds specified limits. Such systems contain manual overrides that the operator can activate in the event of an alarm condition. Shutdown and turnaround can create unanticipated and unexpected conditions in a process unit during start-up. The operator may override the alarm condition. This practice only may occur in accordance with accepted practice and proper procedure. Of considerable concern in such situations is the possibility that real and serious conditions are being overridden, and thereby ignored. These could occur when lines remain blanked off or capped or are reconnected incorrectly, valves remain closed or are reset incorrectly, or misconnection or damage to sensing systems used in the space has occurred. These systems include spark and fire and explosion sensing and suppression systems, temperature and pressure sensors, level sensors, flow sensors, and so on.

WORK REVIEW

Review of work undertaken in the space provides a valuable learning opportunity, as well as closure to the loop. This can provide answers to questions, such as, what can be done to make this worker safer? This question is highly appropriate, as assessment and procedures should evolve as experience and knowledge increase. The process work review is required under some jurisdictions (Canada Gazette 1992). One of the duties of the Qualified Person is to prepare a report about the work that occurred and to submit it to management. Where the report documents problems, management is required to provide it to the Safety and Health Committee for review and consultation.

SUMMARY

The focus of this chapter is entry and work in confined spaces. Confined spaces are highly varied in size, shape, and hazardous conditions. As a result, precautions and work activities must be considered on an individual basis. Certain aspects are common to many operations and jurisdictions.

Work in confined spaces often is subject to time pressure. This pressure applies as much to the Qualified Person responsible for on-site safety as to other supervisory and technical personnel. Time pressure can become especially intense when work has fallen behind some predetermined schedule or when the work was precipitated by an emergency.

Legislation in many jurisdictions and consensus standards have mandated the function of the Qualified Person. The Qualified Person is responsible for the safe conduct of work at the site. This aspect of work involving confined spaces could cause conflict with the objectives of site management, whose focus is timely completion of the work. Support from top management is essential to securing the role of the Qualified Person in the context of the operation.

The Qualified Person completes the entry documents. Entry documents communicate to workers and supervision about conditions in the space and measures needed to ensure the safe conduct of work. Entry documents provide a summary of the hazard assessment.

REFERENCES

American Gas Association: *Purging Principles and Practice*, 2nd ed. Arlington, VA: American Gas Association, 1975. 180 pp.

American Conference of Governmental Industrial Hygienists: *1995–1996 Threshold Limit Values (TLVs™) for Chemical Substances and Physical Agents and Biological Exposure Indices (BEIs™)*. Cincinnati, OH: American Conference of Governmental Hygienists, 1995. 138 pp.

American National Standards Institute: *Safety Requirements for Confined Spaces* (ANSI Z117.1-1995). Des Plaines IL: American Society of Safety Engineers/American National Standards Institute, 1995. 32 pp.

American Petroleum Institute: *Safe Operating Guidelines for Vacuum Trucks in Petroleum Service* (API Pub. 2219). Washington, D.C.: American Petroleum Institute, 1986. 7 pp.

American Petroleum Institute: *Guidelines for Work in Inert Confined Spaces in the Petroleum Industry* (API Pub. 2217A). Washington, D.C.: American Petroleum Institute, 1987. 7 pp.

American Petroleum Institute: *Safe Welding and Cutting Practices in Refineries, Gasoline Plants and Petrochemical Plants*, 5th ed. (API Pub. No. 2009). Washington, D.C.: American Petroleum Institute, 1988a. 6 pp.

American Petroleum Institute: *Ignition Hazards Involved in Abrasive Blasting of Atmospheric Storage Tanks in Hydrocarbon Service*, 2nd ed. (API Pub. 2027). Washington, D.C.: American Petroleum Institute, 1988b. 4 pp.

American Petroleum Institute: *Spark Ignition Properties of Hand Tools,* 3rd ed. (API Pub. No. 2214). Washington, D.C.: American Petroleum Institute, 1989. 5 pp.

American Petroleum Institute: *Protection Against Ignitions Arising out of Static, Lightning and Stray Currents,* 5th ed. (RP 2003). Washington, D.C.: American Petroleum Institute, 1991a. 39 pp.

American Petroleum Institute: *Cleaning Mobile Tanks in Flammable or Combustible Liquid Service,* 6th ed. (API Pub. 2013). Washington, D.C.: American Petroleum Institute, 1991b. 6 pp.

American Petroleum Institute: *Preparing Tank Bottoms for Hot Work,* 4th ed. (API Pub. 2207). Washington, D.C.: American Petroleum Institute, 1991c. 4 pp.

American Petroleum Institute: *Interior Lining of Underground Storage Tanks,* 3rd ed. (RP 1631). Washington, D.C.: American Petroleum Institute, 1992. 10 pp.

American Petroleum Institute: *Safe Entry and Cleaning of Petroleum Storage Tanks* (API Std. 2015, 5th ed.). Washington, D.C.: American Petroleum Institute, 1993. 60 pp.

American Society for Testing and Materials: *Standard Practice for Confined Area Entry* (ASTM D 4276-84, reapproved 1989). Philadelphia, PA: American Society for Testing and Materials, 1984. 3 pp.

American Welding Society: *Recommended Safe Practices for the Preparation for Welding and Cutting of Containers That Have Held Hazardous Substances* (ANSI/AWS F4.1-88). Miami, FL: American National Standards Institute/American Welding Society, 1988. 4 pp.

Aschoff, J.: Circadian rhythms: interference with and dependence on work-rest cycles. In *The Twenty-Four Hour Workday: Proceedings of a Symposium on Variations in Work-Sleep Schedules,* Johnson, L.C., D.I. Tepas, W.P. Colquhoun, and M.J. Colligan (Eds.) (DHHS/NIOSH Pub. No. 81-127). Washington, D.C.: Government Printing Office, 1981.

Ayres, J.W.: *Decontamination of Nuclear Reactors and Equipment.* New York: Ronald Press, 1970.

Bernstein, T.: Electrical systems, terminology and components — relationship to electrical and lightning accidents and fires. In *Electrical Hazards and Accidents, Their Causes and Prevention,* Greenwald, E.K. (Ed.). New York: Van Nostrand Reinhold, 1991. pp. 1–27.

Bier, J.W. and T.K. Sawyer: Amoebae isolated from laboratory eyewash stations. *Curr. Microbiol. 20:* 349–350 (1990).

Bjerner, B., A. Holme, and A. Swenson: Diurnal variations in mental performance: a study of three-shift workers. *Br. J. Ind. Med. 12:* 103–110 (1955).

Brief, R.S. and R.A. Scala: Occupational exposure limits for novel work schedules. *Am. Ind. Hyg. Assoc. J. 36:* 467–471 (1975).

Brief, R.S. and R.A. Scala: Occupational health aspects of unusual work schedules: a review of Exxon's experiences. *Am. Ind. Hyg. Assoc. J. 47:* 199–202 (1986).

Budnick, L.D., S.E. Lerman, T.L. Baker, H. Jones, and C.A. Czeiler: Sleep and alertness in a 12-hour rotating shift work environment. *J. Occup. Med. 36:* 1295-1300 (1994).

Canada Gazette: Part XI, Confined Spaces, *Canada Gazette, Part II. 126*: (17 September 1992) pp. 3863–3876.

Cember, H.: *Introduction to Health Physics.* Oxford, U.K.: Pergamon Press, 1969. 420 pp.

Champagne, R.: A deluge can save the eyes. *Occup. Health Safety 63:* 69–74 (1994).

Colquhoun, W.P. (Ed.): *Biological Rhythms and Human Performance.* London and New York: Academic Press, 1971.

Colligan, M.J. and D.I. Tepas: The stress of hours of work. *Am. Ind. Hyg. Assoc. J. 47:* 686–695 (1986).

Coplen, M.K.: Does shiftwork shift risks? *Safety Health:* June, 1988. pp. 40–45.

Gaunt, J.A.: Relationship between overtime and safety. *Prof. Safety. 25:* 11–15 (1980).

Heinrich, H.W.: *Industrial Accident Prevention.* New York: McGraw-Hill, 1931.

Hockey, G.R.J. and W.P. Colquhoun: Diurnal variation in human performance: a review. In *Aspects of Human Efficiency: Diurnal Rhythm and Loss of Sleep,* W.P. Colquhoun (Ed.). London: The English Universities Press, 1972. pp. 1–23.

Ibbetson, T.A.: Listening in. *Occup. Health Safety 63:* 77–83 (1994).

Jagger, D.: When every second counts. *Occup. Health Safety Cda. 4:* 80–90 (1988).

Kleitman, N.: *Sleep and Wakefulness.* Chicago: University of Chicago Press, 1963.

Lees, F.P.: *Loss Prevention in the Process Industries,* vol. 2. London: Butterworths, 1980. pp. 707–720.

Luce, G.G.: Biological Rhythms in Psychiatry and Medicine (U.S. Public Health Service Pub. No. 2088). Washington, D.C.: Government Printing Office, 1970.

Mine Safety and Health Administration: Think "Quicksand": Accidents Around Bins, Hoppers and Stockpiles, Slide and Accident Abstract Program. Arlington, VA: U.S. Department of Labor, Mine Safety and Health Administration, National Mine Health and Safety Academy, 1988.

National Fire Protection Association: *NFPA 327: Standard Procedures for Cleaning or Safeguarding Small Tanks and Containers* (1975 ed.). Quincy, MA: National Fire Protection Association, 1975. 15 pp.

National Fire Protection Association: *NFPA 306: Control of Gas Hazards on Vessels* (1993 ed.). Quincy, MA: National Fire Protection Association, 1993. 15 pp.

National Institute for Occupational Safety and Health: Occupational Exposure Sampling Strategy Manual, by Leidel, N.A., K.A. Busch, and J.R. Lynch (DHEW/PHS/CDC/NIOSH Pub. No. 77-173). Cincinnati, OH: National Institute for Occupational Safety and Health, 1977. 132 pp.

National Institute for Occupational Safety and Health: Criteria for a Recommended Standard — Working in Confined Spaces (DHEW/PHS/CDC/NIOSH Pub. No. 80-106). Cincinnati, OH: National Institute for Occupational Safety and Health, 1979. 68 pp.

National Institute for Occupational Safety and Health: Worker Deaths in Confined Spaces (DHHS/PHS/CDC/NIOSH Pub. No. 94-103). Cincinnati, OH: National Institute for Occupational Safety and Health, 1994. 273 pp.

National Safety Council: *Entry into Grain Bins and Food Tanks*, No. 12304-0663. Itasca IL: National Safety Council, 1987.

Occupational Safety and Health Administration: Selected Occupational Fatalities Related to Fire and/or Explosion in Confined Work Spaces as Found in OSHA Fatality/Catastrophe Investigations. Washington, D.C.: U.S. Department of Labor, Occupational Safety and Health Administration (U.S. DOL/OSHA), 1982a. 76 pp.

Occupational Safety and Health Administration: Selected Occupational Fatalities Related to Lockout/Tagout Problems as Found in Reports of OSHA Fatality/Catastrophe Investigations. Washington, D.C.: U.S. Department of Labor, Occupational Safety and Health Administration (U.S. DOL/OSHA), 1982b. 113 pp.

Occupational Safety and Health Administration: Selected Occupational Fatalities Related to Grain Handling as Found in Reports of OSHA Fatality/Catastrophe Investigations. Washington, D.C.: U.S. Department of Labor, Occupational Safety and Health Administration (U.S. DOL/OSHA), 1983. 150 pp.

Occupational Safety and Health Administration: Selected Occupational Fatalities Related to Toxic and Asphyxiating Atmospheres in Confined Work Spaces as Found in Reports of OSHA Fatality/Catastrophe Investigations. Washington, D.C.: U.S. Department of Labor, Occupational Safety and Health Administration (U.S. DOL/OSHA), 1985. 230 pp.

Occupational Safety and Health Administration: Selected Occupational Fatalities Related to Welding and Cutting as Found in Reports of OSHA Fatality/Catastrophe Investigations. Washington, D.C.: U.S. Department of Labor, Occupational Safety and Health Administration (U.S. DOL/OSHA), 1988. 225 pp.

Occupational Safety and Health Administration: Selected Occupational Fatalities Related to Ship Building and Repairing as Found in Reports of OSHA Fatality/Catastrophe Investigations. Washington, D.C.: U.S. Department of Labor, Occupational Safety and Health Administration (U.S. DOL/OSHA), 1990. 195 pp.

OSHA: Permit-Required Confined Spaces for General Industry; Final Rule," *Fed. Regist. 58*: 9 (14 January 1993). pp. 4462–4563.

OSHA: "Confined and Enclosed Spaces and Other Dangerous Atmospheres in Shipyard Employment; Final Rule," *Fed. Regist. 59*: 141 (25 July 1994). pp. 37816–37863.

Paszko-Kolva, C., H. Yamamoto, H. Shahamat, T.K. Sawyer, G. Morris, and R.R. Colwell: Isolation of amoebae and *Pseudomonas* and *Legionella* spp. from eyewash stations. *Appl. Environ. Microbiol. 57:* 163–167 (1991).

Paszko-Kolva, C., T.K. Sawyer, and G. Gardner: Laboratory emergency eyewash station contaminants. *Am. Environ. Lab. 7:* 18–20 (1995).

Shapiro, J.: *Radiation Protection, A Guide for Scientists and Physicians*, 2nd. ed. Cambridge, MA: Harvard University Press, 1981. 480 pp.

Standards Australia: *Safe Working in a Confined Space* (AS 2865-1995). North Sydney, New South Wales: Standards Association of Australia, 1995. 42 pp.

Tyndall, R.L., M.M. Lyle, and K.S. Ironside: The presence of free-living amoebae in portable and stationary eye wash stations. *Am. Ind. Hyg. Assoc. J. 48:* 933–934 (1987).

Wallin, L. and I. Wright: Psychosocial aspects of the work environment: a group approach. *J. Occup. Med. 28:* 384–393 (1986).

10 Instrumentation and Testing*

Robert E. Henderson

CONTENTS

* Portions of this chapter have been previously published in *Corporate Health and Safety: Managing Environmental Issues in the Workplace,* George, E., Ed., Ergonomics, Inc., Southampton, PA, 1996. With permission.

INTRODUCTION

Studies published by the National Institute for Occupational Safety and Health (NIOSH) and the Occupational Safety and Health Administration (OSHA) have found that the root cause of many fatal confined space accidents is a hazardous atmosphere associated with the space. The hazardous atmosphere may develop prior to the entry, or may be associated with work activity (NIOSH 1979,

1994; OSHA 1982, 1985, 1988, 1990). A high proportion of these accidents were characterized by failure to test prior to the entry, or during the conduct of work. These findings strongly suggest that atmospheric testing prior to and during entry should be a major component in a hazard management program.

Miners were among the first to become aware of the need for a device to detect hazardous gases (AIHA 1980). The atmosphere in mines can be subject to a variety of hazardous conditions. Toxic gases encountered in this environment include carbon dioxide, carbon monoxide, nitrogen oxides, sulfur dioxide, and others. The atmosphere in mines can also become oxygen deficient. In some circumstances methane may be present in explosive concentrations. Since methane has no warning properties, a fully explosive concentration could accumulate before a worker would realize the potential risk. Any source of ignition, including the original miner's lamp, could readily cause an explosion. The first combustible gas indicator, the Davy lamp, therefore, provided a significant step forward in mine safety. The visible characteristics of the flame of the Davy lamp informed the experienced user about more than just the presence of methane. Variations and refinements of the original Davy lamp design are still used today in some programs.

Carbon monoxide was a particularly important concern to miners. Again, the absence of warning properties meant that miners could be exposed to lethal concentrations without their knowledge. The usage of small animals, birds, or the famous "miner's canary" was a poor substitute for a quantifiable method for the measurement of this hazard. The colorimetric indicator tube designed to measure carbon monoxide became available shortly after the turn of the century, and soon saw wide usage in mines and other environments subject to contamination by this hazard. Soon to follow was the percent oxygen indicator tube.

Atmospheric hazards associated with enclosed spaces on ships has been another spur to the development of modern atmospheric testing devices. In 1926, a string of oil tanker and ship tank explosions led the Standard Oil Company of California to sponsor the research and development of a direct reading indicator for explosive gas. As a result of this sponsorship, Oliver W. Johnson developed and in 1927 introduced a portable, explosive gas indicator based on the catalytic oxidation of flammable gas on a platinum filament wire. Seventy years later, the sensors used in the majority of today's confined space gas detectors continue to be based (with many modern refinements) on this basic detection principle. In the 1960s, development of the first generation of electrochemical oxygen sensors allowed the incorporation of oxygen measurement into real time direct reading portable instruments. Today's practitioner has the choice of indicator tubes, dosimeters, and portable real-time instruments based on a wide variety of detection principles. Indeed, one of the great challenges facing the practitioner is which detection technique to use when testing a particular environment!

Management of air quality during entry and work in confined space reflects two basic requirements: proper assessment of existing or potential atmospheric hazards, and strategies to eliminate, control, or maintain safe atmospheric conditions for potentially affected workers. Before developing any strategy, assessment of all potential atmospheric hazards which may be related to the confined space or entry procedures is required.

Traditional monitoring strategies have focused on the detection of only those chemicals or conditions suspected to be present. This approach is most useful when there is a high level of knowledge about the nature of the hazard. In many cases this is the most appropriate approach. A tightly focused measurement technique usually provides the most accurate readings and is least subject to sources of error, such as cross-interference from other substances simultaneously present. For example, a substance-specific electrochemical sensor designed for detection of hydrogen sulfide shows very little response to most other contaminants. The major pitfall in the use of highly specific measurement techniques is the potential for overlooking or missing an existing hazard.

In some confined spaces, such as sewers, the level of knowledge concerning hazards that may be present is relatively low. A sewer is a very large, interconnected confined space. Achieving complete control, and rendering the atmosphere safe throughout an entire sewer system prior to entry generally is not possible. This means, in most cases, that only the atmosphere in the area where entrants work is ventilated and monitored. While the most commonly encountered hazards (oxygen deficiency, methane, or hydrogen sulfide) are highly predictable, at any moment (with the next flush, as it were) an unforeseeable hazard can suddenly appear. In this case, a more broadly responding detection technology may be more appropriate.

Readings from nonspecific measurement techniques pose interpretive problems. The metal oxide semiconductor (MOS) sensor used in some survey instruments targeted towards sewer entry applications is so widely responsive that a positive reading is impossible to interpret.

The key to resolving this dilemma is to anticipate hazards that could be present, and then to develop a strategy for their assessment and measurement. In other words, when developing a monitoring program, it is critical to ensure that the initial hazard assessment and measurement procedures are broad enough to ascertain all the potential hazards which may be associated with the confined space.

In many cases, the atmospheric hazards associated with a particular confined space are easy to identify. The literature quickly reveals that fatal accidents in confined spaces that involve atmospheric hazards result from a limited number of atmospheric conditions. The most commonly encountered hazardous conditions involve oxygen (deficiency or enrichment), combustible gases and vapors, and toxic contaminants, most prominently, carbon monoxide, and hydrogen sulfide. While a multitude of other toxic contaminants can be present in the confined space atmosphere, CO and H_2S produce the preponderant majority of injuries and fatalities.

The primary focus of this chapter is detection techniques and methods used to measure the most commonly encountered hazards associated with confined space entry. The majority of portable atmospheric monitors manufactured for this application use similar technology and focus on the detection of these most commonly encountered hazards. The differences between one brand of instrument and another can be quite subtle. An understanding about the basic limitations of these designs is essential when considering their use in a monitoring program. This chapter will explain some of the differences, as well as present the advantages and disadvantages of one monitoring technique over another. A secondary focus of the chapter is an examination of other detection techniques. These, while not widely used in confined space monitoring procedures, may have relevance against specific contaminants.

ATMOSPHERIC HAZARDS ASSOCIATED WITH CONFINED SPACES

General

The three basic categories of potential atmospheric hazard that can occur in or be associated with confined spaces include: oxygen level (deficiency or enrichment), ignitable gases, vapors, or particulates, and toxic contaminants. The potential for the development of these hazardous conditions is affected by the following:

- Physical nature of the space
- Work being performed
- Processes associated with the space
- Products used or produced in conjunction with the space
- Natural processes (such as decomposition, fermentation, ripening, etc.) which occur in or are associated with the space
- External sources of contamination

Hazard assessment should always consider conditions and activities in other areas and their potential impact on the atmosphere in the space.

In some situations, the source of the atmospheric hazard may be remote from the space. For example, at many refineries, pulp mills, and other industrial locations, the potential for a sudden catastrophic release of toxic or explosive gas is always present. A sudden release in one area of the plant could quickly spread downwind to other areas. Development of an appropriate monitoring strategy depends on recognition of this potential.

Aerosol Hazards

General

Aerosols are suspensions of fine solid particles or liquid droplets in air. Categories of aerosols include:

- Fumes: formed when metal or other materials vaporize by high heat (as in welding) and recondense to form ultrafine particles
- Mists and fogs: suspensions of tiny droplets of liquids
- Dusts: particles produced by the breakdown of solid materials; dusts include fibrous particles that have longitudinal geometry and particles that have regular or irregular compact geometry.

Ignitable dusts, mists, and fogs are a special concern. Grain, coal, nitrated fertilizers, and other solid materials, as well as many liquids form ignitable mixtures when present in sufficient concentration in air.

OSHA provided a useful "rule of thumb" in the preamble to 29 CFR 1910.146, "Permit required confined spaces," for estimating hazardous conditions involving suspensions of dust (OSHA 1993). OSHA indicated that a hazardous condition exists when ignitable dust obscures vision at a distance of 1.5 m (5 ft).

Activities that disturb settled materials or that introduce materials into the air will have a profound influence on the amount of aerosol present in the air column of the space. Work activities associated with the generation of aerosols should include air monitoring procedures for assessing potentially hazardous conditions. Additional monitoring equipment could be required for assessing buildup of vapors and oxygen levels. The equipment used under these conditions must be classified or approved for use in the types and severity of hazardous conditions which may be present in the space.

MEASUREMENT OF AEROSOLS

Aerosol measurement, in particular, solid particulates, most frequently involves the use of portable, battery-powered, sampling pumps. These pumps are designed to move a precisely maintained volume of air through a collecting mechanism such as a filter, cassette, or impactor over a specific interval of time. Depending on the collection technique, entrained or collected particles may be counted, weighed, or chemically analyzed to determine a concentration in workplace air.

Particulate survey techniques involving use of a sampling pump normally require laboratory analysis. This requirement normally eliminates their use in "real-time" measurement of particulate levels during work activity in confined spaces. Particulate monitors are available for assessing particulate burden in "real time." Direct-reading particulate monitors include clean room particle counters, fibrous aerosol counters, piezobalances, and nephelometers. (The latter are used to measure the total or respirable fraction of particulate contaminants in the air.)

The nephelometer is the type of real-time dust monitor most commonly used during work activity in confined spaces. These devices utilize a calibrated light source (usually in the near infrared range) that radiates through a sensing chamber. Particulates in the sensing chamber reflect the incident light. The amount of backscattering by the particulates is proportional to the mass of airborne material in the sensing chamber. Some nephelometers contain a built-in mechanical pump. A pump enables active sampling of the atmosphere through the sensing chamber. Other instruments utilize natural air currents and simple diffusion to move dust through the sensing chamber. Hand-held units provide the ability to read particulate concentrations directly in units of mg/m^3. One of the most widespread applications of these instruments is real-time dust monitoring in underground coal mines. Other potential applications involving confined spaces include bag-houses, grain elevators, remediation sites, pharmaceutical plants, and other locations containing toxic or explosive dusts.

The nature of the particulate hazard defines the measurement technique that is most appropriate. For example, due to the physical nature and high moisture content of wood smoke, assessment is usually made through the use of a real-time nephelometer. Asbestos sampling protocols, on the other hand, involve use of sampling pumps and after-the-fact laboratory analysis of fibers trapped on the surface of sampling filters, or use of a real-time fibrous aerosol monitor (FAM). Unless there is an urgent requirement to determine particulate concentrations on a real-time basis, the cost effective procedure is usually the after-the-fact laboratory analysis. In addition, the fibrous aerosol monitor may require considerable time at low counting levels to obtain reliable data. Ignitable particulates present a special case where the assessment must occur on a real-time basis.

OXYGEN DEFICIENCY AND ENRICHMENT

The concentration of oxygen in confined spaces is a concern from two stand-points. Too little oxygen can cause asphyxiation. Excessively high (or "enriched") levels of oxygen above normal oxygen concentrations dramatically promote or accelerate combustion and other chemical processes. The concentration of oxygen in normal air is approximately 20.9%. The balance (over 78%) consists primarily of nitrogen. The remaining fraction includes small amounts of water vapor, carbon dioxide, and argon as well as traces of other gases. The pressure of the atmosphere, and hence the number of molecules per unit of volume decreases with increasing altitude.

Most standards on confined spaces and regulatory agencies, such as OSHA, currently define oxygen deficiency by concentration. The usual benchmark is 19.5% by volume. This is also the default low oxygen alarm setpoint used by most instrument manufacturers. Some jurisdictions also define hazardous oxygen level on the basis of partial atmospheric pressure rather than concentration (Workers' Compensation Board 1998). In this case, an oxygen pressure less than 16.3 kPa or 122 mmHg partial pressure is deemed to represent a hazardous condition. (A partial pressure of 16.3 kPa of oxygen is equivalent to a concentration of 16.3% at sea level.) The reasoning behind this choice is that the body responds to atmospheric partial pressure of oxygen, rather than concentration.

The differences inherent in these approaches, as well as the effects of partial atmospheric pressure vs. percent by volume concentration on the function of sensors used to measure oxygen will be discussed at length later in this chapter. Some instrument manufacturers round oxygen readings upward in a properly calibrated instrument to 21.0%, while other manufacturers (the majority) round readings downward to 20.9%.

The definition of oxygen enrichment varies in standards and regulatory documents. OSHA chose 23.5% as the concentration for oxygen enrichment in the 1910.146 "Permit required confined spaces" Standard for general industry, while specifying 22% in the 1915-Subpart B Standard for shipyard employment (OSHA 1993, 1994). The latter value is consistent with nonmandatory recommendations from groups such as the National Fire Protection Association (NFPA) (NFPA 1993). The most conservative standards specify 22% as the concentration above which the atmosphere is deemed to be hazardous due to oxygen enrichment. For this reason, an increasing number of instrument designs that include a high oxygen alarm use 22% as the default high alarm set-point.

An important consideration for use of monitoring instruments in environments containing high concentrations of oxygen is the testing used to classify or approve the instrument for use in hazardous locations. These protocols usually do not include testing for intrinsic safety at elevated concentrations of oxygen. For this reason, labeling on the instrument will include a prohibition against use in oxygen concentrations that exceed testing or design parameters. A typical warning found in the owner's manual of a confined space monitoring instrument might state:

> Intrinsic safety is based on tests conducted in explosive gas/air (21% oxygen) mixtures only. This instrument should not be used in atmospheres where oxygen concentrations exceed 23.5%.

The user should always consult the owner's manual or contact the manufacturer directly to verify design limitations before using any instrument in highly oxygen enriched atmospheres.

Any oxygen concentration other than 20.9% (or 21% in the case of instrument designs that round the reading upward) indicates an abnormal condition. A less than normal concentration of oxygen by definition indicates a greater than normal concentration of some other component or the presence of a contaminant in the atmosphere being sampled. Even when the oxygen concentration does not constitute a statutory hazardous condition, the user should determine the cause of the abnormal reading prior to entering a confined space.

The safest approach is to initiate entry only when monitoring has determined that a "fresh air" oxygen concentration exists in the space. The only exceptions to this approach are circumstances where the cause of the abnormal oxygen concentration is known precisely, and where entry in these situations is explicitly permitted by written procedures.

Causes of Oxygen Deficiency

Confined spaces are particularly prone to the development or containment of oxygen deficient atmospheres. The contents and boundary surface conditions associated with confined spaces make them particularly prone to the development of oxygen deficiencies. Causes of oxygen deficiency include microbial action, displacement, combustion, oxidation, and absorption and adsorption. Work activity is also an important cause. Use of solvents, paints, degreasers, or hot work can affect oxygen levels.

Microbial Action

The microbial decomposition of organic material proceeds by a number of metabolic pathways. Aerobic decomposition is one of the most efficient (or rapid) of these biochemical pathways. Aerobic microbes consume oxygen and produce carbon dioxide as their chief atmospheric metabolite. One of the first detectable changes in the atmosphere of a confined space due to microbial action is a reduction in the oxygen concentration and increase in carbon dioxide. Microbial action can rapidly consume the oxygen present in the atmosphere of a confined space. This process requires surprisingly little organic debris. In an example personally witnessed by the author, a small, dry vault (approximately 2.2 m^3 or 80 $feet^3$) containing a potable water valve with no visible signs of rusting or corrosion, was found to have an oxygen concentration of 5% at the lowest point in the vault. The vault had been routinely sampled in the past, and had no prior documented history of oxygen deficiency. Investigation revealed that the only abnormal condition was the desiccated corpse of a small rodent in the bottom of the vault. We believe that microbial decomposition of this small amount of organic material was sufficient to produce this highly oxygen-deficient condition.

Displacement

The leading cause of oxygen deficiency in many industrial settings is displacement of the normal atmosphere by other gases and vapors. Gases and vapors tend to disperse quickly and spread evenly

TABLE 10.1
Oxygen Concentration vs. Depth Below Surface

Depth (in feet) below surface	Oxygen concentration (%)
5	20.5
7	20.0
9	14.0
11	6.5
13	4.0

in a horizontal direction. On the other hand, gases and vapors tend to form very distinct density-dependent layers in the vertical direction. Gases that are less dense than air tend to rise, while gases that are denser than air tend to sink. Denser than air gases tend to behave like invisible liquids. They can travel considerable distances along boundary surfaces, seeking the lowest level. This could be a pit, vault trench, excavation, or other subgrade structure. Denser than air gases frequently implicated in the generation of oxygen deficiency include argon, nitrogen, and propane. Denser than air vapors include gasoline, as well as many other solvent vapors.

Sometimes displacement is intentional, as when storage areas, curing ovens, or vessels are inerted or flooded with gases and vapors. This is a deliberate strategy to displace atmospheric oxygen. Although deliberate inerting with stack gas, carbon dioxide (dry ice vapor), or nitrogen is a common industrial procedure, most oxygen deficiencies are accidental.

Testing for oxygen or any other atmospheric hazard that may be present must occur at all vertical levels between the highest and lowest point in the confined space. Investigations of fatal accidents due to oxygen deficiency frequently have documented nearly normal concentrations of oxygen near the entrance. At the same time, a lethal oxygen deficiency existed near the bottom of the space.

Table 10.1 illustrates this problem from a fatal accident investigation cited by NIOSH (1994).

Testing by the investigators determined that the oxygen concentration dropped from 20.0 to 4% within a vertical distance of 6 feet (2 m). Sampling only at the point of entry — had it occurred — would not have identified the profound oxygen deficiency that existed near the bottom of the manhole. Evaluation of conditions in a confined space must always include testing at all vertical levels prior to entry.

Oxidation

Rusting is an oxidative process that both requires and consumes oxygen. In the presence of moisture, rusting consumes oxygen by means of the following reaction:

$$2Fe + 3O_2 \rightarrow 2Fe_2O_3 \tag{10.1}$$

In confining environments such as ship compartments, empty water tanks, and other containers with exposed metal surfaces, rusting alone may be sufficient to produce a lethal oxygen deficiency.

Combustion

Combustion requires and consumes oxygen. Deliberate introduction of an internal combustion engine into a confined space or hot work of any kind should be viewed with considerable caution. Not only do combustion processes consume oxygen, they also generate toxic products, including carbon monoxides, oxides of nitrogen, and sulfur oxides.

Absorption/Adsorption

Some substances are capable of directly absorbing or adsorbing oxygen from the atmosphere. Probably the most common adsorbent is activated carbon. Wet activated carbon filter beds will adsorb oxygen directly from the atmosphere until fully saturated. Similarly, wet uncured concrete will absorb oxygen from the air.

Oxygen Detection

General Discussion

Most portable or survey instruments utilize fuel-cell type oxygen sensors. Fuel-cell oxygen sensors contain the following parts:

- Diffusion barrier
- Sensing electrode (cathode) made from a noble metal such as gold or platinum
- Working electrode (anode) made from a base metal, such as lead or zinc
- Basic electrolyte (such as a solution of potassium hydroxide)
- Many designs additionally include an external moisture barrier or filter

Most currently available oxygen sensors used in portable instruments have working electrodes made from lead.

Oxygen diffusing into the sensor is reduced to hydroxyl ions at the cathode:

$$O_2 + 2H_2O + 4e^- \rightarrow 4OH^- \tag{10.2}$$

Hydroxide ions in turn oxidize the lead (or zinc) anode:

$$2Pb + 4OH^- \rightarrow 2PbO + 2H_2O + 4e^- \tag{10.3}$$

This yields an overall cell reaction of:

$$2Pb + O_2 \rightarrow 2PbO \tag{10.4}$$

Fuel-cell oxygen sensors generate electrical current. The amount is proportional to the amount of oxygen consumed. Oxygen detecting instruments simply monitor current output from the sensor. Of course, this is a simplification of the way oxygen sensing instruments are designed.

The electrolyte may be buffered, or may consist of a solution (such as potassium acetate) which is less prone to poisoning or being ruined by gases such as carbon dioxide. The electrolyte also may be a semisolid gel, rather than a liquid. Use of a gel electrolyte improves low temperature performance as well as reducing the potential for leakage or drying out.

The working electrode seldom is a simple chunk of lead, since this surface must be available and accessible for contact with oxygen molecules before the electrochemical reaction can occur. The working electrode usually consists of lead wool or some other form of the metal that provides a good surface area-to-volume ratio.

Other design features can include temperature-compensating thermistor/resistors, internal membranes, and current collectors. The latter serve to convert the current output to a voltage. Of course, there are always tradeoffs in design choices. A design that improves cold temperature performance may at the same time slow the speed of response, current output, sensor life, or other performance parameters. Numerous variations on these general design options are used to optimize the performance of an oxygen sensor for use in a specific product.

One of the most important design constraints is the buildup of lead oxide that develops in the capsule over the life of the sensor. As lead is converted into lead oxide, an increasing fraction of the volume of the sensor capsule is occupied by solid material. If the sensor design does not include a provision for this increase, it will eventually rupture. The latter could lead to leakage of electrolyte into the (usually) delicate and expensive electronics.

The working electrode in a fuel-cell oxygen sensor is consumed over time. In the cell reaction discussed above, when all available surface area of the lead (Pb) anode is converted to lead oxide (PbO), electrochemical activity ceases and current output falls to zero. At this point, the sensor must be rebuilt or replaced. Fuel-cell sensors are designed to last no more than 1 to 2 years.

In most instruments, even during inactivity, the sensor continues to generate current and is used up. Some instrument designs prevent the flow of electricity by breaking the circuit. This blocks the electrochemical reaction as long as the instrument is turned off and increases the effective life span of the sensor. The drawback to this design is the need for a lengthy restabilization period (sometimes several minutes) when the instrument is turned back on. During restabilization, current once again flows in the circuit. The reason for the lengthy stabilization period is that oxygen that diffused into the sensor over time accumulates in the electrolyte. The only means to remove the oxygen is the electrochemical reaction that converts it to lead oxide. The accumulated oxygen must be consumed before the sensor can provide accurate measurements. Effectively, this results in a "counting down" process when the instrument is turned on. Readings tend to start high, then slowly decrease to a stable value. Instruments having this design must not be zeroed or calibrated until full stabilization has occurred.

The temperature of the atmosphere influences the output from oxygen sensors. The warmer the atmosphere, the faster the electrochemical reaction proceeds. For this reason, oxygen sensors usually include a temperature-compensating load resistor to hold current output steady. Microprocessor-based designs usually provide additional signal correction in software to further improve accuracy.

Cold temperatures are a major factor that limits performance. The freezing temperature of electrolyte mixtures commonly used in some oxygen sensors is as high as –20°C (–5°F). Once the electrolyte freezes, electrical output falls to zero. The gelled electrolytes used in some oxygen sensors show much better cold temperature performance (Figure 10.1).

There are two fundamental variations in designs of fuel-cell oxygen sensors, "partial atmospheric pressure" and "capillary pore" type designs. These variations reflect the mechanism by which oxygen diffuses into the sensor.

Dalton's law states that the total pressure exerted by a mixture of gases is the sum of the partial pressures of the constituents. The partial pressure of oxygen is that fraction of the total pressure due to oxygen.

Partial-pressure oxygen sensors rely on the partial pressure (or p_{O_2}) of oxygen to drive molecules through the diffusion barrier into the sensor. As long as p_{O_2} remains constant, current output is a reliable indicator of oxygen concentration. Shifts in barometric pressure, altitude, or other conditions that affects atmospheric pressure will cause a systemic change in sensor output. To illustrate this problem, consider a sensor calibrated at sea level where atmospheric pressure is 760 mmHg. Now consider the same sensor at an elevation of 3,000 m (10,000 ft). Although at both elevations the air contains 20.9% oxygen, at 3,000 m the total atmospheric pressure is only 530 mmHg. Since there is less force driving oxygen molecules through the diffusion barrier into the sensor, current output is significantly lower. Partial-pressure sensors offer a considerably greater area for passage of oxygen into the sensor than does the capillary pore design. As a result, partial-pressure sensors respond more quickly to changes in oxygen concentration at a given partial pressure. It should be noted that several instrument designs which include partial-atmospheric oxygen sensors include software to correct sensor readings for barometric pressure fluctuation. In the case of these designs, oxygen readings will not be affected by changes in ambient pressure within design parameters.

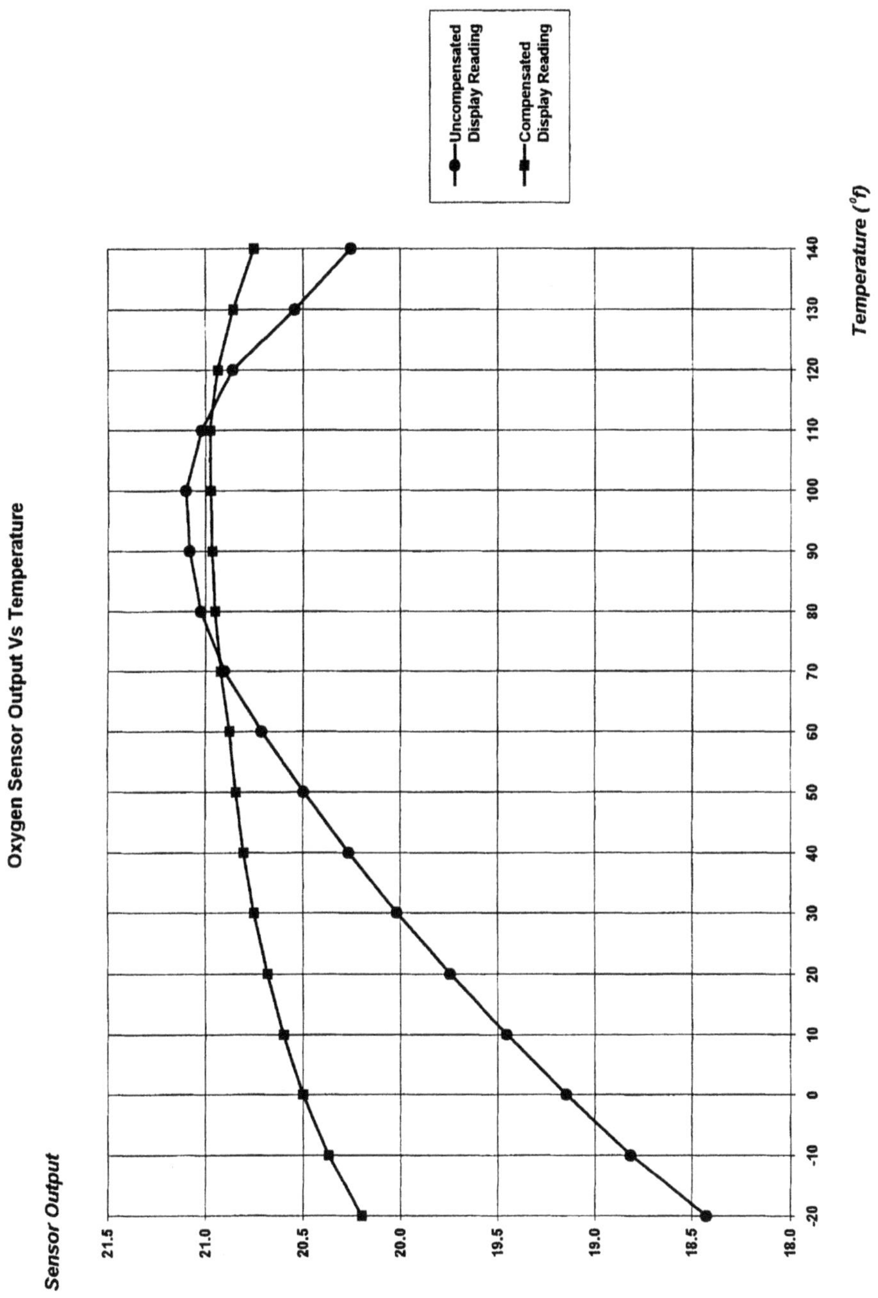

FIGURE 10.1 Raw (uncompensated) oxygen sensor output vs. temperature-compensated output for an oxygen sensor normalized to 70°F (21°C). (Courtesy of Biosystems, Middletown, CT.)

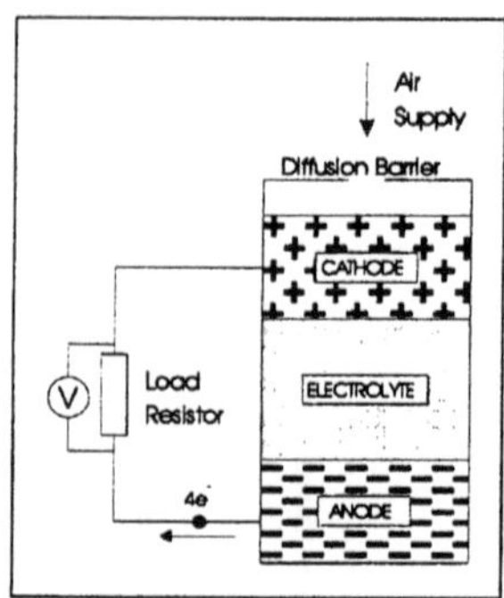

FIGURE 10.2 Diagram of a capillary-pore oxygen sensor. The amount of current produced by an oxygen sensor is proportional to the amount of oxygen that enters the sensor. Capillary-pore sensors include a narrow diameter tube through which oxygen diffuses into the sensor. (From *Product Data Handbook,* Vol. 1, *Safety,* City Technology, Ltd., Portsmouth, England, 1997. With permission.)

Capillary-pore oxygen sensors utilize a narrow diameter tube through which oxygen diffuses into the sensor (Figure 10.2). Oxygen is drawn into the sensor by capillary action in much the same way that water or fluid is drawn up into the fibers of a paper towel. Capillary-pore sensors are much less influenced by changes in pressure than partial-pressure oxygen sensor designs. Although rapidly changing pressure leads to a change in sensor output, as soon as the diffusion barrier capillary has stabilized at the new pressure, the output will return to the previous level. Because the volume of atmosphere contained in the diffusion barrier capillary is very small, stabilization at the new pressure is usually achieved within 10 to 30 seconds. This effect can be seen clearly when a properly calibrated capillary-pore oxygen sensor is taken on board a commercial jet. Initially the sensor reads 20.9%. As the jet takes off and begins to gain altitude, the p_{O_2} drops, causing a drop in oxygen sensor readings. When the jet reaches cruising altitude (actually, when cabin pressure is stabilized at normal operational levels) the readings return to 20.9%. As the jet begins its descent prior to landing, p_{O_2} increases, causing a rise in readings. As soon as the sensor is back at ground level (and the cabin once again depressurized) readings will return to 20.9%.

Capillary-pore designs are potentially vulnerable to blockage of the capillary. For this reason capillary-pore sensors include an external moisture barrier to prevent the pore from being blocked or plugged by solid materials, water, or other fluids.

Most instruments designed for measuring oxygen in confined spaces display readings in units of percent by volume concentration, regardless of the type of sensor that is employed. Readings are usually in increments of ±0.1% O_2, with a full instrument range on the order of 0 to 25% oxygen. A few designs are capable of displaying readings in partial atmospheric pressure (kilopascals or millimeters mercury) rather than concentration.

Capillary-pore sensors are altitude and weather independent. Physiologically based oxygen deficiency or enrichment may not be obvious to users of this type of equipment because concentration would remain constant at all altitudes of normal use. Partial-pressure sensors are pressure (and therefore, altitude and weather) sensitive. Calibration of these instruments at sea level and use at higher altitude, or calibration at higher barometric pressure than present during actual use, could cause underestimation of concentration. Even more likely, changes in pressure can easily lead to false high or low alarms. In some cases, fluctuation in pressure due to forced air ventilation or maintenance of a pressure regime within an industrial site can be the cause of fluctuation significant enough to cause an alarm. As an example, a slight negative pressure is maintained in many nuclear power-generating station containment buildings. In the case of one generating station known to the author, the partial-atmospheric oxygen sensors installed at the facility went into low oxygen alarm every time the pressure system cycled. The operators eventually sought and received a variance to set the low oxygen alarms at 18% rather than 19.5% to reduce the number of false alarms. Conversely, calibration to 20.9% at the surface of a mine could overestimate the concentration present at operating depth where total pressure is higher. An ambient condition easily could be falsely identified as oxygen deficient or oxygen enriched by oxygen-testing instruments containing sensors that are pressure-sensitive. On the other hand, instruments containing this type of sensor are excellent for estimating adverse conditions when the sensor is located in stable ambient pressure.

Mechanisms of Sensor Failure

Some oxygen sensors are affected by prolonged exposure to acid gases, such as carbon dioxide. The degree to which the sensor is affected is a consequence of design. Some designs can be damaged by relatively low level exposure to CO_2. Most oxygen sensors should not be used continuously in atmospheres containing more than 25% CO_2. In some cases, prolonged exposure to the acid gas damages the basic (alkaline) sensor electrolyte. In other situations, high concentrations of acid gas produce a current flux that alters the normal expected output of the sensor at a given concentration of oxygen.

Use in extreme conditions can affect data provided by the instrument. Extreme conditions are any which exceed published operating specifications for the instrument. Users should always consult the owner's manual or contact the manufacturer directly before using an instrument in any unusual or extreme environment. This includes operation in cold or excessively hot temperatures. Monitoring conditions that should trigger queries about appropriateness for use include:

- Inerted atmospheres
- Corrosive atmospheres
- Atmospheres containing high concentrations of combustible gas
- Atmospheres which contain high concentrations of other known contaminants
- Excessively hot or cold temperatures
- Excessively humid, wet, or dirty conditions

While the instrument may not pose an ignition or intrinsic safety hazard, the accuracy of readings may be adversely affected during use outside of design parameters.

As indicated previously, the working electrode in a fuel-cell sensor is consumed over time. When all available surface area of the lead (Pb) anode is converted to lead oxide (PbO_2), electrochemical activity ceases and current output falls to zero. This is the type of sensor failure most frequently experienced by users. This failure is fail-safe, since the oxygen deficiency alarm activates whenever current output falls below a minimum level. At this point the instrument will alarm continuously until the sensor is replaced. Hence, no one is likely to attempt to use the instrument to obtain readings in this condition.

Capillary-pore oxygen sensors include a narrow diameter tube through which oxygen diffuses into the sensor (Figure 10.3). Partial blockage of the pore would decrease the amount of oxygen drawn into the sensor. This would lead to a drop in current output. When used in a normal atmosphere, this would be interpreted by the instrument as an oxygen-deficient condition, a false low. When used in an oxygen-enriched atmosphere under this condition, the instrument could read a false acceptable. Most designs include an external moisture barrier or filter to minimize the chances for this type of failure.

Some conditions can lead to an erroneously higher, rather than lower, electrical output from the sensor. This potentially is a very dangerous type of malfunction, since it is not easily detectable by the user.

FIGURE 10.3 Single-sensor oxygen monitor. (Courtesy of Biosystems, Inc. Middletown, CT.)

Damage to the sensor housing, such as a crack or opening that creates a new channel through which oxygen can diffuse, can introduce more oxygen than would normally be able to enter. This can sometimes cause erroneously high readings.

Bubbles in the electrolyte also can cause erroneously high sensor output. When a sensor containing bubbles is rapidly taken from a warm area to a very cold area, the difference in temperature can cause sudden contraction in the bubbles. This decreases the pressure inside the sensor capsule and draws more oxygen into the capillary tube than normally would occur. The result is an increase in current output and a false high reading. This type of situation can occur when an instrument stabilized to indoor conditions is taken outside and exposed to winter temperatures. This type of anomaly is not permanent. Once the bubbles stabilize to the new temperature, and the pressure inside the sensor capsule equalizes with that outside, output will decrease to normal levels.

Zero-adjusting or calibrating while the output is erroneously high creates a serious problem. Later on, when the sensor output has dropped back to normal levels, readings will no longer be accurate. Fortunately, the preventive actions that are needed are simple! Verify the accuracy of the instrument by exposing it to known concentration test gas before any daily period of use! This advice applies to all the sensors in the instrument, not just the oxygen sensor. This is by far the most prudent approach.

Most calibration gas mixtures used to verify the accuracy of the oxygen sensor in a confined space instrument use a concentration that is less than 20.9%. The purpose for this is to activate the appropriate alarms when the sensor is exposed to the gas mixture. This means that if the sensor is working properly, readings should drop to the concentration indicated on the label of the calibration gas cylinder and the low O_2 alarm should activate. When calibration gas is not available in the field, ensure that the oxygen sensor reads 20.9% in fresh air, then exhale onto the sensor. Readings should decrease (in many cases to a low enough level to activate the low oxygen alarm), then recover. Readings which fail to decrease, or which require an abnormally long time to recover fully, may indicate a problem with the sensor. Table 10.2 summarizes failure mechanisms in oxygen sensors.

TABLE 10.2
Failure Mechanisms in Oxygen Sensors

Failure modes which lead to lower current output:
All available surface area of Pb anode converted to PbO_2
Electrolyte poisoned by exposure to contaminants
Electrolyte leakage
Desiccation
Blockage of capillary pore
Failure modes which lead to higher current output:
Cracking or damage to sensor housing or capillary pore
Bubbles in electrolyte

Ignitable Gases and Vapors

The ignitable gases and vapors encountered in confined spaces arise from a number of sources. These can include microbial decomposition, displacement of the atmosphere originally contained in the space by ignitable gases and vapors, residuals from previous uses of the space, or emissions from work activity. Sources from previous uses of the space include:

- Vaporization of residual contents (liquids and sludges)
- Products from chemical processes
- Desorption from structural materials

Desorption from vessel walls or other structural elements is a special concern. Desorbed vapors create a number of potential hazards, ranging from oxygen displacement to toxic contamination far in excess of exposure limits. Desorption of substances that form ignitable mixtures from the inner walls of vessels is a particular concern. As an example, during storage, liquid propane is absorbed into the porous walls of the tank in which it is being held. Following drainage, propane continues to desorb into the atmosphere of the tank.

Many work activities involve the use of organic solvents in a manner that creates highly concentrated mixtures in air. A prime example is spray painting. Spray painting creates a suspension of droplets of volatile liquid, as well as large wetted surfaces on structures. Both are sources of vapor.

A wide variety of ignitable gases and vapors may be encountered during initial evaluation of conditions in a confined space and work activity following initial entry. When present in sufficient concentration, gases and vapors of many substances become ignitable. Following contact with energy provided by suitable ignition sources, ignition can occur. Ignition sources present in confined spaces can include hot work activity, sparking tools, lighting, power tools, electrical equipment, or even static electricity.

In order for an atmosphere to be ignitable (i.e., capable of the propagation of flame away from the source of ignition when ignited), four conditions must be met. The atmosphere must contain adequate oxygen, adequate fuel, a source of ignition, and sufficient molecular energy to sustain the fire chain reaction. These four conditions are frequently diagrammed as the "fire tetrahedron." If any side of the tetrahedron is missing or incomplete or insubstantial, combustion will not occur.

Any gas or vapor capable of forming an ignitable mixture when mixed with air or oxygen will ignite at some inherent minimum concentration, provided that the other conditions in the fire tetrahedron are met. An ignitable mixture contains a flammable or combustible substance. The temperature at which there is sufficient vapor from a flammable substance is less than 38°C (100°F). The temperature at which there is sufficient vapor from a combustible substance is between 38°C (100°F) and 93°C (200°F). The minimum concentration at which a mixture is ignitable is the lower flammable limit or LFL. A mixture that will burn also can be made to explode. The term "lower

TABLE 10.3
Examples of Flammability Limits

Substance	LFL/LEL	UFL/UEL
Acetone	2.6	12.8
Acetylene	2.5	100
Ammonia	16	25
Carbon monoxide	12.5	74
Ethylene oxide	3	100
Hydrogen	4	75
Hydrogen sulfide	4.3	46
Methane	5	15
Propane	2.2	9.5

From National Fire Protection Association, *Fire Hazard Properties of Flammable Liquids, Gases, and Volatile Solids,* NFPA, Boston, 1977.

explosive limit," or LEL, often is used interchangeably with LFL. While these terms are not equivalent in strictest terms, both will be used interchangeably here to avoid confusion. Below the LFL/LEL the ratio of gas/vapor to oxygen is too low for combustion to occur. Stated in other words, the mixture is "too lean" to burn.

Most (but not all) ignitable gases/vapors also have an upper limit of concentration beyond which ignition will not occur. The upper flammable limit, or UFL, is the maximum concentration of gas/vapor in air that will support combustion. The term "upper explosive limit," or UEL, is often used synonymously with UFL. This convention will be used here. Above the UFL/UEL the ratio of gas/vapor to oxygen is too high for the fire reaction to propagate. Stated in other words, the mixture is "too rich" to burn. The difference in concentration between the LFL/LEL and UFL/UEL is the flammable range. Gas/vapor concentrations within the flammable range will burn or explode provided that the other conditions required in the fire tetrahedron are met.

The flammable range varies widely between individual gases and vapors. This partly results from the convention of expressing LFL/LEL and UFL/UEL in percent units rather than in g/m^3 (grams per cubic meter). When expressed in the latter units, the LFL/LEL for many substances are similar, averaging around 40 to 50 g/m^3. Table 10.3 provides flammability limits for some commonly encountered substances.

Flammability limits commonly listed in tables are determined at ambient temperatures and pressures, and at standard atmospheric concentrations of oxygen. Moderate oxygen enrichment exerts a profound effect on the flammability range by dramatically promoting and accelerating combustion.

Flammable/combustible gas and vapor detecting instruments usually read in "percent LEL" rather than "percent by volume" (Figure 10.4). This distinction is extremely important! To illustrate, consider an environment in which an instrument produces a reading of 3% by volume. If the exact composition of the gas/vapor or mixture producing the reading is known, ignitability of the atmosphere can be determined. On the other hand, if the exact composition of the gas/vapor or mixture producing the reading is not known, ignitability of the atmosphere cannot be determined. If the reading is due to methane (since the LEL for methane is 5% by volume), the concentration is less than the LFL/LEL. If the reading is due to propane (since the LEL for propane is 2.2% by volume), the concentration is above the LEL, and a source of ignition could cause a fire or explosion.

Most instruments read from 0 to 100% LEL. The reason for this is that consensus and regulatory standards use a percent value of the LFL/LEL to impose a margin of safety in hazard management. The most common limits are 5 or 10% LFL/LEL. The default alarm setpoint on many instruments is 10% of LFL/LEL.

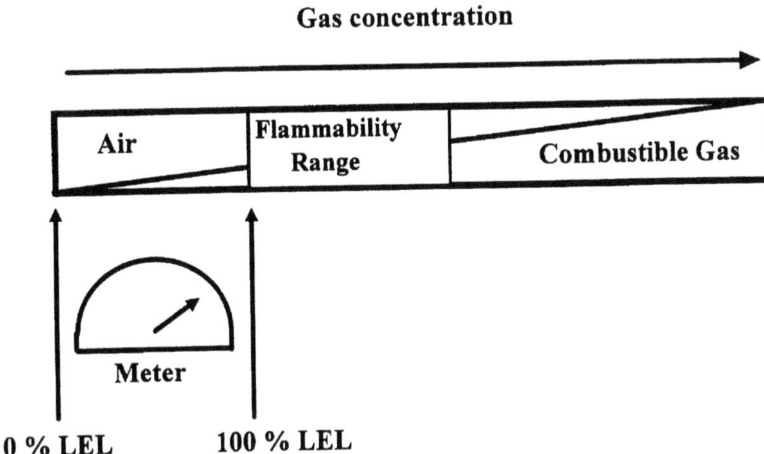

FIGURE 10.4 Flammable/combustible gas sensors read in percent LEL. (From George, E. et al. (Eds.), *Corporate Health and Safety, Managing Environmental Issues in the Workplace*. Copyright 1996 by Ergonomics, Inc., Southampton, PA. With permission.)

A fire hazard should always be deemed to exist whenever readings exceed 10% LFL/LEL. This is the least conservative (or highest acceptable) alarm set-point for instruments used for monitoring flammable/combustible gases and mixtures in confined spaces. An important consideration about the set-point of 10% LFL/LEL is that many circumstances warrant a more conservative, lower alarm set-point. The presence of any detectable concentration of flammable/combustible gas in the confined space indicates the existence of an abnormal condition. The only completely safe concentration of combustible gas in a confined space is 0% LFL/LEL.

Most evaluation for ignitable gases and vapors occurs with instruments designed to detect the widest possible variety of mixtures. Evaluations should consider the size of the source, release or emission rate, the distance from the source to the point where ignition could occur, and work activity.

Some types of instruments do read concentration in percent by volume flammable/combustible. The most notable example is the methanometer approved for use in MSHA-regulated mines. (MSHA is the Mine Safety and Health Administration.) Readings are always stated in units of percent by volume of methane. Monitoring activities related to "gassy" mines fall under the scope of MSHA regulations. These indicate explicitly the amount of methane that is permissible. A reading of 5% methane unambiguously indicates to the instrument operator that the atmosphere is 100% explosive!

Instruments designed to measure high ranges of flammable/combustible concentrations also read concentration in percent by volume. While the primary purpose of these instruments is to read concentrations higher than the LFL/LEL, when used in lower concentration ranges, concentrations are frequently still provided in percent by volume concentrations. For example, a typical flammable/combustible gas detector used to measure natural gas provides readings on a scale of 0 to 100% by volume, regardless of whether the concentration exceeds the LFL/LEL. At the other extreme, some instruments have a low range scale that reads in parts per million (ppm) of combustible gas. It should be noted that some designs include an auto-ranging feature, which displays readings in increments that are appropriate to the concentration encountered. In the case of these designs, the same instrument may display readings in ppm, percent LFL/LEL, or percent by volume.

Role of Flash Point in Monitoring of Ignitable Gases and Vapors

Vapors are the gaseous state of substances that are either liquids or solids at room temperature. Vaporization or evaporation rate, the rate at which the change from liquid or solid to vapor occurs,

TABLE 10.4
Examples of Flash Points

Substance	Flash Point °C	Flash Point °F
Gasoline (aviation grade)[a]	–46	–50
Acetone	–20	–4
Methyl ethyl ketone	–9	16
Ethanol (96%)	17	62
Diesel oil (#1-D)[a]	38	100

[a] Approximate minimum temperatures.
From National Fire Protection Association, *Fire Hazard Properties of Flammable Liquids, Gases, and Volatile Solids,* NFPA, Boston, 1977.

is a key property in consideration of formation of ignitable mixtures. Vaporization is a function of temperature. Increasing the temperature of the liquid increases the rate and amount of vapor that is produced. Conversely, cooling the atmosphere decreases the amount of vapor produced and may condense vapors back to liquid.

In order for combustion to occur, the vapor of the substance must be present in the atmosphere. As a general rule, it's the vapor, not the liquid that burns. The temperature at which sufficient vapor is present for combustion is a key concept in fire protection. This applies especially to confined spaces where boundary surfaces reduce or eliminate the influence of air currents and the wind in vapor dispersion. Flash point is the minimum temperature at which a liquid gives off enough vapor to form an ignitable concentration. The flash point is the temperature at which the LFL/LEL first occurs. The flash point is an inherent property of the substance.

Table 10.4 lists the flash point for a few common substances.

The practitioner must consider the flash point of liquids which may be present in the workplace as part of the monitoring strategy.

Diesel oil or turpentine and other substances that have higher flash points may not be detectable at normal room temperature with a flammable/combustible gas indicator that reads in percent LEL. The detector cannot detect until the substance is present in the atmosphere as a vapor at some minimum level.

An extremely important caveat regarding the assessment process is temperature of the substance. Increasing the temperature of the liquid after the initial test can dramatically alter the amount of vapor in the atmosphere. This could occur in various ways:

- Solar heating on surfaces of the structure
- General work activity
- Spot heating during hot work, such as cutting, grinding, welding, gouging, drilling, and so on

Increasing the temperature sufficiently could provide sufficient vapor for the composition of the atmosphere to enter the flammable range. Lack of attention to this situation has caused many fires and explosions during work activity in confined spaces and during work on the exterior of "empty" containers. Testing must occur under the conditions of work. Testing before work begins in the morning when a structure is cool may not predict the hazard that can arise later in the work shift. For example, At 10°C (50°F), ethanol does not produce a sufficient amount of vapor for ignition. At 21°C (70°F), vapor generation is sufficient to produce an ignitable mixture.

A common concern of individuals attempting to monitor vapor from high-flash liquids, such as diesel fuel, is detection by nose but not by the instrument. The person knows the substance is present because it is clearly nose detectable; yet the combustible gas monitor shows no response in the percent LEL range.

Several factors might contribute to this situation. First, the instrument should be directly calibrated to the substance being measured. An instrument calibrated with methane may not be sensitive to vapor from diesel fuel. The fittings, hoses, or tubing used to convey the sample from the environment to the instrument may absorb the vapor. In this case, the vapor may never reach the sensor. Readings would be strongly depressed. There may also be a temperature-related effect. In winter, atmosphere in the space is often warmer than the external environment where the instrument and operator are located. While the atmosphere inside the space may be warm enough for the diesel fuel to exist as a vapor, the vapor may cool sufficiently while being ducted through the sample tubing to recondense into a liquid. The sensor detects only vapors.

Another important issue is the resolution of the instrument. An instrument that reads in percent LEL, with readings incremented in 1% steps, cannot resolve changes in concentration smaller than ±1% of LFL/LEL. To illustrate, consider a combustible vapor which has an LFL/LEL of 0.7% (7,000 ppm) (One percent is 10,000 ppm). A properly calibrated instrument will only be able to resolve changes that are greater than 70 ppm. Although an individual might be able to smell the substance at 20 ppm, this would be below the detection minimum for the instrument. The instrument reading in this circumstance would probably be zero!

An important point to stress in circumstances where the vapor cannot be detected with an instrument that reads in percent LEL is that this is not an indicator of lack of hazard. An instrument that is capable of resolving vapors into the ppm range may be more appropriate. On the other hand, if what is needed is a determination of ignitability, a properly configured and calibrated instrument that reads in percent LEL will provide that information.

Testing at all levels in a confined space during hazard assessment is critical. Gases and vapors that are less dense than air tend to rise to the top of a structure, while denser-than-air gases and vapors tend to sink. In confining environments, this can lead to stratification of the gases into density-dependent layers. Typical low density gases that can form flammable mixtures include hydrogen, methane, and ammonia. Typical denser-than-air contaminants that can form ignitable mixtures include propane, hydrogen sulfide and gasoline, and many commonly used organic solvents.

Catalytic (Hot Bead) Sensors

General Discussion

Instruments for monitoring ignitable mixtures most frequently use catalytic (hot bead) sensors. Sensors of this type are frequently referred to as pellistors. While there are numerous variations, the underlying detection principle has not changed for the better part of a century. The hot bead sensor is a miniature calorimeter that contains two coils of fine platinum wire that are coated with a ceramic or porous alumina material to form refractory beads. The beads are wired into opposing arms of a balanced Wheatstone bridge electrical circuit (Figure 10.5). One bead is additionally treated with a platinum or palladium-based material that allows catalyzed combustion to occur on the treated surface of the "active" (or detector) bead (Moseley and Tofield, 1987). It should be noted that the porous or sintered nature of the bead means that the available surface is large compared to the diameter of the bead (Figures 10.6, 10.7). The catalyst is not consumed during combustion. Combustion occurs at concentrations far below the LFL/LEL. Trace amounts of gas/vapor in the air surrounding the sensor will oxidize catalytically on the surface of the bead. The "reference" (or compensator) bead in the circuit lacks the catalytic outer coating, but in other respects exactly resembles the active bead.

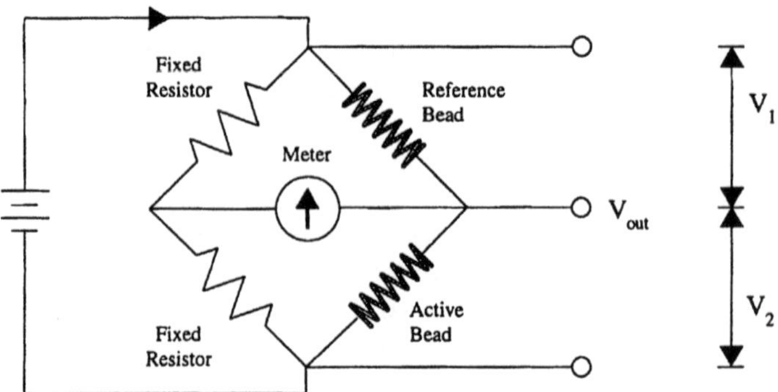

FIGURE 10.5 Wheatstone bridge electrical circuit. (From George, E. et al. (Eds.), *Corporate Health and Safety, Managing Environmental Issues in the Workplace.* Copyright 1996 by Ergonomics, Inc., Southampton, PA. With permission.)

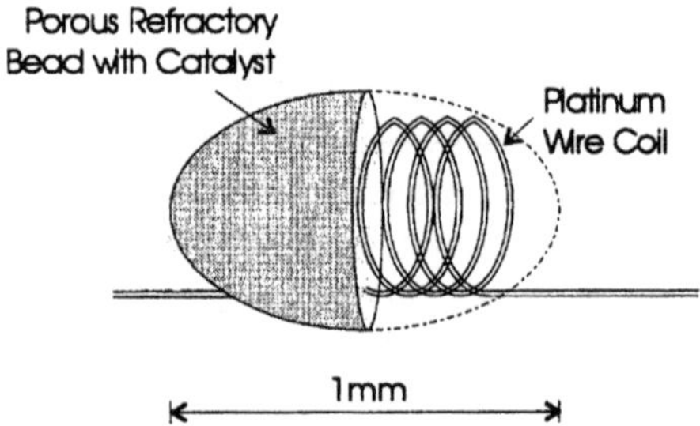

FIGURE 10.6 Schematic drawing of a catalytic bead. (From *Product Data Handbook,* Vol. 1, *Safety,* City Technology, Ltd., Portsmouth, England, 1997. With permission.)

A voltage applied across the active and reference beads causes them to heat. Heating is necessary for catalytic oxidation to occur. The temperature required may be as high as 500°C, or in some cases, even higher (City Technology, 1997). In normal air, the Wheatstone bridge circuit is balanced; that is, $V_1 = V_2$ and the voltage output (V_{out}) is zero. If ignitable gas/vapor is present, oxidation will heat the active bead to a higher temperature. The temperature of the untreated reference bead is unaffected by the presence of gas. Because the two beads are strung on opposite arms of the circuit, the difference in temperature between the beads is registered by the instrument as a change in electrical resistance. Under these conditions, $V_2 > V_1$ and V_{out} is proportional to the amount of oxidation that occurred.

Heating the beads to normal operating temperature requires power from the instrument battery. The amount of power required is a serious constraint on the battery life of the instrument. Recent sensor designs have attempted to reduce the amount of power required by operating the sensor at a lower temperature. While this approach may result in longer battery life, it may also result in the sensor being easier to poison or inhibit, since contaminants which might have been volatilized at a higher temperature can more easily accumulate on the surface of the bead. It is particularly important to verify the calibration of low-power combustible sensors by exposure to known concentration test gas on a regular basis.

Sensors used to measure combustible gas in the ppm range are usually operated at a higher temperature. Operation at the higher temperature can improve the ability of the sensor to oxidize volatile organic compounds and certain other classes of difficult to detect substances which may not be measurable by means of a low-power sensor. For instance, most low-power sensors may not be used to measure halogenated hydrocarbons such as methylene chloride. Halogenated hydrocarbons are

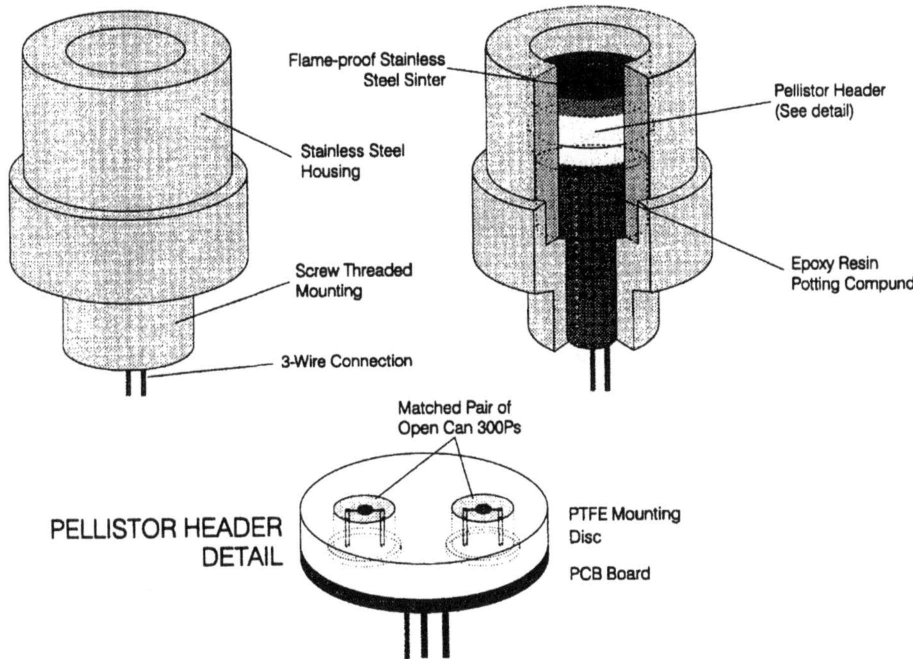

FIGURE 10.7 Schematic drawing of combustible sensor construction. (From *Product Data Handbook,* Vol. 1, *Safety,* City Technology, Ltd., Portsmouth, England, 1997. With permission.)

absorbed or form compounds which are absorbed by the catalyst, thus (at least temporarily) reducing or inhibiting the activity of the sensor. On the other hand, some high-power combustible gas sensors are capable of being used to measure halogenated hydrocarbons such as methylene chloride or trichloroethylene. Consult the owner's manual or contact the manufacturer directly to verify which contaminants may be successfully measured by the sensor prior to use!

A variation on the two-bead (active bead/reference bead) theme is the single-bead pellistor design. This design utilizes a thermocouple rather than a second bead to provide temperature compensation.

There are numerous other design differences between one brand or model of combustible sensor and another. Each design has been optimized for use in a specific instrument or application. Design differences may be found in the composition of the catalyst, coiling of the filament wire used in the beads, diameter of the filament wire, size and available surface area of the beads, power consumption, resistance to poisoning, and applicability for use in the ppm range. In other words, there may be significant differences in detection capability, power consumption, and general robustness of one design vs. another.

An important consideration in use and interpretation of results from instruments equipped with a combustible gas sensor of this type is the concentration of oxygen in the environment being monitored. Catalytic (hot-bead) sensors require at least 8 to 10% oxygen by volume to detect accurately. A combustible sensor in a 100% gas or vapor environment will produce a reading of 0% LEL. This is the reason that testing protocols for evaluating confined spaces specify measuring oxygen first and then flammable/combustible gas/vapor. For this reason, confined space instruments that contain hot-bead sensors should also include a sensor for measuring oxygen. If the instrument being used does not include an oxygen sensor, be especially cautious when interpreting results. A rapid up-scale reading followed by a declining or erratic reading may indicate that the environment contains insufficient oxygen for the sensor to read accurately. (It may also indicate a gas concentration beyond the upper scale limit for the sensor, the presence of a contaminant which has caused a sudden inhibition or loss of sensitivity in the sensor, or other condition which prevents the sensor or instrument from obtaining proper readings.) The minimum amount of oxygen that must be present for the sensor to detect accurately is a function of design. Capabilities vary from one manufacturer to another. Users who anticipate using their instruments in potentially oxygen deficient environments should contact the manufacturer for assistance.

Catalytic hot-bead sensors respond to a wide range of ignitable gases and vapors. The amount of heat produced by the combustion of a particular gas/vapor on the active bead will reflect the heat of combustion for that substance. Heat of combustion varies from one substance to another. For this reason, readings vary between equivalent concentrations of different combustible gases. Remember that the instrument reads electrical units that depend on change in resistance and not concentration units. The amount of heat provided by oxidation of the molecule on the active bead surface actually is inversely proportional to the heat of combustion for that gas. This occurs because of differences in molecular interaction with the catalytic surface. In general, the larger the size of the molecule, the greater the heat of combustion. On the other hand, the smaller the molecule, the more readily it is able to penetrate the sintered surface of the bead, and interact with the catalyst in the oxidation reaction.

Calibration of these instruments is an important issue. A combustible gas sensor may be calibrated to any number of different gases or vapors. Where possible, the user should calibrate the instrument using the substance of interest. Calibration is a two-step procedure. In the first step, the instrument is exposed to contaminant-free "fresh" air (that is, air which contains 20.9% oxygen and no combustible gas), turned on, and allowed to warm up fully. The combustible sensor should read zero. If necessary, the combustible sensor is adjusted to read zero. Instrument manuals and other support materials usually refer to this step as the "fresh air zero." If the instrument cannot be taken to an area where the air is known to be fresh, "zero air" from a calibration gas cylinder should be used as an alternative source of contaminant-free air.

The second step is to expose the sensor to known concentration calibration gas, and (if necessary) adjust the readings to match the concentration. This is called making a "span adjustment." A "span adjustment" sets the sensitivity of the sensor to a specific gas. Always follow the manufacturer's instructions when calibrating or adjusting the instrument. The type and concentration of calibration gas, the flow rate used to introduce gas to the sensors, and the adapters and fittings used during calibration all may affect the accuracy of the calibration procedure. Never use methods or materials that differ from those described by the manufacturer. Use of incorrect flow rates, fittings, concentration of calibration gas, or materials that are incompatible with the gas being used to calibrate the sensor can have a profound affect the accuracy of readings.

The response of a flammable/combustible sensor to an equivalent LEL concentration of gas varies from one substance to another. This is a natural outcome from calibration of these instruments in percent of LFL/LEL, rather than in units of g/m^3. Hence, a 50% LFL/LEL concentration of methane does not produce the same reading as a 50% LFL/LEL concentration of propane. Instruments used only for monitoring a single substance should be calibrated with that substance. An instrument calibrated to a particular substance will be accurate within performance specifications of its design.

Figure 10.8 shows the response of a typical LEL/LFL-reading instrument to several substances. Note that in the case of the gas which was used to calibrate the instrument (the calibration standard), a concentration of 50% LFL/LEL produces a meter response (reading) of 50% LEL in a properly calibrated instrument. Figure 10.8 also illustrates what occurs when an instrument is used to monitor substances other than the one to which it was calibrated. The diagram shows the relative response of the instrument to several different substances.

Note that the response to the substance to which the instrument was calibrated is still accurate. For the other substances, the responses do not match. In the case of some substances the readings are always too high. The result from this is that the instrument overreacts to conditions and alarms

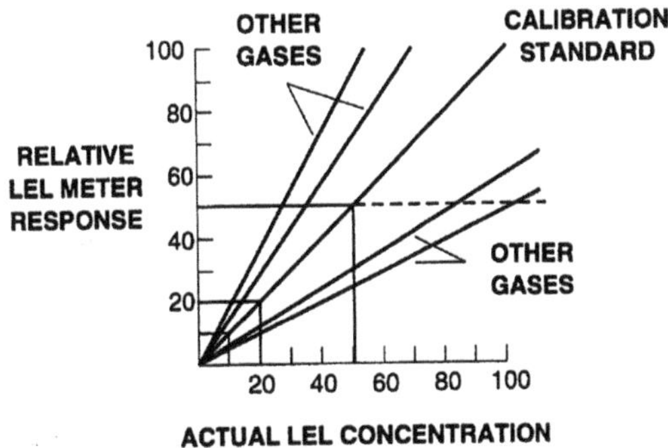

FIGURE 10.8 Relative response curves. (From George, E. et al. (Eds.), *Corporate Health and Safety, Managing Environmental Issues in the Workplace.* Copyright 1996 by Ergonomics, Inc., Southampton, PA. With permission.)

prematurely. This type of error usually is not serious. The most likely result is that workers evacuate the affected area sooner than legally required.

Substances that produce lower relative readings than the calibration standard can create a potentially dangerous error. In Figure 10.8, the worst case only produces a meter reading of 50% LEL, when the actual concentration is 100% LFL/LEL. If the instrument is set to alarm when the display reads 50% LFL/LEL, the alarm would sound simultaneously with a possible fire or explosion. The amount of relative error decreases the lower the alarm point is set. If the instrument is set to alarm when the display reads 20% LEL, a 50% LFL/LEL concentration of the same gas is enough to cause an alarm. If the alarm point is set to 10% LFL/LEL, the differences due to relative response of the sensor are minimized.

Most regulatory Standards, such as OSHA 1910.146, and protocols such as ANSI Z117.1-1995, use 10% LFL/LEL as the threshold concentration above which a hazardous condition exists (OSHA 1993, 1994; ANSI 1995). Many instruments use 10% LEL as the default combustible gas alarm setting. In fact, 10% LEL is the highest or least conservative alarm setpoint which may be used under most regulations and guidelines. This upper limit should not be used as an alarm setting without considerable thought. In its Compliance Directive (CPL 2.100) for Confined Space Entry, OSHA suggests (1995) that when entry is made according to the "Alternate entry procedures" specified in paragraph (c)(5)(ii) of 1910.146, a take action threshold of no higher than 5% LEL should be used to terminate entry and initiate evacuation procedures.

Where alarm set-points and action thresholds are concerned, the primary focus must be to enable work stoppage and safe exit. In some cases where continuous monitoring is occurring, 10% LEL might prove to be a reasonable action level. In other circumstances, the action level is the minimum detection threshold for the instrument being used to monitor for contaminant. Remember that the presence of any measurable ignitable gas/vapor indicates a potential problem.

The curves provided in Figure 10.8 are simplified examples. The response of a flammable/combustible sensor is linear over a wide range, but flattens out near the top of its effective range. Standard catalytic (hot bead) sensors are not usually designed for use in concentrations that exceed the LFL/LEL for the substance being measured. Special techniques must be utilized in order to use catalytic type hot bead sensors in high-range applications.

Relative Calibration

The accuracy of combustible gas readings will be maximized when the instrument is calibrated using the gas or vapor that will actually be monitored. When this is not possible or when the substance is an unknown, the user should select an alarm set-point of 10% LEL or less.

TABLE 10.5
Relative Response of a Flammable/Combustible Sensor

Combustible gas/vapor	Relative response when sensor is calibrated on pentane	Relative response when sensor is calibrated on propane	Relative response when sensor is calibrated on methane
Hydrogen	2.2	1.7	1.1
Methane	2.0	1.5	1.0
Propane	1.3	1.0	0.65
n-Butane	1.2	0.9	0.6
n-Pentane	1.0	0.75	0.5
n-Hexane	0.9	0.7	0.45
n-Octane	0.8	0.6	0.4
Methanol	2.3	1.75	1.15
Ethanol	1.6	1.2	0.8
Isopropyl alcohol	1.4	1.05	0.7
Acetone	1.4	1.05	0.7
Ammonia	2.6	2.0	1.3
Toluene	0.7	0.5	0.35
Gasoline (unleaded)	1.2	0.9	0.6

Courtesy of Biosystems, Inc., Middletown, CT.

Table 10.5 lists the relative response of a typical flammable/combustible sensor when calibrated to one gas/vapor then exposed to another. Note the difference in the relative responses when the instrument is calibrated to propane or pentane, rather than methane.

It is important to note that the values included in Table 10.5 are provided as a general example of how this information is typically conveyed, and should not be used as the basis for actual calculations. The sensors used in a particular instrument may or may not have values similar to those shown in Table 10.5. Even later generations of the same model sensor may exhibit different relative response ratios if the manufacturer has made modifications to the design. Users should consult the owner's manual or contact the manufacturer of the instrument they will be using to verify the correct values to use when making calculations based on relative response.

As an illustration, consider a detector calibrated on methane that is then used to monitor ethanol. From Table 10.5 it can be seen that when calibrated on methane, the sensor shows a relative response to ethanol of 0.8. In other words, the readings will be 20% lower than actual.

Some manufacturers provide a table of correction factors rather than relative response ratios for the gas being measured. The correction factor is the reciprocal of the relative response. In the case of our example, the correction factor would be calculated as: 1/0.8 = 1.25. Multiplying the instrument reading by the correction factor for ethanol (as determined above) provides the true concentration. Given a correction factor for ethanol of 1.25, and an instrument reading of 40% LEL, the true concentration would be calculated as:

$$\underset{\text{Instrument reading}}{40\%\ \text{LFL/LEL}} \times \underset{\text{Correction factor}}{1.25} = \underset{\text{True concentration}}{50\%\ \text{LFL/LEL}} \qquad (10.5)$$

Note that the correction factor for ethanol is different when the instrument is calibrated on propane. In the case of a propane calibrated instrument, instrument readings for ethanol will be higher than actual. Given that the correction factor for ethanol in this case is 1/1.2 = 0.83, when the instrument reads 40% LEL, the true concentration for ethanol is 33% LFL/LEL.

TABLE 10.6
Combustible Sensor Poisons and Inhibitors

Combustible sensor poisons
Lead-containing compounds (especially tetraethyl lead)
Sulfur-containing compounds
Silicones
Phosphates and phosphorus-containing substances
Combustible sensor inhibitors
Hydrogen sulfide
Halogenated hydrocarbons (Freons®, trichloroethylene, methylene chloride, etc.)

$$\underset{\text{Instrument reading}}{40\%\ \text{LEL}} \times \underset{\text{Correction factor}}{0.83} = \underset{\text{True concentration}}{33\%\ \text{LFL/LEL}} \tag{10.6}$$

The closer the relative response to 1.0, the closer to actual the reading becomes. To illustrate, consider a sensor that is calibrated to propane and then exposed to acetone. The response ratio (1.05) is so close to unity that for all intents and purposes any error is trivial.

Prediction of concentration based on theoretical relative response or correction factors deserves caution, since the relative response varies from sensor to sensor. In addition, response ratios can change over the life of a sensor. If the substance measured is identified incorrectly, or the wrong correction factor is used, significant inaccuracy in the calculation could occur. Also, this approach is not suitable for mixtures. It is also very important to understand the method used by a manufacturer to communicate this information. Some manufacturers communicate this information in the form of tables or graphs of relative response. Others provide this information in the form of tables of correction factors. It is critical to understand which method is used by the manufacturer before attempting to calculate true concentration based on relative response!

Follow the manufacturer's instructions when selecting the substance to which the instrument will be calibrated. When not sure what substances might be encountered, the best course usually is to use a calibration mixture that provides a broad sensor response. Calibration using other substances should occur in situations where these are predominantly present, or where the relative response closely approximates that of the substance to be measured.

The data in Table 10.5 indicate that, when an instrument containing this sensor is calibrated with methane, readings for most other substances on the list are dangerously low. On the other hand, when calibrated with pentane, readings for other substances are excessively high. When calibrated with propane, most of the substances on the list produce readings that are close to or slightly higher than actual. For many applications, propane (or a mixture which provides a similar level of sensitivity) is the substance that is most appropriate for calibrating this sensor. Remember that manufacturers that use sensors with different characteristics may offer substantially different advice. Always follow the manufacturer's instructions when deciding which gas to use for calibration.

Sensor Poisons and Inhibitors

The atmosphere in which an instrument is used can have an effect on the sensors. Poisoning or degraded performance can occur when sensors are exposed to certain substances. Some commonly encountered substances that degrade LEL/LFL sensor performance are listed in Table 10.6.

In the case of some substances, the compounds decompose on the catalyst and form a solid barrier over the catalyst surface. Exposure to substances of this type leads to irreversible loss of sensitivity. A single exposure to a high concentration of a silicone-containing substance can destroy

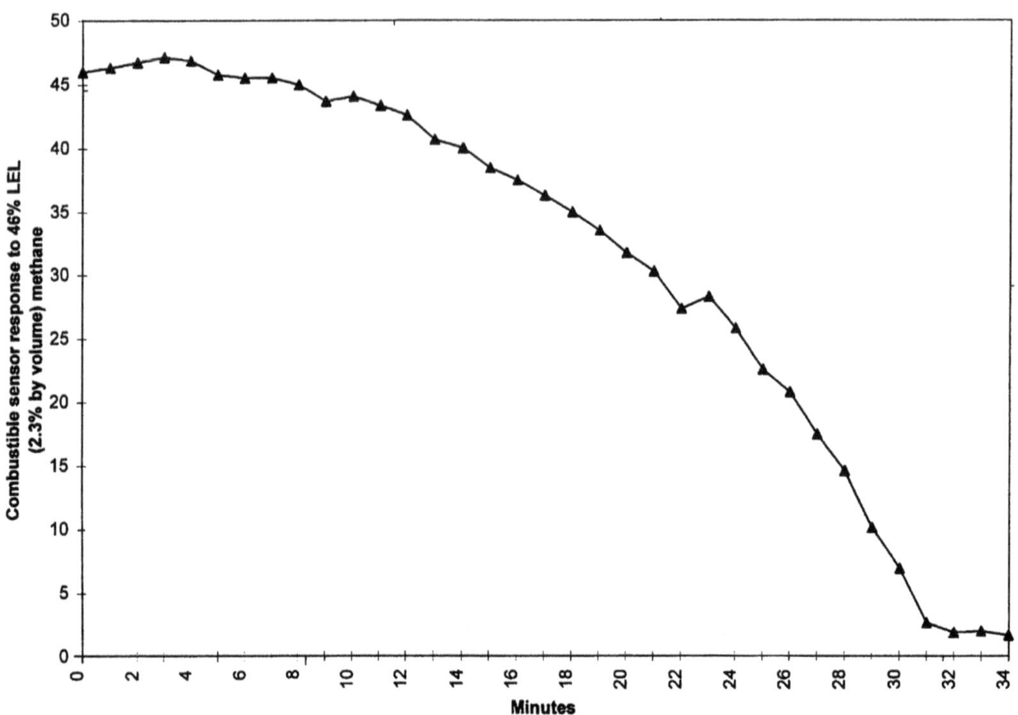

FIGURE 10.9 Loss of sensitivity in a combustible sensor exposed to 20 ppm hexamethyl-disiloxane in atmosphere containing 46% LEL (2.3% by volume) methane in air.

the sensor almost immediately. Figure 10.9 shows the consequences of combustible sensor exposure to a particularly virulent silicon containing sensor poison, hexamethyl-disiloxane (HMDS). The sensor tested was effectively rendered unable to detect combustible gas within a half hour by exposure to a 20 ppm concentration of HMDS.

Other substances are absorbed or form compounds which are temporarily absorbed by the catalyst, inhibiting normal reaction. In the case of these substances, the inhibition is usually temporary, and the sensor may substantially recover after a period of operation in fresh air (City Technology 1997; Moseley and Tofield 1987). Exposure to high concentrations of halogenated hydrocarbons can inhibit sensor performance in this way. Exposure to halogenated solvents causes accumulation of halogen molecules on the surface of the catalyst. Running the instrument while the sensor is located in fresh air tends to "cook off" much of the accumulated contamination. Nevertheless, recovery seldom is complete. Usually some permanent loss of sensitivity is a consequence from any exposure to any sensor poison or inhibitor.

Some substances (such as hydrogen sulfide) may function in both ways to degrade performance. Loss of sensitivity usually is dose dependent. A single very high exposure to hydrogen sulfide may produce an immediate irreversible loss in sensitivity. On the other hand, chronic exposure to low levels of hydrogen sulfide may require years to cause a significant loss of sensitivity.

The accuracy of flammable/combustible sensors can also be affected by exposure to high concentrations of ignitable mixtures. Excessive heating of the active bead can volatilize the catalyst coating. This could cause a partial or total loss of sensitivity. Excessive heating also can cause a break to develop in the filament or circuit wire of the sensor. Exposure to a very high concentration of ignitable gas or vapor (with concurrently low concentrations of oxygen) can lead to deposition of carbon black within the sintered surface of the active bead. Accumulation of carbon black within the bead can cause splitting to occur. This causes a mechanical break in the circuit or significantly alters the sensitivity and stability of the sensor.

To minimize the potential for damage or loss of sensitivity to the sensor, some instruments "alarm latch" whenever the concentration exceeds 100% LEL. (This concentration usually is not high enough to damage the sensor permanently.) Under these conditions, the instrument will indicate an over-limit condition, and audible and visual alarms will sound continuously. In addition, power to the sensor is cut to prevent damage. Until the over-limit alarms are cleared by manually resetting the instrument, the combustible sensor remains unpowered.

This logic is utilized by a number of manufacturers that have met requirements for classification for intrinsic safety by the Canadian Standards Association (CSA) under their standard for combustible gas detection instruments (CSA C-22.2 No. 152-M1984). This testing protocol includes a "methane flood" test which evaluates performance of the instrument when exposed to a high concentration of methane. The instrument is turned on, calibrated, and placed for 8 h in a test chamber containing 80% by volume of methane. This exposure is followed immediately by a test to verify accuracy when the instrument is exposed to 50% LEL of methane. Without the logic discussed above, most flammable/combustible sensors would be quickly destroyed by exposure to 80% by volume of methane.

Loss of Sensitivity to Methane

Age and usage affect sensitivity of flammable/combustible sensors. Chronic exposure to low levels of poisons or inhibitors acts cumulatively. This usually means that the sensitivity must be increased when calibration occurs. In the extreme, the sensor may require replacement. This again demonstrates that regular calibration is essential to the safe use of these instruments.

For most combustible (hot-bead) sensors, if sensitivity is lost due to poisoning, it tends to be lost first to methane. This means that a partially poisoned sensor might still respond accurately to propane, while showing a significantly reduced response to methane. This introduces a significant concern when choosing the substance to calibrate a flammable/combustible sensor. While sensitivity to propane or pentane may be all that is needed, use of propane or pentane as the only calibrant may lead the user to overlook a loss of sensitivity to methane. This could potentially be very dangerous, since methane is by far the most commonly encountered of all flammable/combustible gases associated with confined space entry.

Three methods exist for determining a loss of sensitivity to methane. The first is to calibrate the instrument using the calibrant that provides the best level of sensitivity (for instance, pentane or propane) and then expose the sensor to a known concentration of methane. The relative response factor for methane can then be used to verify whether there has been loss of sensitivity. This approach increases the time needed to calibrate the instrument and complicates the logistics. Another problem is what to do if there has been a loss of sensitivity to methane.

The second approach is to calibrate the instrument directly to methane. An instrument "spanned" to methane will continue to detect methane accurately, even when loss of sensitivity develops. Spanning the instrument during calibration simply makes up for any loss in sensitivity. As discussed, when the sensor is calibrated with methane, readings for most other substances tend to be dangerously low.

The third approach is to calibrate using methane at a concentration that produces a level of sensitivity equivalent to that provided by another calibrant (for instance, propane, pentane, or hexane). Several manufacturers have begun to make use of these "equivalent" or "simulant" calibration mixtures.

For the sensor described in Table 10.5, consider the methane mixture needed to calibrate to a propane level of sensitivity. The LFL/LEL of propane is 2.2% by volume. In a properly calibrated instrument, a concentration of 1.1% propane would produce a reading of 50% LEL. A concentration of 1.62% methane produces the same response. This is exactly the reading that should be shown in an instrument which has been calibrated for a propane level of sensitivity. Other concentrations of methane may be used to simulate other calibration gases such as pentane, hexane, or even substances (such as jet fuel vapor) which are not easily packaged in field-portable cylinders. Since the calibration is based on methane, any loss of sensitivity to methane will result in over-spanning the sensor. Readings for substances other than methane will be a little higher than actual.

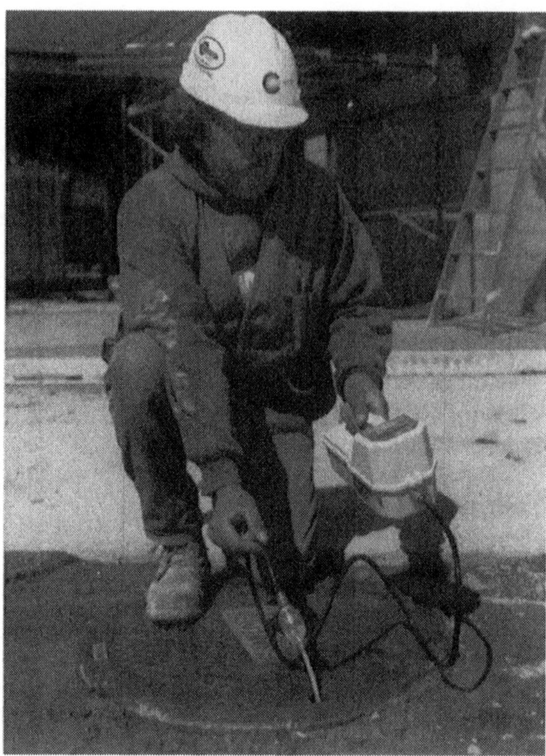

FIGURE 10.10 Multirange (% LEL/PPM hydrocarbon) flammable/combustible gas monitor. (Courtesy of GasTech, Inc., Newark, CA.)

Calibration verifies that sensors remain accurate. If exposure to test gas indicates a loss of sensitivity, the instrument needs adjustment. If the sensors cannot be properly adjusted, they must be replaced before any further use of the instrument. This is an essential part of ownership.

Low Range Hydrocarbon Detectors

Although the primary hazard of most flammable/combustible gases and vapors is fire and explosion, they can pose other hazards. Denser-than-air gases and vapors can displace oxygen in confining environments. In many circumstances, even when the concentration is less than 10% LFL/LEL, a toxic hazard exists. To illustrate, ethanol (or grain alcohol) has an LEL of 3.3% or 33,000 ppm. A 10% LFL/LEL concentration is equivalent to 3,300 PPM. At this level, a significant toxic hazard exists, since the Threshold Limit Value — Time-Weighted Average is only 1,000 ppm (ACGIH 1997).

Flammable/combustible gas and vapor instruments that read in the percent LEL range are designed to monitor contamination in the flammable range. In some instruments, output from the Wheatstone bridge is electronically multiplied and corrected to produce a reading in various ppm ranges (Figure 10.10). Typical ranges are:

- 0 to 10,000 ppm (closely equivalent for many substances to the flammable range)
- 0 to 1,000 ppm
- 0 to 100 ppm

Obtaining a stable signal from a flammable/combustible sensor that is operated in the ppm range is not a trivial engineering challenge. The combustible sensors used in these designs generally have large beads that require considerable power. The sensors are usually operated at higher temperatures in the ppm range than in the percent LEL range. These instruments usually include integral sample draw pumps to improve stability.

High Range Flammable/Combustible Instruments

Standard catalytic (hot-bead) sensors require at least 8 to 10% by volume of oxygen to detect accurately. In addition, extremely high concentrations of gas can heat the bead so hot that it becomes cracked or damaged, or suffers a loss of sensitivity due to vaporization of the catalyst. Different approaches are required to measure concentrations that exceed the LFL/LEL.

Thermal Conductivity Sensors

A thermal conductivity sensor measures flammable/combustible mixtures in the 0 to 100% by volume range. The sensor contains two coils of fine wire that are coated with a ceramic material to form beads. The beads are strung onto the opposite arms of a balanced Wheatstone bridge circuit. Neither bead receives a catalyst coating. Instead, the reference bead is isolated from the air being monitored in a sealed chamber. The active bead is exposed to the air which contains the gas/vapor mixture. Power is provided to the sensor to heat the beads to operating temperature. Detection depends on the "air-conditioning" effect of high concentrations of gas on the active bead. If a flammable/combustible mixture is present, the active bead will dissipate heat more efficiently than the reference bead. Once again, the difference in temperature between the two beads is proportional to the amount of flammable/combustible present. Since the two beads are strung on the opposite arms of a Wheatstone bridge, the difference in temperature between the beads is perceived by the instrument as a change in electrical resistance.

Oxygen Displacement

Several brands of flammable/combustible instruments include a high range mode which allows calculation of combustible gas, based on the amount of oxygen which has been displaced by the combustible gas. As combustible gas is introduced into an environment being monitored, more and more of the oxygen is displaced by combustible gas. Readings from an oxygen sensor are used to calculate the combustible gas concentration. Readings are generally given in percent-by-volume concentration, with a range of 0 to 100% combustible gas. Again, it is critical to reiterate the difference between readings displayed in percent LEL vs. those displayed as percent by volume. Methane has an LFL/LEL of 5% by volume. A reading of 5% by volume is equivalent to a reading of 100% LEL. In either case, the mixture would be fully explosive.

For maximum accuracy, the sensor should be calibrated to the specific combustible gas that will be monitored. In fact, the displacement algorithm may be highly specific to a particular flammable/combustible gas/vapor. For instance, some manufacturers explicitly limit use of this type of high range mode to testing for methane or natural gas. Users should check with the manufacturer before using the instrument to monitor for any flammable/combustible gas/vapor other than those explicitly identified by the manufacturer.

Dilution Fittings

As discussed, the accuracy of standard hot bead sensors is affected when they are used in highly oxygen-deficient atmospheres. Below 8 to 10% by volume (depending on the specific design), the sensor does not have sufficient oxygen to function properly.

A dilution fitting is a sample draw adapter that allows use of a standard hot-bead sensor to obtain direct readings from oxygen-deficient atmospheres. The adapter includes a dilution orifice designed to mix the gas sample with an equal volume of fresh air. Since fresh air contains 20.9% oxygen, the sample would contain at least 10% oxygen. At this level, the sensor will read accurately.

An important consequence of diluting the sample with fresh air is that the amount of flammable/combustible gas/vapor in the sample also is diluted. Since the adaptor provides a 50:50 dilution, the combustible and toxic gas readings must be doubled to obtain the true concentrations.

The adapter should be removed as soon as dilution sampling is completed. Leaving the dilution adapter in place during normal operation can lead to potentially dangerous misinterpretation of test results.

Make sure to locate the instrument in fresh air at all times while the dilution orifice is being used. Only fresh air containing 20.9% oxygen should be used to dilute the sample. If the dilution adapter is located in an oxygen-deficient or otherwise contaminated atmosphere, proper sample dilution will not occur, and accurate readings will not be obtained.

The amount of air drawn into the dilution orifice is affected by the length and inner diameter of the sample draw hose. It is also affected by altitude and the flow rate of the mechanical pump contained in the instrument. Each adapter should be individually calibrated while attached to the monitor and sample probe assembly that will be used during sampling.

Dilution orifices make possible sampling for flammable/combustible gas/vapor from environments which could not be monitored otherwise. Improper use of dilution orifices can lead to inaccurate readings. These have the potential for being the basis of flawed decisions, a major cause of accidents. Manufacturers are very concerned about the potential for misuse of dilution adapters. Users must clearly understand the limitations before making use of this accessory.

Metal Oxide Semiconductor Sensors

A type of sensor occasionally used for measurement of combustible gases and vapors in the percent LFL/LEL range is the metal oxide semiconductor (or MOS) sensor. MOS sensors are also used for monitoring at toxic levels. The MOS sensing element consists of a metal oxide semiconductor such as tin dioxide (SnO_2) on a sintered alumina ceramic bead contained within a flame arrestor. In clean air, electrical conductivity is low. Contact with reducing gases, such as carbon monoxide, or flammable/combustible gases increases conductivity. Sensitivity of the sensing element to a particular gas/vapor is alterable by changing the temperature. MOS sensors are designed to respond to the widest possible range of toxic and flammable/combustible gases/vapors. The most frequent usage of MOS sensors is ppm range hydrocarbon or toxic gas detection. A more detailed discussion of the MOS principle of detection, as well as specific advantages and disadvantages of this type of sensor, is given in a later section of this chapter.

TOXIC GASES AND VAPORS

Many toxic substances are commonly encountered during preparation of confined spaces for entry. Exposures to entirely unrelated substances can occur as a result of work activity. These exposures often are short-term and can occur because of sudden releases of material.

Airborne toxic substances typically are classified on the basis of their ability to produce physiological effects on exposed workers. Toxic substances tend to produce symptoms in two time frames: acute and chronic. Standards for workplace exposure provide a benchmark for protecting individuals against these effects. Confined spaces and the nature of work activity that occurs in them create the potential for accidental exposures above recognized standards. These overexposures can be fatal.

Hydrogen sulfide (H_2S) is a good example of an acutely toxic substance which is immediately lethal at relatively low concentrations. Exposure to 1,000 ppm produces rapid paralysis of the respiratory system, cardiac arrest, and death within minutes. Carbon monoxide (CO) also can act rapidly at high concentrations (1,000 ppm), although not as rapidly as hydrogen sulfide. Similarly, oxygen-deficient atmospheres (<10%) also can cause death in the same time frame. As discussed in the introduction, manufacturers responded to the needs of industry by producing portable instruments capable of providing rapid warning about these conditions.

Another need for instrumentation and monitoring equipment is the lower levels of exposure that occur during normal work activity. Work activity often is short in duration. Determining exposures rapidly is extremely important, since results from these studies form the basis for corrective actions. The atmosphere in a confined space during work activity can be highly complex and difficult to assess. The nature of the toxic contaminants present will also affect the choice of instruments used to monitor the confined space atmosphere.

Another influence on selection of equipment for monitoring is the type of limit applicable to the exposure. Many Standards and permissible exposure protocols define exposure limits in the following ways:

- An 8-h time-weighted average (TWA)
- A short-term exposure limit (STEL) calculated as a 15-min time-weighted average
- A ceiling (CEILING) not to be exceeded

Exposure limits for gases and vapors are usually given in units of parts per million (ppm) and mg/m^3. Exposure limits for liquid and solid aerosols are given in units of mg/m^3.

The TWA concept is based on a simple average of worker exposure during an 8-h day. The TWA concept permits excursions above the TWA limit only as long as they do not exceed the STEL or CEILING, and are compensated by equivalent excursions below the limit.

The regulatory TWA is a calculation using sampling results projected to a full eight hours. Time not measured is projected as zero exposure in many jurisdictions.

The method used by an instrument manufacturer to calculate and display TWA values can be a source of real confusion to users. Some instruments include a TWA calculation that is simply the average of the measured concentration over the monitoring interval, and not projected to an 8-h time frame. Other manufacturers project the accumulated exposure over an 8-h period. How to calculate the TWA when the work shift or monitoring interval exceeds 8 h is another source of potential confusion. Some jurisdictions specify that when the work shift or exposure interval exceeds eight hours, the summed exposure over time continues to be divided by 8 to calculate the TWA. Others specify that the monitoring interval used to calculate the TWA must not exceed eight hours. Manufacturers are equally divided! Some instrument designs calculate TWA for shifts which exceed 8 h by dividing by 8. Other designs calculate TWA only on the basis of the most recently completed 8 h. Once again, the user must understand the method used by the instrument manufacturer to calculate the TWA value. If the owner's manual is vague on this point, call the manufacturer to establish the method used. Then call the regulatory authority to determine how they require this information to be interpreted.

Many instruments include the capability of displaying a "peak" or "ceiling" concentration which represents the highest (or in the case of oxygen, highest and lowest) concentration noted by the instrument during the monitoring interval. The specific nature of the values used by the instrument for this purpose varies between designs.

Sensors respond continuously to changes in the atmosphere being monitored. This information is constantly sampled by the instrument, and used to update the readings on the instrument display. Most instruments integrate sensor output over some period of time (for instance, 1 second) before using it to update the displayed readings. Instrument designs may also use filtering algorithms or further integrate readings to reduce or remove anomalous short-term spikes or "glitches" which might otherwise lead to false positive or negative (downscale) alarms. This information is used to calculate "running" time-history calculations such as CEILING, STEL, and TWA. The amplitude of the "peak" value displayed by the instrument will be a function of the exposure concentration, the speed of sensor response, and how frequently readings are updated. Instrument designs which update information more frequently may display higher readings when challenged by a brief high-level exposure, than designs which integrate readings over a slightly longer time frame.

The best means for keeping track of TWA, STEL, and CEILING calculations on a real-time basis is a microprocessor-based instrument. Microprocessor-based instruments are capable of solving these complex time-history equations, and updating the instrument memory many times per minute. In addition, most include independent alarm points for TWA, STEL, and CEILING, and are capable of determining automatically when an alarm condition exists by the most conservative method of calculation.

Measurement of Toxic Gases and Vapors

Of all the areas of atmospheric assessment, toxic gas measurement has available the most numerous detection technologies. While this may seem like an embarrassment of riches, it is not. Atmospheres in confined spaces, especially during work activity, can be highly complex. No individual technology may be able to resolve a contaminant unambiguously from the others, at high speed and at reasonable cost. The method of choice could be a compromise.

Badge-Type Dosimeters

One of the most useful technologies, the badge-type passive dosimeter, is also one of the simplest to use. The most common varieties are small devices designed to clip onto the shirt collar or lapel as close as possible to the breathing zone. The badges contain one or more layers of a material capable of absorbing the substance of interest. Badges designed to measure a specific contaminant frequently contain external filters or sorbent media that remove interferents. Badges used as organic vapor detectors usually contain one or more layers of an activated carbon material capable of absorbing a wide variety of organic substances.

Gases and vapors in the atmosphere move by passive diffusion into the badge, where they are absorbed into or adsorbed onto the sorbent media. The badge is worn for a specific interval of time, then sealed in a container and sent to a laboratory for analysis. Breakthrough is characterized by the presence of contaminant in the secondary layer. Breakthrough indicates that the primary layer may have been saturated or exposed to such a high concentration of contaminant that the primary sorbent layer was incapable of absorbing all of it. Under these circumstances, prudence would dictate use of dosimeters run concurrently or consecutively. (The most accurate results will be obtained from an interval of exposure just slightly shorter than the time at which breakthrough occurs.)

This technology offers good discrimination between closely related molecular species such as benzene, toluene, and xylene. The precision of rigorous laboratory analysis coupled with the fact that several hours of exposure can be integrated to produce the final time-weighted average exposure makes for highly accurate detection. The chief limitations of passive dosimeter badges are that readings are not obtainable in real-time, and the lack of laboratory analytical protocols. Protocols are available only for a limited range of contaminants.

For a number of important applications, dosimeter badges represent the most practical and cost-effective approach for assessing conditions. This is especially true for contaminants for which no other practical direct reading measurement technique is available. Work activity in confined spaces can occur in highly confined conditions. Workers could experience extreme discomfort and difficulty during the wearing of bulky and heavy sampling equipment. In addition, this could pose a snagging or other safety hazard. Dosimeter badges are small, lightweight, and unobtrusive.

Sorbent Tube Sampling

Sorbent tube sampling utilizes the same concepts as dosimeter badges: absorption or adsorption of contaminants into or onto a sorbent medium. Sorbent tube sampling offers a number of additional refinements. The sorbent is loaded into sealed glass tubes that resemble short straws. A number of sorbent media, each with different properties, are available. (The medium is selected to optimize absorbence or adsorbence for a specific contaminant.) To sample, the ends are broken off and the tube is inserted into a manifold. Instead of relying on passive diffusion, this methodology utilizes a precision low-flow sampling pump to pull a calibrated volume of air through the tube over a specific interval of time. Some manifolds can accommodate several tubes simultaneously, each with an independently adjusted flow rate for concurrent sampling. On completion of sampling, the tubes are removed from the manifold, capped, and sent to a laboratory for analysis.

Sorbent tube sampling is a powerful tool for assessing conditions. As with passive dosimeters, the chief limitation is that readings are not obtainable in real-time. Laboratory analytical protocols are available only for a select group of contaminants. Sorbent tube sampling requires a pump and tubing. For personal samples, this equipment must be worn/carried somewhere in the clothing of the subject. This equipment could become quite bothersome during work in tight quarters.

Colorimetric Measurement Techniques

Colorimetric measurement techniques utilize reagents that undergo a color change when exposed to the specific contaminants they are designed to detect. Colorimetric detection is a real-time measurement technique. It can be used both for broad range screening, and for analysis of specific contaminants.

Colorimetric Detector Tubes

Colorimetric detector tubes resemble short glass straws packed with fine grain silica gel, activated alumina or other medium (Figure 10.11). The tube contents are impregnated with the reagents that undergo a color change reaction when exposed to specific contaminants. To use, the ends are broken off and the tube is inserted into a hand pump designed to pull a calibrated volume of air. Contaminants in the air that can react with the reagent produce a color change progressively along the tube as the air passes through. The length of the color change is proportional to the amount of contaminant present. The outside of the tube usually includes a measurement scale for estimating concentration.

The colorimetric reaction used for the detection of carbon monoxide provides a good example of this type of chemistry. In this reaction, carbon monoxide reduces iodine pentoxide to liberate iodine, which is brown in color:

$$CO + I_2O_5 + H_2S_2O_7 \rightarrow I_2 + CO_2 \tag{10.7}$$

Various types of pumps are used to pull the sample through the tube (Figure 10.12). Types of pumps include hand-operated bellows, syringe-type piston pumps, and mechanically assisted versions. An important point to stress is that pumps and tubes from different manufacturers are not interchangeable.

The chief advantages of colorimetric tubes are their ability to provide real-time test results for a wide range of contaminants. All that is necessary to test for a new contaminant is to change the tube. In addition, identifying an unknown contaminant is often possible by using a specific sequence of "yes/no" indicator tubes.

The chief limitations of colorimetric detector tubes are cross-sensitivity to contaminants other than the ones being measured and inability to provide dynamic results for concentrations that vary with time. Another potential problem is measurement of contaminants which are discontinuously present. Each sample is a "snap-shot" of concentration within a movie containing hundreds of frames. This limits the utility of colorimetric tube measurement when concentrations are potentially subject to rapid change. In addition, interpretation of test results can be strongly affected by the "crispness" of the leading edge of the stain, ambient humidity, and age of the tube.

Automated Colorimetric Measurement Systems

One of the most interesting recent developments in colorimetric measurement technology is a product that integrates colorimetric reagent technology with electronic and photoanalyzer technology. This integrated system recognizes the substance that the colorimetric "chip" is designed to

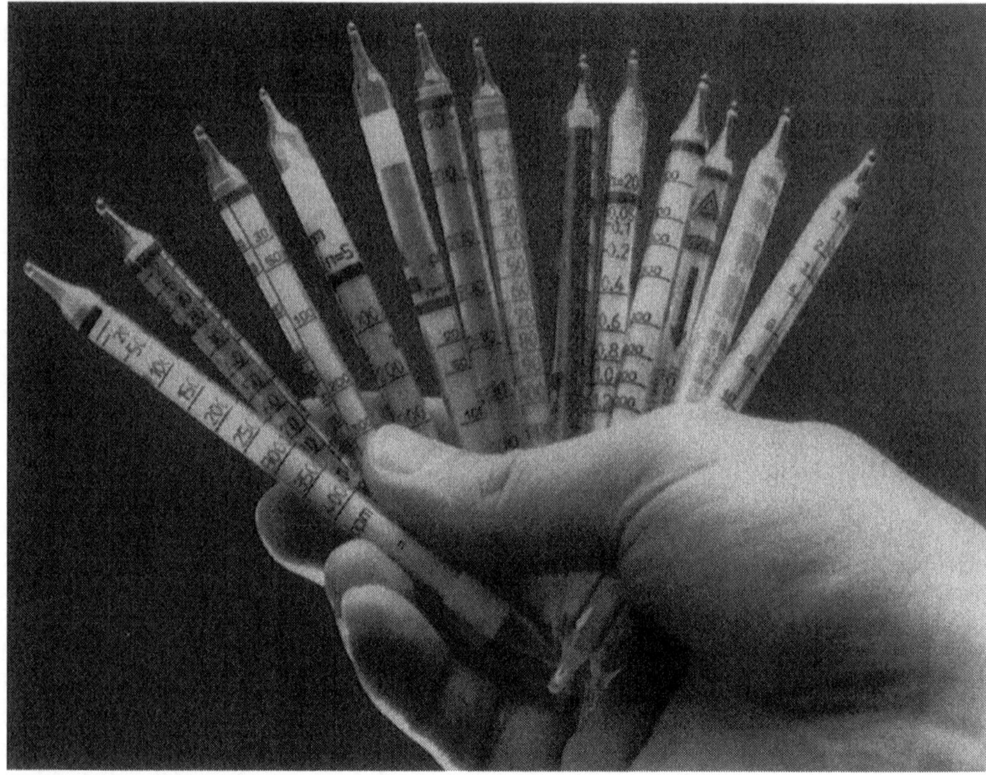

FIGURE 10.11 Colorimetric tubes showing graduated measurement scales. (Courtesy of MSA, Inc., Pittsburgh, PA.)

measure, and draws the appropriate volume of air through the sensing capillary by means of a calibrated automatic pump. It then automatically provides a digital reading by means of an integral photoanalyzer. This sampling system dramatically increases the sensitivity and accuracy of the colorimetric measurement technique. In some cases, this unit can accurately register concentrations of vapor in 0.01-ppm increments. While interfering contaminants and other environmental conditions still affect the accuracy of readings, the improvement in precision and accuracy provided by the new integrated systems is nothing less than revolutionary when compared to products of the past.

Colorimetric Badges and Dosimeter Tubes

The same kind of colorimetric chemistry is also used in passive badge and tube-type dosimeters. These products are worn in the same manner as other types of dosimeters. Contaminants are absorbed into media containing the same reagents as used in colorimetric tubes. As the contaminant is absorbed into the badge, the color becomes more and more intense or darker. Colorimetric badges generally include a chart for comparing color intensity to estimate concentration.

In another variation, contaminants passively diffuse into a dosimeter tube. Dosimeter tubes closely resemble standard colorimetric detector tubes. The length of stain once again is proportional to the concentration of contaminant. The main difference between dosimeter tubes and standard colorimetric detector tubes is the use of diffusion rather than a pump to pull contaminants through the tube. Another important difference is that dosimeter tubes provide a time-weighted average, rather than instantaneous concentration for the contaminant being measured.

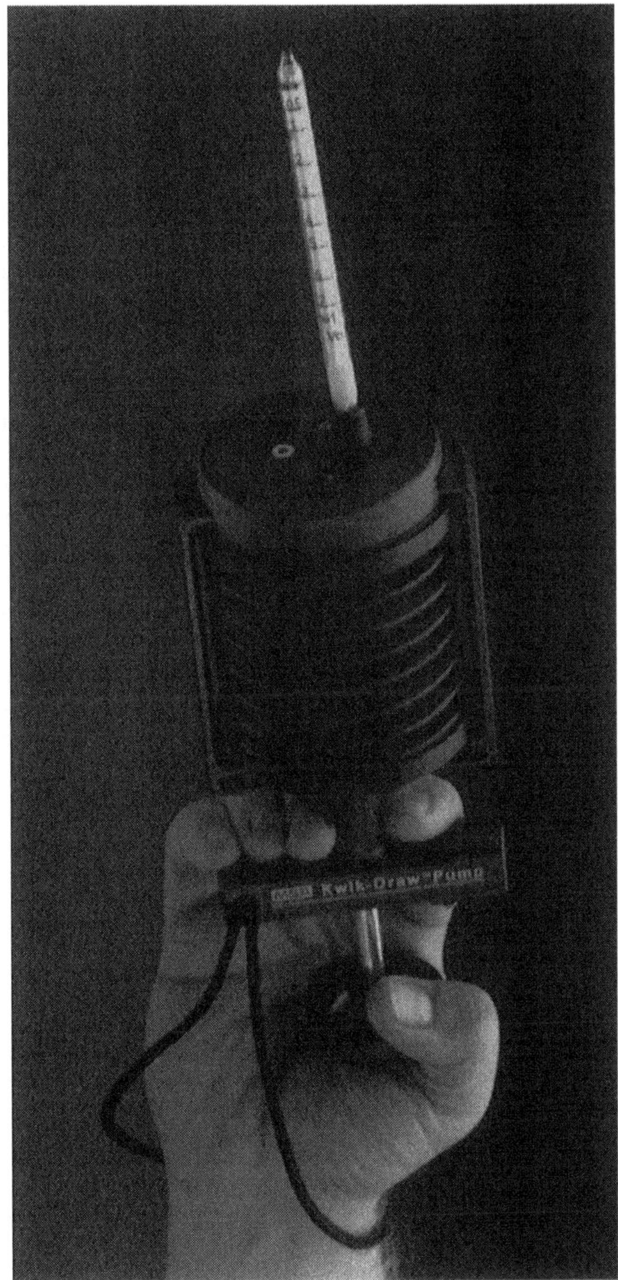

FIGURE 10.12 Colorimetric tube-sampling pump. (Courtesy of MSA, Inc., Pittsburgh, PA.)

Paper Tape and Metal Foil Devices

A variation on the colorimetric badge is the paper tape. Colorimetric reagents are impregnated into a substrate that is wound into a roll. The tape slowly unwinds and passes through a sensing chamber, where it is exposed to the atmosphere being sampled. The instrument includes a precision analyzer that detects and quantifies the color change. Depending on the contaminant, paper tape analyzers can detect accurately into the part-per-billion range.

Paper tape monitors most commonly are used in permanently installed systems. The most common application for field portable instruments is detection of isocyanates and other contaminants that have extremely low exposure limits.

A conceptually similar device uses a gold-film coil. Resistance changes in the presence of mercury or hydrogen sulfide. The increase in resistance provides a highly accurate reading on a digital meter.

The chief drawbacks of paper tape and other cassette-type monitors is the comparative delicacy of the electronics, large size, and overall expensiveness of the systems. For these reasons, they have limited utility in evaluating hazardous conditions in confined spaces. On the other hand, they provide an option where a specific contaminant is present and no other detection technique is available.

Electrochemical Sensors

Electrochemical sensors are available for monitoring a wide variety of toxic contaminants encountered in confined spaces during preentry preparations and work activity. They are used in instruments from many manufacturers. Substance-specific electrochemical sensors are available for a limited number of contaminants (less than 20). Ideally, they show little or no response to interferents. Broad-range electrochemical sensors are designed to provide a measurable response to a wide array of contaminants. Most electrochemical sensors fall somewhere between the two extremes in terms of specificity. Electrochemical sensors are compact, require little power, exhibit excellent linearity and repeatability, and generally have long life span.

Multisensor confined space monitors generally contain an oxygen sensor, a flammable/combustible sensor, and one to three additional electrochemical sensors. Single-sensor instruments equipped with electrochemical toxic sensors are also available for use in situations where a single toxic hazard is present.

Substance-specific electrochemical sensors contain the following parts:

- A diffusion barrier that is porous to gas but nonporous to liquid
- Reservoir of acid electrolyte (usually sulfuric or phosphoric acid)
- Sensing electrode
- Counter electrode
- Reference electrode (in three electrode designs)
- Some sensors additionally include filters designed to remove interferents which would otherwise react with the sensor. Filters may be inboard (integral) components of the sensor, or may be external to the sensor proper.

The sensing electrode catalyzes a specific reaction. Depending on the sensor, the substance being measured is either oxidized or reduced at the surface of the sensing electrode. This reaction causes the potential of the sensing electrode to rise or fall relative to that of the counter electrode. The two-electrode detection principle presupposes that the potential of the counter electrode remains constant. In reality, reactions at the surface of each electrode cause them to polarize. This significantly limits the concentration of contaminant they can measure (Moseley et al. 1991).

In three electrode designs, what actually is measured is the difference between the sensing electrode and reference electrode. Since the reference electrode is shielded from any reaction, it maintains at constant potential. This provides a true point of comparison. In this design, the change in potential of the sensing electrode is due solely to the concentration of the reactant gas or vapor (City Technology 1997).

The current generated by the sensor is proportional to the amount of reactant present. The amount of current generated per ppm of reactant is constant over a wide concentration range. This consistency in output over a wide concentration range provides the exceptional linearity of electrochemical sensors. To illustrate, a City Technology 4 Series carbon monoxide sensor produces

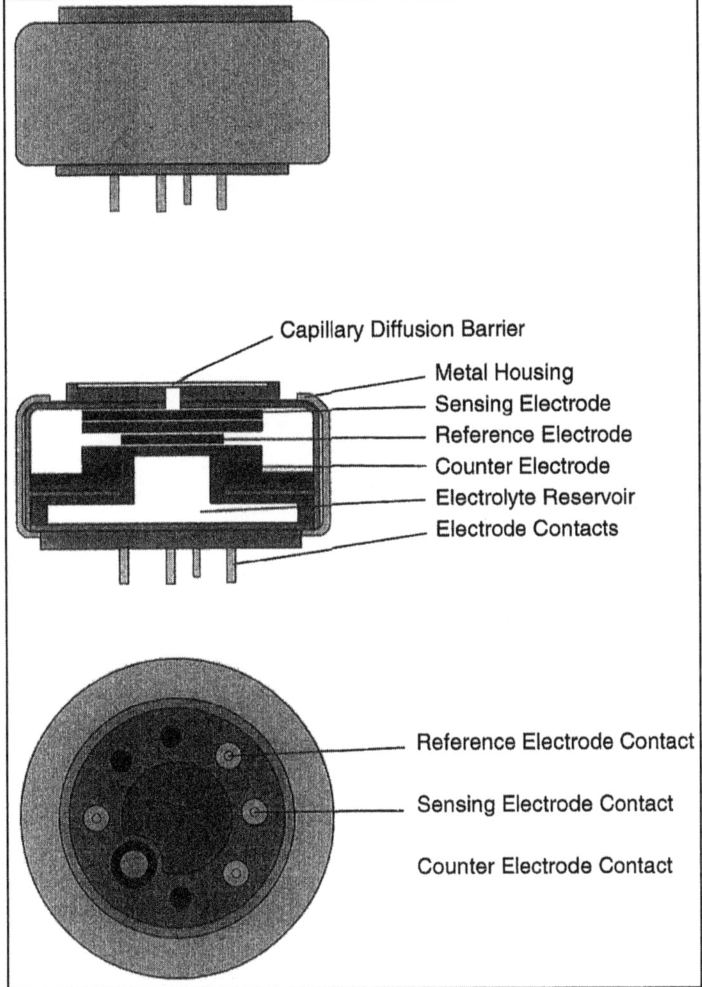

FIGURE 10.13 City Technology 4 Series carbon monoxide sensor. (From George, E. et al. (Eds.), *Corporate Health and Safety, Managing Environmental Issues in the Workplace.* Copyright 1996 by Ergonomics, Inc., Southampton, PA. With permission.)

an output signal of 0.07 ± 0.015 μA/ppm over a nominal range of 0 to 500 ppm (Figures 10.13 and 10.14). Anywhere within this range the output will be highly linear. The sensor is actually capable of discontinuous usage to concentrations as high as 1,500 ppm with only minor loss of linearity.

The oxidation of carbon monoxide in an electrochemical sensor provides a good example of the detection mechanism used in a nonconsuming electrochemical sensor design. Here carbon monoxide is oxidized at the sensing electrode:

$$CO + H_2O \rightarrow CO_2 + 2H^+ + 2e^- \tag{10.8}$$

The counter electrode balances out the reaction at the sensing electrode by reducing oxygen from the air to water:

$$1/2\,O_2 + 2H^+ + 2e^- \rightarrow H_2O \tag{10.9}$$

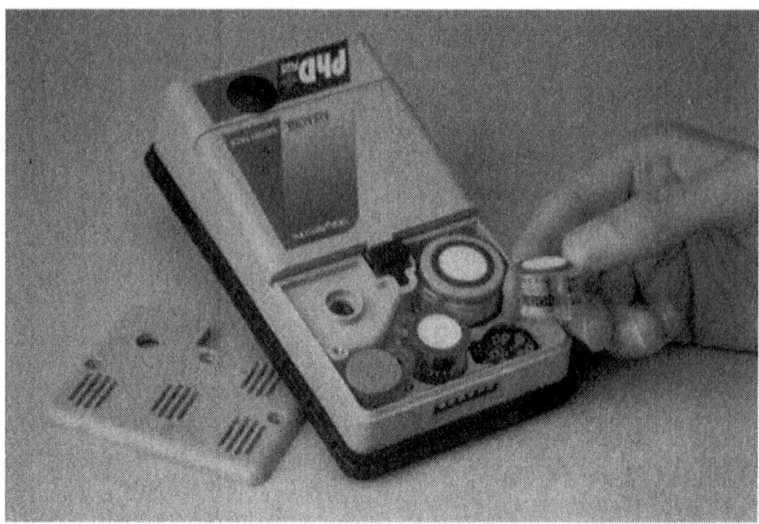

FIGURE 10.14 Portable direct reading instrument with sensor compartment cover removed, showing oxygen, electrochemical toxic, and catalytic bead (pellistor) type sensors installed. (Courtesy of Biosystems, Inc., Middletown, CT.)

The only materials consumed during the detection reaction are the molecule of carbon monoxide, power from the battery of the instrument, and oxygen. This is the reason that nonconsuming electrochemical sensors have such long life spans. The life span of the sensor is not affected by exposure to the contaminant that it measures. The sensor is a catalyst for the reaction. No part of the sensor is consumed during the detection reaction. As long as the sensor is located in an atmosphere containing even trace amounts of oxygen and moisture, the sensor will be able to replenish itself directly from the atmosphere.

Similar nonconsuming reactions are utilized in the electrochemical detection of a variety of reactant gases, including chlorine, ethylene oxide, hydrogen, hydrogen sulfide, nitrogen dioxide, ozone, phosphine, and sulfur dioxide.

Some operating environments render electrochemical sensors of this design impossible to use. For instance, a nonconsuming electrochemical hydrogen sulfide sensor would not normally be usable for monitoring for H_2S in a natural gas pipeline containing no oxygen. Once all of the oxygen available in the electrolyte was consumed, the sensor would then fail to respond to hydrogen sulfide. When re-exposed to an oxygen-containing atmosphere, the sensor would regain its ability to detect H_2S.

The fact that the electrolyte provides a supply of oxygen means that for short periods, nonconsuming sensors can detect the contaminant they are designed to measure, even in the absence of oxygen. This is fortuitous, since many reactive gases (such as chlorine) have very short shelf-lives when packaged in calibration mixtures that include oxygen. Gas mixtures used to calibrate sensors for highly reactive gases, such as chlorine, usually contain no oxygen. For example, a typical calibration gas mixture used to calibrate a chlorine sensor contains 5 ppm of chlorine in nitrogen.

A bias voltage is sometimes applied to the counter electrode. This helps drive the detection reaction for a specific contaminant. Biased sensor designs enable detection of less electrochemically active gases such as hydrogen chloride and nitric oxide.

Several other contaminants (such as ammonia and hydrogen cyanide) are detectable by less straightforward reactions that consume parts of the sensor. For example, the gold-sensing electrode of the hydrogen cyanide sensor is consumed during the reaction:

$$2HCN + Au \rightarrow HAu(CN)_2 + H^+ + e^- \tag{10.10}$$

In the case of the ammonia sensor, the electrolyte is consumed. The life span of the ammonia sensor is directly related to its exposure to NH_3. A City Technology 7AM ammonia sensor has a life span of 1 year when continuously exposed to 2 ppm of ammonia. (This is sometimes expressed as a "2-ppm-year" life span.) While the sensor will last a full year in 2 ppm ammonia, it will last only 6 months when exposed to 4 ppm or 3 months at 8 ppm. This type of sensor should be used only when the normal ambient background concentration of ammonia is sufficiently low to allow a reasonable operational life. For example, this type of sensor should not be used at a poultry farm or nitrate fertilizer plant where ambient concentrations of ammonia may be as high as 20 to 30 ppm.

Electrochemical sensors are stable, long-lasting, require very little power, and are capable of resolution (depending on the sensor and contaminant) to ± 0.1 ppm. Electrochemical sensors are normally usable over a wide range of temperature, in some cases from –40 to 50°C (–40 to 120°F). However, the uncorrected sensor output may be strongly influenced by changes in temperature. For this reason these instruments generally include temperature compensating software and/or hardware.

The chief limitation of electrochemical sensors is the effect of interferents. Substance-specific sensors are designed to respond only to the gases they are supposed to measure. The higher the specificity of the sensor, the less likely that the sensor will be affected by other gases. For example, a substance-specific carbon monoxide sensor is deliberately designed not to respond to other gases, such as hydrogen sulfide or methane. Electrochemical sensors frequently include an external filter. The size and composition of the filter are determined by the type and expected concentration of the interferents to be removed.

While inclusion of a filter is frequently able to increase specificity, removal of a filter may be used to broaden response to a wider variety of gases. To illustrate, the primary difference between a substance-specific carbon monoxide sensor and one capable of detecting both carbon monoxide and hydrogen sulfide is the external filter. The external filter removes hydrogen sulfide from the atmosphere that diffuses into the sensor. Carbon monoxide sensors that do not include an external filter are frequently marketed as "dual purpose" sensors for the simultaneous detection of both CO and H_2S.

Dual purpose CO/H_2S sensors can be calibrated for either gas. A properly calibrated sensor will be exactly accurate (within the design specification of the instrument) for the gas to which it is calibrated. If the sensor is calibrated to carbon monoxide and then exposed to 35 ppm of carbon monoxide, the reading will be 35 ppm. The sensor will show a relative response to other gases. When calibrated on carbon monoxide, the relative response of the sensor to hydrogen sulfide is about 3.5:1.0. This means that a concentration of about 10 ppm of hydrogen sulfide would produce a reading of 10×3.5 or 35 ppm. This is a very convenient relative response, since many jurisdictions use 10 ppm as a TWA for hydrogen sulfide and 35 ppm for carbon monoxide. This means that the instrument will alarm any time the concentration of either carbon monoxide or hydrogen sulfide exceeds this threshold.

Since the sensor responds to both gases simultaneously, the user cannot determine which is producing the reading. That means the user cannot determine which hazard is present or at what specific concentrations. On the other hand, the instrument will provide an immediate indication whenever concentrations exceed the alarm setpoint. These sensors also respond in varying degree to a variety of other toxic contaminants, including gasoline, alcohols, hydrogen, acetylene, ethylene, toluene, nitric oxide, and sulfur dioxide. Where the atmosphere contains a complex mixture of contaminants, this type of sensor could lead to constant alarms that have no meaning for a low setpoint. These sensors cannot discriminate between hazards or tell them apart when more than one contaminant is present at the same time.

A new variant on the dual-purpose carbon monoxide/hydrogen sulfide electrochemical sensor is the 4-electrode "COSH" design. These sensors contain a second sensing electrode. Each sensing electrode provides an independent substance-specific signal to the microprocessor and can be individually calibrated.

When a contaminant is suspected or known to be present, the best approach is to use an instrument containing a substance-specific sensor. Regulations generally support this concept. The

TABLE 10.7
Interferent Effect on Sensor Output

Interfering gas	Effect on the output of a reducing gas sensor (for instance, effect on a hydrogen sulfide sensor)	Effect on the output of an oxidizing gas sensor (for instance, effect on a chlorine sensor)
Reducing gas	Increase output	Decrease output
Oxidizing gas	Decrease output	Increase output

OSHA Standard for general industry explicitly requires the use of a direct-reading instrument containing substance-specific sensors whenever a particular toxic hazard is known to be present (OSHA 1993). For instance, when chlorine is known or suspected to be present, one of the toxic sensors selected should be chlorine-specific and calibrated directly with this gas. A broad-range sensor unable to discriminate between chlorine and other contaminants should not be used for this purpose.

Even though care has been taken to reduce cross-sensitivity in substance-specific designs, interferences still exist. In some cases, the interfering effect is positive; this results in readings that are higher than actual. In other cases the interference is negative; this produces readings that are lower than actual. Reducing gases, such as hydrogen sulfide and carbon monoxide, are oxidized at the sensing electrode. Oxidizing gases, such as chlorine, chlorine dioxide, nitrogen dioxide, and ozone, are reduced at the sensing electrode. The effect of the interfering gas (increase or decrease in output signal) depends on the gas to be measured (reducing or oxidizing) and the interferent (reducing or oxidizing).

Table 10.7 summarizes these relationships.

Table 10.8 lists the interferences of electrochemical toxic sensors used in a typical instrument. Depending on the nature of the reaction with the sensor, the interferent either increases (positive cross-sensitivity) or decreases the signal (negative cross-sensitivity). Each entry represents the response of the sensor to 100 ppm of interferent, thus providing a percentage sensitivity.

The user must understand clearly the effects of potential interferents on the output from sensors used to monitor the atmosphere during preentry preparation and work activities involving confined spaces. To illustrate, the chlorine sensor listed in Table 10.8 will show a reading of approximately –0.3 ppm when exposed to 10 ppm of H_2S. A 10-ppm concentration of H_2S could "cancel out" a 0.3-ppm concentration of chlorine. This situation could occur at pulp and paper mills.

Metal Oxide Semiconductor Sensors

MOS sensors may be used for monitoring at toxic levels in the ppm range, as well as flammable/combustibles in the percent range. As discussed previously, MOS-sensing elements consist of a metal oxide semiconductor, such as tin dioxide (SnO_2), on a sintered alumina ceramic bead contained within a flame arrestor (Figure 10.15). In clean air the electrical conductivity is low. Contact with reducing gases, such as carbon monoxide or combustible gases, increases conductivity. Sensitivity to a particular gas is alterable by changing the temperature of the sensing element.

MOS sensors are broad-range devices. They are designed to respond to the widest possible range of toxic and flammable/combustible gases and vapors. These include vapors from halogenated solvents and other contaminants that are difficult to detect by other means. This nonspecificity can be advantageous in situations where unknown toxic gases may be present, and a simple go/no go determination is sufficient.

Sensitivity of the sensing element to a particular gas is mathematically predictable. A commonly used strategy is to preprogram the instrument with a number of theoretical specific response curves. If the exact nature of the contaminant is known, readings of the sensor can be adjusted to reflect

TABLE 10.8
Cross Sensitivity of Selected City Technology Brand "4 Series" Toxic Sensors When Installed in a Biosystems PhD Ultra Confined Space Gas Detector

Interfering gas	Sensor type						
	Carbon monoxide (CO)	Hydrogen sulfide (H_2S)	CO Plus (Cal to CO)	CO Plus (Cal to H_2S)	Sulfur dioxide (SO_2)	Chlorine	Hydrogen cyanide
CO	100	<10	100	25	0	0	<5
H_2S	<10	100	~ 350	100	0	~ −33	~ 1,100
SO_2	<10	~ 20	~ 65	~ 15	100	0	~ 395
NO	<30	<0	~ 25	~ 6	0	0	0
NO_2	<15	~ −20	~ −60	- 15	~ −120	120	~ −120
Cl_2	<10	~ −20	~ −40	~ −10	<5	100	~ −140
H_2	<60	<5	<60	<15	0	0	0
HCN	<15	0	~ 40	~ 10	<50	0	100
HCl	<3	0	~ 5	~ 1	0	0	~ 65
NH_3	0	0			0	0	0
Ethylene	~ 50	0			0	0	0

Courtesy of Biosystems, Inc., Middletown, CT.

the expected sensitivity of the sensor to the contaminant being measured. MOS sensors offer the ability to detect low (0 to 100 ppm) concentrations of toxic gases over a wide temperature range.

The chief limitations of this kind of sensor are the difficulty in interpreting readings, the potential for false alarms, and the effects of humidity on the sensor. As humidity increases, sensor output increases. As humidity decreases to very low levels, sensor output may fall to zero, even in the presence of a contaminant. In addition, the user must exercise caution in assumptions about contaminants presumed to be present. The user may erroneously use the response curve for a contaminant that is highly detectable by the sensor, yet is not present. The instrument may have low sensitivity to the contaminant that is actually present. The result would be erroneously low readings.

Another issue affecting the use of MOS sensors is the relatively narrow linear response to many common contaminants. As long as the concentration of the contaminant lies within the linear range of the sensor, the output is sufficiently accurate to allow for quantifiable readings. If the concentration lies outside the linear range of the sensor, the effective usage of the sensor is limited to a "yes/no" indication about the presence of the contaminant. That is, the instrument would be unable to provide a quantifiable measurement of the actual concentration present. Some instruments limit the type of information provided by the MOS sensor to a bar graph or other semiquantified output. The MOS sensor in these instruments usually is included as a fourth or fifth channel of detection.

Ionization Detectors

The aim of initial hazard assessment should be determination of overall or total contaminant levels. Further monitoring with more sophisticated or narrowly focused analyzers then refines initial testing results. One of the most useful real-time survey techniques for surveying volatile organic contaminants (VOCs) utilizes detection using ionization. A source of energy removes an electron from neutrally charged target molecules. The electrically charged fragments are called ions. Ionization detectors collect the charged particles on charged plates. This produces a flow of electrical current proportional to the concentration of contaminant.

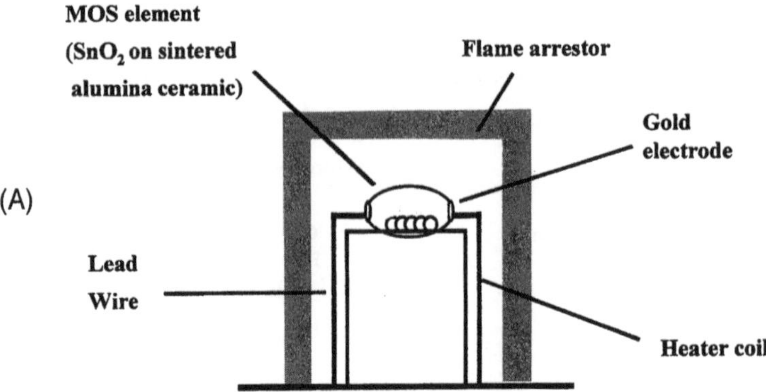

FIGURE 10.15 A. Metal oxide semiconductor (MOS) sensor. (From George, E. et al. (Eds.), *Corporate Health and Safety, Managing Environmental Issues in the Workplace.* Copyright 1996 by Ergonomics, Inc., Southampton, PA. With permission.) B. Hand-held photoionization detector. (Courtesy of MSA, Inc., Pittsburgh, PA.)

The ionization process may be written as:

$$R \rightarrow R^{+} + e^{-} \qquad (10.11)$$

The amount of energy needed to remove an electron from a particular molecule is the ionization potential (or IP). Ionization potentials are usually expressed in units of electron volts (eV).

Photoionization detectors (or PIDs) use ultraviolet light as the source of energy for ionization. Flame ionization detectors (or FIDs) use a hydrogen flame. Regardless of the source, the energy must be greater than the IP in order for an ionization detector to be able to detect a particular

TABLE 10.9
Ionization Potentials for Selected Chemical Substances

Substance	Ionization potential (eV)
Carbon monoxide	14.01
Carbon dioxide	13.77
Hydrogen cyanide	13.60
Methane	12.98
Hydrogen chloride	12.74
Water	12.59
Oxygen	12.08
Chlorine	11.48
Propane	11.07
Hydrogen sulfide	10.46
n-Hexane	10.18
Ammonia	10.18
Acetone	9.69
Benzene	9.24

substance. Ionization detectors are nonspecific, and therefore, should always be used in concert with other methods of detection. This is especially important when contaminants not detected by these units are potentially present. Table 10.9 lists the ionization potentials for selected chemical substances.

The fact that water, oxygen, carbon dioxide, and other gases naturally present in the atmosphere can be ionized puts an upper limit on the useful energetics of the source of energy. The source therefore must not ionize substances in the normal atmosphere as well as the contaminants under investigation.

Instruments containing PID and FID detectors are complementary, rather than competitive. The type of detector chosen should reflect the contaminant(s) being measured. Since neither technique is equally good at detecting all ionizable contaminants, prudence would stress the use of both types of instruments. Several manufacturers now offer dual PID/FID analyzers in the same instrument.

Ionization detectors are nonspecific. They provide a broad-range indication about all detectable molecules in the atmosphere being monitored. Also, the response is always relative to the substance used during calibration. The only time that the instrument actually reads true is when the calibrant and the substance being monitored are the same. Otherwise, readings may be higher or lower than the true concentration. This can be especially confusing when several different detectable molecular species are simultaneously present. Because of differing sensitivity to each, monitoring to establish total VOCs could produce a meaningless result. Adding further to the potential for confusion is the lack of linearity of ionization detectors over their detection ranges. These limitations lead to the bottom-line requirement to treat readings from ionization detectors as qualitative, despite the ability of these instruments to operate in the low part-per-million or even part-per-billion range. Results indicating the presence of contaminants should be elucidated further through the use of more precise analytical techniques.

Ionization detectors can only detect certain gases and vapors. Neither type of detector used alone or together is capable of detecting all potential contaminants. Nonvolatile liquids and solids, particulates, and many toxic gases and vapors cannot be detected at all. Because of the limitations of ionization detectors, a reading of zero should not be interpreted as a guarantee about the absence of contamination.

Photoionization Detectors (PID)

Photoionization detectors use high energy ultraviolet (UV) light from a lamp housed within the detector to provide the energy for ionization. The energy provided by the lamp is determined by the wavelength of the UV light being produced, and is measured in electron volts (eV). Several different lamps are available for use. The choice of a lamp governs the type of compound that will be detectable. Absorption of a photon of ultraviolet radiation energetic enough to ionize a molecule (RH) initiates the process of photoionization.

$$RH + h\upsilon \rightarrow RH^{+} + e^{-} \tag{10.12}$$

The quantity, $h\upsilon$, represents a photon having an energy greater than or equal to the ionization potential of species RH. In general, the smaller the molecule, the tighter the electrons are bound and the higher the ionization potential. The larger the molecule and the more double or triple bonds, the lower the IP. The ions are collected in an ionization chamber that is adjacent to the lamp. The ion chamber contains an accelerating electrode (biased positively) and a collecting electrode where the current is measured. The current measured (after amplification) is proportional to concentration.

The energy of the photons produced by the UV lamp determines whether a specific chemical is detectable. The energy must be higher than the ionization potential of the contaminant in order for detection to occur. Lamps are available in a number of output energies, including 9.5, 10.0, 10.2, 10.6, 11.7, and 11.8 eV (depending on manufacturer). Most manufacturers allow for the use of several lamps in the same detector. The lower the energy of the lamp, the lower the number of chemicals to which the detector will respond. The higher the energy of the lamp, the wider the range of detectable contaminants.

Higher energy lamps are subject to more physical limitations. In general, the higher the lamp energy, the shorter will be the service life. PID lamps generally consist of a glass body filled with an elemental gas such as oxygen, nitrogen, hydrogen, or argon at low pressure. Electric current or radio waves are used to excite the gas to produce the UV light. The UV light is focused in a tight beam directed through a small window in the body of the lamp. The highest energy lamps (11.7 and 11.8 eV) contain a window made from lithium fluoride. Lithium fluoride is easily degraded by the absorption of water vapor as well as exposure to UV light produced by the lamp. In addition, although the energy of the UV produced by high capacity lamps allows the ionization of a wider range of substances, the amount is less than that produced by lamps of lower capacity. This means that the higher capacity lamps produce both a weaker ionization current and have an increased tendency towards drift. The materials used in the windows of 9.5, 10.0, 10.2, and 10.6 eV lamps (magnesium fluoride or calcium fluoride) do not tend to absorb water and are not degraded by exposure to UV. As a result, they generally have much longer service lives.

PID instruments are nonspecific. As a result, equivalent concentrations of gases other than the one used to calibrate the instrument may not produce equivalent readings. PID readings are always relative to the gas that was used to calibrate the detector. Calibration is based on ion current sensed at the detector in response to a known concentration of a known gas or vapor. Instrument response to gases other than the one used for calibration are relative in nature. A reading of 10 ppm only indicates that the ion current experienced by the detector is equivalent to that produced by a 10-ppm concentration of calibrant. The amount of contaminant needed to produce this current may be larger or smaller than the concentration of the gas used to calibrate the instrument. Since PID readings are always relative to the calibrant, they should be recorded as ppm-calibration gas equivalent units, or PID units, never as true concentrations.

Most manufacturers furnish tables containing correction or "sensitivity" factors for normalizing readings when the contaminant being measured is known. Many PIDs provide the ability to key-in the sensitivity factors. This provides real-time readings for the contaminant being measured. Normalized readings still are subject to the limitations discussed above. Concentrations based on calculations using relative response ratios deserve caution. In actual practice, the relative response varies somewhat from one detector to another, and may also be affected by temperature and humidity. Use of the incorrect correction factor could lead to serious errors in assessment of conditions.

Photoionization detectors are affected by a number of environmental factors. The most serious is humidity. Although water vapor (IP = 12.59 eV) is not ionized by PID detector lamps, water vapor does serve to deflect, scatter, and absorb UV light in the ionization chamber. High concentrations of measurable contaminants will produce a reading, even in humid conditions. High humidity may reduce or even block the ability of the instrument to register low concentrations of contaminant. Nonionizing gases and vapors (contaminants with IPs higher than the capacity of the detector lamp) also scatter or block UV light in much the same manner as water vapor. Nonionizing contaminants such as methane can profoundly reduce the sensitivity of the detector. One study reported that 0.5% by volume (or 10% LEL) of methane reduced the sensitivity of the PID detector to toluene by 30%.

Dust and particulates can reduce sensitivity of the detector, similar to water vapor and methane. In addition, deposition of particulates on the lamp window reduces energy transmitted to the sample. Condensation of water vapor or other hot gases and vapors on the window has a similar effect. For this reason, the user should clean the PID lamp regularly. Many manufacturers recommend cleaning the lamp before every use. In the past, photoionization detectors have tended to be bulky, temperamental, and expensive. This has significantly limited their use during work involving confined space. This has changed dramatically over the last few years. The photoionization instruments available now are highly compact, hand-held units.

Flame Ionization Detectors (FID)

Flame ionization detectors use a hydrogen flame as the source of ionization energy. These instruments are able to detect nearly all organic compounds. In clean air, the hydrogen flame is free from ions and nonconducting. Organic contaminants containing either a carbon–hydrogen or carbon–carbon bond break down, according to the following reaction, when exposed to the hydrogen flame in the ionization chamber:

$$RH + O \rightarrow RHO^+ + e^- + CO_2 + H_2O \tag{10.13}$$

Positively charged carbon-containing ions are collected on a negatively charged plate. The ion current is proportional to the hydrocarbon concentration.

A major difference between FID and PID detection is the amount of variance in sensitivity from one organic substance to another. The amount of energy (or ionization potential) necessary to dislodge an electron to create a charged fragment determines the sensitivity of a PID detector to a specific molecule. The shape, size, and specific chemical bonds present within the molecule determine the IP. IP varies widely from one substance to the next. For these reasons, the sensitivity of a PID detector also varies widely from one substance to the next.

FID sensitivity does not vary as significantly from one organic substance to another because the amount of energy necessary to break specific carbon–carbon or carbon–hydrogen bonds is relatively constant. For this reason, FID sensitivity is more generalized, and varies much less significantly between one hydrocarbon and another. Another very important difference between the two detection techniques is that FIDs can detect methane, while PIDs usually cannot. FIDs are generally the preferred instrument for detection of methane and other saturated alkanes, as well as unsaturated hydrocarbons and alkenes at part-per-million levels. Hydroxyl (^-OH) or chloride (^-Cl) functional groups tend to reduce sensitivity, while inorganic contaminants such as chlorine, ammonia, or hydrogen cyanide are not detected.

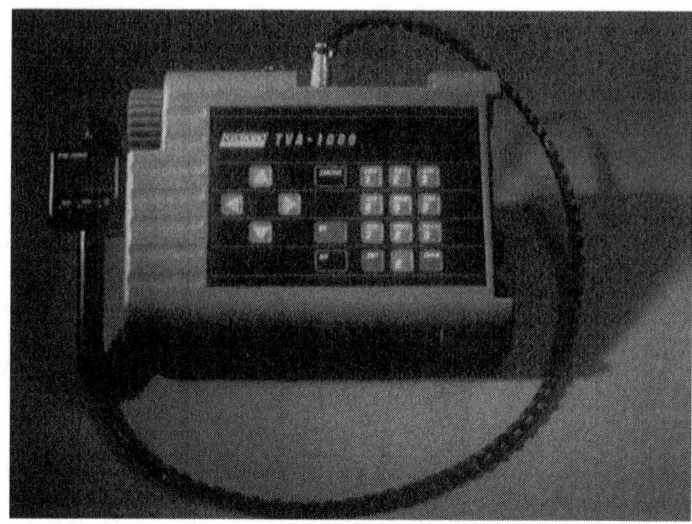

FIGURE 10.16 Combination PID/FID organic vapor analyzer. (Courtesy of Foxboro, Inc., Foxboro, MA.)

One of the most important considerations concerning use of an FID is the high pressure (16 MPa or 2,300 lb/in.^{2}) cylinder of hydrogen required by the instrument. Both refilling and storage may represent a significant issue to the user. An additional consequence of the use of a live hydrogen flame is that some currently available FID detectors are not classified as to intrinsic safety. They cannot be operated in a hazardous location that may be subject to accumulation of flammable/combustible gas and vapor. Ensure that any FID that is to be operated in a potentially hazardous location has been classified as to intrinsic safety!

In the past, FIDs (like PIDs) have tended to be bulky, temperamental, and expensive. This has changed dramatically over the last few years. Today's FIDs by comparison are much less bulky, and substantially easier to operate. In addition, the pairing of FID and PID technology in the same instrument provides a capability that is not easily duplicated by other monitoring techniques (Figure 10.16). While the primary use of this technology continues to be centered on initial hazard assessment, there are increasing circumstances where use of an FID or combination PID/FID during routine monitoring makes sense (Figure 10.17).

Infrared Detectors

In the past, infrared-based instruments tended to be bulky and expensive spectrophotometers. They required a very high level of operator expertise to obtain accurate readings. Incorporation of modern electronics and signal processing capabilities into infrared-based equipment has significantly altered the reality behind these perceptions. Fixed-path length, substance-specific instruments are becoming available for an ever-widening variety of contaminants. Instruments are available for carbon dioxide, Freons®, ammonia, and methane, as well as for low-range hydrocarbon detection in the ppm range.

Molecules may be conceptualized as balls (atoms) held together by flexible springs (bonds) that can vibrate (stretch, bend, or rotate) in three dimensions. Each molecule has certain fixed modes in which this vibratory motion can occur. Vibrational modes are dictated by the nature of the specific bonds that hold the molecule together. The larger the molecule, the greater the number of modes of movement. Each mode represents vibrational motion at a specific frequency. The modes are always the same for a specific molecule. Chemical bonds absorb infrared radiation. The bond continues to vibrate at the same frequency but with greater amplitude after the transfer of energy. For infrared energy to be absorbed (that is, for vibrational energy to be transferred to the molecule), the frequency must match the frequency of the mode of vibration.

FIGURE 10.17 Combination PID/FID organic vapor analyzer in field use. (Courtesy of Foxboro, Inc., Foxboro, MA.)

Specific molecules absorb infrared radiation at precise frequencies. When infrared radiation passes through a sensing chamber containing a specific contaminant, only those frequencies that match one of the vibration modes are absorbed. The rest of the light is transmitted through the chamber without hindrance. The presence of a particular chemical group within a molecule thus gives rise to characteristic absorption bands. Since most chemical compounds absorb at a number of different frequencies, IR absorbance can provide a "fingerprint" for use in identification of unknown contaminants.

There are two methods of presenting the data: transmittance or absorbance. The two terms are related by the following equation:

$$\text{Absorbance} = \log\left(1/\text{transmittance}\right) \quad (10.14)$$

The thermopile detector in a traditional IR spectrophotometer actually measures transmittance. These data are usually substituted into the above equation for presentation as absorbance. (Any error in the transmittance values substituted into this equation is magnified in the calculation of absorbance.)

Infrared absorbance spectra can provide both qualitative and quantitative information. The position (frequencies) and relative strengths of the absorbance maxima provide a qualitative indication about the nature of the contaminants. Peak areas and peak height of the absorbance maxima are proportional to the concentration of the absorbing molecules. When the analyzer is calibrated to a known concentration of the contaminant to be measured, these quantities provide the basis for determining concentration.

Interpretation of infrared absorbance spectra can be complex when multiple contaminants are present. Spectral bands may overlap or obscure each other. Water and carbon dioxide molecules in ambient air also absorb infrared radiation. This absorption can obscure what otherwise might be useful absorbance maxima for a given compound. Only polar molecules are able to absorb infrared radiation. This is a function of the net change in electric dipole moment during vibrational motion. Monatomic (single-atom) gases such as helium (He), argon (Ar), and mercury vapor, as well as nonpolar diatomic (two-atom) gases such as oxygen (O_2), nitrogen (N_2), and chlorine (Cl_2) do not absorb infrared light and are not detectable by this means.

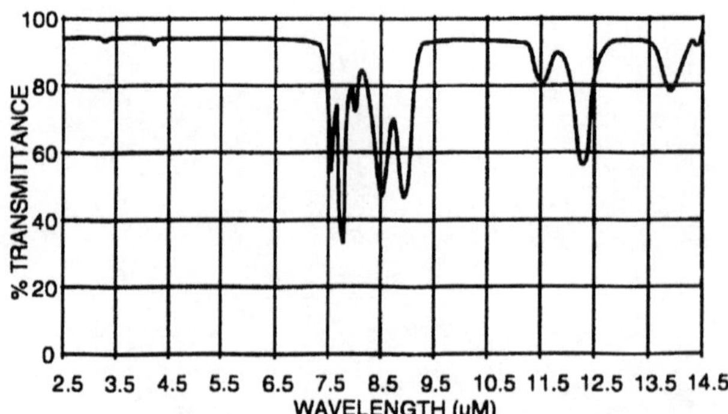

FIGURE 10.18 Infrared absorbance of halothane. Halothane is 2-bromo-2-chloro-1,1,1-trifluoroethane, a common anesthetic. (Courtesy of Foxboro, Inc., Foxboro, MA.)

Typical applications include monitoring for the presence of chlorofluorocarbons, fumigants such as ethylene dibromide and Vikane®, anesthetic gases such as nitrous oxide (N_2O) and halothane (Figure 10.18), sterilants such as ethylene oxide, formaldehyde, hydrogen peroxide, halogenated solvents such as methylene chloride, organic solvents such as toluene, carbon dioxide, carbon monoxide, as well as many other organic compounds (Figure 10.19).

IR Spectrophotometers

Field IR spectrophotometers contain a broad-range (2.5 to 14.5 µm) IR source and a variable path-length gas cell. The longer the optical pathlength, the more sensitive the response. Varying the path-length provides a dynamic monitoring range from sub part-per-million to percent concentrations. The air sample is drawn into the sensing chamber by a pump, where it is exposed to the infrared radiation. A solid state (thermopile) detector measures absorbance. Depending on the model, absorbance data are printed by an internal printer or downloaded to an external printer, datalogger, or personal computer.

IR spectrophotometers can scan the sample. This exposes the sample to the full range of infrared source to determine where absorbances occur. If the contaminant is known, the absorbance frequency is used. A number of models facilitate qualitative identification by providing a library of absorbance spectra in the instrument memory. The software compares these to field results to provide "best fit" identification of unknown contaminants. Field IR spectrophotometers are most useful where the contaminant is suspected or known to be present (Figure 10.20; Figure 10.21). The less that is known about contaminant that may be present, the more challenging the interpretation of results becomes.

Substance-Specific Infrared Instruments

Substance-specific instruments filter the IR source to provide a narrow range of frequencies. Dedicated substance-specific detectors exist for carbon dioxide, halogenated hydrocarbons such as Freons®, methane, and other gases with good absorbance characteristics. Substance-specific IR detectors cannot be used as a "broad range" survey device. The greatest strength of these instruments is their ability to monitor relatively nonreactive contaminants, such as carbon dioxide. IR-based carbon dioxide detectors have been particularly useful for investigating building-related problems. They also are useful for evaluating conditions in confined spaces. This is potentially very important, given the emerging role of CO_2 in fatal accidents. IR-based instruments also are also proving to be very useful for the monitoring of methane in hostile environments. IR-based detectors are not affected by high concentrations of sulfides, silicones, or other substances that would quickly destroy a hot-bead catalytic sensor.

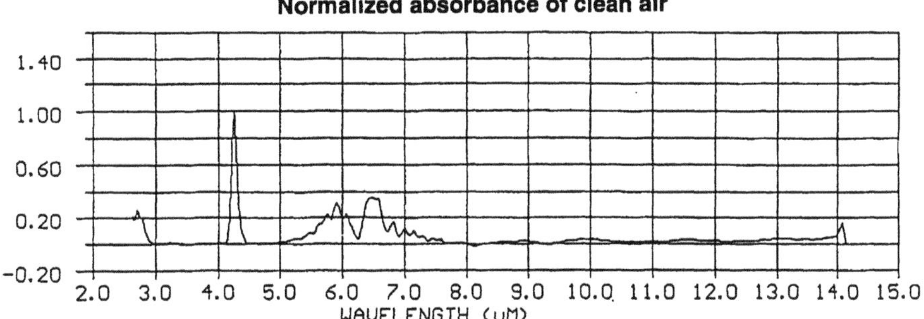

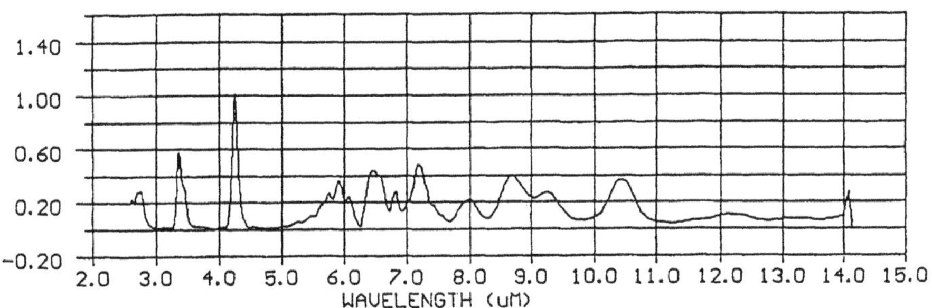

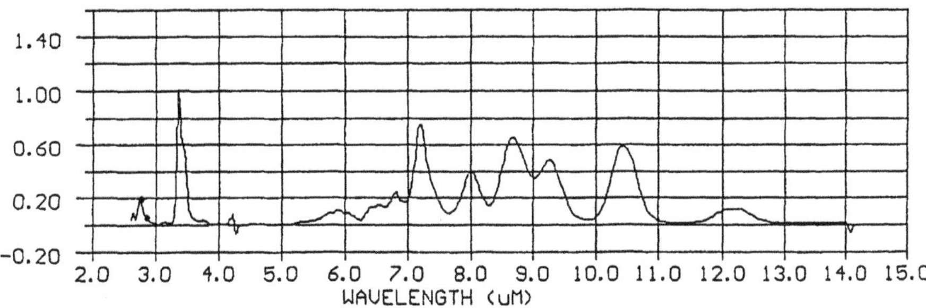

FIGURE 10.19 Infrared absorbance of isopropyl alcohol in fresh air. (Courtesy of Foxboro, Inc., Foxboro, MA.)

Photoacoustic Analyzers

The increase in vibration due to absorbance of IR energy by a molecule is a very short-lived effect. The molecule quickly transfers the energy in the form of heat to adjacent molecules. (The effect is similar to the way a microwave oven heats food. Water molecules in the food absorb microwave radiation, then quickly transfer the absorbed energy to nearby molecules, thus heating the food.) The transfer of absorbed energy has the effect of heating the atmosphere and, when the sensing chamber is sealed, increasing the pressure as well.

The photoacoustic analyzer measures fluctuating pressure changes due to the absorbance and transfer of IR energy using high-precision condenser microphones. The photoacoustic analyzer is accurate to the part-per-billion range. The photoacoustic analyzer measures absorbance directly, as opposed to calculation from percent transmittance. As a result, this instrument is very precise. Increasingly sophisticated software allows photoacoustic analyzers to monitor simultaneously up to 15 contaminants, and through cross-compensation analysis of interferences, to provide specific concentrations for each.

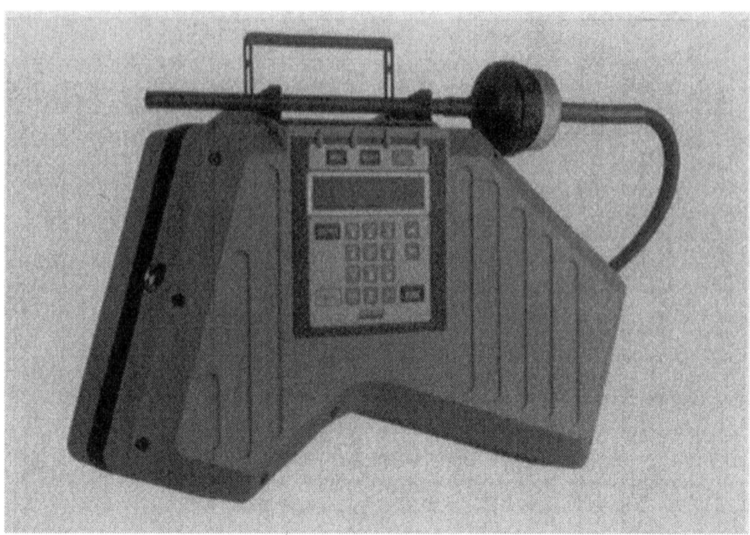

FIGURE 10.20 Field portable infrared ambient air analyzer. (Courtesy of Foxboro, Inc., Foxboro, MA.)

FIGURE 10.21 Infrared ambient air analyzer in use. (Courtesy of Foxboro, Inc., Foxboro, MA.)

Fourier Transform Infrared (FTIR) Analyzers

Field-portable Fourier transform infrared (or FTIR) analyzers represent some of the most exciting new direct-reading instruments to emerge over the last few years. FTIR analyzers compare IR absorbance in reference (fresh) air to absorbance in contaminated air using a Michaelson interferometer. Light from the IR source is split into two beams. One passes through clean reference air. The other passes through the atmosphere containing contaminants. In "monostatic" systems, a remotely located reflector returns the beam to the detector. In "bistatic" systems, both the IR source and source optics are remote from the detector. A parabolic mirror (up to 20 in. in diameter) is used to gather the returning light. The Michaelson interferometer produces an "interferogram" which (by means of Fourier transform analysis) mathematically generates the IR absorbance chart.

Unlike classic IR spectrophotometers, which use an internal pump to draw a sample from a highly localized point source through the sensing chamber, FTIR systems are capable of analysis over very long optical paths. In some cases, the reflector or IR source can be located 0.4 km (0.25 miles) or more from the detector. This makes FTIR systems ideal for assessing plumes and fugitive emissions.

Gas Chromatography

Gas chromatography describes a number of techniques used to separate mixtures of gases and vapors into specific components. All exploit the differential movement of the mixture of gases through a solid or viscous liquid material. The sample is injected into a carrier gas (the mobile phase), which "pulls" the sample through the stationary phase where separation occurs. The stationary phase is frequently called the "separation column." Different molecules in a mixture have differing affinities for the substances in the column and the carrier.

Contaminant molecules having low affinity for the stationary phase and high affinity for the carrier move through the column at the same rate as the carrier. The higher the affinity for the column material, the slower the contaminants move through it. Because contaminants move through the column at different rates, they eventually separate into discrete bands. A suitable detector positioned downstream from the column senses the emergent bands of molecules.

Columns are available in a variety of lengths and can be strung together. A variety of materials are used in the stationary phase. "Packed" columns are solidly filled with material. The columns are usually coiled and may be encapsulated for use in isothermal ovens. "Capillary" columns (0.32 or 0.53 mm ID) have a hollow core. The stationary phase is coated onto the inside wall. The type and length of column selected should be a function of the characteristics of the molecules to be separated.

Once separated, a variety of detection techniques are used to characterize and/or quantify individual components. These include PID, FID, thermal conductivity, electron capture (ECD), and argon ionization, as well as a number of others. Use of separation columns precludes use of the detector as a real-time monitor. Columns provide snapshots of concentrations in a discrete sample, rather than dynamic real-time measurement. The sample must be obtained either automatically by the analyzer, or as a grab sample that is then injected into the GC. Time is required for contaminants to move through the length of the column to the detector.

The carrier gas is a function of manufacturer preference and the nature of the detector. Common carrier gases include "ultra zero air," which contains less than ±0.1 ppm total hydrocarbon (used with PID) and argon (used with argon ionization detectors), as well as nitrogen and helium. Common types of stationary phase media include liquid paraffin, silicone oils, squalene, and apiezon greases for the separation of nonpolar molecules; alkylaryl sulfonate and dinonyl phthalate for separation of intermediate polar molecules; and dimethylsulfolane and polyethylene glycols for the separation of strongly polar molecules. The manufacturer of the detector will supply or specify the appropriate columns, media, and carrier to be used with the analyzer.

Fixed Detection Systems

Proper assessment of existing or potential atmospheric hazards is essential for management of air quality in confined spaces. There are two basic approaches to atmospheric monitoring: portable gas detectors assigned to workers who enter the affected area, or fixed detection systems. Fixed detection systems (Figure 10.22) are permanently installed in the affected area and function 24 hours per day. An approach for addressing ongoing atmospheric hazards in frequently entered confined spaces is a continuously operational fixed-point detection system. Alarms activate whenever conditions become unsafe.

Fixed systems can activate ventilation fans, security notification equipment, such as auto-dialers, as well as external alarm lights and sirens. Fixed detection systems can also be used in process control to activate or deactivate a specific process on the basis of a detected contaminant or condition.

Another important function provided by fixed systems is demonstrating compliance with regulatory requirements. Records from fixed systems can show compliance with OSHA or EPA limits,

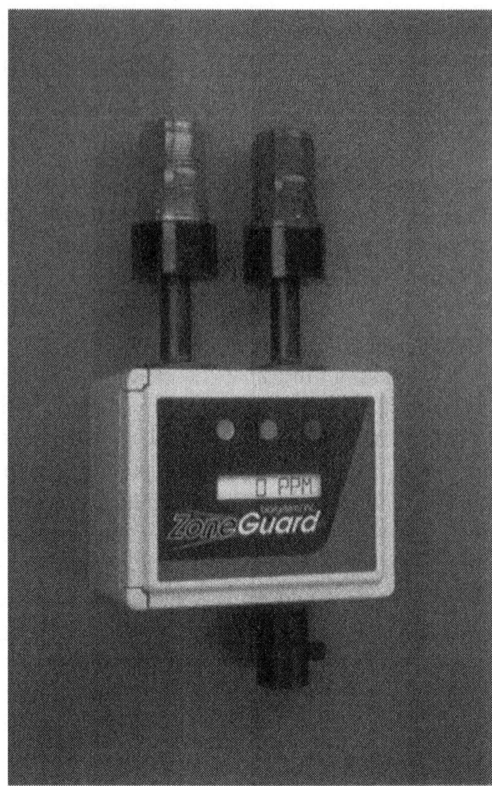

FIGURE 10.22 Single point wall-mounted (fixed) detection system for measurement of carbon monoxide. (Courtesy of Biosystems, Inc., Middletown, CT.)

for example. Although some detection techniques are more commonly used than others, systems are available with virtually every detection technology discussed in this chapter. The hazard to be measured determines the type of detector.

In many circumstances, a permanently operational or fixed-gas detection system in the confined space can provide better protection than portable instruments used on an infrequent basis to determine whether atmospheric conditions are safe for entry.

Hazard Management

Once a specific hazard has been identified and quantified, the most prudent management stratagem is to eliminate it. Regulatory limits, such as OSHA Permissible Exposure Levels, are maximum concentrations to which an unprotected worker may be exposed while on the job. Likewise, Threshold Limit Values (TLVs) provided by the American Conference of Governmental Industrial Hygienists (ACGIH) are recommendations for maximum worker exposure. Above these concentrations the air is deemed to be hazardous. Unprotected workers should not be allowed to remain in those conditions for any reason.

Sometimes the best approach involves the combined use of both fixed and portable monitors. Deciding which approach to use is not always a trivial exercise. The following questions provide a guide for making this decision.

What kind of atmospheric hazard is potentially present?

The kind of hazard profoundly influences the type of warning that is needed. The urgency to warn against rapidly acting hazards, such as hydrogen sulfide or oxygen deficiency, may be greater (depending on the concentrations involved) than that needed for slower acting agents, such as carbon monoxide. Understanding the hazards posed by specific contaminants and conditions that can occur is critical to structuring an appropriate monitoring program.

What is the source of the hazard?

This question brings more questions for which answers are needed. Is the source of the hazard readily identifiable? Is the hazard associated with the work performed in the confined space or near it? Is microbial action involved? What chemical products are used in or near the space? What industrial processes are occurring? Can sources of contaminants that are remote from the confined space present an additional risk under emergency circumstances?

Are the hazards potentially present all of the time, or only when procedures, activities, and/or products associated with the confined space entry are in use?

Some confined spaces are entered on a very frequent basis. For example, some pits in a commercial garage are routinely entered and occupied by mechanics. Hazards known to be chronically present in areas where workers routinely enter without special precautions should be monitored on a continuous basis. This environment would benefit from the installation of a permanently operational fixed-gas detection system.

What is the physical nature of the area affected?

Again, this question prompts more questions. Is the entire facility affected, or only the confined space? Are the areas in which the confined space is located out-of-doors and subject to good ventilation? Is the confined space located indoors or localized in confining areas that prevent rapid dispersal of contaminants? Fixed detection systems are ideal for 24-hour-a-day "sentry" applications.

How much time is required for workers to leave the affected area in safety?

Are the areas immediately outside the confined space congested with equipment, machinery, or other obstacles to safe evacuation? Monitoring programs should provide workers adequate time to "self rescue" from the confined space, as well as areas outside the space also potentially affected in an emergency. Workers require adequate warning before hazardous conditions become life threatening so that they have sufficient time to evacuate the affected area in safety. Fixed detection systems are frequently used to provide the alarm for evacuation of an affected area.

Must the affected area be maintained safe for continuous worker occupancy?

Is the confined space permanently secured against unauthorized entry, or is the area one which is routinely entered by workers without special precautions being taken? Atmospheric hazards in areas routinely entered on an uncontrolled basis should be monitored on a continuous basis.

What is the level of control over worker activities in the affected area?

Is entry into the confined space strictly controlled and limited to specially trained personnel? (It should be!) The lower the level of control over worker activities, the more desirable a continuously operational fixed detection system becomes.

What is the level of training of potentially affected workers?

One of the advantages of fixed detection systems is that workers entering the monitored area usually are not involved in the day-to-day operation of the system. Addition of a fixed detection system coupled with other engineering controls, such as permanently installed ventilation, may actually allow the reclassification of the confined space as an environment outside the scope of regulatory requirements for an entry. What workers entering the area must do is follow company procedures in the event an alarm sounds.

What are the trade-offs in cost?

Equipping workers individually with gas/vapor detectors can become expensive. On the other hand, a permanently installed system that provides general monitoring in an area and equivalent protection can be by far the most cost-effective approach to ensuring worker safety.

Criteria for Instrument Selection

Atmospheric testing is used to assess conditions, determine potential hazards, and establish potential exposure levels. Atmospheric monitoring is used to ensure that conditions remain safe (nonhazardous) for all potentially affected workers. A number of factors should be considered when deciding which instrument to purchase or specify for use in a particular situation:

- Atmospheric hazards
- Monitoring environment
- Sampling strategy
- Level of user sophistication
- Requirements for recordkeeping
- Other performance criteria defined by the characteristics of the job to be performed.

One thing is certain: the purchaser will face a very large number of options and choices. While the instruments most widely used in assessing conditions in confined spaces share many similarities, there are still many important differences between designs. The purchaser should consider the requirements of the entire program before buying an instrument, not after.

Most confined space instruments currently in service include a fuel-cell type oxygen sensor, a catalytic (hot bead) flammable/combustible sensor, and one or two electrochemical sensors for detecting specific toxic gases. A minority of detectors additionally include either an MOS sensor or a photoionization detector for broad-range monitoring purposes. Some instruments also incorporate an infrared sensor for carbon dioxide or methane. This section will explore some of the advantages and disadvantages associated with choices, configurations, and options available for these instruments.

Sample-Draw vs. Diffusion

In normal operation, most confined space instruments are worn on the belt, used with a shoulder strap or chest harness, or held by hand. Once turned on, these devices operate continuously until the battery is exhausted. Most instruments utilize natural air currents to bring the atmosphere being sampled to the instrument. Gases then pass to the sensors by diffusion through holes, vents, or apertures in the instrument housing or cover of the sensor compartment. Normal air movements are sufficiently energetic to bring the sample to the sensors. Most sensors react rapidly to changes in the concentration. This type of operation monitors only the atmosphere that immediately surrounds the detector.

Sample-draw kits enable diffusion-type instruments to sample from remote locations. Two types of sample-drawing kits are available. In each case, the sample is drawn in through a probe assembly and ducted through a length of hose to the instrument. One type uses a hand-operated squeeze-bulb to draw the sample through the hose. The other uses a battery-operated continuous mechanical pump (Figure 10.23). In some designs, the pump attaches to the instrument and pulls the sample from the probe assembly back to the sensors. In other designs a "pistol"-type pump located at the end of the sample hose pushes the air to the location of the sensors. Still other designs contain an integral pump that operates continuously whenever the pump is turned on. Since the pump operates continuously, this type of instrument operates only in the sample-draw mode. Each configuration has both advantages and disadvantages.

Drawbacks of Diffusion Operation

The chief drawback associated with diffusion operation is the inability to sample at locations remote from the instrument. Short of lowering the instrument on a string, there is no way to test the lowest point in a manhole while standing outside the space. A particularly dangerous misuse of diffusion-type instruments occurs when users test only at the point of entry because that is the only point that can be reached. They then base the initial decision about entry on this dangerously incomplete information.

FIGURE 10.23 When atmospheric monitors are operated in the diffusion mode, gases diffuse through holes, vents, or apertures in the housing or sensor compartment cover to reach the sensors. (Courtesy of Biosystems, Inc., Middletown, CT.)

They then enter the space while wearing or carrying the instrument and expose themselves and it to conditions further into the space. This puts the entrant in danger where conditions are life threatening, since the individual then must escape from the space after the instrument alarms. Further, conditions below the level of the instrument may go unmeasured, unless the entrant remembers to lower it. Sampling at all levels in a vertical space is critically important because of potential stratification of gases. This is the information needed before making a determination to proceed.

Fortunately, every leading manufacturer of diffusion-type instruments offers a sample-draw kit for use with their product. Availability of such a kit should be an important consideration when purchasing an instrument. Another is the type of sample-draw unit: squeeze bulb or motorized pump. Potential buyers of squeeze-bulb products should consider usage. Fatigue could become a serious issue. Also, the number of compressions must be calculated to ensure that the sensors sample an atmosphere that is representative of conditions in the space.

Generally, preentry sampling will involve use of the remote sampling kit, while monitoring after entry will utilize diffusion operation.

Drawbacks of Sample-Draw Operation

Several cautions apply to the use of sample-draw kits or other modes of sample-draw operation. The most important is leakage. Components in the sample-draw system upstream from the pump are under negative pressure. In-leakage would dilute the sample with the atmosphere entrained at the point of the leak. Under some circumstances, the atmosphere reaching the sensors is not merely diluted, but entirely replaced by leakage into the system.

Diaphragms in mechanical pumps are notorious for stiffening, deteriorating, or abrading over time or with use. A leak in a pump diaphragm can result in a unit that appears to be performing normally, based on visual observation, yet is incapable of drawing at the correct flow rate. The best way to guard against the potential for leakage is to test the sample-draw system prior to every use.

To test the integrity of manually aspirated, squeeze-bulb type sample-draw kits, attach the instrument to the sample-draw assembly, squeeze the bulb, block the sample probe inlet, and note whether the bulb remains deflated (Figure 10.24). If there are no leaks in the system, the bulb remains deflated until the blockage is removed.

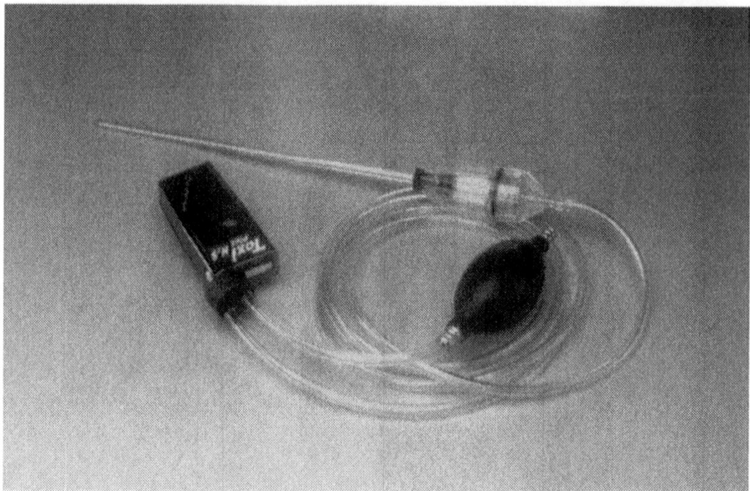

FIGURE 10.24 Hand-aspirated (squeeze bulb) sample-draw kit for a single sensor monitor. (Courtesy of Biosystems, Inc., Middletown, CT.)

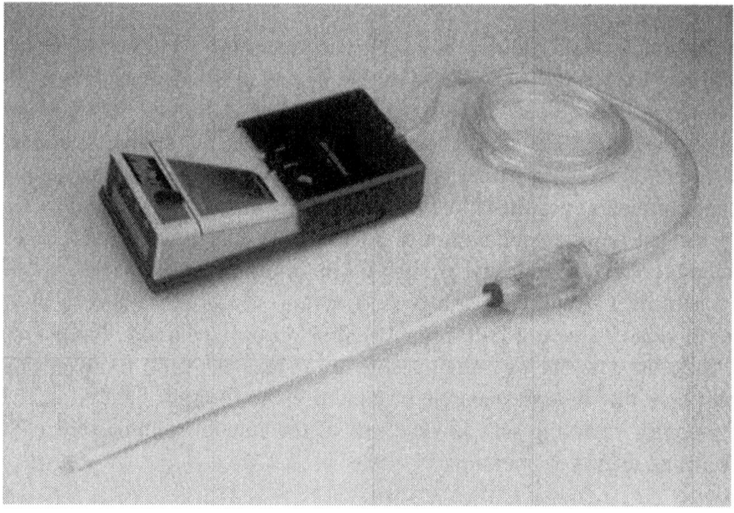

FIGURE 10.25 Motorized sample-draw pump attached to a confined space monitoring instrument. (Courtesy of Biosystems, Inc., Middletown, CT.)

Most mechanical or motorized pumps (Figure 10.25) utilized in monitoring instruments contain a low flow alarm. To test these systems, attach the pump and sample-draw assembly to the instrument and block the end of the probe with a finger. If there are no leaks in the system, the low flow alarm should activate. This method also applies to the sample draw system of instruments containing built-in sample-draw pumps.

Important safeguards in sample-draw systems are the probe assembly and filters attached to the sample line. Probe assemblies usually include filters or traps. These prevent entry of particulate and/or liquid contaminants into the system. Clean air prolongs the life of the sensors by preventing buildup of particulates on membranes. Damage to the pump can occur from particulate abrasion. Damage to the sensors or instrument electronics can result from fluids sucked into the sample-draw system.

A concern about operation of sample-draw systems is time lag. This is the time taken by the pump to move the air from the inlet of the probe to the sensors. This adds to the time required by the sensor to respond to the contaminant. Remember, response can only begin only after the sample reaches the sensors. If the T-90 (or time required for the sensor to reach 90% of final response) is 45 seconds, and the time taken for the sample to reach the sensor also is 45 s, the amount of time needed to reach T-90 output will be 90 s. The longer the sample-draw hose, the greater the amount of time needed for the sample to reach the sensors.

Time lag due to line length highlights another concern about sample-draw operation: absorptive and adsorptive losses. These result from interaction with the material of construction of the sample hose. Absorptive and adsorptive losses reflect the amount of time the sample spends in the sample hose. Highly reactive gases, such as chlorine, are nearly impossible to measure quantitatively without significant losses due to interaction with components in the sample-draw system. Although detection of gross concentrations is possible, measuring chlorine in the 0.1-ppm increments needed to assess exposure is very difficult when the sample is obtained through a sample-draw system. Diffusion sampling (when feasible) is usually the better approach for measurement of highly reactive gases. Fuel mixtures such as diesel oil or "JP-8" also have an affinity for certain types of sample-draw tubing.

Another issue is sample temperature. Confined spaces are frequently warmer than the area where the instrument is likely to be located. In the case of high-flash combustible liquids, the difference in temperature is sometimes sufficient to cause condensation of vapor into liquid in the sample hose. Since the sensors are designed to detect gases and vapors, and not liquid concentrations, condensation can sharply depress the reading.

Sensor Selection

Confined space monitors are available with space for two, three, four, five, or even six sensors. Make sure the instrument chosen for a specific application can accommodate the needed number and type of sensors. While the unit need not be equipped with all the sensors in order to function, the critical issue in the purchasing decision is future need. An instrument that lacks the capability of changing or adding additional sensors if this need is likely in the future would represent a mistake. Field configurability also is highly desirable. Adding an additional sensor should not be difficult. Most leading designs allow the user simply to plug in the new sensor, and make the setup choices necessary to let the instrument know that a change has been made. A number of instrument designs include automatic sensor recognition. This makes the process of adding or changing sensors even easier to accomplish.

One of the major issues facing the gas detector purchaser is the choice of toxic sensors. If the confined space to be monitored is characterized by the known or potential presence of a specific toxic contaminant, the best and safest approach is a substance-specific sensor. The sensor, of course, requires direct calibration by a known concentration of that substance. If the level of knowledge about hazards likely to be encountered in the space is low, a broadly responding sensor may be the better approach.

OSHA wrestled extensively with this issue in the Preamble to 29 CFR 1910.46 — permit required confined spaces (OSHA 1993). According to the most recent (May 1994) of Appendix E (Sewer System Entry), broad-range sensors are best suited for initial use where actual or potential contaminants have not been identified. Appendix E further discusses the fact that such sensors only indicate exceedence of a hazard threshold of a class of chemicals. Therefore, substance-specific sensors are best suited for use where actual and potential contaminants have been identified. However, Appendix E concludes that the atmosphere in sewers can change unpredictably and that substance-specific devices may not detect new potentially lethal hazards. The employer must decide, based on knowledge and experience, the best type of testing instrument for a specific entry operation. The bottom line is that if hydrogen sulfide or carbon monoxide are known to be present, the monitoring instrument should contain sensors that allow for the direct and quantifiable measurement of these contaminants.

Classification for Intrinsic Safety

Instruments purchased for use in a confined space, hazardous location, or other environment characterized by the potential presence of flammable or explosive gases, should carry a "Classification for Intrinsic Safety." A Classification for Intrinsic Safety usually is not an obligatory requirement on instruments purchased or sold for these applications. It ought to be!

Devices classified as "Intrinsically Safe" prevent explosions in hazardous locations by employing electrical designs that eliminate the possibility of ignition. This generally involves adding protective components in series with energy storage devices. The purpose for the protective components is to reduce the risk of ignition due to spark or increased surface temperature of components. Design elements may also include flame arrestors or other components to locally contain an explosion in the event that there is ignition. Combustible sensors contain an integral flame arrestor for this purpose. Classification for Intrinsic Safety is based on performance of the instrument when tested in a specific flammable atmosphere.

The Classification for Intrinsic Safety received by the instrument references the severity of the explosive hazard of the flammable atmosphere in which the testing occurred (NEC, 1995). For example, many confined space instruments are Classified for use in Class I, Division 1, Groups A, B, C, and D Hazardous Locations. This means that the testing was conducted in a "Group A" atmosphere (an atmosphere containing an explosive mixture of acetylene). An instrument capable of passing tests in a Group A atmosphere, by definition is "Intrinsically Safe" for use in Groups B, C, and D atmospheres.

Group B atmospheres contain hydrogen, fuel, and combustible process gases containing more than 30% hydrogen by volume, or gases or vapors of equivalent hazard, such as butadiene, ethylene oxide, propylene oxide, and acrolein. Group C atmospheres contain gases or vapors such as ethyl ether, ethylene, or gases or vapors of equivalent hazard. Group D atmospheres contain gases or vapors such as acetone, ammonia, benzene, butane, cyclopropane, ethanol, gasoline, hexane, methanol, methane, natural gas, naphtha, propane, or gases or vapors of equivalent hazard. Many confined space instruments also carry a Classification for use in Class II, Groups E, F, and G Hazardous Locations. These groups refer to combustible or explosible dusts. Group E atmospheres contain combustible metal dusts, including aluminum, magnesium, and their commercial alloys, or other combustible dusts whose particle size, abrasiveness, and conductivity present similar hazards in the use of electrical equipment. Group F atmospheres contain combustible carbonaceous dusts, including carbon black, charcoal, coal, or dusts that have been sensitized by other materials so that they present an explosion hazard. Group G atmospheres contain combustible dusts not included in Group E or F, including flour, grain, wood, plastic, and chemicals.

Any instrument purchased for use as a confined space gas detector should carry the logo of the testing laboratory that conducted the evaluation, as well as the specific Hazardous Location Groups for which the Classification applies. For instance, the label might read:

> Classified by Underwriters Laboratories, Inc.® and Canadian Standards Association as to Intrinsic Safety for use in Hazardous Locations Class I, Division 1, Groups A, B, C, and D.

Users should beware of statements on the label or literature accompanying the gas detector that indicate the instrument is "intrinsically safe by design." If the logo of a Nationally Recognized Testing Laboratory (such as Underwriters Laboratories, Canadian Standards Association, Factory Mutual, Intertek Testing Services, etc.) is not prominently displayed on the instrument label, the instrument probably has not been submitted for third-party evaluation.

Evaluation methods used to determine Intrinsic Safety may differ from one testing laboratory to another. To illustrate, the Canadian Standards Association (CSA) tests for the Intrinsic Safety of the design of the instrument. CSA also evaluates the accuracy and stability of flammable/combustible sensor readings. Testing conducted by other testing laboratories may or may not include this additional performance evaluation.

ISO Registration

Manufacturers increasingly promote themselves on having attained registration for one or another of the ISO (International Standards Organization) Quality Standards. Attaining ISO registration is no easy feat. ISO-registered companies have submitted their quality systems to a rigorous third-party evaluation, and have met the exacting performance criteria contained in the ISO standard to which they are registered. Being an ISO-registered company, however, does not automatically guarantee the excellence of the products that the company sells. An ISO-registered company is free to produce mediocre products, as long as they produce them in a highly consistent manner. Fortunately, this is seldom the case. The ISO registration process ensures that manufacturers take a careful look at the quality of the products they design, make, or market, and that they do their utmost to increase customer satisfaction. The bottom line is that ISO registration is an excellent indicator about the way that a company does business.

Batteries

Batteries, while not the heart of a portable instrument, are close to it. Batteries determine the size and weight of an instrument. They also determine service life. The flexibility provided by the instrument manufacturer determines whether the user in a small town can buy emergency replacements that will keep the instrument running.

Confined space instruments may be equipped with either disposable alkaline or rechargeable batteries, or both. The primary advantage of rechargeable batteries is overall cost effectiveness. Frequent (or daily) replacement of disposable batteries can be very expensive. While alkaline batteries may not be the most cost-effective approach, having the ability to use them "in a pinch" is a strong design advantage. Several instrument designs offer interchangeable NiCd (nickel cadmium) and alkaline battery packs.

Types of rechargeable batteries used in these instruments include lead–acid, NiCd (nickel cadmium), or NiMH (nickel metal hydride) designs. The primary advantage of lead–acid batteries is their ability to be left continuously on the battery charger whenever the instrument is out of service. Lead–acid batteries are highly resistant to damage due to overcharging. The primary disadvantage of lead–acid batteries is damage sustained when allowed to drain or discharge completely. NiCd (nickel cadmium) batteries are the most widely used rechargeable battery. Although NiCd batteries usually are not damaged by deep-discharging, they lose capacity when not "exercised" by being allowed to discharge fully before being put back on the charger. Many users refer to this loss of capacity as "developing a memory."

The memory effect is attributable not to lack of "exercise," but to the damage due to heating of the battery cells during prolonged overcharging. The charger continues to pump current into the battery even after charging is complete. This heating causes structural changes in the battery that result in sharply reduced capacity. The battery pack then may be able to provide only a few minutes of power to the instrument instead of the normal 8 or 10 hours of continuous operation. Sometimes "cycling" (charging, then draining the battery a number of times) may partially restore the lost capacity. Usually, once damage has occurred, the only way to restore the full operational capability of the instrument is to replace the battery.

Over the last few years, battery manufacturers, as well as manufacturers of battery-charging systems have made major improvements in design. Today's "smart" battery chargers contain electronics for assessing the condition of the NiCd battery pack during charging, and to drop from a "fast" charge rate to a "trickle" the moment charging is complete. The "trickle" charging rate is too low to produce

FIGURE 10.26 Confined space monitor with removable battery pack. (Courtesy of Biosystems, Inc., Middletown, CT.)

damage from heating. As result, instruments containing NiCd batteries can be recharged in a very short period of time, left on the charger for long periods of time without damage. Batteries in these instruments do not require discharging or exercising before being placed back on the charger.

Durability

Instruments designed for use in confined space monitoring programs (Figure 10.26) must be durable. Unfortunately, many designs are less robust than they may appear on the surface. When considering whether an instrument is tough enough to take the abuse it is likely receive, consider asking the following questions:

1. **Has the instrument been tested, and is it protected and constructed in a way to minimize the effects of radio frequency interference (RFI) and electromagnetic interference (EMI)?** This is a very important consideration in light of the effect RFI can have on the output of an unshielded electrochemical sensor.
2. **What are the effects of high and low temperatures on the design?**
3. **How water resistant is the design?** Is the design water resistant, or vulnerable to damage when used in the rain or dropped in a puddle? If the manufacturer says the design is water-proof, ask for their definition.
4. **Does it feel flimsy or provide unstable readings when picked up and turned on?** Don't underestimate the amount of information the "feel" of an instrument can sometimes provide.

Datalogging vs. Nondatalogging Capability

Datalogging capability is available in many single and multisensor instruments. In the past, successfully downloading datalogged information from the instrument to a computer sometimes required a high degree of operational expertise. Today the procedure is nearly automatic. Datalogging instruments usually are set up to retain monitoring information whenever turned on. The utility of this information for compliance and recordkeeping purposes is obvious.

The information also can serve other extremely useful functions. The first is to provide information about accidents or unusual occurrences. The datalogged information provides a record about conditions that occurred. This saves the need to attempt a reconstruction of the event using costly and time-consuming activities. The datalogged information becomes a source of information that can assist in preventing recurrences. The datalogged information also verifies whether the instrument was used correctly at the time the event occurred.

The time history of exposure indicates whether CEILING limits (floor limit in the case of oxygen) have been exceeded during work activity. The output can indicate in considerable detail about exceedences exactly when they occurred, how long they lasted, and the dynamics of events that surrounded them. This information is extremely important for assessing the significance to put on the exceedences. This capability is not available on nondatalogging instruments.

Datalogging as well as nondatalogging confined space instruments frequently include the capability to solve complex time-history calculations, such as those used to compute CEILING, STEL, and TWA values. Datalogging instruments are able to retain this information after the instrument is turned off, or after the initiation of a new sampling interval. Instruments lacking datalogging capability cannot do this.

In order to create a report or display the recorded data, the datalogging instrument periodically takes a "snapshot" of the values being sensed at any given moment, and later uses these to develop a graph or tabular report of stored monitoring data. The period of time between the "snapshots" is the datalogging interval. The instrument logs only one data "snapshot" per datalogging interval. The user generally is able to select the datalogging interval that provides the optimum level of information about the contaminants being measured. For instance, if concentrations are prone to change very rapidly, the user might select a datalogging interval of one second. If concentrations are not subject to rapid change, the user might select a longer interval. The longer the datalogging interval, the less the resolution provided by the graph or tabular report. To illustrate, a datalogging interval of 1 hour will not show short-term spikes in concentrations that last only a second or two.

Values used by the instrument for the data "snapshot" vary from manufacturer to manufacturer. Some manufacturers log the average value over the datalogging interval. Others take the more conservative approach and log the highest value noted by the instrument during the datalogging interval. This sometimes can cause confusion during interpretation of monitoring results. To illustrate the problem that can arise, consider a datalogging instrument that logs the highest value during the datalogging interval and an interval of 1 hour. If 59 min out of the 60 were spent with an average concentration of 1 ppm, and 1 minute was spent at 10 ppm, the tabular report would show that the concentration for that entire datalogging interval was 10 ppm.

Another question frequently asked by instrument users concerns the effect of changing the datalogging interval on running time-history calculations such as CEILING, STEL, and TWA. The answer is that the running calculations will not be affected by changing the datalogging interval. The microprocessor continuously updates these calculations many times per minute. The datalogging interval simply specifies how often the instrument stores a "snapshot" of the current readings for the purposes of generating a printed report or database file of test results.

Included Accessories

Another important consideration is accessories that are included in the purchase price for the instrument. If the instrument includes a rechargeable battery, does the price include a battery charger? Do the accessories include a sample-draw kit or motorized pump? Carrying case? Training video? Calibration materials? Necessary accessories that are not included in the purchase price can considerably add to cost and user frustration.

Warranty

A warranty commensurate with the trust the manufacturer places in the design should accompany a quality instrument. Many manufacturers now offer a lifetime warranty on the instrument, with a 1-year or 2-year warranty on the sensors.

Operability

Probably the most important factor of all in the selection of an instrument is ease of operation. If the person on the shop floor is unable to use the instrument because of unnecessary complexity, difficulty in calibration, or operation — they won't.

Operability also affects the person who selected the instrument and who first must learn how to use it. This individual is the most likely to become familiar with that other part of the equipment package: the manual and associated documentation. These materials should be inclusive enough to cover the routine and the not-so-routine questions in a manner that the average technical person can comprehend without difficulty.

Instrument Performance Specifications

Specifications published by manufacturers to clarify instrument performance are a valuable tool for would-be purchasers. Unfortunately, it sometimes takes a practiced eye to interpret specifications when comparing one instrument design to another. The most significant problem is the terminology used by a particular manufacturer. While some terms are straightforward and fully accepted throughout the industry, other terms have specific meanings to a particular manufacturer. A good example is published concentration ranges. Electrochemical toxic sensors have a "nominal range" in which they may be continuously used without harm or damage, and over which they are capable of accurate readings. Electrochemical sensors can be used discontinuously or for short periods of time above the nominal range, as long as they are not exposed to conditions which exceed the "maximum overload" concentration. While prolonged exposure to concentrations above the nominal range may saturate the electrolyte or create other conditions which prevent the sensor from obtaining accurate readings, short exposures should not do any long-term damage to the sensor. Exposure to concentrations above the maximum overload concentration may permanently harm the sensor.

Some manufacturers differentiate between nominal and maximum overload concentrations. Others list the maximum overload concentration as the upper range limit. Still others take the far more conservative nominal range as the upper range limit. They all may be using the same sensor obtained by the same manufacturer in their designs! In addition, instrument electronics and performance characteristics may also limit performance. That means the same sensor installed in a different brand of instrument may provide substantially different performance.

Another issue is trade-offs made to optimize certain performance characteristics. For instance, partial atmospheric pressure oxygen sensors usually offer slightly faster response than capillary pore sensor designs. Judging only on the basis of the sensor's T-90 (the time from initial exposure for the sensor to reach 90% of its final stable reading), the p_{O2} design might be the better choice. On the other hand, characteristics such as a longer operational life, insensitivity to changes in the ambient or barometric pressure, and smaller size may more than make up for a slightly slower response.

Product liability is a major concern to manufacturers. Conservative firms tend to minimize performance capabilities in written specifications, preferring to err on the side of caution. In some cases, this leads manufacturers to dramatically understate (at least in print) the true capabilities of their design. On the other hand, less conservative manufacturers can occasionally exaggerate. The best advice is to conduct a field trial prior to purchase! There is no substitute for hands-on experience when it comes to making a purchase decision. Commonly used terms in instrument specifications include:

Accuracy: The percentile agreement between the instrument reading and the true concentration. Accuracy may be expressed as a function of the full-scale reading, a percentage of the actual reading, or a specific unitary value. For instance, consider a carbon monoxide instrument with a full scale deflection of 0 to 500 ppm, which is exposed to a concentration of 50 ppm calibration gas. If the published accuracy of the instrument is ±10% of the actual reading, the expected reading in a properly calibrated instrument should fall within 45 to 55 ppm. On the other hand, if the accuracy is ±10% of the full scale deflection, the reading could fall anywhere within 0 to 100 ppm and still be within tolerance! The most straightforward approach is to use a unitary value. A unitary approach to the accuracy of a carbon monoxide sensor might be ±2 ppm over a range of 0 to 250 ppm. Of course, any published accuracy is specific to the design!

Resolution: Lowest concentration of the substance being measured which can be reliably detected by the instrument.

Increments of measurement: The least significant measurement unit used to display readings. The increment of measurement can sometimes exceed the resolution of the instrument. For instance, some confined-space instruments can provide toxic readings in either 1.0- or 0.1-ppm increments. A sensor which shows poor resolution below 1.0 ppm step-change should not be used to obtain 0.1-ppm readings.

Repeatability: The maximum percentage variation between repeated, independent readings on a sensor (using a gas mixture within the nominal range and under identical conditions).

Linearity: The measure of how well the concentration response curve of an instrument fits the equation for a straight line. It's important to note that it is not necessary for the sensor output to be exactly linear in order to provide linear readings. As long as the sensor output is mathematically predictable, instrument electronics may be able to linearize the readings.

Linear range: That portion of the concentration range over which the instrument's concentration response matches (or approximates) a straight line.

Noise: Random fluctuation in signal which is independent of the concentrations being measured.

Drift: Slow or long-term changes in the instrument reading which are not caused by immediate changes in the concentration of the substance being measured.

Response time: The time from initial exposure for the sensor to reach its final stable reading. Response time is usually given as a T-90 (time to 90% of final stable reading) or T-95 (time to 95% of final stable reading) value. Make sure the specification is clear as to the method used!

Recovery time: The time necessary for the sensor to recover after exposure to a step-change in concentration (Figure 10.27).

Alarm Settings

A primary use of confined space instruments is to alert workers about the need to take action when conditions become unsafe. Generally, that action is to leave the confined space immediately, and to return only after further testing determines that the area once again is safe for occupancy.

A very important consideration in the use of instruments is the set-point for the alarm. Alarm set-points should be set conservatively. Alarms should activate before the onset of toxicological effects that might reduce the worker's ability to self-rescue. Setting the alarms at the point above which a hazardous atmospheric condition exists may not be adequately conservative. Workers must have enough time to get to a position of safety before conditions become so hazardous that the ability to self-rescue is impaired.

As an example, OSHA specified 10% of LEL as the concentration above which the atmosphere is deemed to be hazardous (OSHA 1993). This actually is the maximum allowable concentration at which the alarm can be set in this jurisdiction. OSHA used 5% of LEL as an action limit in the compliance document (OSHA 1995).

Possibilities for alarm settings for other gases and vapors are the TWA, STEL, CEILING, AVERAGE, EXCURSION limit, or some fraction of these values. The TWA is the most conservative

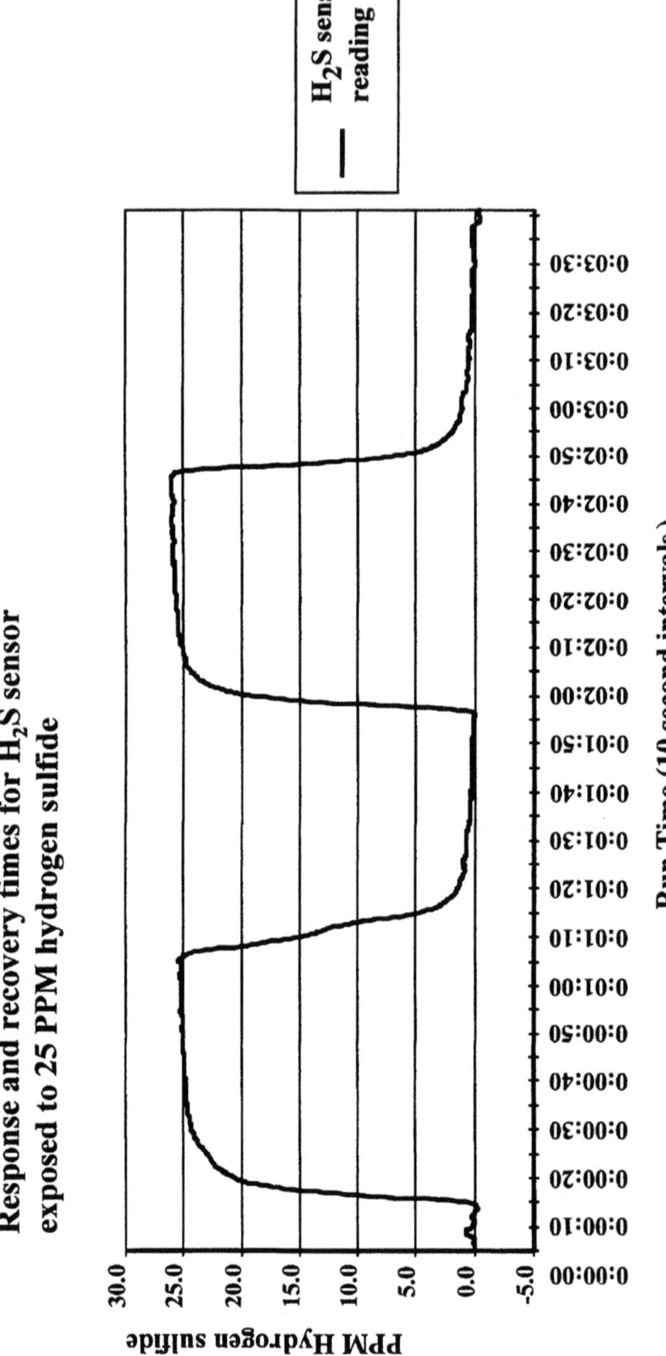

FIGURE 10.27 Electrochemical sensor output showing response and recovery times for hydrogen sulfide sensor installed in a typical confined space instrument. (Courtesy of Biosystems, Inc., Middletown, CT.)

set-point where a STEL or an EXCURSION limit also exists. Setting the set-point at the TWA would ensure that the TWA never would be exceeded during normal work activity provided that the alarm did not sound. The TWA could be exceeded as a result of evacuation during an alarm condition. Setting the alarm level higher could lead to overexposures during work activity because the alarm would not sound until the TWA was exceeded. Setting the alarm to a level below the TWA could lead to needless work stoppages.

Substances with a CEILING designation, or a floor (oxygen, for example), pose a difficult dilemma. Where the concentration changes rapidly as in gas-shielded welding, the alarm condition could develop and clear in less than one minute.

Atmospheric conditions in confined spaces can change rapidly. Conditions can easily go from safe to hazardous in a matter of moments. Datalogged information about normal and abnormal conditions becomes an invaluable resource for understanding what is occurring in the space. This can form the basis for an informed decision about a reasonable alarm setting. Other factors to consider include:

- Distance from the work area to the position of safety
- Rapidity of increase in concentration of contaminant that triggers alarms
- Effects of overexposure

If there is the slightest doubt, use a more conservative setting. Use "factory default" settings only if they are appropriate. Finally, be alert to changes in the job or environment that may require changes in monitoring procedures.

Calibration

All instruments require maintenance and calibration. The only way to achieve assured results is to follow the manufacturer's instructions. Regardless of the use of the instrument, accuracy must be verified on an ongoing basis. This is absolutely important with instruments designed for use in confined spaces, given their use: detection and measurement of potentially life-threatening atmospheric conditions.

The atmospheric conditions that lead to accidents and fatalities during work in confined spaces often provide no sensible warning. The only way to ensure conditions are safe is to use an accurate atmospheric monitor. The only way to know that the readings are accurate is to expose the instrument to a known concentration of test gas. This verifies that readings on the display and those that trigger the alarms are correct. Accuracy is very important. Hexane is fully flammable at 1.1% by volume. A flammable/combustible alarm set to 10% of LEL must activate when the concentration reaches 0.1% by volume. This is a very low concentration from the standpoint of the catalytic hot bead sensors used for this purpose. Toxic gases require action from the instrument at even lower concentrations. In the case of chlorine, the TLV is 0.5 ppm. This is an extremely low concentration to register.

The atmosphere in which the instrument is used can have a profound effect on the sensors or detector element. PID windows become degraded or dirty. Electrochemical and hot-bead sensors can be poisoned or suffer degraded performance when exposed to certain substances. The kinds of conditions that affect the accuracy of sensors vary from one type to the next. For example, oxygen sensors can be affected by prolonged exposure to "acid" gases such as carbon dioxide. Some electrochemical toxic sensors actually consume themselves as they function. Catalytic hot-bead sensors are affected by exposure to silicones, the tetraethyl lead in "leaded" gasoline, chlorinated solvents, hydrogen sulfide, and high (fully flammable) mixtures.

Calibration verifies that the instrument is accurate. If exposure to test gas indicates a loss of sensitivity, the instrument needs to be adjusted. The important thing to understand is that without challenging the instrument with a known concentration of contaminant, there is no way of knowing whether adjustment is needed. For all of these reasons, the most prudent course of action is to verify the accuracy of the instrument by exposure to a known concentration before each day's use.

Calibration should be simple and straightforward. Calibration usually is a two-step procedure. First, the instrument is taken to clean air or exposed to contaminant-free "zero air" from a cylinder. If the readings differ from zero, a "zero" adjustment is required. The second step is to expose the sensor to a known concentration of contaminant. If the readings are the same, the instrument requires no further adjustment. Where a discrepancy exists, the instrument should be "span"-adjusted before further use.

The user should follow the manufacturer's instructions carefully. Incorrect flows produced by an improperly set regulator or improvised fittings can produce inaccurate readings. Adjustment because of these readings could lead to dangerous incorrect readings.

Bump Test

Most manufacturers strongly recommend verifying the calibration of their confined space instruments with a known concentration test gas before use. The "bump" test is very simple and takes only a few seconds to accomplish. Most manufacturers agree that it is not necessary to make a calibration adjustment unless readings are off by more than some percentage of the expected value. While advice may differ between manufacturers, several suggest using ±10% as the criterion for determining whether adjustment is required.

Lengthening the Interval between Calibration Checks

One of the most frequent questions a manufacturer of confined space-targeted instruments hears from its customers is whether there are circumstances that would allow lengthening the period between calibration checks. There has never been a consensus among manufacturers about the approach to take to this vexing issue. One of the industry associations to which many gas-detection manufacturers belong is the Industrial Safety Equipment Association (ISEA). In May 1996, the ISEA published a protocol to clarify the minimum conditions for lengthening the interval between calibration checks for direct-reading portable gas monitors used in confined spaces.

The ISEA protocol begins by clarifying the difference between a functional (bump) test and a full calibration:

a. A functional (bump) test is defined as a means of verifying calibration by using a known concentration of test gas to demonstrate that an instrument's response to the test gas is within acceptable limits (Figure 10.28).
b. A full calibration is defined as the adjustment of an instrument's response to match a desired value compared to a known concentration of test gas.

The protocol goes on to recommend the frequency for verification of calibration:

a. A functional (bump) test or full calibration of direct reading portable gas monitors should be made before each day's use in accordance with the manufacturer's instructions, using an appropriate test gas.
b. Any instrument which fails a functional (bump) test must be adjusted by means of a full calibration procedure before further use.
c. If environmental conditions which could affect instrument performance are suspected to be present, such as sensor poisons, then verification of calibration should be made on a more frequent basis.

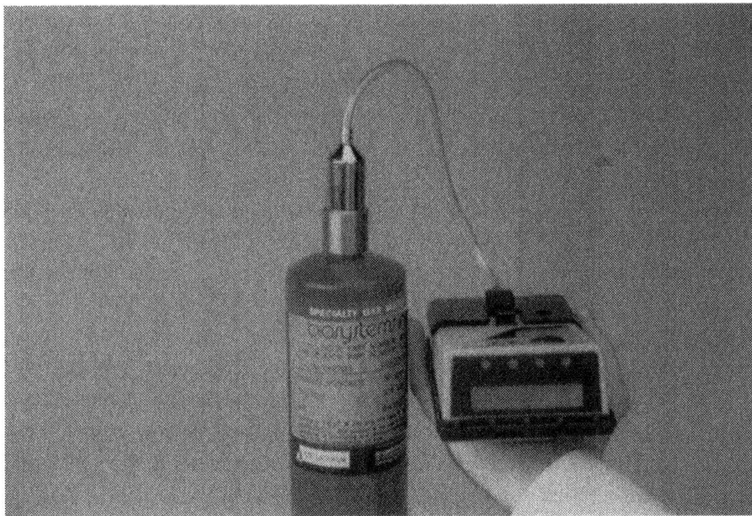

FIGURE 10.28 Confined space monitor being exposed to a known concentration calibration gas. (Courtesy of Biosystems, Inc., Middletown, CT.)

The protocol then goes on to identify the specific minimum circumstance under which the interval between verification checks can be lengthened. If conditions do not permit daily testing of the gas detector to verify calibration, the ISEA protocol permits less frequent verification of calibration only if the following criteria are met:

a. During a period of initial use of at least 10 days in the intended atmosphere, calibration is verified daily to be sure there is nothing in the atmosphere which is poisoning the sensor(s). The period of initial use must be of sufficient duration to ensure that the sensors are exposed to all conditions which might have an adverse effect on the sensors.
b. If the tests demonstrate that it is not necessary to make adjustments, then the time interval between checks may be lengthened, but should not exceed 30 days.
c. The history of the instrument since the last verification can be determined by assigning one instrument to one worker, or by establishing a user tracking system such as an equipment use log.

The ISEA protocol has been accepted by many confined space instrument manufacturers. The user should verify with the manufacturer that the procedure intended for use is in accordance with instructions from the manufacturer of the specific instrument before commencing a calibration schedule. Remember that any conditions, incidents, experiences, or exposure to contaminants that might adversely affect the calibration should trigger immediate reverification before further use. Most important, if there is any doubt about the calibration of the sensors, expose them to a test gas of known concentration before further use.

Confined-Space Monitoring Sequence

Regulatory and other standards on confined space entry require all "permit required" confined-space entry procedures to be performed in accordance with a written program centered on the issuance and use of an entry permit. Confined space regulations explicitly define monitoring requirements. The OSHA Standard on confined spaces for general industry (1993) requires testing of all permit spaces for oxygen, combustible gas, and any toxic gases potentially present. Testing must occur prior to entry, as well as on a continuing basis until the entry is completed. Besides indicating the kind of hazards for which the space must be tested, the Standard also specifies the sequence of testing: oxygen, followed by flammable/combustibles, followed by toxic substances. The reason behind this sequence is that flammable/combustible sensors cannot detect in the absence of oxygen. The oxygen concentration should always be noted prior to obtaining readings from a catalytic hot-bead sensor.

When testing confined spaces prior to entry, note that two out of three explosions involving confined spaces occur at the time the space is first disturbed (NIOSH, 1979). This usually is before the entrants have actually entered the space. For this reason, the OSHA Standard on general industry stresses that the space should not be opened or disturbed until it has proven safe to do.

The best means for assessing conditions in an undisturbed space is to use a remote sample kit to obtain an air sample through holes in the lid or hatch of the space. If holes are not available, the space should be opened cautiously just enough to obtain a sample. Gases and vapors tend to form vertical density-dependent layers in the atmosphere. Gases (such as methane) and vapors that are less dense than air tend to rise and accumulate near the roof of the space. Gases (such as hydrogen sulfide) and vapors that are denser than air tend to sink and accumulate near the lowest point in the space. Once initial readings indicate that work can proceed in safety, additional readings should be obtained for all vertical levels between the entrance and the deepest point in the bottom of the space. Unless all areas of the space are sampled, the existence of a dangerous layer or pocket may not be noted.

Remember that monitoring is only one aspect of confined space entry procedures. Another very important element is use of forced air ventilation. Even if all tests are within allowable limits, the confined space atmosphere should still be purged with a fresh air ventilator blower before entry. The atmosphere should be tested again after the initial purge ventilation. Monitoring and ventilation go hand in hand. The safest course of action is to continuously ventilate and continuously monitor for the entire duration of the entry. Monitoring determines that the atmosphere is safe, while ventilation keeps it that way. Confined space monitors are designed to be operated continuously; take advantage of it!

SUMMARY

Instruments and monitoring are central to assessing conditions in confined spaces and maintaining them safe for entry and work activity. Practitioners of today are fortunate to have available a broad range of technologies for pursuing this end.

REFERENCES

American Conference of Governmental Industrial Hygienists: *1997 TLVs and BEIs, Threshold Limit Values for Chemical Substances and Physical Agents, Biological Exposure Indices.* Cincinnati, OH: American Conference of Governmental Industrial Hygienists, 1997. 148 pp.

American Industrial Hygiene Association: *Manual of Recommended Practice for Combustible Gas Indicators and Portable Direct Reading Hydrocarbon Detectors.* Akron, OH: American Industrial Hygiene Association, 1980. 56 pp.

American National Standards Institute (ANSI): *Standard No. Z117.1-1995, Safety Requirements for Confined Spaces.* Des Plaines, IL.: ANSI, 1995.

City Technology Ltd.: *Product Data Handbook,* vol. 1: *Safety.* Portsmouth, England: City Technology Ltd., 1997.

George, Eric (Ed.), Boissevain, A. L., Henderson, R. E., Claybaugh, D. J., Massey, C., Morton, J. R., Britain, B. J., and Bethel, B.B.: *Corporate Health and Safety: Managing Environmental Issues in the Workplace.* Southampton, PA: Ergonomics, Inc. 1996. www.ergonomicsusa.com.

Moseley, P. T. and Tofield, B. C.: *Solid State Gas Sensors.* Bristol, England: IOP Publishing Ltd., 1987.

Moseley, P. T., Morris, J. O. W., and Williams, D. E.: *Techniques and Mechanisms in Gas Sensing.* Bristol, England: IOP Publishing, 1991.

National Fire Protection Association: NFPA 306: *Control of Gas Hazards on Vessels* (1993 ed.). Quincy, MA: National Fire Protection Association, 1993. 15 pp.

National Fire Protection Association: *Fire Hazard Properties of Flammable Liquids, Gases, and Volatile Solids.* Boston: NFPA, 1977.

National Institute for Occupational Safety and Health: Criteria for a Recommended Standard — Working in Confined Spaces (DHEW/PHS/CDC/NIOSH Pub. No. 80-106). Cincinnati, OH: National Institute for Occupational Safety and Health, 1979. 68 pp.

National Institute for Occupational Safety and Health: Worker Deaths in Confined Spaces (DHHS/PHS/CDC/NIOSH Pub. No. 94-103). Cincinnati, OH: National Institute for Occupational Safety and Health, 1994. 273 pp.

Nyquist, J. E., Wilson, D. L., Norman, L. A., and Gammage, R. B.: Decreased sensitivity of photoionization total organic vapor detectors in the presence of methane. *Am. Ind. Hygiene Assoc. J. 51(6):* 326–330 (1990).

Occupational Safety and Health Administration: Selected Occupational Fatalities Related to Fire and/or Explosion in Confined Work Spaces as Found in OSHA Fatality/Catastrophe Investigations. Washington, D.C.: U.S. Department of Labor, Occupational Safety and Health Administration (U.S. DOL/OSHA), 1982. 76 pp.

Occupational Safety and Health Administration: Selected Occupational Fatalities Related to Toxic and Asphyxiating Atmospheres in Confined Work Spaces as Found in Reports of OSHA Fatality/Catastrophe Investigations. Washington, D.C.: U.S. Department of Labor, Occupational Safety and Health Administration (U.S. DOL/OSHA), 1985. 230 pp.

Occupational Safety and Health Administration: Selected Occupational Fatalities Related to Welding and Cutting as Found in Reports of OSHA Fatality/Catastrophe Investigations. Washington, D.C.: U.S. Department of Labor, Occupational Safety and Health Administration (U.S. DOL/OSHA), 1988. 225 pp.

Occupational Safety and Health Administration: Selected Occupational Fatalities Related to Ship Building and Repairing as Found in Reports of OSHA Fatality/Catastrophe Investigations. Washington, D.C.: U.S. Department of Labor, Occupational Safety and Health Administration (U.S. DOL/OSHA), 1990. 195 pp.

Occupational Safety and Health Administration: Permit-Required Confined Spaces for General Industry; Final Rule, *Fed. Regist. 58:* 9 (14 January 1993). pp. 4462–4563.

OSHA: "Air contaminants," *Fed. Regist. 54:* 36767 (Sept. 5, 1989); 54 FR 41244 (Oct. 6, 1989); 55 FR 3724 (Feb. 5, 1990); 55 FR 12819 (Apr 6, 1990); 55 FR 19259 (May 9, 1990); 55 FR 46950 (Nov. 8, 1990); 57 FR 29204 (July 1, 1992); 57 FR 42388 (Sept. 14, 1992); 58 FR 35340 (June 30, 1993); 61 FR 56746 (Nov. 4, 1996); 62 FR 42018 (August 4, 1997).

OSHA: Confined and Enclosed Spaces and Other Dangerous Atmospheres in Shipyard Employment; Final Rule, *Fed. Regist. 59:* 141 (25 July 1994). pp. 37816–37863.

OSHA: Compliance Directive (CPL 2.100) for Confined Space Entry. Washington, D.C.: U.S. Department of Labor, Occupational Safety and Health Administration (U.S. DOL/OSHA), May 5, 1990.

Sensidyne Inc.: *Gastec Precision Gas Detector System Manual (Blue Book).* Clearwater, FL: Sensidyine, 1994.

Weast, Robert C., Ed.: *Handbook of Chemistry and Physics.* Cleveland, OH. CRC Press, 1977.

Workers' Compensation Board: Occupational Health and Safety Regulation, Core Requirements (BC Regulation 296/97). Richmond, BC: Workers' Compensation Board of BC, 1998.

11 Ventilation for Work in Confined Spaces

CONTENTS

INTRODUCTION

Confined spaces can be inherently more hazardous places in which to work than conventional workspaces. This observation was amply illustrated during investigations of fatal accidents performed by NIOSH and OSHA (NIOSH 1979, 1994; OSHA 1982, 1985, 1988, 1990). There were many reasons for the antecedent conditions that led to these accidents. The most serious hazards resulted from atmospheric contamination: oxygen deficiency or enrichment, toxic gases and vapors, and flammable and explosive atmospheres. The presence of the atmospheric hazard, combined with one or more serious procedural deficiencies, inadequate or improper ventilation, created the conditions antecedent to accidents in these workspaces. In almost all of the fatal accidents documented by NIOSH and OSHA, entrants failed to ensure ventilation of the space prior to entry or during work. This means failure to remove a contaminated atmosphere prior to entry or to prevent its occurrence during the conduct of work. The reasons for this lack of action have not been elucidated. However, it would seem reasonable to presume that this was coincident with the lack of recognition of the grave danger posed by these situations.

One of the outcomes from the hazard assessment performed as part of the planning process for entry and work is to determine the need for ventilation. Ventilation is part of the assurance that the space will be safe for entry and will remain safe for work. This process results from identification of potential or actual contaminants and recognition of the toxicological and flammability hazards that they could represent.

Ventilation is a critical element in hazard control in these workplaces. This position was advocated a number of years ago in the industrial hygiene literature (Garrison and McFee 1988). Also, the role of ventilation as a critical element in hazard control is recognized in technical standards and legislation on confined spaces (ANSI 1995; OSHA 1993, 1994; Standards Australia 1995).

To illustrate the strength of this commitment, the OSHA Standard on confined spaces in general industry (1993) mandated the use of continuous forced ventilation in lieu of other measures for control of atmospheric hazards. OSHA determined that spaces in which all hazards have been eliminated could be reclassified as nonpermit spaces for as long as the hazards remain eliminated. (A "permit space" in OSHA parlance has or could have an atmospheric, engulfment, or configuration hazard, or other recognized serious hazard.)

Ventilation controls hazards; it does not remove them. To take advantage of this provision, the employer must be able to prove that the only hazard in the space is atmospheric and that continuous forced ventilation will maintain the permit space safe for entry. Reclassification must be rigorous. This involves completion of a dated and signed certificate. Affected employees must be informed about the reclassification.

According to OSHA, testing the atmosphere inside the space and providing adequate continuous ventilation sometimes can eliminate the hazardous atmosphere to produce the equivalent of a nonpermit space. The work to be performed must not introduce any additional hazards, such as toxic substances or hot work. A permit is not required for entry into permit spaces whose hazards can be controlled by ventilation alone.

Control of hazards, however, is not the same as elimination of hazards. Alternate ventilation procedures do not turn a permit-required confined space into a nonpermit space. Instead, they allow certain permit-required spaces to be entered without the need for a comprehensive program, including a fully comprehensive entry permit. These spaces, therefore, remain classified as permit-required, even though a less comprehensive "written certification" would suffice instead of the complete permit (Anonymous 1993). In effect, this creates two categories of permit-required confined space (Easter 1993). The first covers those spaces that pose not only atmospheric, but also other hazardous conditions. The second category applies to spaces that contain or potentially contain only atmospheric hazards. This would necessitate an alternate and less extensive program involving training, atmospheric testing, forced air ventilation, and an abbreviated permit and notification program. An attendant would not be required.

A confined space containing a hazardous environment would require initial entry to ascertain conditions and to apply any necessary measures to prepare it for further work. Such measures could include cleaning. Initial entry would require all precautions, as appropriate. Once cleaning and other preparatory activities were complete, the only remaining hazard would be atmospheric. The space then would revert to the condition discussed above.

Awareness of the need for ventilation has widespread recognition in general industry. However, widespread this awareness under normal operations, this does not appear to have made the transition to confined spaces. The unnecessary accidents and the hundreds of injuries and deaths each year, worldwide, continue to provide ample indication about existence of this problem. Fundamental concepts seem to be forgotten, neglected, or otherwise not applied during entry and work involving confined spaces.

Perhaps this occurs because this work is typically part of "maintenance." Maintenance is the most dangerous, least well documented and difficult part of the work cycle. This is inherent in the nature of maintenance. Entry into confined spaces is not routine work. They often are not anticipated, and/or must be performed under time pressure. Because entries into confined spaces are "special," for whatever the reason, they often fail to receive the attention given to the routine activity associated with production.

Occupational health and safety professionals potentially fall into the same trap as everyone else: failure to recognize the potential for atmospheric confinement by boundary surfaces. This can create danger in the midst of otherwise innocuous surroundings. Safety practitioners have died in the confined space accidents documented by NIOSH and OSHA (NIOSH 1994; OSHA 1985).

Awareness about the dangers of confined spaces and strategies for responding to them is not part of the training or experience of many professionals in this field. This information must be acquired outside the regular channels.

The purposes for discussion in this chapter are twofold: to emphasize the importance of ventilation during work in confined spaces and to provide guidance on how to use it effectively. The guiding principle is simple: if in doubt, ventilate. Providing ventilation when it is not needed is far better than failing to provide ventilation when it is. Experience has proven this many times.

STRATEGIES FOR HAZARD CONTROL IN CONFINED SPACES

Confined spaces form a unique class of work environment. They often comprise self-contained workspaces within a larger workplace. Control measures applicable to the hazards in these workspaces incorporate two principal strategies: administrative measures and engineering measures. These differ from each other in significant and fundamental ways. However, both strategies are essential for ensuring worker safety in confined spaces.

Administrative measures involve procedures and actions to protect the individual. The objective behind administrative measures is to manage the risk from hazards through control of the actions of people. This means ensuring use of diagnostic and protective equipment to shield the person against conditions in the workspace. Typically, administrative measures do not alter conditions. Rather, they mandate a response to them. That is, they are reactive. In the context of confined spaces, administrative measures include written programs and procedures, hazard assessments, entry and work permits, entry attendants, atmospheric testing, personal protective equipment, medical surveillance, training and audits.

The most serious limitation arising from the use of administrative measures is human error. These include errors of omission and commission, and failure to develop and implement proper procedures. Also, in no way do the actions included under administrative measures reduce the hazardous condition.

Engineering controls strive to eliminate or reduce hazards by modifying conditions in the workspace. In the context of confined spaces, engineering controls can include modifications to conditions inside the space or modifications to eliminate the need to enter. Either approach reduces or eliminates the hazardous condition. A distinct advantage offered by engineering controls is reduced potential for human error. For this reason, this strategy should be much more reliable than administrative controls, and therefore, should be the preferred approach. However, engineering controls cannot always eliminate or sufficiently reduce hazardous conditions. This is especially true in the context of confined spaces. Hence, administrative controls still are required.

Options for engineering measures to control hazards associated with confined spaces are quite limited. One option is substitution of less hazardous substances. Another is eliminating the need for entry by use of equipment that can accomplish the task. A third approach is ventilation. Ventilation often can reduce and maintain levels of airborne contaminants inside the space during entry and work at acceptable values.

Use of administrative measures as an alternative to engineering controls requires great care. They are not equivalent. Atmospheric testing and/or respiratory protective devices do not provide greater control over exposure to atmospheric hazards. If they fail, hazardous conditions continue to exist. Administrative measures may not be viable substitutes for ventilation.

The purpose for the discussion presented here is to emphasize consideration for engineering controls, of which ventilation is a primary option. The intent is not to advocate reduction of the use of administrative controls. These play an important role in entry safety. Rather, the intent is to caution against adoption of an "either/or" approach. The favored approach instead is complimentary use of both types of controls.

The guidelines developed in this chapter support the following propositions:

- Mechanical ventilation usually increases the safety of the entry when potentially hazardous airborne contaminants are present.
- Utilizing mechanical equipment for ventilation of confined spaces is neither extraordinarily difficult, time-consuming, nor expensive.
- Mechanical ventilation should be provided for most entry situations.
- Utilization of mechanical ventilation should be a standard part of entry practices and procedures, unless a technically defensible decision not to ventilate is made and documented. Documentation must include technical justification.

The authors hope that this discussion will assist individuals responsible for entry safety by enabling ventilation design and utilization in these workspaces. Ventilation can reduce atmospheric hazards in confined spaces, thus providing a safer environment in which to work. Most other precautions, although important, do not alter the environment to reduce the hazard. With the availability of portable air movers designed for use in even the remotest of locations, the means for ensuring adequate ventilation in almost all confined spaces now clearly exist.

VENTILATION OF CONFINED SPACES IN CLOSED SYSTEMS

GENERAL

In many situations involving closed systems, entry must occur periodically, yet the contents of the vessel or structure must not make contact with gases in the atmosphere. The need to avoid contact can arise from concern about generating flammable mixtures, reactivity between the contents and gases in the atmosphere, or high human or environmental toxicity.

Process operations use a number of techniques to prepare closed systems for opening and entry. Preparatory steps also must prevent contact between the contents and the air. Techniques include draining, flushing with cleaning solutions to rid the vessel of solid or liquid residue, lancing, displacement, and purging. None of these techniques necessitates opening the space. Ensuring that the space can be prepared for entry without contact between the atmosphere and its contents obviously requires prior planning and preparation, including installation of appropriate equipment. Protocols for pursuing these activities are contained in technical standards and guidelines (AGA 1975; NFPA 1992, 1993; API 1993).

Lancing/Blowdown

Lancing uses gases under high pressure to remove deposits of liquid, sludges, and solid material from surfaces in the structure. The lance is a tubular pipe-like device that is manipulated externally from the space. Lancing can proceed while process equipment continues to operate. For example, lancing to remove buildup of soot from tubes in heat exchangers and surfaces can occur while the boiler is operating. Material dislodged by the lance moves through process equipment to the appropriate collection point. Compressed gases used for lancing must be compatible with the process and substances with which they will be in contact.

Displacement

Displacement is a ventilation technique that removes gases and vapors from the interior of the space on the basis of density (AGA 1975). Through careful positioning of the injection point and control of injection rate, the displacing gas and the resident gas will form stratified layers. Under controlled conditions, the volume of gas needed to displace the contents is 1.5 to 2.5 times the volume of the container. This is the most efficient means of removal of gas from a container, both in volume and time. The displacing gas is introduced from the direction of its buoyant tendency in air. For example, the buoyant gas would be introduced from the top of the structure to displace the contents downward. A subsident gas would be introduced from the bottom of the structure to displace the contents upward. Introduction from the direction of buoyancy counters the inclination to migrate through the gas to be displaced. Introduced from the bottom of the structure, hydrogen would migrate through the contents toward the roof of the structure. This would create unacceptable turbulence and mixing between the incoming and resident gas.

Hydrogen or nitrogen can be used to displace denser gases and vapors downward. Hydrogen is a flammable gas and therefore must be displaced from the vessel or structure prior to opening and ventilating with air. Hydrogen is used as a feedstock in some processes and as an insulator in some electrical equipment (synchronous condenser) to maintain a reducing atmosphere.

Carbon dioxide is effective for upward displacement of atmospheres containing less dense gases, such as methane and ethane.

Displacement by stratification can pose a serious hazard when used inappropriately or without follow-up ventilation, since the displacing gas may remain in the space. Residual displacing gas, such as nitrogen or carbon dioxide, would constitute a lethal condition, since the atmosphere would be unbreathable. Nitrogen especially could remain in the structure after opening to air due to its neutral buoyancy. Carbon dioxide could remain in low spots (OSHA 1996). Residual gases in structures have caused fatal accidents (OSHA 1985).

Purging

Purging is a ventilation process involving gases other than air (AGA 1975). These gases generally include "inert" process gases such as steam, nitrogen, carbon dioxide, and helium. Purging ventilates or flushes the space, while avoiding contact between the contents and the atmosphere. Purging could be used to remove residual liquid by evaporation and other gases or vapors from the space by dilution or displacement. Purging also can remove volatiles that may be emitted by sludges or residues or from the structure itself. Air is not the agent of choice for ventilating at this stage in preparation, especially when the structure or vessel contains vapors that could form ignitable mixtures or that could react with components of the atmosphere.

Steam offers several distinct advantages over other "inert" gases as a purging agent. While steam itself is a hazardous substance, it condenses to a nonhazardous substance, water. Steam heats surfaces and contents in the space, thereby increasing evaporation of residual materials. The condensate can be drained from the space. Condensate will contain some of the material from the interior of the space. Steaming can be utilized for cleaning by lancing, as well as purging. Steaming could continue as long as condensate escaping from the vessel shows any sign of impurity.

A major disadvantage of the use of steam in closed vessels is potential collapse due to reduction in pressure as cooling occurs. Maintaining internal pressure by introduction of other purge gases is essential under these circumstances.

Steam purging and cleaning must be used with considerable care, due to the potential for buildup of static electricity. This is an especial concern where ignitables are present (Lees 1980). This situation is especially problematic where ignitable vapors can off-gas from residues in the vessel or volatilize from the purging agent.

From the perspective of entry preparation, establishment of an inert atmosphere usually is not the goal in itself, but merely a step in the process of providing a safe atmosphere for entry and work. An inert atmosphere would be lethal, and could be entered only after taking the most stringent of precautions, including complete reliance on air-supplying respiratory protective equipment.

Inerting

In some situations, entry into a space containing a deliberately inerted atmosphere may be required. Inerting would be necessary when safety could not be achieved or assured by other means of control. Inerting of spaces adjacent to the one in which work is to occur also could be required when the work could generate a hazardous condition in those spaces. When inerting is used, the composition of the atmosphere in the affected spaces must be monitored and maintained under careful control.

Inerting involves creating or maintaining an inert atmosphere in the space at all times during critical aspects of the work cycle. Therefore, the "inertness," meaning low level of oxygen, must be maintained using the appropriate gas. The inert gas is chosen for appropriate qualities of buoyancy or subsidence relative to the geometry of the space in which it is to be used.

NFPA considers a gas that will burn in the normal concentration of oxygen in air to be flammable. Gases that do not burn in any concentration of air or oxygen are considered by NFPA to be nonflammable. Some of these gases support combustion of other substances, while others suppress combustion. Gases that suppress combustion are called inert gases. By this definition, under normal conditions, inert gases include: carbon dioxide, nitrogen and sulfur dioxide, as well as the chemically inert gases, helium, neon, argon, krypton, and xenon. However, both carbon dioxide and nitrogen will support combustion of some metals under specific circumstances (Lemoff 1991).

Depletion of oxygen by dilution is a technique exploited in fire and explosion protection. There is a minimum concentration of oxygen below which flame will not propagate (Bodurtha 1980, Zabetakis 1965, Wysocki 1991). Flammability is suppressed by addition of inert gases, such as nitrogen, carbon dioxide, or steam (Zabetakis 1965). Oxygen levels must be maintained at or below 8.0% or 50% of the amount required to support combustion, whichever is less (NFPA 1992).

The relationship between composition of air–gas or air–vapor mixtures and flammability is described by the flammability diagram. Inerting is a direct application of the flammability diagram and the stoichiometry of combustion. This material appeared in Chapter 4 on Fire and Explosion.

Inert gases reduce the height of the ignitability envelope. In general, carbon dioxide reduces the height of the flammability envelope more than nitrogen per unit of volume. Many ignitable mixtures in air can be rendered nonignitable by adding about 28% of carbon dioxide compared to 42% of nitrogen.

Addition of an inert gas also may be required in spaces in which ignitable dusts are present (transfer lines, hoppers, silos, and baghouses). An inert gas can render a dust cloud nonexplosive. The main gases used for this purpose are nitrogen, carbon dioxide, and flue gas. In selecting a suitable inert gas, care must be taken to ensure that it is not reactive with the dust. Certain metallic dusts, for example, can react with carbon dioxide or nitrogen under specific conditions. Helium and argon would be suitable inerts in such situations (Schwab 1991).

The inert gas must reduce the oxygen concentration below the minimum needed to support combustion. Inerting gases vary in their ability to suppress combustion. For example, inerting using carbon dioxide requires reduction of oxygen to 11%, and 8% using nitrogen (Dept. of the Interior 1964). The appropriate margin of safety depends on the dust and conditions, but normally would be a further reduction of at least 2%. Thus, the system should maintain the oxygen concentration below 9% when the minimum needed to support combustion is 11%. Concentrations reported in the literature normally are measured at ambient temperature. The appropriate concentration of inert gas for protection against ignition at high temperatures is considerably higher. This also requires confirmatory testing (Lees 1980).

Minimum ignition energy and ignition temperature of ignitable dusts increase as oxygen concentration decreases. Inerting down to the limiting oxygen concentration of 11% may not be essential, if assurance is available that a strong source of ignition is not present. Partial inerting by carbon dioxide and/or the water vapor generated in a dust dryer may be entirely satisfactory (Bodurtha 1980).

Use of inerting as a mode of control requires careful planning and implementation. Dead spots and points of in-leakage of air must be considered and addressed. Multiple points of injection may be required. Reliance on inerting requires monitoring of oxygen level and procedures to stop work when malfunction occurs.

The use of steam and carbon dioxide for inerting requires great care and consideration where ignitable substances are present because of the potential for buildup of static electrical charges. The electrostatic hazard from steam arises from the formation of liquid water droplets through condensation. Strong electrification can occur at the orifice during escape of steam containing droplets of condensed water. Steam is not recommended for inerting in process or marine environments (NFPA 1993, API 1993).

A similar concern exists with carbon dioxide. When escaping under high pressure through an orifice as a liquid, carbon dioxide immediately changes to gas and solid (snow). This process can result in accumulation of static charge on the discharge device and the receiving container. Carbon dioxide should not be used for the rapid inerting of flammable atmospheres by injection under high pressure for this reason (Scarbrough 1991).

Other inert gases containing solid or liquid impurities also can produce strong electrification on discharge from an orifice. Bombardment of a conductive body by gas contaminated with dust, mist, scale, and metallic oxides can produce strong electrification of conductive fittings that are not bonded and grounded. Gas contaminated in this manner should not be used for purging or cleaning.

Static generated in this manner also can charge a nearby conductor that is insulated from ground (Lees 1980). Conductors insulated from ground can include tramp metal located inside vessels, wire netting around lagging, and insulated metal containers. Charge generation is related to velocity of impaction. Grounded probes should be avoided to minimize the probability of discharge.

VENTILATION OF CONFINED SPACES IN OPEN SYSTEMS

GENERAL

Published literature on the subject of confined space entry, ranging from journal articles to legislated standards, almost always specifies the need for "adequate ventilation." After all, poor natural ventilation is one of the parameters for defining a confined space (Garrison and McFee 1986).

An important problem with phrases such as "adequate ventilation" is that they have become so overworked as to be almost meaningless, and therefore easily misunderstand or ignored. If the word "meaningless" seems too strong, the leeway for interpretation surely certainly is too great. Guidelines and standards for ventilating confined spaces based on actual research involving these workspaces are conspicuous by their absence (Garrison 1990). (For additional information on guidelines, standards, and legislated requirements, see Appendix A.) To some extent, this is a consequence of the multitude of environments encompassed within the meaning of the term, confined space.

The ideal process for addressing ventilation needs in confined spaces is a consequential relationship between the findings of basic research and practical experience, and legislated requirements. In this case, the reverse has occurred. Legislated requirements have incorporated guidelines published for use in normal industrial environments (For additional information on guidelines, standards, and legislated requirements applicable in the U.S., see Appendix I). The appropriateness of this approach is open to question. The basic research needed to understand the intricacies of ventilating these environments is still in its infancy.

Opening to atmosphere is one of the main events in the preparation of many confined spaces for entry. For closed systems, this is the first contact between the internal atmosphere and residual contents, and air in the external surroundings. For spaces that normally are open or vented to atmosphere, this may represent contact with air that is uncontaminated compared to what routinely exists in the space.

Ventilation describes the condition in a space brought about by the inward movement of uncontaminated air and outward movement of contaminated air (ACGIH 1995). The purpose for ventilation in confined spaces is to ensure the quality of air, so that workers can perform their duties, preferably unencumbered by respirators. Ventilation air also can be tempered to provide cooling or heating, as appropriate, thereby improving comfort.

The two periods in the operational cycle of the confined space during which ventilation is critical are preparation of the undisturbed space prior to initial entry and work activity. The intent of ventilation during preentry preparation is to ensure a breathable atmosphere and hospitable conditions. Ventilation could be essential, since a contaminated atmosphere could remain from preparatory activities undertaken prior to opening the space. Also, there is the question about the existence and extent of contamination not known prior to opening the space or not detected by testing.

The intent of ventilation during work activity is to maintain conditions at an acceptable level. Residual contents could continue to cause contamination problems throughout the work period. In addition, some work activities produce hazardous contaminants at significant rates of emission. Under these conditions, continuous ventilation also would be necessary.

An additional consideration in defining ventilation requirements is sudden release of large amounts of pressurized gas, volatile liquids, or aerosols. These emissions may not be controllable even by continuous ventilation.

Is mechanical ventilation necessary in all confined spaces? Should natural ventilation be the primary method of choice? Should mechanical ventilation be employed only when natural ventilation is not effective? People responsible for entry safety have debated these and related questions since the earliest efforts to formalize procedures for entry into confined spaces. There are no hard and fast "yes" or "no" answers. The question about when and how to ventilate deserves critical analysis, probably by more than one knowledgeable individual. This becomes part of the consideration of acceptable risk in addition to technical, legal, administrative, and other considerations.

Effective ventilation for confined spaces "boils down" to considering and responding to all of the factors applicable to the situation. These should include:

- Configuration of the space
- Sources of emission
- Nature of contaminants
- Ventilation options and equipment
- Advantages, limitations, and constraints of each approach

Selecting the best alternative could become a difficult and involved process. It may require substantial advanced planning and purchase of special equipment. In some cases, the way in which work is performed may need significant alteration to accommodate the ventilation system. In assessing the impact that these changes may require on the way of "doing business," it is important to keep in mind the purpose of ventilation in confined spaces: prevention of loss of life.

There is no "cookbook" approach for ventilating confined spaces — far from it. Each situation should include consideration for the advantages and disadvantages of different alternatives. The optimum means to flush the space during preentry preparation may not be the same as for controlling contaminant emissions during work activity. As far as possible, the chosen strategy should be tested prior to use in any potentially hazardous situation. Preparation (equipment, procedures, training) is the critical element to assure successful implementation of ventilation control in confined spaces.

There are two main approaches for ventilating confined spaces: supply air to the space and remove air from the space. These approaches can be used alone and in different combinations. Since no two situations are the same, no single ventilation strategy is appropriate for all cases (Figure 11.1).

Ventilation Involving Hazardous Atmospheres

A major concern about ventilating confined spaces is that the action of ventilation itself must not create a hazard. Hazardous conditions certainly could arise readily enough under some circumstances during ventilation both inside and outside the space. The most critical period is when the space is first opened to air during preentry preparation. Up to that point, the hazard potential posed by the atmosphere in the space was low, since it was contained and undisturbed. Opening the space and starting to ventilate represents considerable potential for escalation of hazard.

The potential for development of a hazardous condition is determined by the physical and chemical properties and ignitability of the atmosphere trapped within the confined space or generated by its residual contents. In addition, the atmosphere that exists in the space prior to opening to air could differ completely from that resulting from work activity. These characteristics are extremely important, since they determine the approach to be taken and design parameters for the ventilation system. The absence of information about the confined atmosphere unnecessarily complicates these considerations. To illustrate, the absence of information could mean that compatibility between the confined atmosphere and the equipment used in ventilation could become an important consideration.

Atmospheric contamination in a confined space can be characterized by a matrix of descriptive elements. Table 11.1 lists descriptive elements for characterizing the atmosphere in the space. The atmosphere in the undisturbed space should be considered separately from contaminants generated during work activity. Some of the elements in Table 11.1 are stated explicitly in the hazard assessment for the undisturbed space and work activities, while others are implicit. Extremes in the descriptors within a descriptive element and certain combinations between descriptors from

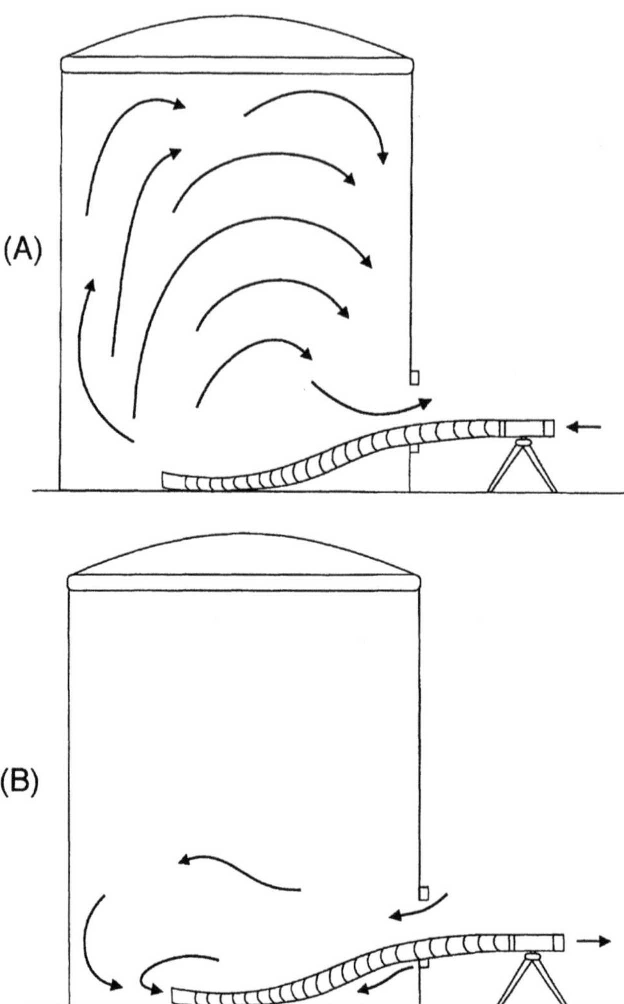

FIGURE 11.1 Portable systems used for ventilating confined spaces. (A) Supply Mode. (B) Exhaust Mode. The choice of mode reflects complex variables concerning the space, the hazardous substance, and the nature of the work. Actual airflow patterns within the structure require elucidation. (Courtesy of Tuthill Corporation, Coppus Portable Ventilation Division, Millbury, MA.)

different descriptive elements signal the need for caution, perhaps considerable. The synthesis from this analysis dictates the feasibility and potential safety of a particular ventilation strategy.

Each circumstance creates its own combination of descriptive properties. Combined together, the descriptors form phrases that provide an indication about conditions that could exist prior to entry or that could develop during work activity. Some signal the need for greater caution than others. For this reason, decisions regarding ventilation require careful consideration.

Atmospheres or sources characterized by certain combinations of descriptors can pose highly hazardous conditions during ventilation. Recognition of this potential in specific situations is essential for preventing disaster. The possibilities raised by these combinations further emphasize the need to obtain chemical, physical, and ignitability data prior to opening and ventilating a space. The preceding considerations are especially important to organizations that rely solely on preentry ventilation, for addressing atmospheric hazards in confined spaces.

TABLE 11.1
Properties of Undisturbed Atmosphere/Contaminant Source

Identity of substances — undisturbed atmosphere	Flammability characteristics
Atmospheric contaminant	Above upper flammable limit
Residual substance	Within flammable range
Process to be undertaken	Below lower flammable limit
Activity to be undertaken	Chemical reactivity
Identity of substances — work activity	Low
Residual substance	Medium
Process to be undertaken	High
Activity to be undertaken	Acidity/alkalinity
Physical state	Corrosive acidic
Gas or vapor	Low pH
Liquid aerosol (fog or mist)	Neutral
Solid aerosol (dust, powdered solid, fume, fiber)	High pH
Liquid	Corrosive alkaline (caustic)
Solid	Oxygen concentration
Thermal characteristics	Low
Cooler than internal atmosphere or external atmosphere	Medium
Same temperature as internal atmosphere or external atmosphere	Normal
Warmer than internal atmosphere or external atmosphere	Enriched
Buoyancy	Toxicity
Positive (less dense than air at ambient temperature)	Low
Neutral (same density as air at ambient temperature)	Medium
Negative or subsident (more dense than air at ambient temperature)	High
Distribution/generation	
Stratified	
Well-mixed	

Ventilation can create a hazard where one did not exist. There are numerous examples of problems that can arise from poorly executed ventilation strategies. The following examples apply to ventilation of the undisturbed space.

Introduction of air into an atmosphere that is above the upper flammable limit dilutes some of the confined atmosphere into the flammable range. A flammable atmosphere can develop inside the space or in the external surroundings as dilution and transfer occur. The presence of an ignition source, such as static electrical buildup involving the ventilation apparatus, could cause a fire or explosion.

A cloud of gas or vapor that is denser than the surrounding air can stratify in low spots in the space. A dense gas/vapor and air mixture can develop through a combination of specific gravity, concentration, and temperature of the substances that form the cloud. A layer formed by stratification at the bottom of the space could be extremely difficult to disperse. (The opposite of a stratified condition is a well-mixed, evenly distributed atmosphere. The latter situation is not necessarily less of a concern to ventilation strategy than the stratified layer.)

Displacement of an oxygen-enriched atmosphere from a confined space can enrich the air in the area surrounding the discharge of the air mover. The presence of combustible substances, including the materials of construction of the ventilating system and an ignition source such as a gasoline engine, arcing electrical component, or source of static electricity could provide the conditions for a fire or explosion.

Similarly, transfer of a highly toxic substance to the exterior of the space can create a serious exposure hazard in the area of the discharge.

Cooling below ambient temperature can reduce the buoyancy of gases such as ammonia and methane that otherwise would rise to the highest point in the space. Under these conditions, such gases could exhibit subsidence and sink to the bottom of the space or low spots in the exterior surroundings. Warming during ventilation could immediately increase the buoyancy, possibly leading to development of a potentially flammable/explosive mixture.

Reactivity with materials used in construction of the ventilation equipment is a concern when the atmosphere contains concentrated acid or alkaline corrosives or chemically reactive or unstable substances. Destruction of the ventilation equipment through chemical attack could result.

Work activity in the space is more likely to have deleterious impact on components of the ventilation system than to create a hazard where none existed. Hot plumes and hot surfaces can melt or burn through components, thereby damaging or destroying the equipment. Spray application of coatings produces liquid aerosols that can coat the interior of the ventilation system. This can destroy fans and other critical components.

Ventilation Modes

Confined spaces usually do not possess ventilation systems. They usually are not intended for occupancy. Creature comforts and safety during occupancy usually were not a consideration in the minds of the designers of these structures, because occupancy was neither expected nor intended. Hence, entry and work in these environments should be considered an aberration. Those responsible for establishing safe conditions of work in these environments sometimes must overcome considerable inertia.

There are relatively few ways to induce air motion. Utilizing these in the ventilation of confined spaces can be limited by constraints imposed by the environment in which the work is to occur and the size and complexity of the space. The modes of moving air in confined spaces involve the following:

- Natural ventilation
    - Convective air motion
    - Wind
- Mechanical ventilation
    - Air mover at opening
    - Portable ventilation system
- Compressed air

Successful strategy for ventilating confined spaces should be considered a matter of recognizing and harnessing what is available and supplementing with what is needed to establish and maintain safe conditions of work.

Natural Ventilation

Some confined spaces, such as open-ended tunnels and sewers, can receive significant natural ventilation. This results from internal geometry, location relative to surface airflows, configuration, and/or process. Natural ventilation utilizes pressure differentials arising from wind and/or convection to cause air movement. Optimum natural ventilation occurs when the space is opened as much as possible and oriented in directions that take advantage of natural airflows, thereby minimizing static pressure losses that restrict air movement and contaminant dilution. In general, for spaces with high roofspaces and high heat release, the ventilation rate is established by the heat release. For low spaces with low heat release, wind is the factor that determines ventilation characteristics (Goodfellow and Smith 1982).

TABLE 11.2
Convective Air Motion in Confined Spaces

- Source is gas, vapor, or fume
- Worker breathing zone does not enter contaminant plume
- Plume discharges from highest point and does not mix with air in the space or stratifies above work zone or height of normal access
- Downward fumigation cannot occur
- Air enters the space below the source of contaminants
- Convective flow begins at or before time of generation of contaminants
- Contaminant has low toxicity or flammability
- Discharge directly to the environment can occur without removal from the airstream
- Dispersion into the air of the workspace above the source can occur without consequence from corrosive or other hazardous chemical property

Convective Air Motion

Convection has been utilized in the ventilation of countless buildings and structures for centuries. Convection is upward motion of air resulting from heating. The decrease in density caused by absorption of energy produces the upward motion. The rise of heated air occurs until the increase in buoyancy is lost due to cooling by the surroundings. Cooler air enters the space to replace that lost by convection.

Convective air motion results from the transfer of energy by internal and external sources to the air in the space. The energy may be provided uniformly to all air in the space or, more typically, to a specific localized zone. Internal sources of energy include processes and equipment that emit radiant energy and provide additional energy to the air through conduction. Convective airflows develop around these sources. Heat sources present in a confined space could include welding arcs, oxyfuel cutting equipment, and many kinds of process and electrical equipment. The most obvious source of energy external to the space is the sun. Energy from the sun usually heats only some of the surfaces that form the boundaries of the structure within which the space is located.

Convective air motion can be exploited for ventilation of confined spaces under circumstances, as summarized in Table 11.2.

Convective airflow can be utilized in an effective manner in some spaces under some conditions. For example, convective air motion can occur in large tanks that have roof-level vents and low level access portals. Contamination from a process, such as welding, would rise to the roofspace.

Of considerable concern in these situations is the rate of escape of contaminated air through openings in the roof. Insufficient escape results in accumulation in the roofspace above the source and fumigation downward into the airspace below. Prominent examples include accumulation of smoke in the roofspace of a building during a fire or welding processes. In both cases, contaminated air rises to the roofspace and spills downward when venting is insufficient relative to production. This situation becomes acutely more hazardous within the small structures that typify many confined spaces.

Venting the roofspace of buildings has been the subject of NFPA standards since 1902 (Heskestad 1991). This source provides detail about venting theory for calculating vent areas in roofs. These calculations require mass and heat generation rates. These are unlikely to be available for sources that would be present during work that occurs in confined spaces. To some extent, trial and error would seem to dictate the suitability of existing venting and exhaust capability of these structures.

Emswiler in 1926 proposed equations to explain natural airflow in buildings. These remain current today (Emswiler 1926). When the air inside a structure is heated to a temperature greater than that of the surrounding air, there results a tendency for flow to occur. Cool air enters at low openings and warm air escapes from upper ones. External heat sources, such as the sun, can influence the ability of convective airflow to ventilate structures to a large extent. This is especially important with metal structures that absorb energy from bottom to top and reradiate it into the interior.

According to Emswiler, the head causing flow at any opening is proportional to the vertical distance of that opening from the "neutral zone" in the structure. The position of the neutral zone is governed by the relative amount of opening provided at the top and bottom of the structure and the temperature difference between inside and outside air at different levels. The internal air pressure at the neutral zone is the same as that outside the building. The amount of air that may pass by a given opening is proportional to the square root of the vertical distance of that opening from the neutral zone. A stack, permanent or temporary, can increase the vertical distance and therefore the amount of air that can pass, provided that flow resistance is small. An opening at the neutral zone is totally ineffective in passing air. The most effective openings are those located at the extremes of height.

Emswiler provided the following equations.

$$h = \frac{z \times D}{T} \tag{11.1}$$

h = the head measured in feet of air
z = difference in height between the opening and the neutral plane in ft
D = difference in temperature between warm air inside the structure and cool air outside, $t_{inside} - t_{outside}$, in °F
T = the average absolute temperature in °R

$$T = \frac{t_{inside} + t_{outside}}{2} + 460 \tag{11.2}$$

Resistance to flow through an opening is given by:

$$R = \frac{1}{2 \times g \times A^2 \times C^2} \tag{11.3}$$

R = resistance of a given opening
g = acceleration due to gravity, 32 ft/s^2
A = clear area of the opening in ft^2
C = a coefficient expressing individual peculiarities of the opening. C has a value between 0.60 to 0.70.

Flow through a particular opening is given by:

$$Q = \frac{\sqrt{z \times D}}{R \times T} \tag{11.4}$$

Q = volumetric flow rate in ft^3/s

The final relationship is that the sum of all flows in equals the sum of all flows out ($Q_{in} = Q_{out}$). Inflow occurs below the neutral zone; outflow occurs above the neutral zone. The height, z, of the neutral zone in the structure is not known. It must be determined by calculation using successive approximation starting with a presumed height. The correct height is such that $Q_{in} = Q_{out}$. Once the height of the neutral zone is determined, the net inflow or outflow, the convective ventilation rate for the structure, is known for a given temperature difference.

TABLE 11.3
Wind Flow in Confined Spaces

- Stable weather conditions
- Linear path between inlet and outlet
- Plane of the inlet oriented perpendicular to the direction of the wind
- Sources and work locations localized in the space
- Source is gas, vapor, or fume
- The space has horizontal orientation
- Worker breathing zone does not enter contaminant plume
- Workers remain upstream from flow of contaminated air
- Discharge directly to the environment can occur without removal from the airstream
- Dispersion into the air of the workspace downstream from the source can occur without consequence from corrosive or other hazardous chemical property

These calculations assume that the temperature inside the structure is uniform. Where temperatures differ with height, the calculations can be performed using exact measured values for each height within the structure.

Wind

Wind can have considerable influence on ventilation of confined spaces, such as horizontal tunnels. Wind pressure exerted at the portal results in flow of a plug of air through the structure. On a steady basis, the velocity of this flow easily can reach 4 m/s (800 ft/min). Flows of this magnitude can provide considerable benefit in the ventilation of confined spaces. This, of course, is highly dependent on the velocity and direction of the wind and weather conditions. Reversal of normal direction can occur as day changes to night. The transition period and air inversions can lead to stagnation in the structure.

Wind sometimes can be concentrated and directed into openings in spaces, using sails or other such structures. As in other applications of wind energy, the results are variable. Use of this approach for ventilating confined spaces also may not be legal.

Wind can be exploited for ventilation of confined spaces under some circumstances, as indicated in Table 11.3.

Wind-induced airflow in a structure may be an essential component in ventilation, especially of large structures, since fans may be incapable of providing sufficient airflow. Wind-induced airflow is susceptible to dead spots, such as chambers off the axis of flow and short-circuiting.

Wind load on structures causes the pressure around and inside the structure to rise or fall (Liu 1979). Pressure increases above ambient on the front or windward side of the structure and decreases below ambient on the windward side, above the roof, and adjacent to the ends that parallel the flow. The maximum pressure rise on the upwind side of a structure is given by the following equation.

$$\Delta P = 0.5 \times \rho \times v^2 \tag{11.5}$$

ΔP = increase in pressure above ambient in Pa (N/m^2)
ρ = density of air in kg/m^3
v = wind velocity in m/s

Since wind pressure depends on the square of the air speed, increasing the air speed by a factor of 2 increases wind pressure by a factor of 4. Wind pressure on the windward surface decreases toward the edges of the wall. Wind pressure also decreases when the angle of contact is less than 90° from the windward surface.

An important percentage of the damage and destruction caused to structures by wind results from excessive internal pressure combined with suction on the roof and leeward sides. Internal pressure in a structure is uniform, except for segregated rooms that each have a different pressure. Interior pressure rises to as high as the maximum external pressure when an opening exists on the windward side, but not the leeward. With openings on both sides, the wind will enter where external pressure is high and exit where it is low. The larger the openings, the stronger the air motion. With more than one opening on windward and leeward side, the internal pressure and airflow will be a function of the respective areas and the corresponding external pressure.

Advantages and Disadvantages of Natural Ventilation

Natural ventilation offers the obvious advantage of avoiding the need for costly mechanical equipment. What equipment is required can be set up relatively easily. Natural ventilation also is not subject to mechanical failure. If natural ventilation can provide significant and reliable air movement, then, legal caveats to the contrary, this mode may be satisfactory for entry purposes.

Natural ventilation has several significant disadvantages. Principal among them is lack of predictability and flexibility. Airflow can change unpredictably without warning. Changes in wind speed and direction can significantly alter ventilation inside a space. Assuring a specified airflow rate at all times is not possible. Natural ventilation provides few options for altering airflow rate and direction. This may not be compatible with changing conditions inside the confined space. Airflow provided by natural sources may not be readily utilized in areas having high localized concentration of contaminant.

Under some conditions, contaminated air from the surroundings could reenter the space due to reentrainment. Air velocity may be insufficient to produce effective mixing and flushing of contaminants. This is especially likely for confined spaces having few and small openings. Natural ventilation usually cannot meet specific criteria for volumetric flow rate and velocity.

The risks associated with reliance on natural ventilation to control hazardous air contaminants are usually too great for its use in routine situations as the sole mode of ventilation. Mechanical ventilation should be the preferred choice when hazardous airborne contaminants are present. Natural ventilation should be utilized to the extent possible to supplement mechanical ventilation.

Mechanical Ventilation

In most situations requiring ventilation, mechanical ventilation provides the greatest margin of safety. Compared to the advantages, the disadvantages are minimal. Mechanical ventilation absolutely is required for confined spaces that have little or no natural ventilation where potentially hazardous contamination is produced.

Mechanical ventilation provides greater control and flexibility in directing air movement than natural ventilation. Air movers can supply fresh air directly to work areas and/or remove contaminated air from the space. A specified rate of ventilation can be achieved. Also, airflow patterns can accommodate to changing circumstances in the space. (See Appendix J for discussion about portable ventilation systems.)

The temporary and intermittent nature of work involving confined spaces usually dictates that permanently installed ventilation systems are not feasible, practicable, or sensible. This results in the need to use portable equipment along with creative, inventive, and novel strategies. Ventilation of small workspaces using portable equipment poses challenges of one type. Ventilation of large confined spaces using portable equipment poses completely different ones.

There are four main techniques for ventilating spaces using mechanical equipment: supply air to the space, remove air from the space, a combination of the two, and source extraction using local

FIGURE 11.2 Examples of air moving equipment used in portable ventilation systems. These include electrical-, fluid-, and air- and steam-powered units. (Courtesy of Tuthill Corporation, Coppus Portable Ventilation Division, Millbury, MA.)

exhaust. The preceding options are routinely employed in the permanent installations used to ventilate large spaces, such as buildings and processes. They also form the basis for ventilating confined spaces. Mechanical ventilation of confined spaces involves use of portable equipment to induce airflow. Fans and venturi-type eductors are the most common types of equipment (Figure 11.2).

A major concern about use of mechanical equipment for ventilating confined spaces is operational failure from loss of power or mechanical breakdown. This could leave entrants little or no protection in situations in which this equipment provides the first line of defense. Other disadvantages include costs of the equipment and the labor for set-up and removal. This equipment definitely encumbers work activities in many circumstances because it competes for surface area in access/egress portals. It also can cause elevated noise levels inside the confined space. In situations involving flammable vapors or dusts, mechanical ventilating equipment must be rated for use in hazardous locations or distanced sufficiently from access/egress portals so as not to act as an ignition source.

Compressed Air

Compressed breathable air from a plant supply system or high pressure cylinders is an option for supplying air into a confined space. Because of the pressure of compression, this air expands into the available volume. Air compressed to 100 lb/in.^{2}, for example, will expand into a volume approximately 7 times the delivery volume.

Compressed air can be delivered through hoses of small diameter. This offers a distinct advantage where the access/egress portal is small. The quality of air used for this purpose, of course, must meet standards applicable to compressed breathing air (CGA 1989; CSA 1985). Discharge of compressed air from hoses also is very noisy. An array of silencers on the discharge can overcome this problem (Silvent 1993).

Supplying Air to a Confined Space

General Discussion

The first option for ventilating confined spaces is to supply air to the space. Supplying fresh air probably is the most common approach to ventilating these workspaces. The principal objective is to establish reliable airflow throughout the space. The freshest air should be provided to the breathing zone in work areas.

Supplying air to a structure slightly pressurizes the atmosphere inside the boundary surfaces, leading to passive outflow. Outflow occurs through any opening that borders the pressurized zone. Supplying air to the space thus displaces air and dilutes contaminants in the atmosphere trapped inside. Removal of contaminants can occur by dilution or displacement.

The mechanics of supplying air to confined spaces can occur in several ways:

- Natural ventilation
    - Wind
    - Convection
- Mechanical ventilation
    - Compressed air
    - Air mover at opening
    - Portable ventilation system

These concepts have been discussed in a previous section.

One option for supplying air to a confined space is to position the air mover at an opening. The unit draws air from the surrounding area and discharges into the space. This configuration may be all that is possible with some types of air movers or situations. This arrangement offers the convenience of rapid set-up and removal. On the other hand, short-circuiting and reentrainment are possible occurrences. Air escaping from the space through the same opening may be reentrained by the air mover. Further, the discharge may fail to ventilate areas off the main axis of flow that are confined by structural boundary surfaces within the space.

A portable ventilation system typically uses a fan as the air mover. Air provided by the fan usually is transported through a flexible duct to the discharge point within the space. In some situations, an intake duct is used upstream to draw supply air from an uncontaminated location. The discharge duct passes through an opening into the space. The discharge from the duct can be directed to the work area.

Supplying air to a confined space during occupancy offers several advantages. First, this technique can deliver fresh air directly to the breathing zone. By using appropriate hardware and movement, this approach can provide fresh air to every work location within the space. The supply mode provides rapid dilution of contaminated air in the area of the discharge, due largely to turbulent mixing produced by the discharge "jet." The discharge jet is highly efficient at projecting air. The velocity of the jet decays slowly, decreasing to 1/10th that in the fan at approximately 30 duct diameters (ACGIH 1995).

TABLE 11.4
Dilution Ventilation — Preentry Preparation

- Release of small quantities of contaminant at uniform, predictable rate from residual sources
- Source is gas or vapor, not particulate
- Air in the space mixes well and rapidly; the space has simple geometry and no dead spots
- Contaminant has low toxicity or flammability
- Discharge directly to the environment can occur without removal from the airstream
- Dispersion into air of the workspace can occur without consequence from corrosive or other hazardous chemical property

Data from McDermott 1976, Burgess et al. 1989, and Heinsohn 1991.

The supply mode can provide convective and evaporative cooling during work involving hot conditions. Heated air also can be provided under cold-weather conditions. Volumetric flow rates also can be designed to meet specific criteria.

Inherent in the supply mode are several disadvantages. The large flow can disperse settled dust and evaporate liquids at an undesirable rate. Contaminants are discharged at opening(s) in the space, including the access/egress portal. Under conditions of poor dispersion outside the access/egress portal, this could result in unacceptable levels of exposure of support personnel. Accumulation of contaminants in the space also can occur when the source of supply air is contaminated. Contamination of supply air can occur through reentrainment of previously discharged air. This mode of ventilation of confined spaces is especially sensitive to short-circuiting. High localized concentrations can occur due to inadequate mixing. This can occur when a single supply of air is used in the presence of multiple sources of contamination or when the space is very complex.

Undesired and possibly even hazardous convective/evaporative cooling can occur when the supply air has low ambient temperature. This is a major problem under severe winter conditions.

Dilution Ventilation

The premise implicit in the use of dilution ventilation is rapid mixing of clean air supplied to the workspace with contaminated air. A ventilation system is considered to be operating in dilution mode when the concentration of contaminants in the discharge is not significantly higher than what is present in the air of the space (Mutchler 1973).

Dilution ventilation is widely exploited in the ventilation of office buildings and some workspaces. It is applicable as a method of ventilation in some situations involving confined spaces. Table 11.4 summarizes application of dilution ventilation during preentry preparation. Dilution ventilation during preentry preparation can produce completely acceptable flushing of the space, provided that concerns raised in Table 11.1 have been addressed. Sources of contamination could be inactive and all that is required is flushing. However, sources of contamination still could be active, as in a space whose surfaces are wetted by residual liquids or that contains pooled liquid or sludge. In the latter case, the ability of the supply to control conditions in the space would depend on emission rate and ventilation rate. Insufficient or incapable supply could dictate the need for further preparatory work, such as cleaning, prior to attempts at ventilation. The appropriateness and success of this approach can be ascertained by air sampling during a prework inspection. This would assure that the supply is sufficient to produce efficient ventilation.

Dilution ventilation is the most common method used to ventilate confined spaces during work activity. Limitations associated with dilution ventilation in normal applications are more significant in confined spaces. Table 11.5 summarizes application of dilution ventilation during work activity. Success in the application of dilution ventilation as a control measure during work activities is sensitive to the factors listed in Table 11.5. The difficulty in applying the concept during work

TABLE 11.5
Dilution Ventilation — Work Activity

- Release of small quantities of contaminant at uniform, predictable rate
- Sources are distributed throughout the space
- Source is gas or vapor, not particulate
- Air in the space mixes well and rapidly; the space has simple geometry and no dead spots
- Contaminant has low toxicity or flammability
- Discharge directly to the environment can occur without removal from the airstream
- Dispersion into air of the workspace can occur without consequence from corrosive or other hazardous chemical property

Data from McDermott 1976, Burgess et al. 1989, and Heinsohn 1991.

activities is generating enough airflow in large confined spaces or spaces with complex internal geometry and dead spots. The largest confined spaces include storage tanks in the range of 1,000,000 gal (4,000 m^3). Railway tunnels can be even larger, depending on length. Use of dilution ventilation to control emissions during work activities can necessitate large volumetric airflow rates. These could be difficult, if not impossible, to achieve with equipment presently available, especially in remote locations where utilities and equipment are difficult to provide.

Dilution of contaminated air generated by the work activity prior to entry into the breathing zone could be difficult to achieve when work activity, and hence the point of origin of the contaminant, requires mobility. Dilution ventilation never provides complete and immediate uniform mixing of air in the space except when high ventilation rates are achievable. Systems used for ventilating confined spaces may have only a single outlet as compared to multiple points of distribution in normal settings. Hence, localized high concentrations can develop when source-to-worker-to-ventilator geometry is poor. Estimating contaminant production rates accurately is very difficult, even in normal circumstances. Hence, this uncertainty increases reliance on the "K factor." The K factor, which ranges from 1 to 10, describes the effectiveness of the ventilation (ACGIH 1995). Since there are no published data on appropriate K factors to use in the dilution equation, selecting a value of K greater than normally used is appropriate in dilution calculations for confined spaces.

Dilution ventilation has been the mode of choice where large distributed sources are present. These could include surfaces wetted by residues, sludges, pooled liquid, and application of coatings. Recent work with models has suggested that control of emissions from large distributed sources using dilution ventilation could be very difficult to achieve. Increasing airflow increases the rate of evaporation from wetted surfaces by removing vapor that has accumulated (Bjerre 1989, Jayjock 1994).

Dilution ventilation does not provide good control over localized sources or spurious (fugitive) emissions. Contaminant concentrations may be unacceptably high during such episodes. Discharge from the supply should be directed to the greatest possible extent to ensure flow of contaminated air away from the breathing zone. This can become counterproductive in the preceding circumstances or where highly toxic substances are involved. Flow away from one individual may direct contamination into the breathing zone of other workers. Also, the induced flow is not an efficient means for removing contaminants from the space.

Dilution ventilation is not considered appropriate for control of emissions from particulate sources. Airflows present during dilution ventilation generally are low. This means that large particles are not transported efficiently from the point of generation into the air volume and could deposit on surfaces.

Displacement Ventilation

Displacement ventilation (plug flow) is a variant of dilution ventilation. Both focus on the supply of air to the space. Whereas dilution ventilation relies on rapid mixing of contaminants to produce

TABLE 11.6
Displacement Ventilation — Preentry Preparation

- Release of small quantities at uniform rate
- Source is gas, vapor, or fume
- The space has linear path between supply and discharge and no dead spots
- Contaminant has low toxicity or flammability
- Discharge directly to the environment can occur without removal from the airstream
- Dispersion into the air of the workspace downstream from the source can occur without consequence from corrosive or other hazardous chemical property
- Source emissions stratify at a height above working height

a uniform mixture, displacement ventilation exploits uniformity of flow and lack of mixing. The premise behind displacement ventilation is to move contaminated air toward the point of discharge without mixing with supply air. The most recognizable examples of displacement ventilation are automotive and other large spray booths in which an individual works. Air from the supply moves past the breathing zone and removes contaminants before they can cause exposure. Displacement by wind and convective airflow removes exhaust gases from railway tunnels after the passage of a train. Displacement by convection carries contaminants, such as welding plumes, into the roof-space of a building. Displacement occurs naturally in other spaces, such as sewers that normally are open to the atmosphere.

Displacement has been utilized in the ventilation of industrial buildings for many years. In this approach, cool air is introduced near floor level with low momentum; heated air is removed at roof level. This approach relies on heat generated by sources in the building to provide positive buoyancy to contaminated air. Recent studies have shown that this approach is more efficient in distributing fresh air from supply sources into the zone of occupancy than dilution ventilation (Breum et al. 1989, Breum and Ørhede 1994).

Displacement ventilation is potentially suited to applications involving confined spaces. Table 11.6 summarizes applications to preentry preparation. The importance of a particular term in Table 11.6 reflects the characteristics of the confined space, the context of the application, and the nature of the contaminant. Displacement ventilation during preentry preparation can provide completely acceptable results, provided that concerns raised in Table 11.1 have been addressed. Displacement of air from the space can be achieved using both natural and mechanical modes of ventilation. These can be applied simultaneously and can compliment one another.

Sources of contamination could be inactive, in which case, all that is required is flushing. However, sources of contamination still could be active, as in a space whose surfaces are wetted by residual liquids or that contains pooled liquid or sludges. In the latter case, the ability of the airflow to control conditions in the space would depend on emission rate and ventilation rate. Insufficient removal could dictate the need for further preparatory work, such as cleaning, prior to attempts at ventilation.

A ventilation regime that utilizes plug flow is susceptible to deadspots in the space. These can occur near the air supply and in side chambers. Hence, the location of the air supply is critical to ensuring displacement in these areas. In addition, this ventilation regime is potentially susceptible to short-circuiting. In order to be effective, the discharge from the space must linearly located as far as possible from the supply. All areas in which occupancy will occur must receive flow in a direct path from supply to discharge. The appropriateness and success of this approach can be ascertained by air sampling during a prework inspection. This would assure that the supply is sufficient to produce efficient ventilation.

Displacement ventilation also has potential for application during work activity. Table 11.7 summarizes criteria for application of displacement ventilation during work activity. The importance

TABLE 11.7
Displacement Ventilation — Work Activity

- Release of small quantities at uniform rate
- Sources and work locations localized in the space
- Source is gas, vapor, or fume
- The space has linear path between supply and discharge and no dead spots
- Worker maintains position upstream from sources
- Geometry between the source and the breathing zone permits airflow to remove contaminants prior to mixing and dispersion
- Contaminant has low toxicity or flammability
- Discharge directly to the environment can occur without removal from the airstream
- Dispersion into the air of the workspace downstream from the source can occur without consequence from corrosive or other hazardous chemical property
- Source emissions stratify at a height above working height

of a particular term in Table 11.7 reflects the characteristics of the confined space, the context of the application, and the nature of the contaminant. Successful application of displacement ventilation as a control measure during work activities is sensitive to the factors listed in Table 11.7.

The most critical requirement for successful application is the worker-to-source-to-flow geometry. This is not a factor during preentry preparation, as the space is not occupied. Displacement ventilation also can be highly successful when emissions can be stratified and isolated in a zone above that occupied by workers. For example, exhaust from vehicles and equipment can be directed into the roofspace of a tunnel. Flow provided by wind and supplemented by roof-level fans discharging horizontally in the same direction in a push-push-push mode can minimize downward fumigation, thus establishing highly effective separation. This approach would be unsuccessful in a situation where vertical stratification does not occur because of mixing, and worksites are spaced along the length of the tunnel. Contaminated air from one worksite becomes the source of breathing air in work locations downstream.

Use of displacement ventilation to control emissions during work activities can necessitate large volumetric airflow rates. These could be difficult, if not impossible, to achieve with the portable equipment presently available. This could pose serious problems in remote locations.

Displacement ventilation is not considered appropriate for control of emissions from sources of large particulates. Airflows during displacement ventilation generally are low. This means that large particles are not transported efficiently from the point of generation into the air volume and could deposit on surfaces.

Removing Air from a Confined Space

Removing air from the confined space (general exhaust ventilation) is another ventilation option. This requires use of mechanical equipment. Removing air from the interior of a structure creates a slight negative pressure. This induces inflow through openings. This approach dilutes contaminated air by drawing in fresh air. Ventilation equipment used in the exhaust mode can be configured in the following ways:

- Air mover at opening
- Portable ventilation system

Positioning the air mover at the opening of the space is the only available option in some circumstances. This configuration offers the advantage of rapid removal of associated equipment from the portal. However, short-circuiting and lack of collection from affected areas are major

TABLE 11.8
General Exhaust Ventilation — Preentry Preparation

- Release of small quantities at uniform rate
- Source is gas, vapor, or fume
- The space has no dead spots
- Contaminant has low flammability
- Removal of contaminant can occur before discharge to the workspace or external environment
- Dispersion into the air of the workspace can be prevented to avoid consequence from corrosive or other hazardous chemical property

concerns. Air entering the space from the exterior through the same opening could supply the air mover. This would reduce the amount of air removed from the interior of the space. Capture by fans is considerably less effective than discharge. Velocity of air on the intake side of a fan decreases to 1/10th that inside at a distance of one duct diameter (ACGIH 1995).

A portable ventilation system is set up in reverse to ducted supply. In this configuration the air mover draws from the space through a duct. In analogous fashion, a duct can be connected on the discharge to move exhaust to a considerable distance from the source. The air mover should not be positioned inside the space. This reduces the possibility of recontaminating air in the space due to leakage on the pressure side of the air mover.

General Exhaust Ventilation

General exhaust ventilation is based on the premise of controlling the flow of air outward from the confined space. Air is removed through a collector system that usually has only a single point of access. General exhaust ventilation utilizes mechanical ventilation. The equipment discharges at the air mover or at a distance through use of discharge duct.

General exhaust ventilation is used for ventilating confined spaces. Table 11.8 summarizes criteria for application of general exhaust ventilation during preentry preparation. The importance of a particular term in Table 11.8 reflects the characteristics of the confined space, the context of the application, and the nature of the contaminant.

Removing air from a confined space offers several advantages. This mode is collective, rather than dispersive. Contaminants are collected from within the space. They can be removed from the airflow or treated and discharged remote from the space and its immediate surroundings. Openings in the space remain under negative pressure during the process. Clean air flows into the space; contaminants exit only under controlled discharge. This minimizes exposure of support personnel working around the space to the atmosphere in the space or contaminants generated from residual contents.

Purging time is likely to be slow compared to supply modes. Capture is considerably less effective in inducing air motion than discharge.

General exhaust ventilation could be susceptible to short-circuiting and dead spots. This is true mainly for confined spaces having multiple openings and partitions and chambers. Placement of the intake relative to openings in the structure is critical to ensuring that incoming air mixes thoroughly with that in the space. The suitability of the placement should be determined by testing with a smoke tube. Reasonable mixing can be achieved, provided that the relationship of the intake to the duct or air mover and the openings through which air enters the space receives appropriate attention.

General exhaust ventilation also has potential for application during work activity. Table 11.9 summarizes criteria for application of general exhaust ventilation during work activity. The importance of a particular term in Table 11.9 reflects the characteristics of the confined space, the context

TABLE 11.9
General Exhaust Ventilation — Work Activity

- Release of small quantities at uniform rate
- Single source per exhaust collector
- Source is gas, vapor, or fume
- Geometry between the source, the collector, and the breathing zone permits airflow to remove contaminants prior to mixing and dispersion
- Removal of contaminant can occur before discharge to the workspace or external environment
- Dispersion into the air of the workspace can be prevented to avoid consequence from corrosive or other hazardous chemical property

TABLE 11.10
Local Exhaust Ventilation — Work Activity

- Sources are discrete, point sources
- Collector is available for each source
- Collector can be positioned to remove contaminants before mixing and dispersion
- Removal of contaminant can occur before discharge to the workspace or external environment
- Dispersion into the air of the workspace downstream from the source can be avoided to prevent consequence from corrosive or other hazardous chemical property

of the application, and the nature of the contaminant. Successful application of general exhaust ventilation as a control measure during work activities is sensitive to the factors listed in Table 11.9.

The most critical requirement for successful application is the worker-to-source-to-flow geometry. This is not a factor during preentry preparation, as the space is not occupied. General exhaust ventilation would have little effect on emissions from distributed sources or from large surfaces. The exhaust mode usually produces less effective air mixing than does supply. As a result, occupants can sustain exposure to contaminants from sources not under the influence of the collector prior to their removal from the space. This is most likely to occur when multiple sources are present. High localized levels of contamination also are possible. Discharge of contaminated air from an air mover located at the access/egress portal could create unacceptable levels in this area. This can be rectified by use of a discharge duct or air-cleaning device.

General exhaust ventilation can control emissions from sources of large particulates, since airflows could be high in the area near the source.

Airflow velocities and undesired cooling effects produced by drafts of outside air typically will be less in general exhaust mode than in supply mode. Exhaust ventilation also can be designed to meet specific criteria for volumetric flow rate.

Local Exhaust Ventilation

Local exhaust ventilation utilizes airflow to entrain and thus to capture and remove airborne contaminants at or near their source(s) of emission. Equipment used for local exhaust includes an intake collector, duct, and an air mover. Local exhaust typically involves the movement of substantially less air, but at higher velocity than general exhaust ventilation. The local exhaust system could be a self-contained unit that collects and discharges in the local area.

Local exhaust ventilation can provide effective control against contaminants generated by work activity. This can accommodate to both intermittent and continuous release. Table 11.10 summarizes criteria for application of local exhaust ventilation during work activity. Local exhaust can provide effective protection against rapid emission of contaminants. Volumetric flow rates typically are less than those required for other modes of ventilation. A smaller duct is needed. This is highly desirable when only a small opening is available at the portal for passage of utilities. Similarly, requirements for make-up air also are reduced. Local exhaust can be used in conjunction with a supply system. Local exhaust also is nondispersive to contaminants. Once collected by the local exhaust system, contaminants can be removed from the space without contaminating the air in the remainder. This offers the opportunity to use aircleaning devices to remove contaminants from the exhaust airflow, as appropriate. Contaminants need not enter the breathing zone of any of the workers in the space.

While local exhaust ventilation offers significant advantages, it also has disadvantages. Contaminant removal from the space will not be effective unless the collector is positioned in an appropriate manner near the source. This requires commitment and finesse by the user of the equipment. The user must move the collector as the source of contamination moves to ensure continuing performance.

Contaminants not recovered at the source will enter the general airspace and possibly the breathing zone. Contaminants in the general airspace would be very difficult to recover. The low volumetric flow rate provided by a local exhaust system would provide low dilution efficiency compared to the other modes of ventilation. Hence, the effectiveness of a local exhaust system is highly dependent on the initiative and skill of the user.

Contaminant discharge from the air mover or duct also can be a problem since the airflow can be highly concentrated. Location of the discharge requires planning to ensure that reentrainment or localized contamination will not occur. Local exhaust is only effective when replacement air is readily available at the zone of collection. This can become a problem when replacement air is not supplied to the confined space. Efficiency of collection can decrease.

Local exhaust ventilation is not applicable where large distributed sources are present. These include surfaces wetted by residual liquids and sludges and pooled liquids.

Simultaneous Supply and Exhaust

Another option for ventilating confined spaces is a combination of the preceding two: simultaneous supply and exhaust. This concept is used routinely in building ventilation systems. Simultaneous supply and exhaust can be adapted to confined spaces, with some modifications.

Many confined spaces have restricted access/egress portal(s). This limitation could restrict the use of multiple air movers at these openings or the use of more than one duct. A supply and return system can be configured using compressed air transported in small diameter, high pressure hoses as the supply and general or local exhaust as the return. This design reduces the cross-sectional area occupied by ducts. Also, the high pressure supply line could be situated inside the exhaust duct. This configuration is likely to produce considerable noise inside the confined space. Arrays of silencers can reduce the noise produced at the discharge from the compressed air line.

This design offers the flexibility to locate the supply in the work area near the breathing zone. Also, the intake of the exhaust can be located near the source of emission to offer a level of control.

Comments raised in the preceding sections apply to the individual components in this system. Because of the nature of the system, controlling airflow rates is more of a concern. Oversupply will pressurize the space, leading to outflow through openings. This could be undesirable when highly toxic contaminants are involved, as mentioned previously. Undersupply will lead to inflow through openings. This also could be disadvantageous when the quality of air near the openings is questionable.

LABORATORY STUDIES OF VENTILATION OF CONFINED SPACES

Garrison and co-workers performed several studies dedicated to determining the characteristics of ventilation in confined spaces (Garrison et al. 1989; Garrison and Erig 1989, 1991; Garrison et al.

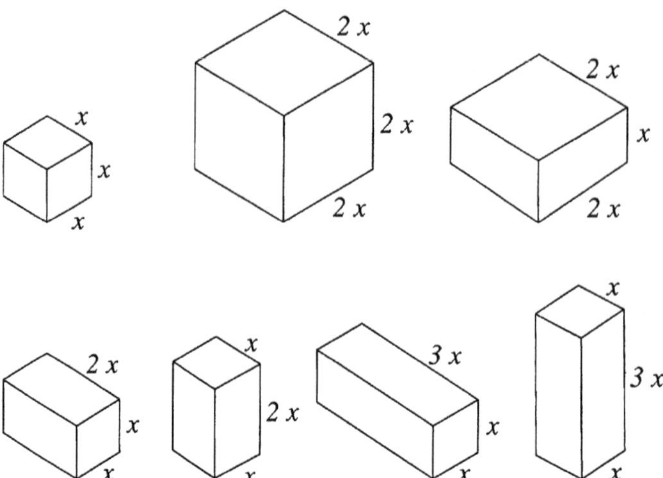

FIGURE 11.3 Model shapes used in studies of ventilation of confined spaces. These models derive from a cube having unit dimensions. (Adapted from Garrison and Erig 1989.)

1991; Garrison 1991; Garrison et al. 1993). These focused on fundamental parameters: shape, size, orientation, mode of ventilation, and contaminant properties.

The tests involved cubic and rectangular box models. The dimensions of all of the models were related by multiples of the side of basic cubic Model A, measuring 0.61 m (2.0 ft) on each side. For example, the walls of vertical, rectangular box Model E2 were three times the length of the basic cube. Models were constructed to examine the effect of size and vertical and horizontal orientation. These shapes were considered representative of the geometry found in many actual confined spaces, such as pits, vaults, and equipment enclosures. Of course, the shape and complexity of many other confined spaces vary considerably from the ones chosen for study during these experiments (Figure 11.3).

Models used in these studies had a single opening in the top near one of the corners. Diameter of the opening was 15 cm (6.0 in.). The tube used to supply or remove air from the space measured 5.1 cm (2.0 in.) inside diameter and 6.1 cm (2.4 in.) outside diameter. This configuration was intended to simulate an access/egress portal containing a ventilation duct. The dimensions used here would correspond to an actual opening of 61 cm (24 in.) in diameter and a duct having inside diameter of 20 cm (8 in.). The models had wooden sides and a front of clear acrylic plastic.

These studies examined atmospheric conditions involving concentrated and dilute air–gas mixtures. Differences in concentration and density of the atmosphere confined in a space can affect ventilation requirements in significant ways.

Concentrated air–gas mixtures displace the normal atmosphere. This produces an oxygen-deficient condition. Oxygen deficiency occurs when the oxygen concentration decreases below levels required for human respiration or combustion of flammable materials. In these experiments, the starting concentration of oxygen was set to approximately 10%. This was ascertained by measurement in the case of nitrogen and by calculation to 50% of the volume of the space in the case of the denser-than-air gases. In the former case, the nitrogen was presumed to distribute uniformly because of equivalence of density. In the latter case, the distribution pattern was not postulated *a priori.*

The ratio of the molecular weight (MW) of the contaminant compared to that of air (MW = 29) at the same temperature and pressure, is the gas specific gravity (air = 1). This quantity provides an indication about buoyancy characteristics of the air–gas mixture. The molecular weight of gases with positive buoyancy is less than 29; the corresponding gas specific gravity is less than unity. Gases with neutral buoyancy have molecular weights around 29. Subsident gases have molecular weights greater than 29 and gas specific gravity greater than unity.

Nitrogen and carbon monoxide (MW = 28) have gas specific gravity of 0.98. These gases are neutrally buoyant in air. Introduced into a confined space in high concentration, they would be expected to displace air uniformly. Other gases and air–vapor mixtures are significantly less dense or more dense than air. Under appropriate circumstances, these would be expected to stratify (form a layer) at the top or bottom of the space. Dense gases studied in these experiments included carbon dioxide (MW = 44), HC_{22} or chlorodifluoromethane (MW = 87), and sulfur hexafluoride (MW = 146). The gas specific gravities are 1.5, 3.0, and 5.0, respectively.

Many contaminants form highly toxic mixtures even at very low concentration. At these concentrations, these contaminants disperse uniformly in air. Isobutylene (MW = 56) was used in these experiments to represent toxic contaminants that are present in low concentration.

Each test utilized mechanical ventilation to reduce contaminant concentration in the test atmosphere. Concentration (O_2 in % or isobutylene in ppm) was recorded as a function of time (min). The choice of substance tested was dictated to some extent by availability and sensitivity of real-time measuring equipment and an attempt to maintain realism with field conditions. Oxygen was measured using electrochemical sensors, and isobutylene using a gas chromatograph equipped with a photoionization detector.

The atmosphere was static at the beginning of the test. Hence, these tests only reflect ventilation during preentry preparation, rather than during work activity when contaminants could be continuously generated. Contaminant was introduced from the bottom of the test chamber. Samples were obtained from four locations at different elevations in different vertical quadrants (oxygen deficiency caused by nitrogen and dilute contaminants), and different elevations on the vertical central axis (dense contaminants).

Ventilation options studied in these experiments paralleled those available under normal circumstances: supply air to the space or remove air from it. Additional variables studied were the height of the discharge/intake of the supply/exhaust duct in the space and volumetric flow rate. The height of the opening in the duct was expressed as a percentage (%H) of the height from the bottom of the model. Volumetric flow rates were expressed as air changes per hour (ac/h). (An air change is the volume of air equivalent to the volume of the space.) These choices of units permitted ready comparison between models of different volume and shape, even though the airflow rate differed in each.

Table 11.11 summarizes findings obtained from studies of laboratory models.

Experimental data from oxygen-deficient situations were regressed against an exponential model of the form in the following equation.

$$\%\ O_2 = 21 - (21 - B)e^{-Ct} \tag{11.6}$$

B = the initial O_2 concentration
C = the rate of oxygen recovery (oxygen recovery time constant)

Values of C were obtained for all test data and are tabulated elsewhere (Garrison et al. 1989, Garrison and Erig 1989, 1991). The larger the value of C, the more effective the ventilation of a particular model and combination of parameters (volumetric flow rate, height of the discharge/exhaust duct, and density of the gas in the space). These values can be used in predictive calculations, subject to the usual caveats about extrapolation from models to real-world situations.

FIELD STUDIES OF VENTILATION OF CONFINED SPACES

The industrial hygiene literature contains little information about conditions in confined spaces or ventilation applications in them.

Stratification or layering is presumed to be able to occur during formation of the contaminated atmosphere. Stratification is known to occur from meteorological studies involving large air masses

TABLE 11.11
Summary of Finding from Studies of Laboratory Models

- Mechanical ventilation can be very effective in reducing potential airborne hazards in confined spaces. The most easily ventilated zone is the region dominated by air motion induced at the end of the supply/exhaust duct. The most difficult zone to ventilate is furthest from the opening in the space and the end of the supply/exhaust duct (Garrison et al. 1989). This applies, regardless of the shape of the space (Garrison and Erig 1989).
- Supply mode generally is more effective than exhaust mode. This is due partly to the dominance of the discharge jet emitted from the end of the supply duct. This jet causes considerably more turbulent mixing than is observed during the exhaust process. Supply ventilation produces very rapid contaminant reduction at locations aligned with the outlet of the supply duct (Garrison et al. 1989, 1991). Supply mode is especially effective in rectangular box models in which the long axis is parallel to the supply duct. The effect is less prominent in rectangular boxes in which the long axis is perpendicular to the supply duct (Garrison and Erig 1989). This observation also applies to denser-than-air mixtures. The difference in effectiveness between supply mode and exhaust mode increases with increasing density of the mixture and at low elevations (Garrison and Erig 1991). (This observation has important applications to ventilation in normal situations prior to entry and in accident situations where the supply duct can be directed into the region near the victim's breathing zone until help can arrive. This also may provide the basis for ventilation in large spaces where complete ventilation cannot occur readily and where contaminant generation is localized.)
- Positioning the outlet of the ventilation duct near the bottom of the space is generally more effective than at higher elevations for both supply and exhaust modes (Garrison and Erig 1989). This indicates that dilution improves by locating the source of supply and the discharge as far from each other as possible. An inflection for marked deterioration in ventilation effectiveness occurs at duct height around 70% H in supply mode and 80% H in exhaust mode. Supply mode is affected mostly at ventilation rates below 20 ac/h. Exhaust mode is affected even at very high ventilation rates. Considerable deterioration in ventilation effectiveness occurs at higher position (increased percent H); some improvement occurs at lower position (Garrison et al. 1989).

  Height of the end of the duct is especially critical when denser-than-air gases are present (Garrison and Erig 1991). These studies found that these gases can stratify at the bottom of the space. Starting oxygen concentrations were considerably less than what were measured during experiments using nitrogen, a gas of neutral buoyancy. This indicates lack of mixing between the air in the space and the contaminant prior to the start of ventilation. Ventilation was most effective when the end of the duct was located at lower rather than higher elevations. This applied in both supply and exhaust mode. Ventilation effectiveness depended on density of the gas/vapor trapped in the space. That is, ventilation effectiveness at low elevations increased with increasing density of trapped gas/vapor.
- Increasing volumetric flow rate decreases ventilation time. However, this effect is limited and not proportional to changes in flow rate. An inflection in ventilation effectiveness exists around 20 ac/h for supply mode and 30 ac/h for exhaust mode. Little improvement in ventilation effectiveness occurs at higher ventilation rates, but marked deterioration occurs at lower ventilation rates. This occurs regardless of the height of the opening of the supply/exhaust duct (Garrison et al. 1989, Garrison and Erig 1991, Garrison et al. 1991).
- Shape has significant impact on ventilation effectiveness (Garrison and Erig 1989, 1991). This is particularly true for vertical noncubical configurations and contaminants that are denser than air. Ventilation in spaces oriented horizontally is less effective than in identical spaces that are oriented vertically. This applies to both supply and exhaust modes.
- Geometric and airflow similarity produce similar ventilation characteristics in identical configurations of different size (Garrison and Erig 1989). This applies regardless of the physical properties of the contaminant (Garrison and Erig 1991).
- Ventilation effectiveness decreases with increasing gas density (Garrison and Erig 1991). This effect is related to stratification at the bottom of the space. This effect is most pronounced for the lowest 10% H elevation in the models.
- Ventilation effectiveness for contaminants present at IDLH (immediately dangerous to life or health) levels is similar to recovery from oxygen deficiency caused by neutrally buoyant contaminants, such as nitrogen (Garrison et al. 1991).

(Perkins 1974). Geographic formations, such as mountains that surround a basin (Los Angeles area), create conditions for stratification that traps contaminants. Air masses that form over the air in the basin prevent flushing. Stratification also can occur in valleys. Dispersion is likely to occur on prolonged standing, except when the stratified layer is disturbed by substantial cross-axis air motion (Briggs et al. 1990).

Groves and Ellwood (1989) demonstrated stratification in sealed vertical farm silos during anaerobic conversion of green crops to silage. They measured conditions in the airspace at varying heights above the silage. Silage production evolves carbon dioxide, nitric oxide, and nitrogen dioxide, and consumes oxygen. The carbon dioxide formed a layer adjacent to the silage. This displaced the atmosphere to produce an oxygen-deficient condition. These studies showed rapid development of a stratified layer of gases, mostly carbon dioxide initially and later nitric oxide, above the last load of material. The stratified layer of gases was about 1 to 2 m thick and highly oxygen deficient. It developed rapidly within 12 h after the addition of crop materials to the top of silage remaining in the silo. This buildup occurred regardless of whether the silo was sealed or the top hatches were open. Dispersion to form uniform concentrations with depth does occur in sealed silos, but the process requires months.

Melcher et al. (1987) reported on the kinetics and mechanism that caused oxygen deficiency to develop in a magnesium chloride brine evaporator. The vessel had been washed, vented, and drained, but remained closed for several days prior to the entry. The oxygen level in the vessel decreased exponentially from normal levels to 1% in 24 h. Investigation showed that surface scale (magnetite) depletes oxygen at a rate more than 10-fold that for bare metal. The tests showed that formation of ferric oxide (Fe_2O_3) over the first 24 h was not a factor. The tests also showed formation on highly scaled surfaces of Iowaite, a magnesium–iron hydroxy complex that contains a high proportion of oxygen. The scale retains a large amount of $MgCl_2$ solution. The scale becomes black, indicating formation of magnetite, Fe_3O_4, the primary corrosion product. This study also showed that other salt solutions, such as KCl, NaCl, $CaBr_2$, and $MgSO_4$, in conjunction with high humidity also can cause hazardous oxygen depletion.

The most extensive source of information on conditions that can develop in confined spaces is the NIOSH summary of accident investigations under the Fatal Accident Circumstances and Epidemiology (FACE) Program (NIOSH 1994). The purpose for FACE was to investigate fatal accidents in detail, using a multidisciplinary team. Despite delays in assessing conditions following the accidents, results obtained by the FACE investigators provide confirmation about the apparent stratification demonstrated in other studies.

Brief et al. (1961) reported on ventilation practices for towers during welding and cutting. These authors related the concentration of welding fume to tower size and ventilation, and provided guidance on ventilation practices.

Bowes et al. (1993) used a tracer gas (sulfur hexafluoride) to determine the effectiveness of ventilation of two petrochemical storage vessels. The tracer was metered into air supplied to the vessel and measured in the discharge. These studies indicated that air in the vessels did not undergo rapid mixing. Furthermore, mixing factors calculated from experimental observations were less consistent than expected, again an indication about the heterogeneous nature of actual ventilation.

These authors considered the implications of various effects that can occur during ventilation: stratification, shunting (short-circuiting), and displacement. Stratification results when incoming air mixes well with air in some parts of the space, but not with that in others. Shunting (short-circuiting) results when some of the incoming air passes directly to the exhaust without mixing with air in the space. Displacement occurs when incoming air displaces air in the space without significant mixing. None of these effects in itself may be disadvantageous, since the intent of ventilation is to protect the occupants in an effective and predictable manner. They may, in fact, provide potential opportunity. The authors comment that shunting results from improper selection of inlet and exhaust locations. However, the reality usually is that little choice about these locations exists (Figure 11.4).

Results from tests using real-world tanks indicate that thorough mixing can occur, as can potentially deleterious effects, such as stratification and shunting. Occurrence of these conditions is sensitive to volumetric flow rate.

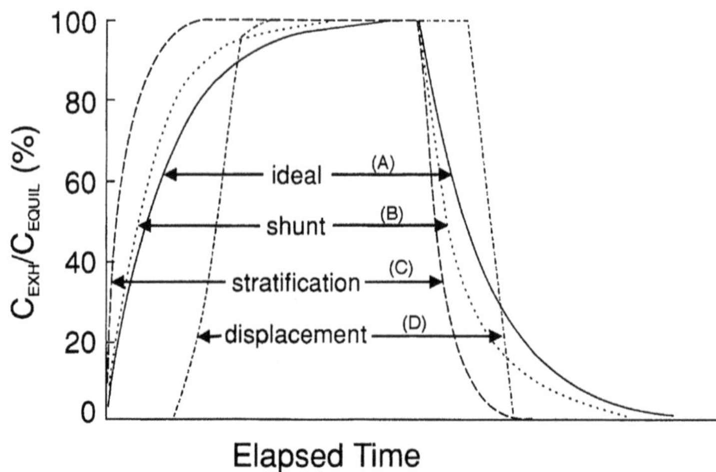

FIGURE 11.4 Theoretical behavior of tracer gas during ventilation of a structure with uncontaminated air (no contaminant generation). C_{EXH} is concentration of tracer in exhausted air. C_{EQUIL} is equilibrium concentration of tracer in the structure. Curve A shows complete and immediate mixing of the uncontaminated air with the air in the structure. Curve C shows stratification, meaning that air mixes well in some parts of the space and poorly in others. Curve B shows shunting or short-circuiting, meaning that some fraction of the incoming air leaves the structure without mixing with the atmosphere in the space. Curve D shows displacement of air from the space by the incoming air. (Adapted from Bowes et al. 1993.)

Wiegand and Dunne (1996) examined the removal of radon from underground structures in the British telecommunications system. These authors used nitrous oxide in controlled studies, as well as real-time measurements of radon progeny. These studies examined natural ventilation by wind motion across the top of the open manhole, as well as ventilation assisted by a baffle. The latter involved a plastic sheet that bisected the opening and extended from 0.75 m above the top of the manhole to 0.5 m above the bottom. The latter increased the natural ventilation rate by a factor of 2 to 10, depending on the geometry of the structure. The volume of these structures ranged from 4 to 20 m^3, but typically were 10 m^3. The concentration of nitrous oxide in these structures decreased by a factor of 10 in 5 to 9 min, using the assisted form of natural ventilation.

The assisted ventilation technique utilized in these studies involving supply and exhaust through the same flue has received considerable interest in the U.K (Stewart 1991, Anonymous 1995). This technology is an application of the air scoop/wind tower used for 2,000 years in the Middle East to create a downwash and upflow of air in structures.

CONSIDERATIONS AFFECTING VENTILATION OF CONFINED SPACES

Confined spaces encompass an unlimited variety of configurations having many different sizes and shapes. The number, size, and location of openings also can vary widely. Contaminants can be dilute or concentrated. The cloud that is trapped within the boundaries of the space may be more dense, less dense, or equal in density to the surrounding air. The contaminants may be highly toxic or low in toxicity. They may be present as an ignitable or nonignitable mixture. They may be highly reactive or low in reactivity. These issues have been discussed previously in this chapter. The following discussion summarizes the issues that dictate the approach that is needed for ventilating a space.

CONFIGURATION

Openings are particularly critical to the effectiveness of ventilation of confined spaces. A confined space having a single, relatively small opening should cause "red flags" to wave in the minds of individuals concerned with ventilation. A single small opening means that all ingress and egress by personnel, ventilation, visual observation, communication, and equipment and utilities must occur by this route. In spaces having more than one opening, consideration should be given to opening all available access points to enable ventilation to be optimized. Spaces should not be designed with a single opening, if at all possible. Also, internal surfaces, such as baffles, bulkheads, equipment, and so on, can greatly affect ventilation. The influence of such features on ventilation effectiveness is difficult to predict. As a result, large variations in ventilation easily can occur at different locations inside a confined space.

Study of even a small fraction of the possible configurations obviously is not possible. Studies of specific operations, thus far, have occurred only in refinery towers and petroleum storage tanks (Brief et al. 1961, Bowes et al. 1993). Study of model confined spaces in the laboratory provides a means of gaining predictive data in an efficient manner. These studies showed for cubic and rectangular box models that characteristics of ventilation and contaminant reduction are similar in geometrically similar models of different sizes (Garrison and Erig 1989, 1991). They also showed that variations in shape can cause significant differences in ventilation characteristics. This is especially true when oxygen deficiency is caused by denser-than-air contaminants (Garrison and Erig 1991).

Undoubtedly, ventilation will be difficult to implement in many geometric shapes. Some examples include long interconnected utility and railway tunnels, ovens, furnaces, degreasers, silos, large storage structures, ship and barge infrastructures, hydroelectric generating structures, and boilers, among many others. For these environments, the best practice is to ventilate as well as can reasonably be accomplished, using experience as a guide. This, at least, will provide better air quality near (if not at) the work area, even though some locations may be ventilated less effectively than others.

Contaminant Characteristics

Ventilation usually can provide an effective means for reducing the hazards presented by three main categories of contaminants found in confined spaces: toxic, flammable or explosive, and oxygen deficiency or enrichment. In some specific situations ventilation cannot be used and entry cannot be avoided. The latter could arise with complex process equipment into which entry was never intended. Other situations arise from conditions in which contact between oxygen and contents in the space must be prevented. This might be necessary to prevent oxidation of process components/chemicals or to prevent fires. (One such situation is a reaction vessel containing a pyrophoric catalyst.) If ventilation is not possible, the decision not to ventilate is easy.

Selection of the fundamental method (supplying air to the space or removing air from it) for ventilating a space is a function of the demands of a particular situation. The critical factor is whether the atmosphere in the space can safely be allowed to disperse in the external surroundings. This determines whether the atmosphere in the space can be permitted to escape without control or whether control must occur. The need to control or prevent contact between the atmosphere in the space and support personnel or the surroundings means that air must flow into the space. Ignitable atmospheres deserve special consideration and attention. Neither method of approach is totally satisfactory. The specific design of the ventilation system depends largely upon the configuration (shape, size, openings). If denser-than-air contaminants are present in high concentration, then locating the ventilation duct near the bottom of the space is particularly important (Garrison and Erig 1991).

Contaminant characteristics can have significant impact on the duration of ventilation required prior to entry and on the need for continuous ventilation during entry. The approach that suits preentry preparation may not suit work activity.

Ventilation time for a given configuration and ventilation design is a function of the nature of the contamination and the initial and the final desired concentration; that is, the extent of contaminant reduction required. Recovery from oxygen deficiency (e.g., 10 to 21% O_2) can be relatively rapid — perhaps only a few minutes when effective dilution is occurring (Garrison et al. 1989, Garrison and Erig 1989). Recovery can require hours under conditions of ineffective ventilation when contaminants have stratified in the bottom of the space (Garrison and Erig 1991).

Correction of oxygen deficiency caused by a nontoxic contaminant may require reduction to a residual concentration of 1 to 2% (10,000 to 20,000 ppm). Ventilation to alleviate fire or explosion hazards requires reduction in concentration to roughly 10% of lower explosive limit, approximately 0.1 to 0.2% by volume (1,000 to 2,000 ppm). Further reduction is suggested, owing to inefficiencies in ventilating and testing in these environments. This can require additional time, possibly several times that needed for eliminating oxygen deficiency. Ventilation to control toxic contaminants to within safe levels (1 to 10 ppm, or lower) could require substantially more time than for other situations.

In theory, continuous ventilation is not needed once oxygen deficiency (caused by nontoxic contaminants) has been eliminated, so long as emission of contaminants from residual sources or production during work activity is unlikely to occur. However, the gap between theory and reality in the area of confined spaces is large and unpredictable. Continuous ventilation is the most prudent course of action. This, similarly, is the case where ignitable atmospheres may develop.

Continuous ventilation would be advisable and could be necessary if significant emissions were to occur following initial preparation for entry. Continuous ventilation probably would be needed to assure safety when toxic contaminants remain or could be released inside a confined space. This applies even when the rate of release is small, and especially so when acceptable levels are very low (1 to 10 ppm or less).

Volumetric Flow Rate

Specification of volumetric flow rate for ventilation of a confined space is a difficult task at best. There are no consistent, well-accepted methods by which to make recommendations of this kind. Ventilation requirements during preentry preparation could differ from those during work activity. There may also be practical limitations, owing to the volume of the space and the size and capacity of available openings and air-moving devices. Unfortunately, the following guidelines cannot prescribe concise recommendations for volumetric flow rate.

The literature contains few studies that provide the basis for guidelines on volumetric flow rate for specific operations. (See Appendix A for further information about guidelines, standards, and legislated requirements.) Some organizations have developed procedures for ventilating confined spaces that include volumetric flow rate and ventilation time. These procedures, however, generally are proprietary to the organizations that prepared them. Simple charts or "nomograms" also have been developed to aid in decision making. Nomograms can be useful tools, but they must be used cautiously. Criteria utilized in developing nomograms for use in specific situations may not be transferrable to others involving different configurations and contaminants.

Mathematical equations for calculating parameters of dilution and local exhaust ventilation in conventional workplaces are available in text and reference sources (Burgess et al. 1989, Heinsohn 1991, ACGIH 1995). Some estimate the volumetric flow rate needed to maintain a desired steady-state contaminant concentration by dilution ventilation. Others estimate time-dependent ventilation needed to achieve a specified amount of contaminant reduction. They are not directly applicable to the ventilation of confined spaces.

Proper use of dilution equations requires reasonable estimates for rate of contaminant generation. These are difficult to obtain at the best of times and are even more so for confined spaces because of the interaction of boundary surfaces. Even more difficult to estimate is the "K factor," the constant intended to account for the relative effectiveness of air mixing within the space (ACGIH 1995). The "K factor" serves as a safety factor to address uncertainties and special hazards associated with exposure to airborne contaminants. Application notes generally advise against using these equations (or dilution ventilation) when highly toxic contaminants are involved. However, the realities of confined spaces as a work environment, in conjunction with the limitations of portable ventilation equipment, force compromises, innovations, and creativity compared to normal situations. In many entry situations, dilution ventilation may be the only option.

Sources sometimes refer to "air changes/time," a measure of the total volume of a space divided by some period of time. This quantity is used as a measure of volumetric flow rate. This quantity usually has the units "air changes"/hour (ac/h) or "air changes"/min (ac/min). The following equation provides the basis for this calculation.

$$\text{“Air changes”/h} = \frac{\text{Airflow rate}\left(\text{ft}^3/\text{min}\right)\times 60\ \left(\text{min/h}\right)}{\text{Volume}\left(\text{ft}^3\right)} \tag{11.7}$$

"Air changes" are useful as a concept because they are somewhat intuitive, easy to calculate, and independent of volume. This means that they can apply to confined spaces of different sizes. However, "air change" is a derived parameter. An air change is not a measurable quantity, as is flow rate.

Contaminant dilution by ventilation is a continuous process involving the simultaneous supply of fresh air, mixing of fresh air with air resident in the space, and removal of surplus air. In a space containing a fixed starting concentration of contaminant, the impact of every air change on the status quo is different. Greatest reduction in contaminant level occurs shortly after ventilation begins in the first few air changes. Subsequent reduction occurs more slowly. As a term to describe ventilation of a space, this concept is something of a misnomer, since it implies complete displacement of contaminated air with fresh replacement air.

The utility of the "air change" concept rests with specification of total volume of air that must be provided to the space to ensure removal of contamination with reasonable confidence. This necessitates calculation of volume. In many spaces, such as utility vaults, where the volume is not large, the overall dimensions represent a reasonable "working" volume for purposes of specifying ventilation.

Laboratory studies of oxygen deficiency in models used ac/h as a parameter to compare between spaces of different size and geometry (Garrison et al. 1989; Garrison and Erig 1989, 1991). These studies found that ventilation efficiency decreased significantly for flow rates below 20 ac/h. Ventilation efficiency increased with increasing airflow to about 60 ac/h. Above this rate, improvement diminished significantly. These results suggest that ventilation rates between 20 and 60 ac/h are most effective. These studies also found longer ventilation at a lower rate to be more effective than shorter ventilation at a higher rate. For example, 20 ac/h for 30 min produced more effective dilution than 60 ac/h for 10 min. Coincidentally, NIOSH recommended a ventilation rate of 20 ac/h for confined spaces in construction projects (NIOSH 1985).

Specific recommendations for ventilation based on the number and rate of air changes for the many variations in confined spaces are not possible. In many situations, 10 air changes (e.g., 20 ac/h for 30 min) will provide sufficient ventilation. However, this likely would be unsatisfactory in other situations involving poor mixing, replenishment of contaminants at a significant rate, and/or contaminants that have stratified in the bottom of the space. Recommendations that are too specific also could create hazardous situations during ventilation to control flammable vapors and contaminants having high toxicity.

Table 11.12 provides general guidelines for ventilation of confined spaces prior to entry. As stressed previously, these are offered without confirmation that they are, in fact, the best recommendations.

For many spaces, such as large tanks and tunnels, calculating the appropriate volume for ventilation purposes based on "air changes" is difficult, if not impossible. Providing ventilation of 20 ac/h, based on the use of portable equipment, could be technically impossible. In these situations, the best action is to consider air movement (velocity) and dilution in proximity to the work area. Whenever feasible, maintaining velocity at 100 ft/min away from the breathing zone is a desirable

TABLE 11.12
General Guidelines for Preentry Ventilation

- 10 Air changes for 10- to 100-fold (i.e., 10% to 1% of initial levels) reduction, assuming good air mixing and negligible release
- 20 to 30 Air changes for 10- to 100-fold reduction, assuming poor mixing or significant contaminant release
- 30 to 60 Air changes for 10- to 100-fold reduction, assuming poor mixing and significant contaminant release
- 60 to 100 Air changes, in cases of known or possibly unknown pockets of negligible air movement and high rates of contaminant release. Ventilation alone may not offer adequate control in these situations.

target. The flow should deliver fresh air to the breathing zone and remove contaminants from it. Long distances of travel, even at substantial velocities, could result in relatively long residence time and possible buildup of contaminants to hazardous levels.

Measurement of contamination levels in the air of the space prior to starting to ventilate and in air supplied to and discharged from the space during ventilation provides a means of evaluating ventilation effectiveness. This can occur without need for entry. Low level of contaminant in discharged air relative to that in the space suggests that the ventilation is ineffective. More purging and/or a higher ventilation rate could be needed. High level in supply air could mean reentrainment and recirculation of contaminants. Injection of tracer gas into the space and monitoring of concentration in the exhaust provides another means of ascertaining ventilation effectiveness. Depending on the tracer chosen, this procedure could require relatively sophisticated equipment. Little work has been done in this area. Research could prove to be very useful.

Preliminary development of computer models with applications for ventilating confined spaces has occurred. Some created for other purposes, such as assessing indoor air quality, may be directly adaptable to confined spaces (Park and Garrison, 1990; Haberlin and Heinsohn 1993; U.S. EPA 1989, 1991a, 1991b; U.S. Dept. of Commerce 1990, 1991). These models provide additional tools for predicting effectiveness of ventilation strategies. This goal is worthy, but difficult to achieve. Unfortunately, the day of computer-based ventilation design for confined spaces is not near.

AIR MOVING EQUIPMENT

Almost any equipment capable of moving air can be used for ventilating confined spaces. The marketplace today offers many different examples of such equipment. Selection of the most appropriate air mover depends on the demands and constraints of a particular situation. For example, the requirements of some situations may necessitate carrying or hoisting air movers to locations from which they be used. (For more information on portable ventilation systems, refer to Appendix J.)

There are two principal types of fans: axial and centrifugal. **Axial fans** offer compactness and relatively large capacity. They are particularly good for general supply and exhaust applications. However, their ability to overcome resistance (static pressure loss) in ducts and small openings is limited. Noise levels sometimes are high. **Centrifugal fans** offer quieter operation and lower sensitivity to airflow resistance. They often are used as air movers for general supply and exhaust and also for local exhaust ventilation. Centrifugal fans are appropriate whenever duct is used.

The source of power available or allowable in a particular situation will influence selection of equipment. If ignition sources must be avoided, then an air mover powered by compressed air or steam, or explosion-proof electrical equipment is necessary. Gasoline- or propane-powered equipment may be appropriate in remote locations without utilities.

Venturi-type eductors, variously called air horns, air ejectors, or air eductors, are used in situations not well-suited to fans. Air- or steam-driven units are available. Eductors offer the advantages of no moving parts and not needing electricity or fuel for power. They can fit into relatively small openings. They also can move air carrying abrasive or corrosive contaminants that otherwise could damage fans not protected by prefiltration.

Venturi-type eductors can generate very high levels of noise. In addition, operational problems can result from variable air/steam pressure, clogged nozzles, and condensate drips. Venturi-type eductors are not capable of moving large quantities of air.

Compressed air sometimes is used for supplying air to confined spaces. It is useful primarily to meet needs for supplementary air and for emergencies. Compressed air has limited flow rate and is subject to possible interruptions. It is inappropriate for use in large spaces. Compressed air used for this purpose must meet standards for breathing air.

Duct used for transporting ventilation air typically is flexible and rugged. These qualities are particularly important for survival and repositioning. A fitting for use in circular manways that minimizes the space required by the duct is a recent innovation. It has round ends for attaching the duct. The part that passes through the manhole has a crescent-shaped cross section. These units are made from rigid durable plastic.

A shroud that seals the edges of the manway and the air mover or duct positioned in this opening will greatly assist in optimizing ventilation effectiveness. A shroud will prevent reentrainment and "short-circuiting" in the vicinity of the opening. Seemingly minor leakage in these locations can be particularly deleterious to performance. Shroud design depends upon specifics of the opening and air mover. Cut-outs may be needed as portals for cables and hoses. These also must be sealed. Typically, the decision to use shrouds necessitates careful planning and preparation.

A combined manway/fan mount is especially useful for optimizing ventilation efficiency in confined spaces having only one opening. A manway/fan mount is an enclosure that has three openings. One connects to the opening of the space. The second mounts the fan or duct, and the third is intended for passage by personnel. The opening for personnel passage also has a door. The manway/fan mount permits ventilation without blocking the opening of the space by the fan.

Movable partitions also can be used for optimizing ventilation effectiveness by directing airflow inside the space.

Another option that may be available is to make additional openings. These can enhance ventilation by controlling airflow into and out of the space.

Audible and/or visual alarms are an important part of the portable ventilation system. These indicate mechanical or other failure. These are essential when ventilation of the space is critical to preventing development of a hazardous situation. Warning is needed when a hazardous situation could arise quickly after loss of ventilation in the space. "Sail" switches located in the duct immediately adjacent to the blower are a simple and effective device for triggering an alarm. On-line monitoring of ventilation air also may be necessary when supply air could become contaminated by discharge from the space or from contaminants in the surroundings.

Additional surfaces, such as barriers, can improve natural ventilation under certain conditions. If wind flow is desirable, additional surfaces can act as sails to "scoop" airflow into the space. Although not "equipment" per se, one of the most effective ways to improve natural and mechanical ventilation is to provide as many openings as possible. Additional openings facilitate air movement and provide flexibility for different ventilation strategies. An example of this approach is the opening of the covers of sewer manholes upstream and downstream from the point of entry.

VENTILATION TESTING AND EVALUATION

Testing and evaluating ventilation effectiveness is critical to ensuring control of conditions. This is as important for confined spaces as for other workplaces. The reason for ventilation testing in confined spaces is simple: to determine pathways and rates of air movement within, and in some situations, external to the space. Reliance on calculations and manufacturers' specifications of performance is no substitute for on-site, in-place tests. There are too many possibilities for degradation of performance. They are no more readily apparent in confined spaces than in other workplace environments. Interaction between flowing air and boundary surfaces is even more consequential in confined spaces than in normal workplace environments. Testing for the presence of atmospheric contaminants does not eliminate the need to test ventilation effectiveness. This applies to spaces ventilated by both natural and mechanical means.

Ideally, a Qualified Person should evaluate ventilation effectiveness. However, in the more usual case, only unsupervised field crews are present at a site. Evaluation by these personnel as a minimum should confirm delivery of fresh air from an uncontaminated source to the work area in the confined space. There is a critical need for transference of knowledge from the Qualified Person to these individuals in such situations. Determining how to position air movers and duct and to eliminate sources of degradation to obtain optimum performance requires investigative enquiry and the benefit of experience. This knowledge is not intuitive. The Qualified Person, having the benefit of training in evaluating and testing industrial ventilation, logically should lead this enquiry and train others.

On-site evaluation of ventilation effectiveness should begin with pathway analysis. This is especially important, since there are many opportunities for degradation of performance. Air in these and other situations always takes the path of least resistance. This could result in a short-circuit at the portal, reentrainment outside the space, or flow along internal surfaces remote from the source of contaminants. Seemingly minor leaks could produce major deterioration of performance. The flow of air is not normally visible and therefore must be visualized by means of visual tracers, such as "smoke tubes." Smoke tubes are the minimum essential tools. They visualize airflow through openings and general patterns inside the space. They indicate whether airflow is being provided as intended and expected, and how it is being removed. An additional valuable function is to identify leakage paths. Leaks usually are undetectable other than through this method of enquiry. Pathway analysis and leak detection should be considered the minimum tasks in this process.

More formal evaluation of ventilation effectiveness would include estimation of the rate of delivery of fresh air and possibly the rate of contaminant evolution. Procedures for evaluating ventilation effectiveness are the same as those used in normal circumstances. Testing protocols utilize conventional devices, such as swinging-vane and thermal anemometers and "smoke" tubes. Velocity measurements at openings (in conjunction with cross-sectional area) permit calculation of general ventilation rate. Measurements inside the space should determine velocity in key locations. In sewers and manholes with connecting influent and effluent pipes, the atmosphere is not static. In these cases, elaborate testing of a mechanical ventilation system would provide limited benefit.

Tracer gas released into the space and monitored at openings and specific interior locations may assist in evaluating ventilation effectiveness. This technique could be particularly useful in large spaces and/or those having complex shapes and internal surfaces. No formal test protocol has been developed yet. Such a technique could lead to meaningful and enforceable criteria for confirming the adequacy of ventilation in these environments (Bowes et al. 1993).

In situations where physical stress is a concern, measurement of temperature and humidity also may be appropriate.

Table 11.13 summarizes considerations for ventilation testing and evaluation. Data obtained from ventilation testing can be compared directly to applicable standards. (Refer to Appendix I for further information on guidelines, standards, and legislated requirements in the U.S. applicable to confined spaces.) Volumetric flow rates can be estimated from measured velocities and cross-sectional area of openings or the interior the space. In most instances, however, these calculations have limited use, since specific guidelines for ventilation do not exist. Data generated from determination of airflow direction and velocity should be evaluated by qualified individuals in a manner consistent with good industrial hygiene practice.

TABLE 11.13
Guidelines for Ventilation Testing in Confined Spaces

Location	Action
Exterior	Trace path of air to intake of ventilation duct and opening of the space to determine that short-circuiting and entrainment of contaminated air from other sources are not occurring
	Trace path of air exhausted from the opening of the space or the ventilation duct to ensure that short-circuiting and undesirable accumulation are not occurring
	Determine leakage paths at openings of the space
	Determine effects of wind (or cross-draft) speed and direction; this is particularly important when natural ventilation is used
Openings	Determine direction of and path of airflow and average velocity
Interior	Determine airflow, particularly in the breathing zone where people are or will be working and in location of contaminant sources
	Evaluate general airflow patterns, existence of dead spaces, and "short-circuiting"
	Evaluate capture and effectiveness of local exhaust ventilation

TABLE 11.14
Topics of Instruction — Ventilation Training

- Atmospheric confinement by containers and structures
- Characteristics of atmospheres confined by containers and structures
- Hazards posed by airborne contaminants
- Ventilation control strategies, including advantages and disadvantages of different ventilation alternatives
- Factors that degrade performance of portable ventilation systems (short-circuiting, reentrainment, and accumulation inside and outside the space)
- Hazards posed by ventilation of hazardous atmospheres contained in confined spaces
- Hazards posed by ventilation equipment (ignition sources, noise)
- Evaluation of ventilation needs in a particular situation, including selection of an approach over alternatives
- Set-up and operation of ventilation equipment, including customization for individual situations
- Storage of ventilation equipment on site
- Inspection and maintenance practices
- Use of testing equipment to evaluate ventilation effectiveness
- Procedure and drills in event of equipment malfunction and emergency situations
- Practical exercises and drills

TRAINING

Individuals involved in ventilating confined spaces should receive thorough training. This should include persons responsible for planning and approving entry, performing and attending at entries, maintaining ventilation equipment, and performing rescues. The previous section has emphasized the experiential nature of the knowledge that is needed for successful ventilation of confined spaces. This is not found in textbooks and must be acquired through on-the-job training and experience. Development of this knowledge base must involve individuals who have had formal training in industrial ventilation, as this is a direct application of these principles.

Table 11.14 lists subjects that should be addressed during training. The relevance of particular topics depends on individual responsibilities.

PROBLEMS RELATED TO VENTILATION OF CONFINED SPACES

As with other measures dedicated to solving problems, their use creates new ones. Portable ventilation systems are no exception to this observation. The use of portable systems for ventilating confined spaces introduces the potential to cause difficult problems during entry preparation and work activity. They are inherent to the approach of using this equipment and difficult to solve. Some are mentioned here without (unfortunately) well-established solutions. Other potential problems inherent to ventilation were mentioned earlier in this chapter.

Air movement, particularly involving cold, damp outside air, can cause unwanted, potentially hazardous cooling. This is particulary true in the supply mode. Ventilation using cold air can create cold, turbulent, high-velocity drafts. The exhaust mode typically involves lower velocities. Local exhaust, if feasible, provides the best alternative for minimizing cold drafts. Equipment is available for heating air in portable ventilation systems. Heating by combustion is probably the most energy-efficient method, but there is potential for exposure to carbon monoxide and carbon dioxide. The limited volume of air in some spaces increases the potential for serious exposures. Continuous monitoring of carbon monoxide during operation of this equipment is highly recommended. Electric heating, if feasible, could avoid this problem.

Heated outside air also can be very dry. The low humidity could cause problems due to accumulation of static electrical charge. Humidification could become necessary. This would add significantly to the complexity of the ventilation system. Antistatic devices, such as "ion generators," may be a feasible alternative for reducing static electricity.

Confined spaces having a single portal are especially problematic. The presence of the ventilation duct and other equipment in the opening can interfere with ingress, egress, and visibility. This is an especial concern where emergency egress may be required. Use of special duct fittings, which provide better "fit" in the opening and minimize obstruction, can help to resolve this problem. Making a second opening for ventilation may be necessary in some situations.

Fans, especially axial fans, can be quite noisy. Other devices, such as eductors ("horns"), also can be very noisy. Centrifugal fans, typically, are the quietest air movers.

SUMMARY

This chapter has examined the question of ventilation of confined spaces. The formal study of this complex topic presently is in its infancy. Ventilating confined spaces is different from ventilating normal work environments, although the overall principles are the same. For one thing, these environments are usually smaller and are enclosed to an extent that is unknown normally. Interaction between boundary surfaces and moving air can considerably complicate ventilation dynamics. Also, controlling the direction of airflow is critical to preventing reentrainment and short-circuiting.

There is no substitute for careful and deliberate consideration of ventilation in each situation. The experience and judgment of qualified professionals are important inputs for assuring safe outcomes. Currently there are few specific standards and guidelines available for these situations.

Mechanical ventilation is preferable to natural ventilation whenever significant uncertainty or lack of knowledge exists regarding the quality of air inside a confined space. Uncertainty and lack of knowledge regarding ventilation have played a major role in accidents involving confined spaces.

Ventilation can eliminate and control hazardous atmospheric conditions in confined spaces. Administrative controls cannot. Mechanical ventilation should be used for entry involving potential exposure to hazardous air contaminants, regardless of the lack of established guidelines or specific regulatory requirements. Most situations can accommodate some method of ventilation without causing serious operational difficulties.

There is no "cookbook" for designing ventilation for these environments. There probably never will be. The relationship between geometry and boundary surfaces, contaminant density, position of access/egress portals, volumetric flow rate, contaminant emission rate, and location of the air supply/exhaust within the space is very complex. Computer models at some point may be able to manipulate these and other parameters to predict ventilation performance. Empirical models based on experimental data are inherently limited in scope and likely would become prohibitively complicated. Analytical models based on theoretical fluid mechanics have a greater chance for success. Enormous challenges face developers of workable, user-friendly, widely applicable programs.

There is an undeniable need for more practical knowledge, based on actual field experience, about ventilating confined spaces.

ACKNOWLEDGMENTS

Laboratory studies described here were sponsored by the National Institute for Occupational Safety and Health.

REFERENCES

American Conference of Governmental Industrial Hygienists: *Industrial Ventilation — A Manual of Recommended Practice*, 22nd ed. Cincinnati, OH: ACGIH, Inc., 1995.

American Gas Association: *Purging Principles and Practice*, 2nd ed. Arlington, VA: American Gas Association, 1975. 180 pp.

American National Standards Institute: *Safety Requirements for Confined Spaces* (ANSI Z117.1-1995). Des Plaines IL: American Society of Safety Engineers/American National Standards Institute, 1995. 32 pp.

American Petroleum Institute: *Safe Entry and Cleaning of Petroleum Storage Tanks* (API Pub. 2015, 5th ed.). Washington, D.C.: American Petroleum Institute, 1993. 60 pp.

Anonymous: Reply to "Confined spaces clarification" by Gary Easter. *Occup. Haz.*, May 1993. p. 9.

Anonymous: *Windcatcher.* High Wycombe, Bucks, U.K.: HP12 3TD, Monodraught, Ltd., 1995. 10 pp. [booklet]

Bjerre, A.: Assessing exposure to solvent vapour during the application of paints, etc. — model calculations vs. common sense. *Ann. Occup. Hyg. 33:* 507–517 (1989).

Bodurtha, F.T.: *Industrial Explosion Prevention and Protection.* New York: McGraw-Hill, 1980. 167 pp.

Bowes, S.M., E.G. Mason, and M. Corn: Confined space ventilation: tracer gas analysis of mixing characteristics. *Am. Ind. Hyg. Assoc. J. 54:* 639–646 (1993).

Breum, N.O., F. Helbo, and O. Laustesen: Dilution versus displacement ventilation — an intervention study. *Ann. Occup. Hyg. 33:* 321–329 (1989).

Breum, N.O. and E. Ørhede: Dilution versus displacement ventilation — environmental conditions in a garment sewing plant. *Am. Ind. Hyg. Assoc. J. 55:* 140–148 (1994).

Brief, R.S., L.W. Raymond, W.H. Meyer, and J.D. Yoder: Better ventilation for close-quarter work spaces. *Air Cond. Heat & Vent. 58*: 74–88 (1961).

Briggs, G.A., R.S. Thompson, and W.H. Snyder: Dense gas removal from a valley by crosswinds. *J. Haz. Mater. 24:* 1–38 (1990).

Burgess, W.A., M.J. Ellenbecker, and R.D. Treitman: *Ventilation for Control of the Workplace Environment.* New York: John Wiley & Sons. 1989. pp. 87–104.

Canadian Standards Association: *Compressed Breathing Air and Systems* (CAN3-Z180.1-M85). Rexdale, ON: Canadian Standards Association, 1985. 32 pp.

Compressed Gas Association Inc.: *American National Standard Commodity Specification for Air* (Pamphlet G-7.1) Arlington, VA: Compressed Gas Association Inc, 1989. [Pamphlet]

Easter, G.: "Confined Spaces Clarification." Letter to the Editor. *Occup. Hazards 55:* May 1993. p. 9.

Emswiler, J.E.: The neutral zone in ventilation. *J. Am. Soc. Heat. Vent. Eng. 32:* 1–16 (1926).

Garrison, R.P. and D.R. McFee.: Confined spaces — a case for ventilation. *Am. Ind. Hyg. Assoc. J. 47*: A708–A714 (1986).

Garrison, R.P., R. Nabar, and M. Erig: Ventilation to eliminate oxygen deficiency in confined spaces — Part I: a cubical model. *Appl. Ind. Hyg. 4*: 1–11 (1989).

Garrison, R.P. and M. Erig: Ventilation to eliminate oxygen deficiency in confined spaces — Part II: Noncubical models. *Appl. Ind. Hyg. 4*: 260–268 (1989).

Garrison, R.P.: Ventilation for Work in Confined Spaces. Testimony on the Proposed OSHA Standard for Permit Entry Confined Spaces (54 CFR 24080), delivered at a hearing in Chicago, Feb. 2, 1990. Washington, D.C.: U.S. Dept. of Labor/OSHA, Technical Data Center, Docket Office (1990).

Garrison, R.P. and M. Erig: Ventilation to eliminate oxygen deficiency in confined spaces — Part III: Heavier-than-air characteristics. *Appl. Occup. Environ. Hyg. 6:* 131–140 (1991).

Garrison, R.P., K. Lee, and C. Park: Contaminant reduction by ventilation in a confined space model — toxic concentrations versus oxygen deficiency. *Am. Ind. Hyg. Assoc. J. 52:* 542–546 (1991).

Garrison, R.P.: Ventilation for Work in Confined Spaces. Project Report, R01-OHO2329. Washington, D.C.: National Institute for Occupational Safety and Health (1991).

Garrison, R.P., M. Erig, and C. Park: Testing and modelling of ventilation for confined work places. In *Ventilation '91: The 3rd Int. Symp. Ventilation for Contaminant Control.* Hughes, R.T., H.D. Goodfellow, and G.S. Rajhans (Eds.). Cincinnati, OH: American Conference of Governmental Industrial Hygienists, 1993.

Goodfellow, H.D. and J.W. Smith: Industrial ventilation — a review and update. *Am. Ind. Hyg. Assoc. J. 43:* 175–184 (1982).

Groves, J.A. and P.A. Ellwood: Gases in forage tower silos. *Ann. Occup. Hyg. 33:* 519–535 (1989).

Haberlin, G.M. and R.J. Heinsohn: Predicting solvent concentrations from coating the inside of bulk fuel storage tanks. *Am. Ind. Hyg. Assoc. J. 54:* 1–9 (1993).

Heinsohn, R.J.: *Industrial Ventilation: Engineering Principles.* New York: John Wiley & Sons, 1991. 699 pp.

Heskestad, G.: Venting practices. In *Fire Protection Handbook,* 17th ed., Cote, A.E. and J.L. Linville (Eds.). Quincy, MA: National Fire Protection Association, 1991. pp. 6-101 to 6-112.

Jayjock, M.A.: Back pressure modeling of indoor air concentrations from volatilizing sources. *Am. Ind. Hyg. Assoc. J. 55:* 230–235 (1994).

Jones, G.W.: Inflammation limits and their practical application in hazardous industrial operations. *Chem. Revs. 22:* 1–26 (1938).

Lees, F.P.: *Loss Prevention in the Process Industries,* vol. 1. London: Butterworths, 1980. pp. 477–634.

Lemoff, T.C.: Gases. In *Fire Protection Handbook*, 17th ed. Cote, A.E. and J.L. Linville (Eds.). Quincy, MA: National Fire Protection Association, 1991. pp. 3-63 to 3-82.

Liu, H.: Understanding wind load on plant buildings. *Plant Eng.* June 14, 1979. pp. 187–191.

McDermott, H.J.: *Handbook of Ventilation for Contamination Control.* Ann Arbor MI: Ann Arbor Science Publishers, 1976. pp. 1–54.

Melcher, R.G., C.E. Crowder, J.C. Tou, and D.I. Townsend: Oxygen depletion in corroded steel vessels. *Am. Ind. Hyg. J. 48:* 608–612 (1987).

Mutchler, J.E.: Principles of ventilation. In *The Industrial Environment — Its Evaluation and Control.* Washington, D.C.: Government Printing Office (DHHS/PHS/CDC/NIOSH), 1973. pp. 573–582.

National Fire Protection Association: *NFPA 69: Explosion Prevention Systems* (1992 ed.). Quincy, MA: National Fire Protection Association, 1992.

National Fire Protection Association: *NFPA 306: Control of Gas Hazards on Vessels* (1993 ed.). Quincy, MA: National Fire Protection Association, 1993. 15 pp.

National Institute for Occupational Safety and Health: Criteria for a Recommended Standard — Working in Confined Spaces (DHEW/PHS/CDC/NIOSH Pub. No. 80-106). Cincinnati, OH: National Institute for Occupational Safety and Health, 1979. 68 pp.

National Institute for Occupational Safety and Health: Safety and Health in Confined Workplaces for the Construction Industry — A Training Resource Manual. Washington, D.C.: Government Printing Office, 1985. pp. 127–135.

National Institute for Occupational Safety and Health: Worker Deaths in Confined Spaces (DHHS/PHS/CDCP/NIOSH Pub. No. 94-103). Cincinnati, OH: National Institute for Occupational Safety and Health, 1994. 273 pp.

Occupational Safety and Health Administration: Selected Occupational Fatalities Related to Fire and/or Explosion in Confined Work Spaces as Found in OSHA Fatality/Catastrophe Investigations. Washington, D.C.: U.S. Department of Labor, Occupational Safety and Health Administration (U.S. DOL/OSHA), 1982.

Occupational Safety and Health Administration: Selected Occupational Fatalities Related to Toxic and Asphyxiating Atmospheres in Confined Work Spaces as Found in Reports of OSHA Fatality/Catastrophe Investigations. Washington, D.C.: U.S. Department of Labor, Occupational Safety and Health Administration (U.S. DOL/OSHA), 1985.

Occupational Safety and Health Administration: Selected Occupational Fatalities Related to Welding and Cutting as Found in Reports of OSHA Fatality/Catastrophe Investigations. Washington, D.C.: U.S. Department of Labor, Occupational Safety and Health Administration (U.S. DOL/OSHA), 1988.

Occupational Safety and Health Administration: Selected Occupational Fatalities Related to Ship Building and Repairing as Found in Reports of OSHA Fatality/Catastrophe Investigations. Washington, D.C.: U.S. Department of Labor, Occupational Safety and Health Administration (U.S. DOL/OSHA), 1990.

OSHA: Permit-Required Confined Spaces for General Industry; Final Rule, *Fed. Regist. 58*: 9 (14 January 1993). pp. 4462–4563.

OSHA: Confined and Enclosed Spaces and Other Dangerous Atmospheres in Shipyard Employment; Final Rule, *Fed. Regist. 59*: 141 (25 July 1994). pp. 37816–37863.

Occupational Safety and Health Administration: OSHA Alerts Workers and Employers to Hazards in Transferring Carbon Dioxide (USDL: 96-242). Washington, D.C.: U.S. Department of Labor, Occupational Safety and Health Administration (U.S. DOL/OSHA), 1996. [news release]

Park, C. and R.P. Garrison: Multicellular model for contaminant dispersion and ventilation effectiveness with application for oxygen deficiency in a confined space. *Am. Ind. Hyg. Assoc. J. 51*: 70–78 (1990).

Perkins, H.C.: *Air Pollution*. New York: McGraw-Hill, 1974. 407 pp.

Scarbrough, D.R.: Control of electrostatic ignition sources. In *Fire Protection Handbook*, 17th ed. Cote, A.E. and J.L. Linville (Eds.). Quincy, MA: National Fire Protection Association, 1991. pp. 2-284 to 2-292.

Schwab, R.F.: Dusts. In *Fire Protection Handbook*, 17th ed. Cote, A.E. and J.L. Linville (Eds.). Quincy, MA: National Fire Protection Association, 1991. pp. 3-133 to 3-142.

Silvent AB: *Blowing Nozzles for Improved Environment.* Göteborgsvägen 99 Sweden: Silvent AB, 1993. [Product Literature]

Standards Australia: *Safe Working in a Confined Space* (AS 2865-1995). North Sydney, N.S.W.: Standards Association of Australia, 1995. 42 pp.

Stewart, S.J.: Boilers in Balanced Compartments: Some background to the new provisions of BS 6644: 1990. *Clean Air 21:* 21–26 (1991).

U.S. Department of Commerce: A Numerical Method for Calculating Indoor Airflows Using a Turbulence Model, by T. Kurabuchi, J.B. Fang, and R.A. Grot (Pub. No. NISTIR 89-4211). Gaithersburg, MD: U.S. Department of Commerce/National Institute of Standards and Technology, 1990.

U.S. Department of Commerce: User Manual NBSAVIS. CONTAM88. A User Interface for Air Movement and Contaminant Dispersal Analysis in Multizone Buildings, by R.A. Grot (Pub. No. NISTIR 4585). Gaithersburg, MD: U.S. Department of Commerce/National Institute of Standards and Technology/Building and Fire Research Laboratory, 1991.

U.S. Department of the Interior: Preventing Ignition of Dust Dispersions by Inerting, by J. Nagy, H.G. Dorsett, and M. Jacobson (Report of Investigation, 6543). Pittsburgh, PA: U.S. Department of the Interior/Bureau of Mines, 1964b.

U.S. Environmental Protection Agency: Project Summary. Indoor Air Quality Model Version 1.0, by L.E. Sparks (EPA/600/S8-88/097). Research Triangle Park, NC: U.S. Environmental Protection Agency/Air and Energy Engineering Research Laboratory, 1989.

U.S. Environmental Protection Agency: EXPOSURE Version 2. A Computer Model for Analyzing the Effects of Indoor Air Pollutant Sources on Individual Exposure, by L.E. Sparks (EPA-600/8-91-013). Research Triangle Park, NC: U.S. Environmental Protection Agency/Air and Energy Engineering Research Laboratory, 1991a.

U.S. Environmental Protection Agency: Indoor Air Quality Simulator for Personal Computers, by P.A. Lawless and M.K. Owen (EPA/600/S8-91-014). Research Triangle Park, NC: U.S. Environmental Protection Agency/Air and Energy Engineering Research Laboratory, 1991b.

Wiegand, K. and S.P. Dunne: Radon in the workplace — a study of occupational exposure in BT (British Telecommunications) underground structures. *Ann. Occup. Hyg. 40:* 569–581 (1996).

Wysocki, T.J.: Carbon dioxide and application systems. In *Fire Protection Handbook*, 17th ed. Cote, A.E. and J.L. Linville (Eds.). Quincy, MA: National Fire Protection Association, 1991. pp. 5-232 to 5-240.

Zabetakis, M.G.: Flammability Characteristics of Combustible Gases and Vapors (Bull. 627). Washington, D.C.: U.S. Department of the Interior/Bureau of Mines, 1965. 121 pp.

12 Respiratory and Other Personal Protective Equipment

CONTENTS

INTRODUCTION

Work that occurs in confined spaces often is highly hazardous, short in duration, and performed during irregular hours, typically the off-shifts. The first line of defense against workplace hazards in these situations usually is personal protection, supplemented by portable ventilating equipment. Respiratory and other personal protective equipment may be required to assure safety during entry and work. Reliance on personal protective equipment results from the nature of the work and the impracticality of other means of control.

Selection of personal protective equipment for use in confined spaces can become a considerable challenge because of the breadth of hazardous conditions that can occur. Compounding the problem, a small opening may be the only means of access/egress. Rapid egress through such an opening by persons encumbered by bulky and heavy ensembles of personal protective equipment under emergency conditions could be seriously compromised. In addition, the situation could require work to occur during extremes of hot or cold weather. Hot weather creates potential for heat-induced injury. Cold weather increases potential for cold-induced injury and reduces the flexibility of fabrics and gloves. Reduced flexibility decreases dexterity and impedes movement. These situations can compound the risk of injury caused by hazardous conditions in the space. In addition, the wearing of protective equipment causes serious psychological stress in some people.

The selection of personal protective equipment for use in these workspaces must reflect conditions against which protection is required and address the necessity for the protected person to work in potentially confining surroundings. This equipment should not compound the hazards of work. This problem must be considered carefully during hazard assessment and planning. Discovering these problems during entry and work activity is too late.

ISSUES INVOLVING USE OF RESPIRATORS IN CONFINED SPACES

GENERAL

Data examined in the first chapter regarding fatal accidents in confined spaces indicated that respirators either were not used, or were used incorrectly or inappropriately (OSHA 1985, 1990; NIOSH 1994). The faulty decision making that caused these tragedies resulted from lack of knowledge and training of persons involved in the work activity.

Most of the information available to practitioners focuses on the technical aspects of respirator use. As a result, the practitioner may not become knowledgeable about the strengths and shortcomings of these products. The latter information appears in research reports and notices. This information provides insight into technical realities concerning use of respirators under less than ideal circumstances. The latter could have serious implications regarding their use in confined spaces, especially where life-threatening circumstances may be present or could arise.

Respirators function by isolating the face and the respiratory system from the contaminated atmosphere. This occurs through formation of a seal between the face and the facepiece, and provision of breathing air from a known source: either an air supply or a purification device. The seal prevents entry of contaminated air from the external environment. Fit-testing procedures provide an estimate of the effectiveness of this seal. The fit is degraded by a number of factors associated with the human face and conditions of use:

- Facial hair (beards, sideburns, bangs, hair overhanging the face)
- Temple pieces of glasses

- Protruding bone structure in the jaw or cheek
- Improper attachment to the face
- Incorrect facepiece size
- Damaged or deteriorated facepiece or sealing surface
- Deteriorated or improperly seating valves

Respirator designs incorporate decisions about a number of variables. These represent the judgment of manufacturers in pursuit of the ultimate in design. Areas of variation include:

- Shape of the sealing surface
- Material of construction
- Thickness of material of construction
- Mode of collapse against the face during securement
- Leakage paths into the facepiece
- Movement of air inside the facepiece

Respirator manufacturers usually produce both full- and half-facepiece models. Parts from one manufacturer usually are not interchangeable with those from another. Interchangeability may exist within a particular brand.

Ventilatory Demand

Respiratory protection comes at a physiological cost. In a review of the subject, Raven et al. (1979) concluded that the primary physiological cost is altered pulmonary response to work. This results from the increase in resistance to inspired and expired flow produced by the respirator. The magnitude of acceptable resistance has been the topic of considerable research. Respirators and other personal protective devices must be compatible with the needs and capabilities of human physiology.

In a review of imposed ventilatory resistance, Myhre (1982) noted that the main effects observed by others were reduction in airflow velocity and an increase in duration of the impeded phase. This was accompanied by reduction of respiratory rate and increase in both tidal volume and expiratory reserve. The net result was reduced pulmonary ventilation (Zechman et al. 1957). This was accompanied by increased alveolar carbon dioxide tension and decreased oxygen tension. Although the respiratory pump responds to resistance in breathing through proportional increases in respiratory work, hypoventilation persists. Increase in alveolar CO_2 and decrease in alveolar O_2 still occur. This is believed to represent a balance between respiratory control and biological economy. The rise in alveolar CO_2 is considered to be the primary stimulating factor supporting the increase in respiratory effort.

The extent to which work performance is compromised by breathing resistance is influenced by both the resulting changes in alveolar gas exchange and the subjective response to stress. Consequently, threshold values for initial detection of discomfort, and maximal tolerable level of resistance differ markedly. The question of tolerable limits of resistance for people wearing respiratory protection thus reflects imposed impedance to inspiration and/or expiration, and increased alveolar CO_2 and reduced O_2 tensions.

Almost all of the work on respiratory demand has occurred under laboratory situations involving treadmills and bicycle ergometers. Myhre (1980) reported on one of the few real-world situations: a simulated emergency evacuation of a gantry tower in a rocket launch complex. Ventilatory requirements during this task were estimated to be approximately 65% of total aerobic capacity. Associated with this level of work was a ventilatory requirement of 51 L/min with predicted peak inspiratory flow of 180 L/min.

Myhre repeated this work while subjects wore a currently approved escape SCBA (delivery rate of 25 L/min). Escape occurred with considerable difficulty under conditions that could seriously compromise the wearer's safety. Most obvious hazards included condensation that almost totally obstructed forward and downward vision, dizziness, and collapse of the hood with each inspiration.

Kamon et al. (1983) reported on another real-world task: an "escape" from an underground mine. The escape path was a passageway that permitted walking and running erect or stooped, duckwalking, and crawling. The miners traveled at different speeds, for each mode of locomotion. Judged by subjective evaluation and the competitive spirit that developed, these results are believed to represent the best effort to travel the escape route as quickly as possible. Compared to treadmill tests, the average effort of the "escape" was 64% and the peak effort, 70%, of the miners' aerobic capacity.

Myhre et al. (1979) reported on respiratory demands of firefighters wearing SCBAs (self-contained breathing apparatus) under heavy work (up to 80% of aerobic capacity). All workloads were performed without complaint when the subjects wore the SCBA on their backs, but did not wear the facepiece (free breathing). Wearing of the facepiece with its accompanying breathing resistance imposed varying degrees of discomfort. This was tolerable at lower work levels. However, breathing resistance imposed by the facepiece during exercise at 80% of aerobic capacity often caused extreme discomfort. The sensation of suffocation was experienced frequently. Under the latter conditions; only by exercising considerable self-restraint were some of the subjects able to resist pulling off the facepiece during the final minute of exercise. At the 80% workload (120 L/min), peak pressure change inside the facepiece (18 cmH_2O) approached the limits of tolerance. This work suggested that a pressure swing of 18 cmH_2O represents the near maximal level of tolerance to breathing resistance for an individual performing 10 min of work at 80% of aerobic capacity. Further, a pressure swing of 30 cmH_2O imposes a feeling of suffocation that is tolerable only for a minute or two in an emergency situation.

Raven et al. (1981) compared the cardiorespiratory response of subjects with normal lung function and exercise tolerance against those with moderate and severe impairment. Subjects worked at rates up to 63% of maximum capacity while wearing a full-facepiece, supplied-air respirator operated in "demand" mode. Physiologically and subjectively, the response of the normal and moderately impaired subjects to respirator wear was the same. However, work at 63% of maximum workload for the moderately impaired was equivalent to work at 50% for normal subjects. Significant differences in the peak flow/pressure ratio of the severely impaired compared to the normals and moderately impaired were found.

Little is known about the impact of respirator use during work when the wearer already has pulmonary impairment. Much of this loss can occur through aging (>25 years), smoking habits, environmental pollutants, and chronic disease (asthma, bronchitis, emphysema and fibrosis).

Raven et al. (1982) repeated the previous investigation using an air-line, full-facepiece respirator operated in "pressure-demand" mode. Physiologically and subjectively, the response of the normal and moderately impaired subjects to respirator wear up to 80% of their maximum aerobic capacity were the same. Pressure swings inside the facepiece exceeded 24 cmH_2O. Half of the subjects were unable to finish 10 min of work at the 80% work rate. Hence, the presumption of increased protection and reduced inspiratory resistance as a result of pressurizing the facepiece is seriously questioned. Further, the greater the ventilatory demand placed on the respirator due to increasing workload, the more "demand"-like the response became.

The data suggest that the critical factors concerning lung impairment and work performance with a respirator are workload and its relationship to lung function. If the impairment is insufficient to limit a person's ability to perform a specific task, use of a respirator will not compound this. Apparently, the operation of the respirator in "demand" or "pressure-demand" mode has no influence on this finding.

Hodous et al. (1986) examined the response of individuals with restrictive lung disease to respirator use under conditions of mild exercise. Restrictive lung disease was considered to occur when either diffusing capacity or total lung capacity is less than 80% of the predicted value. Participants breathed through a single combination cartridge, such as would be used in a half- or full-facepiece air-purifying respirator. Under the conditions of these tests, the data suggested that the stress induced by the breathing resistance used was minor compared to that of the exercise.

Dahlbäck and Novak (1983) investigated potential in-leakage problems with pressure-demand units during exercise at different workloads. Sulfur hexafluoride (SF_6) was used to measure facepiece penetration. Inward leakage was detected at inhalation peak flows of 300 L/min. One subject achieved an inhalation peak flow around 450 L/min. These results argue for testing of pressure-demand systems at peak flows exceeding 300 L/min to ascertain ability to maintain positive pressure in the facepiece during heavy work.

In a subsequent article, Dahlbäck and Balldin (1984) investigated the pressure-demand question, using subjects who worked to exhaustion on a bicycle ergometer. The tests employed four different breathing regimes: two facepieces with positive pressure (0.5 kPa and 0.8 kPa) and two mouthpieces without positive pressure, one control, and one with high resistance. No significant difference was detected in endurance time, heart rate, oxygen consumption, perceived exertion, and blood lactate. These authors also raised concerns about swings in pressure in the facepiece.

Wilson et al. (1989) studied use of a pressure-demand full-facepiece respirator during exercise up to maximum workload. Negative pressures obtained in the facepiece at these workloads imply that respiratory protection may be compromised. Current respirator designs provide respiratory protection for 50% of the population at exercise intensities up to 50% of maximum capacity. These authors recommended redesign of pressure-demand respirators to provide adequate protection for 99% of the general population up to maximal exercise levels.

Burgess and Crutchfield (1995) investigated the consequences of negative facepiece pressures on firefighters working at high levels of exertion. They found, for pressure-demand respirators, that facepiece leakage requires both negative pressure excursions and poor fit. All of the subjects overbreathed their respirators to some extent. The highest average excursion, 0.3 cm (0.1 in.) of water, occurred up to 5.75% of the time. Facepiece fit was excellent for most firefighters, including the subjects. Only 1 of the 51 firefighters fitted for a large facepiece obtained a protection factor less than 10,000. Application of the maximum measured negative-pressure excursion during testing to the worst leakage rate measured during fit-testing resulted in a calculated protection factor of 4,000. During the interval when the respirator was unable to supply the required inhalation rate, the facepiece would develop a pressure less than ambient.

Jones (1991) examined the physiological cost of wearing a disposable respirator at workloads ranging from light to heavy. As work intensity increased, an increase in breathing resistance occurred. Peak resistances ranged to ±2 cm water. Respirator use significantly increased respiratory and heart rates at different workloads. Temperature in the dead space between the respirator and the face averaged 7.5°C higher than in the surroundings. This author concluded that use of a disposable respirator is associated with significant physiological costs, especially at heavy workloads.

Leakage around the seal of a facepiece could permit entry of contaminated atmosphere. The extent of contamination of breathing air will depend on the level of exertion, magnitude and duration of excursions, and the quality of facepiece fit. The preceding studies generally involved fit individuals who were carefully supervised during the course of the experiment. This profile, however, does not describe the average user, nor the conditions of use. Misuse or abuse of respirators is much more likely under real-world conditions.

The situation could be especially serious in confined spaces where use of respiratory protection could be essential because of atmospheric conditions. Overbreathing or other cause of inward leakage could be a factor in rescue situations involving highly stressed individuals who are not fully conversant with use of this equipment. These factors continue to put respirator users in these circumstances at higher risk than where use of a respirator is a precaution rather than a necessity.

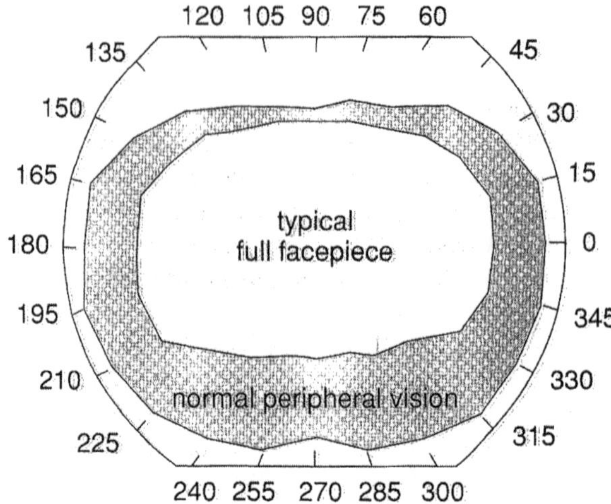

FIGURE 12.1 Loss of visual field caused by a typical full-facepiece respirator. Reduction in visual field can increase potential for accidents. (Adapted from Zelnick et al. 1994.)

VISION

One of the most important criteria for safety when wearing full facepiece respirators is visual field. Yet, visual field is not a consideration in national standards for respiratory or visual protection (OSHA 1991; ANSI 1979; NFPA 1991; CEN 1989). The European Standard (CEN 136) utilizes mannequins as an experimental model for addressing the issue of visual field. The range of the human visual field and its implications for safety during use of a full-face respirator cannot be adequately assessed using a mannequin. Zelnick et al. (1994) examined this question using Goldmann perimetry and the Esterman functional binocular field. Goldmann perimetry and the Esterman functional binocular field are recognized methods for assessing visual impairment in the workplace (AMA 1990).

Results indicated that a full-facepiece respirator considerably restricts vision in all four directions — up, down, and to the sides. The extent of the restriction is innate to the design of individual products, as well as individual differences in visual capability (Figures 12.1, 12.2).

The significance attached to loss of specific portions of the visual field reflects the importance placed on them for the conduct of normal activities. This would be most evident for occupations or activities that have specific requirements in the visual field — overhead, floor level, or side-to-side. The types of injuries that can occur due to loss of visual field are difficult to predict or to measure because they are so varied.

Differences in visual field among subjects wearing full-face respirators support the premise that individual testing should be considered (Zelnick et al. 1994). This would seem particularly useful for individuals engaged in jobs requiring preservation of specific areas of the visual field.

Johnson et al. (1994) examined the role of visual acuity on task performance by wearers of full-facepiece respirators. Clarity of the lens generally is not an issue during performance of tasks requiring visual acuity. Visual acuity can limit performance in some circumstances. This can result from condensation, deposition of particulates and sprays, flow onto the lens by liquids or solids, and scratching and crazing due to normal wear and tear. Also, a spectacle insert containing prescription lenses may not be immediately available.

These authors scratched the lenses with steel wool to produce specific visual characteristics. Results indicated that identification of letters on a computer monitor is considerably more sensitive to degradation of visual acuity than tasks requiring hand–eye coordination.

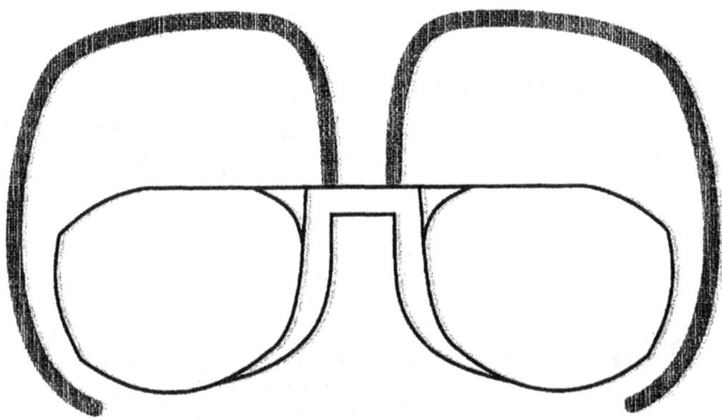

FIGURE 12.2 Spectacle kit for full-facepiece respirator. Spectacle kits utilize suction cups and wire springs formed into the shape of the inner facepiece to hold the front frame of the glasses in position. Some products can be mounted permanently by cementing into place. Optical placement of the lenses is 13 to 15 mm from the eye. Visual improvement occurs only along the path directly through the lenses and not on the periphery. The wearer must turn the head side to side to view these objects. Users requiring extreme visual correction may experience considerable difficulty with these products.

Results from these tests suggest that restriction of visual field could be a factor in the safe performance of work in some confined spaces. Many confined spaces are dimensionally small or contain obstructions that prevent normal unhindered movement. Also, these spaces may contain hot, sharp, or pointed surfaces or other cause of traumatic injury. Reliance on visual information for maintaining appropriate distance or correct orientation is total in these circumstances. The loss of information caused by restricted or deteriorated visual field demonstrated in these experiments highlights the essential role of vision in the safe conduct of work.

Heat Load

An additional consideration arising from use of respirators is the thermal load generated in the process.

James et al. (1984) determined physiological responses and perceived strain on unacclimatized male subjects during respirator use (half- and full-facepiece, cartridge-type, air-purifying respirators) combined with exposure to heat. The subjects walked on a treadmill for 1 hour in a controlled environmental chamber at different energy expenditure levels and thermal loads.

Wearing of the full-facepiece respirator added significant physiological strain to that caused by the heat and workloads used in the study. Increases occurred in heart rate, minute ventilation, oxygen consumption, energy expenditure, and oral temperature. Sweat rate was unaffected. Reduced tolerance to moderate or higher levels of work under hot conditions while wearing a respirator also occurred. The greater reduction in tolerance caused by the full-facepiece respirator was attributed to the greater facial area or dead space enclosed by the facepiece, or both.

Gwosdow et al. (1989) examined the physiological and subjective responses of sedentary subjects wearing half-facepiece respirators over a wide range of room and respirator air conditions. Respirator air temperatures less than 33°C always were comfortable and 100% acceptable. Respirator air temperatures above 33°C, or higher humidity levels, decreased acceptability. Acceptability of the respirator environment decreased as lip temperature increased above 34.5°C, or when respirator dew-point temperature increased above 20°C. Increased respirator air temperature and humidity often made breathing seem "slightly hard." Respirator conditions influenced judgment about the acceptability of the surrounding thermal environment.

DuBois et al. (1990) reached similar conclusions for subjects wearing half-facepiece respirators.

These results suggest that prevention of excessive physiological strain from respirator use at moderate or high workloads at hot job sites could necessitate more rest breaks or limits to work time. These concepts could be especially important at confined spaces. Many of these structures absorb thermal energy from the sun. The walls and interior space can become considerably warmer than the surrounding air. Also, heat generated by work activity, such as welding, will dissipate in the structure and airspace.

Psychological Factors

The main focus of interest about respirator use has been physiological stress and strain. However, respirator use imposes considerable psychological burdens on some users. Shepard (1962) has reported that physiological problems associated with respirator use were minor compared to psychological responses in some users. Morgan (1983a) reviewed the literature on psychological factors and reported that general agreement suggests that "wearer comfort" is a crucial consideration in the selection of respirators. Douglas et al. commented that everyone who wears a respirator experiences some discomfort (ERDA 1976). The distress associated with work in an extreme environment is exacerbated by wearing a respirator. Vision is restricted to some extent with all devices. This can lead to "tunnel vision" or narrowing of the visual field (Landers 1980). This response can, in turn, lead to anxiety, claustrophobia, and hyperventilation in some individuals.

About 10% of the subjects in any given respirator study are unable to perceive changes in breathing, nor are they capable of accurately judging exercise intensity (Morgan 1983a). Approximately 10% experience an excessive amount of discomfort while wearing respirators. Ten percent may represent a conservative estimate of the proportion of individuals who experience discomfort while wearing respirators. Results of research in the field of respirator performance undoubtedly are attenuated due to volunteerism. Indeed, most of the subjects are healthy, male volunteers.

Morgan and Raven (1985) reported on the effectiveness of a psychological test for trait anxiety as a predictor of respiratory distress during heavy physical work performed while wearing a respirator.

The Spielberger trait anxiety scale is a standardized psychological inventory designed to measure proneness to distress (Spielberger et al. 1969). This was administered prior to treadmill tests. The State-Trait Anxiety Inventory (STAI) contains separate scales for reporting state and trait anxiety. The state-anxiety scale evaluates anxiety at a particular time. The trait-anxiety scale measures differences in anxiety proneness.

Some of the participants performed the exercise tasks while wearing an SCBA operated in demand mode. The remainder of the group used an SCBA operated in pressure-demand mode. Subjects classified the reason for terminating the test as respiratory or nonrespiratory. Five of the six individuals predicted to have respiratory distress based upon trait anxiety scores responded as expected. Thirty-eight of the 39 subjects predicted not to experience distress also responded as expected. Objective measures of trait anxiety according to these results can identify individuals most likely to experience distress while wearing a respirator and performing heavy physical exercise.

Johnson et al. (1995) followed up the work of Morgan and Raven (1985). Their study was designed to provide quantitative information about the effect of anxiety level on work performance during respirator use. Subjects were tested for trait anxiety using the Spielberger inventory and then worked on a treadmill at 80 to 85% of their maximum heart rates until reaching voluntary end point (Spielberger et al. 1969). While experimental variability prevented many of the results from achieving statistical significance, performance times with the respirator averaged less than without. Even when they could adjust the rate of work, anxious subjects experienced more discomfort, performed for shorter times, and accomplished less work than their counterparts who had low anxiety.

Jaraiedi et al. (1994) examined the impact of wearing a half-facepiece dual cartridge respirator on productivity in mentally stressing tasks. Sedentary subjects inspected printed circuit boards for defective traces. Neither speed nor accuracy was significantly affected by wearing of the respirator. Individuality was the only factor that consistently affected both speed and accuracy of performance.

Caretti et al. (1995) examined the impact of wearing a full-facepiece respirator on computer-controlled tasks in sedentary subjects. Tasks provided measures of simple and choice reaction time, serial pattern matching, lexical discrimination, visual selective attention, rapid visual scanning, and form discrimination. Subjects were tested for trait anxiety levels using the Spielberger inventory prior to the start of testing (Spielberger et al. 1983). Anxiety levels also were assessed at the start of the day and periodically during the tests. Reaction time and response accuracy did not differ significantly with respirator use and did not change over time. Mean decision-making times were significantly faster during respirator use. This improvement probably reflects increased arousal and focused attention through exclusion of peripheral visual stimuli. Further, the results suggest that respirator wear over a relatively short period in the absence of other stressors should not inhibit cognitive function.

Hyperventilation Syndrome

Hyperventilation syndrome is a potentially serious condition that affects some respirator users (Morgan 1983b). Published research on this syndrome dates back at least a century. Symptoms include breathlessness or dyspnea with effort, paresthesia (tingling), trembling, tachycardia, tetany, carpopedal spasms, and convulsions. Hyperventilation syndrome occurs in some individuals for no apparent reason. Hyperventilation, and the resulting physiological changes, are associated with decrements in psychomotor performance along with increased error rates.

Symptoms that characterize hyperventilation syndrome can be readily produced in certain "types" of individuals through overbreathing or introduction of a CO_2 challenge. The symptoms, once produced, can be quickly reversed by rebreathing of air expired into a bag. Some individuals or "types" appear to be especially sensitive to the effects of overbreathing and/or CO_2 loading.

Identification of these individuals is extremely important, since they appear to be at greater risk of attack during tasks requiring vigorous exercise and the wearing of respirators. An attack occurring during work in a confined space could have serious implications on the safety of these individuals, as well as their co-workers. Identification of susceptible individuals should occur through routine medical evaluation.

Operational Problems

Respirators that are reusable require cleaning after use, regardless of whether they are personal or communal. Little exists in the published literature to describe problems that can arise from servicing of respirators. One of the few accounts was provided by McDermott and Hermens (1980). These authors described equipment for routine field testing of self-contained and airline breathing equipment. The equipment was used communally, and cleaned, serviced, and tested prior to reuse.

These authors identified the following problems:

- **Sticking valves:** Valves can stick closed after washing. This can occur due to formation of a water film/soap/detergent film between the seat and the valve.
- **Regulator blockage:** Reduced airflow through the valve controlling compressed airflow into the regulator can occur because of corrosion or dust in the valve.
- **Deteriorated sealing surface:** Aging or deterioration due to chemical or physical degradation can impair the sealing surface.
- **Damaged facepiece:** Cracking due to aging, or chemical or physical attack, or puncture due to improper handling can shorten service life of the facepiece.
- **Lens damage:** Crazing or scratching due to improper handling, or chemical or physical degradation can shorten service life of the lens.

- **Exhalation valve leakage:** Exhalation valve leakage or noisy regulator operation can occur due to an imbalance in operating pressure. As an illustration, a regulator operating near the maximum mask static pressure may open an exhalation valve set to open at a slightly lower static pressure, thus causing slight leakage.

None of the problems mentioned here significantly reduced the protection provided by the equipment. Many were identified when workers donned the equipment for use.

Beckett and Billings (1985) detailed additional factors that impair work performance during respirator use. These comments are highly relevant to work occurring in confined spaces. All respirators interfere with speech, hearing, vision, or mobility, or some combination of these functions. Tight-fitting, full-face respirators commonly fog in cold, ambient environments. Expiration valves can freeze and remain open or closed. Powered air-purifying respirators produce noise and may temporarily impair hearing or interfere with communications. The communication requirements of some jobs may render proper use of certain respirators unsafe.

Space limitations of the work environment are frequently overlooked. Enclosed spaces with small entrances may not permit entry while wearing an SCBA in the prescribed manner. Users have entered small openings wearing only the facepiece and attempted to put on the tank once inside. This practice has brought fatal consequences to both entrants and would-be rescuers. This equipment adds considerable weight to the individual. The bulk of the equipment plus the extra weight considerably reduces mobility and reserve strength in cramped quarters. Risk of injury to unfit users can increase substantially.

The airline respirator equipped with full facepiece operated in pressure-demand mode does offer the benefit of lighter weight and less bulk (Figure 12.3). In the U.S., the airline respirator equipped with a full facepiece and operated in pressure-demand mode plus emergency escape cylinder of compressed air can be used where the atmosphere is assessed to be IDLH (immediately dangerous to life or health) (NIOSH 1987). The length of the airline hose can limit use of air-supplied respirators (Figure 12.4).

Certain combinations of equipment may prove incompatible for performance of required tasks. For example, the use of a full-face respirator with vapor-barrier clothing in a hot environment may be exceedingly uncomfortable for sustained tasks. Some forms of hearing aids and hearing and eye protection are incompatible with certain respiratory devices. Tight-fitting, full facepieces can deform the face in a way that causes contact lenses to fall out. These also are incompatible with the temple pieces of regular eyeglasses. Denture wearers, those with facial hair, and those with past surgery of the face and neck are less likely to obtain proper fit with this kind of respirator. Tight-fitting facepieces can exacerbate skin diseases that affect the face and neck. For example, pseudofolliculitis barbi is exacerbated by shaving. In most affected males, shaving greatly increases the difficulty of obtaining a proper fit with a tight fitting respirator.

Additional factors include chronic cough productive of sputum, and inability to tolerate the discomfort produced by the device. These can result in such frequent interruptions of protection that the respirator loses its utility. The need for regular oral or inhaled medication also could lead to unacceptable loss of protection.

Hudnall et al. (1993) reviewed fatal accidents involving airline respirators that were incorrectly or inadvertently connected to inert gas sources. From 1984 to 1988 the Occupational Safety and Health Administration (OSHA) investigated 10 incidents (11 fatalities) involving inadvertent connection of airline respirators to inert gas supplies. Seven deaths resulted from connection to a line that normally carried inert gas. Four deaths resulted from leakage or backflow of inert gas into a line that normally carried breathable air. Ten of the deaths were attributed to nitrogen and one to argon. The circumstances of the 11 deaths indicated that coupling compatibility and supervisory oversight were major factors in the accidents.

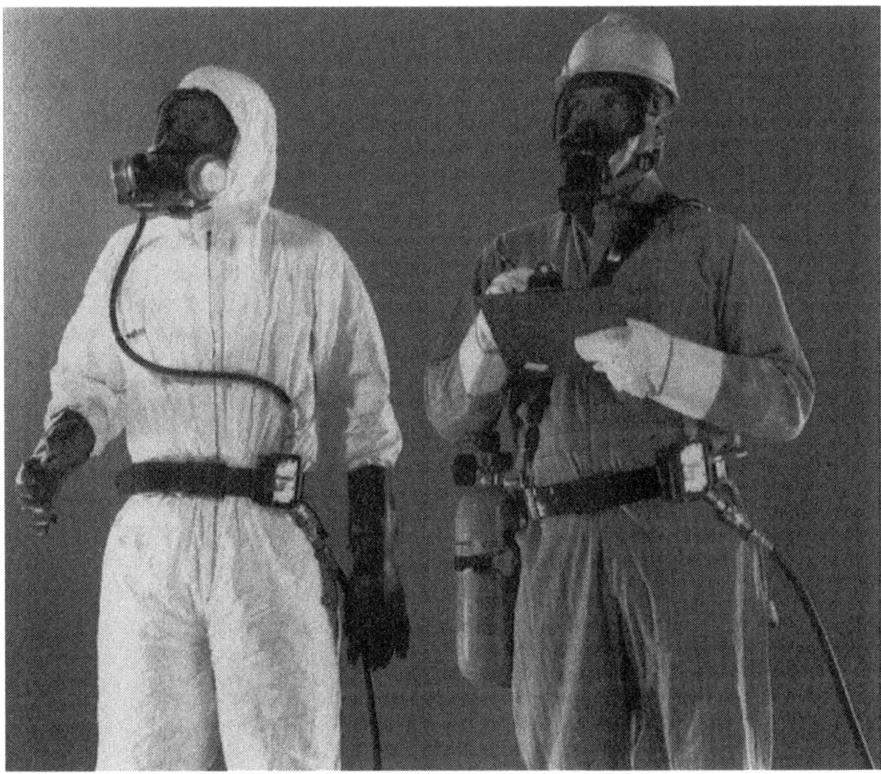

FIGURE 12.3 Respirator flexibility. Respirator manufacturers are now offering considerably greater flexibility in respirator configurations. Flexibility to meet the needs of the situation, rather than rigidity, is needed, especially in the environment of the confined space. The product in this illustration can be configured in 16 different ways as a supplied-air respirator. (Courtesy of Mine Safety Appliances, Pittsburgh, PA.)

FIGURE 12.4 Compressed breathing air purification unit. The unit pictured has a three-stage filtration system and an on-line monitor for carbon monoxide. The first stage removes bulk water and particulates. The second stage coalesces oil droplets and removes ultrafine particulates. The third stage removes organic vapors. All three stages provide end-of-service indication. Some units contain a carbon monoxide to carbon dioxide converter. Location of the air intake and maintenance of the compressor are especially important where the unit has only the on-line CO monitor. (Courtesy of Air Systems International, Inc., Chesapeake, VA.)

A supplied-air respirator, whether configured with hood, helmet, coverall, or facepiece, must have a hose with terminal detachable couplings. In the U.S., breathable air supplied to these respirators is required to be Grade D or higher (NIOSH 1995). In Canada, the corresponding requirement for compressed breathing air is Grade F (Compressed Gas Association 1989, CSA 1985). In 1980, an estimated 513,000 supplied-air respirators were in use in the U.S. (Toulmin and Moran 1982). The potential for misadventure during use of this type of equipment, therefore, is considerable.

Inert gases such as argon, helium, and nitrogen are widely used in industrial settings. (Strictly speaking, nitrogen is not inert.) They are used as fire suppression blankets with flammables, to operate pneumatic equipment, and to prevent oxidation in industrial processes (Compressed Gas Association 1990).

A nonbreathable inert gas provides little warning. The victim would be unconscious in about 12 seconds (Miller and Mazur 1984). Except in the case of persons with severe lung disease, the sensation of breathlessness is driven primarily by the level of carbon dioxide in arterial blood, rather than the level of oxygen. When air in the lungs is replaced by an inert gas, carbon dioxide is still being removed from the blood and exhaled; hence, there is little sensation of "breathlessness." The breathing receptors are fooled because there is no clear indication that something is amiss. Blackout occurs quickly, without warning.

Uniqueness of couplings and inability to misconnect to inappropriate sources are major strategies contained in regulations and consensus standards on respiratory protection for preventing these accidents (CSA 1985; OSHA 1991; ANSI 1992a; NIOSH 1995).

Face-to-Facepiece Seal

The effect of facial hair on the quality of fit obtained from a tight-fitting respirator has been and continues to be a controversial subject. Stobbe et al. (1988) summarized the results of fourteen studies on fit-testing and facial hair. All but two of the studies found that face-seal leakage increased from 20- to 1,000-fold in the presence of facial hair. In addition, leakage variability increased when facial hair was present. In the two studies, one using positive-pressure SCBAs and the other completed in the workplace, no statistically significant differences in leakage were found.

The results of the controlled studies generally were consistent. Leakage increased as the length of facial hair increased. Fit variability was greater in the presence of a beard. These authors offered the following recommendations:

Negative-pressure respirator: When a health hazard exists or is likely to exist, facial hair along or in the seal area should not be permitted — i.e., no beards.

Positive-pressure respirator: For modern, full-facepiece positive-pressure respirators, the answer remains ambiguous and elusive. Good practice dictates that facial hair should not be present.

Convenient use of respirators: There is no simple answer to the question posed by respirator use to control odors, tastes, etc. Respirator use under these conditions will not increase exposure, but could lead to sloppy practices and administrative problems.

Fitting

Gross and Horstman (1990) examined the problem of fitting a mixed-gender workforce. A mixed-gender workforce likely contains a wide variety of facial shapes and sizes. Due to the large number of facial variations, no line of facepieces from a single manufacturer will fit all workers. Quantitative fit tests indicated that 95% of men and women could receive an adequate fit when provided a selection from two randomly chosen brands of half-facepiece respirators. That is, an employer should make available at least two brands of respirator for trial.

TABLE 12.1
Dead Space in Respirator Facepieces

Facepiece type	Dead space (mL)
Quarter	190
Half	270
Full (no nosecup)	1,250

After Hinds and Bellin, 1993.

Facepiece Dead Volume

The volume of conducting air passages in the respiratory system is approximately 150 mL. This dead space has no role in gaseous exchange (Comroe 1965). The facepiece of a respirator traps an additional volume of air. This volume includes the last air exhaled from the lungs and the first air that is rebreathed. Table 12.1 lists dead space volumes for different facepieces.

Hinds and Bellin determined these values for respirators used during their experiments by means of a mannequin. For a given product line, the volume of the dead space depends on several factors:

- Presence or absence of a nosecup
- Design of an individual manufacturer
- Shape of the wearer's face
- Activity level

The physiological effects of dead space are attributable mainly to the rebreathing of accumulated carbon dioxide and reduced oxygen from the previously exhaled breath (Johnson and Cummings, 1975). An individual responds to increasing dead space by increased minute volume. This increase is almost fully due to increased tidal volume without a change in rate. Stannard and Russ (1948) provided the following equation to relate the increase in minute volume to added dead space.

$$(\text{Increase in minute volume}) = 12 \times (\text{added dead space}) \qquad (12.1)$$

The units of increase in respiratory minute volume are L/min; units of added dead space are L. Although tidal volume increases in an amount equal to 50 to 90% of the added dead space, the increase during exercise is not sufficient to restore alveolar partial pressure of CO_2 to levels that existed before dead space was added. The maximum dead volume considered to be without effect on pulmonary ventilation is 100 mL.

The response to dead space appears to be related to the resting CO_2 responsiveness of the individual (Jones et al. 1971). This implies increased sensitivity to dead space in the heat (Hey et al. 1966). Stannard and Russ (1948) recommended on the basis of constant or nearly normal alveolar CO_2 and O_2 partial pressures that the dead space in respirators not exceed 400 mL under exercise conditions. Functional dead space is related to, but not necessarily equal to, the physical space enclosed by the facepiece. Functional dead space depends on the amount of mixing that occurs during the breathing process. Designing for laminar flow inside the facepiece is one technique for minimizing functional dead space (Johnson and Cummings 1975). This has the effect of reducing mask resistance, as well as effective dead volume. Dead space decreases in importance during exercise as tidal volume increases (Silverman et al. 1951).

TABLE 12.2
Criteria for CO_2 in the Facepiece of SCBAs

Service time	Maximum conc in inspired air %
Not more than 30 min	2.5
1 h	2.0
2 h	1.5
3 h	1.0
4 h	1.0

Data from NIOSH 1972 and 1995.

Laminar flow can be optimized by straightening flow paths (Johnson and Cummings 1975). This means locating the exhalation valve and air inlet opposite the mouth. Introducing air at the top of the lens provides the best means of minimizing lens fogging. If operation at low temperature is not a concern, confining valves to the outer surface, rather than inside the facepiece, would dampen turbulence. Valves in the nosecup can introduce turbulence downstream from their location. Inhalation valves located before the filters should have minimal effect on dead volume unless they leak in the reverse direction. Exhalation valves should be located at the end of a somewhat tubular approach to minimize discontinuity effects.

Jones et al. (1971) demonstrated that increases in ventilatory and alveolar partial pressures of CO_2 depend on workload and the volume of added dead space. The magnitude of response was related to the ventilatory response of the subject to CO_2 measured at rest. Craig et al. (1970) investigated levels of inspiratory resistance ranging from 1.5 to 15.5 $cmH_2O/L/s$ and expiratory resistances of 2.0 and 3.9 $cmH_2O/L/s$ along with levels of inspired CO_2 ranging from 1.1 to 4.5%. When CO_2 was present above 3%, a consistent reduction in exercise endurance was noted. Significant hyperventilation was observed at low inspiratory resistance.

Love et al. (1979) reported about the effect of inspired CO_2 (2 to 5%) on miners working for 30 min at 2 L/min oxygen uptake while breathing through an inspiratory resistance of 10 cmH_2O at 100 L/min. Results were similar to those obtained by Jones et al (1971). Several subjects were unable to finish the test when inspired CO_2 was 4% or greater. Reasons for stopping included headache and dyspnea. Several subjects reported breathlessness. Inspired levels of CO_2 ranging from 2 to 5% caused a relative hyperpnea of 30 to 70% with a primary increase in respiratory rate and only a slight increase in tidal volume. When alveolar CO_2 tension exceeded 40 mmHg, headache and dyspnea were prevalent. Older subjects were more prone to stop the test at 5% CO_2. Thus, with the increased respiratory resistance, inhaled CO_2 was not well tolerated during work when its concentration exceeded 3%.

Regulatory testing protocols for the facepieces of SCBAs provide some background about this matter (NIOSH 1972, 1995). (The more recent protocol, 42 CFR 84, retains the requirements of 30 CFR Part 11.) There are no corresponding requirements for half or full facepieces of other types of respirators. The protocol for testing full-facepiece units requires a breathing machine operated at a rate of 14.5 respirations per minute with a minute volume of 10.5 L/min. Air exhaled into the facepiece must contain 5% (50,000 ppm) CO_2 during the test. By comparison, the concentration of oxygen in the alveolar airspaces is approximately 53,000 ppm (partial pressure approximately 40 mmHg at sea level) (Comroe 1965). To pass the test, inhaled air must satisfy the criteria stated in Table 12.2. The preceding performance criteria for the facepieces of SCBAs also provide an indication about capabilities of air-purifying units. Facepieces used in both units are essentially similar. The preceding test specifies a low workload.

The draft European standard for valved air-purifying half facepieces recommends 1.0% as a maximum average concentration of CO_2 in inhalation air (CEN 1991). The European test uses a dummy head and air supplied from a breathing machine set to 25 cycles/min at a delivery rate of 2.0 L/cycle. Exhaled air has a concentration of 5% of carbon dioxide.

Use of a nosecup in a full facepiece has some interesting implications. Originally the nosecup was intended to reduce condensation and the attendant fogging on the lens. Development of fog indicates circulation of exhaled humid air from the nose and mouth into the airspace beside the lens. Use of the nosecup can reduce the dead space in the facepiece by 40%.

The volume of air trapped in the facepiece has considerable implications when respirators are used in atmospheres having partial pressures of oxygen lower than normal. This situation is especially important in the following circumstances: high altitude, and oxygen-deficient atmospheres at normal altitudes. NIOSH (1976) initially recommended against the wearing of air-purifying respirators at high altitudes. This caveat is especially important for individuals acclimatized to sea level who work at high altitude.

An analogous situation is use of an air-purifying respirator in an oxygen-deficient atmosphere that contains additional contaminants. In these circumstances, the wearer rebreathes exhaled oxygen-deficient air trapped in the dead space of the facepiece prior to inhaling oxygen-deficient air from the workspace. The data presented in Table 12.2 provides some elaboration.

Hinds and Bellin (1993) published iterative equations that can be used to determine the impact of dead space on rebreathing and exposure. Results from one calculation become input to the next until equilibration occurs. The impact of dead space is minimized with increasing workload as depth and frequency of breathing increase (Harber 1984). Equilibrium for half facepieces usually is reached after three breaths (Hinds and Bellin 1993).

The dead space of a full-facepiece, air-purifying respirator can contain as much as 2.5% carbon dioxide, according to the performance criteria for SCBAs that are essentially similar in design. Rebreathed air would contain a lower level of oxygen than inhaled air. Under these conditions, both the carbon dioxide and the lower level of oxygen would stimulate breathing (Comroe 1965). This stimulation would tend to increase consumption of contaminated air and further exacerbate the situation. Use of an air-purifying respirator in an oxygen-deficient atmosphere could accelerate the onset of hypoxia.

Carbon dioxide, with a Threshold Limit Value — Time-Weighted Average (TLV-TWA) of 5,000 ppm, is considered to be a toxic gas (ACGIH 1991). Yet, protocols for respirator performance permit exposure to as much as 25,000 ppm of CO_2 in rebreathed air. This could increase to almost 30,000 ppm when CO_2 is present as a contaminant. In an atmosphere containing sub-TLV levels of carbon dioxide, use of an air-purifying respirator likely would enhance the toxic effects of this gas.

A nosecup in the facepiece under such conditions would be slightly beneficial since the dead volume would decrease.

Mixed Exposures

Simultaneous exposure to mixtures of chemical substances is a fact of life in many circumstances. These are the conditions against which protection through air-purifying respirators containing cartridges often is sought. Yet, the data that form the basis for the certification protocols were based on exposure to single substances. Very little yet is known about the performance of these devices under these real-world conditions (Moyer 1983). Moyer documents the complexity of the sorption process and the many unknowns that exist.

For example, conditions of use may be considerably warmer than test conditions (25°C). An increase in temperature would be expected to have an adverse effect on retention. However, this area remains to be explored.

Conditions of use may be considerably more humid than test conditions (50 ± 5% relative humidity). High humidity considerably shortens service life for substances that are hydrophobic. The converse is true for substances that are hydrophilic.

The contaminated atmosphere could contain a complex mixture of gases, vapors, aerosols, and particulates. Breakdown and pyrolysis products from resins considerably complicate this picture. Interaction between these substances and the sorbents used in cartridges are largely unknown.

A major consideration in such circumstances is the end-of-service-life indicator. Conventional approaches include color change in an indicator, odor following breakthrough, and increasing breathing resistance. An electronic device also has been developed (Maclay et al. 1991). However, this technology presently is in its infancy.

What determines service life of a cartridge where multiple contaminants are present is breakthrough by the most volatile contaminant (Lara et al. 1993). Less volatile compounds displace previously adsorbed more volatile compounds.

Usage Factor

A critical presumption in calculating protection offered by respiratory protection is usage. Usage normally is presumed to occur at all times during the exposure. Achieving this compliance in real-world situations can be very difficult. This is especially problematic where workers do not readily perceive the hazard or consider the situation serious enough to warrant use of respiratory protection. The problem of achieving fully cooperative behavior was highlighted in the literature (Cooper 1962, Morgan 1983a). The surest enticement to ensure compliance in respirator use at all times is the knowledge that this decision is clearly a choice between life or death. Again, the key to this decision is knowledge about when the exposure is occurring.

Exposure under conditions of use and nonuse can be calculated from the following equation.

$$\text{Exposure Factor} = (\%\ \text{Nonwear Time}) \times \frac{(\text{exposure level})}{(\text{Exposure Standard})} + \frac{(\%\ \text{Wear Time})}{(\text{Protection Factor})} \times \frac{(\text{Exposure Level})}{(\text{Exposure Standard})} \tag{12.2}$$

Equation 12.2 is a variant of the standard equation used to calculate time-weighted average where exposure occurs at different concentrations. In this equation, exposure level and Exposure Level may be the same or different. To illustrate, using 10 as the protection factor for a half-facepiece respirator, wear time of 95% during exposure at the same level as during nonwear time, and an exposure level of 10 times the exposure standard, the exposure factor would be 145%. Very brief lack of use can cause significant overexposure.

During work in confined spaces, exposure level could be considerably higher than would occur in open surroundings during the same work. This results from the geometric relationship between the structure, the source, and worker. Examples to consider include welding and spray or brush application of coatings. This situation considerably enhances the problem posed by nonwear time. The only options to ensure protection would be to decrease exposure level, and Exposure Level and nonwear time or to increase the protection factor of the respirator.

Contact Lenses and Respirator Use

In order to receive a satisfactory seal from a full-facepiece respirator, protrusions such as the temple pieces on glasses are not considered acceptable. The only options available to individuals needing visual correction are inserts for the facepiece that contain prescription lenses or the wearing of contact lenses. Not everyone can obtain satisfactory performance from the use of a vision kit, as alignment is critical in some prescriptions (Cullen 1993). Wearers of bifocals likewise would experience considerable difficulty in use of a vision kit. (See Figure 12.2.)

Wearing of contact lenses inside a full facepiece has generated considerable controversy. On the one hand, they offer the wearer equal opportunity to use any full-facepiece respirator at a moment's notice. On the other hand, contact lenses are a barrier to the normal function of the cornea for which due consideration must be given.

Wearers of contact lenses are prone to less than ideal conditions. Airborne particulates can cause serious irritation. Very dry compressed breathing air used for winter conditions could dry out the lenses. Contaminated air that enters the facepiece could interact with the lenses.

DaRosa and Weaver surveyed U.S. firefighters about use of contact lenses (LLNL 1986). Despite the threat of legal sanctions against them and their employers, hundreds of firefighters (29% of respondents) admitted to wearing contact lenses during daily use of SCBAs. Of these, only six encountered problems serious enough to require removal of the facepiece. No firefighter was injured or died as a result of wearing contact lenses. These findings and follow-up on comments led the authors to conclude that contact lenses were not significantly more hazardous than insert glasses. A survey performed by the U.S. Department of Energy (1986) obtained similar results. OSHA subsequently took the position (1988) that the wearing of soft and gas permeable contact lenses would constitute only a de minimis violation of its standards. ANSI also adopted the position in the respiratory protection standard (1992a) that contact lenses were no more hazardous than insert glasses.

The contact lens question extends beyond the safety aspects mentioned here. The question also extends to medical and human rights issues. This question will be explored in more depth in Chapter 14.

Operation in Cold Weather

Cold weather imposes considerable difficulties on the use of respiratory protective equipment. Facepieces are uncomfortable under such conditions. The elastomers and polymers in sealing surfaces and lenses become brittle and less flexible. Condensation can occur on lenses. Valves and regulators can stick due to ice formation.

Air supply systems, both piped and self-contained, are especially vulnerable under such conditions. The Canadian Standards Association addressed this problem in one of its standards (CSA 1985). The focus of this standard is the assemblage of components that comprises a complete breathing air system from the air intake of the compressor to the point of end use. This standard includes specifications for the compressor, air receiver, pipelines, and fittings. An Appendix to the standard recommends acceptable levels of moisture at typical operational pressures for both self-contained and piped systems. The intent of these recommendations is to prevent freeze up and internal corrosion during cold weather operations.

Disinfection and Cleaning

Disinfection and cleaning of respirators have received little attention in the literature. With awareness about HIV and hepatitis viruses and the reappearance of tuberculosis, disinfection and cleaning have become important considerations. Respirators often are cleaned together in a common bench-level facility or by users who have had little training. The status quo could be perceived as capable of causing spread of disease. This situation carries serious implications for users of respiratory equipment and administrators of respiratory protection programs.

A number of terms in common English usage have specific microbiological significance (Block, 1977). An **antiseptic** prevents or arrests the growth or action of microorganisms, either by inhibiting their activity or destroying them. The mode of action may depend on temperature, duration of contact, pH, nature of the organism, presence of organic matter, and other factors. **Disinfection** is the process of destroying disease or other microorganisms, but not ordinary bacterial spores. Disinfectants kill the growing forms, but not necessarily the resistant spore forms of bacteria. Some disinfectants kill spores of specific organisms and inactivate specific viruses. **Sanitization** is disinfection to levels considered safe according to public health criteria (for example, 99.9% or more). A product that is a disinfectant at one dilution could be a sanitizer at another. **Sterilization** is the act or process of destroying or eliminating all forms of life, especially microorganisms. Sterilization conveys an absolute meaning, not a relative one.

Disinfection and sterilization are broad subjects of considerable complexity (Hugo 1971, Block 1977). Active agents include: alcohols, aldehydes, phenols and substituted phenols, halogens, salicylanilides and carbanilides, quaternary ammonium salts, antibiotics, and other compounds. Tests for efficacy are provided in legislation and guidance documents from technical sources.

Regulatory authorities and standard-setting agencies require regular cleaning and disinfection of respirators, but have yet to define the meaning of these terms (Eisenberg 1992). A nonmandatory appendix in the ANSI and other standards on respiratory protection provides guidance on cleaning and sanitizing respirator parts (ANSI 1992a; CSA 1993). The procedures and recommendations have appeared in several editions of these standards, but carry no references to enable verification of efficacy. Without antecedents to enable verification of the efficacy of these recommendations, these practices could receive serious challenges from the operational level.

Respirator cleaning and sanitization is a balancing act. The process must destroy microorganisms, yet, at the same time, not damage the facepiece. The outcome of the process is not entirely predictable, since soil remaining on the facepiece following initial cleaning can consume the aggressive capability of the disinfectant/sanitizer. Hardness of the water also may be a factor. Soil can remain on surfaces not accessed by casual cleaning methods. Products used for cleaning and sanitization can affect the performance and longevity of respirator parts. Oxidizers, such as hypochlorites, can attack the elastomers used in the facepiece and valves. Alcohols can cause disintegration of the facepiece and valves and also can pose an exposure problem to the individual performing the cleaning. Iodine can leave an unpleasant after-odor. High heat can prematurely age elastomers in the facepiece and valves.

An added concern is incomplete rinsing and the presence of residues from the process. Quaternary ammonium salts and other residuals that can remain on the surface of the respirator can irritate the skin and possibly cause allergic sensitization.

Respirator Certification Standards

Current certification protocols in the U.S. for respirators reflect the work of Silverman et al (1945). Respiratory airflow measurements and breathing waveforms published by these authors formed the basis for machine testing in respirator certification. Specific breathing machines with the capability to provide different waveforms and ventilation volumes have been used in nonhuman certification tests since 1950 (Wilson et al. 1989). Human testing also occurs. Metabolic and respiratory requirements have been elucidated for selected work tasks during these tests (Kamon et al. 1975).

The protocols developed for certification testing were not designed to meet the challenge of maximal exercise or the extremes of ventilation and inspired pressures and flows (Wilson et al. 1989). To complicate matters further, the average height of the population has increased since the standards were created (DHEW 1980). These population statistics are important to the respirator certification process, since lung volumes vary directly with body weight (Åstrand and Rodahl 1977).

Silverman et al. (1945) concluded that a limit on internal respiratory work appears to be the best basis for stating tolerable limits of resistance. They investigated the effects of breathing against resistance while working for 15 min at various rates on a bicycle ergometer (Silverman et al. 1943, 1945, 1951). Work loads ranged from 0 to 1,660 kg-m/min; respiratory resistances ranged from 0.6 to 10.6 cmH_2O at a measured flow rate of 85 L/min. Below workload of 1,107 kg-m/min, increases in resistance did not affect tidal volume or respiratory rate, although the greater resistance increased the inspiratory phase of the respiratory cycle. Most subjects were able to tolerate the increased resistance, provided that total external respiratory work did not exceed 2.5 kg-m/min at low workloads and 13.3 kg-m/min at high workloads.

In the U.S., regulatory standard 30 CFR Part 11 has controlled respirator development and performance criteria since 1972 (NIOSH 1972). This standard recently was superceded by 42 CFR 84 (1995).

International Standards

The signing of the General Agreement on Tariffs and Trade (GATT) and the formation of the World Trade Organization highlights one direction in which the respirator market may head. The purpose of these agreements is to remove tariff and nontariff barriers to trade. To the point of this writing, NIOSH approval has been the benchmark applied by many jurisdictions for acceptance of respiratory protective equipment. In the future, NIOSH approval could be challenged as a nontariff barrier to trade. NIOSH is one of several standard-setting agencies worldwide that have created performance standards for respirators (Lara and Smith 1994). These standards take different approaches to the same problem. In some situations, these standards permit use of a cartridge-based respirator where NIOSH requires a supplied-air respirator or SCBA. Current standards on performance criteria for respiratory protection exist in Europe, Japan, Australia/New Zealand, and the U.S. European and Australia/New Zealand standards share many common features.

RESPIRATORY PROTECTION PROGRAMS

Respiratory protection programs form the subject of legislation and consensus standards, as well as excellent technical resources. Legislative standards provide the operating framework for a jurisdiction. In the U.S., for example, the legislative framework is 29 CFR 1910.134 (OSHA 1991). Consensus standards, such as the ANSI, CSA, and CEN (European) standards elaborate on requirements for the selection, use, and care of respiratory protective devices and ancillary equipment (ANSI 1992a; CSA 1993; CEN 1990). Technical resources include the NIOSH guides and publications from organizations such as the American Industrial Hygiene Association (AIHA) (NIOSH 1976, 1987; Colton et al. 1991). The 1976 edition of the NIOSH guide contains information that still is highly relevant to this subject. The more recent edition provides comprehensive information about programs and selected topics in respiratory protection.

A respiratory protection program applied to operations in confined spaces must contain the following elements:

- Written standard operating procedures
- Selection based on anticipated hazards and levels
- Instruction and training of users
- Cleaning and disinfection facilities
- Secure storage
- Inspection and maintenance
- Monitoring of the work area to verify selection
- Evaluation of the program
- Medical evaluation of users
- Respirator selection based on certification by an appropriate agency

SKIN PROTECTION

General

The skin is the most prominent interface between the body and the environment. The skin also is the largest organ of the body at 1.8 m^2 and comprises 15% of its weight (Carruthers 1962). The skin consists of three major layers: epidermis (including follicles and sweat glands), dermis, and subcutaneous fat (Tucker and Key 1983). The epidermis prevents entry of water-soluble substances into lower layers. Contact with organic solvents and use of strong detergents can disrupt this defense. The dermis acts as the envelope for the internal organs. A large shear force, such as causes a scrape, removes the epidermis and upper dermis. The subcutaneous fat acts as an insulating layer and cushion.

The skin is second only to the lung as the normal route of entry for substances into the body. The skin is highly susceptible to traumatic injury from physical contact. The skin also is highly susceptible to injury from exposure to the sun and other sources of radiant energy. The skin is the major target for attack by insects and animals and contact with poisonous plants. The skin is susceptible to damage from exposure to hot and cold air, as well as hot and cold surfaces. As a result, protection of the skin is equal in importance to protection of the respiratory system.

The skin is a route of entry for many substances (Bird 1981). This can occur through permeation and destruction.

Permeation is a multistage process in which the substance passes through intact skin without causing destruction. The first stage is contact with the outer layers of the epidermis. The next step is diffusion to regions having lower concentration. This could lead to storage in the fat or transport in the blood (Snodgrass et al. 1982). The uppermost layers of the epidermis (stratum corneum) are believed to provide the major barrier for almost all substances (Schaefer et al. 1982). Polar substances apparently diffuse through the outer surface of protein filaments of the hydrated stratum corneum. Nonpolar molecules diffuse through the nonaqueous lipid matrix (Blank and Scheuplein 1969). Differences in water content of the stratum corneum profoundly influence permeation, particularly of polar substances.

Destruction can occur following attack by corrosive substances. Corrosives cause ulceration. Severe ulceration is designated as a chemical burn (Emmett 1991). Some substances, such as caustics and fluorides, continue to act after initial injury has occurred. Continued action can include penetration deeper into tissues. Preventing further injury necessitates rapid and thorough removal of the causative agent in most cases. This is the basis of the recommendation for prolonged flushing in running water. This recommendation, however, is not universally applicable. The MSDS for the substance of interest must be consulted.

Skin Protective Products

The main protective agents applied to the skin are insect repellents, sunscreens, and barrier creams. None of these products is without hazard.

Spaul et al. (1985) examined potential physiological costs from use of protective lotions (insect repellent and sunscreen) in hot environments. Oil- and alcohol-based lotions were applied thinly onto most body surfaces (approximately 80% coverage). Subjects exercised on a bicycle ergometer at 50% of maximum capacity in an environmental chamber at 43.5°C and 20% relative humidity. Significant differences in rectal temperatures occurred after 30 min under these conditions. A compensatory increase in sweat production was not observed. The skin was visibly wetter during tests with the protective lotions. The latter may have reduced the sweat evaporation rate.

Insect Repellents

The main ingredient in insect repellents is *N,N*-diethyl-meta-toluamide (DEET) (Anonymous 1993). DEET remains the most effective repellent despite years of tests of other compounds. DEET has been available since the 1940s and is used virtually without negative impact (Maibach and Johnson, 1975). DEET provides effective repellency at low application density (16 $\mu g/m^2$) (Maibach et al. 1974).

Snodgrass et al. (1982) evaluated skin penetration and deposition of ^{14}C-labeled DEET in three animal species following a single application. Potential for transplacental transfer and fetal accumulation were determined following repeated application using pregnant rabbits. Significant dermal absorption occurred in all species, primarily during the first several hours. Absorption was essentially complete after 3 days. No bioaccumulation was evident after 7 days following the single application. Bioaccumulation also was not evident in pregnant rabbits or fetuses following repeated application. These authors concluded that DEET should not pose a dermatotoxic hazard when applied to the skin, and that absorption should be less than 10% of the applied dose.

Sunscreens

Sunscreens are formulated to provide protection against UV (ultraviolet) energy radiated in sunlight. The sun is only one of many sources of UV that can occur in the workplace. Other sources include: welding arcs, lamps used for nondestructive testing involving fluorescent dyes and resin curing, mercury vapor lamps, unshielded electrical arcs, insect attractant lamps, and ozone generators. Sunscreens may provide protection against exposure from these sources.

Damage to the skin produced by UV energy is measured through production of erythema (reddening) (Anonymous 1991, Parent 1993). The minimum erythemal dose (MED) is the shortest exposure to sunlight needed to induce a barely perceptible redness of the skin within 24 h. The MED value varies by skin type, latitude and altitude, reflectivity of surroundings, time of year, and time of day. The MED for fair skin is as small as 1/4 of the value for black or very dark brown skin. Radiant energy from the sun is considerably more intense near the equator than at higher latitudes. The time to reach the MED near the equator is about half that at latitudes of 40 to 44°. The time to reach the MED at 1,500 m (5,000 ft) is about 20% less than at sea level. Reflectivity of snow and water also decreases the time to reach the MED. Intensity of the sun is greatest in the late spring and early summer and between 10:00 and 15:00.

The sun protection factor (SPF) is the measure of effectiveness of a sunscreen. SPF is the ratio of the MED of protected skin to the MED of unprotected skin. To illustrate, an SPF of 15 means that erythema will develop in protected skin after 15 times the length of exposure required for the unprotected skin. SPF applies only to energy in the B-region (290 to 320 nm) of the ultraviolet spectrum (UVB) and does not apply to energy in the A- or C-regions. Some formulations offer protection in the A- and B-regions. SPF originally was developed over concerns that UVB energy burns the skin and poses a cancer risk.

SPF is determined under ideal conditions in the laboratory. Use of a sunscreen may not provide protection at the level advertised. One deleterious factor is moisture (Anonymous 1991, Parent 1993). Immersion in water or sweating can reduce protection by 50%. Application density is another variable. Application density is difficult to control. As a result, assumed protection depends on consistency in application onto exposed areas and application in the manner presumed during tests for efficacy.

Sunscreens function by absorbing UV energy. This differs from the action of blocking agents. The latter physically block the path of the energy. Most commercial products use a blend of chemical absorbers. Each substance absorbs energy most strongly in a particular region of the UV spectrum. Overall absorption for the formulation is the sum of the absorptions at a particular wavelength by all of the ingredients.

Absorption of energy raises molecules of the absorber from the ground state to a higher level energy state (March 1985). Chemical bonds or groups of chemical bonds that absorb energy are called "chromophores." A molecule in an excited state must lose this energy in some manner. This can occur through emission as UV or visible energy or through chemical processes. The strength of covalent chemical bonds is similar in magnitude to the energy of visible and UV radiation. As a result, one possible outcome from energy absorption is bond breakage. Bond breakage can result in production of two smaller molecules, molecular rearrangement, reaction with other molecules, or production of free radicals. Free radicals are highly reactive species. Another possible process is transfer of the energy to another molecule.

Some individuals are abnormally sensitive to the effects of UV energy. Most commonly, the cause of this photosensitivity is the presence of an exogenous substance on the skin that absorbs

UV energy and initiates the response (Emmett 1987). Hence, the decision to provide sunscreens for routine prolonged use deserves careful consideration. These substances frequently are drugs. Phototoxic and photoallergenic reactions can occur. Phototoxicity refers to nonimmunologically increased reactivity of a target tissue induced by UV or visible energy. These components are related by a dose–response curve. Photoallergenic reactions involve the immune system. Substances that bind with skin protein have the potential to cause allergic sensitization (Hjorth 1987).

Not every substance sensitizes. Substances that cause contact allergic dermatitis are chemically simple. Molecular structure provides no basis for predicting the allergenic potential of a substance. Molecular weight rarely exceeds 500. Cross-sensitivity can develop between chemically related molecules. The substance first must combine with skin proteins to form a full antigen before the sensitization process can begin. A similar conjugate must form on subsequent exposure before the inflammatory reaction can develop. The photoallergen is formed only in the presence of the radiation. Presently known photoallergens include substances used as sunscreens, for example, 6-methyl coumarin and glyceryl *p*-aminobenzoic acid.

Protective Creams, Lotions, and Ointments

Protective creams, lotions, and ointments form a thin barrier on the surface of the skin. This can be extremely effective in some circumstances where full dexterity in the hands and fingers is required. These products can provide protection where other approaches, such as use of gloves, are unworkable or ineffective.

Protective creams and lotions include: vanishing creams, water- and solvent-repellent products, and specialty products (Birmingham, 1981). The vanishing creams contain soap. The soap remains in the pores and on the surface of the skin and permits ready removal of soil once wetted by water. Water-repellent products contain lanolin, beeswax, petrolatum, silicone, or related substances. These help prevent contact with water-soluble substances. Solvent-repellent (oil-repellent) products utilize ointment or dry repellent films. One specialty product offers protection against poison ivy.

Protective products should be applied only to clean skin. Otherwise, dirt, soil, and possibly hazardous substances can become incorporated into the coating. The coating should be removed and reapplied at least twice per shift. Product selection is important, since in some instances, inappropriate selection will enhance penetration by the contaminant, rather than prevent it (Eydt 1983).

Chemical Protective Clothing and Gloves

Discussion

Almost ubiquitous on sites containing or contaminated by chemical products is chemical protective clothing. The purpose of these garments is to form a barrier between the skin and contaminated surfaces. The nature of the barrier that is required is governed by the nature and properties of the contaminant and mode of contact with the surface. Contaminants may be present as gas or vapor, liquid or sludge, or solid. Substances that are gaseous require gas-tight fabrics and construction. The challenges posed by these substances are considerably greater than posed by liquids, sludges, or solids.

Chemical protective clothing and gloves are fabricated from supported and nonsupported polymer films and fabrics (Figure 12.5). Fabrics include woven and nonwoven, coated and uncoated, breathable and nonbreathable, and single layer and multilayer materials. The possible combinations for a fabric are almost endless.

The original intent of chemical protective clothing was to provide a barrier against gross contamination arising from splash or immersion (Perkins et al. 1987). Entry of gases or vapors from the atmosphere surrounding the garment was not a consideration. Concern about preventing exposure to carcinogens changed the focus of design and protection criteria, as well as testing protocols (Perkins 1987). Other developments that created interest in barrier properties were emergency response and hazardous waste sites. In both cases, little information about the chemicals might be known or available. This did little to assuage the need to ensure protection.

Passage of contaminant across the barrier provided by the fabric occurs by the following mechanisms: degradation, penetration, and permeation. This can occur in gross or molecular quantities.

Degradation refers to change in physical properties of the fabric that result from contact and interaction with the substance. These changes include swelling (determined by weight gain) and dissolving or outright destruction of the material.

Penetration refers to bulk flow across the barrier. Penetration could involve liquids, as well as gases and vapors. Penetration occurs across defects such as pinholes, and through damage resulting from punctures, cuts, abrasions, and creases. Penetration also can occur across paths created during construction of the garment: needle holes, thread, and misaligned fabric in seams; gaps in closures and zipper tape; and vents, pockets, and openings.

Permeation is a molecular process. Polymers and elastomers used in chemical protective clothing and gloves therefore are chosen to minimize this process. Permeation occurs in three stages (Hansen and Skaarup 1967). The first stage, sorption, involves contact with and dissolution into the membrane at the surface of the fabric. The second part, diffusion, occurs in response to the concentration gradient that develops within the fabric. Diffusion can occur by both Fickian and non-Fickian processes. Non-Fickian processes result from interactions between the permeant and the polymer or additives. These interactions may be important enough to affect quantitative measures of permeation. The third part of the process, desorption, occurs on the inside of the barrier. Desorption can lead to evaporation into the airspace or dissolution into sweat on the skin. Wetted skin is an important consideration regarding entry into the body. Permeation mainly is a function of solubility and diffusivity (Perkins and Tippit 1985) (Figure 12.6).

Work in a space containing hazardous residual contents, or hazardous material to be removed (such as asbestos or lead-containing paint dust) creates the potential for contamination of persons and equipment. Where contamination can occur, the means to prevent contamination and to decontaminate also must occur. Otherwise, the potential for transfer of contamination offsite becomes a major concern.

Contamination is unwanted material that adheres to a surface. Surfacial contamination may be fixed or loose. Fixed material is removable only with great difficulty.

Personal contamination occurs on the exterior surfaces of the body — hair, skin, and eyes. Personal contamination also may be fixed or loose. Permeation through boundary surfaces of the skin and lungs can occur in some instances. Personal contamination was not mentioned in the fatal accidents documented in the NIOSH, OSHA, and MSHA reports (NIOSH 1994; OSHA 1985, 1990). The acute toxicity of atmospheric hazards and traumatic action of other hazardous conditions were the sole modes of action in these accidents.

While not a recognized causative agent in fatal accidents, contamination of workers and equipment at the site is a potential concern. Contamination of personnel and equipment can result from normal operations or breakdown of protective or control measures (Cember 1969). Prevention of personal contamination is the function of respiratory and skin protection, the latter in the form of clothing, gloves, and boots.

Contamination during work in confined spaces generally takes the form of loose residual solids, liquids, sludges and slurries, and solid and liquid aerosols. These can transfer to the skin and surfaces of clothing. Depending on circumstances, contamination can be radioactive or nonradioactive. Fixed contamination, unless radioactive, does not pose the level of concern of loose contamination.

Radioactive contamination has received attention for many years. It is detectable relatively easily by unsophisticated instruments. The most common instrument for monitoring beta–gamma contamination is a Geiger–Mueller counter (2.5-cm or 1-in. tube) equipped with a thin end-window (equivalent to 0.03 mm of unit density material). Shielding against beta radiation enables detection

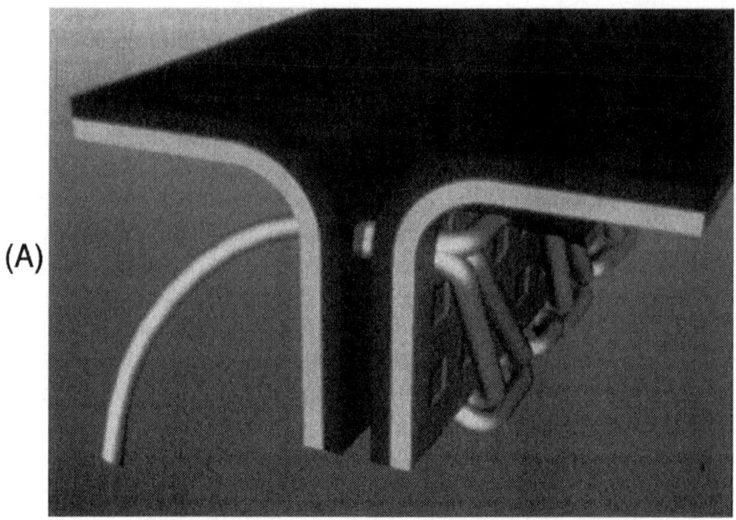

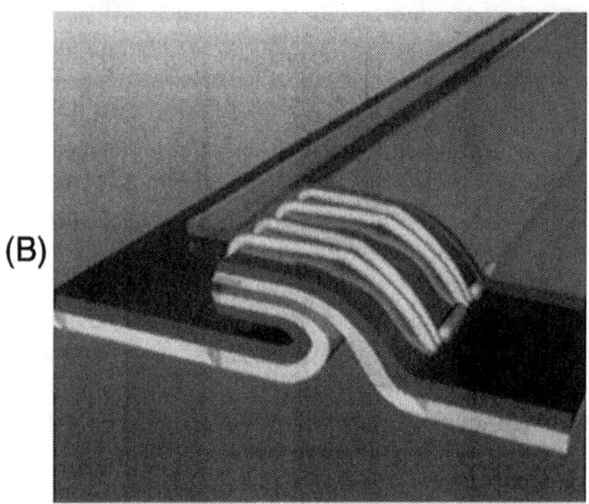

FIGURE 12.5 Seam construction techniques used in protective clothing. (A) Serged seam. (B) Bound seam. (C) NSR® seam. (D) Taped seam. The type of seam influences the level of protection afforded against penetration by liquids. Serged seams are at the low end, bound and NSR® seams in the middle, and taped seams at the high end of protection. (Courtesy of Kappler Safety Group, Guntersville, AL.)

of the gamma component. The difference between the total and the shielded value is the beta count. Alpha contamination is best detected using a proportional counter containing a thin end-window. Small areas of contamination on skin and clothing are easily identifiable using these techniques. As a result, permissible limits for loose and fixed surface radioactive contamination were established (Cember 1969, Shapiro 1981).

Loose contamination is determined by means of a wipe (swipe) test. This involves rubbing loose material from a predetermined area (usually 100 cm^2) and determining the level of activity by beta–gamma counter containing a shielded G-M tube or liquid scintillation method.

Chemical contamination is difficult to detect, unless the contamination fluoresces while irradiated by UV energy (Ness 1994). There are few limits for residual chemical contamination. Detection of surface contamination has focused on:

- Amines
- Highly toxic metals and dusts

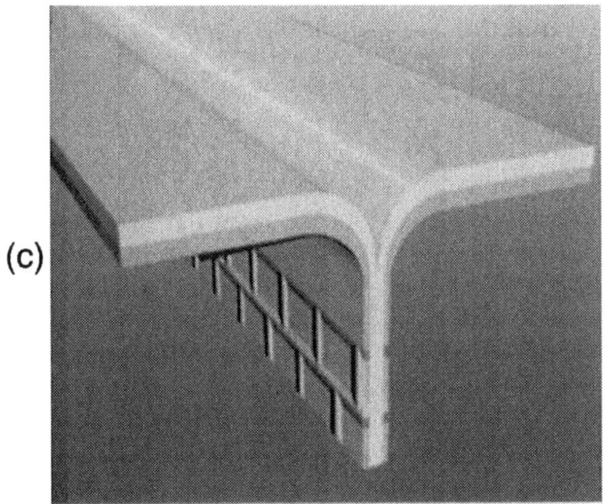

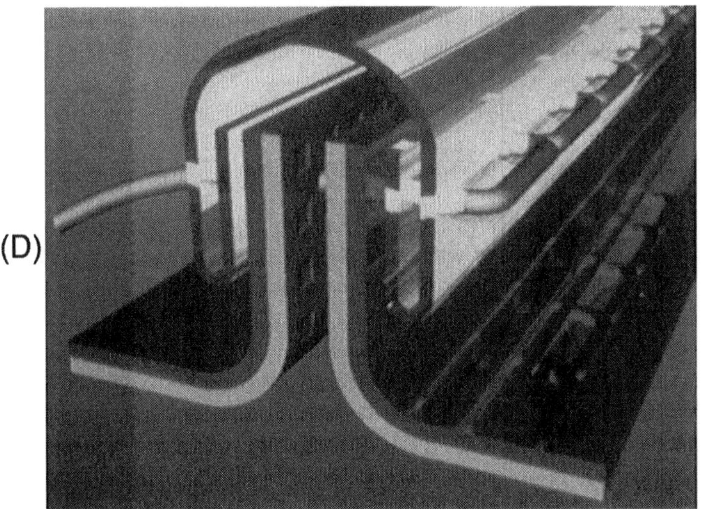

FIGURE 12.5 (continued)

- Polychlorinated biphenyls, furans, and dioxins
- Pesticides
- Microorganisms

These agents potentially can damage or pass through intact skin.

The hazard posed by a chemically contaminated surface is difficult to assess. The most probable mechanism for personal contamination is inhalation following suspension into the air or ingestion following skin contamination.

A well-known technique for assessing loose surface contamination on surfaces is wipe (swipe) sampling. Wipe (swipe) sampling is inexact (Lichtenwalner 1992, McArthur 1992, Caplan 1993, Ness 1994). Even in health physics, where most of the pioneering work occurred, no useful relationship between surface and airborne levels of contamination was established. Swipe (wipe) sampling holds some potential for estimating ingestion and skin absorption, and the need for better housekeeping. Wipe samples can serve as an indicator of "as clean as practicable." This criterion can apply to surfaces of structures as well as protective clothing and equipment.

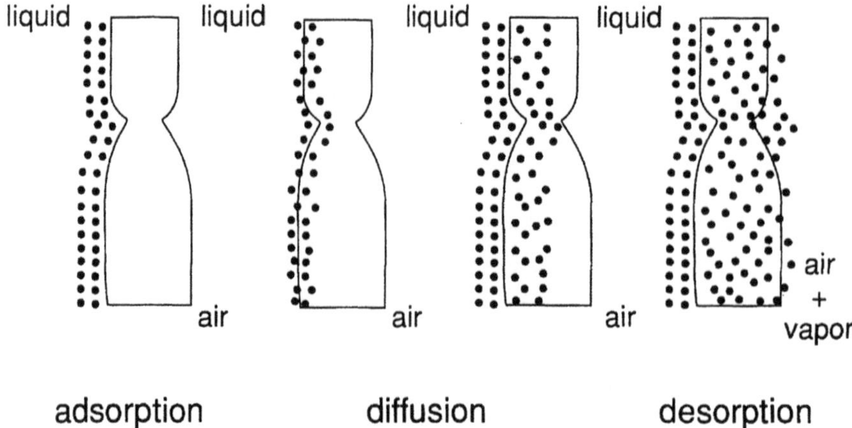

FIGURE 12.6 The permeation process. Permeation involves three describable steps: adsorption onto the surface, diffusion through the material, and desorption from the opposite surface. Most rapid permeation occurs through irregular thin spots created during manufacturing and creases created during use. (Adapted from Stull and White 1992.)

Performance Testing

Establishing performance capabilities of chemical protective clothing and gloves prior to field use is a major concern. Testing protocols and performance standards that have become available over the past decade have provided the means to evaluate performance against anticipated need. Most of the protocols are best suited for use by manufacturers or possibly by large-scale purchasers of equipment.

Table 12.3 summarizes tests that are used for evaluating performance of fabrics and garments. The tests may apply to the complete garment or to materials of construction of individual components (ANSI 1991a; ASTM 1990; CSA 1984; NFPA 1988, 1990a,b,c). (See also Figures 12.7 and 12.8.)

While formal protocols now are available for a wide variety of tests, the meaning of the data generated is not generally appreciated. There likely will be a lag in generation of both data and the means of interpreting the data.

The best-known test in this group, ASTM F 739 (permeation resistance), was developed in 1979 (Stull 1993). The intent of the test is to determine when breakthrough of the challenge substance has occurred. The difficulty with the test has been that measurement of breakthrough depends on the test protocol, and the limit of detection of the analytical instrument. The assumption that no chemical passes through the material until breakthrough is not supportable, since breakthrough is relative to the capability of the measuring equipment. In actual fact, breakthrough may occur almost immediately after the test begins and contact occurs. Since breakthrough time can depend on the test sensitivity, comparing fabrics on the basis of breakthrough times measured at different sensitivities might lead to an erroneous conclusion. To overcome this problem, ASTM F 739 now specifies a normalized breakthrough time when the permeation rate equals 0.1 $\mu g/cm^2/min$.

Critical to the overall performance of a chemical protective garment is the performance and integrity of nonfabric entities: visors, gloves, and boots; exhaust and inflation valves; closures; and seams. Gloves and boots, in particular, are most likely to come into contact with the challenge agent. Materials whose properties differ markedly from the fabric could seriously limit performance of the ensemble. Exhaust valves, for example, utilize materials found in SCBAs. These include neoprene, silicone, and natural rubbers and polycarbonate. Seams are sealed either by heat or radio frequency techniques or by attaching a strip of adhesive-backed material.

TABLE 12.3
Performance Tests for Chemical Protective Clothing

Purpose	Test/Description
Vapor protection	**ASTM F 1052, NFPA 1991:** loss of internal pressure with time following inflation indicates leakage
Liquid protection	**ASTM F 1359, NFPA 1992, 1993:** leakage into the suit of surfactant-amended water from sprays indicates potential lack of splash resistance
Permeation resistance	**ASTM F 739:** continuous contact of chemical with the test surface (fabric, seams, or closures) provides worst-case estimate of permeation parameters — breakthrough time and permeation rate
Penetration resistance	**ASTM F 903:** contact of liquid chemical with the test surface under applied pressure simulates splash of pressurized liquid
Test chemicals	**ASTM F 1001, NFPA 1991, 1992:** provide guidance on selection of liquid and gaseous chemicals for testing of test surfaces for permeation and penetration resistance
Postabrasion barrier	**ASTM D 4157:** test prepares samples for permeation or penetration resistance testing, following simulated wear caused by crawling on an asphalt surface
Postflexing barrier	**ASTM F 392:** test prepares samples for permeation or penetration resistance testing, following simulated wear caused by flexing, twisting, and compression
Burst strength	**ASTM D 751, ASTM D 3786:** test measures force needed to burst material
Tear resistance	**ASTM D 4533:** test measures effort needed to continue a tear across a fabric
Puncture/tear resistance	**ASTM D 2582:** test measures resistance to snagging and tearing of material by protruding nail or other sharp object
Seam/closure strength	**ASTM D 751, D 1682:** test measures resistance of seam/closure or fabric to pulling force
Puncture resistance	**ASTM F 1342:** test measures resistance of fabric to puncture caused by a protruding nail
Temperature effects	**ASTM D 747, ASTM D 2136:** determines how temperature changes affect stiffness characteristics of material
Burning characteristics	**ASTM F 1358:** determines ease of ignition in a flame and self-extinguishment on removal from the flame
Static retention	**NFPA 1991:** determines buildup and retention of static charge on material
Light transmission	**ASTM D 1003:** test expresses clarity of visor material as measured by light transmission and scattering (haze)
Scratch resistance	**ASTM D 1044:** test measures scratch resistance of visor material
Cut resistance	**ASTM F 1790:** test measures resistance of test surface to slicing action
Puncture resistance	**CSA Z195-M:** test measures resistance of sole/heel of footwear to penetration by sharp objects
Abrasion resistance	**ASTM D 1630:** test measures durability of sole/heel in protective footwear
Slip resistance	**ASTM F 489:** test measures traction of footwear based on coefficient of friction
Impact resistance	**ANSI Z41 (1.4):** test provides a measure of protection to toes in footwear following impact and compression of toe box

Sources: Stull 1992 and 1995; Stull and White 1992.

Selection Strategies

Selection of chemical protective clothing and gloves seemingly can become a complex process. This is especially so where mixtures and carcinogens may be present (Perkins 1988). A fabric may perform well against one chemical in a mixture, but poorly against another. No fabric provides absolute protection against all challenges. In aid of simplification, the U.S. Environmental Protection Agency/Coast Guard/National Institute for Occupational Safety and Health jointly published guidelines for the ranking of chemical protective clothing (NIOSH 1985). The intent for the guidelines was application to cleanup of sites containing hazardous waste. They since have become institutionalized and applied in many different contexts.

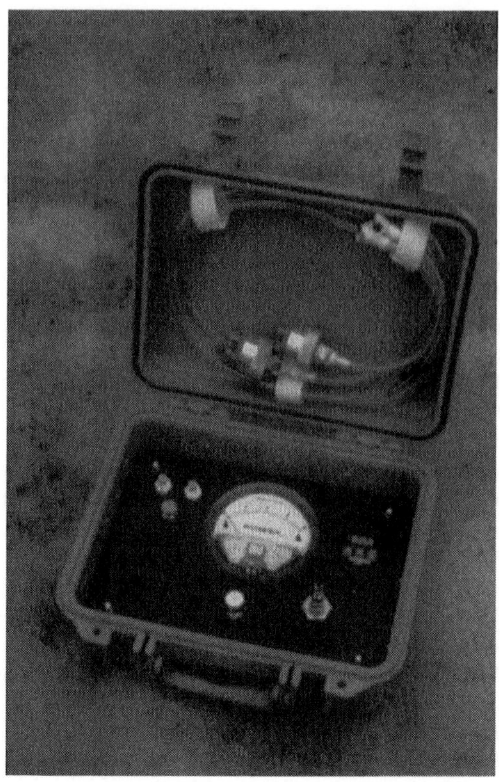

FIGURE 12.7 Pressure integrity test kit. This kit provides the means for determining pressure integrity of gas-tight suits according to the requirements of ASTM F1052. (Courtesy of Kappler Safety Group, Guntersville, AL.)

These guidelines described in general terms the type of boots, gloves, clothing, or ensemble that would be required for specific applications (Figure 12.9). However, the guidelines did not elaborate on performance requirements and capabilities. As a result, what the guidelines actually describe are the following conditions: complete encapsulation, splash protection, and normal work clothing (Stull and White 1992). Complete encapsulation means that gases or vapors, as well as liquids, liquid aerosols, or particulates from the hostile environment must not be permitted to gain entry into the suit by any route. Splash protection means that the ensemble protects mainly against contact with gross liquids or flowable solids. Vapors and gases and possibly liquid aerosols and particulates can enter through vents and incompletely closed zones of overlap. Entry of liquid can occur through overlapping flaps, and closures, such as zippers and Velcro®. Wicking also may be a problem.

The National Fire Protection Association updated performance standards for chemical protective garments (NFPA 1990a,b,c). Performance dictates design of the ensemble. Each standard sets minimum levels of performance for the overall garment, materials, seams, closures, and other components (Stull and White 1992). NFPA 1991 describes a vapor-tight, permeation-resistant garment that provides a completely encapsulated environment. NFPA 1992 and NFPA 1993 set standards of performance of penetration resistance and liquid tightness. Suits meeting this standard are not vapor tight. A suit that is vapor tight and meets NFPA 1991 also is liquid tight and provides splash protection.

The NFPA concepts provide the basis for selecting chemical protective clothing (Stull 1992). The first step involves identifying and characterizing the chemicals of concern, along with worker activity, and modes and duration of possible contact. The next step is to distinguish whether exposure is likely to occur due to vapor or liquid or both. Vapor is a concern in irritation and passage through

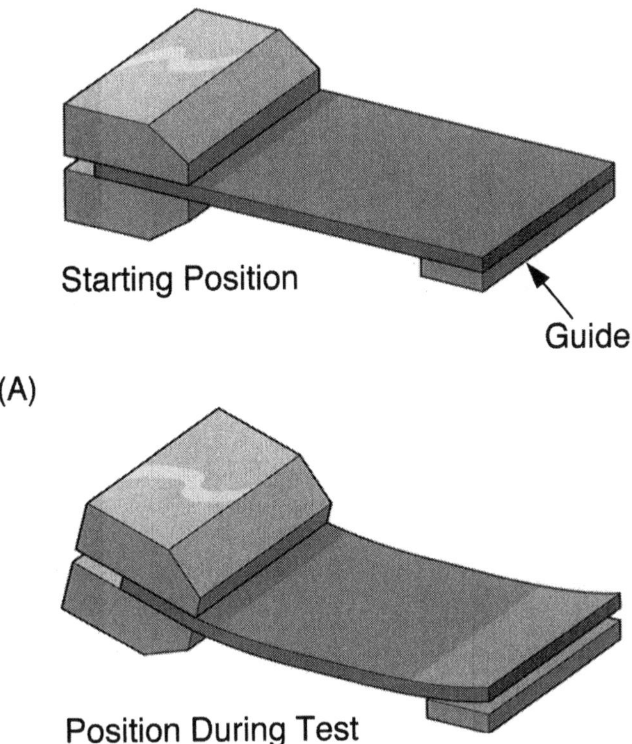

FIGURE 12.8 Protective apparel testing procedures. (A) Cold temperature stiffness. These tests follow ASTM D747 and ASTM D2136. ASTM D747 determines the effort needed to bend a strip of fabric. ASTM D2136 determines whether a fabric will crack when bent at low temperature. (B) Burst strength. These tests follow ASTM D3786 and ASTM D751 to measure pressure required to burst the fabric. (C) Tearing strength. This test follows ASTM D4533 to determine the effort needed to continue a tear across the fabric. (D) Puncture and tear resistance. This test follows ASTM D2582 to determine the effort needed to puncture and then tear a fabric. (E) Permeation resistance. This test follows ASTM F739 to determine resistance to permeation of fabrics by liquids. (F) Penetration resistance. This test determines resistance to penetration of the fabric by a liquid at a specified pressure according to ASTM F903. (G) Tensile/breaking strength. This test follows ASTM D1682-64 and ASTM D751 to determine the effort needed to pull apart the fabric and the elongation that occurred at breakage. (Courtesy of DuPont Protective Apparel, Wilmington, DE.)

the skin. The next step is to locate data on chemical resistance (permeation and penetration) for the chemical(s) of concern. Test data should be relevant to the conditions of interest. Lack of data may necessitate custom testing. Additional concerns include dexterity, abrasion and cut, puncture and tear resistance, and utility in cold conditions.

A starting point in screening potential candidate fabrics is an immersion screening test (Stampfer et al. 1988). Weight and volume changes in the fabric when immersed in the challenge liquid provide the basis for predicting breakthrough times less than 1 h with 90% certainty. Using discriminant analysis, probabilities for breakthrough times can be estimated. This approach can form an initial approach for decisions about additional testing.

Sizing recently has become an issue in garment selection. The historical approach was to stock only very large sizes and to use duct tape to overlap and contain surplus material for smaller individuals. Surplus fabric could constitute a snagging and mobility hazard. For other individuals, the largest size somehow was too small. Adams and Keyserling (1995) examined the impact of coverall size on work performance. Undersized, heavy-weight garments cause deleterious effects on most types of motion. These authors argue that oversizing offers possible benefits.

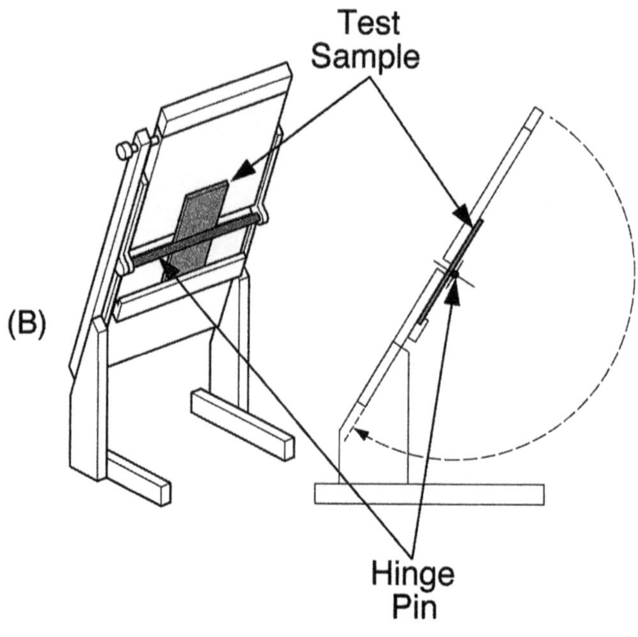

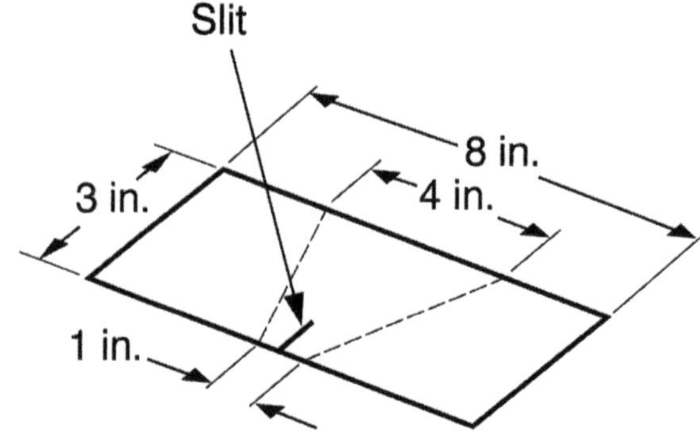

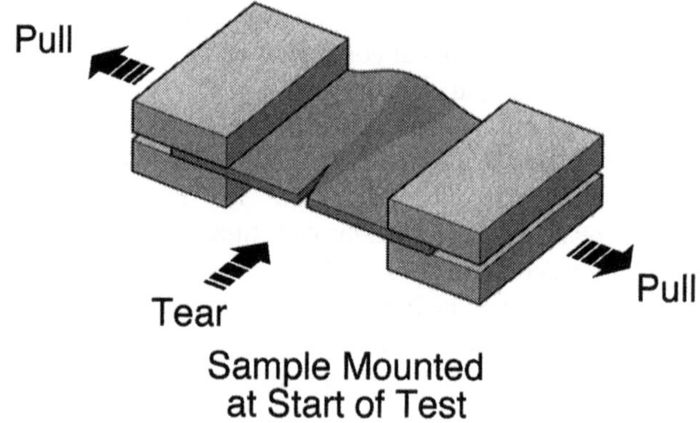

FIGURE 12.8 (continued)

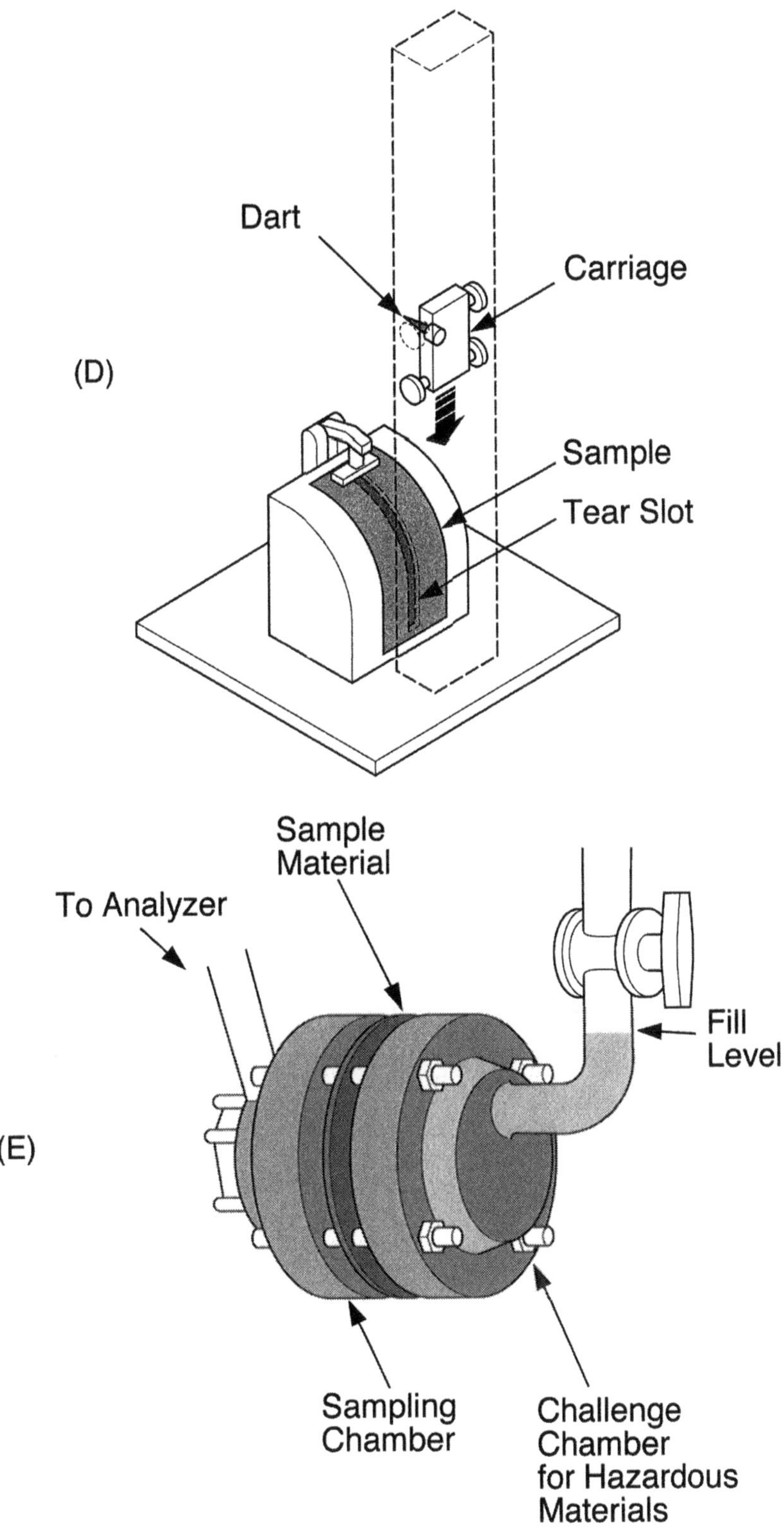

FIGURE 12.8 (continued)

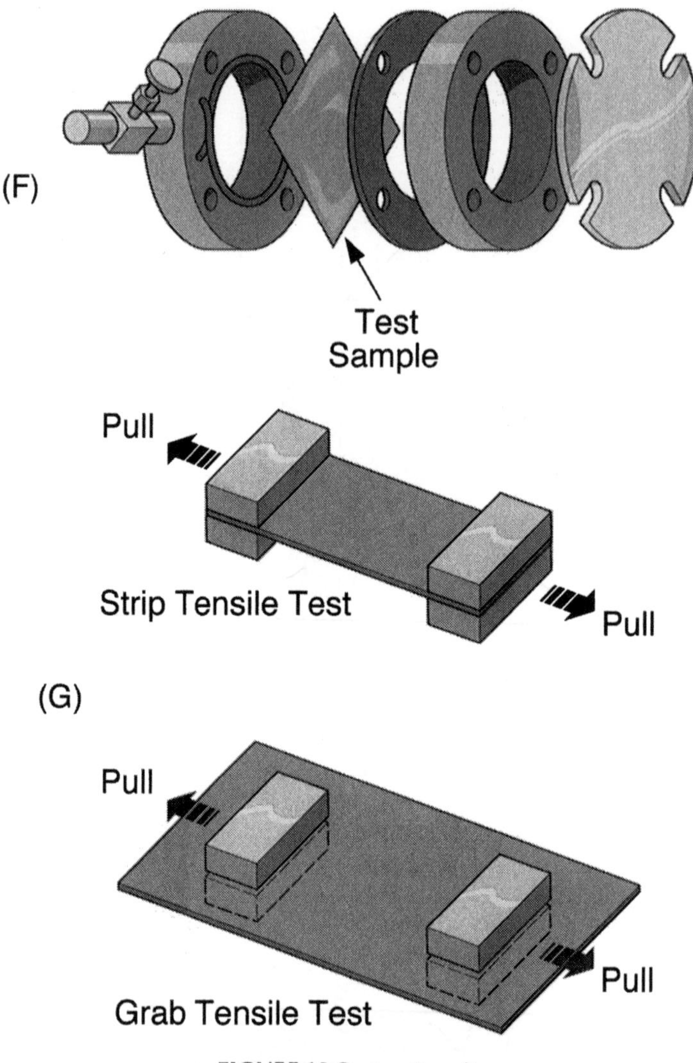

FIGURE 12.8 (continued)

To alleviate concerns about fitting, the American National Standards Institute (ANSI) and the Industrial Safety Equipment Association (ISEA) produced a standard for producing protective coveralls in standard sizes (ANSI/ISEA 1993).

Decontamination

Decontamination is the process of removal of contamination from surfaces. Contamination results from chemical and physical binding forces. Decontamination, therefore, also involves chemical and physical methods. The nuclear industry was confronted by the contamination problem long before this became an issue in the chemical industry. As a result, considerable information on this subject is available (Ayres 1970). The extent and thoroughness needed in the process reflects the level of acceptable residual contamination and the value attached to the item.

A common sense, logical approach to decontaminating skin and other surfaces is provided in the Radiological Health Handbook (BRH 1970, Shapiro 1981). Decontamination of the skin begins with a wash using mild soap and detergent and water. This is followed by progressively more aggressive methods, as necessary. These progress from abrasive soaps to pastes made from detergents and other

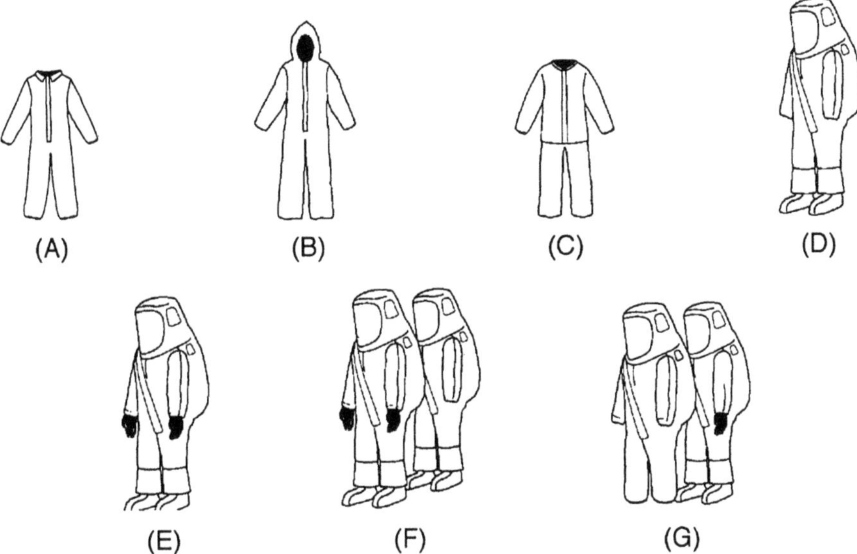

FIGURE 12.9 Styles of protective barrier clothing. (A) Coverall. (B) Hooded coverall. (C) Splash suit. These styles are best suited in applications where skin contamination or skin protection are not serious issues. The hooded coverall can act as an outer disposable garment. (D) Full coverage suit. This type of suit can enclose a self-contained breathing apparatus (SCBA) and provide a reasonably high level of skin protection. (E) Fully encapsulating suit. This suit provides a fully enclosed internal atmosphere. This is achieved through sealed seams, exhaust valves, gas-tight zippers, double storm flaps, attached sock boots with boot flaps, and sealed gloves. (F) NFPA 1991-certified totally encapsulating ensemble. (G) NFPA 1992-certified totally encapsulating ensemble. The ensembles shown here in outline consist of an aluminized overcover and an inner totally encapsulating suit. (Courtesy of Kappler Safety Group, Guntersville, AL.)

cleaners to acid solutions containing oxidizing and complexing agents. The focus of the approach is to preserve the structure and integrity of the skin.

Decontamination of surfaces is more complex. This depends on characteristics (texture, finish, etc.) of the surface to be decontaminated. Decontamination begins with vacuum cleaning. This removes from further consideration and concentrates easily accessed material. This approach minimizes the amount of residual material that could be dislodged and spread by more aggressive methods. Vacuum cleaning is followed progressively by water washing (low and high pressure), and steam and detergent cleaning. This is followed by treatment with complexing agents, organic solvents, acids and acid mixtures, caustics, and abrasive blasting, as required.

The rehabilitation of contaminated sites and structures has focused attention on the processes by which contamination occurs (Bonem and Borah 1995). Contaminants are believed to migrate through pores and microscopic voids in materials and to bond chemically or electrostatically to the substrate. Recent decontamination efforts have focused on solvent technologies that reopen pores and capillary pathways and penetrate into pores and break bonds, and bind or encapsulate the contaminant.

Protective equipment becomes contaminated. Its sole function is to intercept contamination prior to contact with underclothing or the skin (Cember 1969). Bearing this in mind can form the strategy on which selection and planning decisions are made.

The extent to which decontamination becomes an issue during the work phase depends on strategies selected during planning. Decontamination, therefore, is both a planning and an operational issue. The first issue is the nature of the contamination. This is determined by the nature of residual materials contained in the space and work activity. Contamination usually results from contact with surfaces containing residual or newly applied materials. Contact also can occur with aerosols generated during spraying or similar activity.

The nature of the activity and expected contact and properties of contaminants should determine the need and criteria for selection of protective clothing and equipment, and for decontamination. The need for decontamination can be minimized where the space has been cleaned prior to entry for work activity and where contamination is readily visible. Precleaning can considerably reduce operating time.

The need for decontamination can be simplified by the use of sacrificial layers (Perkins 1988). The outer is an inexpensive sacrificial suit. That is, the suit is intended for disposal, rather than decontamination. The seams and weak points on the sacrificial suit can be reinforced using duct tape prior to use to prolong its survival. Lightweight suits tear easily under some conditions. The purpose for the inner layer(s) is to provide a secondary or even tertiary barrier in the event of damage to the outer. The strategy may call for an innermost layer, a more expensive suit that provides a more resistant barrier, albeit at higher cost. The basis for this strategy is that the outer layer (layers, as appropriate) are sacrificed to protect the inner. The outer layers become coated with grime and dirt. The innermost layer remains clean. Another option is to use inner suits manufactured from fabrics having different chemical resistance. This strategy could be useful where mixtures are present.

Special consideration must be given to accident situations (Shapiro 1981). The objectives of remedial action in these situations are to:

- Minimize entry of contaminants into the body of the victim by any route, including the wounds
- Prevent spread of contamination from the area of the accident
- Remove contamination from the victim
- Begin decontamination of the area, as necessary

This approach can apply, regardless of the type of contamination, radiological or chemical. Accidents in confined spaces involving radiological contamination likely would involve small quantities. The situation involving chemical contamination is likely to be similar, unless substances having acute toxicity are involved. Normal decontamination procedures can lead to needless delay and spread of contamination. Removal of contaminated clothing away from the wound and decontamination of the wound should occur under the direction of a physician. This would involve use of mild soap or detergent and tepid water.

In the U.S., the conditions of work involving confined spaces where hazardous materials are present may necessitate adherence to legislation on hazardous waste and emergency response (OSHA 1989).

Some of the first procedures for decontamination of chemical protective clothing were proposed by NIOSH (1985). These utilized a series of wash and rinse tubs utilizing water and possibly detergent or bleach. This approach is intended to remove surface contamination only. No evaluation of the field effectiveness of this technique has occurred. Greatest effectiveness should occur with polar compounds. Its effectiveness in removing nonpolar, viscous organic materials is questionable. Redistribution onto a larger area of surface could occur (Perkins 1991).

Perkins et al. (1987) utilized several decontamination techniques on butyl rubber fabric. Butyl is widely used in chemical protective suits and is moderately priced. The contamination process simulated a splash. The contaminant was spread evenly and permitted to dry. This is consistent with the real-world process. Decontamination procedures included hanging in a moving airstream at room and elevated temperatures, and washing with Freon® 113 or detergent. Hanging in a heated airstream was almost completely effective in removing residual contaminants from the fabric. Washing in Freon® or detergent was considerably less effective. In the case of spillage of small amounts of solvents having substantially different solubility properties from the fabric, special decontamination techniques are not required. Air-drying probably is sufficient. Formal decontamination should be considered when spillage involves substances that have solubility properties that are similar to the fabric.

Vahdat and Delaney (1989) examined the impact of air-drying at normal and elevated temperatures and detergent-washing on new and decontaminated materials using seven polymer/chemical pairs. Test samples were obtained from the palm and gauntlet of gloves, and from sheets of material. Samples were wetted by contaminant until steady-state permeation was achieved. The results paralleled those obtained by Perkins et al. (1987) for most polymer/solvent combinations. In addition, the decontaminated materials exhibited permeation parameters similar to new material.

In reviewing the literature on decontamination of chemical protective clothing, Perkins (1991) concluded that no easy answers exist. The truism that "like dissolves like" and the observation that permeation rate is directly related to concentration gradient are strategic starting points. Water and detergents are effective against polar substances. Freon® washing was not completely effective. In addition, Freon® may leach additives from the fabric. Freon® and other organic washing solvents permeate into the fabric. Exposure to heated air appears to be effective, provided that temperature is controlled to prevent damage to the fabric.

Perkins offered the following recommendations:

- Remove surface contamination using detergents and hot water whenever possible. Use a prewash or solvent having polarity similar to the contaminants on soiled spots.
- Where permeation is suspected, store the garment or glove fully extended in a ventilated environment at 50 to 60°C for 24 h.
- Where contaminants have low toxicity, machine wash using detergent at 40 to 50°C. A prewash additive and bleach also may be required.
- Where contaminants have high toxicity, utilize laboratory tests to determine the effectiveness of the process.

Heat Stress

The trend in many circumstances is to encapsulate the individual completely in a barrier formed by a respirator, a chemical protective suit, gloves, and boots. The interfaces between these layers usually are tightly sealed by a layer of duct tape. Enclosing an individual in this manner prevents normal body cooling through evaporation of sweat, convection, and thermal radiation. Also, the individual may be required to wear an SCBA and to perform heavy work at the same time. This combination is highly conducive to causing physical exhaustion and heat-induced injury. (See Figure 12.10.)

Smolander et al. (1985) determined the physiological cost of wearing an encapsulating, gas-impermeable suit and a self-contained breathing apparatus during simulated repair and rescue tasks in a chemical plant. Total weight of the equipment was 27 kg. The tests were performed outdoors under cool conditions (ambient temperature 2.0°C (35.6°F), wind velocity 0 to 4 m/s). Total work time averaged 37 min. During tasks of search, handling vents, and sawing and replacing bolts, mean (±SE) heart rates were 146 ± 2, 148 ± 2, and 147 ± 5 beats/min, respectively. Mean rectal temperature increased 0.81°C during the whole work period. Weight loss due to sweat averaged 300 g. These authors concluded that typical tasks with gas protective clothing caused marked physiological strain among subjects in average physical condition. This occurred despite the fact that thermal strain was relatively low because of cool weather.

Paull and Rosenthal (1987) determined the impact of wearing chemical resistant suits (PVC-coated Tyvek®) and full-facepiece respirators equipped with chemical cartridges during tasks of equal workload. Tasks included sampling the contents of wastewater lagoons during summer conditions. This combination imposed a heat stress burden equivalent to 6 to 11°C (11 to 20°F) on the ambient wet bulb globe temperature (WBGT) index. A similar result was obtained by calculating the WBGT from the microclimate inside the protective suit. Mean body temperature

FIGURE 12.10 Cooling vest. This product uses phase change of material contained in pouches in pockets to provide cooling to the wearer of encapsulating clothing. Other products use ice and circulating water to provide cooling. (Courtesy of Kappler Safety Group, Guntersville, AL.)

failed to reach equilibrium during these experiments. This is an indication that the body was continuing to store heat. Body core temperature would reach an unacceptable level under continuance of these experiments. These authors conclude that significant risk of heat injury is present during work under these conditions.

White and Hodous (1987) examined the impact of different types of protective clothing and SCBAs at workloads ranging up to 60% of maximum. The test environment was thermally neutral. Mean tolerance time at the highest workload was 4 min for firefighter turnout gear and 13 min for a protective ensemble. These authors concluded that the wearing of protective clothing and the SCBA causes significant stress to the cardiorespiratory and thermoregulatory systems, even at low workload. The most stressful response was produced by the firefighter turnout gear, which was the heaviest ensemble (23 kg total weight).

In a follow-up article, White and Hodous (1988) examined the effect of barrier liners made from Gore-Tex® and neoprene in firefighter turnout gear on work performance. The barrier liner prevents entry of moisture from the environment into the clothing. Gore-Tex® is permeable to vapor but not liquid. Neoprene is impermeable to both vapor and liquid. Workload ranged to 71% of maximum. The authors concluded that the benefit from the use of a vapor-permeable liner in this application was minimal.

White et al. (1989) extended these studies to garments and respiratory protection commonly used in asbestos abatement. Subjects wore Tyvek® disposable coveralls and different types of respiratory protection: a full-facepiece respirator equipped with HEPA (high efficiency particulate air) filters, a pressure-demand, full-facepiece supplied-air respirator (including nosecup and emergency filter) and a pressure-demand full-facepiece SCBA. Workload was 24% of average maximum capacity. Environmental conditions were 34°C and 50% relative humidity. Results suggest that the heavier, more protective respirators produce only minimal additional physiological and subjective stress at the workloads anticipated to occur during this type of work.

Dessureault et al. (1995) examined the impact of wearing an SCBA and an encapsulating garment at constant workload under different environmental conditions. Participants wore a Saranex®-coated Tyvek® totally encapsulating suit and rode a bicycle ergometer at a rate of 255 W (220 kcal/h). This corresponds to light work with the body (ACGIH 1994). These authors found that exposure of encapsulated workers to environmental conditions higher than 32.5°C dry bulb and 40°C globe temperature would be very hazardous. Dry bulb temperatures above 30.5°C have a major influence on heat strain. The critical range of temperature between no harm and maximum permissible heat stress level is very narrow. These authors created a model for predicting heat strain that incorporates globe and dry bulb temperatures and remote measurement of heart rate or resting heart level and estimated work rate.

Payne et al. (1994) examined the difference between fully encapsulating suits in which the SCBA is worn inside against those in which the SCBA is worn externally. In the former case, air discharged from the respirator enters the cavity of the suit and discharges to the surroundings. In the latter, waste air is discharged directly to the surroundings from the facepiece. The suit remains unventilated. Work was performed in an environmental chamber at WBGT up to 23.8°C. Subjects performed tasks simulating work at the site of a chemical accident: drum rolling and carrying, walking, and hose dragging. The unventilated suit significantly retarded heat dissipation. The ventilated suit acts as a bellows. Air discharged from the respirator has some moisture capacity and can participate in evaporative cooling. These authors noted that the SCBA tank became quite cold during use and may have acted as a heat sink. Mean body temperature failed to reach equilibrium during these experiments. This is an indication that the body was continuing to store heat. Body core temperature would reach an unacceptable level under continuance of these experiments.

Solomon et al. (1994) examined the impact of repeated days of light work at moderate temperatures on subjects wearing encapsulating suits and full-facepiece air-purifying respirators. Studies reported in the literature normally determined only short-term impacts. Subjects cycled in an environment of 29°C dry bulb and 22°C wet bulb. These authors observed no general trend in physiological response over the 4 days of this study.

Standard heat stress indices do not apply when heavy or impermeable protective clothing is being worn (ACGIH 1994). Goldman (1985) noted that a tolerance limit based solely on deep-body temperature or heart rate does not address the problem of convergence between deep core and skin temperatures. Skin temperature normally is several degrees cooler than core temperature. This enables heat loss. Protective clothing interferes with heat loss from the skin, primarily by interfering with evaporative cooling of sweat. As a result, skin temperature increases. Blood has a decreased capacity for transferring heat produced in the core to the skin. Conditions that produce these effects can cause rapid deterioration in performance of work and consequent onset of heat-induced injury.

PHYSICAL PROTECTIVE EQUIPMENT

Physical protective equipment is clothing and other equipment that provides protection against physical agents and traumatic injury.

Head Protection

Overhead obstructions are an ever-present hazard in many confined spaces. As confined spaces often are not intended for occupancy, little attention is given regarding placement of equipment, and supporting and other structures. Head injury, while not the most frequently incurred work injury, is among the most serious (Srachta 1986a). Statistics compiled by the National Safety

Council in the U.S. reported about 110,000 head injuries per year in the 1980s. Approximately 8% of electrocutions were caused by head contact (National Safety Council 1984). The relative importance of head injuries in accidents involving confined spaces is not known, since statistics are not available.

Hardhats function to soften the blow from contact with an obstacle and to spread the force over a larger area. National standards define levels of protection and performance that are required from this equipment (ANSI 1996). Classes A, B, and C offer protection against the force of impact from falling objects. Class A offers additional protection against electric shock arising from contact with exposed low voltage conductors. Class B offers additional protection against electrical shock arising from contact with exposed high voltage conductors.

Eye and Face Protection

Eye and face protection are important considerations for work occurring in confined spaces. Unlike identical work that may occur in the surroundings, work that occurs in the space is innately more hazardous because of the confining effect of geometry. Projectiles can ricochet off surfaces and cause injury by this indirect path. In the event of such an occurrence, there may be little that the worker can do to avoid the path of projectiles. This situation puts considerable reliance on the capabilities of the protective equipment.

Statistics on eye injuries during accidents occurring in confined spaces are unavailable. Statistics on eye injuries in industry indicate that the vast majority occur through scratches and abrasions caused by metal items and airborne particulates (Bureau of Labor Statistics 1980).

Eye and face protection is designed to offer protection against a broad range of specific hazards (Srachta 1986b, Mohr 1986). These include: flying objects, airborne particulates and aerosols, molten metal, liquid chemical products, and radiant energy. The nature of the hazard determines the type of eye protection that is required.

The term "eye and face protection" describes a broad range of protective equipment:

- Safety glasses (corrective and noncorrective, tinted and nontinted lenses)
- Goggles (eyecup or rigid cover)
- Faceshields
- Welding helmets and shields

National standards define levels of protection and performance required from this equipment (ANSI 1989, 1996; CAN/CSA 1988). These depend on function and can include: impact resistance, flammability and burning rate, optical properties, and transmittance of radiant energy (ultraviolet, visible, and infrared). The applicable standard may recommend policies for good practice and type of protection for specific situations.

Within product lines, options for selection are available (Mohr 1986). Polycarbonate is considered best overall for safety lenses. Although more resistant to scratching and offering superior optical properties, glass is less impact resistant and may fail current performance requirements. Polycarbonate scratches considerably more easily than glass. Plastic formulations, while second in strength to polycarbonate, scratch more easily and are prone to warping under high heat. Fitting of safety glasses is important. Eye size ranges from 48 to 58 mm. Eye size is related to head width and position of the eyes behind the lenses. The pupils should be located behind the center of the lenses. For this reason, a range of sizes should be provided.

Sideshields considerably improve the protective benefit of safety glasses. The optical performance of off-axis (side and top) shielding is extremely important where the lenses provide protection against radiant energy. Off-axis shielding should provide at least the level of protection offered by the lenses.

Faceshields are not recommended as the sole means for protection where impact is a consideration (National Safety Council 1988). The faceshield is used in conjunction with safety glasses and cup or cover goggles. Faceshields are intended to protect the face and neck from hazards other than direct impact.

A serious concern in the use of personal protective equipment is lack of coordinated use. Some activities could involve the simultaneous use of a hardhat, hearing protection (especially muffs), safety glasses or tight-fitting goggles, cover goggles, a faceshield, and a half-facepiece respirator. Coordinating the specification of this equipment is extremely important in order to avoid the situation where the individual cannot function without extreme difficulty because of the protective equipment. The potential outcome from this situation is loss of motivation and potential refusal to wear the equipment.

Hand and Torso Protection

Injuries to the torso rank first and those to the hand, second, in accident statistics (Lahey 1986). Corresponding statistics for accidents that occur in confined spaces are unavailable.

Hand and torso protection is necessary in some circumstances because the confining aspects of the space can dramatically increase the level of hazard posed by physical conditions. Personal confinement restricts avoidance and escape from sources of energy and other hazardous conditions. These can include: hot and cold surfaces, molten metal, flame and sparks, impact caused by moving parts on equipment, projectiles, cuts and punctures, and abrasion.

Products manufactured for hand and torso protection utilize the characteristics of the fabric to afford protection. Fabrics include natural materials, such as leather and wool, synthetic fabrics, coated and metallized fabrics, and chemically treated fabrics.

Foot Protection

Statistics indicate that about 10% of lost-time injuries involve the feet, toes, and ankles (Cleveland 1984). Most of the injuries occurred due to striking by a foreign object. The balance resulted primarily from rolling an object onto the foot or stepping onto a sharp object (Bureau of Labor Statistics 1981).

National standards define levels of protection and performance required from this equipment (ANSI 1991a; CSA 1984). These include impact and compression resistance of the toe box, puncture and corrosion resistance of the sole, and resistance to application of high voltage. Performance of the toe box to impact determines the capability of the footwear. Electrical resistance deteriorates rapidly in wet conditions or where conductive surfaces are present.

Specific requirements for safety footwear include conductive or nonsparking, and foundry or molding (Anonymous 1986). Conductive or nonsparking footwear is designed to bleed off static charge. Nonferrous parts are used. Conductive footwear is used in certain hazardous locations. Foundry footwear is used where splashes of molten metal can occur. Preventing contact between the hot metal and the skin is essential to ensuring protection.

Safety footwear is available in lined or unlined versions. With the steel present in many products, boots can be especially susceptible to heat loss in cold weather. "Felt-pacs" (safety boots containing removable felt or foam liners) are manufactured for this application. Special sole materials, such as wood or cork, are used where hot surfaces are present.

Fall Arrest and Fall Protection

Falls from heights are a leading cause of fatal traumatic injury. (See Figure 12.11.) For the period 1980 to 1985, falls caused nearly 10% of deaths at work in the U.S. for which a cause was identified (Bobick et al. 1990; NIOSH 1991). Falls from heights also are a major cause of nonfatal accidents. Statistics on fall injuries in nonfatal accidents that occurred in confined spaces are not available.

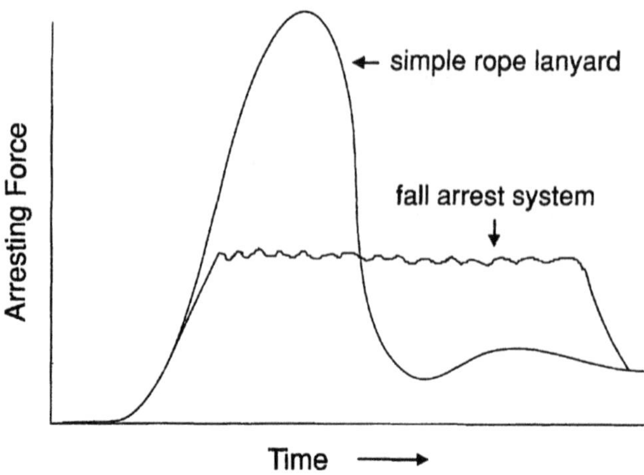

FIGURE 12.11 Moderation of arresting forces in a fall by energy absorption. (Modified from Crawford 1991.)

Prevention of falls from heights is one aspect of avoiding injury. This can occur through permanent and temporary structures, such as railings and barriers to passage, as well as covers fitted over openings. Lack of consideration during design about future occupancy and work render impracticable the installation of temporary structures in many confined spaces. In other circumstances, the work occurs at heights above the entry point into the structure. These situations put reliance onto fall-protective equipment.

National standards define levels of performance required from this equipment (ANSI 1991b, 1992b). The ANSI standards provide representative comment about performance required from this equipment.

ANSI A10-14-1991 was produced for use in construction and demolition. This standard permits use of body belts and full-body harness. This standard limits free fall to 1.5 m (5 ft). The faller cannot come into contact with any surface. The body belt/harness system must withstand a tensile test of 22.2 kN (5,000 lb_f). Anchorage for vertical lifelines shall have anchorage strength at least 22.2 kN (5,000 lb_f). Arresting force must not exceed 8.0 kN (1,800 lb_f).

ANSI Z359.1-1992 was produced for use in general industry. This standard permits use only of full-body harness in personal fall-arrest systems. Arresting force of the harness must not exceed 8.0 kN (1,800 lb_f). Arresting force of the harness-energy absorber combination shall not exceed 4.0 kN (900 lb_f).

Regulations determine application in particular jurisdictions. In the U.S., the OSHA Standard for general industry requires use of a full-body harness and availability of a mechanical device for retrieval from a permit-required confined space (OSHA 1993). The OSHA Standard for fall protection in the construction industry permits the continuing use of body belts in fall-arrest systems (other than positioning or ladder safety devices) until the end of 1997 (OSHA 1994). The current trend is to require use of the full-body harness to minimize the potential for injury.

The laws of motion provide the basis for interpreting the meaning of this information. The first equation provides the means to calculate final velocity during a free fall of a known distance.

$$(\text{final velocity})^2 = (\text{initial velocity})^2 + 2 \times (\text{acceleration}) \times (\text{displacement}) \tag{12.3}$$

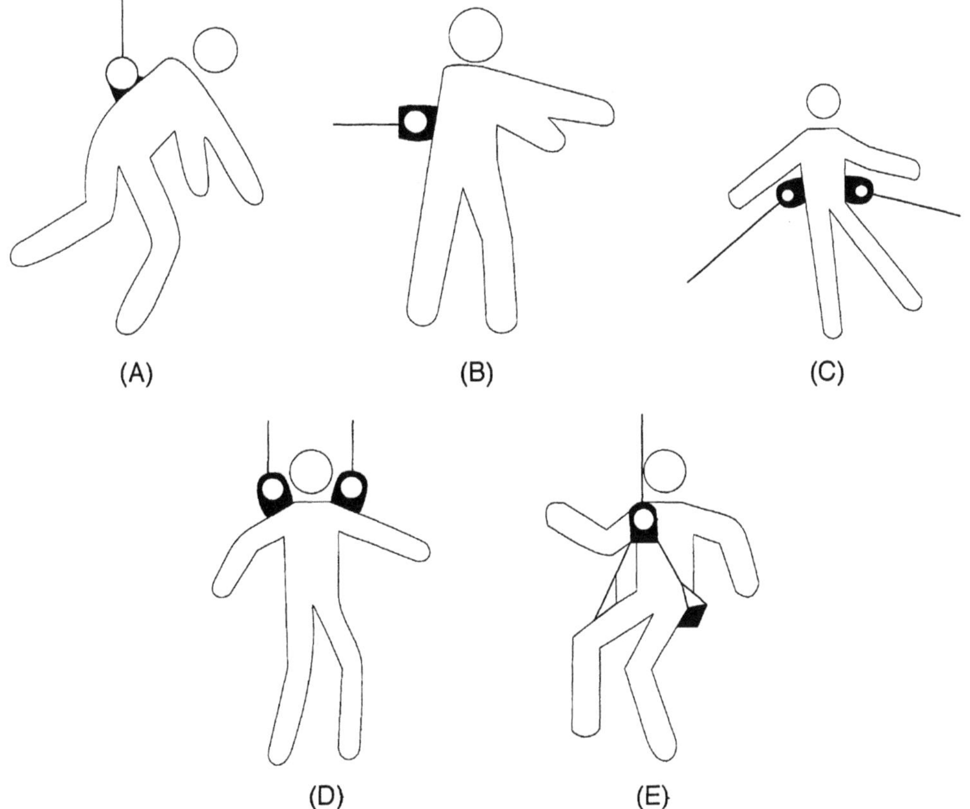

FIGURE 12.12 Functional attachment points on harnesses. (A) Fall arrest. (B) Fall prevention. (C) Positioning. (D) Retrieval. (E) Suspension. The function of the harness dictates its design and points for attachment. Few of the harnesses presently available are designed to perform all functions. (Modified from Crawford 1991.)

In this equation, initial velocity is zero. Acceleration due to gravity is 9.8 m/s^2 or 32 ft/s^2. Displacement is the vertical distance of the fall.

The force exerted by the falling body at the time of impact is given by the following equation.

$$\text{force} = \text{mass} \times \text{acceleration} \tag{12.4}$$

In this equation, the units of force are newtons (N) or lb_f. The retarding force due to air resistance is negligible. A 100-kg individual in free fall would exert a force of 100 kg × 9.8 m/s^2 = 980 kgm/s^2 = 980 N.

Implementation of any fall-arrest system requires careful consideration about several concerns. The worker must not work above the point of attachment of the lanyard. This would increase the distance of the free fall prior to arrest by the fall-arrest system. Also, a fall off-axis from the point of securement could lead to a pendular motion.

Fall protection is functionally divisible into several categories: fall arrest, positioning at height, suspension at height, restraint, and retrieval. Any of these categories could apply to work that occurs in confined spaces. Fall arrest stops the fall once in progress and acts to minimize its potential impact on the individual. Positioning equipment holds the individual in place and permits hands-free work. A fall-arrest system should be utilized simultaneously whenever possible as a backup. Suspension systems are designed for lowering and suspending the worker to a desired working height, and then permitting hands-free operation. The bosun's chair is a typical example. A fall-arrest system should be utilized simultaneously as a back-up, since the suspension system is not designed to arrest a free fall. Restraint prevents the user from attaining a position from which a free fall could occur. Retrieval equipment is used for lifting a worker through a vertical distance (Figure 12.12).

(A)

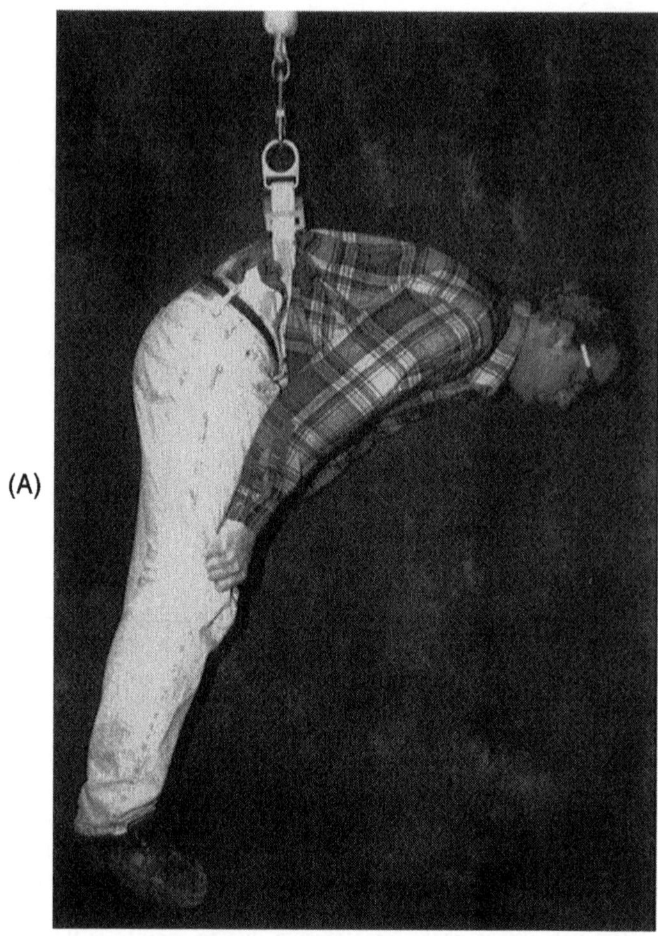

FIGURE 12.13 (A) Suspension from a waist-level restraint belt. Suspension from the belt causes the torso to bend in half at the waist. This causes strain expressed in the face (redness in forehead and protruded veins) and in the hands (whiteness). Considerable damage to internal organs can occur during a fall arrested in this manner. (B) Suspension from a fall-arrest harness. The harness spreads the weight during a fall. A suspended person assumes a partially seated position. (Courtesy of DBI/Sala, D B Industries, Inc., Red Wing, MN.)

The basis of a fall protection system is the ensemble worn by the worker, the arresting mechanism, and the point of securement. The ensemble worn by the worker traditionally has included a belt or a partial or fully supportive harness. The level of protection offered by this equipment depends on its ability to distribute the force created during the fall.

The belt includes a single D ring to which the tether is attached (Figure 12.13). In the event of a fall, the weight and force generated by the motion are localized in the plane around the waist. When suspended, the body pivots about this point. This motion threatens the abdominal organs, in particular the liver and spleen. In actual situations, the victim could remain suspended by the arresting equipment for a considerable period prior to rescue. Breathing is difficult when the weight of the body is suspended from a rear-mounted D ring. The victim also could be injured and unable to participate in the rescue operation. Prolonged suspension could exacerbate the injury (Noel et al. 1991, Brinkley 1991). These authors concluded following static suspension tests that protection should include complete body harnesses. Few knowledgeable individuals today would willingly wear a body belt in place of a full-body harness as part of a fall-arrest system (Amphoux 1991, Arden 1992).

(B)

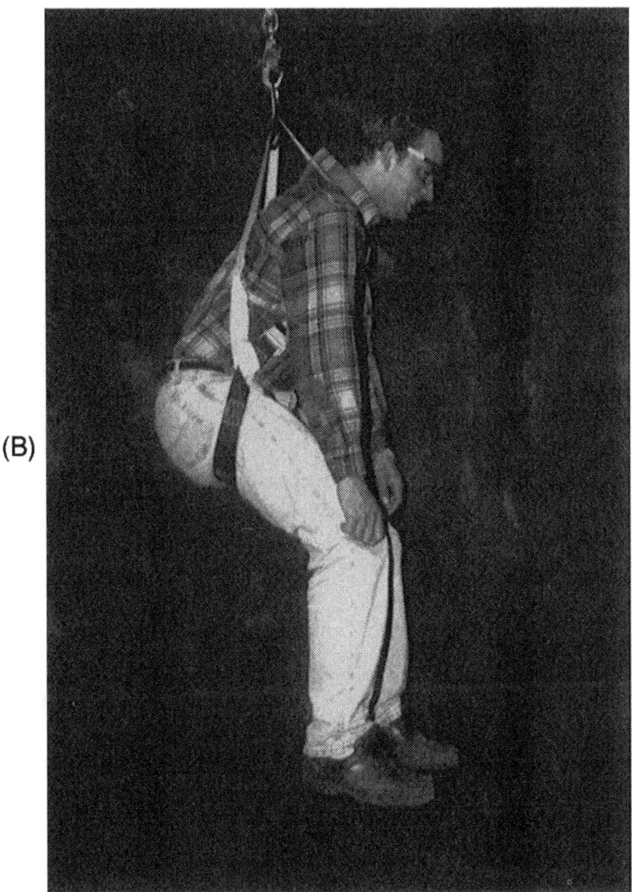

FIGURE 12.13 (continued)

The full-body harness distributes the forces created in the fall throughout the body (Figure 12.14). Attachment points may include a single D ring at the base of the shoulders or center of the chest, or two D rings near the top front or rear of the shoulders or front of the hips. Location of the D rings reflects the intended application for the harness. These should never be misconnected. Otherwise, considerable injury could occur during a fall. Modern designs for fall-arrest or positioning, full-body harnesses are specialized for specific applications (Figures 12.15, 12.16).

Two fall-arresting mechanisms are utilized: the shock-absorbing lanyard and self-retracting lifeline. Both are located between the worker and the point of securement. Neither prevents the onset of the fall. Both mitigate the injury potential of the fall by minimizing energy transfer to the body of the worker or arresting the progress of the fall.

The shock-absorbing lanyard is constructed to elongate during the fall (Figure 12.17). The elongation absorbs the energy of the fall. The elongation occurs through progressive destruction of stitching or stretching of the fabric. The lanyard and other components are not reusable after a fall. The shock of the fall decreases the strength of all parts of the system. Elongation also increases the fall distance.

The self-retracting lifeline acts like a seatbelt retractor (Figure 12.18). Line can be extracted from the tether unit through slow withdrawal. Rapid withdrawal causes the arresting mechanism to lock. Slight intentional slippage may occur prior to locking as a means of damping. The locking action occurs during a fall or rapid motion. This can arrest a fall prior to uncontrolled onset. Some units also

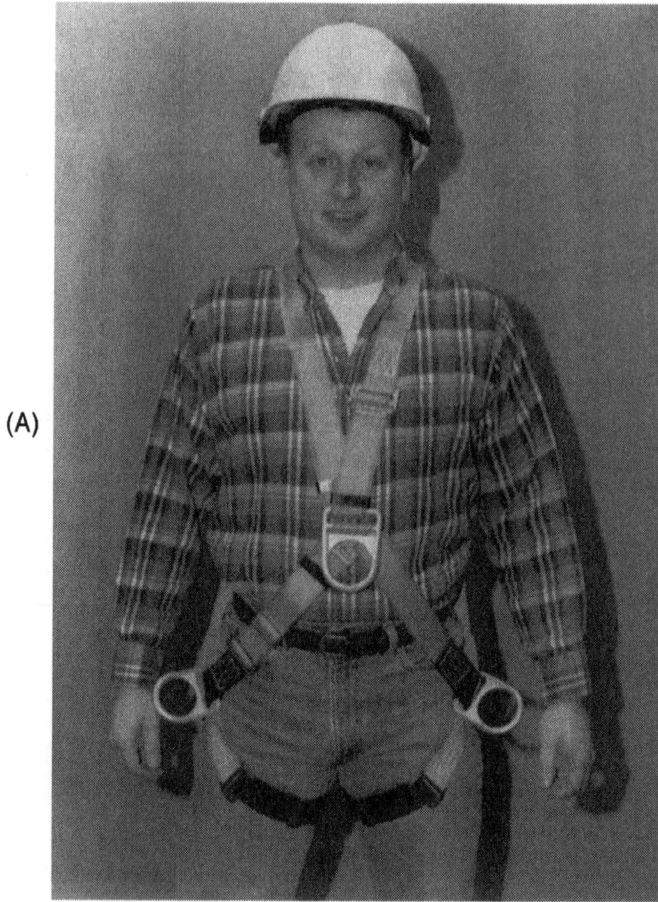

FIGURE 12.14 Full body harnesses. (A) Cross-over style. This model features a front D ring for work positioning or ladder climbing, a back D ring for fall arrest or restraint, and side D rings for restraint or work positioning. (B) Vest-type. This model features a back D ring for fall arrest or restraint and side D rings for restraint or work positioning. (Courtesy of DBI/Sala, D B Industries, Inc., Red Wing, MN.)

can be used to lift/lower/retrieve individuals. The retractor unit contains three subassemblies: the retractor, and locking and braking mechanisms. Some units also incorporate a rescue winch/controlled descent mechanism. Backup fall arrest may be required when the winch is used to lift an individual. This depends on the regulatory outlook of a particular jurisdiction (Ellis 1994). See Figures 12.18, 12.19, and 12.20.

Fully predictable action from the tether unit is required during events whose outcome has potential life or death consequences. As would be expected, maintenance of the tether unit is critical to ensure competent performance during this critical moment (McQuarrie 1993). Utilization of these units could occur indoors and outdoors. Indoor use likely occurs in the roofspace of the building. The roofspace is the location where the most contaminated air tends to accumulate. Outdoor use could expose the unit to all kinds of weather conditions.

While designed for rugged use, these units require periodic testing and recertification. Parts require cleaning and replacement. Warning indicator pins require resetting. Braking subassemblies may fail. This can occur through "aging" of components held under tension or entry of foreign substances, such as grease, oil, dirt, or ice, that reduce frictional properties of the brake pads. The ratchet mechanism in the locking subassembly is subject to wear and tear through routine extension and retraction of the cable. The retractor subassembly is subject to failure. This can occur through

(B)

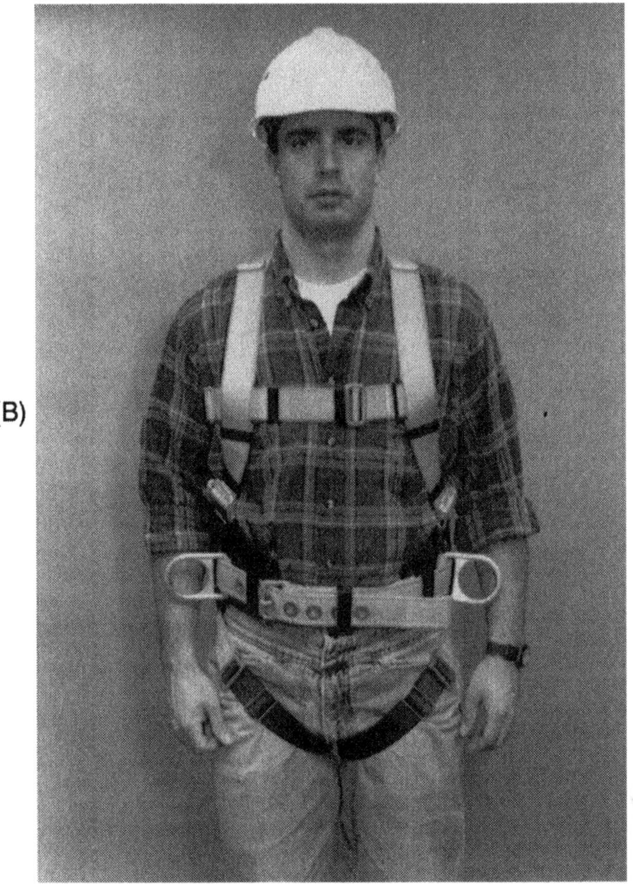

FIGURE 12.14 (continued)

aging of the spring that retracts the line or through entry of a foreign object that blocks motion. Last, the wire rope or other lifeline is subject to failure. This can occur through breakage of wire strands or through weakness caused by chemical attack.

The last crucial factor in use of fall protection is the anchorage point (Scheeler 1995). Anchorage must be able to sustain a load of 22 kN (5,000 lb) at a single tie-off point for one individual. This value may vary according to the jurisdiction. The difficulty in complying with the specification is assuring that a particular attachment will meet this criterion. The layman is severely restricted in making this judgment. Certification by a structural engineer may be required. Strength of attachments under consideration may not be known. Also, structural members may have sustained weakness from corrosion through exposure to hostile atmospheric conditions or from other causes. The Canadian Standards Association recently published a guide for implementing fall protection (CSA 1996).

The anchorage should limit vertical free fall to the shortest possible distance. This minimizes the force imposed on the fall protection system and minimizes the chance that the fall-arrest mechanism will engage. An anchorage point directly overhead of the working area minimizes the chance of swing-fall injury caused by a swinging motion after fall arrest. See Figures 12.21, 12.22, and 12.23.

The currently permitted free-fall distance is 1.5 to 1.8 m (5 to 6 feet), or more, depending on the jurisdiction. Slack in the lanyard could permit the free fall to extend another 1.5 m prior to engaging the fall-arrest mechanism. The self-retracting unit minimizes free-fall distance by maintaining tension on the lifeline and locking during rapid change. This action minimizes the instability that occurs at the beginning of a fall and may permit the individual to recover balance.

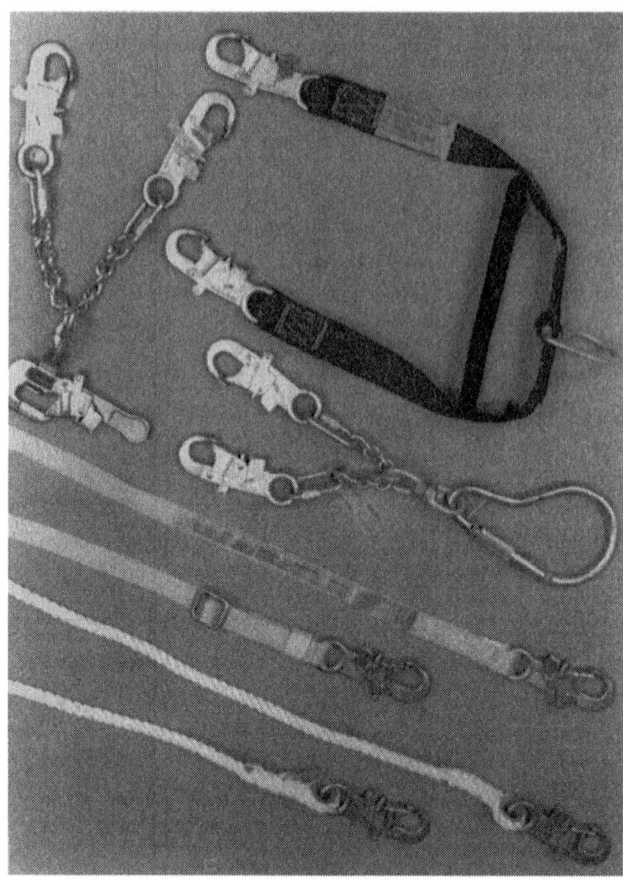

FIGURE 12.15 Lanyards and load spreaders. These devices do not provide fall arrest. (Courtesy of DBI/Sala, D B Industries, Inc., Red Wing, MN.)

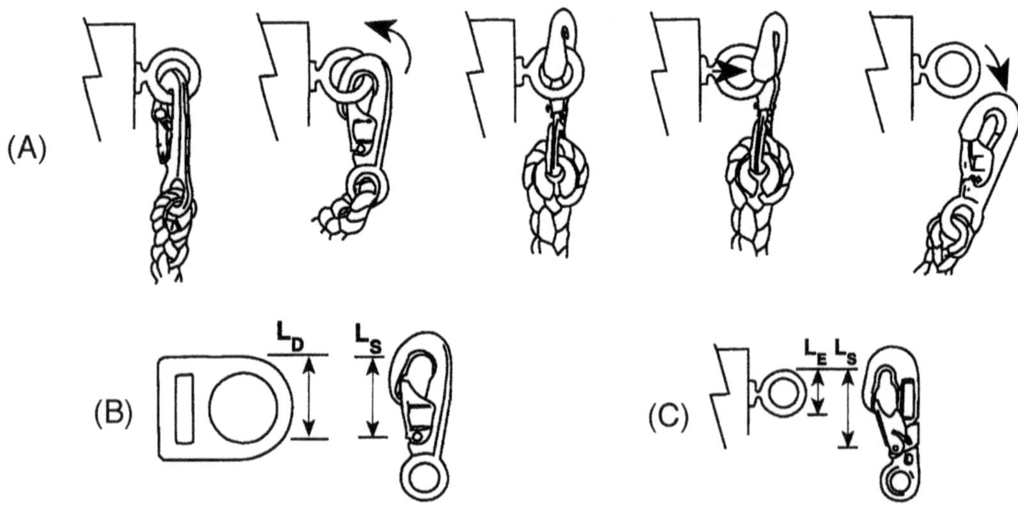

FIGURE 12.16 Snap hook roll-out. (A to C) Roll-out is the disconnection of the snap hook from the eye, D ring, or other incompatible attachment. Disconnection of the snap hook occurs through a rolling (twisting) motion. (Courtesy of Surety Manufacturing & Testing Ltd. © Surety Manufacturing & Testing Ltd., Edmonton, AB.)

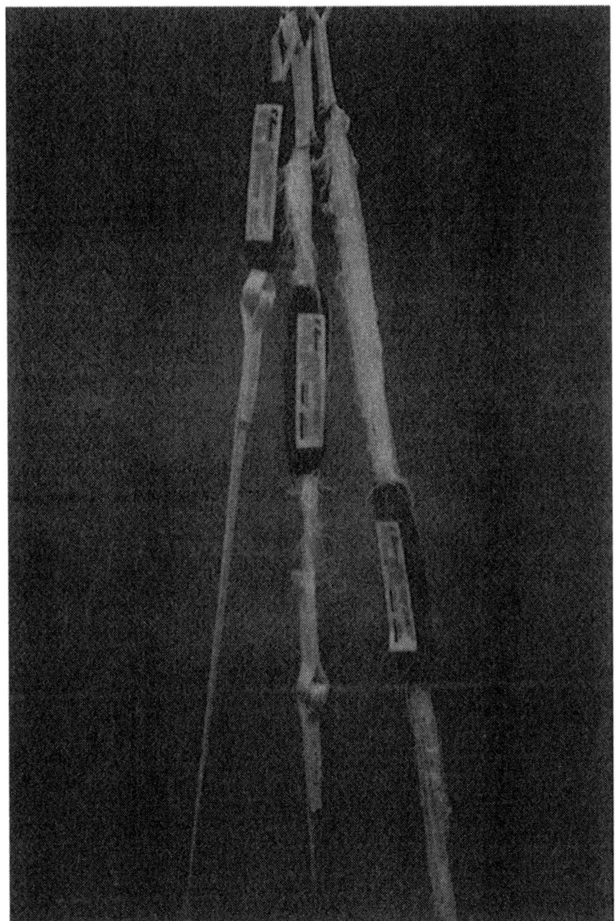

FIGURE 12.17 Shock-absorbing lanyard. (A) Unused lanyard. (B) Partially elongated lanyard. (C) Fully elongated lanyard. The shock absorber absorbs energy through controlled destruction of the interior structure. Lanyards that have become partially or fully elongated must be removed from service and destroyed. (Courtesy of DBI/Sala, D B Industries, Inc., Red Wing, MN.)

Fall protection is a complex subject. Implementation of a fall-protection system demands knowledge and competence from the specifying individual.

SUMMARY

The nature of work activity that occurs in confined spaces fosters greater reliance on personal protective equipment than would occur in production situations. The conditions of work may create extremes of personal confinement and reduced mobility and dexterity, as well as other hazardous conditions that are not encountered in regular surroundings.

Utilization of personal protective equipment in these applications can compound the impact of individual hazardous conditions. Use of respiratory protection plus encapsulating garments, combined with heavy physical labor or work activity that produces heat, greatly increases the risk of heat-induced injury. Burdening workers with uncoordinated layers of personal protective equipment can greatly decrease user acceptance and increase risk of injury.

(A)

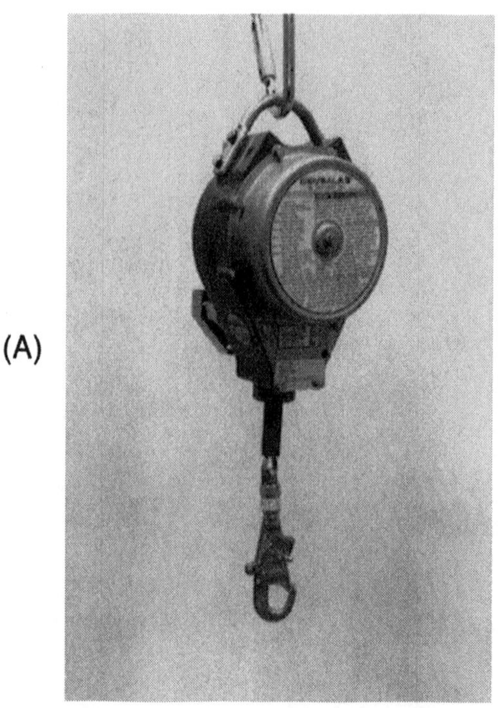

(B)

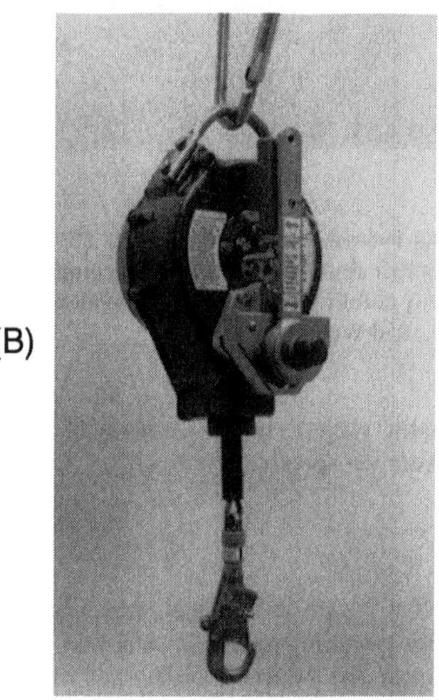

FIGURE 12.18 Self-retracting lifeline with emergency retrieval winch. (A) Front view (B) Back view. Self-retracting units can provide fall arrest. Equipment that has experienced fall arrest must be removed from service and recertified for use. (Courtesy of DBI/Sala, D B Industries, Inc., Red Wing, MN.)

(A)

(B)

FIGURE 12.19 Extrication devices. (A) Tripod (B) Monopod. Note that the height of the tripod or support structure should be sufficient to permit raising of the torso of an injured worker. An injured worker may require immobilization prior to raising. (Courtesy of DBI/Sala, D B Industries, Inc., Red Wing, MN.)

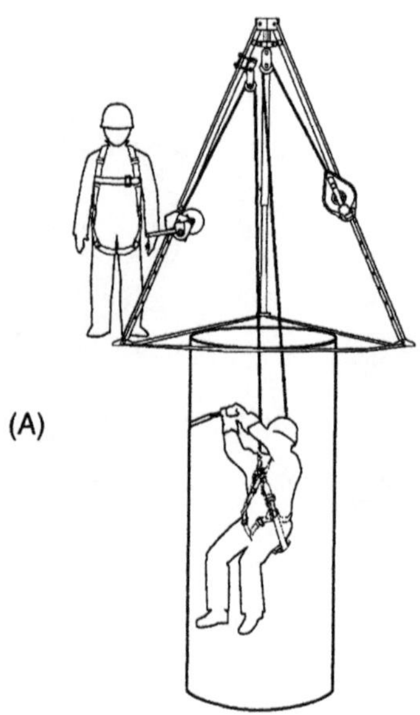

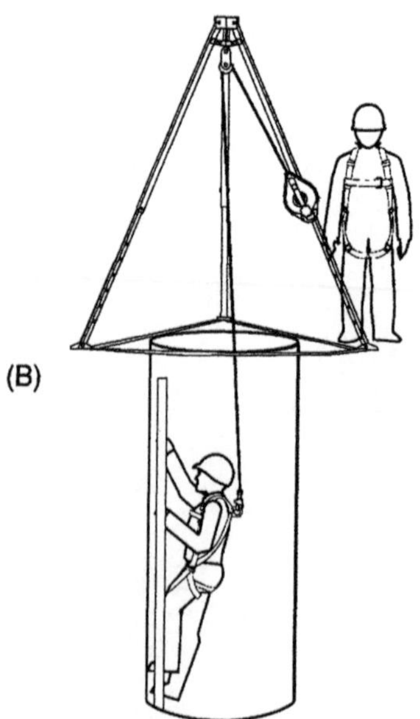

FIGURE 12.20 Work configurations. (A) Suspension/fall arrest. In this configuration the winch raises and lowers the worker, while the self-retracting lifeline provides fall arrest. (B) Fall arrest/retrieval. The self-retracting lifeline provides both fall arrest and retrieval functions. (Courtesy of Surety Manufacturing & Testing Ltd. © Surety Manufacturing & Testing Ltd., Edmonton, AB.)

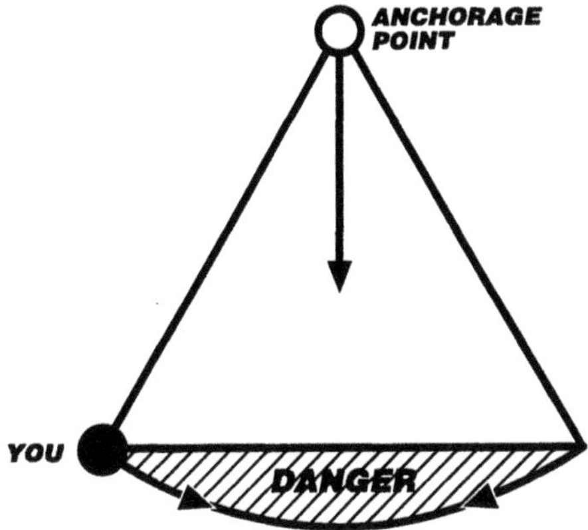

FIGURE 12.21 Swing fall hazard calculation. Anchored falls that are not completely vertical contain a horizontal component. Injury can occur following impact with objects and surfaces during the swing. The geometric relationship between the individual and the anchorage point defines the space that requires protection from contact. (Courtesy of Surety Manufacturing & Testing Ltd. © Surety Manufacturing & Testing Ltd., Edmonton, AB.)

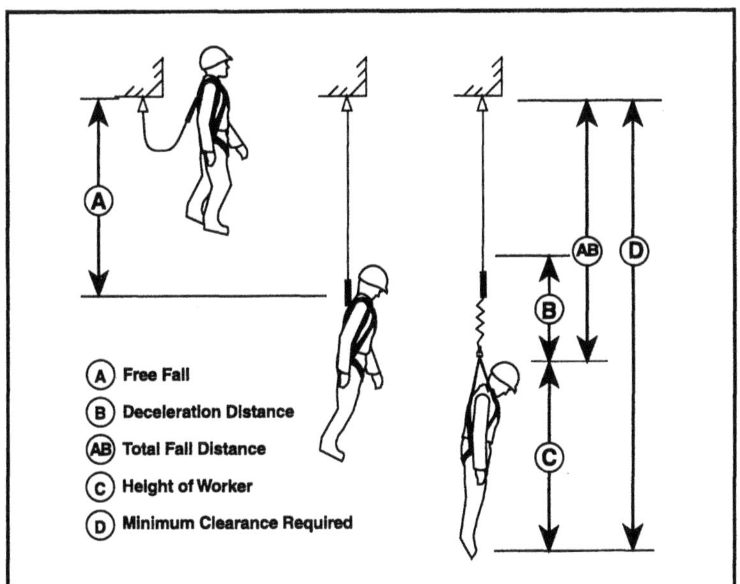

FIGURE 12.22 Fall distance calculation — fixed anchorage. The minimum clearance required below the anchorage is the sum of the free fall distance, the deceleration distance of the arresting device, the height of the wearer of the harness, and a safety factor. Any distance less than this could result in injury following impact with a lower surface. (Courtesy of Surety Manufacturing & Testing Ltd. © Surety Manufacturing & Testing Ltd., Edmonton, AB.)

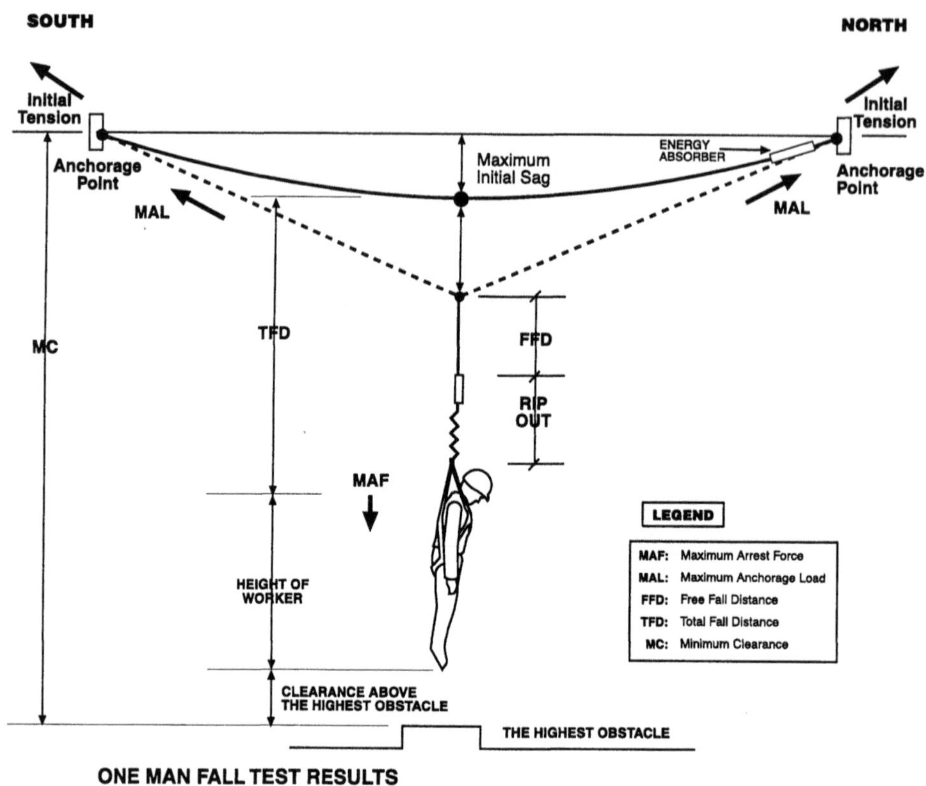

FIGURE 12.23 Fall distance calculation — horizontal lifeline. The minimum clearance required below the horizontal lifeline is the sum of the maximum initial sag in the lifeline, the total fall distance, the height of the wearer of the harness and a safety factor. The total fall distance is the free fall distance plus the deceleration distance of the arresting device. Any distance less than this could result in injury following impact with a lower surface. (Courtesy of Surety Manufacturing & Testing Ltd. © Surety Manufacturing & Testing Ltd., Edmonton, AB.)

REFERENCES

Adams, P.S. and W.M. Keyserling: Effect of size and fabric weight of protective coveralls on range of gross body motions. *Am. Ind. Hyg. Assoc. J. 56:* 333–340 (1995).

American Conference of Governmental Industrial Hygienists: *Documentation of the Threshold Limit Values and Biological Exposure Indices*, 6th. ed. Cincinnati, OH: American Conference of Governmental Industrial Hygienists, 1991. pp. 222–223.

American Conference of Governmental Industrial Hygienists: *1994–1995 Threshold Limit Values for Chemical Substances and Physical Agents and Biological Exposure Indices.* Cincinnati, OH: American Conference of Governmental Industrial Hygienists, 1994. 119 pp.

American Medical Association: *Guides to the Evaluation of Permanent Impairment.* 3rd ed. Chicago IL: American Medical Association, 1990. pp. 164–166.

American National Standards Institute: *American National Standard Practice for Occupational and Educational Eye and Face Protection* (Z87.1-1989). New York: American National Standards Institute, 1989.

American National Standards Institute: *Safety-Toe Footwear* (Z41-1991). New York: American National Standards Institute, 1991a.

American National Standards Institute: *Requirements for Safety Belts, Harnesses, Lanyards and Lifelines for Construction* (ANSI A10.14-1991). New York: American National Standards Institute, 1991b.

American National Standards Institute: *Practices for Respiratory Protection* (ANSI Z88.2-1992). New York: American National Standards Institute, 1992a.

American National Standards Institute: *Safety Requirements for Personal Fall Arrest Systems, Subsystems and Components* (ANSI Z359.1-1992). New York: American National Standards Institute, 1992b.

American National Standards Institute/Industrial Safety Equipment Association: *American National Standard for Limited-Use and Disposable Coveralls — Size and Labelling Requirements* (ANSI/ISEA 101-1993). Arlington, VA: Industrial Safety Equipment Association, 1993.

American National Standards Institute: *Protective Headwear for Industrial Workers* (Z89.1-R1996). New York: American National Standards Institute, 1996.

American Society for Testing and Materials: *1990 Annual Book of ASTM Standards.* Philadelphia, PA: American Society for Testing and Materials, 1990.

Amphoux, M.: Physiopathological aspects of personal equipment for protection against falls. In *Fundamentals of Fall Protection*, Sulowski, A.C. (Ed.). Toronto, ON: International Society for Fall Protection, 1991. pp. 33–48.

Anonymous: Protective footwear: fit the foot and the hazard. *Occup. Health Safety Canada 2:* No. 3 (1986). pp. 45–48.

Anonymous: Sunscreens. *Consumer Reports:* June (1991). pp. 400–406.

Anonymous: Bug off! How to repel biting insects. *Consumer Reports:* July (1993). pp. 451–454.

Arden, P.: Fall-protection rules still in the air. *Safety Health*: August (1992). pp. 34–38.

Åstrand, P.O. and K. Rodahl: *Textbook of Work Physiology*, 2nd ed. New York: McGraw-Hill, 1977. pp. 350–352.

Ayres, J.W.: *Decontamination of Nuclear Reactors and Equipment.* New York: Ronald Press, 1970.

Beckett and Billings: Individual factors in the choice of respiratory protective devices. *Am. Ind. Hyg. Assoc. J. 46:* 274–276 (1985).

Bird, M.G.: Industrial solvents: some factors affecting their passage into and through the skin. *Ann. Occup. Hyg. 24:* 235–244 (1981).

Birmingham, D.J (Ed.): *The Prevention of Occupational Skin Diseases.* NY: The Soap and Detergent Association, 1981. 56 pp.

Blank, I.H. and R.J. Scheuplein: Transport into and within the skin. *Br. J. Dermatol. 81* (Suppl. 4): 4–10 (1969).

Block, S.S.: Definition of terms. In *Disinfection, Sterilization and Preservation*, Block, S.S. (Ed.), 2nd ed. Philadelphia, PA: Lea & Febiger, 1977.

Bobick, T.G., P.G. Schnitzer, and R.L. Stanevich: Investigation of selected occupational fatalities caused by falls from elevations. In: *Advances in Industrial Ergonomics and Safety*, vol. 2, Das, B. (Ed.) London, U.K.: Taylor & Francis, 1990. pp. 527–534.

Bonem, M.W. and R.E. Borah: Extraction chemistry goes deep to grab PCBs. *Environ. Prot.*: December, 1995. [reprint].

Brinkley, J.W.: Experimental studies of fall protection equipment. In *Fundamentals of Fall Protection*, Sulowski, A.C (Ed.). Toronto, ON: International Society for Fall Protection, 1991. pp. 139–154.

Bureau of Labor Statistics: Accidents Involving Eye Injuries (Report 597). Washington, D.C.: U.S. Department of Labor, 1980.

Bureau of Labor Statistics: Accidents Involving Foot Injuries (Report 626). Washington, D.C.: U.S. Department of Labor, 1981. 22 pp.

Bureau of Radiological Health: Radiological Health Handbook, rev. ed. (U.S.DHEW/PHS/FDA/BRH). Rockville, MD, 1970. pp. 194–203.

Burgess, J.L. and C.D. Crutchfield: Quantitative respirator fit tests of Tucson firefighters and measurement of negative pressure excursions during exertion. *Appl. Occup. Environ. Hyg. 10:* 29–36 (1995).

Canadian Standards Association: *Protective Footwear (Z195-M1984).* Rexdale, ON: Canadian Standards Association, 1984.

Canadian Standards Association: *Compressed Breathing Air and Systems* (CAN3-Z180.1-M85). Rexdale, ON: Canadian Standards Association, 1985. 32 pp.

Canadian Standards Association: *Industrial Eye and Face Protectors* (CAN/CSA-Z94.3-M88). Rexdale, ON: Canadian Standards Association, 1988. 37 pp.

Canadian Standards Association: *Selection, Use and Care of Respirators* (Z94.4-93). Rexdale, ON: Canadian Standards Association, 1993. 103 pp.

Canadian Standards Association: *A Guide to Fall Protection.* Rexdale, ON: Canadian Standards Association, 1996. 76 pp.

Caplan, K.J.: Significance of wipe samples. *Am. Ind. Hyg. Assoc. J. 54:* 70–75 (1993).

Caretti, D.M., L.A. Bay-Hansen, and W.D. Kuhlmann: Cognitive performance during respirator wear in the absence of other stressors. *Am. Ind. Hyg. Assoc. J. 56:* 776–781 (1995).

Carruthers, C.: *Biochemistry of Skin in Health and Disease.* Springfield, IL: Charles C Thomas, 1962.

Cember, H.: *Introduction to Health Physics.* Oxford, U.K.: Pergamon Press, 1969. 420 pp.

Cleveland, R.J.: Factors that influence safety shoe usage. *Prof. Safety. 29:* 26–29 (1984).

Colton, C.E., L.R. Birkner, and L.M. Brosseau (Eds.): *Respiratory Protection: A Manual and Guideline,* 2nd ed. Akron, OH: American Industrial Hygiene Association, 1991. 130 pp.

Comité Européen de Normalisation: *Respiratory Protective Devices: Full-Face Masks: Requirements, Testing, Marking (EN 136:1989E/90/35487).* Brussels, Belgium: European Committee for Standardization, 1989. pp. 1–51.

Comité Européen de Normalisation: *Guidelines for Selection and Use of Respiratory Protective Devices* (CEN/TC 79/SG1 N150). Brussels, Belgium: Central Secretariat, 1990. 107 pp.

Comité Européen de Normalisation: *Respiratory Protective Devices; Valved Filtering Half Masks to Protect Against Gases or Gases and Particles; Requirements, Testing, Marking* (prEN 405:1991 E). Brussels, Belgium: Central Secretariat, 1991. 44 pp.

Compressed Gas Association: *Commodity Specification for Air* (ANSI/CGA G-7.1). Arlington, VA: Compressed Gas Association, 1989.

Compressed Gas Association: *Handbook of Compressed Gases,* 3rd ed. New York: Van Nostrand Reinhold, 1990.

Comroe, J.H.: *Physiology of Respiration.* Chicago: Year Book Medical Publishers, 1965.

Cooper, E.A.: The work of ventilating respirators. In *Design and Use of Respirators,* Davies, C.N. (Ed.). New York: Macmillan, Company, 1962. pp. 67–75.

Craig, F.N., W.V. Blevins, and E.G. Cummings: Exhausting work limited by external resistance and inhalation of carbon dioxide. *J. Appl. Physiol. 29:* 847 (1970).

Crawford, H.: Who's afraid of fall factor two? In *Fundamentals of Fall Protection,* Sulowski, A.C. (Ed.). Toronto, ON: International Society for Fall Protection, 1991. pp. 391–406.

Cullen, A.P.: Contact lenses in the work environment. In *Environmental Vision: Interactions of the Eye, Vision and the Environment,* Pitts, D.G. and R.N. Kleinstein (Eds.). Boston: Butterworth-Heinemann, 1993. pp. 315–331.

Dahlbäck, G.O. and L. Novak: Do pressure-demand breathing systems safeguard against inward leakage? *Am. Ind. Hyg. Assoc. J. 44:* 383–387 (1983).

Dahlbäck, G.O. and U.I. Balldin: Physiological effects of pressure demand masks during heavy exercise. *Am. Ind. Hyg. Assoc. J. 45:* 177–181 (1984).

Department of Health, Education and Welfare: *Monthly Vital Statistics Report.* Washington, D.C.: U.S. Government Printing Office, 1980. pp. 51–52.

Dessureault, P.C., R.B. Konzen, N.C. Ellis, and D. Imbeau: Heat strain assessment for workers using an encapsulating garment and a self-contained breathing apparatus. *Appl. Occup. Environ. Hyg. 10:* 200–208 (1995).

DuBois, A.B., Z.F. Harb, and S.H. Fox: Thermal discomfort of respiratory protective devices. *Am. Ind. Hyg. Assoc. J. 51:* 550–554 (1990).

Eisenberg, P.: Do you make these common mistakes in maintaining your respirator masks? *Occup. Health Safety 61:* 43–44 (1992).

Ellis, J.N.: Suspenseful Suspension. *Occup. Health Safety 63:* 39–46 (1994).

Emmett, E.A.: Photobiologic effects. In *Occupational and Industrial Dermatology,* 2nd ed., Maibach, H.I. (Ed.). Chicago, IL: Year Book Medical Publishers, 1987. pp. 94–104.

Emmett, E.A.: Toxic responses of the skin. In *Casarett and Doull's Toxicology, The Basic Science of Poisons,* 4th ed., Amdur, M.O., J. Doull, and C.D. Klaassen (Eds.). New York: Pergamon Press, 1991. pp. 463–483.

Energy Research and Development Administration: *Respirator Manual* (LA-6370-M), by Douglas, D.D., A.L. Hack, B.J. Held, and W.H. Revoir. Washington, D.C.: Energy Research and Development Administration, Division of Safety, Standards and Compliance, 1976.

Eydt, J.J: Developing a Skin Hygiene Program for Industry. *Can. Occup. Safety.* September/October, 1983. 3 pp. [reprint]

Goldman, R.F.: Heat stress in industrial protective encapsulating garments. In *Protecting Personnel at Hazardous Waste Sites*, Levine, S.P. and W.F. Martin (Eds.). Boston: Butterworth-Ann Arbor Science, 1985. pp. 215–266.

Gross, S.F. and S.W. Horstman: Half-mask respirator selection for a mixed worker group. *Appl. Occup. Environ. Hyg. 5:* 229–235 (1990).

Gwosdow, A.R., R. Nielsen, L.G. Berglund, A.B. duBois, and P.G. Tremml: Effect of thermal conditions on the acceptability of respiratory protective devices on humans at rest. *Am. Ind. Hyg. Assoc. J. 50:* 188–195 (1989).

Hansen, C.M. and K. Skaarup: Independent calculation of the parameter components. *J. Paint Technol. 39:* 511 (1967).

Harber, P.: Medical evaluation for respirator use. *J. Occup. Med. 26:* 496–502 (1984).

Hey, E.N., B.B. Lloyd, D.J.C. Cunningham, M.G.M. Jukes, and D.P.G. Bolton: Effects of various respiratory stimuli on the depth and frequency of breathing in man. *Resp. Physiol. 1:* 193 (1966).

Hinds, W.C. and P. Bellin: Effect of respirator dead space and lung retention on exposure estimates. *Am. Ind. Hyg. Assoc. J. 54:* 711–722 (1993).

Hjorth, N.: The Allergens. In *Occupational and Industrial Dermatology*, 2nd ed., Maibach, H.I. (Ed.). Chicago, IL: Year Book Medical Publishers, 1987. pp. 22–27.

Hodous, T.K., C. Boyles, and J. Hankison: Effects of industrial respirator wear during exercise in subjects with restrictive lung disease. *Am. Ind. Hyg. Assoc. J. 47:* 176–180 (1986).

Hudnall, J.B., A. Saruda, and D.L. Campbell: Deaths involving airline respirators connected to inert gas sources. *Am. Ind. Hyg. Assoc. J. 54:* 32–35 (1993).

Hugo, W.B (Ed.): *Inhibition and Destruction of the Microbial Cell.* London, U.K.: Academic Press, 1971.

James, R., F. Dukes-Dobos, and R. Smith: Effects of respirators under heat/work conditions. *Am. Ind. Hyg. Assoc. J. 45:* 399–404 (1984).

Jaraiedi, M., W.H. Iskander, W.R. Myers, and R. Giorcelli Martin: Effects of respirator use on workers' productivity in a mentally stressing task. *Am. Ind. Hyg. Assoc. J. 55:* 418–424 (1994).

Johnson, A.T. and E.G. Cummings: Mask design considerations. *Am. Ind. Hyg. Assoc. J. 36:* 220–228 (1975).

Johnson, A.T., C.R. Dooly, and E.Y. Brown: Task performance with visual acuity while wearing a respirator mask. *Am. Ind. Hyg. Assoc. J. 55:* 818–822 (1994).

Johnson, A.T., C.R. Dooly, C.A. Blanchard, and E.Y. Brown: Influence of anxiety level on work performance with and without a respirator mask. *Am. Ind. Hyg. Assoc. J. 56:* 858–865 (1995).

Jones, N.L., G.B. Levine, D.G. Robertson, and S.W. Epstein: Effect of added dead space on the pulmonary response to exercise. *Respiration 28:* 389 (1971).

Jones, J.G.: Physiological cost of wearing a disposable respirator. *Am. Ind. Hyg. Assoc. J. 52:* 219–225 (1991).

Kamon, E., T. Bernard, and R. Stein: Steady-state respiratory responses to tasks used in federal testing of self-contained breathing apparatus. *Am. Ind. Hyg. Assoc. J. 36:* 886–896 (1975).

Kamon, E., D. Doyle, and J. Kovac: Oxygen cost of an escape from an underground coal mine. *Am. Ind. Hyg. Assoc. J. 44:* 552–555 (1983).

Lahey, J.W.: Hand protection: it's a basic challenge. *Natl. Safety Health News:* March (1986). pp. 60–62.

Landers, D.M.: The arousal-performance relationship revisited. *Res. Quart. Exercise Sport 51:* 77–90 (1980).

Lara, J., Y.E. Hoon, and J.H. Nelson: The test of time: take two. *Occup. Health Safety Canada 9: Buyers' Guide,* 1993. pp. 16–24.

Lara, J. and S. Smith: Comparing standards in a global economy. *Occup. Health Safety Canada 10: Buyers' Guide,* 1994. pp. 14–23.

Lawrence Livermore National Laboratory: *Is It Safe to Wear Contact Lenses with a Full Facepiece Respirator?,* by da Rosa, R.A. and C.S. Weaver (UCRL-53653). Livermore, CA: Lawrence Livermore National Laboratory, 1986.

Lichtenwalner, C.P.: Evaluation of wipe sampling procedures and elemental surface contamination. *Am. Ind. Hyg. Assoc. J. 53:* 657–659 (1992).

Love, R.G., D.C.F. Muir, K.F. Sweetland, R.A. Bentley, and O.G. Griffin: Tolerance and ventilatory response to inhaled CO_2 during exercise and with inspiratory resistive loading. *Ann. Occup. Hyg. 22:* 43–53 (1979).

Maclay, G.J., C. Yue, M.W. Findlay, and J.R. Stetter: A prototype active end-of-service-life indicator for respirator cartridges. *Appl. Occup. Environ. Hyg. 6:* 677–682 (1991).

Maibach, H.I., A.A. Khan, and W.A. Akers: Use of insect repellents for maximum efficacy. *Arch. Dermatol. 109:* 32–35 (1974).

Maibach, H.I. and H.L. Johnson: Contact urticaria syndrome. *Arch. Dermatol. 111:* 726–730 (1975).

March, J.: *Advanced Organic Chemistry, Reactions, Mechanisms and Structure,* 3rd ed. New York: John Wiley & Sons, 1985. pp. 202–217.

McArthur, B.: Dermal measurement and wipe sampling methods: a review. *Appl. Occup. Environ. Hyg. 7:* 599–606 (1992).

McDermott, H.J. and G.A. Hermens: System for routine testing of self-contained and airline breathing equipment. *Am. Ind. Hyg. Assoc. J. 41:* 489–493 (1980).

McQuarrie, R.H.: Toeing the line on maintenance. *Occup. Health Safety Canada 9: Buyers' Guide* (1993). pp. 38–42.

Miller, T.M. and P.O. Mazur: Oxygen deficiency hazards associated with liquid gas streams: derivation of program controls. *Am. Ind. Hyg. Assoc. J. 45:* 293–298 (1984).

Mohr, A.: Safety spectacles and goggles: a "standard" approach to buying. *Occup. Health Safety Canada 2:* 41–48 (1986).

Morgan, W.P.: Psychological problems associated with the wearing of industrial respirators: a review. *Am. Ind. Hyg. Assoc. J. 44:* 671–676 (1983a).

Morgan, W.P.: Hyperventilation syndrome: a review. *Am. Ind. Hyg. Assoc. J. 44:* 685–689 (1983b).

Morgan, W.P. and P.B. Raven: Prediction of distress for individuals wearing industrial respirators. *Am. Ind. Hyg. Assoc. J. 46:* 363–368 (1985).

Moyer, E.: Review of influential factors affecting the performance of organic vapor air-purifying respirator cartridges. *Am. Ind. Hyg. Assoc. J. 44:* 46–51 (1983).

Myhre, L.G., L.G. Holden, R.D. Baumgardner, and F.W. Tucker: Physiological Limits of Firefighters (ESL-TR-79-06). Alexandria, VA: Defense Technical Information Center, 1979. 77 pp.

Myhre, L.G.: Field Study Determination of Ventilatory Requirements of Men Rapidly Evacuating a Space Launch Complex (Report SAM-TR-80-43). Brooks Air Force Base, TX: USAF School of Aerospace Medicine, Aerospace Medical Division (AFSC), 1980. 21 pp.

Myhre, L.G.: Imposed ventilatory resistance. In *Oxygen Transport to Human Tissues,* Loeppky, J.A. and M.L. Riesdel (Eds.). Amsterdam, The Netherlands: Elsevier North Holland, 1982. pp. 267–278.

National Fire Protection Association: *Gloves for Structural Fire Fighting (NFPA 1973).* Quincy, MA: National Fire Protection Association, 1988.

National Fire Protection Association: *Standard on Vapor-Protective Suits for Hazardous Chemical Emergencies (NFPA 1991).* Quincy, MA: National Fire Protection Association, 1990a.

National Fire Protection Association: *Standard on Liquid Splash-Protective Suits for Hazardous Chemical Emergencies (NFPA 1992).* Quincy, MA: National Fire Protection Association, 1990b.

National Fire Protection Association: *Standard on Support Function Protective Garments for Hazardous Chemicals Operations (NFPA 1993).* Quincy, MA: National Fire Protection Association, 1990c.

National Fire Protection Association: *Standard on Fire Department Occupational Safety and Health Program* (NFPA 1500). Quincy, MA: National Fire Protection Association, 1991. pp. 1500-6 to 1500-14.

NIOSH: Respiratory Protective Devices; Tests for Permissibility; Fees, *Fed. Regist. 37:* 59 (25 March 1972) pp. 6244–6271.

National Institute for Occupational Safety and Health: A Guide to Industrial Respiratory Protection, by Pritchard, J.A. (DHEW [NIOSH] Pub. No. 76-189). Cincinnati, OH: DHEW/PHS/CDC/NIOSH, 1976. 150 pp.

National Institute for Occupational Safety and Health/Occupational Safety and Health Administration/U.S. Coast Guard/Environmental Protection Agency: Occupational Safety and Health Guidance Manual for Hazardous Waste Site Activities, (DHHS [NIOSH] Pub. No. 85-115). Washington, D.C.: U.S. Government Printing Office, 1985.

National Institute for Occupational Safety and Health: NIOSH Guide to Industrial Respiratory Protection, by Bollinger, N.J. and R.H. Schutz (DHHS [NIOSH] Pub. No. 87-116). Cincinnati, OH: DHHS/PHS/CDC/NIOSH, 1987. 296 pp.

National Institute for Occupational Safety and Health: National Traumatic Occupational Fatality (NTOF) Database. Morgantown, WV: U.S. Department of Health and Human Services, Public Health Service, Centers for Disease Control, National Institute for Occupational Safety and Health, Division of Safety Research, 1991.

National Institute for Occupational Safety and Health: Worker Deaths in Confined Spaces (DHHS/PHS/CDC/NIOSH Pub. No. 94-103). Cincinnati, OH: National Institute for Occupational Safety and Health, 1994. 273 pp.

NIOSH: Respiratory Protective Devices; Final Rules and Notice, *Fed. Regist. 60:* 110 (8 June 1995) pp. 30336–30404.

National Safety Council: *Safety Hats* (Data Sheet, 1-561-Rev. 84). Chicago, IL: National Safety Council, 1984. 4 pp.

National Safety Council: *Accident Prevention Manual for Industrial Operations, Administration and Programs*, 9th ed. Chicago IL: National Safety Council, 1988. pp. 331–367.

Ness, S.A.: *Surface and Dermal Monitoring for Toxic Exposures.* NY: Van Nostrand Reinhold, 1994. 561 pp.

Noel, G., M.G. Ardouin, P. Archer, M. Amphoux, and A. Sevin: Some aspects of fall protection equipment employed in construction and public works industries. In *Fundamentals of Fall Protection*, Sulowski, A.C (Ed.). Toronto, ON: International Society for Fall Protection, 1991. pp. 1–32.

Occupational Safety and Health Administration: Selected Occupational Fatalities Related to Toxic and Asphyxiating Atmospheres in Confined Work Spaces as Found in Reports of OSHA Fatality/Catastrophe Investigations. Washington, D.C.: U.S. Department of Labor, Occupational Safety and Health Administration (U.S. DOL/OSHA), 1985. 230 pp.

Occupational Safety and Health Administration: "Contact Lenses Used With Respirators (29 CFR 1910.34 (e) (5) (ii))," by T. Shepich through L. Carey. Washington, D.C.: Occupational Safety and Health Administration, February 8, 1988. [Memo]

OSHA: Hazardous Waste Operations and Emergency Response; Final Rule, *Fed. Regist. 54:* 42 (6 March 1989) pp. 9294 ff.

Occupational Safety and Health Administration: Selected Occupational Fatalities Related to Ship Building and Repairing as Found in Reports of OSHA Fatality/Catastrophe Investigations. Washington, D.C.: U.S. Department of Labor, Occupational Safety and Health Administration (U.S. DOL/OSHA), 1990. 195 pp.

OSHA: Occupational Safety and Health Standards, *Code of Federal Regulations*, Title 29, Part 1910, Subpart I (.134). 1991, pp. 402–406.

OSHA: Permit-Required Confined Spaces for General Industry; Final Rule, *Fed. Regist. 58*: 9 (14 January 1993). pp. 4462–4563.

OSHA: Safety Standards for Fall Protection in the Construction Industry; Final Rule, *Fed. Regist. 59*: (9 August 1994).

Parent, M.: Guarding against the sun. *Occup. Health Safety Canada 9: Buyers' Guide,* 1993. pp. 50–54.

Paull, J.M. and Rosenthal, F.S.: Heat strain and heat stress for workers wearing protective suits at a hazardous waste site. *Am. Ind. Hyg. Assoc. J. 48:* 458–463 (1987).

Payne, W.R., B. Portier, I. Fairweather, S. Zhou, and R. Snow: Thermoregulatory response to wearing encapsulated protective clothing during simulated work in various thermal environments. *Am. Ind. Hyg. Assoc. J. 55:* 529–536 (1994).

Perkins, J.L. and A.D. Tippit: Use of three-dimensional solubility parameter to predict glove permeation. *Am. Ind. Hyg. Assoc. J. 46:* 455–459 (1985).

Perkins, J.L., J.S. Johnson, P.M. Swearengen, C.P. Sackett, and S.C. Weaver: Residual spilled solvents in butyl protective clothing and usefulness of decontamination procedures. *Appl. Ind. Hyg. 2:* 179–182 (1987).

Perkins, J.L.: Chemical protective clothing: I. Selection and use. *Appl. Ind. Hyg. 2:* 222–230 (1987).

Perkins, J.L.: Chemical protective clothing: II. Program considerations. *Appl. Ind. Hyg. 3:* 1–4 (1988).

Perkins, J.L.: Decontamination of protective clothing. *Appl. Occup. Environ. Hyg. 6:* 29–35 (1991).

Raven, P.B., A.T. Dodson, and T.O. Davis: Physiological consequences of wearing industrial respirators. *Am. Ind. Hyg. Assoc. J. 40:* 517–534 (1979).

Raven, P.B., A.W. Jackson, K. Page, R.F. Moss, O. Bradley, and B. Skaggs: Physiological response of mild pulmonary impaired subjects while using a "demand" respirator during rest and work. *Am. Ind. Hyg. Assoc. J. 42:* 247–257 (1981).

Raven, P.B., O. Bradley, D. Rohm-Young, F.L. McClure, and B. Skaggs: Physiological response to "pressure-demand" respirator wear. *Am. Ind. Hyg. Assoc. J. 43:* 773–781 (1982).

Schaefer, H., A. Zesch, and G. Stuttgen: *Skin Permeability.* Berlin: Springer-Verlag, 1982.

Scheeler, E.: Do's and don'ts of anchorage points. *Occup. Health Safety Canada 11:* Buyers' Guide (1995). pp. 34–39.

Shapiro, J.: *Radiation Protection, A Guide for Scientists and Physicians*, 2nd ed. Cambridge, MA: Harvard University Press, 1981. 480 pp.

Shepard, R.J.: Ergonomics of the respirator. In *Design and Use of Respirators*, Davies, C.N., Ed. New York: Macmillan, 1962. pp. 51–66.

Silverman, L., R.C Lee, G. Lee, K.R. Drinker, and T.M. Carpenter: Fundamental Factors in the Design of Protective Respiratory Equipment: Inspiratory Air Flow Measurements on Human Subjects With and Without Resistance (Report # 1222). Washington, D.C.: Office of Scientific Research and Development, U.S. War Research Agency, 1943.

Silverman, L., G. Lee, A.R. Yancey, L. Amory, L.J. Barney, and R.C. Lee: Fundamental Factors in the Design of Protective Respiratory Equipment: A Study and an Evaluation of Inspiratory and Expiratory Resistances for Protective Equipment (Report # 5339). Washington, D.C.: Office of Scientific Research and Development, U.S. War Research Agency, 1945. pp. 48–60.

Silverman, L., G. Lee, T. Platkin, L.A. Sawyers, and A.R. Yancey: Air flow measurements on human subjects with and without respiratory resistance at several work rates. *A.M.A Arch. Ind. Hyg. J. 3:* 461 (1951).

Smolander, J., V. Louhevaara, and O. Korhonen: Physiological strain in work with gas protective clothing at low ambient temperature. *Am. Ind. Hyg. Assoc. J. 46:* 720–723 (1985).

Snodgrass, H.L., D.C. Nelson, and M.H. Weeks: Dermal penetration and potential for placental transfer of the insect repellent, N,N-diethyl-m-toluamide. *Am. Ind. Hyg. Assoc. J. 43:* 747–753 (1982).

Solomon, J., P. Bishop, S. Bomalaski, J. Beaird, and J. Kime: Responses to repeated days of light work at moderate temperatures in protective clothing. *Am. Ind. Hyg. Assoc. J. 55:* 16–19 (1994).

Spaul, W.A., J.A. Boatman, S.W. Emling, H.G. Dirks, S.B. Flohr, W.H. Crocker, and M.A. Glazeski: Reduced tolerance for heat stress environments caused by protective lotions. *Am. Ind. Hyg. Assoc. J. 46:* 460–462 (1985).

Spielberger, C.D., R.L. Gorsuch, and R.E. Lushene: *The State-Trait Anxiety Inventory Manual.* Palo Alto, CA: Consulting Psychologist Press, 1969.

Spielberger, C.D., R.L. Gorsuch, and R.E. Lushene: *The State-Trait Anxiety Inventory.* Palo Alto, CA: Consulting Psychologist Press, 1983.

Srachta, B.J.: What you should know about head protection. *Natl. Safety Health News:* March (1986a). pp. 47–49.

Srachta, B.J.: What you should know about eye and face protection. *Natl. Safety Health News:* March (1986b). pp. 50–52.

Stampfer, J.F., R.J. Beckman, and S.P. Berardinelli: Using immersion test data to screen chemical protective clothing. *Am. Ind. Hyg. Assoc. J. 49:* 579–583 (1988).

Stannard, J.N. and E.M. Russ: Estimation of critical dead space in respiratory protective devices. *J. Appl. Physiol. 1:* 326 (1948).

Stobbe, T.J., R.A. daRoza, and M.A. Watkins: Facial hair and respirator fit: a review of the literature. *Am. Ind. Hyg. Assoc. J. 49:* 199–204 (1988).

Stull, J.O.: Chemical protective clothing. *Occup. Health Safety 61:* November (1992). pp. 49–52.

Stull, J.O. and D.F. White: Review of overall integrity and material performance tests for the selection of chemical protective clothing. *Am. Ind. Hyg. Assoc. J. 53:* 455–462 (1992).

Stull, J.: Using, applying chemical permeation data. *Safety Prot. Fabrics 2:* 10–11 (1993).

Stull, J.O.: Performance-based selection of chemical protective clothing. *Occup. Hazards.* January, 1995. pp. 47–51.

Toulmin, L. and J. Moran: *Preliminary Survey of Existing Data and Economic Overview of the Respirator Industry.* NIOSH Contract no. 21-81-1102. Washington, D.C.: The Granville Corporation, 1982.

Tucker, S.B. and M.M. Key: Occupational skin disease. In *Environmental and Occupational Medicine*, Rom, W.N. (Ed.). Boston: Little, Brown and Company, 1982.

U.S. Department of Energy: Amendment of the Occupational Safety and Health Administration (OSHA) Prohibition on Wearing Contact Lenses in Contaminated Atmospheres With Full Facepiece Respirators, by M.L. Walker. Washington, D.C.: Department of Energy, September 23, 1986. [Memo]

Vahdat, N. and R. Delaney: Decontamination of Chemical Protective Clothing. *Am. Ind. Hyg. Assoc. J. 50:* 152–156 (1989).

White, M.K. and T.K. Hodous: Reduced work tolerance associated with wearing protective clothing and respirators. *Am. Ind. Hyg. Assoc. J. 48:* 304–310 (1987).

White, M.K. and T.K. Hodous: Physiological responses to the wearing of firefighter's turnout gear with neoprene and GORE-TEX® barrier liners. *Am. Ind. Hyg. Assoc. J. 49:* 523–530 (1988).

White, M.K., T.K. Hodous, and J.B. Hudnall: Physiological and subjective responses to working in disposable protective coveralls and respirators commonly used by the asbestos abatement industry. *Am. Ind. Hyg. Assoc. J. 50:* 313–319 (1989).

Wilson, J.R., P.R. Raven, W.P. Morgan, S.A. Zinkgraf, R.G. Garmon, and A.W. Jackson: Effects of pressure-demand respirator wear on physiological and perceptual variables during progressive exercise to maximal levels. *Am. Ind. Hyg. Assoc. J. 50: 85–94 (1989).*

Zechman, F., F.G. Hall, and W.E. Hull: Effects of graded resistance to tracheal air flow in man. *J. Appl. Physiol. 10:* 356–362 (1957).

Zelnick, S.D., R.T. McKay, and J.E. Lockey: Visual field loss while wearing full-face respiratory protection. *Am. Ind. Hyg. Assoc. J. 55:* 315–321 (1994).

13 Training

CONTENTS

INTRODUCTION

Among the most important deficiencies identified in the investigative reports published by NIOSH, OSHA, and MSHA about fatal accidents in confined spaces was lack of knowledge (NIOSH 1979, 1994; OSHA 1982a, 1982b, 1983, 1985, 1988, 1990; MSHA 1988). Entrants, their superiors, and would-be rescuers knew little, if anything, about such fundamentally important concepts as:

- Hazardous conditions that could be present
- Atmospheric confinement
- Physical, chemical, or toxicological properties of atmospheric contaminants
- Methods of testing for atmospheric contaminants
- Portable ventilation systems
- Deenergization, deactivation, isolation, and lockout
- Use and limitations of personal protective equipment
- Strategies for emergency response
- Stabilization of victims

Each of the preceding deficiencies alone is sufficient to cause a fatal accident. Therefore, what distinguishes an inadequate response from an adequate one could differentiate a fatal accident from a nonfatal one. There is little doubt, however, that a workforce knowledgeable in the topics mentioned above could have markedly influenced the outcome of these situations.

Senior management must understand and acknowledge their obligations in this area and follow this through by a commitment that is clearly and unambiguously communicated to personnel at all levels of the organization. Chapter 7 and Appendix E discuss programs for hazard management. The most definite evidence of this commitment is a policy statement regarding entry and work involving confined spaces. (For an example of such a statement, consult Chapter 7.) The more clearly expressed the commitment of senior management, the greater the potential for a successful program.

Insufficient or inadequate training instantly can negate the most carefully thought-out of policies and work practices backed up by the finest of equipment. Hence, the importance and thoroughness of training of personnel involved with confined spaces and confined atmospheres cannot be overstated.

Training is an investment to ensure the well-being of an organization and its people. On-the-job training is expensive. Loss of production during training is extremely costly. In the industrial world, the wealth of an organization clearly rests on the output of those who produce the product. The worth to an organization of the productive output of these individuals easily could exceed $1,000 per hour. This perhaps explains the reluctance of organizations to provide training to persons at these levels. However, the loss to an organization of any of these people through an accident is staggering.

Perception of benefit, such as avoidance of injury or death, is a powerful motivator. People must perceive benefit before accepting the need to expend the effort required to complete formal learning. Formal learning represents change, a challenge to an existing status quo. Change in the working environment is intimidating for many. Formal learning requires considerable effort, effort expended through concentration and assumption of some risk of failure. Formal learning can be a difficult process for adults who, believing that they have lost their learning skills, are afraid to try.

General guidelines on training about confined spaces should include the following considerations:

- Employees must clearly understand the seriousness of the hazards posed by confined spaces and confined atmospheres. In addition, all persons must respect the policy that entry into confined spaces is controlled and forbidden to anyone not having received specialized training.
- Any employee involved in any aspect of an operation that involves confined spaces must receive training in that area.
- Retraining must occur on a scheduled basis to ensure continuing competence.
- The entry process must contain controls to ensure that the preceding components cannot be overlooked or circumvented.

The preceding concepts underlie consensus standards and regulations on confined spaces. A bona fide program therefore must include at least these elements.

The overall purpose for training programs in this area is easy to comprehend: everyone wants to prevent injury or loss of life. The overriding question then becomes what is needed to achieve this goal. What is needed are defined outcomes. These establish the scope of the program for both management and participants. Defining these outcomes is equivalent to the process mentioned above for minimizing the magnitude and extent of the problem. They provide points of achievement against which success can be measured. They also provide points of focus for determining the curriculum needed to achieve them. The following end points should be considered as a measure of success for an individual in any program:

- Can persons involved with confined spaces and confined atmospheres identify and describe hazardous conditions known or likely to be present or to develop?
- Can persons involved with confined spaces and confined atmospheres demonstrate measures to protect their health and safety against these hazards?
- Can they demonstrate appropriate actions during an emergency situation?
- Is there a system in place to assess hazards and to manage entry and work involving these environments?

Individual organizations face different training problems. For this reason, their training strategies will vary somewhat. General considerations that will influence these decisions should include:

- Number of confined spaces requiring entry
- Number of persons affected
- Assessment of hazardous conditions
- Geographic coverage by the organization

From the perspective of financial management, there is an incentive to minimize each of the preceding factors. Through minimization, management can exert greater control over conditions, provide more extensive training to a smaller group, and thereby reduce the risk of performing the work. This strategy can provide other benefits in the area of training. Minimization reduces the complexity and extent of training needed to ensure competence.

The extent to which an organization is spread geographically will exert a major influence over the strategy for training. Organizations with many small operating centers that extend across a large geographic area represent one extreme. Their need is to develop competent local expertise. At the opposite extreme are organizations that have a single location and a small number of confined spaces. The problems faced by other organizations fit somewhere in between. Obviously, considerable scope is available for approaching a particular problem.

The technical structure dedicated to addressing the problem posed by confined spaces should contain an expert, such as an industrial hygienist or a safety professional, or both, if available. This person can provide technical assistance at different levels of this hierarchy. (Appendix B discusses the background issues involved in developing the broad base of expertise needed to respond to the technical challenges of confined spaces.)

In all cases, the training strategy probably will utilize a multilevel approach. The level of training provided to persons at different levels should reflect their need to know. Most people in organizations have no involvement with confined spaces other than being aware of warning signs and barriers used to isolate them from other workspaces. These persons need to know about the general dangers that they pose. This is important to their safety, since casual passersby, such as these persons, on occasion have assumed the role of would-be rescuer during accident situations.

The outcome from these well-intentioned, sometimes tragic actions reflects lack of awareness about dangers present in these workspaces.

Other persons work in these environments. They have a considerably greater need for information than casual passersby. The person at the highest level of the technical hierarchy who assesses the hazards and proposed activities and authorizes entry has the greatest need for training. In so doing, this person assumes moral and legal responsibility for the safety of others. The need for technical knowledge by this individual is considerably greater than that of persons carrying out the task.

Educational materials on the subject of confined spaces generally attempt to foster recognition through use of familiar structures, such as silos, hoppers, sewers, and tanks, that are readily recognized as confined spaces. The strategy inherent in this approach is extrapolation from the specifics of the example to the general, the multitude of structures and situations in the industrial environment. This approach requires the learner to recognize and extract hazard elements from the examples and then to extrapolate these to unfamiliar situations. Breakdown in the process occurs because this approach does not identify all hazard elements that constitute the problem. The gap between the familiar and recognizable and the unrecognized and uncharacterized has proven often to be too great to bridge, as demonstrated by the statistics in confined space incidents. A new approach is needed.

Successful implementation of workplace training is difficult to achieve. Often training must be redone because it was insufficient or not done properly the first time. This chapter provides information about learning research performed by experts in the field of education and incorporates these ideas into a practical approach to training about confined spaces. Consensus and regulatory standards, such as the OSHA Standards for general industry and shipyard employment, impose a considerable training burden (OSHA 1993, 1994). Training related to recognizing, evaluating, and controlling potentially life-threatening situations, as can occur in confined spaces, is a very serious undertaking. The importance of proper training cannot be overstated. Conducting a confined space training program is a major challenge, one that many feel less than able to perform (Zeimet and Van Ast 1996).

The purpose for conducting training should be understood clearly from the outset. The intent may be to satisfy one or more of the following outcomes:

- To convey information
- To produce a desired behavior
- To modify existing behavior
- To provide new skills

Understanding the intent of the training is critical, as this should directly influence methods used to achieve it. The overall success of the program reflects long-term utilization of information or modification of behavior as set out in objectives.

ADULTS AS LEARNERS

Before progressing further to detailed aspects about the training program, discussion about adults as learners is imperative. This subject is too important to relegate to a less prominent position in a chapter on training.

The process of learning has been studied in both adults and children. Fundamental differences between the modes utilized by the two groups led to coining of the terms "pedagogy" and, more recently, "andragogy." While pedagogy is the art and science of teaching children, andragogy describes the process of adult learning. Knowles (1980) defines andragogy "as the art and science of helping adults to learn." He espouses four important attributes about adult learners:

- Though dependent in some situations, the adult learner strives to be self-directed. This dependence is only temporary.
- Personal experience is a rich resource for learning. Learning from experience has personal meaning that can be shared with others.
- Readiness to learn is correlated to the need to know in order to cope with a real-life task.
- Learning is a process to gain increased competence. Applying knowledge and skills to performance is necessary.

Andragogy holds that adults learn best when the learning process anticipates and incorporates their experiences. The design of workplace training, therefore, must reflect past experiences and skills. The trainer should be experienced in the subject and employ a collaborative, informal, problem-solving, hands-on approach. To the extent possible, training should occur at the worksite and directly relate to its processes and activities.

Though Knowles propounds andragogy, he is not so rigid as to reject other approaches. He recommends use of either andragogical or pedagogical strategies to suit the needs of the situation, regardless of the learner's age. In other words, if an adult learner needs some hand-holding to learn a task, then by all means provide it. If a child demonstrates self-direction in a learning situation, then by all means allow him/her to do so.

HOW PEOPLE LEARN

Background

"But they were trained. They *knew* what to do." Yet, the accident happened just the same. The people were victims of a recognized workplace hazard.

These words may sound familiar. Sometimes we have difficulty making sense of these accidents. Various factors contribute to these misfortunes. Perhaps the training program was not as effective as believed. Perhaps a key element was missing. Perhaps people lacked motivation to apply the learning to the situation. Perhaps they couldn't bridge the gap between the classroom and the real world.

One of the main aspects of training involves transfer of information from the trainer to the learner (Figure 13.1). This transfer is a multistep process. Persons not directly involved in this field have a need to appreciate the complexities of the process. The process involves:

- Providing information to the learner
- Assimilation of the information by the learner
- Application of the information by the learner
- Feedback from the instructor
- Modification of the action, where necessary

A breakdown in any one of these steps can obliterate the entire process. Without careful consideration of this important subject, success of the program could be jeopardized. Economic and other realities dictate that training should be conducted in the most efficient manner possible. Efficiency of learning may not coincide with the least expensive mode of delivery. Inexpensive delivery that sustains no ongoing result represents wasted money.

Training in the area of confined spaces encompasses many topics. The complexity of information is high. The quantity of information is large. The urgency to communicate is high. Together, these factors represent a totally undesirable combination. This situation places extreme demands on the instructional designer to create an effective learning package and on the learner to acquire new knowledge and skills.

Retention	Activity	Sensory Involvement	
10%	reading	• visual	passive
20%	listening	• auditory	passive
30%	looking at pictures	• visual	passive
50%	watching a demonstration	• auditory and visual	passive
70%	exchanging ideas	• auditory and visual	active
90%	performing	• auditory and visual	active

FIGURE 13.1 The learning process. Learning is a complex endeavor that involves the senses. The greater the number of senses and the greater the involvement in the process, the more rapid and effective is the learning. (Adapted from Grund 1995).

Most of us, for example, know the rules of the road. Yet, when driving a car, we are not always motivated enough to follow them consistently. Past experience leads us to believe that taking risks usually does not result in a penalty, in this case a ticket for speeding. There also are those persons who never learn from their mistakes. Motivation and learning are very personal experiences. Collectively, we share much in common, but we also have individual idiosyncrasies. The keys to successful training are to incorporate principles of learning and motivation and to acknowledge and address individual idiosyncrasies.

The tragedy experienced by others, unfortunately, is one of the most powerful motivators of the need to accept training in the area of confined spaces. A fatal accident or serious nonfatal accident usually is a sufficiently powerful motivator. However, the infrequent occurrence of serious accidents rapidly dilutes the sense of urgency to learn. Unfortunately, in some quarters there may still exist the perception that accidents involving confined spaces and confined atmospheres are part of the risk of doing the job, and not the result of preventable oversights.

People are complex animals. Whereas less developed animals learn in well-defined, predictable fashions, people learn in many ways. A point of considerable importance is that unlike other animals, people recognize when they are being manipulated. Over the years, psychologists and educators have studied how people learn. No one has yet devised a universal approach that incorporates all facets of human learning. What appeals to one learner may not appeal to another. What appeals to one person at a particular point in time may not appeal to the same person at a later moment. Training, therefore, must address and accommodate to the needs of learners at various points in time.

The following sections describe approaches to learning that have resulted from studies by researchers. Each has its merits and applicability in the appropriate setting. None of these approaches will work at all times for all people and should not be relied upon exclusively. Successful trainers utilize different features of each of these approaches in the appropriate context at the appropriate moment.

Sensory Stimulation Theory

Since our earliest conscious moments, we have used our senses to learn about the environment. Most of us possess the most-utilized of the senses, namely: sight, hearing, touch, taste, and smell.

Some persons lack one or more of the senses at birth. Others lose them through accident or disease. People who have lost a sense often compensate by relying more diligently on signals provided by their remaining senses.

People routinely and subconsciously utilize their senses as they interact with the environment. For example, we sometimes use the sense of smell to distinguish one chemical substance from another. However, relying on the sense of smell to warn about the identity or presence of chemical substances is potentially very dangerous. The fallibility of reliance on the sense of smell was readily apparent from the outcomes of many fatal accidents that occurred in confined spaces (OSHA 1985; NIOSH 1994).

Another sense sometimes utilized is taste. Characteristic taste sometimes indicates the presence of a chemical substance or exposure to it. However, only a poorly informed person would use taste as a means of identifying substances encountered in the workplace. In normal experience, people use the sense of touch considerably more than taste or smell. Touch is used for many enquiries: to locate position, to determine texture, to judge temperature, to detect corrosivity, to identify objects by shape, and so on. By far the most important senses are sight and hearing. We use these continuously for processing information. In fact, almost all education and training is based on applying these two senses.

The sensory-stimulation approach to learning focuses on arousing and utilizing the senses (Chambers and Sprecher 1983). As expected, the greatest emphasis is placed on sight, followed by hearing. Experience has shown that people retain more information that they see and hear than what they only see or only hear. Training will be most effective when the person can simultaneously see and hear the information. In fact, the more senses aroused during training, the more successful the learning experience will be.

For training to be most effective the learner must participate actively in the process. Training based solely on stimulation of the senses is passive. Usual examples of training approaches that exploit sensory stimulation include presentations utilizing films, videos, CD-ROMs or slides, and demonstrations of procedure. An immediate follow-up activity requiring the learner to apply the concept or technique presented during the instruction will result in a more effective learning experience. Subsequent repetitions of the activity by the learner will add further reinforcement to the learning.

A video on confined space hazards can play an introductory role in the training process. Follow-up discussion of the most relevant points as they relate to circumstances particular to the location lends more meaning to the video. Additional follow-up using actual hands-on activity during a simulation provides greater enhancement. Some training companies utilize simulators to create realistic simulations of confined spaces. Trainees follow up the audiovisual presentation with an authentic drill and practice. This instructional approach can be highly effective, since trainees must use all of their senses to participate actively in the process.

Skinner's Reinforcement Theory

B.F. Skinner, a world-renowned psychologist in the study of learning theory, taught at Harvard University. He studied learning behavior in animals. Skinner's reinforcement theory is widely applied in various areas of animal training and to some extent, in the education and training of humans.

Skinner showed that animals will do tasks for a reward (Chambers and Sprecher 1983). They learn that a particular action is followed by a reward and that other actions are not. Once one step is mastered, a reward is given only when progress is made toward mastering the next step in the sequence. By this technique, the trainer communicates a complex sequence of desired or required actions step by step. This approach minimizes failure, since learning one step is considerably easier than attempting to learn the entire task at once.

Skinner's reinforcement theory is routinely applied to the training of animals. Obedience training of pets utilizes this approach. Other examples include the training of dolphins, killer whales, and circus animals. Perhaps the ultimate expressions of this approach are the unusual routines performed by animals in television commercials.

Some of Skinner's ideas also have been applied to the human animal. Consider, for example, the use of incentives in safety programs. Achieving a defined standard of performance results in acquisition of a known reward. Other motivators include paychecks and reduction in the price of goods in a store during a sale.

Practice is an important part of learning a skill. For example, learning to type or play a musical instrument requires practice. The course of mastering the patterns for the fingers involves self-administered reinforcement.

Reinforcement theory utilizes several additional principles when applied to humans (Laird 1985).

- Rewards must be **individualized**. What works with one person will not necessarily work with another. This reflects the individuality of people.
- The second principle is **immediacy**. The reward must follow immediately after the desired action. A compliment given freely at the right moment goes a long way toward reinforcing a desired result. Genuine praise given immediately after a correct response can be extremely effective during training in a difficult task or to sustain interest.
- Rewards must have **strength** to have an effect on learning. A person must feel that the reward is worth the effort.

Skinner also found interesting variations on the reward theme (Chambers and Sprecher 1983). An important challenge is to be able to motivate animals that have learned a task. Skinner found that animals would continue to perform a task, even when rewards tapered off. In fact, the greatest continuation of performance occurred when the animal could not predict the point at which the reward would be given.

The implication of these observations is that there is no need to provide continuous rewards once the desired action is mastered. A reward provided at **variable** points in the performance is sufficient to maintain maximum continuance (Chambers and Sprecher 1983). Many students work diligently to receive an occasional bit of praise or a smile from a well-liked teacher. A paycheck received regularly is a powerful form of reward. Yet, people eagerly work and perform in organizations that pay less than top dollar. A simple smile and a few sincere, complimentary words now and then, for a task done well, are invaluable encouragement for many. Drop these rewards, and the good performance discontinues. The paycheck then becomes the only incentive.

Perhaps the most blatant exploitation of this observed behavior is the psychology applied to gambling. Some people become conditioned to putting an endless stream of coins into slot machines. Periodically, the machine rewards a lucky person. The lure of the game is the belief that the player will be the lucky person. For some people, this belief becomes an obsession. These people likely gamble at things far beyond the gaming table.

Up to this point, discussion has focused only on **positive reinforcement** as a means of providing a reward. Positive reinforcement utilizes pleasurable rewards. **Negative reinforcement** is feedback that has unpleasant consequences (Chambers and Sprecher 1983). Negative reinforcement may involve physical pain or mental abuse. Negative reinforcement includes criticism, half-praise, sarcasm, and frowns and other disapproving nonverbal gestures. People respond negatively to negative reinforcement. Negative reinforcement can slow the learning process or stop it completely. As indicated by preceding comments, trainers must be aware that their actions and comments can control the behavioral responses of trainees and, ultimately, the success of the learning process.

Persons in a training and supervisory role can easily slip into a negative perspective. This can become so pervasive that they are not even aware that it is happening. The trainer must remain diligent to avoid this trap.

Social Learning Theory

Social learning theory espouses that people learn through observation or modeling (Bandura 1977). Modeling studies have shown that people can describe a new task quite well, simply by observing models perform the task. With appropriate inducements or rewards, they often are able to perform the task accurately on the first try out. Observational learning, then, occurs **before** performing. Reinforcement, though not necessary, plays a facilitative role. That is, anticipation of reinforcement can be highly effective. A situation where trainees are informed about the benefits or rewards of a particular way of doing something **before** observing the modeled behavior can strengthen learning. However, the trainees must value the ensuing benefits. This can induce trainees to be more attentive to the model and to motivate them to rehearse or practice the behavior. Whether or not trainees choose to follow the modeled behavior is dependent upon its consequences. For example, anticipation of a monetary bonus for doing a task in a particular way will likely reinforce learning. On the other hand, if one anticipates a negative consequence, such as a dock in pay, one likely will not follow the modeled behavior.

Social learning theory has relevant implications for a confined spaces training program. An old adage says it all: "do as I do," rather than, "do as I say." Corporations must ensure that safety "role models" practice what they preach. Incentives for continual safe practices could reinforce the desired behavior. Incentives can be safety awards, baseball hats, coffee mugs, televisions, Thanksgiving turkeys, etc.

Cognitive Learning Theory

Cognitive learning theory builds on the tenets of the sensory perception and reinforcement theories. It supplements the concepts of rewards and stimulation of the senses with thinking, that is, information processing, and memory utilization (Chambers and Sprecher, 1983). Cognitive learning theory forms the basis for activities conducted in the formal educational system.

Our minds are complex processors of information. These actions are facilitated in part by memory. Memory is a biochemical process of information storage. Memory itself is a complex subject. Conscious memory enables recollection of the here and now. This is a form of short-term memory that lasts only about 20 seconds unless a person makes a deliberate effort to retain it for a longer period (Gagné et al. 1992).

Short-term memory can hold only about four different items simultaneously. This, in turn, affects the difficulty of learning various tasks. Long-term memory develops from short-term memory through repetition and practice.

Long-term memory is the information that we retain for many years. Recollection of a key item of information can trigger recall of a past event from long-term memory into short-term memory. The key item could be a taste, smell, word, color, sound, or some other cue that has no particular meaning in the present context. These memories can be pleasant, unpleasant, or neutral.

Combining new information with recalled items appears to result in new learning. Short-term memory of this form often is referred to as working memory.

INSTRUCTIONAL STRATEGIES

Discussion

Anyone is capable of learning. Learning can occur at all ages. Babies and small children learn complex tasks and assimilate voluminous amounts of information. The elderly also are effective learners. Learning occurs at all levels of human intelligence. Mentally challenged people learn many skills to various degrees of efficiency. In all cases the factors common to successful learning include:

- Instructional design
- Empathy with the learner
- Motivators
- Time
- Practice
- The teachable moment (Havighurst 1961)
- Adaptable instructor

Psychological studies have shown that humans can process about seven bits of information at a time. The seven-digit telephone number is a practical application of this observation. This application exploits the ability of people to remember enough information to perform an operation, that is, to direct the connection from one telephone to another. This innovation provided enough flexibility to enable automation of many functions in the telephone system.

A critical area in process control is the management of information provided by sensory inputs. A computerized system can manage considerably more inputs than a human manager. Information overload can become a serious problem to the human operator of complex equipment. Controlling the flow of information presented for processing and decision making by the human operator is critical to preventing this situation. Overload could lead to accidents, the causes being the inability to assimilate and act on all of the information or to process it rapidly enough. To accommodate to the capabilities of a human operator, the information manager in the process controller must screen out extraneous information and present only that needed to make a particular decision.

Training presents similar problems. The challenge is to enable a person to master a task, yet not to overload the mind with extraneous information. Perhaps the best example of a training session that leads to overload is the average seminar. Seminars usually provide a large amount of information during a short period of time. Usually a person stands at the front of the room and talks throughout the presentation. The presentation also may incorporate overhead transparencies or slides, and occasionally, a movie or video. The learner usually is expected to sit passively during the process. The amount of information retained in short-term memory is questionable. Long-term retention likely is very small. Unless the instructional designer has provided some means for revitalizing the learning, this technique has very limited effectiveness as a training vehicle.

The preceding example illustrates the challenge faced by persons who occupy the training role: how to communicate information to the learner in a way that enables long-term retention. Usually the instructor attempts to convey far more information than can be handled by the learner. One strategy for addressing this problem is to separate information according to hierarchy of need (Mills 1977). This concept stratifies information into three layers, according to need to know:

- Must know
- Should know
- Nice to know

In the limited amount of time available for training, ensuring that the learner has assimilated and can utilize information listed as **must know** makes the most sense.

Another strategy for learning is to present the information as single complete ideas or a group of ideas (Jay 1983). The learner is permitted to assimilate the material before presenting the next concept. This is the essence of the grade concept in a school.

Educational strategies practiced in the elementary school system utilize the concepts of cognitive learning theory extensively. Information is presented in small amounts that are spread over a prolonged period. Repetition is a key element in this approach. The teacher of very young students in the primary grades acts as a guide. Guided hands-on activities are used to advance concepts. Through the advanced grades, the communication of information and acquisition of skills relies less on hands-on activity and more on the use of words and images. The increased reliance on the abstract rather than the concrete can create problems for students.

Reliance on abstract word-based concepts as the basis for teaching and learning may be an important factor in early leaving from high schools. Early leaving is a characteristic of many persons in lower levels of the workforce. Strategies for learning that apply to the university graduate have no relevance to laborers. The trainer must recognize these needs and accommodate to them. Data presented in the first chapter suggest that people interacting with confined workspaces have a broad spectrum of education. Job title, while not wholly reflective of educational attainment, ranged from unskilled laborer to skilled tradesman to supervisor to superintendent and technical personnel. Educating higher levels of supervision and technical personnel is just as important as educating lower levels. The data indicated that victims who were would-be rescuers were more likely to be higher in the organizational hierarchy.

As mentioned in a previous section, motivation to learn is a major impediment to successful workplace training. Unlike the self-motivated mature student who attends night school, many workers lack the motivation to learn at work. The challenge facing the trainer is to reach out to these individuals by inducing the teachable moment (Havighurst 1961). The teachable moment is an instant in time in which the mind of the participant opens to accept information. Knowledge about the audience and empathy toward their understanding of the world around them are essential. Inducing the teachable moment taxes the best of instructors. The demonstrable urgency in this training, the need to ensure competency in addressing the hazards of confined spaces, may assist in inducing the teachable moment. The trainee's life could depend on applying this information.

Instructional strategies are the paths that guide the trainer to develop successful learning activities. These models also are useful resources for addressing unexpected situations. However, any strategy for developing skills is extensive and complex, and time-consuming. All of these are incompatible with the economic realities of the production-oriented workplace. Use of one-dimensional, linear approaches such as lectures, while seemingly time- and cost-effective, are likely to be met with failure. Lack of retention and inability to make the bridge from the abstract world in the training room to the concrete one in the workplace are major obstacles to success.

Events of Instruction

Gagné states that both internal and external factors contribute to learning (Gagné 1985). External factors refer to external stimuli. Internal factors refer to previous learning. External factors can influence internal learning processes. Collectively, Gagné calls these factors "conditions of learning." Gagné et al. (1981) devised a practical approach for applying cognitive learning theory. This approach incorporates nine steps known as the **events of instruction**. The events of instruction are external to the learner. They are designed into the instructional strategy to support internal learning processes. Incorporating the events of instruction into training sessions produces effective results. This approach is methodical. The events of instruction focus on the learner throughout the process. They also provide a point of reference for evaluating the instructional merit of individual lessons or the whole programs.

Table 13.1 presents the "events of instruction," along with examples to illustrate their application. Most steps follow in sequence, but these may vary somewhat depending on the situation.

TABLE 13.1
Events of Instruction

1. Gain the learner's attention.
Example: Show something visual — diagram, slide, overhead transparency.
Gaining the person's attention at the beginning of a session, rather than at some later point, is very important. The challenge of maintaining attention then arises. Keller's motivational strategies (described in a later section) provide devices for maintaining interest (Keller 1983).
In a similar way, Grabowski and Aggen (1984) stress that assisting the learner to ignore both internal and external "noise" is necessary in order to achieve sensory processing. That is, obtaining the learner's full and undivided attention is essential for optimizing this cognitive experience.
Organized structure and logical progression in the material are essential for efficient sensory processing. Presenting large amounts of data in small, logical segments enables processing efficiency.

2. Inform the learner about the purpose for the lesson.
Example: Indicate clearly what the trainee will be able to do as a result of this training. Sometimes this type of statement appears in promotional material either as a money-back or no-risk guarantee.
This step sets up internal expectancy by providing the trainee with an end point for completion.

3. Review prior learning.
Example: Utilize prior experience in examples.
New learning is based on prior mastery. Learning of mathematics provides a good example for discussion. A person learns first to count, then to add and subtract, and later to perform higher level operations. Reviewing previous work before proceeding with new concepts narrows and bridges the gap between the old and the new. The process of review facilitates the transition.

4. Present new information clearly.
Example: Use illustrations wherever possible to explain concepts.
Confusion is frustrating. This is one of the impediments to learning that pushes people to quit. Always present new information as clearly as possible. Though well-described verbal illustrations may suffice, visual material provides essential support by stimulating other senses. The greater the number of senses invoked, the greater will be the reinforcement of the concept and success and satisfaction obtained through the process of learning.

5. Guide the learner through the instruction.
Example: Cues such as color and arrows provide direction. Cues should not be used excessively, however. Too small a number may cause ambiguity; too many will distract attention.
Providing the learner with clear instructions about how to proceed is critical. This is especially true when the process is self-guided. Providing clear instructions is partly the function of a live instructor during classroom sessions. Distance learning poses greater challenges to both instructors and learners. Instructors are not able to receive immediate visual and audible feedback as would occur during live sessions. Students must possess considerably greater skills of patience and concentration in order to maintain interest and attention.
In order to transfer the information to short-term memory, the brain must process the material. Since the brain can process only small amounts of information at a time, cuing the learner about the most important features of the material and "aides memoires" is essential.

6. Involve the learner actively in the process.
Example: Provide practice exercises.
Learning requires active repetition. However, the learner needs the practice, rather than the trainer. This is a trap into which inexperienced trainers can fall. They become very proficient at the skill, while learners watch passively. A demonstration, for example, has greatest value when the student tries to replicate and repeat the activity on demand.

7. Inform the learner about progress.
Example: Provide constructive feedback.
Learning is most effective when the trainee receives feedback that sustains and prolongs the teachable moment. People need feedback about their progress, and encouragement and help when they are experiencing difficulty.

8. Assess performance.
Example: Assign a final exam or project or demonstration of competence.
Assessment indicates how well the learner has assimilated the information and can apply new skills to meet some required level of performance or competence. Assessment also identifies areas needing corrective action.

TABLE 13.1 (continued)
Events of Instruction

9. Provide practice by applying learning to new situations.
Example: After practicing using a simulation, assign a real-world task and monitor the situation.
The ability to apply knowledge and skills beyond the protected environment of the classroom is a critical point in any training program. This is especially the case with work involving confined spaces. Making this transition should be an integral part of the instructional design.
After completing the session, provide the trainee with realistic opportunities to apply the learning. This is essential. On-the-job activity must include monitoring and feedback. The value in learning something new lies in the application. Without this component, a person rapidly loses the newly acquired skill. Refresher training is often necessary to ensure continuity.
Grabowski and Aggen (1984) point out that assimilation and analogical reasoning are cognitive processes that enable more resistant storage in long-term memory. Assimilation is the integration of the new information with related existing knowledge. Analogical reasoning ties new concepts to other concepts using structural relationships.

Elaboration Model

Training often involves large amounts of information. This information could cover one large topic, several interrelated topics, or a large number of small unrelated ones. This situation poses potential problems, because the information could rapidly overwhelm the learner. One approach for addressing this problem is the **elaboration model** (Reigeluth et al. 1980). The elaboration model incorporates the following strategies:

- Sequencing the information
- Showing the relationship between details
- Moving from the simple to the complex
- Presenting periodic summaries

In support of this approach, Mills (1977) provides the following additional strategies. These apply to all learning situations, including technical instruction. Instruction should proceed from:

- The known to the unknown
- The simple to the complex
- The concrete to the abstract
- The particular to the general
- Observations to reasoning
- The whole to the parts, and back to the whole

Movies often utilize these techniques. The camera first pans across a large expanse, then zooms onto a selected area, and finally stops on an aspect of intended focus. This technique initially provides the overall picture to the audience and then minute detail about a selected part or situation. Applied to training, this technique first provides the learner with the overall view, and then focuses on selected topics. The learner can see the overall context of the instruction and hence its purpose.

To illustrate the use of the elaboration model, consider induction training for new employees at a large complex. The complex may contain processes that require emergency action and possible evacuation. The new employee must be made aware of these processes and their interrelationships and evacuation routes, as quickly as possible. The first step in the introduction might be to indicate areas of concern and routes of evacuation on a layout of the plant. This also would include alarm signals. A tour of the plant indicating key landmarks would reinforce what was seen on the diagram. This provides the person with the overall picture.

Training at the next level would include detail about specific process operations. After receiving this information, the trainee would be guided to the area in which he or she will work. As much as possible, the person should apply information already learned about the plant and its layout to find the way.

MOTIVATING LEARNERS

One of the most important elements in the process of learning is motivation. Motivation is the spark that initiates the action to achieve a result. Motivation can arise from an external stimulus, such as a paycheck, or from an undefined internal source. Motivation is an intrinsic factor in determining how a person will respond to a challenge.

For instructors, motivating people to learn can be a difficult task. What motivates one person does not necessarily motivate another. Worse still, what motivates one person may alienate another. Also, a particular motivator for an individual may function only during certain circumstances. The trainer must be sensitive to and respond to the different moods and personalities of people. This sensitivity is essential for optimizing the learning process.

Instructional design has been the focus of considerable attention and publication. Motivational strategies, by contrast, have received little attention. J.M. Keller, a professor and consultant to schools, corporations, and the military, has developed strategies for motivating learners (Keller 1983). These strategies have proven to be very effective and are described under the following categories:

- Interest
- Relevance
- Expectancy
- Satisfaction

Interest arouses and sustains the learner's curiosity. Skilled instructors use various devices to create interest:

- Humorous anecdotes
- Challenges to common perception
- Extensions of behavior of individuals in the group
- Items from the popular news media

The trainer must establish the **relevance** of the material within the learner's innermost sphere of needs and motives. Individual learners perceive relevance differently. Some are motivated by the need to achieve, others by the need to gain power. Still others are motivated by the need to belong.

Expectancy describes strategies for developing confidence in the learner. A learner who is confident of success is likely to pursue learning for its own sake.

Satisfaction addresses strategies for managing rewards. As mentioned previously, people are motivated by rewards. The learner must accept the merit and relevance of the reward.

Table 13.2(a to d) outlines Keller's strategies for motivation.

In a similar vein, Malone (1983) suggests stimulating curiosity to an optimal level. That is, the instructional design should incorporate an element of novelty and surprise. Auditory and visual effects attract attention and arouse sensory curiosity. Constructive feedback containing an element of surprise also stimulates cognitive curiosity. The material should fit with the student's base of existing knowledge — neither too difficult or too elementary (Table 13.2b). (This is a tall order for trainers having only commercially available products from which to select.)

In additional thoughts on this topic, Neher and Hauser (1982) emphasize the need for consistency in the new information with the values and interests of the learner. Carlson (1980) and Carlisle

TABLE 13.2a
Strategies for Motivating Learners — Interest

- Refer to new, unusual, out-of-place, conflicting, and strange events to generate curiosity and attention.
- Describe personal experiences related to the situation.
- Provide an opportunity to learn more about a situation than a person already knows.
- Use familiar ideas to explain new concepts. (By way of example, the approach utilized throughout this chapter was to use familiar ideas and concepts to introduce or explain new material.)
- Utilize the experiences of people in the group regarding a situation to create the starting point for proceeding in a new direction. By asking the right questions, the trainer can stimulate the curiosity of the learner.

TABLE 13.2b
Strategies for Motivating Learners — Relevance

- Provide the learner with opportunities for success by breaking the subject into small pieces. These make the tasks more manageable. This strategy will appeal especially to people who are afraid of failure and those who strive to achieve high levels of success.
- Provide the learner with opportunities to make choices, to assume responsibility, and to influence the direction of the training. This will appeal to people who are motivated by power.
- Put the learner at ease. Avoid situations that expose the learner to the risk of failure. To a greater or lesser extent, each person has a need to sense belonging. Therefore, avoid situations that expose the learner to the possibility of failure, embarrassment, or humiliation.

TABLE 13.2c
Strategies for Motivating Learners — Expectancy

- The oft-used expression, nothing succeeds like success, is no understatement when applied to training. Establishing a situation in which the learner expects to succeed and can succeed at all times optimizes learning potential.
- Inform the learner about requirements for success.
- Use methods that permit the learner to define the completion of a step. For example, permitting the learner to decide when to move on to the next step provides a modicum of control over the learning process.
- If possible, indicate to the learner the connection between personal effort and success.

TABLE 13.2d
Strategies for Motivating Learners — Satisfaction

- As mentioned earlier, people are motivated by financial rewards (paycheck) and equally as important, praise. Legitimate praise is highly effective.
- Praise given at unexpected points in the training process is likely to be more effective than that given in a predictable pattern.
- Positive comments about progress are important. Even when little progress has been made, accentuate the positive. Trainers should avoid negative comments, half-hearted praise, threats, and so on. At the same time, evaluation should be treated in a low-key manner. Fear of evaluation can be a block to progress. A learner who fears evaluation is unlikely to be relaxed and may not perform to fullest potential.
- Positive feedback given at the end of the job influences the quantity and quality of work produced by a person. This reward is highly motivating when offered, for example, by the crowd at a hockey game after a goal has been scored.
- Corrective feedback given at critical points in the learning affects the quality of the result. The most useful time to provide corrective feedback usually is just before the learner repeats or practices the task.

(1983) both stress the importance of informing the learner about the reason for teaching the material. Adult learners have a need to understand the relevance of the information and what will be gained from the experience. This concept is particularly important in training on confined spaces. Trainees must fully understand the relevance of this topic. One day their lives could be at risk (Table 13.2c).

Fear of failure is thought to be the basis of negative motivation. This may result in subsequent avoidance of learning (North 1984). Neher and Hauser (1982) have identified four conditions — autonomy, utility, integration, and security — that are necessary to overcome these fears. Autonomy refers to self-initiated action to gain access to resources — information, communication, or instructional materials. Utility refers to flexibility in the learning program that will permit the user to branch to areas that provide remedial assistance. Integration enables the adult learner to set the pace of learning, and to receive help and review, as needed. Security refers to a risk-free environment in which the adult can learn without threat of humiliation or embarrassment (Table 13.2d).

Carlisle (1983) emphasizes that the way in which the information is presented will influence motivation. Inaccurate, poorly presented instruction will not sustain interest. Subject matter expertise, obviously, is an extremely important attribute in the trainer. No one wants to participate in a confusing, difficult-to-follow, or inaccurate learning situation.

Some authors suggest individualizing the instruction. This process would necessitate assessment of the learning styles and capabilities of individuals (Neher and Hauser 1982, Snow 1978). Snow maintains that enough is known about individual differences in aptitude and learning to justify this approach. In this way, different instructional approaches aimed at the same objective could be utilized to meet the needs of individual learners. This approach is unworkable in the industrial setting.

INSTRUCTIONAL AND PROGRAM DESIGN

Designing a training program on confined spaces requires considerable effort and skill. The subject matter is both complex and extensive. Under ideal circumstances, a team containing several well-qualified professionals would prepare the program. At a minimum, members of this team would include:

- An industrial hygienist who acts as a subject-matter expert
- A safety professional who acts as a subject-matter expert
- An instructional designer who converts the subject matter into a teachable format (possibly with the aid of a graphics artist who illustrates the concepts)
- A trainer, an expert in communicating information to learners

Reality usually dictates that fewer persons will actually participate in the process than indicated here. Of necessity, one person may act in all of the roles. Equally as important as input from well-qualified professionals is that the training program utilize a systematic approach to instruction.

Training for work associated with confined spaces entails procedural training related to tasks to be performed during preparation, entry, and work, and education about the subject of confined spaces and peripheral topics.

PERFORMANCE-BASED TRAINING

Performance-based training is a systematic approach to the process of training. Blank (1982) refers to this approach as competency-based training, but notes that this concept appears under about a dozen different names. Regardless of the name, the approach focuses most closely on procedural development. This model provides the basis for setting up well-planned programs. The focus of performance-based training is to analyze existing activities and to synthesize better procedures.

TABLE 13.3
Comparison between Performance-Based and Traditional Training

Performance-based approach	Traditional approach
What the student learns, is based only on competencies or tasks essential to performing a job successfully	The material is covered using textbooks and other reference sources that often are unrelated to the job
Performance-based programs require careful planning and organization. The result is quality student-centered instruction. Students move at their own pace and repeat parts of the program or move ahead, as appropriate. Feedback is provided periodically, so that students can correct themselves as they move through the program.	The traditional approach tends to be instructor-centered. The pace is controlled by the instructor, little if any feedback is provided, and the learning activities rely on instructor delivery. Common modes of delivery include lectures, live demonstrations, and discussions.
Performance-based programs allow each student to master a task before moving on to the next.	The traditional approach uses a fixed time slot. All students spend equal time on a unit and must move ahead, whether ready or not. For some this could be either too soon or not soon enough.
Before receiving credit, the performance-based method requires a student to perform according to a predetermined standard.	The traditional method relies on written tests that compare an individual's score to a group norm. Students move onto the next unit regardless of the level of performance. Some may have actually failed the unit or passed marginally.

After Blank 1982.

The analysis and synthesis of procedures follow a formal process. Other aspects of the process provide guidance for creating educational programs.

Blank (1982) describes the basic characteristics of this approach and compares them to those of the traditional approach (Table 13.3). Blank also makes the following points about learning and the performance-based method. Anything worth teaching is worth learning. This means that failure by a student is as much the instructor's problem as the student's. Failure should not exist. Producing only a handful of excellent achievers out of a group is not satisfactory, either. Any student is capable of mastering a task at a proficient level, given enough time and instruction of sufficient quality. Ability to learn does not predict how well a student learns a task. That is, ability may reflect amount of time needed to learn a task, but not the actual amount learned.

Differences in levels of mastery are caused by errors in the training program, rather than individual differences. Three factors tend to influence amount of learning: the number of necessary learning prerequisites, feelings and attitudes about learning, and quality and length of instruction. Learning ability, rate of learning, and motivation to learn more become comparable for all learners when favorable learning conditions exist.

The performance-based approach focuses more on the instructional process, rather than characteristics of individual learners. That is, rather than blaming failure on such things as age, ethnic group, and lack of interest, the instruction is examined critically and corrected systematically. The kind and quality of instruction that a student experiences during training is the most important factor in the teaching–learning process. This implies well-planned instruction using a systematic approach.

The following sections outline the various processes of the performance-based approach that could be applied to the creation of a confined space training program. These include: task analysis, creation of standard procedures, setting of performance objectives, creating test(s), characterizing learners, selecting training methods, materials and the training environment, evaluation and revision, implementation, and recordkeeping.

Task Analysis

Jobs usually consist of a number of complex activities. People usually fail to appreciate this complexity, not giving it much heed. This complexity becomes most evident when training of replacement workers is required. At this point, everyone concerned senses how much learning must occur, but does not know how to expedite the process in an efficient and effective manner. For training purposes, splitting complex activities into simpler parts is crucial to satisfying these needs.

Each part of a training program contains one or more complete units of work (Blank 1982). Units associated with work involving confined spaces would include identification, assessment of hazardous conditions, preentry preparation, testing and entry, and work activity. A unit consists of discrete procedures. A procedure has a starting point, contains a sequence of logical steps, and ends in a specific result. Procedures associated with assessing the hazards of a particular confined space include: determining present and previous contents, determining hazards associated with them, determining potential interaction with the activity to be performed, determining actions necessary to isolate the confined space from supply and discharge lines, de-energizing and deactivating equipment, determining the internal condition of the space, determining potential accidents that may occur despite other precautions, determining emergency response, and so on.

Task analysis identifies and examines the steps intrinsic in tasks or skills that form the steps in a procedure. The intended outcome is to describe tasks that result in the most efficient performance of the procedure with the minimum associated risk of injury, disease, or environmental impact. In this discussion, the terms "task(s)" and "skill(s)" are used interchangeably.

The first step involves observation of the manner in which a number of experienced people perform the present procedure. The steps are recorded. Different people may perform a task differently. Subtle differences could be very important. The best method combines a logical sequence and includes all critical steps. Videotaping provides a useful means for storing this information for future analysis. If observing people in action is not possible, discussion may be appropriate. Though not as effective, discussion has its merits. This process should examine all normally encountered scenarios.

The next step is to identify procedural problems and health and safety hazards that may arise during performance of the task.

Procedural problems refer to inefficiencies and illogical actions in procedures as performed by individuals. Examining each step in the procedure provides insight to determine whether better ways exist for accomplishing the same action.

Health and safety hazards refer to hazards indigenous to the confined space and to the work to be performed. The hazard of the procedure may arise as a consequence of entering the space. The task to be carried out may present no hazard. On the other hand, the work itself may create the hazard. Examining each step in the task is critical to determining the nature and level of hazard. The next step is to identify alternatives. The best alternative entails the least personal and environmental risk consistent with completing the action. All of the present practices may pose risks judged to be unacceptable. In this case, reevaluation of the entire procedure should occur and alternatives be explored.

The preceding techniques identify critical steps, procedural problems, and hazards. The next step in the process is to specify preventive measures to address the procedural problems and hazards. Preventive measures include engineering controls, work practices, and personal protective equipment.

Devising Standard Procedures

Information compiled during the task analysis forms the basis for creating the standard procedure. The process is to optimize the sequence of steps to produce best results and to integrate preventive measures for controlling associated hazards. Each step must be meaningful to the process, neither too simple nor too complex. A clearly written, easy-to-follow procedure is extremely valuable. Thus, a standard procedure represents a synthesis from the best points identified during the task analysis.

An important part of the standard procedure is a clearly stated title that describes the contents. The most suitable title is the task itself. This provides rapid access for future reference. An integral component of the standard procedure is the performance objective that specifies requirements to be achieved by the person.

Setting Performance Objectives

A performance objective is similar to a road map. It guides a person between a starting point and a destination. A performance objective guides the person developing the training package, the person using it, and the person receiving it.

Stated formally, a performance objective contains three components. First is what the trainee will be able to do after completing the training. The "what" is always an observable action. Second are the conditions under which the trainee will be able to do this. The third part contains the indicators of performance, or how well the person will be able to perform the task. This part of the objective must be measurable.

Example: Using this chapter as a reference, readers will be able to apply the concepts in their own training programs to increase skills and productivity and reduce accidents.

The criterion for success is better training. The indicators are reduced training time and increased skills, and ultimately, fewer accidents. (Well-trained people have fewer accidents than poorly trained people and those just learning. Thus, trainees who learn more thoroughly and quickly are less likely to have accidents.) The condition in this objective is, "Using this chapter as a reference." The "what" is, "will be able to apply" (observable). The "how well" is, producing "better training" (measured by reduced training time, increased skills, and fewer accidents).

The overall performance objective of a standard procedure describes the result of training. It defines the complete task to be achieved, and states the outcome of the steps in the procedure.

Devising Tests

The next step is to devise a means to evaluate or measure the ability of the learner to fulfill the performance objective. A common practice in industry when conducting training is to assemble a package from materials available commercially and to supplement this with fill-in that bridges the gap to the actual situation. The test is then produced from this material. Unfortunately, this approach emphasizes the content of the training materials, rather than site-specific aspects. Usual commercially available training materials include CD-ROMs, interactive videodisk, video tapes, films, slide-tape presentations, and books. These always involve generalization. They often contain unnecessary information and provide information inapplicable to the specific circumstance and omit relevant information. A better strategy is to construct the test(s) before selecting training methods and materials. In this way, the test focuses on the standard procedure and its inherent performance objective, and site-specific needs.

The standard procedure is a valuable tool for devising a test. Restated into questions, the steps form a performance checklist. A performance checklist can apply to any hands-on procedure.

Assessing knowledge-based information such as technical facts requires written or verbal tests. The best written or verbal tests follow an objective format. That is, the answer is either correct or incorrect. Multiple choice and true/false questions are the most susceptible to correct guessing. The short answer format is least susceptible to ambiguity and is strongly advocated. Acceptable formats include: fill-in-the-blank, definitions, short explanations, location of positions, matching, describing, and constructing objects (Dick and Carey, 1985). Testing is a complex subject that demands respect. Poorly created tests can be a major source of frustration to learners. For additional information on this important subject, the text by Gronlund (1993) is highly recommended.

Characterizing Learners

Common practice during preparation of training programs is to create the training program and then to present it to any and all trainees. This approach is content oriented and fails to examine the needs of trainees as learners.

Many adults regard adulthood as the period in their lives in which the need to learn ceases. In fact, many persons apparently are relieved to reach this stage of life for this reason. Many individuals left school prematurely because of a lack of interest in the subject matter or because of the fear of failure. Despite these views, adulthood is a period of learning. Learning usually occurs through choice, rather than compunction. The more obvious vehicles of information or sources that influence behavior include:

- The media
  - Newspapers
  - Magazines
  - Television
  - Radio
  - Movies
- Comments of peers
- Social groups
- Lessons of life

Information provided through these vehicles consists for the most part of short bursts. The rate of repetition during commercial messages is often very high. The same message is repeated frequently in order to promote retention. The process of learning is casual, subtle and often long-term. There is no formal indication that it is happening. There are no exams or assignments or fears of failure. A person can tune out or turn off the source and restart it at any time without penalty.

As a place of learning, the workplace provides a direct contrast to what occurs in other venues. The workplace is the closest reminder of the formal institutional setting represented by the school. On-the-job training represents an imposed form of information transfer and behavior specification or modification. Usually this is conducted by the supervisor, a person usually ill-equipped to carry out this function. Many people are distinctly distrustful and sometimes outwardly hostile toward anything that originates from this source. This is the framework within which on-the-job training must compete for the attention of adult learners. This is a distinct contrast to the casual atmosphere in which learning occurs off-the-job.

A major hurdle facing trainers today is illiteracy in the workplace. Illiteracy is an inability to read, write, calculate, or solve problems efficiently. There are two classes of illiteracy, functional and borderline. Functional illiteracy is an inability to cope at the simplest level. A person who is borderline illiterate survives in today's society, but with great difficulty. One study reported that illiteracy affects one in six employees in all fields (Goddard, 1987). Some organizations address the problem of illiteracy by screening before hiring. However, this approach does not solve the problem for those previously hired, nor does it address illiteracy in the broader context.

An international study on literacy conducted by the Organization for Economic Cooperation and Development (OECD) measured prose, document, and quantitative literacy levels. The levels ranged from level 1 to level 5 in increasing difficulty. Table 13.4 briefly summarizes the results for Canada and the U.S.

Recognizing the symptoms of illiteracy is the first roadblock. An illiterate person is unlikely to come forward to admit to this problem. Admitting to being illiterate in a society that prides itself on its high level of literacy is much too embarrassing. One approach is to observe how workers respond to written instructions. A climate of trust and support is crucial. These people need to feel encouragement, not shame, for addressing and correcting this problem. Specialized adult reading

TABLE 13.4
OECD Literacy Study

		Percent			
Prose scale:		**Level 1**	**Level 2**	**Level 3**	**Level 4/5**
Canada	Employed	11.5	24.7	37.5	26.4
	Unemployed	32.6	23.2	35.7	8.6
	Student	11.5	22.7	39.7	26.1
	Other	29.0	31.0	25.2	14.8
United States	Employed	15.0	26.2	34.0	24.7
	Unemployed	31.5	26.8	32.2	9.6
	Student	27.1	25.0	34.4	13.5
	Other	32.2	25.9	28.5	13.5
Document Scale:		**Level 1**	**Level 2**	**Level 3**	**Level 4/5**
Canada	Employed	11.9	24.0	34.5	29.6
	Unemployed	30.4	29.4	23.1	17.1
	Student	8.1	26.0	31.9	33.9
	Other	38.0	24.8	27.5	9.7
United States	Employed	17.8	25.5	34.0	22.7
	Unemployed	35.7	26.5	24.6	13.1
	Student	23.8	25.7	33.7	16.8
	Other	37.3	29.8	25.2	7.6
Quantitative scale:		**Level 1**	**Level 2**	**Level 3**	**Level 4/5**
Canada	Employed	11.4	25.0	36.0	27.6
	Unemployed	32.9	30.6	27.2	9.3
	Student	7.5	26.6	45.3	20.6
	Other	32.5	27.3	29.5	10.7
United States	Employed	15.9	24.5	32.5	27.1
	Unemployed	37.2	23.7	26.7	12.4
	Student	25.5	27.3	36.6	10.6
	Other	30.5	28.8	28.9	11.8

After OECD 1995.

programs are available. Public libraries and educational authorities can provide information on what is available. Many organizations provide paid time off and worksite classrooms to facilitate this process. Programs operate through several vehicles: paid time off, worksite classrooms, and volunteer tutors.

Eradication of workplace illiteracy is feasible (Kenter 1987). Kenter noted that all too often the problem only becomes evident because of an incident. This situation is avoidable through screening of new employees and determining the level of literacy required to do the job.

Use of visual aids during training de-emphasizes the need for word comprehension. The cliché, "a picture is worth a thousand words," states the situation succinctly. The buddy system is another option worth considering. People help each other along the path to comprehension.

Customizing the instruction to suit the needs of the group, or better still for individuals, can be especially helpful. The well-experienced or older worker may resist retraining out of fear, hostility, or sense of inadequacy. Beginners require more extensive training than experienced workers. To have the greatest possibility for success, the training method must respond to both needs. The more information that is available about the learners, the better the design can reflect them.

As mentioned previously, tailoring instruction to the learning and motivational needs of individual learners is a very difficult task. What motivates one person may not motivate another. Many factors contribute to this situation: age, experience, social background, educational level, and personality. These may influence features such as reinforcement, choices, control, skill levels, and content and style of presentation (e.g., graphics or film, audio, or print).

Note: Collecting personal information about individual workers could infringe on privacy and human rights. Becoming sensitive to these and other issues is extremely important. When acquired by a person in a training role, this information must be treated in strictest confidence.

Training Methods and Materials

The function of training is to present and to transfer information. Information transferred during training sessions covers two main categories: fact-based or skill-based. Several methods have evolved for information transfer. Table 13.5 summarizes the strengths and weaknesses of common training methods.

TABLE 13.5
Training Methods

Lecture

- Presents information quickly
- Audiovisuals needed to reinforce verbal concepts
- Learning involves only listening and note-taking
- Requires expertise in public speaking to be most effective
- No interaction between the learner and the material
- Written outline and/or summary makes note-taking easier

Assigned reading

- Works best when relevant information is identified
- No assurance that material will be read
- Relies on individual initiative and reading ability
- Without follow-up, there is little incentive to read the material
- Requires much effort to process and remember information

Discussion (led by trainer)

- Clear focus on training objective required
- Negative viewpoints can interfere with flow
- Requires strong leadership to maintain direction and pace
- Easy to lose involvement by some members of the group

Socratic method

- Question and answer format
- Either trainer or trainee can ask a question
- Stimulates thinking
- Clarifies information
- Very time-consuming
- Limited number of participants
- Pace is controlled by least knowledgeable participant
- Onlookers may be discouraged
- Best in one-on-one situations

Demonstration

- Allows learner to see concept in action
- Information must be presented sequentially
- Learner should be able to try out skill as early as possible
- May involve chaining, e.g., step 1, then step 1 and 2, etc.
- Depending on complexity, skill will not be learned without adequate practice and feedback

TABLE 13.5 (continued)
Training Methods

- Suits small groups when performed in person
- Staging is critical to ensure that information is shown as intended

Simulation/simulator

- Uses models that are exact replicas of work environments
- Used for training airline pilots, locomotive engineers, process operators
- May include role playing
- Presents various conditions of situation
- Involves decision making at critical points
- Presents outcomes of choices
- Permits mistakes to be made without consequences
- Simulator is costly
- Can accommodate only small number at a time

Workshop

- Small group interaction
- Focuses on building skills
- Narrow theme possible

Performance-based approach

- Information presented (e.g., reading, demonstration)
- Trainee practices the skill until mastery achieved
- Each trainee receives individual feedback
- Learning is self-paced — trainee moves to next task only when ready

The depth of involvement by the trainee in the learning process varies from one method to another. From previous discussion, methods that involve the trainee most actively are the most effective. Unfortunately, methods that work best under controlled conditions may not be practicable in a particular workplace situation. Where possible, use the activity-based features from several methods to involve trainees actively in the learning process.

Another choice facing instructors is training materials. These range from traditional paper- and print-based materials to media that reflect the electronic age in which we live. Table 13.6 summarizes the strengths and weaknesses of common training materials.

Selection of training methods and materials should reflect the learning style and needs of the group. For example, introducing computer-based technology to a group who is not already computer literate could be highly intimidating.

Training Environment

In an ideal situation, the facility has a training area that can accommodate learners and training methods and materials. Unfortunately, the conditions under which training actually occurs often fall far short of this ideal. Consider, for example, the following locations.

- **Production area:** The supervisor trains while actively supervising others. The atmosphere is noisy and chaotic.
- **Coffee room:** This area is available only when other workers are not taking a break or eating their lunch. In some situations, break and lunch periods require considerable time due to multiple sittings.
- **First-aid room:** The area is small and out of the way. It contains expensive medical equipment used for administering first aid and cannot be left unattended. The room cannot accommodate usual audio-visual equipment and is suitable only for one-on-one discussion.

TABLE 13.6
Training Materials

Books
- Very effective when well designed and illustrated
- Can enable active participation
- Can provide continuing resource
- Individual scheduling
- No human interaction; requires effective instructor
- Presentation of concepts must match the audience
- Difficult to provide detail relevant to specifics of an operation
- Expensive to produce in small quantities

Slides
- Custom package can incorporate specifics of an operation
- Passive
- No human interaction; requires effective instructor
- Requires script
- Requires darkened room for effective viewing
- More effective when combined with audio
- Requires high level of interest
- Production expensive

Films
- Custom package can incorporate specifics of an operation
- Passive
- No human interaction; requires effective instructor
- Requires darkened room for effective viewing
- Requires high level of interest
- Production and maintenance expensive

Videotapes
- Custom production possible for specific situations using consumer equipment (**Caution:** consider conditions of use)
- Involvement by local people possible; production values must be set to high standard
- Passive
- No human interaction; requires effective instructor
- Requires high level of interest
- Playback requires adequate television sets for audience
- Commercial production expensive

Computer-based training: software only, videodisk, CD-ROM
- Highly interactive instruction
- Mastery learning
- Self-paced instruction
- High throughput for fact-based learning
- Delivery time compressed as much as 4:1 compared to traditional methods
- Instruction available on demand
- Graphics, animation, video, sound
- Program development and upgrading can be very expensive
- Human instructor necessary for support
- Learner must be highly motivated
- Rewards for performance must be chosen carefully to avoid cynicism
- Presentation of concepts must match the audience
- Difficult to provide detail relevant to specifics of an operation
- Expensive to produce for small quantities of information

TABLE 13.7
Optimizing the Training Environment

- Adequately-sized room to accommodate normal group and associated materials and equipment
- Seating facing parallel to and facing away from windows
- Drapes on windows, or windowless — to minimize distraction
- Isolation from noise and interruptions
- Neatly organized area
- No equipment, material, or supporting columns to obstruct view of the front of the room
- Comfortable chairs and nonreflective work surfaces
- Round or square tables — promote interaction between participants.
- Projection screen located at front of the room above heads
- Adequate lighting level — at least 1000 lux
- Good ventilation — at least 0.6 m^3/min (20 ft^3/min) of outside air per person
- Smoking and other distractions prohibited
- Temperature controlled to comfortable levels

After Laird 1985.

- **Dedicated classroom:** This area, possibly a former storage area, is located beside a noisy section of the plant. There is constant noise from production equipment. The room contains no ventilation. By midafternoon, trainees are yawning and sleepy.
- **Boardroom:** The opulent surroundings and exhortations not to dirty the plush furniture intimidate blue-collar workers.

Training is possible in the less than perfect locations that exist within a facility. The conditions of the training environment are very important. Surroundings profoundly affect mood and behavior and ability to learn. Analyzing conditions and making improvements wherever possible will improve the situation. Inadequate facilities impede the training process. Laird (1985) compiled a list of characteristics for optimizing the training environment. Table 13.7 incorporates his findings, as well as other characteristics that should be considered.

IMPLEMENTING THE PROGRAM

PRELIMINARIES

Prior to implementing the program, determining the entry knowledge of the group is a useful exercise (Dick and Carey 1985). This provides a benchmark that establishes the extent of training that is necessary. Training an employee in a procedure already learned and correctly performed wastes resources. On the other hand, not retraining a person who is making mistakes could be very costly. Performance against a checklist provides a basis for evaluation.

Other items to consider before implementing the training program include:

- Time scheduling
- Number of trainees per session
- Availability of training environment, materials, and equipment

The time of day or even the day of the week can affect the outcome of training. People generally are most receptive in the morning, when their minds are fresh. Performance falls off, however, just before and after lunch, and at the end of the day and the workweek. Avoid these time frames where possible.

Attention spans vary. Attention span for groups reaches a peak at 10 minutes and decreases thereafter. Changing the approach can revitalize the session. A rule of thumb is to change strategy after every 20 minutes. Some examples of variations include changing the pace, using a different prop, or changing to a different training method. Where possible, the training should fit this time frame (Mills, 1977).

The number of trainees per group is very important. Where possible, keep the group small. The maximum workable size depends on a number of factors: nature of the training, complexity of the task, the training environment, and so on. The more difficult the task, the smaller the group should be in order to optimize interaction. Education sessions can accommodate larger groups.

Grouping learners according to ability provides the trainer with the opportunity to conduct training at the narrow level that suits the learners' needs. Within groups, pairing using the buddy system is a productive strategy. In the case of a trainee for whom English is a second language, the buddy should be one who has good command of English.

Often, the trainer does not have the luxury of being acquainted with the group prior to the start of the session. Whenever possible, collect information about the group in advance. This could come from management or other sources. However, be prepared for expression of biases about individuals or the group itself. Individuals and groups may act differently from predictions in the presence of a stranger who brings a compelling message.

Observing the group for 5 min prior to starting the session can provide invaluable information about the characteristics of its members. Certain individuals, the bold, stand out immediately. Recognize them and respond accordingly. Involving these individuals immediately by gaining their attention and appealing to their interests can forge a rapid link to the group. Knowing the audience and anticipating what appeals to them is critical to involving them in the learning process. With practice and experience, a trainer will be able to adapt quickly to the needs of the group and its individual learners.

Inform workers that testing will occur. Then give them adequate time to prepare. Schedule tests so that trainees will have the best possible opportunity for success. Avoid late afternoon at the end of the workweek. This is a time when participants are especially likely to be tired.

Evaluation and Revision

Input from a group is a valuable resource for improving the efforts of individuals, especially trainers. A training program is no exception: the better the input from qualified people, the better the result. Major resources available within organizations to review training programs include management, other occupational health and safety professionals, the occupational health and safety committee or representative, and union representatives. These groups can provide valuable insight about various aspects of the proposed training package. This strategy will work most effectively when the designer guides the reviewer to specific considerations: technical content, readability, logical flow, and so on. The ideal reviewer is familiar with both the characteristics of the trainees and the subject matter. Feedback from any of these sources could be extremely valuable. Use this information to revise the program before implementing it.

Evaluating and revising the program during the implementation phase is extremely important. Training often requires fine-tuning. Feedback from trainees is the major form of two-way communication available to the trainer. This area is often overlooked. Making notes during and after the session provides subjective feedback about the outcome of the session. Sensitivity to verbal and nonverbal messages is extremely important. Recognizing weak and strong points in existing materials and new ideas that arise during the sessions provides the basis for revising and improving the program.

Feedback from written evaluations is another valuable input. A guided questionnaire provided at the end of the session can be an extremely useful source of information. Trainees may feel more at ease expressing themselves in an anonymous questionnaire. Guiding the responder in these surveys is important. The trainer needs specific information to identify points of weakness, rather than vague comment. Keep the questions to the point.

The performance test given after training also provides evaluation information. If trainees are successful, then the program is deemed effective. If trainees are unsuccessful, revision of the program is necessary.

Long-term evaluation also is an extremely important, yet underused tool. Determining the effectiveness of training, days, weeks, or even months later is worth considering. This can help to establish when refresher training should occur. Training must occur frequently enough to ensure capable performance by employees. Refresher training may include updates on process changes, as well as reinforcement of the main course. The suggested period for retraining for trained and qualified workers is annually. Rescue teams should practice at least once every 3 months.

Recordkeeping

Recordkeeping is a fundamental aspect of any program. Records ensure continuity as the persons in a function change. Records are invaluable for helping to reduce the probability that something will go wrong. Recordkeeping establishes the credibility of the program by proving performance, competence, and compliance. The main areas requiring recordkeeping within the context of training include:

- Dates of presentation and roster of attendance at presentations of general level information
- Contents of program
- Date of certification and recertification in CPR (Cardiopulmonary Resuscitation) and first aid
- Dates of refresher training
- Records of qualification as Qualified Persons

Training about Confined Spaces and Confined Hazardous Atmospheres

Confined spaces include complex and diverse environments with equally complex and diverse potential hazards. From the perspective of training this diversity poses a considerable challenge. Virtually everyone in an organization who has access to production areas is vulnerable to dangers posed by confined spaces. Four groups normally are associated with confined spaces: those who assess the hazards and authorize and supervise entry, those who perform work, management and technical personnel, and passers-by. The need for training in some organizations is widespread and can affect virtually all levels.

All persons potentially affected by confined spaces have a need to know about certain basics: recognition and awareness of hazards, and policies and procedures respecting these workspaces. Beyond the basics needed by casual passers-by, the level of knowledge needed by members of other groups increases according to circumstances intrinsic to their positions. Accordingly, to satisfy this need, training should be structured into three ascending levels:

- General level
- Entrant/external support personnel
- Qualified Person

This terminology is generic. It reflects levels of knowledge that could be required across functional job titles. For example, tradespeople, technical personnel, and management all could require training to the level of entrant. "Entrant" describes a common base of knowledge that is needed, regardless of the reason for entry, rather than a job title.

All persons having any reason to enter a confined space should receive education and training to the level of entrant. Trained persons must be fully acquainted with hazards of confined spaces and control measures.

"Qualified Person" is a creation of consensus and regulatory standards. A Qualified Person is described as both trained and experienced in all aspects of work involving confined spaces and confined hazardous atmospheres. The Qualified Person is the key technical individual in the hazard management program. The Qualified Person provides local expertise for assessing hazards and ensuring that work practices are satisfactory to ensure safety. (The Qualified Person is the subject of Appendix B.)

While consensus and regulatory standards specify the duties of trained individuals in global terms, they do not specify what these entail in macro and micro terms. Organizations that must comply with these requirements are provided little, if any substantive assistance. Smith and Ragan (1993) have identified the steps in a needs assessment related to developing an instructional program.

Establish performances: These are the outcomes that must occur to fulfill the requirement, or as Mills stated (1977), the "must knows."

Establish benchmarks: These form the status quo. This process involves an investigation to determine the qualifications or entry level of the trainees. Entry level was discussed in the section on performance-based training. Methods to determine entry level include written tests, performance tests, and observations in simulated situations.

Describe the gap between "What is" and "What should be": Smith and Ragan stress the importance of this step. They mention that assumptions based on formal education or lack thereof should not occur. This step can provide the basis to avoid unnecessary and costly training.

Prioritize action: This step determines the sequence for addressing the difference between the status quo and what is required. This judgment entails a risk assessment to determine which gaps in knowledge/training are the most consequential.

Table 13.8 compiles macro and micro tasks that could be required from individuals participating in a confined space program. Some of the tasks may not require additional training. If required in their entirety, these tasks could necessitate extensive study of a very broad area of knowledge. The need exists, therefore, to identify only those tasks that fit within the context of a particular organization. This approach fulfills the need to identify actions required without overwhelming participants or the organization.

Table 13.8 borrows heavily from the organization used in the OSHA Standards (OSHA 1993, 1994). The OSHA Standards and their preambles currently provide the most comprehensive list of potential tasks. Table 13.8 also contains programs that were not mentioned in the OSHA Standards. Table 13.8 organizes tasks according to the technical structure found in actual organizations, rather than the mission-oriented approach used by OSHA. Also, Table 13.8 refers to the term "confined space" generically, rather than in the specific manner used by OSHA. The tasks outlined in Table 13.8 could be achieved by one person or could involve a number of individuals. Overlap between tasks provided by individuals also is possible.

For clarity, each section in Table 13.8 begins with a possible title for a program of study/training and indicates possible candidates for the program. Once a needs analysis is conducted, each of the required tasks would be subjected to a task analysis followed by development of standard procedures, and the other processes of developing a performance-based program.

The end of this chapter contains sample forms for use in the development of a performance-based program.

TABLE 13.8
Training Needs for Confined Spaces

Program: Administrative Aspects of Confined Spaces
For whom: Managers, administrators, and/or anyone responsible for overseeing the legal, financial, and organizational effects on the organization
Tasks:

- Assess impact of legislation on the organization
- Assess status of present program, i.e., compare against legal and other requirements and identify deficiencies
- Obtain management support and commitment for necessary changes
- Develop policy to express management support
- Assess capability of existing organizational structure to satisfy legal requirements and identify deficiencies
- Specify training needs
- Assess personnel needs
- Assess ability of internal/external resources to meet training needs
- Determine need for additional equipment and technical resources
- Devise recordkeeping systems
- Assess emergency response needs and capabilities
- Determine external interfaces and liaise with external resources
- Obtain funding

Program: Hazard Recognition, Assessment, Management and Control
For whom: Technical Advisors (Industrial Hygienist, Safety Professional, Certified Marine Chemist, Safety Engineer)
Tasks:

- Identify actual and potential confined spaces
- Identify existing and potential hazards associated with each space
- Assess severity of existing and potential hazards
- Assess capability of equipment and resources available in the marketplace
- Recommend equipment for purchase
- Evaluate the impact of exposure to actual and potential hazards considering:
  - Number and characteristics of employees affected by the exposure
  - Magnitude of hazards considered through energy release, toxicity and quantity of chemicals involved
  - Likelihood of exposure
  - Seriousness of exposure
  - Likelihood and consequences of change in conditions or activities not considered initially
  - Strategies for control
- Develop and review written procedures that specify measures and precautions needed for safe entry, work, and egress
- Identify conditions that would change the hazard status assigned to a particular confined space and require reevaluation
- Determine need and frequency for periodic reassessment of hazards
- Establish guidelines for assigning the status of a space
- Establish a system of entry permits
- Specify protective equipment required during entry and work
- Specify training requirements
- Assess specific needs for emergency response
- Develop emergency procedures
- Train in entry procedures and conditions that would prohibit entry
- Audit outcome of entries and revise procedures, as necessary

Program: Safety and Health Committee Consultation
For whom: Safety and Health Committee
Tasks:

- Develop consultation process with Safety and Health Committee on all aspects of confined space program, including:
  - Hazard assessments by Qualified Person
  - Procedures developed for entry and work
  - Results of compliance monitoring by Qualified Person
  - Selection of protective and other equipment
  - Selection of rescue equipment

TABLE 13.8 (continued)
Training Needs for Confined Spaces

Program: Confined Spaces — General Level
For whom: All Employees
Tasks:

- Be informed about the existence and location of confined spaces
- Recognize confined spaces through labels and other means of identification
- Understand differentiation between permit and nonpermit spaces
- Understand the nature of hazards posed by confined spaces
- Be trained in the understanding, knowledge, and skills needed to perform duties safely

Program: Confined Spaces — Entrant Level
For whom: Anyone who could enter the space (Authorized Entrants)
Tasks:

- Know hazards that could be present during entry
- Understand mode of exposure to hazardous substances and agents
- Recognize signs or symptoms of overexposure to hazardous substances and agents
- Understand consequences of exposure to hazardous conditions
- Understand particulars of a permit space and the reason for entry
- Be aware of necessary protective equipment and its limitations
- Be able to use properly equipment necessary for safe entry
- Communicate with the attendant in a uniform manner to indicate status
- Alert the attendant about any indication of exposure to a dangerous or about any prohibited condition
- Alert the attendant when self-initiating evacuation
- Recognize conditions under which exit must occur (orders from entry supervisor or attendant, warning sign or symptom of exposure to a hazardous substance, prohibited condition, sounding of the evacuation alarm)
- Be familiar with procedure for emergency egress

Program: Confined Spaces — External Support for Entrants
For whom: Support Personnel (Attendants)
Tasks:

- Know the hazards that may be present during entry
- Understand mode of exposure to hazardous substances and agents
- Recognize signs or symptoms of overexposure to hazardous substances and agents
- Understand consequences of overexposure to the hazardous conditions
- Recognize possible behavioral effects resulting from overexposure
- Know and watch for hazards that may be present inside the space
- Monitor the status of authorized entrants
- Warn unauthorized employees to stay out of the space
- Inform unauthorized entrants to leave the space immediately
- Inform authorized entrants that unauthorized persons have entered the space
- Assist in evacuating entrants
- Monitor activities inside and outside the space to determine whether entrants can remain inside safely
- Order evacuation of entrants during existence of a prohibited condition, behavioral effects of hazardous exposure, a situation outside the space that could endanger the entrants or whenever the attendant is unable to perform required duties
- Summon rescue services in the event of an emergency
- Know effective work rules
- Remain outside the permit space during entry operations until relieved by another attendant
- Provide information to the rescue service
- Maintain attention on entrants
- Perform only duties that do not detract from the primary function of monitoring and protecting authorized entrants
- Be familiar with procedure for emergency egress

Program: Confined Spaces — Entry Authorization
For whom: Operational Supervisors (Entry Supervisor)
Tasks:

- Recognize that a particular workspace is a confined space
- Know the hazards that may be present during entry

TABLE 13.8 (continued) Training Needs for Confined Spaces

- Understand mode of exposure to hazardous substances and agents
- Recognize signs or symptoms of overexposure to hazardous substances and agents
- Understand consequences of overexposure to the hazardous conditions
- Recognize possible behavioral effects resulting from overexposure
- Ensure systematic review of hazards prior to each entry
- Systematically assess hazards prior to issuing permit considering:
    - Proposed occupancy
    - Work to be performed
    - Possible methods for performing work
    - Physical and chemical hazards and associated risks
    - The actual method proposed and associated equipment
    - Impact of proposed operations on conditions in the space
    - Soundness and security of the structure
    - Need for illumination
    - Substances last contained in the space
    - Steps needed to bring the space to atmospheric pressure
    - Atmospheric testing
    - Medical fitness of entrants
    - Training of entrants
    - Instruction in work procedures
    - Availability and adequacy of personal protective equipment
    - Adequacy of warning signs
    - Special considerations for hot work
    - Fire control
    - Other special considerations
    - Emergency and rescue procedures
    - Arrangements for rescue, first aid, and resuscitation
- Specify tests necessary to determine whether entrants likely would be exposed
- Specify changes in conditions or the nature of the work that would terminate the permit
- Specify protective equipment and tools to be used by entrants and emergency equipment required by rescuers
- Verify the appropriateness of entries on the permit before signing and authorizing entry to begin
- Verify the existence of acceptable entry conditions
- Verify availability of rescue services and operability of the means for obtaining them
- Authorize entry through signing and dating the permit
- Explain procedures to entrants and obtain acknowledgment through signature
- Remove unauthorized persons from the space
- Determine, whenever responsibility is transferred and at intervals as dictated by the hazards and operations performed in the space, that operations within the space remain consistent with the terms of the entry permit and that acceptable conditions are maintained
- Terminate entry operations and cancel the permit, when necessary
- Be familiar with procedure for emergency egress

Program: Testing/Inspection

For whom: Testing/Inspection Personnel

Tasks:

- Be knowledgeable about use and calibration of testing instruments
- Be knowledgeable about limitations of testing instruments and information provided
- Interpret results
- Just prior to entry and during occupancy perform tests to verify:
    - Exposure to chemical agents singly or in combination is less than legal limit/guideline
    - Concentration of flammables is less than legal (safe) limit
    - Percentage of oxygen is within legal (safe) limits
    - Liquid in which entrants could drown has been removed
    - Free-flowing solid in which entrants could become entrapped has been removed
    - Entry of liquid, free-flowing solid, or hazardous substance is prevented by a secure means of disconnection or fitting of blank flanges
    - Disconnection from power sources, real or residual, and locking out of electrical and mechanical equipment that could present a hazard due to release of stored energy

TABLE 13.8 (continued)
Training Needs for Confined Spaces

- Perform atmospheric testing prior to entry or reentry
- Perform testing to assess sufficiency of the ventilation system
- Perform testing throughout the period of occupancy
- Prepare a signed and dated report outlining work performed and results obtained from the tests

Program: Rescue Training
For whom: Rescue Services
Tasks:

- Acquire training in performance of duties
- Demonstrate proficiency in use of personal protective and rescue equipment
- Know hazards that may be present during entry
- Understand mode of exposure to hazardous substances and agents
- Recognize signs or symptoms of exposure to hazardous substances and agents
- Understand consequences of exposure to the hazardous conditions
- Practice making rescues from the actual or representative permit spaces at least once every 12 months. Practice rescue may involve manikins, dummies, or actual persons.
- Obtain training in basic first aid and current certification
- Obtain training in cardiopulmonary resuscitation and current certification

As presented in Table 13.8, the tasks fit a hierarchical structure. They also have temporal and sequential components. Once tasks required within an organizational context are identified and sorted, roles for preexisting personnel can be assigned and the need for new personnel identified.

The tasks identified in Table 13.8 are specific to individual organizations. The only way to provide training that addresses the specifics of an organization is a custom-tailored program. This could occur in large organizations that have the resources to create such a program or utilize the services of consultants. More likely, organizations will attempt to utilize training programs and materials available commercially or attempt to utilize academic training. The latter situation possibly would involve application of concepts in specific disciplines in novel ways.

An important consideration about many of the tasks identified in Table 13.8 is that a short phrase or even a single word can indicate a complete course of study. The problem in a confined space could require a synthesis from a broad range of knowledge in a single technical area.

Table 13.9 identifies possible areas of study that could be applicable to confined spaces. Individual situations could require proficiency in any or all of these topics.

The list, while not expected to be all-inclusive, provides a reasonable starting point for outlining areas of study that may be needed.

Short courses may be available for many of the topics listed in Table 13.3. Trade organizations could be vital resources of supplemental information. Safety associations can provide assistance sectorally. Governments and government agencies also are excellent sources of information.

Expecting a single individual within an organization to possess or develop the requisite working knowledge in all of the areas that could be needed is unreasonable and unrealistic. Development of specialized knowledge within members of a group is a realistic solution to this dilemma. Each could assist the others in specific aspects of a situation. With time the knowledge base of the entire group would broaden and that of individuals would begin to merge and overlap.

TABLE 13.9
Potential Competencies Needed in Qualified Persons

Process- and Structure-Related Hazards

- Hydraulic systems and control circuits
- Pneumatic systems and control circuits
- Mechanical systems and control circuits
- Process/machine control and feedback circuits
- Current electricity and electrical hazards
- Static electricity and static electrical hazards
- Bulk liquid and solid materials handling
- Compressed and cryogenic gases
- Process control
- Chemical incompatibility and instability
- Fire and explosion behavior of matter
- Radiation sources in processes and radiation protection
- Structural integrity

Protective/Preventive Measures

- Deactivation, de-energization, and isolation practices and procedures
- Crane and rigging safety
- Fall arrest/fall protection
- Extrication and retrieval techniques
- Fire prevention/fire protection
- Machine guarding
- Noninvasive cleaning and servicing techniques
- Selection, calibration, operation, and limitations of monitoring instruments
- Application of portable ventilation systems
- Selection and placement of protective equipment
- Isolation techniques
- Purging techniques
- Inerting techniques
- Selection, use, maintenance, and limitations of personal protection
- Decontamination techniques
- Cardiopulmonary resuscitation
- Emergency first aid
- Emergency and contingency planning

SUMMARY

Training is a fundamental element in the strategy for addressing the hazards posed by confined spaces and confined hazardous atmospheres. This chapter has outlined fundamental concepts used in education and training. These concepts include sound principles upon which a training program can be based. Successful training utilizes principles of instructional design and implementation. This approach minimizes inefficiencies created by haphazard or inappropriate approaches.

This chapter provides a curriculum of study for persons involved with confined spaces. This curriculum considers three levels of involvement: casual, persons working in confined spaces, and persons assessing the hazards of such work and authorizing entry and work by others.

SAMPLE FORMS FOR PERFORMANCE-BASED PROGRAM

Form 1

Task Analysis Worksheet		
Task	Job Title	Department
Participants		
Analyzed By — Date	Approved By — Date	
Steps in Procedure	Problems/Hazards	Preventive Measures

Form 2

Standard Procedure		
Task	Job Title	Department
Prepared By — Date	Approved By — Date	
Performance Objective		
Equipment and/or Materials	Protective Equipment	
Steps in Procedure		

Form 3

Performance Checklist

Task	Job Title	Department

Employee Name	Number	Name of Tester	Test Date

Step	Performance		Comments
	OK	No	

Form 4

Worker Characteristics

Job Title and Number of Workers
Range of Experience
Education Range
Literacy Levels
English Fluency
Cultural Background
Age Range and Interests
Other

Form 5

Training Methods and Materials

Task	Job Title	Department

Prepared By	Date	Approved By	Date

Performance Objective

Equipment and/or Materials	Protective Equipment

Step	Methods and Materials	Time

Form 6

Training Record

Employee Name	Number	Job Title	Department

Training Received	Training Date	Test Date	Results	Trainer Signature	Employee Signature	Scheduled Retraining

REFERENCES

Bandura, A. *Social Learning Theory*. Englewood Cliffs, NJ: Prentice-Hall, 1977. pp. 35, 37–38.

Blank, W.E.: *Handbook for Developing Competency-Based Training Programs*. Englewood Cliffs, NJ: Prentice-Hall, 1982. 378 pp.

Carlisle, K.E.: Handling hostile training audiences: motivation and attitude change. *Perform. Instruc. 22:* 26–28 (1983).

Carlson, N.A.: General principles of learning and motivation. *Teach. Excep. Child. 12:* 60–62 (1980).

Chambers, J.A. and J.M. Sprecher: *Computer-Assisted Instruction: Its Use in the Classroom*. Englewood Cliffs, NJ: Prentice-Hall, 1983. 232 pp.

Dick, W. and L. Carey: *The Systematic Design of Instruction*, 2nd ed. Glenview, IL: Scott, Foresman & Company, 1985. 277 pp.

Gagné, R.M., W. Wager, and A. Rojas: Planning and authoring computer-assisted instruction lessons. *Educ. Technol. 21:* 17–26 (1981).

Gagné, R.M. *The Conditions of Learning*, 4th ed. New York: Holt, Rinehart and Winston, 1985.

Gagné, R.M., L.J. Briggs, and W.W. Wager: *Principles of Instructional Design*, 4th ed. Orlando, FL: Harcourt Brace Jovanovich College Publishers, 1992. pp. 9–11.

Goddard, R.W.: The crisis in workplace literacy. *Person. J.:* December (1987). pp. 73–81.

Grabowski, B. and W. Aggen: Computers for interactive learning. *Instruc. Innovator 29:* 27–30 (1984).

Gronlund, N.E.: *How to Make Achievement Tests and Assessments*, 5th. Ed. Boston, MA: Allyn and Bacon, 1993. 181 pp.

Grund, E.V.: Lockout/tagout: the process of controlling hazardous energy. Itasca, IL: National Safety Council, 1995. 429 pp.

Havighurst, R.J.: *Developmental Tasks and Education*. New York: David McKay, 1961. p. 2.

Jay, T.B.: The cognitive approach to computer courseware design and evaluation. *Educat. Technol. 23:* 22–26 (1983).

Keller, J.M.: Motivational design of instruction. In *Instructional-Design Theories and Models: An Overview of Their Current Status,* Reigeluth, C.M. (Ed.). Hillsdale, NJ: Lawrence Erlbaum Associates, 1983.

Kenter, P.: Workplace illiteracy: the three RRR'S. *Occup. Health Safety Canada 3(4):* 25–66 (1987).

Knowles, M.S.: *The Modern Practice of Adult Education*. New York: Cambridge, Adult Education Company, 1980. pp. 40–45.

Laird, D.: *Approaches to Training and Development*, 2nd ed. Reading, MA: Addison-Wesley, 1985. pp. 116–202.

Malone, T.W.: Guidelines for designing educational computer programs. *Childhood Educ. 59:* 241–247 (1983).

Mills, H.R.: *Teaching and Training. A Handbook for Instructors*, 3rd ed. New York: John Wiley & Sons, 1977. pp. 14–20.

Mine Safety and Health Administration: Think "Quicksand": Accidents Around Bins, Hoppers and Stockpiles, Slide and Accident Abstract Program. Arlington, VA: U.S. Department of Labor, Mine Safety and Health Administration, National Mine Health and Safety Academy, 1988.

National Institute for Occupational Safety and Health: Criteria for a Recommended Standard — Working in Confined Spaces (DHEW/PHS/CDC/NIOSH Pub. No. 80-106). Cincinnati, OH: National Institute for Occupational Safety and Health, 1979. 68 pp.

National Institute for Occupational Safety and Health: Worker Deaths in Confined Spaces (DHHS/PHS/CDC/NIOSH Pub. No. 94-103). Cincinnati, OH: National Institute for Occupational Safety and Health, 1994. 273 pp.

Neher, W.R. and L. Hauser III.: How Computers Can Help Adults Overcome the Fear of Learning. *Training/HRD:* February (1982) pp. 48–50.

North, S.L.: The Psychology of Instructional Design for Interactive Videodisc. *Videodisc Optic. Disk 4:* 230–232 (1984).

Occupational Safety and Health Administration: Selected Occupational Fatalities Related to Fire and/or Explosion in Confined Work Spaces as Found in OSHA Fatality/Catastrophe Investigations. Washington, D.C.: U.S. Department of Labor, Occupational Safety and Health Administration (U.S. DOL/OSHA), 1982a. 76 pp.

Occupational Safety and Health Administration: Selected Occupational Fatalities Related to Lockout/Tagout Problems as Found in Reports of OSHA Fatality/Catastrophe Investigations. Washington, D.C.: U.S. Department of Labor, Occupational Safety and Health Administration (U.S. DOL/OSHA), 1982b. 113 pp.

Occupational Safety and Health Administration: Selected Occupational Fatalities Related to Grain Handling as Found in Reports of OSHA Fatality/Catastrophe Investigations. Washington, D.C.: U.S. Department of Labor, Occupational Safety and Health Administration (U.S. DOL/OSHA), 1983. 150 pp.

Occupational Safety and Health Administration: Selected Occupational Fatalities Related to Toxic and Asphyxiating Atmospheres in Confined Work Spaces as Found in Reports of OSHA Fatality/Catastrophe Investigations. Washington, D.C.: U.S. Department of Labor, Occupational Safety and Health Administration (U.S. DOL/OSHA), 1985. 230 pp.

Occupational Safety and Health Administration: Selected Occupational Fatalities Related to Welding and Cutting as Found in Reports of OSHA Fatality/Catastrophe Investigations. Washington, D.C.: U.S. Department of Labor, Occupational Safety and Health Administration (U.S. DOL/OSHA), 1988. 225 pp.

Occupational Safety and Health Administration: Selected Occupational Fatalities Related to Ship Building and Repairing as Found in Reports of OSHA Fatality/Catastrophe Investigations. Washington, D.C.: U.S. Department of Labor, Occupational Safety and Health Administration (U.S. DOL/OSHA), 1990. 195 pp.

OSHA: Permit-Required Confined Spaces for General Industry; Final Rule, *Fed. Regist. 58*: 9 (14 January 1993). pp. 4462–4563.

OSHA: Confined and Enclosed Spaces and Other Dangerous Atmospheres in Shipyard Employment; Final Rule, *Fed. Regist. 59*: 141 (25 July 1994). pp. 37816–37863.

Organization for Economic Cooperation and Development: Literacy, Economy and Society: Results of the First International Adult Literacy Survey. Paris, France: Secretary-General of the Organisation for Economic Co-operation and Development and Statistics Canada, 1995. Errata & p.29

Reigeluth, C.M., M.D. Merrill, B.G. Wilson, and R.T. Spiller: The elaboration theory of instruction: a model for sequencing and synthesizing instruction. *Instruc. Sci. 9:* 195–219 (1980).

Smith, P.L. and T.J. Ragan: *Instructional Design.* New York: Macmillan, 1993. pp. 27–32.

Snow, R.E.: Individual differences and instructional design. *J. Instruc. Devel. 1:* 23–26 (1977–1978).

Zeimet, D.E. and J. Van Ast: Lessons learned in confined space training. *Appl. Occup. Environ. Hyg. 11:* 108–116 (1996).

14 Medical Aspects of Work Involving Confined Spaces

CONTENTS

INTRODUCTION

Confined spaces pose a considerable challenge for the occupational physician. The physician normally does not have access to these workspaces, as compared to those that are readily observable during a plant tour. Previous chapters indicated that asphyxiation and intoxication are the main causes of fatal accidents where atmospheric hazards are present Engulfment is the most frequent cause of accidents involving nonatmospheric hazards. These represent accident conditions to which physicians must respond during emergency situations.

Occupational medicine is a discipline in its own right. There is good reason for this. The physician who treats work-related injury and disease must have knowledge about human exposure to chemicals and chemical processes, and physical agents. This is beyond the scope of knowledge of physicians in family practice and the specialities. In addition to atmospheric and physical hazards, the physician also must consider the physical and psychological demands of work in confined spaces. Many of these workspaces are small. Entry and exit are often limited. These restrictions can increase the burden on the musculoskeletal system. The limited range of motion increases the need for flexibility, strength, and reserve energy. Physical exhaustion can occur much more rapidly in these situations. These workspaces also can be psychologically overpowering to some individuals.

Phobias that express themselves during work in confined spaces may never appear under normal conditions.

Physicians encounter employees who work in confined spaces in several types of circumstances. First are emergency situations where workers have been injured. These can occur through exposure to hazardous atmospheric conditions or physical agents or trauma. Among the emergency situations that physicians must address are traumatic injuries and intoxications. Traumatic injuries or fractures are similar to those resulting from other emergency situations, such as car accidents. First aid attentdants and rescuers at the jobsite may have initiated treatment to stabilize the victim and to enable transport to the hospital. The physician in the emergency or operating room will be called upon to take over and provide care.

Patients overexposed to hazardoius atmosphere can present a variety of medical conditions unusual to many physicians. They may arrive at the emergency room in a comatose state. Overexposure to many solvents produces the same response. The Material Safety Data Sheet or questioning of fellow workers provide clues to the identity and effects of the hazardous substance. These may or may not be correct.

In some circumstances, chemical reaction has occurred and generated a new compound. Knowledge of chemical processes and possible reactions, therefore, is essential for identifying the substance causing the intoxication and for initiating treatment of the medical condition triggered by the overexposure.

Preparation for emergency care could occur through an on-site medical facility staffed by physicians and nurses or an on-site first aid facility staffed by a first aid attendant. These facilities could be located in all possible surroundings from large cities to remote sites. The needs of facilities in each location vary. Emergency first aid is a topic of discussion in Chapter 15 on emergency preparedness and response.

The second involvement of physicians with confined spaces is assessment of fitness of individuals to work in these environments. In order to be able to assess the ability of individuals to perform taks, the physician needs a quantitative assessment of abilities required from them. The physician should be provided the opportunity to visit the worksite. A first-hand inspection enables a better understanding of the physical demands of the work. The employer should provide a description of work tasks and schedules, shift length and frequency of change. This should also include requirements for use or personal protective equipment.

If evaluation of working conditions has not occurred, the physician may request assistance from industrial hygienists and safety professionals. Effective occupational health programs have always been considered a partnership between medical professionals and industrial hygienists and safety professionals (Dinman 1973, Zenz 1988). Nowhere is this partnership more important than in work involving confined spaces. The hygienist and safety professional act as the eyes and ears for the physician, since they have access to these workspaces. Since hazardous conditions during work in confined spaces can range from innocuous to life threatening, there is little leeway for error in the performance of work, especially in the latter situations. Lack of leeway for error in the performance of work indicates the need for individuals who are unlikely to pose a risk to themselves and others. One way to ensure this is prescreening of individuals to identify physiological and psychological limitations relative to demands of the work. A parallel to this is the protective management of individuals with disease conditions, so that workplace conditions do not exacerbate the situation. Allowing individuals who are a risk to themselves and others to work in these situations needlessly compounds the risk of accident and tragedy.

The third role of the physician in addressing problems involving confined spaces is medical surveillance. A medical surveillance determines exposure to hazardous substances and some physical agents using specific tests. These tests usually require body fluids or materials. Medical surveillance usually occurs during the course of work; however, this also can occur prior to placement. Medical surveillance can determine that overexposure did not occur, and therefore, that protective and preventive measures were successful. Medical surveillance tailored to specific circumstances is known as a hazard-oriented medical examination. Such a program reflects legal requirements, recommendations made by agencies such as NIOSH (National Institute for Occupational Safety and Health), and medical and scientific justifications. The surveillance program should reflect "good practice" and comply with all legal, ethical, and regulatory requirements.

OCCUPATIONAL MEDICAL PROGRAMS

To date, there has been no discussion in the literature about occupational medical programs for work involving confined spaces. Medical surveillance programs are available for general industrial activity and hazardous waste operations (NIOSH/OSHA/USCG/EPA 1985, EPA 1992, Zenz 1988, Cooper and Zavon 1994). Similarities exist between the latter and work involving confined spaces. Workers potentially exposed to some hazardous conditions are required to wear encapsulating suits and supplied air respiratory protection. Conditions inside these suits can be extremely stressful. Work in these environments can pose considerable risk of injury because of hazardous mechanical conditions. Confined spaces compound these hazards because of confinement by the boundary surfaces. Restricted access and egress in some spaces can complicate first aid treatment and extrication of an injured victim.

Adaptation of programs created for other purposes may be possible directly or by extrapolation. Caution is warranted, since work involving confined spaces often occurs over a brief time frame and can be highly varied. Also, many of the hazardous conditions result from rapidly acting toxic agents or from agents that can cause traumatic injury. Medical surveillance programs designed for long-term operations would be inappropriate in this context.

Table 14.1 lists components that should appear in a medical surveillance program. The specifics of an individual situation could lead to modifications, additions, and greater weighting to individual components. As with other components of a health and safety program, the effectiveness of a medical surveillance program depends on active worker involvement and management commitment.

Table 14.2 provides information about clinical tests frequently requested by occupational physicians.

Cell counts provide an important clue to diagnosis of blood diseases and abnormalities (Fairbanks 1990, Hutton 1990, George 1990). Deviations from normal limits are indicative of damage or disease conditions. Normal range is defined arbitrarily as the mean ±2 standard deviations. This definition also defines 2% of the population as anemic (abnormally low hemoglobin level) and 2% as polycythemic (abnormally high hemoglobin level) (Fairbanks 1990). Microscopic examination can indicate the presence of disease or other conditions.

Liver function tests are performed on enzymes present in blood (Kaplan 1990). As used in occupational medicine, these tests identify and distinguish among different types of liver disorders and gauge the extent of damage. However, normal results can be obtained in cases of severe liver disease. In addition, abnormal results can occur in cases of disease not affecting the liver. The liver contains thousands of enzymes, some of which are present in blood. There is no known reason for the presence of liver enzymes in blood. Elevated levels are believed to reflect damage to liver cells.

Urine tests provide indications about kidney and other functions (Forland 1990). Color can provide a useful diagnostic clue about disease in other systems, as well as food and drug consumption. Elevated specific gravity reflects the presence of high molecular weight species, a possible indicator of kidney damage. Elevated pH can indicate kidney damage as well as urinary infection. Elevated protein in urine is indicative of primary renal disease or renal involvement in systemic illness. Elevated glucose can indicate renal or pancreatic disease. The appearance of bile in urine is a potential indication of liver damage or disease. The presence of acetone and other ketone bodies in urine is evidence of faulty metabolism of fatty acids in the liver and other tissues. Blood or blood pigments in urine are indicative of potential kidney damage. Centrifuged sediment can include

TABLE 14.1
Suggested Components of a Medical Surveillance Program

Component	Elements
Preemployment screening	Medical history
	Occupational history
	Physical examination
	Determination of fitness to work
	Baseline monitoring
Periodic medical examination	Annual update of medical and occupational history
	Annual physical examination
	Routine medical tests
	Specific testing based on examination results, exposures, job class, tasks
Termination examination	Update of medical and occupational history
	Routine and specific medical tests
Emergency treatment	Emergency first aid on site
	Liaison with hospital/emergency medical center
	Arrangement for decontamination of victim
	Arrangement for transport to hospital/emergency medical center
	Arrangement for transfer of medical records to next care provider
Nonemergency treatment	Case-by-case mechanism
Recordkeeping	Confidential, secure system for keeping records of individuals and reports and records of accidents, injuries, and illnesses
Program review	Apply recognized procedures to audit effectiveness of program

From NIOSH/OSHA/USCG/EPA 1985, Zenz 1988, and EPA 1992.

TABLE 14.2
Clinical Tests Performed Frequently

Organ	Body fluid	Test
Blood	Blood	Complete cell count including white and red cells and platelets; hemoglobin level; hematocrit or packed cell volume and red cell indices. Test shows status of blood-forming function.
Liver	Blood	Total protein, albumin, globulin, bilirubin. Test shows general functional status of liver.
		Alkaline phosphatase. Test shows obstruction of bile production and release.
		Gamma glutamyl transpeptidase; lactic dehydrogenase; serum glutamic-oxaloacetic transaminase; serum glutamic-pyruvic transaminase. Tests show cell injury.
Kidney	Blood	Blood urea nitrogen; creatinine; uric acid. Tests show general functional status of kidneys.
	Urine	Color; appearance; specific gravity; pH; glucose; protein; bile; acetone; occult blood; centrifuged sediment. Test shows functional status of kidneys.
Eyes	None	Vision test. Test measures refraction, depth perception, and color vision to ensure ability to read instruments and printed materials and to respond appropriately to color-coded labels and signals.
Ears	None	Audiometric test. Test measures ability to hear pure tones at standard frequencies to ensure ability to respond to warnings.
Lung	None	Chest X-ray; pulmonary function test. Tests determine functional status of lungs and respiratory system and ability to tolerate use of respiratory protection.
Heart	None	Electrocardiogram. Test determines functional status of heart while resting and under stress of exercise.

After NIOSH/OSHA/USCG/EPA 1985.

blood cells, bacteria, and other invaders of the urinary tract, cell fragments, crystals, and fat globules. Presence of blood cells is indicative of kidney disease or infection.

Confidentiality, trust, and short- and long-term security of information are critical to the success of these programs. Many people are reluctant to provide any information about themselves to any medical service provider that has a relationship with management. This is an unfortunate consequence of past attempts to gain and misuse of confidential information. Also, samples of blood and urine obtained for routine clinical tests can become the feedstock for drug screening. Use of body fluids and excreta in any unauthorized manner is prohibited on human rights grounds in some jurisdictions.

ASSESSMENT OF FITNESS

Discussion

Physicians have the power to declare individuals fit or unfit to work in a particular situation. Exercising that power carries considerable responsibility. On the one hand, rejecting a suitable individual could cause financial hardship because of wrongful deprivation of the opportunity for employment. On the other hand, accepting an unsuitable individual could lead to risk of serious injury or loss of life for that person or someone else. This situation has obvious implications with human rights and accommodation to disabilities.

The physician has the need for a systematic process for making these decisions in a manner that is consistent and fair, and technically defensible. In order to do this, the physician requires detailed information about the physiological and psychological demands of the position, status of the individual, and technical benchmarks.

The physician must be able to describe the requirements of the work activity in quantitative terms. The starting point for this process is a comprehensive description about the nature of the work activity. Of course, this information usually is not available. Usually all that is available to the physician is a job title and at best a vague job description. Neither of these is particularly helpful for recognizing the required physiological and psychological demands of the position. As a result, the physician must rely on intuition and information gleaned from other sources.

As an illustration of the problem, consider an emergency service provider at a site containing confined spaces. The incumbent weighed more than 125 kg (275 lb) and appeared to be considerably overweight for his height. He was 53 years of age. This individual perished during an attempt to rescue another person who was down in an oxygen-deficient confined space. The entrant, who was later extricated, survived the episode without harm.

The victim was a full-time first aid attendant. This position was sedentary and involved little walking around the site. The position required the individual to remain in a centralized location to await injured parties and to attend to their needs.

On rare occasions, such as the one described, an accident could occur in one of the many confined spaces present on the site. At this instant, the first aid attendant assumed the role of emergency services provider. The demands of the position are fundamentally different from those of the normal routine. However, the job title, "emergency services provider," says nothing about the physiological and psychological demands of the position.

The closest analogous job description to the emergency services provider is the firefighter or hazmat responder. These individuals routinely work in confined spaces under all kinds of conditions. An in-depth examination of these jobs would indicate the following demands:

- Able to carry 25 kg (55 lb) of equipment for at least 1 hour
- Able to utilize an SCBA (self-contained breathing apparatus)
- Able to work in a fully encapsulating suit

- Able to climb up and down ladders while wearing the equipment
- Able to lift and carry at least 35 kg beyond requirements of the extra equipment
- Able to move through small openings inside the framework of a structure
- Able to work under extreme emotional duress
- Able to work in awkward postures

By contrast to the first aid attendant, the role of emergency services provider is highly strenuous and demanding. Without knowing about the demands of the position, the physician, therefore, is completely unable to anticipate the capabilities required from individuals who fulfill this role.

Hazard assessments produced for work activities to be performed in the confined space provide a valuable resource for facilitating the mental image of the conditions under which individuals must work. The physician, of course, is unlikely to be able to be present to witness these activities. The hazard assessment can assist by providing information about the nature of ergonomic, physiological, and psychological demands. Also, a difficulty scale could be incorporated into the hazard assessment or derived from it.

Work requirements form the basis for assessing capabilities that are required in individuals. The work activity described above is a composite of elemental tasks and actions. These include types and ranges of motion, repetition, duration, forces, dexterity, psychological stability, etc. Considering the demands of the job as a composite of elemental tasks and actions provides a means for simplification and measurement.

Individual elemental tasks and actions require output of physiological and psychological energy. The literature provides little assistance for assessing the demands of the broad variety of such tasks. However, the literature does contain benchmarks of performance required for specific tasks and jobs. Analogy to specific circumstances may be possible.

Physiological and psychological status of the individual is an intrinsic quantity that can be determined through test protocols. Standardized tests that mimic the demands of the discrete tasks provide the closest possible indication about the ability of an individual to perform capably.

Technical benchmarks provide a basis for decision making through comparison of individual abilities with the required performance. These benchmarks reflect research into various aspects of human performance. Some of this research was performed to address requirements of specific activities, for example, military or mission-oriented tasks.

The outcome from this process is a decision about fitness to perform the work activity. The physician could assess the person as fit to perform the activity or limited by minor or significant impairment (Zenz 1988).

Intermittent impairment could occur under predictable or unpredictable conditions. Intermittent impairment includes: uncontrolled or inadequately treated epilepsy and diabetes. Either of these conditions could be a major factor in the decision about fitness to work in a confined space.

Stable impairment refers to conditions unlikely to progress or worsen through job activity. Stable impairments include: amputation, ankylosis (fixed joints), spasticity, blindness, deafness, cerebral palsy, residual effects of polio, and so on. Some of these conditions could be major factors in the decision about fitness to work in a confined space.

Impairments worsened by work conditions include chronic skin diseases, liver disease, and asthma. These impairments could be stable or unstable. Some of these conditions could be major factors in the decision about fitness to work in a confined space.

Progressive conditions worsen with time. These include: chronic obstructive lung disease (for example, emphysema), congestive heart disease, and arthritic diseases. Some of these conditions could be major factors in the decision about fitness to work in a confined space.

Accommodating Individuals

In recent years, considerable emphasis has been put onto accommodation of the workplace to suit the needs of individuals. Confined spaces conflict with this concept, since in many cases the configuration cannot be changed, and the individual must accommodate to the space. This can be very difficult, and potentially impossible for some individuals. For example, some spaces contain small openings. These delineate who can and cannot enter. The size of openings is regulated by codes for pressure vessels. Changing the size to accommodate other individuals is not possible. Similarly, work in some confined spaces requires individuals who are very strong and can work in awkward postures. In neither case is the average person necessarily suitable to perform the task. Females are shorter and less strong than men, on average.

The average person is not a unique individual. There are many "average persons." What constitutes an average person, however, differs with the homogeneity of the group through factors such as age, gender, and race. Age affects other characteristics of the average person as the group ages. Some races are larger than others, as automobile designers have discovered.

Aging is a natural and gradual phenomenon (Courau 1983). The effects of aging usually become apparent between 40 to 45 as compensatory systems fail (Figure 14.1). One worker in three in the developed countries and one in four in the developing countries is beyond the age of 45 (ILO 1977).

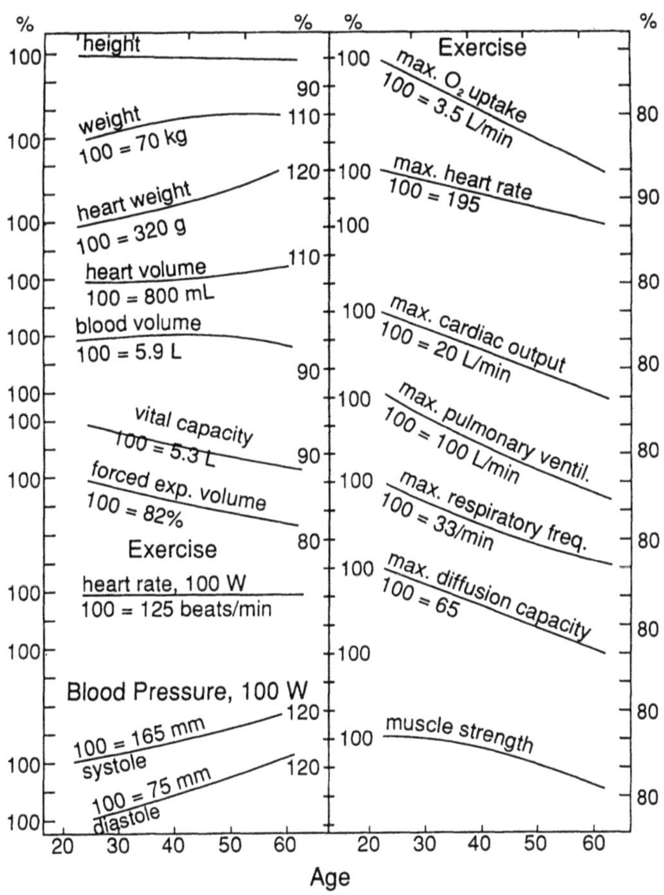

FIGURE 14.1 Effect of aging on parameters of interest in workplace physiology. (Adapted from Astrand and Rodahl 1986.)

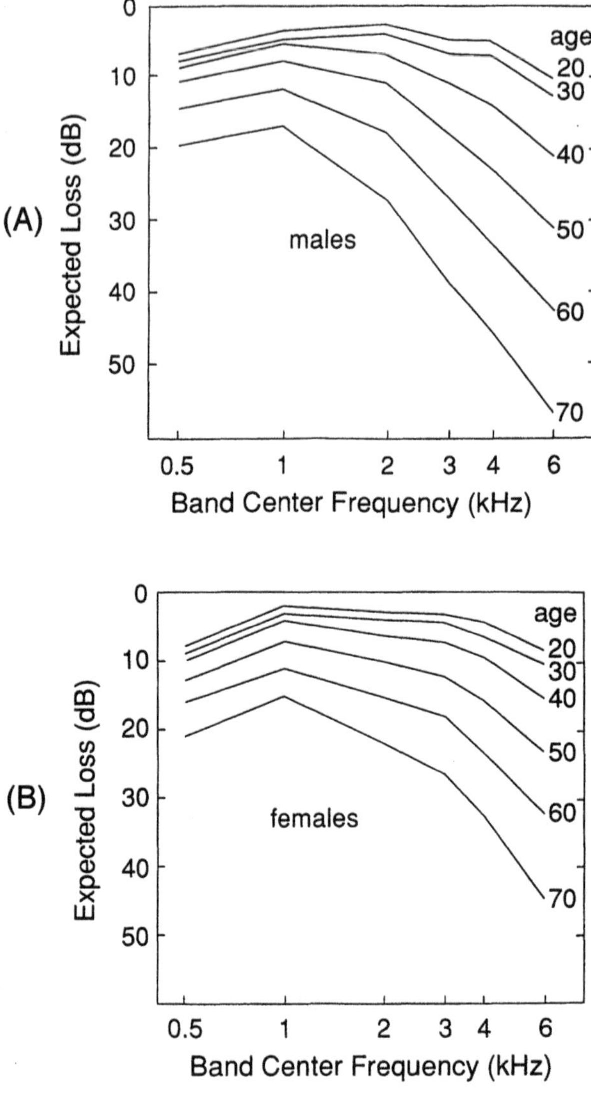

FIGURE 14.2 Decline of hearing acuity at different frequencies with age. (A) Male. (B) Female. (Adapted from NIOSH 1972.)

Muscular strength reaches its maximum around the 25th year and declines thereafter. Yet, the impact of fatigue does not appear to be greater in the middle aged than in the young. The ability to tolerate maximum effort declines with age. Older workers are less able to tolerate the heat. Manual skill deteriorates after the 20th year. Physical agility and muscular coordination decline. Balance is less precise.

Visual functions (accommodation, acuity, field of vision, adaptation to darkness, and resistance to dazzle) and auditory functions may change with age (Figures 14.2 through 14.4). However, conpensatory appliances such as spectacles and/or contact lenses can often be used to counteract any of these changes.

Cerebral functions peak between 25 to 30 years and then decline. These are a complex mix of crystallized functions involving the sum of personal knowledge and fluid functions, such as immediate memory, power of concentration, and rapidity of comprehension and reasoning. Some capabilities improve with age, while others decline.

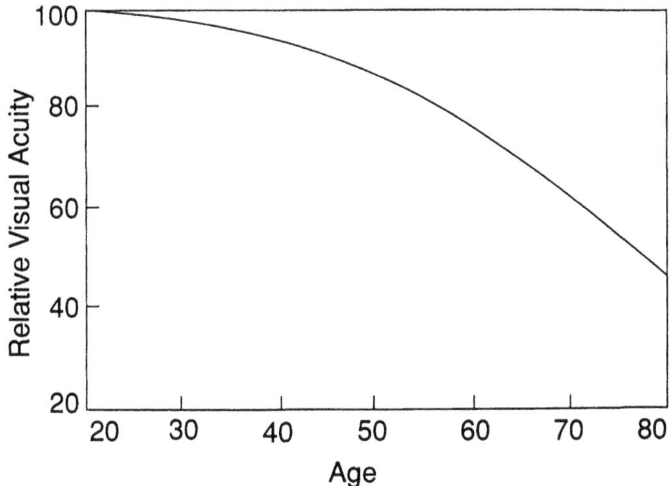

FIGURE 14.3 Decline of visual acuity with age. (Adapted from Zerbe and Hofstetter 1957–1958.)

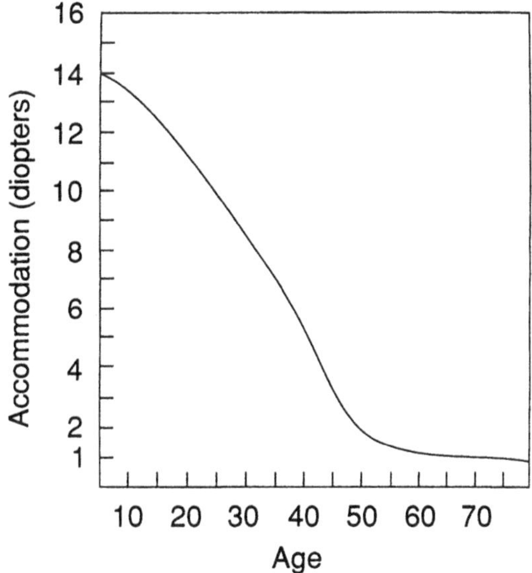

FIGURE 14.4 Decline of visual accommodation with age. (Adapted from Kaufman and Christensen 1981.)

Work capacity remains relatively the same with increasing age. What declines is the ability to maintain the pace of work. Reaction time needed to process information to make decisions increases.

Increasing age of a stable workforce can have a significant impact on performance capabilities. Despite the limitations imposed by growing older, the demographics of fatal accidents involving confined spaces provided in Chapter 1 suggest that older workers are no more at risk than other groups. This would strongly argue that the knowledge and experience of this group are far too important to lose through lack of accommodation to their declining physical capabilities.

Legislation in recent years has attempted to protect workers whose capabilities are less than some ideal of perfection. One example is the Americans With Disabilities Act (Equal Employment Opportunity Commission 1992). The substance of this type of legislation is that workers who can perform the essential functions of a position (with or without accommodation) without substantial risk or threat to themselves or others should not be restricted in performing their duties or denied the opportunity.

As discussed above, most people are incapable of performing work that demands superhuman capabilities. Even those who can perform at this level lose this capability as they age. In order to enable individuals to continue to work as they age, or to permit wider access of work to both genders or to enable the physically and mentally disabled to work, accommodation is necessary. Accommodation to the needs of older workers has occurred for many years. Older workers are too valuable a resource to lose for many industries.

Yet, some activities require superior performance (Anderson 1989). Superior performance requirements could lead to exclusion of many individuals and virtually all women. This would lead to disproportional representation. Disproportional representation is acceptable, provided that it is job related. Job-relatedness can be established through strength and endurance testing. Tests chosen for the battery must bear high similarity to the job, be predictive of risk of injury and performance, have minimum bias, and be replicatable.

The critical problem for the physician and the organization is to define, as precisely as possible, the physical requirements of the work activity. Accommodation to individual situations should occur where possible. However, confined spaces deserve special consideration, especially where hazardous conditions cannot be eliminated or controlled to a high probability. In these situations, rapid evacuation may be required. This could require strength and agility, as well as the ability to pass through small openings while wearing considerable protective equipment.

Fitness to Use Respiratory Protection

Respiratory protection is one of the main lines of defence against the hazardous conditions that can be present in confined spaces. Respiratory and other personal protective equipment are the subject of Chapter 12. Use of respiratory protection alone or combined with impervious encapsulating protective suits can cause physiological and psychological difficulties for many individuals.

Spirometry

Spirometry is one of a number of techniques used to assess pulmonary function (Comroe et al. 1962, Shigeoka 1983, Cotes 1983). Spirometry often measures forced vital capacity (FVC) and forced expiratory volume (FEV). FVC is the volume of air exhaled with maximum forced expiration following maximum inspiration. FEV is the volume of air exhaled during performance of the FVC. FEV is based on a time frame. $FEV_{1.0\,s}$ is the forced expiratory volume in 1 second.

The ratio of $FEV_{1.0\,s}/FVC$ corrects for volume (Shigeoka 1983). A low FVC and normal ratio indicates a restrictive breathing pattern. The lower the FVC compared to normal values, the more severe the restriction. A normal FVC and low ratio indicates an obstructive pattern. The smaller the ratio, the more severe is the obstruction. Normal FVC and normal ratio indicates normal condition. Low FVC and low ratio indicates advanced obstruction or combined obstruction and restriction. TLC (total lung capacity) in restrictive disorders is less than predicted. The lower the measured value, the greater is the restriction.

Patients with asthma exhibit an obstructive type of impairment (Cotes 1983). This condition may produce severe obstruction during an attack. Asthma may be detectable only following exercise or administration of a bronchoconstrictor. Bronchodilator therapy reverses the condition, except during some acute episodes.

By its nature as an experimental science, spirometry is subject to variations in technique and interpretation. The American Thoracic Society (ATS) provides recommendations for standardization of these techniques (American Thoracic Society 1987). Spirometry is an effort-dependent maneuvre. Successful assessment requires careful subject instruction, understanding, coordination, and cooperation. The ATS provides recommendations for equipment, conduct of tests, and interpretation of results.

Pulmonary function tests, such as spirometry, provide objective documentation of respiratory function (Shigeoka 1983). However, results from pulmonary function tests cannot be attributed to a specific condition. For example, low vital capacity can result from pulmonary interstitial fibrosis or emphysema. The latter occupy opposite ends of the spectrum of pathophysiology.

Constraints in Respirator Usage

Successful respirator use is contingent on selection, fitting, training, and the ability to tolerate the facepiece and resistance to breathing during normal work requirements (Raven et al. 1979). Much of the evolution of respiratory protection has been directed to enabling healthy young males to work at high workloads without physiological consequences (Beckett and Billings 1985). Demographically, however, in terms of pulmonary function the worker population ranges from superior to little or no loss to moderate to severe loss (Raven et al. 1979). Much of this loss is attributable to age (>25 years), smoking habits, environmental pollutants and chronic disease entities (asthma, bronchitis, emphysema, and fibrosis).

Negative pressure respirators limit maximal breathing in normal young males at extreme workload (Deno et al. 1981). Inspiratory resistance is the most important of the various types: threshold, inspiratory, expiratory, and dead space. Examples of workloads at which limitations might occur are the most strenuous activities of fire fighting or mine rescue. The combination of breathing resistance and dead space during maximal exercise in normal workers alters the duration and timing ratio of inspiration and expiration, as well as peak mouth pressures and inspiratory flow rate (Harber et al. 1982). Normal subjects respond to the added load during moderate exercise by increasing respiratory effort and decreasing total minute ventilation (Ceretelli et al. 1973). At lower workload, the inspiratory resistance of commonly used, negative-pressure, air-purifying respirators increases breathing resistance by a small amount.

The ability to tolerate commonly used negative pressure respirators varies among individuals. Part of the variability may be related to individual variation in perception of respiratory load. The greater the perceived load, the more likely the respirator will feel uncomfortable or produce the sensation of dyspnea (shortness of breath). Older subjects show decreased sensitivity of perception of inspiratory resistance compared to younger subjects (Gottfried et al. 1981). Asthmatics show thresholds of perception similar to normal subjects. Persons with chronic obstructive lung disease show a higher threshold to the perception of inspiratory resistance than normal (Tack et al. 1982). Whether these differences are important for the magnitude of resistance in commonly used respirators is not known.

Hodous et al. (1983) studied physiologic and subjective responses to exercise with a combination of inspiratory and expiratory resistance in males up to 45 years of age. The magnitude of the resistance was similar to that in negative pressure respirators. Responses of normals were compared to those with chronic obstructive pulmonary disease. Using progressive treadmill exercise up to 10 min duration and making comparisons at moderate exercise intensity, they found relatively little difference in subjective and physiologic response between controls, mildly obstructed, and moderately obstructed.

In normal adults, the major stress to the cardiovascular system occurs from the added weight of equipment, such as the self-contained breathing apparatus. The cardiovascular demands of the task are increased by a factor at least equal to the metabolic cost of carrying the added weight through the task. There is little information available on the effects of respirators on those with cardiovascular disease who are, without respiratory protection, able to perform their jobs without symptoms. It is not known whether breathing with inspiratory resistance might increase myocardial oxygen demand enough to induce angina at an otherwise tolerated exercise level.

Psychological Factors

Respirator use imposes considerable psychological burdens on some users. Shepard (1962) reported that physiological problems associated with respirator use were minor compared to psychological responses in some users. The distress associated with work in an extreme environment is exacerbated by wearing a respirator. Vision may be restricted to some extent with all devices. This can lead to narrowing of the visual field ("tunnel vision") (Landers 1980). This can, in turn, lead to anxiety, claustrophobia, and hyperventilation in some individuals.

About 10% of the subjects in any respirator study are unable to perceive changes in breathing, nor are they capable of accurately judging exercise intensity (Morgan 1983a). Approximately 10% of users experience excessive discomfort while wearing respirators; 10% may represent a conservative estimate.

Morgan and Raven reported on the effectiveness of a psychological test for trait anxiety as a predictor of respiratory distress during heavy physical work performed while wearing a respirator (Morgan and Raven 1985).

The Spielberger trait anxiety scale is a standardized psychological inventory designed to measure proneness to distress (Spielberger et al. 1983). This was administered prior to treadmill tests. The State–Trait Anxiety Inventory (STAI) contains separate scales for reporting state and trait anxiety. The state–anxiety scale evaluates anxiety at a particular time. The trait–anxiety scale measures differences in anxiety proneness.

Five of the six individuals predicted to have respiratory distress based upon trait anxiety scores responded as expected. Of the 39 subjects predicted not to experience distress, 38 also responded as expected. Objective measures of trait anxiety according to these results can identify individuals most likely to experience distress while wearing a respirator and performing heavy physical exercise.

Johnson et al. (1995) followed up the work of Morgan and Raven (1985). Even when they could adjust the rate of work, anxious subjects experienced more discomfort, performed for shorter times, and accomplished less work than their counterparts who had low anxiety.

Hyperventilation Syndrome

Hyperventilation syndrome is a potentially serious condition that affects some respirator users (Morgan 1983b). Published research on this syndrome dates back at least a century. Symptoms include breathlessness or dyspnea with effort, paresthesia (tingling), trembling, tachycardia, tetany, carpopedal spasms, and convulsions. Hyperventilation syndrome occurs in some individuals for no apparent reason. Hyperventilation, and the resulting physiological changes, are associated with decrements in psychomotor performance along with increased error rates.

Symptoms that characterize hyperventilation syndrome can be readily produced in certain "types" of individuals through overbreathing or introduction of a CO_2 challenge. The symptoms, once produced, can be quickly reversed by rebreathing of air expired into a bag. Some individuals or "types" appear to be especially sensitive to the effects of overbreathing and/or CO_2 loading.

Identification of these individuals is extremely important, since they appear to be at greater risk of attack during tasks requiring vigorous exercise and the wearing of respirators. An attack occurring during work in a confined space could have serious implications on the safety of these individuals, as well as their co-workers. Identification of susceptible individuals should occur through routine medical evaluation.

Contact Lenses

Use of contact lenses by individuals in general industry has generated considerable controversy (Nejmeh 1982, Stein and Slatt 1984, Rosenstock 1986, Randolph and Zavon 1987). Nejmeh (1982) reported on a detailed survey performed through the National Safety Council. This indicated highly variable policies and practices throughout the chemical sector that participated in the survey. In the three instances involving injury, the contact lenses were believed to have protected the eye. Stein and Slatt (1984) documented follow-up on reports of injury reportedly exacerbated by contact lenses and found no supporting evidence. These authors described measures for safe use of contact lenses. More recently, Randolph and Zavon (1987) reported on a survey of organizations. These authors found that the decision not to permit wearing of contact lenses appeared to be perceptual, rather than fact-based. Literature reviewed by these authors indicated that contact lenses were protective during injury situations and simulations using animal tests (Rengstorff and Black 1974). In another part of the article, Randolph and Zavon provided guidelines for the use of contact lenses by workers in industry.

Cullen (1993, 1994) has reviewed this question extensively. He noted numerous instances in which contact lenses have provided protection and mitigated against more serious injury. The extent of protection depends on the nature of the hazard and the type of lens worn. Some types of contact lens provide added protection to wearers of eyeglasses against chemical splashes, dust, flying particles, and nonionizing radiation (arc flashes). Also, depending on type, contact lenses do not trap chemicals under or in them. Contact lenses offer added benefit over eyeglasses in many occupational circumstances (Cullen 1993).

In order to receive a satisfactory seal from a full-facepiece respirator, protrusions such as the temple pieces on glasses are not considered acceptable. The only options available to individuals needing visual correction are inserts for the facepiece that contain prescription lenses or the wearing of contact lenses. Not everyone can obtain a satisfactory outcome from the use of a vision kit, as alignment is critical in some prescriptions. Wearers of bifocals, likewise, would experience considerable difficulty in use of a vision kit.

Wearing of contact lenses inside a full facepiece has generated considerable controversy. On the one hand, they offer the wearer equal opportunity to use any full-facepiece respirator at a moment's notice. On the other hand, contact lenses are a barrier to the normal function of the cornea, for which due consideration must be given.

Wearers of contact lenses are prone to less than ideal conditions. Airborne particulates can cause serious irritation. Very dry compressed breathing air used for winter conditions could dry out the lenses. Contaminated air that enters the facepiece could interact with the lenses.

DaRosa and Weaver surveyed U.S. firefighters about use of contact lenses (LLNL 1986). Despite the threat of legal sanctions against them and their employers, hundreds of firefighters (29% of respondents) admitted to wearing contact lenses during daily use of SCBAs. Of these, only six encountered problems serious enough to require removal of the facepiece. No firefighter was injured or died as a result of wearing contact lenses. These findings and follow-up on comments led the authors to conclude that contact lenses were not significantly more hazardous than insert glasses. A survey performed by the U.S. Department of Energy obtained similar results (Department of Energy 1986). OSHA (1988) subsequently took the position that the wearing of soft and gas-permeable contact lenses would constitute only a de minimis violation of its standards. ANSI (1992) also adopted the position in the respiratory protection standard that contact lenses were no more hazardous than insert glasses.

Prohibition of contact lens use in the context of current knowledge about safety in their use puts the physician into conflict with human rights legislation, such as the Americans With Disabilities Act (Equal Employment Opportunity Commission 1992, Blais 1994). Some people must wear contact lenses to obtain best visual performance or efficiency. An example is an individual with monocular aphakia who must wear contact lenses to obtain binocular vision.

Respirator Assessment Protocols

ANSI Z88.6-1984 provides guidance to physicians performing medical assessments of personnel required to use respirators (ANSI 1984). ANSI Z88.6 incorporates a medical history, medical examination, and spirometric and exercise stress testing. The medical history is comprehensive and begins with present and previous disease conditions, with emphasis on known cardiovascular and respiratory diseases. It solicits information about psychological problems, including claustrophobia, previous experience with respirator use, usage of medication, deformities or abnormalities that could interfere with respirator use, previous occupations, and tolerance of heated air. Disqualifying conditions lead to one of two outcomes: specific limitations on use or no use under any circumstances.

Disqualifying conditions include: facial deformities and facial hair, use of prescription eyeglasses or contact lenses, hearing impairment, respiratory diseases, cardiovascular diseases, endocrine disorders, neurological disability, medications, and psychological conditions. The impact of facial deformities can be determined through a fit test. Facial hair is a variable quantity that is not easily quantified. Requirements for visual correction and the requirements of the visual task and design of the facepiece determine the practicality of use of contact lenses and eyeglass frames that fit into the facepiece. Persons wearing respirators must be able to hear alarm systems and communications. Respirator use with a perforated eardrum is potentially contraindicated, since the contaminated atmosphere can enter the body through this route. This could be alleviated somewhat through use of nonvented, deeply seated earplugs.

Disability from respiratory disease can be assessed from personal history, chest X-ray and spirometry, as appropriate. Disqualifying cardiovascular diseases include: symptomatic coronary artery disease or arrhythmias, history of recent myocardial infarction, or premature ventricular contraction with elevated pulse rate. Additional potential factors include uncontrolled hypertension or related symptoms and use of blood pressure or cardiovascular medications. Sudden loss of consciousness or response capability shall be grounds for prohibition due to endocrinal disorders. Disqualifying neurological disorders affect the ability to perform coordinated movements or affect response and consciousness.

Disqualifying concerns about medication refer to performance, judgment or reliability deficiencies caused by use of prescription and nonprescription drugs. Disqualifying psychological conditions cause judgment or reliability deficiencies and claustrophobia.

Spirometric testing utilizes the FVC and $FEV_{1\,s}$ as indicators of status for at least incidental respirator use. An FVC less than 80% or $FEV_{1\,s}$ less than 70% of age and gender predictive values is cause for consideration of restriction. For individuals using respirators extensively during heavy or strenuous work, the $MVV_{0.25}$ (maximum voluntary ventilation for 0.25 min or 15 s) is the recommended parameter. The $MVV_{0.25}$ test is performed at a breathing frequency of 40 breaths per minute in a standing position. Tidal volume of each breath should approximate 2/3 of vital capacity and be corrected to body temperature and pressure saturated (BTPS). Maximum tolerable steady-state exercise ventilation is approximately 64% of the $MVV_{0.25}$. Exercise ventilation can be sustained indefinitely at 50% of the $MVV_{0.25}$.

Use of the $MVV_{0.25}$ is an outcome from the work of Raven and colleagues (Raven et al. 1981a). Wilson and Raven (1989) showed that the $MVV_{0.25}$ with a respirator was the best single item predictor of maximal exercise performance, both with and without the respirator. The justification for the single item predictor is that standard pulmonary measures are highly correlated with each other (Weil 1973). The primary effect of respirators is change in normal pulmonary function (Raven et al. 1981b, 1982). These changes were related directly to increased resistance to inspired and expired flow.

Fitness to Work under Hot Conditions

A consideration accompanying the use of respirators, chemical protective clothing, and the combination of both is the thermal load generated in the process.

James et al. (1984) determined that wearing a full-facepiece respirator added significant physiological strain to that caused by heat and workload on unacclimatized male subjects. Increases occurred in heart rate, minute ventilation, oxygen consumption, energy expenditure, and oral temperature. Sweat rate was unaffected. Reduced tolerance to moderate or higher levels of work under hot conditions while wearing a respirator also occurred. The greater reduction in tolerance caused by the full-facepiece respirator was attributed to the greater facial area or dead space enclosed by the facepiece, or both.

Gwosdow et al. (1989) found that respirator air temperature less than 33°C always was comfortable and 100% acceptable to sedentary subjects over a wide range of room and respirator air conditions. Respirator air temperatures above 33°C or higher humidity levels decreased acceptability. Acceptability of the respirator environment decreased as lip temperature increased above 34.5°C, or when respirator dew-point temperature increased above 20°C. Increased respirator air temperature and humidity often made breathing seem "slightly hard." Respirator conditions influenced judgment about the acceptability of the surrounding thermal environment. DuBois et al. (1990) reached similar conclusions for subjects wearing half-facepiece respirators.

The ensemble in many circumstances is complete encapsulation in a barrier formed by a respirator, a chemical protective suit, gloves, and boots. The interfaces between these layers usually are tightly sealed by a layer of duct tape. Enclosing an individual in this manner prevents normal body cooling through evaporation of sweat, convection, and thermal radiation. Also, the individual may be required to wear an SCBA and to perform heavy work at the same time. This combination is highly conducive to causing physical exhaustion and heat-induced injury.

Smolander et al. (1985) determined the physiological cost of wearing an encapsulating, gas-impermeable suit and a self-contained breathing apparatus during simulated repair and rescue tasks in a chemical plant. These authors concluded that typical tasks performed in gas-protective clothing caused marked physiological strain among subjects in average physical condition.

Paull and Rosenthal (1987) determined the impact of wearing chemical-resistant suits (PVC-coated Tyvek®) and full-facepiece respirators equipped with chemical cartridges during tasks of equal workload. Tasks included sampling the contents of wastewater lagoons during summer conditions. This combination imposed a heat stress burden equivalent to 6 to 11°C (11 to 20°F) on the ambient wet bulb globe temperature (WBGT) index. Mean body temperature failed to reach equilibrium during these experiments. This indicates that the body was continuing to store heat. Collapse due to heat exhaustion can occur at deep-body temperatures as low as 38°C (100°F) under conditions of convergence of skin and core temperatures. These authors concluded that significant risk of heat injury is present during work under these conditions.

White and Hodous (1987) examined the impact of different types of protective clothing and SCBAs at workloads ranging up to 60% of maximum. The test environment was thermally neutral. Mean tolerance time at the highest workload was 4 min for firefighter turnout gear and 13 min for a protective ensemble. These authors concluded that the wearing of protective clothing and an SCBA causes significant stress to the cardiorespiratory and thermoregulatory systems, even at low workload. The most stressful response was produced by the heaviest ensemble, the firefighter turnout gear (23 kg total weight).

Dessureault et al. (1995) examined the impact of wearing an SCBA and an encapsulating garment at constant workload under different environmental conditions. Participants wore a Saranex®-coated Tyvek® totally encapsulating suit and rode a bicycle ergometer at a rate of 255 W (220 kcal/h). This corresponds to light work with the body (ACGIH 1994). These authors found that exposure of encapsulated workers to environmental conditions higher than 32.5°C dry bulb and 40°C globe temperature would be very hazardous. Dry bulb temperatures above 30.5°C have a major influence on heat strain. The critical range of temperature between no harm and maximum permissible heat stress level is very narrow.

Standard heat stress indices do not apply when heavy or impermeable protective clothing is being worn (ACGIH 1994). Goldman (1985) noted that a tolerance limit based solely on deep-body temperature or heart rate does not address the problem of convergence between deep core and skin temperatures. Skin temperature normally is several degrees cooler than core temperature. This enables heat loss. Protective clothing interferes with heat loss from the skin, primarily by interfering with evaporative cooling of sweat. As a result, skin temperature increases. Blood has a decreased capacity for transferring heat produced in the core to the skin. Conditions that produce these effects can cause rapid deterioration in performance of work and consequent onset of heat-induced injury.

The National Institute for Occupational Safety and Health (NIOSH) has published a major report on heat-related conditions (1986). NIOSH commented that an environment that is acceptable for acclimatized workers can pose a health threat to the unacclimatized. Heat acclimatization represents a dynamic state of conditioning rather than a permanent change in the physiology of the individual. The proof of heat acclimatization is the ability to work in a hot environment without excessive physiologic heat strain. Acclimatization to one level of heat stress does not ensure acclimatization to a higher level. NIOSH points out that up to 5% of workers cannot adequately acclimatize to heat stress.

NIOSH recommends direct involvement by physicians in efforts to protect workers exposed to potentially hazardous levels of heat stress in the workplace. NIOSH views the role of the physician in preplacement evaluation, supervision during initial days of exposure to the hot environment, and detection of evidence of heat-induced illness.

The preplacement physical examination should include a detailed medical and occupational history. The latter should identify episodes of heat-related disorders and evidence of successful adaptation on previous occasions. This should include prescribed and over-the-counter medications, and use of alcohol and other drugs. Of particular concern are medications and drugs that affect cardiac output, electrolyte balance, renal function, sweating capacity, or function of the autonomic nervous system. These include: diuretics, antihypertensives, sedatives, antispasmodics, anticoagulants, psychotropics, anticholinergics, and drugs that alter thirst or sweating.

Direct evaluation should include a physical examination, clinical chemistry, and blood pressure. Additional detail may be warranted in specific circumstances, including: evidence of heart, circulatory, or lung disease; use of medications; and hypertension. Skin disease or injury can impair sweat production. Obesity is a possible factor.

Kenney et al. (1986) described a simple exercise test for assessing heat tolerance. This test is intended to be a reliable, scientifically based predictor of physiological performance in environments capable of producing high heat stress. In this test, the subject pedals on an ergometer or performs other energy-consuming exercise (bench stepping) at a constant workload of 600 kcal/h in a neutral environment at normal room temperature. Clothing consists of shorts, tee-shirt, socks, sneakers, one-piece vapor-barrier suit (Tyvek®), rubber gloves, and boots. The head is left uncovered. The subject attempts to maintain the work output as long as possible to a maximum of 20 min. At volitional end point or after 20 min, the subject enters the recovery period. The subject continues moving, but under no load. This is intended to prevent pooling of blood in the lower extremities. Heart rate is monitored after 5 min in the recovery mode.

The physiological effectors of thermoregulation are vasodilation, and production and evaporation of eccrine sweat. Vasodilation transfers deep body heat to the skin for dissipation via convection and radiant exchange. In high heat stress conditions, vasodilation occurs at the expense of adequate venous return. To maintain cardiac output, heart rate increases proportionally to decreasing stroke volume. If sweating is eliminated as a viable means of heat dissipation by the vapor-barrier suit, heat load should be reflected by increase in heart rate alone.

Under the conditions of the test, a large portion of the heat load is supported by the peripheral circulation and is measurable as an increase in heart rate. A lower heart rate indicates a greater ability to compensate for the heat load and lower stress on the cardiovascular system.

These investigators noted that most subjects reached a psychological barrier prior to reaching a physiological maximum. The physiological limit was based on rectal temperature of 38.5°C or heart rate of 80% of the age-adjusted maximum ($0.8 \times [220 - \text{age}]$). There was a significant linear relationship between the physiological limit and the heart rate after 5 min of recovery. The physiological limit was unrelated statistically to fitness, age, or body fat.

Heat stress is a difficult concept to manage. In many geographic areas, outdoor temperatures high enough to cause heat stress occur for brief periods. Acclimatization may barely have occurred prior to the appearance of moderating temperatures. The structure in which the work is to occur could be subjected to high thermal load from the sun, while exterior temperatures are moderate.

Interior temperatures and humidities could be considerably higher than those outside the structure. Indoor work in temperature-controlled buildings is considerably more predictable.

Biomechanical Fitness

Many work activities in confined spaces involve lifting and awkward postures. The ergonomics literature and textbooks are replete with concerns about such use of body motions. Yet, the reality in many confined spaces is that work activity occurs under considerably less than ideal biomechanical conditions. Many spaces are not conducive to introduction of equipment to assist in lifting, pushing, pulling, rotating, and other biomechanical motions. Designing the job to fit the person is not a viable option in these situations. The work usually is infrequent and short in duration. Many confined spaces are not designed for occupancy, and the work that must occur is not anticipated in the design. This puts serious practical constraints on the ability of many people to perform the required tasks.

The bases for assessment of an individual are the requirements of the job. The realities of work involving individual confined spaces severely limit the assessment process. Protocols available in the literature focus on *a posteriori* analysis. There is little information available for attempting *a priori* analysis. The closest analogy to work that occurs in confined spaces is construction. Construction presents many ergonomic hazards (Schneider and Susi 1994). Schneider and Susi have detailed qualitatively the ergonomic hazards encountered during construction of a multistoried building. These could provide a basis for understanding the rigors of analogous work that occurs in confined spaces.

The protocol for assessing strength requirements normally begins with observation of the work activity in person or by videotaping to enable division into subtasks (Ayoub et al. 1987, Anderson 1989). Within each subtask are the biomechanical elements of stooping, lifting, holding, lowering, carrying, grasping, twisting, and so on. Within each biomechanical element are quantifiers of range, frequency, duration, weights, and forces. Establishing these parameters is especially important for tasks identified as physically demanding. The next step is to describe tests of strength and stamina that simulate the biomechanical elements identified as physically demanding. Ayoub et al. commented on the difficulty of performing this work. Self-administered questionnaires are unreliable. Supervisors tend to misperceive the actual demands of a task. Verification can be very time-consuming.

Snook provided one of the first approaches to the assessment of lifting tasks (Snook 1978, Snook and Cirello 1991). The premise behind this approach was the belief that individuals can determine loads they can lift, lower, carry, push, and pull throughout the day with minimum risk of injury. Injury in this case means back injury. Demands at other body joints were not assessed. Snook provides tables of maximal acceptable weight vs. percentile of the population and frequency of the motion. These data are strictly statistical and provide no physiological, anatomical, or biomechanical rationale.

The NIOSH lifting guides provide a protocol for assessing lifting tasks (NIOSH 1981, Waters et al. 1993). The guides utilize the terms "action limit" (AL) and "maximum permissible limit" (MPL). Action limit and maximum permissible limit are computed values. The basis for these limits is the compressive force on the L5/S1 (L = lumbar; S = sacral) disc in the lower back. The load that can be handled safely is affected by postural demands, and output and workrate. The guides consider only injury to the back and do not address demands on other body parts. The guides address only limited range of motion.

Another approach for observing whether a method of performing a task is potentially dangerous to the lower back is based on the moment of force (Chaffin and Andersson 1991). The moment of force is the product of the force times the perpendicular distance to a point, axle, or axis. The moment of force, therefore, is the tendency of a force applied to an object to cause it to rotate about a point, axle, or axis. The moment of force is the single most important biomechanical concept underlying human motion. This has direct bearing on how and why tissues respond to external forces. The body must counter moments of force produced by external forces, otherwise rotation will occur in the direction of the imbalance. Moments provide a means of analyzing forces acting on the body and counter forces produced by the tissues of joints and ligaments of the spine to resist them.

A fourth approach to assessing the impact of manual work on the lower back is based on abdominal pressure (Davis and Stubbs 1977). These authors found that forces acting on the lower back are closely related to pressures generated in the abdominal cavity. Significantly higher incidence of back pain occurs in workers whose intraabdominal pressure exceeds 100 mmHg. This value formed the basis of the European approach to limits for safe lifting (Materials Handling Research Unit 1980). The limits incorporate a 10% safety factor in abdominal pressure.

Worker selection for particular jobs by assessing muscular strength and mobility and aerobic capacity has considerable merit (Kroemer 1989, Dukes-Dobos 1989). Chaffin et al. have shown that job demands that exceeded isometric strength were a factor in the occurrence of musculoskeletal injuries (NIOSH 1977). Workers whose job demands exceeded their isometric strength had more severe and three times as many injuries as in the converse situation. Chaffin and Andersson (1991) provide information about the range of measuring tools used in biomechanical assessments.

Static techniques require the individual to exert isometric strength against an external measuring instrument. The protocol of Caldwell et al. (1974) forms the basis for many researchers. The strong points offered by the isometric (static) techniques are simplicity and ease of measurement and interpretation. The same factors, however, also are the limitations of this approach, since predictive power is low unless the tasks evaluate whole-body response (Anderson 1989, Kroemer 1989). Whole-body tests evaluate the ability of the individual to perform the task using whatever posture is most suitable. This loads the entire musculoskeletal system and not isolated muscle groups. Local muscle testing in constrained postures is particularly problematic for individuals who may be able to perform the task in a posture that compensates for some handicap.

Dynamic techniques are considered more relevant to materials handling problems (Kroemer 1989). These involve moving limbs or trunk against imposed torques or lifting or moving weights between fixed points. Dynamic techniques offer the potential to mimic actual situations, and the opportunity for continuous monitoring of torque and exertion. The problems with dynamic techniques include the complexity of the analysis and mimicking the conditions and dynamics of actual motions. These techniques assess only lifting or lowering ability.

Previous discussion has considered *a posteriori* approaches to assessment of biomechanical fitness. *A priori* approaches are limited to making analogies to types of work for which the results of assessment are available. In some cases, these assessments have led to development of protocols for testing the capabilities of individuals. One example is the protocol used to screen applicants to the fire, hazmat, and rescue services. Some types of work activity performed in confined spaces are similar to that performed by these groups. However, work performed by fire, hazmat, and rescue services likely represents an extreme of the type of work normally performed in confined spaces.

Many fire departments require applicants to pass medical examinations and physical test protocols. NFPA (National Fire Protection Association) 1001 specifies minimum performance requirements for entry into the fire service (NFPA 1992). Similar screening tests exist for police and ambulance personnel, as well as military recruits. The physical test protocols involve lifting, pulling, pushing, rotating, and other biomechanical motions. These involve work against predetermined loads and the clock.

These tests are not without controversy. In order to be defensible against challenges from human rights groups, they must reflect the rigors of the position. Some of these protocols may provide the basis for screening tests for work activity in confined spaces. In considering these protocols for use in screening individuals for work in confined spaces, several caveats must be highlighted. First, test protocols used in the fire, hazmat, and rescue services reflect one extreme of the demands of normal work in confined spaces. Also, these protocols is used to screen applicants for career service, not persons already employed. Many persons who enter confined spaces are highly skilled trades and technical people. Their training has screened them for performing this work. Overly rigid screening tests, while intending to reduce the potential for injury during this work, could exclude the individuals who will perform it.

EXPOSURE ASSESSMENT

The third major role of the industrial physician in work involving confined spaces is assessment of exposure to intoxicants during work activity. This could occur as a routine assessment or assessment of accidental exposure. The purpose for routine assessment is to ensure the protection offered by protective equipment and procedures, as well as compliance in their use.

Some substances exhibit complex routes of entry. The second part of this role is assessment of accidental exposures. In addition to inhalation, passage across the skin following contact with sprays or mists or by splashes or immersion can occur with some substances. Ingestion also is possible in some circumstances. Accidental exposure could result from failure of protective measures or failure to use protective measures. The possibility of exposure to substances not anticipated during the hazard assessment for which existing protective measures are inadequate also exists.

Assessment of exposure by monitoring of biological fluids and excreta is an after-the-fact measure. Medical surveillance cannot undo the situation that led to the need to obtain the sample. This situation could have involved an accidental overexposure. Medical surveillance can only indicate severity of the exposure and possible damage. In this respect, medical monitoring shows its weakest side. This activity should be followed only if it yields useful information.

Medical surveillance can provide information useful for treating an overexposure. For example, carbon monoxide in exhaled breath or carboxyhemoglobin in blood can provide important information regarding treatment following exposure to exhaust gases from engines. It is important to bear in mind that these episodes can occur just as easily in remote field locations as in large cities. In the former case, sophisticated equipment for assessing exposure may not be available, yet rapid response is required.

The question about information is important. Individuals may seek assistance following a suspected exposure. Suspicion of exposure may occur following detection of odor. Detection of odor carries no health significance. Individuals with a high level of distrust may seek assistance privately through a family physician if thwarted in a request to a physician employed through management. A family physician may have had no training in this area and may sample inappropriately and have no basis on which to interpret the results.

Main sources of information for consideration about medical monitoring include the MSDS and Documentation of the Threshold Limit Values and Biological Limit Values (ACGIH 1991). These resources provide information about the immediate, intermediate, and delayed effects of acute exposure. This information in itself may be sufficient to allay concerns about a presumed (over) exposure.

Medical surveillance is based on toxicokinetic and toxicodynamic principles (Droz 1989). Toxicokinetics is the verbal and mathematical description of what the body does to the chemical. Toxicodynamics is the verbal and mathematical description of what the chemical does to the body. The determinant is the substance that is measured. The determinant can be the original substance to which exposure occurred or a metabolite or a reversible biochemical change induced by the substance. The determinant should be unique to the substance under consideration, otherwise exposure attributed to one substance could have occurred because of another or a combination of both. Based on the determinant, the biological specimen chosen, and the time of sampling, the measurement will indicate either the intensity of a recent exposure, a daily average, or a chronic cumulative exposure. Hence, an isolated sample taken without regard to these parameters can have little (if any) utility in establishing the nature of an exposure.

The situation that prompts the request for surveillance could follow multiple exposures over a period of time or a single exposure. Models for toxicokinetics consider the substance to enter and migrate between "compartments" (Rowland and Tozer 1989, Klaassen and Rozman 1991). Compartments can include benign organs, target organs, and storage. Storage can occur in plasma proteins, liver, kidney, fat, and bone. Toxicokinetic behavior can be simple or complex, depending on the toxicological behavior of the substance. The latter, therefore, influences the nature of the protocol for biological monitoring. Biological monitoring can utilize body fluids, excreta, and exhaled breath as the matrix on which to perform analytical tests. Blood, urine, and exhaled breath are the most common. The significance of the information provided by a sample is hindered by the complexities of this subject (Droz 1989). These include: physiological and health status, environmental and dietary sources, and life style sources.

Some determinants do not accumulate in the body. Sampling is critical with regard to exposure and postexposure periods. Some determinants accumulate during a period of exposure and are eliminated during periods of nonexposure, such as weekends and vacation. In these cases, sampling must consider previous exposure and accumulation and the new exposure and accumulation. For some determinants, timing of the sampling is not critical, since accumulation and elimination occur over a prolonged period. The documentation provided with the biological exposure indices (BEIs) specifies sampling times relative to exposure (ACGIH 1991). NIOSH also provides recommendations for medical monitoring of exposures, but provides no information about the time frame (NIOSH/OSHA/USCG/EPA 1985).

A major problem with these tests is the interpretation of the results. ACGIH points out that variability associated with the human condition and individual response could lead to a value that exceeds the BEI. In the view of ACGIH, an exceedence does not necessarily signify an increased health risk (ACGIH 1991). Interpretation of these data is a serious issue (Cooper and Zavon 1994). Bland reassurances to individuals who have received exceedences are unlikely to be met with acceptance. These easily could lead to refusals to work or shutdown of the site through wildcat strike. Part of the response to this problem is baseline and follow-up monitoring. Of course, the only time to perform baseline monitoring is **prior** to the start of the work.

The opposite problem is hypersusceptibility. A hypersusceptible individual is unlikely to be reassured by a result that is within the normal range when difficulty is being experienced. Hypersusceptibility usually is a condition for exclusion (Cooper and Zavon 1994).

SUMMARY

Successful programs in occupational health and safety represent a team effort between medical professionals and industrial hygienists and safety professionals. The latter evaluate conditions in workplaces and provide firsthand input to the occupational physician. This input is extremely valuable, since the physician often does not have the opportunity to visit the workplace, yet bears important responsibility for decisions about worker health.

Occupational medical programs for work involving confined spaces include the following main outcomes: planning and preparing for emergency situations, assessment of health status and fitness to use respiratory and other protective equipment under the conditions of work, and surveillance to determine the consequences of exposure to hazardous conditions.

Hazardous conditions that occur in confined spaces can produce traumatic injury and symptoms of acute overexposure to chemical substances and physical agents. The confining geometry in many spaces and inability to use engineered solutions for handling materials and equipment put considerable stress on human capabilities. These include: lifting, lowering, holding, carrying, pushing, pulling, crouching, bending, and twisting, as well as other motions. Persons assessed as fit to work in confined spaces should not pose a threat to themselves or to fellow workers.

REFERENCES

American Conference of Governmental Industrial Hygienists: *Documentation of the Threshold Limit Values and Biological Exposure Indices*, 6th ed. Cincinnati, OH: American Conference of Governmental Industrial Hygienists, 1991.

American Conference of Governmental Industrial Hygienists: *1994–1995 Threshold Limit Values for Chemical Substances and Physical Agents and Biological Exposure Indices.* Cincinnati, OH: American Conference of Governmental Industrial Hygienists, 1994. 119 pp.

American National Standards Institute: *American National Standard for Respiratory Protection — Respirator Use — Physical Qualifications for Personnel* (ANSI Z88.6-1984). New York: American National Standards Institute, 1984. 15 pp.

American National Standards Institute: *Practices for Respiratory Protection* (ANSI Z88.2-1992). New York: ANSI, 1992.

American Thoracic Society (Medical Section of the American Lung Association): Standardization of Spirometry — 1987 Update. *Am. Rev. Respir. Dis. 136:* 1285–1298 (1987).

Anderson, C.K.: Strength and endurance testing for preemployment placement. In *Manual Material Handling: Understanding and Preventing Back Trauma.* Akron, OH: American Industrial Hygiene Association, 1989. pp. 73–78.

Ayoub, M.M., B.C. Jiang, J.L. Smith, J.L. Selan, and J.W. McDaniel: Establishing a physical criterion for assigning personnel to U.S. Air Force jobs. *Am. Ind. Hyg. Assoc. J. 48:* 464–470 (1987).

Beckett and Billings: Individual factors in the choice of respiratory protective devices. *Am. Ind. Hyg. Assoc. J. 46:* 274–276 (1985).

Blais, B.R.: Contact lenses in the chemical industry. *J. Occup. Med. 36:* 706–707 (1994). [Letter to the Editor]

Caldwell, L.S., D.B. Chaffin, F.N. Dukes-Dubot, K.H.E. Kroemer, L.L. Laubach, S.H. Snook, and D.E. Wasserman: Proposed standard procedure for static muscle strength testing. *Am. Ind. Hyg. Assoc. J. 35:* 201–206 (1974).

Ceretelli, P., R.S. Sikand, and L.E. Farhi: Effects of increased airways resistance during steady-state exercise. *J. Appl. Physiol. 35:* 361–366 (1973).

Chaffin, D.B. and G.B.J. Andersson: *Occupational Biomechanics*, 2nd ed. New York: John Wiley & Sons, 1991. 518 pp.

Comroe J.H., Jr., R.E. Forster II, A.B. DuBois, W.A. Briscoe, and E. Carlsen: *The Lung, Clinical Physiology and Pulmonary Function Tests*, 2nd ed. Chicago: Year Book Medical Publishers, 1962. 390 pp.

Cooper, W.C. and M.R. Zavon: Health surveillance programs in industry. In *Patty's Industrial Hygiene*, 3rd ed., vol. 3, Part A, *Theory and Rationale of Industrial Hygiene Practice: The Work Environment*, Harris, R.L., L.J. Cralley, and L.V Cralley (Eds.). New York: John Wiley & Sons, 1994. pp. 605–626.

Cotes, J.E.: Lung function tests. In *Encyclopaedia of Occupational Health and Safety*, 3rd. rev. ed., vol. 2, L–Z, Parmeggiani, L. (Ed.). Geneva: International Labor Organisation. 1983. pp. 1250–1256.

Courau, P.J.: Older workers. In *Encyclopedia of Occupational Health and Safety*, 3rd rev. ed., vol. 2, Parmeggiani, L (Ed.). Geneva: International Labor Organisation, 1983. pp. 1565–1570.

Cullen, A.P.: Contact lenses in the work environment. In *Environmental Vision: Interactions of the Eye, Vision and the Environment*, Pitts, D.G. and R.N. Kleinstein (Eds.). Boston: Butterworth-Heinemann, 1993. pp. 315–329.

Cullen, A.P.: Contact lenses in the workplace. In *Eye Injury Prevention in Industry*, 2nd ed., McRae, E. and M. Grimm (Eds.). Edmonton, AB: Alberta Labor, Occupational Health and Safety Division, 1994. pp. 26–29.

Davis, P.R. and D.A. Stubbs: Safe levels of manual force for young males, 1. *Appl. Ergonom. 8:* 141–150 (1977).

Deno, N.S., E. Kamon, and D.M. Kiser: Physiological responses to resistance breathing during short and prolonged exercise. *Am. Ind. Hyg. Assoc. J. 42:* 616–622 (1981).

Dessureault, P.C., R.B. Konzen, N.C. Ellis, and D. Imbeau: Heat strain assessment for workers using an encapsulating garment and a self-contained breathing apparatus. *Appl. Occup. Environ. Hyg. 10:* 200–208 (1995).

Dinman, B.D.: Medical aspects of the occupational environment. In *The Industrial Environment — Its Evaluation & Control* (DHHS/PHS/CDC/NIOSH). Washington, D.C.: Government Printing Office, 1973. pp. 197–205.

Droz, P.O.: Biological monitoring. I: Sources of variability in human response to chemical exposure. *Appl. Indust. Hyg. 4:* F-21 to F-24 (1989).

DuBois, A.B., Z.F. Harb, and S.H. Fox: Thermal discomfort of respiratory protective devices. *Am. Ind. Hyg. Assoc. J. 51:* 550–554 (1990).

Dukes-Dobot, F.N.: The physician's role as a team member in preventing low back injury. In *Manual Material Handling: Understanding and Preventing Back Trauma.* Akron, OH: American Industrial Hygiene Association, 1989. pp. 51–54.

Environmental Protection Agency: Standard Operating Safety Guides (Pub. 9285.1-03). Washington, D.C.: U.S. Environmental Protection Agency/Office of Emergency and Remedial Response, 1992. pp. 78–89.

Equal Employment Opportunity Commission: A Technical Assistance Manual on the Employment Provisions (Title I) of the Americans With Disabilities Act 1990. Washington, D.C.: U.S. Government Printing Office, 1992.

Fairbanks, V.F.: Evaluation of erythrocytes. In *Internal Medicine*, 3rd ed., Stein, J.H. (Ed.). Boston: Little, Brown and Company, 1990. pp. 977–983.

Forland, M.: Urinalysis. In *Internal Medicine*, 3rd ed., Stein, J.H. (Ed.). Boston: Little, Brown and Company, 1990. pp. 762–768.

George, J.N.: Evaluation of hemostasis and thrombosis. In *Internal Medicine*, 3rd ed., Stein, J.H. (Ed.). Boston: Little, Brown and Company, 1990. pp. 992–996.

Goldman, R.F.: Heat stress in industrial protective encapsulating garments. In *Protecting Personnel at Hazardous Waste Sites*, Levine, S.P. and W.F. Martin (Eds.). Boston: Butterworth-Ann Arbor Science, 1985. pp. 215–266.

Gottfried, S., M. Altose, S. Kelsen, and N. Cherniack: Perception of changes in airflow resistance in obstructive pulmonary disorders. *Am. Rev. Respir. Dis. 124:* 566–570 (1981).

Gwosdow, A.R., R. Nielsen, L.G. Berglund, A.B. duBois, and P.G. Tremml: Effect of thermal conditions on the acceptability of respiratory protective devices on humans at rest. *Am. Ind. Hyg. Assoc. J. 50:* 188–195 (1989).

Harber, P., J. Tamimie, A. Bhattacharya, and M. Barber: Physiologic effect of respirator dead space and resistance loading. *J. Occup. Med. 24:* 681–684 (1982).

Hodous, T., L. Petsonk, C. Boyles, J. Hankinson, and H. Amandus: Effects of added resistance to breathing during exercise in obstructive lung disease. *Am. Rev. Respir. Dis. 128:* 943–948 (1983).

Hutton, J.J.: Evaluation of leukocytes. In *Internal Medicine*, 3rd ed., Stein, J.H. (Ed.). Boston: Little, Brown and Company, 1990. pp. 983–986.

International Labor Office: *Labor Force Estimates and Projections, 1950–2000. Vol. V: World Summary,* 2nd ed., Geneva: International Labor Office, 1977. 101 pp.

James, R., F. Dukes-Dobos, and R. Smith: Effects of respirators under heat/work conditions. *Am. Ind. Hyg. Assoc. J. 45:* 399–404 (1984).

Johnson, A.T., C.R. Dooly, C.A. Blanchard, and E.Y. Brown: Influence of anxiety level on work performance with and without a respirator mask. *Am. Ind. Hyg. Assoc. J. 56:* 858–865 (1995).

Kaplan, M.M.: Evaluation of hepatobiliary diseases. In *Internal Medicine*, 3rd ed., Stein, J.H. (Ed.). Boston: Little, Brown and Company, 1990. pp. 439–446.

Kaufman, J.E. and J.F. Christensen (Eds.): *IES Lighting Handbook, Reference Volume,* 6th ed. New York: Illuminating Engineering Society of North America, 1981. pp. 3-1 to 3-27.

Kenney, W.L., D.A. Lewis, R.K. Anderson, and E. Kamon: A simple exercise test for the prediction of relative heat tolerance. *Am. Ind. Hyg. J. 47:* 203–206 (1986).

Klaassen, C.D. and K. Rozman: Absorption, distribution, and excretion of toxicants. In *Casarett and Doull's Toxicology, The Basic Science of Poisons*, 4th ed., Amdur, M.O., J. Doull, and C.D. Klaassen (Eds.). New York: Pergamon Press, 1991. pp. 50–87.

Kroemer, K.H.E.: Models, methods and techniques for personnel testing and selection. In *Manual Material Handling: Understanding and Preventing Back Trauma.* Akron, OH: American Industrial Hygiene Association, 1989. pp. 65–71.

Landers, D.M.: The arousal-performance relationship revisited. *Res. Quart. Exercise Sport 51:* 77–90 (1980).

Lawrence Livermore National Laboratory: *Is It Safe to Wear Contact Lenses With a Full Facepiece Respirator?,* by da Rosa, R.A. and C.S. Weaver (UCRL-53653). Livermore, CA: Lawrence Livermore National Laboratory, 1986.

Materials Handling Research Unit: *Force Limits in Manual Work.* Guildford, U.K.: IPC Science and Technology Press, 1980. 25 pp.

Morgan, W.P.: Psychological problems associated with the wearing of industrial respirators: a review. *Am. Ind. Hyg. Assoc. J. 44:* 671–676 (1983a).

Morgan, W.P.: Hyperventilation syndrome: a review. *Am. Ind. Hyg. Assoc. J. 44:* 685–689 (1983b).

Morgan, W.P. and P.B. Raven: Prediction of distress for individuals wearing industrial respirators. *Am. Ind. Hyg. Assoc. J. 46:* 363–368 (1985).

National Fire Protection Association: *NFPA 1001: Fire Fighting Professional Qualifications* (1992 ed.). Quincy, MA: National Fire Protection Association, 1992. 20 pp.

National Institute for Occupational Safety and Health: Occupational Exposure to Noise. Criteria for a Recommended Standard (DHEW (NIOSH) Pub. No. 73-11001). Cincinnati, OH: DHEW/PHS/CEC/NIOSH, 1972.

National Institute for Occupational Safety and Health: Preemployment Strength Testing, by Chaffin, D.B., G.D. Herrin, W.M. Keyserling, and J.A. Foulke (DHEW(NIOSH) Pub. No. 77-163). Cincinnati, OH: DHHS/PHS/CDC/NIOSH, 1977.

National Institute for Occupational Safety and Health: Work Practices Guide for Manual Lifting (DHHS(NIOSH) Pub. No. 81-122). Cincinnati, OH: DHHS/PHS/CDC/NIOSH, 1981. 91 pp.

National Institute for Occupational Safety and Health, Occupational Safety and Health Administration, U.S. Coast Guard, U.S. Environmental Protection Agency: Occupational Safety and Health Guidance Manual for Hazardous Waste Site Activities (DHHS(NIOSH) Pub. No. 85-115). Cincinnati, OH: DHHS/PHS/CDC/NIOSH, 1985. pp. 5-1 to 5-10.

National Institute for Occupational Safety and Health: Criteria for a Recommended Standard.... Occupational Exposure to Hot Environments, Revised Criteria 1986 (DHHS(NIOSH) Pub. No. 86-113). Cincinnati, OH: DHHS/PHS/CDC/NIOSH, 1986. 140 pp.

Nejmeh, G., Jr.: Contact lens, to keep them in or to keep them out, on the job — that is the question! *Natl. Safety News:* June, 1982. pp. 58–61.

Occupational Safety and Health Administration: "Contact Lenses Used With Respirators (29 CFR 1910.34 (e) (5) (ii))," by T. Shepich through L. Carey. Washington, D.C.: Occupational Safety and Health Administration, February 8, 1988. [Memo]

Paull, J.M. and Rosenthal, F.S.: Heat Strain and heat stress for workers wearing protective suits at a hazardous waste site. *Am. Ind. Hyg. Assoc. J. 48:* 458–463 (1987).

Randolph, S.A. and M.R. Zavon: Guidelines for contact lens use in industry. *J. Occup. Med. 29:* 237–242 (1987).

Raven, P.B., A.T. Dodson, and T.O. Davis: The physiological consequences of wearing industrial respirators. *Am. Ind. Hyg. Assoc. J. 40:* 517–534 (1979).

Raven, P.B., R.F. Moss, K. Page, R. Garmon, and B. Skaggs: Clinical pulmonary function and industrial respirator wear. *Am. Ind. Hyg. Assoc. J. 42:* 897–903 (1981a).

Raven, P.B., A.W. Jackson, K. Page, R.F. Moss, O. Bradley, and B. Skaggs: The physiological response of mild pulmonary impaired subjects while using a "demand" respirator during rest and work. *Am. Ind. Hyg. Assoc. J. 42:* 247–257 (1981b).

Raven, P.B., O. Bradley, D. Rohm-Young, F.L. McClure, and B. Skaggs: Physiological response to "pressure-demand" respirator wear. *Am. Ind. Hyg. Assoc. J. 43:* 773–781 (1982).

Rengstorff, R.H. and C.J. Black: Eye protection from contact lenses. *J. Am. Optom. Assoc. 45:* 270–276 (1974).

Rosenstock, R.: Contact lenses: are they safe in industry? *Prof. Safety. 31:* January, 1986. pp. 18–21.

Rowland, M. and T.N. Tozer: *Clinical Pharmacokinetics: Concepts and Applications*, 2nd ed. Philadelphia: Lea and Febiger, 1989. 541 pp.

Schneider, S. and P. Susi: Ergonomics and construction: a review of potential hazards in new construction. *Am. Ind. Hyg. Assoc. J. 55:* 635–649 (1994).

Shepard, R.J.: Ergonomics of the respirator. In *Design and Use of Respirators*, Davies, C.N. (Ed.). New York: Macmillan, 1962. pp. 51–66.

Shigeoka, J.W.: Pulmonary function testing. In *Environmental and Occupational Medicine.* Boston: Little, Brown and Company, 1983. pp. 99–112.

Smolander, J., V. Louhevaara, and O. Korhonen: Physiological strain in work with gas protective clothing at low ambient temperature. *Am. Ind. Hyg. Assoc. J. 46:* 720–723 (1985).

Snook, S.H.: The design of manual handling tasks. *Ergonometrics 21:* 963–985 (1978).

Snook, S.H. and V.M Ciriello: The design of manual handling tasks: revised tables of maximum acceptable weights and forces. *Ergonometrics 34:* 1197–1213 (1991).

Spielberger, C.D., R.L. Gorsuch, and R.E. Lushene: *The State-Trait Anxiety Inventory Manual.* Palo Alto, CA: Consulting Psychologist Press, 1983.

Stein, H.A. and B.J. Slatt: *Contact Lenses in Industry* (Ref. No. B01732). Toronto, ON: Industrial Accident Prevention Association of Ontario, 1984. 18 pp.

Tack, M., M.D. Altose, and N. Cherniack: Effect of aging on the perception of resistive ventilatory loads.*Am. Rev. Respir. Dis. 126:* 463–467 (1982).

U.S. Department of Energy: "Amendment of the Occupational Safety and Health Administration (OSHA) Prohibition on Wearing Contact Lenses in Contaminated Atmospheres With Full Facepiece Respirators," by M.L. Walker. Washington, D.C.: Department of Energy, September 23, 1986. [Memo]

Waters, T.R., V. Putz-Anderson, A. Garg, and L.J. Fine: Revised NIOSH equation for the design and evaluation of manual lifting tasks. *Ergonometrics 36:* 749–776 (1993).

Weil, H.: Pulmonary function testing in industry. *J. Occup. Med. 15:* 693–699 (1973).

White, M.K. and T.K. Hodous: Reduced work tolerance associated with wearing protective clothing and respirators. *Am. Ind. Hyg. Assoc. J. 48:* 304–310 (1987).

Wilson, J.R. and P.B. Raven: Clinical pulmonary function tests as predictors of work performance during respirator wear. *Am. Ind. Hyg. Assoc. J. 50:* 51–57 (1989).

Zenz, C.: The occupational physician. In *Fundamentals of Industrial Hygiene,* 3rd ed., Plog, B.A. (Ed.). Chicago, IL: National Safety Council, 1988. pp. 607–636.

Zerbe, L.B. and H.W. Hofstetter: Prevalence of 20/20 with best previous and no lens correction.*J. Am. Optical Assoc. 29:* 772 (1957–1958).

15 Emergency Preparedness and Response

CONTENTS

INTRODUCTION

Most of the material written about confined spaces has concerned itself with descriptions of accidents. Descriptions of accidents form the theme of entire reports produced by NIOSH, OSHA, and MSHA (NIOSH 1979, 1994; OSHA 1982a, 1982b, 1983, 1985, 1988, 1990; MSHA 1988). In many accidents involving atmospheric hazards, the tragedy occurred because of the actions of would-be rescuers. Fellow workers or supervisory or management personnel entered the confined space to retrieve the stricken worker. The manner in which this occurred indicated lack of prior consideration about the consequences of the action to the would-be rescuer. These inappropriate actions led to further casualties that often included the would-be rescuer, as well as the initial victim. In some circumstances, the would-be rescuer was the only victim. These situations provide ample evidence to indicate that lack or inadequacy of preparation for emergencies and inappropriateness or inadequacy of response were major factors in the outcome.

The first part of this book focused on an examination of the dynamics and intrinsic elements of accidents that occur in confined spaces. Understanding as much as is possible about these events is essential to enabling the synthesis of strategies for prevention. The second part of this book systematically focuses on assessing and eliminating, or at least controlling, hazardous conditions that could be present in confined spaces during work activity. The outcome from the hazard assessment is the identification of hazardous conditions that could persist despite implementation of precautionary measures. Residual hazardous conditions could be intrinsic to the space or could be caused by the work activity. Residual hazardous conditions, therefore, are potential causes of injury to persons who enter and work in the space.

This chapter focuses on emergency preparedness and emergency response. Emergency preparedness describes the process of determining actions that enable response to occur in a competent manner during an emergency situation. Emergency response describes the actions taken to mitigate the effects of the emergency once in progress.

EMERGENCY PREPAREDNESS

The adequacy of response to accident situations is a direct reflection of the preparation made for them. Emergency preparedness seems straightforward enough when approached in the literal sense; however, experience time and again has shown that this is a complex undertaking. The shortcomings of this endeavor often become apparent only following an event. More fortuitously, they become apparent during a drill to simulate a realistic situation when consequences of the shortcoming are minimal.

The first consideration in preparing for accidents that could occur in a confined space is to recognize the multidimensional nature of the problem. Emergency preparedness involves more than technical considerations. Accident situations are complex events. Accidents provoke response from agencies and parties affected either directly or indirectly by the situation. The situation potentially exposes the organization to inquiry by outside agencies. Interested and affected parties can include:

- The victim(s) and family(ies)
- Persons involved in the activity at the time of the accident
- Casual bystanders
- Supervisors; technical and management personnel
- Emergency response organizations (police, fire, medical services)
- The news media
- Government agencies

Each of the preceding groups will view the accident situation from a particular perspective. While the nature of activity directed toward assisting the victim is completely technical, the importance of the perceptual aspects of the situation cannot be stressed strongly enough.

As the owner of the equipment/structure that is involved in the emergency, the organization has the responsibility of responding in a manner that is perceived as being, as well as is, technically proficient and competent. In order to satisfy these requirements, the response requires tangible evidence of prior planning, and preparation.

Needs Assessment

Emergency preparedness can cover very broad situations and circumstances. The process begins with a needs assessment. That is, what are the needs that the preparation must address? Unless the needs are understood clearly, the rest of the process could become somewhat haphazard at best. At worst, the organization could suffer a fatality because of lack of preparation to respond appropriately. This application of needs assessment focuses on the nature of accidents that can occur in confined spaces. As indicated previously, confined spaces can occur anywhere, and can have any size and configuration. The nature and extent of preparation by an organization should reflect these considerations.

Post-Accident Conditions

The major reason for which emergency response at a confined space is required is to mitigate injuries that have occurred as the result of an accident. The circumstances for which an emergency response could be required could be almost infinite, as the nature of possible combinations of work activity and hazardous conditions involving confined spaces would indicate. Simplifying these possibilities to workable limits is the rationale for the concept of control that forms the basis of the second part of this book. The concept of control rests on establishing and responding to the following concerns:

- What hazardous conditions can exist or can develop in the confined space?
- What must be done to eliminate or at least control these conditions?
- That having been accomplished, what still can happen during an accident situation?
- What injuries are likely to occur during an accident situation?
- What preparations are necessary to minimize the impact of these accidents and injuries?

The emergency response section of the hazard assessment in Chapter 8 provides the starting point for considering what can cause an accident situation. Recognition and anticipation are the basis for the planning and preparation needed to respond to these situations.

Post-accident conditions in the space have a controlling impact on the nature of the response. These affect both the speed and effectiveness of actions taken to mitigate the situation. Delay in attending to an injured victim is a critical factor in loss of life following traumatic injury.

The most desirable situation is no hazardous conditions. This means that atmospheric and other types of hazardous conditions have been eliminated or at least controlled during preentry preparation, and that the work activity does not generate uncontrolled hazardous conditions. The victim can breathe the atmosphere in the space without difficulty and the emergency service provider can enter without concern for respiratory protection. The main focus of the emergency response then becomes addressing the needs of the injured person.

Hazardous conditions could exist in the space post-accident. These may have existed prior to the accident or may have been created by it. This situation considerably complicates and delays the process of attending to the victim. Entry cannot occur until the situation has been assessed. The cause of the hazardous condition must be determined and corrected before the emergency service provider can enter the space — either that, or the emergency service provider must don high-level protective equipment in anticipation of hazardous conditions that could be present. Another option is to extricate the victim, where possible.

The need to control post-accident hazardous conditions shifts emphasis away from attending to the needs of the victim to the safety of the emergency service provider. Delay in attending to an injured victim decreases the probability for survival. On the other hand, rapid extrication of an injured person can exacerbate the existing injury and can cause further injury. This, in itself, could decrease the chance of survival. Extrication, as currently depicted, shows the subject being lifted from a manhole in the head-bent-forward position. This posture, coupled with the uncontrolled motion caused by movement around obstacles in the space, could subject the victim of a neck/spinal injury to further injury, including possible permanent paralysis. Similarly, the OSHA and NIOSH reports amply illustrate the consequences of entry into a hazardous space by an unprotected emergency service provider (OSHA 1985; NIOSH 1994).

In some circumstances, the work occurs in hazardous conditions immediately dangerous to life and health. This term usually refers to atmospheric conditions. However, many other hazardous conditions are possible. The emergency service provider must wear high-level personal protective equipment in order to enter and work in the space. This could include both atmosphere-supplying respiratory protection and an encapsulating suit. An accident that occurs under these conditions could be extremely difficult to mitigate. First, the victim also would be wearing high-level protective equipment. This could obscure the cause of injury and render diagnosis very difficult. Preparing the victim for extrication could be very difficult, as would be the process of extrication because of the need for use of high-level protective equipment at all times. This situation should be avoided, if at all possible.

Accidents sometimes occur following unsanctioned entry into confined spaces. The accident situations detailed in the OSHA, MSHA, and NIOSH reports indicate that unsanctioned entry constitutes a recognizable concern (OSHA 1982b, 1985; MSHA 1988; NIOSH 1994). Preparation for accidents involving these episodes takes even the best-prepared organization back to square one. Minimizing the occurrence of these episodes is absolutely essential. Education and training and worksite security provide the best starting-point in this regard.

Assessment of Location

The geographic location where the work occurs has a major influence on the considerations that are required to formulate an emergency response. Locations can range from central city to remote locales. The needs for preparation differ by location.

Central city locations usually are close to specialized hospitals. A specialized hospital can provide treatment for extensive burns, shock, orthopedic injury, and so on. One possible problem with using a central city hospital as an emergency treatment center is traffic. Central city traffic can become "gridlocked" during certain times of the day. Valuable time could be lost in this manner. This question should be investigated with representatives from the hospital and local ambulance services.

Suburban city locations usually are close to general hospitals. These hospitals usually are large enough to provide adequate emergency treatment. As necessary, a patient could be transferred to a specialized hospital, once stabilized.

Small city locations often are close to small general hospitals. A small town may have its own hospital, but a trip to a small city hospital could be required. The question about small hospitals is whether they operate at all times, and whether they are equipped to handle trauma cases.

Urban and remote locations require long trips to small or large city hospitals. Locations in these locales pose accessibility problems depending on weather and condition of roads. The most rapid means of transporting the patient to hospital likely will be airlift by helicopter or fixed-wing aircraft. These modes of transport and use of hospitals under these circumstances require prior arrangement.

Traffic and industrial activity around the site could be additional factors in evacuation of a victim to hospital by road. For example, a site isolated from transport routes by a busy railway

TABLE 15.1
Summary of Occurrence of Disabling Work Injuries

Type of accident	Occurrence %
Overexertion	31.3
Struck by or against	24.0
Falls	17.1
Bodily reaction	7.6
Caught in, between, or under	5.2
Contact with radiation, caustics, etc.	3.1
Rubbed or abraded	2.0
Contact with temperature extremes	2.0
Other (nonclassifiable)	4.6

After National Safety Council 1993.

track where train marshaling occurs requires special attention. Prior arrangement with the railway company and emergency service providers is essential to ensure that passage of emergency vehicles can occur on demand. This situation also could occur on roads containing a bridge that raises to permit passage of ships or where extensive construction has reduced passage to a single lane.

Assessment of Potential Injuries

An endeavor that could be beneficial to this process is to determine the nature of injuries that could be expected. Statistics on nonfatal accidents in confined spaces are not available. Sources of information on general workplace accidents provide a valuable starting point for consideration. If preparation has been thorough, these injuries likely would be the same as would occur in any workspace during the same activity. This question must be considered carefully, because the confining aspects of the structure and its internal geometry could markedly increase the severity of the accident.

Disabling injuries arising from workplace accidents are tabulated according to specific categories. Table 15.1 provides the type of accident by category in decreasing order of occurrence. "Overexertion," "struck by or against," and "falls" were by far the most common categories of accident across industrial categories. Motor vehicle accidents (3.1%) were not included in the preceding list as these are unlikely to have a role in confined space accidents.

Table 15.2 provides information about affected body parts. Possible additional sources of information are accident statistics compiled by an organization or trade group, workers' compensation boards, or insurance companies. The latter may make available statistics about rate groupings. Together these can provide a picture about what could be expected in a specific situation. The following sections provide examples of the depth of information that is obtainable on specific body parts.

Eye Injuries

The statistics in Table 15.2 indicate that certain types of accidents, for example, involving the eye and, head occur in low frequency. Despite the low occurrence indicated by the statistics, eye and head injuries can cause highly significant outcomes. Sight is extremely important for performance of most workplace tasks. Loss or impairment could have devastating consequences for an individual.

Chapter 12 provides information on respiratory and other personal protective equipment. The intent of personal protective equipment, of course, is to prevent occurrence of the injuries mentioned in Table 15.2. However, these can and still do occur, despite the use of protective eyewear and faceshields.

TABLE 15.2
Summary of Occurrence of Work Injuries by Body Part

Body part	Occurrence %
Eyes	4
Head (excluding eyes)	4
Neck	2
Arms	12
Hands	5
Fingers	11
Back	24
Trunk	11
Legs	13
Feet	4
Toes	1
Body systems	3
Multiple	10

After National Safety Council 1993.

TABLE 15.3
Summary of Occurrence of Disabling Eye Injuries by Type

Type of injury	Occurrence %
Scratch and abrasion	61.0
Cut, laceration, puncture	9.0
Other diseases of the eye	8.0
Chemical burn	8.0
Radiant energy effect	7.0
Thermal burn	3.0
Other injury	5.0

From Bureau of Labor Statistics 1980.

More detailed statistics on work-related, disabling eye injuries are available. Table 15.3 summarizes statistics for type of eye injury. These statistics, although dated, provide considerable insight about what can be expected during work in confined spaces. The overwhelming injury, scratch and abrasion, often results from the presence of a foreign object on the surfaces of the eye. Cuts, lacerations, punctures, and chemical burns are considered severe in that they have the greatest potential to produce visual impairment (Bromberg and Hirschfelder 1984).

Table 15.4 and Table 15.5 provide additional information about work-related, disabling eye injuries. Table 15.4 lists causative agents in the eye injuries summarized in Table 15.3. Metal items and unidentified particulates were the causative agents in more than half of the injuries. More than 95% of the injuries caused by radiant energy sources occurred due to welder's flash.

TABLE 15.4
Causative Agents in Disabling Eye Injuries

Causative agent	Occurrence %
Metal items	30.0
Particles (not specified)	22.0
Chemical substances	12.0
Radiant energy sources and equipment	7.0
Wood items	4.0
Mineral items, nonmetallic	3.0
Soaps and detergents	2.0
Glass items	2.0
Other	19.0

From Bureau of Labor Statistics 1980.

TABLE 15.5
Type of Accident Involved in Disabling Eye Injuries

Accident type	Occurrence %
Rubbed or abraded	48.0
Struck by	23.0
Contact with radiant energy, chemicals	23.0
Contact with temperature extremes	3.0
Struck against	2.0
Other	2.0

From Bureau of Labor Statistics 1980.

Table 15.5 summarizes the type of accident that produces disabling eye injuries. More than half of the "struck by" accidents and 97% of the "rubbed or abraded by" accidents resulted in scratches and abrasions (Bromberg and Hirschfelder 1984).

These accidents reflected either lack of use of eye protection (60% of injuries) or eye protection inappropriate to the circumstances (Bureau of Labor Statistics 1980). In the latter circumstances, the eyewear failed to enclose the eye sufficiently to stop entry by the particle or chemical splash. In other words, the particle went around the protector. Almost 25% of those injured while wearing glasses did not use eyewear rated for use in industrial applications. Reasons for not wearing proper eye protection ranged from perceived lack of need, to not normally used, to not practical, to not required or not available.

Consideration about the nature of work activity that is performed in a confined space with the previous statistics provides some indication about the nature of eye injuries that can be expected. These injuries should be considered even more likely to occur due to the influence of the confining geometry of the structure. Prevention of or at least minimizing the potential for occurrence of these injuries is the best strategy. This requires selection of appropriate eye and face protection, training in its use, awareness about the need for strict compliance, and enforcement. An emergency should not be precipitated just because some individual decided that eye protection was not needed. For additional information about the eye and eye injuries, see Chapters 12 and 14.

TABLE 15.6
Summary of Occurrence of Disabling Foot Injuries by Type

Type of injury	Occurrence %
Toes	57
Toes only	45
Toes/other parts of foot	13
Metatarsal	38
Metatarsal only	23
Metatarsal/other parts of foot	15
Sole	19
Sole only	12
Sole/other parts of foot	7
Heel	6
Heel only	2
Heel/other parts of foot	3
Ankle/other parts of foot	5

From Bureau of Labor Statics 1981.

Skin Injury

The skin is at risk from chemical contact and burns. This can occur during traumatic incidents and failure of protective clothing. This also could occur when protective clothing is not worn, or is inappropriate for the circumstances.

The skin is more resistant to damage than the eye, owing to the layer of dead cells on its surface. This layer is considered sacrificial. There is not the urgency to remove contamination from the skin that exists with the eye. However, significant damage can occur if action is not taken quickly.

Skin disorders account for about 40% of reported occupational illnesses (Lahey 1986). Dermatitis (inflammation of the skin) can result from direct contact with mechanical agents, and physical, chemical, and biological agents. About 75% of occupational dermatitis results from contact with primary irritant chemicals.

Foot Injuries

Another example of injuries for which historical data exist are those involving the foot. About 10% of lost-time reportable injuries involve the feet, toes, and ankles (Cleveland 1984). Occupational groups reporting at least 3% or more of foot injuries include carpenters, mechanics, repairmen, assemblers, welders, flame cutters, machine operators, construction laborers, freight and materials handlers, warehouse workers, and general laborers.

Detailed statistics on work-related foot injuries are available. Table 15.6 summarizes statistics for type of foot injury. These statistics, although dated, provide considerable insight about what can be expected during work occurring in confined spaces. The toes, metatarsal, and sole are involved in the overwhelming majority of injuries.

Table 15.7 summarizes the type of accident that caused the foot injury. The vast majority of accidents occurred through stepping onto a sharp object or being struck by a falling object or one that rolled onto or over the foot.

More than 75% of the injured workers were not wearing safety footwear at the time of the accident (Bureau of Labor Statistics 1981). Reasons for not wearing foot protection ranged from

TABLE 15.7
Type of Accident Involved in Foot Injuries

Accident type	Occurrence %
Stepped on sharp object	16
Struck by falling object	58
Object rolled onto or over the foot	13
Squeezed between two surfaces	5
Struck foot against object	2
Other	6

From Bureau of Labor Statistics 1981.

lack of requirement by the employer to lack of instruction in foot protection, to lack of policy, to requirement for individual purchase, to perceived impracticality.

Considering the nature of work activity that is performed in confined spaces with the previous statistics provides some indication about the nature of foot injuries that can be expected. Preventing or at least minimizing the potential for occurrence of these injuries is the best strategy. This requires selection of appropriate foot protection, training in its use, and awareness about the need for strict compliance, and enforcement. An emergency should not be precipitated just because some individual decided that foot protection was not needed.

The "Golden Hour"

An important concept in provision of emergency medical services is the "golden hour" (Workers Compensation Board 1990). The concept of the golden hour arose from analysis of records of morbidity and mortality. Use of specially trained medics and rapid transport to field trauma centers dramatically reduced morbidity and mortality during the Korean and Viet Nam wars. The basis for the concept is that medical service providers have about 1 hour to initiate treatment of an injured victim in order to maximize the potential for recovery or survival. The earlier within the hour that appropriate treatment begins, the greater the potential for success.

The following truisms apply to accidents involving traumatic injury:

- Stabilization of patients suffering major trauma cannot adequately occur in field locations
- Many trauma victims die on the way to the operating room
- Trauma care demands efficient use of time

The victim therefore must be transferred to the hospital operating room as quickly as possible.

The condition that forms the basis for the concept of the "golden hour" is shock. Shock is the state of inadequate perfusion of the cells of the body. Perfusion is the flow of blood to and from the cells. Proper cellular function requires adequate perfusion, since this supplies oxygen and nutrients and removes wastes. The outcome from inadequate perfusion occurs in the following sequence of ever-increasing consequence: cessation of cellular function → cell death → tissue death → organ death → body death. The intent of rapid intervention as early as possible within the "golden hour" is to interrupt this sequence, preferably before cessation of cellular function occurs.

The term "shock" describes several conditions having different origins. "Hypovolemic shock" refers to the condition caused by loss of blood through bleeding or loss of fluid through burns. "Cardiogenic shock" results from damage to the heart caused by heart attack or direct traumatic injury. "Anaphylactic shock," "bacteremic (septic) shock," and "neurogenic (spinal) shock" result in excessive dilation of blood vessels. The normal volume of blood is insufficient to fill dilated blood vessels. Anaphylactic shock results from severe allergic reaction. Agents that cause anaphylactic shock include insect stings, antibiotics, seafood, and nuts, as well as other agents. Bacteremic shock results from the action of bacterial toxins on blood vessels. Neurogenic shock occurs in rare cases following a spinal injury that causes complete paralysis.

One of the important responses of the body to shock is vasoconstriction — contraction of blood vessels that supply nonvital organs — to provide blood flow to the brain and heart. This deprives the nonvital organs of blood needed to maintain metabolic activity. If not alleviated rapidly, this situation could lead to cellular death in these organs.

The "golden hour" has important implications for the provision of care following accidents that occur in confined spaces. The victim may be trapped in the structure of the space with injuries of unknown seriousness and also may require extrication through a small opening. This situation is compounded by occurrence of accidents in remote areas where transport over long distances to a hospital will be required. This means that action to stabilize the victim to prepare for the trip to the hospital must begin as soon as possible after occurrence of the accident is known. The realities of accidents that occur in confined spaces therefore put extra emphasis on planning and preparation.

Preparation for Traumatic Accidents

Emergency preparedness is contingent on having the appropriate equipment (and the associated training) that is needed for the emergency when it occurs. The equipment is broad in scope and covers some familiar and some not so familiar conditions. The circumstances of the location have a major influence in selection of equipment. Location of the site can range from the middle of a city where medical help is readily available within minutes to a remote area where medical help may not be available without hours of transportation. Medical equipment maintained at the site could overlap with that provided by the external emergency service provider. Selection of what is appropriate reflects judgment of the emergency service provider who will be present at the site or could reflect requirements of the jurisdiction.

Eye Contamination

Where chemical contamination of the eye by acids, caustics, and solvents still can occur despite other precautions, immediate flushing by water or specially formulated eye-flushing products is essential. The first 10 to 15 seconds are considered critical to saving the sight of the individual (Jagger 1988, Anonymous 1991, Champagne 1994). A 3-min flush that begins 10 s after contact is superior to a 15-min flush that begins within the first 10 min after contamination. This is especially important where chemical splash has occurred. The minimum length of time recommended for flushing the eyes is 15 min. Most medical experts recommend 20 to 30 min for optimum cleansing. The material safety data sheet (MSDS) for the product should be consulted for exact guidance on treatment of eye injuries, since some chemicals are incompatible with water.

Caustics are more consequential than acids. Acids tend to form a barrier of precipitated protein and do not penetrate further into tissue. Hence, initial appearance is a good indicator of ultimate damage. Caustics penetrate into tissue as long as contact occurs. The end result is scarring.

ANSI Standard Z358.1 provides minimum standards of design and performance for emergency eyewash stations (ANSI 1990). ANSI Z358.1 describes eyewashes that can flush both eyes simultaneously at a minimum delivery rate of 1.5 L/min (0.4 U.S. gal/min) for 15 min. An eye/face wash should have a minimum delivery rate of 11 L/min (3 U.S. gal/min). The preferred flushing solution is a preserved, buffered saline solution. Installation should occur in an accessible location no further than 30 m (100 ft) from the hazard. Where strong acids or caustics are present, installation should occur within 3 m (10 ft) of the source. Eyewash units that meet the ANSI standard can be either plumbed-in or self-contained.

Growth of microorganisms in stored water in plumbed-in and portable eyewashes is a serious concern (Tyndall et al. 1987; Bier and Sawyer 1990; Paszko-Kolva et al. 1991, 1995; Bowman et al. 1996). These organisms could cause difficult-to-treat disease conditions in the damaged eye. This could compound the consequences of the eye injury. For this and other reasons, ANSI recommends weekly checking and or testing and flushing out the lines of plumbed-in units.

Discovery of microbial contamination prompted development of flushing solutions containing preservatives (Anonymous 1991). Coupled with this was the observation that prolonged exposure to tap water during flushing causes damage to corneal cells due to lack of isotonicity. Buffered saline solutions containing preservatives also must be replaced periodically. Replacement should occur at least every 6 months. More frequent replacement could be required where high bacterial levels could be present in the workplace environment.

Adherence to the ANSI standard could be very difficult in situations where the workspace is a considerable distance from the entrance to the confined space. In situations where rapid evacuation cannot occur, provision of emergency medical supplies at the workspace must be considered. "Personal" eyewash devices that do not meet the ANSI standard are available in the marketplace. Some are small enough to be incorporated into a kit that could be carried by the individual. Larger units easily could be incorporated into a rescue kit carried by a first aid attendant. These units also are suitable for flushing foreign bodies from the eye.

The eye lavage is a cup-like device that connects by tubing to a supply of sterile saline solution. When activated, the saline flows across the surfaces of the eye to remove residual substances. This product can act as a adjunct to an eyewash to provide treatment during transport to a medical facility.

Use of water-based products in cold weather complicates matters considerably. This could be the reality in remote areas. Cold water in the eye is extremely uncomfortable. Freeze-up is unlikely to occur in products containing balanced saline solution. Freeze-up, however, could occur in products containing only water. Temperature control in these situations, therefore, is critical.

Skin Contamination

The preferred means for removing contamination from the skin is to flush with copious amounts of water. The MSDS for the substance of interest should be consulted for exact recommendations, as compatibility problems may exist.

ANSI Standard 358.1-1990 provides minimum standards for deluge showers (ANSI 1990). A deluge shower must provide an overhead or angled flow of water over the body at a rate of 135 L/min (30 U.S. gal/min) for a minimum of 15 min. Some shower booths provide water from the sides and top, using nozzles in the walls and ceiling (Jagger 1988). The spray from a deluge shower could be used to extinguish a clothing fire.

Deluge equipment also includes hand-held drench hoses. These, however, do not provide the high rate of flow of the fixed emergency showers. This limits their utility to treatment of contaminants in the eye and minor skin contamination. Operation requires a free hand to manipulate the nozzle. Pressurized self-contained units are available. However, these units do not provide the required flow rate for deluge equipment.

Delivery temperature of the water is a major concern. The ANSI standard recommends delivery temperatures ranging from 15 to 35°C (60 to 95°F). The shock of exposure of an injured person to very hot or very cold water could compound the situation. Water standing in long runs of uninsulated pipe that are exposed to the sun could become hot enough to cause first degree burns. Maintaining comfortable temperature during use in cold weather is a difficult problem, especially in remote locations where permanent facilities are not available.

Water that is cool can prove beneficial where thermal burn has occurred. Cooling the skin removes the hot sensation.

FIGURE 15.1 Face masks. The removable cap covers the opening to the self-inflating face seal. (Courtesy of Ambu Inc., Linthicum, MD.)

FIGURE 15.2 Single patient use resuscitators (adult and infant). The adult version has a stroke volume of 2.6 L. The handle provides support for maintaining a steady grip and uniform compression rate. These models have an oxygen reservoir tube. Other models contain a bag reservoir. (Courtesy of Ambu Inc., Linthicum, MD.)

Resuscitator/Inhalator

The resuscitator and inhalator are used for airway maintenance (Figures 15.1 and 15.2). The resuscitator is used with individuals who are not breathing. The inhalator is used with an individual who is breathing. Both units utilize a mask that fits over the nose and mouth. A nose-only version also is available.

The simplest resuscitators consist of a mask and one-way valve assembly. These are used primarily for cardiopulmonary resuscitation. The mask prevents skin-to-skin contact. The one-way valve and discharge hose prevent entry of breath exhaled by the victim into the rescuer's breathing zone. Some units contain an absorbing filter that provides a physical barrier in the air passing between the rescuer and the victim.

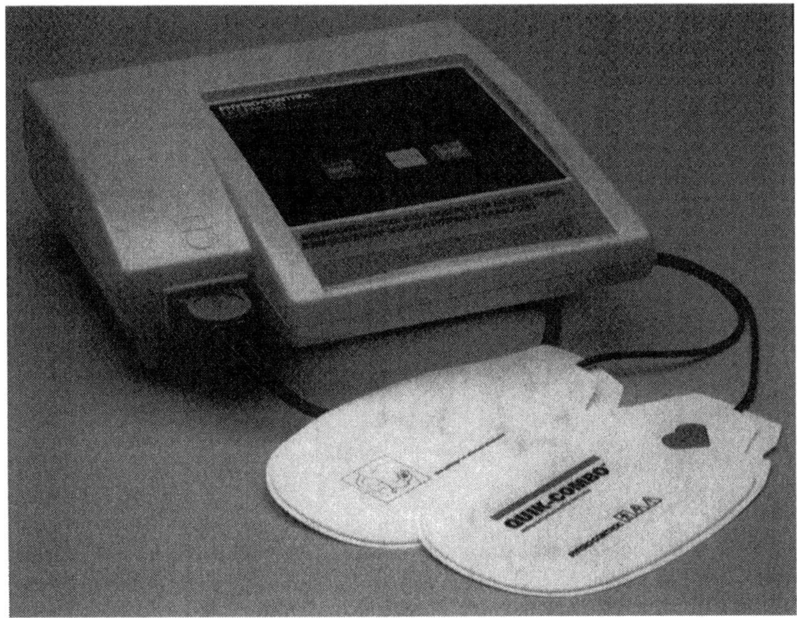

FIGURE 15.3 Defibrillator. This automated unit assists the operator through a series of screen and synthesized voice prompts. The unit analyzes the condition of the patient and prompts when to apply the electrical shock and cardiopulmonary resuscitation. (Courtesy of Physio-Control Corporation, Redmond, WA.)

Basic resuscitators utilize a hand-operated squeeze bulb to deliver air or oxygen through the mask. The oxygen-delivering units utilize a cylinder of compressed oxygen and demand regulator. Automated units also are available. These can preset delivery of batches of known tidal volume.

Inhalators can provide oxygen by constant flow or demand mode.

Defibrillator

Uncoordinated firing by ventricular pacemaker sites can produce no coordinated contraction and ineffective pumping of blood (Abel 1982). This is visualized as a totally chaotic pattern in the electrocardiogram. The only effective means to treat this condition is to depolarize the entire heart muscle (defibrillation) and to restart it from a single pacemaker site. The defibrillator is the medical device that corrects cardiac arrhythmia (White 1995) (Figure 15.3). The defibrillator was inspired by electrical accidents involving linemen. Original defibrillation techniques used AC voltages. Later applications utilized high voltage direct current pulses of opposite polarity.

Immobilization

Immobilization is the treatment for fractures, bone injuries, and breakage (Wilkerson, 1985). Almost all bone and joint injuries pose major problems. Damage to other structures can result in permanent disability or death. Accompanying damage can occur to blood vessels and nerves. This can lead to hemorrhage, gangrene, or paralysis. Additional measures may be required for open fractures and bleeding. Immobilization prevents further damage to surrounding tissue by bone ends and reduces pain and decreases the risk of shock. For a fracture to be effectively immobilized, the joints above and below the fracture also must be immobilized. The immobilizing splint must permit blood circulation and not cause nerve damage.

Fractures of the spine in the back and neck are always accompanied by the possibility of injury to the spinal cord. The higher the level at which damage occurs, the greater is the risk of damage

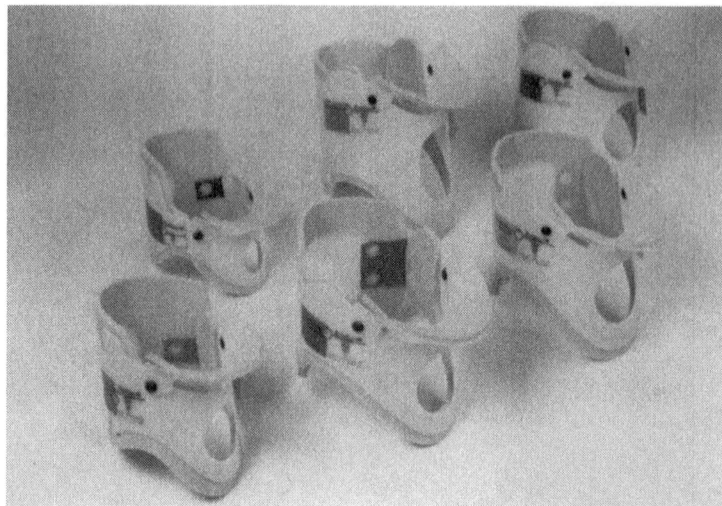

FIGURE 15.4 Immobilization devices: extrication collars. These units are color-coded according to neck length. Collar diameter is adjustable to neck size. This unit contains a flip chin piece that allows for intubation and airway management. (Courtesy of Ambu Inc., Linthicum, MD.)

FIGURE 15.5 Immobilization devices: splints — vacuum splints. This set includes an arm and leg splint and vacuum pump. (Courtesy of Ambu Inc., Linthicum, MD.)

to the nervous system and the consequence of the injury. Paralysis resulting from an injury to the spinal cord usually is permanent.

Splints are manufactured in various configurations. The full-body splint/litter/spineboard/stretcher can provide horizontal, as well as lateral immobilization of the spine. The head also is immobilized as part of the ensemble. The traditional unit is the rigid backboard. Some units have a split-leg design that permits each leg to be anchored independently. Many units are designed to fit the immobilized victim through the small openings of some confined spaces. Upper body immobilization is available in some products. Some combine a partial body splint for cervical immobilization and a harness for vertical lifting/extrication. Traction splints provide local immobilization of limbs. The cervical collar provides support to the head, chin and base of the skull (Figures 15.4 and 15.5).

The stretcher is an important component in the movement of immobilized patients (Figure 15.6). Some types provide a semicylindrical supporting structure around the body. One design uses a plastic pad that is secured in a curled position around the victim. The opened pad is secured around the patient in the axis opposite to its natural direction of curl. This orientation takes advantage of the natural rigidity in the plastic.

Pneumatic Antishock Garments

Pneumatic antishock garments apply air pressure circumferentially against the lower limbs, similar to the cuff in a blood pressure apparatus. The intent of this externally applied pressure is to mitigate against fluid loss into the lower extremities during hypovolemic shock. These garments are not without controversy (Chang et al. 1995). However, most of the study has occurred in large metropolitan centers where transport times are short, rather than rural or remote areas.

Dressings and Bandages

Burn dressings that contain gels have high heat capacity and form a protective, soft barrier over the wound. Covering the wound is extremely important to prevent fluid loss and to prevent bacterial invasion. Bacterial infection is a major cause of death in burn patients.

Cold compresses stimulate circulation and help to reduce pain and swelling. Products include ice bags that hold ice cubes and gel packs, as well as products that become cold through chemical reaction of the contents. The ice and gel packs require storage in a freezer until use.

Hot compresses contain chemicals that generate heat through chemical reaction.

Asthma

Asthma affects about 3 to 5% of the adult population (Mossesso and Yee 1996). Asthma is a disease of the small and medium airways. Asthma results from an uncontrolled immune response. The latter produces three primary reactions: inflammation of the bronchial wall, increased mucous production, and bronchospasm. The most obvious functional effect produced by asthma is the inability to exhale completely. Triggers of an asthma attack are wide and varied. These can include infections, allergens, drugs, physical and emotional or psychological stress, cold air, and exposure to certain chemicals.

Management of an asthmatic patient prior to hospital treatment is essential. This could require use of oxygen and a bronchodilatory agent.

Heat-Related Conditions

Heat-related conditions include dehydration, heat cramps, heat exhaustion, and heat stroke. Severe cases could lead to unconsciousness. An important consideration for work in confined spaces is absorption of solar energy by surfaces of the structure. The interior temperature can be considerably higher than that of surrounding air.

Management of a heat-related condition begins with stopping physical activity and forced cooling (Murphy and Tkach 1996).

This could require use of icepacks, as in the case of heat stroke. Electrolyte replacement either orally or intravenously also is a priority. While cooling is essential, this must occur at a controlled rate to avoid shivering and shutdown of peripheral circulation. Overrapid cooling can cause blood to shift to the core and overload the heart, causing pulmonary edema.

Heat stroke is one of the true medical emergencies (Wilkerson 1985). Delay in seeking treatment increases the potential for residual disability. The complications from heat stroke are numerous and often severe: kidney failure, liver failure, blood clotting abnormalities, gastrointestinal ulceration with bleeding, heart damage, biochemical alterations, and brain damage. For additional information on this subject, see Chapters 12 and 14.

(A)

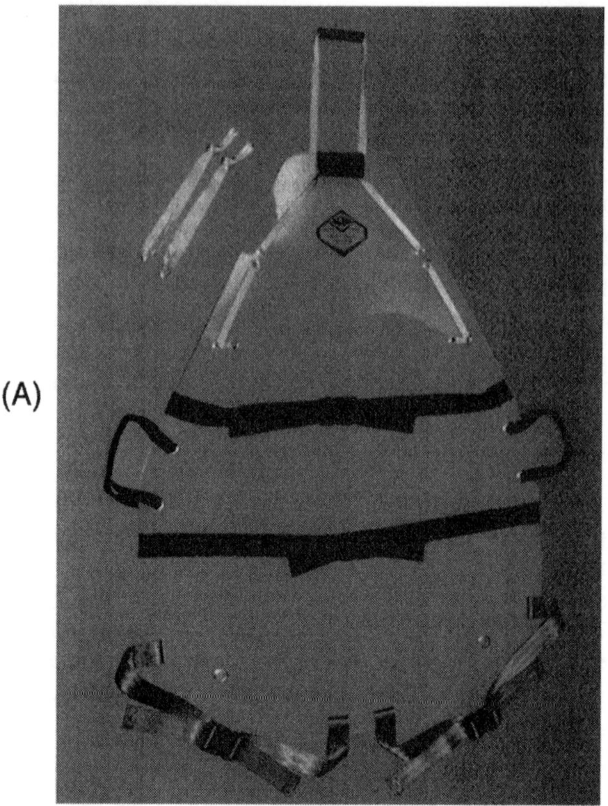

(B)

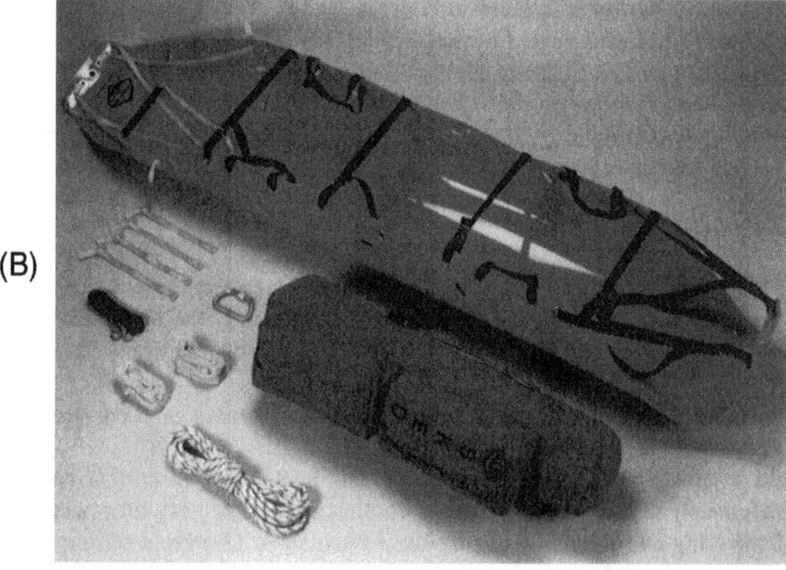

FIGURE 15.6 Immobilization devices: stretchers: (A) Roll-up half-length stretcher. (B) Roll-up full-length stretcher. These stretchers roll into a compact cylinder for transport. They derive their rigidity from utilization of the reverse curl in the plastic material. The half-length stretcher provides upper body immobilization, while permitting the legs to be bent at the hips (less than 90°). This can facilitate movement of the immobilized person around corners in convoluted spaces. This movement would be impossible with a full-length stretcher. (Courtesy of Skedco Incorporated, Portland, OR.)

Cold-Related Conditions

The most common cold-related conditions are hypothermia, frostbite, and immersion foot (Wilkerson 1985). Hypothermia is a decrease in the core temperature of the entire body. Hypothermia impairs muscular and cerebral functions. Frostbite is freezing of tissue. Immersion foot produces similar damage but does not involve freezing of the tissues.

Treatment of mild hypothermia involves decreasing heat loss and increasing heat production. The former can be achieved by adding clothing and taking shelter. The latter occurs through exercise and intake of high calorie, rapidly metabolized foods.

Treatment of severe hypothermia is considerably more complicated and requires considerable care. The victim is not capable of generating the heat or performing the vigorous exercise needed to correct the condition. Rapid rewarming through application of external sources of heat is hazardous, since this produces rapid modification of internal adaptive biochemical conditions. Rewarming of internal organs must occur first. Also, ventricular fibrillation of the heart could occur. This could be triggered by minor movements. Severe hypothermia requires hospitalization.

The preferred treatment for frostbite is rapid rewarming in a water bath. This minimizes tissue damage. Temperature should be maintained at 38 to 42°C (100 to 108°F). This should occur in a hospital setting. Healing requires weeks to months.

Immersion of the extremities in cold water for a prolonged period causes vasoconstriction. This deprives the area of sufficient blood supply. Permanent damage to nervous tissue can result from prolonged exposure.

Anaphylactic Shock

Anaphylactic shock follows the (Type I) allergic reaction produced by insect bites and stings, food, drugs, and other allergens (Fortenberry et al. 1995). Each year 50 to 150 people in the U.S. die from insect-sting anaphylaxis. Most die within 30 min of the initial injury. Prompt administration of epinephrine, therefore, is essential to prevent the anaphylactic reaction. Improvement is observable within 2 to 25 min of administration of epinephrine by subcutaneous injection. Modern products utilize a hypodermic autoinjector.

Snake Bite

About 45,000 bites by all species of snake occur annually in the U.S. (Perez and Sumner 1979). Of these, 15 to 20% result from poisonous snakes. Fortunately, the overwhelming majority of victims reach the hospital within 2.5 hours. Snake habitat includes the undisturbed settings that characterize some confined spaces. Disturbance that occurs during entry can lead to conditions conducive to snake arousal. The consequences of not treating a bite by a poisonous snake range from nothing to death. While death is a rare outcome, painful and crippling injuries are not.

Snake venoms cause both local and systemic effects, including serious effects on the blood and blood vessels, nervous system, and heart. There is considerable disagreement about treatment options. Hospitalization is considered an essential first step. Follow-up treatment is best rendered in the hospital setting.

Animal Contact

The danger of infection following the bite by any animal is high. This results from introduction of mouth-borne bacteria into the wound during transfer of saliva (Wilkerson 1985). Bite wounds require treatment with copious quantities of soap and water, followed by an antimicrobial wash.

Rabies is a viral infection that is transmitted in the saliva by infected animals. Following extensive vaccination of domestic pets, the main transmitters of the disease to humans now primarily include skunks, foxes, bats, and raccoons. The virus can be transmitted by biting, licking, and

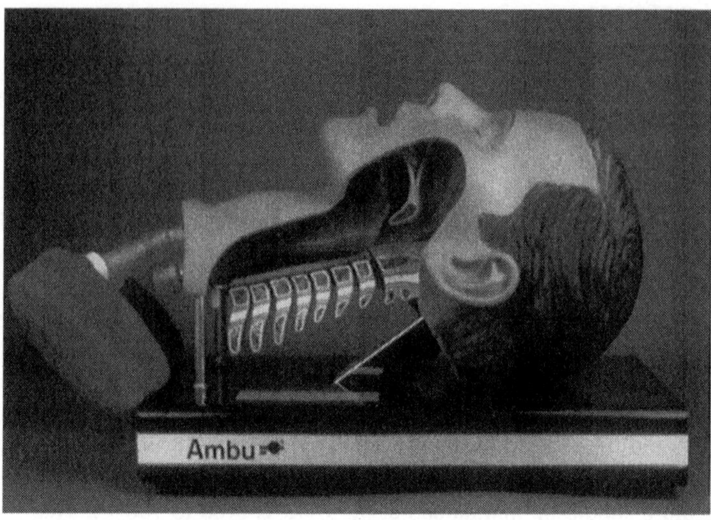

FIGURE 15.7 Nasotracheal intubation. This device used for training emergency medical service providers in nasotracheal intubation indicates the finesse required in performing this procedure. Care is required to avoid ventilating the stomach. (Courtesy of Ambu Inc., Linthicum, MD.)

breathing aerosolized forms. The head of the animal is required to determine whether rabies is present. If the animal that caused the contact cannot be captured for study, rabies should be assumed and treatment begun. This could require hospitalization.

A more recently recognized problem is hantavirus (Centers for Disease Control 1993). Hantavirus is carried by rodents and possibly other animals. The virus is spread through excreta — urine and feces. Hantavirus pulmonary syndrome strikes otherwise healthy adults. Infection of humans occurs through inhalation of aerosols containing the virus, open wounds, ingestion and rodent bites. Some confined spaces are natural habitat for these animals. Contact with the virus can occur regardless of recommendations for prevention, since the victim may be unaware about the hazard.

Symptoms following infection initially are flu-like, including fever, chills, and muscle aches. These progress to pulmonary edema (lungs filled with fluid). Immediate hospitalization should be considered following contact with contaminated areas or when a rodent bite has occurred.

Some biting spiders are capable of causing serious illness (Wilkerson 1985). Some confined spaces provide habitat for these organisms. A spider bite can cause painful muscle spasms and prostration for several days. Dizziness, nausea, and vomiting also are common. Hospitalization is recommended.

Scorpions that produce lethal bites are found in some parts of the southern U.S. and Mexico. Some confined spaces provide habitat for these organisms. Victims should receive hospital treatment as quickly as possible. Death can result from sudden high blood pressure.

Nasotracheal Intubation

Nasotracheal intubation is the passage of an endotracheal tube into the nostril, through the throat and into the trachea (Click 1996). The nasotracheal intubator is a piece of plastic tubing to which are attached adapters, one at each end (Figure 18.7). This device is suited to maintaining the airways of patients with suspected spinal injuries, facial fractures and jaw dislocations.

Specialized Treatments

Some chemical products that cause skin injury require unique treatment. Fluoride in hydrofluoric acid continues to penetrate until immobilized by calcium. As a result, serious burns can result long after treatment by normal agents has ceased. Successful treatment requires use of specialized agents that the receiving hospital may not routinely stock. The emergency service provider should alert the hospital to the nature of injuries requiring specialized treatment and provide supplies with the patient.

Universal Precautions

The latter half of the 1980s brought major changes in procedures for stranger-to-stranger interactions involving health-care, emergency medical services, and public safety workers. These changes resulted from concerns about transmission of the human immunodeficiency virus (HIV) and the hepatitis viruses, principally hepatitis B, and tuberculosis. The risk of infection of health-care and public-safety workers by these viruses is unknown (NIOSH 1989). The regulatory response in the U.S. to these concerns was the OSHA Bloodborne Pathogens Standard (OSHA 1991).

The approach taken to concerns about infection is to consider all body fluids as potentially infectious and hazardous, and to prevent contact with them (Centers for Disease Control 1987). Precautions include the use of impervious gloves, masks, and respirators; protective eyewear; faceshields; and impervious gowns. Infection prevention protocols require use of these products in a manner to prevent infection of one patient with fluid generated by another. Resuscitators should be used to avoid potential contact with blood or body fluids. Persons should receive vaccination against hepatitis B virus. Additional precautions include use of puncture-resistant, leak-proof containers to collect used needles, syringes and other sharp objects that may have contacted blood and body fluids. Decontamination and cleaning and disinfection of hands and skin, clothing, and surfaces that may have had contact with blood and body fluids also must occur.

Infection protection devices include a wide range of products. Suction devices are used to collect body fluids, including blood, other body fluids, and vomitus. These devices can be hand-, pneumatically, or electrically powered.

Surface disinfectants, decontaminants, and cleaners are essential to eliminate bacterial and viral contamination. A number of terms in common English usage have specific microbiological significance (Block, 1977). An **antiseptic** prevents or arrests the growth or action of microorganisms, either by inhibiting their activity or destroying them. The mode of action may depend on temperature, duration of contact, pH, nature of the organism, presence of organic matter, and other factors. **Disinfection** is the process of destroying disease or other microorganisms, but not ordinary bacterial spores. Disinfectants kill the growing forms, but not necessarily the resistant spore forms of bacteria. Some disinfectants kill spores of specific organisms and inactivate specific viruses. **Sanitization** is disinfection to levels considered safe according to public health criteria (for example, 99.9% or more). A product that is a disinfectant at one dilution could be a sanitizer at another. **Sterilization** is the act or process of destroying or eliminating all forms of life, especially microorganisms. Sterilization conveys an absolute meaning, not a relative one.

Disinfection and sterilization are broad subjects of considerable complexity (Hugo 1971, Block 1977). Active agents include: alcohols, aldehydes, phenols and substituted phenols, halogens, salicylanilides and carbanilides, quaternary ammonium salts, antibiotics, and other compounds. Tests for efficacy are provided in legislation and guidance documents from technical sources.

Air-filtering respirators that meet the new NIOSH N95 filtration requirements are essential for use against communicable diseases, such as tuberculosis (NIOSH 1995). Tuberculosis is caused by *Mycobacterium tuberculosis*. N95 is the required level of filtration against tuberculosis according to CDC (Centers for Disease Control) guidelines.

Latex Allergy

Latex rubber traditionally has been the favored material in the manufacture of surgical gloves and many other products used in emergency medical services (McKenna et al. 1996). Rapid increase in use of latex products occurred because of concerns about controlling transfer of blood-borne pathogens and other infectious diseases. This led to the concomitant increase in appearance of allergy to latex products. Proteins in the sap from the rubber tree from which latex is extracted currently are believed to be responsible for the sensitization (Hamann 1994).

The incidence of latex allergy in the general population is believed to be 7% (Voelker 1995). The incidence could range from 10 to 17% in health-care providers (Charous 1994). In most cases, the response is a Type IV reaction that leads primarily to skin reactions: sores, drying, cracking, hives, itching, and redness. Onset can range from hours to days. A small fraction of latex-sensitive individuals exhibit a Type I reaction that involves anaphylactic shock. Onset can occur in minutes to hours. Death can occur in extreme cases.

This situation can affect emergency preparedness, especially in remote areas where transport to hospital may be required. A considerable number of products used in emergency services contain latex (McKenna et al. 1996). Alternatives are available for some of these products. Prudence would suggest use of latex-free products where possible. So widespread is the use of latex in medical products, that some sensitized individuals have assembled their own emergency medical kits. They may, as well, refuse to accept medical treatment and transport unless latex-containing materials have been removed from the emergency response vehicle.

FIRE PROTECTION

The potential for fire is a serious concern during hot work and work involving ignitable materials. Preventive and protective measures are absolutely essential for preventing the incidence of fire and explosion and minimizing the impact of their occurrence. Fire preventive and protective measures could be considered from the perspective of either work activity or emergency preparedness. Either way, these measures should be regarded as a seamless part of hazard management, rather than an add-on.

Fire prevention and fire protection form an integral part of NFPA 306, the hazard management system for confined spaces on ships (NFPA 1993). NFPA 306 was adopted into the OSHA Standard on confined spaces in shipyard employment (OSHA 1994). The Standard permits assessment only by a Certified Marine Chemist in some situations. This is formal recognition of the extensive practical training in fire prevention and fire protection that Certified Marine Chemists receive as part of the certification process. Fire prevention and protection are not stressed as heavily in other standards on confined spaces (ANSI 1995; OSHA 1993).

NFPA 306 directs the Certified Marine Chemist to examine the space in which the work is to occur as well as adjacent spaces.

Adjacent spaces make contact with the space under consideration through shared common walls in all directions. Work activity in the space under consideration can affect conditions in the adjacent space and vice versa. This means that attention to fire prevention and fire protection must occur on both sides of the common wall.

Table 15.8 lists fire preventive and protective measures for the structure in which hot work is to occur and measures that apply during use of ignitable materials.

Detection of fire should prompt evacuation from the confined space. Workers normally are not trained or equipped to fight fires. Also, the confining geometry of the space could enable the smoke to accumulate rapidly. Toxic substances in the smoke normally are considered to be the main cause of death of victims of fire (Hartzell 1991). These include asphyxiants, narcotics, irritants, and intoxicants that produce specific health effects. The hazard assessment normally does not consider fire fighting as part of work activity.

Prolonged combustion requires a high temperature and generation of heat at least as fast as it is dissipated (Friedman 1991). Extinguishment occurs by removing the oxidizer (usually oxygen), cooling the fuel or the gaseous combustion zone, or terminating the free radical chain reaction.

TABLE 15.8
Fire Preventive and Protective Measures for Work Activity

Preventive Measures
Displace the atmosphere by filling with inert gas (inerting)
Fill with water
Remove residual materials and wash with detergents
Steam-clean
Remove coatings and combustible materials
Wet combustible materials surrounding the work area
Cover susceptible surfaces with wetted fire-resistant fabrics
Cover susceptible horizontal surfaces with wetted sand
Eliminate sources of ignition (smoking materials, lighters, matches, arcing tools, hot surfaces)
Ventilate to prevent accumulation of vapors and vapor/mist combinations to levels in the flammable range
Protective Measures
Post a "fire watch"
Provide fire fighting equipment (hoses, fire extinguishers)

Removing the oxidizer is possible only in limited amounts, since this would affect an unprotected firefighter.

Portable fire extinguishers can provide a rapid means to attack the fire during its initial stages. These units are small enough that they can be positioned inside the space during the work activity. Use of an extinguisher can permit evacuation of the space to occur under less urgent conditions.

Extinguishing agents that are usable in confined spaces are somewhat limited. For example, water used to fight a fire in a ship could cause the vessel to capsize. Water can be used casually for this purpose only in limited quantities. Use of water in portable units in cold weather could be contraindicated, due to potential for freezing. Water also could be contraindicated where electrical equipment is being used, since exposed live conductors could result during the fire. Water use is contraindicated where certain metals, such as aluminum, magnesium, and zinc are present, due to reaction to evolve hydrogen and their ability to melt at low temperatures. Water is contraindicated where many flammable liquids are present, since most flammable liquids are insoluble and float on water. For liquids that are more dense than water, a layer of water on the hot liquid could lead to an explosive condition. An explosion could spread the hot liquid and, consequently, the fire.

Water-based foams are used against fires involving flammable liquids. The foam floats on the surface of the liquid. This produces a continuous layer that excludes air and cools the liquid. The foam must cover the entire surface of the liquid. It also disintegrates with time and must be applied in sufficient quantity to maintain the cover blanket for the needed time.

Dry chemical agents act on the fire by a mechanism believed to involve absorption of heat and the formation of volatile species that react with free radicals, thus interrupting the chain reaction. Effectiveness is related to particle size. Ammonium phosphate melts and forms a glassy coating on the burning surface. Dry chemical extinguishants, especially ammonium phosphate, are corrosive to metal surfaces to varying extent. These substances can damage sensitive equipment, and may never be satisfactorily removed following the fire. Dry chemical extinguishants are safe for use in electrical fires.

Inert gas extinguishants, principally carbon dioxide, extinguish fires by diluting the oxidizer (oxygen). Inert gases suitable for extinguishing fires include carbon dioxide, nitrogen, and steam, and the chemically inert gases, helium, neon, and argon. Reignition of deep-seated fires can occur following dispersal of the extinguishant. The required addition of carbon dioxide or other inert reduces the concentration of oxygen to levels that are oxygen deficient for humans. The structure of the confined space can retain the extinguishant if ventilation is poor, since the gas is discharged at low temperatures. Stratification at the bottom of the structure should be expected unless ventilation has caused dispersion to occur. Discharge of fire suppression systems containing carbon dioxide have produced fatalities.

Winter use (temperatures below –18°C or 0°F) can affect operation and performance of extinguishers containing carbon dioxide (Wysocki 1991). "Winterization" of cylinders may be required.

Halogenated extinguishants (halons) combine with free radicals created in the combustion process (Friedman 1991). This combination stops combustion. Low concentrations of some types of halons are required. Thus, oxygen in the space would not be depleted to serious levels. The halogenated extinguishants currently in use are being phased out, as these substances have been linked to ozone depletion. Replacement products do not exist at this time. Reignition is a potential concern following dispersal of the extinguishant.

EMERGENCY PREPAREDNESS SYSTEM

The first important consideration in formalizing emergency preparedness is to create a system under which this will occur. As mentioned above, the fatal accidents documented by OSHA (1985) and NIOSH (1994) clearly indicated the consequences of lack of formal emergency preparedness systems. Emergency preparedness systems combine organization with people and equipment. These determine the means through organization, people, and equipment to satisfy the need for emergency response.

Emergency preparedness has become an important topic in site management in recent years. As a result, consensus standards have become available for formalizing the process. One such example is published by the Canadian Standards Association (CAN/CSA 1995). CAN/CSA-Z731-95 provides a framework suitable for planning responsive action to emergencies that occur in confined spaces.

The emergency preparedness plan is the backbone of the emergency preparedness system. The emergency preparedness plan for responding to accidents that occur in confined spaces should contain the following elements:

- Needs assessment
- Corporate policy statement
- Chain of command
- Roles and responsibilities
- Reporting requirements
- Administrative details
- Response procedures
- Internal and external network (communication/resources)
- Training requirements
- Auditing and revision

The detail required to address individual elements reflects the nature of potential emergencies for which preparedness and response are required. The emergency preparedness plan must be completed prior to the event in order to have any opportunity for success.

The needs assessment was considered in the previous sections. The extent of technical preparation for these accidents reflects the following determinants:

- Nature of potential accidents and injuries
- Post-accident conditions in the space
- Entry for assessment of injury and preparation for extrication
- Extrication and preparation for transport
- Transport to hospital

Each of the preceding requires detailed evaluation and follow-up.

The corporate policy statement could be a subset of the general safety policy statement or that prepared for the confined space program. The policy statement also could be part of an organizational policy statement on emergency preparedness. A corporate policy statement specifically directed to confined spaces could enable decision making during an extremely stressful time.

The chain of command and accompanying roles and responsibilities establish who will report to whom and where and when and how. Establishing the chain of command unambiguously is essential since reporting relationships during an emergency could change significantly from those during normal activities. This requires an organization chart and description of responsibility for each of the roles.

External emergency service providers are accustomed to interacting through formal incident command structures (National Fire Academy 1983). External responders and internal decision makers must be able to collaborate on decisions that have potential life–death outcomes. Input to the decision from the organization must clearly reflect the well-being of the victim and not other concerns.

Emergency preparedness should occur in parallel with the other aspects of the confined space program, since the participants could be the same or they could be different. Defining the structure is critical, as this establishes roles for the participants and the nature of interactions that will occur. The last consideration is extremely important, since the individual who will assume control during an incident may not have supervisory or management authority under normal circumstances.

Fulfilling the roles of the incident command structure requires knowledge and training. This becomes most obvious when the command structure is disrupted through changes forced by vacations, illness, downsizing, retirement, and so on. Replacements should be able to move seamlessly into positions vacated by the incumbents.

At large sites, members of the incident command structure could be present routinely. They could easily maintain contact through two-way radios that are routinely used, or through cellular telephones. This situation becomes considerably more complicated at remote sites where a small number of persons are working. In these circumstances, the same people may be required to fill the roles of site workers and emergency service providers. The site may be remote or inaccessible, so that external emergency service providers would be unavailable except following prolonged travel. During an emergency involving a small crew, external expert assistance initially could be available only through radio communication. Crews working under these circumstances must be as self-sufficient as possible.

Multishift operations can create problems for the emergency preparedness plan. Hazardous operations often are performed during the "off-shifts." The intent behind this strategy was to expose the least number of people to the hazardous condition. Of course, the key people identified in the emergency preparedness plan also are not present. This situation can create serious complications, since the individuals who otherwise would provide leadership are not present at the site. The only alternatives that are available are 24-hour call-out of key individuals or creation of incident command teams to cover each shift or a hybrid structure. The second alternative is probably the least disruptive. However, this can encounter serious difficulties, since off-shift operations usually are less fully staffed than the main shift.

The key people likely would attend an emergency during the off-shift. However, this necessitates an after-hours call-out system. Delays in availability of key members of the incident command structure are inevitable. Regardless, in the interest of the needs of the victim the response must begin as quickly as possible and proceed competently at all times. First aid coverage must be present, whether required by law or not, so that the initial actions can occur quickly. Also, a key decision maker must be present on the site.

Consensus standards and regulations on confined spaces increasingly are specifying the role for emergency service providers. For example, the OSHA Standard on confined spaces in general industry contains a number of requirements for emergency rescue services (OSHA 1993). Table 15.9 summarizes requirements for these groups. Other considerations may override the preceding specific requirement for work involving confined spaces.

TABLE 15.9
Summary of Training Requirements for Emergency Service Providers

Training in performing duties
Training in use of personal protective and rescue equipment
Know hazards that may be faced during entry
Understand mode of exposure
Recognize signs or symptoms of exposure
Understand consequences of exposure to the hazards
Practice rescue from the actual or representative permit spaces at least once every 12 months. Practice rescue may involve manikins, dummies or actual persons.
Obtain training in basic first aid and maintain current certification
Obtain training in cardiopulmonary resuscitation and maintain current certification

Employers using in-plant services or providing services to other employers must equip and train members of the emergency response team properly. An employer that utilizes a rescue service must inform the contract employer about hazards likely to be encountered. This is essential, so that the outside team can equip, train and conduct itself appropriately. Further, the employer must provide access to the contract rescue service, so that they may develop an appropriate rescue plan and practice in permit spaces similar to those from which a rescue may be required.

In many communities, the fire service has expanded its traditional role to include response to a range of emergency conditions. These include hazardous materials incidents, traffic accidents, workplace accidents, and so on. Fire services have responded on many occasions to requests for assistance involving confined spaces. Tragically, members of the fire service, both full-time and volunteer, have numbered disproportionately among the victims of fatal accidents involving confined spaces. The best current approach for these groups, as reflected in the intent of the OSHA Standard, is to visit the site. This provides an opportunity to become familiar with its layout and emergency response facilities, and the confined spaces and hazardous conditions that could be present post-accident. Also, they can offer consultative advice about obvious deficiencies in the status quo that may be unknown to management.

OSHA was concerned about the ability of rescue services to perform under the pressures of actual situations. For this reason, the Standard requires practice sessions at least annually, using spaces that are representative of those actually entered. This is especially important for ensuring that equipment will fit through openings of portals or that encumbered members of the rescue team can move throughout the space and perform their function. Successful rescues in actual circumstances are equivalent to practice sessions. Unsatisfactory performance of a rescue or practice session indicates the need for further development.

The OSHA Standard made no reference to provision of emergency medical services. The focus of the Standard is extrication. Some jurisdictions, such as the Province of British Columbia, specify requirements for provision of emergency first aid (Workers' Compensation Board 1994). These regulations describe different levels of qualifications for first aid attendants and specify qualifications required for different worksites. The latter are based on accident experience. The regulations also describe requirements for first aid facilities and transport vehicles. The regulations indicate a hierarchy of authority in first aid. The first aid attendant has authority until relieved by a person having a higher level of qualification. There is no mention of unusual situations, such as provision of emergency services in confined spaces in these regulations. The only potential accommodation to emergency first aid involving confined spaces is the delay in transport to hospital caused by the need to extricate a victim. This delay could increase the level of hazard classification built into the regulations.

The emergency preparedness plan for accidents involving confined spaces should be considerably more inclusive than the technical aspects mentioned thus far. Emergency preparedness should include senior management of the organization. As mentioned previously, an emergency on the site could involve intervention by varied outside agencies whose interests are disparate. These could include police, news media, and officials from different government agencies. (Involvement by government agencies could vary by jurisdiction. Some government agencies require contact when a serious accident has occurred.) Senior management or their representatives are the public face of the organization. Preparation for this role is essential for ensuring a caring, concerned and competent response to enquiries for information.

An emergency should set into motion a carefully orchestrated logical sequence of actions. Documentation in the form of a standard procedure is essential to ensure that each step is followed in sequence and acknowledged. This sequence could be initiated by a call to a central listening post, for example, security or reception or directly to the emergency response organization. The listening post can contact key individuals and then wait for instructions. The key individuals should include emergency first aid and health and safety.

The next step in the process could involve an assessment of the situation by a key individual and the setting into motion of calls to internal and external emergency service providers. There are other possible variations on this theme, depending on the circumstances of individual situations. The fundamental issue is that someone will make a decision to initiate a full emergency response.

The decision to initiate a full emergency response means that the call-out network will be activated. The call-out network is the group of individuals internally and externally who have a role in the response or a need to know that an emergency response is occurring.

The strengths of the emergency preparedness plan lie in its flawless execution. The deficiencies of the emergency preparedness plan lie in its flawed execution. The emergency preparedness plan is a living document. It requires periodic revision and upgrading in order to remain effective.

EMERGENCY RESPONSE

Emergency preparedness and emergency response are akin to painting a house. Almost all of the work lies in the preparation. If the preparation is comprehensive and has occurred competently, then the painting goes smoothly and the work has permanence. Similarly, in emergency work, the response will proceed smoothly with a satisfactory outcome if the preparation was competent and comprehensive. That is, the time to prepare for an emergency is before it happens, rather than during or after the event. The ultimate test of the adequacy of the process of planning and preparation is the ability to respond in an efficient and effective manner.

An emergency is a complex, rapidly changing situation in which the information available to act upon is minimal (Bryan 1991). Elements common to all emergencies are the unpredictable and the unknown. The motivator is urgency. Combining these elements produces a situation that is extremely stressful for all concerned. Resources needed for addressing the situation must be readily at hand. First responders must be thoroughly trained in the use of equipment that is used in these situations. The goal of emergency response is to stabilize the situation. The more efficient the expenditure of time and use of resources during this activity, the more effective the actual and perceived response.

Two groups of individuals are affected by emergencies: those whose lives and health are directly affected, and bystanders and emergency service providers. The first group is caught within the envelope of the life threatening situation. The progression of the emergency for them is a conscious awareness that the envelope of safety is closing in around them. Once the envelope collapses, they could perish. Escape means life and health. Escape is a time-dependent phenomenon. Additional participants in an emergency are bystanders and emergency service providers. The health and safety of these individuals are not threatened by the emergency, yet they are aware that their actions can have an impact on the outcome for the others.

Research on fire emergencies provides valuable information about behavior of individuals trapped in burning buildings whose health and safety are threatened (Bryan 1991). Generalizing beyond fire situations, behavior during emergency situations reflects a logical attempt to address a complex, rapidly changing situation for which little information is available. What is appropriate as a response at one moment may be inappropriate at the next. Research into fire situations suggests that some people act altruistically and heroically. Some individuals deliberately remain in the building or reenter to fight the fire or attempt to rescue persons known or believed to remain inside (Bryan 1991). These actions are undertaken in a deliberate and purposeful manner. Firefighting by occupants appears most prevalent when the individual is economically or emotionally involved in the outcome.

Behaviors reported in fire situations are consistent with what has occurred in fatal accidents involving confined spaces (OSHA 1985; NIOSH 1994). These accidents involved only hazardous atmospheric conditions and not other causative agents. Persons not overcome by the hazardous condition generally evacuated the space. Once outside, one or more individuals from this group or individuals not directly involved in the proceedings sometimes reentered the space to attempt rescue.

Individuals use a multistep process in the attempt to structure and evaluate situational threats (Withey 1962). These include: recognition, validation, definition, evaluation, commitment, and reassessment.

Recognition occurs when the individual perceives through cues that a threat exists. The cues may be continuous or discontinuous, constant or changing. They depend on the parameters of a particular situation. Individuals recognize these cues, based on past experience, most probable source, and repetitive training.

Validation is the process of attempting to confirm the existence or the seriousness of the hazardous condition. When the cues are sufficiently ambiguous, the individual will attempt to obtain additional information. This can delay the initiation of avoidance or protective actions. Minimizing the duration of this activity is a function of training in recognition skills.

Definition is an attempt to put meaning to the threat by describing it in qualitative terms and assessing a time context. This involves assembling all of the inputs received about the threat and synthesizing a model from them.

Evaluation is the process used to establish a response to the threat. This is the decision making involved in the "fight or flight" syndrome. The individual seeks a way to reduce stress and anxiety levels through this decision. The decision may be to mimic the actions of others facing similar choices or to choose an individual course of action. The decision is influenced by decision making exhibited by the group. Cultural influences and training to assume a role also can act as determinants. Training to assume a role is considered to be a key determinant in the process of making a decision. Training can reduce the anxiety level associated with a particular course of action. The outcome of the decision is never totally predictable, nor are the actions of individuals trained in these situations, regardless of the level and repetition of training.

Commitment refers to the mechanisms used by the individual to initiate the action decided on during the evaluation. Success leads to reduction of the stress and anxiety intrinsic to the situation. Lack of success leads to reassessment.

Reassessment is a highly stressful part of the process, as it represents tacit acknowledgment that actions taken during the commitment phase have failed to produce the expected results. Actions chosen during reassessment tend to be less well considered as the anxiety level increases. The potential for injury increases considerably, as the risk inherent in the action increases.

The first consideration during emergency response must be the welfare of the people who are directly or indirectly involved. Accidents are highly stressful events. An accident affects members of a group in different ways. In this context, members of the group can include individuals directly involved in the accident, bystanders, and emergency services personnel (Wilkerson 1985). About 1/8 to 1/4 of the members of the group react effectively. They often are excited and too busy to worry. About 3/4 of the members of the group are stunned and bewildered, showing no emotion. They usually are inactive, indecisive, and docile, and display physiological reactions of fear. The latter include sweating, palpitations, tunnel vision, or dry mouth. The remaining 1/8 to 1/4 of the members of the group react with grossly inappropriate behaviors. These can include paralyzing anxiety, hysteria, or psychotic reactions.

At the time of the accident and for several hours thereafter, some victims are stunned, confused, and paralyzed with anxiety. Some are vigilant and cool. Some are emotionally aggressive and may display anger toward themselves or others, or a sense of relief. All of these symptoms are normal. All of these individuals need listening and reassurance and acceptance about the normalcy of their responses. Involving capable victims at this time to provide assistance to others is an important reinforcer of self-esteem. These actions are essential to minimize the long-term impact of the event on these individuals.

The technical aspects of emergency response include:

- Recognizing the occurrence of the accident
- Initiating the emergency response by notifying appropriate participants
- Identifying the nature of the accident
- Assessing the impact of the event
- Minimizing the impact by taking measures appropriate to the circumstances
- Stabilizing the situation to prevent additional damage

Recognition of the onset of the accident should occur rapidly if working relationships are normal and communications systems are satisfactory. This is one important benefit from the "buddy" system. The "buddy," if not also involved in the accident, should sound the alarm that starts the response operation. Any situation where working alone is occurring must provide an equivalent basis for communication of the need for help.

The alarm could be a yell to others in the vicinity for help, sounding of a horn, or a call on the radio. In any case, a prearranged protocol for the alarm can save valuable time during the response. The alarm should indicate the nature of the problem, number of victims, and the location. This assists recipients of the message who must initiate further action. This is especially important when time is crucial or where difficulties in gaining access exist.

In some situations, an attendant will be involved in the process. The attendant monitors the progress of work in the space and assists the entrants in their needs. The attendant may be stationed at the entry to the space or at some other location. While the attendant may perform other duties, these must not impede the ability of this individual to monitor work in the space. Where work involves many individuals, the attendant may monitor radio or other communication. The attendant could be a computerized communication system. The electronic attendant sends a signal to a receiver/sender carried by the worker. The worker must respond in a specific manner within a specified time-frame to indicate well-being; otherwise, an alarm sounds to initiate the emergency response.

The attendant or other authorized or Qualified Person, such as the entry supervisor, may order evacuation of unaffected workers from the space. This offers the advantage of determining the whereabouts and status of the other workers. This is essential where conditions inside the space following the accident are uncertain. Persons working in the space, but not directly affected by the accident, could decide to reenter in an effort to assist the rescue effort. This could involve removal of debris and obstacles that could hamper rescue efforts. Permitting reentry is a judgment call reflecting information known about the nature of the accident and the conditions in the space. Reentry should occur only under assured conditions of safety; otherwise, secondary accidents could occur and cause additional victims. This could necessitate shutdown and lockout of electrical

FIGURE 15.8 Extrication equipment. (A, B) Equipment for extricating victims immobilized in stretchers should be high enough to provide sufficient clearance for complete extrication. Note the protection provided to the victim's head and neck provided by the top curl of plastic in the stretcher. (Courtesy of Skedco, Inc., Portland, OR.)

circuits, equipment, and gas supplies for welding, as well as other processes. Accident summaries for fatal accidents published by NIOSH, OSHA, and MSHA indicate that derivative accidents are likely only where hazardous atmospheric conditions are present (NIOSH 1979, 1994; OSHA 1982a, 1982b, 1983, 1985, 1988, 1990; MSHA 1988). In most circumstances, the best course of action would be to evacuate nonaffected persons from the space.

In some circumstances, retrieval equipment may be present at the entrance to the space. This equipment provides rapid removal capabilities (Figure 15.8). It is especially useful under some circumstances, such as entry that must occur in the presence or suspected presence of hazardous atmospheres. However, rapid action using this or other mechanical equipment to extricate or retrieve a victim who has a back or neck injury could exacerbate this state and potentially cause needless permanent paralysis. Management of the emergency response is essential, so that the rescue occurs as quickly as possible, in a manner that minimizes further injury to the victim.

A parallel situation occurs in serious automobile accidents. Older model automobiles held together by bolts tended to break open during the impact of the accident. This seriously compromised the safety of the occupants who often were flung from the automobile. Newer methods of welded construction have improved the integrity of the passenger compartment. At the same time, this created new problems. Deformation of the structure prevents opening of doors and may reduce the size of window spaces. The victim can be trapped inside and may have a serious neck or spinal injury. Gasoline may have leaked onto the ground and could pose a fire hazard. Traditional methods, such as cutting, using oxyacetylene torches and metal saws, to free the victim are contraindicated because of the potential for ignition of the fuel.

FIGURE 15.8 (continued)

The "jaws of life" have revolutionized extrication in these circumstances. The jaws of life use hydraulics to separate parts of the car body. This provides rapid access to the victim without creating a fire/explosion hazard and at the same time minimizes the chance of causing further injury. This technology is especially useful where the automobile has left the road and the site is inaccessible to fire hoses.

While there are some technical similarities between accidents involving automobile accidents and those involving confined spaces, there are important differences. Rescue operations provided by fire and emergency medical services occur through a formal management structure. This structure functions day after day and, as a result, attains some efficiency. This is much less likely to occur in the industrial context, at the very least, because of the unusual nature of an accident situation.

Yet, efficiency is essential for success during emergency response. Success is measured in terms of minimizing the impact of the accident on the victim and other affected individuals, and ultimately, the organization. Success, therefore, demands efficiency. Efficiency results from management of the situation. The manager of the situation must be able to act decisively to ensure activation of external agencies, and assessment and preparation for transport of the victim. An effective emergency response manager knows from previous contact under nonthreatening conditions which of the bystanders will be capable of providing assistance. This involves assessment of personality, as well as recognition of the influence of cultural and racial heritage.

Assessment, extrication, critical interventions, and preparation for transport of the victim must occur in a systematic and efficient manner. The fact that the accident occurred in a confined space can complicate this requirement. Some of the fundamental questions to which emergency service providers need answers should have been addressed during the sounding of the initial alarm.

Additional questions for which answers are needed include the nature of hazardous conditions in the space, need for special equipment, and need for assistance from other trained individuals.

Assessment of the scene for hazardous conditions can occur through the actions of qualified or authorized persons, such as the entry supervisor, testing personnel, industrial hygienist, or safety professional. Hazardous conditions should be anticipated prior to the entry of emergency service personnel. The approach taken in this book is to minimize hazardous conditions prior to entry and start of work activity. If this approach is followed, residual hazardous conditions and hazardous conditions resulting from the work activity prior to the accident should be known and minimized. This approach also should enable anticipation of hazardous conditions resulting from the accident. Establishing safe conditions could necessitate shutdown and lockout of electrical circuits, equipment, and gas supplies for welding and other processes if this did not occur previously. Accident summaries for fatal accidents published by NIOSH, OSHA, and MSHA indicate that derivative accidents are most likely where hazardous atmospheric conditions are present (NIOSH 1979, 1994; OSHA 1982a, 1982b, 1983, 1985, 1988, 1990; MSHA 1988).

In some circumstances, elimination or control of hazardous conditions is not possible. This could happen, for example, when the entry and work occur in response to an emergency situation. Entry of the emergency service provider could not occur in safety until the resulting hazardous conditions were identified and assessed and appropriate protective measures were adopted. This necessity, unfortunately, lengthens the period before assistance to the victim can begin.

Once assessment of conditions inside the space has occurred and appropriate measures have been taken, the emergency service provider can enter to appraise the situation and condition of the victim. Protocols established by the Workers' Compensation Board of British Columbia provide an example of an approach for emergency service providers (Workers' Compensation Board 1991, ABC Emergency Care Training School 1991). The first aid attendant should be activated by the call for help and could be the first emergency service provider to arrive at the scene. The first aid attendant brings a trauma pack to the scene. The trauma pack contains medical equipment used for assessing the condition of the victim. The trauma pack should include:

- Gloves
- Assorted airways
- Pocket mask
- Oxygen/bag-valve mask with reservoir
- Suctioning device
- Assorted hard collars
- Spine board + straps + padding
- First-aid kit, including pressure dressings, scissors and tape

The trauma pack should include **everything** needed for initial assessment of the condition of the victim.

The first action of the emergency service provider once gaining access to the space is to assess the situation. Verbalizing observations and comments provides reinforcement for the learning process undertaken by emergency service provider, as well as informing others about conditions. This is critical, so that mobilization of additional resources can occur. Verbalizing into a radio provides a means for recording the observations and comments. This information may be needed during a subsequent accident investigation or other follow-up action. Cross-training in both emergency first aid and safety is an ideal combination for an emergency service provider, since attributes from both areas are needed in these individuals.

An important component of the assessment is to determine the number of victims. The number of victims is important, since this will influence priorities and order of providing attention. The actual number may not have been available to observers outside the space. In the confusion of the accident and the evacuation, not all entrants or other workers necessarily could be documented. This situation easily could arise where the number of workers at a site changes continuously or where contractors and subcontractors may be present.

Recognition of conditions in the space and their relationship to the accident can provide insight about the nature of injuries suffered by the patient. Injury results from transfer of energy from the surroundings to all or part of the body. Energy transfer can occur during a fall from a height, rollover of equipment, contact with a moving object, crushing, shearing, or other violent action. Key areas of enquiry are the magnitude and direction of the force and areas of the body to which it was applied. Suspected or apparent serious injury is the basis for assigning the victim to the category of rapid transport to hospital. Certain types of injury are more likely than others to cause shock. These include falls from heights greater than 6.5 m (20 ft), rapid deceleration, contact with objects moving faster than 30 km/h (19 mi/h), severe crush injuries, major fractures, and bleeding. Treatment administered to these individuals is intended for critical intervention only. Intervention in this situation other than preparation for transport would be cervical spinal control and insertion of an airway.

The first action performed by the first aid attendant is the primary survey (Figure 15.9). This requires about 2 minutes. The focus of the primary survey is to determine the level of consciousness, and the status of the airway, breathing, and circulation. The primary survey utilizes simple observational tools to elicit this information. Circulation must be checked frequently for signs of developing shock. Initial stability could deteriorate into circulatory shock. All trauma patients who have a decreased level of consciousness should be assumed to have a cervical spinal injury and treated accordingly.

The secondary survey is performed later. Its function is a rapid and complete assessment of the patient. The basics of the secondary survey include: reassessment of vital signs, details of the accident and injury, medications, allergies, and a head-to-toe examination.

In some accidents involving confined spaces, the victim falls into the structure or becomes trapped by fallen structure or debris and thus becomes inaccessible to rescuers. Assessing, stabilizing, and preparing for transport becomes a race against time. With an injured person trapped inside, the risk of onset of traumatic shock becomes a real possibility. Freeing the victim so that preparation for transport can begin is utmost in importance. This requires efficiency to minimize the delay between notification of the accident and beginning of the response. Freeing the victim may be possible by clearing debris or the structure. In some cases this could require demolition of the internal and/or external structure. These actions could create new hazardous conditions, especially where ignitable and/or toxic materials are present and were not considered in the original hazard assessment and preparation for entry and work. Emergency actions of this kind should be undertaken only with the greatest of care.

In some cases, the victim cannot be freed readily by emergency demolition. These situations pose the greatest of difficulty to emergency service providers. A victim who is not endangered possibly can wait for further action. A victim who has entered shock or potentially could do so poses a considerable moral dilemma. Saving the life of the individual could require amputation of a limb or movement of the body that potentially could cause permanent paralysis. On the other hand, failing to act in a decisive manner in these types of situations could cost a life.

RESTORATION AND REHABILITATION

Restoration and rehabilitation begin when the emergency has ended. That is, the situation has been stabilized. In some circumstances, the restorative and rehabilitative phase may begin immediately. In these instances, a division between the emergency and restorative and rehabilitative phases represents only an administrative distinction. However, in many other circumstances, this discontinuity provides a valuable interlude to shift emphasis and approach.

PATIENT ASSESSMENT CHART
Prevention Division

Patient's Name			Social Insurance Number	
Last First Middle				
Address			Phone	
Occupation	Marital Status		Date of Birth (Day Month Year)	Age
Employer's Name	Location of Project			Phone
Employer's Address				
Time/Date of Accident a.m. p.m. Day Month Year	Time/Date To Attendant a.m. p.m. Day Month Year			

OXYGEN ADMINISTRATION	N/A	AIRWAY
Method		Flow Rate
Started		Discontinued

CONTINUOUS MONITORING CHART

Time										
Level Of Conc.		Eyes								
		Motor								
		Verbal								
Resp		Rate								
		Character								
Pulse		Rate								
		Character								
Pupils	L	Size								
		React								
	R	Size								
		React								
Skin		Colour								
		Temp								
Capillary Refill										
Body Temperature										

Pupil Size (mm)
8
7
6
5
4
3
2
1

	Eye Opening	Motor Response	Verbal Response
Coma Scale	4 Spontaneous 3 To Voice 2 To Pain 1 None	6 Obeys Commands 5 Localizes to Pain 4 Withdraws to Pain 3 Decorticate Response (Flexion) 2 Decerebrate Response (Extensor) 1 No Response	5 Normal 4 Confused but Coherent 3 Simple Inappropriate Words 2 Incomprehensible Speech 1 No Response

55M60 (R4/94)

FIGURE 15.9 (A and B) Patient assessment chart. This chart provides a systematic approach for evaluating the status and condition of potential victims of an industrial accident. (Courtesy of Workers' Compensation Board of British Columbia, Prevention Division, Richmond, BC.)

An accident can have a devastating impact on individuals directly or peripherally involved in the event. Training never can duplicate the actual situation. What is missing is the psychological/emotional trauma. This can affect victims and bystanders, as well as on-site (and external) emergency service providers. Emotional and psychological responses are a normal response to traumatic accidents (Wilkerson 1985). These responses are similar to bereavement or grief. Grief

PHYSICAL ASSESSMENT

History and/or Chief Complaint ______________________

POSITION FOUND IN	NATURE OF INJURY		LOCATION OF INJURY	L	R		L	R	SYMPTOMS		SIGNS
Standing	Amputation	Exposure (Low Temp)	Scalp			Knee			Pain (Rest)	Loss of Sensation Partial	External Bleeding
Sitting	Asphyxia	Heat Stress Exhaustion	Skull			Lower Leg			Pain on Movement	Loss of Sensation Complete	Bleeding from Orifices
Semi Recl.	Burn (Heat)	Irritation of Joints	Brain			Ankle			Pain on Breathing	Vision Impaired	Skin Bruised
Supine	Burn (Chemical)	Poisoning	Ear			Foot			Pain on Swallowing	Vision Absent	Excessive Persp.
Semi Prone	Concussion	Radiation/ Sunburn	Eye			Toes			Feels Cold	Agitation	Vomitting
On Side	Infectious Disease	Abrasions	Face			Shoulder			Feels Hot	Hyperactivity	Diarrhea
Prone	Bruise	Strains/Sprains	Neck			Upper arm			Feels Weak	Other	Shivering
Hanging	Laceration	Multiple Fractures	Chest			Elbow			Nauseous		Convulsions
Enmeshed	Dermatitis	Occupational Disease	Abdomen			Forearm			Dizziness		Odor on Breath
Other	Dislocation	Other	Hip			Wrist					Abnormal Behaviour
	Electric Shock		Back			Hand					Deformity
	Fracture		Thigh			Fingers					Swelling
											Discolouration
											Point Tenderness
											Other

TREATMENT AND DISPOSITION

PATIENT POSITIONED	TREATMENT		TRANSPORTED ON	TRANSPORTED IN	DISPOSITION	
Standing	Airways Cleared	Limb Immob	Ambulance Cot	Helicopter	Returned To Work	Personal Residence
Sitting	Art. Vent.	Trunk Immob	Spine Board	Airplane	Light Duty	Doctors Clinic
Semi-sitting	Vent. Assisted	Cervical Collar	Basket Stretcher	EHSC Ambulance	Off For Day	Acute Care Hospital
Semi-recumbent	CPR	Traction	Furley Stretcher	Company Vehicle	Refused Medical Attention	Other
Supine	Cleansed Wound	Joint Manipulation	Other	Own Vehicle	Refused Transport	
¾ Prone	Dressed Wound	Other		Taxi		
Other	Cold Applied			Other		

Additional Comments ______________________

Name of First Aid Attendant *(Please Print)* ______________________

FIGURE 15.9 (continued)

is the cycle of emotional and behavioral reactions set in motion by major loss. The cycle of grief contains the major elements of protest, despair, and detachment. This cycle can be lengthy (6 to 18 months) and painful.

The protest phase is characterized by shock, denial, anger, and guilt. Physical symptoms of nausea, loss of appetite, and sleep disturbance, as well as other changes, can accompany the psychological/emotional ones. This phase could be especially difficult for individuals who were present at the scene and permitted or enabled the accident to happen. This phase could continue for days or weeks.

The second phase, despair, is characterized by anguish, grief, and depression. Individuals in this phase exhibit slowed thinking and action. As a consequence, they are potentially at risk of involvement in other accidents. This phase could continue for weeks or months.

Apathy, indifference, and the desire to withdraw and to give up characterize the third phase, detachment. During the third stage, the victim begins the return to normalcy by accepting ownership for the problems arising from the accident and attempting to correct them. Emotional responses are less intense than during the second stage. Some individuals require care from a therapist to overcome these difficulties.

Most individuals suffer the effects of grief immediately after the accident. For some the final effects may not occur until 6 to 12 months after the incident. The impact of this situation could last a lifetime. Abnormal reaction to grief is called the delayed stress syndrome or post-traumatic stress disorder. The manifestations of this disorder include psychological, psychosomatic, and physical illness. Professional counseling is needed.

Emergency service providers have the same potential needs as the victims and therefore could require assistance. This is especially true for emergency service providers who are internal to the organization. These individuals normally could perform other jobs completely unrelated to this work. As a result, the accident easily could be as traumatic as, if not more so, for the victims. Internal emergency service providers, who may be volunteers, may sense the harsh glare of the spotlight during rescue activities, and may hold themselves accountable for errors, omissions, and deficiencies in their actions.

Vigorous, stress-relieving exercise for emergency service providers is recommended within 24 hours after the end of the accident. A mandatory debriefing should occur within 48 hours. A supportive, noncritical atmosphere is essential. A debriefing also could be beneficial for individuals who were present at the scene of the accident. Support is especially important for any feelings of negativity experienced by each individual. Prompt debriefing is important to prevent entrenchment of these feelings.

These manifestations of grief indicate that an industrial accident could produce a catastrophic impact on the operation of an organization. Individuals who were directly or indirectly involved could suffer recrimination and postincident stress-related reactions for a prolonged period. Postincident stress counseling should be considered an essential service to enable these individuals to return to normalcy. The need for postincident stress counseling is being recognized by large organizations, but this service is not readily available to individuals in the community.

SUMMARY

This chapter is concerned with preparing for and responding to accidents that can happen in confined spaces. Hazard assessment performed in a previous chapter should indicate the nature of accidents that are believed possible. Analysis of injury statistics for nonfatal accidents provides the basis for predicting the nature of injuries that could result.

Preparation for accidents should consider the nature of possible accidents and injuries, as well as location of the site. The needs for preparation are somewhat different for remote and inaccessible sites compared to those located in cities and urban areas.

The overriding consideration in emergency first aid is to prevent the onset of shock in the victim. This dictates the nature of actions that should and can be attempted at the accident scene. These actions reflect the conflict between the need to transport the victim to a hospital operating room as rapidly as possible and the desire to avoid causing further injury.

A traumatic accident and emergency response form a highly stressful experience for everyone: the victim, bystanders, and emergency service providers. The emotional needs of everyone should be addressed as part of rehabilitation and restoration.

REFERENCES

ABC Industrial Emergency Care Training School Inc.: *Industrial First Aid, A Study Guide.* Vancouver, BC: ABC Industrial Emergency Care Training School, 1991. pp. 7–21.

Abel, F.L.: Heart and circulation. In *Basic Physiology for the Health Sciences,* 2nd ed., Selkurt, E.E. (Ed.). Boston: Little, Brown and Company, 1982. pp. 272–322.

American National Standards Institute: *American National Standard for Emergency Eyewash and Shower Equipment* (Z358.1-1990). New York: American National Standards Institute, 1990.

American National Standards Institute: *Safety Requirements for Confined Spaces* (ANSI Z117.1-1995). Des Plaines, IL: American Society of Safety Engineers/American National Standards Institute, 1995. 32 pp.

Anonymous: *Emergency Eyewash Handbook.* Arlington Heights, IL: Fendall Company, 1991. 37 pp.

Bier, J.W. and T.K. Sawyer: Amoebae Isolated from Laboratory Eyewash Stations. *Curr. Microbiol. 20:* 349–350 (1990).

Block, S.S.: Definition of terms. In *Disinfection, Sterilization and Preservation,* Block, S.S. (Ed.), 2nd ed. Philadelphia, PA: Lea & Febiger, 1977.

Bowman, E.K., A.A. Vass, R. Mackowski, B.A. Owen, and R.L. Tyndall: Quantitation of free-living amoebae and bacterial populations in eyewash stations relative to flushing frequency. *Am. Ind. Hyg. Assoc. J. 57:* 626–633 (1996).

Bromberg, J. and D. Hirschfelder: Workplace eye injuries: the problem and solution. *Prof. Safety:* June (1984). pp. 15–20.

Bryan, J.L.: Human behaviour and fire. In *Fire Protection Handbook,* 17th ed. Cote, A.E. and J.L. Linville (Eds.). Quincy, MA: National Fire Protection Association, 1991. pp. 7-3 to 7-17.

Bureau of Labor Statistics: Accidents Involving Eye Injuries (Report 597). Washington, D.C.: U.S. Department of Labor, 1980.

Bureau of Labor Statistics: Accidents Involving Foot Injuries (Report 626). Washington, D.C.: U.S. Department of Labor, 1981. 22 pp.

Canadian Standards Association: *Emergency Planning for Industry: Major Industrial Emergencies, A National Standard of Canada* (CAN/CSA-Z731-95). Rexdale, ON: Canadian Standards Association, 1995.

Centers for Disease Control: Recommendations for prevention of HIV transmission in health-care settings. *Morbid. Mortal. Weekly Rep. 36(2S):* August 21, 1987. 17 pp.

Centers for Disease Control: Hantavirus Infection — Southwest. Recommendations for Risk Reduction (MMWR July 30/1993/vol. 42/No. RR-11. Atlanta, GA: DHHS/PHS/CDC, 1993. 13 pp.

Champagne, R.: A deluge can save the eyes. *Occup. Health Safety 63:* October, 1994. pp. 69–74.

Chang, F.C., P.B. Harrison, R.R. Beech et al.: PASG: does it help in the management of traumatic shock? *J. Trauma 39:* 453–456 (1995).

Charous, L.: The puzzle of latex allergy: some answers, still more questions. *Ann. Allergy 70:* 177–180 (1994).

Cleveland, R.J.: Factors that influence safety shoe usage. *Prof. Safety. 29:* August, 1984. pp. 26–29.

Click, M.: Airway another way: blind nasotracheal intubation. *J. Emerg. Med. Serv. 21(2):* 58–63 (1996).

Fortenberry, J.E., J. Lane, and M. Shalit: Use of epinephrine for anaphylaxis by emergency medical technicians in a wilderness setting. *Ann. Emerg. Med. 25:* 785–787 (1995).

Friedman, R.: Theory of fire extinguishment. In *Fire Protection Handbook,* 17th ed. Cote, A.E. and J.L. Linville (Eds.). Quincy, MA: National Fire Protection Association, 1991. pp. 1-72 to 1-82.

Hamann, C.: Latex hypersensitivity: an update. *Allergy Proc. 15:* 17–20 (1994).

Hartzell, G.E.: Combustion products and their effects on life safety. In *Fire Protection Handbook,* 17th ed. Cote, A.E. and J.L. Linville (Eds.). Quincy, MA: National Fire Protection Association, 1991. pp. 3-3 to 3-14.

Hugo, W.B (Ed.): *Inhibition and Destruction of the Microbial Cell.* London, U.K.: Academic Press, 1971.

Jagger, D.: When every second counts. *Occup. Health Safety Can. 4:* 80–90 (1988).

Lahey, J.W.: Dermatosis under your skin? *Natl. Health Safety News:* March, 1986. pp. 80–83.

McKenna, K., J. Hamilton, and K. Walsh: Latex allergy, the dark side of infection protection. *J. Emerg. Med. Serv. 21(4):* 59–64 (1996).

Mine Safety and Health Administration: Think "Quicksand": Accidents Around Bins, Hoppers and Stockpiles, Slide and Accident Abstract Program. Arlington, VA: U.S. Department of Labor, Mine Safety and Health Administration, National Mine Health and Safety Academy, 1988.

Mossesso, V.N. and M. Yee: Asthma. *J. Emerg. Med. Serv. 21(4):* 65–73 (1996).

Murphy, P.M., III and T. Tkach: The heat is on: treating heat-related emergencies. *J. Emerg. Med. Serv. 21(6):* 54–62 (1996).

National Fire Academy: *Incident Command System.* Stillwater, OK: Fire Protection, Oklahoma State University, 1983. 220 pp.

National Fire Protection Association: *NFPA 306: Control of Gas Hazards on Vessels* (1993 ed.). Quincy, MA: National Fire Protection Association, 1993. 15 pp.

National Institute for Occupational Safety and Health: Criteria for a Recommended Standard — Working in Confined Spaces (DHEW/PHS/CDC/NIOSH Pub. No. 80-106). Cincinnati, OH: National Institute for Occupational Safety and Health, 1979. 68 pp.

National Institute for Occupational Safety and Health: Guidelines for the Prevention of Transmission of Human Immunodeficiency Virus and Hepatitis B Virus to Health-Care and Public-Safety Workers (DHHS(NIOSH) Pub. No. 89-107). Atlanta, GA: DHHS/PHS/CDC/NIOSH, 1989.

National Institute for Occupational Safety and Health: Worker Deaths in Confined Spaces (DHHS/PHS/CDC/NIOSH Pub. No. 94-103). Cincinnati, OH: National Institute for Occupational Safety and Health, 1994. 273 pp.

NIOSH: Respiratory Protective Devices; Final Rules and Notice, *Fed. Regist. 60:* 110 (8 June 1995) pp. 30336-30404.

National Safety Council: *Accident Facts,* 1993 ed. Itasca, IL: National Safety Council, 1993. pp. 36–38.

Occupational Safety and Health Administration: Selected Occupational Fatalities Related to Fire and/or Explosion in Confined Work Spaces as Found in OSHA Fatality/Catastrophe Investigations. Washington, D.C.: U.S. Department of Labor, Occupational Safety and Health Administration (U.S. DOL/OSHA), 1982a. 76 pp.

Occupational Safety and Health Administration: Selected Occupational Fatalities Related to Lockout/Tagout Problems as Found in Reports of OSHA Fatality/Catastrophe Investigations. Washington, D.C.: U.S. Department of Labor, Occupational Safety and Health Administration (U.S. DOL/OSHA), 1982b. 113 pp.

Occupational Safety and Health Administration: Selected Occupational Fatalities Related to Grain Handling as Found in Reports of OSHA Fatality/Catastrophe Investigations. Washington, D.C.: U.S. Department of Labor, Occupational Safety and Health Administration (U.S. DOL/OSHA), 1983. 150 pp.

Occupational Safety and Health Administration: Selected Occupational Fatalities Related to Toxic and Asphyxiating Atmospheres in Confined Work Spaces as Found in Reports of OSHA Fatality/Catastrophe Investigations. Washington, D.C.: U.S. Department of Labor, Occupational Safety and Health Administration (U.S. DOL/OSHA), 1985. 230 pp.

Occupational Safety and Health Administration: Selected Occupational Fatalities Related to Welding and Cutting as Found in Reports of OSHA Fatality/Catastrophe Investigations. Washington, D.C.: U.S. Department of Labor, Occupational Safety and Health Administration (U.S. DOL/OSHA), 1988. 225 pp.

Occupational Safety and Health Administration: Selected Occupational Fatalities Related to Ship Building and Repairing as Found in Reports of OSHA Fatality/Catastrophe Investigations. Washington, D.C.: U.S. Department of Labor, Occupational Safety and Health Administration (U.S. DOL/OSHA), 1990. 195 pp.

OSHA: Occupational Exposure to Bloodborne Pathogens; Final Rule, *Fed. Regist. 56*: 235 (6 December 1991). pp. 64175–64182.

OSHA: Permit-Required Confined Spaces for General Industry; Final Rule, *Fed. Regist. 58*: 9 (14 January 1993). pp. 4462–4563.

OSHA: Confined and Enclosed Spaces and Other Dangerous Atmospheres in Shipyard Employment; Final Rule, *Fed. Regist. 59*: 141 (25 July 1994). pp. 37816–37863.

Paszko-Kolva, C., H. Yamamoto, H. Shahamat, T.K. Sawyer, G. Morris, and R.R. Colwell: Isolation of amoebae and *Pseudomonas* and *Legionella* spp. from eyewash stations. *Appl. Environ. Microbiol. 57:* 163–167 (1991).

Paszko-Kolva, C., T.K. Sawyer, and G. Gardner: Laboratory emergency eyewash station contaminants. *Am. Environ. Lab. 7:* March, 1995. pp. 18–20.

Perez, L. and D. Sumner: Snakes, snakebites, first aid and treatment. In *The Whole Hiker's Handbook,* Kemsley, W., Jr. (Ed.). New York: William Morrow and Company, 1979. pp. 111–117.

Tyndall, R.L., M.M. Lyle, and K.S. Ironside: The presence of free-living amoebae in portable and stationary eye wash stations. *Am. Ind. Hyg. Assoc. J. 48:* 933–934 (1987).

Voelker, R.: Latex-induced asthma among health care workers. *J. Am. Med. Assoc. 273:* 764 (1995).

White, R.D.: Shocking history: the first portable defibrillator. *J. Emerg. Med. Serv. 20(10):* 41–45 (1995).

Wilkerson, J.A (Ed.): *Medicine for Mountaineering*, 3rd ed. Seattle, WA: The Mountaineers, 1985. 438 pp.

Withey, S.B.: Reaction to uncertain threat. In *Man and Society in Disaster*, Baker, G.W. and D.W. Chapman (Eds.). New York: Basic Books, 1962.

Workers' Compensation Board of British Columbia: *Industrial First Aid, A Reference and Training Manual*, Dresser, A.A., R.L. Hazelton, D.G. Hunt, D.H. Schreiber, T.D.B. Smith, and S. Wilson (Eds.). Richmond, BC: Workers' Compensation Board of B.C., 1990. 417 pp.

Workers' Compensation Board of British Columbia: *Occupational First Aid Regulations.* Richmond, BC: Workers' Compensation Board of B.C., 1994. 62 pp.

Wysocki, T.J.: Carbon dioxide and application systems. In *Fire Protection Handbook*, 17th ed. Cote, A.E. and J.L. Linville (Eds.). Quincy, MA: National Fire Protection Association, 1991. pp. 5-232 to 5-240.

Appendix A: Standards, Guidelines, and Regulations

CONTENTS

INTRODUCTION

Various groups — standard-setting organizations, industry and trade associations, and regulatory agencies — have sought to produce a management framework for work involving confined spaces. Inherent in each of these attempts is recognition of the special concerns arising from entry and work in these workspaces. Each provides a philosophy for minimizing the risk of performing work in these environments. Creating a workable management framework is very difficult. Confined spaces encompass an infinite number of workspaces, environments, and combinations of hazards. Also, hazards associated with confined spaces cover a broad range of severity ranging from the inconsequential to the life threatening. An important point also worth stressing is the difficulty in raising the level of knowledge of entrants about the nature of the hazards and the technical expertise needed for assessing and controlling them.

By necessity, approaches for addressing hazards posed by confined spaces/confined atmospheres must be performance based, rather than specification based. Performance-based strategies specify end results. They provide key outcomes, but few, if any, details about the process for attaining them. Inherent in this type of approach is the need for highly knowledgeable, highly skilled, and highly creative resource persons. Those charged with implementing this process must be able to analyze the needs of an organization and to synthesize a strategy that best suits them. This requirement can be highly intimidating. There are few resources to assist in the process. Building a confined space program on a performance-based framework requires high level, broadly based expertise. This type of expertise is difficult to obtain; yet, the very real hazards present in these environments have no respect for this dearth. The strength of performance-based standards is their flexibility and the opportunities that they provide for creative solutions that uniquely suit the needs of the organization.

Another challenge in working with performance-based standards is the need for real-world strategies that fit within the larger operational context of an organization. No single legal statute can be permitted to monopolize the attention of health and safety resources within an organization. This forces compromise. Compromise tends to limit coverage of the standard only to environments that contain major hazards. The nature of the compromises applied in various standards is a subject of discussion in the remainder of this Appendix.

Specification-based standards contain detailed requirements for action. While these may suit the characteristics of a particular hazard or reflect the needs of a specific industry, they may be completely inapplicable to others. The specific requirements outlined in a specification-based standard can provide comfort zones to the less knowledgeable. However, the narrowness of tolerance in this type of standard can severely restrict creative solutions and certainly cannot accommodate all hazards and situations. A comprehensive specification standard would prescribe actions for all situations. This would require intricate detail and would be cumbersome in the extreme.

Specification-based standards are not practical for confined spaces, given the broad spectrum of workspaces, environments, and hazards that they contain. However, performance-based standards require consideration for the lack of knowledge in the many who will be required to implement them. This view certainly is reflected in the work produced by groups and agencies reviewed in this Appendix.

Strategies for managing entry and work in confined spaces reviewed here have a number of similarities. Generally, the response starts following recognition of the presence of confined spaces in the physical surroundings of an organization. Encoding the cue for that critical first aspect in the process — recognition — has proven elusive in all standards. Without recognition of the relationship between the characteristics of the workspace and the potential for development of hazardous conditions, none of the requirements of any of the standards will ever be undertaken.

An agent common to the implementation strategies of all of the approaches reviewed here is the "qualified person." Rarely, however, are the qualities required in the Qualified Person ever defined precisely. The typical description for a Qualified Person is an individual who, by education, training, and experience, is qualified to perform the required duties. Some legislation identifies generic professional descriptions, such as industrial hygienist, safety professional, or marine chemist. The "Qualified Person" is the subject of Appendix B. Qualified Persons are required to perform duties requiring varying levels of knowledge and skill. Some approaches vary the level of qualification according to the requirements of the duty to be performed. This provides flexibility, so that the duties of the Qualified Person can be spread among many individuals, each performing duties to the level of a particular requirement.

Another theme common to almost all approaches is top-down management style. Management, acting through the Qualified Person, identifies and assesses the hazard, and prescribes work practices and procedures, and control and emergency measures. That is, management determines the conditions of work and specifies how the work is to be performed. The people who will perform the work often have no role in the process other than following instructions.

A variation on this approach mandates active involvement of the health and safety committee in the process. In this approach, management engages a "Qualified Person," who assesses the situation, determines actual and potential hazards, and recommends actions to be taken. Management presents the hazard assessment to the committee and then prepares entry and work practices, and control and emergency measures. These are presented to the committee for review and discussion.

Standards generally contain requirements for hazard identification and assessment; specification of work practices and procedures; control measures; training and education; recordkeeping; and communication and emergency measures.

While the strategies reviewed here for managing entry and work involving confined spaces have many similarities, they also have some significant differences. The starting point of divergence is the meaning of the term "confined space." **Confined space** has a different meaning to the creators of many of the standards. This lack of agreement obviously has major implications, since minor variations in the definition can produce considerable differences in interpretation and application. These differences may be academic, since in many jurisdictions the definition is codified in the law of the land. A definition that is too restrictive would be incapable of encompassing the diversity of workspaces and situations in which accidents have happened. Table A.1 presents definitions contained in standards reviewed in this Appendix. These are broken down by concept. This provides the opportunity for comparison and highlights the difficulty in describing in concise language the meaning of the term, confined space.

Several of the standards reviewed here introduce strategies for breaking the problem into manageable pieces. One approach is to divide confined spaces by severity into serious and nonserious environments. This approach utilizes as divisors terms such as "enclosed space" and "confined space," classes of hazard, duration of entry, and nonpermit- and permit-required to differentiate these environments. The benefit inherent in this approach is that each workspace or situation receives the level of attention that it deserves. This enables limited organizational resources to be apportioned to the extent needed. The potential limitations in this approach lie in the suitability of the classification and its appropriateness during the passage of time. A change in severity of hazard or development of a new hazard could lead to inadequacy of response during subsequent entry and work. Assignment to a particular category and actions that are taken because of this are only as good as the skill of the person who performs this task, the constancy of conditions, and other vagaries of the real world.

RESEARCH STANDARDS: NATIONAL INSTITUTE FOR OCCUPATIONAL SAFETY AND HEALTH

In 1979, the National Institute for Occupational Safety and Health (NIOSH) published the Criteria Document entitled "Working in Confined Spaces" (NIOSH 1979). This comprehensive reference formed the basis for much of what followed in the development of the literature in this field. The Criteria Document described the results of a NIOSH-sponsored analysis of confined space accidents. It also reviewed the literature of the time.

NIOSH published the Criteria Document in response to concerns about the approach taken in the ANSI Standard on confined spaces (ANSI Z117.1-1977) and the OSHA General Industry Safety and Health Standards (ANSI 1977). These concerns focused on training, recommendations for personal protective equipment, acceptance of tagging as a substitute for lockout, and use of lifelines. Many of these concerns since have been resolved through updates in OSHA legislation. The purpose of the Criteria Document was to inform management and workers about work practices needed to address the hazards in confined spaces.

TABLE A.1
Definitions for the Term "Confined Space"

Characteristic	Description
Consensus Standards	
American National Standards Institute, 1977	
Geometric characteristics	Enclosures with limited means of access and egress, such as storage tanks, open-topped spaces more than 4 feet in depth, sewers
Occupancy	
Ventilation	Poor natural ventilation
Atmospheric confinement	
Other hazards	
American National Standards Institute, 1989	
Geometric characteristics	Completely enclosed space Restricted entry and exit meaning physical impediment of the body Isolation of occupants from external rescuers
Occupancy	Human occupancy not primary function
Ventilation	
Atmospheric confinement	Atmospheric contamination by toxic or flammable atmospheres or oxygen deficiency or excess
Other hazards	Potential or known hazards Engulfment hazard Physical hazards Biological hazards Mechanical hazards Unexpected entry of liquids, gases or solids during occupancy
American National Standards Institute, 1995	
Geometric characteristics	Completely enclosed space Restricted entry and exit meaning physical impediment of the body Isolation of occupants from external rescuers Person can enter bodily
Occupancy	Human occupancy not primary function
Ventilation	
Atmospheric confinement	Atmospheric contamination by toxic or flammable or explosive atmospheres or oxygen deficiency or enrichment
Other hazards	Potential or known hazards Engulfment hazard Physical agents Biological hazards Mechanical hazards Unexpected entry of liquids, gases or solids during occupancy
National Institute for Occupational Safety and Health, 1979	
Geometric characteristics	Limited openings for entry and exit
Occupancy	Not intended for continuous employee occupancy
Ventilation	Unfavorable natural ventilation
Atmospheric confinement	Geometry that could contain or produce dangerous air contaminants including, but not limited to, oxygen deficiency, explosive or flammable atmospheres, and/or concentrations of toxic substances
Other hazards	
American Society for Testing and Materials, 1984	
Geometric characteristics	Uses descriptive examples, such as vapor degreasers, cold cleaning tanks, storage vessels, tank cars and trucks, van trailers, ships or barges, pits or sumps, and unventilated rooms to define confined areas

TABLE A.1 (continued)
Definitions for the Term "Confined Space"

Occupancy	
Ventilation	Lack of ventilation
Atmospheric confinement	
Other hazards	
Standards Association of Australia, 1986	
Geometric characteristics	A space of any volume having restricted means for entry and exit
	Includes, but are not limited to, storage tanks, tank cars, process vessels, boilers, silos, other tank-like compartments having only a manhole for entry, open-topped spaces of more than 1.5 m in depth, such as pits or degreasers, which are not subject to good natural ventilation, pipes, sewers, tunnels, shafts, ducts and similar structures, shipboard spaces entered through a small hatchway or manhole, cargo tanks, cellular double bottom tanks, duct keels, cofferdams, ballast and oil tanks, and void spaces, but not including cargo holds
Occupancy	Not intended as a regular workplace
Ventilation	May have inadequate ventilation
Atmospheric confinement	May have an atmosphere which is either contaminated or oxygen deficient
Other hazards	Accidental operation of machinery, services
	Performance of nonroutine tasks
	Space is at atmospheric pressure during occupancy
Legal Statutes	
Occupational Safety and Health Administration, 1970	
Geometric characteristics	Enclosed workspaces having limited means of entry and exit
Occupancy	Not intended for continuous employee occupancy
Ventilation	May include open-topped spaces not subject to good natural ventilation
Atmospheric confinement	May be subject to accumulation of toxic or flammable contaminants and where an oxygen-deficient atmosphere may develop
Other hazards	
Occupational Safety and Health Administration, 1993 — General Industry	
Geometric characteristics	Large enough and so configured that an employee can bodily enter and perform assigned work
	Limited or restricted means for entry or exit
Occupancy	Not intended for continuous employee occupancy
Ventilation	
Atmospheric confinement	
Other hazards	
Occupational Safety and Health Administration, 1994 — Shipyard Employment	
Geometric characteristics	Large enough and so configured that an employee can bodily enter and perform assigned work
	Limited or restricted means for entry or exit
Occupancy	Not intended for continuous employee occupancy
Ventilation	
Atmospheric confinement	
Other hazards	
Mine Safety and Health Administration, 1989	
Geometric characteristics	Restricted openings for entry and exit
Occupancy	Not intended for continuous occupancy

TABLE A.1 (continued)
Definitions for the Term "Confined Space"

Ventilation	
Atmospheric confinement	Unfavorable atmosphere that could contain or produce dangerous air contaminants
Other hazards	
Labor Canada — Canada Occupational Safety and Health Regulations 1986	
Geometric characteristics	Enclosed or partially enclosed space
	Restricted means of access and egress
Occupancy	Not designed or intended for human occupancy except for performing work
Ventilation	
Atmospheric confinement	May become hazardous to an employee entering
	Presence of a hazardous atmosphere
Other hazards	Design
	Construction
	Location
	Materials or substances in it
	Any other condition relating to it
Labor Canada — Marine Occupational Safety and Health Regulations 1987	
Geometric characteristics	Storage tank, ballast tank, pump room, coffer dam, other enclosure excluding a hold
Occupancy	Not designed or intended for human occupancy except for performing work
Ventilation	Poor ventilation
Atmospheric confinement	Oxygen deficiency
	Presence of airborne hazardous substance
Other hazards	Design
	Construction
	Location
	Materials or substances in it
	Any other condition relating to it

The Criteria Document recommended broad-based measures for addressing the health and safety hazards associated with these workspaces. Of greatest interest to the industrial hygienist are aspects related to adverse atmospheric conditions. These included toxic and/or flammable substances as well as oxygen-enriched and oxygen-deficient atmospheres. Measures to address adverse atmospheric conditions formed the major focus of the Criteria Document.

The Criteria Document provides a rigorous approach to addressing the problem. The first step is appointment of a Qualified Person. A Qualified Person is designated by the employer in writing as someone capable by education and/or specialized training of anticipating, recognizing, and evaluating employee exposure to hazardous substances or other unsafe conditions in a confined space. Also, the Qualified Person shall be capable of specifying the necessary control and/or protective action to ensure worker safety. The Qualified Person obviously is the key to making the process work. As in other approaches, NIOSH did not specify the nature of the education or specialized training that would be required. While the NIOSH concept would put considerable responsibility onto the Qualified Person, it did not address the parallel need for authority.

The next step in the process is to identify and classify confined spaces within the organization. NIOSH established the following system for classifying the space.

- **Class A** confined spaces are immediately dangerous to life and health (IDLH) due to oxygen deficiency, flammable or explosive atmosphere, or the presence of toxic agents.
- **Class B** confined spaces have the potential to cause injury and illness if preventive measures are not taken. They involve hazardous conditions that are not immediately life threatening. (An example is airborne contaminants at concentrations above the Permissible Exposure Limit, but below the IDLH.)
- **Class C** confined spaces represent situations with much lower hazard potential. No specific modification of work procedures is required.

Classification is required before precautions specific to the workspace are recommended. The most hazardous condition that could occur during entry, working in, or exit from the confined space determines the classification. A work practice that could increase existing hazards or generate additional ones, would necessitate reevaluation to determine whether a change in classification would be warranted.

The Qualified Person would issue a permit prior to entry into any confined space. A permit is an authorization and approval in writing for entry and the performance of work. It certifies that the Qualified Person has evaluated and considered all existing or potential hazards. It specifies the location and type of work to be performed, establishes conditions for performing the work, and specifies necessary control and/or protective action to ensure worker safety.

The Criteria Document also specifies additional actions which are organized under the following headings:

- Entry and rescue
- Medical
- Training
- Testing and monitoring
- Labeling and posting
- Safety equipment and clothing
- Work practices

The following discussion highlights concepts mentioned in these sections.

Rescue procedures would be specific to each entry. A Class A or B confined space would require a trained standby person. The standby would be prepared for entry with equipment including a fully charged, positive-pressure, self-contained breathing apparatus (SCBA). Additional duties for the standby person would include procedures for ensuring the safe conduct of work and readiness for response in the event of an emergency. Specific duties would include maintaining unobstructed lifelines and communication with all workers within the confined space, and initiating the emergency response by summoning rescue personnel.

Workers entering Class A or B confined spaces would require a preplacement physical examination. This would address the rigors of the environment and demands posed by respiratory protection.

Training would be considered as complete only when the supervisor, or safety or training officer, judged that the employee had attained a level of proficiency acceptable for entry and work. The trainee's judgment of the adequacy of the training also would be considered.

Entry would be prohibited until evaluation of atmospheric conditions inside the space had occurred. This evaluation would occur from outside the space. Atmospheric tests would include oxygen content, flammability, and presence and level of toxic materials. Class A confined spaces would require continuous monitoring. The Qualified Person would determine monitoring requirements for Class B and C spaces.

Entry for any type of hot work would be prohibited when the concentration of flammable gases/vapors exceeded 10% of the lower flammable limit (LFL). A fire watch would be required for entry into any confined space containing an atmosphere classified as a Class II or Class III hazardous location according to Article 500 Sections 5 and 6 of the U.S. National Electrical Code (NFPA 70 –1993). In such areas, surface dust and fibers would be removed and no hot work initiated until the airborne particulate level was below 10% of the LFL for the material.

The percentage of oxygen required during entry and work would be at least 19.5% and not greater than 25% at 760 mmHg. When contaminants in the atmosphere could not be kept within permissible exposure levels, an approved respirator would be required.

Posting of warning signs printed in English and the predominant language of non-English-reading workers would be required at entrances to any confined space.

The Criteria Document specified purging and ventilating for achieving and maintaining environmental control within confined spaces. Conditions encountered or produced during the work would dictate the method to be used. Airflow measurements would be made before each workshift to ensure a safe environmental level. Continuous general ventilation would be required where toxic atmospheres are produced as part of a work procedure or could develop due to the nature of the confined space. Local exhaust ventilation would be required when general ventilation is ineffective due to restrictions in the confined space or when high concentrations of contaminant occur in the breathing zone of the worker.

Class A or B confined spaces would require complete isolation from all other systems by physical disconnection, double block and bleed valves, or the blanking of lines. In continuous systems, where complete isolation is not possible, specific written safety procedures could be used. In addition to blanking, pumps and compressors serving lines entering the confined space would be locked out to prevent accidental activation.

Isolation of the confined space to prevent accidental activation of equipment also would be required. This would be achieved by locking out circuit breakers and/or disconnects in the open (off) position. Mechanical isolation of moving parts would be required by disconnecting linkages, removing drive belts or chains, and blocking to prevent accidental rotation.

The Criteria Document also assigns to the Qualified Person the role of assessing work practices and procedures, and processes to be employed. These include, for example, procedures, such as cleaning and steaming, and use of tools and equipment.

The last section in the Criteria Document concerned recordkeeping. Recordkeeping would include records of training including safety drills, inspections, tests, and maintenance.

The Criteria Document leaves decision making to the judgment of the Qualified Person. While this may be one of the strengths of the concept proposed by NIOSH, it also is one of the weaknesses. The abilities of the Qualified Person are critical to the success of the concept. The training and experience needed to become a Qualified Person are difficult commodities to acquire.

Another weakness of the concept proposed by NIOSH was the definition of the confined space. The NIOSH definition does not encompass workspaces that can confine hazardous atmospheres. This is a result of the concept of occupancy and enclosure. The definition also fails to include physical hazards such as engulfment, mechanical and process hazards, and unstable interior conditions.

The third weakness of the concept, the classification scheme, simultaneously is one of its strengths. Some form of classification is a necessary real-world compromise to the problem of determining how to allocate scarce resources. However, rapid change in conditions could negate the precautions taken within the envelope of a particular hazard class. Of course, this criticism would apply to any classification scheme.

In addition to the Criteria Document, NIOSH published several other documents on the hazards of confined spaces. A comprehensive training manual for the construction industry was published in 1985 (NIOSH 1985). This document received considerable input from sources outside NIOSH. It is a broadly based resource for constructing a training program in this area. While intended for training construction workers, it is important to emphasize that construction includes both new construction and renovation and rehabilitation of existing structures. Tradespeople were prominently represented in the fatal accidents analyzed by OSHA and NIOSH and discussed in the first and second chapters. However, while directed to this audience, the concerns mentioned in this document are equally applicable to other work in confined workspaces.

In 1986 NIOSH issued "Request for Assistance in Preventing Occupational Fatalities in Confined Spaces" (NIOSH 1986a). The substance of this ALERT was concern that fatal accidents in confined spaces were continuing to occur, despite the existence of readily available resources and strategies to address the problem. NIOSH mentioned that over 60% of the fatalities in confined space accidents occurred among "would-be rescuers." Also, previously highlighted deficiencies in the management and conduct of work were recurring elements in these accidents. Workers and management failed to recognize confined spaces and to act prudently in preparing for entry and occupancy. The ALERT also cited the lack of specific regulatory rules in the U.S. to address these workspaces.

Another ALERT, entitled "Preventing Fatalities Due to Fires and Explosions in Oxygen-Limiting Silos" also was published in 1986. This ALERT was directed to fire situations involving oxygen-limiting silos in agricultural operations. Decaying organic matter can generate flammable atmospheres under anaerobic conditions. Leakage of small amounts of air into these silos can allow smoldering to occur through spontaneous combustion. Smoldering can result in accumulation of combustible gases. Agricultural dust made airborne during fire fighting can become the fuel for a dust explosion. While this information is not directly applicable to the normal industrial sector, it does highlight the complexity of processes that can occur in confined spaces and the highly technical nature of the measures needed to control them (NIOSH 1986b).

In 1987, NIOSH published "A Guide to Safety in Confined Spaces" (NIOSH 1987). This booklet is oriented toward workers. It explains the concepts of confined space hazards and control measures in concise terms, using simple drawings and text.

NIOSH issued an ALERT during 1988 that addressed engulfment caused by flowable bulk materials (NIOSH 1988). The substance of this ALERT was NIOSH concern about the continuing occurrence of fatal accidents involving flowable bulk solids and caked materials, despite the availability of resources to prevent them. Flowable bulk solids have the ability to cause engulfment. Caked materials that form on the surfaces of structures can dislodge suddenly and unpredictably and cause traumatic injury. NIOSH also expressed concern about the lack of specific U.S. regulatory involvement to prevent these accidents.

An ALERT published in 1989 warned about vapor hazards from halogenated substances (NIOSH 1989). At that time, NIOSH noticed an increase in fatal accidents attributed to inhalation of vapors from chlorofluorocarbons. The hazard from these products results from especially high vapor pressures and lack of sufficient sensory discomfort that otherwise would act as a warning property.

The latest ALERT was published in 1990. "Request for Assistance in Preventing Deaths of Farm Workers in Manure Pits" documented the continuing occurrence of fatal accidents in manure pits or tanks (NIOSH 1990). Manure pits or tanks potentially can confine gases such as methane, hydrogen sulfide, carbon dioxide, ammonia, and oxygen-deficient atmospheres that result from anaerobic digestion. NIOSH also expressed concern about the lack of specific U.S. regulatory requirements involving these workspaces.

During the same period, NIOSH initiated the Fatality Assessment and Control Evaluation (FACE) program to investigate fatal accidents (NIOSH 1994). The purpose of FACE was to identify factors that increase the risk of work-related fatal injury. While previous studies utilized only written reports or forms as the source of information, FACE included on-site visits. These visits provided the opportunity for firsthand observation and investigation of the worksite by a team of specialists and interviews with witnesses. The intent of the team approach was to provide continuity and consistency from one investigation to another and the opportunity for more precise reconstruction of accidents. FACE currently focuses on accidents involving falls from heights, contact with sources of electrical energy, entry into confined spaces, and contact with machinery.

FACE investigated 423 accidents involving the deaths of 480 workers from 1983 to 1993. Of these, 70 involving confined spaces caused 109 deaths. Multiple deaths occurred in 25 of these accidents.

CONSENSUS STANDARDS

American National Standards Institute

The American National Standards Institute (ANSI) published a consensus standard entitled "Safety Requirements for Working in Tanks and Other Confined Spaces" in 1977 (ANSI 1977). This standard was revised and updated in 1989 and again in 1995 (ANSI 1989, 1995). The revision in 1989 emphasized hazard recognition, hazard evaluation, training, and several other aspects not included in the 1977 version. The latter include classifying confined spaces on the basis of potential to pose hazards, relations with contractors, and provision of rescue and emergency services. A logic diagram for the process also is included. By necessity, this ANSI Standard is performance oriented. The revision in 1995 added finesse to the previous version and also brought it into line with the OSHA Standard on confined spaces for general industry (OSHA 1993).

The first step in following the ANSI Standard would be to appoint a Qualified Person. According to ANSI, a Qualified Person is knowledgeable in the operation to be performed and competent by virtue of training, education, and experience to judge the hazards involved and specify controls and/or protective measures. Acting on behalf of management, the Qualified Person would identify confined spaces present in the operation, equipment, and premises. The next step would be to identify hazards associated with each confined space. According to ANSI, hazards could include:

- Atmospheric contamination by toxic or flammable atmospheres or oxygen deficiency or enrichment
- Physical agents
- Biological hazards
- Mechanical hazards
- Unexpected entry of liquids, gases or solids during occupancy
- Isolation of occupants from external rescuers

The identification process would include consideration for past and current uses of the confined space, as well as physical characteristics and location. Additional considerations would include properties of previous and present contents, potential role of coatings used to line the space, and processes to be used during entry. The ANSI Standard also would require consideration of spaces and operations adjacent to the one under evaluation.

The next phase would require involve evaluation of the impact of exposure to the real and potential hazards. This evaluation would consider:

- Number and characteristics of employees affected by the exposure
- Magnitude of the hazard examined through energy release, toxicity, and quantity of chemicals involved
- Likelihood of exposure
- Seriousness of the outcome of the exposure
- Likelihood and consequences of change in conditions/activities not considered initially
- Strategies for control
- Specific needs for emergency response due to characteristics of the confined space

Completion of the hazard evaluation puts the Qualified Person at a critical point in the application of the Standard: deciding about the magnitude of actual and potential hazards. The ANSI Standard utilizes two classes — nonpermit confined space and permit-required confined space — to regulate the extent of the response.

For nonpermit confined spaces the Standard would require development of written procedures that specify measures and precautions to ensure safe entry and work. These could include some of the measures needed for entry into a permit-required confined space. These procedures also would identify conditions that would revoke the status as a nonpermit confined space and require reevaluation. Persons working in nonpermit confined spaces would need training in entry procedures and conditions that would prohibit entry. A condition for entry into a nonpermit confined space would include successful control of atmospheric hazards within acceptable limits, as indicated by testing performed by the Qualified Person. Testing could be waived if the space is ventilated properly before and during entry, provided that this means of control is shown to be satisfactory through formal hazard identification and evaluation.

For permit-required confined spaces, the Standard would require issuance of a permit prior to all entries. The intent for the permit is to ensure systematic review of hazards prior to each entry, communication of this information to potential occupants, and acceptance by them of requirements to be followed. The Qualified Person would complete the permit. Changes in conditions in the confined space or the nature of the work would terminate the permit.

The ANSI Standard also discusses other topics involved in safe entry and work in confined spaces. These include:

- Atmospheric testing
- Standby attendant, entrant and entry supervisor/leader
- Isolation and lockout/tagout
- Ventilation
- Cleaning/decontamination
- Personal protective equipment
- Safeguards
- Warning signs and symbols
- Emergency response
- Training
- Medical suitability
- Contractors

The following discussion highlights concepts mentioned under the preceding headings.

The Qualified Person would be responsible for atmospheric testing. Testing would be performed prior to entry, usually for oxygen, combustible gases, and toxic substances. In each case, prior to entry or reentry, the ventilation system would be shut down. Testing also would be performed to assess the adequacy of the ventilation system. Testing would be required throughout the period of occupancy.

Acceptable levels for oxygen would be 19.5 to 23.5%. The acceptable level for flammables would be less than 10% of the lower explosive limit (LEL) or lower flammable limit (LFL). Acceptable levels for toxic agents would be less than relevant exposure limits. Failure to meet these limits would terminate the entry and necessitate corrective action or reassessment.

A major distinguishing factor of a permit-required confined space is the requirement for one or more standby attendants. The attendant maintains two-way communication with the occupants, provides assistance, monitors conditions, and initiates evacuation and emergency procedures.

The current edition of the ANSI standard added sections on the additional classifications: entrant and entry supervisor/leader. These sections describe duties and requirements for performance of these individuals. This addition brings the standard into line with requirements of the OSHA Standard on confined spaces in general industry (OSHA 1993).

Isolation and lockout/tagout would be required prior to entry. The intent is to secure, relieve, disconnect, and restrain potentially hazardous energy sources. This would prevent unexpected energization, start-up or release of stored energy. Isolation could occur by one or more of blinding, depressurization and disconnection, removal of contents, or use of double block and bleed valves. Selection of the appropriate method would include consideration of its impact on the entry process. The Standard also recognizes the need for additional procedures and precautions to address situations that could not be handled routinely. Lockout/tagout would be implemented according to principles stated in ANSI Z244.1-1992(R) on lockout and tagout of energy sources (ANSI 1992).

According to the Standard, ventilation is to occur until atmospheric contaminants are within the acceptable range of concentration. Ventilation normally would include a preentry purge of several air changes and continuous ventilation during occupancy. Continuous ventilation during occupancy would be required only if the concentration of contaminants could increase beyond the acceptable range. Natural ventilation also could be acceptable. Alternate methods would be acceptable where ventilation is not possible or feasible.

The Standard requires cleaning/decontamination of the confined space to the extent possible prior to entry. The intent of this provision is to minimize exposure during entry and work to the substances being removed. The Standard cautions about the need for careful selection of agents used for cleaning.

Personal protective equipment would be selected according to the respective ANSI Standards and legal statutes.

The Standard also includes a requirement to consider equipment associated with safe work. This would begin with consideration of the entry/exit portal. Retrieval equipment or methods would be a requirement for entry/exit involving a permit-required confined space, except in exceptional circumstances. Fall protection and fall arrest would be required to prevent persons from falling into the opening of the space or during the work, as appropriate.

Warning signs would be required to indicate entrances to confined spaces where inadvertent entry could occur.

The Standard requires a formal plan of emergency response. This would include assessment of equipment and methods needed to effect a rescue, communications aspects, and practice drills. Rescuers could include off-site personnel. The Standard specifies the requirement for rescuers to use self-contained breathing apparatus (SCBA) or combination Type C airline/SCBA during rescue attempts. This would apply unless a hazardous atmosphere was shown not to be present.

Training would be performance oriented in aspects related to the duties of an individual and work in these environments. The Qualified Person periodically would assess the effectiveness of training.

Medical suitability would be assessed through physical and psychological testing. Suitability to wear respiratory protection would be assessed according to ANSI Z86.6-1984 "Physical Qualifications for Respirator Use" (ANSI 1984).

The issue of contractors who perform work in confined spaces on behalf of an employer also is raised in the standard (ANSI 1989). The current version of the standard considerably increased emphasis on contractors (ANSI 1995). Again, this is to bring the ANSI standard into line with the requirements of the OSHA Standard on confined spaces in general industry (OSHA 1993). Employers would be required to inform contractors about known or potential hazards associated with the confined space, as well as the classification assigned by the host employer and precautionary measures taken. This would leave the burden of hazard identification and assessment with the employer. Also, the employer and contractor jointly would establish emergency response and rescue procedures.

The Qualified Person would determine the need for periodic reassessment of hazards and the decision about the status of the space. This would apply to nonpermit and permit-required confined spaces.

The approach taken in the ANSI standard of two rather than three levels of assignment creates an either/or situation. The additional measures needed to comply with the requirements of the permit-required confined space clearly direct organizations toward the lesser approach. This emphasis could put considerable personal and organizational pressure onto the Qualified Person.

The ANSI standard provides no guidance to the Qualified Person on how to decide about the status of a space, other than listing factors to be considered. While this approach may be consistent with performance-oriented concepts, this provides no help to individuals having limited training and experience. While the skills needed to make decisions about confined spaces may be acquired relatively easily, the background knowledge and experience needed to recognize and assess potential and real hazards is extensive.

American Society for Testing and Materials

The American Society for Testing and Materials (ASTM) has published a standard for entry and work in confined spaces "Standard Practice for Confined Area Entry" (ASTM D 4276-84) (ASTM 1984). This standard is particularly applicable to entry into confined spaces associated with the use of halogenated organic solvents. Vapor degreasers, cold cleaning tanks, storage vessels, tank cars and trucks, van trailers, ships or barges, pits or sumps, and unventilated rooms are examples of confined spaces mentioned in this standard. As with the other standards, ASTM D 4276 preceded the OSHA standard. As a result, this standard may require revision to make it compliant with U.S. regulatory requirements. However, the present version does provide insight into the historical development of this subject.

ASTM D 4276 defines "confined area" by way of examples and the presence or potential presence of chemicals or vapors or lack of ventilation to define the hazard.

ASTM D 4276 begins with a brief description of the risks of working near chlorinated solvents, especially in enclosed areas. The main focus of the standard is procedural. This approach utilizes the entry permit, which would be completed by the responsible supervisor or other Qualified Person. The nature of qualification needed for completing the entry permit is not mentioned in the Standard. ASTM D 4276 includes a sample of an entry permit. The Standard and the permit indicate that a single test is sufficient for atmospheric contaminants. The Standard recommends positive ventilation through use of fans, portavents, or air movers, although natural drafts considered adequate are acceptable. The Standard requires use of a safety harness and lifeline. Self-contained breathing apparatus or supplied air respirators are required when the "toxicity level" exceeds OSHA limits.

The Standard also recommends other protective measures, such as an external observer and use of low voltage electrical equipment when flammable vapors could be present. Labeling of the entrance to the space also is suggested.

One unusual aspect of this standard is recommendation of test equipment. This is especially useful for halogenated substances that could be present.

National Fire Protection Association

The National Fire Protection Association (NFPA) was an early leader in addressing the hazards of entry and work in confined spaces. This leadership has been expressed through the issuance of standards and sponsorship and administration of the Marine Chemist program. The primary focus of NFPA standards as implied by the name has been prevention and control of fire and explosion hazards. However, concern about health hazards is inherent in current versions of NFPA 306. NFPA 306 derived from a comprehensive series of standards intended to stop fires and explosions in the marine environment. These were originally adopted in 1922 (Keller 1982, NFPA 1993).

NFPA also sponsors and administers the certification program for Marine Chemists. Marine Chemists can obtain certification through a program of defined academic achievement and practical experience and formal examination. The Marine Chemist Qualification Board is developed and maintained by an NFPA technical committee that represents government and industry. Certificated Marine Chemists remain bound by the stipulations of NFPA 306. These govern the conditions under which a "gas free" certificate may be issued (Keller 1982, NFPA 1993).

NFPA 306 applies to confined space hazards on vessels carrying or burning as fuel flammable or combustible liquids, flammable compressed gases, chemicals in bulk, and other products capable of creating hazardous conditions. It also applies to vessels under construction, or being repaired, altered, or scrapped. NFPA 306 specifically applies to those spaces on vessels that are subject to concentrations of combustible, flammable, and toxic liquids, vapors, gases, and chemicals. It also applies to spaces that may not contain sufficient oxygen to permit safe entry.

NFPA 306 describes conditions required prior to entry into a space or the start of work. It applies to cold work, application or removal of protective coatings, and work involving riveting, welding, burning, or similar fire-producing operations.

An important concept introduced in NFPA 306 is that of the "adjacent space." Adjacent spaces are those spaces extending in all directions that make contact with the space under consideration. Adjacent spaces include diagonals and corners, as well as spaces located above, below, and to the side of the space under consideration.

In following NFPA 306, the Marine Chemist issues a certificate that describes conditions found in the space at the time of the inspection and precautions required for entry and work. In determining the condition of the space, the Marine Chemist considers:

- The three previous cargoes
- Nature and extent of the work to be performed
- Starting time and duration of the work
- Results of tests of compartment or spaces, cargo and vent lines at manifolds and accessible openings, and cargo heating coils
- Tagging and securing of cargo valves in prescribed areas in a manner to prevent accidental opening and operation

The certificate issued by the Marine Chemist carries several possible standard safety designations. These outline conditions necessary for complying with safe performance of the work to be undertaken. NFPA 306 also contains directions for preparing the vessel for the issuance of the certificate by the Marine Chemist.

NFPA 328 focuses on control of hazards posed by flammable and combustible liquids and gases in manholes, sewers, underground vaults and similar underground structures (NFPA 1992). The latter include sanitary sewers, storm drains, water lines, fuel gas distribution systems, electric light and power conduits, telephone and telegraph communication lines, street lighting conduits, police and fire signal systems, traffic signal lines, refrigeration service lines, steam lines, and petroleum pipelines.

This Standard then describes the nature of the problem that has resulted from enclosure of utilities into underground corridors. This includes potential ignition hazards, as well as a accumulation of substances that can enter these structures. The Standard then discusses protective practices that should be employed as part of entry and work procedures. These include use of testing instruments, ventilation, flushing with water, and inspection.

American Petroleum Institute

The American Petroleum Institute (API) has published several consensus guidelines for work involving tanks that have contained petroleum products. These cover various aspects of work on the exterior of containers, as well as entry and work in confined spaces. Specific topics addressed in these guidelines include:

- API RP 1631: interior lining of tanks (API 1992)
- API RP 2003: protection against static electricity and stray currents (API 1991a)
- API 2214: protection against spark ignition from hand tools (API 1989)
- API RP 2013 and API 2015: cleaning of mobile and stationary tanks (API 1991b, API 1993)
- API 2009: welding and cutting practices in refineries, gasoline plants, and petrochemical plants (API 1988a)
- API 2207: preparing tank bottoms for hot work (API 1991c)
- API 2217A: work in ventilated and inerted confined spaces (API 1987)
- API 2027: ignition hazards arising from abrasive blasting (API 1988b)
- API 2219: safe operation of vacuum trucks in petroleum service (API 1986)

As these guidelines were prepared for use primarily in the U.S., publication of the OSHA confined space Standard for general industry has forced reevaluation of some of the concepts in order to ensure consistency.

These guidelines are performance based. The foundation for successful implementation is the knowledgeable, competent person who is immediately available at the worksite. The guidelines rely on administrative controls including:

- Entry and work permits
- Atmospheric testing before and during entry
- Training of workers and supervision
- Supervision by qualified individuals at the worksite
- Use of safeguards to control hazards
- An emergency plan of action
- Compliance with company policy and government regulations

As part of the process of achieving the intended outcome, the guidelines begin by providing information about hazards likely to be present or encountered, and relevant precautionary measures. The next aspect is detailed, real-world information needed for planning, executing and auditing the progress of the work. This is based on operating experience needed for preparing for entry and executing the task and ensuring control over potential hazards.

The approach taken by API in the previous version of API 2015 led to three possible outcomes to categorizing hazards (API 2015-1991d). These included:

- **Category I — Entry not Permitted**
  Category I describes conditions of excessive flammable vapors (20% or more of the lower flammable limit), low oxygen level (less than 16%), excessive hydrogen sulfide (greater than 100 ppm), airborne levels of toxic substances exceeding levels set by the employer and government, or unsafe physical condition of the tank.
- **Category II — Entry Permitted with Respiratory Equipment**
  Category II would require use of full facepiece, positive-pressure respiratory protective equipment. Also, an entry permit must be issued. Category II describes the following inclusive conditions. Flammable vapors must be less than 20% of the lower flammable limit; oxygen level 16% or more; hydrogen sulfide concentration less than 100 ppm; airborne levels of toxic substances within levels set by the employer and government; and the physical condition of the tank must be safe for entry.
- **Category III — Entry Permitted without Respiratory Equipment**
  Category III describes the following inclusive conditions. Also, an entry permit also would be required. Flammable vapors must be 10% or less than the lower flammable

limit; oxygen level 19.5% or more; and toxic vapors and dusts must be within limits established by the employer and government. Also, the tank must not have contained leaded gasoline or, if it did, it must have been cleaned and concentration of organic lead in air was less than 0.075 mg/m^3 of air, and the physical condition of the tank safe for entry.

The information provided in these guidelines for hazard recognition, assessment, and control is an extremely important contributor to the literature of confined space entry. The real-world, in-depth nature of this information emphatically emphasizes the breadth of knowledge needed by the Qualified Person. The Qualified Person involved with entry and work involving containers that hold or have held hydrocarbons definitely needs more technical knowledge than merely how to assess atmospheric conditions.

Standards Association of Australia

The Standards Association of Australia is an example of an organization outside North America that has published a standard on confined spaces (AS 2865-1986) (Standards Association of Australia 1986). The purpose of the Standard was to assist in preventing injuries and fatalities in confined spaces. This is assured by ensuring that confined spaces are made safe for entry, as well as highlighting hazards and relevant safe work practices.

This Standard was updated by a draft proposed standard published in 1992 (National Occupational Health and Safety Commission 1992). The joint draft national Standard would require the taking of adequate steps to eliminate or control hazards. It also would require training and instruction in the nature of hazards and precautions to be followed for persons involved in the entry of a confined space. The joint draft Standard primarily upgrades AS 2865-1986 by providing additional appendices on atmospheric monitoring, and provision and use of protective and rescue equipment.

The joint draft Standard utilized the definition for confined space provided in AS 2865-1986. For the purposes of this Standard, a person whose upper body or head is within the confined space is considered to have entered.

The first step in implementing this Standard would be for the employer to identify confined spaces within the organization. This would include posting a sign and securing the space. This also would include determining whether proposed work could be performed without entering the space. In the event that such work could be undertaken from outside the space, preventing entry by physical means then would be required. The joint draft Standard and its predecessor provide an extensive list of examples of confined spaces.

In the event that work would involve entry, a written assessment performed by a competent person would be required. A competent person has the combination of training, education and experience, and knowledge and skills needed to perform the specified task. The employer is required to ensure that the background of the individual deemed to be the competent person meets these qualifications. How this assurance is to occur is not specified. The hazard assessment would consider the work to be performed and possible methods for doing it, hazards and associated risks, the actual method proposed and associated equipment, and emergency and rescue procedures. The hazard assessment would be performed prior to issuance of any permit to enter.

The joint draft Standard provides assistance for performing the hazard assessment. The section on hazard assessment lists major and minor hazards associated with work in confined spaces and various factors to consider. These factors include:

- Arrangements for rescue, first aid, and resuscitation
- Proposed occupancy
- Persons required outside the space to assist entrants, maintain equipment, ensure communication and surveillance of the occupants, and to initiate emergency response

- Impact of proposed operations on conditions in the space
- Soundness and security of the structure
- Need for illumination
- Substances last contained in the space
- Steps needed to bring the confined space to atmospheric pressure
- Atmospheric testing
- Hazards that may be encountered
- Medical fitness of entrants
- Training of entrants
- Instruction in work procedures
- Availability and adequacy of personal protective equipment
- Adequacy of warning signs
- Special considerations for hot work
- Fire control
- Other special considerations

Appendices provide guidelines for specific precautions and methods of work for particular tasks.

The next part of the joint draft Standard considers testing prior to entry. The joint draft Standard and its predecessor specified an upper limit for flammable contaminants of 5% of the LEL for entry. If the level exceeds 5% during entry, continuous monitoring would be required. Vacating would be required if the level exceeded 10% of the LEL at any time during occupancy. The joint draft Standard also proposes acceptable levels of oxygen during entry and occupancy. Acceptable levels of flammable contaminants and oxygen were issues in the joint draft Standard on which discussion was requested. The document mentions the various levels that exist in U.S. documents. The joint draft Standard would require the use of appropriate personal protective equipment where these levels could not be ensured. The document also would require compliance with legislated exposure standards. Retesting at specified intervals would be required.

The joint draft Standard would require each entry to be treated as a separate task and assessed accordingly.

The joint draft Standard lists the hierarchy of control measures to be taken to assure safe entry and occupancy in the following priority:

- Elimination of the need for entry
- Substitution of less hazardous products for tasks during entry and occupancy
- Isolation of the space from external hazards and deactivation of external and internal hazardous energy sources
- Engineering controls
- Adoption of safe work practices
- Use of personal protective equipment

The joint draft Standard would require employers to apply this hierarchy of controls as much as possible. This application would involve reviews to ensure the adequacy of the controls. These would occur regularly, or as indicated by the assessment report, in the event of significant change and for each entry.

The joint draft Standard mentions preparatory measures needed to ensure effective isolation. These would include prevention of accidental introduction of materials and deenergization and lockout or tagout of equipment. The document also discusses methods and other considerations involved in achieving isolation.

The joint draft Standard also discusses the use of ventilation to establish and maintain a safe breathing atmosphere. Ventilation must continue throughout the period of occupancy as a safeguard against unexpected release of contaminants. The joint draft Standard condones use of natural, forced, or mechanical ventilation, as appropriate. Where reliance on mechanical ventilation to maintain a safe breathing atmosphere is necessary, techniques for assurance of supply, such as continuous monitoring and tagging of control circuits to minimize accidental shutdown, would be required. The joint draft Standard also raises concerns about entrainment of contaminants from sources such as engine exhaust into the ventilation system. The document cautions that oxygen or compressed air containing more than 21% oxygen shall not be used to ventilate the space.

The joint draft Standard would require use of a document of approval — a written permit — prior to the start of work. In issuing the permit, the employer must be satisfied that the work can be performed safely and without risk to health. The permit would include precautions or instructions for the safe execution of the work. The authorized person supervising the work and employees performing it would sign the permit to indicate comprehension. The permit would require revalidation following a change in supervisor, significant break in continuity of work, or change in conditions.

Standby person(s) would be required outside the confined space at all times. The standby would maintain communication with entrants. The joint draft Standard offers flexibility in the type and manner of communication device.

Additional sections briefly discuss the following topics:

- Posting of warning signs
- Safety equipment for personal protection, first aid, fire protection, and rescue
- Respiratory protective equipment
- Safety harnesses, lines, and extraction equipment
- Electrical equipment
- Rescue and first aid
- Recordkeeping

The joint draft Standard puts considerable emphasis on training for all persons involved with confined spaces. As a minimum, training should include: hazards of confined spaces; assessment procedures; control measures and the selection; and use, fit, and maintenance of safety equipment. Training should be provided to employees in all aspects of the process of entry and work, and emergency response and rescue. The curriculum should match the needs and responsibilities of each participant in the process. The joint draft Standard provides a list of suggested topics. Training should occur until the employer is satisfied that each person is competent to a reasonable level of attainment. The training should be evaluated and reviewed with relevant employees and their representatives.

National Safety Council

The National Safety Council has published a series of Occupational Safety and Health Data Sheets that were created by various industrial groups within the Council (NSC 1985, 1986, 1987a, 1987b). The information contained in these Data Sheets reflects the experience base of the group that prepared them. The Data Sheets relevant to confined spaces include:

- Confined Space Entry Control System for R & D Operations, No. 12304-0704 (1985)
- Inspecting and Cleaning Pipes and Sewers, No. 12304-0577 (1986)
- Atmospheres in Sub-Surface Structures and Sewers, No. 12304-0550 (1987)
- Entry into Grain Bins & Food Tanks, No. 12304-0663 (1987)

Generally, these Data Sheets are performance oriented. They begin by describing hazards associated with the environment on which the specific Data Sheet focuses. At this point they list measures known to be effective in controlling the hazards. These Data Sheets provide valuable information, since hazards in specific environments vary considerably. A measure that may be acceptable in one environment, such as use of compressed air as a source of ventilating air, can be absolutely contraindicated in another due to potential for buildup of static electricity.

The Data Sheets provide very brief information about any subject within a particular topic. They are written for an audience that is aware of the hazards and controls employed against them, yet needs reminding. The reader is left to work out the specifics of addressing a hazard in the context of a particular operation.

REGULATORY STANDARDS

U.S. Department of Labor

Two federal government agencies in the U.S. Department of Labor, the Occupational Safety and Health Administration (OSHA) and the Mine Safety and Health Administration (MSHA), have produced regulations or proposed standards that address safety in confined spaces.

Occupational Safety and Health Administration

The Occupational Safety and Health Administration (OSHA) regulates confined spaces in more than 50 different sections of 29 CFR (Code of Federal Regulations) 1910, 1915, 1916, 1917, 1918, and 1926 (NIOSH 1979). Until recently, none of these Standards (General Industry, Maritime, or Construction Industry) provided comprehensive coverage to address all of the hazards associated with entry and work involving confined spaces. Most of the Standards address very limited applications or specific situations.

Permit-required confined spaces for general industry

In 1993, OSHA issued the confined space Standard (Permit-Required Confined Spaces for General Industry) (OSHA 1993). Prior to 1993, the "general duty" clause was cited. In its rule making, OSHA stated that it felt that none of the existing standards adequately protected persons working in confined spaces from the atmospheric, mechanical, and other hazards. Further, OSHA expressed the view that effective protection could be attained only through a comprehensive confined space entry program. Improvising or following "traditional methods" could not address the additional hazards created by confinement, limited access, and restricted airflow. OSHA noted that inadequate practices led to more frequent fatal accidents than predicted by standard statistical models. This affirmed concerns expressed by NIOSH in the Criteria Document (NIOSH 1979).

OSHA first published an Advanced Notice of Proposed Rulemaking (ANPR) for a comprehensive confined space standard in 1975. ANPRs also were issued in 1979 and 1980. A proposed Standard was issued in June, 1989 after a series of delays (OSHA 1989a). The final rule for this Standard (Permit-Required Confined Spaces for General Industry) was published in the *Federal Register* on January 14, 1993 as 29 CFR 1910.146 (OSHA 1993). In taking so long to publish its Standard, OSHA had the benefit of hindsight for the review of standards, guidelines, and legislation published in other domains, both domestic and international. An outcome of the insight gained from this review was the decision to increase the emphasis on nonatmospheric hazards compared to that found in existing standards, such as the NIOSH Criteria Document and ANSI Z117.1-1989 (NIOSH 1979; ANSI 1989).

The intent of the OSHA Standard is to provide comprehensive protection from exposure to all types of confined space hazards, including mechanical and physical hazards. OSHA intends to achieve this goal through direct linkage to the Lockout/Tagout Standard, 29 CFR 1910.147 (which is reviewed briefly in a subsequent section) (OSHA 1989b). As a result, employers must evaluate mechanical hazards found in permit-required confined spaces and take all steps necessary to protect entrants.

OSHA expects the Confined Space Standard to affect more than 1.6 million workers who enter 4.8 million "permit-required confined spaces" at over 230,000 worksites throughout the U.S. Further, OSHA predicts that 80 to 90% of the 10,000 injuries and 54 fatalities that occur annually in confined spaces in the U.S. will be prevented as a result of implementation of this rule (OSHA 1993).

The Standard applies to confined spaces in general industry. Major sectors that are affected include: electric and gas utilities, manufacturing, transportation, agricultural services, and wholesale trade. Agriculture, construction, and shipyard employment industries are excluded.

The process of setting regulatory standards in the United States is among the most rigorous in the world. Deliberations for the confined space standard involved participation from many interested parties. The OSHA Standard is a highly sophisticated document. It introduces concepts not found in the approaches taken by other groups. For this reason, considerable space in this Appendix will be devoted to examining this statute. The most informative part of the document is the preamble. In this section OSHA discussed the Standard, paragraph by paragraph, and provided justification for its decision making.

OSHA defines a "confined space" as a space that:

- Has adequate size and configuration for employee entry, and
- Has limited means of access or egress, and
- Is not designed for continuous employee occupancy

The preceding characteristics are common to all confined spaces. OSHA took the view that isolating characteristics that make some types of confined spaces hazardous from those that do not would simplify addressing the hazards posed by these spaces.

OSHA recognized that the Standard would not cover small spaces that could accommodate part, but not all, of the body. While recognizing that fatal accidents or exposures can occur within such an enclosed area, their view was that this Standard was not intended to address all locations that pose atmospheric hazard, only those that permit entry of the entire body. OSHA believes that small spaces could be readily ventilated to prevent accumulation of a hazardous atmosphere. This discussion highlights an important consideration in this subject. How small an enclosed space reasonably should be considered as a confined space for the purposes of legislation? Where should the line of demarcation occur?

Entry is deemed to occur when any part of the body breaks the plane of the entry portal. That is, entry can occur either head-first or feet-first. This accommodates situations where only part of the body actually enters the space. This also recognizes hazards from respiratory and other types of hazards, such as mechanical hazards. (The term "entry only" covers entry into confined spaces that are permit-required.)

Limited means of access or egress was not intended to include a doorway or other portal through which a person can walk. This excludes many types of structures whose geometry can cause atmospheric accumulation.

Continuous employee occupancy is another important concept. OSHA focused on the design of the space, rather than its primary function, when choosing the words in the definition. In OSHA's view, design of the space for occupancy renders irrelevant its primary function. This is what determines whether a human can occupy the space under normal operating conditions. The condition of the space at the time of entry, rather than its ultimate use, is an important distinction. This distinction has important ramifications to structures under construction. While the finished structure may not pose a hazard to occupants because of ventilation, the geometry of the partly assembled structure may pose a hazard because of processes used during fabrication.

OSHA uses the terms "nonpermit confined space (nonpermit space)" and "permit-required confined space (permit space)" to distinguish the level of hazard. The strength of this approach is to permit allocation of resources to situations that pose serious and potentially life-threatening hazards. The weakness in this approach is that changes in the seriousness of a particular situation can be overlooked. OSHA has commented that the hazard element that differentiates permit spaces from nonpermit spaces may vary in nature. A permit space is a confined space that has certain characteristics that make entry hazardous without special precautionary measures.

A "nonpermit confined space (nonpermit space)" is a confined space that does not contain or, with respect to atmospheric hazards, have the potential to contain any hazard capable of causing death or serious physical harm. Examples provided by OSHA include vented vaults, motor control cabinets, and dropped ceilings. The element that minimizes concern about hazards in these environments is natural or mechanical ventilation. This prevents accumulation of hazardous atmospheric conditions. They do not present risk of engulfment. However, there must be some proof that they do not present other serious hazards.

OSHA chose the term "permit-required confined space (permit space)" to identify confined spaces that pose or could pose health and safety problems. A "permit-required confined space (permit space)" is a confined space that presents or has a potential to present one or more of the following characteristics:

- An atmospheric hazard
- An engulfment hazard
- A configuration hazard
- Any other recognized serious hazard

According to OSHA, poor natural ventilation is not a necessary condition for a confined space to be a permit space under this definition. The question of poor natural ventilation is relevant only when a space is a confined space or is not a confined space. OSHA takes the view that the most important characteristic regarding atmospheric hazard is the content of air inside the space. Even under well-ventilated conditions, certain areas within a space may be able to accumulate a hazardous atmosphere.

A contentious issue not carried forth in the final rule was the concept of the low-hazard permit space. The intent of the concept was to allow entry into confined spaces that pose low hazard without requiring the full precautions needed for a permit space. The concept was based on the premise that either the space posed a low level of risk or that its hazards were controlled. This concept also would have permitted differentiating entry requirements based on perceived level of risk. This situation was addressed in the final rule by requiring increasingly more stringent measures as the level of risk increases. That is, spaces posing the least level of risk above the threshold that defines the permit-required space necessitate the least stringent approach. Confined spaces below the threshold require the least level of effort.

A "hazardous atmosphere" as defined by OSHA is an atmosphere that may expose employees to the risk of death, incapacitation, impairment of ability to self-rescue (unaided escape from a permit space), injury, or acute illness. A hazardous atmosphere can arise from one or more of the following causes:

- Flammable gas, vapor, or mist in excess of 10% of the lower flammable limit (LFL)
- Airborne combustible dust at a concentration that exceeds the LFL
- Oxygen concentration less than 19.5% or greater than 23.5%
- Concentration of any toxic substance exceeding the OSHA PEL
- Any other atmospheric condition recognized as Immediately Dangerous to Life or Health (IDLH)

A rule of thumb regarding many combustible dusts is that a concentration that obscures vision at a distance of 5 feet (1.5 m) or less is explosible.

Regarding toxic substances, a concentration that is not capable of causing death, incapacitation, impairment of the ability to self-rescue, injury, or acute illness due to its health effects is not covered in this provision. An atmosphere containing a substance at a concentration exceeding a Permissible Exposure Limit (PEL) intended solely to prevent long-term adverse health effects is not considered to be hazardous on this basis alone.

Use of the term "IDLH" differs from that chosen by NIOSH. IDLH here means a condition that poses an immediate or delayed threat to life, or that would cause irreversible adverse health effects, or that would interfere with the ability to escape unaided from a permit space.

Engulfment can include surrounding and effective capture of a person by a liquid or finely divided (flowable) solid. This means any solid or liquid that can flow into a confined space and can drown or suffocate a person.

The Standard provided no indication about the meaning of the fourth criterion, "any other recognized serious hazard." As a regulatory document, the meaning of this term will evolve through inquiry from interested persons through appropriate channels. OSHA stated the fourth criterion in the broadest possible terms in order to embrace serious hazards and to afford protection to affected employees (OSHA 1996). According to this view, these hazards could include conditions that necessitate preventive or protective actions such as the wearing of personal protective equipment. On the other hand, they would not include conditions such as poor communications in a nonpermit space. Other conditions would require case-by-case evaluation. These would include biological hazards, slippery surfaces and tripping hazards, noise and vibration, and illumination.

The first actions required of employers in implementing the Standard are to identify permit-required confined spaces in their workplaces and to determine whether their employees enter them. If this is occurring, the employer then is required to close off the space, and to retain a contractor to provide the service performed in the space or to implement a permit space program. The employer need not evaluate the hazards posed by the space until a reasonable time before entry. This leeway was intended to address concerns about assessing all permit spaces at one point in time when entry might occur many months or even years later.

Moreover, employers must notify employees about the presence and location of the permit spaces. This can occur by posting danger signs or by any other equally effective means that indicate the danger posed by the permit space. The latter provision alleviates the need to post a sign on all known permit spaces. Posting signs on all permit spaces in a process operation, for example, would be extremely difficult. Many industrial complexes have thousands of permit spaces that on occasion require entry. On the other hand, OSHA noted that lack of recognition of the danger was a deficiency during many fatal accidents. Signs tend to be ignored. Also, a space that has no sign could become hazardous due to a change in conditions. Similarly, a space that has a sign could become nonhazardous. Awareness training can replace signs, provided that employees are informed about the location of confined spaces.

The Standard identifies persons allowed to work in the vicinity of permit spaces.

Authorized entrants are persons authorized by the employer to enter a permit space. Authorized entrants can work in the permit space, test the atmosphere, supervise the entry or provide rescue services.

The **entry supervisor** is responsible for evaluating conditions in and around a permit space where entry is planned. The entry supervisor authorizes entry to begin or allows entry already underway to continue and oversees entry operations. This means taking necessary measures to protect personnel from hazards in the permit space and prohibiting or terminating entry, as required. The OSHA Standard clearly sets out who is in charge. The entry supervisor must determine at appropriate intervals whether conditions within the space remain safe and to cancel the permit whenever conditions are otherwise. The entry supervisor and the person who supervises the actual work to be undertaken inside the permit space could be one and the same or they could be different.

The **attendant** monitors authorized entrants and performs all duties assigned by the permit program. The attendant monitors activities inside and outside the space to determine whether entrants can safely remain inside. The attendant orders occupants to leave the space when evidence of an uncontrolled hazard exists. The attendant watches for conditions not envisioned during preparation of the permit and for other prohibited conditions. The attendant works from a location outside the permit space. This could include a control room that allows remote monitoring of entrants in one or more permit spaces through television monitors, or by electronic monitors or radio communications.

Testing personnel determine the condition of atmospheric hazards in the confined space.

Rescue and emergency services respond to accidents and other emergency situations involving permit spaces.

The Standard requires a written "permit-required confined space program (permit space program)" where entry of permit spaces occurs. The general purpose for the program is to regulate entry into permit spaces. It also provides accountability for auditing purposes and should reduce the opportunity for error. The permit space program includes provisions for:

- Identifying, evaluating, and controlling hazards
- Preparing, issuing, and implementing written entry permits
- Procedures to prevent unauthorized entries
- Informing and training workers
- Providing equipment needed for safe entry (including testing, monitoring, communication, rescue, and personal protection equipment)
- Rescue procedures
- Informing contractors
- Other measures

Review of the permit space program is required periodically. Revisions should reflect changes in hazards. OSHA approached this requirement by requiring review any time conditions in the workplace indicate that existing procedures provide inadequate protection. Also, employers are required to review the permit program within 1 year from entry. The annual review provides the opportunity to correct deficiencies not previously discovered.

An important part of the permit space program is the permit system. The permit system is the infrastructure involved in preparing and issuing permits for entry and for returning the permit space to service following termination of entry. OSHA separated requirements for the permit system from the content of the permit. The intent was to emphasize the inclusive nature of the system from the preparation of the permit. OSHA provided examples of permit space programs and permits in appendices to the Standard.

The permit authorizes entry into permit spaces and documents measures required to ensure that employees can safely enter and work in the permit space. OSHA believes that the action of preparing the permit will assist in determining suitability for entry. Also, the permit provides a concise summary of the entry procedure. This can be valuable to persons involved in the entry and in auditing the process. OSHA specifies that the permit must document measures required by the Standard. All critical information must appear on the permit. This critical information includes:

- Identification of the space to be entered
- Purpose for the entry
- Date and authorized duration of the entry permit (this may not be actual time, but may be completion of the task)
- List of authorized entrants (this may occur through a systematic tracking system)
- Current entry supervisor
- Hazards of the space
- Specific measures for isolating the permit space or controlling hazards prior to entry

- Acceptable entry conditions
    - Atmospheric conditions
    - Energy control measures
- Measures for obtaining acceptable entry conditions
- Initial and periodic test results corresponding to the specified entry conditions plus signature or initials of the tester and an indication about when the tests were performed
- Rescue and emergency services that can be summoned and the means for summoning them
- Communication procedures to be used by the attendant and authorized entrants during entry operations
- Equipment to be provided for compliance with requirements of the Standard
    - Personal protective equipment
    - Testing equipment
    - Communications equipment
    - Alarm systems
    - Rescue equipment
    - Other equipment
- Existence of any additional permits, such as hot work permits

The permit must be prepared after establishing safe entry conditions and prior to the entry. The entry supervisor must sign the permit. The permit must be available for review by all entrants at the time of entry. The duration of the permit is the period of time needed for completion of the assigned task or job. As long as acceptable entry conditions are present, employees can safely enter and perform work in the permit space. In OSHA's view, limiting the duration of the permit to an arbitrary length of time, such as a single shift would not reduce the risk of entry, provided that other requirements in the Standard were implemented. The entry supervisor cancels the permit. Since entry supervisor is a transferrable title, someone other than the author of the permit may terminate entry and cancel the permit.

In recognition of concerns for employers who are not knowledgeable about confined spaces, OSHA provided an indication of measures that could be needed to control hazards in permit spaces. These include:

- Specifying acceptable conditions for entry (for example, 10% of the LFL of a gas)
- Specifying conditions for evacuation (for example, 50% of the level of flammable or toxic substances that would constitute a hazardous atmosphere
- Isolation techniques
- Purging and ventilating
- Security measures to prevent unauthorized entry and to protect entrants
- Testing and monitoring

The Standard clearly requires evaluation of conditions in permit spaces before, and as necessary, during entry. The latter evaluation is necessary to determine that acceptable conditions are being maintained during the course of operations. For spaces such as sewers that limit isolation because they are part of a continuous system, both preentry and continuous monitoring are required. The type of testing depends on the type of hazard.

Testing is to occur in the following sequence: oxygen, combustible gases, and toxic gases and vapors. OSHA discussed testing in an appendix of the Standard.

A controversial aspect of the Standard is the permissible use of continuous forced ventilation in lieu of other measures. To do this, the employer must be able to prove that the only hazard is atmospheric and that continuous forced ventilation will maintain the permit space safe for entry. According to OSHA, testing the atmosphere inside the space and providing adequate continuous ventilation sometimes can eliminate the hazardous atmosphere to produce the equivalent of a nonpermit space. The work to be performed must not introduce any additional hazards such as toxic substances or hot work. A permit is not required for entry into permit spaces whose hazards can be controlled by ventilation alone.

Control of hazards, however, is not the same as elimination of hazards. Alternate ventilation procedures do not turn a permit-required confined space into a nonpermit space. Instead, they allow certain permit-required spaces to be entered without the need for a comprehensive program, including a fully comprehensive entry permit. These spaces, therefore, remain classified as permit-required, even though a less comprehensive "written certification" would suffice instead of the complete permit (Anonymous 1993). In effect, this creates two categories of permit-required confined space (Easter 1993). The first covers those spaces that pose not only atmospheric, but also other hazardous conditions. The second category applies to spaces that contain or potentially contain only atmospheric hazards. This would necessitate an alternate and less extensive program involving training, atmospheric testing, forced air ventilation, and an abbreviated permit and notification program. An attendant would not be required.

A confined space containing a hazardous environment would require initial entry to ascertain conditions and to apply any necessary measures to prepare it for further work. Such measures could include cleaning. Initial entry would require all precautions, as appropriate. Once cleaning and other preparatory activities were complete, the only remaining hazard would be atmospheric. The space then would revert to the condition discussed above.

OSHA determined that spaces in which all hazards have been eliminated can be reclassified as nonpermit spaces for as long as the hazards remain eliminated. Ventilation controls hazards; it does not eliminate them. Reclassification must be rigorous. This involves completion of a dated and signed certificate. Affected employees must be informed about the reclassification.

The OSHA Standard differs from ANSI Z117.1-1995 by separating nonpermit spaces into two groups — permit spaces with atmospheric hazards controlled by ventilation alone, and permit spaces that have been reclassified as nonpermit spaces because the hazards have been eliminated (ANSI 1995).

The issue of training was a major concern identified by OSHA during its deliberations. Concern about lack of training or insufficient or inconsistent or inappropriate training led OSHA to put considerable emphasis on training in the Standard.

Training is required before an employee is assigned to duties. Training also is required prior to a change in assigned duties. Training is required when there is a hazard in a permit space about which the employee previously has not been trained. Training is required when the employer has reason to believe that there are deviations from entry procedures or inadequacies in knowledge or use of procedures.

The employer must certify that the training that is provided is adequate to the need. The certificate must contain the name of the person trained, the signature or initials of the person who provided the instruction, and dates. On-the-job training by a qualified individual is acceptable. As a first step, the process of certification involves determining areas of proficiency required in tasks to be performed.

The Standard does not provide specific training requirements for each category of individual involved with permit required confined spaces. The Standard does, however, specify duties. Training must match the duties of each job class involved in entry and work procedures. OSHA specifically requires training in the understanding, knowledge and skills necessary for the safe performance of duties. Table A.2 indicates criteria that must be considered. These are based on comments provided in the preamble to the Standard.

OSHA did not require intervention by certified professionals, such as Certified Safety Professionals, Certified Industrial Hygienists, or Certified Marine Chemists in evaluation of conditions or development of control measures for use in permit spaces. OSHA argued that professional certification is not an automatic assurance of competence. On the other hand, lack of certification

TABLE A.2
Training Criteria Mentioned in the OSHA Standard

All Employees

- Be informed about the existence and location of permit-required confined spaces
- Recognize confined spaces
- Understand difference between permit and nonpermit spaces
- Recognize hazards posed by permit spaces
- Be trained in the understanding, knowledge and skills needed to perform their duties safely

Authorized Entrants

- Know hazards that may be faced during entry
- Understand mode of exposure
- Recognize signs or symptoms of exposure
- Understand consequences of exposure to the hazards
- Understand particulars of a permit space and the reason for entry
- Be aware of necessary protective equipment and its limitations
- Be able to use equipment necessary for safe entry properly
- Communicate with the attendant in a uniform manner to indicate individual status
- Alert the attendant about any sign or symptom of exposure to a dangerous condition or about any prohibited condition
- Alert the attendant when self-initiating evacuation from the permit space
- Recognize conditions under which exit from the permit space must occur (orders from entry supervisor or attendant, warning sign or symptom of exposure to a hazardous substance, prohibited condition, sounding of the evacuation alarm)

Attendants

- Know hazards that may be present during entry
- Understand mode of exposure
- Recognize signs or symptoms of exposure
- Understand consequences of exposure to the hazards
- Recognize possible behavioral effects resulting from exposure
- Know and watch for hazards that may be present inside the space
- Monitor the status of authorized entrants
- Warn unauthorized employees to stay out of the space
- Inform unauthorized entrants that they must leave the space immediately
- Inform authorized entrants that unauthorized persons have entered the space
- Assist in evacuating entrants
- Monitor activities inside and outside the space to determine whether entrants can safely remain inside
- Order evacuation of entrants from the permit space during existence of a prohibited condition, behavioral effects of hazard exposure, a situation outside the space that could endanger the entrants, or whenever the attendant is unable to perform required duties
- Summon rescue services in the event of an emergency
- Know effective work rules
- Remain outside the permit space during entry operations until relieved by another attendant
- Provide information to the rescue service
- Maintain attention on entrants in the permit space
- Perform only duties that do not detract from the primary function of monitoring and protecting authorized entrants

Entry Supervisors

- Know the hazards that may be present during entry
- Understand mode of exposure
- Recognize signs or symptoms of exposure
- Understand consequences of exposure to the hazards
- Recognize possible behavioral effects resulting from exposure
- Verify the appropriateness of entry on the permit before signing and authorizing entry to begin
- Verify the existence of acceptable entry conditions

TABLE A.2 (continued)
Training Criteria Mentioned in the OSHA Standard

- Verify availability of rescue and emergency services and that the means for obtaining them are operable
- Authorize entry through signature on the permit
- Remove unauthorized persons from the space
- Determine, whenever responsibility for a permit space is transferred and at intervals as dictated by the hazards and operations performed in the space, that operations within the space remain consistent with the terms of the entry permit and that acceptable conditions are maintained
- Terminate entry operations when necessary
- Terminate entry operations and cancel the permit

Atmospheric Testers
- Use and calibration of testing instruments
- Limitations of instruments being used
- Interpretation of results

In-Plant Rescuers
- Training in performing duties
- Training in use of personal protective and rescue equipment
- Know hazards that may be faced during entry
- Understand mode of exposure
- Recognize signs or symptoms of exposure
- Understand consequences of exposure to the hazards
- Practice rescue from the actual or representative permit spaces at least once every 12 months. Practice rescue may involve manikins, dummies, or actual persons.
- Obtain training in basic first aid and maintain current certification
- Obtain training in cardiopulmonary resuscitation and maintain current certification

is not an automatic assurance of lack of competence. OSHA did acknowledge the extensive experience required by Marine Chemists in performance of their profession. The understanding, knowledge, and skills of other professionals were recognized for their generalized nature. Specific experience with the type of permit space found in the workplace also is required.

Employee involvement in the process of complying with the Standard was a controversial issue. OSHA decided that employees could be involved to the extent of reviewing the permit space program and employee training. Further involvement was disallowed because of concerns regarding disputes over contentious issues and their resolution.

The Standard also addresses the problem of contractors hired to perform work inside the confined space. A contractor usually is unfamiliar with the host workplace and its particular hazards. Host employers are responsible for coordinating entry and work involving contractors in permit spaces. The host employer must enable the contractor to develop and implement permit space programs that satisfy the Standard. This includes informing the contractor about the permit spaces and their attendant hazards, about permit space procedures, coordinating entry operations with the contractor, and debriefing the contractor at the conclusion of entry operations. OSHA requires contractors to obtain any available information concerning permit space hazards and entry operations from the host employer.

OSHA was very concerned about security around confined spaces, having recognized the danger to which unauthorized entrants subject themselves and others. OSHA mentioned its concern about the overly casual attitude that, in its view, led to many of the permit space accidents.

An unusual aspect of the Standard is the attention that it requires toward concluding the entry. The intent of this concern is to ensure orderly return of the permit space to service. OSHA is concerned about the orderly conduct of entry operations from start to finish.

Rescue and emergency response are difficult territory to address. Few organizations possess the resources to provide this service in-house. On the other hand, the fire service which increasingly is becoming involved in this type of work, undertakes these actions under highly nonideal conditions. Rapid action is needed to sustain life. Yet, rapid action is completely incompatible with the need for caution on the part of the rescuers in order to ascertain conditions. Ascertaining conditions is essential so as not to put the rescuers into the same situation as the potential victim whose plight already is known.

Considerable input was received in the discussion phase regarding the need to respond to an accident involving an oxygen deficiency within four minutes. Some of the commenters acknowledged that even in-plant response teams could not provide service within that time frame. Outside services would require even longer. OSHA took the view that it was not reasonable to require employers to develop the capability of responding to these emergencies within four minutes. OSHA also was concerned about the assumption of unnecessary risk by rescuers attempting to respond within too short a period and prior to assessing the situation properly. For this reason, OSHA stressed the need for nonentry rescue methods and assuring the safety of rescuers, rather than response time. OSHA stressed the preventive nature of the Standard, rather than focusing on response time and rescue of incapacitated entrants.

OSHA approached the problem of rescue by both in-plant rescue teams and other providers of this service through designation of the term "rescue service." Employers may use on-site or off-site rescue services. OSHA expects that rescue could involve either entry into the permit space by the rescuers or extraction of the entrant using retrieval equipment from outside.

Employers using in-plant services or providing services to other employers must equip and train members of the emergency response team properly. An employer that utilizes a rescue service must inform that employer about hazards likely to be encountered. This is essential, so that the outside team can equip, train, and conduct itself appropriately. Also, the employer must provide access to the rescue service, so that they may develop an appropriate rescue plan and practice in permit spaces similar to those from which a rescue may be required.

OSHA was concerned about the ability of rescue services to perform under the pressures of actual situations. For this reason the Standard requires practice sessions at least annually, using spaces that are representative of those actually entered. This is especially important for determining that equipment will fit through openings of portals or that encumbered members of the rescue team can move throughout the space and perform their function. Successful rescues in actual circumstances are equivalent to practice sessions. Unsatisfactory performance of a rescue or practice session indicates the need for further development.

In keeping with the emphasis on nonentry rescue, the Standard requires the use of retrieval systems or methods unless these would increase the overall risk of entry or would not contribute to the rescue. OSHA will use the following guidelines in making this determination:

- A permit space with obstructions or turns that prevent pull on the retrieval line
- A permit space from which an employee being rescued using the retrieval system would be injured
- A permit space that was entered by an entrant using an air-supplied respirator does not require the use of a retrieval system if the retrieval line could not be controlled so as to prevent entanglement hazards with the air line

The entrant is required to wear a chest or full-body harness with the retrieval line attached to minimize the profile that would be presented during retrieval. Wristlets also may be used, when appropriate. The retrieval device must be attached to a fixed point, so that retrieval can begin as soon as possible. A mechanical device is required for vertical drops exceeding 5 feet (1.3 m).

An additional issue relating to rescue is provision of an MSDS or other similar written information to the treating medical facility.

Confined and Enclosed Spaces ... in Shipyard Employment

The Occupational Safety and Health Administration issued a second standard addressing confined spaces in 1994. This standard covers confined spaces in maritime industry. This standard modernizes existing statutes dating from 1971, and extends ship-based concepts to confined spaces in the remainder of the shipyard. This extension removes the overlap of jurisdiction with the standard for general industry, discussed in the previous section.

In proposing to modify regulations regarding entry into confined spaces in the maritime industry, OSHA started with the observation that shipbuilding and repairing had the highest incidence of lost workdays for injury of any industry (OSHA 1994). OSHA believes that 20 deaths in shipyard and boatbuilding industries between 1983 and 1992 resulted from accidents in confined spaces (OSHA 1990).

The shipyard standard, 29 CFR Part 1915, applies to all aspects of shipbuilding, ship repair, shipbreaking, and the premises in which these activities occur. OSHA believes that the hazards in land-side confined spaces in the shipyard are similar to those found in vessels and vessel sections. The shipyard standard thus provides an interesting opportunity to compare the application of shipyard strategies to land-side confined spaces.

The focus of the shipyard standard clearly is atmospheric conditions. The standard relies on ventilation to prevent overexposure of employees to atmospheric hazards. This is accomplished through a built-in system of testing, visual inspection, and application of ventilation principles. The intent of the standard is to prevent or eliminate hazardous conditions.

The standard requires assessment of conditions at any time deemed appropriate during the cycle of entry and work. Conditions must be determined prior to entry of any worker. Where conditions are unacceptable, they must be corrected to within acceptable limits. At this point, workers may enter the space to perform work tasks. Since the work itself can be the source of the hazard, the process of evaluation and correction must continue at all stages. When conditions exceed acceptable limits during work, work must stop and workers must vacate the space. Work can resume once conditions are restored to within acceptable limits. Subsequent discussion will provide additional detail about these concepts.

The process prior to entry by workers begins with visual inspection and testing. A visual inspection is a physical survey of the space, its contents, and surroundings. The purpose of the inspection is to identify such hazards as restricted accessibility, cargo residues, unguarded machinery, and piping or electrical hazards. Spaces that contain or have contained combustible or flammable liquids or gases and spaces adjacent to them specifically must be inspected. Visual inspection is a necessary adjunct to atmospheric testing. This is especially important where products, such as diesel fuel, may be present. Diesel fuel has a high flash point and low vapor pressure. At ambient temperatures, diesel fuel produces insufficient vapors for detection by the combustible gas indicator.

The presence of residual combustible liquid in a confined space would be expected to produce a reading on a combustible gas detector. Lack of such a reading could indicate a defect in the instrument or presence of a high boiling liquid of low volatility. The presence of residual material or hazardous concentration of vapor or gas indicates the need for follow-up. The visual inspection provides the Qualified Person the means to pursue additional enquiry.

Testing also must occur as part of hazard assessment. Tests must occur close enough to the time of first entry for the results to reflect the conditions to be encountered during entry. To meet this requirement, testing almost always will be performed just prior to the start of work.

The Standard also establishes requirements for periodic retesting. Tests must occur as often as necessary to ensure that the required atmospheric conditions are maintained. When a change that could alter conditions in a tested space occurs, work must stop, and the space be vacated and retested. OSHA argues that the need for testing is directly related to the potential for change to occur within the space. Any change that could affect the designation of a space as "Safe for Workers" necessitates reinspection, retesting, and recertification.

TABLE A.3
Testing and Assessment of Confined Spaces

Situation	Authorized agent
Preentry	
Assessment	Competent person Certified Industrial Hygienist Certified Marine Chemist
Follow-up	Certified Industrial Hygienist
Toxic condition	Certified Marine Chemist
Flammable condition	
Oxygen-deficiency	
Oxygen enrichment	
Work to be performed	
Cold work	Competent person Certified Marine Chemist Certified Industrial Hygienist
Hot work	
Nonflammable condition	Competent person Certified Marine Chemist Certified Industrial Hygienist
Flammable condition	Certified Marine Chemist Coast Guard authorized person

The shipyard Standard uses a two-tiered approach for hazard assessment and control. The foundation of this approach is the shipyard **competent person**. The shipyard competent person is designated by the employer. The Standard directs the competent person to complete the initial evaluation, and as appropriate, to implement the first level of control. The competent person can determine that the atmosphere exceeds a regulatory limit and can recommend installation of ventilation to reduce the atmospheric concentration to within acceptable limits. If ventilation fails to control the level or other aspects of the situation exceed their capabilities, the competent person must seek advice from an NFPA (National Fire Protection Association) Certified Marine Chemist, a U.S. Coast Guard authorized person, or Certified Industrial Hygienist. The latter then will provide assistance in the development and implementation of control measures.

The shipyard standard authorizes specific actions by each of the preceding experts, as summarized in Table A.3.

Where hot work is to occur in a space that contains or contained flammable liquids or gases, or in a space that is adjacent to such a space, the evaluation and certification must be performed only by a Certified Marine Chemist or Coast Guard authorized person. (This is the only evaluation and certification duty permitted to the Coast Guard authorized person.) In other situations where hot work is to occur, the competent person can test and certify the space.

Interpretation and communication of results of inspection and testing are additional strategies underlying the shipyard Standard. Results of tests made by the competent person and outside experts must be recorded by location, time, and date and posted in the immediate vicinity of the affected operation before work starts. They also must be retained for a further 3 months. Along with results of testing is the interpretation or certification of conditions in the space. A certificate must be issued, either by the competent person or the outside expert. The certificate issued by the outside expert can contain instructions to be followed by the shipyard competent person and management. All spaces, therefore, must be certificated prior to entry. The certificate contains one or more of the designations contained in the following Table A.4. Spaces that are Not Safe for Workers or Not Safe for Workers — Enter with Restrictions must be posted with a sign in order to prevent

TABLE A.4
Designations Contained in Certificates

Designation	Justification/possible cause
Safe for Workers	Oxygen content at least 19.5% and less than 22%
	Flammable vapors less than 10% of LEL
	Toxic materials in the atmosphere associated with cargo, fuel, tank coatings, or inerting media are within permissible limits at time of inspection
	Residues or materials associated with authorized work will not produce uncontrolled release of toxic materials under existing atmospheric conditions while maintained as directed
Enter with Restrictions	A space where engineering controls, personal protective equipment, clothing, and time limitations are specified in the certificate to permit safe entry for work
Not Safe for Workers	A space that an employee may not enter because the conditions do not meet the criteria for Safe for Workers
	IDLH atmosphere (no PEL exists)
	Atmosphere exceeds PEL
	Oxygen deficiency (less than 19.5%)
	Oxygen enrichment (22% or greater)
Not Safe for Hot Work	Concentration of flammable vapors or gases equals or exceeds 10% LEL in the space or adjacent space where hot work is to be performed
Not Safe for Workers — Not Safe for Hot Work	Oxygen enrichment (22% or greater)

inadvertent entry. Spaces containing or adjacent to spaces containing flammable gases or vapors at concentrations at or above 10% of their lower explosive limit also must be posted.

Spaces that previously held flammable or toxic liquids must be cleaned prior to being assessed "safe for entry without restrictions." First, residues must be removed and then the space cleaned. The atmosphere must be tested for flammability and retested as often as necessary throughout the course of the work to ensure that the concentrations remain within the safe range. These tests are additional to those required prior to entry. The Standard requires ignition sources to be controlled or eliminated during cold work in these spaces in order to limit further the possibility of explosion or fire.

Pipelines that carry hazardous materials must be blocked or flushed and cleaned to prevent discharge into the space. The space must be tested periodically to ensure that safe working conditions are maintained. Work must cease and the space be vacated when conditions change and the space no longer meets the criteria specified in the certificate.

Thus far, discussion has indicated that entry by workers can occur when testing indicates airborne concentrations are within acceptable limits. When concentrations are not within acceptable limits, the space must undergo corrective action, starting with installation of ventilation.

Ventilation can be used to correct a preexisting problem. Once conditions have been restored to acceptable levels, ventilation is no longer required. Ventilation also can be used to control a preexisting problem during the work period. The third possible application of ventilation is to control atmospheric hazards created by the work to be performed.

The intent of ventilation is not simply to maintain a local work zone inside a space within acceptable limits, while other areas exceed them. OSHA requires ventilation to maintain conditions in the entire compartment within acceptable limits. Ventilating only a portion of a compartment means that pockets of a hazardous atmosphere can exist or develop or remain during the work period.

A space maintained by ventilation within acceptable limits does not require labeling. Thus, the label applied as a result of the initial evaluation could be removed following successful control by ventilation. This strategy seems at odds with some of the other labeling requirements presented in Table A.4. This adds to the burden of training, since everyone working in or around the space must

TABLE A.5
Permissible and Hazardous Conditions in Confined Spaces

Condition	Permissible	Dangerous
Oxygen	19.5% ≤ conc. <22%	<19.5% ≥22%
Flammable/combustible	<10% LEL	≥10% LEL
Toxic	≤PEL ≤IDLH	>PEL >IDLH

Note: In this table, PEL is Permissible Exposure Limit. LEL is Lower Explosive Limit. In the case of toxic substances, IDLH is used where a PEL does not exist. In this standard, IDLH (Immediately Dangerous to Life or Health) means an atmosphere that poses an immediate threat to life or that is likely to result in immediate severe health effects.

recognize the presence, purpose, and vital function of ventilating equipment. This becomes critical when ventilation fails or is inoperative or has been deliberately moved or tampered with. Labeling spaces that require functional ventilation to maintain control of atmospheric hazards seems a logical necessity.

In spaces whose atmospheric conditions are maintained by ventilation, testing during work is required. OSHA uses the term "as often as necessary" to describe the nature of this requirement. Ventilation is required when the concentration does not remain below the applicable standard on its own accord. Similarly, testing and ventilating also are required to ensure that the concentration of toxic, corrosive, or irritating vapors remains within permissible limits or less than the IDLH, where these do not exist.

Factors that could affect the frequency of testing include:

- Temperature
- Work being performed
- Length of elapsed time
- Unattended space
- Work breaks
- Ballasting or trimming

Another requirement for testing when ventilation is used to control the atmosphere in a space is to assess the discharge for accumulation of contaminants. When this accumulation exceeds hazardous limits, work must stop and evacuation must occur until the vapors have dissipated or been removed.

The presence of an initial atmospheric condition that lies outside permissible limits requires action. OSHA has decided that workers shall not be permitted to work in dangerous atmospheres. A dangerous atmosphere may expose employees to risk of death, incapacitation, impairment of the ability to self-rescue (escape unaided from a confined space), injury, or acute illness. A dangerous atmosphere has a composition that lies outside permissible limits. Permissible conditions and dangerous conditions are presented in Table A.5.

Having decided that workers cannot work in hazardous atmospheres, OSHA permitted entry in several exceptional situations. These include entry:

- For testing and visual inspection
- For installation of ventilating or other control equipment
- To perform rescue

In all cases, entry is expected to be short in duration. Entrants must wear appropriate respiratory protection and other personal protective equipment. The atmosphere must be monitored continuously for the contaminant. Also, an attendant must be present to assist the entrant.

Considerations vary according to specific atmospheric conditions. The following discussion considers each briefly.

OSHA indicates that the following spaces may contain oxygen-deficient atmospheres:

- Spaces that have been sealed, coated, and closed up, or painted and lack ventilation
- Spaces and adjacent spaces that contain or have contained combustible or flammable liquids or gases
- Spaces and adjacent spaces that contain or have contained liquids, gases, or solids that are toxic, corrosive, or irritating
- Spaces that have been fumigated
- Spaces containing materials or residues that could create an oxygen-deficient atmosphere (for example, cargo spaces containing scrap iron, fresh fruit, molasses, vegetable drying oils)

OSHA permits entry into oxygen-deficient or oxygen-enriched spaces designated Not Safe for Workers, as indicated above, but only under tightly controlled conditions. Entrants must wear appropriate respiratory protection and other personal protective equipment. The atmosphere must be monitored continuously for oxygen.

A parallel question concerns spaces in which the atmosphere exceeds 10% of the LEL. An atmosphere whose composition is within the flammable/explosive range potentially can be ignited. The specifics under which OSHA permits entry into such spaces include:

- No ignition sources may be present — only explosion-proof, self-contained portable lamps can be used
- The atmosphere must be monitored continuously
- Flammable vapors and gases at or above their UEL must be maintained in this condition
- Respiratory protection and other personal protection must be used
- Signs indicating that ignition sources are prohibited must be prominently displayed at the entrance to the space and in adjacent spaces and in the open area adjacent to those spaces

Ventilation equipment and components that are used in these spaces must not be capable of generating a static electrical charge of sufficient energy to create a source of ignition. They also must be bonded electrically to the structure of the vessel or vessel section or grounded in the case of land-side spaces to prevent unintentional discharge of static charge. Fans must have nonsparking blades. Portable duct must be made from nonsparking material.

The third type of dangerous atmosphere contains toxic substances. Toxic atmospheres can arise from gases and liquids or solids that are toxic, corrosive or irritant. The Standard references the PEL, or if a PEL does not exist, the IDLH. A space containing a substance above its PEL or IDLH can be entered for emergency rescue or installation of ventilation provided that:

- The atmosphere is monitored continuously
- Respiratory protection and other personal protection are used

Where ventilation cannot maintain concentration within the required limits, following intervention by the Certified Marine Chemist or Certified Industrial Hygienist, the space may be listed as "Enter with Restrictions." If intervention produces a means to control the hazard, the space becomes "Safe for Workers."

Regarding cleaning and other cold work that occurs after permission for entry is granted, the Standard specifies that spaces must be tested as often as necessary and ventilated as required. Liquid residues in tanks shall be removed as thoroughly as practicable before manual cleaning starts. Also, spills or releases of flammable, combustible, toxic, corrosive, irritant, or fumigant materials must be cleaned up as work progresses.

Hot work in or on spaces and adjacent spaces that contain or have contained combustible or flammable liquids or gases, related piping and accessories receives special attention in the shipyard Standard. Hot work is any activity including riveting, welding, burning, cutting, use of powder-actuated tools, or similar fire-producing operations. Grinding, drilling, or similar spark-producing operations also are considered hot work, except when isolated physically from any atmosphere containing more than 10% LEL.

Certain spaces must be certified by a Certified Marine Chemist or Coast Guard authorized person as "Safe for Hot Work" prior to the start of work. The competent person also may approve hot work in certain circumstances.

Circumstances requiring certification by a Certified Marine Chemist or Coast Guard authorized person include:

- Work within, on, or immediately adjacent to spaces that contain or have contained combustible or flammable liquids or gases
- Work within, on, or immediately adjacent to fuel tanks that contain or have last contained fuel
- Work on pipelines, heating coils, pump fittings, or other accessories connected to spaces that contain or have last contained fuel
- Exception: within spaces adjacent to spaces in which the flammable gas or liquid has a flash point less than 66°C (150°F) and the distance between such spaces and the work is greater than 7.5 m (25 ft)

The certificate issued by the Certified Marine Chemist or Coast Guard authorized person requires the competent person to recheck the space to ensure that conditions do not change. If a change occurs, the competent person must stop the work, and recall the Certified Marine Chemist to recertify the space.

The competent person also may issue a certificate for hot work. This may occur in the following locations:

- Dry cargo holds
- Bilges
- Engine room and boiler for which certification by the Marine Chemist or Coast Guard authorized person is not required
- Vessel and vessel sections for which certification by the Marine Chemist or Coast Guard authorized person is not required
- Land-side spaces

When the concentration of flammable vapors or gases equals or exceeds 10% LEL in the space or adjacent space where hot work is to be performed, the space shall be labeled "Not Safe for Hot Work" and ventilated to reduce the concentration to within acceptable limits.

Training is an important component of the shipyard Standard. Training applies to shipyard competent persons who enter the space for evaluation and testing, to workers who enter the space for the purpose of performing work and to persons involved in rescue. The employer must certify that entrants have been trained before they can enter these spaces. This certification must include the name of the trainee, trainer, and date(s).

Knowledge-based requirements for competent persons include:

- Legislative requirements
- Surface preparation and preservation
- Welding, cutting, and heating
- Tools and related equipment
- Location and designation of spaces where work is to be performed
- Calibration and use of test equipment — oxygen indicators, combustible gas indicators, carbon dioxide indicators
- Decision making to determine whether further assessment should occur by a higher level technical individual
- Understanding and carrying out instructions and other information provided by the higher level technical individual
- Ability to maintain records

Training for each person entering the space must occur prior to entry and whenever a change in operations or duties occurs. Specific hazards and hazardous conditions addressed by OSHA in the Standard include:

- Recognition of characteristics of the space
- Anticipation and awareness of hazards that could be encountered during entry
- Recognition of the signs, symptoms, or other adverse health effects that could be caused by exposure to hazards
- Understanding the signs and reactions following exposure to hazards
- Knowledge about how to use personal protective equipment
- As necessary, awareness of the presence and proper use of barriers or other devices needed to protect entrants from hazards
- Conditions for exiting the space:
  - Order of the employer
  - Sounding of an evacuation signal
  - Perception of danger or threat to safety
- Self-rescue techniques
- Recognition of implications of signs used to indicate safe and hazardous conditions

The last major strategy on which the shipyard standard is built is rescue. Inevitably, workers will continue to be injured in these workspaces. The requirement to establish a breathable atmosphere prior to the start of work should de-emphasize the importance of rescue. However, rescue from the space following an accident during the initial inspection and installation of ventilation still could be required.

Employers must provide for rescue. Rescue can occur using a team from within the shipyard or an outside rescue service. Persons on shipyard rescue teams must be provided with and trained in use of the personal protective equipment that they would require. This includes respirators and rescue equipment. Rescues must be attempted only in accordance with safe entry procedures. The shipyard team must practice these skills at least once per year. Practice can include use of manikins and simulations of the physical facilities in which rescue may be needed. At least one member of the team must maintain certification in first aid and cardiopulmonary resuscitation.

When outside services are to be used, the employer must establish contact and inform them about hazards and hazardous conditions that might be encountered. Also, the Standard encourages employers to evaluate the skills of the outside rescue service and to determine what in-house hazard information would be most helpful to them. To follow this through, OSHA requires transfer of hazard information between employers. This transfer must include information on hazards, safety rules, and emergency procedures. The purpose for this interchange is to ensure that knowledge about hazards that could change daily will be communicated to people who could be affected by them.

The quality of decision making by the competent person is critical to the success of this Standard. OSHA stresses that the competent person must be able to recognize the limitations of training and capability and the need for input from higher level technical resources. Yet, OSHA specified only performance outcomes for these individuals and declined to specify a curriculum of formal study. That is to say, a person whose background education and training are not specified in law is to be held fully responsible for technical actions and for deciding when to call in higher level assistance. The inconsistency in this policy is that the individuals who are to be called in as advisors are the products of rigorous programs of education, training, and certification.

Many of the intervenors to the standard-setting process were Certified Marine Chemists. These individuals often work alone or as part of small enterprises. Participation in a standard-setting exercise required considerable effort and potential sacrifice on their part. They voiced many comments of concern regarding the need for high level technical input in decision making regarding hazard management in these spaces. It is certainly true that this group had a vested interest in the outcome of these deliberations. Yet, there was little input from individuals whom OSHA labels shipyard competent persons and in whom they have vested so much responsibility. In effect, OSHA has downgraded the role of professionals in the implementation of the revised Standard and not heeded their statements of concern.

Another area of concern is the manner in which ventilation is used as a control measure. Where testing determines that the concentration exceeds some standard or permissible limit, ventilation must be used as the control measure. A space maintained by ventilation within acceptable limits does not require labeling. Thus, the label applied as a result of the initial evaluation could be removed following successful control by ventilation. This strategy seems at odds with some of the other labeling requirements presented in Table A.4. This adds to the burden of training, since everyone working in or around the space must recognize the presence, purpose, and vital function of ventilating equipment. This becomes critical when ventilation fails or is inoperative or has been deliberately moved or tampered with. Labeling spaces that require functional ventilation to maintain control of atmospheric hazards seems a logical necessity.

Comparison between General Industry and Shipyard Standards

The shipyard Standard, 29 CFR Part 1915, utilizes fundamentally different strategies for managing the hazards in confined spaces compared to the general industry Standard, 29 CFR Part 1910.146. These differences reflect the rapidly changing presence, configuration and design, and hazards of the confined spaces managed in the maritime industry, especially shipbuilding and shipbreaking (Willwerth 1994). Encounters with confined spaces in the maritime industry occur daily, rather than occasionally. In one extreme example, workers involved in construction of an aircraft carrier could enter 1,000 confined spaces per day.

Work in confined spaces in a shipyard is highly complex and can involve application of many industrial processes. Processes whose hazards are highly interactive could be utilized in close proximity. To illustrate, poorly planned hot work on one side of a bulkhead could cause a fire during spray painting on the other. Complicating things further, these activities and the hazards inherent in them must be managed within the confines of tight production schedules. Once a ship is drydocked, no matter how limited the repair, the entire vessel is out of service and not productive. The pressure to do the job correctly and quickly the first time is especially acute in the shipyard.

Structure- or location-specific concepts that are applicable in one industrial sector are not applicable to another, according to this rationale. The marine program is expert- and prevention-centered. An expert is readily at hand to address and assess concerns. Also, both management and the workforce are trained in recognition of hazards associated with work in confined spaces. Consultation by management and the workforce in hazard control with the expert is a routine part of hazard management in this environment. The expert routinely inspects, evaluates, and certifies the status of confined spaces and specifies controls for the hazards of work to be performed. The emphasis in this approach is to discover and control hazards before work begins.

There is a fundamental difference in approach between the shipyard and general industry Standards. OSHA considers the shipyard Standard to be a more restrictive subset of the general industry Standard and even more comprehensive in its coverage of hazardous atmospheres (OSHA, 1994). Further, its approach to inspection, testing and ventilating spaces has become a part of routine work activity in the shipyard. OSHA comments that confined space entry and work in shipyards is routine; hazardous atmospheres are common, and the work activity introduces or creates hazards. Furthermore, the spaces are complex and may be nested inside one another.

The general industry Standard requires employers to identify confined spaces in their workplaces and to determine whether they are permit-required or nonpermit-required. Similarly, they must reevaluate nonpermit spaces following changes that might increase hazards to entrants above the threshold of concern. By contrast, the shipyard Standard treats all confined spaces and other spaces that might contain a hazardous atmosphere equally. This means that all spaces to be entered require initial testing and visual inspection, possibly followed by continuous ventilation and further testing.

Another requirement of the general industry Standard is the implementation of a permit system. The employer must document by means of a permit the completion of measures required for safe entry. The permit must be completed before entry is allowed into a permit-required confined space. OSHA argues that the requirements of the shipyard Standard constitute an informal permit system. That is, it requires evaluation, tracking, and control measures that are equivalent in protection. The only spaces that employees are permitted to enter are those certified as Safe for Workers. Entry is not permitted until that certification is granted. Shipyard employees are trained to remain outside spaces that do not have this certification.

The general industry Standard requires the presence of at least one attendant outside a permit space while entry operations are occurring. The attendant monitors entrants and conditions inside and outside the space, prevents entry of unauthorized persons, and summons rescue services in emergencies. OSHA rejected the attendant concept in the shipyard Standard, arguing that the requirements for the space to be completely safe prior to entry and work provide equivalent protection. OSHA also commented that some permit-required spaces can be made safe for entry without the need for a permit or attendant. The shipyard Standard does require the presence of the attendant during initial inspection and testing where a hazardous atmosphere is present.

The general industry Standard included serious safety hazards, such as engulfment and the internal geometry of the space, in decision making about whether it is permit-required. OSHA did not incorporate these considerations into the shipyard Standard, arguing that the shipyard competent person, in making the initial visual inspection of the space, determines whether conditions warrant application of other standards. The competent person would be expected to alert the employer to this situation.

Control of Hazardous Energy Source (Lockout/Tagout)

A companion to the confined spaces Standards is the Control of Hazardous Energy Source (Lockout/Tagout) (OSHA 1989). This Standard is intended to protect employees from hazardous energy sources while they are servicing or performing maintenance on machines and equipment. Hazardous energy sources involve the unexpected start-up of equipment or release of stored energy. The Lockout/Tagout Standard identifies practices and procedures necessary to shut down and lock out or tag out machines and equipment. In general, this Standard requires the turning off, disconnecting from the energy source and the locking out or tagging out of the energy-isolating device prior to starting servicing or maintenance. Locking out clearly is the preferred method of action for complying with this Standard. The Standard supplements existing statutes by requiring development and utilization of written procedures, training of employees, and periodic auditing to ensure effective use of the procedures.

The Lockout/Tagout Standard applies to general industry employment. The Standard clearly applies during work involved in the preparation of permit spaces for entry and during occupancy, since these activities are not part of normal production operations. Machinery must be isolated and inoperative before any employee performs service or maintenance. This is to ensure that unexpected energization, start-up, or release of stored energy cannot occur.

Central to application of the Standard is establishment of the energy control program. This program must include:

- Documented energy control procedures
- Employee training program
- Periodic auditing to ensure effective use of the procedures

The energy control procedures must identify information that authorized employees must know in order to control hazardous energy during servicing or maintenance. These require a comprehensive disciplined approach according to procedural elements contained in the Standard. Also, they also must address multiple energy sources that can affect a machine or piece of equipment.

The primary tool for providing protection under the Standard is the energy-isolating device. This mechanism prevents transmission or release of energy. In practical terms, this means start-up of the machine or equipment or re-energization of equipment. These devices must be singularly identified and the only devices used for controlling hazardous energy. Also, they must be durable, substantial, identifiable, and standardized by color, shape, or size.

Employee training must be effective and substantial. Initial and periodic training are required. The employer must document the training and certify that it has been provided to all employees affected by the Standard.

Periodic inspection must occur at least annually during actual application of a procedure. Results must be documented and certified. A review of employee responsibilities under the energy control program must occur.

Mine Safety and Health Administration

OSHA's sister agency, the Mine Safety and Health Administration (MSHA), issued an Advanced Notice of Proposed Rulemaking for a confined space Standard in 1991 (MSHA 1991). This action followed a less comprehensive proposed rule issued in 1989.

MSHA concern focused on atmospheric hazards, including toxic gases and vapors and oxygen deficiency. Additionally, MSHA specifically focused on engulfment and entrapment. Their research indicated that the vast majority of fatal accidents in mining are caused by engulfment and entrapment. MSHA estimates that 45 metal and nonmetal mining fatalities and 33 coal mining fatalities occurring between 1980 and 1991 involved confined spaces. They indicate that fatal injuries from accidents involving confined spaces in mining result from many causes including: asphyxiation, engulfment, entrapment, burn and explosion, electrical shock, toxic substances, and mechanical and physical hazards.

State Regulations

Several states have enacted their own confined space Standards. These include California, Kentucky, Maryland, Michigan, New Jersey, Pennsylvania, Virginia, and Washington (OSHA 1993). OSHA will allow states with "approved plans" to issue and enforce more stringent confined space requirements than those outlined in the OSHA Standard.

REGULATIONS IN OTHER COUNTRIES: CANADA

Canada has federal, provincial, and territorial jurisdictions. Legislation that addresses confined space hazards exists within 12 of the 13 jurisdictions in this country. Most of this legislation involves a step-by-step procedure that includes testing and ventilation or the use of respiratory protection. Other common requirements between the jurisdictions include: written procedures, communication, standby attendants, isolation, and rescue procedures.

Among the more detailed of the legal requirements in Canada are those contained in Part II of the Canada Labour Code. The Canada Labour Code applies primarily to operations that cross provincial boundaries. The Canada Labour Code contains a number of regulations, including the Canada Occupational Safety and Health Regulations and the Marine Occupational Safety and Health Regulations. Part XI of the Canada Occupational Safety and Health (COSH) Regulations and Part IX of the Marine Occupational Safety and Health Regulations (MOSH) cover confined spaces (Canada Gazette 1987, 1992).

These regulations provide an illustration of variations on the theme of the approach taken by OSHA.

Canada Occupational Safety and Health Regulations

Part XI of the COSH regulations was revised recently and therefore represents the more recent approach to this topic. The regulation allows consideration of individual confined spaces and groups of workspaces as a class by virtue of similarity of hazards. The approach used in the COSH regulations begins with recognition that a particular workspace is a confined space. At this point the employer must appoint a Qualified Person. The Qualified Person assesses the physical and chemical hazards to which an entrant could be exposed, and specifies tests necessary to determine whether the entrant likely would be exposed. The Qualified Person then provides the results of the assessment in a signed and dated report to the employer. The employer then establishes procedures for entry, work, and egress. The employer also must establish a system of entry permits. Also, the employer must specify protective equipment and tools to be used by entrants and emergency equipment required by rescuers.

One important feature of the Canadian approach to workplace health and safety is the statutory involvement of the Safety and Health Committee or representative at all stages in the process. The employer must inform and consult with the committee or representative about all aspects of the situation.

Just prior to entry and during occupancy of the confined space, the Qualified Person performs tests to verify compliance with the following specifications:

- Exposure to chemical agents singly or in combination is less than specifications in the 1985–1986 TLVs (the present legal reference)
- The concentration of flammables is less than 10% of the lower explosive limit
- Percentage of oxygen is not less than 18% or more than 23% at normal pressure
- Liquid in which the entrant could drown has been removed
- Free-flowing solid in which the entrant could become entrapped has been removed
- Entry of liquid, free-flowing solid, or hazardous substance is prevented by a secure means of disconnection or fitting of blank flanges
- Disconnection from power sources, real or residual, and locking out of electrical and mechanical equipment that may present a hazard

At the conclusion of the work, the Qualified Person prepares a signed and dated report outlining the work performed and results obtained from the tests. Where the report indicates that the occupants have been in danger, the employer must revise the procedures.

In circumstances where conditions in the space or the nature of the work preclude compliance with the preceding specifications, the employer must establish emergency and evacuation procedures and provide other measures to ensure communication and immediate assistance to occupants. A Qualified Person trained in entry and rescue procedures must attend outside the confined space. Also, additional trained help must be immediately available. The Qualified Person specifies equipment required and procedures to be followed in a signed and dated report to the employer. Also, the Qualified Person explains the procedures to entrants and obtains their acknowledgment through signature.

Performance of hot work introduces additional constraints. The concentration of an explosive or flammable substance must not exceed 10% of the lower explosive limit. The concentration of oxygen must not exceed 23%. Where the confined space contains hazardous concentrations of flammable or explosive materials, the employer must provide a Qualified Person to maintain a fire watch in the area outside the space. The hazards of substances produced as a result of the hot work must be addressed through ventilation or the use of personal protective equipment.

Ventilation can be used to control the concentration of chemical substances or the level of oxygen in the confined space. Measures are required to ensure proper function of this equipment and to alarm in the event of failure.

This Part contains additional requirements including training and recordkeeping.

Marine Occupational Safety and Health Regulations

Part IX of the MOSH Regulations addresses confined spaces. An initial difference in this older approach is the lack of consideration of classes of confined spaces. Part IX introduces the involvement of the Qualified Person, in this case a marine chemist or other Qualified Person, just prior to entry. "Marine chemist" in Canadian usage is a defined term. The definition is less stringent than that used by the National Fire Protection Association in the Marine Chemist program in the U.S. The Qualified Person verifies by tests the existence of conditions mentioned above in the COSH regulations and then prepares a signed report that includes an evaluation of the hazards. The Qualified Person then identifies normal and emergency procedures to be followed during entry and occupancy or prepares them where they do not exist. In the latter case, the Qualified Person also would specify protective equipment.

Entrants are to be informed about the contents of the report and to acknowledge understanding.

Other parts of the regulation are comparable to corresponding sections in the COSH regulation. The COSH version also contains clarifications and strengthened requirements for training and recordkeeping.

The fundamental differences between the COSH and MOSH regulations are the trend to full participation by the Safety and Health Committee, breaking the process into more segments and increasing the involvement of the Qualified Person. The opportunity for involvement of the Safety and Health Committee was not stated explicitly in the older MOSH regulation. Increasing this involvement is part of a trend toward strengthening the internal responsibility system in Canadian workplaces. In the MOSH regulation, the Qualified Person did not have to follow-up the entry process by monitoring and preparing a report that outlined the success of the procedures. This extra segment in the process now provides an audit and feedback mechanism.

SUMMARY

This Appendix examined standards and guidelines produced by various groups — standard-setting organizations, industry and trade associations, and regulatory agencies. These standards attempt to provide a management framework for the safe conduct of work in confined spaces. Inherent in each of these attempts is recognition of the special concerns arising from these workspaces. Creating a workable management framework is very difficult. Confined spaces encompass an infinite number of workspaces, environments, and combinations of hazards. Also, hazards associated with confined spaces cover a broad range of severity ranging from the inconsequential to the life threatening.

Most standards, guidelines, and regulations only address the most complex and extreme hazards in confined spaces. These situations account for the most serious reported accidents, as reflected primarily in fatal accidents. Overexposures and injuries occurring in less hazardous situations usually are not reported, and therefore controls for these types of hazards are not addressed as comprehensively.

By necessity, approaches for addressing hazards posed by confined spaces must be performance based, rather than specification based. A key agent common in all of the approaches reviewed here is the Qualified Person. The Qualified Person is charged with very high levels of responsibility, but usually no authority. Rarely, however, is the term, "qualified person," ever defined. Some legislation identifies generic professionals, such as industrial hygienist, safety professional, or marine chemist.

Almost all standards operate through a top-down style of management. Management, acting through the Qualified Person, identifies and assesses the situation, and prescribes work practices and procedures, and control and emergency measures. That is, management determines the conditions of work and specifies how the work is to be performed. The people who will perform the work have no role in the process other than following instructions. One variation on this theme is the active involvement of the Safety and Health Committee mandated in some jurisdictions.

Most of the standards contain hazard classification systems. These are used to assign resources appropriate to the level of concern. These systems differ from standard to standard.

REFERENCES

American National Standards Institute: *Safety Requirements for Working in Tanks and Other Confined Spaces* (ANSI Z117.1-1977). New York: American National Standards Institute, 1977. 24 pp.

American National Standards Institute: *American National Standard for Respiratory Protection — Respirator Use — Physical Qualifications for Personnel* (ANSI Z88.6-1984). New York: American National Standards Institute, 1984. 15 pp.

American National Standards Institute: *Safety Requirements for Confined Spaces* (ANSI Z117.1-1989). Des Plaines, IL: American Society of Safety Engineers/American National Standards Institute, 1989. 24 pp.

American National Standards Institute: *Safety Requirements for Lockout/Tagout of Energy Sources* (ANSI Z244.1-1992(R)). New York: American National Standards Institute, 1992.

American National Standards Institute: *Safety Requirements for Confined Spaces* (ANSI Z117.1-1995). Des Plaines, IL: American Society of Safety Engineers/American National Standards Institute, 1995. 32 pp.

American Petroleum Institute: *Safe Operating Guidelines for Vacuum Trucks in Petroleum Service* (API Pub. 2219). Washington, D.C.: American Petroleum Institute, 1986. 7 pp.

American Petroleum Institute: *Guidelines for Work in Inert Confined Spaces in the Petroleum Industry* (API Pub. 2217A). Washington, D.C.: American Petroleum Institute, 1987. 7 pp.

American Petroleum Institute: *Safe Welding and Cutting Practices in Refineries, Gasoline Plants and Petrochemical Plants*, 5th ed. (API Pub. No. 2009). Washington, D.C.: American Petroleum Institute, 1988a. 6 pp.

American Petroleum Institute: *Ignition Hazards Involved in Abrasive Blasting of Atmospheric Storage Tanks in Hydrocarbon Service*, 2nd ed. (API Pub. 2027). Washington, D.C.: American Petroleum Institute, 1988b. 4 pp.

American Petroleum Institute: *Spark Ignition Properties of Hand Tools*, 3rd ed. (API Publ. 2214). Washington, D.C.: American Petroleum Institute, 1989. 5 pp.

American Petroleum Institute: *Protection against Ignitions Arising Out of Static, Lightning, and Stray Currents*, 5th ed. (API Publ. RP2003). Washington, D.C.: American Petroleum Institute, 1991a. 39 pp.

American Petroleum Institute: *Cleaning Mobile Tanks in Flammable or Combustible Liquid Service*, 6th ed (API Pub. 2013). Washington, D.C.: American Petroleum Institute, 1991b. 6 pp.

American Petroleum Institute: *Preparing Tank Bottoms for Hot Work*, 4th ed (API Pub. 2207). Washington, D.C.: American Petroleum Institute, 1991c. 4 pp.

American Petroleum Institute: *Safe Entry and Cleaning of Petroleum Storage Tanks,* 4th ed. (API Pub. 2015). Washington, D.C.: American Petroleum Institute, 1991d. 26 pp.

American Petroleum Institute: *Interior Lining of Underground Storage Tanks,* 3rd ed. (RP 1631). Washington, D.C.: American Petroleum Institute, 1992. 10 pp.

American Petroleum Institute: *Safe Entry and Cleaning of Petroleum Storage Tanks,* 5th ed. (API Pub. 2015). Washington, D.C.: American Petroleum Institute, 1993. 60 pp.

American Society for Testing and Materials: *Standard Practice for Confined Area Entry* (ASTM D 4276-84, reapproved 1989). Philadelphia, PA: American Society for Testing and Materials, 1984. 3 pp.

Anonymous: Reply to "Confined Spaces Clarification" by Gary Easter. *Occup. Haz.* May 1993 p. 9.

Canada Gazette: "Part IX, Confined Spaces," *Canada Gazette, Part II. 121*: 1337–1364 (1987).

Canada Gazette: "Part XI, Confined Spaces," *Canada Gazette, Part II. 126*: 3863–3876 (1992).

Easter, G.: "Confined Spaces Clarification." Letter to the Editor. *Occup. Haz.* May 1993 p. 9.

Keller, C.: Firesafety in the shipyard. *Fire J.*: July 1982. pp. 60–66.

National Fire Protection Association: *NFPA 306: Control of Gas Hazards on Vessels* (1993 ed.). Quincy, MA: National Fire Protection Association, 1993. 15 pp.

National Fire Protection Association: *NFPA 328: Control of Flammable and Combustible Liquids and Gases in Manholes, Sewers, and Similar Underground Structures* (1992 ed.). Quincy, MA: National Fire Protection Association, 1992. 11 pp.

National Fire Protection Association: *NFPA 70: National Electrical Code* (1993 ed.). Quincy, MA: National Fire Protection Association, 1993.

National Institute for Occupational Safety and Health: Criteria for a Recommended Standard — Working in Confined Spaces (DHEW/PHS/CDC/NIOSH Pub. No. 80-106). Cincinnati, OH: National Institute for Occupational Safety and Health, 1979. 68 pp.

National Institute for Occupational Safety and Health: Safety and Health in Confined Workspaces — for the Construction Industry — A Training Resource Manual (DHEW/PHS/CDC/NIOSH, Division of Training and Manpower Development). Cincinnati, OH: National Institute for Occupational Safety and Health, 1985. 197 pp.

National Institute for Occupational Safety and Health: Alert: Request for Assistance in Preventing Occupational Fatalities in Confined Spaces (DHHS/PHS/CDC/NIOSH Pub. No. 86-110). Cincinnati, OH: National Institute for Occupational Safety and Health, 1986a. 9 pp.

National Institute for Occupational Safety and Health: Alert: Preventing Fatalities Due to Fires and Explosions in Oxygen-Limiting Silos (DHHS/PHS/CDC/NIOSH Pub. No. 86-118). Cincinnati, OH: National Institute for Occupational Safety and Health, 1986b. 6 pp.

National Institute for Occupational Safety and Health: *A Guide to Safety in Confined Spaces* (DHHS/PHS/CDC/NIOSH Pub. No. 87-113). Cincinnati, OH: National Institute for Occupational Safety and Health, 1987. 20 pp.

National Institute for Occupational Safety and Health: Alert: Preventing Entrapment and Suffocation Caused by the Unstable Surfaces of Stored Grain and Other Materials (DHHS/PHS/CDC/NIOSH Pub. No. 88-102). Cincinnati, OH: National Institute for Occupational Safety and Health, 1988. 8 pp.

National Institute for Occupational Safety and Health: Alert: Preventing Death from Excessive Exposure to Chlorofluorocarbon 113 (DHHS/PHS/CDC/NIOSH Pub. No. 89-109). Cincinnati, OH: National Institute for Occupational Safety and Health, 1989. 12 pp.

National Institute for Occupational Safety and Health: Alert: Request for Assistance in Preventing Deaths of Farm Workers in Manure Pits (DHHS/PHS/CDC/NIOSH Pub. No. 90-103). Cincinnati, OH: National Institute for Occupational Safety and Health, 1990. 7 pp.

National Institute for Occupational Safety and Health: Worker Deaths in Confined Spaces (DHHS/PHS/CDC/NIOSH Pub. No. 94-103). Cincinnati, OH: National Institute for Occupational Safety and Health, 1994. 273 pp.

National Occupational Health and Safety Commission and Standards Australia: Joint Draft National Standard for Planning and Work for Confined Spaces (BS92/20113 Cat. No. 92 0995 8). Canberra, ACT: Australian Government Publishing Services, 1992. 55 pp.

National Safety Council: *Confined Space Entry Control System for R & D Operations,* No. 12304-0704. Itasca, IL: National Safety Council, 1985.

National Safety Council: *Inspecting and Cleaning Pipes and Sewers,* No. 12304-0577. Itasca, IL: National Safety Council, 1986.

National Safety Council: *Atmospheres in Sub-Surface Structures and Sewers*, No. 12304-0550. Itasca, IL: National Safety Council, 1987a.

National Safety Council: *Entry into Grain Bins & Food Tanks*, No. 12304-0663. Itasca, IL: National Safety Council, 1987a.

Occupational Safety and Health Administration: Selected Occupational Fatalities Related to Ship Building and Repairing as Found in Reports of OSHA Fatality/Catastrophe Investigations. Washington, D.C.: U.S. Department of Labor, Occupational Safety and Health Administration (U.S. DOL/OSHA), 1990. 195 pp.

Occupational Safety and Health Administration: Letter to Mary C. DeVany in Reply to Her Letter of Enquiry of June 13, 1996 by Richard S. Terrill through Ronald T. Tsunehara. Seattle, WA: Occupational Safety and Health Administration, July 1, 1996. [letter]

OSHA: Permit-Required Confined Spaces; Notice of Proposed Rulemaking, *Fed. Regist. 54*: 106 (5 June 1989). pp. 24080–24109 (1989a)

OSHA: Control of Hazardous Energy Sources (Lockout/Tagout); Final Rule, *Fed. Regist. 54*: 169 (1 September 1989). pp. 36644–36696 (1989b)

OSHA: Confined Spaces; Advance Notice of Proposed Rulemaking, *Fed. Regist. 56*: 250 (30 December 1991). pp. 67364–67366.

OSHA: Permit-Required Confined Spaces for General Industry; Final Rule, *Fed. Regist. 58*: 9 (14 January 1993). pp. 4462–4563.

OSHA: Confined and Enclosed Spaces and Other Dangerous Atmospheres in Shipyard Employment; Final Rule, *Fed. Regist. 59*: 141 (25 July 1994). pp. 37816–37863.

Standards Association of Australia: *Safe Working in a Confined Space* (AS 2865-1986). North Sydney, NSW: 1986. 19 pp.

Willwerth, E.: Maritime Confined Spaces. *Occup. Health Safety 63:* 39–44 (1994).

Appendix B: The Qualified Person

CONTENTS

INTRODUCTION

The current approach of standard-setting organizations and legislators toward managing the conduct of work in confined spaces is performance based, rather than specification based. A performance-based approach specifies end results or key outcomes, but provides few, if any, details about how to attain them. On the other hand, the specification-based approach dictates the action required in a particular situation. Providing sufficient detail to cover all situations is impossible. The critical element that is fundamental to the success of the performance-based approach is the Qualified Person. The Qualified Person is expected to have the knowledge, skills, and experience needed in a particular situation to recognize what must be done, and when and how. The Qualified Person thus personifies the encyclopedic information that would be needed in the equivalent specification-based document.

While variously defined, a relatively comprehensive description for the Qualified Person is someone designated by the employer as capable by education and/or specialized training of anticipating, recognizing, and evaluating exposure to hazardous substances or other unsafe conditions. Also, the Qualified Person must be capable of specifying control and/or protective action to ensure worker safety (Standards Assoc. of Australia 1986).

In the context of confined spaces, this definition places an extraordinary technical and moral burden onto any individual who assumes the role of Qualified Person. Appreciation about the level of knowledge, skill, and performance required from the Qualified Person is extremely important to all concerned: these individuals, their employers, and ultimately the persons whose safety and well-being falls under their judgment and care.

The level of performance required from the Qualified Person leaves little room for error. As noted in Chapters 1 and 2, mistakes that have minor consequences under normal circumstances can cause fatal outcomes in confined spaces. The Qualified Person must simultaneously focus an inordinate amount of knowledge, skill and experience toward hazard assessment and decision making regarding these workspaces.

The aspects of performance from the Qualified Person that are critically important are the quality and level of decision making. The safety of the Qualified Person and that of others could depend totally on these intangibles. Quality and level of decision making both have similar characteristics. They depend not only on noncontrollable factors, such as the inherent capability of the individual, but also on controllable ones, such as knowledge, skill, and experience.

Performance-based standards and legal statutes on confined spaces presently focus only on outcomes. An outcome of critical importance is the classification of a workspace. The ANSI Standard (1985), for example, provides no guidance on how to make this decision, other than a list of factors to be considered. While, in theory, this approach provides maximum flexibility, this really benefits only the expert. The reality is that this approach provides no help to real-world decision makers. The background knowledge and experience needed to recognize, assess, and devise measures to control potential and real hazards is extensive. This never should be underestimated.

Other activities in human endeavor offer parallels to the challenges posed by this situation and remedies created to address them. In medical and fire fighting situations, people totally depend on the knowledge, skill and training of the care provider. Society refuses to allow these individuals to operate without strict guidelines covering necessary education and skills development. In the case of medicine, these criteria have evolved over the centuries and are put into practice through medical schools and disciplinary colleges. This ensures development and provision of a consistent level of expertise. Such is not yet the case with the Qualified Person. While standards and legal statutes generally describe performance outcomes required from the Qualified Person, they provide no guidance for practitioners on how to attain the level of proficiency implicit in them.

At the present time, the meaning of the term "Qualified Person," as used in context of confined spaces, differs according to who provides the interpretation. To some, a supervisor having no training in health and safety could become qualified after attending a 5-day, 3-day, or even a 2-day course. To others, "qualified" means completing a rigorous process of certification in a professional discipline in the safety/industrial hygiene/marine chemist/engineering field. To even others, "qualified" means attainment of the preceding certification(s) plus additional study.

In the preamble to the confined space Standard for general industry, the U.S. Department of Labor took the view that certification in safety, industrial hygiene, engineering, and marine chemistry in and of itself did not qualify these professionals as Qualified Persons (OSHA 1993). This is the clearest expression yet by a jurisdiction about the highly complex nature of confined spaces as working environments and the depth of knowledge needed to address them. By extension, this also implies the extent of knowledge, skills, and experience needed in the Qualified Person. OSHA also argued that professional certification alone is not an automatic assurance of competence. On the other hand, lack of certification is not an automatic assurance of lack of competence. OSHA did acknowledge the extensive experience required by Certified Marine Chemists in attainment of their professional designation as an example of what is required. The understanding, knowledge, and skills of other professionals was recognized for its generalized nature. Specific experience with the type of permit space found in the workplace also is required in OSHA's view.

Another view about the concept of qualified is expressed in terms of amateurs and experts (Wilson 1994). Wilson holds that the amateur has superficial knowledge or knowledge at the introductory level. The amateur has no idea that the subject is much more complex and/or isn't interested in learning more about it. The amateur operates at the level of instinct, impulse, and bureaucracy and upon a few basics or general knowledge of rules. More seriously, the amateur falsely thinks and represents that the superficial level of knowledge is adequate and does not realize or refuses to acknowledge the deficiency in this position. The unprofessional or apprentice is the next level of attainment. What distinguishes the unprofessional from the amateur is recognition about the deficiency of knowledge and expression of desire to learn. During this phase, the individual gains competence in individual segments of the overall picture. The professional has attained journeyman proficiency in most of the tasks, enough to sustain an operation. What separates the professional from the expert is the "big picture." The expert can handle almost any type of situation with confidence and competence.

The incongruities mentioned in the preceding discussion deserve clarification and resolution. Individuals who assume the burden of Qualified Persons and their employers must know what is required from them. This is needed as soon as possible in order for training needs to be defined and programs created.

In order for performance-based standards and statutes to produce the intended results and not to become well-intentioned failures, nurturing the Qualified Person must become a priority. The realities of the real-world workplace are very blunt and unforgiving.

A practical working definition for "qualified" as applied to confined spaces in this context might well be the following:

> **"Qualified" is a state of attainment such that no one will suffer illness, injury or death as a result of deficiency of judgment brought about by deficiency of knowledge.**

This is a bottom-line requirement. The knowledge, experience, assessment skills, and judgment of the Qualified Person will make the difference between success and injury or loss of life. This is a sobering leveler of perceptions about qualifications and abilities. Responsible individuals will be humbled or perhaps even terrified by the fact that commitment of their words and signatures to paper could make this difference.

DESCRIBING THE QUALIFIED PERSON

Assuming that the view of the U.S. Department of Labor expressed in the preamble to the confined space Standard for general industry is correct, where should people start? Without a starting description of academic or technical background, how can anyone say what a Qualified Person should look like? Further, how can anyone decide what to do to become a Qualified Person? For that matter, how should organizations respond to the need to acquire or develop these highly knowledgeable, highly skilled, and highly creative resource persons?

A starting point in this process is recognition that there are two types of interests: personal and organizational.

Probably the most important starting point for developing individuals to become Qualified Persons is to determine what is needed to perform the role. That is, what must be known in order to comprehend, assess, and make decisions about all of the elements that together form the scope of work involving the confined spaces within an organization? For the individual, this means determining all that is possible to know about the confined spaces and the activities that occur within them. That is, so that there are no unknowns. In the course of doing this, the individual hopefully can determine what is not known about these situations and therefore where additional information or study is needed. What one does not know is by far the more important commodity.

The difficulty in determining what one does not know is comprehending how to identify the deficiency. If one does not understand the work process and recognize the importance or consequence of some aspect, then one cannot identify deficiencies in knowledge or skill. Without knowing fully what knowledge and skills are required, one cannot identify what one does not know. This is a classic chicken and egg problem.

One approach for beginning this process is to identify the universe of performances required from Qualified Persons. Standards and legal statutes on confined spaces are useful beginning sources of information about performances and performance outcomes. Performance outcomes vary somewhat from document to document. What is specified in one in some detail may appear only as a word or phrase in another. Lack of inclusion or lack of detail does not necessarily mean lack of need for the performance. A performance outcome missing from one document that appears in another may be essential to the process. The omission simply may reflect the skill of one group or the detail orientation of another. Once again, all that matters with performance-based standards and legal statutes is outcome. Lack of inclusion of a performance outcome merely may complicate the path for attaining the required end result. This should not be interpreted as lack of need for the performance.

Table B.1 lists performances identified in various sources as essential to the safe conduct of work involving confined spaces. Not all performances necessarily apply to every situation that may impact a particular organization. Knowledgeable individuals should be able to recognize those that apply. For those who are less sure, starting from a broad, generic perspective offers greater assurance than underestimating the need.

Standards and legislative statutes use two approaches for defining the role of the Qualified Person. The first specifies actions that must be fulfilled. This approach does not identify specific job titles for the Qualified Person, nor does it imply that the same individual is required to perform all of the functions. This approach provides maximum flexibility for integrating the confined space program within existing organizational structures. However, it does force organizational discipline, since the requirements must be clearly and unambiguously assigned to ensure that all are completed during the course of entry and work in the confined space.

The other approach involves mission-oriented job titles and specification of performances required from them. This is the approach taken in the OSHA Standard for general industry (OSHA 1993). The focus of this approach is a hierarchical structure that is directed solely to entry and work in the confined space (in OSHA parlance, a permit-required confined space). The difficulty with the approach used by OSHA is that it deviates markedly from the operational structure found in organizations. The latter structure is oriented toward the orderly completion of work tasks. Under the OSHA scheme, the person supervising work tasks may not necessarily supervise the entry. An individual with specialized expertise, in safety or industrial hygiene, could be required to supervise tasks related solely to entry. This approach could create serious jurisdictional problems. On the other hand, supplementing the skills of functional supervisors with those needed in entry supervisors could be an intimidating task.

Table B.1 borrows heavily from the organization used in the OSHA Standard for general industry (OSHA 1993). The OSHA Standard currently offers the most comprehensive list of performance outcomes. Table B.1 also contains performances that were not mentioned in the OSHA Standard. Table B.1 organizes performances and performance outcomes according to the technical structure found in actual organizations, rather than the mission-oriented approach used by OSHA. Also, Table B.1 uses the term "confined space" in the generic sense, rather than the specific one used in the OSHA Standard. The performances outlined in Table B.1 could be performed by one person or could involve a number of individuals. Overlap between performances provided by individuals also is possible.

TABLE B.1
Performances Required from Qualified Persons

Managerial

- Assess impact of legislation on the organization
- Assess status of present program, compare against legal and other requirements, and identify deficiencies
- Obtain management support and commitment for necessary changes
- Express management support in a policy and generalized statement
- Assess capability of existing organizational structure to satisfy legal requirements and identify deficiencies
- Specify training needs
- Determine personnel needs
- Determine ability of internal/external resources to meet training needs
- Determine need for additional equipment and technical resources
- Devise recordkeeping systems
- Assess emergency response needs and capabilities
- Determine external interfaces and liaise with external resources
- Obtain funding

TABLE B.1 (continued)
Performances Required from Qualified Persons

Technical

- Identify actual and potential confined spaces
- Identify existing and potential hazards associated with each space
- Assess severity of existing and potential hazards
- Assess capability of equipment and resources available in the marketplace
- Recommend equipment for purchase
- Evaluate the impact of exposure to actual and potential hazards considering:
    - Number and characteristics of employees affected by the exposure
    - Magnitude of hazards considered through energy release, toxicity, and quantity of chemicals involved
    - Likelihood of exposure
    - Seriousness of exposure
    - Likelihood and consequences of change in conditions or activities not considered initially
    - Strategies for control
- Develop and review written procedures that specify measures and precautions needed for safe entry, work, and egress
- Identify conditions that would change the hazard status assigned to a particular confined space and require reevaluation
- Determine need and frequency for periodic reassessment of hazards
- Establish guidelines for assigning the status of a space
- Establish a system of entry permits
- Specify protective equipment required during entry and work
- Specify training requirements
- Assess specific needs for emergency response
- Develop emergency procedures
- Train in entry procedures and conditions that would prohibit entry
- Audit outcome of entries and revise procedures, as necessary

Safety and Health Committee

- Develop consultation process with Safety and Health Committee on all aspects of confined space program, including:
    - Hazard assessments by Qualified Person
    - Procedures developed for entry and work
    - Results of compliance monitoring by Qualified Person
    - Selection of protective and other equipment
    - Selection of rescue equipment

All Employees

- Be informed about the existence and location of confined spaces
- Recognize confined spaces through labels and other means of identification
- Understand difference between permit and nonpermit spaces (as appropriate)
- Understand the nature of hazards posed by confined spaces
- Be trained in the understanding, knowledge and skills needed to perform duties safely

Authorized Entrants

- Know hazards that may be faced during entry
- Understand mode of exposure to hazardous substances and agents
- Recognize signs or symptoms of overexposure to hazardous substances and agents
- Understand consequences of exposure to hazardous conditions
- Understand particulars of a permit space and the reason for entry
- Be aware of necessary protective equipment and its limitations
- Be able to use properly equipment necessary for safe entry
- Communicate with the attendant in a uniform manner to indicate status
- Alert the attendant about any indication of exposure to a dangerous condition, or about any prohibited condition
- Alert the attendant when self-initiating evacuation
- Recognize conditions under which exit must occur (orders from entry supervisor or attendant, warning sign or symptom of exposure to a hazardous substance, prohibited condition, sounding of the evacuation alarm)
- Be familiar with procedure for emergency egress

TABLE B.1 (continued)
Performances Required from Qualified Persons

Attendants

- Know the hazards that may be present during entry
- Understand mode of exposure to hazardous substances and agents
- Recognize signs or symptoms of overexposure to hazardous substances and agents
- Understand consequences of exposure to the hazardous conditions
- Recognize possible behavioral effects resulting from overexposure
- Know and watch for hazards that may be present inside the space
- Monitor the status of authorized entrants
- Warn unauthorized employees to stay out of the space
- Inform unauthorized entrants to leave the space immediately
- Inform authorized entrants that unauthorized persons have entered the space
- Assist in evacuating entrants
- Monitor activities inside and outside the space to determine whether entrants can remain inside safely
- Order evacuation of entrants during existence of a prohibited condition, behavioral effects of hazardous exposure, a situation outside the space that could endanger the entrants, or whenever the attendant is unable to perform required duties
- Summon rescue services in the event of an emergency
- Know effective work rules
- Remain outside the permit space during entry operations until relieved by another attendant
- Provide information to the rescue service
- Maintain attention on entrants
- Perform only duties that do not detract from the primary function of monitoring and protecting authorized entrants
- Be familiar with procedure for emergency egress

Operational Supervision

- Recognize that a particular workspace is a confined space
- Know the hazards that may be present during entry
- Understand mode of exposure to hazardous substances and agents
- Recognize signs or symptoms of overexposure to hazardous substances and agents
- Understand consequences of overexposure to the hazardous conditions
- Recognize possible behavioral effects resulting from overexposure
- Ensure systematic review of hazards prior to each entry
- Systematically assess hazards prior to issuing permit considering:
    - Proposed occupancy
    - Work to be performed
    - Possible methods for performing work
    - Physical and chemical hazards and associated risks
    - The actual method proposed and associated equipment
    - Impact of proposed operations on conditions in the space
    - Soundness and security of the structure
    - Need for illumination
    - Substances last contained in the space
    - Steps needed to bring the space to atmospheric pressure
    - Atmospheric testing
    - Medical fitness of entrants
    - Training of entrants
    - Instruction in work procedures
    - Availability and adequacy of personal protective equipment
    - Adequacy of warning signs
    - Special considerations for hot work
    - Fire control
    - Other special considerations
    - Emergency and rescue procedures
    - Arrangements for rescue, first aid, and resuscitation
- Specify tests necessary to determine whether entrants likely would be exposed
- Specify changes in conditions or the nature of the work that would terminate the permit
- Specify protective equipment and tools to be used by entrants and emergency equipment required by rescuers

TABLE B.1 (continued)
Performances Required from Qualified Persons

- Verify the appropriateness of entries on the permit before signing and authorizing entry to begin
- Verify the existence of acceptable entry conditions
- Verify availability of rescue services and operability of the means for obtaining them
- Authorize entry through signing and dating the permit
- Explain procedures to entrants and obtain acknowledgment through signature
- Remove unauthorized persons from the space
- Determine whenever responsibility is transferred and at intervals as dictated by the hazards and operations performed in the space, that operations within the space remain consistent with the terms of the entry permit and that acceptable conditions are maintained
- Terminate entry operations and cancel the permit, when necessary
- Be familiar with procedure for emergency egress

Testing/Inspection

- Be knowledgeable about use and calibration of testing instruments
- Be knowledgeable about limitations of testing instruments and information provided
- Interpret results
- Just prior to entry and during occupancy perform tests to verify:
  - Exposure to chemical agents singly or in combination is less than legal (safe) limit
  - Concentration of flammables is less than legal (safe) limit
  - Percentage of oxygen is within legal (safe) limits
  - Liquid in which entrants could drown has been removed
  - Free-flowing solid in which entrants could become entrapped has been removed
  - Entry of liquid, free-flowing solid, or hazardous substance is prevented by a secure means of disconnection or fitting of blank flanges
  - Disconnection from power sources, real or residual, and locking out of electrical and mechanical equipment that could present a hazard due to release of stored energy
- Perform atmospheric testing prior to entry or reentry
- Perform testing to assess sufficiency of the ventilation system
- Perform testing throughout the period of occupancy
- Prepare a signed and dated report outlining work performed and results obtained from the tests

Rescue Services

- Acquire training in performance of duties
- Demonstrate proficiency in use of personal protective and rescue equipment
- Know hazards that may be present during entry
- Understand mode of exposure to hazardous substances and agents
- Recognize signs or symptoms of exposure to hazardous substances and agents
- Understand consequences of exposure to the hazardous conditions
- Practice making rescues from the actual or representative permit spaces at least once every 12 months. Practice rescue may involve manikins, dummies, or actual persons.
- Obtain training in basic first aid and current certification
- Obtain training in cardiopulmonary resuscitation and current certification

Standards Assoc. of Australia 1986; Nat. Occ. Health & Safety Comm. & Stds. Australia 1992; ANSI 1992, 1995; OSHA 1989, 1993, 1994; NIOSH 1979; Canada Gazette 1987, 1992.

As presented in Table B.1, the performances fit a hierarchical structure. They also have temporal and sequential components. Once tasks required within an organizational context are identified and sorted, roles for preexisting personnel can be assigned and the need for new personnel identified.

In any organization there is a need for Qualified Person(s) to act at the following levels:

- Conceptual: to assess organizational need and to create the program
- Managerial: to create policies and procedures within the program

- Technical: to assess needs associated with individual confined spaces, to select instruments and equipment, to train others
- Operational: to support on-site activity preceding and during entry and work

The reality in many organizations is that one individual will become the Qualified Person. This individual may need to develop skills at all levels and be able to apply them on a continuing basis. This will be a difficult endeavor, given the level of knowledge and skill that are required. Organizations caught in this situation may best be served through the services of consultants who can develop a program to fit their specific needs.

Larger organizations likely will have available several Qualified Persons. The presence of more than one person offers the opportunity for specialization. In this way, no one necessarily must become proficient in all of the skills.

QUALIFYING THE QUALIFIED PERSON

The preceding analysis identifies an extensive number of performances and performance outcomes. The effort and action needed to master a performance must not be underestimated. A single word or word phrase can signify a complex training program. Thus far, this discussion has attempted to separate consideration about the individual from the organization. In reality, both are intimately intertwined and cannot be separated. Hence, qualifying the individual also involves qualifying the organization.

Within any large organization, a hierarchy of individuals with capabilities applicable to this problem likely will develop. This is a natural outcome from the gradation of skills, knowledge, and decision making resident in the available technical resources. Critical to the outcome of the process in the organization is the key individual. The key individual accepts ownership for and organizes, implements, and manages the confined space program. The key individual must possess or develop a broader base of knowledge and skills than the other participants. More importantly, the key individual also must possess the analytical skill to recognize what is needed, and the vision to determine how to achieve it.

The critical starting point in the process for the key individual is to define competencies that are required in the program and to assess capabilities that exist. The process starts with recognition of competencies, the knowledge and skills that are needed to satisfy the meaning of the term, "Qualified Person," within the context of needs of a particular organization. The difficulty is to identify what participants should know, but presently do not. This presumably is the basis for comments made by OSHA regarding the relevance of professional designations and the level of competence demanded in the OSHA Standard (OSHA 1993).

Assessment of competencies is a difficult task. One possible approach is to prepare an inventory of needed skills to compare against those presently available. Performances and performance outcomes identified in Table B.1 provide a starting point. Another possible starting point for assessment is a detailed resume of relevant academic education and post academic courses and work experience. Again, the competencies outlined in this document can be compared against performances and performance outcomes listed in Table B.1.

The need for certain types of competencies in high-level Qualified Persons is very clear cut. Anticipatory knowledge and recognition skills are key attributes. The direction and development of the confined space program totally depend on the ability to anticipate and to recognize hazards against which action is required. Recognizing and addressing hazards after entry is underway that should have been self-evident represents a loss of control in the process. This situation exposes entrants to unnecessary risk. It also indicates a deficiency in knowledge or skill of the Qualified Person responsible for the entry. Further, this also could indicate failure of the program to ensure the ability of Qualified Person to anticipate and recognize.

Table B.1 provides an outline of competencies that could be needed in the Qualified Person. If required in their entirety, this would necessitate extensive study of a very wide area of knowledge. The need exists, therefore, to identify only those that are appropriate within the context of the organization. This approach would fulfill the need to upgrade skills and knowledge, yet would not overwhelm participants or the organization.

One means for obtaining the training in the competencies outlined in Table B.1 is specialized short courses or academic study. Table B.2 identifies possible areas of study. Individual situations could require knowledge at varying levels of proficiency in any or all of these topics.

TABLE B.2
Possible Areas to Develop Competencies in Qualified Persons

Process- and Structure-Related Hazards

- Hydraulic systems and control circuits
- Pneumatic systems and control circuits
- Mechanical systems and control circuits
- Process/machine control and feedback circuits
- Current electricity and electrical hazards
- Static electricity and static electrical hazards
- Bulk liquid and solid materials handling
- Compressed and cryogenic gases
- Process hazards
- Chemical incompatibility and instability
- Fire and explosion properties of matter
- Radiation sources and protection
- Structural soundness (integrity)
- Cleaning methods

Work-Related Hazards

- Work processes (welding, spray painting, etc.)
- Heat and cold stress
- Biomechanical/ergonomic hazards
- Biological hazards
- Noise
- Laser
- Chemical/material compatibility/incompatibility

Protective/Preventive Measures

- Practices and procedures for lockout
- Crane and rigging safety
- Fall arrest/fall protection
- Extrication and retrieval techniques
- Fire prevention/fire protection
- Machine guarding
- Noninvasive cleaning and servicing techniques
- Selection, calibration, operation, and limitations of monitoring instruments
- Portable ventilation fundamentals
- Selection and placement of protective equipment
- Isolation techniques
- Purging techniques
- Inerting techniques
- Selection, use, maintenance, and limitations of personal protection
- Decontamination techniques
- Cardiopulmonary resuscitation
- Emergency first aid
- Contingency planning
- Communication methods

The preceding list, while not expected to be all-inclusive, provides a reasonable starting point for outlining competencies that may be needed. One of the most important questions facing individuals attempting to become Qualified Persons is determining what they should learn. The next is determining sources of training. Short courses are unavailable for many of the competencies listed in Table B.2.

Publications from trade organizations could be vital resources of supplemental information. The American Petroleum Institute, for example, publishes standards on hydrocarbon safety. These identify hazards and preventive actions applicable to the handling of all quantities of these products. Similarly, the National Fire Protection Association publishes standards on fire and explosion protection and prevention. The information in these publications has widespread application across industry. Safety associations can provide assistance sectorally. Governments and government agencies also are excellent sources of information.

Expecting a single individual within an organization to possess or develop the requisite working knowledge in all of the areas that could be needed is both unreasonable and unrealistic. Developing specific expertise by each member within a group is a realistic solution to this dilemma. Each then could assist the others in specific aspects of a situation. With time the knowledge base of the entire group would broaden and that of individuals would begin to merge and overlap.

ORGANIZATIONAL ASPECTS

To this point, discussion about the Qualified Person has focused on development of the skills and knowledge base of the individual. However, the need for involvement by the organization in this process should be apparent. An important caveat about this role must be acknowledged from the outset of this discussion. That is, the process of developing the skills of individuals is relevant to organizations only if the physical plant contains confined spaces and only if they are entered. On the other hand, if confined spaces are not present or are not entered, there is no need to proceed.

Another critical point in participation by the organization is recognition by management of the need to become involved in the process. Typically, this recognition occurs for the following reasons:

- A near miss or fatal accident
- Action initiated by the Safety and Health Committee, by an individual or the union
- An order following inspection by a regulator
- Introduction of legislation
- Perception of hazard

The preceding list is hierarchical. The least likely reasons for involvement in this pursuit by many, especially smaller organizations, are the last two.

The weakness in the process of organizational response is that it is totally contingent on the actions of individuals. The ability of individuals to further the process on behalf of the organization depends on their level of awareness and training. As should be apparent, the actions of the organization are closely intertwined with those taken by the key individual who undertakes to develop and implement the program.

Discovery of one or more confined spaces within an organization or activation caused by one of the other factors mentioned above could set into motion an avalanche of activity. Once the avalanche begins, everything must happen yesterday. Fully Qualified Persons instantly must appear and address and fulfill legal requirements. That is, everything must happen at once.

Real-world realities, however, make things considerably different from the preceding situation. Time for making an organized approach to the problem does not exist. Individuals fully qualified to address the matter may not exist within the organization. The number of persons qualified to address requirements in any of the standards, but especially the OSHA Standard, is insufficient to fulfill the need. Skills development is a time-consuming, difficult exercise. This burden in many cases is added to the regular pressures of performing a day's work. Those charged with implementing the process must be able to analyze needs of the organization and to synthesize a strategy that minimizes inefficiency. This process is likely to be intimidating and highly stressful for all concerned.

Addressing the need of an organization is best started through an audit. An audit could identify needed performances and performance outcomes against those indicated in Table B.1. A previously implemented program provides an invaluable starting point. A comparison of present operating practices with current requirements will indicate the extent of change that is required. Previous work also could indicate reasons for entry, nature of work performed, experience of entrants, protective equipment, ventilating equipment, testing equipment, practices and procedures, emergency response procedures, training, recordkeeping, calibration and maintenance of equipment, and so on. The audit also should determine the operational structure that regulates entry. That is, who controls the process of hazard assessment and entry, and how this is done.

The present program may be structured and operate in a manner consistent with current requirements. All that may be needed to bring the program into line with present requirements is minor revision. Of course, the audit also could indicate that major improvement is needed. Even in the latter case, the existence of a structure even with flaws is a better position from which to build than no structure.

Another possible benefit from an audit is an indication about how to develop a program that maximizes use of existing resources within the organization. A large organization, for example, may be able to create several programs, each containing a subset of the competencies listed in Tables B.1 and B.2.

The ability to create subsets would simplify the need to train all Qualified Persons to the same level in all skills. This is consistent with the wording in standards and legal statutes on confined spaces. These impose a variety of performances requiring action from a Qualified Person. Fulfilling these requires varying levels of knowledge and skill. These documents do not necessarily require the same individual to perform all actions. To a varying extent, these documents permit approaches that utilize the capability of individuals. This means that the duties of the Qualified Person can be spread among many individuals, each performing at the level of a particular requirement.

The key to the success of the process is successful utilization of the knowledge and skills base of the senior Qualified Person. Also, the key individual, having created the big picture, then can focus on identifying and developing the skills and knowledge needed by the others.

The starting point for organizations needing, but not already possessing, a confined spaces program again is an inventory of skills and knowledge requisite in individuals participating in the future structure. The next step is to establish the level of skills, knowledge, and experience resident in individuals within the existing structure. This establishes the difference between what is needed and what is presently available. Matching the attributes of existing personnel to the demands of the future structure will minimize the learning curve.

Additional information on training is contained in Chapter 13.

SUMMARY

This Appendix has examined the concept, role, and development of the Qualified Person in the conduct of work involving confined spaces. The Qualified Person is an individual who by education, training and experience is capable of fulfilling the moral, legal, and practical requirements and responsibilities of the position. The philosophy underlying the Qualified Person is that this individual will make the appropriate decision in a particular circumstance, guided by performance outcomes from standards and statutes. This approach puts responsibility for safe conduct of work onto the judgment of a person rather than specifications provided by a regulatory authority. The Qualified Person thus is a creation of standard-setting and legislative agencies. The education, training, and experience of real-world practitioners rarely match what is required from a Qualified Person.

Developing qualifications for a Qualified Person is a complex and difficult activity. Initially, this requires identification of performances that will or could be required from the Qualified Person. These performances indicate areas of study from which performance-oriented actions must result. Individual performances may be simple or complex.

The key components in the success of the organizational effort are the knowledge, skills, and experience of the most knowledgeable person involved in the program. Where several persons are involved, skills can be matched to required outcomes.

REFERENCES

American National Standards Institute: *Safety Requirements for Lockout/Tagout of Energy Sources* (ANSI Z244.1-R1992). New York: American National Standards Institute, 1992.

American National Standards Institute: *Safety Requirements for Confined Spaces* (ANSI Z117.1-1995). Des Plaines, IL: American Society of Safety Engineers/American National Standards Institute, 1995. 32 pp.

Canada Gazette: Part XI, Confined Spaces, *Canada Gazette, Part II. 126*: (17 September 1992) pp. 3863-3876.

Canada Gazette: Part IX, Confined Spaces, *Canada Gazette, Part II. 121*: (26 March 1987) pp. 1337–1364.

National Institute for Occupational Safety and Health: Criteria for a Recommended Standard — Working in Confined Spaces (DHEW/PHS/CDC/NIOSH Pub. No. 80-106). Cincinnati, OH: National Institute for Occupational Safety and Health, 1979. 68 pp.

National Occupational Health and Safety Commission and Standards Australia: Joint Draft National Standard for Planning and Work for Confined Spaces (BS92/20113 Cat. No. 92 0995 8). Canberra, ACT: Australian Government Publishing Services, 1992. 55 pp.

OSHA: Control of Hazardous Energy Source (Lockout/Tagout); Final Rule, *Fed. Regist. 54*: 169 (1 September 1989). pp. 36644–36696.

OSHA: Permit-Required Confined Spaces for General Industry; Final Rule, *Fed. Regist. 58*: 9 (14 January 1993). pp. 4462–4563.

OSHA: Confined and Enclosed Spaces and Other Dangerous Atmospheres in Shipyard Employment; Final Rule, *Fed. Regist. 59*: 141 (25 July 1994). pp. 37816–37863.

Standards Association of Australia: *Safe Working in a Confined Space* (AS 2865-1986). North Sydney, NSW: 1986. 19 pp.

Wilson, R.G.: Turn your servicemen into experts. *The Cleaner*. June 1994.

Appendix C: Standards for Oxygen Deficiency and Enrichment

CONTENTS

INTRODUCTION

Probably the biggest controversy involving confined spaces is the acceptable limit for atmospheres deficient or enriched in oxygen. This controversy has arisen, in part, because oxygen is essential for life, and because people can adapt in both the short and the long term to situations where oxygen levels are both greater and less than they are at sea level. Sea level, of course, is merely a convenient height of reference. There is no particular significance to this altitude, as people live and work at attitudes far below and far above this height.

Complicating things further is the fact that oxygen levels can be measured in units of concentration and partial pressure. Oxygen concentration remains constant within normal habitable altitudes. This results from the relative constancy of composition of the atmosphere (Moran and Morgan 1989). Total atmospheric pressure, and by implication, the pressure of oxygen, vary according to altitude and weather pattern. Also, oxygen sensors used in safety instruments measure either concentration or partial pressure, depending on their design. Sensors that measure concentration are unaffected by altitude. Those that measure partial pressure are affected by altitude and changes in barometric pressure.

ATMOSPHERIC CONSIDERATIONS

The gas laws of physical chemistry provide an important basis for discussing this problem (McQuarrie and Rock 1984). Boyle's law states that the volume of a gas is inversely proportional to the pressure exerted by it. Charles' law states that the volume of a gas is proportional to the absolute temperature. Avogadro hypothesized that equal volumes of gases at the same temperature and pressure contain equal numbers of molecules. These relationships are used routinely to convert between the concentration of a gas and the pressure it exerts against the walls of its containment. Dalton's law of partial pressures states that the total pressure exerted by an atmosphere of mixed gases is the sum of the pressures (partial pressures) exerted by individual constituents. That is, each gas within the containment acts independently of the others. In the case of the normal atmosphere:

$$P_{Tot} = P_A + P_B + P_C + \dots + P_N \tag{C.1}$$

P_{Tot} = Total pressure
P_A = Partial pressure of component A

A useful application of these relationships is calculation of the partial pressure of a gas in an atmosphere.

$$P_A = (\text{concentration fraction}) \times P_{Tot} \tag{C.2}$$

For example, the partial pressure of oxygen in the standard atmosphere at sea level is:

$$P_{Oxygen} = \frac{209{,}500 \text{ ppm}}{1{,}000{,}000 \text{ ppm}} \times 760 \text{ mmHg} = 159 \text{ mmHg} \tag{C.3}$$

In this calculation, the units of total pressure define the units of partial pressure. Following are the partial pressures (in mmHg) of gases in the normal atmosphere at sea level, assuming dry air (after Weast 1968).

Nitrogen:	(0.78090) (760) =	593
Oxygen:	(0.20950) (760) =	159
Argon:	(0.00934) (760) =	7.1
Carbon dioxide:	(0.00034) (760) =	0.3
Other:	(0.00079) (760) =	0.6
Water vapor:	(0.00000) (760) =	0.0
Total	(1.00000) (760) =	760

The pressure of the real atmosphere varies from place to place, from day to day, and from hour to hour. Peak-to-trough interdaily variations at the same location easily can equal 60 mmHg.

Long-term study and averaging of atmospheric conditions has led to development of the standard atmosphere. The latter stands for discussion as a model of the actual atmosphere. The standard atmosphere is averaged for all latitudes and seasons. The standard atmosphere references a standard temperature (15°C) and pressure (1013.25 mb [millibars] or 101.325 kPa [kiloPascals] or 760 mmHg) at sea level (Moran and Morgan 1989).

Air density changes as the atmosphere is heated or cooled because the volume is not constrained. That is, atmospheric air is able to expand or contract. By contrast, the density of air confined within a container remains constant as temperature increases or decreases, since the volume does not change. When heated, the pressure of atmospheric air decreases due to expansion, as intermolecular distance increases. Decreasing air density lowers the pressure. Thus, given equal volumes of atmospheric air at different temperatures, the warmer air will be less dense and will exert less pressure. A hot air balloon rises for this reason.

Water vapor also decreases the density of air. The average molecular weight of dry air is 29; the molecular weight of water is 18. Water molecules take the place of molecules of the other gases

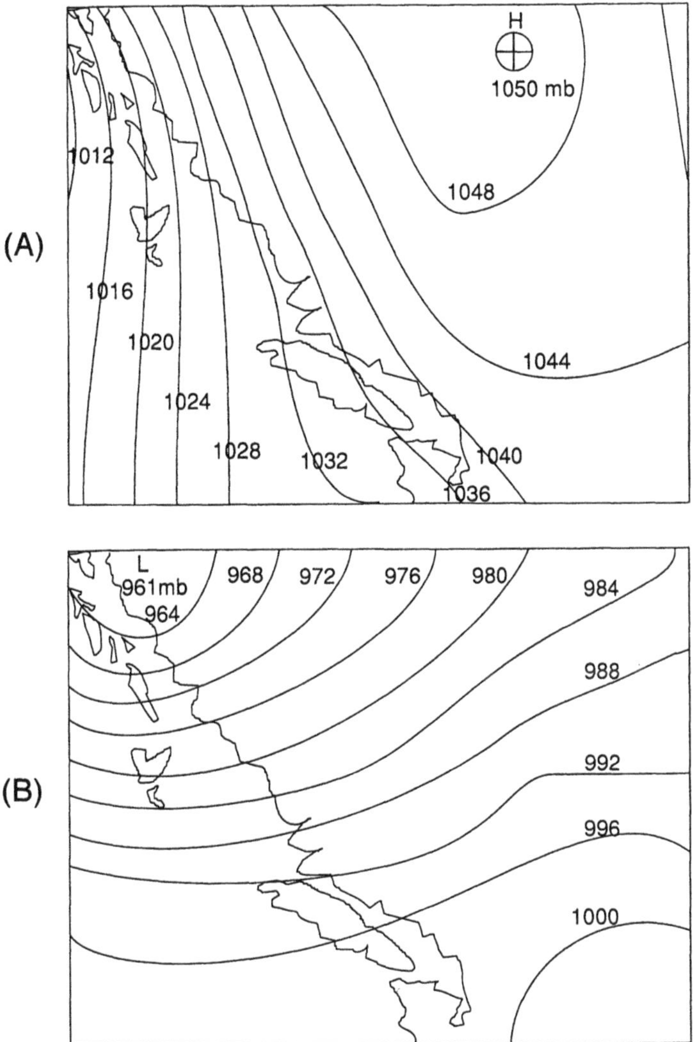

FIGURE C.1 Variations in atmospheric pressure with location. (A) December 17, 1996; (B) December 5, 1996. The weather maps illustrate the considerable variations that can exist in atmospheric pressure by location. (The units of the numbers are millibars (mb). Normal atmospheric pressure at sea level is 1,013 mb). Major weather systems usually move eastward at 800 to 1200 km/d. These variations can affect calibration of pressure-sensitive instruments for measuring oxygen, as well as considerations about oxygen deficiency. (Used by permission of Environment Canada "The Source™," www.WeatherOffice.com.).

(principally nitrogen and oxygen). As the concentration of water vapor increases, the net effect is to decrease the density, and hence the weight and pressure.

In summary, cold dry air masses produce higher surface pressure than warm humid air masses. A warm dry air mass produces higher pressure than a humid air mass at the same temperature. The usual range of air pressure at sea level is 728 to 788 mmHg (970 to 1,050 mb) (Figure C.1). The record low of 653 mmHg (870 mb) was measured in the eye of a typhoon; the record high of 813 mmHg (1,084 mb) was measured in winter in Siberia. (Readings are corrected to sea level.)

Water vapor obeys Dalton's law in the same manner as other gases in the atmosphere. Water vapor pressure is directly proportional to the concentration of water vapor in air. Table C.1 provides the saturation vapor pressures for water vapor at temperatures normally encountered in the workplace.

TABLE C.1
Variation of Saturation Vapor Pressure of Water with Temperature

Temp.	Saturation Vapor Pressure			
	Over water		Over ice	
°C	mb	mmHg	mb	mmHg
–40			0.13	0.10
–30			0.38	0.28
–25			0.64	0.48
–20			1.03	0.77
–15			1.65	1.24
–10			2.60	1.95
–5			4.01	3.01
0	6.11	4.58		
5	8.72	6.54		
10	12.3	9.21		
15	17.1	12.8		
20	23.4	17.5		
25	31.7	23.8		
30	42.4	31.8		
35	56.3	42.2		
40	73.8	55.3		
45	95.9	71.9		
50	123	92.2		

After Weast 1968 and Moran and Morgan 1989.

The relationship between temperature and saturation vapor pressure is based on vaporization rate, rather than "water-holding capacity." Water vapor merely coexists in the airspace with other gases and vapors.

At any time the partial pressure of water vapor in the atmosphere usually is considerably less than saturation values presented in Table C.1. A more likely value is about 50% of saturation. This level would be half of the values contained in Table C.1. Thus, the usual contribution of water vapor to atmospheric pressure within the range of temperatures normally encountered in the workplace would be about 10 mmHg.

Real-world measurements of barometric pressure are routinely converted to equivalent values at sea level for comparative purposes. This requires the addition of correction factors incorporating the temperature of the air column (assumed) from the station to sea level, altitude of the station, and the value of the reading. The sea level equivalent pressure of component "A," based on an actual reading is:

$$P_{atm,\ sea\ level} = P_{reading} + \text{correction} \tag{C.4}$$

$$CF_A = \frac{\left(P_{atm,\ sea\ level} - P_{water}\right)}{P_{atm,\ sea\ level}} \times CF_{A,\ std\ atm} \tag{C.5}$$

$$P_A = \left(CF_A\right) \times P_{atm,\ sea\ level} \tag{C.6}$$

TABLE C.2
Variation of Standard Atmosphere with Altitude

Height		Atmospheric pressure		Oxygen (re: sea level)	
ft	m	mmHg	mb	Pressure (mmHg)	Concentration (%)
-8,200	–2,500	1,014.0	1,352	212.1	28.0
-6,560	–2,000	958.4	1,278	200.5	26.4
-4,920	–1,500	905.2	1,207	189.4	25.0
-3,281	–1,000	854.5	1,139	179.0	23.6
-1,640	–500	806.1	1,075	168.9	22.2
-1,000	–305	787.9	1,050	165.1	21.7
- 500	–152	774.0	1,032	162.2	21.3
0	0	760.0	1,013	159.2	20.9
1,000	305	733.6	978	153.7	20.2
2,000	610	708.2	944	148.4	19.5
3,000	915	683.3	911	143.3	18.9
4,000	1,220	659.2	879	138.1	18.2
5,000	1,525	636.1	848	133.3	17.5
6,000	1,830	613.5	818	128.5	16.9
7,000	2,135	591.6	789	123.9	16.3
8,000	2,440	570.3	760	119.5	15.7
9,000	2,745	549.7	733	115.2	15.2
10,000	3,050	529.7	706	111.0	14.6
11,000	3,355	510.1	680	106.9	14.1
12,000	3,660	491.3	655	102.9	13.5
13,000	3,965	473.1	631	99.1	13.0
14,000	4,270	455.5	607	95.4	12.6
15,000	4,575	438.3	584	91.8	12.1
16,000	4,880	421.8	562	88.4	11.6
17,000	5,185	405.5	541	85.0	11.2
18,000	5,490	390.0	520	81.7	10.8
19,000	5,795	375.0	500	78.6	10.3
20,000	6,100	360.3	480	75.5	9.9
21,000	6,405	346.1	461	72.4	9.5

After Bolz and Tuve 1970 and Lide 1991.

The concentration fraction of A in the real-world reading will be slightly smaller than that in the standard atmosphere, due to the presence of water vapor. The most pronounced example of this effect occurs in the interior of the respiratory system where saturation conditions (47 mmHg) of water vapor at body temperature (37°C) exist.

Combining Equations C.5 and C.6 simplifies to the following:

$$P_A = \left(P_{atm,\ sea\ level} - P_{water}\right) \times CF_{A,\ std\ atm} \tag{C.7}$$

This is the more familiar form of the equation, as found in reference sources (ANSI 1980).

The relationship between normally accessible depths and altitudes and standard atmospheric pressure is presented in Table C.2. In this table, the partial pressure of oxygen was calculated by multiplying total pressure by the mole fraction of oxygen in dry atmospheric air (0.2095). The effect of water vapor was not considered in this calculation.

The values in Table C.2 for $-1,524 \le H < 11,000$ m are given approximately by the following equation (Lide 1991):

$$H = 44,331.514 - 11,880.516\ (1.3332237\ P)^{0.190\ 263\ 2} \tag{C.8}$$

P = pressure in mmHg

PHYSIOLOGICAL CONSIDERATIONS

OXYGEN DEFICIENCY (HYPOXIA)

The underlying basis for concern about low levels of oxygen is physiologically based. Oxygen, of course, is essential for survival. The physiological expression of oxygen deficiency is hypoxia. Hypoxia can be caused by asphyxiants. Asphyxiants include both simple asphyxiants and chemical asphyxiants. Simple asphyxiants are physiologically inert; that is, they do not interact biochemically within the body (Dinman 1978). They produce their toxic effect by diluting or displacing the normal atmosphere. The resultant partial pressure of oxygen in inspired air is insufficient to maintain oxygen saturation of hemoglobin high enough for normal tissue respiration. Simple asphyxiants include: acetylene, argon, ethane, ethylene, helium, hydrogen, methane, nitrogen, neon, propane, and propylene (ACGIH 1994).

A dry atmosphere at sea level contains 20.9% oxygen at a partial pressure of 159 mmHg. The presence of saturated water vapor (and residual air) in the lung reduces this value to about 100 mmHg in the alveolar spaces. At this partial pressure, hemoglobin saturates to 95% with oxygen. (See Figure 3.4.) As the oxygen concentration or atmospheric partial pressure are reduced, hemoglobin saturation decreases. At alveolar oxygen partial pressure of 60 mmHg, hemoglobin saturation reduces to 90%. The atmospheric partial pressure of oxygen corresponding to this alveolar partial pressure is about 120 mmHg. At this point, most physiologists agree that symptoms of oxygen deficiency become evident (NIOSH 1976).

Under normal conditions, the alveolar–capillary pressure gradient is approximately 60 mmHg. This causes rapid transfer of oxygen from the alveolar airspace into the fluid of the capillary. The partial pressure of O_2 in blood and that in the alveolar airspaces equilibrate before the end of travel through the pulmonary capillary. In an oxygen-deficient atmosphere containing, for example, 14% oxygen, the initial pressure gradient may be only 25 mmHg. Because of the shallower pressure gradient, oxygen transfer occurs at a slower rate. A measurable pressure gradient exists between oxygen in the alveolar airspace and blood at the end of the capillary. Under this condition, equilibration fails to occur. Hence, a decrease in the partial pressure of O_2 in inspired air leads to a decrease in arterial partial pressure and a decrease in saturation of hemoglobin (Comroe et al. 1962).

The areas of the body considered most sensitive to oxygen deprivation are the brain and the myocardium (heart muscle). Simple asphyxiants do not directly suppress cardiac output. Brain cells perish in 3 to 5 min under conditions of severe hypoxia. Cerebral hypoxia occurs when the partial pressure of inspired oxygen is lowered to 60 to 70 mmHg (Comroe et al. 1962). Damage sustained by oxygen-sensitive tissues is not reversible upon restoration of the atmosphere (Ayres et al. 1969, Davis 1985).

Table C.3 summarizes the effects of acute exposure to oxygen deficient atmospheres as normally presented.

The rate of onset of the symptoms presented in this table depends on many factors, including breathing rate, work rate, temperature, emotional stress, age, and individual susceptibility. These factors can exacerbate the effects of an oxygen-deficient atmosphere.

Altitude introduces an additional complicating factor. The body responds to partial pressure of oxygen, rather than concentration. Total atmospheric pressure, and hence the partial pressure of

TABLE C.3
Effects of Acute Exposure to Oxygen-Deficient Atmosphere

Effect	Atmospheric (dry air, sea level) Concentration (%)	Pressure (mmHg)
No symptoms, healthy adults	16 to 20.9	122 to 159
Increased heart and breathing rate, some incoordination, increased breathing volume, impaired attention and thinking	16	122
Abnormal fatigue upon exertion, emotional upset, faulty coordination, impaired judgment	14	107
Very poor judgment and coordination, impaired respiration; tunnel vision; vigorous movement may cause permanent heart damage; nausea and vomiting	12	91
Nausea, vomiting, lethargic movements, perhaps unconsciousness, inability to perform vigorous movement or loss of all movement, unconsciousness followed by death	<10	<76
Convulsions, then shortness of breath, the cardiac standstill, spasmodic breathing, death in minutes	<6	<46
Unconsciousness after 1 or 2 breaths	<4	<30

After NIOSH 1976, Miller and Mazur 1984, and CSA 1993.

oxygen both decrease with increasing altitude. Alveolar oxygen partial pressure of 60 mmHg corresponds to atmospheric oxygen partial pressure at 3,000 meters (10,000 ft). Altitudes exceeding this height are normally considered to be oxygen-deficient for individuals acclimatized to sea level (Davis 1985). At these altitudes, less oxygen depression in a workspace atmosphere is required to produce an oxygen-deficient condition. Also, a greater percentage of oxygen is required in supplied breathing air to prevent physiological oxygen deficiency. For example, at 10,000 m (33,000 feet), an atmosphere containing 100% oxygen is needed (NIOSH 1976).

Oxygen Enrichment (Hyperoxia)

Oxygen enrichment is the condition resulting when the partial pressure of oxygen exceeds that found under normal ambient conditions. Normal ambient conditions can include workspaces, such as deep mines, whose workings occur at depths considerably below sea level. At partial pressures considerably greater than those found in normal atmospheres, oxygen exerts both acute and chronic toxic effects.

Hyperoxia has little impact on hemoglobin saturation. Increasing alveolar partial pressure beyond normal values increases hemoglobin saturation insignificantly. This outcome results from the dynamics of the saturation process as reflected in the saturation/partial pressure curve (Bouhuys 1974).

Table C.4 indicates the toxic activity of oxygen at elevated partial pressures.

At partial pressures exceeding 400 mmHg, oxygen produces respiratory irritation. In hyperbaric atmospheres exceeding 2,280 mmHg, oxygen produces nervous signs and symptoms that culminate in convulsive seizures. Oxygen toxicity is exerted in the lungs, central nervous system and the eyes, although it is probably toxic to all organs at sufficient concentration (Piantadosi 1991). Generally, the rate of onset is a hyperbolic function of the inspired partial pressure (Clark and Lambertson 1971a, b). Sensitivity of the central nervous system to the toxic effects of oxygen is considerably greater than the that of the pulmonary system. Tolerance to elevated partial pressures of pure oxygen atmospheres ranges from several minutes to 2 hours. Toxic action of hyperbaric oxygen atmospheres

TABLE C.4
Toxic Action of Oxygen

Atmospheric pressure (mmHg)		
Total	**Oxygen**	**Comments**
760	159	Sea level
	400	Respiratory irritation
	760	Throat irritation; no systemic effects provided that exposure is brief
	1,520	Tracheal irritation, slight burning on inhalation; tolerance increased when periods of oxygen interspersed with air; reduced vital capacity develops
	>1,520	Signs and symptoms of oxygen poisoning: tingling of fingers and toes, visual disturbances, acoustic hallucinations, confusion, muscle twitch, nausea, vertigo, possible convulsions
	>2,280	Nervous signs and symptoms: twitching, vertigo, anxiety, paresthesia in toes and fingers, nausea, convulsive seizures

Data from Yarborough 1947, Donald 1947, Dukes-Dobos and Badger 1977, and Behnke 1978.

is greatly enhanced by exercise and elevated levels of carbon dioxide (Yarborough 1947). This translates into reduced tolerance time. Individual tolerance varies widely (Donald 1947).

Oxygen toxicity is expressed through production of reactive intermediates, such as the superoxide anion O_2^- and the hydroxyl radical (OH) (Freeman and Crapo 1982). The superoxide anion is highly reactive toward biological molecules. Normally, these species are removed by enzymatic action and reaction by free radical scavengers, such as reduced glutathione. During hyperoxia, production of reactive oxygen metabolites greatly increases and may exceed capacity of scavengers to remove them. Tissue injury and subsequent effects in both brain and lungs appear to be related to increased metabolism (Mayevsky 1984).

Another consideration about oxygen enrichment that is extremely important is the increased ignitability of clothing and other combustible materials, including the skin (OSHA 1985). OSHA documented fatal accidents in which oxygen enrichment occurred through inadvertent or deliberate release of pressurized gas. The resulting fires indicated that risk of combustibility is greatly enhanced under these conditions, even at normal atmospheric pressure.

The fire hazard in an oxygen-enriched atmosphere is significantly greater than that in a normal atmosphere (Frankel 1991). This is due in part to the reduction in minimum energy needed for ignition and the greater rate of flame spread. That is, combustible materials ignite more easily and burn more rapidly in an oxygen-enriched atmosphere. Generally, ignition energy decreases with increasing oxygen concentration, and flame spread rate increases with increasing atmospheric pressure. Almost all materials will burn in a pure oxygen environment. This situation can seriously challenge presumptions about safety in selection of materials for use in oxygen service.

The effects of exposure of substances, fabrics, and polymers to an enriched oxygen atmosphere on ignitability are summarized in Table C.5. Lubricants and hydraulic fluids are the most sensitive of the types of substances for which information is available. In the case of lubricants, this sensitivity changes from oxygen-deficiency through normal concentrations through oxygen-enrichment. The lowest of the tested partial pressures corresponded to a concentration of 31% oxygen relative to the sea level dry atmosphere.

CONSENSUS AND REGULATORY STANDARDS

Like limits set for toxic substances, those for oxygen reflect laboratory-based studies and human experience. The standard-setting process does not involve documented judgment of expert committees

TABLE C.5
Effect of Oxygen-Enrichment on Combustibility/Flammability

Pressure (mmHg)		
Oxygen	**Total**	**Comments**
160	760	Normal atmosphere, sea level, dry air
Range	760	Decrease in autoignition temperature of hydraulic fluids with increase in partial pressure of oxygen
Range	760	Decrease in autoignition temperature of lubricants with increase in partial pressure of oxygen from less than normal through 760 mmHg
236	760	Increase in combustibility in oxygen/nitrogen mixture of materials (fabrics, paper, polymers) that did not burn in normal atmosphere
258	760	Considerable increase in flame spread rate in combustible materials (fabrics and polymers)
319	760	Decrease in ignition temperature of combustible fabrics and sheeting
760	760	Slight decrease in autoignition temperature of most hydrocarbon fuels, solvents, and anesthetic gases; broadening of flammable range by increase in upper flammable limit

Data from Hugget et al. 1965, Johnson and Woods 1966, Kuchta et al. 1967, Kuchta and Cato 1968, and Frankel 1991.

in toxicology or human physiology in the same manner as other toxic agents. For example, there is no TLV for oxygen. The only mention of oxygen by ACGIH in the current edition of the TLV booklet is contained in the preamble (ACGIH 1994). Further, there is no documentation for oxygen in the current Documentation volumes (ACGIH 1991).

Table C.6 summarizes values published in various regulatory and other standards. It provides limits set by some of the regulators and committees that produce exposure standards. The consensus expressed through more recent standards indicates decreased tolerance for oxygen deficiency and enrichment. Permissible limits for work activity usually are expressed as percent of oxygen by volume in air.

Of the preceding standards, the only ones that provide discussion and rationale for decisions and direction are the OSHA final rules (OSHA 1993, 1994). The latter provide invaluable insight into concerns and comments raised by intervenors to the process. However, the necessary discussion is far from complete.

In the preamble to the Standard for general industry, OSHA (1993) recommended 19.5% as an acceptable lower limit. OSHA stated its belief that concentrations less than 19.5% would be oxygen deficient. No technical evidence about oxygen deficiency that could serve as the basis for informed discussion about this subject was provided by either OSHA or any of the intervenors to the process.

Commenters representing the ANSI Z88.2 Committee on respiratory protection argued that 19.5% as a lower limit was too high. They argued for a lower limit, namely, 12.5%, using as the rationale that no respiratory protection was needed at 16% oxygen. The value, 12.5%, should in their view be considered as immediately dangerous to life and health. The direction of reasoning shown by the ANSI Z88.2 Committee was considerably different from that taken by the ANSI Z117.1 Committee on confined spaces (ANSI 1989). The difference in approach taken by these Committees indicates the depth of the controversy that surrounds this issue.

In making its selection of 19.5% for the lower limit, OSHA stated its heavy reliance on the judgment of the ANSI Z117.1 Committee on confined spaces and the NIOSH Respirator Decision Logic (ANSI 1989; NIOSH 1987).

The OSHA Standard for shipyard employment raised the minimum acceptable level from the 16.5% contained in the previous version of the rule to 19.5% (OSHA 1994). Intervenors to this process included a high proportion of individuals and groups with direct, on-going experience in assessment and control of oxygen-deficient atmospheres. This group was highly supportive of the

TABLE C.6
Concentration Limits for Oxygen Deficiency and Enrichment

Source	Concentration limit (%)	Comments
ANSI Z117.1-1977	18 ≤ C	Achievable through ventilation; applies to cold and hot work; excess oxygen not permitted, but level not specified
ANSI Z117.1-1989	19.5 ≤ C ≤ 23.5	Achievable through controls or provision of appropriate personal protective equipment
ANSI Z117.1-1995	19.5 ≤ C ≤ 23.5	Achievable through controls or provision of appropriate personal protective equipment
ANSI Z88.2-1980	14 < C < 19.5	Concentrations less 19.5% or the governing legal minimum and greater than that which is immediately dangerous to life and health (approximately 14% in dry standard atmosphere at sea level) are oxygen deficient — not immediately dangerous to life or health
ANSI Z88.2-1992	C ≤ 16	Sea level, dry normal atmosphere; atmosphere-supplying respirator required under some conditions of work; oxygen-deficient atmosphere, not immediately dangerous to life or health
ANSI Z88.2-1992	C ≤ 12.5	Sea level, dry normal atmosphere; self-contained breathing apparatus required; oxygen-deficient atmosphere, immediately dangerous to life or health
CSA Z94.4-M1982	C ≤ 14	Air at normal atmospheric pressure, air having normal oxygen content at altitudes above 3.66 km; immediately dangerous to life or health
CSA Z94.4-93	C ≤ 14	Air at normal atmospheric pressure, air having normal oxygen content at altitudes above 3.66 km; immediately dangerous to life or health
NIOSH	19.5 ≤ C ≤ 21.4	No modification of work procedures
	16.1 to 19.4	Ventilation and protective measures required
	21.5 to 25	
	≤16, >25	Ventilation and protective measures required
NFPA 306	19.5 ≤ C ≤ 22	Safe for Workers and Safe for Hot Work
API 2015	<16	Entry not permitted
	≥16	Entry permitted with respiratory protection (positive pressure air supplying respirator)
	≥19.5	Entry permitted with other type of respiratory protection, as needed
AS 2865-1986	18 ≤ C	Evaluation can occur in spaces where higher levels are present; supplied air respiratory protective equipment required; excess oxygen not permitted, but level not specified
AS 2865-Draft	18 ≤ C ≤ 21	
CEN/TC79/SGI1/N161	17 ≤ C	
OSHA 1993	19.5 ≤ C ≤ 23.5	Within these limits the atmosphere is nonhazardous and the space would be a nonpermit space in the absence of other hazards
OSHA 1994	19.5 ≤ C < 22	Safe for Workers and Safe for Hot Work
Canada — Federal	18 ≤ C ≤ 23	
Canada — Ontario	18 ≤ C ≤ 23	

Compiled from ANSI 1977, 1980, 1989, 1992, 1995; CSA 1982, 1993; NIOSH 1979; Human Resources Development Canada 1994; Ontario Ministry of Labor 1984; NFPA 1993; API 1991; Standards Association of Australia 1986; National Occupational Health & Safety Commission & Standards Australia 1992; CEN 1988; OSHA 1993, 1994.

change. However, neither OSHA nor the intervenors provided any technical documentation in support of this change. The Standard on shipyard employment represents a special case in confined spaces, since affected sites are located at sea level or low altitudes.

OSHA (1993) recommended 22% as an acceptable upper limit for oxygen in the Standard for general industry. In so doing, OSHA stated its belief that concentrations greater than 22% would be oxygen enriched and would pose a hazard of fire and explosion. No technical evidence about oxygen enrichment was provided by OSHA or any of the intervenors to the process.

Some commenters argued that 22% as an upper limit was too low. They argued for a higher limit, as high as 25 or 26%, using as the rationale that an ambient condition, such as that in a deep mine or that produced by barometric pressure, easily could be falsely identified as oxygen enriched.

In making its selection of 23.5%, OSHA stated its heavy reliance on the judgment of the ANSI Z117.1 Committee on confined spaces (ANSI 1989). This value also was used in other OSHA standards.

In the Standard for shipyard employment, OSHA adopted 22% as the maximum permissible oxygen level on the basis of the criterion used in NFPA 306 (1993). Neither OSHA nor NFPA provided any supporting documentation regarding oxygen enrichment on which technical discussion could occur.

The second approach to setting of limits for acceptable levels of oxygen utilizes partial pressure. Standards from various sources are summarized in Table C.7. Limits provided here are contained in the source documents and not converted from concentration units.

The limit of greatest interest from the perspective of acute exposures is the IDLH (Immediately Dangerous to Life or Health). IDLH originally was defined in the NIOSH Standards Completion Project for the purposes of selecting respiratory protection (NIOSH 1990). Under the NIOSH usage of the term, IDLH is the presumed minimum concentration from which a person could escape in 30 min in the event of respirator failure without experiencing any escape-impairing or irreversible health effects. Since then, IDLH has been adopted as an acronym by other groups. Each has modified the original meaning. Under the ANSI usage of the term, IDLH means an atmosphere that poses an immediate hazard to life or produces immediate, irreversible, debilitating health effects (ANSI 1980, 1992). Under the CSA usage of the term, IDLH means an atmosphere where the concentration of oxygen could cause a person without respiratory protection to be fatally injured or to suffer immediate irreversible or incapacitating health effects (CSA 1980, 1993).

The rationale used by ANSI and CSA in their definitions of IDLH was to select the partial pressure of oxygen that would produce 90% saturation of hemoglobin in alveolar blood (ANSI 1980; CSA 1982, 1993). In the current standard on respiratory protection, the ANSI Z88.2 Committee chose 83% saturation for the IDLH (ANSI 1992). These saturation values are the minimum below which symptoms of oxygen deficiency are believed to become noticeable. Neither make reference to time frame of exposure, since onset of this condition would not necessarily be immediate.

DISCUSSION

Selection of acceptable limits for exposure to oxygen is one of the most difficult and controversial of decisions. As indicated in previous discussion, oxygen level is influenced by:

- Barometric pressure (weather)
- Humidity
- Altitude
- Atmospheric composition

Limits based on concentration are altitude independent, since concentration remains constant in normal air at the depths and altitudes normally accessible without respiratory protection. These limits, therefore, only reflect atmospheric composition caused by local contamination. However, the body responds to the partial pressure of oxygen and not to concentration. Limits based on partial pressure are altitude, barometric pressure (weather) and composition dependent. Saturation of hemoglobin

TABLE C.7
Partial Pressure Limits for Oxygen Deficiency and Enrichment

Source	Atmospheric pressure limit (mmHg)	Comments
ACGIH	P ≤ 135	Minimum partial pressure without need for respiratory protection; normal atmospheric pressure
NIOSH	P ≤ 122	Immediately dangerous to life, normal atmospheric pressure, sea level
	P < 132	Oxygen deficiency, normal atmospheric pressure, sea level
	122 ≤ P ≤ 147	Dangerous, but not immediately life threatening; respiratory protection determined by Qualified Person; normal atmospheric pressure, sea level
	148 ≤ P ≤ 163	No modification of work procedures, normal atmospheric pressure, sea level
	163 ≤ P < 190	Dangerous, but not immediately life threatening; respiratory protection determined by Qualified Person; normal atmospheric pressure, sea level
	P >190	Oxygen-enriched atmosphere, normal pressure, sea level, immediately dangerous to life
ANSI Z88.2-1980	P ≤ 106	Atmospheric partial pressure of oxygen in dry air at sea level corresponding to partial pressure of 100 mmHg in freshly inspired air in the upper portion of the lung that is saturated with water vapor at 37°C; immediately dangerous to life or health
ANSI Z88.2-1992	P ≤ 95	Dry atmosphere, sea level, immediately dangerous to life or health; may occur through any combination of reduction in oxygen content or altitude
	95 < P ≤ 122	Oxygen deficient — not immediately dangerous to life or health; may occur through any combination of reduction in oxygen content or altitude
CSA Z94.4-M1982	P ≤ 106	Atmospheric partial pressure of oxygen in dry air at sea level corresponding to partial pressure of 100 mmHg in freshly inspired air in the upper portion of the lung that is saturated with water vapor at 37°C; immediately dangerous to life or health
CSA Z94.4-93	P ≤ 106	Atmospheric partial pressure of oxygen in dry air at sea level corresponding to partial pressure of oxygen in inspired air in the upper respiratory passages falls to 13.3 kPa (100 mmHg) or less; immediately dangerous to life or health

Compiled from ACGIH 1994; NIOSH 1979; ANSI 1980, 1992; and CSA 1982, 1993.

depends only on partial pressure of oxygen in the local atmosphere. Therefore, altitude, barometric pressure, and local contamination all combine inseparably to affect performance and safety.

Previous discussion has identified the individual elements that influence atmospheric pressure, as well as the physiological basis for response to oxygen level. This discussion highlighted the dichotomy between concentration- and pressure-based limits for assessing oxygen level in consensus and regulatory standards. Table C.8 summarizes this information in order to provide the basis for further discussion.

Humidity levels usually vary from hour to hour and day to day in a location. As indicated in Table C.1, 10 mmHg would be a conservative value for water vapor pressure under most situations. The contribution of water vapor to total atmospheric pressure is small. This contribution easily could become lost in fluctuations of barometric pressure. Hence, the contribution of humidity on total pressure is so small as to be ignorable under most conditions.

Based on standard total atmospheric pressure associated with the standard (dry) atmosphere and 19.5% as a regulatory limit, oxygen deficiency would occur at altitudes as low as 2,000 ft (610 m), based on standard atmospheric pressure. However, atmospheric pressure varies continuously in a location from hour to hour and day to day.

TABLE C.8
Altitude and Weather Effects on Atmospheric Conditions

Condition (re: sea level, dry atmosphere)	Pressure (mmHg)	Oxygen concentration (%)	Altitude equivalent	
			ft	m
Atmospheric				
Dry atmosphere	760			
Water vapor pressure	≈10			
Typical high pressure	+28			
Record high pressure	+53			
Typical low pressure	–32			
Record low pressure	–107			
Geographic				
Highest mine in Andes (Chile)	357	9.8	20,262	6,176
La Paz (Bolivia)	493	13.6	11,916	3,632
Bogota (Colombia)	555	15.3	8,724	2,659
Mexico City (Mexico)	581	16.0	7,487	2,282
Denver (CO)	630	17.4	5,280	1,609
Deep mine (Sudbury, ON)	950	26.2	–6,317	–1,925
Dead Sea (Jordan)	799	22.0	–1,337	–408
Qattara Depression (Egypt)	772	21.3	–440	–134
Death Valley (CA)	768	21.2	–282	–86
Regulatory limits				
16% oxygen	580		7,500	2,286
18% oxygen	653		4,288	1,307
19.5% oxygen	707		2,000	610
22% oxygen	798		–1,394	–425
23.5% oxygen level	853		–2,871	–875

Data from Kemball 1985, McIntyre 1987, and Anonymous 1991.

During a period of typical high pressure at this altitude, the atmospheric pressure easily could increase from 707 mmHg to 707 + 28 = 735 mmHg. This would increase the oxygen concentration from 19.5% to 735/760 × 20.95% = 20.3%, relative to sea level. Under this weather condition, this location would not be oxygen deficient. During a period of typical low pressure at this altitude, the atmospheric pressure would decrease from 707 mmHg to 707 – 32 = 675 mmHg. This would decrease the oxygen concentration to 675/760 × 20.95% = 18.6%, relative to sea level. Under this weather condition, oxygen deficiency would be more severe compared to the standard atmosphere. An unusual low would further exacerbate the oxygen-deficient condition.

It could be argued that the true altitude at which oxygen deficiency should be considered to occur would be that whose normal low pressure would be 707 mmHg. This would correspond to a standard atmospheric pressure of 707 + 32 = 739 mmHg and an altitude of 795 ft (242 m).

The second approach would be more protective, since this would minimize the chance that oxygen deficiency could occur due simply to variation in atmospheric pressure. Adoption of this approach would permit use of an instrument that is altitude and weather independent.

The preceding is one approach to this problem. Another is to incorporate all losses in partial pressure of oxygen. These would include elevation, weather pattern, and local atmospheric contamination caused by the work to be performed. This approach would require an instrument responsive to both altitude and weather conditions. Calibration could occur at sea level during an average day. Using this approach, hazard assessment could be subjected to day-to-day variability, as weather conditions and local contamination experienced change. This would be an especial concern at altitudes where partial pressures would be close to oxygen-deficient conditions and where weather conditions alone could necessitate control measures.

Instruments designed for measuring oxygen in confined spaces display in units of concentration. Some sensors function independently of altitude and variations in partial pressure of oxygen due to weather. Physiologically based oxygen deficiency or enrichment may not be obvious to users of this type of equipment, because concentration would remain constant at all altitudes of normal use. The other type of sensor is sensitive to partial pressure of oxygen (and therefore, altitude and weather). Calibration of this type of instrument at sea level for use at higher altitude, or calibration at higher barometric pressure than present during actual use, could cause underestimation of concentration. An ambient condition easily could be falsely identified as oxygen deficient by an oxygen-testing instrument containing a sensor that is sensitive to partial pressure. On the other hand, an instrument containing this type of sensor provides the best potential for estimating adverse conditions, since the body responds to partial pressure of oxygen, not concentration.

The question of oxygen deficiency is more complex even than discussed thus far. Demands of today's industrialized society also must be considered. Individuals acclimatize to the conditions of a particular altitude. Acclimatization brings about physiological change that occurs over a period of time. However, many people routinely work at altitudes considerably different from that to which they are acclimatized. Travel to a worksite could entail a flight in a commercial airplane whose cabin is pressurized to 8,000 ft (2,438 m). This corresponds to 570 mmHg total pressure or 15.7% oxygen, relative to sea level, dry atmosphere (Bancroft 1971). For a person traveling during work time, or for the flight crew, this level technically represents an occupational exposure to an oxygen-deficient atmosphere. For long flights, this exposure can occur for most of the workday. For individuals acclimatized to sea level, travel to a location at a higher altitude, coupled with work in an office tower or stay in a high-rise hotel, could constitute exposure to an oxygen-deficient atmosphere. The mere act of moving from ground level to a worksite in a high-rise building in a geographic location at a higher altitude could lead to exposure to an oxygen-deficient condition.

The problem of limits for oxygen deficiency was mentioned in the original NIOSH Guide to Respiratory Protection (NIOSH 1976). This document indicated that oxygen deficiency could develop through decrease in oxygen content or through increase in altitude. Altitudes greater than 10,000 ft (3,050 m) at the time of writing of the NIOSH report were considered to be oxygen deficient. However, as mentioned in the NIOSH report, workers at altitudes of 10,000 ft (3,050 m) routinely used air-purifying respirators without apparent difficulty.

As the report optimistically commented, this problem was under study, and eventually "oxygen-deficient atmosphere" will be redefined to eliminate the present discrepancies and account for the effect of altitude. The irony in this statement is only partly apparent. The NIOSH report listed the standards of the day for oxygen deficiency. By comparison, as indicated by the data in Table C.6, the permissible limit for oxygen deficiency has increased, that is, has become more stringent, thus making resolution of the discrepancy even more difficult to achieve.

The former ANSI Standard on respiratory protection, ANSI Z88.2-1980, only obliquely approached the question of altitude (ANSI 1980). This occurred in example problems provided for clarification in use of the equation used for calculating oxygen partial pressure. The corresponding CSA Standard, CSA Z94.4-M1982, utilized the same criteria as the ANSI Standard, but introduced an altitude limit of 3.66 km (12,000 ft) for oxygen deficiency (CSA 1982). Standard atmospheric pressure at this altitude would be 491 mmHg. Corresponding concentration of oxygen would be 13.5% relative to sea level.

The current CSA standard on respiratory protection, CSA Z94.4-93, did not vary from the previous version in this respect (CSA 1993). By contrast, the current ANSI standard on respiratory protection, Z88.2-1992, addressed this question head-on (ANSI 1992).

The direction taken by ANSI Committee Z88.2 was not shared by ANSI Committee Z117.1 on confined spaces or by regulatory agencies such as OSHA (ANSI 1989; OSHA 1993, 1994).

The approach taken by ANSI Z117.1 and OSHA on this question provided no recognition about the altitude question nor a means to resolve it.

The current ANSI Standard on respiratory protection has embraced the concept of partial pressure rather than concentration for resolving the question of oxygen deficiency (ANSI 1992). By taking this approach, the ANSI Z88.2 Committee created a basis for technical dialogue about this question. ANSI Z88.2 established 95 mmHg as the partial pressure of oxygen in air for the IDLH. This could be reached at an altitude of 14,000 ft (4,267 m), in an atmosphere containing 12.5% oxygen at sea level, or some combination of altitude and oxygen deficiency.

The Standard also provides additional important information. At and above 10,000 ft (3,000 m), an ordinary supplied-air respirator or SCBA provides oxygen at a partial pressure less than 121 mmHg, even though the concentration is 20.9%. This would be equivalent to a gas mixture that provides 16% oxygen at sea level. In cases where a supplied-air respirator or SCBA would be required at these altitudes, the gas mixture must contain at least 23% oxygen at 10,000 ft and 27% at 14,000 ft, relative to sea level. This situation would necessitate a specially designed respirator or rebreather. Compressed air tanks containing these mixtures could not be used at sea level due to the enriched atmospheres.

Data provided in Table C.8 indicate that people live and work over a wide range of altitudes. Healthy people live long and active lives at high altitudes where arterial saturation ranges from 85 to 95%. Few patients with cardiopulmonary disease have arterial oxygen saturation less than 85%. The lower limit of arterial oxygen saturation compatible with moderately active existence depends on the abruptness with which hypoxemia develops, compensatory mechanisms, and other limiting factors in the disease process. Hemoglobin saturation in persons with congenital heart disease may be less than 80% without causing disability. On the other hand, an asthmatic may sustain adequate alveolar gas exchange and arterial saturation only by extreme effort. Persons with emphysema may experience disability, despite the fact that arterial saturation is 90 to 95% (Comroe et al. 1962).

Thus far, discussion has considered oxygen-deficient atmospheres. Oxygen-enriched atmospheres also pose considerable hazard, for the reason of enhanced ignitability. In the average circumstance, the hazard of enrichment seems to be related to use of oxygen in flame-cutting processes, namely, oxyfuel equipment. Oxygen could be present in confined spaces where it is generated for use as a process gas. Some work occurs under pressurized atmospheres. In these situations, pressurization increases the partial pressure of oxygen above that found under normal circumstances.

Barometric pressure increases with depth. The Dead Sea, the lowest surface feature on earth is situated at –1,337 ft (–408 m). Partial pressure of oxygen at this depth corresponds to a concentration of 22% at sea level. Some deep mines exceed 1 mile (1.6 km) in depth below sea level. Oxygen partial pressure at these depths easily could exceed the partial pressure of 23.5% used in many regulatory limits.

Based on standard total atmospheric pressure associated with the standard (dry) atmosphere and 23.5% as a legal limit, oxygen enrichment would occur at a depth of –2,871 ft (–875 m), based on standard atmospheric pressure. However, atmospheric pressure varies continuously in a location from hour to hour and day to day.

During a period of typical low pressure at this depth, the atmospheric pressure easily could decrease from 853 mmHg to $853 - 32 = 821$ mmHg. This would decrease the oxygen concentration from 23.5% to $821/760 \times 20.95\% = 22.6\%$, relative to sea level. Under this weather condition, this depth would not be oxygen enriched. During a period of typical high pressure at this depth, the atmospheric pressure would increase from 853 mmHg to $853 + 28 = 881$ mmHg. This would decrease the oxygen concentration to $881/760 \times 20.95\% = 24.3\%$, relative to sea level. Under this weather condition, oxygen enrichment would be more severe compared to the standard atmosphere. An unusual high pressure system would further exacerbate the condition of oxygen enrichment.

It could be argued that the true depth at which oxygen enrichment should be considered to occur would be that whose normal high pressure would be 853 mmHg. This would correspond to a standard atmospheric pressure of 853 – 28 = 825 mmHg and a depth of –2,276 ft (–694 m).

The second approach would be more protective in the regulatory sense, since this would minimize the chance that oxygen enrichment could occur due simply to variation in atmospheric pressure. Adoption of this approach would permit use of an instrument that is altitude and weather independent.

The preceding is one approach to this problem. Another is to incorporate all gains in partial pressure of oxygen. These would include depth, weather pattern, and local atmospheric contamination caused by the work to be performed. This approach would require an instrument responsive to both altitude and weather conditions. Calibration would occur at sea level during an average day. Using this approach, hazard assessment could be subjected to day-to-day variability, as weather conditions and local contamination changed. This would be an especial concern at depths where partial pressures would be close to oxygen-enriched conditions and where weather conditions alone could necessitate control measures.

Sensors that are altitude and weather independent may not be capable of identifying problems, because concentration would remain constant at all depths of normal use. The other sensor type is pressure (and therefore, altitude and weather) sensitive. Calibration of these instruments at sea level and use at depth, or calibration at lower barometric pressure than present during actual use, could cause overestimation of concentration. An ambient condition easily could be falsely identified as oxygen enriched. On the other hand, instruments containing this type of sensor provide the best facility for estimating adverse conditions, since the body responds to partial pressure of oxygen, not concentration.

Oxygen deficiency and enrichment create a fundamental dilemma for the practicing industrial hygienist. One the one hand, exposure to an atmosphere containing a narrow range of concentration is essential for survival. Controlling or eliminating exposure through active intervention is not an option. Conversely, not permitting exposure beyond normal atmospheric concentrations, regardless of the wider range of permissible concentration is a conservative strategy. On the other hand, hygienists sanction exposure of workers to contaminants at nonzero concentrations. The question is whether oxygen deserves special consideration because of our relationship with it and the acute nature of action under hazardous conditions.

There is another dimension concerning oxygen level that is not considered in the discussion about physiologically based oxygen deficiency. Physiologically based studies occur under controlled conditions where the composition of the atmosphere is known. This is not the case in the workplace. There is a real danger that reliance on physiologically based arguments concerning the effects of hypoxia on the human metabolism could miss the real basis for the concern about oxygen deficient conditions.

One reason for monitoring oxygen deficiency is to allow workers to note a decline in oxygen concentration well before the onset of physiological symptoms. This provides sufficient time for workers to evacuate the space, not merely to hear the alarm moments before they begin to feel the effects of hypoxia. While it is true that most healthy individuals have no trouble in working in an environment of 18% oxygen by volume (at sea level), this ignores the benefit provided by having a take action level of 19.5% or even higher.

The other reason for monitoring oxygen deficiency is the question of atmospheric dilution or displacement. In the workplace, any measured concentration less than 20.9% indicates the presence of something that is not normally present and should not be there. While there are working environments where the nature of that "something else" is fully understood, these are few and far between.

Selection of 19.5% as a limit has occurred in some consensus and regulatory standards since at least the early 1970s. There is evidence to suggest that this was the highest concentration at which a decrease in oxygen concentration was measurable using the field equipment of the time, such as colorimetric detector tubes. That is, this value was a function of the resolution and stability limitations of the measurement technique, rather than physiological effects. Today, of course, the situation is much different. Most instruments are capable of resolving within ±0.1% in stable atmospheric pressure. This means that the potential for false alarms for set-points higher than 19.5% is greatly reduced. Given the accuracy and stability of today's instruments, there is no reason not to use an alarm set-point even more conservative than 19.5% when the instrument is operated in stable atmospheric pressure. In situations where the atmospheric pressure is not stable, instruments containing capillary pore sensors or barometric pressure compensation will provide stable readings.

In summary, while physiologically based limits reflect the way in which the body responds to an oxygen-deficient condition, they do not consider the effect of the substance(s) that dilute(s) or displace(s) the atmosphere. The reality is that codifying use of physiologically accurate, but less conservative thresholds, based on partial pressure of oxygen would have the following effects:

- Enabling entrants to remain in the space until much closer to the onset of physiological symptoms caused by oxygen deficiency and possibly intoxication by other contaminants.
- Enabling entrants to ignore incipient effects of oxygen deficiency and possibly intoxication by other contaminants at an early stage. After all, entrants do not tend to watch the display of their monitoring instruments. Their focus is performing the job. They depend on the alarm to indicate when evacuation must occur. This approach carries the risk in noisy environments that entrants may not hear the alarm.
- Reducing the likelihood that workers can evacuate under their own power during an emergency because of oxygen deficiency and possibly intoxication by other contaminants or a combination of both.

Standards based on concentration rather than partial pressure of oxygen provide a means of addressing the presence of contaminants in the atmosphere. This, in combination with a set- point higher than 19.5%, seems to provide the best approach to resolving this issue.

REFERENCES

American Conference of Governmental Industrial Hygienists: *Documentation of the Threshold Limit Values and Biological Exposure Indices,* 6th ed., vol. 1. Cincinnati, OH: American Conference of Governmental Industrial Hygienists, 1991. pp. vii-xv.

American Conference of Governmental Industrial Hygienists: *1994–1995 Threshold Limit Values for Chemical Substances and Physical Agents and Biological Exposure Indices*. Cincinnati, OH: American Conference of Governmental Industrial Hygienists, 1994. 119 pp.

American National Standards Institute: *Safety Requirements for Working in Tanks and Other Confined Spaces* (ANSI Z117.1-1977). New York: ANSI, 1977.

American National Standards Institute: *Practices for Respiratory Protection* (ANSI Z88.2-1980). New York: ANSI, 1980. 38 pp.

American National Standards Institute: *Safety Requirements for Confined Spaces* (ANSI Z117.1-1989). Des Plaines, IL: American Society of Safety Engineers/American National Standards Institute, 1989. 24 pp.

American National Standards Institute: *Practices for Respiratory Protection* (ANSI Z88.2-1992). New York: ANSI, 1992.

American National Standards Institute: *Safety Requirements for Confined Spaces* (ANSI Z117.1-1995). Des Plaines, IL: American Society of Safety Engineers/American National Standards Institute, 1995. 32 pp.

American Petroleum Institute: *Safe Entry and Cleaning of Petroleum Storage Tanks* (API Std. 2015, 4th ed.). Washington, D.C.: American Petroleum Institute, 1991. 26 pp.

Anonymous: *CMJ's (Canadian Mining Journal) 1991 Mining Sourcebook.* Toronto, ON: Southam Mining Group, 1991. pp. 31–34.

Ayres, S.M., H.S. Mueller, J.J. Gregory, S. Gianelli, Jr. and J.L. Penny: Systemic and myocardial hemodynamic responses to relatively small concentrations of carboxyhemoglobin (COHb). *Arch. Environ. Health 18*: 699–704 (1969).

Bancroft, R.W.: Pressure cabins and rapid decompression. In *Aerospace Medicine,* 2nd ed. Baltimore, MD: Williams & Wilkins, 1971. pp. 337–363.

Behnke, A.R., Jr.: Physiological effects of abnormal atmospheric pressures. In *Patty's Industrial Hygiene*, 3rd. rev. ed., vol. 1: *General Principles*. New York: John Wiley & Sons, 1978. pp. 237–274.

Bolz, R.E. and G.L. Tuve (Eds.).: *Handbook of Tables for Applied Engineering Science*. Boca Raton, FL: CRC Press, 1970. p. 534.

Bouhuys, A.: *Breathing; Physiology, Environment and Lung Disease*. New York: Grune & Stratton, 1974. pp. 25–233.

Canadian Standards Association: *Z94.4-M1982 Selection, Care, and Use of Respirators*. Rexdale, ON: Canadian Standards Association, 1982. 55 pp.

Canadian Standards Association: *Z94.4-93 Selection, Use, and Care of Respirators*. Rexdale, ON: Canadian Standards Association, 1993. 103 pp.

Clark, J.M. and C.J. Lambertsen: Pulmonary oxygen toxicity: a review. *Pharmacol. Rev. 23*: 37–133 (1971a).

Clark, J.M. and C.J. Lambertsen: Rate of development of pulmonary O_2 toxicity in man during O_2 breathing at 2.0 ATA. *J. Appl. Physiol. 30*: 739–752 (1971b).

Comité Européen de Normalisation: *Guidelines for Selection and Use of Respiratory Protective Devices* (CEN Report CEN/TC79/SG1 N150). Brussels: Central Secretariat, 1991. 107 pp.

Comroe J.H., Jr., R.E. Forster II, A.B. DuBois, W.A. Briscoe, and E. Carlsen: *The Lung, Clinical Physiology and Pulmonary Function Tests*. 2nd ed. Chicago: Year Book Medical Publishers, 1962. 390 pp.

Davis, J.C.: Abnormal pressure. In *Patty's Industrial Hygiene,* vol. 3, 2nd. ed. *3B Biological Responses*. New York: John Wiley & Sons, 1985. pp. 431–449.

Dinman, B.D. The mode of entry and action of toxic materials. In *Patty's Industrial Hygiene,* vol. 1, *General Principles*, 3rd rev. ed. G.D. Clayton and F.E. Clayton (Eds.). New York: John Wiley & Sons, 1978. pp. 135–164.

Donald, K.W.: Oxygen poisoning in man. *Br. Med. J. 1:* 717–722 (1947).

Dukes-Dobos, F.N. and D.W. Badger: Atmospheric Variations. In *Occupational Diseases: A Guide to Their Recognition,* rev. ed. DHEW (NIOSH) Pub. No. 77-181. Washington, D.C.: U.S. Government Printing Office (DHEW/PHS/CDC/NIOSH), 1977. pp. 497–520.

Frankel, G.J.: Oxygen-enriched atmospheres. In *Fire Protection Handbook,* 17th. ed. Cote, A.E. and J.L. Linville (Eds.). Quincy, MA: The National Fire Protection Association, 1991. pp. 3-160 to 3-169.

Freeman, B.A. and J.D. Crapo: Free radicals and tissue injury. *Lab. Invest. 47*: 412–426 (1982).

Hugget, C. et al.: The Effects of 100% Oxygen at Reduced Pressure on the Ignitability and Combustibility of Materials (SAM-TR-65-78). Brooks Air Force Base, TX: 1965.

Human Resources Development Canada: *Consolidated Working Copy: Canada Labor Code, Part II*. Ottawa, ON: Canada Communication Group — Publishing, 1994. pp. X-7 and X-8.

Johnson, J.E. and F.J. Woods: Flammability in Unusual Atmospheres. Part 1 — Preliminary Studies of Materials in Hyperbaric Atmospheres Containing Oxygen, Nitrogen, and/or Helium (NRL Report 6470). Washington, D.C.: Naval Research Laboratory, 1966.

Kemball, W.G (Ed.): *The Canadian Oxford School Atlas,* 5th ed. Toronto, ON: Oxford University Press (Canada), 1985. 128 pp.

Kuchta, J.M. et al.: Flammability of Materials in Hyperbaric Atmospheres (USDI/BM Final Report 4016). Pittsburgh, PA: Explosive Research Center, 1967.

Kuchta, J.M. and R.J. Cato: Review of Ignition and Flammability Properties of Lubricants (Technical Report AFAPL-TR-67-126). Wright Air Force Base, OH: Air Force Aero Propulsion Laboratory, 1968.

Lide, D.R (Ed.): *Handbook of Chemistry and Physics,* 72nd ed. Boca Raton, FL: CRC Press, 1991. pp. 14-11 to 14-14.

Mayevsky, A.: Brain oxygen toxicity. In *Proc. of the Eighth Symp. on Underwater Physiology.* Bachrach, A.J. and M.M. Matzen (Eds.). Bethesda, MD: Undersea Medical Society, 1984. pp. 69–89.

McIntyre, L.: The high Andes. South America's islands in the sky. *Natl. Geogr. Mag. 171:* 422–460 (1987).

McQuarrie, D.A. and P.A. Rock: *General Chemistry,* 2nd. ed. New York: W.H. Freeman, 1984. pp. 126–154.

Miller, T.M. and P.O. Mazur: Oxygen deficiency hazards associated with liquefied gas systems: derivation of a program of controls. *Am. Ind. Hyg. Assoc. J. 45:* 293–298 (1984).

Moran, J.M. and M.D. Morgan: *Meteorology: The Atmosphere and The Science of Weather,* 2nd ed. New York: Macmillan, 1989. pp. 105–151.

National Fire Protection Association: *NFPA 306: Control of Gas Hazards on Vessels* (1993 edition). Quincy, MA: National Fire Protection Association, 1993. 15 pp.

National Institute for Occupational Safety and Health: A Guide to Industrial Respiratory Protection, by J.A. Pritchard (DHEW [NIOSH] Pub. No. 76-189). Cincinnati, OH: DHEW/PHS/CDC/NIOSH, 1976. pp. 15–24.

National Institute for Occupational Safety and Health: Criteria for a Recommended Standard — Working in Confined Spaces (DHEW/PHS/CDC/NIOSH Pub. No. 80-106). Cincinnati, OH: DHEW(NIOSH), 1979. 68 pp.

National Institute for Occupational Safety and Health: NIOSH Respirator Decision Logic (DHHS [NIOSH] Pub. No. 87-108). Cincinnati, OH: DHHS/PHS/CDC/NIOSH, 1987. 55 pp.

National Institute for Occupational Safety and Health: NIOSH Pocket Guide to Chemical Hazards (DHHS [NIOSH] Pub. No. 90-117). Cincinnati, OH: DHHS/PHS/CDC/NIOSH, 1990. p. 5.

National Occupational Health and Safety Commission and Standards Australia: Joint Draft National Standard for Planning and Work for Confined Spaces (BS92/20113 Cat. No. 92 0995 8). Canberra, ACT: Australian Government Publishing Services, 1992. 55 pp.

Occupational Safety and Health Administration: Selected Occupational Fatalities Related to Toxic and Asphyxiating Atmospheres in Confined Work Spaces as Found in Reports of OSHA Fatality/Catastrophe Investigations. Washington, D.C.: U.S. Department of Labor, Occupational Safety and Health Administration (U.S. DOL/OSHA), 1985. 230 pp.

Ontario Ministry of Labor: Occupational Health and Safety Act and Regulations for Industrial Establishments: Occupational Health and Safety Act, Revised Statutes of Ontario, 1980, Chapter 321. Toronto, ON: Ontario Ministry of Labor, 1984. p. 84.

OSHA: Permit-Required Confined Spaces for General Industry; Final Rule, *Fed. Regist. 58*: 9 (14 January 1993). pp. 4462–4563.

OSHA: Confined and Enclosed Spaces and Other Dangerous Atmospheres in Shipyard Employment; Final Rule, *Fed. Regist. 59*: 141 (25 July 1994). pp. 37816–37863.

Piantadosi, C.A.: Physiological effects of altered barometric pressure. In *Patty's Industrial Hygiene*, 4th ed., vol. 1, Part A. New York: John Wiley & Sons, 1991. pp. 329–359.

Standards Association of Australia: *Safe Working in a Confined Space* (AS 2865-1986). North Sydney, NSW: 1986. 19 pp.

Weast, R.C (Ed.): *Handbook of Chemistry and Physics*, 49th ed. Boca Raton, FL: CRC Press, 1968.

Yarborough, O.D., W. Welham, E.S. Brinton, and A.R. Behnke: *Experimental Diving Unit Report No. 1*. Washington, D.C., 1947.

Appendix D: Ignitability Limits

One of the controversies involving confined spaces is the acceptable limit for work in an atmosphere containing ignitable substances. Unlike limits set for toxic substances, these limits do not result from laboratory-based studies. Rather, they reflect experience and judgment.

Regulated ignitability limits are expressed as a percent of the Lower Flammable (Explosive) Limit in air. Strictly speaking, Lower Flammable Limit (LFL) and Lower Explosive Limit (LEL) are not identical, although they have come to be used interchangeably. (Flammability refers to propagation of flame, whereas explosibility refers to ability to deflagrate or detonate.) Lower flammable limits usually are expressed as a percent in air. When expressed in units of mass/volume, such as g/m^3, the LFLs for many hydrocarbons are relatively constant, in the range of 45 to 55 g/m^3 (U.S. Dept. of the Interior 1965). Addition of oxygen or other substituents increases these values, in some cases, considerably.

The permissible limit for work activity contained in regulatory and other standards is expressed as percent of LFL or LEL. Thus, the actual unit used in standards is percent of percent concentration. Table D.1 summarizes values published in various regulatory and other standards. Table D.1 provides examples of limits set by regulators and authors of consensus standards. The consensus expressed through more recent standards indiçates decreased tolerance for atmospheric contamination.

Of the preceding standards, the only ones that record discussion and rationale for decisions and direction are the OSHA final rules (OSHA 1993, 1994). The latter provide invaluable insight into concerns and comments raised by intervenors to the process.

In the preamble to the Standard for general industry, OSHA provided no rationale for its choice of 10% of the LFL as an acceptable limit. OSHA stated its reliance on the value chosen by NIOSH and ANSI in their respective standards (NIOSH 1979; ANSI 1989). No technical evidence in favor of this choice was offered in either of these antecedents. Also, neither OSHA nor any of the intervenors offered any technical evidence about fire or conditions that occur in confined spaces that could serve as the basis for informed discussion about this subject.

Some commenters argued for a higher ignitability limit, namely, 20% of the LFL, using as the argument that this was the set-point for the alarm in the instruments that they owned. OSHA rejected this argument by stating that the choice of 10% of LFL instead of 20% would be more protective.

The OSHA Standard for shipyard employment provided greater insight into this question (OSHA 1994). Intervenors to this process included a high proportion of individuals and groups with direct, on-going experience in assessment and control of flammable atmospheres.

One of the issues raised during the process was the acceptability of ventilating a portion of a compartment containing a potentially flammable atmosphere. This approach would permit work to occur in the locally ventilated area, so long as the concentration of flammables would be maintained at less than 10% of LEL. Ventilation of only a portion of a compartment would permit pockets of atmosphere containing unknown levels of contaminant to remain. These could create significant

TABLE D.1
Summary of Flammability Limits

Source	Limits (% of LFL)	Comments
ANSI Z117.1-1977	<10	Achievable through ventilation; applies to cold and hot work
ANSI Z117.1-1989	<10	Achievable through controls or provision of appropriate personal protective equipment
ANSI Z117.1-1995	<10	Achievable through controls or provision of appropriate personal protective equipment
NIOSH	≤10	No modification of work procedures
	10 to 19	Ventilation and protective measures
	≥20	Ventilation and protective measures
Canada — Federal	<50	No sources of ignition present; protective measures required
	<10	Where source of ignition may cause ignition
Canada — Ontario	<50	Cleaning or inspection permitted, provided that sources of ignition are not present or created
	<10	Cold work permitted
NFPA 306	<10	Safe for Workers and Safe for Hot Work
API 2015	>20	Entry not permitted
	<20	Entry permitted with respiratory protection (positive pressure air-supplying respirator)
	≤10	Entry permitted with other type of respiratory protection, as needed
AS 2865-1986	≤5	Evaluation can occur in spaces where higher levels are present; supplied-air respiratory protective equipment required
AS 2865-Draft	≤5	Use of discrete monitoring equipment
	≤10	Continuous monitoring used
OSHA 1993	<10	Within this limit the atmosphere is nonhazardous and the space would be a nonpermit space in the absence of other recognized serious safety and health hazards
OSHA 1994	<10	Safe for Workers and Safe for Hot Work

From ANSI 1977, 1989, 1995; NIOSH 1979; Human Resources Development Canada 1994; Ontario Ministry of Labor 1984; NFPA 1993; API 1991; Standards Association of Australia 1986; National Occupational Health & Safety Commission & Standards Australia 1992; OSHA 1993, 1994.

fire potential. Also, there is no ready assurance that all sources of ignition could be eliminated. OSHA chose not to permit local ventilation of large spaces containing flammable atmospheres in the updated standard. OSHA instead requires ventilation of the entire compartment to maintain a level less than 10% of LEL. This requirement also applies to spaces adjacent to the one in which the work is to occur.

One well-experienced marine chemist challenged as too high the choice of 10% of LEL for the ignitable limit during the performance of work. He commented that in preparing thousands of certificates the level of flammables was 1 to 2% of LEL at most. OSHA justified the choice of 10% LEL on the following grounds:

- It was adopted from a national consensus standard NFPA 306, Appendix A, which reflects "current practices and sampling technology"
- The shipyard industry has followed this standard for 23 years
- OSHA believes that it provides a sufficient margin of safety

Neither OSHA nor any of the other standard-setters has supported the choice of 10% of LEL with technical data. The mere allowance of such a value in published standards does not mean that practitioners actually adhere to it. They may, in fact, utilize a much more conservative level, such as the 1 to 2% of LEL mentioned by the intervenor.

In accordance with this standard, OSHA will permit entry into compartments containing greater than 10% of LFL under certain circumstances. Having decided that workers cannot work in hazardous atmospheres, OSHA permitted entry in several exceptional situations:

- Testing and visual inspection
- Installation of ventilating or other control equipment
- To perform rescue

In all cases the duration of entry is expected to be short. Entrants must wear appropriate respiratory protection and other personal protective equipment. The atmosphere must be monitored continuously for the contaminant. Also, an attendant must be present to assist the entrant. Additional specifics under which OSHA permits entry into such spaces include:

- No ignition sources may be present — only explosion-proof, self-contained portable lamps can be used
- Flammable vapors and gases at or above their upper explosive limit must be maintained in this condition
- Signs indicating that ignition sources are prohibited must be prominently displayed at the entrance to the space and in adjacent spaces and in the open area adjacent to those spaces

Draft Australian Standard AS 2865, the upgrade of AS 2865-1986, sought comment on the level to set for ignitables (National Occupational Health and Safety Commission 1992). This draft offered two options for consideration. The first would require evacuation when the concentration reached 5% of LEL. This was based on use of instruments capable of taking only discrete grab samples. The other proposal would require evacuation when the concentration measured on a direct-reading instrument reached 10% of LEL.

The appropriateness of the level selected for incorporation into standards should reflect technical discussion about how ignitable atmospheres arise and the role of amplifying factors, such as geometry, and controlling factors, such as ventilation.

These standards all permit use of ventilation as a controlling factor. The two situations of interest here are conditions existing prior to the start of work and conditions generated as a result of the work activity.

Possible characteristics of sources that require control by ventilation in a space otherwise prepared for work include:

- High volatility
- Temperature above ambient
- Large surface area
- Passive generation of contaminant
    - Desorption
    - Vaporization
    - Sublimation

These characteristics should be recognized during hazard assessment. The source of contamination is contained within the space and the need to remove the source should be self-evident. Spaces containing such sources should undergo cleaning and flushing prior to any attempt at control by ventilation. Ventilation in this circumstance should be seen as the step of last resort. The need for ventilation to maintain levels below 10% of LFL should indicate a major source of emission. In fact, the concentration of vapors could approach the LFL close to the source.

Work activity can introduce sources requiring control by ventilation. Possible characteristics of these sources include:

- High volatility
- Temperature above ambient
- Large surface area
- Active generation of contaminant
  - Sprays
  - Jets
  - Turbulent mixing
  - Aeration
  - Chemical reaction
  - Mechanical action
- Multiple phases (e.g., liquid droplets and vapor)
- Mobile source
- Multiple sources

In any of the standards, the method of detection to alert about problems is the flammability tester. In order to provide proper input about the potential for hazard, measurement must occur in appropriate locations with sufficient frequency.

The question about location for obtaining samples has not received discussion in the literature or in the standard-setting process. To illustrate, samples can be obtained in several ways.

One method of sampling is to measure the composition in the airspace normally encountered by entrants. Provided that the ventilation is sufficient to induce turbulent and thorough mixing, these measurements would provide an indication about the average concentration in the airspace. Individual measurements would provide input about the thoroughness of mixing induced by the ventilation in each location. This approach would provide input about sources actively generating contamination, as in spraying, use of jets, aeration of liquid, and turbulent mixing produced by mechanical action.

The flammability tester could underestimate true fire potential when both vapor and liquid droplets are present, since these instruments are designed to measure only vapor. Each droplet provides a surface for evaporation. Vaporization from droplets in the presence of vapor could be extremely important where combustible liquids are involved. The combination of liquid droplets plus vapor still could be ignitable regardless of any measurement provided by the instrument. The in-line dust filter and fire suppression screen could prevent the droplets from reaching the sensor chamber. Even if the droplets could reach the sensor chamber, there is nothing to indicate that the instrument could provide an accurate measure of what actually is present.

Another option is to position the intake of the instrument near the source of the emission, such as a wetted surface. Concentration of vapor is considerably higher near a wetted surface than in air further away. Measurement at this position could produce readings greater than the standard. This could be a more reliable indicator of hazard when a wetted surface is the sole cause of the airborne contamination.

A wetted surface could pose a problem when ignition sources inadvertently are brought into the confined space. Both wetted surfaces and sprays could pose serious problems when static electricity accumulates and discharges at an inopportune moment.

As discussed, detection of the hazard depends on the position of the intake of the instrument relative to the position of the source. This is critical to detection of ignitability hazards, compared to oxygen deficiency. The individual is likely to carry the instrument, especially the current generation. An oxygen deficiency has immediate implications to the individual. The instrument provides a rapid warning about the environment in which the person is working. An ignitability

TABLE D.2
Comparison of Flammability Limits and Oxygen Levels

Substance	LFL (%)	Tenth LFL (%)	Net oxygen (%)
Air			20.9
Acetylene	2.5	0.3	20.6
Ethane	3.0	0.3	20.6
Ethylene	2.7	0.3	20.6
Hydrogen	4.0	0.4	20.5
Methane	5.0	0.5	20.4
Propane	2.1	0.2	20.7
Propylene	2.4	0.2	20.7

Source: ACGIH 1994 and U.S. Dept. of the Interior 1965.

hazard, about which rapid warning is needed with equal urgency, could develop distant from where the person and the intake of the instrument are located.

An additional concern is timing of the samples. Previous generations of instruments could only provide grab samples — discrete frames in a movie containing many discrete frames. There was no way of knowing exactly when a measurement should occur in order to capture the worst case, highest level. Without knowing that instant, the choice of time would, at best, reflect only a very good guess. While the current generation of instruments provides real-time capabilities, there is no way of knowing whether the intake is positioned opportunely at precisely the right moment to capture the worst-case situation.

Fugitive emissions and catastrophic releases represent another concern in selection of the appropriate value for ignitability limits. The boundary surfaces of a workspace define the limits of motion of contaminants, hence the extent to which atmospheric confinement and accumulation can occur. The geometry of ordinary workspaces imposes less stringent boundary conditions than those in confined workspaces. An excursion in a confined space could produce rapid increase in concentration. This easily could go unnoticed when discrete sampling is employed.

A final concern about the choice of ignitablity limit is the relationship of the value to respirator requirements dictated by the value of Occupational Exposure Standards. A few of the flammable substances encountered in industry are considered simple asphyxiants. Table D.2 provides LFL and compares tenth LFL values against atmospheric oxygen (20.9%) for flammable, simple asphyxiants. The presence of a flammable, simple asphyxiant at 10% of its LEL would have little impact on the level of oxygen. Under these conditions, the atmosphere would not be considered oxygen deficient by any standard measure (NIOSH 1979; ACGIH 1994; ANSI 1989; OSHA 1993, 1994; Standards Australia 1986; Canada Gazette 1992).

Use of 10% of LEL as a practical operating limit could create conflict where other flammable gases and vapors are present. Occupational Exposure Standards, such as the Threshold Limit Values, have decreased steadily during the preceding decade. Where use of air-supplying respirators is permitted, the Maximum Use Concentration is governed by multiplication of the Occupational Exposure Standard by the Assigned Protection Factor for the respirator (Colton et al. 1991). That is, MUC = OES × APR. Normally accepted real-world exposure is somewhat less than this calculated value. Most organizations attempt to maintain exposures as low as possible relative to the Occupational Exposure Standard. This concern also would apply during the wearing of air-purifying respirators.

TABLE D.3
Comparison of Flammability Limits and Exposure Standards

Substance	LFL (%)	Tenth LFL (ppm)	Occupational exposure standard (ppm)	Ratio
Acetone	2.5	2,500	750	3
Benzene	1.3	1,300	10	130
(proposed)			0.3	4,333
Butane	1.6	1,600	800	2
N-butanol	1.4	1,400	C 50	28
Carbon disulfide	1.3	1,300	10	130
Carbon monoxide	1.3	1,300	25	52
Cyclohexane	1.3	1,300	300	4
Ethyl acetate	2.0	2,000	400	5
Ethyl alcohol	3.3	3,300	1,000	3
Ethyl benzene	1.0	1,000	100	10
Ethyl ether	1.9	1,900	400	5
Gasoline	1.3	1,300	300	4
n-Heptane	1.1	1,100	400	3
n-Hexane	1.1	1,100	50	22
Isopropyl alcohol	2.0	2,000	400	5
Methyl alcohol	6.0	6,000	200	30
2-hexanone (MnBK)	1.2	1,200	5	240
Methyl chloroform	7.5	7,500	350	21
Styrene	1.1	1,100	50	22
Toluene	1.2	1,200	50	24
Turpentine	0.8	800	100	8
VM&P naphtha	0.9	900	300	3
Xylene	1.0	1,000	100	10

Data from Alliance of American Insurers 1987, ACGIH 1994, and U.S. Dept. of the Interior 1965.

Table D.3 compares LFL and tenth LFL values with TLVs for commonly encountered substances. Occupational Exposure Standard as used in this table refers to the current Threshold Limit Value. One tenth LFL expressed in parts per million is the value in percent multiplied by 1,000. Ratio is calculated by dividing 1/10 of LEL by the Occupational Exposure Standard, in this case the TLV. In effect, the ratio is the minimum protection factor needed in the facepiece. This results from setting 1/10 of the LEL as the concentration outside the respirator and the Occupational Exposure Standard as the permissible concentration inside the respirator. A half-facepiece air-purifying respirator carries an Assigned Protection Factor of 10, and a full-facepiece, 50 (assuming that cartridges are available for the contaminant of interest) (NIOSH 1987; ANSI 1992). Different standards worldwide for respiratory protection take different views regarding limitations of air-purifying respirators (Lara and Smith 1995).

Results from Table D.3 indicate that a full-facepiece respirator would be needed for protection against many of the contaminants when controlled to 10% of LEL by ventilation. For many of these substances, this could mean reliance on supplied-air respirators, depending on the dictates of a particular jurisdiction. Reducing the acceptable working concentration to 1% of LEL or less would substantially reduce this requirement, except for key substances, such as benzene. Benzene is a minor component in many solvents and mixed products. Concern about overexposure to benzene, assuming reduction of the TLV to 0.3 ppm as a time-weighted average, could dictate the approach to ventilation and acceptable airborne contamination, rather than concern about ignitability.

Last, consideration in setting a standard should consider safety under normally anticipated conditions, as well as extraordinary ones. Preceding discussion would suggest that the lower the standard for ignitables, the better the assurance of safety, regardless of legal requirements.

There is considerable evidence to argue for the lowest practicable alarm level as the criterion for terminating the entry. That is, let the limitations of instrument design set the threshold and not perception of what works. The design and performance of today's instruments are considerably improved over those that existed when 10% of LFL (LEL) was selected as a feasible limit.

Another factor that argues for lowering of the limit is the manner in which air sampling occurs. During real-world sampling, the position of the detector rarely is near the position of the source of the vapor. Normal practice is to sample the air in the space at different heights and sometimes different horizontal positions prior to entry. These samples provide at best an average concentration, depending on air movement at the time. During work activity, the instrument, if brought into the space, usually is set in some convenient position or sometimes worn by a worker. The suitability of these monitoring locations relative to the source of emission and potential sources of ignition for providing adequate warning is debatable. These realities argue for an alarm set-point that is as low as possible in order to optimize worker safety. In order to reach the sensor, vapor must migrate considerable distance from the source. Concentration obviously is much higher at the source than at the worker position, this being the usual closest position for the instrument.

Worker safety has several meanings in this discussion. While the normal function served by these instruments is fire safety, a low set-point may be able to assist in other areas. A low set-point can provide information about exposure to substances for which real-time monitoring is otherwise not available. For a substance having an LFL of 1.0% or 10,000 ppm, the minimum concentration readable on the scale of the instrument is 1% of that value or 100 ppm for a properly calibrated instrument. This technique, in conjunction with a datalogging instrument, could be useful with some contaminants for assessing the need for respiratory protection and also to sound an evacuation alarm.

The factor that limits the minimum set-point for the alarm is the need to minimize the occurrence of false (positive) alarms. This in turn is affected by accuracy and precision of the signal, lack of specificity and response to contaminants that differ from the calibrant.

Relative response of the sensor varies between flammable gases and vapors. In many cases the user has no way of knowing what is present, and, therefore, how to interpret the reading. The instrument reading could underestimate the true concentration of contaminant. To illustrate, consider a substance that has a relative response of 35% compared to the calibrant. This could easily occur in a pulp mill where an instrument calibrated to methane is exposed to turpentines. In this circumstance, a true concentration of 58% of LFL (LEL) would produce a reading of only 20%. A high take action level increases the potential for underestimation. On the other hand, a low take action level minimizes the effect of relative response on readings at this value. For example, at a take action level of 5% of LFL (LEL), even in this extreme example, a true concentration of only 14.5% would activate the alarm.

Other practical considerations about the minimum value for the alarm set point include zero-line bounce, drift, and digital zero clamp. Zero line bounce occurs when the span is increased to accommodate substances of low sensitivity, such as turpentine in the preceding example. Zero line bounce manifests as an unstable zero value that might display in successive readings as ..., 1, -1, 0, -1, 0, 1, ... High span settings, as in this example, also cause drift of the zero value. Some instruments incorporate a "digital zero clamp". "Digital zero clamp" refers to a deliberate inability to display nonzero readings. To illustrate, a true reading of 1 or 2% might read zero on these instruments.

An alarm set point of 5% of LFL (LEL) provides high confidence that false (positive) alarms will not occur. An alarm set point of 2 or 3% would be more protective, but at higher risk of false (positive) alarms.

REFERENCES

Alliance of American Insurers: *Handbook of Organic Industrial Solvents,* 6th ed. Schaumberg, IL: Alliance of American Insurers, 1987. 135 pp.

American Conference of Governmental Industrial Hygienists: *1994–1995 Threshold Limit Values for Chemical Substances and Physical Agents and Biological Exposure Indices.* Cincinnati, OH: American Conference of Governmental Industrial Hygienists, 1994. 119 pp.

American National Standards Institute: *Safety Requirements for Working in Tanks and Other Confined Spaces* (ANSI Z117.1-1977). New York: ANSI, 1977.

American National Standards Institute: *Safety Requirements for Confined Spaces* (ANSI Z117.1-1989). Des Plaines, IL: American Society of Safety Engineers/American National Standards Institute, 1989. 24 pp.

American National Standards Institute: *ANSI Z88.2 (1992) Practices for Respiratory Protection.* New York: American National Standards Institute, 1992.

American National Standards Institute: *Safety Requirements for Confined Spaces* (ANSI Z117.1-1995). Des Plaines, IL: American Society of Safety Engineers/American National Standards Institute, 1995. 32 pp.

American Petroleum Institute: *Safe Entry and Cleaning of Petroleum Storage Tanks,* 4th ed. (API Std. 2015). Washington, D.C.: American Petroleum Institute, 1991. 26 pp.

Colton, C.E., L.R. Birkner, and L.M. Brosseau (Eds.): *Respiratory Protection: A Manual and Guideline,* 2nd ed. Akron, OH: American Industrial Hygiene Association, 1991. 130 pp.

Human Resources Development Canada: *Consolidated Working Copy: Canada Labour Code, Part II.* Ottawa, ON: Canada Communication Group — Publishing, 1994. pp. X-7 and X-8.

Lara, J. and S. Smith: Comparing standards in a global economy. *Occup. Health Safety Canada — Buyers Guide 1995* pp. 14–23.

National Fire Protection Association: *NFPA 306: Control of Gas Hazards on Vessels* (1993 ed.). Quincy, MA: National Fire Protection Association, 1993. 15 pp.

National Institute for Occupational Safety and Health: Criteria for a Recommended Standard — Working in Confined Spaces (DHEW/PHS/CDC/NIOSH Pub. No. 80-106). Cincinnati, OH: DHEW(NIOSH), 1979. 68 pp.

National Institute for Occupational Safety and Health: Guide to Industrial Respiratory Protection (DHHS [NIOSH] Pub. No. 87-116). Cincinnati, OH: DHHS/PHS/CDC/NIOSH, 1987. 296 pp.

National Occupational Health and Safety Commission and Standards Australia: Joint Draft National Standard for Planning and Work for Confined Spaces (BS92/20113 Cat. No. 92 0995 8). Canberra, ACT: Australian Government Publishing Services, 1992. 55 pp.

Ontario Ministry of Labor: Occupational Health and Safety Act and Regulations for Industrial Establishments: Occupational Health and Safety Act, Revised Statutes of Ontario, 1980, Chapter 321. Toronto, ON: Ontario Ministry of Labor, 1984. p. 84.

OSHA: Permit-Required Confined Spaces for General Industry; Final Rule, *Fed. Regist. 58*: 9 (14 January 1993). pp. 4462–4563.

OSHA: Confined and Enclosed Spaces and Other Dangerous Atmospheres in Shipyard Employment; Final Rule, *Fed. Regist. 59*: 141 (25 July 1994). pp. 37816–37863.

Standards Association of Australia: *Safe Working in a Confined Space* (AS 2865-1986). North Sydney, NSW: 1986. 19 pp.

U.S. Department of the Interior: Flammability Characteristics of Combustible Gases and Vapors, by M.G. Zabetakis (Bull. 627). Washington, D.C.: U.S. Department of the Interior/Bureau of Mines, 1965. 121 pp.

Appendix E: Model Confined Space Program

CONTENTS

PREAMBLE

This Appendix describes an organizational program for work involving confined workspaces. This program applies to a moderate-sized organization that has sufficient resources to employ an industrial hygienist, as well as safety officers. The organization is assumed to be widely distributed geographically. This program assumes that these and other high level technical resources act as advisors to on-site decision makers. In this program, the high level advisors create the hazard assessment following a visit to the site. This forms the basis for preparation of entry documents (entry permit or certificate) by supervisors at the site. On-site supervisors are expected to contact the author of the hazard assessment when the situation varies from what is described in the documents. This approach is consistent with general requirements in standards and legislation, even though variations in terminology may exist. It also enables scarce corporate resources to be utilized in an efficient manner.

INTRODUCTION

Confined workspaces pose unique challenges to the safe conduct of work. Structures within this Organization contain a number of confined spaces. Although individual operating departments have addressed this problem for some time, experience indicates the need for a consistent Organization-wide approach. The approach to be followed by this Organization will meet or exceed all government requirements.

Research indicates that the hazardous conditions that can arise in confined spaces include one or more of the following:

- Personal confinement
- Unstable interior condition
- Flowable solid materials or residual liquids or sludges

- Release of energy through uncontrolled or unpredicted motion or action of equipment
- Atmospheric confinement
    Toxic substances
    Oxygen deficiency/enrichment
    Flammable/combustible atmosphere
- Chemical, physical, biological, ergonomic, mechanical, hydraulic, pneumatic, process, and safety hazards

The meaning of individual terms may be defined in the legislative framework of this jurisdiction. This definition represents 36 or more potentially hazardous conditions. Many could exist or could develop simultaneously and independently of each other. One important ramification from this definition is that the term "confined space" encompasses a range of hazardous conditions. Concern about individual hazardous conditions ranges from minimal to life threatening.

Research has shown that confined spaces and structures that can cause atmospheric confinement pose greater health and safety hazards than normal workspaces.

This document outlines the approach to be followed within this Organization for hazard management in confined spaces and on structures that confine hazardous atmospheres. It contains six main sections: education and training, recognition and identification, hazard assessment, entry and hazard control, emergency preparedness and response, and recordkeeping. This program identifies participants and assigns responsibility.

In carrying out this policy, the Organization will take all reasonable measures to ensure competent evaluation and control of hazards arising from work involving confined spaces.

PARTICIPANTS AND ACCOUNTABILITIES

Participants involved in implementing and executing this program include:

- Health and Safety Manager
- Members of the health and safety committee
- Industrial Hygienist/Safety Professional
- Operations and mechanical personnel
- Qualified Person
- Entrant
- Attendant

The Health and Safety Manager is responsible for the administration and operation of this program. The Manager is accountable to senior management.

Members of the Health and Safety Committee assist in the identification and description of confined spaces at the site. The Health and Safety Committee reviews the program and offers input into its direction.

The Industrial Hygienist/Safety Professional acts as a technical expert in the recognition, assessment, and elimination or control of hazardous conditions that occur in confined spaces. They collaborate to prepare hazard assessments for confined spaces. These individuals are accountable to the Health and Safety Manager. The Industrial Hygienist and Safety Professional both have certifications in their respective disciplines. These certifications are obtainable only by individuals who have suitable levels of education and professional experience. Both certifications are governed by strict codes of ethics. Operations and mechanical personnel act as experts in operation of equipment in the space and activities that must occur. They prepare procedures for the entrants who perform the work.

The Qualified Person is responsible for on-site evaluation of hazardous conditions and preparation of entry documents. They utilize hazard assessments prepared by the Industrial Hygienist/Safety

Professional as guidance documents. The Qualified Person provides expertise to on-site management and supervision. The Qualified Person has education and training in the recognition, assessment and elimination or control of hazards in confined spaces. The Qualified Person may have other duties. The Qualified Person is accountable to site management.

Entrants work in the confined space or in proximity to it. Entrants receive education and training about hazardous conditions and protective measures.

The Attendant monitors the course of work in the space and assists the entrants.

The titles "Qualified Person" and "Entrant" do not exist as regular workplace titles. They receive their existence from this program. These titles apply in addition to regular job titles.

EDUCATION AND TRAINING

The most important resources for assuring safe conditions for work involving confined spaces and confined atmospheres are well-informed and well-trained people.

The Organization will make general information about the hazards of confined spaces available to all interested employees. This will be provided primarily through safety meetings. Specialized training will be provided to persons directly affected by confined spaces.

Responsibilities for Education and Training

1. All persons working in confined workspaces shall receive education and training to at least the level of entrant.
   Action: Manager/Supervisor
2. Appropriate persons within this group shall be selected to receive education and training to become Qualified Persons.
   Action: Manager/Supervisor
3. Persons nominated to become entrants and Qualified Persons shall be provided the opportunity to receive this training.
   Action: Manager/Supervisor
4. A curriculum for the education and training program shall be developed. Training shall be developed using the performance-based model.
   Action: Industrial Hygienist/Safety Professional
5. Records of education and training for each Qualified Person and entrant shall be maintained.
   Action: Manager/Supervisor
6. Education and training of Qualified Persons and entrants shall be valid for one year. An opportunity for retraining for maintaining this certification shall be made available.
   Action: Manager/Supervisor

RECOGNITION AND IDENTIFICATION

The Organization will create and maintain an inventory of confined spaces. The inventory will be used for determining priorities in the program and for labeling purposes.

Responsibilities for Recognition and Identification

1. All confined spaces shall be identified and characterized. This information shall be maintained in a written (computerized) inventory and made available to operating departments.
   Action: Members of Health and Safety Committee/Qualified Person/Health and Safety Manager

2. The inventory shall be updated at least annually and shall include all current changes.
   Action: Members of Health and Safety Committee/Qualified Person/Health and Safety Manager
3. Where reasonable and practicable, the entrance to the confined space shall be marked by a warning sign, and some means of restricting entry.
   Action: Members of Health and Safety Committee/Qualified Person/Manager/Supervisor

HAZARD ASSESSMENT

The purpose of hazard assessment is to document and evaluate hazardous conditions that could be present or could develop during work in a confined space. Hazard assessment considers the present and previous history of the space, including uses and processes. Hazard assessment also considers work to be performed in the space.

Responsibilities for Hazard Assessment

1. Hazard assessment shall precede entry into confined spaces. Requests for hazard assessment shall be forwarded to the Industrial Hygienist and Safety Professional. These shall include a detailed description of work to be performed, including procedures. The request shall be submitted 1 month prior to the start of the project.
   Action: Manager/Supervisor
2. Hazard assessment shall proceed as expeditiously as possible. Hazards likely to be present shall be identified and evaluated. The assessment shall determine the safety of entry and the proposed work.
   Action: Industrial Hygienist/Safety Professional/Operations and Mechanical Personnel
3. A confined space shall not be entered prior to completion of on-site entry documents. The basis for the entry document is the hazard assessment.
   Action Manager/Supervisor/Qualified Person
4. Instruments and methodologies capable of assessing conditions shall be investigated and recommended to the Organization. Where reasonable and practicable, emphasis shall be placed on monitoring instruments that provide continuous, real-time output.
   Action: Industrial Hygienist/Safety Professional
5. Instruments suitable for assessing conditions shall be purchased in numbers sufficient to satisfy the need.
   Action: Manager/Supervisor
6. A facility for calibration and repair of monitoring instruments shall be established and sustained.
   Action: Manager/Supervisor
7. All testing equipment shall be maintained and calibrated according to manufacturer's recommendations.
   Action: Manager/Supervisor/Qualified Person
8. A written log of performance, calibration and service shall be maintained for each piece of testing equipment.
   Action: Manager/Supervisor/Qualified Person
9. Equipment suitable for controlling exposure to hazardous conditions shall be investigated and recommended to the Organization. This shall include ventilation equipment and personal protective equipment.
   Action: Industrial Hygienist/Safety Professional

ENTRY AND HAZARD CONTROL

Control can be achieved in a number of different ways as outlined in this section.

Engineering control occurs through use of equipment, such as portable air supply and exhaust ventilation units, to control or eliminate hazards. In some cases, specially engineered equipment can eliminate the need for entry to perform a particular task. **Procedural control** utilizes procedures to control hazards by regulating the potential for exposure to them. **Personal protective equipment** is equipment, such as hard hats, faceshields, respirators, clothing, gloves, and footwear. Personal protective equipment provides the only feasible means of control in many situations. Personal protective equipment must be chosen carefully, since this can impose considerable burden on the wearer. The burden imposed by personal protective equipment itself may constitute a safety hazard. **Administrative control** occurs through hazard assessment by an on-site Qualified Person. The Qualified Person assesses conditions at time of entry and authorizes work to occur under specified environmental conditions.

Responsibilities for Entry and Hazard Control

1. Persons entering and working in confined spaces shall have training to at least the level of entrant.
   Action: Manager/Supervisor
2. All persons working in the confined space shall sign the entry documents to acknowledge understanding of terms, conditions and requirements.
   Action: Qualified Person/Entrants
3. The Qualified Person shall define the conditions under which work may proceed. This includes specification of control measures. The Qualified Person also shall determine conditions under which work must be halted.
   Action: Manager/Supervisor
4. Testing shall be conducted throughout the confined space during the prework assessment.
   Action: Qualified Person
5. Where reasonable and practicable, the atmosphere inside the confined space shall be tested continuously during entry and work.
   Action: Qualified Person/Entrant
6. If the conditions stated in the entry document cannot be followed, the Qualified Person shall be notified immediately. Entry and work shall not proceed until clarification has been obtained from the Qualified Person.
   Action: Manager/Supervisor
7. If conditions specified in the entry document are not being followed, work shall be stopped and the supervisor notified about the noncompliance.
   Action: Qualified Person
8. If corrective actions required for compliance with conditions stated in the entry documents are not taken prior to continuing the work or the deficient action occurs again, the work shall be stopped and the manager notified verbally and in writing.
   Action: Qualified Person
9. Following evaluation by the Qualified Person of the hazards likely to be present in confined spaces, ventilating and personal protective equipment shall be obtained in sufficient quantity to address the need.
   Action: Manager/Supervisor
10. A sample of compressed breathing air shall be submitted for testing every 6 months, or sooner, if fault in the compressor or purification system is suspected.
    Action: Qualified Person

11. If the quality of the atmosphere or other hazardous condition in the confined space cannot be guaranteed by testing and control measures, it shall be declared Immediately Dangerous to Life and Health (IDLH).
    Action: Qualified Person

EMERGENCY PREPAREDNESS AND RESPONSE

Preceding sections of this document address routine situations. Emergency preparedness is an important action for lessening the consequences of accidents that can occur in these environments. Planning for emergency response involves anticipating situations reasonably likely to occur and taking the steps needed to lessen their impact.

Responsibilities for Emergency Preparedness and Response

1. Accidents still considered possible after taking control measures shall be identified during hazard assessment. Specialized equipment needed for accident situations shall be investigated and recommended to the Organization.
   Action: Industrial Hygienist/Safety Professional
2. Equipment needed to address accident situations shall be specified in entry documents.
   Action: Qualified Person
3. Equipment needed for emergency response shall be acquired in sufficient quantity to satisfy the anticipated need.
   Action: Manager/Supervisor
4. Equipment needed for emergency response shall be readily available on the site at the time of entry and work activity.
   Action: Manager/Supervisor
5. A written plan for emergency response shall be prepared for the operating unit.
   Action: Qualified Person
6. Emergency response personnel shall be appointed and trained. (These persons also may function in other voluntary emergency response activities such as the fire crew.)
   Action: Manager/Supervisor
7. Where emergency response personnel are not available within the Organization, suitable external sources of assistance shall be identified and contacted. The external response group shall be informed fully about possible accident situations and assistance that might be required.
   Action: Manager/Supervisor/Qualified Person
8. Persons working in confined spaces and attendants shall be provided with the means to communicate with emergency response providers.
   Action: Manager/Supervisor
9. The emergency response plan shall be tested through drills. Testing should occur at least once per year.
   Action: Manager/Supervisor/Qualified Person

RECORDKEEPING

Records ensure continuity as functional staff changes. Records provide invaluable links to the lessons of the past. They help to reduce the probability that something will go wrong in the present and future. Recordkeeping establishes the credibility of the confined space program by proving performance, competence and compliance. The main areas in which records are kept include:

- Performance, maintenance, and calibration of instruments
- Training of qualified and trained persons
- Hazard assessments
- Entry documents (permits/certificates)
- Work procedures
- Monitoring results
- Records of qualifications
- Inventory of confined spaces
- Inspection of safety equipment (e.g., harnesses, fall arrest, retrieval equipment, etc.)

Responsibilities for Recordkeeping

1. Written records shall be maintained for all aspects of work involving confined spaces.
   Action: Manager/Supervisor
2. Records of hazard assessments, entry procedures, work procedures and entry documents shall be maintained.
   Action: Industrial Hygienist/Safety Officer/Qualified Person

Appendix F: Guided Questions for Identifying and Assessing Hazardous Conditions in Confined Spaces

CONTENTS

INTRODUCTION

Recognition is the key first step in addressing the hazardous conditions associated with confined workspaces and structures capable of atmospheric confinement. While certain conditions elicit instant recognition from all observers, other equally valid circumstances produce no response. This can occur from lack of knowledge, lack of sensory warning, lack of experience with the problem, or other factors.

The following questions provide a guide for identifying and assessing hazardous conditions that define the status of a particular workspace and work to be performed. The user must recognize that this list may not be exhaustive.

IDENTIFICATION AND FUNCTION

- Identify the space.
- What is the present function of the space?
- What were previous functions of the space?

ADMINISTRATIVE PARTICULARS

- What is the reason for entering the space?
- How frequently is the space entered?
- Can the space be modified to eliminate the need for entry? If not, why not?
- Describe the work to be performed.
- Can the work be modified to eliminate the need for entry? If not, why not?
- Is this space similar to others presently considered to be confined spaces?
- Who will perform work in this space?
- Are contractors involved?
- How many persons will be present at the site?
- What trades do they represent?
- What training in confined spaces have the entrants had?
- At what time of the day is the work to be performed?
- Will the work be performed during the normal workweek or on weekends or holidays?

EXTERIOR CHARACTERISTICS

- Describe the exterior surroundings of the space.
- Is the space above ground?
- Is the space below ground or the normal floor level?
- How many sides of the space are enclosed?
- Are openings into the space normally open or closed?
- Does air circulate into and out of the space?
- Is the space normally occupied by workers?
- Is protective equipment required during normal occupancy? If yes, identify what is required and state the reason.
- Do beams from radioactive gauges pass through the space? If yes, state precautions to be taken to shut off this equipment prior to entry.
- Describe access (entry) or egress (exit) portals (doorway, hatchway, manhole, etc.).
- How many entry/exit portals are there?
- What is the size of the entry/exit portal?
- Do openings provide easy access to the interior of the space and ready egress from it during emergency?
- Will equipment or service cables and hoses partially block the opening?

ADJACENT SPACES AND PROCESSES

- List and describe all processes and activities occurring in the immediate vicinity outside the space.
- Does any process involve substances that are:
  Toxic
  Flammable
  Explosive
  Compressed gases
- Could any substance produced in an adjacent process enter the workspace? If yes, explain.
- Do adjacent processes involve sources of physical energy such as:
  Ionizing radiation
  Nonionizing radiation
  Lasers

- Are heat sources adjacent to the space?
- Can sources of energy adjacent to the workspace interfere with activity to be carried out in the space?
- Is the space an isolated structure?
- Describe structures that share common walls with the confined space.
- Describe the function of structures attached to the space.
- What processes occur in these structures?
- Will the process in the adjacent structure continue to occur during work in the space?
- What contents are stored in these structures?
- Will these contents remain in the adjacent structure during the work to be undertaken in the space?
- Determine the hazardous properties of the contents of the adjacent structure from the Material Safety Data Sheet (MSDS).
- Can the work to be undertaken in the space destroy the integrity of the wall separating it from the adjacent structure?
- What are the possible outcomes from this occurrence?
- Do these outcomes include the possibility of contamination of the atmosphere in the space by toxic substances or fire or explosion? If yes, what precautions will be taken to ensure that these outcomes will not occur?
- Can the work to be undertaken in the space transfer energy into the adjacent structure?
- Can the contents of the adjacent structure or the process transfer energy into the space?
- What are the possible outcomes from the transfer of energy?
- Do these outcomes include physiological stresses or the possibility of fire or explosion? If yes, what precautions will be taken to ensure that these outcomes will not occur?

INTERIOR CHARACTERISTICS

- Describe the interior of the space.
  - Materials of construction
  - Floor area
  - Height
  - Volume
- Is the interior structure of the space stable? If no, what precautions are required to ensure the stability of the structure?
- Can the interior geometry restrict movement of persons working in the space?
- Is visual contact from the entry point to all workspaces obstructed? If yes, describe measures to maintain contact with entrants.
- Are workspaces in the space accessible by unobstructed path from the entry/exit?
- Is the space partitioned off into subspaces (compartments or other rooms)?
- How is entry gained into each subspace?
- Describe the subspace(s). List their dimensions.
- Will the dimensions or geometry of the subspaces or internal equipment confine the movements of persons working in them?
- Does the elevation of the subspace differ from that of the main space?
- Does this difference in elevation pose a fall hazard? If so, describe measures taken to prevent a fall from height.
- Is airflow into/from subspaces obstructed?
- Are workspaces within the subspace accessible by unobstructed path from the entry?
- Is visual contact with workspaces in the subspace from the entry point obstructed? If yes, describe measures to maintain contact with entrants.

- Will service cables and hoses supplying equipment pass through the opening leading to each subspace?
- Will service cables and hoses passing through openings hinder access and ready egress during emergency?
- How will communications between persons working inside the space and those working outside be maintained?
- Where electrical/electronic equipment is to be used, is this rated for safety in an explosive atmosphere?
- Is the interior of the space wet or dry?

EQUIPMENT HAZARDS

- What equipment does the space contain (mixers, augers, agitators, etc.)?
- Will this equipment remain in place during the work to be done?
- What is the function of this equipment ?
- What energy sources are required to power this equipment (electrical, pneumatic, hydraulic, etc.)?
- Describe the action of each energy source.
- Describe how each energy source is controlled inside the equipment.
- Does this equipment store energy (pressure, vacuum, potential energy, kinetic, etc.) as part of its operation?
- How is this energy released?
- From where is this equipment activated?
- How is this equipment controlled externally?
- How many controls are there?
- Where are control devices located?
- How is this equipment de-energized?
- Where are shut-off devices located?
- Can this equipment be started by other systems?
- What precautions will be taken to prevent the equipment from starting up during entry or work?
- What precautions will be taken to ensure that energy sources connected to the equipment cannot energize the equipment or release energy or contents into the space?
- What precautions will be taken to ensure that energy stored in the equipment will be relieved prior to entry?
- Does the space contain level sensing devices? If yes, describe.
- What type of energy is used in the process?
- Does this energy exceed exposure limits? What controls will be applied to prevent exposure?

PROCESS HAZARDS

- What was the function of the space at time of entry?
- Identify present and previous contents.
- If the space is used for a process, describe the process fully.
- What materials enter the space through supply lines?
- What materials are produced in the space?
- What materials leave the space through outlet lines or process vents?
- Identify the number, size, and purpose of each supply line.
- Identify the number, size, and purpose of each outlet line.

- If the space can be isolated completely from supply and outlet lines to ensure that no substance can enter during the work to be undertaken, describe measures to be taken to prevent entry of any material (solid, liquid, or gas) into the space.
- If the space cannot be isolated from supply and outlet lines, describe precautions to be taken to prevent entry of process chemicals or waste products?
- Will contents be drained from the space prior to entry? State measures to remove material from low points in the space and associated lines, valves, and pumps.
- Will the interior of the space be washed prior to entry? Describe the washing process, indicating all substances to be used.
- Is the space to be purged? Describe the purging process, indicating all substances to be used.
- Can the washing or purging agent remain in isolated areas of the space or in low spots in the space or associated equipment?
- What materials could remain in the space following draining and purging?
- Will the space be ventilated following purging? Describe how and for how long.
- Describe how adequacy of ventilating is to be determined.
- What will be the interior pressure of the space prior to opening for entry? If other than atmospheric, describe precautionary measures during equalization procedure.
- Do utility lines pass through the space (e.g., steam, gas, electrical, hot water, compressed air, nitrogen, hydrogen, other compressed gas, etc.)?
- Will these lines be active at the time of entry? If yes, describe protective measures to prevent exposure in the event of accidental failure of this equipment during occupancy. Describe precautions to ensure that contact with live electrical conductors cannot occur.
- Does the process that occurs in the space involve energy from laser, ultraviolet, visible, infrared, microwave, or radio frequency source? If yes, describe precautions to deactivate the source prior to entry.

BULK MATERIAL HAZARDS

- Does the space contain flowable solid materials or liquids? If yes, describe.
- What quantities?
- What is the depth of the flowable solid or liquid?
- Can a bridge form in the solid material?
- Can caking on walls occur?
- What is the purpose for entering this space?
- What precautions are planned to prevent engulfment or immersion?
- Does the bulk material extend above the working level?
- Describe precautions to prevent the material from falling onto entrants?

FIRE/EXPLOSION HAZARDS

- Does the space normally contain pure oxygen or an oxygen-enriched atmosphere? At what pressure?
- Can original contents or any material used to wash, purge or vent the space burn?
- Will ignition sources be used inside or near the space?
- State precautions to be taken to prevent fire.
- How easily does the bulk material become airborne as dust?
- Will work activity cause the production of airborne dust from this material?
- Does movement of the bulk material (solid or liquid) generate static electricity?
- Can the work to be performed produce static electricity?
- State precautions to be taken to control generation of static electricity.

ATMOSPHERIC HAZARDS

- Can the boundary surfaces of the space trap residual contents, substances used for washing, or purge gases or contaminants produced during the activity?
- What is the expected behavior of airborne contaminants produced by these substances? That is, will they rise in air, remain stagnant, or settle? Consider temperature and density of ambient air and the substance.
- Describe the properties of the substances mentioned in the preceding questions?
  - Physical
  - Chemical
  - Toxicological
  - Flammability/combustibility
  - Exposure standards
    - Threshold Limit Value
    - Concentrations Immediately Dangerous to Life or Health
- Describe precautions to be taken to prevent overexposure to these substances.
- Can the atmosphere inside the space be or become oxygen deficient prior to or during entry? Why?
- Describe precautions to be taken to prevent exposure to an oxygen-deficient atmosphere.
- Does the task require or produce oxygen? For what purpose? Describe precautionary measures.
- Does the task require use of compressed gases? Identify which and state the purpose.
- Does the task generate gas? Which gas and what is the rate of production?
- Describe precautions to be taken to prevent exposure to gases produced or used during the task.
- Does the task require use of chemical products?
- Identify hazards associated with each substance from the MSDS.
- Does the process generate other atmospheric hazards: vapors, dusts, fumes, mists, or smoke?
- Describe precautions to be taken to prevent exposure to atmospheric hazards produced during the task.
- Can any of the substances used in a task or produced during a task burn?
- Describe measures to prevent fire.
- Can any tasks produce static electricity?
- Describe steps to control generation of static electricity.
- Will plant compressed air be used in the space for any purpose (purging, ventilating, powering equipment, or breathing air)?
- Is the intake of the compressor protected against contamination by vehicle exhausts, process gases, or vapors or solvents in storage?
- Does the system that supplies compressed air for breathing contain an in-line monitor for carbon monoxide, oil, moisture, and particulate traps and an audible alarm?
- Is the purity of compressed air to be used in the space assured by regular testing?
- Does the interior of the space contain hot or cold surfaces, or are hot or cold surfaces produced during work activity? Describe precautionary measures to protect against contact.

ACTIVITY HAZARDS

- Describe tasks to be performed during the activity.
- Does the activity involve sources of physical energy such as:
  - Noise
  - Ionizing radiation

Nonionizing radiation
- Ultraviolet
- Infrared
- Microwave
- Laser
- Radio frequency

- Can these forms of energy escape into the workspace?
- Describe protective measures to be taken.
- Do residual materials or materials associated with work activity pose a skin contact hazard? Describe protective measures for work in the space. Describe decontamination measures for personnel and equipment upon leaving the space.
- Does the activity involve sources of heat?
- Can buildup of heat occur?
- Does the activity require physical exertion?
- What measures will be taken to control heat stress?
- Does the plant contain utility supply lines that provide process gases and liquids?
- Do hoses that attach to supply lines have common fittings?
- Describe precautions to be taken to prevent mistaken interconnection of hoses to supply lines?
- Will the interior of the space be slippery due to residual materials in the space? Describe precautions against accidents resulting from slippery surfaces.
- Will the interior of the space contain electrically conductive liquids? How will protection from contact with live electrical circuits be ensured?
- How will illumination be provided in the space? If flammables/combustibles are used during work activity or remain in the space as residual contents, describe measures to prevent fire/explosion.
- How much noise is produced by equipment to be used in and around the space? Describe precautions to be taken to prevent this noise from interfering with normal communications and recognition of warning alarms.

VENTILATION

- Is the space normally ventilated regardless of occupancy?
- Is ventilation provided by natural means (wind or convection)?
- Is the space ventilated by mechanical means (fan or other air mover)?
- Does the space have supply ventilation?
- Could the ventilation fail? Under what circumstances?
- What would be the effect on airborne contaminants if ventilation ceased?
- How will the space be ventilated during occupancy?
- Is compressed air used to ventilate the space tested to ensure compliance with standards for compressed breathing air?
- Is the dew point of the air sufficiently low to prevent freeze-up of regulators, valves, or other equipment in the system?
- Does the system contain a breathing air purification unit?
- Is the intake for the compressor in such a location as to prevent entry of contaminants from vehicle exhaust, solvent storage, process emissions, or excessive humidity?
- Can the system provide an adequate flow to ventilate the space?
- Does the nozzle incorporate noise-reducing technology?
- What is the source of intake air of the air mover used to ventilate the space?
- Is the intake of the air mover in such a location as to prevent entry of contaminants from vehicle exhaust, solvent storage, process emissions, or excessive humidity?
- Can the blower provide an adequate volumetric flow to ventilate the space?

EMERGENCY PREPAREDNESS AND RESPONSE

- What extraordinary situations could occur despite implementation of precautionary measures?
- Rank the accidents in descending order of likelihood.
- Are atmospheric hazards likely to be involved? If yes, indicate why.
- Could the victim be removed from the space rapidly without the risk of further injury?
- What conditions would prevent rapid extrication of a victim?
- Could first responders enter the space without undue risk of personal injury or overexposure to atmospheric contaminants?
- What actions could be taken to stabilize a victim in the space prior to arrival of community emergency services?
- What equipment should be readily available for first responders?
- What training do first responders need in order to address accidents in these situations?
- What liaison has been established with local community emergency services?
- Do the contents of the space or the process to be undertaken involve potential exposure to substances (such as hydrofluoric acid) for which specialized treatment is required?
- Are community emergency services aware of the possible need for such therapeutic agents?
- Is someone at the site designated to take charge in an emergency? If so, who?
- What qualifies this person to take charge? Indicate training in emergency response.
- Does the organization have an emergency response plan? If yes, describe.
- Has this plan been tested? If yes, describe.
- Were deficiencies identified?
- Were they corrected?
- If the work is to be performed during off-shifts or weekends, what precautions will be taken to ensure adequate level of emergency and rescue services?

Appendix G: Deactivation, De-Energization, Isolation, and Lockout

CONTENTS

INTRODUCTION

Lockout/tagout is a major theme in the control of hazardous conditions in confined spaces. Lockout/tagout is actually one part of a much bigger whole: deactivation, de-energization, isolation, and lockout.

An interesting way to introduce these concepts is an analogy, a story about an accident at a zoo. This involved a tiger and its keeper. Normally the tiger remained in the cage and the keeper remained outside, as one would expect. People in this role often develop close attachments to those in their care. This is a normal expression of human emotional needs, as seen with the keeping of pets. Pets include domesticated animals, such as dogs and cats, who have developed a reason in their psyche to have a relationship with humans. "Pets" also include animals, such as raccoons, monkeys, elephants, bears, and lions that normally avoid contact with humans. The relationship that we have with all animals is tenuous. Their behavior is not totally predictable at all times. This unpredictability increases, especially with carnivores that will kill humans in the wild. That is, there is always an element of uncertainty in the relationship between humans and animals. There is always danger of attack and injury.

Regarding the keeper at the zoo, this individual developed a belief that he had developed a special relationship with the tiger. So powerful was the strength of this belief that he felt that he could enter the cage to pay the tiger a "social visit". Unbeknownst to him, the tiger was unaware of these feelings and the role that he was expected to play, and acted accordingly. This episode had a tragic outcome.

The cage is a protective enclosure whose function is to keep the tiger inside and visitors outside. To normal, aware individuals, the hazardous condition is evident and receives the appropriate level of respect. However, we continue to see reports in the newspaper about individuals who fail to accord this hazardous condition the respect due. They receive mauling injuries from attempting to feed the animals, as well as from disrespecting protective (for humans) and isolating (for animals) moats and trenches.

During normal servicing of the animal's needs, the keeper must enter the tiger's cage to clean the floor and to provide fresh bedding. Entry is impossible unless the tiger is not present or is otherwise incapacitated by tranquilizer dart. Normally this means luring the animal through a doorway into a holding area. The doorway is closed securely and locked. So long as the tiger remains in the holding area, the cage is safe to enter and occupy.

There is no level of protective equipment that would enable the keeper to coexist in the cage with the tiger. The power inherent in the strength, speed, and agility of the tiger and sharpness of its teeth would rapidly overpower any personal protective device normally available. (Divers who photograph sharks routinely express concern about the power of these animals and their potential ability to overcome the protection afforded by the cage within which they work.)

The only breakdown in the procedure in the zoo would be the ability of the tiger to break down the door or for the keeper to fail to close the door or for someone to open the door while the keeper was in the cage. The latter could occur accidentally or through malicious intent. The potential for breakdown of the door is controllable through engineering design and choice of materials. The atmosphere in the cage may be corrosive to the metals used in hinges, screws, and locks. This corrosion could lead to eventual failure. This situation is addressable through maintenance. Control of opening and closing and locking the door is addressable through procedures and training.

Much of industry and industrial equipment is concerned with the storage, transfer and transport, and use or conversion of substances and commodities. The common denominator in all of these activities and processes is energy. Energy commonly encountered and utilized in industry includes potential (deformational, gravitational, and chemical), kinetic (linear and rotational), electromagnetic, and thermal. Energy cannot be created or destroyed, but can be converted from one form to another. While the outcome of an industrial endeavor involves a tangible item or substance, every action or process involves energy.

Containment of energetic processes and equipment is an important function of many industrial structures. The boundary surfaces prevent contact with equipment that has moving parts or other sources of hazardous energy. A variation on this theme is a space whose boundary surfaces confine materials, as well as the moving parts or energy source. This is the case with a mixer containing liquid or solid materials.

These situations are much like the tiger in the cage. That is, the energy sources have no respect for the presence of human occupants. Under no conditions can humans coexist in safety with the hazardous conditions created by aggressive and unpredictable energy sources. For this reason, entry into these structures before the energy is discharged and maintained at nonhazardous levels represents a pre-accident condition. However, under appropriate conditions, occupancy can occur without undue risk. The remainder of the time occupancy is highly hazardous. The key to eliminating or at least controlling hazardous conditions in these situations (the tiger's cage and the examples above) is to create a system for the systematic analysis of energy flow.

Even today there are situations in which workers enter and work in confined spaces containing "tigers", albeit a sleeping one perhaps. One example is the silo containing bridged or hung-up solid materials. The task may be to break up the bridge or to remove the adherent material. A parallel example is the unshored trench. Sudden release of gravitational energy leads to collapse. Shoring and the accompanying requirements to use it form the basis to address this problem in the latter circumstance.

Unexpected and rapid release or conversion of energy is a major theme in the accident situations described in Chapter 2. A typical example is the slumping of bulk materials. In almost an instant the material moves from an unstable configuration to a stable one. In some situations, the hazardous condition developed prior to entry and the start of work. In other situations, the hazardous condition resulted from the work activity. In either case, transformation of energy occurred during work activity. Victims experienced traumatic injury from the rapid release or conversion of energy. This occurred regardless of whether the accident resulted from engulfment, entanglement, entrapment, or mechanical, process, or other safety hazards.

Accident summaries in the OSHA, MSHA, and NIOSH reports repeatedly demonstrated the importance of unpredictability as a factor in these accidents (NIOSH 1979, 1994; OSHA 1982a, 1982b, 1983, 1985, 1988, 1990; MSHA 1988). What was predictable and familiar through many repetitions and iterations, suddenly and inexplicably became unpredictable. Following the accident, conditions returned to "normal." Addressing the unpredictable causes of these accidents in a proactive manner requires a systematic approach.

STANDARDS AND GUIDELINES

In 1982 the American National Standards Institute (ANSI) published a standard (ANSI Z244.1-1982) on lockout/tagout of hazardous energy sources (ANSI 1982). This standard recently was reaffirmed (ANSI 1992). ANSI Z244.1-R1992 outlines minimum requirements for enabling the safe conduct of work involving hazardous energy sources. These requirements include policies covering training, communications, planning and implementation of procedures for hazard assessment and control, and work activity. Hazard assessment includes identification and assessment of energy sources. Hazard control includes identification of energy-isolating devices and protective materials and hardware, procedures for de-energizing and start-up, and verification of control. ANSI Z244.1-1982 also includes special considerations about shift overlap, group lockout, authorization for application, and removal of locks and tags and other issues.

In 1983, NIOSH published guidelines for controlling hazardous energy sources during maintenance and servicing (NIOSH 1983). These were developed using methodology from systems analysis. The document provided results from a worldwide search of the literature on maintenance hazards and control techniques and review of 300 accidents. NIOSH attributed cause of these accidents to:

- Failure to control hazardous energy
- Inadequate energy isolation
- Failure to dissipate residual energy
- Accidental reactivation of hazardous energy

NIOSH argued that accidents having these identifiable causes are preventable, provided that effective energy control techniques or procedures are employed.

According to NIOSH, control can occur through de-energization or when sources of hazardous energy are present. In both cases, identification of hazardous energy sources must occur as a first step. During de-energization, each source of hazardous energy must be controlled and secured and tested to verify that control has occurred. Control techniques include isolation, blockage, and dissipation. Securement means that accidental or deliberate reenergization cannot occur. Verification through testing is necessary to prove that energy flow cannot occur and to confirm the effectiveness of control techniques. Documentation of procedures must occur. Personnel who design and implement these measures must obtain qualifications through education, training, and experience.

When sources of hazardous energy are present, control is considerably more complex. This process requires identification of energy sources and documented procedures for establishing control. Verification to establish effectiveness of control measures is essential. Personnel who design and implement these measures must obtain qualifications through education, training, and experience.

In 1989, OSHA enacted a Standard (29 CFR 1910.147) on control of hazardous energy sources (OSHA 1989). The scope of this Standard is the servicing and maintenance of machines and equipment in which unexpected energization or start-up or release of stored energy could cause injury. Sources of energy covered in the Standard include electrical, mechanical, hydraulic, pneumatic, chemical, thermal, or other energy. The Standard uses the term energy isolating device to describe a mechanical device that physically prevents the transmission or release of energy. Energy isolating device include, but are not limited to (Office of General Industry Compliance Assistance 1991):

- A manually operated electrical circuit breaker.
- A disconnect switch.
- A manually operated switch by which the conductors of a circuit can be disconnected from all ungrounded supply connectors. No pole can be operated independently.
- A line valve.
- A block.
- Any similar device used to block or isolate energy.

The preamble to the OSHA standard provides useful insight into lockout and tagout of hazardous energy sources and the regulatory process. OSHA believes that failure to control hazardous energy sources accounts for nearly 10% of serious accidents and 7% of fatal injuries in general industry. Shutting down a machine or equipment is only a partial solution to the problem. Unanticipated movement of a component or of the material being handled can occur due to the release or conversion of energy. OSHA believes that the most effective method to prevent injury from these sources is to dissipate or minimize remaining potential energy or to utilize restraining devices to prevent movement. Preventing inadvertent reactivation of deactivated and de-energized equipment also is essential. OSHA also provided a summary of results from several reports on lockout-related accidents. OSHA concluded that the major factors in lockout-related accidents are:

- Servicing equipment while operating
- Failure to ensure power off
- Inadvertent activation
- Lack of procedure
- Lack of training

An important point made by OSHA and commenters to the Standard is that locks and tags themselves do not control hazardous energy. Control is achieved by isolation of equipment from the energy source and following procedures for deactivation and de-energization. Attachment of locks and tags should occur only after control is achieved.

The monograph by Grund (1995) on lockout and tagout provides additional detail about regulations and standards on this subject.

DEACTIVATION AND DE-ENERGIZATION

Previous discussion has indicated that the sudden release of energy is the cause of many accidents involving nonatmospheric hazards. In order to create safe conditions of work around equipment and structures in which energy storage, transfer, or conversion occurs, deactivation and de-energization must occur. Deactivation and de-energization possibly can occur through disconnection of the source of energy/power. However, experience has shown that obtaining verifiable and predictable deactivation and de-energization of equipment can be considerably more complicated than simple disconnection from the source of energy/power.

The logical starting point for determining how to deactivate and de-energize a system is to ascertain how energy is stored, utilized, and transformed in equipment and how this is controlled. This process begins at energy inputs and ends at sinks and discharges. Between these end points, storage, transfer, and transformation can occur.

In the zoo example in the Introduction, as in many other real-world situations, the hazardous condition is apparent and so also is the strategy for elimination and control. Locating the tiger in a separate room behind a door that can withstand any level of physical attack is an obvious strategy. Once behind the door, the tiger cannot re-enter the cage until permitted to do so by the keeper. This approach is satisfactory, so long as the door remains closed in some predictable and reliable manner.

In many other real-world situations, hazardous conditions and strategies for their elimination or control are not so obvious. Many machines that have enterable spaces receive inputs from utilities: electricity, water, steam, natural or fuel gas, compressed air, vacuum and possibly gases, such as hydrogen, helium, argon, carbon dioxide, nitrogen and oxygen. Process units can have any of these, as well as supply and drain lines and materials storage. It is not unusual for design drawings to be unavailable for equipment, or even when they are, to provide little assistance in hazard assessment and control. Essentially what maintenance personnel are facing in these circumstances is a black box that has some recognizable inputs and outputs. This lack of information originates in design.

Even today, current textbooks do not recognize the ramifications of design decisions on servicing. The text by Juvinall and Marshek (1991), for example, refers to safety primarily in product design. The concepts of deactivating, de-energizing, isolating, and lockout for servicing are not mentioned.

The manufacturer is the logical resource for determining energy flow within equipment. The manufacturer has designed the equipment to perform according to some specification. This should include calculation of forces, or at the very least, trial-and-error determination of performance of subsystems that contribute to output from the entire unit. The manufacturer also knows precautionary measures needed for manipulating the components during manufacture.

Some manufacturers have addressed the question of disassembly of their equipment for servicing. They indicate points for securement and procedures for de-energization, isolation, and lockout. The manufacturer may be unable to provide information for several reasons. The focus of manufacturers is to create equipment that performs an intended function at a stated level of performance. There is no emphasis on design to enable dismantlement in order to gain access for service. Many manufacturers lack the resources or the expertise to perform this type of analysis. All but the largest manufacturers buy off-the-shelf components produced by other companies and assemble these into specialized configurations. Also, the manufacturer may no longer be in business or manufacturing the equipment in question.

The ultimate responsibility for resolving this issue rests with the owner and operator of the equipment. Depending on the complexity of the equipment, resolving this issue may require input from a group representing different backgrounds. Potential participants include safety and industrial hygiene, and engineering, operations and maintenance personnel who work with the equipment. Collectively this group provides the best functional expertise for addressing this problem. Gaps in representation could hinder the functioning of this group.

Safety and industrial hygiene professionals provide generalized knowledge about the anticipation, recognition, evaluation, and control of hazardous conditions. This includes the requirements and implications or regulations and standards.

Engineering, operations, and maintenance personnel from the department that "owns" the equipment know what it does, how it functions, and problems that have occurred in its operations. They have the knowledge about electrical, hydraulic and pneumatic, mechanical, utility and process and control systems. Knowledge about control systems and feedback circuits is especially important where integrated systems and processes are involved. An action that occurs in one part of a machine or one process unit in an integrated chemical plant can instigate unpredicted and unanticipated effects in another. This knowledge provides the basis for discussion about energy flow and possible sources of stored energy. One formalized technique for doing this is energy barrier trace analysis (Wiggins 1997).

Engineering, operations and maintenance personnel from the department that "owns" the equipment also are the most likely to enter the space to inspect or to perform work. Maintenance personnel who are tradespeople have hands-on knowledge about equipment inside the confined space and also about energy sources and control equipment (electrical breakers, for example) that are external to the space.

Once the flow and control of flow of energy within equipment are understood, the next step is to determine how to deactivate and de-energize in a predictable, reliable, and verifiable manner. Determining the answer to this question could require experimentation with systems and subsystems within the equipment. These studies may indicate that while complete deactivation can occur, complete de-energization may not be possible. If this is the case, prevention of release, translation, and conversion of residual stored energy must occur. To accomplish this could require immobilization of movable parts in a position from which damage cannot occur, or devising novel means to enable release of stored energy.

There is a need to express these concepts in diagrammatic form. There is an especial need to integrate pneumatic, hydraulic, electrical, mechanical, utility, and process components. Models can provide a means of describing the input, conversion, storage, transfer and transport, and release of energy, as well as describing the action of systems that activate, control and deactivate these systems. Models also can show how an action in one system can cause a seemingly unrelated and unpredicted action to occur in another.

The network model provides a graphic representation of paths or routes by which the preceding processes can occur (Watts 1991). Links (also called arcs or branches) join nodes (actions, events, or outcomes). Links describe how one event leads to another. Often the link is the answer to a question posed about an intermediate event or outcome.

The simplest network contains only terminal nodes (initiating event and final outcome). More complex networks contain terminal and intermediate nodes. The tree is a network that contains branches. Branches outline alternate paths between initiating events and final outcomes. Branches usually reflect possible answers to questions posed to create the links.

Fault and event trees and cause–consequence diagrams provide the basis for *a priori* analysis of the events and actions that occur in equipment. These models are used to analyze all of the events and combinations of events that can lead to an undesired outcome, in this case the unexpected and uncontrolled release of energy by or within equipment (Lees 1980, National Mine Health and Safety Academy 1991).

The first step in the process involving the fault tree is to identify the undesired event or outcome. The next is to regress from the event to determine how this could happen. This regression identifies primary causes and determines how they interact to produce the undesired event. The same question, asked for each of the primary causes, leads to the events that led to them. The process finishes when all primary causes and the events leading to them are identified and linked.

Whereas the approach taken with the fault tree is to begin with the undesirable event and work backwards, the approach with the event tree is to start with initiating events and move forward. An event tree begins with the definition of the primary event and considers consequences and paths that emanate from this point. Each of the secondary events has a path for success and a path for failure. The event tree follows only one path.

The cause–consequence diagram is a hybrid of the other approaches. The cause–consequence diagram starts with a critical event, usually a fault or deviation. Development occurs in both directions to obtain both the causal events and outcomes and logical relations between them.

Logic gates link events at individual levels in the tree. These combinations produce events at the next higher level. That is, the logic gate indicates the relationship between primary causes that produce the undesired event. The "and" and "or" gates are the most commonly used logic gates. For the "and" gate to apply, all events connected through it must exist or occur simultaneously. For the "or" gate to apply, the event will occur if one or any combination of events exists. Events include output events, normal, and independent events, as well as undeveloped events. Output events include the final undesired event, as well as those intermediate within the tree that are analyzed further. The output of one intermediate event becomes the input to the next. Normal events occur naturally within system operation. As such, they do not require further analysis. Independent events do not depend on other parts of the system for their occurrence and represent a clearly defined failure of some component. An undeveloped event is not totally developed due to lack of information or significance.

Following an understanding of the system obtained from observation and other sources of information, the next step is to construct the appropriate tree using the concepts discussed here. Steps following this involve evaluating the likelihood of events in the tree and to controlling the hazards identified.

ISOLATION AND LOCKOUT

Isolation and lockout are essential activities prior to entry and the performance of work in many types of confined spaces. Isolation is the severing of pathways or connections between sources of energy (external or internal) or supply or drain lines or utilities and the space. The OSHA Standard on hazardous energy sources (1989) considers isolation to be prevention of transmission or release of energy. The OSHA Standard on confined spaces in general industry (OSHA 1993) considers isolation to be the process by which a permit space is removed from service and completely protected against the release of energy and material into the space. According to OSHA this occurs by such means as blanking or blinding; misaligning or removing sections of lines, pipes, or ducts; a double block and bleed system; lockout or tagout of all sources of energy; or blocking or disconnecting all mechanical linkages. Isolation precedes lockout.

Lockout is the placement of a lockout device on an energy-isolating device. Lockout cannot occur without an established, verified procedure that ensures that the energy-isolating device and the equipment being controlled cannot be operated until the lockout device is removed. The energy-isolating device is a mechanical device that physically prevents the transmission or release of energy (OSHA 1989). Another view of lockout is the mechanical interruption and neutralization to a zero energy state of all energy sources leading to and within a piece of equipment or machinery (UAW-GM National Joint Committee on Health and Safety 1985).

The concept that has evolved over the years is a unique lock for each person affected by the lockout. The lock should identify the owner through a photograph or substantially attached tag (Figure G.1).

The lock should be a key lock, rather than a combination type. Lockout devices, such as chains and scissors, should be case-hardened steel so that they cannot be easily removed (Figure G.2). Some scissor-type devices are made from white metal that is easily broken (UAW-GM National Joint Committee on Health and Safety 1985). Considerable development of lockout devices has occurred during recent years (Figure G.3).

An additional facet of the lockout paradigm is that no one should have a key to the personal lock used by another individual. This includes supervisory personnel.

Isolation and lockout are important practices for controlling hazardous conditions. In the U.S., the OSHA Standard on confined spaces in general industry (OSHA 1993) provides a mechanism for reclassification from permit-required to non-permit-required. Reclassification can occur following elimination of all serious safety and health hazards. This then begs the question whether successful isolation and lockout of all hazardous energy sources enables reclassification (provided that no other serious safety or health hazard exists). To reiterate, lockout is the placement of a lockout device on an energy isolating device. Isolating and locking out a hazardous energy source

(A)

(B)

(C)

FIGURE G.1 Personalization. (A) Photographic identification. This personalized lock contains a photograph of the owner. (Courtesy of Idesco Corporation, New York, NY.) (B, C) Color-coded identity tags are affixed to the lock. Locks of different shape, size and texture can provide additional cues about identity. (Courtesy of Panduit Corporation, Tinley Park, IL.)

prevents its activation. This approach is successful, so long as everyone respects the sanctity of lockout device. However, all that is required to reactivate the hazard is to defeat the lockout device and to re-engage the isolating device. This situation has occurred.

According to OSHA, isolation and locking out using their suggested methods completely protect against the release of energy and material into the permit space. The resolution of this question rests with whether complete protection is the same as or equivalent to elimination of the hazard. While this question may reflect semantics, it is underlain by an important issue of risk assessment. Any safety measure, no matter how stringent, is defeatable by a determined individual. An obvious difference in safety measures is the extent of difficulty required to defeat them. Another matter that deserves consideration is reliability of isolating and lockout devices. A third matter is the risk of instituting high level precautionary measures. Any measure involving alteration of the status quo exposes individuals to potentially hazardous conditions. The real cost of implementing higher-level

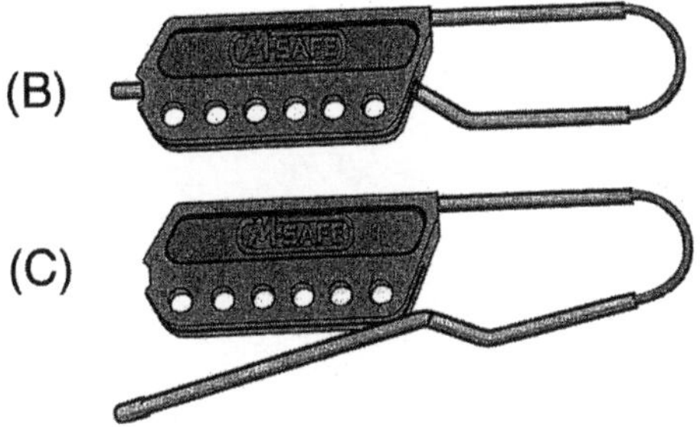

FIGURE G.2 Attachment hardware. (A) Hasps: scissor and fold-over type. (B, C) bend-in type. (D) Straight and angled padlock eyes. (E) Chain. Materials of construction should be case-hardened steel or other suitable material to prevent tampering and easy removal. (A, D, E: Courtesy of Panduit Corporation, Tinley Park, IL; B, C: courtesy of Arkon, Inc., Ville d'Anjou, QC.)

precautionary measures for one group of workers can outweigh the perceived benefit received by another

Organizations affected by the need to lock out hazardous energy sources should prepare a complete written lockout program. This recommendation regularly occurs in lockout/tagout standards, regulations, and codes of practice. Statutes, such as the U.S. Lockout/Tagout Standard (OSHA 1989) require annual training on specific methods and procedures involved with locking out hazardous energy sources.

The order of the following discussion of energy sources reflects the hierarchy suggested by the UAW-GM National Joint Committee on Health and Safety (1985). Each section applies to a simple system containing a single energy source, as well as the hierarchy.

Mechanical Systems

Mechanical devices change the magnitude, direction or intensity of forces, or the speed resulting from them. Mechanical devices can be as simple as the screwdriver, wrench or hammer, and as complex as a passenger aircraft. In the industrial context, a machine usually is an assembly of subsystems, each of which can contain simpler subsystems composed of simple mechanical devices. Simple mechanical devices and the subsystems found within industrial equipment can store, trans-

(D)

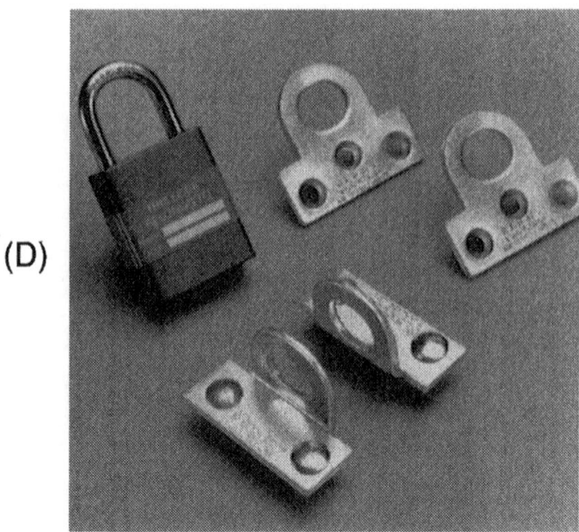

(E)

FIGURE G.2 (continued)

form and transfer energy. Mechanical action occurs through the motion of levers in straight-line directions and rotating parts (Olivo and Olivo 1984).

The sole function of enclosures (actually large guard structures) around many types of equipment is to prevent entry and detrimental contact. That is, by the presence of the enclosure, the manufacturer or installer is stating implicitly that entry for cleaning, servicing, modification, or other purpose cannot occur safely while the equipment is energized and operational. An electrical isolation and lockout can achieve the goal of preventing start-up and operation of equipment, such as a motor, that transfers energy to mechanical subsystems. The same is true for a fluid power system that operates within a machine. However, these lockouts do not ensure the existence of a zero-energy state within the equipment. Deactivation, de-energization and isolation of mechanical subsystems within the machine can require the immobilization of freely moving parts following run-down or those that store gravitational energy, or removal of tension in springs. These requirements have evolved through the unfortunate experience of many individuals.

Momentum is the tendency of a moving body to remain in motion. In mechanical subsystems, momentum is most likely in oscillating and rotating components. Momentum should cease soon after deactivation of external energy sources, unless an internal source of stored energy is present. Parts capable of free motion during operation of a machine are potentially capable of movement following run-down. This applies to lever subsystems and rotating equipment. Lever subsystems can amplify minor movement into major motion in large parts of the machine. Motion in large parts having considerable mass can strike or entrap. Freely rotating parts can strike or fail to support body weight through misplaced footing or other action that presumes immobility. These subsystems require immobilization to prevent rotational and translational motion.

(A)

(B)

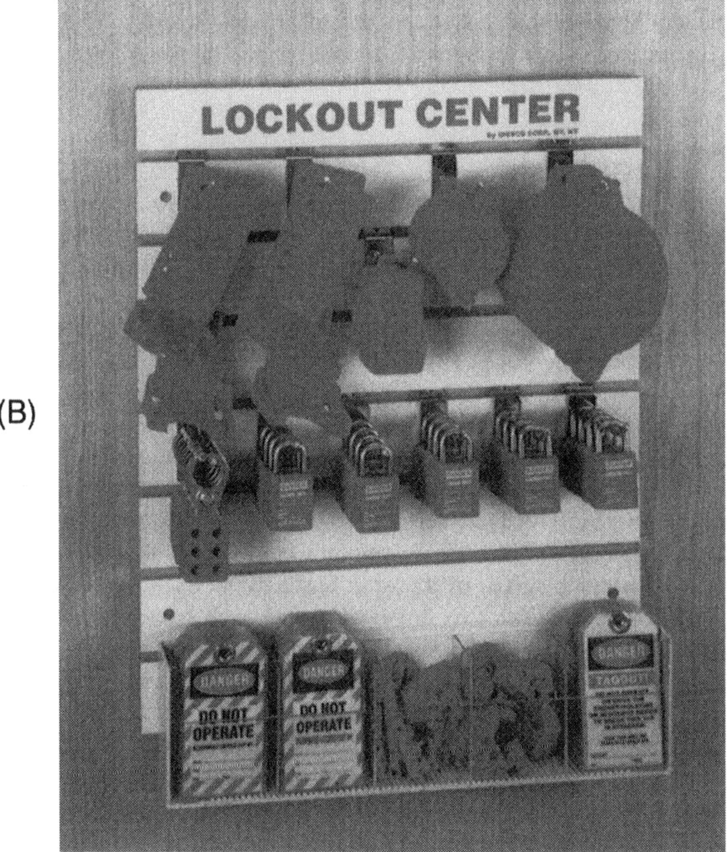

FIGURE G.3 (A, B) Lockout organizers. Lockout organizers can provide a detailed display of status of use of personalized and shared locks. (A: Courtesy of Panduit Corporation, Tinley Park, IL; B: courtesy of Idesco Corporation, New York, NY.)

Other mechanical subsystems store gravitational energy. Gravity is the most widespread mechanism of energy storage in machines. Parts in machines that move as part of normal operation, such as levers, knives, hammers, and many others, can stop in a position from which downward motion can occur under appropriate conditions. As well, loss of anchorage during dismantlement can lead to pendular motion of many types of functional and structural parts of machines. This can release suddenly or gradually, depending on the circumstance. Sudden conversion of gravitational to kinetic energy leading to unexpected motion and the potential for injury has occurred on numerous occasions during adjustment while workers were within the confines of equipment (NIOSH 1983).

Springs are another type of energy storage device. Springs are elastic materials. They exert forces or torques, and absorb and store energy which is released later. Release of energy stored in a spring can occur rapidly or slowly. Rapid release of energy in conjunction with the compounding action of lever systems, can produce unexpected movement in machine parts. Release of gravitational energy stored in movable parts and compressional, tensional or torsional energy stored in springs requires in-depth knowledge of the operation of the equipment. Maintaining the released member in a neutral position could require securement by means of blocks and clamps (UAW-GM National Joint Committee on Health and Safety, 1985).

Blocks are metal structures that are inserted in the path of moving parts (Figure G.4). The function of blocks is to prevent drift in machine parts. Drift can occur in the horizontal, as well as the vertical direction. Some blocking systems also incorporate wedges. Wedges fitted into place with the block prevent even slight movement. Blocks are not intended to stop the impact from a cycling machine.

Electrical Systems

Electrical equipment situated outside the confined space (for example, stirrer motors) will require deactivation, de-energization, isolation, and locking out. This equipment is less obvious than pumps and pump motors located inside. Similarly, these requirements affect electrical circuits in equipment located inside and outside the space.

Lockout devices prevent closure of the circuit from the open (nonenergized) position to closed without removing the lock (Figures G.5 through G.11). Pulling the fuse and locking closed the panel and pulling the disconnect and locking it in the open position are other possibilities.

Electrical circuits are complex and may be multi-phasic. Phase refers to the time-behavior of the sinusoidal wave of electrical energy flowing in a circuit (Nelson 1983). Single-phase service is used for lighting and small motors. Three phase service is utilized where heavy motor loads exist. Three phase service involves three separate conductors and sinusoidal waves, each 120° out of phase. Provision of energy in this manner smooths out the irregularities in flow. Three phase current provides three separate surges of power in half a cycle, each separated by one sixth of a cycle. Single phase current provides two surges of power, each separated by half a cycle. Three phase induction motors are reversible. Reversal of direction occurs through interchanging any two of the three supply leads. For this reason, care when reconnecting supply leads to a three-phase motor is essential to ensure that reversal has not occurred.

An additional necessity with complex circuits is to trace the path to the power (energy) source(s). This activity should ascertain whether shutting down part of the system will eliminate the hazard, as desired, or create an additional one. Some circuits contain uninterruptable power (energy) sources. These include stand-by generators, circuits that receive feed from more than one supply, and storage batteries. Surety of supply in some installations is so critical that they receive electrical feed from two separate generating stations. Following normal deactivation, deactivation and lockout procedures in these situations provides no assurance that the circuit cannot reenergize. Investigative diligence is required at all times.

Control systems pose additional complications in electrical circuits. Some systems contain multiple start/stop controls. Computer-controlled systems and remotely located, radio-controlled

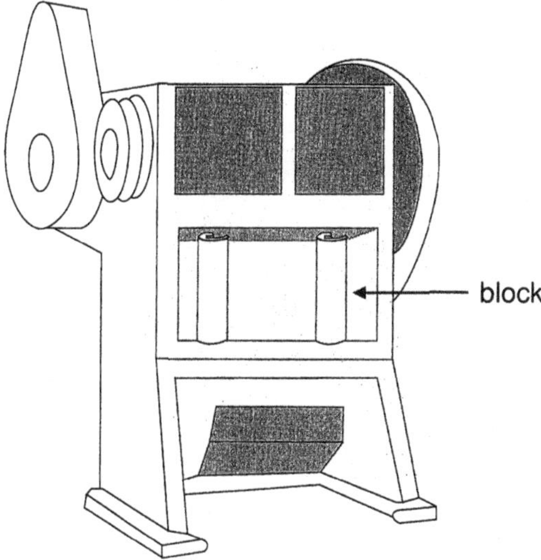

FIGURE G.4 Blocking device. The blocking device is used to stop machine motion due to gravity. The block is only effective when the machine is not operational. (Adapted from National Safety Council 1993.)

FIGURE G.5 Electrical lockout. In this representation of an electrical lockout, each individual puts a lock through the holes in the scissor device (if used). A minimum of one lock prevents the switch from being activated. (Courtesy of Panduit Corporation, Tinley Park, IL.)

systems may be present and not readily apparent. The potential for problems from these systems can be especially difficult to recognize because the low voltages involved can activate high voltage equipment. Activation of computer- and radio-controlled circuits can occur without warning.

The action of deactivating, de-energizing, isolating and locking out electrical systems can create hazards. In some circumstances, dust from the ambient atmosphere has entered and accumulated in electrical disconnects. Shutting down a circuit under load by tripping the disconnect has caused dust explosions because of arcing inside the electrical box. For this reason, deactivation of electrical equipment should occur through the local ON/OFF control switch. The next step is to shut off the disconnect/breaker. To avoid injury from a possible explosion, the person shutting off the disconnect should face away from the panel (UAW-GM National Joint Committee on Health and Safety, 1985).

(A)

(B)

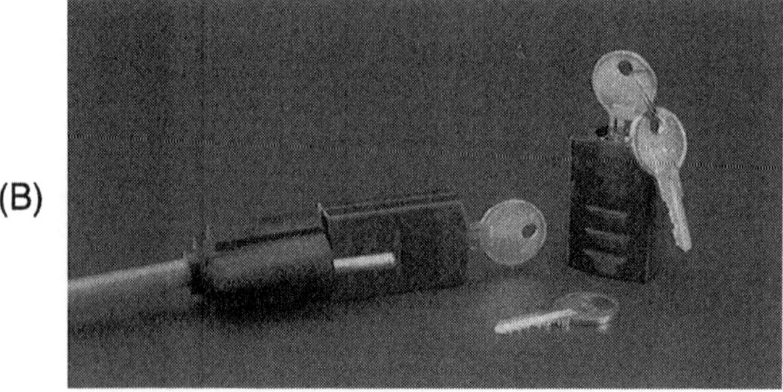

(C)

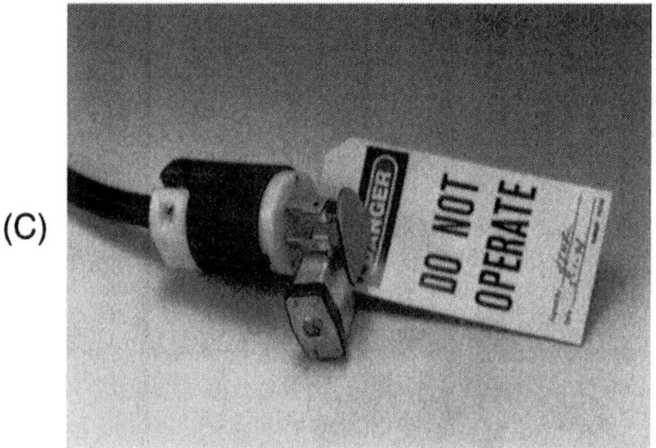

FIGURE G.6 Plug shields. The shackle of the lock fits through the eye of the pin, which fits through the holes in the prongs of the plug. The shield prevents the plug from being pushed into the receptacle. (A, B) 110 Volt domestic type. (A: Courtesy of Panduit Corporation, Tinley Park, IL; B: courtesy of W.H. Brady, Richmond Hill, ON.) (C) Industrial type. (Courtesy of Panduit Corporation, Tinley Park, IL.)

(A)

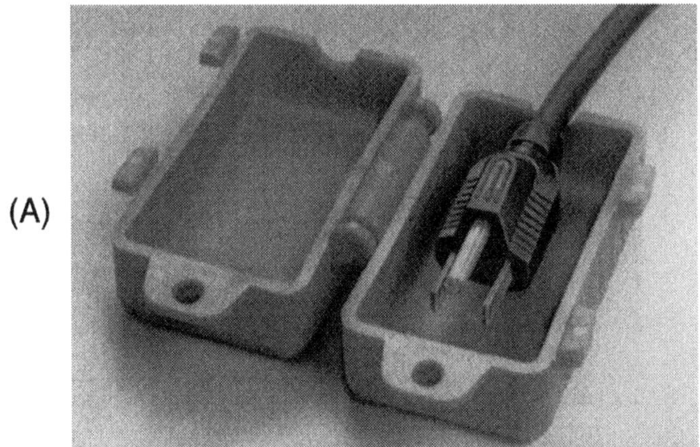

(B)

FIGURE G.7 Plug enclosures (A–D). Depending on design, plug boxes can accommodate plugs of various sizes. (A: Courtesy of Panduit Corporation, Tinley Park, IL; B–D: courtesy of W.H. Brady, Richmond Hill, ON.)

Testing the circuit downstream from the lockout is absolutely necessary in order to ensure that deactivation has occurred. Testing should involve activating ON/OFF controls, and as appropriate, use of diagnostic equipment. Deactivation and de-energization must provide an absolute guarantee of results. Too many accidents have occurred because of disconnects that failed in the ON position, breakers that were mislabeled, or unknown redundancy in the electrical supply. Input from an individual with broad knowledge of electrical control equipment during development of lockout procedures, therefore, is absolutely essential. Many organizations permit only journeymen electricians to perform electrical lockout for this reason.

Fluid Power (Pneumatic, Hydraulic, and Vacuum) Systems

Fluid power systems include hydraulic, pneumatic, and vacuum systems. Isolation and lockout of these systems must occur as part of preparation for entry. Pneumatic systems sometimes are difficult to distinguish from hydraulic systems. Distinguishing components of pneumatic systems include air filters, regulators, lubricators, and noise silencers/mufflers. Distinguishing components of hydraulic systems include flow control, electric motor and hydraulic pump, cylinders, and fluid reservoir. Piping and fittings in pneumatic and hydraulic systems are similar. However, hydraulic systems may contain steel-braided hose and extra heavy pipe and fittings (UAW-GM National Committee on Health and Safety 1985).

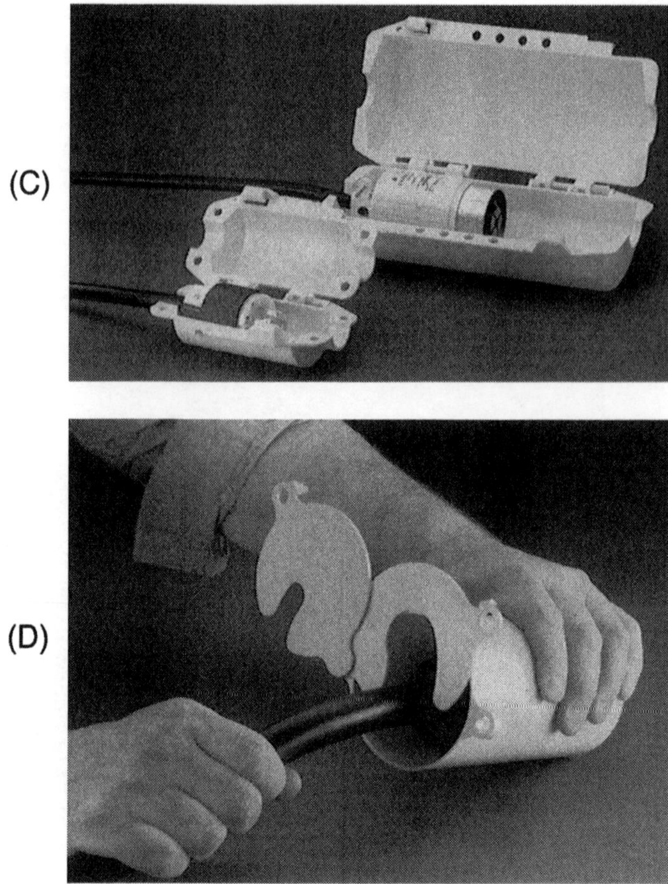

FIGURE G.7 (continued)

Tracing lines back to their source and examining component parts will assist in this process. The investigator should be aware that backup systems could exist in a location remote from the main unit. Also, this investigation should determine whether a hazard could develop as a result of shutting down the supply. For example, pressure in the system could be essential for supporting some component. Support from a block could be insufficient to maintain safe positioning. Leaving the system energized could be essential. If this is the case, safety of the work could be compromised by failure of the supply system. Decisions about how to proceed in this context require input from knowledgeable individuals.

Lockout of pneumatic systems can occur at shut-off valves (Figure G.12). Testing to determine the effectiveness of the lockout is essential. Tests should include attempts to move valve levers and stems. Leakage through valves locked out by loosely fitting devices could compromise the safety provided by the isolation. Even slight leakage could produce significant consequences. Some safety valves used for isolation bleed pneumatic pressure from the downstream side.

Lockout of hydraulic systems, most of which are driven by electric motor, should occur at the hydraulic pump (UAW-GM National Joint Committee on Health and Safety 1985, Vickers 1992). Lockout in this manner could necessitate shutdown of additional equipment powered by the same pump. Lowering or mechanically securing all suspended loads should occur first. The next step is to exhaust pressure trapped in the system, including the accumulators. A bleeder valve may be present to drain hydraulic fluid from the accumulator and system. Some valves have manual override.

FIGURE G.8 Wall switch lockout. Both designs immobilize the cover plate. (A, B) A setscrew tightens the device onto the toggle of the switch. The shackle of the lock covers the setscrew and prevents access. (Courtesy of Panduit Corporation, Tinley Park, IL.) (C, D) The lockout device is held in position by screws that connect into the light switch. Closing the cover and applying the lock prevents access. (Courtesy of W.H. Brady, Richmond Hill, ON.)

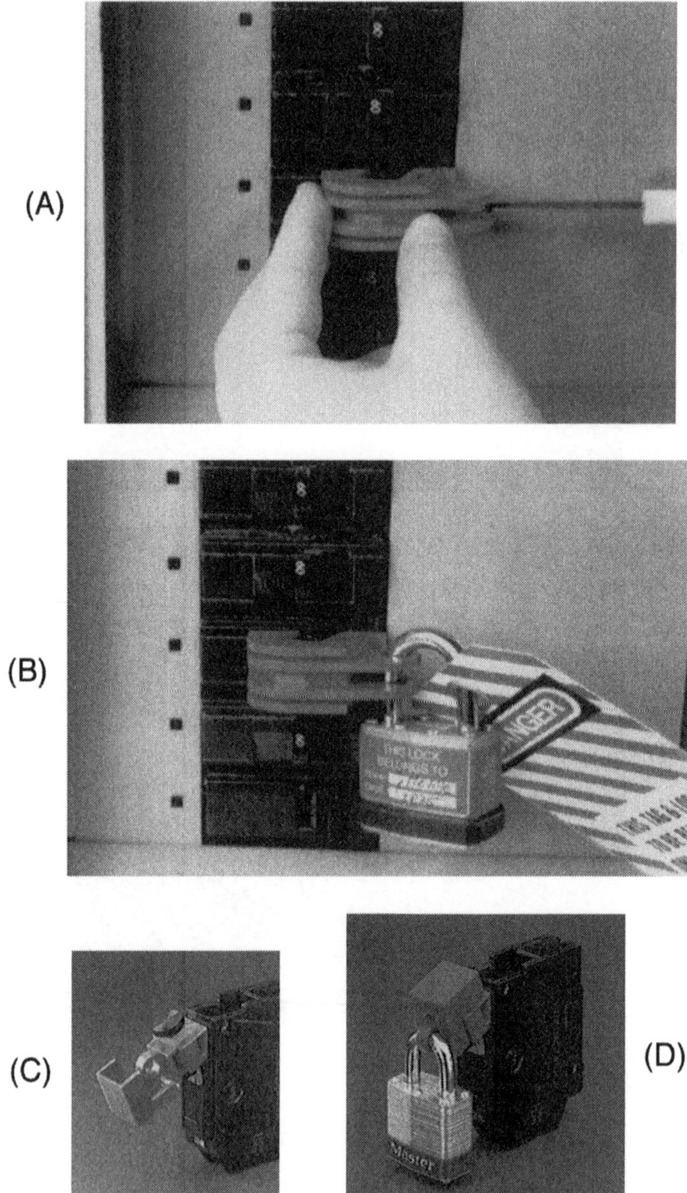

FIGURE G.9 Circuit breaker lockouts. (A, B) A setscrew tightens the device onto the toggle of the switch. The shackle of the lock covers the setscrew and prevents access. (Courtesy of Panduit Corporation, Tinley Park, IL.) (C-I) Snap on, snap off designs fit a variety of breaker designs. (Courtesy of W.H. Brady, Richmond Hill, ON.)

Fluid should be discharged from both ends of intensifiers. Bleeding fluid pressure could cause machine movement. This would indicate that the zero energy state has not been achieved. Blocking, pinning, and physical disconnection also may be required. Relief of fluid pressure also may require opening a fitting. Once the hydraulic system is deactivated and de-energized, the last steps are to isolate and lock out the electrical control system and power supply.

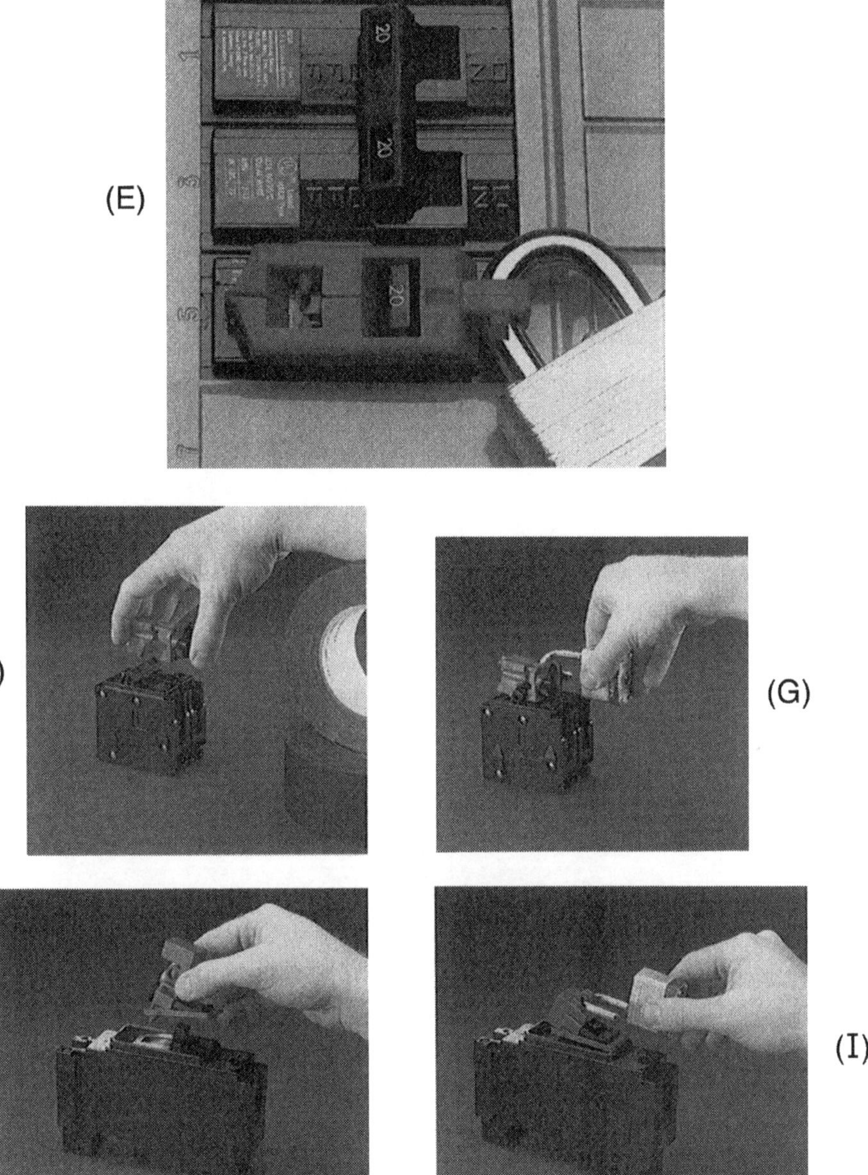

FIGURE G.9 (continued)

Process and Utility Systems

Among the most complex systems to isolate and lock out are the process and utility systems and utility feeds into machines. For the purpose of this discussion utility systems include systems that provide water, steam, natural or fuel gas, and possibly gases, such as hydrogen, helium or argon, carbon dioxide, nitrogen and oxygen. Process units can have any of these, as well as supply and drain lines and materials storage.

Germane to process and utility systems, the OSHA Standard on confined spaces in general industry (OSHA 1993) considers isolation to occur by such means as blanking or blinding; mis-

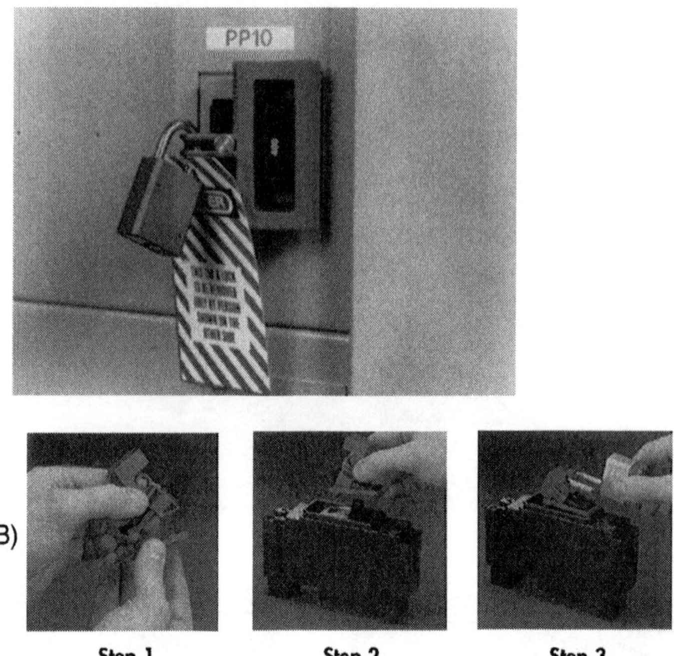

FIGURE G.10 Large circuit breaker lockout. (A) A setscrew tightens the device onto the toggle of the switch. The shackle of the lock covers the setscrew and prevents access. (Courtesy of Panduit Corporation, Tinley Park, IL.) (B) Snap on, snap off design. (Courtesy of W.H. Brady, Richmond Hill, ON.)

FIGURE G.11 Aircraft circuit breaker lockout. This device fits both 2.5 and 5 Volt Texas Instruments Klixon® circuit breakers. (Courtesy of W.H. Brady, Richmond Hill, ON.)

aligning or removing sections of lines, pipes, or ducts; or a double block and bleed system (Figures G.13 to G.18). The Standard considers isolation to be the process by which a permit space is removed from service and completely protected against the release of energy and material into the space. The OSHA Standard on control of hazardous energy sources (Office of General Industry Compliance Assistance 1991) includes line valves and similar devices as energy isolating devices.

Blanking (blinding) involves breaking open the line and inserting a solid plate or cap that completely covers the bore and is capable of withstanding the maximum pressure of the piping without leakage (OSHA 1993; Workers' Compensation Board 1998). A similar approach involves the use of caps and plugs on threaded pipe. The pipe must be depressurized, drained and possibly flushed prior to opening and exposing the contents to air. The piping may require additional support during this process.

Spectacle blanks resemble the wire frame of the front of a pair of eyeglasses. One side is completely closed, while the other is an open ring. Spectacle blanks remain in place in the pipe.

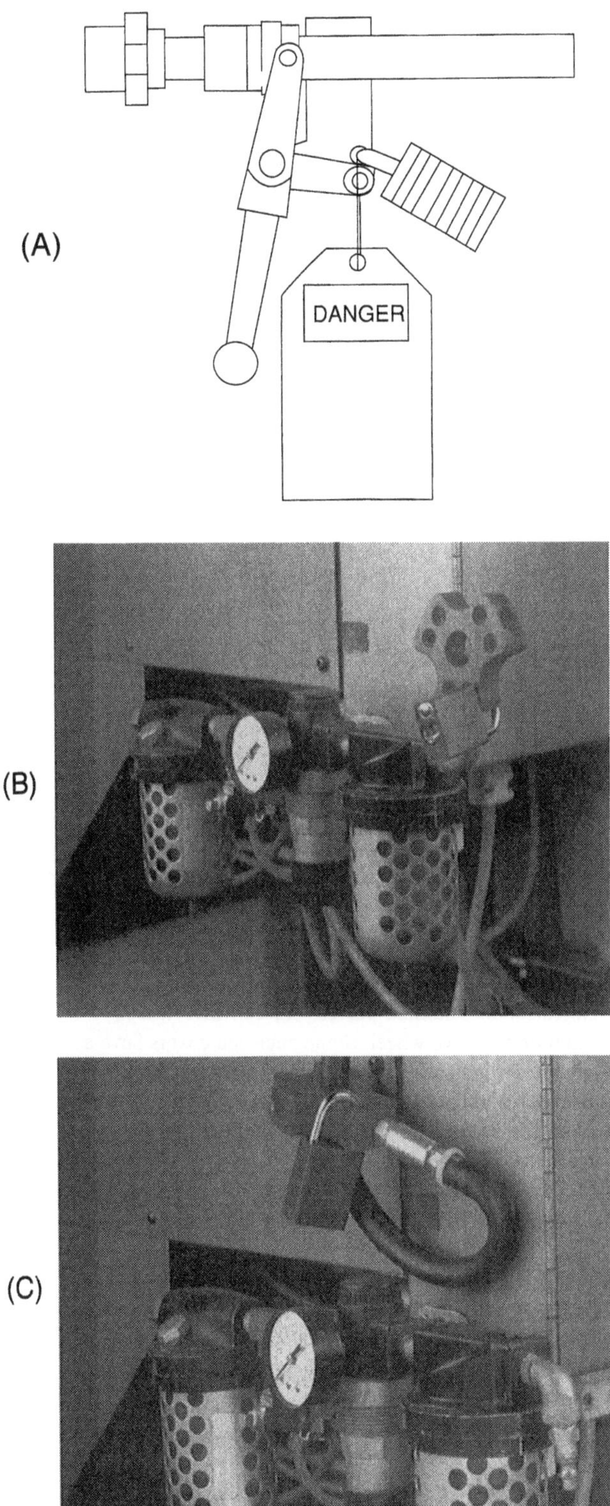

FIGURE G.12 Pneumatic valve lockout. (A) Many valves contain lockable fittings. (Adapted from UAW–GM National Joint Committee on Health and Safety 1985.) (B,C) Lockout devices, (Courtesy of W. H. Brady, Richmond Hill, ON.)

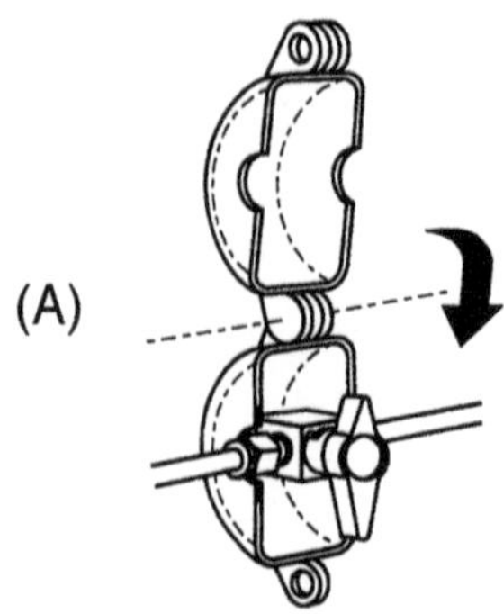

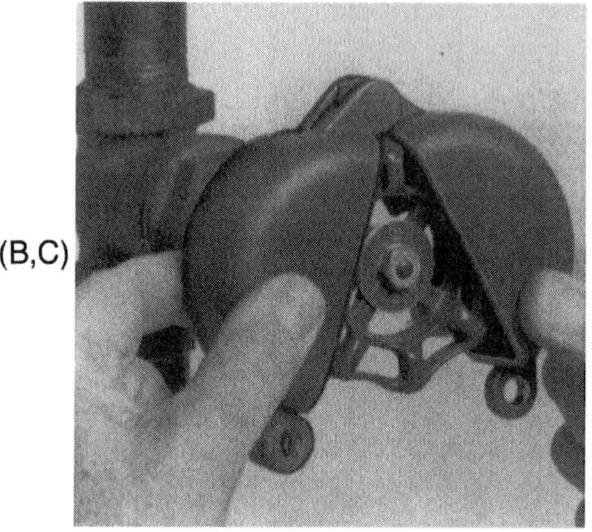

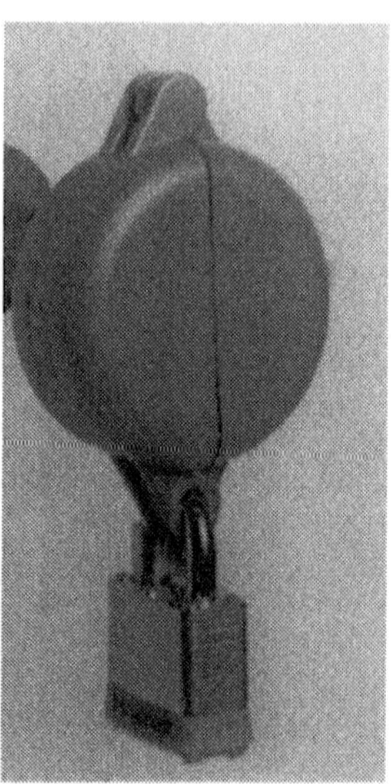

FIGURE G.13 Valve lockout devices. (A) In-line valve enclosure. This unit completely encloses the valve. (B-D) Valve-wheel covers for short-stemmed valves. (E, F) Valve-wheel covers for long-stemmed valves. Valve covers prevent rotation of the valve wheel. Some enclosed covers have a knockout to accommodate the long stem. Colors are available for indicating status of the valve or contents. (G, H) Ball valve lockouts. The body of the device slips over the stem and topside of the handle of the valve. The outer sheath slides over the body of the device and underside of the valve. (I-K) Universal ball valve lockout device. This device can be positioned in various orientations. In control applications, the device can lock a valve in the "on" and throttled positions. (A, D, E: Courtesy of Arkon, Inc., Ville d'Anjou, QC; B, C, G: courtesy of Panduit Corporation, Tinley Park, IL; F, H–K: courtesy of W.H. Brady, Richmond Hill, ON.)

The blank is rotated around a pivot and indicates by position the status of the line. Skillet (paddle) blanks have a handle that protrudes from the flange to indicate their presence in the line. The skillet (paddle) spacer is the open-ringed structure that corresponds to the open ring in the spectacle blank. The handle of the skillet (paddle) spacer has a hole to differentiate it from the skillet (paddle) blank.

Blanks, caps and plugs require clearly identifiable features to indicate their presence in lines. Otherwise, there is the possibility that they may be overlooked during reconnection in preparation for start-up.

Blanks are the subject of API (American Petroleum Institute) Standard 590 (API 1985). API 590 describes characteristics of steel line blanks intended for installation between flanges conforming to ASME/ANSI B16.5 dimensions. These characteristics include size, class, facing, style, plate material, inside diameter and lifting or handling device. ASME/ANSI B16.5-1988 describes dimen-

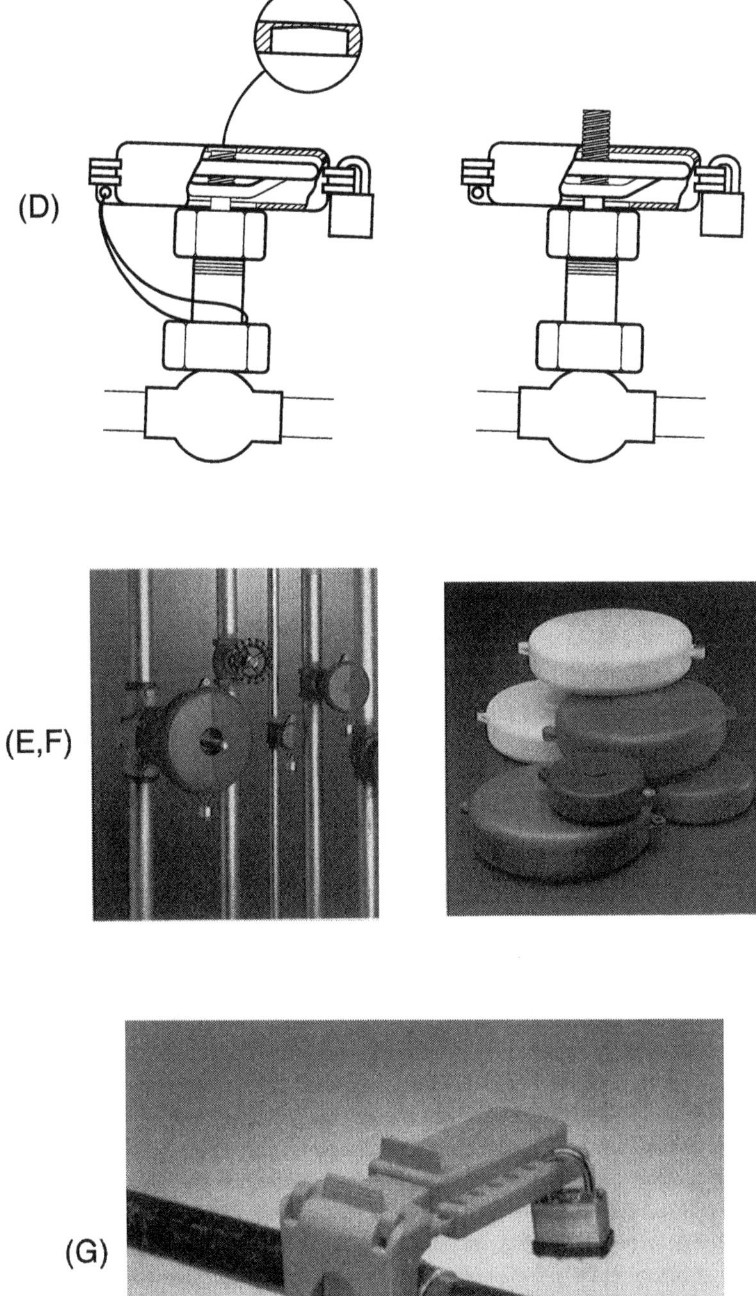

FIGURE G.13 (continued)

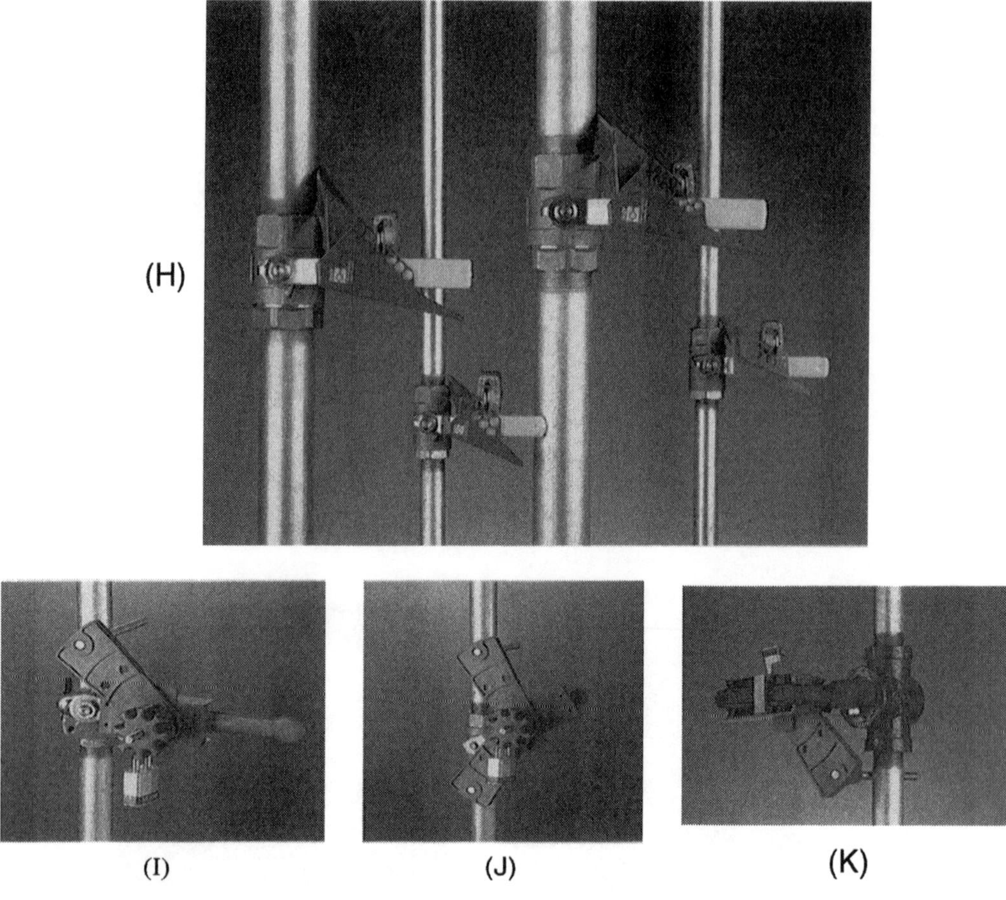

(H)

(I) (J) (K)

FIGURE G.13 (continued)

sions and other characteristics of pipe flanges and flanged fittings (ASME/ANSI 1988). (ASME is the American Society of Mechanical Engineers.)

Leakage from flanges and joints can be significant during shutdown as the system cools and start-up as the system reheats. Leakage of ignitable materials increases the possibility of fire and poses an exposure risk to process operators and maintenance people who service pumps, valves, meters and ancillary equipment. Leakage also means higher levels of atmospheric emissions.

A parallel area of concern is risk associated with opening lines for this or other purposes (Lees 1980). The breaking open of lines connecting process equipment during preparation for isolating a tank, vessel, reactor, and so on, can be extremely hazardous. Many incidents have occurred due to errors in identification of equipment on which work is required. Process plants and even small-scale equipment can seem to be a confusing maze of vessels, equipment and pipe. Oftentimes the plant or equipment, as built, differs from what the drawings indicate. Modifications may never be recorded. Identification of equipment and pipe is particularly important, because breaking open the wrong joint has caused serious accidents.

Workers breaking open lines risk exposure to pressurized liquids and gases. Pressure can remain in a line for any number of reasons, starting with human error. If the line is not pressurized, it may contain residual liquid. Sometimes the worker is splashed or even drenched during the opening of an overhead line. Spilled liquid can pose a fire and possibly thermal burn hazard. Volatile liquids can pose an exposure hazard.

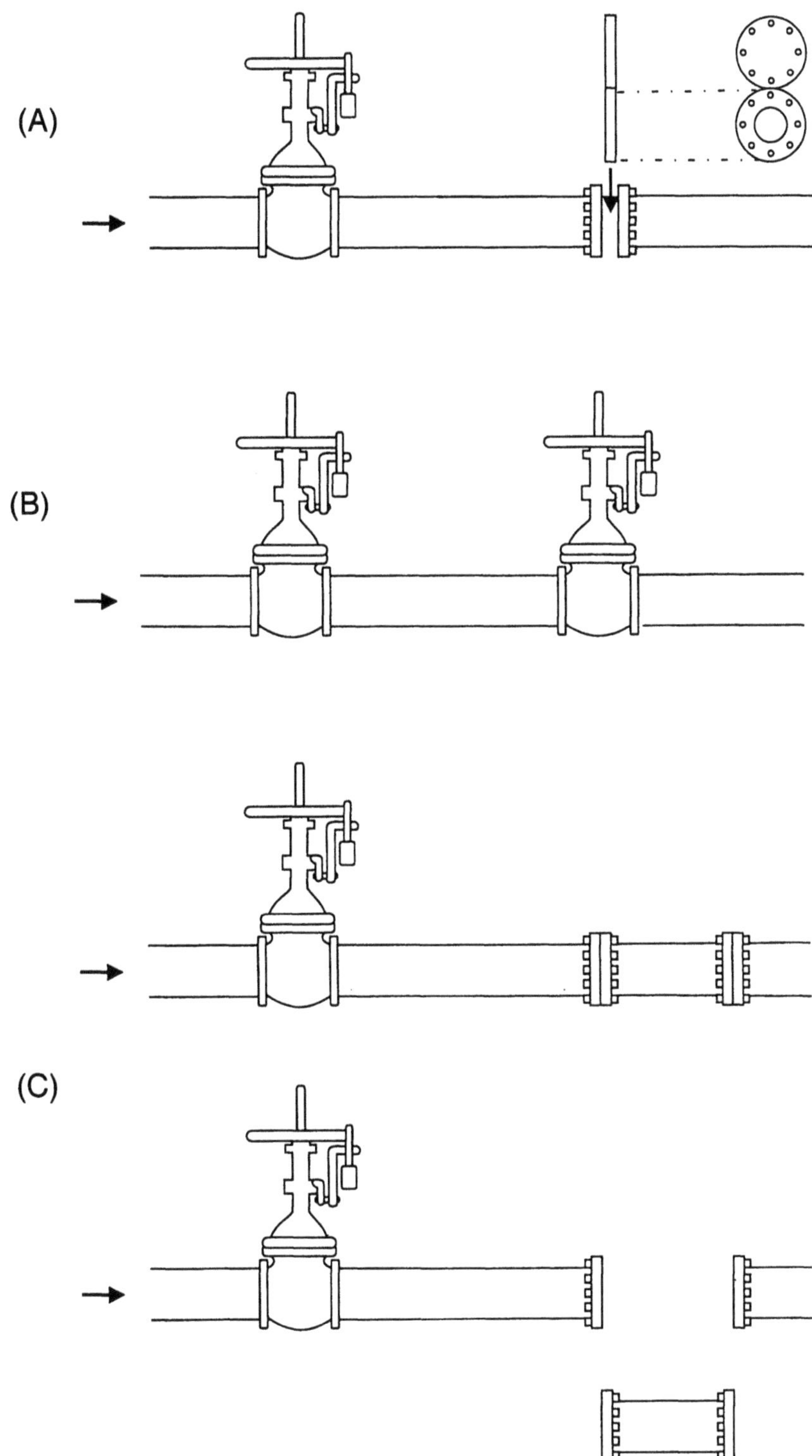

FIGURE G.14 Configurations for isolating process and utility lines. (A) Locked valve and blanked. (B) Double-locked valves. (C) Locked valve and spool piece removed. (D) Locked double block and bleed valves. (Adapted from Lees 1980, Grund 1995.)

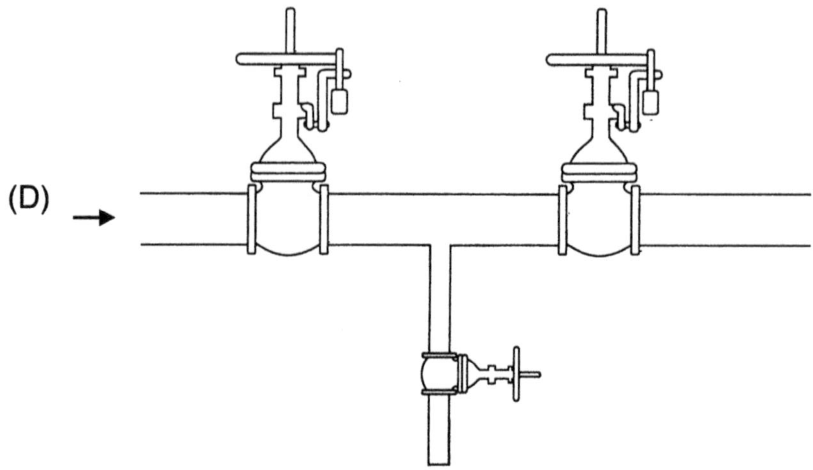

FIGURE G.14 (continued)

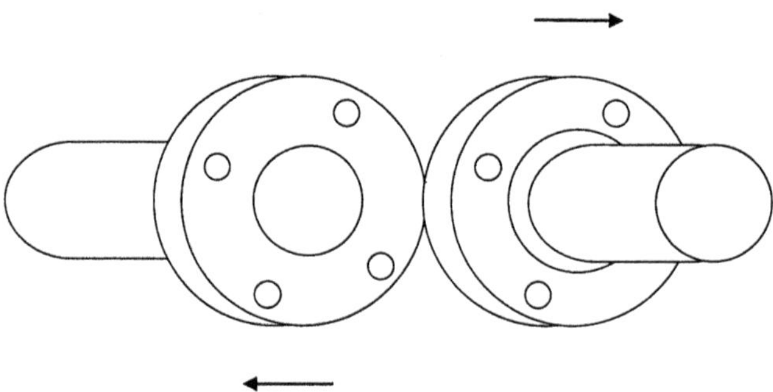

FIGURE G.15 Line breaking at a flange. This technique assures that liquid from the supply side of the line could not continue down the downstream side. Line breaking is most effective when the lines are separated enough to prevent entry of liquid following leakage or unexpected major flow. This method does not prevent movement of vapors between lines. (Adapted from UAW–GM National Joint Committee on Health and Safety 1985.)

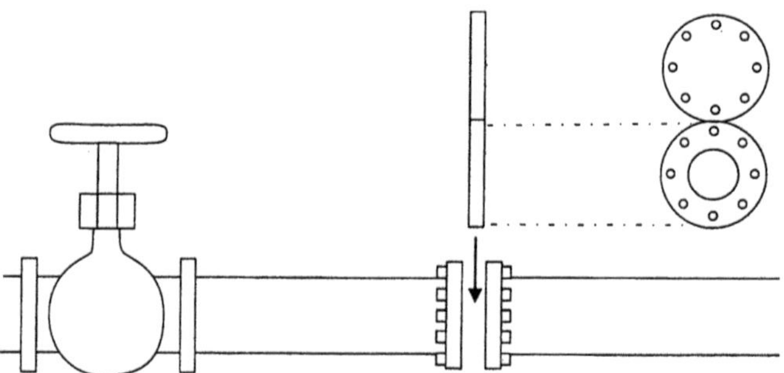

FIGURE G.16 Line blanking. A common sight in process operations is the spectacle blank. Removing all but one of the bolts between the flanges permits rotation of the blank. The position of the blank provides an immediate indication about the status of the line. (Adapted from NIOSH 1987.)

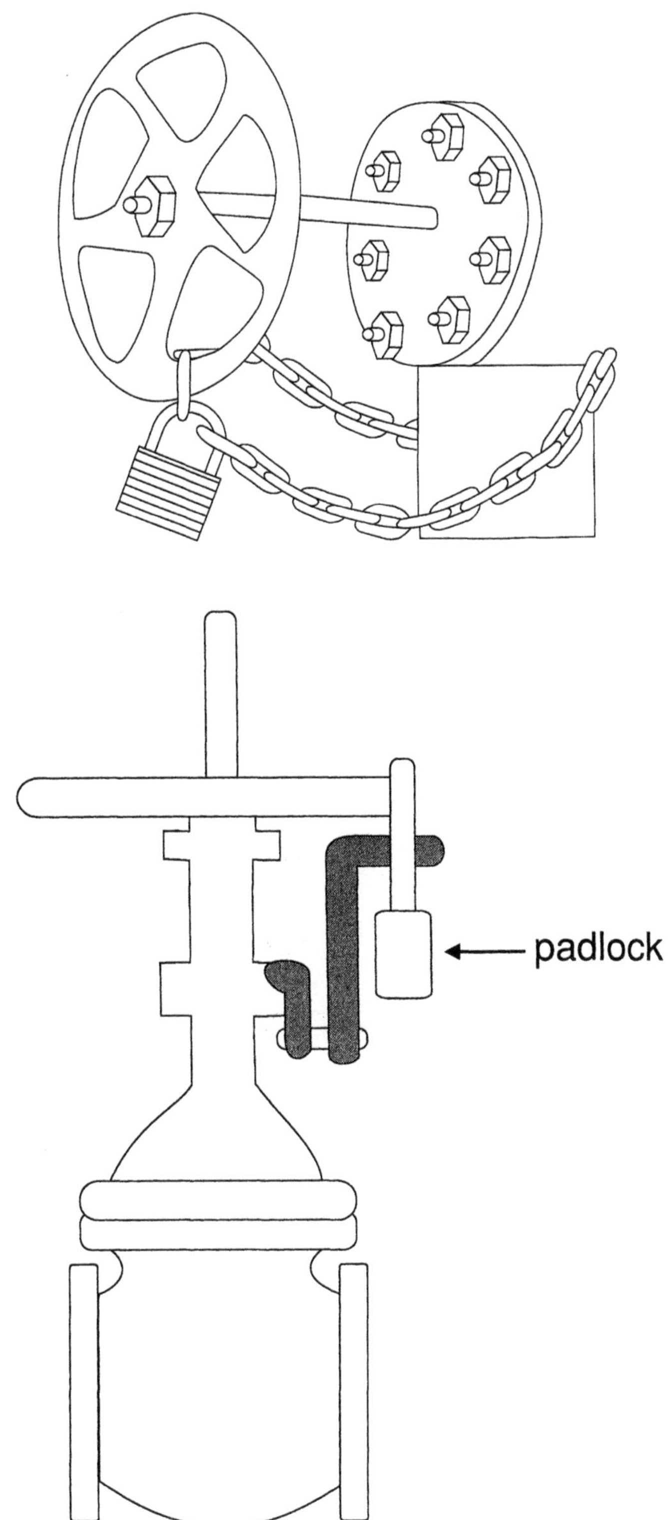

FIGURE G.17 (A) Valve chained and locked closed. (Adapted from UAW-GM National Joint Committee on Health and Safety 1985.) (B) Valve locked closed using adapter. (Adapted from Grund 1995.)

(A)

(B)

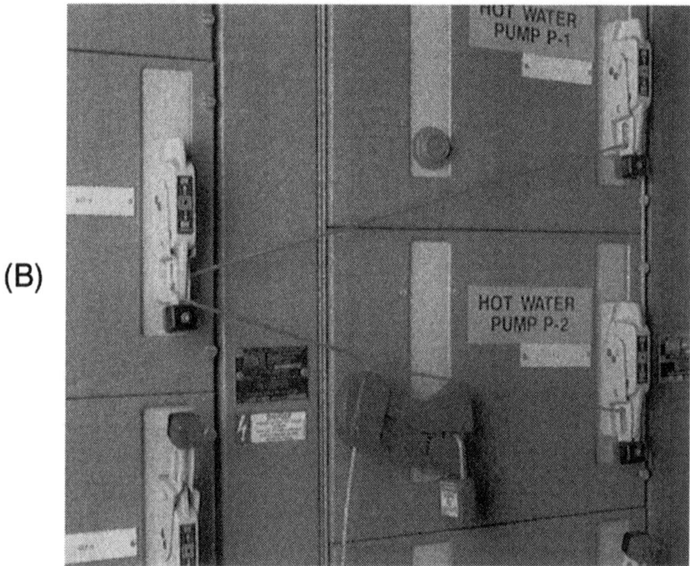

FIGURE G.18 (A, B) Cable lockout devices. Cable lockout devices provide capability for locking out more than one device or type of device. (Courtesy of W.H. Brady, Richmond Hill, ON.)

Another concern about the opening of lines is stress fractures. The action of separating sections of pipe by longitudinal or angular pulling to break the seal exerts a stress on the intact piping. Steam utility systems seem especially vulnerable to the problem of stress fractures.

Most of the process-related accidents occurring in confined spaces reported by OSHA resulted from failure of pressurized systems containing steam (OSHA 1985). Utility steam systems seem especially prone to sudden in-service failure. This may result from the stress of long high temperature, high pressure service and lack of maintenance. Most of the work activity in the accidents studied was unrelated to the process system that failed.

An accident reported by Debban and Eyre (1997) highlights a related problem that can affect steam and other process systems: water hammer. Water hammer is the hammering (banging) noise heard in piping following abrupt change in flow. This can result from the abrupt closing or opening of a valve. It also can result from sudden entry of gases into lines containing condensate. This change in flow exerts heavy stresses that can fracture the system. Subsequent investigation following this accidents showed that slow operation of valves would not prevent this situation. Steam and water cannot be mixed safely in a piping system.

The double block and bleed system utilizes two line valves separated by a short section of pipe. The short section of pipe contains a drain or vent. The two line valves are locked in the closed position; the vent valve is locked open. In this way leakage from the upstream valve encounters an open drain or vent prior to contacting the downstream valve. Valve manufacturers are actively attempting to develop products that accommodate the concepts of this paradigm in a single package.

Use of valves as isolating devices is not without controversy both inside and outside the regulatory community. Valve lockouts must be secure. Use of loose-fitting chain, as shown in Figure G.17 can allow the valve to be partially opened. Chain must be taut (UAW-GM National Joint Committee on Health and Safety 1985). The Workers' Compensation Board of British Columbia recently upgraded regulations for lockout and confined space entry (Workers' Compensation Board 1998). The regulation prohibits the use of line valves for isolation, even when several are closed in series. Double block and bleed systems are permissible only for liquids having low volatility. In other circumstances, the only permissible option is to disconnect, blank or blind piping, unless shown not to be possible.

The diagrams provided here (Figures G.13 to G.18) illustrate the classic paradigm of manually operated valves. Computerization of process control introduces the possibility of using motorized valves and achieving isolation through a software package. Control through software introduces the potential for tampering.

Valve leakage and failure are the major concerns about the use of the double block and bleed approach for controlling exposure to hazardous substances during work in confined spaces. However, valves play a critical role in the other approaches to isolation. For example, a valve(s) upstream from a disconnected section of line could fail. This would lead to leakage from the open-ended piece of pipe. Depending on the volume of the liquid or gas, and physical, chemical and toxicological properties of the chemical, a spill of this nature could have serious consequences.

An issue with the preparation of confined spaces for entry that tends to be overlooked is the nature of the work activity to follow. Many welding and cutting processes require the use of gases. These can include: helium or argon; acetylene, hydrogen or other fuel gases; carbon dioxide, oxygen and others. These gases likely will enter the space through flexible hoses. The latter are subject to damage and cutting. Welding units and valves on guns are subject to leakage. In large projects, custody of gas lines is an important issue. Errors in connection can and do happen. The level of care employed in ensuring safe conditions during the work activity should reflect the level in preparing the space for occupancy.

Other influences are acting in conflict with the classic paradigm used in isolation and lockout of process systems. There is a trend in process system design to move to welded construction. Welded systems leak less than systems containing bolted construction and are more reliable. These designs offer obvious benefits to process safety and reduction of emissions to the environment. This trend has important implications in preparing confined spaces for entry, since the ability to remove sections of pipe conflicts with other design priorities.

In the U.S. process safety is the subject of OSHA Standard 29 CFR 1910.119 (OSHA 1992). The focus of this Standard is maximizing knowledge about processes and the equipment in which they occur for the benefit of control. Part of this focus centers on piping and equipment design for the control of process hazards, such as fire and personal exposure and emissions to the environment. (McGuire Moran 1996).

Preparation for entry is an integral part of the cycle involving confined spaces. While the main focus of this book is work that occurs inside the space, pre-entry preparation can pose considerable hazards. The direction of approach provided by the process safety Standard is worthwhile, even where it is not mandated. One benefit is consideration of downstream consequences of actions

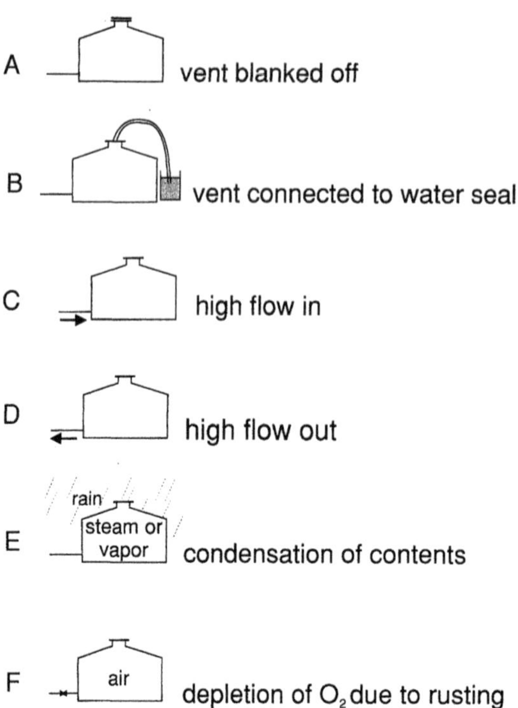

FIGURE G.19 Venting and preparatory conditions that can lead to tank collapse or overpressurization. (A) Blocking or blanking off the vent. This can occur through use of substantial or insubstantial materials. (B) Vent connected to a water seal. (C) High liquid flow into the tank with insufficient venting. This could involve process fluid or cleaning solutions or steam or air. (D) High liquid flow out of the tank with insufficient venting. (E) Decrease in temperature/condensation of contents. (F) Depletion of oxygen due to rusting. (Adapted from Lees 1982.)

(Figure G.19). Regardless of the method chosen for controlling hazardous conditions, management has the responsibility for diligence to ensure that risk is minimized. The means assessing the hazards of the situation, following recognized codes of practice, creating standardized procedures, providing training, maintaining equipment and facilities, and so on.

Multiple Energy Systems

Modern equipment can incorporate several of the systems mentioned here. While these systems could be independent of each other, the likelihood is that they are interconnected. Interconnection can considerably complicate the task of deactivation and de-energization. In some cases, systems and processes in equipment are controlled remotely as a result of activities occurring in other equipment. As a result, testing for deactivation is not as simple as pushing a "start" button. While this is one means of activation, activation by a signal from a microprocessor in a control circuit within the equipment or in a distant control room easily could be another.

Electrical systems are used to operate valves in hydraulic, pneumatic, and process and utility systems (Stewart 1976). Unlike mechanical systems, electrical systems provide limited power for actuation. However, an electrically operated pilot valve can control the activity of a much larger fluid power valve. Electrically operated valves utilize motors and solenoids. Motors operate through gears and gear reducers. The solenoid produces straight-line motion of a plunger through electromagnetic action. The solenoid operates a mechanical operator that in turn operates the valve mechanism.

The solenoid also is used to control circuits in electrical systems (Nelson 1983). The solenoid (called a relay in this application) closes contacts in high voltage/high current applications. The solenoid operates in a low voltage/low current circuit through innocuous wire of small gauge. These circuits can be battery powered and radio controlled. Hence, the control switches touched by the operator do not carry the high voltage supplied to equipment. Relays can introduce time delays ranging from seconds to minutes.

Solenoids/relays can be triggered by push-button switches, limit switches, pressure switches, timers, or the contacts of another solenoid/relay (Vickers 1992). Solenoid/relays can be used for remote control of circuits. A relay that activates or deactivates heavy electrical equipment, such as motors and heaters, is a contactor. The contacts of this device are strong enough to switch heavy current flow.

The proportional valve is a cross between the solenoid and the servo valve (Vickers 1992). Unlike conventional ON/OFF solenoid valves, the working range of the proportional valve includes an infinite number of positions. That is, proportional valves are electronically controlled, variable position fluid power valves. This type of valve can control flow rate and direction, as well as pressure. These valves represent another example of the potential for remote control of fluid power equipment. Because the fluid power output from these valves is proportional to the strength of the electronic input signal, their output is subject to feedback control through microprocessors.

Modern electronic amplifiers are another important part of hydraulic/pneumatic systems (Vickers 1992). Electrohydraulic devices, such as servo valves and pumps, pressure and flow control valves and speed control circuits require high power input electrical signals. Devices such as microprocessors, temperature, pressure, and speed sensors typically produce low power signals. Signal amplification, therefore, is a necessity. Operational amplifiers can provide signal gains as high as 10^6. This means that very weak signals from electrical devices can be accommodated through minor difficulty.

The only way to achieve safe conditions of work in equipment containing complex subsystems is to deactivate, de-energize, isolate, and lock out all energy sources. This must occur in a logical sequence that considers the implications of prior and subsequent actions on the stability of equipment. Applying a logical sequence provides the benefit of consistency of approach. Table G.1 provides the hierarchy recommended by the UAW–GM National Joint Committee on Health and Safety (1985).

TABLE G.1
Suggested Hierarchy of Deactivation and Lockout

Energy source	Comments
Momentum	During shutdown sequence, moving parts in machinery come to a stop
Gravity	Immobilization through pinning and blocking should occur before the primary energy source is turned off
Stored mechanical energy	While sources of stored mechanical energy should receive consideration at this point in the sequence, they could be neutralized later in the procedure; relieving residual and stored energy may require shutting down a primary energy source
Electrical system	Normal operation of the equipment must cease prior to shutting down the electrical supply system; electrical energy may remain stored within electrical equipment — for example, capacitors and transformers
Pneumatic system	Residual stored air pressure existent in accumulators must be released during de-energization
Hydraulic system	Residual stored hydraulic pressure in accumulators must be released during de-energization
Process and utility systems	Deactivated, de-energized, and locked out as needed

From UAW–GM National Joint Committee on Health and Safety 1985.

The hierarchy in Table G.1 provides the basis in a specific situation for the formulation of standardized procedures. As recommended by the UAW-GM National Joint Committee on Health and Safety, the standardized procedure should contain:

- Layout drawing of machine, including location and identification of lockout equipment labels
- Number of locks required for complete lockout
- Written description of location of lockout equipment
- Type of energy source and identification by label, and method used to lock it out
- Reference position locator
- Sequential procedure for lockout
- Information about peripheral areas and equipment affected by the neutralization and lockout of energy sources
- Procedure for verification check

Hazardous conditions still can occur during the implementation of standardized procedures (Lees 1980, Grund 1995). Line-breaking external to the confined space can lead to exposure to residual liquid contents contained in valve casings and pipe. Volatile residual flammable/combustible liquids can pose a serious inhalation and fire hazard. Service-stressed isolation devices could fail.

More elaborate approaches are available for complex mechanical and process systems (Grund 1995). Some use the concept of group lockout/tagout. For more discussion on this topic, refer to the book by Grund.

In discussing standard procedures, emphasis must be put on the fact that a procedure is valid only so long as the system to which it applies has not changed. Change is the only constant in industrial settings, just as in other facets of life. Change or modification of equipment or process can negate the applicability of a standard procedure. Without a mechanism to ensure updating of standard procedures for lockout at the same time as a modification or change in equipment, the effort expended in creating these procedures could be wasted. Worse still, duplication of effort could result following recognition of the need to update a group of standard procedures following an accident caused because of demonstrated inadequacy in one.

SUMMARY

This appendix has highlighted some of the issues in the deactivation, de-energization, isolation, and lockout of equipment and utility and process systems. NIOSH attributed cause of equipment-related accidents to failure to control hazardous energy, inadequate energy isolation, failure to dissipate residual energy and accidental reactivation of hazardous energy.

Depending on the system, this process can be as simple as shutting off the control switch and removing a fuse or shutting down a breaker and locking out access to the electrical panel. On the other hand, deactivation, de-energization, isolation, and lockout of complex, real-world, integrated systems could require considerably more effort.

REFERENCES

American National Standards Institute: *American National Standard for Personnel Protection— Lockout/Tagout of Energy Sources— Minimum Safety Requirements* (Z244.1-1982). New York: American National Standards Institute, 1982.

American National Standards Institute: *American National Standard for Personnel Protection— Lockout/Tagout of Energy Sources— Minimum Safety Requirements* (Z244.1-R1992). New York: American National Standards Institute, 1992.

American Petroleum Institute: *Steel Line Blanks* (API Standard 590). Washington, D.C.: American Petroleum Institute, 1985. 22 pp.

American Society of Mechanical Engineers/American National Standards Institute: *Pipe Flanges and Flanged Fittings* (ASME/ANSI B16.5-1988). New York: American National Standards Institute, 1988.

Debban, H.L. and L.E. Eyre: Steam distribution systems: managing and preventing condensate-induced water hammer. *Prof. Safety.* 42: May 1997. pp. 34-37.

Grund, E.V.: *Lockout/Tagout, The Process of Controlling Hazardous Energy.* Itasca, IL: National Safety Council, 1995. 429 pp.

Juvinall, R.C. and K.M. Marshek: *Fundamentals of Machine Component Design,* 2nd ed. New York: John Wiley & Sons, 1991. 804 pp.

Lees, F.P.: *Loss Prevention in the Process Industries,* vol. 1 and 2. London: Butterworths, 1980. 1346 pp.

McGuire Moran, M.: *OSHA's Process Safety Management Standard, A Proven Written Program for Compliance.* Rockville, MD: Government Institutes, 1996. 235 pp.

Mine Safety and Health Administration: Think "Quicksand": Accidents Around Bins, Hoppers and Stockpiles, Slide and Accident Abstract Program. Arlington, VA: U.S. Department of Labor, Mine Safety and Health Administration, National Mine Health and Safety Academy, 1988.

National Institute for Occupational Safety and Health: Criteria for a Recommended Standard — Working in Confined Spaces (DHEW/PHS/CDC/NIOSH Pub. No. 80-106). Cincinnati, OH: National Institute for Occupational Safety and Health, 1979. 68 pp.

National Institute for Occupational Safety and Health: Guidelines for Controlling Hazardous Energy Sources During Maintenance and Servicing (DHHS(NIOSH) Pub. No. 83-125. Morgantown, WV: DHHS/PHS/CDC/NIOSH, 1983. 72 pp.

National Institute for Occupational Safety and Health: A Guide to Safety in Confined Spaces, by Pettit, T. and H. Linn (DHHS (NIOSH) Pub. No. 87-113). Cincinnati, OH: DHHS/PHS/CDC/NIOSH, 1987. 20 pp.

National Institute for Occupational Safety and Health: Worker Deaths in Confined Spaces (DHHS/PHS/CDC/NIOSH Pub. No. 94-103). Cincinnati, OH: National Institute for Occupational Safety and Health, 1994. 273 pp.

National Mine Health and Safety Academy: Fault Tree Analysis, rev. 1991 (Safety Manual No. 8). Beckley, WV: U.S. Department of Labor/Mine Safety and Health Administration/National Mine Health and Safety Academy, 1991. 33 pp.

National Safety Council: *Safeguarding Concepts Illustrated,* 6th ed., Itasca, IL: National Safety Council, 1993. 100 pp.

Nelson, C.A.: *Millwrights and Mechanics Guide.* New York: Bobbs-Merrill Company, 1983. 1032 pp.

Occupational Safety and Health Administration: Selected Occupational Fatalities Related to Fire and/or Explosion in Confined Work Spaces as Found in OSHA Fatality/Catastrophe Investigations. Washington, D.C.: U.S. Department of Labor, Occupational Safety and Health Administration (U.S. DOL/OSHA), 1982a. 76 pp.

Occupational Safety and Health Administration: Selected Occupational Fatalities Related to Lockout/Tagout Problems as Found in Reports of OSHA Fatality/Catastrophe Investigations. Washington, D.C.: U.S. Department of Labor, Occupational Safety and Health Administration (U.S. DOL/OSHA), 1982b. 113 pp.

Occupational Safety and Health Administration: Selected Occupational Fatalities Related to Grain Handling as Found in Reports of OSHA Fatality/Catastrophe Investigations. Washington, D.C.: U.S. Department of Labor, Occupational Safety and Health Administration (U.S. DOL/OSHA), 1983. 150 pp.

Occupational Safety and Health Administration: Selected Occupational Fatalities Related to Toxic and Asphyxiating Atmospheres in Confined Work Spaces as Found in Reports of OSHA Fatality/Catastrophe Investigations. Washington, D.C.: U.S. Department of Labor, Occupational Safety and Health Administration (U.S. DOL/OSHA), 1985. 230 pp.

Occupational Safety and Health Administration: Selected Occupational Fatalities Related to Welding and Cutting as Found in Reports of OSHA Fatality/Catastrophe Investigations. Washington, D.C.: U.S. Department of Labor, Occupational Safety and Health Administration (U.S. DOL/OSHA), 1988. 225 pp.

OSHA: Control of Hazardous Energy Sources (Lockout/Tagout); Final Rule, *Fed. Regist. 54*: 169 (1 September 1989). pp. 36644–36696 (1989).

Occupational Safety and Health Administration: Selected Occupational Fatalities Related to Ship Building and Repairing as Found in Reports of OSHA Fatality/Catastrophe Investigations. Washington, D.C.: U.S. Department of Labor, Occupational Safety and Health Administration (U.S. DOL/OSHA), 1990. 195 pp.

OSHA: Process Safety Management of Highly Hazardous Chemicals; Explosives and Blasting Agents; Final Rule, *Fed. Regist. 57:* 36 (24 February 1992). pp. 6356–6417.

Office of General Industry Compliance Assistance: Presentation of 29 CFR 1910.147, The Lockout/Tagout Standard, incorporates the amendments and corrections of the Federal Register, dated, September 20, 1990. Washington: U.S. department of Labor, Occupational Safety and Health Administration (U.S. DOL/OSHA), 1991. [presentation]

Olivo, C.T. and T.P. Olivo: *Fundamentals of Applied Physics*, 3rd ed. Albany, NY: Delmar Publishers, 1984. 500 pp.

Stewart, H.L.: *Pneumatics and Hydraulics.* Indianapolis, IN: Theodore Audel, 1976. 486 pp.

UAW–GM National Joint Committee on Health and Safety: *Lockout.* Detroit, MI: UAW–GM Human Resources Center, 1985.

Vickers Training Center: *Vickers Industrial Hydraulics Manual.* Rochester Hills, MI: Vickers, 1992.

Watts, J.M., Jr.: Probabilistic fire models. In *Fire Protection Handbook*, 17th ed. Cote, A.E. and J.L. Linville (Eds.). Quincy, MA: National Fire Protection Association, 1991. pp. 10-93 to 10-98.

Wiggins, Jr., J.H.: Control of hazardous energy. *Compliance Magazine.* 4:May 1997. pp 14-17.

Workers' Compensation Board: *Occupational Health & Safety Regulation, General Hazard Requirements* (Parts 9 and 10). Vancouver, BC: Workers' Compensation Board of British Columbia, 1998.

Appendix H: Hazard Reduction for Work Involving Confined Spaces

CONTENTS

INTRODUCTION

One of the fundamental assumptions upon which the confined space "industry" rests is that entry is a necessary activity. Certainly, this is undeniably true for many of the tasks that must be performed in these workspaces. However, for an ever-increasing number, alternatives that do not necessitate entry have become available. In still other situations, noninvasive techniques can reduce or eliminate the initial level of hazard of subsequent entry and work. Wider adoption of existing techniques and technologies offers considerable opportunity to reduce the level of risk associated with work involving confined spaces and confined hazardous atmospheres.

DESIGN MODIFICATIONS

The design phase is the most opportune time to modify equipment and structures. Relocating a piece of equipment or inserting a wall into a structure can be as simple as making a few keystrokes in a design program. Modification of design to eliminate the presence of or reduce the consequence of entry into confined spaces that occur in structures and equipment offers distinct advantages. Modifications at the design stage are extremely cost-effective.

Considerations for change could include:

- Accessibility to the space
- Mobility within the space
- Means to isolate the space — valves, piping, and fittings
- Means to achieve zero energy state within the space
- Means for lockout of key components
- Hazards present in adjacent spaces
- Locating equipment out of the space

All modifications that eliminate the need for entry or minimize associated risk should be incorporated into the design.

STRUCTURAL AND EQUIPMENT MODIFICATIONS

In many situations, entry occurs because of equipment that is located in the space. That is, a person must enter the space to perform some action or interaction with equipment. Some of the actions and interactions that occur during these entries include:

- Reading meters, gauges
- Adjusting position of valves
- Activating/deactivating controls
- Inspecting equipment

The critical element that motivates these activities is the location of the equipment in the space and not somewhere else. That is, the activity occurs because of location of the equipment. Thus, the action would remain the same, regardless of the location of the equipment. Entries of this type are necessitated because of a decision taken during design to locate the equipment inside the space. Modification of the existing structure or the equipment could eliminate the need for entry except for unusual situations. Relocation or modification of the critical element involved in the entry depends on a number of factors:

- Frequency of the activity for which entry is needed
- Seriousness of the actual or potential hazard
- Ancillary activities undertaken during routine entry

Confined spaces which are entered frequently, in which a high level of actual or potential hazard exists and where no ancillary activity is undertaken, offer the greatest incentive for correction. Some of the options available for eliminating the need for entry include:

- Relocating the task critical element outside the space
- Installing video cameras to provide surveillance of the critical element
- Continuously ventilating the space
- Extending the stem of valves outside the space

- Installing lighting to improve visibility
- Installing monitoring device to provide remote indication of status
- Installing ventilation systems

Each of the preceding is an option that offers its own advantages and disadvantages. Exploring each of these through a disciplined decision analysis system provides the means to make the best choice.

ROBOTICS

The pioneering work in this area occurred in the nuclear industry. The extreme level of hazard that can exist with radioactive materials often prohibits access to the actual worksite. Sometimes this prohibition necessitates considerable distance or use of shielding materials as an intermediary between the worker and the work activity. A less desirable third option is duration of exposure. The latter requires rigid administrative control. Overcoming these necessities led to development of ingenious solutions, such as specialized tools, remote manipulators, and robotic equipment.

Recent developments in the nuclear industry are typified by miniature tools that can operate inside the reactor itself (Janzen, 1995). The pressure tube sampling tool is one example. First it moves within the pressure tube to the desired location. (The Canadian-developed power reactor contains pressure tubes, similar to those found inside old-style, fuel-fired boilers.) Then it isolates a zone, removes water from the sampling location, cuts metallic oxide from the wall, and finally metal from the area from which the oxide was removed. Another tool, named the wet replication tool, creates replicas of flaws in the wall of the pressure tube, using an impressionable material that is positioned in-place by the tool.

More recent applications of these concepts include equipment developed for use in undersea and extraterrestrial exploration. In both situations, the environments are hostile and preclude entry by humans by usual means.

The most familiar of the "civilian" applications of these technologies is the remotely controlled equipment used in police work (Figure H.1). Application of this equipment often occurs during life-threatening situations, such as bomb-disposal and hostage-taking incidents. This equipment includes wheeled or tracked vehicles that incorporate "arms and hands" that can clasp, hold, lift, and rotate objects just like their human counterparts. The difference represented by this equipment, despite emotional attachments sometimes displayed by its operators, is that it is machinery. It is intended for sacrifice should the need arise. A mechanical arm that has been torn or blown off during an explosion or highly contaminated by radioactive or chemical contaminants can be replaced. Its human equivalent cannot.

Robotic equipment recently has been employed in manufacturing in ever-increasing applications. The ideal application for this technology is elimination of environments that are highly hostile to workers. Thus far, more likely applications for this equipment are repetitive manufacturing operations.

The fundamental feature of the preceding applications of this equipment is its ability to operate without difficulty in environments that are hostile and potentially lethal to humans. There should be considerable opportunity for application of these concepts in confined spaces.

An important consideration regarding confined workspaces that are not entered by humans is that the hazard still remains. Without proper care in design, selection, and operation of this equipment, the hazard could create the same result, a fire or explosion, for example. The only difference would be that workers would be outside the space. The consequences could be just as severe.

Presently available equipment has application in confined spaces. This will be described in the following sections. The main applications of this equipment include video and instrumental assessment of conditions and cleaning technologies. Widespread application of even the preceding concepts would dramatically reduce the need for entry.

(A)

(B)

FIGURE H.1 (A and B) Police robot (or remote mobile investigator). The police robot is one of the more familiar applications of robotics. The units pictured here can operate through cable or radio control. The claw on the front arm can provide both a soft grip for picking up objects without damaging them or a stronger grip for heavier objects. The claw also can provide rotational motion. Arm extenders provide the basis for reaching less accessible locations. The larger unit (H.1B) has 10 driving wheels and can be fitted with tracks for extra climbing capability. (Courtesy of Pedsco, (Canada) Ltd., Scarborough, ON.)

EQUIPMENT FOR EXPLORATION AND ASSESSMENT

General

Some of the equipment utilized for exploration and assessment of spaces has been commercially available for over 25 years (Sefton, 1994). The main application of video inspection has been exploration and assessment of conditions in sewers and pipes and air ducts. Originally, the cameras were designed for either underwater applications or inspection tasks in nuclear power plants. The cameras invariably employed vidicon tubes. They are similar in construction to the television picture tube. Both are glass devices. Equipment used for video inspection must be extremely rugged. Vidicon tubes ultimately proved to be unreliable, casualties of the very rough and unknown conditions within pipes. The advent of the charge coupled device (CCD), the solid state analog of the vidicon, improved the reliability of the equipment. The CCD has no glass envelope to shatter and thus essentially is unbreakable. The latest development in this equipment was the introduction of color.

The simplest of the video inspection equipment is a float, cart, skid, or sled on which the camera is mounted (Figure H.2). These units are drawn by winch between two points of access. The procedure necessitates floating a line down the sewer to the downstream point. This is attached to a winch line, which is then drawn back through the sewer and attached to the camera, which in turn is towed back through the pipe to conduct the survey.

More sophisticated equipment includes the snake or push, and self-propelled units. The snake consists of the sensing head and associated illumination, connecting cable, and supporting structure. The user must manipulate the snake through the pipe. The supporting framework provides the resistance and flex needed for pushing the snake forward and forcing it to follow the contours and changes of direction of the structure. The sensing head incorporates the video camera, as well as light-emitting diodes to supply illumination. These units can be pushed 60 to 90 m (200 to 300 feet) through large and small diameter pipes, 5 cm (2 in.) or greater, even around 90° bends. Color and monochrome equipment are available.

Larger, more complex units incorporate low light, high resolution, color cameras and built-in light sources on mounts that tilt and rotate in some cases. These units can operate in pipes as small as 20 cm (8 in.) in diameter. The camera and lighting equipment is mounted on a self-propelled wheeled cart called a tractor or a tracked vehicle. Propulsion sources include electric motors and high pressure water jets.

Safety Hazards and Protective Measures

An environment that is dangerous for a person to enter, due to the possible presence of ignitable vapors, is no less hazardous to equipment that contains sources of ignition. The decision to utilize equipment in place of people to perform the work does not alter or diminish the hazards of a confined space (Sefton 1994). The only difference is that the people remain outside. An explosion triggered by equipment that propels sewer covers into the air or ruptures a tank is potentially no less hazardous to spectators.

Possible interaction between the equipment and the hazardous atmosphere or contents therefore must receive consideration. The voltage in some video inspection systems approaches 300 V. This is necessary to overcome resistive losses incurred over the length of the cable that powers the camera, lights, and in some cases, the tractor.

An additional concern is longitudinal variation in the atmosphere of elongated spaces, such as sewers, in which this equipment operates. The atmosphere in one part of the confined space may have been tested and found acceptable. However, this assessment applies only to the local zone under test. Video inspection systems can travel considerable distances. This travel can include environments where assessment has not occurred. Sewers and storm drains are capable of containing pockets of naturally formed methane gas. Also, conditions may have changed since the test was performed. This can occur because of discharge of an ignitable liquid into the pipe. This would be especially relevant to chemical sewers. During video inspection of sewers, gas detectors often are employed at each manhole before insertion of the camera. Checking at the end points of a long run has little relevance if a discharge of ignitable effluent occurs midway.

(A)

(B)

FIGURE H.2 Video inspection units. (A) Steerable tractor operates in pipes and ducts 40 cm (15 in.) in diameter and larger. The unit pictured is rated for use in hazardous atmospheres. (Courtesy of Pearpoint, Inc., Thousand Palms, CA.). (B–C) Track-mounted systems for inspection of small pipes and ducts and other applications. The latter unit can change shape during operation. (Courtesy of Inuktun Services, Ltd., Cedar, BC.). (D) Telescoping pole-mounted video camera. The camera head in this unit contains the camera, lights, focus, and tilt motor. The camera assembly can tilt through 180° and pan through 360°. (Courtesy of Inuktun Services, Ltd. Cedar, BC.)

With the potential increase in utilization of this equipment and broadening of applications, the urgency to assure safety has increased dramatically. There is a move within this industry to ensure the safety of this equipment in hazardous locations through voluntary testing by Nationally Recognized Testing Laboratories (NRTLs).

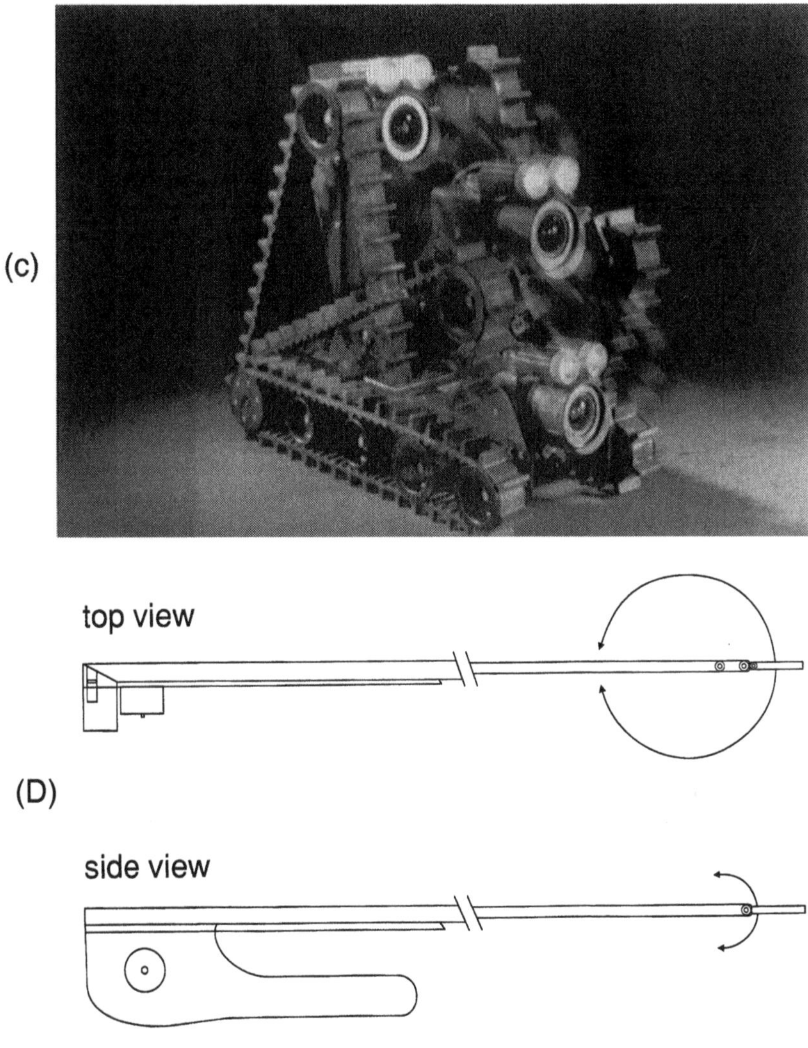

FIGURE H.2 (continued)

CLEANING TECHNOLOGIES

An appreciable amount of the activity that occurs in confined spaces is related to cleaning. Removal of residual materials often is required for its own sake. That is, the presence of adherent material interferes with other activities or processes. To illustrate, material adherent to the interior of a batch reactor could destroy the quality of succeeding production. A batch reactor may require cleaning after each use. In storage silos, material adherent on walls or which has formed bridges interferes with the free flow of the cargo. Cleaning also is required in order to make the space hospitable for performing other work. Removal of residual materials from the interior of a space prior to entry also can dramatically reduce risks unrelated to cleaning that are associated with subsequent work (Figure H.3).

Water-Jet Cleaning

Water-jet cleaning, or hydroblasting, has been used for over 60 years (Summers 1993). Water jetting equipment pressurizes water to levels far above normal tap flow. The jet is used to loosen and

FIGURE H.3 Robotic applications. Track-mounted system containing motorized brush and inspection camera. This unit is used for inspection and cleaning of duct systems. (Courtesy of Inuktun Services, Ltd., Cedar, BC.)

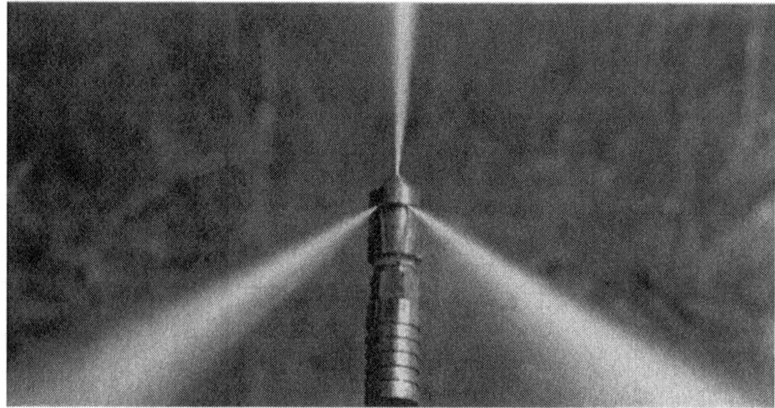

FIGURE H.4 High pressure water jet. This jet of high pressure water can remove rust, scale, chemical residues, and other contaminants from surfaces. High pressure jets can cut concrete and steel. (Photo courtesy of NLB Corporation, Wixom, MI.)

remove residues and intentionally applied coatings. The blast can be applied manually or by remotely controlled delivery systems (Figure H.4). Most of the equipment operates at pressures of 7 to 14 MPa (1,000 to 2,000 lb/in^2) and flows of 20 to 40 L/min (5 to 10 gal/min).

Equipment used for low pressure waterblasting operates at less than 35 MPa (5,000 lb/in^2). This pressure was chosen as an upper limit because considerable energy is required to improve jet performance above this the value. The major application of this equipment is industrial cleaning. High pressure waterblasting involves pressures exceeding 35 MPa (5,000 lb/in^2) at the orifice. High pressure waterblasting is a relatively recent development.

Table H.1 indicates minimum pressures required for various kinds of cleaning operations.

The basic components of all waterblast cleaning systems are the same: a pump, pressure relief system, hose (piping), and nozzle.

TABLE H.1
Pressures Needed for Removal of Embedded Materials

Pressure MPa (psi)	Application
10 (1,500)	Silt, loose debris
20 (3,000)	Light marine fouling, light scale, fuel oil residue, aluminum cores and shells
30 (4,500)	Weak concrete, medium marine growth, sandstone and mudstone, light millscale, limited core removal, loose paint and rust
40 to 70 (6,000 to 10,000)	Concrete in pipes, severe marine fouling, ferrous casting molds, runway rubber, soft limestone, lime scale, burnt oil deposits, medium millscale, petrochemicals
70 to 100 (10,000 to 15,000)	Concrete cutting and removal, most paints, medium limestone, most millscale, silica cores, burnt carbon deposits, heavy clinkers
100 to 200 (15,000 to 30,000)	granites, marble, limestone, marine epoxies, aluminum, lead, rubber, frozen food

The heart of this equipment is the pump. The most common type for waterblasting is a positive displacement plunger pump. A plunger pump has some resemblance to a piston pump (Yared, 1994). The plungers are driven through rotation of a crankshaft (Anonymous 1994a). The plunger, having very small surface area compared to a piston, concentrates the motive force in order to produce higher pressure. Operating pressures range from 35 to 210 MPa (5,000 to 30,000 lb/in^2) with flows of 40 to 2,000 L/min (10 to 500 gal/min). Ultrahigh pressure systems can range as high as 410 MPa (60,000 lb/in^2). Achieving these high pressures typically requires use of intensifiers.

Intensifiers are powered hydraulically or by air. Hydraulically powered intensifiers typically operate in the range of 210 to 410 MPa (30,000 to 60,000 $lb/in.^2$) with flows of 4 to 20 L/min (1 to 5 gal/min) (Anonymous 1994a). Air-powered intensifiers can produce pressures as high as 700 MPa (100,000 $lb/in.^2$) with a flow of 0.4 L/min (0.1 gal/min). An intensifier uses a large, low pressure hydraulically driven piston to drive small high pressure plungers that contact the water. The plunger displaces the water in the high pressure cavity. An intensifier is equivalent in electrical terms to a step-up transformer.

A critical component in these systems is the pressure-relief system (Williams 1995). These systems utilize two designs: an on/off valve or a bypass. Both are located in the gun and are used to control flow. The on/off valve is either open or closed. Pressure in the system is modulated by a nitrogen-charged bladder that controls an internal valve. Expansion or contraction of the bladder maintains constant pressure. The bypass design dumps water to atmosphere through an unrestricted port. The bypass may be hand- or foot-operated. Pressure and flow can be modulated according to the setting of the valve in the bypass. Each design has its respective advantages and disadvantages.

Flexible, wire-reinforced hose is the common way to conduct water from pump to nozzle (Anonymous 1993a). With correct sizing, lengths of several hundred feet are possible. A variety of delivery tools are available: hand lances, shotguns, mole heads, flex lances, and stiff lances. Nozzle tips can produce round, needle, or fan sprays. Nozzles are made from steel, tungsten carbide, ceramic, and sapphire, as well as coated and plated materials.

Heating the water after it leaves the pressure pump is a technique for improving performance when cleaning hydrocarbon and similar dirt coatings. Chemical substances also are used in some cleaning operations. Long-chain polymers enable pressure jets to deliver greater energy over longer distances by increasing cohesive forces. Bubbles increase the area of coverage without requiring an increase in delivery power. Abrasives can be added to the water jet to assist in removing especially difficult coatings. Hard media such as sand, garnet, and slag are used, as well as soluble abrasives, such as sodium bicarbonate (Dorman 1993a, Anonymous 1994b).

FIGURE H.5 Vertical tank cleaning units. The lance on which the self-rotating cleaning head is mounted can telescope to permit vertical positioning within the structure. (Photo courtesy of NLB Corporation, Wixom, MI.)

Applications

The most important application to confined spaces is remote cleaning of containers and structures. These can include tanks, reactor vessels, pipes, towers, and so on. Several manufacturers produce equipment expressly for this purpose (Figure H.5). The underlying concept is that the equipment provides omnidirectional, uniform coverage of the surfaces within the container. This equipment includes a baseplate and attached structure for positioning the cleaning heads. The baseplate containing the cleaning unit bolts to the opening of the manway. The cleaning unit fits inside the structure to be cleaned. Operation is controlled externally.

Omnidirectional cleaning is most effective when the unit is centered in the vessel. Problems occur due to off-center or side-positioned manways and tank internals, such as agitator blades and cooling coils. Positioning devices ensure that the rotating head is properly centered. These provide adjustment along the two coordinate axes.

The secret of the three-dimensional capabilities of current equipment is the rotating cleaning head and rotor arm. Together, these provide 360° omnidirectional cleaning. This configuration is one of the most useful and adaptable tools available to the water jetting industry (Figure H.6).

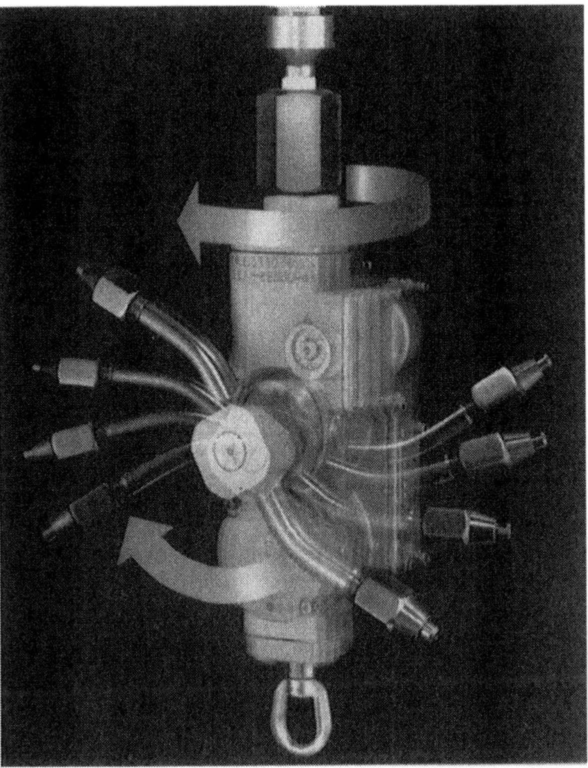

FIGURE H.6 Self-rotating cleaning head. Self-rotating heads provide 360° coverage for cleaning the interior of structures. In this configuration, the water jets spin vertically, while the head rotates horizontally. (Photo courtesy of NLB Corporation, Wixom, MI.)

Features that contribute to the success of these units include variable rotational speed and rugged, compact design. Some designs also offer dual rotor arms. Propulsion for the rotating arm(s) is provided by the reactive force of the discharged water. Although usually used with a rigid lance or telescoping arm, the cleaning head is dynamically balanced for effective operation, even while suspended on a flexible high pressure hose.

A parallel application is equipment designed for cleaning railway tank cars and similarly configured structures (Anonymous 1993b). The three-dimensional rotating cleaning head and rotor arm provide 360° omnidirectional cleaning. Several designs are available for positioning the cleaning head along the main axis of the tank. One uses a curved guide arm that pivots at both ends and moves horizontally along the tank. This unit cleans only in one direction. Another uses a telescoping tube that extends horizontally (Figure H.7). This unit also cleans in one direction along the axis of the tank car. A third design uses a scissor frame that extends horizontally in both directions. Each design can position the cleaning head at any desired location along the main axis of the tank. These units can be powered by compressed air, electrically, or hydraulically.

The improved safety offered by use of this equipment is amply illustrated by the conditions that are experienced during manual tank cleaning:

- Low visibility due to mist
- Difficulty in providing lighting due to potential for shock hazard
- Slippery floor
- High heat in summer
- Risk of injection from direct or reflected path

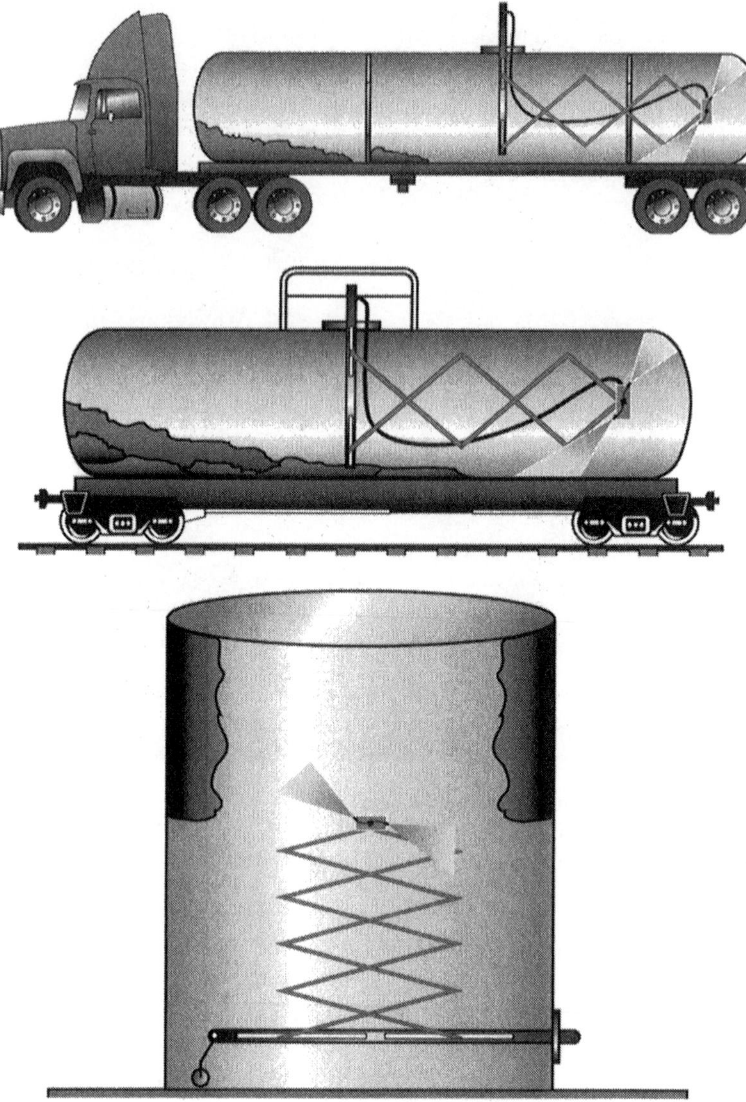

FIGURE H.7 Horizontal tank cleaning units. Articulated arms (as shown) and other designs position the cleaning head along the long axis of horizontal tanks. (Photo courtesy of NLB Corporation, Wixom, MI.)

- Reaction force from spray gun
- Need for more water during manual cleaning
- Potential inhalation hazard from solvent cleaners or chemical additives

Another application of high pressure water jetting is a system that cleans air-preheater baskets in boilers (Anonymous 1992, Dorman 1993a). Air-preheater baskets transfer heat in air-to-air heat exchangers. The baskets sit in a wheel structure that rotates between the clean and dirty sides of the system. They gain heat on the ash-laden dirty side and transfer it to incoming air on the clean side. Ash and soot deposits on the surfaces of the baskets reduce the efficiency of heat transfer and increase pressure drop. Baskets can be cleaned by various techniques: high-volume water washing, steam and air soot blowers, and manual cleaning. These can require dismantling the structure.

High pressure water jets can clean the baskets during normal operation. Supporting equipment consists of a steel support beam and a trolley assembly. A nozzle block is attached to the trolley.

A motorized chain drive positions the nozzle block. A variable-speed drive moves the preheater rotor at a predetermined speed. A computer controller simultaneously adjusts nozzle movement and rotor speed. The support beam spans the radius of the rotor. This may be left in place, or removed and reattached for each cleaning.

Safety Hazards and Protective Measures

Use of pressurized fluid technology creates some serious safety hazards (Vijay, 1993). Serious injury to the hand, the eye, and the abdomen has occurred due to use of high pressure jets. The superficial aspect of the injury from a pressure jet usually is small. This can induce a false sense of security, leading the victim not to seek medical attention. Delay in seeking treatment can be extremely serious. Minor delay can drastically increase the probability of need to amputate the injured limb.

Traumatic injury produced by fluid jets differs from that produced by other types of cutting tools, such as a chainsaw. Injury or damage caused by the chainsaw usually is localized. That caused by a fluid jet can be widespread. Several papers have examined the literature of injuries that can be caused by fluid jets (Katakura and Tsuji 1985, Summers and Viebrock 1988, Vijay 1989). Injury at pressure as low as 0.55 MPa (80 psi) has been reported. At low pressure, the severity of the injury depends on the duration of exposure and site of the impact.

Studies on the cutting of seafood and meat products using high pressure jets indicate that water can penetrate extensively, perpendicular to the plane of cut (Vijay and Brierley, 1983). These observations suggest that, once the fluid jet penetrates through the skin, there is no way of telling how far and how deep it will spread in the underlying tissue. The extent of spread can increase when the jet hits a resistant structure, such as bone. The bleeding or the sensation of pain that occur when the jet penetrates through the epidermis and reaches the dermis (bottom layer of the skin) are only the obvious indicators of injury.

Although a great number of injuries caused by different types of fluid jets have been reported, the mechanism of damage is not well understood (Vijay 1989). Reports in the dental literature and other sources provide the basis for speculation about how streams of fast-moving fluids interact with human flesh.

The site of the impact is usually small. Spread of the fluid in the tissue and the intensity of the ensuing injury is highly dependent on kinetic energy of the jet (proportional to pressure and mass flow rate) and the time of exposure. The amount of fluid entering the body depends on standoff distance (distance of the jet from point of entry). Injury will vary according to the site of injection. Tissue necrosis (death) and breakdown of tissues which are surrounded by healthier ones result from deposition of bacteria and toxins. Direct mechanical damage to tissue and blood vessels also can occur due to the abnormally high shear forces induced by the jet. Release of gas (usually air) along soft tissue planes can cause severe emphysema. Gas-induced infections can be fatal if not treated early by adequate surgical debridement (cleansing of the wound by cutting away dead or infected tissue, foreign matter, etc.) (Vijay, 1989).

Three stages of injury due to fluid injection have been characterized:

Acute stage: There are immediate symptoms of swelling and an increase in interstitial pressure, accompanied by edema. This can cause compression of the arteries or thrombosis making the tissues white and anesthetic.

Intermediate stage: This phase is characterized by the presence of oleogranulomas (nodular tumors formed as a result of the reaction between the foreign fluid and the tissue). Oleogranuloma formation may not occur with water. These can remain within the body without change for a number of years.

Late stage: Breakdown of the oleogranulomas close to the surface of the skin is associated with the appearance of widespread lesions (tumors, etc.) Malignancy also may occur.

TABLE H.2
Personal Protection for Water Jetting

- Head protection
- Full-face shield
- Eye protection — combination of visor and goggles or a full hood with shield
- Body protection — waterproof and/or liquid or chemical resistant suits
- Hand protection — waterproof and/or liquid or chemical resistant gloves
- Foot protection — waterproof boots with steel toe caps + metatarsal guard for jetting gun operators
- Hearing protection — jetting noise easily can exceed 90 dBA
- Respiratory protection — required where toxic substances can be produced as a result of the jetting process

In practice there is no pressure at which fluid jets can be considered totally safe. Every fluid jet must be considered as having the potential to cause human injury. Basic protective clothing and equipment thus are mandatory for any operator. Katakura and Tsuji (1985) tested the effectiveness of PVC clothing as protection against injury. They concluded that PVC clothing may be useful, but there is no guarantee that it can provide full protection against injury.

The best course of action is never to ignore any fluid jet injury, no matter how small it may appear at first sight. Regard an injury caused by a fluid jet as a surgical emergency (Vijay 1989). The victim should report immediately to a hospital to seek the attention of an experienced physician or surgeon, not the inexperienced staff usually found in emergency departments.

When a person enters the hospital within a day of incidence of an injury of this type, there is a 20% chance that amputation of the limb that was affected will be required. When the patient delays 2 days, the probability increases to 60%. When the patient waits a week, loss of the limb that sustained the injection is almost certain.

In response to the recognized safety hazards attendant with use of this equipment, the Water Jet Technology Association has produced a series of manuals of recommended practices for use of high pressure water-jetting equipment (WJTA 1994). This document considers hazards according to range of pressure. Pressure cleaning or cutting refers to pump pressures less than 35 MPa (5,000 lb/in.^{2}). High pressure ranges from 35 to 210 MPa (5,000 lb/in.^{2} to 30,000 lb/in^2). Ultrahigh pressure includes pump pressures exceeding 210 MPa (30,000 lb/in.^{2}). Adequate precautions are required at all pressures. The Recommended Practices emphasize the importance of training for persons who operate or maintain high pressure water-jetting equipment.

An important concern in this industry is freeze-up. Ice in lines can have a devastating effect on the safe use of this equipment.

The Recommended Practices include a list of personal protective equipment (Table H.2). This, of course, is most relevant when the water jet is used in open conditions and contact either by direct path or reflection can occur. Protective equipment may not protect the operator from injury from direct impact by the high pressure water jet. Fluid jets have tremendous penetrating power.

Lancing

Lancing is a technique that employs compressed air or steam. The pressure and flow of the discharge dislodge and break up material. Lancing is most effective on systems during pre-shutdown preparation, especially closed systems that operate under negative pressure. The operator positions the lance through a small opening in the structure near the surface to be cleaned and manipulates the flow and discharge to produce the desired effect. Lancing is used to dislodge ash and soot from tubes and other surfaces in boilers prior to entry. The normal flow of air in the boiler carries away the dislodged material.

Lancing is likely to generate considerable airborne dust. Lancing also can release volatiles trapped in the structure of the material. Some solids can emit vapor through sublimation. Heat provided by steam can accelerate the emission of volatile substances. Airborne dust and volatiles generated in this way can constitute a source of exposure to the operator. This is especially possible when the work is performed through a manhole or other large opening. Lancing must be used with considerable caution in these circumstances because of the possibility for generation and accumulation of static electricity. Static electricity potentially could trigger a vapor explosion.

High viscosity materials and solidified sludges respond to steam lancing. Lancing using steam breaks up aggregated material or makes it more mobile by increasing its temperature. This technique is extremely useful for mobilizing heavy oils and greasy solids, and provides an excellent alternative to entry into spaces and use of physical methods of removal.

Steam lancing may not be suitable when the material reacts with water or contains ignitable substances that are released by heating. Sludges or high boiling mixtures can trap volatiles. The heat provided by the steam can liberate these materials. Volatiles released in this process can constitute a source of exposure to the operator of the lance, when this technique is performed through an open manhole. Also, both air and steam must be used with considerable caution in these circumstances because of the generation of static electricity.

The operator of the lance requires eye, face, and head protection; protective clothing; and hand protection; as well as protective footwear.

Dry Removal Systems

Some entry operations occur for the purpose of removing flowable solid material that has formed bridges or caked onto walls (Figure H.8). In ready-mix equipment, for example, concrete sets in the cavity formed between the fins and wall of the mixing drum. This material must be removed periodically in order to maintain proper operation. The concrete sets into a curved shape and is removed using jackhammers. These can fall from the drum without warning during breakout.

An additional problem with flowable solids is formation of bridges and ratholes. Bridges form when a layer of flowable material sets and underlying material is drained away. Bridges are a well-known hazard in silos and hoppers. Collapse of bridges on many occasions has led to partial or full engulfment of individuals working on them. Ratholes are channels in the bridge through which loose material can continue to pass.

Dry removal systems are a necessary alternative to systems based on high pressure water jets. Dry removal systems have application where the product must not come into contact with water. For example, cement that has caked or bridged in a storage silo is still saleable, so long as it has not been wetted (Hauck 1993). Dry removal systems also have advantages where cold weather is a concern. High pressure water-jetting systems cannot be used where freeze-up is a possibility.

An appreciation of the nature of problems confronting dry removal is helpful for putting the accomplishments of this technology into perspective. Powdered materials can form extremely hard, caked solids. Removal by blasting with dynamite has been tried as a last resort (Anonymous 1989a, Anonymous 1989b).

Cutting (Attrition) Head Systems

Some equipment used in this industry is proprietary. One type of equipment uses cutting heads (Anonymous 1989a, b; Hauck 1993). One type of unit is described as an industrial rototiller (Anonymous 1989a). This equipment can be powered by compressed air, steam, other inert gas, water, or hydraulic fluid (Anonymous 1989a, b). These systems can clean anything from a 10-cm (5 in.) pipe to a 15-m (50 foot) diameter silo. These systems can clean from top to bottom or bottom to top and work on corners and chutes, and downward-pointing cones. They cannot accommodate sludges.

The operator controls the motion of the cutting head through guidelines located outside the space. The opening used to admit the cutting equipment is the only entry into the space. The cutting

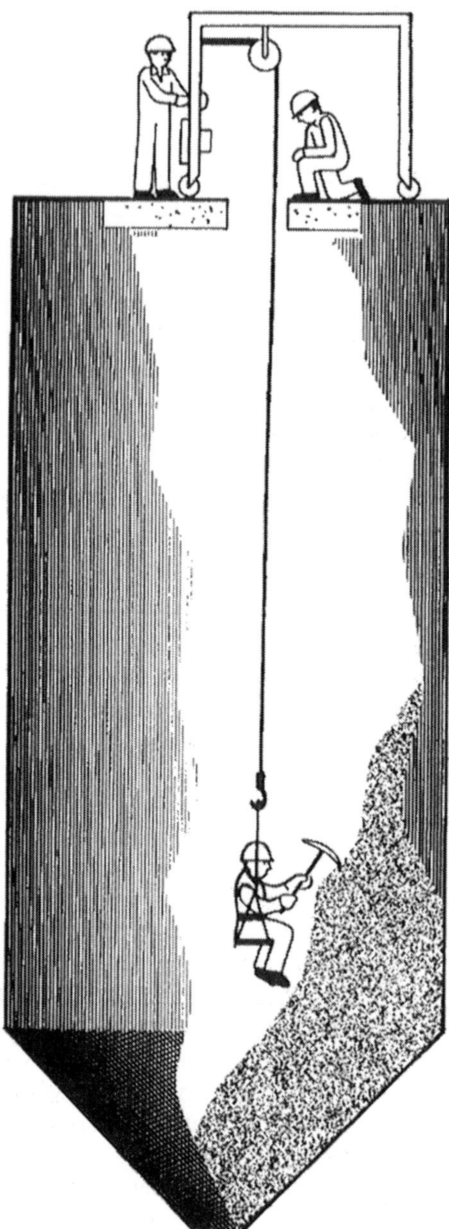

FIGURE H.8 Manual silo cleaning. This labor-dependent approach requires entry of the cleaner into the silo to dislodge the material. This procedure has high hazard potential because of the unpredictable behavior of the flowable solid. An additional hazard can arise due to atmospheric hazards posed by dust and vapor released due to off-gassing. (Courtesy of Northern Vibrator Mfg. Ltd., Georgetown, ON.)

heads can be positioned into a blocked area to cut a path or to remove built-up material from walls. The system can accommodate one or more cutting heads. The cutting heads vary in size, depending on the application. Materials used in the cutting (or attrition) heads include hardened steel, fiberglass–reinforced rubber, acrylonitrile–butadiene–styrene (ABS), or aluminum. Selection for a particular circumstance depends on the hardness and other properties of the material being removed. To illustrate, the preferred material of construction in the attrition head for removing coal is fiberglass-reinforced rubber. Fiberglass-reinforced rubber is sufficiently hard to remove the relatively soft coal without causing sparks that could cause fire. (An additional safety measure is to purge the structure with carbon dioxide.) Cement normally is the hardest of the materials removed by this equipment. This system also is able to remove rust from steel silos.

This technology has been applied to a wide variety of structures utilized in processing and storage of bulk powdered materials. Examples of materials that have been removed include: coal, coke, cement, sand, ore, grain and flour, coffee, cottonseed meal, ladle and furnace linings, powdered inks and pigments, lead oxide, sawdust, clays, powdered adhesive resins, powdered milk, and others. These materials span a broad range of industry. They provide evidence of the many potential applications for this equipment that avoid the need to enter into confined spaces.

Whips

Another type of abrading device is the whip (Hauck, 1993). The whip is a motorized device that is lowered and positioned in the structure (Figure H.9). Attached to the shaft of the motor is a hub to which are attached cord-like flails containing disks and bolts. The flails, disks, and bolts are made from materials compatible with the properties of the material to be abraded. Considerations in selection of these materials include hardness of the deposit and underlying structure, and fire and explosibility hazard of the resulting dust cloud. Chain is used against hard noncombustible materials. On the other hand, plastics are used with soft materials, such as grain or sand, to prevent sparking. The flails and attached disks and bolts strike against the deposit, thereby abrading the material. Compressed air or hydraulics are used to power the drive motor in the central hub. An expandable boom is used to position the abrading head into position.

The whip is effective against deposits on vertical surfaces. Whips cannot be used to clean out corners, tops, bottoms, or cones or large pipes or chutes.

Drills

An adjunct to the abrading equipment is the drill (Figure H.10). The drill is used to create a channel in the solidified material. Two basic designs are available. In the one case, an external hydraulic motor rotates the entire shaft. In the other, the drill motor is located at the drill head. Depending on the needs of the situation, these units can work from above or under the bridged material. The hole created by the drill provides a channel for drainage, as well as operation of the other equipment mentioned.

Safety Hazards and Protective Measures

Cleaning and dislodging powdered bulk solids from inside structures that store and handle them can involve a number of health, safety, and fire hazards. These structures often contain mechanical and pneumatic systems. These must be deactivated, de-energized and locked out to prevent accidental start-up and operation.

Equipment of this type can create potential safety hazards. Primary among these are fire and explosion. Pulverizing the caked materials in the exhaust of compressed air can create a highly concentrated dust cloud and possibly vapors. Without proper design and execution, the equipment and its ancillaries could act as ignition sources. Measures that minimize this potential include:

- Use of steam or other nonflammable gases or hydraulics to power drive motors
- Maintaining an atmosphere in the space of inert gas, such as carbon dioxide or nitrogen
- Use of nonsparking materials for the attrition heads or whip
- Application of bonding and grounding principles in conductive hoses, carbon based or with interior wire and connections
- Low voltage electrical equipment, as appropriate

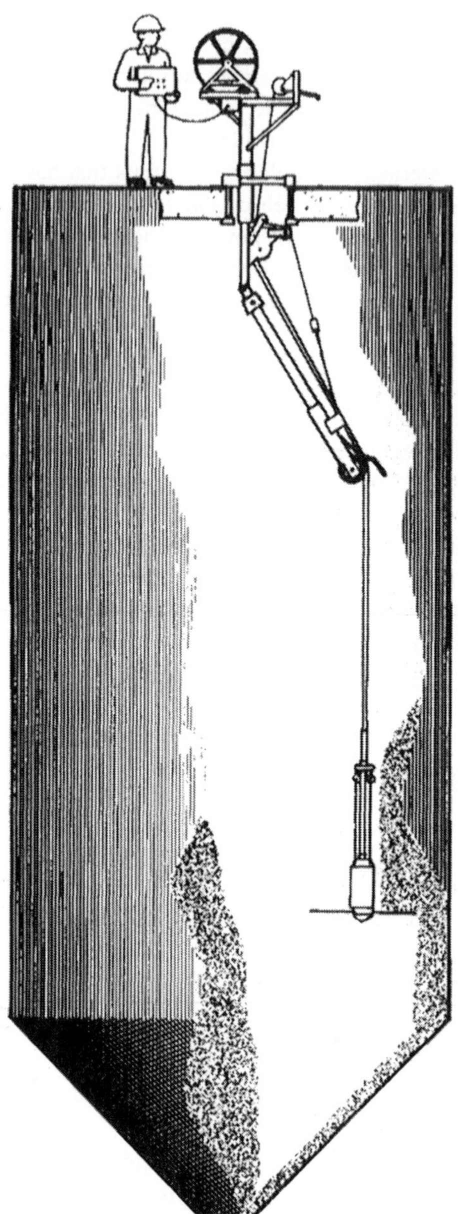

FIGURE H.9 Mechanical whip for remote silo cleaning. Mechanical equipment of this type can use air, steam, or hydraulics for motive power. (Courtesy of Northern Vibrator Mfg. Ltd., Georgetown, ON.)

Emission of highly toxic dust from openings in the structure can occur as part of the escape path of the compressed air used to power equipment. This material could create exposure hazards to individuals working in the area.

Vacuum Loaders

Clean-up using vacuum-loading equipment is an important follow-up to cleaning methods mentioned in previous sections. Vacuum-loading equipment is available skid-, trailer-, or truck-mounted (Figure H.11). For discussion purposes, these units are classified as wet/dry vacuum loaders, and

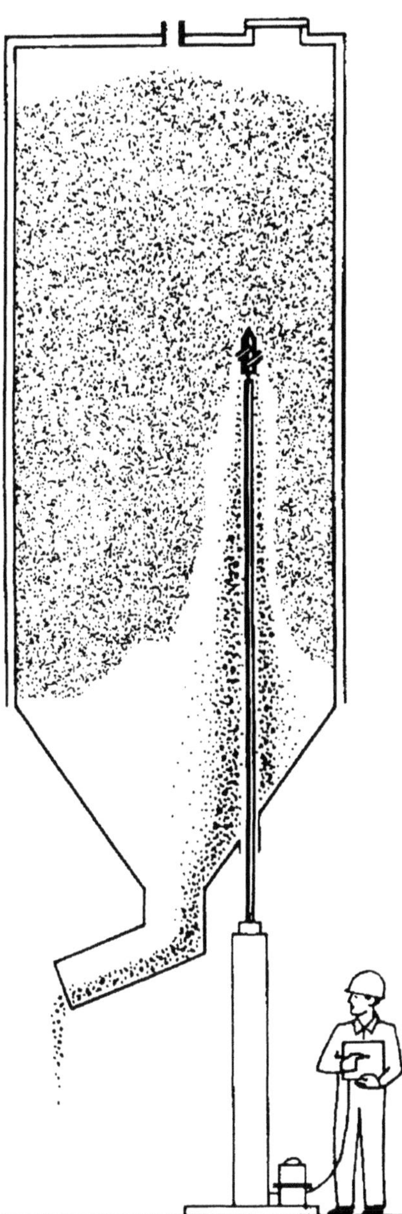

FIGURE H.10 Drill unit. The drill is used to create an opening through the column of coalesced material. The opening then becomes the pilot hole for use of the whip or other cleaning device. (Courtesy of Northern Vibrator Mfg. Ltd., Georgetown, ON.)

vacuum liquid tankers (API 1986). The wet/dry units can remove liquids and sludges, and solid materials simultaneously without modification or adaptation. The vacuum liquid tankers subclassify into units that handle only water-based liquids and sludges and those that can handle all liquids and sludges, especially those containing volatile hydrocarbons. Water-capable units are utilized primarily in sewer cleaning and cleaning wastewater treatment facilities.

An unusual application for a wet/dry vacuum loader was to save the life of an individual who was in process of being engulfed and suffocated in grain (Dorman 1993b).

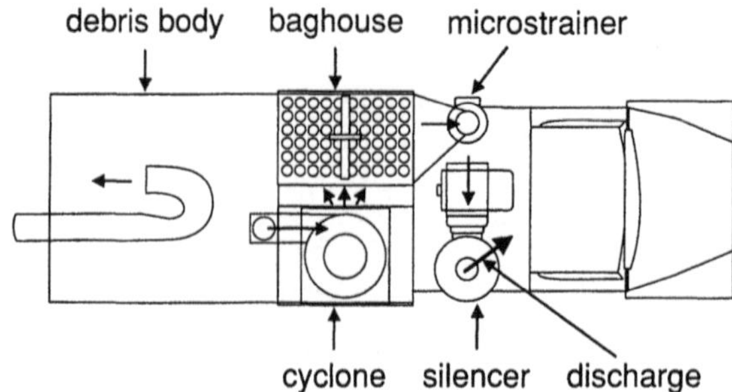

FIGURE H.11 Vacuum-loading truck. Vacuum-loading trucks use a positive displacement vacuum blower to draw air and debris into a collecting tank and then through various separatory devices (cyclone, baghouse, and microstrainer) prior to discharge. These units can handle a wide range of wet and dry materials having a variety of viscosities and sizes. Some units can generate a vacuum of 69 cm (27 in.) Hg. (Courtesy of Guzzler Manufacturing Inc., Birmingham, AL.)

Vacuum-loading equipment consists of some or all of the following parts:

- Suction hose
- Primary separator
- Debris collection tank
- Centrifugal (cyclone) separator(s)
- Baghouse
- Final filter
- Vacuum blower/sludge pump
- Discharge air silencer
- Sludge loading/offloading pump
- Water washdown equipment

This material and the following discussion was obtained from manufacturer's literature and should be considered in its generic context, rather than in the context of specific products.

This equipment can accommodate suction hose or metal tubing to 8 in. (20 cm) diameter. The suction hose contains the vacuum safety relief. This is a flat disc containing beveled edges to which is attached a perpendicular rod. It is located in a "T"-shaped fitting. The "rip cord" attaches to the rod. During an emergency, the operator pulls on the "rip cord." This dislodges the disk from its seat and admits air into the line, thereby reducing the suction.

The primary separator is an optional add-on. This unit consists of a centrifugal (cyclone) separator and associated valves and control equipment. The primary separator intercepts debris in the vacuum line upstream from the vacuum loader. The cyclone separator removes particulates and liquid aerosols from the airstream with 98% efficiency. Intercepted material can be collected into 200-L drums (45 gal U.K./55 gal U.S.), bags, dumptrucks, or roll on/roll off containers. Some units contain a dual discharge. This permits continuous operation.

For nonhazardous materials, the primary separator can dramatically improve productivity of the vacuum-loading equipment, since the debris collection tank does not become filled. In the case of hazardous materials, such as asbestos and lead-containing dust or debris, this equipment minimizes contamination of the collection body in the vacuum loader. A water-injection system is available for cleaning the hopper of the cyclone and eliminating fugitive emissions at the discharge.

The debris collection tank contains structural or air-guided deflectors that cause initial separation of entrained material from the airflow. These also act to minimize resuspension of material already in the tank. Redirection of flow reduces the air velocity in the entrained material. This initial separation reduces particulate loading in air entering the centrifugal (cyclone) separators. A float device prevents overflow of material from the collector body into the downstream air path. Most of the units drain by gravity from a raised position. A cover at the base of the rear panel or the rear panel itself opens to allow escape of collected material. Pneumatic (pressurized), pumped, or augered emptying also are available.

The centrifugal (cyclone) separator(s) in the truck body may occur singly or in multiples. Multiples offer the advantage of smaller diameter and redundancy in the event of damage. These units are rated at 98% efficiency. Liquid and solid debris from the cyclones collects in a storage area located below. This area drains through chutes that empty at the rear of the truck when the debris collection tank is elevated for dumping. A high pressure water injection system is available for cleaning the hopper and chute(s).

Airflow from the cyclones next passes through a baghouse. The baghouse is rated at 99% efficiency for all material larger than 1 μm. Debris that accumulates on the bags is blown off by periodic pulsed blasts of air. Debris from the bags collects in a storage hopper located below. The hopper drains through a chute when the rear of the truck is elevated for dumping. Hatches provide access for washing.

The final filter removes particulates that escaped collection in the baghouse because of leakage in a bag or around a seal, or through resuspension of debris knocked loose from the bags. The final filter protects the positive displacement vacuum blower. HEPA (high efficiency particulate air) filtration is available as an option.

The air mover in wet/dry vacuum loaders is a high capacity rotary lobe positive displacement blower. These units can generate Hg (mercury) vacuum up to 28 in. (94 kPa) of Hg (mercury) vacuum at flow rates up to 6,000 ft^3/min (2.8 m^3/s). A hydraulic sludge pumping system can be used for vacuum loading and off loading of liquids. Loading and off loading can occur simultaneously.

An exhaust air silencer is situated downstream from the blower.

Some units incorporate a high pressure water-jetting unit for washdown. This is essential where hazardous materials are involved and decontamination of the internals of the unit is required.

Safety/Environmental Concerns

Vacuum loaders will accept anything that can fit through the opening at the end of the hose and be drawn along its length into the debris collection body. As a result, this capability simultaneously constitutes the greatest strength and the greatest liability of this equipment. Debris can originate from almost any conceivable source. What enters the collection hose reflects decisions made by the operator of the vacuum loader and the organization requesting the service. An infinite number of complex mixtures and possibly some reaction products could form inside the collection tank when materials from different sources are combined. Strict attention to requirements of environmental and waste management legislation is required.

Operation of this equipment is not regulated directly in the U.S. (Dorman 1993b). As a result, responsibility for safety in its operation and application falls to the user. Manufacturers provide some assistance. The American Petroleum Institute (API) produced a code of practice for the operation of vacuum trucks in petroleum service (API 1986). The foreward to this document prominently distinguishes between pneumatic conveying equipment and liquid vacuum tankers. The former utilize high airflow created by a positive displacement blower to entrain material. According to API, this approach is not suitable for hydrocarbon service. Indeed, manufacturer's literature for wet/dry vacuum loaders refers to crude oil as the sole hydrocarbon liquid that they handle. Liquid vacuum tankers use vacuum created in the collection tank to draw liquids and sludges and pressure to expel them. The source of the pressure/vacuum is a low flow vane pump. This minimizes vapor formation in the airflow above the liquid.

TABLE H.3
Safety Concern with Operation of Vacuum-Loading Equipment

Concern	Comment
Electrical contact	Some units have long extension booms that permit operation over obstacles. Booms of any kind increase potential for contact with powerlines.
Mechanical contact	Vacuum loaders and their optional add-ons are complex units. They contain subsystems that rotate, articulate, operate under high vacuum, and contain spring-loaded and pressurized mechanisms. These subsystems produce or can produce motion during normal operation, failure mode, or during maintenance. Moving parts create potential for crushing, pinching, cutting, amputation, and decapitation.
Exhaust gases	Exhaust gases from diesels contain nitric oxide and nitrogen dioxide, carbon monoxide and carbon dioxide, sulfur dioxide, hydrocarbon vapors, and soot. Operation of these units in enclosed areas can confine exhaust gases. This is especially the case during operation in cold weather when exhaust gases stratify at low levels.
Hot surfaces	Blower casings and associated piping and accessories, and pulleys with slipping drive belts can become hot enough to cause major skin burns.
Coupling failure	Failure at a coupling can lead to whipping action in discharge hose or leakage from suction hose.
Spills	Spills have occurred following accidental overfilling due to failure of shut-off mechanism in the collector tank.
Vacuum cleaning	Wear hard hat, safety glasses, protective clothing, and gloves; respiratory and other protection also may be required.
	Use straight lengths of aluminum or steel tubing or hose containing as few bends as necessary.
Vacuum relief	Valve on truck — release vacuum when engine shuts down while still vacuuming to prevent damage to blower
	In-line emergency relief valve — pull to release (Figure H.12)
Noise	Noise is mentioned as a major operational hazard in trade discussion about this equipment. Noise levels are not available from manufacturers. Operational noise can mask warning signals and other communication.
Decontamination	Interior of debris collection tank and collection hoppers and chutes may require decontamination following contact with hazardous material.
Confined spaces	Debris collection tank, cyclones, baghouse structure, hoppers, and chutes are potential confined spaces. These spaces can trap volatile atmospheric contaminants. Tailgate and access hatches require securement to prevent accidental closing.

The major concern of API with high airflow is formation of ignitable mixtures. High airflow across a surface wetted by a volatile hydrocarbon liquid promotes evaporation. Under suitable conditions of temperature and airflow, ignitable mixtures could form in the airspaces within this equipment. These could include the collector hose, debris collection tank, centrifugal separator, baghouse, collection hoppers, final filter, blower, and exhaust air silencer.

API 2219 documents incidents involving vacuum trucks in which fires have occurred. Diesel engines, despite the limited electrical system compared to gasoline engines, still must be considered a potential source of ignition. Other possible sources include exhaust systems and external wiring. API 2219 points out that shielded ignition systems and muffler-mounted spark arresters should not be considered as a substitute for a gas-free work area. Locating the truck remote from possible sources of vapor is considered an important safety precaution. Dieseling or run-on can occur when high concentrations of flammable vapor enter the air intake of diesel engines. Vapor clouds can develop through discharge of volatiles by the vacuum pump or blower. Venting to a location remote from the truck is considered an important safety precaution.

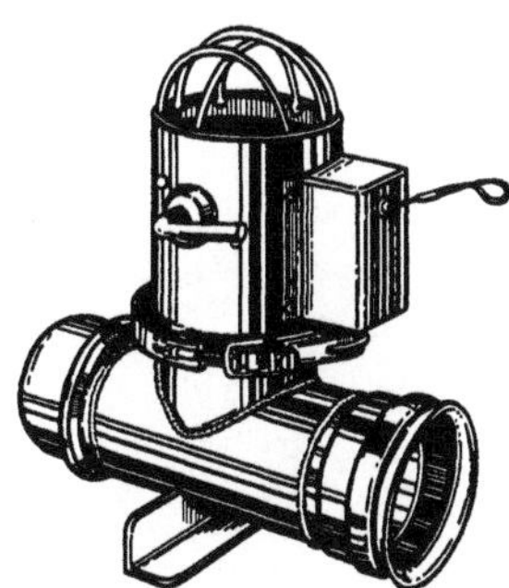

FIGURE H.12 In-line vacuum relief valve. This device is essential for safe use of this equipment. The in-line vacuum relief valve provides a bypass to relieve the vacuum in the lines. (Courtesy of Guzzler Manufacturing Inc., Birmingham, AL.)

The blower used in wet/dry loading units runs hot and must be considered a potential source of ignition to ignitable mixtures both internally and externally. Fire and explosion suppression systems are installed in some of these units as an option. Slippage of drive belts can cause pulleys to run hot.

Bonding and grounding are important safety precautions for ensuring that buildup of static electrical charge does not occur. Continuity testing of the bonding cable is required. API 2219 mentions that petroleum industry experience indicates that static electricity does not present an ignition problem with either conductive or nonconductive hose. Exposed metal in nonconductive hose, such as flanges and couplings, can act as an isolated conductor and can accumulate static charge. A spark can occur if this metal touches or is brought close to ground. API 2219, therefore, recommends grounding all metal parts when any hose is used in applications other than closed systems. An alternative would be to verify by means of electrical testing that a hose is conductive.

Vacuum loaders are complex machines that are operated in complex environments. Table H.3 lists additional concerns associated with operation of this equipment.

Additional safety issues regarding operation of this equipment concern road safety from shifting cargo, braking distance, and cargo dumping.

OTHER APPLICATIONS

An interesting application of robotics in confined spaces is rehabilitation of aging pipe through trenchless technology (Anonymous 1995) (Figure H.13). This technology eliminates the need for digging and replacement and can be utilized *in situ* where pipe has deteriorated and collapse has occurred. One process begins with installation of a thin tube into the pipe to be rehabilitated. The replacement pipe, which initially is collapsed, is drawn through this tube. A "rounding device," drawn through the interior of the new pipe, expands it into the contours of the existing pipe. The new pipe, which was heated prior to being installed, cools and sets into the expanded shape. Sufficient pressure is applied by the rounding device, that openings at connections form dimples in the wall of the new pipe. The dimples are located by a video inspection camera. A robotic cutter then cuts away the material that blocks the opening.

Operation of this equipment requires consideration for hazardous conditions. The cutter, used in the pipe relining equipment, for example, operates at 15,000 rev/min. Used on plastic materials, this could be the source of considerable heat, as well as static electrical charge. Used in an ignitable atmosphere, a fire or explosion could result.

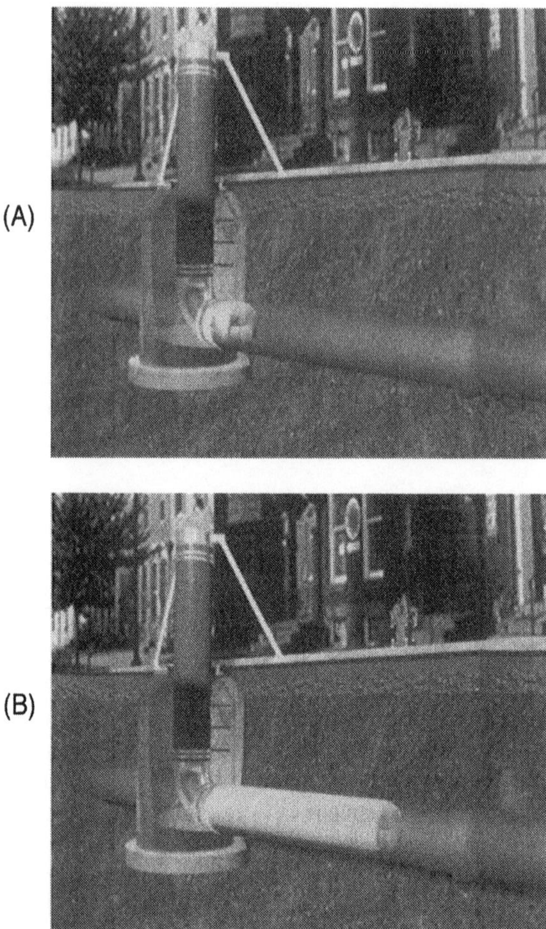

FIGURE H.13 (A-D) Trenchless pipe rehabilitation. Trenchless pipe rehabilitation technologies utilize liners to replace the existing inner surface of the pipe (a pipe within a pipe). The liners are drawn through the pipe and expanded, using several techniques. All of these procedures occur externally to the pipe. The following illustrates a specific process. (A) The liner is inverted into the deteriorated pipe. (B) Water pressure propels the inverting liner through the deteriorated pipe. (C) Hot water circulated through the fully inverted liner sets the thermosetting resin. (D) A robotic unit cuts through the liner to restore service laterals. The restored pipe is then inspected. (Courtesy of Insituform® Technologies, Inc., St. Louis, MO.)

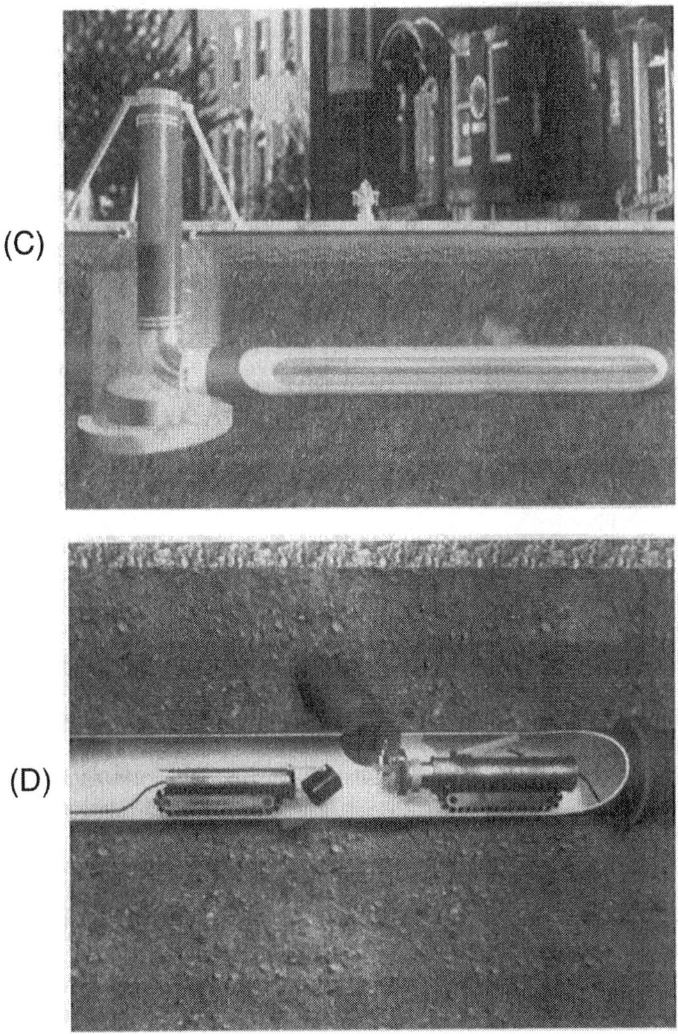

FIGURE H.13 (continued)

REFERENCES

American Petroleum Institute: *Safe Operation of Vacuum Trucks in Petroleum Service* (API Pub. No. 2219). Washington, D.C.: American Petroleum Institute, 1986. 7 pp.

Anonymous: Mechanical moles get to the bottom of silo storage problems. *Powder Bulk Eng. 3:* January, 1989a. 3 pp. [reprint].

Anonymous: Silo cleaning system moves cement terminal's stock. *Powder Bulk Eng. 3:* December, 1989b. pp. 22–25.

Anonymous: New Computerized Air Preheater Cleaning System. Houston, TX: *Butterworth Jetting News* (newsletter published by Butterworth Jetting Systems Inc.): Summer, 1992. 4 pp.

Anonymous: Contractor's program featured at 7th water jet conf. *Ind. Clean. Contractor 1:* October, 1993a. pp. 14–17.

Anonymous: New rail tank car cleaning machines use high pressure water. *Ind. Clean. Contractor 1:* March, 1993b. pp. 34–36.

Anonymous: High pressure pumps, intensifiers and accumulators. *Ind. Clean. Contractor 2:* November/December, 1994a. pp. 18–19.

Anonymous: New baking soda system adds muscle to power washing. *Indust. Clean. Contractor 2:* July/August, 1994b. pp. 22–25.

Anonymous: Pipe replacement technique. *Canadian Heavy Equip. News 10:* September, 1995. p. 6.

Dorman, S.: High pressure water jetting systems. Part I. *Ind. Clean. Contractor 1:* May, 1993a. pp. 9–13.

Dorman, S.: Technological developments in industrial vacuum loaders. Part II. *Ind. Clean. Contractor 1:* May, 1993b. pp. 24–28.

Hauck, R.M.: How to clean your storage vessel quickly and safely. *Powder Bulk Eng. 6:* November, 1993. pp. 47–50.

Katakura, H. and S. Tsuji: A study to avoid the dangers of high speed liquid jets. *Bull. Jpn. Soc. Mech. Eng. 28:* 623–630 (1985).

Janzen, P.: "Remotely Operated Tools in the Canadian Nuclear Industry." 1995 January 11. [Personal communication] Mechanical Equipment Development Branch, Chalk River Laboratories (Atomic Energy of Canada Limited), AECL Research, Chalk River, Ontario, Canada K0J 1J0.

Sefton, A.K.: Explosion proof products from Pearpoint are certified for use in hazardous locations. *The Cleaner:* April 1994.

Summers, D.A. and J. Viebrock: The impact of water jets on human flesh. Paper H4. In *Proc. 9th Int. Symp. on Jet Cutting Technology (Sendai, Japan).* Cranfield, Bedford, England: British Hydromechanics Research Association, 1988. pp. 423–433.

Summers, D.: Historical perspective of fluid jet technology. In *Fluid Jet Technology. Fundamentals and Applications,* 2nd ed. St. Louis, MO: Water Jet Technology Association, 1993. pp. 1–21.

Vijay, M.M. and W.H. Brierley: Feasibility study of cutting some materials of industrial interest with high pressure water jets. In *Proc. 2nd U.S. Water Jet Conf. (Rolla, Missouri).* St. Louis, MO: Water Jet Technology Association, 1983. pp. 289–298.

Vijay, M.M.: A critical examination of the use of water jets for medical applications. Paper No. 42. In *Proc. 5th American Water Jet Conf. August 29 to 31, Toronto, Canada.* St. Louis, MO: Water Jet Technology Association, 1989. pp. 425–448.

Vijay, M.M.: High pressure safety. In *Fluid Jet Technology. Fundamentals and Applications*, 2nd ed. St. Louis, MO: Water Jet Technology Association, 1993. pp. 9.1–9.30.

Water Jet Technology Association: *Recommended Practices for the Use of Manually Operated High Pressure Water Jetting Equipment.* St. Louis, MO: Water Jet Technology Association, 1994. 40 pp.

Williams, D.: Shut off gun versus dump gun. *Ind. Clean. Contractor 3:* January/February, 1995. pp. 18–19.

Yared, D.J.: Contractor waterjetting: equipment and applications. *Ind. Clean. Contractor 2:* July/August, 1994. pp. 10–15.

Appendix I: Standards, Guidelines, and Legislation Affecting Ventilation of Confined Spaces

CONTENTS

INTRODUCTION

This Appendix summarizes ventilation requirements and guidance presented in standards, guidelines, and legislation that affect confined spaces. Few of the published standards or guidelines on confined spaces address ventilation (Garrison and McFee 1986).

Many documents use the term "adequate ventilation" to describe a desired state of attainment. The problem with the term, aside from its vagueness, is that practitioners have no sense of what this means or, more importantly, how to attain it. Does "adequate," in a confined space where airflow is not precisely controllable, have the same meaning as "adequate" in a normal workspace where it is? Does "adequate" mean airflow past the breathing zone of 100 ft/min (0.5 m/s), when the air that is flowing was contaminated upstream? While persons trained in ventilation principles may interpret "adequate" to mean control exposure to less than some standard, how can they assure this in the rapidly changing work environment in the confined space? How is the potential for overexposure to be recognized *a priori*? Would some simple measure, such as air velocity, that can be checked easily, provide assurance of protection? The best resource for ensuring success, that is, providing answers to these questions, is the experience of those who have learned these lessons. This experience surely is describable in concrete terms, rather than being left in the abstract in vague terms such as "adequate ventilation."

Principal sources of information about ventilation of confined spaces are guidelines produced by standard-setting organizations, trade associations, individual organizations, and government Standards. (The term "standard" is used to denote publications from voluntary organizations. The term "Standard" is used to denote legislated rules.) Other sources include textbooks written by ventilation experts and reference publications.

Procedures for ventilating confined spaces *do* exist. Industries such as telecommunications, petroleum refining, and chemical manufacturing addressed this issue years ago. Some companies have developed specific requirements for ventilation. However, this information has not been published or otherwise been made available to wide readership. There is an undeniable need to compile what is known and practiced and to provide means for making this knowledge, which is essentially experience based, widely available.

Description of practices opens them to scrutiny and enquiry, and possibly to criticism. This obviously entails some risk, but at the same time can provide benefit. The risk is that the practices, as described, may not be entirely satisfactory or optimum. The benefit is that improvement and optimization for the benefit of all can occur through sharing them with others.

GUIDELINES AND STANDARDS

Guidelines and standards have resulted from efforts by government agencies, such as NIOSH (National Institute for Occupational Safety and Health), and voluntary and industry groups also produced guidelines. In the U.S., these efforts preceded adoption of legal statutes by many years. These guidelines and standards, while having no force in the U.S. unless cited in legal statutes, have provided important guidance to practitioners in other countries. Also, they provide a historical record and alternate approaches for discussion.

National Institute for Occupational Safety and Health (NIOSH)

The National Institute for Occupational Safety and Health published one of the first comprehensive approaches for entry and work involving confined spaces (NIOSH 1979). This approach required identification of confined spaces and classification into classes (A, B, or C) according to hazard. A "Qualified Person" would determine the class of a particular confined space. Assignment to a particular class depended on results from preentry air monitoring and the occurrence and potential severity of other hazards.

Class A spaces would satisfy one or more of the following conditions: an atmosphere immediately dangerous to life or health, oxygen content 16% or less, flammables/combustibles at 20% or more of the Lower Explosive Limit (Lower Flammable Limit). In Class B spaces, the contaminant level might exceed OSHA Permissible Exposure Limits, oxygen content would lie between 16.1% to 19.4%, and/or flammables/combustibles would be 10% to 19% of the LEL (LFL). Class C spaces would have the potential for generating a hazardous atmosphere. Measured air contaminant levels would be less than OSHA PELs, oxygen content would lie between 19.5% and 21.4%, and flammables/combustibles would be less than 10% of the LEL (LFL).

Mechanical ventilation was required for entry into Class A and B spaces. In Class C spaces, a "Qualified Person" would assess the need for ventilation. Ventilation systems for Class A entry would require audible devices to warn in the event of failure. Airflow measurements on the ventilation system were recommended before entry to confirm performance. Where mechanical ventilation was not part of the operating procedure, at least three periodic air tests were required to determine that contaminant and oxygen concentrations would remain within safe limits.

The NIOSH guidelines required continuous mechanical ventilation under certain conditions. These included production of toxic atmospheres as part of a work operation, such as welding or spray painting, or development of a toxic atmosphere due to the nature of the confined space.

NIOSH recommended use of local exhaust when general dilution ventilation was ineffective. The latter could occur because of complexity of the space or generation of high concentrations of contaminants from localized sources, such as welding and manual cleaning. The NIOSH guidelines acknowledged that ventilation might not be completely effective in eliminating the hazardous atmosphere in some situations.

NIOSH also addressed ventilation in confined spaces in a training guide directed at the construction industry (NIOSH 1985). This document contained a detailed discussion on ventilating confined spaces, including recommendations for volumetric airflow rates and velocity compiled from various sources. It also recommended a minimum volumetric flow rate of 20 air changes per hour. This is one of the few published recommendations on this subject.

American National Standards Institute (ANSI)

The American National Standards Institute has published comprehensive standards for working in confined spaces (ANSI 1977, 1989, 1995). The standard states that ventilation should continue until the atmosphere is acceptable. Acceptable means oxygen levels between 19.5% and 23.5%, flammable/combustible levels below 10% of the LEL (LFL), and/or contaminants below appropriate Threshold Limit Value (TLV) or applicable legal limit.

The ANSI standards also mentioned that stratification of contaminants can occur in some spaces. It cautions about placing the exhaust intake low in the space for contaminants that are denser than air. The concept of contaminant stratification must be addressed with care. Stratification occurs only when conditions in the space are stable and relatively still, and when significant difference in gas/vapor density exists between the contaminant and surrounding air. In stable conditions, gas/vapor stratification is a real concern. Fatalities have occurred despite the use of mechanical ventilation when the end of the duct was positioned too high to ventilate the lower reaches of the space.

The ANSI standard mentions the importance of various parameters in ventilation design and implementation. These include volume of the confined space, blower capacity, and air distribution. It states that purging with fresh air for several "air changes" usually is sufficient. Further, it mentions the need for continuous air supply during entry and occupancy when conditions could change and become unacceptable. According to the standard, natural ventilation is acceptable, provided it can achieve results equivalent to those produced by mechanical ventilation. This is a very high hurdle because natural ventilation cannot possibly outperform mechanical ventilation under most conditions.

Standards Australia

Standards Australia has issued two versions of the standard on confined spaces in 1986 and 1995 (Standards Australia 1986, 1995). The standards require ventilation (natural, forced, or mechanical means) to maintain a safe environment when entrants are not wearing supplied-air respiratory protection. Mechanical equipment that is necessary to maintain safe working conditions shall be monitored continuously during occupancy of the space. Further, controls of this equipment shall be tagged clearly to avoid unauthorized interference. Exhaust facilities shall ensure that air exhausted by this equipment does not create a hazard to persons and equipment.

National Fire Protection Association (NFPA)

Standards applicable to work involving confined spaces are NFPA 306 and NFPA 69 (NFPA 1992, 1993). NFPA 306 sets a general protocol for assessing, and controlling or eliminating hazards. Ventilation is one of the strategies in hazard control. NFPA 69 sets out requirements for purging.

Industry Guidelines

In addition to government agencies and consensus organizations, trade associations have developed specific policies and procedures for confined spaces. Examples include guidelines produced by the American Gas Association (AGA) and American Petroleum Institute (API) (AGA 1975, API 1993).

Industry guidelines have several advantages over regulations from governmental agencies and industry-wide standards. They can be much more detailed and specific to the actual situations encountered by organizations within the group. They also can eliminate the many nonapplicable general considerations found in regulatory standards.

American Gas Association

The American Gas Association published a valuable book on purging, ventilating and inerting (AGA 1975). This book explains principles and practices needed to assure safe working conditions primarily involving fuel gases. This document provides specific guidance on the use of inerting and purging gases in gas plants and pipelines, pipes, holders, tanks, and other facilities. This book illustrates the complexities and subtleties involved in these activities. Although not upgraded since 1975, as testimony to its continuing utility, this book has been reprinted four times. Omitted from the current edition are practices for boilers, furnaces, ovens, and atmosphere generators.

American Petroleum Institute

API 2015 details practices and procedures for preparing, emptying, ventilating, atmospheric testing, cleaning, entry, hot work, and recommissioning in, on, and around atmospheric and low pressure petroleum storage tanks. API 2015 recently underwent a considerable upgrade (API 1993). This standard applies to stationary aboveground storage tanks.

Company Procedures

Company procedures can be even more specific than industry guidelines. They can provide a well-conceived step-by-step protocol. Ventilation needs can be anticipated well in advance of the entry, proper equipment selected and provided, and individual responsibilities assigned. Company procedures and practices must include a training component, since this can profoundly affect the performance of individuals involved in the entry.

LEGISLATIVE REQUIREMENTS: OCCUPATIONAL SAFETY AND HEALTH ADMINISTRATION (OSHA)

Legislation on confined spaces has existed in various jurisdictions and countries for many years. This generally assigns few specific requirements for ventilation aside from limits for flammability, exposure to toxic substances and oxygen that must not be exceeded.

OSHA Standards provide an example of the nature of governmental regulation of work involving confined spaces. These Standards result from consideration of extensive input and are among the most detailed anywhere. As detailed as the standard-setting process actually is, the outcome regarding ventilation in confined spaces is piecemeal, and applicable only to specific, limited activities. OSHA currently has two Standards on confined spaces (OSHA 1993, 1994). A third Standard for the construction industry was proposed at the time of this writing.

The intent of the Standard on confined spaces for general industry was to amalgamate activities involving confined spaces that were covered in other Standards (OSHA 1993). The Standard is performance based, as opposed to specification based. It contains many provisions that are essential for assuring safe work in these environments.

The Standard specifically encourages use of ventilation for controlling atmospheric hazards to levels below regulated limits. Utilizing this approach can considerably reduce other obligations contained in the Standard. While encouraging considerable reliance on ventilation as a control measure, the Standard provides no assistance in its specification and implementation. What is provided are limits for flammability, exposure to toxic substances, and oxygen that must not be exceeded.

This Standard does not address the mechanics of implementing and using ventilation in a meaningful or effective way. Ventilation is mentioned very briefly in connection with written permits, as an item on a checklist of possible controls. Ventilation is listed in the appendices as one of several possible actions for preparation prior to entry. The list mentions an "air blower" as an example of safety equipment.

What requirements on ventilation do exist in the OSHA statutes are scattered throughout the individual volumes of the Maritime/Marine, Construction, and General Industry Standards. These address specific situations, and are not applicable to confined spaces in general. More than 50 different sections in the General Industry Standard address safety in confined spaces (OSHA 1992a). By comparison, the few references to ventilation requirements in the OSHA Standard on confined spaces in general industry are vague and incomplete.

Of the three OSHA Industry Standards, the General Industry Standard covers the greatest number of workers, but has few references to confined spaces (OSHA 1992a). Ventilation requirements during entry into confined spaces typically are vague. For example, the section applicable to open-top tanks containing liquids to clean or alter surface finishes (Section 29 CFR 1910.94(d)(11)(iv)) provides one example. This simply states that before individuals enter for maintenance or clean-up, "the tank shall be ventilated until the hazardous atmosphere is removed." This section also prescribes preentry air monitoring to assess the concentration of suspected toxic materials and the possibility of oxygen deficiency, defined as a concentration less than 19.5%. Mechanical ventilation is required to prevent recurrence of a hazardous atmosphere as long as a worker is in the tank.

Ventilation requirements for work in manholes and unvented vaults (29 CFR 1910.268(o)(2)) apply only to manholes related to telecommunications. They do not apply to construction, or municipal or electric utility work that may occur in the same or similar manholes. This section requires a forced and continuous supply of air to the manhole under specified conditions. The mode of ventilation is not specified, but this presumably means use of a fan. The specified conditions include detection of oxygen deficiency, combustible vapors or toxic contaminants during preentry testing, use of organic solvents or open-flame torches, or entry by contaminants from vehicular traffic or other sources. This section does not specify airflow rates.

The section on welding and torch cutting in confined spaces (29 CFR 1910.252(c)) contains some of the most specific ventilation requirements in government regulations. Continuous mechanical ventilation is mandatory under the following conditions. There is less than 10,000 ft^3 (280 m^3) of space per welder. Also, ceiling height is less than 16 feet (5 m), or structural barriers obstruct natural cross-ventilation. In these situations, a minimum volumetric flow rate of 2,000 ft^3/min (1 m^3/s) per welder is required, except when local exhaust ventilation is provided to remove fumes at their source. Local exhaust ventilation must generate an air velocity of 100 ft/min (0.5 m/s) in the zone of fume generation. Specific volumetric flow rates for a flanged circular inlet are tabulated (29 CFR 1910.252(f)(3)(i)) as a function of distance from the point of welding/cutting and duct diameter. For example, the Standard specifies 150 ft^3/min (0.07 m^3/s) at 4 to 6 in. (10 to 15 cm) using a 3-in. (8 cm) diameter duct. The volumetric flow rate remains the same when toxic fumes, such as lead, cadmium, or zinc are generated or oxygen deficiency may occur. The Standard states simply that "adequate ventilation" must be provided.

The OSHA Standard for work in confined and enclosed spaces in shipyard employment was recently upgraded. This upgrade enables a consolidated approach to all work occurring in confined spaces within the boundaries of the shipyard (OSHA 1994). Ventilation has a major role as a control measure in the application of this Standard. This Standard also specifies specific limits to be achieved by ventilation or other means, but provides no guidance on how to achieve them.

The confined space Standard for shipyard employment requires a competent person to assess atmospheric and other conditions and to determine requirements for safe entry and performance of work. The competent person must personally inspect the space and determine that hazards have been controlled by ventilation or other means. These requirements are performance oriented and depend on the capabilities of the competent person. They do not provide specifics for ventilation. The competent person requires training, the general nature of which is specified in the Standard. Training would include application of ventilation.

Other sections of the Standard acknowledge that certain work activities can rapidly worsen atmospheric conditions. Cleaning, painting, torch cutting, welding, brazing, fumigation, use of forklifts, and operation of internal combustion engines are mentioned as activities that require frequent atmospheric testing. The Standard also describes steps to be taken when hazardous atmospheres exceed PELs (Permissible Exposure Limits). These include halting the activity, evacuating personnel, and mechanically ventilating the space using local exhaust ventilation. Clean, respirable replacement air is specified when exhaust ventilation is used. Also, contaminated exhaust air must be discharged in a manner that does not contaminate the source of breathing air.

The Standard places heavy reliance on mechanical ventilation for protection against hazardous atmospheres in confined spaces. Use of respiratory protection is permitted as the final measure of protection only under narrowly defined criteria. Hence, the role of ventilation is to provide a margin of safety. Respirators represent the last line of defence.

In situations where the ventilation duct would block the only access, then ventilation may be infeasible unless additional openings can be provided. There also are work situations where off-gassing or generation of gases/vapors may occur at such a high rate that they overcome any feasible ventilation system. In these cases, airline or supplied-air respirators may be the only option in addition to mechanical ventilation.

Exceptions to reliance upon respirators do occur in the Maritime/Marine Standards (OSHA 1992b). These occur in Section 29 CFR 1915.136(c), which covers ship repairing, building, and breaking operations, and Section 29 CFR 1918.93(a)(1)(iii), which covers longshoring operations. These do not allow personnel to enter into confined spaces on board ships when the concentration of carbon monoxide exceeds 50 ppm as a ceiling level (building/breaking) or as a time-weighted average (longshoring).

Section 29 CFR 1917.152(f)(3)(i) covers wharves and other marine terminals. This section requires sufficient ventilation during hot work (welding, flame cutting, or riveting) on metals containing lead, cadmium, or chromium, so that the atmospheric concentrations of these fumes do not become "hazardous." These are the only sections in the Maritime/Marine Standards to require an engineering control (ventilation) to take precedence over respiratory protection.

The OSHA Construction Industry Standards also address confined spaces (OSHA 1992c). These Standards apply to all work involving any type of construction, renovation, or demolition. Application of the Standard is limited to specific types of activities.

The Standards provide most comprehensive coverage to construction of tunnels and shafts (Section 29 CFR 1926.800). The Standards specify environmental control measures, such as ventilation, as the primary means for protecting workers. Reliance on respiratory protection is to occur only when environmental controls are not yet developed, or when made necessary by the nature of the work involved (Section 29 CFR 1926.800 (k)(1)(i)). Examples of the latter include welding, sandblasting, and lead burning.

The Standards also specify mechanical ventilation for all work areas of a tunnel (Section 29 CFR 1926.800 (k)(1)(ii)). In the event of failure of mechanical ventilation, workers are not permitted to return until areas subject to accumulation of flammable gases or other suspect air contaminants are tested and found safe (Section 29 CFR 1926.800 (k)(8)). The Standards specify a volumetric flow rate of fresh air not less than 200 cfm per person underground. The linear velocity through the tunnel must be at least 30 ft/min when dusty activities are occurring and/or hazardous concentrations of fumes, vapors, or gases can accumulate (Section 29 CFR 1926.800(k)(2) and (3)). The Standards require evacuation of workers and de-energization of equipment and lights when air exhausted from an underground workplace contains 1.5% (15,000 ppm) or higher of flammable gases/vapors. This restriction remains until ventilation reduces the concentration to less than 1% (Section 29 CFR 1926.800(j)(1)(vii)).

The OSHA regulations concerning ventilation in tunnels and shafts under construction are among the most detailed of any OSHA Standards. They are patterned after the regulations promulgated by the federal Mine Safety and Health Administration for ventilation in underground mines. They are not applicable to other situations.

The Construction Industry Standards contain provisions very similar to those in the Maritime/Marine Standards regarding welding and torch cutting in any type of confined space. They require general mechanical ventilation or local exhaust ventilation whenever hot work is performed, unless the ventilation duct blocks access to the confined space (Section 29 CFR 1926.353(b)(1) and (2)). The ventilation must reduce the concentration of welding fumes to levels below OSHA PELs. Respirators are allowed only where accomplishing adequate control of welding emissions by ventilation alone would be difficult.

The Construction Industry Standards contain one other reference to ventilation for confined spaces. This requires provision of "sufficient ventilation" when space heaters are used. This requirement is to ensure "proper combustion" and to "maintain the health and safety of the workmen" (Section 29 CFR 1926.154(a)(2)).

TEXTBOOKS

Present texts focus on fixed industrial ventilation systems. Confined spaces either receive no mention or simply a definition. Portable ventilating equipment receives no mention (Alden and Kane 1970, McDermott 1976, Burgess et al. 1989, Heinsohn 1991, ACGIH 1995). Persons depending on these sources as resources must recognize the context of information provided by them. Of course, these books provide a wealth of information about air movement and control of exposure.

SUMMARY

Current standards and guidelines almost always leave the decision about use of mechanical ventilation to qualified individuals or those who interpret preentry air-monitoring tests. These documents usually specify legal limits that must not be exceeded, but provide little if any guidance or protocol for achieving these targets. By way of contrast, these documents often do provide assistance with other requirements, such as the form of entry permits, or management programs. In typical fashion with performance standards, this approach works best only with individuals who are well qualified and experienced in hazard control. These individuals are the exception, rather than the norm, in industrial workplaces. A better approach might be to require mechanical ventilation routinely, unless a Qualified Person demonstrates lack of need and to specify performance requirements based on task. This at least would provide some guidance to persons who are not experts in ventilation. Today's portable equipment and power sources also make feasible the widespread use of mechanical ventilation even at the most remote of job sites.

REFERENCES

Alden, J.L. and J.M. Kane: *Design of Industrial Exhaust Systems*, 4th ed. New York: Industrial Press, 1970. 243 pp.

American Conference of Governmental Industrial Hygienists: *Industrial Ventilation — A Manual of Recommended Practice*, 22nd ed. Cincinnati, OH: ACGIH, 1995.

American Gas Association: *Purging Principles and Practice*, 2nd ed. Arlington, VA: American Gas Association, 1975. 180 pp.

American National Standards Institute: *Safety Requirements for Working in Tanks and Other Confined Spaces*. ANSI Z117.1-1977. New York: American National Standards Institute, 1977. 24 pp.

American National Standards Institute: *Safety Requirements for Confined Spaces*. ANSI Z117.1-1989. Des Plaines, IL: American Society of Safety Engineers/American National Standards Institute, 1989. 24 pp.

American National Standards Institute: *Safety Requirements for Confined Spaces*. ANSI Z117.1-1995. Des Plaines, IL: American Society of Safety Engineers/American National Standards Institute, 1995. 32 pp.

American Petroleum Institute: *Safe Entry and Cleaning of Petroleum Storage Tanks* (API Pub. 2015). Washington, D.C.: American Petroleum Institute, 1993.

Burgess, W.A., M.J. Ellenbecker, and R.D. Treitman: *Ventilation for Control of the Workplace Environment*. New York: John Wiley & Sons, 1989. 476 pp.

Garrison, R.P. and D.R. McFee.: Confined Spaces — A Case for Ventilation. *Am. Ind. Hyg. Assoc. J. 47*: A708–A714 (1986).

Heinsohn, R.J.: *Industrial Ventilation: Engineering Principles*. New York: John Wiley & Sons, 1991. 699 pp.

McDermott, H.J.: *Handbook of Ventilation for Contaminant Control*. Ann Arbor, MI: Ann Arbor Science Publishers, 1976. 368 pp.

National Fire Protection Association: *NFPA 69 — Explosion Prevention Systems*, 1992 ed. Quincy, MA: National Fire Protection Association, 1988.

National Fire Protection Association: *NFPA 306 — Control of Gas Hazards on Vessels*, 1993 ed. Quincy, MA: National Fire Protection Association, 1993.

National Institute for Occupational Safety and Health: Criteria for a Recommended Standard — Working in Confined Spaces (DHEW/PHS/CDC/NIOSH Pub. No. 80-106). Cincinnati, OH: National Institute for Occupational Safety and Health, 1979. 68 pp.

National Institute for Occupational Safety and Health: Safety and Health in Confined Workspaces for the Construction Industry (DHEW/PHS/CDC/NIOSH), Division of Training and Manpower Development, Cincinnati, OH: National Institute for Occupational Safety and Health, 1985. pp. 127–135.

Occupational Safety and Health Administration: *Code of Federal Regulations Title 29, Part 1910*. Washington, D.C.: Office of the Federal Register, National Archives and Records Administration, 1992a.

Occupational Safety and Health Administration: *Code of Federal Regulations Title 29, Parts 1915, 1917, and 1918*. Washington, D.C.: Office of the Federal Register, National Archives and Records Administration, 1992b.

Occupational Safety and Health Administration: *Code of Federal Regulations Title 29, Part 1926*. Washington, D.C.: Office of the Federal Register, National Archives and Records Administration, 1992c.

OSHA: Permit-Required Confined Spaces for General Industry; Final Rule, *Fed. Regist. 58*: 9 (14 January 1993). pp. 4462–4563.

OSHA: Confined and Enclosed Spaces and Other Dangerous Atmospheres in Shipyard Employment; Final Rule, *Fed. Regist. 59*: 141 (25 July 1994). pp. 37816–37863.

Standards Australia: *Safe Working in a Confined Space* (AS 2865-1986). North Sydney, NSW: Standards Association of Australia, 1986. 19 pp.

Standards Australia: *Safe Working in a Confined Space* (AS 2865-1995). North Sydney, NSW: Standards Association of Australia, 1995. 42 pp.

Appendix J: Portable Ventilation Systems

CONTENTS

INTRODUCTION

Ventilation is essential for the control of atmospheric hazards in confined spaces. The importance of ventilation as a control technique has long been recognized in standards and guidelines on this subject. The OSHA Standard on confined spaces in general industry, for example, offers considerable incentive to use ventilation to control atmospheric hazards (OSHA 1993). Implicit, of course, in reliance on ventilation to bring about and maintain control of atmospheric hazards are the dual caveats of dependability and reliability. This Appendix will focus on practical aspects of ventilating confined spaces using portable ventilation systems. Discussion will include consideration about the relationship between hazardous atmospheres and ventilating equipment, methodology, general principles, criteria for selecting equipment, and ventilation practices.

PORTABLE VENTILATING EQUIPMENT

An appropriate beginning for this Appendix is a discussion about portable ventilating equipment and portable ventilation systems, and what they can and cannot do. The capabilities of the equipment and its utility in these workspaces set the limits of ventilation effectiveness. Any portable ventilation system must be built around this equipment. Hence, the capabilities of the system, therefore, reflect those of the equipment.

Components produced for use in portable ventilation systems include:

- Air movers
- Duct
- Accessories

In many ways these components reflect their permanently installed cousins. However, there are significant differences. These differences can impact on performance, and hence, success in using this equipment.

AIR MOVERS

Overview

Air Movers are devices that induce air motion. Induction of air motion in a controlled and predictable manner is the critical consideration in ventilation. As a group, air movers used in portable systems should have some or all of the following characteristics:

- Portability
- Compactness
- Output assessed using standardized measuring techniques
- Standardized inlet and outlet sizes
- Ability to be used in hazardous locations (available as an option)
- Connections to enable grounding and bonding to other components (available as an option)
- Common labeling format to indicate applications and limitations
- Low flow detection and alarm (required in some jurisdictions)
- In-line contaminant detection (required in some jurisdictions)

Equipment presently available may not incorporate all of these features. Some are available as options. To an extent, the preceding may seem to be a wish list. A particular application or range of applications within an organization may not warrant some of the features mentioned in this list. However, the opposite situation can happen. Equipment is incapable of meeting the needs of the situation because of lack of features. That is, the equipment could become part of the problem, rather than being part of the solution. To illustrate, rental equipment is used in a broad range of applications. So also is the equipment used by itinerant contractors. Also, while needs and conditions within organizations change, equipment that is used infrequently could remain static. What was suitable for one application at a point in time may no longer be suitable.

TABLE J.1
Comparison of Air-Moving and Air Supply Equipment

Axial fans
- Large volume
- Low static pressure capability
- Low dust loading
- Can operate in reverse direction, although not efficiently
- More compact than centrifugal fans
- Straight-line installation
- Higher efficiency

Centrifugal fans
- Can operate in reverse direction
- Greater ability to cope with uncertain or fluctuating operating conditions than axial fans
- Better access to the motor than axial fans
- Easier to support than axial fans
- Lower noise than axial fans
- Natural ability to adapt to duct requiring 90° turn
- High static pressure capability
- Low volume

Flow amplifiers
- High efficiency
- Placement is critical so as not to propel contaminated air into the space
- Noisy
- Moderate static pressure capability
- Compact

Compressed air
- Can use small diameter supply line
- In-line purification unit necessary to ensure air quality
- Purification unit may limit airflow rate
- Noisy

After Bleier 1946, 1985.

The direction taken in current legislation and standards on confined spaces provides considerable incentive to manufacturers to upgrade present designs. Meeting the unstated but intimated requirements in these documents goes considerably beyond moving air in a predictable way.

Air-moving or air supply equipment commonly used in portable ventilation systems includes axial fans, centrifugal fans, airflow amplification devices and compressed air.

Table J.1 provides a comparison of air-moving or air supply equipment used in ventilating confined spaces.

As indicated in Table J.1, air movers differ in performance capabilities. Fans impart energy to air. Fan-imparted energy is apportioned between static and velocity pressure. Static pressure measures the resistance the fan can overcome in moving air through the ventilation system from intake to discharge. The ventilation system includes the fan, duct, hoods, filters, etc. Velocity pressure is a measure of the velocity imparted to the air by the fan.

Fan performance curves provide a measure of the differences in performance between fans. Just as fingerprints distinguish people from each other, performance curves do the same for individual fans. One type of fan performance curve shows the relationship between volumetric flow rate and static pressure (Figure J.1). The blocked tight pressure is the maximum static pressure generated by the fan; that is, energy is totally converted to static pressure. Said another way, this is the static pressure above which the fan will be unable to pull air through the ventilation system.

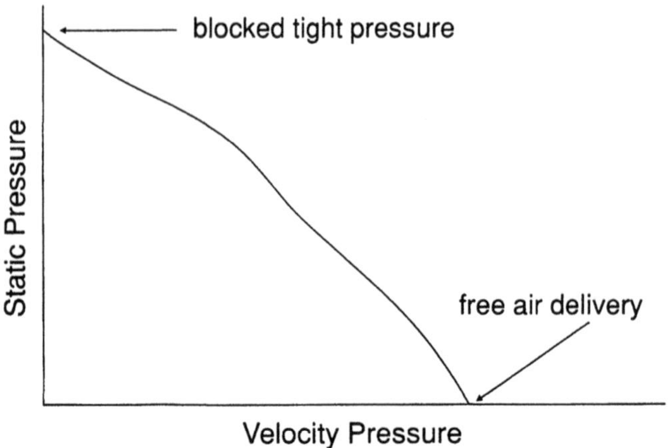

FIGURE J.1 Fan performance curve. This performance curve for a typical fan illustrates the interconversion between static pressure and velocity pressure. (Volumetric flow rate, the usual unit on the horizontal axis, is proportional to the product of the square root of the velocity pressure × area of the opening.)

The free air delivery rate is the maximum volumetric flow rate deliverable by the fan. The free air delivery rate represents the situation where energy is totally converted to velocity pressure.

Volumetric flow rate is expressed in cubic feet per minute (cfm or ft^3/min) or cubic meters per hour (m^3/h). Static pressure is expressed in inches or millimeters of water gauge (in wg or mm wg). Generally, operation at different speeds forms a family of parallel fan curves (Russell et al. 1973). Fan curves are a vital part of the process of selection, since for any given system, a fan will operate at a point on the fan curve at all times.

Noise

Portable ventilation equipment produces noise. Noise produced by this equipment is unavoidable. It arises from the motors, gasoline engines and fluids used as power sources, from internal motion of the air mover, and motion of air through the ventilation system.

Vendor literature does not provide information about noise produced by these products. Sound pressure levels are available in product literature for equipment from only one manufacturer. These data are provided in Table J.2 as representative of noise produced by this equipment. Sound levels were measured at a distance of 5 ft (1.5 m) from the unit during operation at published airflow rates.

Noise at worksites that contain confined spaces could pose a serious problem because of the need to rely on audible warnings produced by testing instruments or other equipment. These warnings easily could be masked by traffic and equipment noise and not be heard. Personnel working outside the space in proximity to this equipment likely would require hearing protection. The noise also could interfere with communication between persons working inside the space and those working outside.

Some air movers can operate at different speeds or provide airflow at different rates, depending on the pressure of compressed gas used as the source of energy. Operating a unit of larger capacity at lower pressure in some cases produces less noise than operation of a smaller air mover at its maximum operating pressure for the same volumetric flow rate.

For a given pressure in the fan casing, the noise level is proportional to the tip speed of the impeller and to the air velocity leaving the wheel. Fan noise is approximately proportional to the pressure developed, regardless of type of blade (McQuiston and Parker 1982).

TABLE J.2
Noise Produced by Portable Ventilating Equipment

Type of unit	Volumetric airflow rate ft³/min	Volumetric airflow rate m³/h	Sound pressure level, dBA
Propeller fan	2,850	4,840	73
(electrical)	8,600	14,600	84
	16,800	28,500	93
Tubeaxial fan			
(electrical)	5,500	9,300	96
Compressed air (reaction)			
Fan A	2,140	3,640	104
Fan B	5,100	8,600	109
Fan C	7,000	11,900	103
	11,000	18,700	106
	16,900	28,700	108
Fan D	14,600	24,800	109
	16,900	28,700	111
Vaneaxial fan	1,500	2,550	90
(electrical)	3,000	5,100	92
Centrifugal fan	560	950	84
(backward curved blades;	815	1,385	87
electrical)	940	1,600	90
	1,600	2,900	91
	1,700	2,900	84
	2,500	4,250	88
	4,100	7,000	94
	7,450	12,700	100
Flow amplifier			
(compressed air)			
Unit A	1,520	2,580	85
	1,700	2,900	88
Unit B	3,980	6,760	89
	4,500	7,650	92
Unit C	5,600	9,500	91
	6,250	10,600	94
Unit D	6,850	11,600	92
	8,000	13,600	95
Compressed air			
Open pipe, 0.75 in diameter, 80 lb/in.²	400	680	110
As above, with silencer			70

Axial Fans

Air flows through the casing parallel to the rotational axis of the impeller in axial fans (Figure J.2). In general, axial fans can move large volumes of air, but generate only relatively low static pressure (Alden and Kane 1970). (The greater the static pressure that the fan can develop, the more effective it is in a broad range of applications.) The fan curve for a typical axial fan is only slightly curved at different airflow rates. This illustrates empirically the ability of this type of fan to develop relatively low static pressure (Figure J.3).

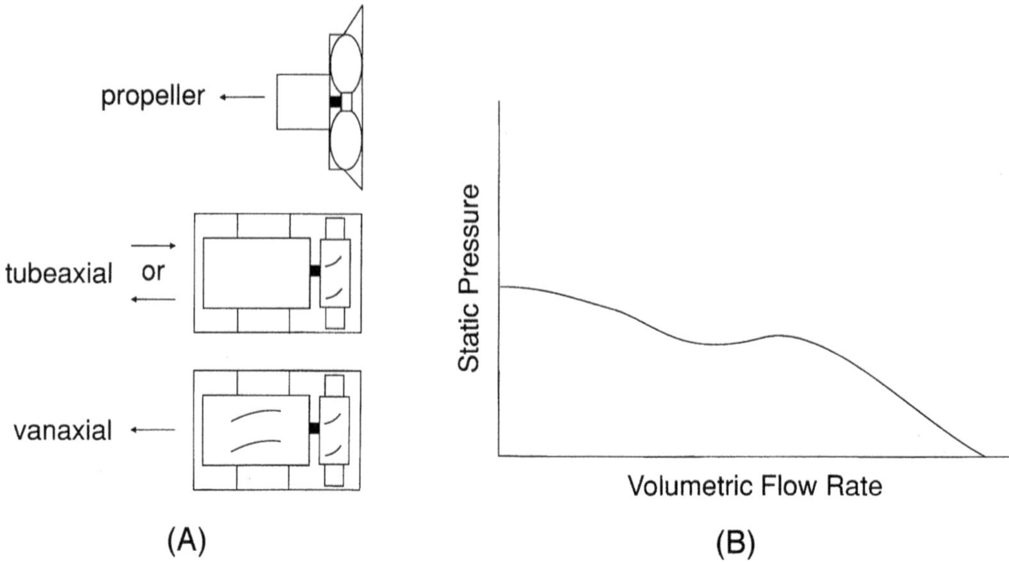

FIGURE J.2 Axial fans. (A) Types. (B) Typical performance curve. (Adapted from ACGIH 1995.)

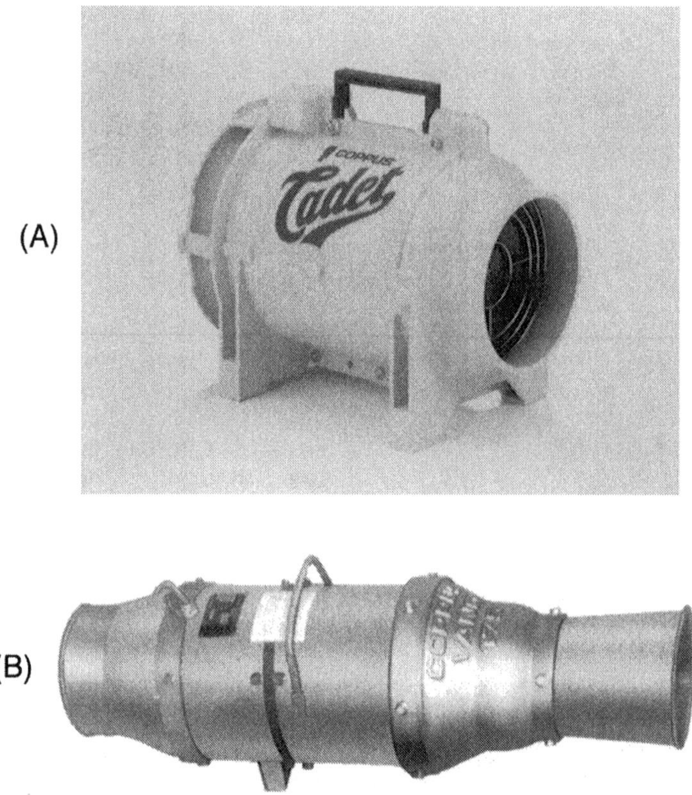

FIGURE J.3 Axial fans used in portable ventilation systems. (A, B) Electrically driven vaneaxial fans. (C) Electrically driven vaneaxial fan configured for use in hazardous locations. (D) Air/steam-driven tubeaxial fan. (E) Water/hydraulic fluid-driven tubeaxial fan. (Courtesy of Tuthill Corporation, Coppus Portable Ventilation Division, Millbury, MA).

(C)

(D)

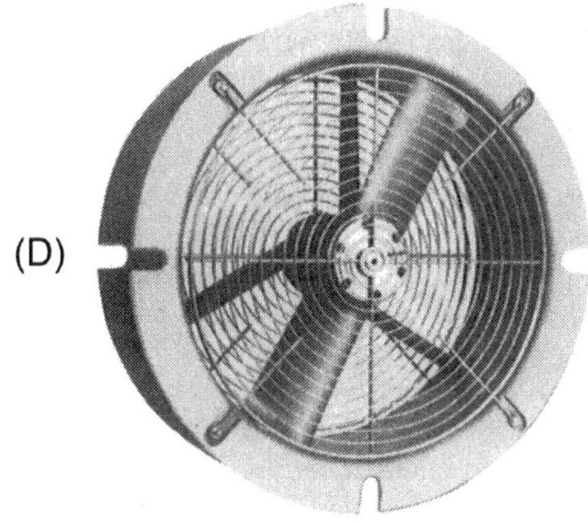

(E)

FIGURE J.3 (continued)

The dip in the fan performance curve represents a region of instability (McDermott 1976). The region between the blocked tight no delivery pressure and the top of the dip (the stall point) is an undesirable operating zone. The stall point is the point at which the flow lines begin to depart from the inlet side of the blade surface and eddies and vortices begin to form. These fans must be operated in the region to the right of the stalling point. This requirement is more stringent for axial fans than centrifugal fans where characteristics of the system fluctuate.

Axial fans will function when rotated in reverse direction. Propeller and tubeaxial fans provide up to 50% of design capacity when operated in reverse. Vaneaxial fans will barely draw when operated in the reverse direction. Tubeaxial and vaneaxial fans are commonly used in confined space applications.

The **propeller fan** is the most common and simplest type of axial fan. The impeller contains two or more blades and rotates on a central shaft within a short casing. Propeller fans are capable of moving large volumes of air, but develop very low static pressures, less than 0.5 in. wg (inches of water gauge). Propeller fans may be direct-drive or belt-driven. This type of fan can be used successfully in applications where static pressure is not an issue. However, the potential for success is low due to intolerance to less than ideal conditions. This type of fan cannot be used reliably with duct.

The **tubeaxial** fan is a variant of the propeller fan. The fan is mounted inside a longer casing. This design provides higher efficiency. Tubeaxial fans can develop higher static pressures, up to 2.5 in. wg. More sophisticated designs have a larger central hub and blades that are shorter in length than those used in propeller fans. The larger hub blanks off the less effective central portion of the impeller through which airflow reenters at higher static pressures. The ability to generate higher static pressure permits the use of duct and other fittings, thus providing greater control over air movement. Tubeaxial fans are considerably bulkier and heavier than propeller fans, owing to the longer housing. This can restrict their use in some circumstances.

The **vaneaxial** fan is an improved version of the tubeaxial fan. The vaneaxial fan has short, stubby airfoil blades and a large central hub. The vaneaxial fan uses guide vanes on either or both sides of the impeller to remove the rotational component from the airflow. This straightening of the airflow produces static pressure regain. Vaneaxial fans are the most efficient of the axial fans. They can produce greater static pressure (up to 8 in. wg) than tubeaxial fans. This increase in performance increases the utility and reliability of this type of fan for ventilating confined spaces where operating conditions can change considerably following minor change in configuration.

Centrifugal Fans

Airflow discharges perpendicular to intake in centrifugal fans (Figure J.4). Centrifugal fans use a wheel-type impeller to induce airflow. In general, centrifugal fans produce lower airflow, but higher static pressure than a comparable axial fan (Alden and Kane 1970). The ability to generate high static pressures is important where long runs of flexible or collapsible duct may be involved. Centrifugal fans have obvious utility for ventilating confined spaces.

The blades in the impeller of a centrifugal fan are oriented in one of three ways — forward-inclined, radially or backward-inclined — relative to the direction of intended rotation. As illustrated in Figure J.5, the slope of the performance curve for a typical centrifugal fan is much steeper than that for an axial fan. This illustrates the enhanced ability to generate high static pressure with relatively small decrease in airflow. Centrifugal fans provide 40 to 50% of rated capacity when operated backwards.

Forward-inclined fans have blades inclined in the direction of rotation. These are the so-called squirrel-cage fans. Forward-inclined fans are the least efficient of the centrifugal fans. Forward-inclined fans typically operate at slower speeds than the other types of centrifugal fans. Because of their forward inclination, the blades accumulate particulate debris and are difficult to keep clean. They also are more sensitive to erosion than blades in the other types of centrifugal fans. The shape of the fan performance curve for a forward inclined fan is similar to that of axial fans. That is,

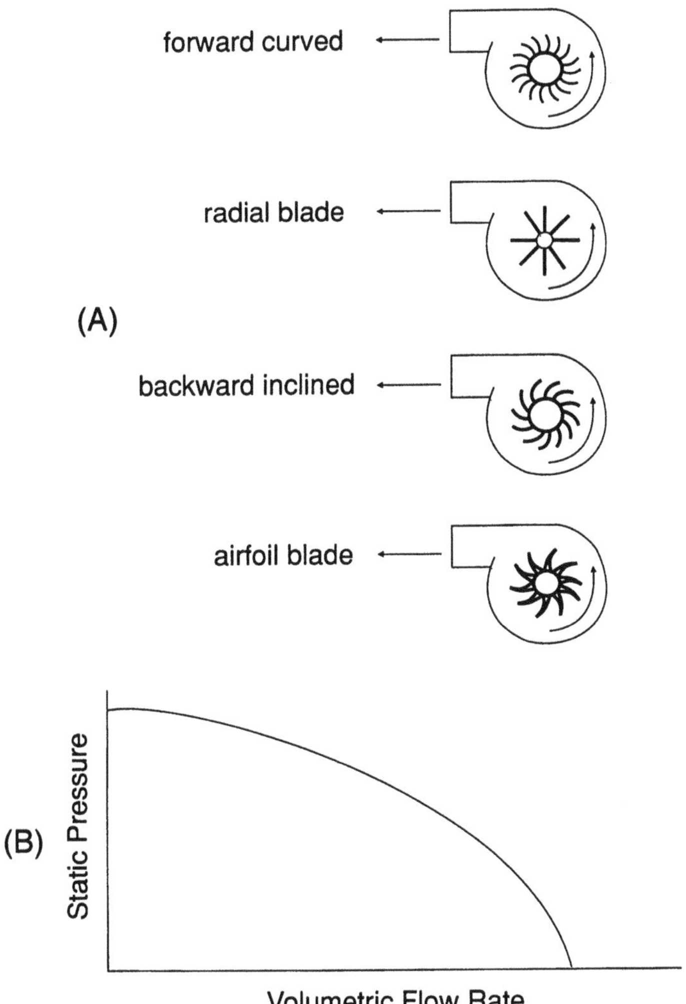

FIGURE J.4 Centrifugal fans. (A) Types. (B) Fan performance curve. (Adapted from ACGIH 1995.)

there is a stalling point and region of instability between this and the blocked tight static pressure. The power requirement for a forward inclined fan is greatest at maximum airflow. This characteristic can overload the motor.

Radial fans have blades that radiate like spokes from a central hub (paddle wheel). Radial fans typically operate at higher speeds and generate greater static pressures than forward-inclined fans. Radial fans are suitable for conveying particulates and condensable mists or vapors because this design resists accumulation and erosion. The efficiency of radial fans is approximately the same as that of forward-inclined fans. There is no region of instability in the fan performance curve. The power requirement for a radial fan also is greatest at maximum airflow. This characteristic can overload the motor.

Backward-inclined fans have blades inclined against the direction of rotation. These are the most efficient of the common centrifugal fans. Backward-inclined fans typically can operate at higher speeds and against greater static pressures than forward-inclined fans. Since the blade configuration in backward-inclined fans is inherently self-cleaning, this type of fan also is suitable for handling dusts and small particles. There is no region of instability in the fan performance curve. Unlike forward-inclined and radial fans, backward-inclined fans cannot overload. Maximum

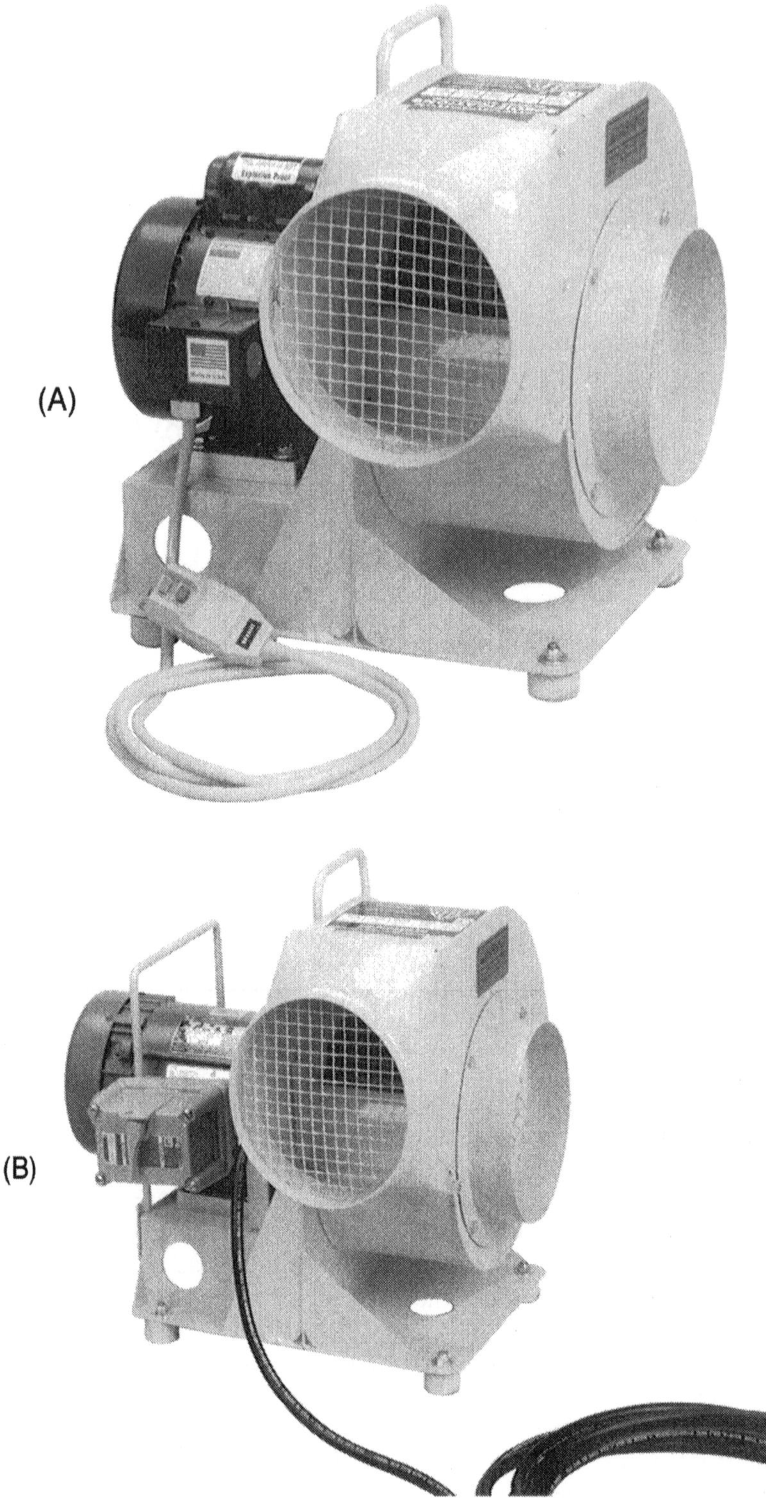

FIGURE J.5 Centrifugal fans used in portable ventilation systems. (A) Electrically driven centrifugal fan. (B) Electrically driven centrifugal fan configured for use in hazardous locations. (C) Pneumatically driven centrifugal fan. (D) Gasoline/propane-powered centrifugal fan (Courtesy of Air Systems® International, Inc., Chesapeake, VA.)

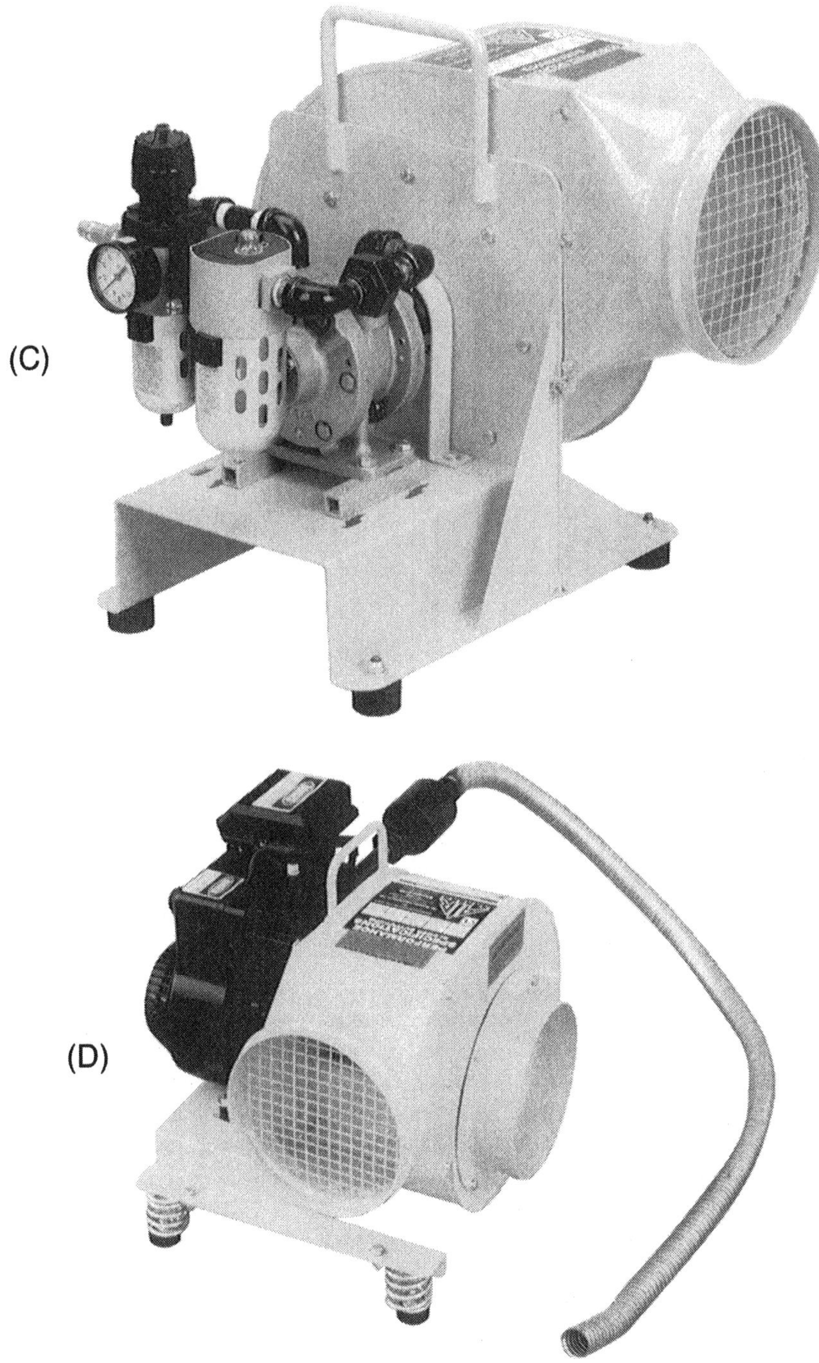

FIGURE J.5 (continued)

power is required when the fan is operating at maximum efficiency. This occurs part way between the blocked-tight and free-delivery points on the performance curve.

Airfoil fans are a refinement of the backward-inclined design. Airfoil fans are the most efficient of the centrifugal designs. The blades have airfoil shape. Airfoil fans can generate higher static pressure and operate at higher speeds than radial and backward-inclined fans.

FIGURE J.6 Flow amplifier (venturi blower). (Courtesy of Air Systems® International, Chesapeake, VA.)

The **inline centrifugal fan** is a hybrid between the axial and centrifugal fan. The inline fan utilizes a backwardly inclined or airfoil centrifugal impeller in an axial configuration. In comparison to other fans of similar size, the inline centrifugal generates static pressures intermediate between axial and centrifugal fans. Similarly, airflow produced by the inline centrifugal is intermediate between that produced by comparable centrifugal and axial fans. Inline centrifugals typically are utilized where compact (inline) installation is required. For this reason, they are ideal for application in portable systems where compactness and high capacity may be critical.

Flow Amplifiers

Flow amplifiers induce air motion by controlled release of pressure in a geometrically optimized configuration (Figure J.6). Flow amplifiers used for ventilating confined spaces are powered by compressed air or steam. The choice of compressed gas in this application is restricted, since the powering flow mixes with the ventilating flow. Operating pressure typically ranges from 40 to 150 psig (275 to 1,030 kPa). Effectiveness of induction of airflow depends on the difference between ambient and operating pressure, as well as geometry. The pressurized gas is introduced into an inlet chamber located in the base. This contains tiny nozzles, holes, or an annular gap. Controlled expansion of the pressurizing gas to ambient pressure induces flow of air into the inlet. This is a classic application of the venturi — flow being greatest where pressure is least.

Flow amplifiers are portable and very efficient. Some models can induce 40 volumes of flow per volume of compressed gas consumed, although 25 to 30 are more typical. The induction ratio (ratio of induced flow to consumed flow) is a measure of efficiency of a particular design. Flow amplifiers have no moving parts. They are well suited for gas-freeing and ventilating, as well as moving particulates and small objects. These units can operate against static pressure as high as 7 in. wg using normally available plant supply air (80 psig). These units must be properly bonded and grounded when used in hazardous locations, in order to dissipate static charge. Properly bonded and grounded, the flow amplifier is ideal for use in hazardous locations where volatile or ignitable vapors may be present.

Compressed Air

Compressed air from plant systems is another possible ventilation resource. There are situations, for example, very small openings, where duct from portable ventilating equipment would limit access. Where normal duct might be impractical due to large size, a small-diameter hose might be acceptable.

Plant air systems often provide compressed air at 80 to 100 psi. Upon release, this air expands into the space in front of the discharge point to a volume five to seven times the delivery volume. This expansion provides the means for diluting or displacing the atmosphere in the space.

Compressed air used for this purpose must meet standards for compressed breathing air (CSA 1985, CGA 1989). Air from plant compressor systems can contain carbon monoxide and other contaminants. Equipment used to purify plant air for breathing purposes or to assure its quality must be used in this application. Delivery rate through the purification system could limit the feasibility of this application.

Use of compressed air in this manner may not be appropriate where ignitable mixtures are present, because of potential generation of static electricity.

Mode of Power

An important consideration in the selection of a portable air mover is the mode of power. Several options are available. These include:

- Electrical — AC or 12 V DC
- Pneumatic — compressed air or steam
- Hydraulic — water or hydraulic fluid
- Engine — gasoline, propane, or diesel

Electrical and pneumatic-powered (compressed air) units are the most frequent choices.

Selection of mode of power typically depends on factors related to the application or the environment. This process should begin with consideration about availability of the power source and safety of operation. Additional considerations should include expected duration of operation and cost of operation. Each mode of power offers its own opportunities and suffers from its own limitations.

Use of electrically powered equipment in remote areas can pose problems due to availability and security of electrical power. Three options are available to address this situation: string power cables from existing supplies, or use a portable electrical generator or an engine-driven air mover. The latter two options are very tempting in these situations. In fact, these situations may be well served by either of them.

Use of portable generators or engine-driven air movers requires extreme care. Fuel vapors and exhaust gases can be ever present around this equipment. Fuel tanks on the engines are small and rapidly emptied compared to the length of the shift. This means that refueling must occur in order to maintain continuity of ventilation. Refueling "on the fly," whether condoned or not or permitted or not, is likely to be a reality. At the least, this practice can lead to spillage of fuel. In the more extreme situation, the spilled fuel could catch fire on the engine of the generator or the air mover. A fire caused by a fuel spill could destroy this equipment, halt ventilation, and require evacuation of the confined space and cessation of work.

Exhaust gases equally are a concern in these situations. Care must be taken to ensure that exhaust gases are not drawn into the intake of the air supply or duct on the negative pressure side of the air mover. An equal concern is entry of contaminated air into the confined space through entry/exit portals or the work area around the space. Exhaust gases can readily accumulate around stationary fuel-powered equipment, especially when located at or below ground level. Obstructions to wind flow, such as natural features or man-made structures or equipment, and lack of wind exacerbate the lack of dispersion. Contaminants in the exhaust cloud around small engines easily can exceed regulated limits. A catalytic converter in the exhaust system can dramatically reduce the problem posed by carbon monoxide.

Air supplied by this equipment could be expected to meet requirements for compressed breathing air (CSA 1985, CGA 1989). This would maintain consistency with breathing air provided by other means. The only means of assuring the quality of breathing air provided by a portable system under these circumstances is to measure levels of carbon monoxide and possibly nitrogen dioxide, as appropriate. This would require use of discrete or more preferably, continuous monitoring instruments containing audible and visual alarms. Some standards require these assurances of performance in portable systems (Standards Association of Australia 1986, National Occupational Health and Safety Commission and Standards Australia 1992).

Another option for remote applications is a unit that can operate from a 12-V DC electrical supply. This unit can utilize the battery in a motor vehicle or a stand-alone 12-V battery as a power source.

Electrically powered units used in hazardous locations require special motors and accessories. These units should satisfy requirements of Class I, Group D and Class II, Groups E, F, and G locations as specified in the National Electrical Code (NFPA 1993). These are the most common levels of performance required from motors and electricals used in hazardous locations.

Concern about ignition sources should dictate consideration about equipment powered by energy sources that are inherently nonsparking — pneumatic (compressed air and steam), and hydraulic (water and hydraulic fluid). Steam must be used with caution since this can be a source of static electricity. Ventilators powered by compressed air and steam are used frequently in large plants and facilities having plumbed-in sources of supply. In situations where compressed air is not readily available, portable compressors sometimes are brought on site to power air-driven units and other equipment. Aboard marine vessels where water is available as a source of fluid power, water-driven units are the preferred choice.

All fluid-powered ventilators must be mechanically bonded and the system grounded in order to dissipate static electrical charges. However, static charges can accumulate as a result of movement of air through *any* ventilator.

Duct

Duct is an essential component in portable ventilation systems. Duct is the conduit for transporting air between the point of collection, the air mover, and point of discharge. Duct made from both rigid and flexible types of materials has application in specific situations involving confined spaces.

Rigid Duct

Rigid metal duct is made from aluminum, and carbon and stainless steels. Plastics also are used. The metal may be flat or corrugated. Duct made from corrugated metal is inherently stronger than that made from flat metal. Metal duct is available in two styles: spiral wound and formed sheet. Spiral wound duct is formed from a long, narrow piece of metal. The seam between edges of the metal follows the spiral and is much longer than the duct itself. Formed sheet duct is made from a rectangular piece of metal that is bent into a cylinder. The length of the seam in a formed sheet duct is the length of the duct. Spiral wound duct is potentially better suited to the needs of portable systems because of its strength. Cuts made to spiral wound duct can be finished better than those made to formed sheet duct.

Metal duct offers obvious advantages of strength, and puncture and burn resistance. In applications involving hot work, for example, welding and cutting, hot slag can burn through less substantive materials, such as fabric. Metal duct used at terminations where hot materials are collected can dramatically improve longevity. Smooth metal also offers the least resistance to airflow. At the same time, the biggest advantage offered by the rigidity of these products also is the biggest limitation. Rigid duct cannot be moved or formed around obstacles without considerable effort. On the other hand, the positive attributes of metal duct and fittings, such as elbows, transitions, hoods, and branch entries, can be exploited where these strengths can be utilized to best advantage. For example, elbows and straight sections can provide structural stability and strain relief to flexible duct. Reducing sag and needless bends reduces dynamic losses, thereby improving airflow.

Another positive attribute of metal duct is malleability — the ability to be formed into a shape other than the starting one. Cylindrical duct may be formed into shapes appropriate to the needs of the situation, such as a slot or flanged hood. A cylindrical duct becomes a slot or flanged hood merely by cutting into the edge, flattening the cylinder, and bending, as needed. Modifying the shape of flexible duct is not readily achievable.

Flexible Duct

Flexible duct is widely used for ventilating confined spaces. Flexible duct usually means wire-wound fabric or metal duct. These products are available from manufacturers of portable ventilating equipment or from third-party manufacturers. A wide variety of fabrics is available — polyester, nylon, fiberglass, polyvinyl chloride (PVC), Teflon®, and polyurethane. Coatings include neoprene, PVC, acrylic, silicone, aluminized polyester/polyaramid, and Teflon®.

A wire helix provides shape, reinforces the fabric, and prevents collapse. The wire usually is located on the outside of the duct where it is isolated by the fabric from contact with the air transported inside. This situation could mean that static electrical charges could accumulate on inner surfaces, unless the fabric is conductive. Static buildup can result during passage of air, suspended particulates, and liquid aerosols. Static charge is not necessarily dissipated by the metal of the helix.

The helical construction of wire-wound fabric duct offers the advantages of longitudinal compressibility for compactness during storage, relatively low weight, strength, and the ability to follow the contours of the "landscape." Fabric products are semidisposable in the event of contamination. Contamination can result from contact between the fabric and substances in the gas phase or from wetting of the surface.

Wire-wound metal foil duct is another option. This type of duct became popular with asbestos-removal projects as the conduit for discharging air movers located inside the contaminated zone. Wire-wound metal foil duct is disposable. It offers most of the advantages of the fabric duct and most of the advantages of the metal. Wire-wound metal foil duct offers the solvent resistance of sheet metal without the rigidity. It also offers the flexibility of fabric duct and resistance of the metal to chemical attack. Wire-wound metal foil duct should be conductive, and if so, should be electrically bondable to the air mover.

Tubular fabric and plastic materials also are available. These materials lack the structural integrity offered by helical-wound products. That is, they collapse when not in use and therefore can be used only on the discharge side of an air mover. This lack of dimensional stability exerts a cost on effectiveness of the ventilation system through increased static pressure drop.

Accessories

Portable ventilation systems lack the sophisticated hardware of permanently installed systems. Nowhere is this more apparent than in the range of accessories that are available to complement the duct and air mover.

Saddle Vent®

An innovative and important product that appeared several years ago was the Saddle Vent®. The product is widely known by this name or variations offered by licencees. The Saddle Vent is a duct that contains a flattened section in the shape of a crescent and cylindrical end fittings. It is made

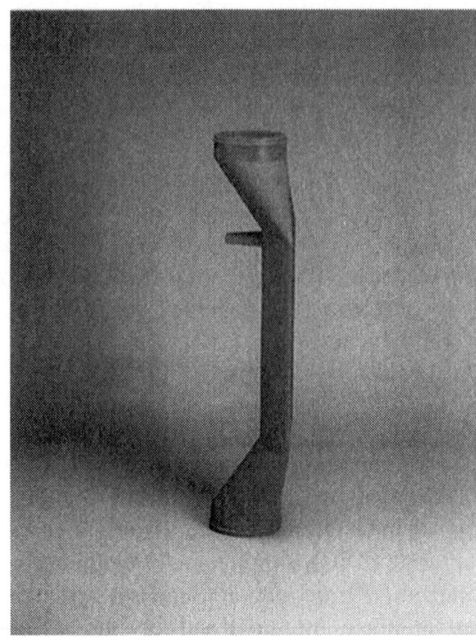

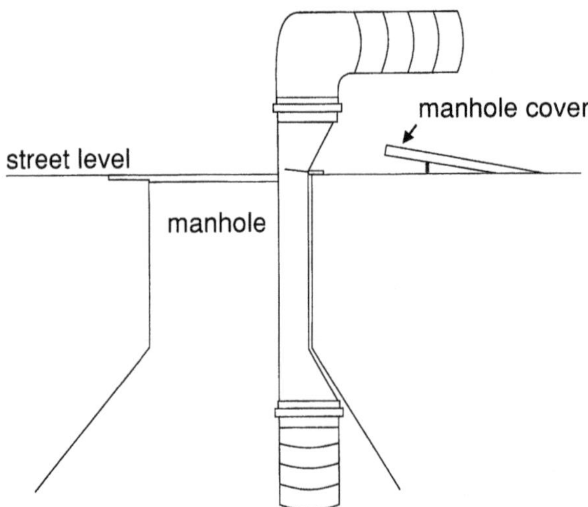

FIGURE J.7 Ventilation accessories: (A) Saddle Vent®. The Saddle Vent® and similar products enable crowded openings to accommodate ventilation duct, as well as transit by occupants. (B) Saddle Vent in position in a manhole. (C) Propane heater. (Courtesy of Air Systems® International, Inc., Chesapeake, VA. Saddle Vent® is a Registered Trademark of Air Systems International.)

from molded, rigid plastic to ensure dimensional stability. It is intended for installation in manholes. The Saddle Vent minimizes the cross section occupied by the duct in the manhole. This makes available extra area in the opening for use by personnel and equipment. An added benefit of the Saddle Vent is the strength inherent in its design. No longer is flattening or puncture, or destruction of the wire reinforcing or fabric wall, a concern following contact with equipment or personnel moved through the opening. The Saddle Vent is self-supporting on the manhole structure. It also provides a support to anchor the duct from the top in its descent into the structure and relieves the strain in the connection to the fan (Figure J.7).

FIGURE J.7 (continued)

Air Heaters

In-line heaters for heating incoming air utilize either electricity or propane as the energy source. The heating elements in the electrically powered unit are located in the direct path of incoming air. The propane-powered unit contains an air-to-air heat exchanger to isolate combustion gases in the heating section from the airflow.

Inherent in the use of fuels for heating is concern about exhaust gases. Care must be taken to ensure that exhaust gases from the heater are not drawn into the intake of the air supply or into duct on the negative pressure side of the air mover. An additional concern is entry into the space through entry/exit portals or the work area around the space. Exhaust gases can readily accumulate around stationary fuel-powered equipment, especially when located at or below ground level. Obstructions to wind flow, such as natural features or man-made structures or equipment and lack of wind can exacerbate the lack of dispersion. Contaminants in exhaust gases around this equipment could exceed regulated limits. Air supplied by this equipment could be expected to meet requirements for compressed breathing air (CSA 1985, CGA 1989). This would maintain consistency with breathing air provided by other means.

The only way to assure the quality of breathing air provided by a portable system under these circumstances is to measure levels of carbon monoxide and carbon dioxide, as appropriate. This would require use of discrete or, more preferably, continuous monitoring instruments containing audible and visual alarms. Some standards require assurance of performance of portable systems (Standards Association of Australia 1986, National Occupational Health and Safety Commission and Standards Australia 1992). The heat exchanger must be inspected periodically to assess its integrity.

An additional concern with air heaters is the hot surface. If sufficiently hot and containing the appropriate geometric configuration, these surfaces could serve as a source of ignition to ignitable atmospheres in the airstream that passes through the unit. This again emphasizes the need to assure the quality of air utilized by the ventilation system.

Inlet Manifold

The inlet manifold provides the means to connect several hoses to a single air mover. This device provides an alternative at the manhole to the Saddle Vent. Several small-diameter hoses can be arranged around the perimeter of the manhole. This arrangement can be less obtrusive than the Saddle Vent or a single large duct. Also, supply or exhaust capability can be spread around more of the space by use of the small hoses. An important point here is that multiple supply outlets or exhaust inlets are available through use of a single air mover.

Exhaust Hoods

Hoods are available for the intake of the exhaust duct. A hood improves collection efficiency over open duct. Open duct is an inefficient means for inducing flow; yet this is used in many situations.

PORTABLE VENTILATION SYSTEMS

The outward manifestations of a ventilation system are its hardware and software components. Ventilation systems also reflect the philosophy of the designer. Ventilation systems most familiar to people are permanent installations. Permanently installed systems usually operate so seamlessly within the normal course of activity that their performance capabilities tend to be taken for granted. However, fundamental differences exist between permanent installations and portable systems, such as are used for ventilating confined spaces. These differences are more profound than the obvious differences in hardware. They extend to the philosophy of design and reflect the real-world constraints within which portable systems are required to operate. Appreciating these differences is critical to achieving success in applications using this equipment.

Ventilation systems used in permanent installations usually contain supply and exhaust subsystems. Rare today is an industrial building that has only exhaust ventilation and no supply subsystem. Both the supply and exhaust subsystems are complex configurations containing many components. Table J.3 lists components that could be found in permanent and portable systems.

Supply and return/exhaust subsystems in buildings occur in many variations. Typically, the supply system mixes air returned from within the building envelope by the return subsystem with air drawn into the system from outside. To maintain balance, some of the air is diverted from the return flow and expelled from the building envelope. Incoming outdoor air is cleaned by filtration, heated or cooled to the desired temperature, and humidified or dehumidified. Supply air is propelled through the distribution system by the supply fan. Duct in the distribution system divides the main flow into the quantities appropriated for particular areas and conducts it to them. Diffusers at the ends of the distribution system provide flow resistance and promote mixing of incoming air with that already in the space.

In office environments, the return/exhaust subsystem typically collects air that escapes from chambers (rooms) due to pressurization. This can involve a ceiling plenum that collects from many sources and exhausts from a single outlet. It also can involve a network of ducts that collect from discrete sources. In either case, motive power provided by the return fan draws the air through the system.

The local exhaust subsystem utilizes discrete extractors to withdraw air from the space and expel it from the building. The local exhaust subsystem may be part of a recirculating system, as described above, or may act as the building exhaust.

These systems are designed to move air to and from multiple locations within a building envelope, in controlled quantities at predictable velocities. Failure of the system to do this leads to complaints from occupants for corrective action. Corrective actions can include:

- Balancing to redirect flow
- Relocating points of delivery

TABLE J.3
Components of Ventilation Systems

Component
Permanent Installations
Supply subsystem
Intake air collector
Intake air duct
Air-cleaning device
Temperature-conditioning device
Humidification device
Dehumidification device
Main return duct (connects return fan to supply fan)
Mixing chamber
Main supply duct (connects supply fan to distribution network)
Distribution duct network
Distribution control chambers
Supply diffusers
Exhaust/return subsystem
Receiving hood/ceiling plenum
Main duct (connects plenum to fan)
Return fan
Building discharge duct
Pressure/volume compensation equipment
Local exhaust subsystem
Collector
Upstream duct
Fan
Downstream duct
Aircleaner
Discharge structure
Portable System
Supply mode
Upstream duct
Heater
Air mover
Saddle Vent
Manifold
Downstream duct
Exhaust mode
Collector
Upstream duct
Saddle Vent
Manifold
Air mover
Downstream duct
Air cleaner

- Relocating points of collection
- Increasing or decreasing flow or velocity
- Changing temperature and humidity

Implicit in being able to facilitate these macro or micro changes is the tremendous flexibility that is designed into building ventilation systems.

Portable ventilation systems, on the other hand, are considerably less sophisticated than the permanent systems described here. This lack of sophistication deserves to be emphasized, so that the extensive nature of the expectations that are placed on portable systems can be appreciated and reconciled with their capabilities. Users accustomed to the high level of performance provided by permanent systems naturally would expect to achieve the same result from a single fan and some flexible duct. Failure to appreciate these realities will lead to less than satisfactory performance from portable systems.

Confined spaces often are isolated geometric structures. They usually have no supply or exhaust subsystem or provision for ventilation. Of its own accord, air normally will not flow through these structures. The portable system must induce this flow. A portable system just as easily can be operated to supply air to the space or to exhaust air from it. This merely is a matter of reversing the orientation of the air mover. In either case, air outside the space is the source for either supply or replacement.

This realization is fundamental to consideration about how to utilize portable systems and to understand what can be expected from them. There is no make-up system that supplies air into the space when the portable system is used as an exhauster. The energy needed to draw air into the space to replace that removed originates from the same fan; that is, the exhaust fan. Similarly, when a portable system is used to supply air to the space, there is no exhaust system to remove the excess air. The energy needed to expel excess air from the space must come from the same fan; that is, the supply fan.

This situation means that air movers used in portable systems must operate simultaneously against losses in static pressure in both modes, supply and exhaust. This situation is compounded by losses from unnecessary sags and bends in flexible duct, and those due to elbows and fittings.

VENTILATING CONFINED SPACES: PLANNING CONSIDERATIONS

Designing a portable ventilation system for a confined space is a complex undertaking. There are many variables to consider. The starting point is to become completely familiar with the physical layout and geometry of the space. This includes, for example, special attention to natural pockets where contaminants could collect, even after purging and preentry ventilation has occurred.

The process of selecting the mode of ventilation and ventilating equipment is fairly subjective. It largely depends on the characteristics of a particular application. Those unfamiliar with the process often are disappointed to learn that universal rules and criteria for making these choices do not exist.

The process should start with consideration about the space as a geometric structure, and about its atmospheric contents and the hazard that they represent. This consideration should occur during the hazard analysis and planning stage. A number of factors should be considered prior to specifying and selecting equipment. These include:

- The characteristics of the preentry atmosphere
- Contaminants generated during work activity
- Air used for ventilation
- Air removed or displaced from the space
- Ventilation mode

The Preentry Atmosphere

The preentry atmosphere could represent the first point of contact between the ventilating equipment and the atmospheric hazard. The chemical and physical characteristics of the preentry atmosphere, therefore, should have a major influence in determining the type of ventilating equipment, mode of power, and materials of construction that are appropriate. The preentry atmosphere could be very different from what is present during occupancy and work. Recognition and careful consideration about the properties of the preentry atmosphere are extremely important in order that the interaction between the ventilating equipment and the atmosphere does not create a hazard. Following are some of the factors that could influence selection of equipment:

- Oxygen level
- Temperature difference
- Ignitable gases or vapors
- Ignitable dusts and mists
- Chemical incompatibility
- Particulate loading
- Toxic substances

The preceding factors presume that the space to be ventilated exists at normal atmospheric pressure. A pressurized or depressurized space could pose hazards due to the sudden equalization during initial opening.

Oxygen level is a fundamental concern. This is mainly true for enriched atmospheres. Oxygen enrichment can result from the presence of oxygen in a process gas stream or dissolved in fluids, as well as use of process chemicals, such as hydrogen peroxide, that decompose to liberate oxygen.

Atmospheres enriched in oxygen (above about 23%) pose especial hazards to ventilating equipment containing combustible materials or combustible metals. Relatively minor enrichment of normal atmospheres seriously enhances combustibility of normal materials (Turner 1987, Frankel 1991). This problem was demonstrated in fatal accidents investigated by OSHA (OSHA 1982). Most materials burn in oxygen. Also, conditions needed for combustion in oxygen-enriched atmospheres are considerably less stringent than those needed normally. Some of the more common sources of ignition in oxygen-enriched atmospheres in which ventilating equipment could be involved include:

- Rubbing friction
- Mechanical impact
- Particle impact against structures during change in direction
- Static electricity
- Current electricity

Materials suitable for use in enriched oxygen service, such as polytetrafluoroethylene (PTFE) and fluorocarbon oils and greases, have high autoignition temperatures and low heats of combustion (Lowrie 1987). Adhesives other than inorganic silicates or phosphates have poor compatibility with oxygen. This concern extends also to the metal used in construction of the ventilator. Aluminum and magnesium alloys, chosen for lightness, have low compatibility with oxygen. Static electricity produced by motion of gases, particles, and liquid aerosols through ducts and fan casings, and arcing in electrical switches and motors are potential sources of ignition.

Use of ventilating equipment under conditions that may include contact with enriched oxygen atmospheres should be undertaken only after the most careful of considerations. These must include discussion with the manufacturer.

Temperature difference between the atmosphere in the space and the atmosphere external to the space is another important concern. Exhausting an atmosphere that is hotter than the air outside the space could lead to rapid cooling in the ventilation system. Material existing as vapor in the warmer environment of the confined space could condense and coat interior surfaces of the exhaust system — duct, fan casing, and fan blades. This condensation also could occur beyond the point of discharge. Depending on the characteristics of the substance, this process could create or enhance a fire hazard. The potential for production of static electrical charges increases with the presence of mist. Also, condensation potentially exposes the ventilating equipment to chemical and physical attack.

The opposite situation involves removal of cooler air from the confined space. Contact with warmer air exterior to the space could involve rapid heating of the contaminated air transported from the space. In this case volume, vapor pressure and vaporization rate increase as the temperature increases. This could alter the vapor-to-liquid ratio in mixtures containing vapor and mist. Increased vaporization also could put a mixture that was nonignitable into the ignitable range, thus potentially increasing the fire hazard.

Ignitable atmospheres could result from the several physical forms of materials that could be present in the atmosphere inside the space. These could include:

- Gases and vapors
- Dusts
- Mists

The atmosphere simultaneously could contain a mixture of several of these forms. Combustible solvents present simultaneously as a mixture of mist and vapor pose considerably greater ignitability hazard than when present solely as vapor, as more mass is present. Finely divided solids suspended in air as dusts pose a potential ignitability and explosibility hazard.

An additional hazard could develop when air is introduced into a space containing an ignitable gas or vapor or dust or mist that initially exists at a concentration above the upper flammable (explosive) limit. As dilution air enters the space, the composition of the mixture will enter the flammable/combustible/explosive range. This occurs regardless of whether outside air is supplied to the space or the atmosphere is exhausted from it. In either case, the ignitable mixture could develop inside the space or in the surroundings where the contents are discharged. It also is conceivable that a warning from a gas-detecting instrument about the status of the initial condition could be missed, due to commotion or noise from fans and engines or other equipment. This could further compound the hazard.

Again, an improperly chosen ventilator or mode of ventilation could become a source of ignition when used in a hazardous atmosphere. Consideration for the application and potential outcomes in a particular location is essential. A situation initially assessed as low hazard could evolve rapidly into a highly hazardous one, following only minor change in conditions. For example, rapid suspension of finely divided, settled material could occur simply by blowing air onto surfaces in the space during lancing or initial ventilation. This could rapidly produce an explosible atmosphere. An improperly chosen ventilator or flexible duct could act as a source of ignition. Motion of particulates in the air stream within the system could generate static electricity.

Selecting equipment suitable for use in hazardous locations requires an understanding of Class, Division, Group, and T-code designations defined by the National Electrical Code and Underwriters' Laboratories (NFPA 70 1993). Special motors and electrical accessories are required for service in environments in which explosive or ignitable gases or vapors and dusts or mists are present, or are likely to become present. These special components are essential to ensure that electrical or thermal energy generated inside the motor enclosure cannot act as a source of ignition.

Use of equipment under these conditions should be undertaken only following the most careful of considerations, including discussion with the manufacturer.

Chemical incompatibility between the materials of construction of the ventilating equipment and the components of the atmosphere in the space is another potential complication in the selection of equipment. Incompatibility could lead to damage and destruction of components, as well as premature failure of the equipment. Components of the portable ventilation system that should receive consideration for incompatibility include:

- Fan casings
- Fan blades
- Bearings
- Duct materials

Some portable systems contain only metal components. However, fabric duct materials are widely used. Fabrics and polymeric coatings are subject to different types of attack than are metal components.

Materials used in construction of process and containment structures are chosen to resist attack by the contents. Portable ventilation systems are intrusions to this environment. Materials used in construction of this equipment should not be presumed *a priori* to be equivalent in resistance to those of the existing structure.

Reaction between the materials of construction and the components of the hazardous atmosphere is not likely to be so vigorous as to be violent or explosive. However, destruction, as in the reaction between concentrated sulfuric acid and polyurethane-coated nylon, could be rapid.

Incompatibility is expressed through solvent action and oxidative processes.

A solvent is a substance that dissolves another substance. Solvent action is a concern during contact between fabric materials and airborne contaminants or residual contents. Organic rather than inorganic solvents are the more important group of solvents in this discussion. Solvent action on the fabrics and coatings used in ducts occurs in the same manner as that affecting chemical protective fabrics and gloves.

Solvent action occurs through penetration, permeation, and degradation. Penetration occurs through seams and needle holes in stitching, and pinholes (Schwope 1983, Stull 1992). Contact between the liquid and the fabric may result in permeation. Permeation is a three-step process. The first step, sorption, involves contact between the liquid and the fabric. The second step, diffusion through the fabric, occurs from the contaminated surface. The third step, desorption, occurs at breakthrough on the other side of the fabric at which time evaporation from this surface begins. Once started, permeation continues in the direction of the concentration gradient from the higher to the lower value. Degradation or destruction of the fabric or polymer also can occur. This involves swelling, once penetration or permeation by the solvent has begun. See Chapter 12 for more information.

Technical publications on chemical protective clothing and gloves and vendors' literature can provide guidance on potential incompatibilities. A full discussion on selecting equipment for use in these situations goes beyond the scope of this Appendix. The resources mentioned in the previous paragraph should be consulted for more detailed discussion. Use of equipment under these conditions should follow careful considerations, including discussion with the manufacturer.

Oxidative processes include oxidation and corrosion (Shackelford 1988, Meyer 1989). Oxidation refers to destructive processes involving attack by atmospheric oxygen. Oxidation is a potential concern for metals in duct and materials of construction of the air mover. For some metals, the oxide coating is tenacious and protective against further attack. For others, the coating tends to crack and is not protective.

Corrosion is the dissolution of metal into an aqueous environment. Corrosion is a physical change, usually deterioration or destruction brought about by a chemical or electrochemical reduction–oxidation (redox) process. Fabrics and polymeric coatings are susceptible to other types of oxidative attack.

Aqueous corrosion is a form of electrochemical attack (Shackelford 1988). A variation in metal ion concentration in aqueous solution above two different regions of a metal surface leads to passage of an electrical current through the metal. The area of low ionic concentration corrodes or "loses" metal into ionic form in the solution. Galvanic corrosion occurs when two dissimilar metals are held in contact in an aqueous environment. The active metal corrodes. In the absence of ionic concentration differences or galvanic coupling, corrosion also can occur by gaseous reduction. Gaseous reduction involves dissolved oxygen.

Corrosives include acids and bases (alkalis or caustics) (Meyer 1989). Acids and bases attack amphoteric metals such as aluminum and zinc that may be present in fan casings and blades. Acidic substances or substances that react with water to form acids are more likely to be able to cause damage. Concentrated sulfuric acid extracts water from organic materials containing hydroxyl (–OH) groups. Fabrics and polymeric coatings used in flexible duct, that contain hydroxyl groups, face potential attack from sulfuric acid. Less concentrated sulfuric acid will attack carbon steel.

Acids, and to some extent bases, can hydrolyze the intermolecular linkages present in natural and synthetic polymers of textiles and coatings.

Corrosives of greatest concern to the materials of construction of portable ventilating equipment are oxidizers and oxidizing agents. These include nitrates, chlorine, bromine, iodine, fluorine, hypochlorites, ferric chloride, peroxides such as hydrogen peroxide, chromic acid and chromates, permanganates, and ozone (Furr 1989). Oxidizers can react with alloys containing active metals such as aluminum, magnesium, and zinc, used in air movers and organic polymers used in duct fabrics and coatings.

Nitric acid is both an acid and an oxidizing agent (Meyer 1989). It attacks certain metals, such as zinc and aluminum, and even steel-producing nitrogen oxides. Nitric acid also oxidizes organic materials. Fabrics and polymeric coatings used in flexible duct may be susceptible to this attack. Perchloric acid is another powerful oxidizing agent. Perchloric acid can attack both metals and organic substances, such as those used in duct fabrics. Duct fabrics that have been soaked in or have become coated by perchloric acid or perchlorates should be regarded as fire and explosion hazards. This situation could be compounded by use of fabric or metal duct on which organic and reactive inorganic materials have deposited during previous use. Chlorosulfuric acid is a third strongly oxidizing acid.

Electrochemical corrosion can occur when reactive metals such as aluminum, magnesium, and zinc found in lightweight alloys come into contact with metallic forms that occur lower in the electrochemical series (Barrow 1966). Some organic compounds also may be sufficiently reactive to cause corrosion of these metals.

Extensive discussion about chemical corrosion is beyond the scope of this Appendix. The reader is directed to references by Shackelford (1988), Meyer (1989), and Breatherick (1990) for further information. This subject is extremely important for the safe use of ventilating equipment, especially in chemical process operations.

Particulate loading of the air being transported from the space is an important concern. Particulates can cause erosion when transported at high velocity. Erosion refers to the physical wastage of components of the portable ventilation system. Erosive action of particulates is most likely to occur on fan blades and casings and elbows in duct.

Systems used to transport particulates must provide sufficient velocity. Otherwise, the particulates will settle onto bottom surfaces of the duct or into irregularities in the surface of the wall formed by spiral wire, seams, folds, and bends.

High particulate loadings could be explosible mixtures. The action of moving the particulates could generate static electrical charges on the particles and interior surfaces of the duct. Also, the fan or drive unit could be a source of ignition.

Use of equipment under these conditions should follow careful consideration, including discussion with the manufacturer.

Toxic substances generally pose less of a concern to equipment than other considerations mentioned in previous discussion. While substances present at levels of concern toxicologically are unlikely to pose an incompatibility problem with ventilating equipment, they could injure persons involved with the ventilation process.

Dusts and mists could deposit onto interior surfaces of the equipment. Residual contents in the space could deposit onto the external surfaces of duct. Exposure to the substances could pose a toxicity problem long after initial ventilation of the space has ceased. Contaminated equipment would require decontamination or disposal.

Removal of toxic material from the confined space could create an atmospheric hazard at the discharge of the duct or fan. This problem also could result from deposition of particulates and mists onto surfaces outside the confined space. In effect, the ventilation system could transfer the toxicity problem from inside the confined space to the outside.

Work Activity

Atmospheric conditions during work activity can be dramatically different from those during preentry preparation. In many cases, the purpose for ventilating during preentry preparation is to remove contaminants that have accumulated over a period of time and are not expected to reappear. That is, following initial ventilation, the space will remain contamination free. Under these conditions the portable ventilation system provides two basic functions: maintaining conditions for comfort or maintaining conditions for safety.

Maintaining conditions for comfort involves maintaining the status quo — that is, the condition present at the time that entry was authorized. Interior surfaces of the space may be contamination free or they may continue to act as sources of contamination. In either case, the movement of air induced by the portable ventilation system is sufficient to maintain the status quo.

Often the work to be undertaken itself is the source of new or additional contamination. The implications of this must be addressed when ventilating equipment is being considered. Work undertaken in the space can be the source of contamination from the following sources:

- Hot work
- Wetted surfaces
- Particulate generation processes
- Liquid aerosolization
- Process emissions

Many of the concerns raised in the previous section about the preentry atmosphere apply in addition to the two mentioned here. For this reason, they will not be repeated. Due to the nature of human needs and the nature of work undertaken in confined spaces, the level of concern regarding atmospheric contaminants during occupancy likely is very different from concerns relating to the preentry atmosphere.

Hot work includes processes such as welding, brazing, burning, cutting, and air-arcing and mechanical activities such as planing, grinding, and drilling. Hot work can lead to degradation of existing coatings on metal surfaces and volatilization of breakdown products. Slag from these processes can burn through fabric surfaces. Fabric duct used as a conduit for heated plumes is at risk from premature destruction. Burning of fabric also must be considered as a potential fire situation.

Wetted surfaces result from the presence of residual liquid and sludges on interior surfaces. They also can result from use of cleaning agents or from application of coatings. Under appropriate circumstances, these liquids may coat the inside and exterior of the duct, and attack and destroy the coating or fabric.

Particulate generation processes are common occurrences in work involving confined spaces. These include hot work processes as mentioned above, chipping, hammering, chiseling, scraping, and so on. These activities can be the source of very high levels of airborne dust.

Liquid aerosolization results from processes that break liquids into small droplets. Spray application of coatings is a major source of liquid aerosols. Liquid aerosols from coating products contain solvents, solids, and plasticized materials. These can deposit on interior and exterior surfaces of the portable ventilation system, including fan casings, fan blades, and duct. Physical and chemical damage, as well as increased combustibility of these components, can result.

Process emissions can occur when pressurized gases are used in confined spaces. These gases can include oxygen and acetylene or propane used in oxyfuel cutting, and shield gases used in welding, burning, and cutting. Additional sources can include combustion gases from engines, as well as fire.

Air Used for Ventilation

Ventilation systems move air. While the system is not concerned about the quality of air handled, the user must be. The quality of air supplied to confined spaces must be assured. While this statement is self-evident, the ability to assure the quality of air supplied for breathing purposes is not. Previous discussion considered causes of contamination associated with operation of equipment in the portable system. The focus of this part of the discussion is the source of breathing air. Assuring the quality of supply is especially important when supply air must be drawn from within the confines of a process operation. A fugitive emission near the intake of a duct used to supply air to a space could lead to disaster. This also could happen during ventilation of the residual atmosphere during preentry preparation. Failure to assure the quality of air resident in the space following ventilation could lead to a disastrous situation.

The same concern holds true for air provided during occupancy. Where there is any doubt about the quality of air to be provided, on-line, real-time testing must be considered the minimum option.

Ventilation Mode

The two approaches utilized for ventilating confined spaces include supplying air to the space and removing air from the space. Supplying air to the space displaces air from the space. The action of providing air to a fixed volume dilutes contaminated air. Removing air from the space causes air from the surroundings to enter the space. Entry of outside air will dilute contaminated air in the space. Placement of the inlet to the system near the source of contamination (source capture) can minimize spread of contamination throughout the airspace.

Important differences exist between supply and removal of air using portable systems. However, one thing is the same. Volumetric flow rate in the duct remains the same, despite differences at the outlet or intake. Air discharges from the end of a duct in a narrow plume as the energy imparted to it by the fan (static and velocity pressure) is expended in the larger air volume. Air collects into a duct or hood from a wide spherical area that surrounds the opening (Figure J.8). This motion can be focused somewhat by the presence of slotted and flanged openings. Velocity on the discharge side will be approximately the same at 30 duct diameters from the outlet as at one duct diameter from the inlet on the suction side. Air discharges through the opening of a duct perpendicular to the face. This results in a high velocity through small cross-sectional area. Air drawn into a duct comes from the zone surrounding the inlet, not just the area perpendicular to the face. The difference in air motion and velocity can have profound impact on effectiveness in controlling contaminants.

This fundamental difference in the characteristics of airflow at the intake and discharge underscores the importance of control in the flow pattern of air.

Dilution Mode

The basis for dilution methods is the forced mixing of uncontaminated air from outside the space with the contaminated air inside (Figure J.8). The action of mixing dilutes the contaminated air, and the excess pressure associated with the inflow expels the diluted mixture from inside the space. The continuous mixing and expulsion lowers the concentration of contaminants to zero unless generation is occurring. In the latter case an equilibrium will develop at a nonzero concentration.

Air exchange and air exchange rate represent a first approximation toward a quantitative description of ventilation of confined spaces. An air exchange is the volume of air equal to the volume of the space. Air exchange rate is the number of air exchanges per hour that a system can provide.

The following equations relate exchange rate (air changes per hour) to the volume of a space and volumetric flow rate:

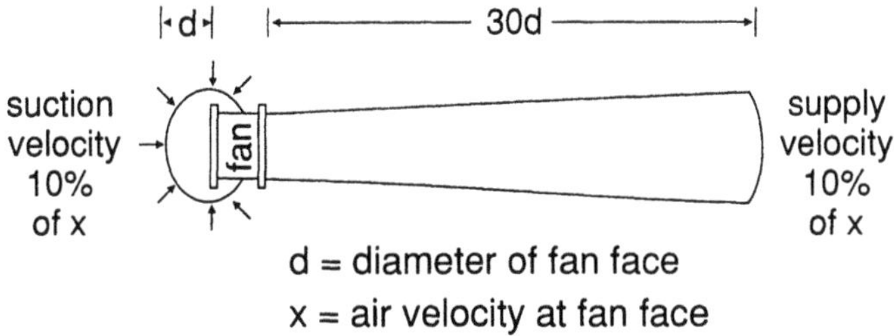

FIGURE J.8. Flow pattern around a bare fan. Use of short sections of duct on the intake and discharge side of the fan will minimize reentrainment of contaminants.

In the **English** system:

$$ER = \frac{Q \times 60}{V} \quad (J.1)$$

ER = ac/h
Q = volumetric flow rate (cfm) or (ft^3/min)
V = volume of space (ft^3)

In the **metric/SI** system:

$$ER = \frac{Q}{V} \quad (J.2)$$

ER = ac/h
Q = volumetric flow rate (m^3/h)
V = volume of space (m^3)

The limitations of the term, air exchange rate, must be stressed. The ventilating air is added gradually over a prolonged period, rather than through bulk exchange. Also, this term considers the confined space as a uniform macro environment and ignores the presence of poorly ventilated areas and the influence of source number and geometry.

A more realistic view of the process of replacing the air in the space with outside air and thereby diluting contaminants is given by the following equation. This describes the general case where contaminant generation is occurring (Burgess et al. 1989). This equation applies regardless of whether air is being supplied to the space or removed from it by the portable system.

$$C_2 = \frac{1}{Q}\left[G - (G - QC_1)e^{-\frac{Q}{V}(t_2 - t_1)}\right] \quad (J.3)$$

C_2 = the concentration at time t_2
C_1 = the concentration at time t_1
G = the generation rate
Q = the airflow rate
V = the volume of the space

This equation presumes that perfect mixing occurs, that G and Q are constant, and that dilution air contains negligible contamination. This equation applies in situations in which a source continually generates contamination. This would correspond to a space that was not cleaned prior to the start of ventilation. The quantity Q/V is the same as air exchanges per unit of time. Hence, this links the intuitive concept to the mathematical one.

There are several special applications of Equation J.3. First is the situation where the source is no longer present at the time of ventilation (G = 0).

$$C_2 = C_1 e^{-\frac{Q(t_2 - t_1)}{V}} \tag{J.4}$$

This could apply during preentry preparation to correct an existing situation or following some work activity in which a source was generating contamination. Setting $C_2 = 0.5\ C_1$ introduces the concept of half-time (half-life). This is the time needed to reduce the concentration to half its starting value.

The second situation involves generation of a contaminant in a space where the starting concentration was zero ($C_1 = 0$). The resulting concentration at any time t is given by:

$$C = \frac{G}{Q}\left(1 - e^{-\frac{Qt}{V}}\right) \tag{J.5}$$

The third situation occurs where generation of contamination continues (G > 0) and ventilation has occurred for a considerable time.

$$C_{max} = \frac{G}{Q} \tag{J.6}$$

Although dilution ventilation is more than adequate for many applications, this method does have limitations. Dilution is especially suited for ventilation during preentry preparation. Considerations about dilution ventilation are summarized in Table J.4.

Local Exhaust (Source Capture) Mode

The basis for the local exhaust or source capture method is capture of contaminants at the point of generation (Figure J.11). This minimizes entry into the air in the space. Source capture requires induction of a controlled flow of air that traverses the point of generation. In the process of capturing and removing the contaminant from the space, outside air is drawn in to replace that exhausted.

Considerations that govern the application of local exhaust ventilation are contained in Table J.5.

DESIGN CRITERIA FOR PORTABLE VENTILATION SYSTEMS

The last part of this Appendix is concerned with criteria for designing portable ventilation systems. Designing a portable system is somewhat different from designing one that is permanently installed. A previous section demonstrated that portable systems are considerably simpler in concept and utilize considerably less hardware. They also are operationally inflexible compared to permanently installed systems.

The approach taken in designing a portable supply system differs from that used in designing a permanent installation. In the former case, many of the parameters, such as capability of the air mover and duct size are set prior to the start of the process. The focus of the process is to determine the capabilities of the system resulting from connecting together various (available) pieces of hardware. By contrast, the approach taken in a permanent installation is to determine performance

TABLE J.4
Application of Dilution Ventilation

- Dilution is best suited in applications where the contaminant level is low. Where contaminant levels are high, the required volumetric flow rate could become impractical (with regard to drive requirements, fan size and weight, and portability).
- Dilution is best suited in applications where the LFL is high. Where LFL is low, the required volumetric flow rate could become impractical (with regard to drive requirements, fan size and weight, and portability).
- Dilution is best suited in applications where the exposure limit (TLV, PEL, or other unit) is high (toxicity is relatively low). Even where contaminant levels are low, the required volumetric flow rate could become impractical (with regard to drive requirements, fan size and weight, and portability). Dilution ventilation rate is inversely proportional to the TLV.
- Clean air used for dilution should be supplied to the breathing zone of occupants and then into the contaminated zone to dilute contaminants. Contaminated air should not pass through the breathing zone of occupants prior to dilution. Otherwise, the application, at best, is only partly successful.
- Occupants must be located far enough away from the source of the contaminant or the level must be low enough, so that overexposure will not occur.
- Dilution ventilation is best suited for applications where the distribution of sources or contaminants within the confined space is uniform.
- When using dilution ventilation, do not create additional problems by introducing air contaminated from other sources into the confined space (Figure J.9).
- When diluting the atmosphere in a space, locate the duct so that all pockets of contaminated air are purged (Figure J.10).

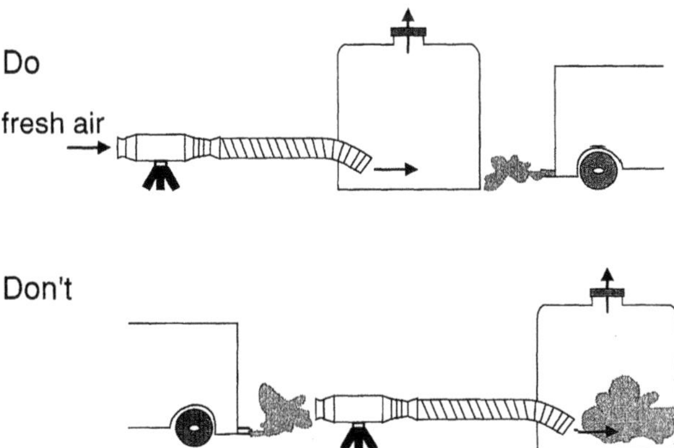

FIGURE J.9 Dilution ventilation practices: air supply. Air used to supply a dilution system should not contaminate the space being ventilated.

characteristics and then to specify hardware that best meets them. The following discussion compares the design approach used for permanent installations with the realities of portable systems (Burgess et al. 1989).

The first step is to choose capture hood or inlet geometry. Portable systems often do not include hoods or other intake devices. A hood or intake device, such as a flanged or bell-mouthed opening would improve performance of the system.

The second step is to calculate required airflow. Required airflow in the usual context is relevant to airflow specified for performance of the capture hood. This concept usually does not apply to portable systems, since hoods usually are not employed. However, results of airflow studies can be utilized here. NIOSH (1985) recommended an air exchange rate of 20 air exchanges per hour for confined spaces in the construction industry. The validity of this recommendation was supported by experimental work using models of confined spaces (Garrison et al. 1989, 1991; Garrison and Erig 1991). During the initial purge of a space during preentry preparation, 20 air changes (theoretically) would reduce concentration of a contaminant to 2×10^{-9} of its original concentration,

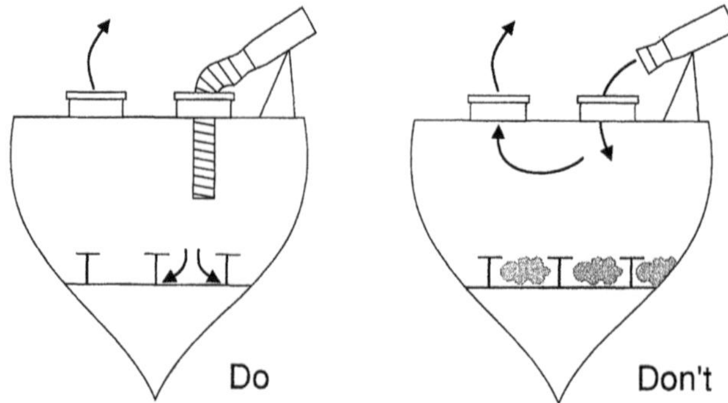

FIGURE J.10 Dilution ventilation practices: thoroughness of mixing. Lack of attention to airflows in the structure can lead to incomplete mixing of residual contaminants. Dilution ventilation relies on turbulence to induce motion in the air contained within a structure.

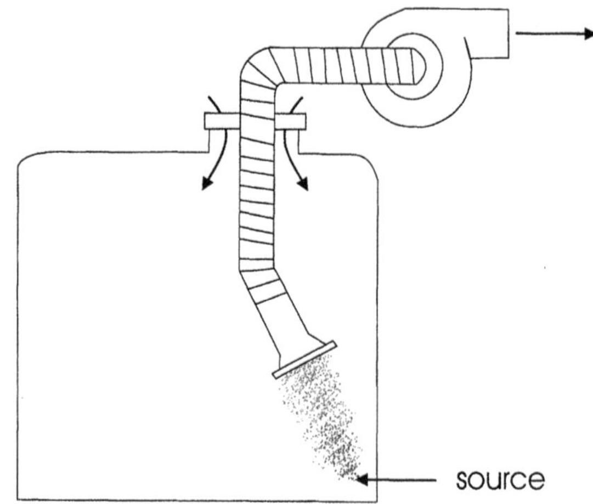

FIGURE J.11 Local exhaust ventilation. Local exhaust ventilation utilizes induction of airflow to remove contaminants at the point of generation.

TABLE J.5
Application of Local Exhaust Ventilation

- Local exhaust is preferred for point sources.
- Utilize the local exhaust method to capture contaminants whenever possible. When treating contaminants generated at a point source, this method not is only more protective, but will require less airflow and consume less energy than the dilution method (Figure J.12).
- When exhausting contaminants from a confined space, locate the blower and duct so that discharged air does not contaminate another area (Figure J.13).
- Locate the hood as close as possible to the source of the contaminants in order to optimize capture efficiency (Figure J.14).

assuming G = 0. Provision of 20 air changes per hour could become the basis for calculating required airflow. This would require the volume of the space and delivery rate of the air mover.

The third step is to specify minimum duct velocity. The fourth step is to specify duct size (diameter). These do not pose an issue where the portable system supplies clean air to the space.

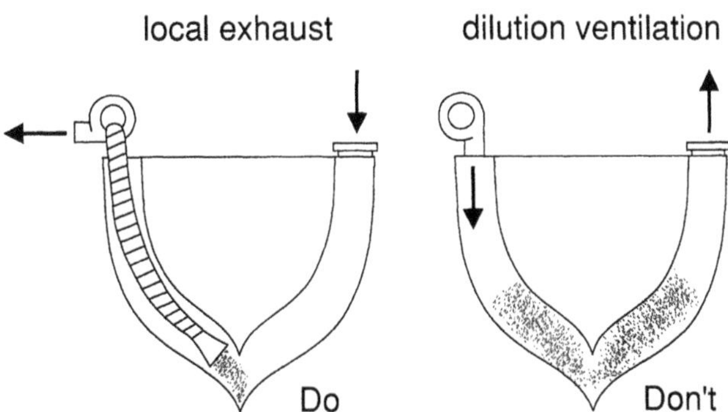

FIGURE J.12 Ventilation mode should fit the circumstances. In this configuration, several workspaces are located along a linear corridor. Local exhaust removes contaminants at source. In dilution mode, the fan pushes contaminated air throughout the workspace. This situation would benefit from use of both supply and local exhaust ventilation, where possible.

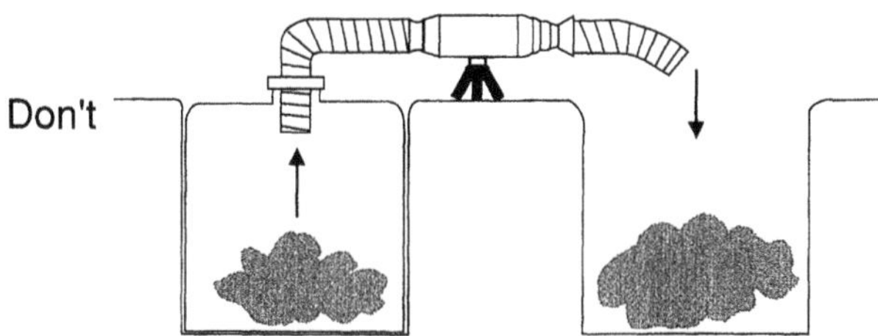

FIGURE J.13 Exhaust ventilation practices: fate of discharged air. Contaminated air discharged from the space should not enter another in which work is occurring.

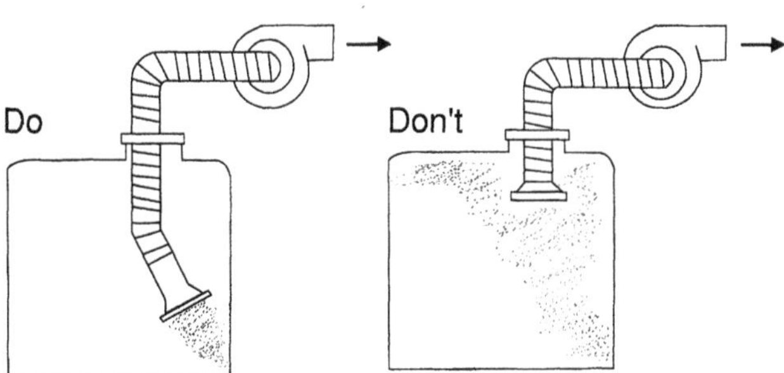

FIGURE J.14 Exhaust ventilation practices: location of collector. Position the collector as closely as possible to the source. Otherwise, the contamination could fill the space.

However, these could become an issue where aerosols (dusts, mists, or fumes) are present in air to be exhausted from a space. Minimum velocities are specified to prevent settling or deposition on the walls of the duct (ACGIH 1995). Systems transporting gases and vapors usually are designed to a standard of 1,500 to 2,000 ft/min (7.6 to 10 m/s). Systems for transporting particulates are designed to a standard of 3,500 to 4,500 ft/min (18 to 23 m/s).

Velocity determines duct diameter and could affect choice of air mover. In the real world of portable systems, neither of the preceding may be a variable. While flexible duct is produced in a range of diameters, individual users are unlikely to maintain stock of enough sizes to enable design flexibility. Also, users are unlikely to purchase a range of air movers for this reason. These constraints need not become serious problems, provided that users purchase duct sizes that would enable adequate velocity for anticipated use with the air mover(s). Users should discuss their needs with the manufacturer or vendor at time of purchase.

The fifth step is to calculate energy losses in the system. Ideally, the layout of the system should minimize duct length, the number and tightness of bends (fixed elbows), and sags in duct. The reality is that the configuration of portable systems often changes continually. Duct may be stretched out at one moment and bunched together at another. The number of bends and sags also could vary from moment to moment as the position of the work area changes. For this reason, values calculated for portable systems should be considered approximate.

Energy losses in systems are characterized as static or dynamic.

Static losses (discussed earlier) associated with the resistive effect of boundary surfaces are also known as frictional losses. Roughness factors are used to account for the variation in surface characteristics of materials of construction — metal, fabric, and plastic; and internal configuration (flat, corrugated, and helical). Roughness factors also influence air velocity. Roughness factors can be calculated through equations, approximated through the use of graphs and charts, or obtained from the duct manufacturer.

Static losses in duct are given by: (ACGIH 1995)

$$SL = \frac{f \times L \times (VP)}{d} \tag{J.7}$$

SL = friction loss in duct (in wg)
f = a dimensionless friction factor
L = the length of duct (ft)
VP = velocity pressure (in. wg); VP is related to air velocity.
d = duct diameter (ft)

The value of "f" can be calculated from the following equations:

$$f = 8\left[\left(\frac{8}{Re}\right)^{12} + (A+B)^{3/2}\right]^{1/12} \tag{J.8}$$

$$A = \left\{-2.457 \times \ln\left[\left(\frac{7}{Re}\right)^{0.9} + \left(\frac{e}{3.7\,D}\right)\right]\right\}^{16} \tag{J.9}$$

$$B = \left(\frac{37{,}530}{Re}\right)^{16} \tag{J.10}$$

e = the absolute surface roughness of the duct (in.)
D = duct diameter (in.)
Re = the Reynolds number. This is calculated from the following equation.

$$Re = \frac{\eta dV}{\mu} \tag{J.11}$$

η = the density of air (lb/ft^3)
μ = the viscosity of air (lb/ft min)

TABLE J.6
Absolute Surface Roughness for Duct Materials

Duct material	Surface roughness (in.)	Constants		
		a	b	c
Galvanized metal	0.006	0.0307	0.533	0.612
Aluminum, stainless steel, PVC plastic	0.0018	0.0425	0.465	0.602
Flexible duct, wires exposed	0.12			
Flexible duct, wires covered, rigid fibrous glass	0.036	0.0311	0.604	0.639
Galvanized steel with spiral seams, 1 to 3 ribs	0.0036			

At standard conditions of 70°F (21°C) and 1 atm pressure and air density of 0.075 lb/ft^3 and viscosity of 7.42×10^{-4} lb_{mass}/(ft min), the equation for calculating Reynolds number reduces to:

$$Re = 101\ d \times V \tag{J.12}$$

ACGIH (1995) offer an alternative method for determining the friction factor, f, using a single equation. This method provides values of f within 5%. This equation for calculating f is specific only to the duct for which the complete data were provided.

$$f = \frac{d \times a \times V^b}{Q^c} \tag{J.13}$$

d = duct diameter (ft)
a = a constant
V = duct velocity (ft/min)
b = a constant
Q = volumetric flow rate (ft^3/min)
c = a constant

Values of a, b, and c are provided in Table J.6. Values of surface roughness, e, are available from tables for various duct materials (Guffey 1992, ACGIH 1995). Those of interest in portable systems are provided in Table J.6. Values of the constants, a, b, and c, are available only for a limited number of materials. Galvanized steel with spiral seams would be expected to perform similarly to galvanized metal, according to this data. Surface roughness has a range, even for new materials.

Static losses in duct vary directly with length, velocity, and the roughness factor, and inversely with diameter. Therefore, for a given fan and power source, frictional losses increase and airflow decreases when the duct is lengthened, or reduced in diameter. Roughness of molded plastics, such as PVC, appears to be less than or at least similar to that of flexible duct. For the purpose of design calculations, treating the roughness of plastic fittings such as tapers, rigid elbows, the Saddle Vent, and so on, as equivalent to flexible duct appears to be a reasonable assumption.

Dynamic losses occur as air passes through structures in the ventilation system. Dynamic losses are given by the following equation:

$$DL = F \times VP \tag{J.14}$$

F = a loss coefficient
VP = duct velocity pressure (in. wg)

Dynamic losses are associated with hoods, elbows, sags, compressed zones in flexible duct, transitions, and changes in area. Losses for rigid structures are provided in reference sources (ACGIH 1995). Durr et al. (1987) derived equations for losses in flexible duct. The loss due to a drooping or sagging section is given by the following equation.

$$DL_{sag} = 31.7 \times VP \times f \times \frac{h}{d} \times e^{\frac{-31}{3d}} \quad (J.15)$$

h = the depth of the sag (consistent units)
l = the length of the sagging section (consistent units)
d = duct diameter (consistent units)

The corresponding loss due to a bend or turn is given in Equation J.16.

$$DL_{bend} = 1.808 \times VP \times \frac{C}{360} \times \left[e^{-\frac{3\times tr}{4}} + 9.956 \times f \times e^{-\frac{tr}{100}} \right] \quad (J.16)$$

c = the angle of bend (degrees)
tr = the turning ratio, the ratio of turning radius (radius of the circle formed by the duct in the bend) to diameter of the duct

Previous discussion has provided the means to determine deteriorative losses of performance in portable systems. The sum of these losses dictates overall performance. The latter usually is all that users of this equipment can observe. Table J.7 provides a comparison of importance of each of these factors. Calculations used in creating this table assumed a duct diameter of 8 in. and flow rate of 1,000 ft³/min through the openings to the air mover. At this flow rate under standard conditions, velocity pressure is 0.51 in. wg, and the friction factor for flexible duct with covered wire is 0.031 by Equation J.13.

The greatest contributor to deterioration in performance from the portable system is duct sag. One or more medium or deep narrow sags could produce serious deterioration in performance of the system. This easily could exceed loss from the duct. Unnecessary sags resulting from careless placement of duct can be eliminated through training and awareness. Sags or bends around obstacles that cannot be eliminated dictate attention to assure performance from the system.

The confined space itself can contribute to static and dynamic losses. This is especially true for spaces having small diameter and limited openings, as well as baffles, piping, machinery, and internal structures through which air must pass. A static loss of 1 in. wg (25 mm wg) due to the structure easily could cause an unexpected delivery loss of 10 to 20% from the air mover. All other things being the same, greater airflow can occur through a large open manway than through a small one. Also, losses would be less when several manways of a small structure are open instead of only one.

Dynamic losses due to structural geometry are nearly impossible to calculate. They depend on the situation. They can be determined through measurement of airflow and pressures, and the cross section of openings, and back calculation to compare known vs. expected performance. The preceding discussion highlights a critical requirement for success when ventilating using portable systems: minimize flow restrictions in the system and the confined space.

The system resistance curve is the mathematical representation of losses imposed by a specific system. This specifies the airflow that will pass through the system for any value of static pressure applied by the air mover. The system resistance curve is specific to a combination of duct, hoods, filters, and so on. Changing parameters of airflow produces a new position on the system resistance curve. Changing configuration of the system through repositioning duct produces a new system

TABLE J.7
Sources of Deterioration of Performance in Portable Systems (Flexible Duct, Covered Wire)

			Loss (in. wg)
Product (flexible duct, covered wire)			
Manufacturer A			1.45/100 ft
Manufacturer B, C			2.8/100 ft
Spiral-wound corrugated metal duct			3.5/100 ft
Sag (flexible duct, covered wire)			
Depth/width	**(h/d**	**l/d)**	
Shallow/narrow	(0.1	1)	0.03
Shallow/medium	(0.1	6)	0.01
Shallow/wide	(0.1	12)	0.00
Medium/narrow	(5	1)	1.67
Medium/medium	(5	6)	0.26
Medium/wide	(5	12)	0.03
Deep/narrow	(10	1)	3.33
Deep/medium	(10	6)	0.51
Deep/wide	(10	12)	0.05
Bend (flexible duct, covered wire)			
(Turn angle	**Turn ratio)**		
(30	1.5)		0.05
(45	1.5)		0.07
(60	1.5)		0.10
(90	1.5)		0.15
(135	1.5)		0.21
(30	7.0)		0.02
(45	7.0)		0.03
(60	7.0)		0.04
(90	7.0)		0.07
(135	7.0)		0.10
(30	13.5)		0.02
(45	13.5)		0.03
(60	13.5)		0.04
(90	13.5)		0.06
(135	13.5)		0.09
Interior of space			unknown

resistance curve. There is a family of system resistance curves. Burgess et al. (1989) shows a family of system resistance curves. The slope of these curves becomes steeper as additional losses are incurred (Figure J.15).

Just as there is a family of system resistance curves, there also is a family of fan static pressure curves, one for each rotational speed. Rotational speed also governs volumetric airflow rate and power requirements through a series of fan laws (May 1970, Russell et al. 1973).

The first fan laws relate changes in volumetric flow rate, pressure (static or total), and horsepower to variations in fan speed.

$$Q_2 = Q_1 \times (N_2/N_1) \tag{J.17}$$

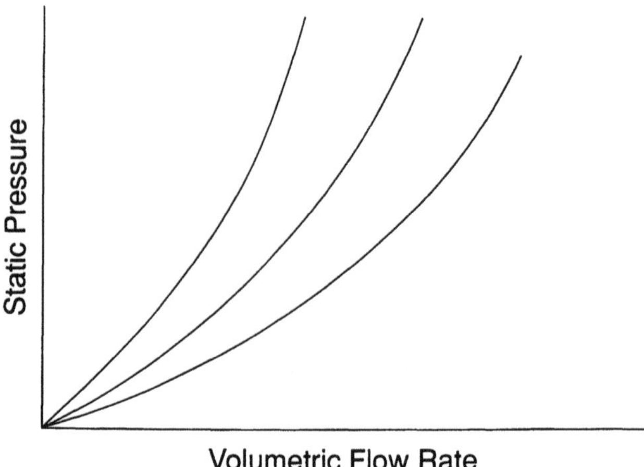

FIGURE J.15 Family of system resistance curves.

$$P_2 = P_1 \times (N_2/N_1)^2 \tag{J.18}$$

$$HP_2 = HP_1 \times (N_2/N_1)^3 \tag{J.19}$$

Q = volumetric flow rate (ft^3/min or m^3/hr)
N = rotational speed (rev/min)
P = pressure (in. wg or mm wg)
HP = power requirement (horsepower or Watts)

As indicated by dimensional analysis, the fan laws can be solved in both English or metric units.

Example: Calculate the performance of an electrically driven ventilator to be operated at 2,850 rev/min instead of 3,450 rev/min. The unit is rated to deliver 400 ft^3/min at 3.00 in wg using a 1/2-horsepower motor operating at 3,450 rev/min.

$$Q_2 = 400\,ft^3/min \times (2850/3450) = 330\,ft^3/min \tag{J.20}$$

$$P_2 = 3.00\text{ in. wg} \times (2850/3450)^2 = 2.05\text{ in. wg} \tag{J.21}$$

$$HP_2 = 0.5\text{ hp} \times (2850/3450)^3 = 0.3\text{ hp} \tag{J.22}$$

The preceding example is relevant only to portable equipment whose speed can be varied.

Operating point is the final concept in the design of portable systems (Figure J.16). Operating point is the intersection of the system resistance curve and the fan curve. Operating point indicates the predicted volumetric flow rate and fan static pressure for the actual system for a particular fan operating at a particular speed. When duct is added to an existing system, the system resistance curve changes, becoming steeper (higher static pressure). The operating point moves up the corresponding fan static pressure curve, as the static pressure increases and airflow decreases.

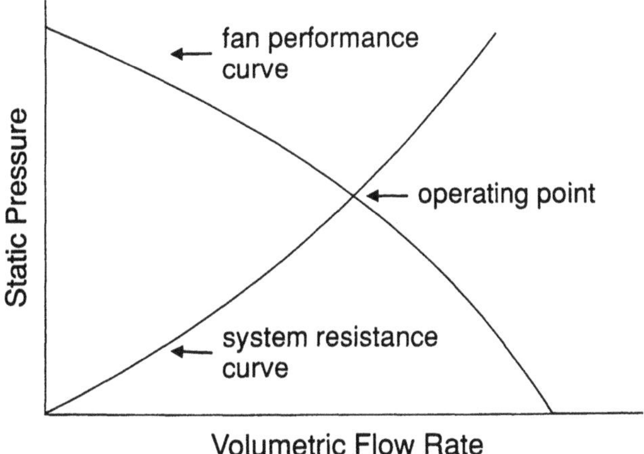

FIGURE J.16 Intersection of the system resistance curve with the fan performance curve.

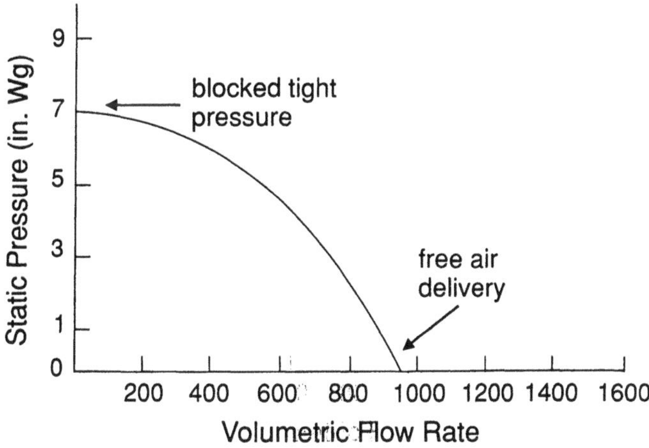

FIGURE J.17 Fan performance curve for a specific fan.

Example: A 1-hp centrifugal ventilator is used to extract vapors from an underground vault using a 30-ft length of 6-in. diameter flexible duct. The duct is connected to the suction side of the fan. One 90° elbow having a centerline radius-to-diameter ratio of 2.0 is employed. The duct is made from polyvinyl chloride (PVC)-impregnated nylon fabric. The duct is supported by an internal wire helix. It has a roughness factor of 1.25. Determine the volumetric flow rate of this system.

The solution to this problem involves locating the operating point for the system. The operating point is the point of intersection of the fan and system resistance curves. The fan curve indicates that the operating point will occur to the left of 960 cfm, the free air delivery point (Figure J.17). The system resistance curve is determined by plotting static pressure losses at various airflows. The following equations will be used to calculate loss:

$$\text{Total loss} = \text{static loss} + \text{dynamic loss} \tag{J.23}$$

$$\text{Static loss} = \text{SL} = \text{F}/100 \times \text{L} \tag{J.24}$$

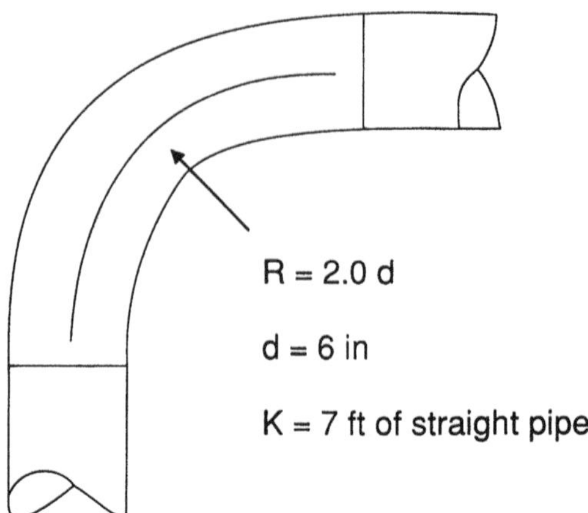

FIGURE J.18 Loss of static pressure in an elbow equivalent to straight pipe.

F = frictional loss per 100 ft of duct
L = duct length in feet

To calculate static loss from either volumetric flow rate or duct velocity, consult reference tables or the equations provided in previous discussion. In this example, the frictional loss factor is 4.4 in. wg/100 ft of duct for 800 cfm flowing through a 6-in. diameter duct.

$$\begin{aligned} SL &= F/100 \times L \\ &= 4.4 \text{ in. wg}/100 \text{ ft} \times 30 \text{ ft} \\ &= 1.32 \text{ in. wg} \end{aligned} \tag{J.25}$$

Dynamic losses incurred at elbows and other fittings can be expressed in terms of equivalent lengths of straight duct.

$$\text{Dynamic loss} = DL = F/100 \times K \tag{J.26}$$

F = frictional loss factor per 100 ft of duct
K = equivalent resistance in linear feet of duct

As shown in Figure J.18, a 6-in. elbow with a centerline radius-to-diameter ratio of 2.0 is equivalent to 7 feet of straight duct. Substituting into the equation for dynamic losses:

$$\begin{aligned} DL &= F/100 \times K \\ &= 4.4 \text{ in. wg}/100 \text{ ft} \times 7 \text{ ft} \\ &= 0.31 \text{ in. wg} \end{aligned} \tag{J.27}$$

$$\begin{aligned} \text{Total loss} &= SL + DL \\ &= 1.32 \text{ in. wg} + 0.31 \text{ in. wg} \\ &= 1.63 \text{ in. wg} \end{aligned} \tag{J.28}$$

TABLE J.8
System Losses

Airflow (cfm)	Friction loss factor (in. wg/100 ft)	Static losses (in. wg)	Dynamic losses (in. wg)	Total losses (in. wg)
800	4.40	1.65	0.39	2.04
700	3.40	1.02	0.23	1.25
600	2.50	0.75	0.18	0.93
500	1.80	0.54	0.12	0.66
400	1.15	0.34	0.08	0.42
300	0.67	0.20	0.05	0.25
200	0.315	0.10	0.02	0.12
100	0.085	0.02	0.01	0.03

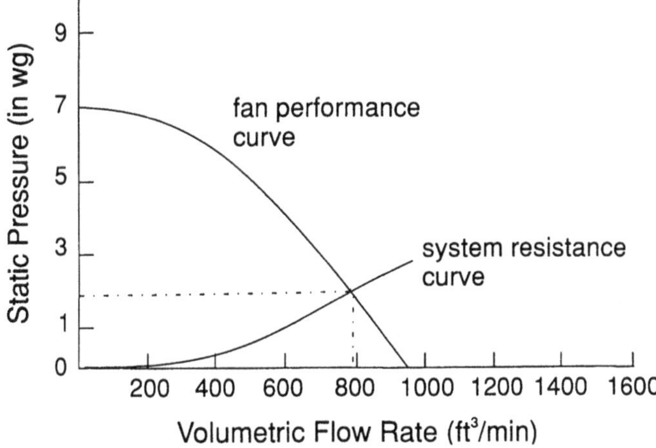

FIGURE J.19 Intersection of the system resistance curve with the fan performance curve for a specific system–fan combination.

Repeating these calculations for airflow rates between 100 and 700 cfm produces data that is listed in Table J.8.

The system resistance curve now can be generated by plotting airflow (cfm or ft^3/min) vs. total losses (in. wg). Intersection of this curve with the fan curve, the operating point, occurs at 790 cfm and 1.95" wg (Figure J.19). Tables and graphs provided in reference books simplify the process of calculating static and dynamic losses. Software packages also are available for this purpose.

SUMMARY

Practical use of ventilation equipment in confined spaces is very much a combination of the art of experience and the science of ventilation. Selection of ventilation equipment that will provide sufficient airflow is application specific. A particular ventilator probably will not be suitable for every application.

This Appendix has explored the considerations needed for successful application of portable ventilating equipment in confined spaces. Selection of appropriate ventilating equipment is a vital part of any confined space entry program. This requires careful thought and consideration.

Table J.9 summarizes in general terms practices appropriate to successful operation of portable systems.

TABLE J.9
General Practices for Use of Portable Ventilation Systems

- When selecting a method for ventilating a confined space, consider whether the gas is denser or less dense than air. Let these natural forces work for you, not against you (Figure J.20).
- In order to optimize airflow and minimize losses, maximize duct diameter, minimize duct length, minimize restrictions to airflow, and avoid elbows (Figure J.21).
- Supply air always must be readily available to the fan in order to prevent stalling. Stalling is a condition where the air mover (especially a propeller fan) rotates, but does not move any air.
- Always use close pitch duct on the suction side of a fan in order to avoid collapse (Figure J.22).
- Air movement through a ventilation system can cause a buildup of static electrical charges under certain conditions. Where a flammable/combustible/explosive or oxygen-enriched atmosphere may be present or may develop, an explosion can result from a static discharge or spark. As a result, fluid-driven ventilators must be mechanically bonded to dissipate the buildup of static charges and all electrically driven units be grounded through electrical circuits (Figure J.23).
- In some situations, placement of ventilators can cause short-circuiting of incoming air. In this condition, air enters and exits without ventilating the confined space. Test to ensure that the performance of the portable ventilation meets expectations.
- Harness allied forces — wind and thermal buoyancy. These can be the greatest of allies or they can be the greatest of enemies to the user of portable ventilation systems. Portable ventilation systems can overcome these forces only with the greatest of difficulty. They can make use of them with the greatest of ease.

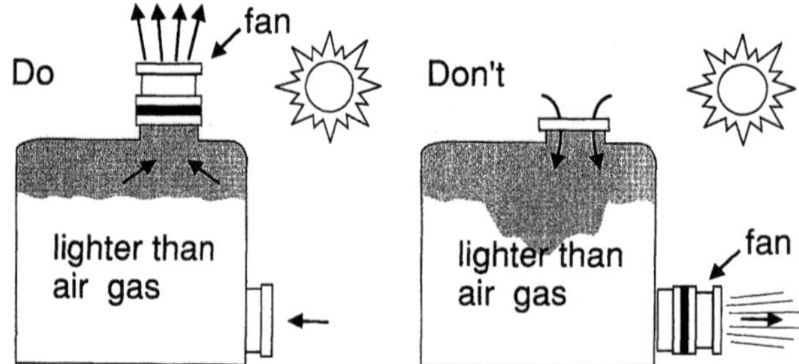

FIGURE J.20 Ventilation practices: natural forces. During ventilation with mechanical equipment, harness natural forces where possible.

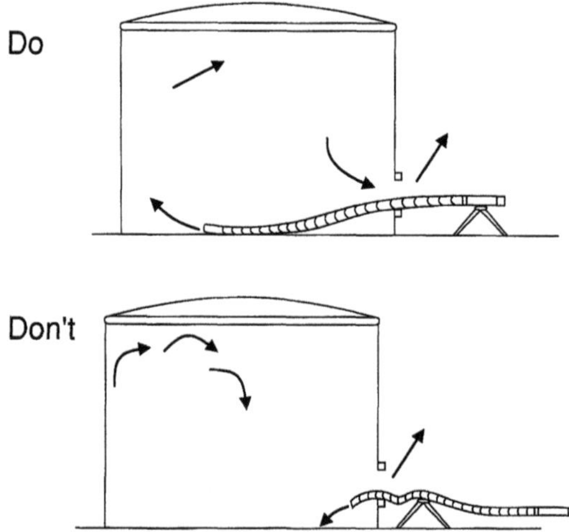

FIGURE J.21 Ventilation practices: bends and sags in duct. Minimizing bends and sags in flexible duct minimizes losses due to static pressure and maximizes airflow.

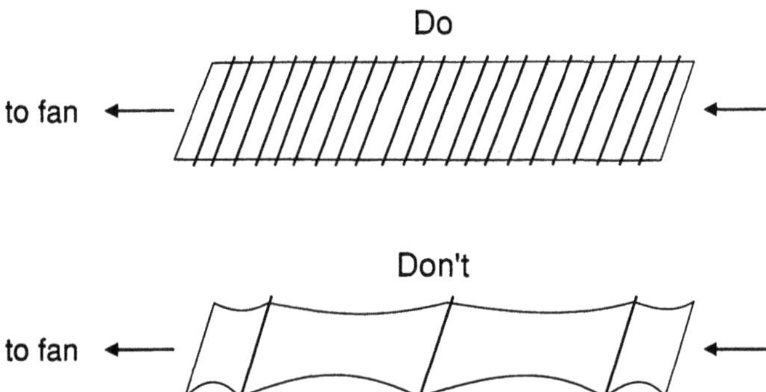

FIGURE J.22 Ventilation practices: duct collapse and contraction. Use close-pitched wire-wrapped duct on the suction side of the fan. Close-pitched wire-wrapped duct retains its shape and length. This minimizes losses due to static pressure and maximizes airflow.

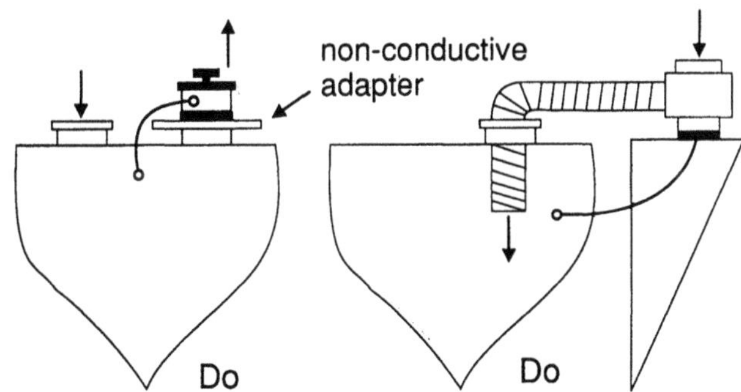

FIGURE J.23 Ventilation practices: bonding. Bonding is an essential preventive measure when ignitable vapors may form. Bonding is an electrical connection between the air mover and the metal of the structure.

REFERENCES

Alden, J.L. and J.M. Kane: *Design of Industrial Exhaust Systems,* 4th ed. New York: Industrial Press, 1970 pp. 184–209.

American Conference of Governmental Industrial Hygienists (ACGIH): *Industrial Ventilation: A Manual of Recommended Practice,* 22nd ed. Cincinnati, OH: American Conference of Governmental Industrial Hygienists, 1995.

Barrow, G.M.: *Physical Chemistry,* 2nd ed. New York: McGraw-Hill, 1966. pp. 712–752.

Bleier, F.P.: Design, performance and selection of axial-flow fans. Reference section. *Heating and Ventilating 43:* 83–94 (1946).

Bleier, F.P.: Fans. In *Handbook of Energy Systems Engineering*, Section 4.2. Leslie C. Wilbur (Ed.). John Wiley & Sons, 1985.

Breatherick, L.: *Breatherick's Handbook of Reactive Chemical Hazards,* 4th ed. London: Butterworths, 1990. 2003 pp.

Burgess, W.A., M.J. Ellenbecker, and R.D. Treitman: *Ventilation for Control of the Work Environment.* New York: Wiley-Interscience, 1989. pp. 229–307.

Canadian Standards Association: *Compressed Breathing Air and Systems (CAN3-Z180.1-M85).* Toronto, ON: Canadian Standards Association, 1985. 32 pp.

Compressed Gas Association: *Commodity Specification for Air (ANSI/CGA G-7.1).* Arlington, VA: Compressed Gas Association, 1989.

Durr, D.E., N.A. Esmen, C. Stanley, Jr., and D.A. Weyel: Pressure drop in flexible ducts. *Appl. Ind. Hyg. 2:* 99–102 (1987).

Frankel, G.J.: Oxygen-enriched atmospheres. In *Fire Protection Handbook*, 17th ed. Cote, A.E. and J.L. Linville (Eds.). Quincy, MA: National Fire Protection Association, 1991. pp. 3-160 to 3-169.

Furr, A.K.: *CRC Handbook of Laboratory Safety*, 3rd ed. Boca Raton, FL: CRC Press, 1989. pp. 268–284.

Garrison, R.P., R. Nabar, and M. Erig: Ventilation to eliminate oxygen deficiency in confined spaces. Part I: A cubical model. *Appl. Ind. Hyg. 4*: 1–11 (1989).

Garrison, R.P., and M. Erig: Ventilation to eliminate oxygen deficiency in confined spaces. Part III: Heavier-than-air characteristics. *Appl. Occup. Environ. Hyg. 6*: 131–140 (1991).

Garrison, R.P., K. Lee, and C. Park: Contaminant reduction by ventilation in a confined space model — toxic concentrations versus oxygen deficiency. *Am. Ind. Hyg. Assoc. J. 52:* 542–546 (1991).

Guffey, S.E.: *Heavent User Manual.* Seattle, WA: S.E. Guffey, 1992.

Lowrie, R.: Materials for oxygen service. *Chemical Engineering.* April 27, 1987. pp. 75–80.

May, J.W.: *The Physics of Air*, 7th ed. Louisville, KY: American Air Filter Company, 1970 p. 50.

McDermott, H.J.: *Handbook of Ventilation for Contamination Control.* Ann Arbor, MI: Ann Arbor Science Publishers, 1976. 368 pp.

McQuiston, F.C. and J.D. Parker: *Heating, Ventilating, and Air Conditioning: Analysis and Design*, 2nd ed. New York: John Wiley & Sons, 1982. pp. 366–432.

Meyer, E.: *Chemistry of Hazardous Materials*, 2nd. ed. Englewood Cliffs, NJ: Prentice Hall, 1989. pp. 204–359.

National Fire Protection Association: *National Electrical Code (1993 Edition).* Quincy, MA: National Fire Protection Association, 1993.

National Institute for Occupational Safety and Health: Safety and Health in Confined Workplaces for the Construction Industry — A Training Resource Manual. Washington, D.C.: Government Printing Office, 1985. pp. 127–135.

National Occupational Health and Safety Commission and Standards Australia: Joint Draft National Standard for Planning and Work for Confined Spaces (BS92/20113 Cat. No. 92 0995 8). Canberra, ACT: Australian Government Publishing Services, 1992. 55 pp.

Occupational Safety and Health Administration: Selected Occupational Fatalities Related to Fire and/or Explosion in Confined Work Spaces as Found in OSHA Fatality/Catastrophe Investigations. Washington, D.C.: U.S. Department of Labor, Occupational Safety and Health Administration (U.S. DOL/OSHA), 1982.

OSHA: Permit-Required Confined Spaces for General Industry; Final Rule, *Fed. Regist. 58*: 9 (14 January 1993). pp. 4462–4563.

Russell, D.K., Q. Keeny, J.E. Mutchler, and G.D. Clayton: Design of ventilation systems. In *The Industrial Environment — Its Evaluation and Control.* Washington, D.C.: Government Printing Office (DHHS/PHS/CDC/NIOSH), 1973. pp. 609–628.

Schwope, A.D., P.P. Costas, J.O. Jackson, and D.J. Weitzman: *Guidelines for the Selection of Chemical Protective Clothing. Volume II: Technical and Reference Manual.* Cincinnati, OH: American Conference of Governmental Industrial Hygienists, 1983. pp. 1–17.

Shackelford, J.F.: *Introduction to Materials Science for Engineers*, 2nd ed. New York: Macmillan, 1988. pp. 605–641.

Stull, J.O.: Chemical protective clothing. *Occup. Health Safety.* November, 1992 pp. 49–52.

Standards Association of Australia: *Safe Working in a Confined Space* (AS 2865-1986). North Sydney, NSW: 1986. 19 pp.

Turner, K.B.: Oxygen safety. *Professional Safety.* January, 1987. pp. 13–16.

Index

A

D

E

F

G

H

J

K

L

N

O

P

S

U

V

W

X

Milton Keynes UK
Ingram Content Group UK Ltd.
UKHW050306111024
449327UK00043B/2003